Culicipedia: Species-group, genus-group and family-group names in Culicidae (Diptera)

Culicipedia: Species-group, genus-group and family-group names in Culicidae (Diptera)

Ralph E. Harbach
Department of Life Sciences
Natural History Museum
Cromwell Road
London
SW7 5BD
UK

CABI

CABI is a trading name of CAB International

CABI
Nosworthy Way
Wallingford
Oxfordshire OX10 8DE
UK

Tel: +44 (0)1491 832111
Fax: +44 (0)1491 833508
E-mail: info@cabi.org
Website: www.cabi.org

CABI
745 Atlantic Avenue
8th Floor
Boston, MA 02111
USA

T: +1 (617)682-9015
E-mail: cabi-nao@cabi.org

A catalogue record for this book is available from the British Library, London, UK.

Library of Congress Cataloging-in-Publication Data

Names: Harbach, Ralph E., author.
Title: Culicipedia : species-group, genus-group and family-group names in Culicidae (Diptera) / Ralph E. Harbach.
Description: Boston, MA : CABI, [2018] | Includes bibliographical references.
Identifiers: LCCN 2018023336 (print) | LCCN 2018028612 (ebook) | ISBN 9781786399069 (ePDF) | ISBN 9781786399076 (ePub) | ISBN 9781786399052 (hbk : alk. paper)
Subjects: LCSH: Mosquitoes--Nomenclature. | Mosquitoes--Classification.
Classification: LCC QL536 (ebook) | LCC QL536 .H347 2018 (print) | DDC 595.77/2--dc23
LC record available at https://lccn.loc.gov/2018023336

ISBN-13: 978 1 78639 905 2

Commissioning editor: Ward Cooper
Editorial assistant: Emma McCann
Production editor: Marta Patiño

Typeset by SPi, Pondicherry, India
Printed and bound in the UK by Bell & Bain Ltd, Glasgow

You, whose name has not been formed by anyone,
Whose proof not even intellect can grasp,
To label you would be a loathsome act,
For you are far removed from genus and species.

Nosiri Khisrav
Persian Poet
(1004–1088)

Contents

Preface

The idea for this volume began with the development of an 'index' for the online Mosquito Taxonomic Inventory (http://mosquito-taxonomic-inventory.info/). This involved the acquisition of all articles in which the names that comprise the nomenclature of the family Culicidae were originally published. As the work progressed, it become apparent that the catalogue of world Culicidae (Knight & Stone, 1977), its supplements (Ward, 1984; Gaffigan & Ward, 1985; Ward, 1992), and consequently the online catalogue of the Walter Reed Biosystematics Unit (http://www.mosquitocatalog.org/), which was copied verbatim from the world catalogue and its supplements, contain many errors and include only species-group, genus-group and currently valid family-group names. The tome presented here is a comprehensive compilation of all scientific names introduced at all levels of classification for Culicidae from the official start of zoological nomenclature on 1 January 1758 until 30 June 2018, with corrections to errors contained in the world catalogue and its supplements. Corrections and additions made to information contained in those publications are too numerous to list individually. In general, however, corrections have been made to literature citations, page numbers of taxon citations, and the names of authors, species and type localities.

Culicid nomenclature may be defined as a lexicon of generally latinized names applied to groups of mosquitoes (actual or perceived monophyletic taxa) that constitute the hierarchical classification of the family. The groups (taxonomic ranks) whose names are regulated by the *International Code of Zoological Nomenclature* (International Commission on Zoological Nomenclature, 1999) are the species, genus and family groups. The species group includes names given to subspecies and species, the genus group includes names of subgenera and genera, and the family group includes the names of tribes, subfamilies and families. Emphasis is given to accurately reproducing the names given to members of these three groups and correctly associating and listing the original literature sources in which they were published.

Much of what has happened in mosquito taxonomy since the start of zoological nomenclature has no value besides historical interest: notwithstanding, the present work is a unique compendium of mosquito taxonomy. The contents and composition, unlike the four catalogues of the Culicidae published during the 20th century (Theobald, 1905e; Edwards, 1932a; Stone *et al.*, 1959; Knight & Stone, 1977), comprise a practical amalgamation of taxonomic history, rules that govern the formation of names, mosquito nomenclature and classification. The tome provides an up-to-date resource for linking names with one another and higher taxonomic groups while providing a concise account of all names (nearly 6000) used in the Culicidae, their origins and taxonomic status.

The need for this tome coincides with a time when few professional traditional taxonomists are left to assist and advise the new generation of molecular taxonomists in the basic principles and practice of mosquito taxonomy. For this reason, this compilation is intended to be a valuable resource for students, researchers, entomologists, librarians and anyone interested in mosquito taxonomy, systematics and nomenclature.

Many colleagues have contributed directly or indirectly to the consummation of this work. I am grateful to all persons who aided in any way, but I would like to particularly thank the following colleagues. First of all, this tome owes its existence to the exhortations of Lorna Culverwell and Simon Rycroft. Their suggestions and encouragement provided the impetus to transform an intended 'index' into this monograph. I am equally grateful to the Library and Archives staff of the Natural History Museum, especially Harriet Campbell Longley, for assistance in locating articles and obtaining numerous elusive publications through the inter-library loan service that aided the preparation of this tome. A huge thank you to my revered colleagues who took time from busy schedules to review and provide advice and comments on various parts of this work: Art Borkent, Lorna Culverwell, Bruce Harrison and Christian Kammerer. Thanks also to David Notten for discussing provisions of the *International Code of Zoological Nomenclature*, and very special thanks to Christian Kammerer for his critical appraisal and scholarly comments on text dealing with the application of Latin and Greek in the formation of names. Chen Bin and Ichiro Miyagi are gratefully acknowledged for providing information about names proposed in Chinese and Japanese works, respectively. My particular thanks to freelance copy editor Val Porter for diligently and thoroughly reviewing the manuscript and meticulously checking the references cited. I hasten to add that any lapses and errors are strictly my responsibility.

Finally, I am grateful to everyone, past and present, whose taxonomic studies have contributed names and information compiled in this volume. The bibliography provides a complete listing of those individuals. However, it would be very remiss of me not to express special gratitude to Kenneth L. Knight, who unfortunately will not see this work, but whose ennobling and inspiring influence started my career as a mosquito taxonomist and enabled the ambitious undertaking and completion of this work.

Ralph E. Harbach
London
June 2018

Introduction

Men and women have been giving names to animals and plants since the dawn of human history. According to the Book of Genesis, God brought every living animal to the man (Adam) to see what he would name them (*Genesis* 1.19). The process of applying single names to living organisms continued until partial use of binomial names emerged in the 17th century CE with the work of Bauhin (1623) and others. The application of binomial names, now known as binomial nomenclature, was not applied consistently until Carl Linnaeus (Caroli von Linné, titled name; Caroli Linnæi, Latin genitive form) introduced Linnaean taxonomy in his multiple editions of *Systema Naturae* published between 1735 and 1772. As all students of biology have or should have learned, the tenth edition of *Systema Naturae* (Linnaeus, 1758) is considered the official starting point of zoological nomenclature, which, with the recognition of two mosquito species, *Culex pipiens* and *Culex bifurcatus*, is also the beginning of mosquito taxonomy.

Linnaeus's meagre contribution to mosquito taxonomy is significant because the family Culicidae Meigen, 1818 is based on the genus *Culex* (Stephens, 1829), for which *Cx. pipiens* Linnaeus, 1758 is the type species (Latreille, 1810). Linnaeus's collection, maintained by the Linnean Society of London, includes two pinned mosquitoes – the first and second species of '*Culex*' described in the tenth edition of *Systema Naturae*. The first is a female which bears a label inscribed '1. pipiens' and the second is a male bearing a label inscribed '2. bifurcatus'. The former is actually a damaged and unidentifiable species of *Aedes* (*Ochlerotatus*) and the latter is a member of the *Anopheles maculipennis* species complex. Implementing provisions of the *International Code of Zoological Nomenclature* (*Code* hereafter) (International Commission on Zoological Nomenclature, 1985), Harbach *et al.* (1985) fixed the identity of *Cx. pipiens* by designating a neotype from a series of specimens collected near Veberöd, Scania, Sweden – Linnaeus's homeland. An alternative action to neotype designation would have been to accept the specimen in the Linnaean Collection as the 'type' of *Cx. pipiens*, but this would have had catastrophic consequences for mosquito taxonomy. The family Culicidae, the subfamily Culicinae and the tribe Culicini would still be based on the genus *Culex*, but *Culex* would be based on an aedine mosquito and comprise species traditionally placed in *Aedes* (*Ochlerotatus*), the taxon traditionally known as *Culex* would take on the next available name, *Heteronycha* Lynch Arribálzaga, and the tribe Culicini would be known as Heteronychini (Harbach, 2012). Other than being an example of practical application of provisions of the *Code*, this paragraph reveals the interconnected vocabulary of culicid nomenclature – 12 names, three species-group, five genus-group and four family-group names. The principal value of taxonomic names is that they denote concepts that can be understood and communicated. An important purpose of this volume is therefore to indicate which names represent currently accepted concepts, which are excluded from current usage and which are deemed to apply to the same concept.

So, how many names have been introduced at all levels of classification within the family Culicidae since the official start of zoological nomenclature on 1 January 1758? How many of those names are currently recognized as valid concepts? How many species-group names have appeared in the scientific literature? How many genus-group names? How many family-group names? How many names represent the same taxon, i.e. how many are junior synonyms? How many of the currently recognized genera and species are derived from personal names, i.e. how many are patronymics? How many are formed from geographical names? All of these questions, as well as others, are answered in this volume. Additionally, effort has been made to provide explanatory information that is interesting, informative and to some degree thought-provoking.

Historical Compilation of Names in Culicidae

Theobald (1905e), *Genera Insectorum* Fascicule 26, recognized seven subfamilies and 60 genera for 464 species of Culicidae (excluding subfamily Corethrinae). Five years later, Theobald (1910c), with the removal of Corethrinae, recognized ten subfamilies, 149 genera, about 1050 extant species and 22 extant varieties. Edwards (1932a), *Genera Insectorum* Fascicule 194, radically revised the classification of Theobald (1905e) to include three subfamilies: the Culicinae (true mosquitoes), Dixinae and Chaoborinae. His subfamily Culicinae (present-day Culicidae) included three tribes: Anophelini with three genera, 166 species and 49 varieties; Megarhinini with one genus, 52 species and three varieties; and Culicini with 31 genera, 1195 species and 70 varieties – a total of 35 genera, 1400 extant species and 135 extant varieties. It is noteworthy that neither Theobald nor Edwards recognized subspecies.

An extensively revised and updated catalogue of the Culicidae of the world was produced by Stone *et al.* (1959). This catalogue omitted Edwards's subfamilies Dixinae and Chaoborinae, which were raised to full family status by Stone (1957a), thus restricting the family Culicidae to Edwards's subfamily Culicinae. The classification set out in the catalogue comprised three subfamilies: Anophelinae, Culicinae and Toxorhynchitinae (Edwards's Megarhinini); and two tribes, Culicini and Sabethini, were recognized within Culicinae. The classification included 31 genera, 95 subgenera, 2426 species, 113 subspecies, 94 varieties, three infrasubspecific forms and 12 *nomina dubia*. In addition to valid names, there were 308 synonyms of genus-group names and 3372 synonyms of species-group names.

Knight & Stone (1977) produced an updated second catalogue of the mosquitoes of the world to include taxonomic acts published prior to the end of 1973. The suprageneric classification utilized in the catalogue only differed from that of Stone *et al.* (1959) in recognizing ten tribes within the subfamily Culicinae (Belkin, 1962). While three genera, 35 subgenera and 1026 extant species were added to the culicid fauna, the number of subspecies and varieties had decreased by three and 30, respectively. An additional infrasubspecific form was recognized and 23 nominal species were added to the list of *nomina dubia*. As a result of these changes, the species-group taxa known at the end of 1973 included 2960 species, 109 subspecies, 64 varieties and four infrasubgeneric forms. Thirty-five names proposed for species were classified as *nomina dubia*.

Three supplements to Knight & Stone's catalogue were published in the final three decades of the 20th century. The first supplement (Knight, 1978b) included one new subgenus, 107 new species and four new subspecies. No other changes were made to the existing classification. The authors of the second supplement (Ward, 1984; Gaffigan & Ward, 1985) did not disclose the total numbers of species, subspecies and varieties recognized following taxonomic acts published between 1978 and 1983. Careful assessment of taxa listed in the supplement reveals that the authors correctly listed three new genera and three new subgenera, but incorrectly determined the number of valid species-group taxa known at the time. The failure of Ward (1984) to fully compensate for losses and gains led Ward (1992) to miscalculate the number of species recognized in the third supplement, which included additions and changes in the taxonomy of mosquitoes published before the end of January 1992. In short, 3120 species (an increase of 53 since the 1978 supplement), 131 subspecies (increase of 18), 52 varieties (decrease of 12), five infrasubspecific forms (increase of one) and 40 *nomina dubia* (increase of five) were recorded in the second supplement. Twelve years later, 144 subgenera (further increase of ten), 3273 species (gain of 153), 125 subspecies (loss of four), 49 varieties (loss of three) and three infrasubspecific forms (loss of two) were recognized based on taxonomic acts recorded in the third supplement (Ward, 1992). Three subfamilies and ten tribes within Culicinae were recognized in all three supplements. Whereas 35 *nomina dubia* were listed in the first supplement, 40 were recorded in the second and third supplements.

Taxonomic additions and losses since January 1992 have resulted in the species-group, genus-group and family-group names compiled in Sections 1, 2 and 3 of this volume, respectively, which correspond to the groups governed by provisions of the *Code*. The names of valid taxa are those recognized in the 'traditional' classification of the family (Knight & Stone, 1977), except Toxorhynchitinae which was reduced to tribal status in subfamily Culicinae (Harbach & Kitching, 1998), and the modified 'traditional classification' of the tribe Aedini of Wilkerson *et al.* (2015), with the exception being a separate section (Section 8) that lists the species- and genus-group names that comprise the phylogenetic classification of the composite genus *Aedes* established by Reinert *et al.* (2009), which some researchers prefer over the modified traditional polyphyletic classification of the genus (Wilkerson *et al.*, 2015).

Arrangement, Format and Definitions

First of all, this publication is not a catalogue. It is principally a lexicon, divided into sections, most of which contain a comprehensive list of taxonomic names. There are three main sections, one each for species-group, genus-group and family-group names. These sections are followed by three sections devoted to currently valid species- and genus-group names formed from personal and geographical names, and a section featuring some interesting and unusual names. These seven sections subsume the traditional classification of the Culicidae, including the composite genus *Aedes* Meigen (Knight & Stone, 1977; Wilkerson *et al*., 2015: S3 Appendix). The sections are followed by a separate listing of names (Section 8) that comprise the multi-generic classification proposed by Reinert *et al*. (2009) for taxa that comprise the traditional polyphyletic *Aedes*. Taxonomic works published in recent years that recognize and use generic-level taxa contained in the classification of Reinert *et al*. justify the inclusion of this section. To illustrate the unstable taxonomic history of the family, a final section provides information and a list, in an appendix, of currently valid species-group names in their original formation and generic placement compared with their formations and generic assignments in the current traditional classification and the classification of Reinert *et al*.

All names of the genus and species groups are *italicized*; the names of valid taxa (including *nomina dubia*) are in **boldface** type. Names are listed in alphabetical order in each section. Each name is followed by its author(s) and year of publication. The **Format** of information provided for the names listed in the main sections is explained in the introductory text of those sections. The following examples provide an overview (preview) of the general format and type of information provided for species-, genus- and family-group names. It is necessary to mention that type species of genus- and family-groups are not indicated. The type species of currently recognized generic- and family-level names can be found in the online Mosquito Taxonomic Inventory (http://mosquito-taxonomic-inventory.info/).

Valid species-group names:

affinis (Edwards, 1913c) (p. 241) – species of *Tripteroides* (*Rachionotomyia*) [*nomen novum* for *caeruleocephala* Theobald, 1910c; original combination: *Rachionotomyia affinis*].
Synonym: *caeruleocephala* Theobald, 1910c.

aequatorianus Levi-Castillo, 1953f (p. 72) – *nomen dubium*, *Toxorhynchites*; status by Stone *et al*., 1959.

Invalid species-group names (synonyms):

adolphoi Neiva, 1908b (p. 457) – synonym of *Anopheles* (*Kerteszia*) *cruzii* Dyar & Knab, 1908 [original combination: *Myzorhynchella adolphoi*]; synonymy by Barretto & Coutinho, 1951.

africanus Theobald, 1909b (p. 8) – synonym of *Aedes* (*Catageiomyia*) *stenoscutus* (Edwards, 1912c) [original combination: *Stenoscutus africanus*]; synonymy by White, 1975. **Note**: this nominal taxon is a junior secondary homonym of *africanus* (Theobald, 1901d).

Valid genus-group names:

Culiseta Felt, 1904 (p. 391c) – f., genus (subgenus) of tribe Culisetini Belkin, 1962.
Synonyms: *Theobaldia* Neveu-Lemaire, 1902a; *Theobaldinella* Blanchard, 1905; *Pseudotheobaldia* Theobald, 1907.

Invalid genus-group names (synonyms):

Empihals Reinert, Harbach & Kitching, 2008 (m., subgenus of *Ochlerotatus* Lynch Arribálzaga, 1891a,b) (p. 105) – synonym of *Ochlerotatus* Lynch Arribálzaga, 1891a,b; synonymy by Wilkerson *et al*., 2015.

Conventional family-group names:

Aedini Neveu-Lemaire, 1902c (coordinate tribal name of Aedeinae Neveu-Lemaire, 1902c) (p. 220) – tribe of subfamily Culicinae Meigen, 1818.

Disused (excluded) family-group names:

Lynchiellina Lahille, 1904 (pp. 79, 81 [11, 13], tableau) – subtribe of Culicinae (= Culicidae Meigen, 1818 as currently defined); proposed by da Costa Lima *et al.* (1962) as a subtribe of Toxorhynchitini Lahille, 1904 (not currently recognized).

Symbols, Abbreviations, Words and Phrases in this Tome

& – and (logogram known as an ampersand)
♀ – female (standard sex symbol derived from astrological symbol denoting planet Venus)
♂ – male (standard sex symbol derived from astrological symbol denoting planet Mars)
< – derived from (e.g. *chionodes* Belkin, the stem *chiono* < Gr. *chion*, *-os*>
Ar. – Arabic
CE – Common Era or Current Era
Code – *International Code of Zoological Nomenclature*
dim. – diminutive
ed. – editor (plural, eds)
e.g. – *exempli gratia* (Latin, for example)
et al. – *et alia* (Latin, and others)
f. – feminine (gender)
Fr. – French
Ger. – German
Gr. – Greek
L. – Latin
m. – masculine (gender)
ME. – Middle English
Mod. – Modern
ML. – Medieval Latin
n. – neuter (gender)
NL. – New Latin
nomen dubium – Latin, doubtful name (plural, *nomina dubia*)
nomen novum – Latin, new name (plural, *nomina nova*)
non – Latin, not (negative participle)
OE. – Old English
p. – page (plural, pp.)
pl. – plural
sic – thus, just as (in full: Latin *sic erat scriptum*, thus was it written)
Sp. – Spanish
stat. nov. – *status novus* (Latin, new status, representing a change in taxonomic rank)

Declension (Inflection) and the Formation of Names

Latin and Greek, and certain other languages, are 'inflected' languages, i.e. languages in which words comprise roots (Stems, see below) and inflections (changes in the form of a word, typically the ending, to express a grammatical function or attribute, such as tense, mood, person, number, case and gender). The stems remain stable and the endings change depending on the function (see Case and Declension, below) of the word in a sentence. The following definitions and information are provided for individuals, like a reviewer of this tome, who are not familiar with the elements of declension (inflection) that govern the formation of names.

Case (grammatical case)

Case is a special grammatical category of a noun, pronoun, adjective, participle or numeral whose value reflects the grammatical function performed by that word in a phrase, clause or sentence. In Latin and Greek, and some other languages, different types of words take different inflected forms depending on what case they are in. Two cases, the nominative and genitive, feature in zoological nomenclature.

Nominative case

The nominative case is the subject case. For most Latin nouns, the form seen in the dictionary is the nominative singular, followed by an ending for the genitive, and the gender of the noun. The word 'nominative' means 'pertaining to the thing designated'.

Genitive case

The genitive case is a descriptive case. The genitive case in Latin is used to indicate any number of relationships that are most frequently and easily translated into English by the preposition 'of': 'state of the union', 'species of the genus'. The common meaning of the genitive is the possessive. Broadly speaking, the genitive is the 'belonged-to' case, the source or origin or owner. In the two preceding examples, the second noun (genitive) owns the first noun.

Declension

The variation (inflection) of the form of a noun, pronoun, adjective, adverb, numeral or article by which its grammatical case (nominative, genitive, dative, etc.), number (singular or plural) and/or gender are identified. Latin has five declensions, but only the first, second and third declensions feature in the formation of names. For inflections (endings) see Latin declensions in brief (below).

First declension

Latin words of the first declension have an invariable stem and are generally of feminine gender. The nominative singular form of this declension consists of the stem and the suffix *-a*, and the genitive singular form is the stem plus *-ae*. There is a small number of masculine exceptions that generally refer to occupations, e.g. inhabitant (m., *incola*, *incolae*) and writer (m., *scriba*, *scribae*).

Second declension

Latin words of the second declension are regularly of masculine gender (ending in *-us*, *-er* and *-ir*) or neuter gender (ending in *-um*), and have a genitive ending in *-i*.

Third declension

The third declension is a category of nouns in Latin and Greek with broadly similar case formation – diverse stems, but similar endings. Words of the third declension comprise consonant stems and *i*-stems. They are of all three genders. Third declension nouns are identifiable by the genitive singular ending *-is*. Third declension nouns in the nominative cannot be identified because they have various forms and spelling, have endings that do not reveal their gender and can be masculine, feminine or neuter.

Latin declensions in brief

1st declension	Singular	Plural
Nominative	*-a*	*-ae*
Genitive	*-ae*	*-arum*

	Singular		Plural	
2nd declension	m.	n.	m.	n.
Nominative	*-us*, *-er*, *-ir*	*-um*	*-i*	*-a*
Genitive	*-i*	*-i*	*-orum*	*-orum*

	Singular		Plural	
3rd declension	m. & f.	n.	m. & f.	n.
Nominative	(various)		*-es*	*-a* (*-ia**)
Genitive	*-is*	*-is*	*-um* (*-ium**)	*-um* (*-ium**)

*For *i*-stem nouns.

Stem (word stem)

The stem is the part of an inflected word that remains (does not change) after the inflected part is removed. The stem of a generic name is the combining form, for example in the formation of family-group names. The stem is derived from the genitive (possessive) case of the name, minus the case ending. For example, the genitive singular of *Culex* is *Culicis* and the stem is *Culici-*, thus Culicidae for the family name, Culicinae for the subfamily name and Culicini for the tribal name.

Classification

The classification of the Culicidae took root in the 18th century (1758–1799)[1] with the naming of 22 species of the genus *Culex*, only ten (in boldface) of which are recognized today (see Knight, 1972): ***aegypti*** Linnaeus, 1762; ***annulatus*** Schrank, 1776; *argenteus* Poiret, 1787 (synonym of *aegypti* Linnaeus); *bifurcatus* Linnaeus, 1758 (synonym of *pipiens* Linnaeus); *bipunctatus* Robineau-Desvoidy, 1827 (synonym of *flavescens* Müller); ***caspius*** Pallas, 1771; *ciliaris* Linnaeus, 1767 (*nomen dubium*, *Aedes*); ***ciliata*** Fabricius, 1794; ***communis*** de Geer, 1776; *fasciatus* Müller, 1764 (synonym of *pipiens* Linnaeus); ***flavescens*** Müller, 1764; ***geniculatus*** Olivier, 1791; ***haemorrhoidalis*** (Fabricius, 1787); ***hyrcanus*** Pallas, 1771; *lutescens* Fabricius, 1775 (synonym of *flavescens* Müller); *molestus* Forskål, 1775 (synonym of *pipiens* Linnaeus); ***rusticus*** Rossi, 1790; *trifurcatus* Fabricius, 1794 (synonym of *pipiens* Linnaeus); *variegatus* Schrank, 1781 (synonym of *flavescens* Müller); ***vulgaris*** Linnaeus, 1792 (*nomen dubium*, *Culex*).

The classification of the family became increasingly more diversified with the discovery of additional species. Seven genera, still recognized today, were introduced during the 19th century (1800–1899), including ***Aedes*** Meigen, 1818; ***Anopheles*** Meigen, 1818; ***Haemagogus*** Williston, 1896; ***Psorophora*** Robineau-Desvoidy, 1827; ***Sabethes*** Robineau-Desvoidy, 1827; and ***Uranotaenia*** Lynch Arribálzaga, 1891a,b. This was followed during the first three decades of the 20th century (1900–1999) by a period of unequalled mosquito taxonomic proliferation as a result of the discovery at the turn of the 19th century that mosquitoes transmit the causal agents of malaria and yellow fever.

The British Museum (Natural History) hosted Fred. V. Theobald in 1899 to examine and classify mosquitoes received from the British Colonies. In the first catalogue of mosquitoes and 'allied insects', Theobald (1905e) recognized seven subfamilies, 60 genera and 462 species of true mosquitoes, and one subfamily for corethrellids and chaoborids. Theobald later excluded the corethrellids and chaoborids from the family Culicidae and recognized, by the end of the first decade of the 20th century, ten subfamilies and 149 genera for about 1050 species of mosquitoes (Theobald, 1910c). As Theobald's generic classification was neither practical nor natural, significant changes were made during the second and third decades that culminated in a much more conservative classification comprising broad genus-group concepts (Edwards, 1932a), which gave birth to the 'traditional' classification of the 20th century (Stone *et al.*, 1959; Belkin, 1962; Knight & Stone, 1977). Edwards (1932a) included dixid and chaoborid midges as subfamilies of Culicidae and classified the 'true mosquitoes' as a separate subfamily, Culicinae, comprising three tribes: Anophelini, Culicini and Megarhinini. Stone (1957a) removed dixids and chaoborids from Culicidae and restricted the family Culicidae to the Culicinae of Edwards (1932a). This brought about changes in family-group taxa that were adopted by Stone *et al.* (1959) in their world catalogue of mosquitoes. This classification recognized subfamilies Anophelinae, Culicinae and Toxorhynchitinae (the Megarhinini of Edwards), and two tribes within Culicinae: the Culicini and Sabethini. Belkin (1962) retained Edwards's subfamilies and divided the Culicinae ('true mosquitoes') into 12 tribes instead of three. He retained Anophelini, Sabethini and Toxorhynchitini and recognized nine tribes in place of Edwards's Culicini. At least some authors (e.g. Belkin *et al.*, 1970) continued to treat dixids and chaoborids as subfamilies of Culicidae until Knight & Stone (1977) again excluded them from the family. This resulted in the recognition of three subfamilies, Anophelinae, Culicinae and Toxorhynchitinae, and the division of the subfamily Culicinae into the ten tribes. Mattingly (1971, 1981) and Service (1993) continued to follow Stone *et al.* (1959) in recognizing only Culicini and Sabethini as tribes of Culicinae, but mosquito taxonomists have otherwise accepted the division of the subfamily into ten tribes: Aedeomyiini, Aedini, Culicini, Culisetini, Ficalbiini, Hodgesiini, Mansoniini, Orthopodomyiini, Sabethini and Uranotaeniini. The only significant change to family-groups within the 'traditional' classification was the downgrading of the subfamily Toxorhynchitinae to tribal status (Harbach & Kitching, 1998; Mitchell *et al.*, 2002). With the addition of Toxorhynchitini, the subfamily Culicinae now includes 11 tribes.

The foregoing only reflects a small portion of the multitude of changes and fluctuations that have occurred in culicid taxonomy since Linnaeus (1758). Taxonomic acts completed to date have resulted in the internal classification of the family that is currently accepted by most taxonomists and mosquito workers, i.e. Culicidae comprises

[1] Strictly speaking, centuries run from 01 to 100, with a new century beginning on the first day of year 01. The practice of considering a century to begin when the terminal two digits change to 00, e.g. 1 January 1800, when 1799 (last year of the 18th century) became 1800 (the first year of the 19th century), is followed here.

two subfamilies, Anophelinae and Culicinae, and the subfamily Culicinae is divided into the 11 tribes, 41 genera and 186 subgenera (Appendix 1).

Looking back at the history of culicid nomenclature, it appears that little change has been made to the 'traditional' generic classification of mosquitoes since the publication of *A Catalog of the Mosquitoes of the World* (Knight & Stone, 1977). Numerous additions to the generic classification of the tribe Aedini by Reinert *et al.* (2009) were largely downgraded to subgenera of *Aedes* by Wilkerson *et al.* (2015) in agreement with the practice, beginning with Edwards (1932a), of recognizing few genera and many subgenera. The Aedini of Knight & Stone included nine genera and 47 subgenera, with 38 of the subgenera contained in *Aedes*. Today, Aedini includes only one additional genus, *Verrallina*, which in 1977 was classified as a subgenus of *Aedes*, and 96 subgenera, 87 of which are subgenera of *Aedes*. In view of the decline in mosquito taxonomists and revisionary studies, it seems unlikely that this classification will be changed much in future years to reflect the evolution of the Culicidae. Lumping numerous small groups of indefinite affinities into a very large group for convenience obscures phylogenetic relationships.

Section 1: Species-group Names

The list below includes all species-group names applied to mosquitoes from Linnaeus (1758). The species group comprises a total of 5390 names (including the three new replacement names proposed below): 3559 valid extant species; 130 extant subspecies; 59 *nomina dubia*; 25 valid fossil species; 76 *nomina nuda*, including one nominal fossil species; three rejected names; one unavailable name; and 1537 synonyms, including two for fossil species and those of the replacement names proposed below. Varietal names are no longer recognized in the family Culicidae, based on implementation by Harbach & Howard (2007) of provisions contained in Article 45 of the *International Code of Zoological Nomenclature* (International Commission on Zoological Nomenclature, 1999) that regulate the rank of species-group names.

Format

The names are listed in alphabetical order, with two or more identical names listed alphabetically by author, and by date where identical names are by the same author. Valid names (senior synonyms), including *nomina dubia*, are in ***boldface italics*** and all synonymous names (junior synonyms), as well as *nomina nuda*, are in *italicized normal type*. Each name is followed by authorship, date of publication, page of citation, taxonomic status, original combination (if applicable) and source of status, e.g. synonymy (if applicable and known). Junior synonyms are listed for valid taxa, in chronological order. Literature citations are provided for synonymies recorded in Knight & Stone (1977) and later publications. However, in the majority of cases where the sources of synonymies are not supported by authorship in Knight & Stone's catalogue, synonymous names are followed by 'synonymy established before Knight & Stone, 1977'.

The endings (terminations) of Latin or latinized adjectival and participial names of valid species and subspecies that are currently included in a genus other than the one in which they were originally placed have been changed to agree in gender with the current generic name, and the authorship-date of those names is enclosed in parentheses to indicate that they have been transferred. This has not been done for synonymic names, which are spelled exactly as originally spelled, or should have been spelled for gender agreement, with the generic name with which they were originally combined.

Formation of Names

Species-group names of mosquitoes include names that are nouns and adjectives. Nounal names may be formed from Latin and Greek names (e.g. animals, plants, shapes, colours, body parts, etc.) or derived from personal names (living and deceased persons and mythical entities), localities (geographical names), environment (habitat names) and other sources (acronymic and cultural names).

In many cases, it is not readily apparent whether a name is a noun or an adjective. Adjectival names may be formed from adjectives or nouns, i.e. nouns translated into their Latin adjectival forms. In addition to Latin nouns, proper names of many individuals, peoples and geographical localities have also been latinized to form adjectival names. In general, species-group names may be placed in one or other of the following categories based on their derivations.

Descriptive names

These names are based on important or prominent anatomical features, usually a feature that distinguishes a species from a related species. Some specific names are essentially brief descriptors of the species, e.g. *Anopheles gigas* (Gr. noun meaning 'giant'), *Anopheles superpictus* (adjective derived from *super*, L. over, above + *pictus*, L. painted, coloured; meaning 'coloured above'), *Culex melanura* (noun derived from *melas*, Gr. black, dark + *oura*, Gr. tail; meaning 'black tail') and *Verrallina cautus* (L. adjective meaning 'careful, wary'). The derivations of some fanciful and imaginative descriptive names are included in Section 7 (see), but the formations of the majority of such names are not dealt with herein.

Personal names

These names, also known as eponyms or patronymics, serve to preserve the memory of some person or fictional entity. These names may be formed as nouns or adjectives. Currently valid species-group names formed from personal names are compiled in Section 4 (see).

Geographical names

These names, also formed as nouns or adjectives, are based on the geographical place, locality, administrative territory or governmental region where a species was first collected, e.g. an island, mountain range, village, town, district, state, province or country. Currently valid species-group names formed from geographical names are compiled in Section 5 (see).

Habitat names

Many specific names of mosquitoes are based on the habitat where the species was first discovered, predominantly where larval forms were collected, e.g. *Aedes bambusae* (noun derived from *Bambusa*, a genus of bamboos, in reference to adults and larvae collected in bamboo forest and bamboo stems, respectively), *Aedes deserticola* (noun derived from *desertum*, L. desert, wasteland + -*cola*, L. dweller or inhabitant), *Culex barrinus* (adjective *barrinus*, L. of elephants, named for larvae found in elephant footprints). Some other such names are included in Section 7 (see).

Acronymic, cultural and fanciful names

These types of species-group names are fewer in number than other types of names. Acronyms used as species names are nouns that essentially equate to arbitrary combinations of letters, e.g. *Anopheles nimpe* (an abbreviation formed from the initial letters of the Vietnamese institution previously known as the National Institute of Malariology, Parasitology and Entomology. Names derived from cultures and peoples are normally nouns, e.g. *Anopheles guarani* (named for the Guarani people of South America), but may also be adjectives, e.g. *Aedes lepchanus* (named for the Lepcha people of India). Examples of many of these unusual names are given, along with their derivations, in Section 7 (see).

Suffix -*cola* versus -*colus*

Use of the Latin suffix -*cola* (masculine and feminine), versus -*colus*, -*cola* and -*colum* (masculine, feminine and neuter, respectively), in forming species-group names is a controversial issue that deserves attention. These suffixes are formed from the Latin verbal stem *col*, which derives from the verb *colo*, meaning to dwell or inhabit. Names ending in the suffix -*cola* are nouns and those ending in the suffixes -*colus*, -*cola* and -*colum* are adjectives. The mosquito nomenclature includes 26 compound species-group names that were originally proposed with the -*cola* ending and 13 originally proposed with the -*colus* ending. Most of the names ending in -*colus* (adjectives) when originally proposed were later emended to end in -*cola* (nouns). Nine names, listed as originally spelled in the catalogue of Stone *et al.* (1959), were emended from -*colus* to -*cola* in Knight & Stone (1977). Only two names, *Aedes rubicolus* and *Culex arboricolus*, were listed in Knight & Stone (1977) as originally proposed. The first was emended to *rubicola* in Reinert (1999b) and the second has remained unchanged. *Culex axillicolus*, published more recently than Knight & Stone, was emended to *axillicola* in Ward (1984). *Anopheles bambusicolus* was emended to *bambusicola* in Knight & Stone (1977) and later changed back to its original spelling by Zavortink (1973), who considered the change to *bambusicola* to be an unjustified emendation.

Obviously names proposed with the -*cola* ending might be interpreted to be either nouns (masculine or feminine) or adjectives (feminine) as they both end in -*a*. There is no doubt that names ending in -*cola* that are combined with masculine genera are nouns in apposition whereas such names in combination with feminine genera may be either nouns or adjectives. Eight species-group names ending in -*cola* were originally placed in feminine genera – *Topomyia* (one), *Uranotaenia* (six) and *Wyeomyia* (one). Although the authors of those names did not indicate whether they were nouns or adjectives, their derivations suggest that most, if not all, were intended to be nouns, and should be treated as such to ensure that their original spellings are maintained if they should be transferred to a masculine genus.

Names formed from the verbal stem *col* are uniformly derived nouns ending in -*cola*, but in modern Latin there are many specific epithets, including *col*, which are definitely established by their authors as nouns and in other cases as adjectives. The question is: should the adjectival ending be regarded as an orthographic error and corrected? Knight & Stone (1977) changed such adjectives to nouns to conform to Latin precedent, to treat adjectives as solecisms or unintentional orthographic errors and to correct them to nouns ending in -*cola* in apposition to the generic name. Are such changes warranted? The answer should be no – if a species name was originally included in a masculine genus and terminated in -*colus*, that is *prima facia* evidence that it was intended to be an adjective, and should have remained an adjective. In agreement with Zavortink (1973), the changed spellings of species-group names originally ending in -*colus* are here regarded as unjustified emendations. Thus, ten names whose endings were changed from -*colus* to -*cola* (and are still combined with masculine generic names) are spelled with their original adjectival endings in the list below. These include ***arboricolus*** Galindo & Mendez, ***arboricolus*** Knight & Rozeboom, ***arboricolus*** Zavortink, ***axillicolus*** Steffan, ***bambusicolus*** Knight & Rozeboom, ***bambusicolus*** Komp, ***bihaicolus*** Dyar & Nuñez Tovar, ***domicolus*** Edwards, ***rabanicolus*** Floch & Abonnenc and ***rhodesiensis rupicolus*** Lewis. The species originally named *Aedes nubicolus* by Laffoon, unjustifiably changed to *nubicola* in Knight & Stone, is now correctly spelled as a feminine adjectival name of the species now known as ***Verrallina nubicola*** (Laffoon). As the derivations of the names formed from the verbal stem *col* are interesting and unusual, those that are currently valid (33) are included in Section 7 with other species-group names that are also intriguing.

Homonyms

Careful examination of the list of species-group names reveals that 135 names are homonyms. Some are junior or senior secondary homonyms of one or more species-group names (names originally placed in different genera) and some are junior or senior primary homonyms (names originally combined with the same generic name). Homonymous names in the list are provided with a **Note** indicating the nature of their homonymy.

When two or more species-group names are considered congeneric, the junior homonym(s) is/are invalid (*Code*, Article 59.1) and must be replaced either by an available and potentially valid synonym, or if not available, by a new substitute name (*Code*, Article 60.1–3). In rare instances, however, prevailing usage must be maintained if the senior homonym has not been used as a valid name after 1899 and the junior homonym has been used as the valid name of the taxon in at least 25 publications by at least ten authors over a span of not less than 10 years during the immediately preceding 50 years (*Code*, Article 23.9.1). There is only one junior homonym among mosquito species-group names that qualifies for retention under this provision of the *Code*: *Aedes* (*Downsiomyia*) *niveus* (Ludlow, 1903) [original combination: *Stegomyia nivea*] is the junior secondary homonym of *niveus* (von Eichwald, 1837), *nomen dubium* in *Aedes* Meigen [original combination: *Culex niveus*]. In the interests of stability, the prevailing usage of the junior homonym of *niveus* (Ludlow, 1903) is maintained based on criteria for reversal of precedence (*Code*, Article 23.9).

New Replacement Names

New replacement names are proposed here for the following three junior secondary homonyms (of presumed valid taxa) that do not qualify for retention of prevailing usage, and for which synonymous names are not available. Replacement names were chosen whose derivations have meanings similar to the original names.

intermedius (Chagas, 1908), recognized as a species of *Anopheles* (*Anopheles*) [original combination: *Cycloleppteron intermedium*], is the junior secondary homonym of *intermedia* Rothwell, 1907 [original combination: *Neocellia intermedia* = *Anopheles* (*Cellia*) *stephensi* Liston, 1901], and thus requires a replacement name; it is also the junior secondary homonym of *intermedius* Shingarev, 1928. Herewith, ***medialis*** *nomen novum* is proposed to replace *intermedius* (Chagas, 1908); henceforth, *Anopheles* (*Anopheles*) *medialis* Harbach. Etymology: *medialis*, L. of or belonging to the middle.

mediopunctatus (Theobald, 1909b), recognized as a subspecies of *Aedes* (*Aedimorphus*) *cumminsii* (Theobald, 1903a) (see Harbach & Howard, 2007) [original combination: *Culicada mediopunctata*], is the junior secondary homonym of *Aedes* (*Stegomyia*) *mediopunctatus* (Theobald, 1905h) [original combination: *Stegomyia mediopunctata*]. Herewith, ***mesostictus*** *nomen novum* is proposed to replace *mediopunctatus* (Theobald, 1909b); henceforth, *Aedes* (*Aedimorphus*) *cumminsii mesostictus* Harbach. Etymology: *mesos*, Gr. middle; *stiktos*, Gr. punctured, dappled, spotted. **Note**: a replacement name is not needed for this taxon in the phylogenetic classification of composite *Aedes* established by Reinert *et al.* (2009) where *mediopunctatus* (Theobald, 1909b) is a subspecies of *Aedimorphus cumminsii* (Theobald, 1903a) (see Section 8).

sylvaticus (Le Goff, Boussès & Brunhes, 2007), recognized as a species of *Aedes* (*Neomelaniconion*) [original combination: *Neomelaniconion sylvaticum*], is the junior secondary homonym of *sylvaticus* Meigen, 1818 [original combination: *Culex sylvaticus* = *Aedes* (*Ochlerotatus*) *communis* (de Geer, 1776)], and thus requires a replacement name; it is also the junior secondary homonym of *sylvaticus* Brunhes, 1983, currently a synonym of *Aedes* (*Paulianius*) *madagascarensis* van Someren, 1949a. Herewith, ***hylaius*** *nomen novum* is proposed to replace *sylvaticum* Le Goff, Boussès & Brunhes, 2007; henceforth, *Aedes* (*Neomelaniconion*) *hylaius* Harbach. Etymology: *hylaios*, Gr. of the wood, forest, savage, wild. **Note**: a replacement name is not needed for this taxon in the phylogenetic classification of composite *Aedes* established by Reinert *et al.* (2009) where *sylvaticus* Le Goff, Boussès & Brunhes is a species of the genus *Neomelaniconion* (see Section 8).

Comprehensive list of Species-group Names in Culicidae

Aa

abadsantosi Baisas, 1946c (p. 25) – synonym of *Aedes* (*Hulecoeteomyia*) *rizali* (Banks, 1906b); synonymy by Knight, 1968.

abascanta Dyar & Knab, 1908 (p. 65) – synonym of *Wyeomyia* (*Wyeomyia*) *medioalbipes* Lutz, 1904; synonymy by Lane, 1953.

abdita (Barraud, 1931b) (p. 614) – species of *Verrallina* (*Harbachius*) [original combination: *Aedes* (*Aedes*) *abditus*].

abdita Peyton, 1977 (p. 125) – species of *Uranotaenia* (*Pseudoficalbia*).

abdominalis Taylor, 1913 (p. 13) – synonym of *Culex* (*Oculeomyia*) *bitaeniorhynchus* Giles, 1901a [original combination: *Culicelsa abdominalis*]; synonymy established before Knight & Stone, 1977.

abebela Dyar & Knab, 1908 (p. 67) – species of *Wyeomyia* (*Wyeomyia*).

aberrans Harrison & Scanlon, 1975 (p. 152) – species of *Anopheles* (*Anopheles*).

abfitchii Felt, 1904 (p. 381) – synonym of *Aedes* (*Ochlerotatus*) *excrucians* (Walker, 1856a) [original combination: *Culex abfitchii*]; synonymy established before Knight & Stone, 1977.

abia Dyar & Knab, 1908 (p. 67) – synonym of *Wyeomyia* (*Wyeomyia*) *mitchellii* (Theobald, 1905d); synonymy by Belkin, 1970a.

ablabes Dyar & Knab, 1908 (p. 66) – synonym of *Wyeomyia* (*Wyeomyia*) *adelpha* Dyar & Knab, 1906a; synonymy by Stone, 1969.

ablechra Dyar & Knab, 1908 (p. 66) – species of *Wyeomyia* (*Wyeomyia*).

ablusa Harbach, 2009 (in Harbach & Howard, 2009) (p. 5) – species of *Chagasia*.

abnormalis Lane, 1936b (p. 189) – species of *Culex* (*Culex*).
Synonym: *scutatus* Rozeboom & Komp, 1948.

abnormalis (Theobald, 1909b) (p. 10) – species of *Aedes* (*Aedimorphus*) [original combination: *Bathosomyia abnormalis*].

abnormalis Theobald, 1910c (p. 512) – synonym of *Uranotaenia* (*Uranotaenia*) *pallidocephala* Theobald, 1908c; synonymy established before Knight & Stone, 1977.

abominator Dyar & Knab, 1909d (p. 257) – species of *Culex* (*Melanoconion*).

abonnenci Clastrier, 1970c (p. 468) – species of *Culex* (*Melanoconion*).

aboriginis Dyar, 1917b (p. 99) – species of *Aedes* (*Ochlerotatus*).

abrachys Dyar & Knab, 1909d (p. 262) – synonym of *Wyeomyia* (*Wyeomyia*) *scotinomus* (Dyar & Knab, 1907c); synonymy established before Knight & Stone, 1977.

abserratus (Felt & Young, 1904) (p. 312) – species of *Aedes* (*Ochlerotatus*) [original combination: *Culex abserratus*].
Synonyms: *centrotus* Howard, Dyar & Knab, 1917; *dysanor* Dyar, 1921e.

absobrinus Felt, 1904 (p. 318) – synonym of *Culiseta* (*Culiseta*) *impatiens* (Walker, 1848) [original combination: *Culex absobrinus*]; synonymy established before Knight & Stone, 1977.

abstrusa Peyton, 1977 (p. 35) – species of *Uranotaenia* (*Pseudoficalbia*).

abyssinicus Edwards, 1941 (p. 325) – subspecies of *Culex* (*Culex*) *andersoni* Edwards, 1914b.

abyssinicus Ribeiro, 1991 (p. 43) – subspecies of *Toxorhynchites* (*Toxorhynchites*) *brevipalpis* Theobald, 1901a.

abyssinicus van Someren, 1945 (p. 79) – synonym of *Culex* (*Oculeomyia*) *aurantapex ellinorae* Ovazza, Hamon & Neri, 1956 [original combination: *Culex* (*Culex*) *aurantapex abyssinicus*]; synonymy by Ovazza *et al.*, 1956.

acaci Baisas, 1946a (p. 30) – species of *Anopheles* (*Anopheles*).

acanthotorynus Komp, 1937a (p. 358) – species of *Anopheles* (*Stethomyia*).

acaudatus (Leicester, 1908) (p. 49) – species of *Toxorhynchites* (*Toxorhynchites*) [original combination: *Teromyia acaudata*].

accelerans Root, 1927c (p. 581) – species of *Culex* (*Aedinus*).

accraensis Theobald, 1909b (p. 19) – synonym of *Culex* (*Culex*) *thalassius* Theobald, 1903a [original combination: *Culicelsa accraensis*]; synonymy established before Knight & Stone, 1977.

acer Walker, 1848 (p. 8) – synonym of *Culex* (*Culex*) *quinquefasciatus* Say, 1823; synonymy established before Knight & Stone, 1977.

achaetae (Leicester, 1908) (p. 257) – species of *Heizmannia* (*Mattinglyia*) [original combination: *Dendromyia achaetae*].

acharistus Root, 1927c (p. 578) – species of *Culex* (*Culex*).

aconitus Dönitz, 1902 (p. 70) – species of *Anopheles* (*Cellia*).
Synonyms: *albirostris* Theobald, 1903a; *brahmacharii* Christophers, 1912.

acrophilus Dyar, 1917d (p. 127) – synonym of *Aedes* (*Ochlerotatus*) *pullatus* (Coquillett, 1904c); synonymy established before Knight & Stone, 1977.

acrostichalis Edwards, 1941 (p. 268) – species of *Culex* (*Eumelanomyia*).

aculeatus Colless, 1965 (p. 286) – species of *Culex* (*Lophoceraomyia*).

aculeatus (Theobald, 1903a) (p. 233) – species of *Aedes* (*Ochlerotatus*) [original combination: *Gilesia aculeata*].

acutipalus Colless, 1965 (p. 299) – species of *Culex* (*Lophoceraomyia*).

acutisentis Arnell, 1973 (p. 58) – species of *Haemagogus* (*Haemagogus*).

adairi Kirkpatrick, 1926 (p. 112) – species of *Culex* (*Lasiosiphon*) [*nomen novum* for *pluvialis* Kirkpatrick, 1925].
Synonyms: *pluvialis* Kirkpatrick, 1925; *kirkpatriki* Edwards, 1926c; *kirkpatricki* Stackelberg, 1927.

adamesi Sirivanakarn & Galindo, 1980 (p. 26) – species of *Culex* (*Melanoconion*).

adami Ferrara & Eouzan, 1974 (p. 261) – species of *Eretmapodites*.

adami Geoffroy, 1971b (p. 273) – species of *Aedes* (*Catageiomyia*).

adami (Hamon & Mouchet, 1955) (p. 860) – species of *Culex* (*Eumelanomyia*) [original combination: *Neoculex adami*].

adelae Baisas, 1938 (p. 200) – synonym of *Culex* (*Culex*) *annulus* Theobald, 1901a; synonymy by Colless, 1957a.

adelpha Dyar & Knab, 1906a (p. 140) – species of *Wyeomyia* (*Wyeomyia*).
Synonyms: *ablabes* Dyar & Knab, 1908; *rolonca* Dyar & Knab, 1910.

adenensis Christophers, 1924c (p. 296) – synonym of *Anopheles* (*Cellia*) *culicifacies* Giles, 1901 [original combination: *Anopheles culicifacies adenensis*]; synonymy by Harrison, 1980.

adenensis Edwards, 1941 (p. 198) – subspecies of *Aedes* (*Aedimorphus*) *hirsutus* (Theobald, 1901a) [original combination: *Aedes* (*Aedimorphus*) *hirsutus* variety *adenensis*]; subspecific status by White, 1975.

adentata van den Assem, 1959 (p. 48) – species of *Tripteroides* (*Rachisoura*).

adersi (Edwards, 1917) (p. 214) – species of *Aedes* (*Diceromyia*) [original combination: *Ochlerotatus* (*Diceromyia*) *adersi*].

adersianus Edwards, 1941 (p. 268) – species of *Culex* (*Eumelanomyia*).

adiei James & Liston, 1911 (p. 89) – synonym of *Anopheles* (*Cellia*) *annularis* van der Wulp, 1884 [original combination: *Nyssorhynchus fuliginosus* variety *adiei*]; synonymy by Christophers, 1933.

adolphoi Neiva, 1908b (p. 457) – synonym of *Anopheles* (*Kerteszia*) *cruzii* Dyar & Knab, 1908 [original combination: *Myzorhynchella adolphoi*]; synonymy by Barretto & Coutinho, 1951.

adusta (Laffoon, 1946) (p. 236) – species of *Verrallina* (*Neomacleaya*) [original combination: *Aedes* (*Aedes*) *adustus*].

advieri Senevet, 1938 (p. 185) – synonym of *Culex* (*Melanoconion*) *atratus* Theobald, 1901c; synonymy by Rozeboom & Komp, 1950a.

aedes Leicester, 1908 (p. 189) – synonym of *Culex* (*Eumelanomyia*) *malayi* (Leicester, 1908); synonymy established before Knight & Stone, 1977.

aegypti (Linnaeus, 1762) (p. 470) – species of *Aedes* (*Stegomyia*) [original combination: *Culex aegypti*].

Synonyms: *argenteus* Poiret, 1787; *fasciatus* Fabricius, 1805; *calopus* Meigen, 1818; *frater* Robineau-Desvoidy, 1827; *mosquito* Robineau-Desvoidy, 1827; *sugens* Wiedemann, 1828; *taeniatus* Wiedemann, 1828; *kounoupi* Brullé, 1833; *toxorhynchus* Macquart, 1838a; *annulitarsis* Macquart, 1846a; *excitans* Walker, 1848; *inexorabilis* Walker, 1848; *viridifrons* Walker, 1848; *exagitans* Walker, 1856a; *insatiabilis* Bigot, 1859; *bancrofti* Skuse, 1889; *elegans* Ficalbi, 1890c; *rossii* Giles, 1899; *luciensis* Theobald, 1901a; *nigeria* Theobald, 1901a; *queenslandensis* Theobald, 1901a; *lamberti* Ventrillon, 1904; *canariensis* Pittaluga, 1905; *persistans* Banks, 1906b; *albopalposus* Becker, 1908; *anguste-alatus* Becker, 1908; *alboannulis* Ludlow, 1911b; *pulcherrima* Taylor, 1919; *atritarsis* Edwards, 1920a.

aegypti Salem, 1938 (p. 1) – synonym of *Anopheles* (*Cellia*) *rhodesiensis rupicolus* Lewis, 1937; synonymy by Edwards, 1941.

aenea Thurman, 1959 (p. 39) – species of *Topomyia* (*Topomyia*).

aenescens Edwards, 1941 (pp. 323, 421) – subspecies of *Culex* (*Culex*) *trifilatus* Edwards, 1914b.

aeneus (Edwards, 1921c) (p. 79) – species of *Tripteroides* (*Tripteroides*) [original combination: *Rachionotomyia aenea*].

aeneus (Evans, 1926) (p. 102) – species of *Toxorhynchites* (*Afrorhynchus*) [original combination: *Megarhinus* (*Toxorhynchites*) *aeneus*].

aenigma Harbach, 2000 (in Harbach & Peyton, 2000) (p. 132) – species of *Onirion*.

aenigmaticus Cerqueira & Costa, 1946 (p. 133) – species of *Aedes* (*Ochlerotatus*).

aequatorialis Levi-Castillo, 1952 (p. 131) – synonym of *Wyeomyia* (*Dodecamyia*) *aphobema* Dyar, 1918e [original combination: *Wyeomyia* (*Wyeomyia*) *aphobema* variety *aequatorialis*]; synonymy by Lane, 1953.

aequatorianna Leví-Castillo, 1953b (p. 14) – species of *Uranotaenia* (*Uranotaenia*).

aequatorianna Leví-Castillo, 1954a (p. 249) – species of *Wyeomyia* (subgenus uncertain, determined by Motta & Lourenço-de-Oliveira, 1995).

aequatorianus Leví-Castillo, 1953f (p. 72) – *nomen dubium*, *Toxorhynchites*; status by Stone *et al.*, 1959.

aerarius McIntosh, 1975 (p. 281) – species of *Aedes* (*Aedimorphus*).

aeritinctus Galindo & Trapido, 1967 (p. 103) – species of *Haemagogus* (*Haemagogus*).

aestivalis Dyar, 1904b (p. 245) – synonym of *Aedes* (*Ochlerotatus*) *sticticus* (Meigen, 1838) [original combination: *Culex aestivalis*]; synonymy established before Knight & Stone, 1977.

aestivus Sirivanakarn, 1977a (p. 50) – species of *Culex* (*Lophoceraomyia*).

aestuans Wiedemann, 1828 (p. 11) – synonym of *Culex* (*Culex*) *quinquefasciatus* Say, 1823; synonymy by Stone, 1958a.

affinis Adams, 1903 (p. 25) – synonym of *Culex* (*Culex*) *thriambus* Dyar, 1921a; synonymy by Strickman, 1988.

affinis (Edwards, 1913c) (p. 241) – species of *Tripteroides* (*Rachionotomyia*) [*nomen novum* for *caeruleocephala* Theobald, 1910c; original combination: *Rachionotomyia affinis*].

Synonym: *caeruleocephala* Theobald, 1910c.

affinis Stephens, 1825 (p. 452) – synonym of *Culiseta* (*Culiseta*) *annulata* (Schrank, 1776) [original combination: *Culex affinis*]; synonymy established before Knight & Stone, 1977.

affinis Theobald, 1909b (p. 25) – synonym of *Aedes* (*Pseudarmigeres*) *argenteoventralis* (Theobald, 1909b) [original combination: *Dendromyia affinis*]; synonymy established before Knight & Stone, 1977.

affirmatus Dyar & Knab, 1906b (p. 164) – synonym of *Haemagogus* (*Haemagogus*) *equinus* Theobald, 1903g; synonymy established before Knight & Stone, 1977.

africana Neveu-Lemaire, 1906 (p. 273) – species of *Aedeomyia* (*Lepiothauma*).

africana Newstead, 1907 (in Newstead *et al.*, 1907) (p. 20) – synonym of *Aedes* (*Catageiomyia*) *tarsalis* (Newstead, 1907) [original combination: *Duttonia africana*]; synonymy established before Knight & Stone, 1977. **Note**: this nominal taxon is the junior secondary homonym of *africanus* (Theobald, 1901d), *africanus* Theobald, 1901d and *africanus* Neveu-Lemaire, 1906; the senior secondary homonym of *africanus* Theobald, 1909b and *africana* Theobald, 1911c.

africana Newstead, 1907 (in Newstead *et al*., 1907) (p. 28) – synonym of *Ficalbia nigra* (Theobald, 1901d) [original combination: *Mimomyia africana*]; synonymy established before Knight & Stone, 1977.

africana (Theobald, 1901d) (p. 12) – species of *Mansonia* (*Mansonioides*) [original combination: *Panoplites africanus*].

africana Theobald, 1911c (p. 151) – synonym of *Aedes* (*Diceromyia*) *furcifer* (Edwards, 1913b) [original combination: *Diceromyia africana*]; synonymy established before Knight & Stone, 1977. **Note**: this nominal taxon is the junior secondary homonym of *africanus* (Theobald, 1901d), *africanus* Theobald, 1901d, *africanus* Neveu-Lemaire, 1906, *africana* Newstead, 1907 and *africanus* Theobald, 1909b.

africanus Neveu-Lemaire, 1906 (p. 271) – synonym of *Aedes* (*Ochlerotatus*) *caspius* (Pallas, 1771) [original combination: *Taeniorhynchus africanus*]; synonymy established before Knight & Stone, 1977. **Note**: this nominal taxon is the junior secondary homonym of *africanus* (Theobald, 1901d) and *africanus* Theobald, 1901d; the senior secondary homonym of *africana* Newstead, 1907, *africanus* Theobald, 1909b and *africana* Theobald, 1911c.

africanus Roque, 1903 (p. 110) – *nomen dubium*, *Anopheles*; status by Stone *et al*., 1959.

africanus Theobald, 1901d (p. 1) – rejected name, suppression by International Commission on Zoological Nomenclature, 1968 [original combination: *Anopheles barbirostris* variety *africanus*].

africanus Theobald, 1901d (p. 2) – synonym of *Aedes* (*Mucidus*) *grahamii* (Theobald, 1909b) [original combination: *Mucidus africanus*]; synonymy established before Knight & Stone, 1977. **Note**: this nominal taxon has page priority over *africanus* (Theobald, 1901d) but is considered to be the junior secondary homonym of that nominal taxon.

africanus (Theobald, 1901d) (p. 3) – species of *Aedes* (*Stegomyia*) [original combination: *Stegomyia africana*]. **Note**: this nominal taxon is the senior secondary homonym of *africanus* Theobald, 1901d, *africanus* Neveu-Lemaire, 1906, *africana* Newstead, 1907, *africanus* Theobald, 1909b and *africana* Theobald, 1911c. Synonym: *dubia* Theobald, 1910c.

africanus Theobald, 1909b (p. 8) – synonym of *Aedes* (*Catageiomyia*) *stenoscutus* (Edwards, 1912c) [original combination: *Stenoscutus africanus*]; synonymy by White, 1975. **Note**: this nominal taxon is a junior secondary homonym of *africanus* (Theobald, 1901d), *africanus* Theobald, 1901d, *africanus* Neveu-Lemaire, 1906 and *africana* Newstead, 1907; the senior secondary homonym of *africana* Theobald, 1911c.

afridii Qutubuddin, 1956 (p. 140) – synonym of *Culex* (*Oculeomyia*) *infula* Theobald, 1901a; synonymy by Sirivanakarn, 1976.

agastyai Tewari & Hiriyan, 1992b (p. 154) – species of *Aedes* (*Tewarius*).

ager Giles, 1901c (p. 196) – synonym of *Culex* (*Oculeomyia*) *bitaeniorhynchus* Giles, 1901a [original combination: *Taeniorhynchus ager*]; synonymy by Edwards, 1913c.

agilis Bigot, 1889 (p. 122) – synonym of *Culex* (*Culex*) *pipiens* Linnaeus, 1758; synonymy established before Knight & Stone, 1977.

agitator Dyar & Knab, 1907b (p. 100) – synonym of *Culex* (*Melanoconion*) *pilosus* (Dyar & Knab, 1906d) [*nomen novum* for *cubensis* Dyar & Knab, 1906d; original combination: *Mochlostyrax agitator*]; synonymy established before Knight & Stone, 1977.

aglischrus Dyar, 1924d (p. 121) – synonym of *Culex* (*Culex*) *maracayensis* Evans, 1923b; synonymy established before Knight & Stone, 1977.

agnostips Dyar & Knab, 1907c (p. 211) – synonym of *Wyeomyia melanocephala* Dyar & Knab, 1906a; synonymy established before Knight & Stone, 1977.

agoggylia Dyar, 1922e (p. 115) – synonym of *Psorophora* (*Psorophora*) *holmbergii* Lynch Arribálzaga, 1891b; synonymy established before Knight & Stone, 1977.

agranensis Singh & Prakash, 2008 (p. 197) – species of *Lutzia* (*Metalutzia*).

agrestis (Barraud, 1931b) (p. 613) – species of *Verrallina* (*Neomacleaya*) [original combination: *Aedes* (*Aedes*) *agrestis*].

agrihanensis Bohart, 1957 (p. 39) – species of *Aedes* (*Stegomyia*).

aguilari Baisas & Feliciano, 1953 (p. 164) – species of *Zeugnomyia*.

agyrtes Dyar & Knab, 1909d (p. 265) – synonym of *Wyeomyia clasoleuca* Dyar & Knab, 1908; synonymy by Lane, 1951.

ahomi Chowdhury, 1929 (p. 986) – species of *Anopheles* (*Anopheles*) [original combination: *Anopheles barbirostris* variety *ahomi*].

aikenii (Aiken & Rowland, 1906) (p. 34) – *nomen dubium*, *Culex* [original combination: *Gnophodeonyia aikenii*]; status by Belkin, 1970b.

aikenii Dyar & Knab, 1908 (p. 61) – synonym of *Culex* (*Culex*) *quinquefasciatus* Say, 1823; synonymy established before Knight & Stone, 1977.

ainshamsi Gad, El Said, Hassan & Shoukry, 1987 (p. 214) – *nomen nudum*, *Anopheles*.

ainshamsi Gad, Harbach & Harrison, 2006 (p. 367) – species of *Anopheles* (*Cellia*).

airosai Lane & Cerqueira, 1942 (p. 593) – species of *Wyeomyia* (*Spilonympha*).

airozai Lane, 1945b (p. 204) – species of *Culex* (*Phenacomyia*).

aitkeni Schick, 1970a (p. 61) – species of *Aedes* (*Protomacleaya*).

aitkenii James, 1903 (in Theobald, 1903a) (p. 22) – species of *Anopheles* (*Anopheles*).

akkeshiensis Tanaka, 1998 (p. 215) – species of *Aedes* (*Ochlerotatus*).

akritos Forattini & Sallum, 1995 (p. 125) – species of *Culex* (*Melanoconion*).

alagoanii Peryassú, 1925 (p. 258) – synonym of *Anopheles* (*Anopheles*) *peryassui* Dyar & Knab, 1908; synonymy established before Knight & Stone, 1977.

alani Forattini, 1965 (p. 143) – species of *Culex* (*Culex*).
alani Lane & Cerqueira, 1957 (p. 244) – species of *Wyeomyia* (*Antunesmyia*) [*nomen novum* for *rooti* Lane & Cerqueira, 1942].
Synonym: *rooti* Lane & Cerqueira, 1942.
alaskaensis (Ludlow, 1906c) (p. 326) – species of *Culiseta* (*Culiseta*) [original combination: *Theobaldia alaskaensis*].
Synonyms: *arctica* Edwards, 1920a; *siberiensis* Ludlow, 1920b.
alba Baker, 1936 (p. 1) – species of *Orthopodomyia*.
alba Theobald, 1901c (p. 262) – species of *Uranotaenia* (*Uranotaenia*) [original combination: *Uranotaenia mashonaensis* variety *alba*]; specific status by Theobald, 1903a.
albertae Dyar, 1920e (p. 115) – synonym of *Aedes* (*Ochlerotatus*) *mercurator* Dyar, 1920a [original combination: *Aedes stimulans albertae*]; synonymy by Wood, 1977.
albertianus Edwards, 1941 (p. 265) – species of *Culex* (*Eumelanomyia*).
albertii Theobald, 1912 (in Schouteden, 1912) (p. 78) – synonym of *Aedes* (*Stegomyia*) *apicoargenteus* (Theobald, 1909b) [original combination: *Kingia albertii*]; synonymy established before Knight & Stone, 1977.
albertoi Anduze, 1943a (p. 193) – synonym of *Culex* (*Culex*) *ousqua* Dyar, 1918c; synonymy by Bram, 1967b.
albertoi Unti, 1941 (p. 9) – species of *Anopheles* (*Nyssorhynchus*) [original combination: *Anopheles* (*Nyssorhynchus*) *strodei* variety *albertoi*]; specific status by Sallum *et al.*, 2010.
albescens Edwards, 1921d (p. 308) – species of *Aedes* (*Rusticoidus*).
albescens Taylor, 1914c (p. 705) – species of *Uranotaenia* (*Uranotaenia*).
albicosta (Chagas, 1908) (in Peryassú, 1908) (p. 220) – species of *Coquillettidia* (*Rhynchotaenia*) [original combination: *Taeniorhynchus albicosta*].
albicosta (Edwards, 1913b) (p. 47) – species of *Aedes* (*Neomelaniconion*) [original combination: *Banksinella luteolateralis* variety *albicosta*]; specific status by ? Edwards, 1941.
albicosta (Lutz, 1904) (in Bourroul, 1904) (pp. 40, 59) – species of *Orthopodomyia* [original combination: *Bancroftia albicosta*].
albifasciatus (Macquart, 1838b) (p. 35) – species of *Aedes* (*Ochlerotatus*) [original combination: *Culex albifasciatus*].
Synonyms: *flavipes* Macquart, 1838b; *annuliferus* Blanchard, 1852; *vittatus* Philippi, 1865; *colonarius* Dyar, 1924g; *philippii* Dyar, 1924g.
albifera (Prado, 1931a) (p. 193) – species of *Coquillettidia* (*Rhynchotaenia*) [original combination: *Mansonia* (*Rhynchotaenia*) *albifera*].
albigenu Enderlein, 1920 (p. 50) – synonym of *Culex* (*Culex*) *ventrilloni* Edwards, 1920b; synonymy established before Knight & Stone, 1977.
albigenu (Lutz, 1908) (in Peryassú, 1908) (p. 155) – species of *Psorophora* (*Janthinosoma*) [original combination: *Janthinosoma albigenu*].
Synonyms: *paraguayensis* Strickland, 1911c; *bruchi* Petrocchi, 1927.
albilabris Edwards, 1925 (p. 258) – species of *Aedes* (*Rampamyia*).
albimanus da Cunha Ramos & Brunhes, 2004 (p. 91) – species of *Uranotaenia* (*Pseudoficalbia*).
albimanus Wiedemann, 1820 (p. 10) – species of *Anopheles* (*Nyssorhynchus*).
Synonyms: *cubensis* Agramonte, 1900; *albipes* Theobald, 1901a; *tarsimaculata* Goeldi, 1905; *gorgasi* Dyar & Knab, 1907c; *bisignatus* Hoffmann, 1938; *trisignatus* Hoffman, 1938.
albinensis Bonne-Wepster & Bonne, 1920 (p. 173) – species of *Culex* (*Melanoconion*).
Synonyms: *maroniensis* Bonne-Wepster & Bonne, 1920; *gordoni* Evans, 1924.
albinervis Edwards, 1929b (p. 338) – species of *Culex* (*Oculeomyia*).
albineus Séguy, 1923 (p. 206) – species of *Aedes* (*Ochlerotatus*).
Synonym: *flavidorsalis* Luh & Lee, 1975.
albino Stoker & Waktoedi Koesoemawinangoen, 1949 (p. 26) – synonym of *Anopheles* (*Cellia*) *vagus* Dönitz, 1902 [original combination: *Anopheles* (*Myzomyia*) *vagus albino*]; synonymy by Reid, 1968.
albinotata da Cunha Ramos & Brunhes, 2004 (p. 220) – *Uranotaenia* (*Pseudoficalbia*).
albionensis MacGregor, 1919 (p. 451) – synonym of *Orthopodomyia pulcripalpis* (Rondani, 1872); synonymy established before Knight & Stone, 1977.
albipes (Edwards, 1922a) (p. 287) – species of *Toxorhynchites* (*Toxorhynchites*) [original combination: *Megarhinus albipes*].
albipes (Giles, 1904a) (p. 366) – *nomen dubium*, *Orthopodomyia* [original combination: *Finlaya albipes*]; status by Knight & Stone, 1977.
albipes Leicester, 1904 (in Theobald, 1904c) (p. 237) – species of *Orthopodomyia*.
albipes Lutz, 1904 (in Bourroul, 1904) (p. 6) – species of *Culex* (*Microculex*).
albipes Peyton, 1977 (p. 64) – species of *Uranotaenia* (*Pseudoficalbia*).
albipes Theobald, 1901a (p. 125) – synonym of *Anopheles* (*Nyssorhynchus*) *albimanus* Wiedemann, 1820; synonymy established before Knight & Stone, 1977.
albipes (Theobald, 1907) (p. 157) – species of *Psorophora* (*Janthinosoma*) [original combination: *Janthinosoma albipes*].
albipes Theobald, 1910a (p. 11) – synonym of *Aedes* (*Stegomyia*) *desmotes* (Giles, 1904a) [original combination: *Stegomyia albipes*]; synonymy established before Knight & Stone, 1977.
albipes Theobald, 1911b (p. 243) – synonym of *Anopheles* (*Cellia*) *theileri* Edwards, 1912 [original combination: *Pyretophorus albipes*]; synonymy by Edwards, 1912d.

albiprivatus Lutz, 1904 (in Bourroul, 1904) (p. 67) – synonym of *Sabethes* (*Sabethes*) *albiprivus* Theobald, 1903a; synonymy established before Knight & Stone, 1977.

albiprivatus Theobald, 1907 (p. 30) – synonym of *Sabethes* (*Sabethes*) *albiprivus* Theobald, 1903a; synonymy established before Knight & Stone, 1977.

albiprivatus Theobald, 1907 (p. 620) – synonym of *Sabethes* (*Sabethinus*) *melanonymphe* Dyar, 1924b; synonymy by Lane & Cerqueira, 1942.

albiprivus Theobald, 1903a (p. 323) – species of *Sabethes* (*Sabethes*).
 Synonyms: *albiprivatus* Lutz, 1904; *albiprivatus* Theobald, 1907.

albiradius (Le Goff, Boussès & Brunhes, 2007) (p. 197) – species of *Aedes* (*Neomelaniconion*) [original combination: *Neomelaniconion albiradius*].

albirostris Macquart, 1850 (p. 10) – rejected name, suppression by International Commission on Zoological Nomenclature, 1972 [original combination: *Culex albirostris*].

albirostris Theobald, 1903a (p. 24) – synonym of *Anopheles* (*Cellia*) *aconitus* Dönitz, 1902 [original combination: *Myzomyia albirostris*]; synonymy established before Knight & Stone, 1977.

albitarsis (Brug, 1939) (p. 91) – subspecies of *Toxorhynchites* (*Toxorhynchites*) *inornatus* (Walker, 1865) [original combination: *Megarhinus inornatus* variety *albitarsis*]; subspecific status by Stone *et al.*, 1959.

albitarsis Gordon & Evans, 1922 (p. 335) – synonym of *Uranotaenia* (*Uranotaenia*) *calosomata* Dyar & Knab, 1907c [original combination: *Uranotaenia calosomata* variety *albitarsis*]; synonymy by Lane, 1953.

albitarsis Ludlow, 1905a (p. 131) – synonym of *Aedes* (*Stegomyia*) *desmotes* (Giles, 1904a) [original combination: *Anisocheleomyia albitarsis*]; synonymy established before Knight & Stone, 1977. **Note**: this nominal taxon is the senior secondary homonym of *albitarsis* Taylor, 1914d.

albitarsis Lynch Arribálzaga, 1878 (p. 150) – species of *Anopheles* (*Nyssorhynchus*).
 Synonyms: *limai* Galvão & Lane, 1937b; *imperfectus* Corrêa & Ramos, 1943.

albitarsis Taylor, 1914d (p. 194) – synonym of *Aedes* (*Patmarksia*) *novalbitarsis* King & Hoogstraal, 1946d [original combination: *Leucomyia albitarsis*]; synonymy established before Knight & Stone, 1977. **Note**: this nominal taxon is the junior secondary homonym of *albitarsis* Ludlow, 1905a.

albitarsis Theobald, 1901c (p. 25) – synonym of *Mimomyia* (*Mimomyia*) *plumosa* (Theobald, 1901a) [original combination: *Culex albitarsis*]; synonymy established before Knight & Stone, 1977.

albiventris Edwards, 1922d (p. 88) – species of *Culex* (*Eumelanomyia*).
 Synonym: *inconspicuosa* Theobald, 1909b.

alboabdominalis Theobald, 1910c (p. 508) – species of *Uranotaenia* (*Uranotaenia*).

alboannulata Taylor, 1912a (p. 31) – synonym of *Aedes* (*Macleaya*) *tremulus* (Theobald, 1903e) [*Danielsia alboannulata*]; synonymy established before Knight & Stone, 1977.

alboannulata (Theobald, 1905a) (p. 54) – species of *Uranotaenia* (*Uranotaenia*) [original combination: *Anisocheleomyia alboannulata*].

alboannulatus Giles, 1904c (p. 38) – *nomen nudum*, *Anopheles* [original combination: *Myzorhynchus alboannulatus*].

alboannulatus (Macquart, 1850) (p. 314) – species of *Aedes* (*Dobrotworskyius*) [original combination: *Culex albo-annulatus*].

alboannulatus Theobald, 1905b (p. 154) – synonym of *Aedes* (*Elpeytonius*) *apicoannulatus* (Edwards, 1912c) [original combination: *Aedimorphus alboannulatus*]; synonymy established before Knight & Stone, 1977.

alboannulis Ludlow, 1911b (p. 132) – synonym of *Aedes* (*Stegomyia*) *aegypti* (Linnaeus, 1762) [original combination: *Duttonia alboannulis*]; synonymy established before Knight & Stone, 1977.

alboapicalis Theobald, 1910c (p. 25) – synonym of *Anopheles* (*Cellia*) *minimus* Theobald, 1901a [original combination: *Myzomyia christophersi alboapicalis*]; synonymy established before Knight & Stone, 1977.

alboapicus Schick, 1970a (p. 50) – species of *Aedes* (*Protomacleaya*).

alboaurata Petrocchi, 1923 (p. 92) – *nomen nudum*, *Psorophora*.

albocaerulea Senevet & Abonnenc, 1939b (p. 259) – synonym of *Wyeomyia argenteorostris* (Bonne-Wepster & Bonne, 1920); synonymy by Lane, 1951.

albocephala da Cunha Ramos & Brunhes, 2004 (p. 334) – species of *Uranotaenia* (*Uranotaenia*).

albocephalus (Theobald, 1903c) (p. 4) – species of *Aedes* (*Polyleptiomyia*) (original combination: *Stegomyia albocephala*].
 Synonym: *seychellensis* Theobald, 1912b.

albocinctus (Barraud, 1924d) (p. 1002) – species of *Aedes* (*Hopkinsius*) [original combination: *Finlaya albocincta*].

albodorsalis Fontenille & Brunhes, 1985 (p. 151) – species of *Aedes* (*Aedimorphus*).

albofasciata Taylor, 1920 (p. 164) – synonym of *Uranotaenia* (*Uranotaenia*) *nivipes* (Theobald, 1905a); synonymy established before Knight & Stone, 1977.

albofimbriatus Giles, 1902 (p. 302) – synonym of *Anopheles* (*Cellia*) *pharoensis* Theobald, 1901a [original combination: *Anopheles pharoensis* variety *albofimbriatus*]; synonymy established before Knight & Stone, 1977.

albolateralis (Theobald, 1908b) (p. 289) – species of *Aedes* (*Downsiomyia*) [original combination: *Stegomyia albolateralis*].

albolineatus Giles, 1901a (p. 609) – synonym of *Culex* (*Culex*) *quinquefasciatus* Say, 1823; synonymy established before Knight & Stone, 1977.

albolineatus (Theobald, 1904c) (p. 77) – species of *Aedes* (*Scutomyia*) [original combination: *Scutomyia albolineata*].

albomaculatus Theobald, 1903a (p. 308) – species of *Haemagogus* (*Haemagogus*).

albomarginatus (Newstead, 1907) (in Newstead *et al.*, 1907) (p. 16) – species of *Aedes* (*Pseudarmigeres*) [original combination: *Stegomyia albomarginata*].

alboniveus Barraud, 1934 (p. 210) – species of *Aedes* (*Downsiomyia*).

albonotatus (Coquillett, 1906a) (p. 183) – species of *Aedes* (*Howardina*) [original combination: *Gymnometopa albonotata*].

albopalposus Becker, 1908 (p. 80) – synonym of *Aedes* (*Stegomyia*) *aegypti* (Linnaeus, 1762) [original combination: *Culex albopalposus*]; synonymy established before Knight & Stone, 1977.

albopictus (Skuse, 1895) (p. 20) – species of *Aedes* (*Stegomyia*) [original combination: *Culex albopictus*].

Synonyms: *samarensis* Ludlow, 1903; *nigritia* Ludlow, 1910; *quasinigritia* Ludlow, 1911b.

albopleura Theobald, 1907 (p. 456) – synonym of *Culex* (*Culiciomyia*) *pallidothorax* Theobald, 1905g; synonymy established before Knight & Stone, 1977.

albopunctata Theobald, 1907 (p. 262) – synonym of *Aedes* (*Fredwardsius*) *vittatus* (Bigot, 1861) [original combination: *Reedomyia albopunctata*]; synonymy established before Knight & Stone, 1977.

albo-punctatus Rondani, 1872 (p. 31) – synonym of *Aedes* (*Dahliana*) *geniculatus* (Olivier, 1791) [original combination: *Culex albo-punctatus*]; synonymy established before Knight & Stone, 1977.

alboscutellatus Lee, 1946 (p. 235) – species of *Tripteroides* (*Tripteroides*).

alboscutellatus (Theobald, 1905f) (p. 80) – species of *Aedes* (*Aedimorphus*) [original combination: *Lepidotomyia alboscutellata*].

Synonyms: *argentinotus* Banks, 1909; *omurensis* Yamada, 1921.

albosquamata Bonne-Wepster & Bonne, 1919 (p. 107) – species of *Wyeomyia* (subgenus uncertain, determined by Motta & Lourenço-de-Oliveira, 1995).

albosternopleura Peters, 1963e (p. 203) – species of *Uranotaenia* (*Uranotaenia*).

albotaeniatus (Leicester, 1904) (in Theobald, 1904c) (p. 111) – species of *Aedes* (*Danielsia*) [original combination: *Danielsia albotaeniata*]. **Note**: this nominal taxon is the senior secondary homonym of *albotaeniatus* Theobald, 1909b.

Synonym: *mikiranus* Edwards, 1922a.

albotaeniatus (Theobald, 1903a) (p. 88) – species of *Anopheles* (*Anopheles*) [original combination: *Myzorhynchus albotaeniatus*].

albotaeniatus Theobald, 1909b (p. 5) – synonym of *Aedes* (*Catageiomyia*) *irritans* (Theobald, 1901d) [original combination: *Aedimorphus albotaeniatus*]; synonymy established before Knight & Stone, 1977. **Note**: this nominal taxon is the junior secondary homonym of *albotaeniatus* (Leicester, 1904).

albothorax (Theobald, 1907) (p. 470) – species of *Aedes* (*Neomelaniconion*) [original combination: *Banksinella luteolateralis* variety *albothorax*]; specific status by ? Edwards, 1941.

alboventralis (Theobald, 1910c) (p. 251) – species of *Aedes* (*Aedimorphus*) [original combination: *Protomacleaya alboventralis*].

albovirgatus Graham, 1910 (p. 264) – synonym of *Culex* (*Culex*) *duttoni* Theobald, 1901d; synonymy established before Knight & Stone, 1977.

albus LaCasse & Yamaguti, 1950 (p. 136) – *nomen nudum*, *Aedes* [original combination: *Aedes* (*Ochlerotatus*) *dorsalis albus*].

albus Leicester, 1908 (p. 148) – synonym of *Culex* (*Culex*) *whitmorei* (Giles, 1904a); synonymy established before Knight & Stone, 1977.

albus Ventrillon, 1904 (in Laveran, 1904) (p. 211) – synonym of *Anopheles* (*Cellia*) *pharoensis* Theobald, 1901a; synonymy established before Knight & Stone, 1977.

alcasidi Huang, 1972 (p. 37) – species of *Aedes* (*Stegomyia*).

alcocki Bonne-Wepster & Bonne, 1920 (p. 171) – species of *Culex* (*Melanoconion*).

aldrichanus Bonne-Wepster & Bonne, 1920 (p. 179) – synonym of *Toxorhynchites* (*Lynchiella*) *bambusicola* (Lutz & Neiva, 1913) [original combination: *Megarhinus aldrichanus*]; synonymy established before Knight & Stone, 1977.

aldrichi Dyar & Knab, 1908 (p. 57) – synonym of *Aedes* (*Ochlerotatus*) *sticticus* (Meigen, 1838); synonymy established before Knight & Stone, 1977.

alektorovi Stackelberg, 1943 (p. 311) – species of *Aedes* (*Bruceharrisonius*); see *kobayashii* Nakata, 1956.

alexandraeschingarevi Shingarev, 1928 (p. 48) – synonym of *Anopheles* (*Anopheles*) *maculipennis* Meigen, 1818; synonymy by Zhelochovtsev, 1937.

alexandrei Gornostaeva, 2005 (p. 485) – species of *Aedes* (*Aedes*).

alfaroi Dyar, 1921b (p. 34) – synonym of *Culex* (*Melanoconion*) *bastagarius* Dyar & Knab, 1906b; synonymy by Komp, 1935.

algeriensis Theobald, 1903a (p. 21) – species of *Anopheles* (*Anopheles*).

Synonym: *lukisii* Christophers, 1916b.

aliciae Duret, 1953 (p. 75) – species of *Culex* (*Melanoconion*).

alienus Colless, 1957a (p. 97) – species of *Culex* (*Culex*).

alineata Schneider, 1913 (p. 37) – synonym of *Aedes* (*Ochlerotatus*) *punctor* (Kirby, 1837) [original combination: *Culicada nemorosa* form *alineata*]; synonymy established before Knight & Stone, 1977.

alinkios Sallum & Hutchings, 2003 (p. 618) – species of *Culex* (*Melanoconion*).

alis Theobald, 1903a (p. 167) – species of *Culex* (*Culex*).

Synonym: *neolitoralis* Bram, 1967b.

alius Lien, 1968d (p. 221) – species of *Aedes* (*Gilesius*) [original combination: *Aedes* (*Finlaya*) *pulchriventer alius*]; specific status by Meng, 1981.

aliyusopi Miyagi & Toma, 2014 (in Miyagi *et al.*, 2014) (p. 28) – species of *Topomyia* (*Topomyia*).

alkatirii Toma, Miyagi & Syafruddin, 1995 (p. 110) – species of *Armigeres* (*Armigeres*).

alleni Turner, 1924 (p. 84) – synonym of *Aedes* (*Protomacleaya*) *zoosophus* Dyar & Knab, 1918b; synonymy by Gjullin, 1946.

allophus (Lutz & Peryassú, 1921) (in Peryassú, 1921a) (p. 70) – *nomen dubium*, *Anopheles* [original combination: *Cellia allopha*]; status by Linthicum, 1988 (inappropriately designated *nomen nudum* by Lourenço-de-Oliveira & Deane, 1984).

allostigma Howard, Dyar & Knab, 1915 (p. 471) – species of *Lutzia* (*Lutzia*).

allotecnon Kumm, Komp & Ruiz, 1940 (p. 417) – species of *Aedes* (*Howardina*).

alocasicola Marks, 1947a (p. 35) – species of *Aedes* (*Finlaya*).

alogistus Dyar, 1918e (p. 126) – species of *Culex* (*Melanoconion*).

Synonym: *megapus* Root, 1927c.

alongensis Venhuis, 1940 (p. 173) – species of *Anopheles* (*Anopheles*).

Synonym: *tonkinensis* Galliard & Ngu, 1940.

alongi Galliard & Ngu, 1947 (as *halongi*) (p. 77) – species of *Aedes* (*Bothaella*); emendation by Galliard & Ngu, 1950.

aloponotum Dyar, 1917b (p. 98) – species of *Aedes* (*Ochlerotatus*).

alorensis Bonne-Wepster & Brug, 1932 (p. 92) – species of *Aedes* (*Stegomyia*) [original combination: *Aedes* (*Stegomyia*) *variegatus* variety *alorensis*]; specific status by Stone & Farner, 1945.

alorensis Sirivanakarn, 1977a (p. 25) – species of *Culex* (*Lophoceraomyia*).

alpha Séguy, 1924 (p. 41) – synonym of *Culex* (*Culex*) *theileri* Theobald, 1903a; synonymy by Harbach, 1988.

alphus Colless, 1965 (p. 283) – species of *Culex* (*Lophoceraomyia*).

alternans (Westwood, 1835) (p. 681) – species of *Aedes* (*Mucidus*) [original combination: *Culex alternans*].

Synonyms: *commovens* Walker, 1856a; *hispidosus* Skuse, 1889; *kermorganti* Laveran, 1901a.

alticola Bonne-Wepster, 1948 (p. 313) – species of *Aedes* (subgenus uncertain).

alticola Galindo, Trapido & Boshell-Manrique, 1956 (in Galindo & Trapido, 1956) (p. 231) – synonym of *Haemagogus* (*Haemagogus*) *mesodentatus* Komp & Kumm, 1938 [original combination: *Haemagogus mesodentatus alticola*]; synonymy by Arnell, 1973.

alticola Martini, 1931a (p. 216) – synonym of *Culex* (*Culex*) *apicinus* Philippi, 1865; synonymy established before Knight & Stone, 1977.

alticola Peters, 1963d (p. 142) – subspecies of *Uranotaenia* (*Uranotaenia*) *novaguinensis* Peters, 1963d.

altiusculus Dyar, 1917b (p. 100) – synonym of *Aedes* (*Ochlerotatus*) *communis* (de Geer, 1776) [original combination: *Culex altiusculus*]; synonymy established before Knight & Stone, 1977.

altivallis Bonne-Wepster, 1948 (p. 311) – species of *Tripteroides* (*Polylepidomyia*).

alvarezi Sutil Oramas, Pulido Florenzano & Amarista Meneses, 1987 (p. 83) – synonym of *Culex* (*Melanoconion*) *spissipes* (Theobald, 1903a); synonymy by Sallum & Forattini, 1996.

amabilis Schick, 1970a (p. 68) – species of *Aedes* (*Protomacleaya*).

amaltheus de Meillon & Lavoipierre, 1944 (p. 52) – species of *Aedes* (*Stegomyia*).

amamiensis Tanaka, Mizusawa & Saugstad, 1979 (p. 312) – subspecies of *Aedes* (*Hulecoeteomyia*) *japonicus* (Theobald, 1901a).

amaniensis van Someren & Hamon, 1964 (p. 78) – species of *Culex* (*Eumelanomyia*).

amaurus Martini, 1929 (p. 135) – synonym of *Anopheles* (*Anopheles*) *claviger* (Meigen, 1804); synonymy by Stone *et al.*, 1959.

amazonensis (Lutz, 1905) (p. 103) – species of *Culex* (*Aedinus*) [original combination: *Aedinus amazonensis*].

Synonyms: *tapena* Dyar, 1919c; *paraplesia* Dyar, 1922g; *hildebrandi* Evans, 1923c.

amazonensis (Theobald, 1901c) (p. 182) – species of *Mansonia* (*Mansonia*) [original combination: *Panoplites amazonensis*].

amazonica Cerqueira, 1960b (p. 1) – species of *Psorophora* (*Janthinosoma*).

amazonica Leví-Castillo, 1954a (p. 247) – species of *Wyeomyia* (subgenus uncertain, determined by Motta & Lourenço-de-Oliveira, 1995).

amazonicus Christophers, 1923b (p. 71) – synonym of *Anopheles* (*Anopheles*) *mattogrossensis* Lutz & Neiva, 1911; synonymy by Shannon, 1933b.

amazonicus Gordon & Evans, 1922 (p. 316) – species of *Sabethes* (*Sabethes*).

Synonyms: *kappleri* Bonne, 1923; *longfieldae* Edwards, 1928a.

ambiguus Dyar & Knab, 1906c (p. 246) – synonym of *Toxorhynchites* (*Lynchiella*) *theobaldi* (Dyar & Knab, 1906c) [*nomen novum* for *ferox* Wiedemann, 1928 (in part)]; synonymy by Lane, 1953.

ambiguus Theobald, 1903a (p. 248) – synonym of *Culex* (*Oculeomyia*) *infula* Theobald, 1901a [original combination: *Grabhamia ambiguus* [*sic*]]; synonymy by Sirivanakarn, 1976.

amboannulatus Theobald, 1913a (p. 320) – synonym of *Culex* (*Maillotia*) *salisburiensis* Theobald, 1901c [original combination: *Culex salisburiensis* variety *amboannulatus*]; synonymy established before Knight & Stone, 1977.

ambodimanga da Cunha Ramos & Brunhes, 2004 (p. 94) – species of *Uranotaenia* (*Pseudoficalbia*).

amboinensis (Doleschall, 1857) (p. 381) – species of *Toxorhynchites* (*Toxorhynchites*) [original combination: *Culex amboinensis*].

Synonyms: *subulifer* Doleschall, 1857; *immisericors* Walker, 1859; *lewaldii* Ludlow, 1904b, *argenteotarsis* Ludlow, 1906d; *grata* Banks, 1906a.

ambreensis Rodhain & Boutonnier, 1983a (p. 826) – species of *Aedes* (*Ochlerotatus*).

ambremontis (Brunhes & Boussès, 2017) (in Brunhes *et al.*, 2017) (p. 356) – species of *Aedes* (*Paulianius*) [original combination: *Paulianius ambremontis*].

ambremontis Brunhes & Hervy, 1995 (p. 21) – species of *Orthopodomyia*.

ameliae Casal, 1967 (p. 29) – species of *Culex* (*Culex*).

americanus (Neveu-Lemaire, 1902b) (p. 23) – *nomen dubium*, *Culex* [original combination: *Aedeomyia americana*].

amesii (Ludlow, 1903) (p. 139) – species of *Aedes* (*Lorrainea*) [original combination: *Stegomyia amesii*].
Synonyms: *fusca* Leicester, 1908; *furvus* Edwards, 1928a.

ambaricus Hunt, Wilkerson & Coetzee, 2013 (in Coetzee *et al.*, 2013) (p. 264) – species of *Anopheles* (*Cellia*).

amictus Edwards, 1921c (p. 71) – species of *Anopheles* (*Cellia*).

amiensis Peters, 1963e (p. 207) – species of *Uranotaenia* (*Uranotaenia*).

amitis Komp, 1936c (p. 333) – species of *Culex* (*Melanoconion*).

ampyx Dyar, 1920i (p. 176) – synonym of *Hodgesia malayi* Leicester, 1908; synonymy established before Knight & Stone, 1977.

amurensis Maslov, 1964 (p. 210) – species of *Culiseta* (*Culicella*) [original combination: *Culiseta silvestris amurensis*].

amutis de Burca, 1943 (p. 233) – synonym of *Anopheles* (*Cellia*) *turkhudi* Liston, 1901; synonymy by Gillies & de Meillon, 1968 (in part).

ananae Knight & Laffoon, 1946 (p. 218) – species of *Aedes* (*Finlaya*).

anarmostus Theobald, 1903a (p. 170) – synonym of *Culex* (*Culex*) *duttoni* Theobald, 1901d; synonymy established before Knight & Stone, 1977.

anastasionis Dyar, 1921k (p. 155) – species of *Haemagogus* (*Haemagogus*).
Synonym: *dominguezi* Duret, 1971b.

anchietai Corrêa & Ramalho, 1968 (p. 115) – species of *Anopheles* (*Anopheles*).

andamanensis Barraud, 1934 (p. 102) – species of *Orthopodomyia* [original combination: *Orthopodomyia anopheloides* variety *andamanensis*].

andamanensis (Edwards, 1922a) (p. 272) – species of *Verrallina* (*Neomacleaya*) [original combination: *Aedes* (*Aedes*) *andamanensis*].
Synonym: *simplex* Theobald, 1908b.

andavakae Doucet, 1950 (p. 44) – species of *Uranotaenia* (*Uranotaenia*).

andersoni Edwards, 1914b (p. 65) – species of *Culex* (*Culex*).

andersoni Edwards, 1926a (p. 112) – species of *Aedes* (*Ochlerotatus*).
Synonyms: *tasmaniensis* Strickland, 1911b; *tasmaniensis* Taylor, 1914b.

andinum Leví-Castillo, 1953c (as *andinus*) (p. 63) – species of *Trichoprosopon*.

andinus Leví-Castillo, 1954a (p. 252) – species of *Limatus*.

andinus Osorno-Mesa, 1944 (p. 170) – species of *Haemagogus* (*Haemagogus*).

andreae Doucet, 1962 (p. 1157) – species of *Uranotaenia* (*Pseudoficalbia*).

andreae Harbach, 2018 (p. 270) – species of *Sabethes* (*Sabethinus*).

andreanus Edwards, 1927a (p. 354) – species of *Culex* (*Eumelanomyia*).

andrewsi Edwards, 1926a (p. 103) – species of *Aedes* (*Stegomyia*) [original combination: *Aedes variegatus* variety *andrewsi*]; specific status by Stone & Farner, 1945.

andricus Root, 1927c (p. 592) – species of *Culex* (*Melanoconion*).

andropus Dyar & Knab, 1908 (p. 68) – synonym of *Wyeomyia* (*Hystatomyia*) *circumcincta* Dyar & Knab, 1907c; synonymy established before Knight & Stone, 1977.

anduzei Cerqueira & Lane, 1944 (p. 219) – species of *Culex* (*Carrollia*).

aneles Dyar & Ludlow, 1922 (p. 63) – synonym of *Culex* (*Melanoconion*) *educator* Dyar & Knab, 1906d; synonymy established before Knight & Stone, 1977.

anggiensis Bonne-Wepster, 1937 (p. 97) – species of *Aedes* (*Patmarksia*).

angkoris Klein, 1977 (p. 123) – species of *Topomyia* (*Topomyia*).

angolae Ribeiro & Ramos, 1974 (p. 94) – subspecies of *Aedes* (*Aedimorphus*) *durbanensis* (Theobald, 1903a).

angolensis da Cunha Ramos, 1985 (p. 74) – species of *Uranotaenia* (*Uranotaenia*).

angolensis da Cunha Ramos & Ribeiro, 1992 (p. 32) – species of *Eretmapodites*.

angolensis Ribeiro, 1992c (p. 188) – species of *Toxorhynchites* (*Afrorhynchus*).

angulatus Theobald, 1901c (p. 324) – synonym of *Culex* (*Culiciomyia*) *viridiventer* Giles, 1901a; synonymy established before Knight & Stone, 1977.

anguste-alatus Becker, 1908 (p. 79) – synonym of *Aedes* (*Stegomyia*) *aegypti* (Linnaeus, 1762) [original combination: *Culex anguste-alatus*]; synonymy established before Knight & Stone, 1977.

angustiplatus Evenhuis & Steffan, 1986 (p. 548) – species of *Toxorhynchites* (*Toxorhynchites*).

angustivittatus Dyar & Knab, 1907a (p. 9) – species of *Aedes* (*Ochlerotatus*).
Synonyms: *argentescens* Dyar & Knab, 1908; *cuneatus* Dyar & Knab, 1908; *traversus* Dyar, 1925g.

angustus Edwards, 1935a (p. 135) – species of *Aedes* (*Stegomyia*).

anhydor Dyar, 1907 (p. 128) – species of *Uranotaenia* (*Pseudoficalbia*).

aningae Motta & Lourenço-de-Oliveira, 2005 (p. 847) – species of *Wyeomyia* (*Spilonympha*).

anips Dyar, 1916a (p. 48) – species of *Culex* (*Melanoconion*).

anisochloros Theobald, 1903c (pl. VI, fig. 4) – synonym of *Anopheles* (*Cellia*) *funestus* Giles, 1900b [original combination: *Anopheles funestus* variety *anisochloros*]; synonymy by Evans, 1938.

ankaratrensis Brunhes & Hervy, 1995 (p. 27) – species of *Orthopodomyia*.

annandalei Barraud, 1926 (p. 343) – species of *Uranotaenia* (*Uranotaenia*).
Synonym: *nanseica* Bohart & Ingram, 1946b.

annandalei Prashad, 1918 (p. 123) – species of *Anopheles* (*Anopheles*).
Synonym: *djajasanensis* Brug, 1926.

annandalei (Theobald, 1910a) (p. 10) – species of *Aedes* (*Stegomyia*) [original combination: *Stegomyia annandalei*].
Synonym: *horishensis* Yamada, 1921.

annettii (Theobald, 1901d) (p. 13) – species of *Coquillettidia* (*Coquillettidia*) [original combination: *Taeniorhynchus annettii*].

annularis van der Wulp, 1884 (p. 249) – species of *Anopheles* (*Cellia*).
Synonyms: *fuliginosus* Giles, 1900a; *jamesii* Liston, 1901; *leucopus* Dönitz, 1901; *lineata* Ludlow, 1908a; *adiei* James & Liston, 1911; *nagpori* James & Liston, 1911.

annulata Leicester, 1908 (p. 174) – species of *Mansonia* (*Mansonioides*) [generic name misspelled as *Masonia*].

annulata (Schrank, 1776) (p. 97) – species of *Culiseta* (*Culiseta*) [original combination: *Culex annulatus*]. **Note**: this nominal taxon is the senior synonym of *annulatus* de Fourcroy, 1785 and *annulatus* Fabricius, 1787.
Synonyms: *annulatus* de Fourcroy, 1785; *annulatus* Fabricius, 1787; *affinis* Stephens, 1825; *nicaensis* Leach, 1825; *ferruginata* Martini, 1924.

annulata Taylor, 1914c (p. 689) – synonym of *Culex* (*Culex*) *sitiens* Wiedemann, 1828 [original combination: *Culicada annulata*]; synonymy established before Knight & Stone, 1977. **Note**: this nominal taxon is the junior secondary homonym of *annulata* Theobald, 1905f, *annulata* Theobald, 1907, *annulata* Theobald, 1913a and *annulata* Taylor, 1914c (based on page priority); the senior secondary homonym of *annulata* Taylor, 1916.

annulata Taylor, 1914c (p. 695) – synonym of *Culex* (*Culex*) *vicinus* (Taylor, 1916) [original combination: *Leucomyia annulata*]; synonymy established before Knight & Stone, 1977. **Note**: this nominal taxon is the junior secondary homonym of *annulata* Theobald, 1905f, *annulata* Theobald, 1907 and *annulata* Theobald, 1913a; the senior secondary homonym of *annulata* Taylor, 1914c (based on page priority) and *annulata* Taylor, 1916.

annulata Taylor, 1916 (p. 571) – synonym of *Culex* (*Lophoceraomyia*) *cubiculi* Marks, 1989 [original combination: *Lophoceratomyia annulata*]; synonymy by Lee *et al.*, 1989b. **Note**: this nominal taxon is the junior secondary homonym of *annulata* Theobald, 1905f, *annulata* Theobald, 1907, *annulata* Theobald, 1913a, *annulata* Taylor, 1914c (p. 689) and *annulata* Taylor, 1914c (p. 695).

annulata Theobald, 1901d (p. 14) – species of *Uranotaenia* (*Pseudoficalbia*).

annulata Theobald, 1905f (p. 98) – synonym of *Culex* (*Oculeomyia*) *squamosus* (Taylor, 1914c) [original combination: *Trichopronomyia annulata*]; synonymy established before Knight & Stone, 1977. **Note**: this nominal taxon is the senior secondary homonym of *annulata* Theobald, 1907, *annulata* Theobald, 1913a, *annulata* Taylor, 1914c (p. 689), *annulata* Taylor, 1914c (p. 695) and *annulata* Taylor, 1916.

annulata Theobald, 1907 (p. 230) – synonym of *Culex* (*Culiciomyia*) *nigropunctatus* Edwards, 1926a [original combination: *Culiciomyia annulata*]; synonymy established before Knight & Stone, 1977. **Note**: this nominal taxon is the junior secondary homonym of *annulata* Theobald, 1905f; the senior secondary homonym of *annulata* Theobald, 1913a, *annulata* Taylor, 1914c (p. 689), *annulata* Taylor, 1914c (p. 695) and *annulata* Taylor, 1916.

annulata Theobald, 1913a (p. 321) – synonym of *Culex* (*Culex*) *theileri* Theobald, 1903a [original combination: *Culex theileri* variety *annulata*]; synonymy established before Knight & Stone, 1977. **Note**: this nominal taxon is the junior secondary homonym of *annulata* Theobald, 1905f and *annulata* Theobald, 1907; the senior secondary homonym of *annulata* Taylor, 1914c (p. 689), *annulata* Taylor, 1914c (p. 695) and *annulata* Taylor, 1916.

annulatus de Fourcroy, 1785 (p. 516) – synonym of *Culiseta* (*Culiseta*) *annulata* (Schrank, 1776) [original combination: *Culex annulatus*]; synonymy established before Knight & Stone, 1977. **Note**: this nominal taxon is the junior synonym of *annulata* (Schrank, 1776); the senior synonym of *annulatus* Fabricius, 1787.

annulatus Fabricius, 1787 (p. 363) – synonym of *Culiseta* (*Culiseta*) *annulata* (Schrank, 1776) [original combination: *Culex annulatus*]; synonymy established before Knight & Stone, 1977. **Note**: this nominal taxon is the junior synonym of *annulata* (Schrank, 1776) and *annulatus* de Fourcroy, 1785.

annulatus Haga, 1930 (p. 380) – species of *Anopheles* (*Cellia*) [original combination: *Anopheles longirostris* variety *annulatus*].

annulifemur Edwards, 1914b (p. 77) – synonym of *Aedes* (*Dendroskusea*) *periskelatus* (Giles, 1902) [original combination: *Ochlerotatus annulifemur*]; synonymy established before Knight & Stone, 1977.

annulifera Ludlow, 1903 (p. 141) – synonym of *Aedes* (*Ochlerotatus*) *vigilax ludlowae* (Blanchard, 1905) [original combination: *Culex annulifera* [*sic*]]. **Note**: this nominal taxon is the junior primary homonym of *annuliferus* Blanchard, 1852.

annulifera (Theobald, 1901c) (p. 183) – species of *Mansonia* (*Mansonioides*) [original combination: *Panoplites annulifera*].
Synonyms: *seguini* Laveran, 1901b; *septemguttata* Theobald, 1907.

annuliferus Blanchard, 1852 (p. 333) – synonym of *Aedes* (*Ochlerotatus*) *albifasciatus* (Macquart, 1838b) [original combination: *Culex annuliferus*]; synonymy by Belkin, 1968b. **Note**: this nominal taxon is the senior primary homonym of *annulifera* Ludlow, 1903.

annulimanus van der Wulp, 1867 (p. 129) – synonym of *Anopheles* (*Anopheles*) *quadrimaculatus* Say, 1824; synonymy established before Knight & Stone, 1977.

annulioris Theobald, 1901a (p. 371) – species of *Culex* (*Oculeomyia*).

Synonym: *gambiensis* Theobald, 1903c.

annulipalpis Lynch Arribálzaga, 1878 (p. 149) – species of *Anopheles* (*Anopheles*).

annulipalpis (Theobald, 1910a) (p. 6) – species of *Armigeres* (*Leicesteria*) [original combination: *Brevirhynchus annulipalpis*].

annulipes (Meigen, 1830b) (p. 241) – species of *Aedes* (*Ochlerotatus*) [original combination: *Culex annulipes*]. **Note**: this nominal taxon is the senior secondary homonym of *annulipes* Taylor, 1914c.

Synonym: *quartus* Martini, 1920a.

annulipes Taylor, 1914c (p. 693) – synonym of *Aedes* (*Ochlerotatus*) *camptorhynchus* (Thomson, 1869) [original combination: *Culicada annulipes*]; synonymy established before Knight & Stone, 1977. **Note**: this nominal taxon is the junior secondary homonym of *annulipes* (Meigen, 1830b).

annulipes Theobald, 1907 (p. 512) – synonym of *Culex* (*Melanoconion*) *taeniopus* Dyar & Knab, 1907b [original combination: *Melanoconion annulipes*]; synonymy by Belkin, 1969.

annulipes Walker, 1856a (p. 433) – species of *Anopheles* (*Cellia*).

Synonyms: *mastersi* Skuse, 1889; *musivus* Skuse, 1889; *perplexus* Taylor, 1943; *derricki* Taylor, 1943; *persimilis* Taylor, 1943.

annulipes Walker, 1856b (p. 6) – synonym of *Mansonia* (*Mansonioides*) *dives* (Schiner, 1868) [original combination: *Culex annulipes*]; synonymy established before Knight & Stone, 1977.

annulirostris Skuse, 1889 (p. 1737) – species of *Culex* (*Culex*). **Note**: this nominal taxon is the senior secondary homonym of *annulirostris* Taylor, 1914c.

Synonyms: *jepsoni* Bahr, 1912; *somerseti* Taylor, 1912b; *consimilis* Taylor, 1913; *simplex* Taylor, 1914c; *palmi* Baisas, 1938.

annulirostris Taylor, 1914c (p. 696) – synonym of *Culex* (*Oculeomyia*) *squamosus* (Taylor, 1914c) [original combination: *Leucomyia annulirostris*]; synonymy established before Knight & Stone, 1977. **Note**: this nominal taxon is the junior secondary homonym of *annulirostris* Skuse, 1889.

annulirostris (Theobald, 1905h) (p. 239) – species of *Aedes* (*Christophersiomyia*) [original combination: *Stegomyia annulirostris*].

annulitarsis (Leicester, 1908) (p. 99) – species of *Armigeres* (*Leicesteria*) [original combination: *Leicesteria annulitarsis*].

annulitarsis Macquart, 1846a (p. 136) – synonym of *Aedes* (*Stegomyia*) *aegypti* (Linnaeus, 1762) [original combination: *Culex annulitarsis*]; synonymy by Mattingly & Bruce-Chwatt, 1954.

annuliventris (Blanchard, 1852) (p. 334) – species of *Culex* (*Culex*) [original combination: *Anopheles annuliventris*].

annuloabdominalis Theobald, 1910c (p. 236) – synonym of *Culex* (*Culiciomyia*) *pallidothorax* Theobald, 1905g [original combination: *Culiciomyia annuloabdominalis*]; synonymy established before Knight & Stone, 1977.

annulus Theobald, 1901a (p. 358) – species of *Culex* (*Culex*).

Synonyms: *pseudoinfula* Theobald, 1911d; *adelae* Baisas, 1938; *permixtus* Hsieh & Liao, 1956; *cheni* Ho, 1963.

anomalophyllus Komp, 1936d (p. 163) – species of *Anopheles* (*Nyssorhynchus*).

anopheloides Brunhes & Razafindrasolo, 1975 (p. 223) – species of *Uranotaenia* (*Uranotaenia*).

anopheloides (Giles, 1903) (in Thomson, 1903) (p. 315) – species of *Orthopodomyia* [original combination: *Mansonia anopheloides*].

Synonyms: *nigritarsis* Leicester, 1908; *maculata* Theobald, 1910a; *maculipes* Theobald, 1910c; *manganus* Baisas, 1946a; *nipponica* LaCasse & Yamaguti, 1948; *lemmonae* Thurman, 1959.

anoplicitus Forattini & Sallum, 1989a (p. 1) – species of *Culex* (*Melanoconion*).

anoplus Komp, 1937b (p. 514) – synonym of *Anopheles* (*Kerteszia*) *homunculus* Komp, 1937b; synonymy by Lane, 1953.

antalahaensis da Cunha Ramos & Brunhes, 2004 (p. 291) – species of *Uranotaenia* (*Pseudoficalbia*).

antanosyorum Rodhain & Boutonnier, 1984 (p. 220) – synonym of *Orthopodomyia vernoni* van Someren, 1949; synonymy by Brunhes & Hervy, 1995.

antennalis Bohart & Farner, 1944 (p. 70) – species of *Tripteroides* (*Tripteroides*).

antennalis Taylor, 1919 (p. 840) – species of *Uranotaenia* (*Uranotaenia*).

antennatus (Becker, 1903) (p. 68) – species of *Culex* (*Culex*) [original combination: *Anopheles antennatus*].

Synonym: *laurenti* Newstead, 1907.

anthropophagus Xu & Feng, 1975 (p. 81) – synonym of *Anopheles* (*Anopheles*) *lesteri* Baisas & Hu, 1936 [original combination: *Anopheles lesteri anthropophagus*]; synonymy by Wilkerson *et al.*, 2003.

antiguae Giles, 1904b (p. 384) – synonym of *Psorophora* (*Grabhamia*) *pygmaea* (Theobald, 1903a) [original combination: *Taeniorhynchus antiguae*]; synonymy established before Knight & Stone, 1977.

antillarum Floch & Abonnenc, 1945a (p. 4) – synonym of *Wyeomyia* (*Wyeomyia*) *grayii* Theobald, 1901c; synonymy by Belkin, 1968b.

antillummagnorum Dyar, 1928 (p. 344) – species of *Culex* (*Micraedes*).

antipodea Dobrotworsky, 1962 (p. 291) – species of *Culiseta* (*Climacura*).

antipodeus (Edwards, 1920a) (p. 132) – species of *Aedes* (*Ochlerotatus*) [original combination: *Ochlerotatus antipodeus*].

antiquus Borkent & Grimaldi, 2004 (p. 883) – fossil species of *Burmaculex* Borkent & Grimaldi, 2004.

antoinetta Dyar & Knab, 1909d (p. 263) – synonym of *Wyeomyia* (*Wyeomyia*) *mitchellii* (Theobald, 1905d); synonymy established before Knight & Stone, 1977.

antuensis Su, Wang & Li, 1978 (p. 439) – synonym of *Aedes* (*Edwardsaedes*) *pingpaensis* Chang, 1965; synonymy by Danilov, 1985.

antunesi Galvão & Franco do Amaral, 1940 (p. 150) – species of *Anopheles* (*Nyssorhynchus*).

antunesi Lane & Guimarães, 1937 (p. 213) – species of *Wyeomyia* (*Phoniomyia*).

antunesi Lane & Whitman, 1943 (p. 389) – species of *Culex* (*Carrollia*).
Synonym: *manaensis* Floch & Fauran, 1955.

anxifer Bigot, 1859 (p. 117) – synonym of *Culex* (*Culex*) *quinquefasciatus* Say, 1823; synonymy established before Knight & Stone, 1977.

aobae Belkin, 1962 (p. 452) – species of *Aedes* (*Stegomyia*).

apeteticus Howard, Dyar & Knab, 1913, 1915 (as *apateticus*) (fig. 91, p. 321, respectively) – synonym of *Culex* (*Melanoconion*) *educator* Dyar & Knab, 1906d; synonymy established before Knight & Stone, 1977.

aphobema Dyar, 1918e (p. 120) – species of *Wyeomyia* (*Dodecamyia*).
Synonyms: *bodkini* Edwards, 1922d; *aequatorialis* Levi-Castillo, 1952.

aphylactus Root, 1927c (p. 584) – species of *Culex* (*Microculex*).

apicalis Adams, 1903 (p. 26) – species of *Culex* (*Neoculex*).

apicalis Theobald, 1903a (p. 171) – synonym of *Psorophora* (*Grabhamia*) *cingulata* (Fabricius, 1805) [original combination: *Culex apicalis*]; synonymy established before Knight & Stone, 1977.

apicalis Theobald, 1903a (p. 298) – species of *Uranotaenia* (*Uranotaenia*).

apicalis Theobald, 1908b (p. 291) – synonym of *Armigeres* (*Leicesteria*) *flavus* (Leicester, 1908) [original combination: *Leicesteria apicalis*]; synonymy established before Knight & Stone, 1977. **Note**: this nominal taxon is the senior secondary homonym of *apicalis* Theobald, 1910a (p. 5) and *apicalis* Theobald, 1910a (p. 7).

apicalis Theobald, 1910a (p. 5) – synonym of *Armigeres* (*Armigeres*) *theobaldi* Barraud, 1934 [original combination: *Desvoidea apicalis*]; synonymy by Barraud, 1934. **Note**: this nominal taxon is the junior secondary homonym of *apicalis* Theobald, 1908b; the senior secondary homonym of *apicalis* Theobald, 1910a (p. 7), based on page priority.

apicalis Theobald, 1910a (p. 7) – synonym of *Armigeres* (*Leicesteria*) *flavus* (Leicester, 1908) [original combination: *Brevirhynchus apicalis*]; synonymy established before Knight & Stone, 1977. **Note**: this nominal taxon is the junior secondary homonym of *apicalis* Theobald, 1908b and *apicalis* Theobald, 1910a (p. 5), the latter based on page priority.

apicimacula Dyar & Knab, 1906a (p. 136) – species of *Anopheles* (*Anopheles*).

apicinus Philippi, 1865 (p. 596) – species of *Culex* (*Culex*).
Synonyms: *debilis* Dyar & Knab, 1914; *escomeli* Brèthes, 1920; *alticola* Martini, 1931a.

apicoannulatus (Edwards, 1912c) (p. 18) – species of *Aedes* (*Elpeytonius*) [*nomen novum* for *alboannulatus* Theobald, 1905b; original combination: *Ochlerotatus apicoannulatus*].
Synonym: *alboannulatus* Theobald, 1905b.

apicoargenteus (Theobald, 1909b) (p. 3) – species of *Aedes* (*Stegomyia*) [original combination: *Stegomyia apicoargentea*].
Synonyms: *albertii* Theobald, 1912; *maculoabdominalis* Theobald, 1913d; *togoensis* Enderlein, 1923; *ziemanni* Enderlein, 1923.

apicosquamata da Cunha Ramos & Brunhes, 2004 (p. 97) – species of *Uranotaenia* (*Pseudoficalbia*).

apicotaeniata Theobald, 1909b (p. 21) – species of *Uranotaenia* (*Pseudoficalbia*).

apicotriangulatus (Theobald, 1910c) (p. 211) – species of *Tripteroides* (*Polylepidomyia*) [original combination: *Mimeteomyia apicotriangulata*].

apoci Marsh, 1933 (p. 193) – species of *Anopheles* (*Cellia*).

apoensis Baisas & Ubaldo-Pagayon, 1953 (p. 68) – species of *Tripteroides* (*Tricholeptomyia*).

apoensis Bohart & Farner, 1944 (p. 69) – species of *Armigeres* (*Armigeres*).

apollo Schick, 1970a (p. 46) – synonym of *Aedes* (*Protomacleaya*) *terrens* (Walker, 1856a); synonymy by Schick, 1970b.

aporonoma Dyar & Knab, 1906d (p. 230) – species of *Wyeomyia* (*Triamyia*).

approximata Peyton, 1977 (p. 36) – species of *Uranotaenia* (*Pseudoficalbia*).

apsarae Klein, 1977 (p. 126) – species of *Topomyia* (*Suaymyia*).
Synonym: *rausai* Miyagi, 1980.

aquacaelestis Curry, 1932 (p. 566) – synonym of *Anopheles* (*Nyssorhynchus*) *oswaldoi* (Peryassú, 1922b) [original combination: *Anopheles* (*Nyssorhynchus*) *tarsimaculatus* variety *aquacaelestis*]; synonymy by Rozeboom, 1941.

aquarius Strickman, 1990 (p. 150) – species of *Culex* (*Culex*).

aquasalis Curry, 1932 (p. 566) – species of *Anopheles* (*Nyssorhynchus*) [original combination: *Anopheles tarsimaculatus* variety *aquasalis*].
Synonyms: *emilianus* Komp, 1941; *guarujaensis* Ramos, 1942; *delta* Anduze, 1948; *guarauno* Anduze, 1948; *deltaorinoquensis* Cova Garcia, Pulido F. & Amarista M., 1977b.

aquilus Graham, 1910 (p. 266) – synonym of *Culex* (*Culex*) *invidiosus* Theobald, 1901d; synonymy established before Knight & Stone, 1977.

arabica Christophers & Chand, 1915 (p. 189) – synonym of *Anopheles* (*Cellia*) *fluviatilis* James, 1902 [original combination: *Anopheles funestus* variety *arabica* [*sic*]]; synonymy by Mattingly & Knight, 1956.

arabica Giles, 1906 (p. 130) – synonym of *Aedes* (*Ochlerotatus*) *caspius* (Pallas, 1771) [original combination: *Mansonia arabica*]; synonymy established before Knight & Stone, 1977. **Note**: this nominal taxon is the senior secondary homonym of *arabicus* Becker, 1910.

arabicus Becker, 1910 (p. 140) – synonym of *Aedes* (*Ochlerotatus*) *caspius* (Pallas, 1771) [original combination: *Culex arabicus*]; synonymy established before Knight & Stone, 1977. **Note**: this nominal taxon is the junior secondary homonym of *arabica* Giles, 1906.

arabiensis Patton, 1905 (p. 625) – species of *Anopheles* (*Cellia*).

arabiensis (Patton, 1905) (species) (p. 633) – subspecies of *Aedes* (*Aedimorphus*) *vexans* (Meigen, 1830b) [original combination: *Culex arabiensis*]; subspecific status by White, 1975.
Synonym: *davidsoni* Ribeiro, da Cunha Ramos, Pires & Capela, 1979.

arabiensis Theobald, 1913a (p. 318) – synonym of *Culex* (*Oculeomyia*) *bitaeniorhynchus* Giles, 1901a [original combination: *Pseudotaeniorhynchus quasigelidus* variety *arabiensis*]; synonymy established before Knight & Stone, 1977.

aranetana Banks, 1906b (p. 1001) – synonym of *Aedes* (*Finlaya*) *flavipennis* (Giles, 1904a) [original combination: *Finlaya aranetana*]; synonymy established before Knight & Stone, 1977.

aranoideoides Dong, Zhou, Gong *et al.* [*sic*], 2004 (p. 187) – *nomen nudum*, *Tripteroides*.

aranoides (Theobald, 1901c) (p. 274) – species of *Tripteroides* (*Rachionotomyia*) [original combination: *Wyeomyia aranoides*].
Synonyms: *fusca* Leicester, 1908; *inornata* Theobald, 1910a; *szechwanensis* Hsu, 1964.

araozi Shannon & del Ponte, 1928 (p. 74) – synonym of *Aedes* (*Ochlerotatus*) *milleri* Dyar, 1922g; synonymy established before Knight & Stone, 1977.

araozi Shannon & del Ponte, 1928 (in Dyar, 1928) (p. 257) – synonym of *Coquillettidia* (*Rhynchotaenia*) *venezuelensis* (Theobald, 1912) [original combination: *Mansonia* (*Rhynchotaenia*) *araozi*]; synonymy by Castro & Bressanello, 1952.

arawak Berlin, 1969b (p. 37) – species of *Culex* (*Micraedes*).

arbieeni Salem, 1938 (p. 20) – species of *Culex* (*Maillotia*).
Synonym: *jenkinsi* Knight, 1953.

arborea Galindo, Carpenter & Trapido, 1951 (p. 86) – species of *Wyeomyia* (*Exallomyia*).

arborealis Bonne-Wepster & Bonne, 1920 (p. 178) – species of *Aedes* (*Howardina*).

arborealis Shannon & del Ponte, 1928 (p. 89) – synonym of *Toxorhynchites* (*Lynchiella*) *guadeloupensis* (Dyar & Knab, 1906c) [original combination: *Megarhinus tucumanus* variety *arborealis*]; synonymy by Lane, 1953.

arboricollis (de Charmoy, 1908) (p. 257) – species of *Orthopodomyia* [original combination: *Culex arboricollis*].

arboricolus Galindo & Mendez, 1961 (p. 3) – species of *Culex* (*Melanoconion*).

arboricolus Knight & Rozeboom, 1946 (p. 88) – species of *Aedes* (*Scutomyia*).

arboricolus Zavortink, 1970b (p. 14) – species of *Anopheles* (*Anopheles*).

arcanus Blanchard, 1905 (p. 303) – synonym of *Aedes* (*Ochlerotatus*) *flavescens* (Müller, 1764) [*nomen novum* for *flavescens* Theobald, 1901a; original combination: *Culex arcanus*]; synonymy established before Knight & Stone, 1977.

archegus Dyar, 1929b (p. 511) – species of *Culex* (*Culex*).

arctica Edwards, 1920a (p. 136) – synonym of *Culiseta* (*Culiseta*) *alaskaensis* (Ludlow, 1906c) [original combination: *Theobaldia arctica*]; synonymy established before Knight & Stone, 1977.

ardensis (Theobald, 1905g) (p. 17) – species of *Anopheles* (*Cellia*) [original combination: *Pyretophorus ardensis*].
Synonym: *pyretophoroides* Theobald, 1907.

arenivaga Marks, 1968 (p. 51) – species of *Culiseta* (subgenus uncertain, determined by the author of the species).

argenteitarsis Brug, 1932 (p. 76) – species of *Aedes* (*Patmarksia*).

argenteiventris (Theobald, 1905f) (p. 118) – species of *Tripteroides* (*Polylepidomyia*) [original combination: *Polylepidomyia argenteiventris*].
Synonyms: *atra* Taylor, 1914d; *brugi* Edwards, 1927b.

argenteolobatus (Gough, 1910) (p. 116) – species of *Anopheles* (*Cellia*) [original combination: *Cellia argenteolobata*].
Synonym: *pseudosquamosa* Newstead & Carter, 1911b.

argenteomaculata Theobald, 1907 (p. 184) – ? synonym of *Aedes* (*Stegomyia*) *gardnerii imitator* Leicester, 1908 [original combination: *Stegomyia argenteomaculata*]; synonymy by Huang, 1977.

argenteopennis Peryassú, 1923a (p. 2) – synonym of *Uranotaenia* (*Uranotaenia*) *nataliae* Lynch Arribálzaga, 1891b; synonymy established before Knight & Stone, 1977.

argenteopunctatus (Theobald, 1901a) (p. 316) – species of *Aedes* (*Catageiomyia*) [original combination: *Stegomyia argenteopunctata*].
Synonym: *quinquepunctata* Theobald, 1913d.

argenteopunctatus (Ventrillon, 1905a) (p. 446) – species of *Culex* (*Culex*) [original combination: *Heptaphlebomyia argenteopunctata*].

argenteorostris (Bonne-Wepster & Bonne, 1920) (p. 167) – species of *Wyeomyia* (subgenus uncertain, determined by Motta & Lourenço-de-Oliveira, 1995) [original combination: *Cleobonnea argenteorostris*].
Synonym: *albocaerulea* Senevet & Abonnenc, 1939b.

argenteoscutellatus Carter & Wijesundara, 1948 (p. 139) – species of *Aedes* (*Aedimorphus*).

argenteotarsis Ludlow, 1906d (p. 367) – synonym of *Toxorhynchites* (*Toxorhynchites*) *amboinensis* (Doleschall, 1857); synonymy by Bohart, 1945.

argenteoumbrosus Theobald, 1907 (p. 461) – synonym of *Culex* (*Microculex*) *imitator* Theobald, 1903a [original combination: *Microculex argenteoumbrosus*]; synonymy established before Knight & Stone, 1977.

argenteoventralis Leicester, 1908 (as *argentoventralis*) (p. 240) – species of *Topomyia* (*Suaymyia*).

argenteoventralis (Theobald, 1909b) (p. 24) – species of *Aedes* (*Pseudarmigeres*) [original combination: *Dendromyia argenteoventralis*].
Synonym: *affinis* Theobald, 1909b.

argentescens Dyar & Knab, 1908 (p. 55) – synonym of *Aedes* (*Ochlerotatus*) *angustivittatus* Dyar & Knab, 1907a; synonymy established before Knight & Stone, 1977.

argenteus Ludlow, 1905a (p. 98) – synonym of *Culex* (*Culex*) *whitmorei* (Giles, 1904a) [original combination: *Taeniorhynchus argenteus*]; synonymy established before Knight & Stone, 1977.

argenteus Poiret, 1787 (p. 245) – synonym of *Aedes* (*Stegomyia*) *aegypti* (Linnaeus, 1762) [original combination: *Culex argenteus*]; synonymy established before Knight & Stone, 1977.

argentina Petrocchi, 1923 (p. 92) – *nomen nudum*, *Uranotaenia*.

argentinotus Banks, 1909 (p. 547) – synonym of *Aedes* (*Aedimorphus*) *alboscutellatus* (Theobald, 1905f) [original combination: *Culex argentinotus*]; synonymy established before Knight & Stone, 1977.

argentinus Brèthes, 1912 (p. 15) – synonym of *Anopheles* (*Anopheles*) *pseudopunctipennis* Theobald, 1901c [original combination: *Proterorhynchus argentinus*]; synonymy established before Knight & Stone, 1977.

argentipleura da Cunha Ramos & Brunhes, 2004 (p. 349) – species of *Uranotaenia* (*Uranotaenia*).

arguellesi Baisas, 1935c (p. 68) – species of *Uranotaenia* (*Uranotaenia*).

argyritarsis Robineau-Desvoidy, 1827 (p. 411) – species of *Anopheles* (*Nyssorhynchus*).
Synonym: *rooti* Brèthes, 1926b.

argyrites Dyar & Nuñez Tovar, 1927a (p. 190) – species of *Aedes* (*Howardina*).

argyromeris Dyar & Ludlow, 1921b (p. 679) – species of *Haemagogus* (*Haemagogus*).
Synonym: *gladiator* Dyar, 1921g.

argyronotum Belkin, 1962 (p. 354) – species of *Aedes* (*Patmarksia*).

argyronotum Edwards, 1928a (p. 283) – synonym of *Sabethes* (*Sabethes*) *belisarioi* Neiva, 1908a; synonymy established before Knight & Stone, 1977.

argyropalpis Leicester, 1908 (p. 242) – species of *Topomyia* (*Topomyia*).
Synonym: *gregoryi* Ludlow, 1911b.

argyropus (Swellengrebel, 1914) (p. 334) – species of *Anopheles* (*Anopheles*) [original combination: *Myzorhynchus argyropus*].

argyropus (Walker, 1848) (p. 2) – species of *Maorigoeldia* [original combination: *Culex argyropus*].

argyrotarsis Leicester, 1908 (p. 214) – species of *Uranotaenia* (*Uranotaenia*).
Synonym: *parangensis* Ludlow, 1909a.

argyrotarsis Lutz, 1904 (in Bourroul, 1904) (p. 77) – *nomen nudum*, *Wyeomyia*.

argyrothorax Bonne-Wepster & Bonne, 1920 (p. 179) – species of *Aedes* (*Protomacleaya*).

argyrura Dyar & Knab, 1908 (p. 70) – synonym of *Wyeomyia* (*Wyeomyia*) *vanduzeei* Dyar & Knab, 1906a; synonymy established before Knight & Stone, 1977.

argyrurus (Edwards, 1934) (in Barraud, 1934) (p. 448) – species of *Udaya* [original combination: *Paraedes argyrurus*].

argyrurus Edwards, 1936 (p. 53) – species of *Eretmapodites*.

arizonensis Bohart, 1949 (p. 341) – species of *Culex* (*Neoculex*).

arnoldi Stephens & Christophers, 1908 (p. 175) – synonym of *Anopheles* (*Cellia*) *squamosus* Theobald, 1901a [original combination: *Cellia arnoldi*]; synonymy established before Knight & Stone, 1977.

arnoulti Grjebine, 1966 (p. 233) – *nomen dubium*, *Anopheles*; status by Brunhes *et al.*, 1998a.

arribalzagae Giles, 1902 (p. 341) – synonym of *Psorophora* (*Janthinosoma*) *discrucians* (Walker, 1856a) [original combination: *Janthinosoma arribalzagae*]; synonymy established before Knight & Stone, 1977.

arribalzagae (Theobald, 1903a) (p. 261) – species of *Coquillettidia* (*Rhynchotaenia*) [original combination: *Taeniorhynchus arribalzagae*].
Synonym: *coticula* Dyar & Knab, 1907b.

artemievi Gordeev, Zvantsov, Goriacheva, Shĕikevich & Ezhov, 2005 (p. 5) – species of *Anopheles* (*Anopheles*).

arthrostigma (Lutz, 1905) (p. 311) – species of *Wyeomyia* (*Wyeomyia*) [original combination: *Dendromyia arthrostigma*].
Synonyms: *bromeliarum* Dyar & Knab, 1906a; *espartana* Dyar & Knab, 1906a; *panamena* Dyar & Knab, 1907c; *drapetes* Dyar & Knab, 1909d.

arthuri Unti, 1941 (p. 9) – species of *Anopheles* (*Nyssorhynchus*) [original combination: *Anopheles* (*Nyssorhynchus*) *strodei* variety *arthuri*]; specific status by Sallum *et al.*, 2010.
Synonym: *artigasi* Unti, 1941.

articularis Philippi, 1865 (p. 596) – species of *Culex* (*Culex*).

articulatus Rondani, 1872 (p. 30) – synonym of *Aedes* (*Aedimorphus*) *vexans* (Meigen, 1830b) [original combination: *Culex articulatus*]; synonymy established before Knight & Stone, 1977.

artigasi Unti, 1941 (p. 9) – synonym of *Anopheles* (*Nyssorhynchus*) *arthuri* Unti, 1941 [original combination: *Anopheles* (*Nyssorhynchus*) *strodei* variety *artigasi*]; synonymy by Galvão, 1943; Sallum *et al.*, 2010.

aruni Sobti, 1968 (p. 482) – species of *Anopheles* (*Cellia*).

arvernensis (Piton, 1936) (pp. 17, 18, 20) – fossil species of *Neoculicites* Evenhuis, 1994 [original combination: *Culicites arvernensis*].

asanumai Sasa, Kano & Takahasi, 1950 (p. 637) – synonym of *Aedes* (*Aedes*) *esoensis* Yamada, 1921; synonymy by Tanaka *et al.*, 1975b.

ascidiicola de Meijere, 1910 (p. 925) – species of *Uranotaenia* (*Pseudoficalbia*).

aseyehae Dyar & Knab, 1915 (p. 112) – synonym of *Culex* (*Culex*) *quinquefasciatus* Say, 1823; synonymy established before Knight & Stone, 1977.

ashworthi Edwards, 1921c (p. 75) – species of *Aedes* (*Halaedes*).

asiaticus Edwards, 1926c (p. 73) – subspecies of *Aedes* (*Ochlerotatus*) *pulcritarsis* (Rondani, 1872) [original combination: *Aedes* (*Ochlerotatus*) *pulchritarsis* [*sic*] variety *asiaticus*]; subspecific status by Monchadskii, 1951.
Synonym: *stegomyina* Stackelberg & Montchadsky, 1926.

asiaticus Leicester, 1903 (p. 291) – species of *Anopheles* (*Anopheles*).
assamensis Bhattacharyya, Tewari, Prakash, Mohapatra & Mahanta, 2004 (p. 115) – species of *Verrallina* (*Neomacleaya*).
assamensis (Theobald, 1908b) (p. 290) – species of *Aedes* (*Phagomyia*) [original combination: *Stegomyia assamensis*].
asteliae Belkin, 1968a (p. 66) – species of *Culex* (*Culex*).
astridianus de Meillon, 1942 (p. 89) – species of *Culex* (*Culex*).
asulleptus (Theobald, 1903a) (p. 315) – species of *Limatus* [original combination: *Dendromyia asullepta*].
Synonym: *methysticus* Dyar & Knab, 1909d.
atacamensis González & Sallum, 2010 (p. 14) – species of *Anopheles* (*Nyssorhynchus*).
atactavittatus Arnell, 1976 (p. 26) – species of *Aedes* (*Ochlerotatus*).
ataeniatus Theobald, 1911b (as *ataeniata*) (p. 261) – synonym of *Culex* (*Culex*) *univittatus* Theobald, 1901c; synonymy by White, 1980.
ater (Daniels, 1908) (p. 265) – species of *Toxorhynchites* (*Toxorhynchites*) [original combination: *Teromyia ater*].
ater Gutsevich, 1955 (p. 321) – synonym of *Aedes* (*Ochlerotatus*) *mercurator* Dyar, 1920a; synonymy by Danilov, 1974.
atheniensis Cardamatis, 1931 (p. 129) – synonym of *Anopheles* (*Cellia*) *superpictus* Grassi, 1899 [original combination: *Pyretophorus atheniensis*]; synonymy by Pandazis, 1935.
atlantica (Edwards, 1932b) (p. 560) – species of *Culiseta* (*Culiseta*) [original combination: *Theobaldia atlantica*].
atlanticus Adames, 1971 (p. 28) – species of *Deinocerites*.
atlanticus Dyar & Knab, 1906d (p. 198) – species of *Aedes* (*Ochlerotatus*).
atra (Lee, 1944b) (p. 220) – species of *Culiseta* (*Culicella*) [original combination: *Theobaldia atra*].
atra Taylor, 1914d (p. 190) – synonym of *Tripteroides* (*Polylepidomyia*) *argenteiventris* (Theobald, 1905f) [original combination: *Stegomyia atra*]; synonymy established before Knight & Stone, 1977.
atra Theobald, 1905f (p. 114) – species of *Uranotaenia* (*Pseudoficalbia*).
Synonym: *nigerrima* Taylor, 1914d.
atracus Colless, 1960 (p. 385) – species of *Culex* (*Lophoceraomyia*).
Synonym: *franclemonti* Belkin, 1962.
atrata Belkin, Heinemann & Page, 1970 (p. 227) – species of *Wyeomyia* (*Wyeomyia*).
atratipes Skuse, 1889 (p. 1755) – species of *Anopheles* (*Anopheles*).
atratulus Edwards, 1922a (p. 287) – synonym of *Culex* (*Lophoceraomyia*) *niger* (Leicester, 1908) [*nomen novum* for *niger* Leicester, 1908, in error]; synonymy established before Knight & Stone, 1977.
atratus Theobald, 1901c (p. 55) – species of *Culex* (*Melanoconion*).
Synonyms: *falsificator* Dyar & Knab, 1909d; *advieri* Senevet, 1938.
atria (Barraud, 1928b) (p. 371) – species of *Verrallina* (*Neomacleaya*) [original combination: *Aedes* (*Aedes*) *atrius*].
atriceps Edwards, 1926a (p. 105) – species of *Culex* (*Culex*).
Synonym: *nigriceps* Buxton, 1927 (*lapsus* for *atriceps*).
atriisimilis (Tanaka & Mizusawa, 1973) (p. 625) – species of *Verrallina* (*Neomacleaya*) [original combination: *Aedes* (*Neomacleaya*) *atriisimilis*].
atripes (Skuse, 1889) (p. 1750) – species of *Tripteroides* (*Polylepidomyia*) [original combination: *Culex atripes*].
atritarsalis (Dobrotworsky, 1954) (p. 73) – subspecies of *Culiseta* (*Neotheobaldia*) *frenchii* (Theobald, 1901c) [original combination: *Theobaldia frenchi atritarsalis*].
atritarsis Edwards, 1920a (p. 129) – synonym of *Aedes* (*Stegomyia*) *aegypti* (Linnaeus, 1762) [original combination: *Stegomyia fasciata* variety *atritarsis*]; synonymy established before Knight & Stone, 1977.
atroapicalis Gillett, 1946 (p. 436) – synonym of *Coquillettidia* (*Coquillettidia*) *fraseri* (Theobald, 1911a) [original combination: *Taeniorhynchus* (*Coquillettidia*) *atroapicalis*]; synonymy by Gillett, 1949.
atropalpus (Coquillett, 1902d) (p. 292) – species of *Aedes* (*Georgecraigius*) [original combination: *Culex atropalpus*].
atroparvus van Thiel, 1927 (p. 389) – species of *Anopheles* (*Anopheles*) [original combination: *Anopheles maculipennis* variety *atroparvus*].
Synonyms: *fallax* Roubaud, 1934; *cambournaci* Roubaud & Treillard, 1936.
atropos Dyar & Knab, 1906b (p. 160) – species of *Anopheles* (*Anopheles*).
Synonym: *ferruginosus* Wiedemann, 1828.
atropotenae Lindtrop, 1924 (p. 40) – synonym of *Anopheles* (*Cellia*) *pulcherrimus* Theobald, 1902a [original combination: *Cellia pulcherrima* variety *atropotenae*]; synonymy established before Knight & Stone, 1977.
aurantapex Edwards, 1914b (p. 74) – species of *Culex* (*Oculeomyia*).
auranticauda Lane, 1992b (p. 301) – species of *Toxorhynchites* (*Toxorhynchites*).
aurantius (Theobald, 1907) (p. 280) – species of *Aedes* (*Mucidus*) [original combination: *Pardomyia aurantia*].
Synonym: *aureostriata* Leicester, 1908.
aurata (Dobrotworsky, 1962) (p. 295) – species of *Coquillettidia* (*Coquillettidia*) [original combination: *Mansonia aurata*].
aurata (Doucet, 1951a) (p. 70) – species of *Mimomyia* (*Ingramia*) [original combination: *Ravanalites auratus*].
auratus Grabham, 1906b (p. 313) – species of *Aedes* (*Ochlerotatus*). **Note**: this nominal taxon is the senior secondary homonym of *auratus* Leicester, 1908.
auratus Leicester, 1908 (p. 153) – synonym of *Aedes* (*Edwardsaedes*) *imprimens* (Walker, 1860) [original combination: *Culex auratus*]; synonymy established

before Knight & Stone, 1977. **Note**: this nominal taxon is the junior secondary homonym of *auratus* Grabham, 1906b.

aurea Brug, 1932 (p. 75) – species of *Heizmannia* (*Heizmannia*).

aurea (Edwards, 1915b) (p. 281) – species of *Coquillettidia* (*Coquillettidia*) [original combination: *Taeniorhynchus aureus*].

aurea (Leicester, 1908) (p. 116) – species of *Mimomyia* (*Mimomyia*) [original combination: *Conopomyia aurea*].

aureoantennata Ferrara, 1973 (p. 199) – species of *Orthopodomyia*.

aureochaeta (Leicester, 1908) (p. 255) – species of *Heizmannia* (*Heizmannia*) [original combination: *Dendromyia aureochaeta*].

aureolineatus Berlin, 1969a (p. 25) – species of *Aedes* (*Howardina*).

aureolineatus (Leicester, 1908) (p. 79) – species of *Armigeres* (*Armigeres*) [original combination: *Desvoidya aureolineata*].

aureonotatus Duret & Barreto, 1956 (p. 91) – species of *Culex* (*Melanoconion*).

aureopunctis Ludlow, 1910 (p. 195) – synonym of *Lutzia* (*Metalutzia*) *halifaxii* Theobald, 1903a [original combination: *Culex aureopunctis*]; synonymy established before Knight & Stone, 1977.

aureosquamiger Theobald, 1907 (p. 73) – synonym of *Anopheles* (*Cellia*) *natalensis* (Hill & Haydon, 1907) [original combination: *Pyretophorus aureosquamiger*]; synonymy established before Knight & Stone, 1977.

aureosquammata (Ludlow, 1909c) (p. 234) – species of *Coquillettidia* (*Coquillettidia*) [original combination: *Taeniorhynchus aureosquammatus*].
Synonym: *pagei* Ludlow, 1910.

aureostriata Grabham, 1906a (p. 171) – synonym of *Aedes* (*Howardina*) *grabhami* Berlin, 1969a; synonymy by Berlin, 1969a. **Note**: this nominal taxon is the junior secondary homonym of *aureostriatus* (Doleschall, 1857) and the senior secondary homonym of *aureostriata* Leicester, 1908.

aureostriata Leicester, 1908 (p. 71) – synonym of *Aedes* (*Mucidus*) *aurantius* (Theobald, 1907) [original combination: *Ekrinomyia aureostriata*]; synonymy established before Knight & Stone, 1977. **Note**: this nominal taxon is the junior secondary homonym of *aureostriatus* (Doleschall, 1857) and *aureostriata* Grabham, 1906a.

aureostriatus (Doleschall, 1857) (p. 385) – species of *Aedes* (*Bruceharrisonius*) [original combination: *Culex aureostriatus*]. **Note**: this nominal taxon is the senior secondary homonym of *aureostriata* Grabham, 1906a and *aureostriata* Leicester, 1908.
Synonym: *leucopleurus* Rozeboom, 1946.

aureoventer (Theobald, 1910a) (p. 27) – species of *Topomyia* (*Topomyia*) [original combination: *Pseudograhamia aureoventer*].

aurescens (Lutz, 1905) (p. 350) – species of *Sabethes* (*Peytonulus*) [original combination: *Sabethinus aurescens*].

aureus Gutsevich, 1955 (species) (p. 320) – subspecies of *Aedes* (*Neomelaniconion*) *lineatopennis* (Ludlow, 1905a); subspecific status by Danilov, 1977.

aureus Lane & Whitman, 1951 (p. 349) – species of *Culex* (*Microculex*).

auriceps Brug, 1939 (p. 96) – species of *Topomyia* (*Suaymyia*).

auridorsum Edwards, 1922d (p. 93) – species of *Aedes* (*Molpemyia*).

aurifer (Coquillett, 1903a) (p. 255) – species of *Aedes* (*Ochlerotatus*) [original combination: *Culex aurifer*].

aurifluus (Edwards, 1921a) (p. 631) – subspecies of *Toxorhynchites* (*Toxorhynchites*) *christophi* (Portschinsky, 1884) [original combination: *Megarhinus aurifluus*].

aurilatus Senevet & Abonnenc, 1939a (p. 94) – synonym of *Culex* (*Melanoconion*) *theobaldi* (Lutz, 1904); synonymy by Forattini & Sallum, 1989b.

aurimargo Edwards, 1922d (p. 94) – species of *Aedes* (*Leptosomatomyia*).
Synonym: *lateralis* Theobald, 1905f.

auripennis Edwards, 1915b (p. 279) – synonym of *Coquillettidia* (*Coquillettidia*) *microannulata* (Theobald, 1911a) [original combination: *Taeniorhynchus auripennis*]; synonymy by Stone *et al*., 1959.

auripes (Edwards, 1935b) (p. 579) – species of *Toxorhynchites* (*Toxorhynchites*) [original combination: *Megarhinus auripes*].

aurirostris (Watson, 1910) (p. 251) – species of *Anopheles* (*Cellia*) [original combination: *Myzomyia aurirostris*].

auritaenia Enderlein, 1920 (p. 49) – synonym of *Culex* (*Oculeomyia*) *poicilipes* (Theobald, 1903a); synonymy established before Knight & Stone, 1977.

aurites (Theobald, 1901d) (p. 13) – species of *Coquillettidia* (*Coquillettidia*) [original combination: *Taeniorhynchus aurites*].

aurites (Theobald, 1907) (p. 216) – species of *Aedes* (*Howardina*) [original combination: *Howardina aurites*].

aurivittatus Cerqueira, 1943 (p. 7) – species of *Aedes* (*Howardina*).

auroides Felt, 1905a (p. 449) – synonym of *Aedes* (*Ochlerotatus*) *punctor* (Kirby, 1837) [original combination: *Culicelsa auroides*]; synonymy established before Knight & Stone, 1977.

auronitens Edwards, 1922a (p. 268) – species of *Aedes* (subgenus uncertain).

aurostriata Banks, 1906b (p. 995) – synonym of *Aedes* (*Belkinius*) *aurotaeniatus* Edwards, 1922a [original combination: *Stegomyia aureostriata*]; synonymy by Edwards, 1922b.

aurotaeniatus Edwards, 1922a (p. 256) – species of *Aedes* (*Belkinius*) [*nomen novum* for *aureostriata* Banks, 1906b].
Synonym: *aureostriata* Banks, 1906b.

aurovenatus Worth, 1960 (p. 312) – species of *Aedes* (*Neomelaniconion*).

austenii (Theobald, 1905b) (p. 102) – species of *Anopheles* (*Cellia*) [original combination: *Pyretophorus austenii*].

austenii Theobald, 1910c (p. 572) – synonym of *Eretmapodites quinquevittatus* Theobald, 1901d; synonymy established before Knight & Stone, 1977.

australicus Dobrotworsky & Drummond, 1953 (p. 143) – species of *Culex* (*Culex*) [original combination: *Culex pipiens australicus*].

australiensis Giles, 1902 (p. 355) – synonym of *Mansonia* (*Mansonioides*) *uniformis* (Theobald, 1901c) [original combination: *Panoplites australiensis*]; synonymy established before Knight & Stone, 1977.

australiensis (Theobald, 1910c) (p. 313) – species of *Aedes* (subgenus uncertain) [original combination: *Leucomyia australiensis*].

australis (Erichson, 1842) (p. 270) – species of *Aedes* (*Halaedes*) [original combination: *Culex australis*].
Synonyms: *crucians* Walker, 1856a; *tasmaniensis* Strickland, 1911b; *concolor* Taylor, 1914c.

australis Strickland, 1911a (p. 133) – synonym of *Aedes* (*Ochlerotatus*) *stricklandi* (Edwards, 1912b) [original combination: *Grabhamia australis*]; synonymy established before Knight & Stone, 1977. **Note**: this nominal taxon is the senior secondary homonym of *australis* Taylor, 1914b.

australis Taylor, 1914b (p. 457) – synonym of *Aedes* (*Macleaya*) *tremulus* (Theobald, 1903e) [original combination: *Aedimorphus australis*]; synonymy established before Knight & Stone, 1977. **Note**: this nominal taxon is the junior secondary homonym of *australis* Strickland, 1911a.

australis Taylor, 1915 (p. 178) – synonym of *Culex* (*Lophoceraomyia*) *hilli* Edwards, 1922d [original combination: *Neomacleaya australis*]; synonymy established before Knight & Stone, 1977.

autocratica Dyar & Knab, 1906d (p. 230) – species of *Wyeomyia* (*Hystatomyia*).

autogenicus Roubaud, 1935b (p. 444) – synonym of *Culex* (*Culex*) *pipiens* Linnaeus, 1758 [original combination: *Culex pipiens autogenicus*]; synonymy with *molestus* Forskål, 1775 by Marshall & Staley, 1937.

automartus Root, 1927c (p. 591) – synonym of *Culex ocellatus* Theobald, 1903a; synonymy by Lane & Whitman, 1943.

autumnalis Weyenbergh, 1882 (p. 23) – synonym of *Culex* (*Culex*) *quinquefasciatus* Say, 1823; synonymy established before Knight & Stone, 1977.

auyantepuiensis Harbach & Navarro, 1996 (p. 207) – species of *Anopheles* (*Kerteszia*).

avianus de Meillon, 1943 (p. 96) – species of *Culex* (*Maillotia*).

avistylus Brug, 1939 (p. 107) – species of *Aedes* (*Finlaya*) [original combination: *Aedes* (*Finlaya*) *flavipennis* variety *avistyla* [*sic*]]; specific status by Stone & Bohart, 1944.

axillicolus Steffan, 1979 (p. 133) – species of *Culex* (*Acallyntrum*).

axitiosus Kulasekera, Knight & Harbach, 1990 (p. 26) – species of *Aedes* (*Downsiomyia*).

ayrozai Unti, 1940a (p. 379) – synonym of *Anopheles* (*Nyssorhynchus*) *evansae* (Brèthes, 1926b) [original combination: *Anopheles* (*Nyssorhynchus*) *oswaldoi* variety *ayrozai*]; synonymy by Faran, 1981.

azaniae Bailly-Choumara, 1960 (p. 532) – species of *Anopheles* (*Cellia*).

azevedoi Ribeiro, 1969 (p. 113) – species of *Anopheles* (*Cellia*).
Synonym: *deaconi* de Meillon & van Eden, 1976.

azoriensis Theobald, 1903a (p. 210) – synonym of *Culex* (*Culex*) *pipiens* Linnaeus, 1758; synonymy established before Knight & Stone, 1977.

azriki Patton, 1905 (p. 632) – synonym of *Anopheles* (*Cellia*) *turkhudi* Liston, 1901; synonymy by F.W. Edwards (in Evans, 1938).

aztecus Hoffmann, 1935 (p. 5) – species of *Anopheles* (*Anopheles*) [original combination: *Aedes maculipennis aztecus*].

azuayus Leví-Castillo, 1954b (p. 264) – synonym of *Culex* (*Culex*) *nigripalpus* Theobald, 1901c; synonymy by Bram, 1967a.

azureosquamata (Bonne-Wepster, 1948) (p. 320) – species of *Verrallina* (*Verrallina*) [original combination: *Aedes* (*Aedes*) *azureosquamatus*].

azurini Basio, 1971 (p. 54) – species of *Armigeres* (*Armigeres*).

azurini Toma, Miyagi & Cabrera, 1984 (p. 172) – species of *Culex* (*Culiciomyia*).

azymus Dyar & Knab, 1906b (p. 169) – species of *Culex* (*Microculex*).

Bb

babahoyensis Levi-Castillo, 1953d (p. 92) – species of *Culex* (*Carrollia*).

bachmanni Petrocchi, 1925 (corrected spelling of Petrochi) (p. 70) – synonym of *Anopheles* (*Nyssorhynchus*) *triannulatus* (Neiva & Pinto, 1922b); synonymy by Faran, 1980.

bacigalupoi Martínez & Prosen, 1958 (p. 17) – synonym of *Orthopodomyia fascipes* (Coquillett, 1906a); synonymy by Zavortink, 1968.

badgeri Dyar, 1924e (p. 125) – synonym of *Culex* (*Culex*) *erythrothorax* Dyar, 1907; synonymy established before Knight & Stone, 1977.

baezai Gater, 1933 (p. 162) – species of *Anopheles* (*Anopheles*).
Synonym: *gateri* Baisas, 1936.

bahama Dyar & Knab, 1906a (p. 138) – species of *Wyeomyia* (*Wyeomyia*).
Synonym: *minor* Dyar & Knab, 1906a.

bahamensis Berlin, 1969a (p. 61) – species of *Aedes* (*Howardina*).

bahamensis Dyar & Knab, 1906d (p. 210) – species of *Culex* (*Culex*).
Synonyms: *eleuthera* Dyar, 1918a; *patersoni* Dyar, 1920b.

bahiensis Duret, 1969b (as *bahiense*) (p. 40) – species of *Culex* (*Melanoconion*).

bahri (Edwards, 1914b) (p. 79) – species of *Culex* (*Culiciomyia*) [original combination: *Culiciomyia bahri*].

baileyi Edwards, 1929b (p. 323) – species of *Anopheles* (*Anopheles*) [original combination: *Anopheles gigas* variety *baileyi*].

bailyi Barraud, 1929a (p. 1062) – species of *Hodgesia*.

bailyi Barraud, 1934 (p. 382) – species of *Culex* (*Culiciomyia*).

baimaii Sallum & Peyton, 2005 (in Sallum *et al*., 2005b) (p. 175) – species of *Anopheles* (*Cellia*).

baisasi Colless, 1957b (p. 137) – species of *Anopheles* (*Cellia*) [original combination: *Anopheles balabacensis baisasi*].

baisasi Knight & Hull, 1951a (p. 197) – species of *Aedes* (*Geoskusea*).

baisasi Sirivanakarn, 1972 (p. 35) – species of *Culex* (*Eumelanomyia*).

baisasi Stone & Thurman, 1958 (p. 240) – species of *Armigeres* (*Armigeres*).

bakeri Theobald, 1911a (p. 19) – synonym of *Coquillettidia* (*Coquillettidia*) *fuscopennata* (Theobald, 1903a) [original combination: *Chrysoconops bakeri*]; synonymy established before Knight & Stone, 1977.

balabacensis Baisas, 1936 (p. 65) – species of *Anopheles* (*Cellia*) [original combination: *Anopheles leucosphyrus* variety *balabacensis*].

balerensis Mendoza, 1947 (p. 173) – species of *Anopheles* (*Anopheles*).

balfouri Theobald, 1904b (p. 82) – species of *Uranotaenia* (*Uranotaenia*).

baltae Porter, 2009 (p. 808) – species of *Wyeomyia* (*Hystatomyia*).

balteatus Dyar & Knab, 1907a (p. 9) – synonym of *Aedes* (*Ochlerotatus*) *tortilis* (Theobald, 1903g); synonymy established before Knight & Stone, 1977.

balteatus Macdonald, 1960 (p. 128) – species of *Armigeres* (*Leicesteria*).

bambiotai Geoffroy, 1987 (p. 101) – species of *Aedes* (*Aedimorphus*).

bamborum Rozeboom & Komp, 1948 (p. 399) – species of *Culex* (*Anoedioporpa*).

bambusa (Yamada, 1917) (p. 61) – species of *Tripteroides* (*Tripteroides*) [original combination: *Rachionotomyia bambusa*].

bambusae Dyar, 1922h (p. 419) – *nomen nudum*, *Culex*.

bambusae Edwards, 1935a (p. 134) – species of *Aedes* (*Stegomyia*).

bambusaihole Dong, Zhou & Dong, 1997a (p. 247) – species of *Topomyia* (*Suaymyia*).

bambusicola da Cunha Ramos & Brunhes, 2004 (p. 225) – species of *Uranotaenia* (*Pseudoficalbia*).

bambusicola (Lutz & Neiva, 1913) (p. 136) – species of *Toxorhynchites* (*Lynchiella*) [original combination: *Megarhinus bambusicola*].
Synonym: *aldrichanus* Bonne-Wepster & Bonne, 1920.

bambusicolus Knight & Rozeboom, 1946 (p. 94) – species of *Aedes* (*Scutomyia*).

bambusicolus Komp, 1937b (p. 515) – species of *Anopheles* (*Kerteszia*).

bananea Wolfs, 1958 (p. 298) – species of *Aedes* (*Diceromyia*).

bancoi Geoffroy, 1987 (p. 103) – species of *Aedes* (*Aedimorphus*).

bancrofti Skuse, 1889 (p. 1740) – synonym of *Aedes* (*Stegomyia*) *aegypti* (Linnaeus, 1762) [original combination: *Culex bancrofti*]; synonymy established before Knight & Stone, 1977.

bancrofti (Taylor, 1914b) (p. 465) – species of *Verrallina* (*Verrallina*) [original combination: *Skusea bancrofti*].
Synonym: *cunninghami* Taylor, 1944.

bancroftianus Edwards, 1921c (p. 74) – species of *Aedes* (*Ochlerotatus*).

bancroftii Giles, 1902 (p. 511) – species of *Anopheles* (*Anopheles*).

bancroftii Theobald, 1901a (p. 367) – synonym of *Culex* (*Culex*) *sitiens* Wiedemann, 1828 [original combination: *Culex annulirostris bancroftii*]; synonymy by E.N. Marks (in Lee *et al*., 1989a).

bandoengensis Brug, 1939 (p. 112) – species of *Culex* (*Lophoceraomyia*) [original combination: *Culex* (*Lophoceratomyia*) *minor* variety *bandoengensis*]; specific status by Colless, 1965.

banksensis Maffi & Tenorio, 1977 (p. 103) – species of *Culex* (*Culex*).

banksi Edwards, 1922a (p. 270) – species of *Aedes* (*Collessius*).

bannaensis Gong & Lu, 1995 (p. 478) – species of *Topomyia* (*Suaymyia*).

baolini Gong, 1989b (p. 220) – species of *Topomyia* (*Topomyia*).

barbarus Dyar & Knab, 1906d (p. 210) – synonym of *Culex* (*Culex*) *quinquefasciatus* Say, 1823; synonymy established before Knight & Stone, 1977.

barberellus Evans, 1932 (p. 94) – species of *Anopheles* (*Cellia*).

barberi Coquillett, 1903b (p. 310) – species of *Anopheles* (*Anopheles*).

barbipes Edwards, 1913b (p. 47) – species of *Toxorhynchites* (*Toxorhynchites*).

barbirostris van der Wulp, 1884 (p. 248) – species of *Anopheles* (*Anopheles*).
Synonyms: *martini* Laveran, 1902a; ? *innominata* Stoker & Waktoedi Koesoemawinangoen, 1949.

barbiventris Brug, 1938 (p. 520) – subspecies of *Anopheles* (*Anopheles*) *bancroftii* Giles, 1902 [original combination: *Anopheles* (*Anopheles*) *bancroftii* variety *barbiventris*]; subspecific status by Harbach & Howard, 2007.

barbumbrosus Strickland & Chowdhury, 1927a (p. 18) – species of *Anopheles* (*Anopheles*) [*nomen novum* for *pallidus* Swellengrebel, 1919].
Synonyms: *pallidus* Swellengrebel, 1919; *pseudumbrosus* Haga, 1930.

barbus Baisas, 1946d (p. 41) – species of *Topomyia* (*Topomyia*).

baresi Cerqueira, 1960a (p. 1) – species of *Haemagogus* (*Haemagogus*).

baria Dyar & Knab, 1908 (p. 69) – synonym of *Wyeomyia* (*Wyeomyia*) *hemisagnosta* Dyar & Knab, 1906d; synonymy established before Knight & Stone, 1977.

barianensis James, 1911 (in James & Liston, 1911) (p. 76) – species of *Anopheles* (*Anopheles*).

barkerii (Theobald, 1907) (p. 623) – *nomen dubium*, *Culex* [original combination: *Philodendromyia barkerii*]; status by Sirivanakarn, 1977a.

barkhuusi Giaquinto-Mira, 1950 (p. 294) – synonym of *Anopheles* (*Cellia*) *sergentii macmahoni* Evans, 1936 [original combination: *Anopheles* (*Myzomyia*) *macmahoni* variety *barkhuusi*]; synonymy by Stone *et al.*, 1959.

barnardi Edwards, 1924a (p. 161) – species of *Aedes* (*Hopkinsius*).

barnesi Belkin, 1953 (p. 332) – species of *Uranotaenia* (*Uranotaenia*).

barraudi Baisas & Ubaldo-Pagayon, 1953 (p. 75) – species of *Tripteroides* (*Tricholeptomyia*).

barraudi Edwards, 1922a (p. 284) – species of *Culex* (*Culex*).

barraudi (Edwards, 1934) (in Barraud, 1934) (p. 447) – species of *Aedes* (*Paraedes*) [original combination: *Paraedes barraudi*].

barretoi Adames, 1971 (p. 40) – species of *Deinocerites*.

barri Rueger, 1958 (p. 34) – synonym of *Aedes* (*Ochlerotatus*) *euedes* Howard, Dyar & Knab, 1913; synonymy by Wood, 1977.

barrinus Bram, 1967a (p. 125) – species of *Culex* (*Culiciomyia*).

basalis Howard, Dyar & Knab, 1913, 1917 (fig. 251, p. 917, respectively) – synonym of *Uranotaenia* (*Uranotaenia*) *coatzacoalcos* Dyar & Knab, 1906d; synonymy established before Knight & Stone, 1977.

basalis Taylor, 1912b (p. 27) – synonym of *Verrallina* (*Verrallina*) *funerea* (Theobald, 1903a) [original combination: *Pseudoskusea basalis*]; synonymy established before Knight & Stone, 1977.

basicinctus Edwards, 1921c (p. 78) – synonym of *Culex* (*Culex*) *vicinus* (Taylor, 1916); synonymy by Stone & Knight, 1958.

basicinctus Edwards, 1922d (p. 96) – synonym of *Culex* (*Oculeomyia*) *starckeae* Stone & Knight, 1958; synonymy by Stone & Knight, 1958.

basilewskyi Leleup, 1957 (p. 69) – subspecies of *Anopheles* (*Cellia*) *garnhami* Edwards, 1930a; elevated from synonymy with *garnhami* by Brunhes *et al.*, 1998b.

basilicus Dyar & Knab, 1906b (p. 169) – synonym of *Culex* (*Phenacomyia*) *corniger* Theobald, 1903a; synonymy established before Knight & Stone, 1977.

basilii Falleroni, 1932 (p. 29) – synonym of *Anopheles* (*Anopheles*) *maculipennis* Meigen, 1818 [original combination: *Anopheles maculipennis basilii*]; synonymy by Hackett & Missiroli, 1935.

bastagarius Dyar & Knab, 1906b (p. 170) – species of *Culex* (*Melanoconion*).
Synonyms: *vapulans* Dyar, 1920d; *alfaroi* Dyar, 1921b; *innominatus* Evans, 1924; *thomasi* Evans, 1924; *cuclyx* Dyar & Shannon, 1924a.

batesi Lane & Cerqueira, 1942 (p. 669) – species of *Sabethes* (*Sabethes*).

batesi Rozeboom & Komp, 1948 (p. 403) – species of *Culex* (*Melanoconion*).

bathana (Dyar, 1928) (in Curry, 1928) (p. 244) – species of *Chagasia* [original combination: *Anopheles* (*Chagasia*) *bathanus*].

beauperthuyi Anduze, 1943b (p. 459) – synonym of *Culex* (*Culex*) *saltanensis* Dyar, 1928; synonymy established before Knight & Stone, 1977.

becki Belkin, 1962 (p. 265) – species of *Culex* (*Lophoceraomyia*).

becki Belkin, 1962 (p. 334) – species of *Aedes* (*Geoskusea*).

bedfordi Edwards, 1936 (p. 50) – species of *Aedes* (*Catageiomyia*).

behningi Martini, 1926 (p. 200) – species of *Aedes* (*Ochlerotatus*).

bejaranoi Duret, 1953, 1954a (pp. 75, 109, respectively) – species of *Culex* (*Melanoconion*).

bejaranoi Martínez, Carcavallo & Prosen, 1960 (p. 50) – species of *Aedes* (*Ochlerotatus*).

bekkui Mogi, 1977 (p. 129) – species of *Aedes* (*Edwardsaedes*).

beklemishevi Denisova, 1955 (p. 58) – synonym of *Aedes* (*Ochlerotatus*) *euedes* Howard, Dyar & Knab, 1913; synonymy by Wood, 1977.

beklemishevi Stegnii & Kabanova, 1976 (p. 198) – species of *Anopheles* (*Anopheles*).

belemensis Duret & Damasceno, 1955 (p. 404) – species of *Culex* (*Anoedioporpa*).

belenrae Rueda, 2005 (p. 2) – species of *Anopheles* (*Anopheles*).

belisarioi Neiva, 1908a (p. 351) – species of *Sabethes* (*Sabethes*).
Synonyms: *schausi* Dyar & Knab, 1908; *goeldii* Howard, Dyar & Knab, 1915; *argyronotum* Edwards, 1928a.

belkini Adames, 1971 (p. 34) – species of *Deinocerites*.

belkini Baisas & Ubaldo-Pagayon, 1953 (p. 80) – species of *Tripteroides* (*Tricholeptomyia*).

belkini Casal & García, 1966 (p. 155) – synonym of *Onirion brucei* (del Ponte & Cerqueira, 1938) [original combination: *Wyeomyia* (*Dendromyia*) *belkini*]; synonymy by Harbach & Peyton, 2000.

belkini Grjebine, 1979 (p. 65) – species of *Uranotaenia* (*Pseudoficalbia*).

belkini Stone & Penn, 1948 (p. 114) – species of *Culex* (*Acallyntrum*).

bellator Dyar & Knab, 1906b (p. 160) – species of *Anopheles* (*Kerteszia*).
Synonym: *bromelicola* Dyar, 1925d.

belleci (Le Goff, Boussès & Brunhes, 2007) (p. 191) – species of *Aedes* (*Neomelaniconion*) [original combination: *Neomelaniconion belleci*].

benarrochi Gabaldon, Cova-Garcia & Lopez, 1941a (p. 3) – species of *Anopheles* (*Nyssorhynchus*).

bengalensis Barraud, 1934 (p. 371) – species of *Culex* (*Lophoceraomyia*) [original combination: *Culex* (*Lophoceraomyia*) *minor* variety *bengalensis*]; specific status by Colless, 1965.

bengalensis Puri, 1930 (p. 955) – species of *Anopheles* (*Anopheles*) [original combination: *Anopheles aitkenii* variety *bengalensis*]; specific status by Reid, 1965.

bengalensis Rosenberg & Evenhuis, 1985 (p. 36) – species of *Toxorhynchites* (*Toxorhynchites*).

benguetensis King, 1931 (p. 753) – subspecies of *Anopheles* (*Anopheles*) *lindesayi* Giles, 1900a [original combination: *Anopheles lindesayi* variety *benguetensis*]; subspecific status by Puri, 1949.

benoiti Wolfs, 1964 (in Benoit, 1964) (p. 309) – species of *Uranotaenia* (*Uranotaenia*).

bentleyi Bentley, 1902 (p. 15) – synonym of *Anopheles* (*Anopheles*) *nigerrimus* Giles, 1900a; synonymy established before Knight & Stone, 1977.

bequaerti Dyar & Shannon, 1925b (p. 39) – species of *Culex* (*Melanoconion*).

bequaerti Wolfs, 1947a (p. 40) – species of *Aedes* (*Neomelaniconion*).

berbericus Roubaud, 1935b (p. 445) – synonym of *Culex* (*Culex*) *pipiens* Linnaeus, 1758 [original combination: *Culex pipiens berbericus*]; synonymy by Knight & Abdel-Malek, 1951.

berestnevi Shingarev, 1926 (p. 48) – synonym of *Anopheles* (*Cellia*) *superpictus* Grassi, 1899 [original combination: *Anopheles* (*Myzomyia*) *superpictus* variety *berestnevi*]; synonymy by Christophers, 1933.

bergerardi Pajot & Geoffroy, 1971 (p. 269) – species of *Aedes* (*Neomelaniconion*).

berghei Vincke & Leleup, 1949 (p. 248) – species of *Anopheles* (*Cellia*).

bergi Belkin, 1962 (p. 258) – species of *Culex* (*Lophoceraomyia*).

bergrothi (Edwards, 1921b) (p. 50) – species of *Culiseta* (*Culiseta*) [original combination: *Theobaldia bergrothi*].
Synonyms: *borealis* Shingarev, 1927; *kogievnikovi* Shingarev, 1927; *kanayamensis* Yamada, 1932.

berlandi Séguy, 1921 (p. 192) – species of *Aedes* (*Ochlerotatus*).
Synonyms: *praeteritus* Séguy, 1923; *longitubus* Cambournac, 1938; *heracleensis* Callot, 1944a.

berlini Schick, 1970a (p. 48) – species of *Aedes* (*Protomacleaya*).

bernardi (Borel, 1926) (p. 111) – *nomen dubium*, *Culex* [original combination: *Lophoceratomyia bernardi*]; status by Sirivanakarn, 1977a.

bernardi (Doucet, 1950) (p. 49) – species of *Mimomyia* (*Ingramia*) [original combination: *Ravenalites bernardi*].

bertii Cova García & Rausseo, 1964 (p. 221) – species of *Uranotaenia* (*Uranotaenia*).

bertrami Schick, 1970a (p. 38) – species of *Aedes* (*Protomacleaya*).

bervoetsi D'Haenens, 1961 (p. 189) – species of *Anopheles* (*Cellia*) [original combination: *Anopheles moucheti bervoetsi*]; specific status by Antonio-Nkondjio *et al.*, 2008.

beta Séguy, 1924 (p. 41) – species of *Culex* (*Culex*).

bevisi (Edwards, 1915b) (p. 275) – species of *Aedes* (*Aedimorphus*) [original combination: *Ochlerotatus bevisi*].

beytouti (Doucet, 1951b) (p. 112) – species of *Mimomyia* (*Ingramia*) [original combination: *Ravenalites beytouti*].

bhayungi Thurman & Thurman, 1958 (p. 186) – species of *Armigeres* (*Armigeres*).

biannulata Theobald, 1907 (p. 263) – synonym of *Aedes* (*Catageiomyia*) *tarsalis* (Newstead, 1907) [original combination: *Reedomyia biannulata*]; synonymy established before Knight & Stone, 1977.

bibulus Dyar, 1920d (p. 74) – synonym of *Culex* (*Melanoconion*) *vaxus* Dyar, 1920d; synonymy by Forattini & Sallum, 1993b.

bicincta da Cunha Ramos & Brunhes, 2004 (p. 237) – species of *Uranotaenia* (*Pseudoficalbia*).

bicki Stone & Penn, 1947 (p. 89) – species of *Culex* (*Acallyntrum*).

bickleyi Forattini, 1965 (p. 146) – species of *Culex* (*Culex*).

bickleyi Thurman, 1959 (p. 14) – species of *Toxorhynchites* (*Toxorhynchites*).

bicolor Leicester, 1908 (p. 225) – species of *Uranotaenia* (*Pseudoficalbia*).
Synonyms: *fusca* Leicester, 1908; *leicesteri* Edwards, 1913c; *kalabahensis* Haga, 1925; *lagunensis* Baisas, 1935c.

bicolor Martini, 1935 (p. 29) – synonym of *Uranotaenia* (*Uranotaenia*) *hystera* Dyar & Knab, 1913; synonymy by Galindo *et al.*, 1954.

bicolor Meigen, 1818 (p. 9) – synonym of *Culex* (*Culex*) *pipiens* Linnaeus, 1758; synonymy established before Knight & Stone, 1977.

bicompressa Lutz, 1928 (p. 33) – synonym of *Wyeomyia* (*Decamyia*) *pseudopecten* Dyar & Knab, 1906a [original combination: *Dendromyia bicompressa*]; synonymy by Lane 1953.

bicornis (Root, 1928) (in Dyar, 1928) (p. 50) – species of *Wyeomyia* (*Nunezia*) [original combination: *Dendromyia* (*Eunicemyia*) *bicornis*].
Synonym: *trujilloi* Pulido F. & Sutil O., 1981.

bicornutus (Theobald, 1910a) (p. 25) – species of *Culex* (*Lophoceraomyia*) [original combination: *Lophoceratomyia bicornuta*].

bicristatus Thurman & Winkler, 1950 (p. 239) – species of *Aedes* (*Rusticoidus*).

bidens Dyar, 1922g (p. 190) – species of *Culex* (*Culex*).

bidentata da Cunha Ramos & Brunhes, 2004 (p. 355) – species of *Uranotaenia* (*Uranotaenia*).

bifasciata da Cunha Ramos & Brunhes, 2004 (p. 102) – species of *Uranotaenia* (*Pseudoficalbia*).

bifoliata Dyar, 1922d (p. 94) – synonym of *Culex* (*Anoedioporpa*) *conservator* Dyar & Knab, 1906d; synonymy established before Knight & Stone, 1977.

bifoliata (King & Hoogstraal, 1947b) (p. 120) – species of *Verrallina* (*Verrallina*) [original combination: *Aedes* (*Aedes*) *bifoliatus*].

bifoliata Osorno-Mesa & Muñoz-Sarmiento, 1948 (p. 105) – *nomen nudum*, *Anopheles* [original combination: *Anopheles pseudopunctipennis* variety *bifoliata* [*sic*]]; status by Harbach & Howard, 2007.

bifoliata Theobald, 1905g (p. 31) – synonym of *Culex* (*Culex*) *duttoni* Theobald, 1901d; synonymy by Edwards, 1941.

bifoliolatus Duret & Barreto, 1956 (p. 96) – species of *Culex* (*Melanoconion*).

bifurcata Dong, Wang & Lu, 1995 (p. 484) – species of *Topomyia* (*Topomyia*).

bifurcatus Linnaeus, 1758 (p. 603) – synonym of *Culex* (*Culex*) *pipiens* Linnaeus, 1758; synonymy established before Knight & Stone, 1977; fixed by Harbach *et al*., 1985.

bigoti (Bellardi, 1862) (p. 3) – species of *Lutzia* (*Lutzia*) [original combination: *Culex bigoti*].
Synonyms: *brasiliae* Dyar, 1923b; *patersoni* Shannon & del Ponte, 1928.

bigotii Theobald, 1901a (p. 135) – synonym of *Anopheles* (*Nyssorhynchus*) *pictipennis* (Philippi, 1865); synonymy established before Knight & Stone, 1977.

bihaicolus Dyar & Nuñez Tovar, 1928 (p. 91) – species of *Culex* (*Carrollia*).

bihamatus Edwards, 1926a (p. 107) – species of *Culex* (*Culex*).

bilineata Theobald, 1909b (p. 21) – species of *Uranotaenia* (*Uranotaenia*).

bilineatus Theobald, 1903a (p. 196) – species of *Culex* (*Culex*); elevated from synonymy with *dolosus* (Lynch Arribálzaga, 1891b) by Laurito *et al*., 2013.

bilobatus Galindo & Blanton, 1954 (p. 236) – synonym of *Culex* (*Melanoconion*) *sardinerae* Fox, 1953; synonymy by Galindo & Blanton, 1955.

bimaculata Leicester, 1908 (p. 226) – species of *Uranotaenia* (*Pseudoficalbia*).

bimaculata Theobald, 1909b (p. 22) – synonym of *Uranotaenia* (*Pseudoficalbia*) *nigromaculata* Edwards, 1941; synonymy established before Knight & Stone, 1977.

bimaculata Theobald, 1910c (p. 393) – synonym of *Lutzia* (*Metalutzia*) *tigripes* de Grandpre & de Charmoy, 1901 [original combination: *Culex tigripes* variety *bimaculata* [*sic*]]; synonymy by Edwards, 1941.

bimaculatus (Coquillett, 1902c) (p. 84) – species of *Aedes* (*Ochlerotatus*) [original combination: *Culex bimaculatus*].
Synonym: *rozeboomi* Vargas, 1941b.

bimaculiala Leicester, 1908 (p. 208) – species of *Uranotaenia* (*Uranotaenia*).

bimaculipes (Theobald, 1905f) (p. 114) – species of *Tripteroides* (*Tripteroides*) [original combination: *Phoniomyia bimaculipes*].
Synonym: *ornata* Taylor, 1914d.

binigrolineatus Knight & Rozeboom, 1945 (p. 289) – species of *Culex* (*Acallyntrum*).

binotatus Belkin, 1950 (as *binotata*) (p. 264) – species of *Tripteroides* (*Tripteroides*).

biocellatus (Taylor, 1914b) (p. 463) – species of *Aedes* (subgenus uncertain) [original combination: *Culex biocellatus*].

biocellatus Theobald, 1903a (p. 224) – synonym of *Culex* (*Culex*) *nigripalpus* Theobald, 1901c; synonymy by Stone *et al*., 1959.

bipartipes Dyar & Knab, 1906a (p. 136) – species of *Sabethes* (*Sabethes*).
Synonym: *chroiopus* Dyar & Knab, 1913.

bipunctata Theobald, 1907 (p. 374) – synonym of *Culex* (*Culex*) *gelidus* Theobald, 1901c [original combination: *Leucomyia gelidus* variety *bipunctata*]; synonymy established before Knight & Stone, 1977.

bipunctata Theobald, 1909b (p. 5) – synonym of *Aedes* (*Catageiomyia*) *tarsalis* (Newstead, 1907) [original combination: *Reedomyia bipunctata*]; synonymy established before Knight & Stone, 1977. **Note**: this nominal taxon is the junior secondary homonym of *bipunctatus* Robineau- Desvoidy, 1827.

bipunctatus Robineau-Desvoidy, 1827 (p. 405) – synonym of *Aedes* (*Ochlerotatus*) *flavescens* (Müller, 1764) [original combination: *Culex bipunctatus*]; synonymy established before Knight & Stone, 1977. **Note**: this nominal taxon is the senior secondary homonym of *bipunctata* Theobald, 1909b.

biroi Theobald, 1905f (p. 82) – synonym of *Culex* (*Culex*) *tritaeniorhynchus* Giles, 1901a; synonymy established before Knight & Stone, 1977.

bironelli Christophers, 1924a (p. 16) – *nomen novum* for *gracilis* Theobald, 1905f [original combination: *Anopheles* (*Bironella*) *bironelli*]; rejected name, suppression by International Commission on Zoological Nomenclature, 1965.

bisanensis Suzuki, Tanimura, Miyagawa & Murata, 1953 (p. 9) – synonym of *Aedes* (*Hulecoeteomyia*) *japonicus* (Theobald, 1901a); synonymy by Sasa & Kamimura, 1971.

biscaynensis Zavortink & O'Meara, 1999 (p. 263) – species of *Culex* (*Micraedes*).

bisignata Grünberg, 1905 (p. 378) – synonym of *Anopheles* (*Cellia*) *funestus* Giles, 1900b [original combination: *Myzomyia bisignata*]; synonymy by Evans, 1938. **Note**: this nominal taxon is the senior secondary homonym of *bisignatus* Hoffmann, 1938.

bisignatus Hoffmann, 1938 (p. 176) – synonym of *Anopheles* (*Nyssorhynchus*) *albimanus* Wiedemann, 1820 [original combination: *Anopheles albimanus* variety *bisignatus*]; synonymy by Ross & Roberts, 1943. **Note**: this nominal taxon is the junior secondary homonym of *bisignata* Grünberg, 1905.

biskraensis Brunhes, 1999 (p. 25) – species of *Aedes* (*Ochlerotatus*).

bisquamatus Lee, 1946 (p. 252) – species of *Tripteroides* (*Rachisoura*).

bisulcatus (Coquillett, 1906a) (p. 185) – species of *Culex* (*Micraedes*) [original combination: *Micraedes bisulcatus*].

bitaeniorhynchus Giles, 1901a (p. 607) – species of *Culex* (*Oculeomyia*).
Synonyms: *ager* Giles, 1901c; *maculipesarabiensis* Patton, 1905; *ethiopicus* Edwards, 1912c; *abdominalis* Taylor, 1913; *karatsuensis* Mochizuki, 1913.

biunguiculatus (Harbach, 2014) (in Harbach & Taai, 2014) (p. 428) – species of *Aedes* (*Nyctomyia*) [original combination; *Nyctomyia biunguiculata*].

blacklocki Evans, 1925a (p. 121) – species of *Aedes* (*Stegomyia*).

blanchardi Surcouf & Gonzalez-Rincones, 1911 (p. 120) – synonym of *Psorophora* (*Psorophora*) *lineata* (von Humboldt, 1819); synonymy established before Knight & Stone, 1977.

bodkini Edwards, 1922d (p. 81) – synonym of *Wyeomyia* (*Dodecamyia*) *aphobema* Dyar, 1918e; synonymy by Lane & Cerqueira, 1942.
bogotanus Arnell, 1976 (p. 22) – species of *Aedes* (*Ochlerotatus*).
boharti Knight & Rozeboom, 1946 (p. 90) – species of *Aedes* (*Scutomyia*).
boharti Brookman & Reeves, 1950 (p. 159) – species of *Culex* (*Neoculex*) [*nomen novum* for *reevesi* Bohart, 1949].
Synonym: *reevesi* Bohart, 1949.
bokorensis Klein & Sirivanakarn, 1970 (p. 587) – species of *Culex* (*Eumelanomyia*).
bolensis Edwards, 1936 (p. 52) – species of *Aedes* (*Neomelaniconion*).
bolii Sirivanakarn, 1968 (p. 100) – species of *Culex* (*Lophoceraomyia*).
boliviensis (Theobald, 1905f) (p. 66) – species of *Anopheles* (*Kerteszia*) [original combination: *Kerteszia boliviensis*].
bonariensis Brèthes, 1916 (p. 213) – synonym of *Culex* (*Culex*) *dolosus* (Lynch Arribálzaga, 1891b); synonymy established before Knight & Stone, 1977.
boneti Gil Collado, 1936 (p. 323) – species of *Aedes* (*Bifidistylus*).
boninensis Bohart, 1957 (p. 75) – species of *Culex* (*Sirivanakarnius*).
bonneae Dyar & Knab, 1919 (p. 3) – species of *Culex* (*Culex*).
bonneae Edwards, 1930b (p. 542) – species of *Mansonia* (*Mansonioides*).
bonneae Mattingly, 1958a (p. 32) – species of *Aedes* (*Paraedes*).
bonneae Root, 1927b (p. 474) – species of *Chagasia*.
bonnei Dyar, 1921k (p. 155) – species of *Culex* (*Carrollia*).
bonnei (Lane & Cerqueira, 1942) (p. 630) – species of *Wyeomyia* (*Phoniomyia*) [original combination: *Phoniomyia bonnei*].
bonneorum da Fonseca & da Silva Ramos, 1940 (as *bonnei*) (p. 386) – synonym of *Anopheles* (*Anopheles*) *costai* da Fonseca & da Silva Ramos, 1940; synonymy and emendation by Sallum *et al*., 1999.
bonneti Belkin, 1962 (p. 501) – species of *Tripteroides* (*Tripteroides*).
bonneti Senevet, 1938 (p. 187) – synonym of *Culex* (*Melanoconion*) *elevator* Dyar & Knab, 1906d; synonymy by Rozeboom & Komp, 1950a.
bonnewepsterae van den Assem, 1958 (p. 205) – synonym of *Mansonia* (*Mansonioides*) *septempunctata* Theobald, 1905f; synonymy by Lee *et al*., 1988b.
borealis Ludlow, 1911a (p. 178) – synonym of *Aedes* (*Ochlerotatus*) *communis* (de Geer, 1776) [original combination: *Culex borealis*]; synonymy established before Knight & Stone, 1977.
borealis Shingarev, 1927 (p. 548) – synonym of *Culiseta* (*Culiseta*) *bergrothi* (Edwards, 1921b) [original combination: *Theobaldia borealis*]; synonymy by Natvig, 1942.
borinqueni Root, 1922 (p. 400) – synonym of *Culex* (*Melanoconion*) *erraticus* (Dyar & Knab, 1906d); synonymy by King & Bradley, 1937.
borneensis McArthur, 1949 (p. 49) – species of *Anopheles* (*Anopheles*) [original combination: *Anopheles aitkeni* variety *borneensis*]; specific status by Reid, 1968.
boscii Robineau-Desvoidy, 1827 (p. 413) – synonym of *Psorophora* (*Psorophora*) *ciliata* (Fabricius, 1794); synonymy established before Knight & Stone, 1977.
boshelli Osorno-Mesa, 1944 (p. 165) – species of *Haemagogus* (*Haemagogus*).
Synonym: *garciai* Leví-Castillo, 1955a.
bosseri Grjebine, 1979 (p. 57) – species of *Uranotaenia* (*Pseudoficalbia*).
bostocki Theobald, 1905g (p. 29) – synonym of *Culex* (*Maillotia*) *salisburiensis* Theobald, 1901c; synonymy established before Knight & Stone, 1977.
bougainvillensis (Belkin, 1962) (p. 294) – species of *Mimomyia* (*Etorleptiomyia*) [original combination: *Ficalbia* (*Etorleptiomyia*) *bougainvillensis*].
bougainvillensis Marks, 1947a (p. 19) – species of *Aedes* (*Finlaya*).
bougainvillensis Steffan, 1979 (p. 138) – species of *Culex* (*Acallyntrum*).
bourrouli (Lutz, 1905) (p. 347) – species of *Wyeomyia* (*Spilonympha*) [original combination: *Dendromyia bourrouli*].
boussesi da Cunha Ramos & Brunhes, 2004 (p. 151) – species of *Uranotaenia* (*Pseudoficalbia*).
boydi Vargas, 1939 (p. 361) – synonym of *Anopheles* (*Anopheles*) *franciscanus* McCracken, 1904; synonymy by Ross & Roberts, 1943.
bozasi Neveu-Lemaire, 1905a (p. 32) – synonym of *Anopheles* (*Cellia*) *pharoensis* Theobald, 1901a [original combination: *Nyssorhynchus bozasi*]; synonymy established before Knight & Stone, 1977.
brachiatus Hutchings & Sallum, 2008 (p. 45) – species of *Culex* (*Melanoconion*).
brachypus Dönitz, 1902 (p. 52) – *nomen dubium*, *Anopheles*; status by Edwards, 1932a (? = *albotaeniatus* Theobald, 1903a).
bracteatus Coquillett, 1906a (p. 184) – synonym of *Aedes* (*Ochlerotatus*) *tortilis* (Theobald, 1903g) [original combination: *Culex bracteatus*]; synonymy established before Knight & Stone, 1977.
bradleyi King, 1939 (p. 468) – species of *Anopheles* (*Anopheles*) [original combination: *Anopheles crucians* variety *bradleyi*]; specific status by King & Bradley, 1941.
brahmacharii Christophers, 1912 (in Brahmachari Bahadur, 1912) (p. 186) – synonym of *Anopheles* (*Cellia*) *aconitus* Dönitz, 1902 [original combination: *Myzomyia brahmacharii*]; synonymy established before Knight & Stone, 1977.
brami Forattini, Rabello & Souza Lopes, 1967 (p. 13) – species of *Culex* (*Culex*).
brasiliae Dyar, 1923b (p. 67) – synonym of *Lutzia* (*Lutzia*) *bigoti* Bellardi, 1862; synonymy by Lane, 1936c.

brayi Knight, 1947b (p. 73) – synonym of *Aedes* (*Christophersiomyia*) *ibis* Barraud, 1931b; synonymy by Abercrombie, 1977.

braziliensis (Chagas, 1907) (p. 18) – species of *Anopheles* (*Nyssorhynchus*) [original combination: *Cellia braziliensis*].
Synonym: *pessoai* Galvão & Lane, 1937a.

braziliensis Gordon & Evans, 1922 (p. 329) – species of *Aedes* (*Protomacleaya*) [original combination: *Aedes* (*Finlaya*) *oswaldi* [*sic*] variety *braziliensis*]; specific status by Belkin, 1968b.

breedensis Muspratt, 1953 (p. 55) – species of *Aedes* (*Ochlerotatus*).

brehmei Knab, 1916 (p. 161) – synonym of *Culex* (*Culex*) *restuans* Theobald, 1901c; synonymy established before Knight & Stone, 1977.

breinli (Taylor, 1914d) (p. 186) – species of *Armigeres* (*Armigeres*) [original combination: *Neosquamomyia breinli*].

breinli Taylor, 1943 (p. 156) – synonym of *Anopheles* (*Cellia*) *novaguinensis* Venhuis, 1933; synonymy by Lee & Woodhill, 1944.

brelandi Zavortink, 1972 (p. 25) – species of *Aedes* (*Protomacleaya*).

brenguesi Brunhes & Ravaonjanahary, 1973 (p. 169) – species of *Culex* (*Eumelanomyia*).

brenguesi Rickenbach & Lombrici, 1975 (p. 39) – species of *Eretmapodites*.

brethesi Dyar, 1919e (p. 86) – species of *Culex* (*Culex*).

brevicellulus Theobald, 1901c (p. 212) – synonym of *Coquillettidia* (*Coquillettidia*) *crassipes* (van der Wulp, 1881) [original combination: *Taeniorhynchus brevicellulus*]; synonymy established before Knight & Stone, 1977.

breviculus Senevet & Abonnenc, 1939a (p. 110) – species of *Culex* (*Tinolestes*).
Synonym: *mojuensis* Duret & Damasceno, 1955.

brevipalpis Brug, 1934 (p. 508) – species of *Tripteroides* (*Tripteroides*).

brevipalpis (Giles, 1902) (p. 384) – species of *Culex* (*Eumelanomyia*) [original combination: *Stegomyia brevipalpis*].
Synonyms: *longipes* Theobald, 1901c; *macropus* Blanchard, 1905; *uniformis* Leicester, 1908; *fidelis* Dyar, 1920i.

brevipalpis Roper, 1914 (p. 141) – species of *Anopheles* (*Anopheles*).

brevipalpis Theobald, 1901a (p. 245) – species of *Toxorhynchites* (*Toxorhynchites*).
Synonyms: *loewi* Giebel, 1862; *marshallii* Theobald, 1903a.

brevipalpus (Theobald, 1905f) (p. 96) – species of *Culex* (*Lophoceraomyia*) [original combination: *Lophoceratomyia brevipalpus*].

brevipes (da Costa Lima, 1931a) (p. 68) – species of *Trichoprosopon* [original combination: *Isostomyia brevipes*].

brevirhynchus Brug, 1934 (p. 503) – species of *Tripteroides* (*Rachisoura*).

brevirostris Edwards, 1915c (p. 284) – synonym of *Uranotaenia* (*Pseudoficalbia*) *moultoni* Edwards, 1914c; synonymy by Peyton, 1972.

brevirostris Reid, 1950 (p. 300) – species of *Anopheles* (*Anopheles*).

brevis Berlin, 1969a (p. 82) – species of *Aedes* (*Howardina*).

brevis Edwards, 1941 (p. 242) – subspecies of *Eretmapodites plioleucus* Edwards, 1941.

breviseta da Cunha Ramos & Brunhes, 2004 (p. 107) – species of *Uranotaenia* (*Pseudoficalbia*).

brevispinosus Bonne-Wepster & Bonne, 1920 (p. 171) – species of *Culex* (*Culex*).

brevitibia (Edwards, 1914c) (p. 125) – species of *Aedes* (*Alanstonea*) [original combination: *Armigeres brevitibia*].

brevivittatus Berlin, 1969a (p. 116) – species of *Aedes* (*Howardina*).

bricenoi Cova García, Pulido, Escalante de Ugueto, Amarista & Mora R., 1987c (p. 65) – species of *Uranotaenia* (*Uranotaenia*).

briseis Dyar, 1925g (p. 216) – species of *Uranotaenia* (*Uranotaenia*).

britanniae Strickland & Chowdhury, 1931 (pp. 3, 12) – subspecies of *Anopheles* (*Anopheles*) *aitkenii* James, 1903 [original combination: *Anopheles aitkenii* variety *britanniae*]; because *britanniae* was proposed expressly for a varietal entity, it has subspecific rank from its original publication (International Commission on Zoological Nomenclature, 1999: *Code*, Article 45.6.4).

britteni Marks & Hodgkin, 1958 (p. 33) – species of *Aedes* (subgenus uncertain).

brittoni Felt, 1905b (p. 79) – synonym of *Culiseta* (*Culicella*) *morsitans* (Theobald, 1901c) [original combination: *Culex brittoni*]; synonymy by Wood *et al.*, 1979.

brohieri Edwards, 1929b (p. 322) – species of *Anopheles* (*Cellia*) [original combination: *Anopheles theileri* variety *brohieri*].
Synonyms: *septentrionalis* Evans, 1934; *masseguini* Hamon, 1954a; *gilroyi* Service, 1960.

bromeliae (Theobald, 1911a) (p. 10) – species of *Aedes* (*Stegomyia*) [original combination: *Stegomyia bromeliae*].

bromeliarum Dyar & Knab, 1906a (p. 138) – synonym of *Wyeomyia* (*Wyeomyia*) *arthrostigma* (Lutz, 1905); synonymy by Lane & Cerqueira, 1942.

bromelicola Dyar, 1925c (p. 27) – synonym of *Anopheles* (*Kerteszia*) *bellator* Dyar & Knab, 1906b [original combination: *Anopheles* (*Dendropaedium*) *bellator* race *bromelicola*]; synonymy established before Knight & Stone, 1977.

broquetii Theobald, 1913c (p. 179) – synonym of *Aedes* (*Ochlerotatus*) *dorsalis* (Meigen, 1830b) [original combination: *Grabhamia broquetii*]; synonymy established before Knight & Stone, 1977.

brottesi Rickenbach, 1967 (p. 374) – species of *Eretmapodites*.

broussesi Edwards, 1929a (species) (p. 83) – subspecies of *Anopheles* (*Cellia*) *rufipes* (Gough, 1910); subspecific status by Rioux, 1961.
Synonyms: *ingrami* Edwards, 1929b; *seneveti* Rioux, 1959; *brucechwatti* Hamon, Taufflieb & Dyemkouma, 1961.

browni Komp, 1936c (p. 326) – species of *Culex* (*Anoedioporpa*).

browni Mattingly, 1955 (in Mattingly & Brown, 1955) (p. 75) – species of *Uranotaenia* (*Pseudoficalbia*).

brownscutumus Dong, Zhou & Dong, 1999 (p. 133) – species of *Aedes* (*Bothaella*).

brucechwatti Hamon, Taufflieb & Dyemkouma, 1961 (p. 25) – synonym of *Anopheles* (*Cellia*) *rufipes broussesi* Edwards, 1929a; synonymy established before Knight & Stone, 1977.

brucei (del Ponte & Cerqueira, 1938) (p. 231) – species of *Onirion* [original combination: *Wyeomyia* (*Dendromyia*) *brucei*].

brucei Service, 1960 (p. 85) – species of *Anopheles* (*Cellia*).

bruchi Petrocchi, 1927 (in Shannon & del Ponte, 1927) (p. 728) – synonym of *Psorophora* (*Janthinosoma*) *albigenu* (Lutz, 1908); synonymy by Stone, 1967.

brugi Edwards, 1924b (p. 371) – synonym of *Aedes* (*Edwardsaedes*) *imprimens* (Walker, 1859); synonymy by Mattingly, 1961.

brugi Edwards, 1927b (p. 356) – synonym of *Tripteroides* (*Polylepidomyia*) *argenteiventris* (Theobald, 1905f) [original combination: *Rachionotomyia brugi*]; synonymy by Lee, 1946 (as synonym of *atra* Taylor, 1914d).

brugi Soesilo & van Slooten, 1931a, 1931b (pp. 1256, 126, respectively) – synonym of *Bironella* (*Neobironella*) *papuae* (Swellengrebel & Swellengrebel de Graaf, 1920a) [original combination: *Bironella papuae* variety *brugi*]; synonymy by Tenorio, 1977.

brumpti Doucet, 1951b (p. 111) – species of *Uranotaenia* (*Pseudoficalbia*).

brumpti Galliard, 1931a (p. 134) – species of *Culex* (*Culex*).

brumpti Hamon & Rickenbach, 1955a (p. 342) – species of *Anopheles* (*Cellia*).

brumpti Neveu-Lemaire, 1905b (p. 9) – synonym of *Aedes* (*Fredwardsius*) *vittatus* (Bigot, 1861) [original combination: *Stegomyia brumpti*]; synonymy established before Knight & Stone, 1977.

brunhesi Grjebine, 1979 (p. 61) – species of *Uranotaenia* (*Pseudoficalbia*).

brunhesi Reinert, 1999a (p. 55) – species of *Aedes* (*Zavortinkius*).

brunhesi Ribeiro, 2004 (p. 250) – species of *Toxorhynchites* (*Afrorhynchus*).

brunnipes (Theobald, 1910c) (p. 64) – species of *Anopheles* (*Cellia*) [original combination: *Nyssorhynchus brunnipes*].

brygooi Brunhes, 1971 (p. 335) – species of *Aedes* (*Zavortinkius*).

brygooi Grjebine, 1986 (p. 119) – species of *Mimomyia* (*Ingramia*).

buenaventura Schick, 1970a (p. 52) – species of *Aedes* (*Protomacleaya*).

bukavuensis Wolfs, 1947b (p. 159) – species of *Culex* (*Culex*).

bulkleyi Causey, 1937a (p. 543) – species of *Anopheles* (*Anopheles*).

bunanoki Sasa & Ishimura, 1951 (p. 103) – synonym of *Aedes* (subgenus uncertain) *oreophilus* (Edwards, 1916); synonymy by Kamimura, 1968.

burgeri Zavortink, 1972 (p. 42) – species of *Aedes* (*Protomacleaya*).

burgosi Baisas, 1946c (p. 27) – synonym of *Aedes* (*Hulecoeteomyia*) *rizali* (Banks, 1906b); synonymy by Knight, 1968.

burjaticus Kuchartshuk, 1973 (p. 98) – species of *Aedes* (*Ochlerotatus*).

burkii Lane, 1936b (p. 194) – synonym of *Uranotaenia* (*Uranotaenia*) *ditaenionota* Prado, 1931b; synonymy by Lane, 1943.

burnetti Belkin, 1962 (p. 363) – species of *Aedes* (*Finlaya*).

burnsi Basio & Reisen, 1971 (p. 58) – species of *Aedes* (*Stegomyia*).

burpengaryensis (Theobald, 1905g) (p. 26) – species of *Aedes* (*Ochlerotatus*) [original combination: *Culicada burpengaryensis*].

busckii (Coquillett, 1906e) (p. 60) – species of *Aedes* (*Howardina*) [original combination: *Stegomyia busckii*].
Synonyms: *thaxteri* Dyar & Knab, 1919; *tracei* Senevet & Quiévreux, 1941.

bustamantei Galvão, 1955 (p. 255) – species of *Anopheles* (*Anopheles*).

butleri (Theobald, 1901c) (p. 230) – species of *Verrallina* (*Verrallina*) [original combination: *Aedes butleri*].
Synonyms: *diurna* Theobald, 1903f; *hatiensis* Carter, 1910b; *umbrosus* Brug, 1924; *carteri* Wijesundara, 1951.

buxtoni Belkin, 1962 (p. 356) – species of *Aedes* (*Patmarksia*).

buxtoni (Edwards, 1923b) (p. 9) – species of *Coquillettidia* (*Coquillettidia*) [original combination: *Taeniorhynchus* (*Coquillettidia*) *buxtoni*].

buxtoni Edwards, 1926a (p. 106) – species of *Culex* (*Lophoceraomyia*) [original combination: *Culex* (*Lophoceratomyia*) *hilli* variety *buxtoni*]; specific status by Belkin, 1962.

buxtoni Service, 1958 (p. 176) – species of *Anopheles* (*Cellia*).

bwamba van Someren, 1950c (p. 67) – subspecies of *Aedes* (*Aedimorphus*) *tricholabis* Edwards, 1941.

bwambae White, 1985 (p. 506) – species of *Anopheles* (*Cellia*).

bwambanus Edwards, 1941 (p. 325) – subspecies of *Culex* (*Culex*) *andersoni* Edwards, 1914b.

Cc

caballus (Theobald, 1912a) (p. 93) – species of *Aedes* (*Ochlerotatus*) [original combination: *Grabhamia caballa*].

cabrerai Cagampang-Ramos & Darsie, 1969 (in Darsie & Cagampang-Ramos, 1969) (p. 34) – synonym of *Anopheles* (*Cellia*) *ludlowae* (Theobald, 1903a) [original combination: *Anopheles ludlowae cabrerai*]; synonymy by Darsie & Cagampang-Ramos, 1977.

cabrerai Miyagi, Toma & Rivera, 1983 (p. 1) – species of *Topomyia* (*Topomyia*).

cachani (Doucet, 1950) (p. 45) – species of *Uranotaenia* (*Pseudoficalbia*) [original combination: *Aedes* (*Skusea*) *cachani*].

cacharanus (Barraud, 1923e) (p. 406) – species of *Aedes* (*Phagomyia*) [original combination: *Finlaya cacharana*].

cacodela Dyar & Knab, 1909d (p. 265) – synonym of *Wyeomyia* (*Decamyia*) *ulocoma* (Theobald, 1903a); synonymy established before Knight & Stone, 1977.

cacophrades Dyar & Knab, 1909d (p. 266) – synonym of *Limatus paraensis* (Theobald, 1903a); synonymy by Lane & Cerqueira, 1942.

cacothius Dyar, 1923a (p. 44) – synonym of *Aedes* (*Ochlerotatus*) *ventrovittis* Dyar, 1916b; synonymy established before Knight & Stone, 1977.

cacozelus Marks, 1963b (p. 45) – species of *Aedes* (*Ochlerotatus*).

caecus (Theobald, 1901a) (p. 413) – species of *Aedes* (*Aedimorphus*) [original combination: *Culex caecus*].

caerulea Theobald, 1908c (p. 267) – synonym of *Uranotaenia* (*Uranotaenia*) *pallidocephala* Theobald, 1908c [original combination: *Uranotaenia pallidocephala caerulea*]; synonymy established before Knight & Stone, 1977.

caeruleocephala Theobald, 1901d (p. 14) – species of *Uranotaenia* (*Uranotaenia*).

caeruleocephala Theobald, 1910c (p. 577) – synonym of *Tripteroides* (*Rachionotomyia*) *affinis* (Edwards, 1913c) [original combination: *Phoniomyia caeruleocephala*]; synonymy established before Knight & Stone, 1977.

caeruleocephalus (Leicester, 1908) (p. 233) – species of *Tripteroides* (*Tripteroides*) [original combination: *Colonemyia caeruleocephala*].

caeruleovittata Ludlow, 1911b (p. 131) – synonym of *Malaya genurostris* Leicester, 1908 [original combination: *Harpagomyia caeruleovittata*]; synonymy by Bohart, 1945.

caeruleus King & Hoogstraal, 1947c (p. 67) – synonym of *Culex* (*Lophoceraomyia*) *digoelensis* Brug, 1932; synonymy by Sirivanakarn, 1973a.

caillardi Rickenbach, Ferrara & Eouzan, 1968 (p. 297) – species of *Eretmapodites*.

cairnsensis (Taylor, 1919) (p. 829) – species of *Culex* (subgenus uncertain, determined by Mattingly & Marks, 1955) [original combination: *Pseudoskusea cairnsensis*].

cairnsensis Taylor, 1919 (p. 837) – synonym of *Culex* (*Lophoceraomyia*) *insequens* Marks, 1989 [original combination: *Lophoceratomyia cairnsensis*]; synonymy by Lee *et al.*, 1989b.

cairnsensis Taylor, 1919 (p. 839) – synonym of *Uranotaenia* (*Uranotaenia*) *lateralis* Ludlow, 1905c; synonymy established before Knight & Stone, 1977.

cairnsensis Taylor, 1919 (p. 842) – species of *Hodgesia*.

calabarensis Edwards, 1941 (p. 265) – species of *Culex* (*Eumelanomyia*).

calabyi Marks, 1963a (p. 204) – species of *Aedes* (*Macleaya*).

calcariae Marks, 1957a (p. 74) – species of *Aedes* (*Ochlerotatus*).

calceatus Edwards, 1924 (in Haworth, 1924) (p. 197) – species of *Aedes* (*Stegomyia*).

calcitrans Robineau-Desvoidy, 1827 (p. 409) – synonym of *Culex* (*Culex*) *pipiens* Linnaeus, 1758; synonymy established before Knight & Stone, 1977.

calderoni Wilkerson, 1991 (p. 25) – species of *Anopheles* (*Anopheles*).

caledonicus (Edwards, 1922d) (p. 100) – species of *Tripteroides* (*Polylepidomyia*) [original combination: *Rachionotomyia caledonica*].

californica Bohart, 1950 (p. 399) – synonym of *Orthopodomyia signifera* (Coquillett, 1896); synonymy by Zavortink, 1968.

caliginosa Philip, 1931 (p. 190) – species of *Uranotaenia* (*Uranotaenia*).

caliginosus de Meillon, 1943 (p. 91) – species of *Anopheles* (*Anopheles*) [original combination: *Anopheles coustani* subspecies *caliginosus*]; specific status by Gillies & de Meillon, 1968.

caliginosus (Graham, 1910) (p. 268) – species of *Aedes* (*Aedimorphus*) [original combination: *Culex caliginosus*].

callithotrys Dyar, 1920a (p. 16) – synonym of *Aedes* (*Ochlerotatus*) *campestris* Dyar & Knab, 1907d; synonymy established before Knight & Stone, 1977.

calloti Rioux & Pech, 1959 (p. 117) – synonym of *Culex* (*Culex*) *pipiens* Linnaeus, 1758 [original combination: *Culex pipiens calloti*]; synonymy by Harbach, 1988.

calopus Meigen, 1818 (p. 3) – synonym of *Aedes* (*Stegomyia*) *aegypti* (Linnaeus, 1762) [original combination: *Culex calopus*]; synonymy established before Knight & Stone, 1977.

calosomata Dyar & Knab, 1907c (p. 200) – species of *Uranotaenia* (*Uranotaenia*).
Synonym: *albitarsis* Gordon & Evans, 1922.

calumnior Belkin, Heinemann & Page, 1970 (p. 169) – species of *Aedes* (*Ochlerotatus*).

calurus Edwards, 1935a (p. 136) – species of *Culex* (*Culex*).

camaronis Ribeiro, 1991 (p. 50) – species of *Toxorhynchites* (*Toxorhynchites*).

cambournaci Hamon & Gandara, 1955 (p. 866) – species of *Culex* (*Culiciomyia*).

cambournaci Roubaud & Treillard, 1936 (p. 730) – synonym of *Anopheles* (*Anopheles*) *atroparvus* van Thiel, 1927 [original combination: *Anopheles maculipennis* biotype *cambournaci*]; synonymy by Bates, 1940.

cameronensis Edwards, 1929b (p. 323) – subspecies of *Anopheles* (*Anopheles*) *lindesayi* Giles, 1900a [original combination: *Anopheles lindesaii* [*sic*] variety *cameronensis*].

cameroni de Meillon & Evans, 1935 (p. 91) – species of *Anopheles* (*Cellia*).

campana Schick, 1970a (p. 79) – species of *Aedes* (*Protomacleaya*).

campestris Dyar & Knab, 1907d (p. 213) – species of *Aedes* (*Ochlerotatus*).

campestris Leicester, 1908 (p. 213) – species of *Uranotaenia* (*Uranotaenia*).
Synonym: *callithotrys* Dyar, 1920a.

campestris Reid, 1962 (p. 15) – species of *Anopheles* (*Anopheles*).

campilunati Carter & Wijesundara, 1948 (p. 143) – species of *Culex* (*Eumelanomyia*).

camposanus Dyar, 1918e (p. 128) – synonym of *Aedes* (*Ochlerotatus*) *scapularis* (Rondani, 1848); synonymy by Arnell, 1976.

camposi Dyar, 1925d (p. 28) – species of *Culex* (*Culex*) [original combination: *Culex coronator* race *camposi*].

camptocomma Dyar, 1924d (p. 120) – synonym of *Wyeomyia* (*Wyeomyia*) *scotinomus* (Dyar & Knab, 1907c); synonymy by Lane & Cerqueira, 1942.

camptorhynchus (Thomson, 1869) (p. 443) – species of *Aedes* (*Ochlerotatus*) [original combination: *Culex camptorhynchus*].
Synonyms: *labeculosus* Coquillett, 1905; *westralis* Strickland, 1911a; *inornata* Strickland, 1911b; *victoriensis* Taylor, 1914b; *annulipes* Taylor, 1914c; *nigra* Taylor, 1914c.

campylostylus (Laffoon, 1946) (p. 240) – species of *Verrallina* (*Neomacleaya*) [original combination: *Aedes* (*Aedes*) *campylostylus*].

canaanensis Lane & Whitman, 1943 (p. 398) – species of *Culex* (*Anoedioporpa*).

canadensis (Theobald, 1901c) (p. 3) – species of *Aedes* (*Ochlerotatus*) [original combination: *Culex canadensis*].
Synonym: *nivitarsis* Coquillett, 1904c.

canariensis Pittaluga, 1905 (p. 432) – synonym of *Aedes* (*Stegomyia*) *aegypti* (Linnaeus, 1762) [original combination: *Stegomyia calopus* variety *canariensis*]; synonymy established before Knight & Stone, 1977.

cancer Leicester, 1908 (p. 215) – synonym of *Uranotaenia* (*Uranotaenia*) *lateralis* Ludlow, 1905c; synonymy established before Knight & Stone, 1977.

cancer Theobald, 1901c (p. 215) – species of *Deinocerites*.
Synonym: *tetraspathus* Dyar & Knab, 1909d.

cancricomes Edwards, 1922a (p. 272) – species of *Aedes* (*Cancraedes*).

candelabrifer Brug, 1939 (p. 99) – species of *Armigeres* (*Armigeres*).

candidiensis Koidzumi, 1924 (p. 98) – synonym of *Anopheles* (*Cellia*) *jeyporiensis* James, 1902; synonymy by Harrison, 1980.

candidipes Edwards, 1912c (p. 42) – synonym of *Uranotaenia* (*Pseudoficalbia*) *nivipous* Theobald, 1912a [*nomen novum* for *nivipous* Theobald, 1912a]; synonymy established before Knight & Stone, 1977.

candidoscutellum Marks, 1947b (p. 1) – species of *Aedes* (subgenus uncertain).

canfieldi Dyar & Knab, 1907c (p. 207) – synonym of *Wyeomyia melanocephala* Dyar & Knab, 1906a [original combination: *Sabethes canfieldi*]; synonymy established before Knight & Stone, 1977.

canorii Floch & Abonnenc, 1945d (p. 1) – species of *Anopheles* (*Stethomyia*).

cantans (Meigen, 1818) (p. 6) – species of *Aedes* (*Ochlerotatus*) [original combination: *Culex cantans*].
Synonyms: *reptans* Meigen, 1804; *waterhousei* Theobald, 1905; *subvexans* Martini, 1922.

cantator (Coquillett, 1903a) (p. 255) – species of *Aedes* (*Ochlerotatus*) [original combination: *Culex cantator*].

capelai da Cunha Ramos, 1993 (p. 235) – species of *Uranotaenia* (*Pseudoficalbia*).

capelai Ribeiro, 1992b (p. 135) – species of *Toxorhynchites* (*Afrorhynchus*).

capensis de Meillon, 1935 (p. 354) – subspecies of *Culex* (*Maillotia*) *salisburiensis* Theobald, 1901c [original combination: *Culex* (*Neoculex*) *salisburiensis* variety *capensis*]; subspecific status by White, 1975.

capensis Edwards, 1924a (p. 162) – species of *Aedes* (*Albuginosus*).

capitis Shannon & del Ponte, 1928 (p. 85) – synonym of *Uranotaenia* (*Uranotaenia*) *nataliae* Lynch Arribálzaga, 1891b; synonymy by Lane, 1953.

capricornii Lutz, 1904 (in Bourroul, 1904) (p. 66) – species of *Haemagogus* (*Haemagogus*).

cara Dyar & Knab, 1909d (p. 264) – synonym of *Wyeomyia* (*Decamyia*) *pseudopecten* Dyar & Knab, 1906a; synonymy established before Knight & Stone, 1977.

caracula Dyar & Núñez Tovar, 1928 (p. 90) – species of *Wyeomyia* (*Wyeomyia*).

caraibeus Howard, Dyar & Knab, 1913, 1915 (fig. 333, p. 257, respectively) – synonym of *Culex* (*Culex*) *nigripalpus* Theobald, 1901c; synonymy established before Knight & Stone, 1977.

carcinicola da Cunha Ramos & Brunhes, 2004 (p. 111) – species of *Uranotaenia* (*Pseudoficalbia*).

carcinophilus Dyar & Knab, 1906d (p. 220) – species of *Culex* (*Melanoconion*).

carcinoxenus de Oliveira Castro, 1932 (p. 98) – species of *Culex* (*Culex*).

cardamatisi Newstead & Carter, 1910 (p. 379) – synonym of *Anopheles* (*Cellia*) *superpictus* Grassi, 1899 [original combination: *Pyretophorus cardamatisi*]; synonymy established before Knight & Stone, 1977.

caribeanus Galindo & Blanton, 1954 (p. 244) – species of *Culex* (*Melanoconion*).

carioca Lane & Whitman, 1951 (p. 358) – species of *Culex* (*Microculex*).

carleti Brunhes & Ravaonjanahary, 1971 (p. 177) – species of *Culex* (*Culex*).

carmenti (Edwards, 1924b) (p. 388) – species of *Verrallina* (*Verrallina*) [original combination: *Aedes* (*Aedes*) *carmenti*].
Synonym: *ceramensis* Brug, 1934.

carmodyae Dyar & Knab, 1906d (p. 210) – synonym of *Culex* (*Culex*) *nigripalpus* Theobald, 1901c; synonymy established before Knight & Stone, 1977.

carnevalei Brunhes, Le Goff & Geoffroy, 1999 (p. 552) – species of *Anopheles* (*Cellia*).

carolinensis Bohart & Ingram, 1946b (p. 30) – species of *Culex* (*Lophoceraomyia*).
caroni Adam, 1961 (p. 714) – species of *Anopheles* (*Cellia*).
carrilloi (Sutil O. & Pulido F., 1978) (p. 199) – species of *Wyeomyia* (*Exallomyia*) [original combination: *Sabethes* (*Sabethes*) *carrilloi*].
carteri Amerasinghe, 1993 (p. 110) – species of *Heizmannia* (*Heizmannia*).
carteri Edwards, 1936 (p. 52) – species of *Aedes* (*Neomelaniconion*).
Synonym: *maculicosta* Edwards, 1936.
carteri Evans & de Meillon, 1933 (p. 276) – species of *Anopheles* (*Cellia*) [original combination: *Anopheles demeilloni* variety *carteri*]; specific status by Gillies & de Meillon, 1968.
carteri Wijesundara, 1951 (p. 176) – synonym of *Verrallina* (*Verrallina*) *butleri* (Theobald, 1901c) [original combination: *Aedes* (*Aedes*) *carteri*]; synonymy by Reinert, 1974.
cartroni Ventrillon, 1905b (p. 429) – synonym of *Culex* (*Culex*) *quinquefasciatus* Say, 1823; synonymy established before Knight & Stone, 1977.
cartroni (Ventrillon, 1906b) (p. 143) – species of *Aedes* (*Skusea*) [original combination: *Stegomyia cartroni*].
casali Schick, 1970b (p. 21) – species of *Aedes* (*Protomacleaya*).
caspius (Pallas, 1771) (p. 475) – species of *Aedes* (*Ochlerotatus*) [original combination: *Culex caspius*].
Synonyms: *punctatus* Meigen, 1804; *siculus* Robineau-Desvoidy, 1827; *maculiventris* Macquart, 1846b; *penicillaris* Rondani, 1872; ? *longisquamosa* Theobald, 1905f; *subtilis* Sergent & Sergent, 1905; *arabica* Giles, 1906; *africanus* Neveu-Lemaire, 1906; *willcocksii* Theobald, 1907; *arabicus* Becker, 1910; *albineus* Séguy, 1923; *epsilonn* Séguy, 1924.
castaneus Sirivanakarn, 1973a (p. 212) – species of *Culex* (*Lophoceraomyia*).
castelli Hamon, 1957 (p. 686) – species of *Culex* (*Culex*).
castor de Meillon & Lavoipierre, 1944 (p. 49) – species of *Culex* (*Eumelanomyia*).
castrensis Edwards, 1922a (p. 285) – species of *Culex* (*Eumelanomyia*).
castroi Casal & García, 1967b (p. 451) – species of *Culex* (*Phytotelmatomyia*).
castroi Lane & Cerqueira, 1942 (p. 514) – species of *Trichoprosopon*.
cataphylla Dyar, 1916b (p. 86) – species of *Aedes* (*Ochlerotatus*).
Synonyms: *prodotes* Dyar, 1917c; *rostochiensis* Martini, 1920b; *pacificensis* Hearle, 1927.
cataractarum Edwards, 1923b (p. 7) – species of *Culex* (*Eumelanomyia*).
Synonym: *tricontus* Delfinado, 1966a.
catasticta Knab, 1909 (p. 387) – species of *Aedeomyia* (*Aedeomyia*).
catesi (Lien, 1968c) (p. 130) – species of *Heizmannia* (*Mattinglyia*) [original combination: *Mattinglyia catesi*].
catharinensis (da Costa Lima, Guitoon & Ferreira, 1962) (p. 230) – species of *Toxorhynchites* (*Ankylorhynchus*) [original combination: *Ankylorhynchus catharinensis*].
caucasicus Stackelberg, 1937 (table following p. 52) – *nomen nudum*, *Anopheles* [original combination: *Anopheles maculipennis maculipennis caucasicus*].
cauchensis Floch & Abonnenc, 1945b (p. 1) – species of *Culex* (*Tinolestes*).
caudatus Clastrier, 1970c (p. 470) – species of *Culex* (*Melanoconion*).
caudelli (Dyar & Knab, 1906d) (p. 224) – species of *Culex* (*Melanoconion*) [original combination: *Mochlostyrax caudelli*].
Synonym: *multispinosus* Bonne-Wepster & Bonne, 1920.
cauta (Barraud, 1928b) (p. 368) – species of *Verrallina* (*Neomacleaya*) [original combination: *Aedes* (*Aedes*) *cautus*].
cavalierii García & Casal, 1967 (p. 436) – species of *Toxorhynchites* (*Lynchiella*).
cavaticus Reinert, 1979 (p. 148) – species of *Aedes* (*Isoaedes*).
cavernicola Mattingly, 1954b (p. 167) – species of *Uranotaenia* (*Pseudoficalbia*).
cavernicola (Rattanarithikul & Harbach, 2007) (in Harbach, Rattanarithikul, Howard *et al.*, 2007) (p. 475) – species of *Aedes* (*Borichinda*) [original combination: *Borichinda cavernicola*].
cavernicolus Abonnenc, 1954, 1956 (pp. 288, 803, respectively) – synonym of *Anopheles* (*Cellia*) *smithii* Theobald, 1905b; synonymy established before Knight & Stone, 1977.
cavernicolus Floch & Abonnenc, 1945c (p. 1) – synonym of *Culex* (*Melanoconion*) *putumayensis* Matheson, 1934; synonymy by Rozeboom & Komp, 1950a.
cayennensis Floch & Abonnenc, 1945b (p. 4) – synonym of *Culex* (*Melanoconion*) *portesi* Senevet & Abonnenc, 1941; synonymy by Floch & Abonnenc, 1947.
cedecei Stone & Hair, 1968 (p. 39) – species of *Culex* (*Melanoconion*).
celaenocephala Dyar & Knab, 1906a (p. 140) – species of *Wyeomyia* (*Wyeomyia*).
Synonyms: *chrysomus* Dyar & Knab, 1907c; *philophone* Dyar & Knab, 1907c; *mataea* Dyar & Knab, 1908; *megalodora* Dyar & Knab, 1908.
celatum Peyton & Harbach, 2000 (in Harbach & Peyton, 2000) (p. 128) – species of *Onirion*.
celebicus Mattingly, 1959 (p. 32) – species of *Aedes* (*Lorrainea*).
celeste Dyar & Nuñez Tovar, 1927a (p. 152) – species of *Haemagogus* (*Haemagogus*).
celidopus Dyar & Shannon, 1925b (p. 41) – synonym of *Anopheles* (*Anopheles*) *peryassui* Dyar & Knab, 1908; synonymy established before Knight & Stone, 1977.
centaurus Theobald, 1905e (p. 16) – *nomen nudum*, *Psorophora*.
centrale Brèthes, 1910 (p. 469) – synonym of *Psorophora* (*Janthinosoma*) *ferox* (von Humboldt, 1819) [original combination: *Janthinosoma centrale*]; synonymy established before Knight & Stone, 1977.
centropunctatus (Theobald, 1913d) (p. 599) – species of *Aedes* (*Aedimorphus*) [original combination: *Culicelsa centropunctata*].

centrotus Howard, Dyar & Knab, 1917 (p. 747) – synonym of *Aedes* (*Ochlerotatus*) *abserratus* (Felt & Young, 1904); synonymy established before Knight & Stone, 1977.
cenus Root, 1927c (p. 590) – synonym of *Culex* (*Melanoconion*) *intrincatus* Brèthes, 1916; synonymy established before Knight & Stone, 1977.
cephasi Edwards, 1923b (p. 8) – synonym of *Tripteroides* (*Polylepidomyia*) *tasmaniensis* (Strickland, 1911b) [original combination: *Rachionotomyia cephasi*]; synonymy established before Knight & Stone, 1977.
ceramensis Brug, 1934 (p. 512) – synonym of *Verrallina* (*Verrallina*) *carmenti* (Edwards, 1924b) [original combination: *Aedes* (*Aedes*) *ceramensis*]; synonymy by Reinert, 1974.
ceramensis Sirivanakarn & Kurihara, 1973 (p. 220) – species of *Culex* (*Culiciomyia*).
cerqueirai (Barreto & Coutinho, 1944) (p. 73) – species of *Mansonia* (*Mansonia*) [original combination: *Taeniorhynchus* (*Taeniorhynchus*) *cerqueirai*].
cerqueirai (Stone, 1944) (p. 339) – species of *Runchomyia* (*Runchomyia*) original combination: *Trichoprosopon* (*Runchomyia*) *cerqueirai*].
cerqueirai Valencia, 1973 (p. 76) – species of *Culex* (*Carrollia*).
cesari del Ponte & Cerqueira, 1938 (p. 226) – species of *Wyeomyia* (subgenus uncertain, determined by Motta & Lourenço-de-Oliveira, 1995).
ceylonensis (Theobald, 1905h) (p. 248) – species of *Tripteroides* (*Rachionotomyia*) [original combination: *Rachionotomyia ceylonensis*].
ceylonica (Edwards, 1917) (p. 221) – species of *Verrallina* (*Neomacleaya*) [original combination: *Aedes ceylonicus*].
Synonym: *lankaensis* Stone & Knight, 1958.
ceylonica Newstead & Carter, 1910 (p. 377) – synonym of *Anopheles* (*Cellia*) *tessellatus* Theobald, 1901a [original combination: *Dactylomyia ceylonica*]; synonymy established before Knight & Stone, 1977.
ceylonica Theobald, 1907 (p. 236) – synonym of *Culex* (*Culiciomyia*) *fragilis* Ludlow, 1903 [original combination: *Culiciomyia ceylonica*]; synonymy established before Knight & Stone, 1977.
ceylonica Theobald, 1910c (p. 391) – synonym of *Aedes* (*Hulecoeteomyia*) *chrysolineatus* (Theobald, 1907) [original combination: *Culex japonicus* variety *ceylonica* [*sic*]]; synonymy by Knight, 1948a.
ceylonica Theobald, 1910c (p. 503) – synonym of *Uranotaenia* (*Uranotaenia*) *lateralis* Ludlow, 1905c; synonymy established before Knight & Stone, 1977.
ceyx (von Heyden, 1870) (p. 252) – fossil species of *Neoculicites* Evenhuis, 1994 [original combination: *Culex ceyx*].
chaetoventralis (Theobald, 1910c) (p. 461) – species of *Culex* (*Neoculex*) [original combination: *Neomelanoconion* [*sic*] *chaetoventralis*].
chagasi (da Costa Lima, 1935a) (p. 464) – species of *Mansonia* (*Mansonia*) [original combination: *Taeniorhynchus* (*Taeniorhynchus*) *chagasi*].
chagasi Galvão, 1941a (p. 92) – synonym of *Anopheles* (*Nyssorhynchus*) *triannulatus* (Neiva & Pinto, 1922b) [original combination: *Anopheles* (*Nyssorhynchus*) *triannulatus* variety *chagasi*]; synonymy by J. Lane (in Boyd, 1949).
chaguanco Casal, García & Fernández, 1968 (p. 217) – species of *Culex* (*Anoedioporpa*).
chaii Miyagi & Toma, 2012 (in Miyagi *et al.*, 2012) (p. 104) – species of *Topomyia* (*Topomyia*).
chalcocephala Dyar & Knab, 1906a (p. 140) – species of *Wyeomyia* (subgenus uncertain, determined by Motta & Lourenço-de-Oliveira, 1995).
Synonym: *luciae* Senevet, Chabelard & Abonnenc, 1942.
chalcocorystes Martini, 1914 (p. 70) – synonym of *Culex* (*Anoedioporpa*) *corrigani* Dyar & Knab, 1907c; synonymy established before Knight & Stone, 1977.
chalcospilans Dyar, 1921g (p. 110) – species of *Haemagogus* (*Haemagogus*).
chamberlaini Ludlow, 1904c (p. 297) – species of *Mimomyia* (*Mimomyia*).
chamboni Cornet, 1968 (p. 398) – species of *Aedes* (*Catageiomyia*).
champerico (Dyar & Knab, 1906a) (p. 134) – species of *Psorophora* (*Janthinosoma*) [original combination: *Janthinosoma champerico*].
Synonym: *pisces* Lassmann, 1944.
chandi Edwards, 1922a (p. 291) – species of *Heizmannia* (*Heizmannia*).
changbaiensis Su & Wang, 1981 (p. 327) – synonym of *Toxorhynchites* (*Toxorhynchites*) *christophi* (Portschinsky, 1884); synonymy by Danilov, 1987a.
changfus Ma, 1981b (p. 65) – synonym of *Anopheles* (*Anopheles*) *sinensis* Wiedemann, 1828; synonymy by Miao *et al.*, 1988 (also see Qu & Zhu, 2008).
changuinolae Galindo & Blanton, 1954 (p. 238) – species of *Culex* (*Melanoconion*).
chaquense Petrocchi, 1923 (p. 91) – *nomen nudum*, *Culex*.
chaquensis Paterson & Shannon, 1927 (pp. 4, 9) – synonym of *Psorophora* (*Janthinosoma*) *lutzii* (Theobald, 1901a); synonymy by Lane, 1953.
charmion Dyar, 1928 (Errata & Addenda) – species of *Wyeomyia* (*Wyeomyia*).
chathamicus (Dumbleton, 1962) (p. 20) – species of *Opifex* (*Nothoskusea*) [original combination: *Aedes* (*Nothoskusea*) *chathamicus*].
chaudoyei Theobald, 1903a (p. 68) – synonym of *Anopheles* (*Cellia*) *multicolor* Cambouliu, 1902 [original combination: *Pyretophorus chaudoyei*]; synonymy established before Knight & Stone, 1977.
chaussieri Edwards, 1923a (p. 397) – species of *Aedes* (*Stegomyia*).
chauveti Brunhes & Rambelo, 1968 (p. 113) – species of *Culex* (*Eumelanomyia*).
cheesmanae Mattingly & Marks, 1955 (p. 166) – species of *Culex* (*Neoculex*).
chelli (Edwards, 1915b) (p. 275) – species of *Aedes* (*Ochlerotatus*) [original combination: *Ochlerotatus chelli*].

chemulpoensis Yamada, 1921 (p. 54) – species of *Aedes* (*Stegomyia*).

chengi Lien, 1968c (p. 137) – species of *Heizmannia* (*Heizmannia*).

cheni Dong, Wang & Lu, 2003 (in Dong, Dong, Zhou *et al.*, 2003) (p. 528) – species of *Culex* (*Culiciomyia*).

cheni Ho, 1963 (p. 368) – synonym of *Culex* (*Culex*) *annulus* Theobald, 1901a; synonymy by Chen, 1980.

cheni Lien, 1968a (p. 2) – species of *Tripteroides* (*Tripteroides*).

cheni Lien, 1968c (p. 135) – synonym of *Heizmannia* (*Heizmannia*) *reidi* Mattingly, 1957c; synonymy by Mattingly, 1970.

cherrapunjiensis (Natarajan, Rajavel & Jambulingam, 2016) (p. 330) – species of *Aedes* (*Hulecoeteomyia*) [original combination: *Hulecoeteomyia cherrapunjiensis*].

chidesteri Dyar, 1921h (p. 117) – species of *Culex* (*Culex*).
Synonyms: *finlayi* Perez Vigueras, 1956; *deanei* Correa & Ramalho, 1959b.

chilensis Blanchard, 1905 (p. 289) – synonym of *Anopheles* (*Nyssorhynchus*) *pictipennis* (Philippi, 1865) [original combination: *Culex chilensis*]; synonymy by Belkin, 1968b (as synonym of *variegatus* Blanchard, 1852).

chionodes Belkin, 1962 (p. 434) – species of *Aedes* (*Christophersiomyia*).

chionotum Zavortink, 1972 (p. 47) – species of *Aedes* (*Protomacleaya*).

chiquitana Pinto, 1932 (p. 291) – synonym of *Psorophora* (*Grabhamia*) *confinnis* (Lynch Arribálzaga, 1891b); synonymy by Lane, 1953.

chiriquiensis Komp, 1936b (p. 156) – species of *Anopheles* (*Anopheles*).

chitae Duret, 1967b (p. 113) – species of *Culex* (*Culex*).

chiungchungensis Hsü, 1963 (p. 229) – synonym of *Culex* (*Lophoceraomyia*) *mammilifer* (Leicester, 1908); synonymy established before Knight & Stone, 1977.

chiyutoi Baisas, 1935d (p. 176) – synonym of *Culex* (*Eumelanomyia*) *foliatus* Brug, 1932; synonymy by Bram, 1967a.

chloropterus (von Humboldt, 1819) (p. 340) – species of *Sabethes* (*Sabethoides*) [original combination: *Culex chloropterus*].
Synonyms: *confusus* Theobald, 1903a; *rangeli* Surcouf & Gonzales-Rincones, 1911; *imperfectus* Bonne-Wepster & Bonne, 1920.

chloroventer Theobald, 1909b (p. 16) – synonym of *Culex* (*Culex*) *invidiosus* Theobald, 1901d; synonymy established before Knight & Stone, 1977.

chocoensis Porter & Wolff, 2004 (p. 3) – species of *Wyeomyia* (*Hystatomyia*).

chodukini Martini, 1929 (p. 141) – species of *Anopheles* (*Anopheles*).

chorleyi Edwards, 1936 (p. 54) – species of *Uranotaenia* (*Uranotaenia*).

chorleyi Edwards, 1941 (p. 330) – species of *Culex* (*Culex*).

chresta Dyar & Knab, 1909d (p. 263) – synonym of *Wyeomyia* (*Wyeomyia*) *scotinomus* (Dyar & Knab, 1907c); synonymy established before Knight & Stone, 1977.

christiani Colless, 1960 (p. 389) – species of *Culex* (*Lophoceraomyia*).

christianus Dyar, 1921i (p. 148) – synonym of *Aedes* (*Stegomyia*) *gardnerii imitator* Leicester, 1908; synonymy by Mattingly, 1965.

christophersi Baisas & Ubaldo-Pagayon, 1953 (p. 86) – species of *Tripteroides* (*Tricholeptomyia*).

christophersi Barraud, 1926 (p. 336) – species of *Uranotaenia* (*Uranotaenia*).

christophersi Edwards, 1922a (p. 267) – species of *Aedes* (*Bruceharrisonius*).

christophersi Theobald, 1902a (p. 378) – synonym of *Anopheles* (*Cellia*) *minimus* Theobald, 1901a; synonymy by Harbach, Garros *et al.*, 2007.

christophersii Theobald, 1907 (p. 453) – synonym of *Culex* (*Culex*) *quinquefasciatus* Say, 1823; synonymy established before Knight & Stone, 1977.

christophi (Portschinsky, 1884) (p. 122) – species of *Toxorhynchites* (*Toxorhynchites*) [original combination: *Megarhina* [*sic*] *christophi*].
Synonym: *changbaiensis* Su & Wang, 1981.

christyi (Newstead & Carter, 1911b) (p. 238) – species of *Anopheles* (*Cellia*) [original combination: *Neocellia christyi*].

chroiopus Dyar & Knab, 1913 (p. 76) – synonym of *Sabethes* (*Sabethes*) *bipartipes* Dyar & Knab, 1906a; synonymy established before Knight & Stone, 1977.

chryselatus Dyar & Knab, 1919 (p. 5) – species of *Culex* (*Microculex*).

chrysocephalus Theobald, 1907 (p. 136) – synonym of *Toxorhynchites* (*Lynchiella*) *solstitialis* (Lutz, 1904) [original combination: *Megarhinus chrysocephalus*]; synonymy established before Knight & Stone, 1977.

chrysochlorus Arnell, 1973 (p. 27) – species of *Haemagogus* (*Haemagogus*).

chrysocorporis Hsieh & Liao, 1956 (p. 126) – subspecies of *Armigeres* (*Armigeres*) *subalbatus* (Coquillett, 1898) [original combination: *Armigeres* (*Armigeres*) *obturbans* variety *chrysocorporis*]; subspecific status by Harbach & Howard, 2007.

chrysogaster Graham, 1909 (p. 157) – species of *Eretmapodites*.

chrysogaster (Taylor, 1927) (p. 67) – subspecies of *Aedes* (*Mucidus*) *aurantius* (Theobald, 1907) [original combination: *Mucidus chrysogaster*].

chrysogona Knab, 1909 (p. 386) – synonym of *Coquillettidia* (*Coquillettidia*) *ochracea* (Theobald, 1903a) [original combination: *Mansonia chrysogona*]; synonymy established before Knight & Stone, 1977.

chrysolineatus (Theobald, 1907) (p. 218) – species of *Aedes* (*Hulecoeteomyia*) [original combination: *Howardina chrysolineata*].
Synonyms: *trilineata* Leicester, 1904; *ceylonica* Theobald, 1910c.

chrysomus Dyar & Knab, 1907c (p. 208) – synonym of *Wyeomyia* (*Wyeomyia*) *celaenocephala* Dyar & Knab, 1906a [original combination: *Phoniomyia chrysomus* [*sic*]]; synonymy established before Knight & Stone, 1977.

chrysonotum Dyar & Knab, 1908 (p. 57) – synonym of *Culex* (*Melanoconion*) *theobaldi* (Lutz, 1904); synonymy by Forattini & Sallum, 1989b.

chrysonotum (Peryassú, 1922a) (p. 117) – species of *Coquillettidia* (*Rhynchotaenia*) [original combination: *Taeniorhynchus chrysonotum*].

chrysoscutus (Theobald, 1910c) (p. 228) – species of *Aedes* (*Paraedes*) [original combination: *Pseudohowardina chrysoscuta*].

chrysosoma (Edwards, 1915b) (p. 281) – species of *Coquillettidia* (*Coquillettidia*) [original combination: *Taeniorhynchus chrysosoma*].

chrysothorax (Newstead & Thomas, 1910) (p. 147) – *nomen dubium*, *Culex* [original combination: *Neomelaniconion chrysothorax*]; status by Forattini & Sallum, 1989b ('species inqueirenda').

chrysothorax (Peryassú, 1908) (p. 244) – *nomen dubium*, *Culex* [original combination: *Melanoconion chrysothorax*]; status by Sirivanakarn, 1983.

chrysothorax Theobald, 1909b (p. 13) – synonym of *Aedes* (*Neomelaniconion*) *taeniarostris* (Theobald, 1909b) [original combination: *Banksinella chrysothorax*]; synonymy by Edwards, 1941.

chubbi Edwards, 1915b (p. 280) – synonym of *Coquillettidia* (*Coquillettidia*) *microannulata* (Theobald, 1911a) [original combination: *Taeniorhynchus chubbi*]; synonymy established before Knight & Stone, 1977.

chungi Lien, 1968b (p. 100) – species of *Aedes* (*Jihlienius*).

chungkiangensis Chang & Chang, 1974 (p. 347) – synonym of *Culex* (*Eumelanomyia*) *foliatus* Brug, 1932; synonymy by Meng & Chen, 1980.

churchillensis Ellis & Brust, 1973 (p. 922) – species of *Aedes* (*Ochlerotatus*).

ciliaris (Linnaeus, 1767) (p. 1002) – *nomen dubium*, *Aedes* [original combination: *Culex ciliaris*]; status established before Knight & Stone, 1977.

ciliata (Fabricius, 1794) (p. 401) – species of *Psorophora* (*Psorophora*) [original combination: *Culex ciliata* [*sic*]].
Synonyms: *cyanopennis* von Humboldt, 1819; *molestus* Wiedemann, 1820; *boscii* Robineau-Desvoidy, 1827; *rubidus* Robineau-Desvoidy, 1827; *tibialis* Robineau-Desvoidy, 1827; *conterrens* Walker 1856a; *lynchi* Brèthes, 1916; *ctites* Dyar, 1918e.

cilipes (Fabricius, 1805) (p. 34) – species of *Psorophora* (*Psorophora*) [original combination: *Culex cilipes*].
Synonyms: *scintillans* Walker, 1848; *iracunda* Dyar & Knab, 1906a.

cinctellus Edwards, 1922a (p. 287) – species of *Culex* (*Lophoceraomyia*) [*nomen novum* for *taeniata* Leicester, 1908].
Synonym: *taeniata* Leicester, 1908.

cinctus (Newstead & Carter, 1910) (p. 381) – species of *Anopheles* (*Cellia*) [original combination: *Cellia cincta*].

cinerellus Edwards, 1922d (p. 90) – species of *Culex* (*Culiciomyia*).

cinereoborealis Felt & Young, 1904 (p. 312) – synonym of *Aedes* (*Rusticoidus*) *provocans* (Walker, 1848) [original combination: *Culex cinereoborealis*]; transferred synonymy by synonymy of senior synonym *trichurus* (Dyar) with *provocans* (Walker) by Wood, 1977.

cinereus Meigen, 1818 (p. 13) – species of *Aedes* (*Aedes*).
Synonyms: *rufus* Gimmerthal, 1845; *nigritulus* Zetterstedt, 1850; *fuscus* Osten Sacken, 1877; *leucopygos* Eysell, 1903; *pallidohirta* Grossbeck, 1905; *hemiteleus* Dyar, 1924h.

cinereus Theobald, 1901a (p. 161) – species of *Anopheles* (*Cellia*).
Synonym: *jehafi* Patton, 1905.

cinereus Theobald, 1901c (p. 58) – species of *Culex* (*Culiciomyia*).
Synonyms: *freetownensis* Theobald, 1901c (in part, syntype ♂ only); *uniformis* Theobald, 1909b.

cingulata (Fabricius, 1805) (p. 36) – species of *Psorophora* (*Grabhamia*) [original combination: *Culex cingulatus*].
Synonyms: *apicalis* Theobald, 1903a; *indoctum* Dyar & Knab, 1906b; *neoapicalis* Theobald, 1910c; *garciai* Leví-Castillo, 1953e.

cingulatus Doleschall, 1856 (p. 405) – synonym of *Culex* (*Culex*) *quinquefasciatus* Say, 1823; synonymy established before Knight & Stone, 1977.

cingulatus (Leicester, 1908) (p. 97) – species of *Armigeres* (*Leicesteria*) [original combination: *Leicesteria cingulata*].

circumcincta Dyar & Knab, 1907c (p. 210) – species of *Wyeomyia* (*Hystatomyia*).
Synonyms: *macrotus* Dyar & Knab, 1907c; *andropus* Dyar & Knab, 1908.

circumflava Cerqueira, 1943 (p. 3) – species of *Psorophora* (*Janthinosoma*).

circumluteolus (Theobald, 1908a) (p. 107) – species of *Aedes* (*Neomelaniconion*) [original combination: *Banksiella* [*sic*] *luteolateralis* variety *circumluteola*]; specific status by ? Edwards, 1941. **Note**: incorrectly listed as *circumluteolum* (Theobald) in the online catalogue of the Walter Reed Biosystematics Unit (http://www.mosquitocatalog.org/).

circumtestacea (Theobald, 1908c) (p. 264) – species of *Ficalbia* [original combination: *Mimomyia circumtestacea*].

circumvolens Walker, 1861 (in Tennent, 1861) (p. 434) – *nomen nudum*, *Culex*.

civinskii Belkin, 1953 (p. 341) – species of *Uranotaenia* (*Uranotaenia*).

claggi Bohart & Farner, 1944 (p. 71) – species of *Tripteroides* (*Tripteroides*).

clara Dyar & Shannon, 1925a (p. 68) – species of *Uranotaenia* (*Uranotaenia*).
Synonym: *delae* Baisas, 1935c.

clarki Evans, 1924 (p. 365) – species of *Culex* (*Melanoconion*).

clarki (Galindo, Carpenter & Trapido, 1953) (p. 531) – species of *Haemagogus* (*Conopostegus*) [original combination: *Aedes* (*Finlaya*) *leucocelaenus clarki*]; specific status by Zavortink, 1972.

clarki Komp, 1943 (p. 197) – synonym of *Anopheles* (*Nyssorhynchus*) *evansae* (Brèthes, 1926b); synonymy by Faran, 1981.

clasoleuca Dyar & Knab, 1908 (p. 68) – species of *Wyeomyia* (subgenus uncertain, determined by Motta & Lourenço-de-Oliveira, 1995).
Synonym: *agyrtes* Dyar & Knab, 1909d.

classicus Dyar, 1920e (p. 113) – synonym of *Aedes* (*Ochlerotatus*) *stimulans* (Walker, 1848) [original combination: *Aedes stimulans classicus*]; synonymy established before Knight & Stone, 1977.

clastrieri Casal & Garcia, 1968b (p. 119) – species of *Culex* (*Aedinus*).

clavata (Barraud, 1931b) (p. 614) – species of *Verrallina* (*Neomacleaya*) [original combination: *Aedes* (*Aedes*) *clavatus*].

claviger (Meigen, 1804) (p. 4) – species of *Anopheles* (*Anopheles*) [original combination: *Culex claviger*].
Synonyms: *villosus* Robineau-Desvoidy, 1827; *grisescens* Stephens, 1828; *turkestani* Shingarev, 1926; *amaurus* Martini, 1929; *habibi* Mulligan & Puri, 1936; *missiroli* del Vecchio, 1939; *pollutus* Torres Cañamares, 1945.

clavipalpus (Theobald, 1908b) (p. 295) – subspecies of *Mimomyia* (*Mimomyia*) *chamberlaini* Ludlow, 1904c [original combination: *Radioculex clavipalpus*]; subspecific status by Mattingly, 1957a.

clavirostris Stone & Bohart, 1944 (p. 212) – synonym of *Mimomyia* (*Etorleptiomyia*) *luzonensis* (Ludlow, 1905a) [original combination: *Aedes* (*Luzonus*) *clavirostris*]; synonymy by Stone, 1945a.

clelandi (Taylor, 1914c) (p. 690) – species of *Aedes* (*Ochlerotatus*) [original combination: *Culicada clelandi*].

cleopatrae Willcocks, 1910 (p. 586) – *nomen nudum*, *Anopheles* [original combination: *Pyretophorus cleopatrae*].

clintoni Taylor, 1946 (p. 211) – species of *Aedes* (*Patmarksia*).

clivis Lanzaro & Eldridge, 1992 (p. 89) – species of *Aedes* (*Ochlerotatus*).

clowi Rozeboom & Knight, 1946 (p. 114) – species of *Anopheles* (*Cellia*).

coatzacoalcos Dyar & Knab, 1906d (p. 186) – species of *Uranotaenia* (*Uranotaenia*).
Synonym: *basalis* Howard, Dyar & Knab, 1913.

cockerelli (Edwards, 1923c) (p. 151) – fossil species of *Mansonia* Blanchard, 1901 [original combination: *Taeniorhynchus cockerelli*].

codiocampa Dyar & Knab, 1907c (p. 209) – species of *Wyeomyia* (*Miamyia*).

coenonus Howard, Dyar & Knab, 1913, 1915 (fig. 38, p. 177, respectively) – species of *Wyeomyia* (*Hystatomyia*).

coerulescens Edwards, 1928a (p. 277) – species of *Culex* (*Lophoceraomyia*).

coeruleus (Brug, 1934) (p. 501) – species of *Toxorhynchites* (*Toxorhynchites*) [original combination: *Megarhinus coeruleus*].

coffini Dyar & Knab, 1906a (p. 134) – synonym of *Psorophora* (*Janthinosoma*) *johnstonii* (Grabham, 1905) [original combination: *Janthinosoma coffini*]; synonymy established before Knight & Stone, 1977.

cogilli Edwards, 1922d (p. 92) – species of *Aedes* (*Phagomyia*).

cohaesa Dönitz, 1903 (p. 233) – synonym of *Anopheles* (*Cellia*) *minimus* Theobald, 1901a [*nomen novum* for *formosaensis I* Tsuzuki, 1902a; original combination: *Anopheles aconitus* variety *cohaesa*]; synonymy validated by Harbach, Garros *et al*., 2007.

coheni Belkin, 1950 (p. 250) – species of *Tripteroides* (*Polylepidomyia*).

colledgei Marks, 1956 (p. 41) – species of *Anopheles* (*Anopheles*).

collessi Grjebine, 1986 (p. 124) – species of *Mimomyia* (*Ingramia*).

collessi Lee, 1946 (p. 271) – species of *Tripteroides* (*Polylepidomyia*).

collessi Mattingly, 1958a (p. 28) – species of *Aedes* (*Paraedes*).

collessi Reid, 1963 (p. 103) – species of *Anopheles* (*Anopheles*).

collessi Sirivanakarn, 1968 (p. 170) – species of *Culex* (*Lophoceraomyia*).

colocasiae Edwards, 1928a (p. 281) – species of *Uranotaenia* (*Pseudoficalbia*).

colombiana Lane, 1945a (p. 142) – species of *Wyeomyia* (*Antunesmyia*).

colombianus Adames, 1971 (p. 64) – species of *Deinocerites*.

colombiensis Dyar, 1924i (p. 184) – synonym of *Culex* (*Melanoconion*) *pilosus* (Dyar & Knab, 1906d); synonymy established before Knight & Stone, 1977.

colon Harris, 1835 (in Hitchcock, 1835) (p. 595) – *nomen nudum*, *Culex*.

colonarius Dyar, 1924g (p. 130) – synonym of *Aedes* (*Ochlerotatus*) *albifasciatus* (Macquart, 1838b); synonymy established before Knight & Stone, 1977.

colsoni Senevet & Quiévreux, 1941 (p. 249) – synonym of *Wyeomyia* (*Wyeomyia*) *grayii* Theobald, 1901c [original combination: *Wyeomyia quasiluteoventralis* variety *colsoni*]; synonymy by Belkin, 1968b.

columbiae (Dyar & Knab, 1906a) (p. 135) – species of *Psorophora* (*Grabhamia*) [original combination: *Janthinosoma columbiae*].
Synonyms: *floridense* Dyar & Knab, 1906a; ? *texanum* Dyar & Knab, 1906a.

coluzzii Coetzee & Wilkerson, 2013 (in Coetzee *et al*., 2013) (p. 249) – species of *Anopheles* (*Cellia*).

coluzzii Rioux, Guilvard & Pasteur, 1998 (p. 357) – species of *Aedes* (*Ochlerotatus*).

comata (Barraud, 1931b) (p. 615) – species of *Verrallina* (*Neomacleaya*) [original combination: *Aedes* (*Aedes*) *comatus*].

comatus Senevet & Abonnenc, 1939a (p. 103) – species of *Culex* (*Melanoconion*).

combesi Doucet, 1950 (p. 39) – species of *Uranotaenia* (*Pseudoficalbia*).

comitatus Arnell, 1976 (p. 66) – species of *Aedes* (*Ochlerotatus*).

comitatus Dyar & Knab, 1909a (p. 35) – synonym of *Culex* (*Culex*) *pipiens pallens* Coquillett, 1898; synonymy established before Knight & Stone, 1977.

commevynensis Bonne-Wepster & Bonne, 1920 (p. 176) – species of *Culex* (*Melanoconion*).
comminutor Dyar, 1920d (p. 70) – species of *Culex* (*Melanoconion*).
commovens Walker, 1856a (p. 432) – synonym of *Aedes* (*Mucidus*) *alternans* (Westwood, 1835); synonymy established before Knight & Stone, 1977.
communis (de Geer, 1776) (p. 316) – species of *Aedes* (*Ochlerotatus*) [original combination: *Culex communis*].
Synonyms: *fasciatus* Meigen, 1804; *nemorosus* Meigen, 1818; *sylvaticus* Meigen, 1818; *obscurus* Meigen, 1830a; *lazarensis* Felt & Young, 1904; *borealis* Ludlow, 1911a; *diplolineata* Schneider, 1913; *altiusculus* Dyar, 1917b; *palmeni* Edwards, 1921b; *prolixus* Dyar, 1922a.
communis (Leicester, 1908) (p. 254) – species of *Heizmannia* (*Heizmannia*) [original combination: *Dendromyia communis*].
comorensis Brunhes, 1977a (p. 136) – species of *Culex* (*Culex*).
comorensis Brunhes, 1977a (p. 145) – species of *Orthopodomyia*.
comorensis Brunhes, Le Goff & Geoffroy, 1997 (p. 178) – species of *Anopheles* (*Cellia*).
comorensis da Cunha Ramos & Brunhes, 2004 (p. 170) – species of *Uranotaenia* (*Pseudoficalbia*).
comosa (Reinert, 1974) (p. 36) – species of *Verrallina* (*Neomacleaya*) [original combination: *Aedes* (*Verrallina*) *comosus*].
completiva Leicester, 1908 (p. 178) – synonym of *Mimomyia* (*Etorleptiomyia*) *luzonensis* (Ludlow, 1905a) [original combination: *Etorleptiomyia completiva*]; synonymy by Stone, 1945a.
complex (Theobald, 1910a) (p. 31) – species of *Heizmannia* (*Heizmannia*) [original combination: *Bolbodeomyia complex*].
Synonym: *stonei* Mattingly, 1957c.
complosa (Dyar, 1928) (p. 74) – species of *Wyeomyia* (*Dendromyia*) [original combination: *Dendromyia* (*Melanolepis*) *complosa*].
compressum Lutz, 1905 (p. 171) – species of *Trichoprosopon*.
Synonym: *trichorryes* Dyar & Knab, 1907c.
compta Senevet & Abonnenc, 1939b (p. 264) – species of *Wyeomyia* (subgenus uncertain, determined by Motta & Lourenço-de-Oliveira, 1995).
conchita Dyar & Knab, 1909d (p. 264) – synonym of *Wyeomyia* (*Wyeomyia*) *vanduzeei* Dyar & Knab, 1906a; synonymy established before Knight & Stone, 1977.
conchobius Edwards, 1941 (p. 231) – subspecies of *Eretmapodites silvestris* Ingram & de Meillon, 1927.
concinnus Lee, 1946 (p. 260) – species of *Tripteroides* (*Rachisoura*).
concinnus Stephens, 1846 (p. 19) – synonym of *Aedes* (*Ochlerotatus*) *sticticus* (Meigen, 1838) [original combination: *Culex concinnus*]; synonymy established before Knight & Stone, 1977.
concolor Edwards, 1938 (in Evans, 1938) (p. 56) – species of *Anopheles* (*Anopheles*).
concolor Robineau-Desvoidy, 1827 (p. 405) – synonym of *Lutzia* (*Metalutzia*) *fuscana* (Wiedemann, 1820) [original combination: *Culex concolor*]; synonymy established before Knight & Stone, 1977.
concolor Taylor, 1914c (p. 700) – synonym of *Aedes* (*Halaedes*) *australis* (Erichson, 1842) [original combination: *Caenocephalus concolor*]; synonymy by Mattingly & Marks, 1955.
condei Ventrillon, 1905a (p. 444) – synonym of *Eretmapodites quinquevittatus* Theobald, 1901d; synonymy established before Knight & Stone, 1977.
conditus Moses, Howard & Harbach, 2000 (p. 992) – species of *Sabethes* (*Sabethoides*).
condolescens Dyar & Knab, 1907a (p. 11) – species of *Aedes* (*Ochlerotatus*).
condylodesmus Grünberg, 1905 (p. 385) – synonym of *Culex* (*Culex*) *duttoni* Theobald, 1901d; synonymy established before Knight & Stone, 1977.
confinnis (Lynch Arribálzaga, 1891b) (p. 149) – species of *Psorophora* (*Grabhamia*) [original combination: *Taeniorhynchus confinnis*].
Synonyms: *scutipunctatus* Lutz & Neiva, 1911; *chiquitana* Pinto, 1932.
confirmatus Lynch Arribálzaga, 1891b (p. 146) – synonym of *Aedes* (*Ochlerotatus*) *scapularis* (Rondani, 1848) [original combination: *Ochlerotatus confirmatus*]; synonymy established before Knight & Stone, 1977.
confundior Komp & Rozeboom, 1951 (p. 132) – species of *Culex* (*Melanoconion*).
confusa Bonne-Wepster, 1951a (p. 67) – species of *Bironella* (*Neobironella*).
confusa (Lutz, 1905) (p. 312) – species of *Wyeomyia* (*Prosopolepis*) [original combination: *Prosopolepis confusus*].
confusa Peyton, 1977 (p. 129) – species of *Uranotaenia* (*Pseudoficalbia*).
confusus Baisas, 1938 (p. 216) – synonym of *Culex* (*Culex*) *mimulus* Edwards, 1915c; synonymy by Delfinado, 1966a.
confusus Edwards, 1915c (p. 283) – species of *Armigeres* (*Armigeres*).
confusus Evans & Leeson, 1935 (p. 40) – species of *Anopheles* (*Cellia*) [original combination: *Anopheles funestus* variety *confusus*]; specific status by Gillies, 1962.
confusus Lee, 1946 (p. 251) – species of *Tripteroides* (*Rachisoura*).
confusus Sirivanakarn, 1968 (p. 105) – synonym of *Culex* (*Lophoceraomyia*) *singuawaensis* Sirivanakarn, 1969; synonymy by Sirivanakarn, 1969.
confusus Theobald, 1903a (p. 328) – synonym of *Sabethes* (*Sabethoides*) *chloropterus* (von Humboldt, 1819) [original combination: *Sabethoides chloropterus*]; synonymy established before Knight & Stone, 1977.
confusus Theobald, 1909b (p. 10) – synonym of *Aedes* (*Catageiomyia*) *irritans* (Theobald, 1901d) [original combination: *Myxosquamus confusus*]; synonymy established before Knight & Stone, 1977.
congolensis Edwards, 1927a (p. 351) – synonym of *Aedes* (*Aedimorphus*) *stenoscutus* (Edwards, 1912c); synonymy by White, 1975.

congolensis Evans, 1923a (p. 91) – synonym of *Culex (Oculeomyia) annulioris consimilis* Newstead, 1907 [original combination: *Culex annulioris* variety *congolensis*]; synonymy by Edwards, 1941.

conjungens Edwards, 1914c (p. 126) – species of *Armigeres (Armigeres)*.

connali Edwards, 1912c (p. 39) – species of *Uranotaenia (Uranotaenia)*.

conopas (von Frauenfeld, 1867) (p. 451) – *nomen dubium, Mansonia* ? [original combination: *Culex conopas*]; status established before Knight & Stone, 1977.

conquensis Torres Cañamares, 1946 (p. 47) – synonym of *Anopheles (Anopheles) marteri* Senevet & Prunnelle, 1927 [original combination: *Anopheles marteri* variety *conquensis*]; synonymy by Shahgudian, 1956.

conradti Grünberg, 1907a (species) (p. 405) – subspecies of *Toxorhynchites (Toxorhynchites) brevipalpis* Theobald, 1901a; subspecific status by Hopkins, 1936.

Synonyms: *schultzei* Enderlein, 1931; *tessmanni* Enderlein, 1931.

conservator Dyar & Knab, 1906d (p. 221) – species of *Culex (Anoedioporpa)*.

Synonyms: *divisior* Dyar & Knab, 1906d; *bifoliata* Dyar, 1922d; *paganus* Evans, 1923b; *surukumensis* Anduze, 1941.

consimilis Newstead, 1907 (in Newstead *et al.*, 1907) (p. 23) – subspecies of *Culex (Oculeomyia) annulioris* Theobald, 1901a [original combination: *Culex tigripes* variety *consimilis*]; subspecific status by Edwards, 1941.

Synonyms: *pseudoannulioris* Theobald, 1909b; *congolensis* Evans, 1923a; *mayumbae* Galliard, 1931b; *major* Edwards, 1935a.

consimilis Taylor, 1913 (p. 8) – synonym of *Culex (Culex) annulirostris* Skuse, 1889 [original combination: *Culicelsa consimilis*]; synonymy established before Knight & Stone, 1977.

consobrinus Robineau-Desvoidy, 1827 (p. 408) – synonym of *Culex (Culex) pipiens* Linnaeus, 1758; synonymy established before Knight & Stone, 1977.

consolator Dyar & Knab, 1906b (p. 169) – species of *Culex (Microculex)*.

Synonym: *trychnus* Root, 1927c.

consonensis (Reinert, 1973c) (p. 252) – species of *Verrallina (Harbachius)* [original combination: *Aedes (Neomacleaya) consonensis*].

conspirator Dyar & Knab, 1906d (p. 217) – species of *Culex (Melanoconion)*.

Synonyms: *holoneus* Dyar, 1921b; *dysmathes* Dyar & Ludlow, 1921a; *merodaemon*, Dyar 1921f; *pasadaemon* Dyar 1921f; *fatuator* Dyar & Shannon, 1924a; *meroneus* Dyar, 1925b; *macaronensis* Dyar & Nuñez Tovar, 1927a; *inducens* Root, 1928.

consternator Dyar & Knab, 1908 (p. 59) – synonym of *Culex (Anoedioporpa) restrictor* Dyar & Knab, 1906d; synonymy established before Knight & Stone, 1977.

contei Duret, 1968c (p. 121) – species of *Culex (Melanoconion)*.

Synonym: *seneveti* Clastrier, 1970c.

conterrens Walker, 1856a (p. 427) – synonym of *Psorophora (Psorophora) ciliata* (Fabricius, 1794) [original combination: *Culex conterrens*]; synonymy established before Knight & Stone, 1977.

contiguus Edwards, 1936 (p. 55) – species of *Aedes (Stegomyia)*.

continentalis Dobrotworsky, 1960a (p. 71) – species of *Aedes (Ochlerotatus)*.

continentalis Dyar & Knab, 1906d (p. 187) – synonym of *Uranotaenia (Uranotaenia) lowii* Theobald, 1901c; synonymy established before Knight & Stone, 1977.

contrahens Walker, 1861 (in Tennent, 1861) (p. 462) – *nomen nudum, Culex*.

contrastata da Cunha Ramos & Brunhes, 2004 (p. 247) – species of *Uranotaenia (Pseudoficalbia)*.

cooki Belkin, 1962 (p. 454) – species of *Aedes (Stegomyia)*.

cooki Root, 1937 (p. 98) – species of *Uranotaenia (Uranotaenia)*.

coonorensis Mattingly, 1981 (p. 30) – species of *Tripteroides (Rachionotomyia)*.

coppenamensis Bonne-Wepster & Bonne, 1920 (p. 173) – species of *Culex (Melanoconion)*.

coquilletti Dyar & Knab, 1906d (p. 187) – synonym of *Uranotaenia (Uranotaenia) sapphirina* (Osten Sacken, 1868); synonymy established before Knight & Stone, 1977.

coquillettii Theobald, 1907 (p. 153) – synonym of *Psorophora (Janthinosoma) ferox* (von Humboldt, 1819) [original combination: *Janthinosoma coquillettii*]; synonymy established before Knight & Stone, 1977.

corbeti Hamon, 1962 (p. 256) – species of *Eretmapodites*.

cordellieri Huang, 1986a (p. 636) – species of *Aedes (Diceromyia)*.

corentynensis Dyar, 1920d (p. 65) – species of *Culex (Melanoconion)*.

corethroides Theobald, 1907 (p. 35) – species of *Anopheles (Anopheles)*.

corneti Huang, 1986b (p. 765) – species of *Aedes (Stegomyia)*.

corniger Theobald, 1903a (p. 173) – species of *Culex (Phenacomyia)*.

Synonyms: *basilicus* Dyar & Knab, 1906b; *hassardii* Grabham, 1906a; *subfuscus* Theobald, 1907; *loquaculus* Dyar & Knab, 1909d; *rigidus* Senevet & Abonnenc, 1939a.

cornuta da Cunha Ramos & Brunhes, 2004 (p. 295) – species of *Uranotaenia (Pseudoficalbia)*.

cornutus Edwards, 1922a (p. 283) – species of *Culex (Oculeomyia)*.

corona Belkin, Heinemann & Page, 1970 (p. 218) – species of *Wyeomyia (Wyeomyia)*.

coronator Dyar & Knab, 1906d (p. 215) – species of *Culex (Culex)*.

Synonym: *mooseri* Vargas & Martinez Palacios, 1954.

corrigani Dyar & Knab, 1907c (p. 203) – species of *Culex (Anoedioporpa)*.

Synonym: *chalcocorystes* Martini, 1914.

corsicanus Edwards, 1928c (p. 166) – synonym of *Anopheles (Anopheles) plumbeus* Stephens, 1828 [original combination: *Anopheles plumbeus* variety *corsicanus*]; synonymy suggested by Bates *et al.*, 1949.

costai da Fonseca & da Silva Ramos, 1940 (p. 385) – species of *Anopheles* (*Anopheles*).
Synonym: *bonneorum* da Fonseca & da Silva Ramos, 1940.

costalimai Coutinho, 1944 (p. 425) – synonym of *Anopheles* (*Anopheles*) *mediopunctatus* Lutz, 1903; synonymy by Lane, 1951.

costalis Giles, 1900b (p. 49) – synonym of *Anopheles* (*Cellia*) *gambiae* Giles, 1902; synonymy established before Knight & Stone, 1977.

costalis Loew, 1866 (p. 55) – *nomen dubium*, *Anopheles*; status by Gillies & de Meillon, 1968.

costaricensis Adames & Hogue, 1970 (p. 12) – species of *Deinocerites*.

coticula Dyar & Knab, 1907b (p. 101) – synonym of *Coquillettidia* (*Rhynchotaenia*) *arribalzagae* (Theobald, 1903a) [original combination: *Taeniorhynchus coticula*]; synonymy established before Knight & Stone, 1977.

cotopaxense Leví-Castillo, 1953c (p. 66) – *nomen dubium*, *Trichoprosopon*; status by Zavortink, 1979.

cottlei Sirivanakarn, 1968 (p. 180) – species of *Culex* (*Lophoceraomyia*).

coulangesi Rodhain & Boutonnier, 1983b (p. 332) – species of *Aedes* (*Paulianius*).

courdurieri Grjebine, 1966 (p. 378) – *nomen dubium*, *Anopheles*; status by Brunhes *et al.*, 1998a.

coursi Doucet, 1949a (species) (p. 144) – subspecies of *Culex* (*Maillotia*) *salisburiensis* Theobald, 1901c; subspecific status by White, 1975.

coustani Laveran, 1900 (p. 109) – species of *Anopheles* (*Anopheles*).
Synonyms: *mauritianus* de Grandpré & de Charmoy, 1901; *similis* Theobald, 1901a.

covagarciai Forattini, 1965 (p. 128) – species of *Culex* (*Culex*).

covagarciai Sutil Oramas & Pulido F., 1974 (p. 25) – species of *Wyeomyia* (subgenus uncertain, determined by Motta & Lourenço-de-Oliveira, 1995).

covelli Barraud, 1929b (p. 265) – species of *Heizmannia* (*Heizmannia*).

cozi Cornet, 1973 (p. 175) – species of *Aedes* (*Cornetius*).

cozumelensis Diaz Nájera, 1966 (p. 334) – species of *Aedes* (*Howardina*).

cracens Sallum & Peyton, 2005 (in Sallum *et al.*, 2005b) (p. 165) – species of *Anopheles* (*Cellia*).

craggi (Barraud, 1923c) (p. 227) – species of *Aedes* (*Stegomyia*) [original combination: *Stegomyia craggi*].
Synonym: *purii* Barraud, 1931a.

crassicomus Colless, 1965 (p. 296) – species of *Culex* (*Lophoceraomyia*).

crassiforceps Edwards, 1927a (p. 352) – species of *Aedes* (*Neomelaniconion*).

crassipes (van der Wulp, 1881) (p. 9) – species of *Coquillettidia* (*Coquillettidia*) [original combination: *Culex crassipes*].
Synonyms: *brevicellulus* Theobald, 1901c; *pygmaeus* Theobald, 1908b; *diaeretus* Dyar, 1920i.

crassistylus Brug, 1934 (p. 516) – species of *Culex* (*Neoculex*).

crawfordi Reid, 1953 (p. 41) – species of *Anopheles* (*Anopheles*).

creole Anduze, 1949b (p. 149) – species of *Culex* (*Melanoconion*).

cretata (Delfinado, 1967) (p. 14) – species of *Verrallina* (*Neomacleaya*) [original combination: *Aedes* (*Neomacleaya*) *cretatus*].

creticus Theobald, 1903a (p. 189) – synonym of *Culex* (*Culex*) *theileri* Theobald, 1903a; synonymy established before Knight & Stone, 1977.

cretinus Edwards, 1921d (p. 325) – species of *Aedes* (*Stegomyia*).
Synonyms: *delta* Séguy, 1924; *lindtropi* Shingarev, 1927.

cricillium Martini, 1932b (p. 99) – synonym of *Anopheles* (*Anopheles*) *hectoris* Giaquinto-Mira, 1931; synonymy by Martini, 1935.

crinicauda Edwards, 1921c (p. 77) – species of *Culex* (*Culex*) [*nomen novum* for *parvus* Taylor, 1912a].
Synonym: *parvus* Taylor, 1912a.

crinifer (Theobald, 1903a) (p. 209) – species of *Aedes* (*Ochlerotatus*) [original combination: *Culex crinifer*].
Synonyms: *lynchii* Brèthes, 1910; *tapinops* Brèthes, 1917; *iguazu* Shannon & del Ponte, 1928.

cristata (Theobald, 1904b) (p. 78) – species of *Coquillettidia* (*Coquillettidia*) [original combination: *Taeniorhynchus cristatus*].
Synonym: *nocturnus* Theobald, 1913d.

cristata Thurman, 1959 (p. 46) – species of *Topomyia* (*Suaymyia*).

cristatus King & Baisas, 1936 (p. 80) – species of *Anopheles* (*Cellia*).

cristipalpis Service, 1977 (p. 233) – species of *Anopheles* (*Cellia*).

cristovaoi Duret, 1968a (p. 14) – species of *Culex* (*Melanoconion*).

croceus Knight & Laffoon, 1946 (p. 213) – species of *Aedes* (*Finlaya*).

crockeri Colless, 1955a (p. 337) – subspecies of *Anopheles* (*Anopheles*) *gigas* Giles, 1901b.

crossi Lien, 1967 (p. 177) – species of *Aedes* (subgenus uncertain).

crowei Sirivanakarn, 1968 (p. 115) – species of *Culex* (*Lophoceraomyia*).

crucians Walker, 1856a (p. 432) – synonym of *Aedes* (*Halaedes*) *australis* (Erichson, 1842) [original combination: *Culex crucians*]; synonymy by Mattingly & Marks, 1955.

crucians Wiedemann, 1828 (p. 12) – species of *Anopheles* (*Anopheles*).

cruzii Dyar & Knab, 1908 (p. 53) – species of *Anopheles* (*Kerteszia*).
Synonyms: *lutzii* Theobald, 1901a; *adolphoi* Neiva, 1908b; *montemor* Corrêa, 1950.

crybda Dyar, 1924i (p. 184) – species of *Culex* (*Melanoconion*).

crypticus Coetzee, 1995 (p. 129) – species of *Anopheles* (*Anopheles*).

ctites Dyar, 1918e (p. 126) – synonym of *Psorophora* (*Psorophora*) *ciliata* (Fabricius, 1794); synonymy established before Knight & Stone, 1977.

cubensis Agramonte, 1900 (p. 460) – synonym of *Anopheles* (*Nyssorhynchus*) *albimanus* Wiedemann, 1820; synonymy established before Knight & Stone, 1977.

cubensis Bigot, 1857 (p. 329) – synonym of *Culex* (*Culex*) *quinquefasciatus* Say, 1823; synonymy established before Knight & Stone, 1977.

cubensis Dyar & Knab, 1906d (p. 225) – synonym of *Culex* (*Melanoconion*) *pilosus* (Dyar & Knab, 1906d) [original combination: *Mochlostyrax cubensis*]; synonymy established before Knight & Stone, 1977.

cubiculi Marks, 1989 (in Lee *et al.*, 1989b) (p. 94) – species of *Culex* (*Lophoceraomyia*) [*nomen novum* for *annulata* Taylor, 1916].
Synonym: *annulata* Taylor, 1916.

cubitatus Colless, 1965 (p. 273) – species of *Culex* (*Lophoceraomyia*).

cuccioi (Belkin, 1962) (p. 414) – species of *Verrallina* (*Verrallina*) [original combination: *Aedes* (*Verrallina*) *cuccioi*].

cuclyx Dyar & Shannon, 1924a (p. 48) – synonym of *Culex* (*Melanoconion*) *bastagarius* Dyar & Knab, 1906b; synonymy by Rozeboom & Komp, 1950a.

cucphuongensis Anonymous (in IMPE, 1987) (pp. 8, 36, 49) – *nomen nudum*, *Anopheles*; unavailable name [authorship cannot be determined from the content of the publication and is therefore deemed to be anonymous; a new name published after 1950 with anonymous authorship is not made available (International Commission on Zoological Nomenclature, 1999: *Code*, Articles 50.1 and 14, respectively)].

cucphuongensis Vu, Nguyen, Tran & Nguyen, 1991 (p. 145) – species of *Anopheles* (*Anopheles*).

culebrae Dyar, 1923b (p. 65) – synonym of *Wyeomyia* (*Wyeomyia*) *melanopus* Dyar, 1919d; synonymy established before Knight & Stone, 1977.

culicifacies Giles, 1901b (p. 197) – species of *Anopheles* (*Cellia*).
Synonyms: *indica* Theobald, 1901a; *listoni* Giles, 1901b; *punjabensis* James, 1911; *adenensis* Christophers, 1924c.

culiciformis Cogill, 1903 (p. 333) – species of *Anopheles* (*Anopheles*).
Synonym: *culiciformis* James & Liston, 1904.

culiciformis James & Liston, 1904 (p. 122) – synonym of *Anopheles* (*Anopheles*) *culiciformis* Cogill, 1903; synonymy by Christophers, 1916a.

culiciformis (Theobald, 1905f) (p. 77) – species of *Aedes* (*Pseudoskusea*) [original combination: *Skusea culiciformis*].

culicinus Edwards, 1922a (p. 271) – species of *Aedes* (*Aedimorphus*).

culicivora Dyar & Knab, 1907c (p. 207) – synonym of *Johnbelkinia ulopus* (Dyar & Knab, 1906a): [original combination: *Lesticocampa culicivora*]; synonymy by Zavortink, 1979.

culionicus Delfinado, 1966a (p. 130) – synonym of *Culex* (*Eumelanomyia*) *hinglungensis* Chu, 1957; synonymy by Sirivanakarn, 1972.

cumminsii (Theobald, 1903a) (p. 214) – species of *Aedes* (*Aedimorphus*) [original combination: *Culex cumminsii*].
Synonyms: *fuscopalpalis* Theobald, 1909b; *daruensis* Evans, 1925a.

cumpstoni Taylor, 1914c (p. 692) – synonym of *Aedes* (*Dobrotworskyius*) *rubrithorax* (Macquart, 1850) [original combination: *Culicada cumpstoni*]; synonymy established before Knight & Stone, 1977.

cunabulanus Edwards, 1924b (p. 378) – species of *Aedes* (*Ochlerotatus*).

cuneatus Dyar & Knab, 1908 (p. 54) – synonym of *Aedes* (*Ochlerotatus*) *angustivittatus* Dyar & Knab, 1907a; synonymy established before Knight & Stone, 1977.

cuneatus Theobald, 1901c (p. 22) – synonym of *Culex* (*Culex*) *gelidus* Theobald, 1901c [original combination: *Culex gelidus* variety *cuneatus*]; synonymy established before Knight & Stone, 1977.

cunninghami Taylor, 1944 (p. 123) – synonym of *Verrallina* (*Verrallina*) *bancrofti* Taylor, 1914b [original combination: *Aedes* (*Aedes*) *cunninghami*]; originally proposed as a replacement name for *bancrofti* Taylor, 1914b, *non bancrofti* Skuse, 1889, when both were included in *Aedes* Meigen.

curiche Adames, 1971 (p. 68) – species of *Deinocerites*.

curopinensis Bonne-Wepster & Bonne, 1920 (p. 177) – synonym of *Culex* (*Melanoconion*) *pilosus* (Dyar & Knab, 1906d); synonymy established before Knight & Stone, 1977.

curriei Coquillett, 1901 (p. 259) – synonym of *Aedes* (*Ochlerotatus*) *dorsalis* (Meigen, 1830b) [original combination: *Culex curriei*]; synonymy by Wood *et al.*, 1979.

curryi Dyar, 1926 (p. 112) – synonym of *Culex* (*Melanoconion*) *elevator* Dyar & Knab, 1906d; synonymy by Rozeboom & Komp, 1950a.

curtipalpis (Edwards, 1914c) (p. 127) – species of *Culex* (*Lophoceraomyia*) [original combination: *Lophoceratomyia curtipalpis*].

curtipes Edwards, 1915c (p. 283) – species of *Aedes* (*Cancraedes*).
Synonym: *miachaetessa* Dyar & Shannon, 1925a.

curvibrachius Angulo, 1993 (in Angulo & Olivares, 1993) (p. 22) – species of *Culex* (*Culex*).

curvirostris Laveran, 1902c (p. 1160) – synonym of *Limatus durhamii* Theobald, 1901c [original combination: *Simondella curvirostris*]; synonymy by Edwards, 1932a.

cuttsi van den Assem, 1959 (p. 46) – species of *Tripteroides* (*Rachisoura*).

cuyabensis Neiva & Pinto, 1923 (p. 235) – synonym of *Anopheles* (*Nyssorhynchus*) *triannulatus* (Neiva & Pinto, 1922b) [original combination: *Cellia cuyabensis*]; synonymy by Pinto, 1939.

cuyanus Duret, 1968e (p. 323) – species of *Culex* (*Culex*).

cyanescens (Coquillett, 1902a) (p. 137) – species of *Psorophora* (*Janthinosoma*) [original combination: *Culex cyanescens*].

Synonyms: *purpurascens* Edwards, 1922d; *tovari* Evans, 1922.

cyaneus (Fabricius, 1805) (p. 35) – species of *Sabethes* (*Sabethes*) [original combination: *Culex cyaneus*].

Synonyms: *locuples* Robineau-Desvoidy, 1827; *remipes* Wiedemann, 1828.

cyanopennis von Humboldt, 1819 (p. 340) – synonym of *Psorophora* (*Psorophora*) *ciliata* (Fabricius, 1794) [original combination: *Culex cyanopennis*]; synonymy established before Knight & Stone, 1977.

cyclocerculus Dyar, 1920a (p. 23) – synonym of *Aedes* (*Ochlerotatus*) *hexodontus* Dyar, 1916b; synonymy established before Knight & Stone, 1977.

cydippis de Meillon, 1931a (p. 318) – species of *Anopheles* (*Cellia*) [original combination: *Anopheles squamosus* variety *cydippis*]; specific status by Lips, 1962.

Synonym: *entebbiensis* Evans, 1938.

cylindricus Theobald, 1903a (p. 202) – species of *Culex* (*Lophoceraomyia*).

cyprioides Danilov & Stupin, 1982 (p. 824) – species of *Aedes* (*Ochlerotatus*).

cyprius Ludlow, 1920b (p. 158) – species of *Aedes* (*Ochlerotatus*).

Synonym: *freyi* Edwards, 1921d.

cyptopus Theobald, 1909b (as *cuptopous*) (p. 25) – species of *Hodgesia*; emendation by Edwards, 1912c.

cyrtolabis (Edwards, 1928a) (p. 273) – species of *Verrallina* (*Neomacleaya*) [original combination: *Aedes* (*Aedes*) *cyrtolabis*].

Dd

daciae Linton, Nicolescu & Harbach, 2004 (in Nicolescu *et al.*, 2004) (p. 530) – species of *Anopheles* (*Anopheles*).

daggyi Stone & Bohart, 1944 (p. 215) – species of *Aedes* (*Geoskusea*).

dahlae (Nielsen, 2009) (p. 123) – species of *Aedes* (*Ochlerotatus*) [original combination: *Ochlerotatus dahlae*].

dahomeensis Holstein, 1951 (pp. 464, 465) – *nomen nudum*, *Anopheles funestus* variety.

dahuricus Danilov, 1987b (p. 35) – species of *Aedes* (*Aedes*).

daisetsuzanus Tanaka, Mizusawa & Saugstad, 1979 (p. 271) – subspecies of *Aedes* (*Ochlerotatus*) *impiger* (Walker, 1848).

daitensis Miyagi & Toma, 1981 (p. 428) – species of *Aedes* (*Stegomyia*).

daliensis (Taylor, 1916) (p. 565) – species of *Aedes* (*Ochlerotatus*) [original combination: *Stegomyia daliensis*].

dalzieli (Theobald, 1910c) (p. 234) – species of *Aedes* (*Aedimorphus*) [original combination: *Culiciomyia dalzieli*].

Synonym: *rhecter* Dyar, 1921d.

damascenoi Duret, 1969c (p. 143) – species of *Culex* (*Anoedioporpa*).

Synonym: *menui* Clastrier, 1971.

damasei Grjebine, 1979 (p. 70) – species of *Uranotaenia* (*Pseudoficalbia*).

damnatorum Scudder, 1890 (p. 582) – fossil species of *Culex* Linnaeus, 1758.

damnosus Say, 1823 (p. 11) – synonym of *Aedes* (*Ochlerotatus*) *taeniorhynchus* (Wiedemann, 1821) [original combination: *Culex damnosus*]; synonymy established before Knight & Stone, 1977.

damzeni (Szadziewski, 1998) (p. 239) – fossil species of *Eoaedes* Harbach & Greenwalt, 2012 [original combination: *Aedes* (*Finlaya*) *damzeni*].

danaraji Ramalingam, 1988 (p. 245) – species of *Topomyia* (*Topomyia*).

danaubento Mochtar & Walandouw, 1934 (p. 932) – subspecies of *Anopheles* (*Anopheles*) *gigas* Giles, 1901b [original combination: *Anopheles gigas* variety *danaubento*]; subspecific status by Harbach & Howard, 2007.

dancalicus Corradetti, 1939b (p. 277) – species of *Anopheles* (*Cellia*).

dandakaranyensis Natarajan, Rajavel & Jambulingam, 2017 (p. 255) – species of *Uranotaenia* (*Pseudoficalbia*).

dangi Anonymous (in IMPE, 1987) (pp. 14, 33, 57) – *nomen nudum*, *Anopheles*; unavailable name [authorship cannot be determined from the content of the publication and is therefore deemed to be anonymous; a new name published after 1950 with anonymous authorship is not made available (International Commission on Zoological Nomenclature, 1999: *Code*, Articles 50.1 and 14, respectively)]; unnecessary synonymy with *Anopheles* (*Anopheles*) *crawfordi* Reid, 1953 by Ngo *et al.*, 2013.

darlingi Root, 1926 (p. 706) – species of *Anopheles* (*Nyssorhynchus*).

Synonym: *paulistensis* Galvão, Lane & Corrêa, 1937.

daruensis Evans, 1925a (p. 123) – synonym of *Aedes* (*Aedimorphus*) *cumminsii mesostictus* Harbach (herein) [original combination: *Aedes* (*Aedimorphus*) *cumminsi* [*sic*] variety *daruensis*]; synonymy with *mediopunctata* Theobald, 1909b by Edwards, 1941.

darwini Taylor, 1914b (p. 458) – synonym of *Aedes* (*Macleaya*) *tremulus* (Theobald, 1903e) [original combination: *Aedimorphus australis* variety *darwini*]; synonymy established before Knight & Stone, 1977.

daryi Schick, 1970a (p. 80) – species of *Aedes* (*Protomacleaya*).

dasyorrhus King & Hoogstraal, 1946c (p. 100) – species of *Aedes* (*Lorrainea*).

daudi Coluzzi, 1958 (p. 69) – species of *Anopheles* (*Cellia*).

daumastocampa Dyar & Knab, 1908 (p. 58) – species of *Culex* (*Microculex*).

daumasturus Dyar & Knab, 1906d (p. 220) – synonym of *Culex* (*Microculex*) *imitator* Theobald, 1903a; synonymy established before Knight & Stone, 1977.

davidi Basio, 1971 (p. 51) – species of *Aedes* (*Aedimorphus*).

davidsoni Ribeiro, da Cunha Ramos, Pires & Capela, 1979 (p. 75) – synonym of *Anopheles* (*Cellia*) *arabiensis*

Patton, 1905 [original combination: *Anopheles* (*Cellia*) *quadriannulatus davidsoni*]; synonymy by Cambournac *et al.*, 1982.

davisi Kumm, 1933 (p. 565) – species of *Culex* (*Microculex*).

davisi Lane, 1943 (p. 158) – species of *Uranotaenia* (*Uranotaenia*).

davisi (Lane & Cerqueira, 1942) (p. 647) – species of *Wyeomyia* (*Phoniomyia*) [original combination: *Phoniomyia davisi*].

davisi Paterson & Shannon, 1927 (p. 5) – synonym of *Anopheles* (*Nyssorhynchus*) *triannulatus* (Neiva & Pinto, 1922b); synonymy by Faran, 1980.

dazhaius Ma, 1981b (p. 65) –synonym of *Anopheles* (*Anopheles*) *lesteri* Baisas & Hu, 1936; synonymy with *anthropophagus* Xu & Feng by Maio *et al.*, 1988 (see Qu & Zhu, 2008).

deaconi de Meillon & van Eeden, 1976 (p. 336) – synonym of *Anopheles* (*Cellia*) *azevedoi* Ribeiro, 1969; synonymy by White, 1980.

deanei Corrêa & Ramalho, 1959b (p. 141) – synonym of *Culex* (*Culex*) *chidesteri* Dyar, 1921h; synonymy established before Knight & Stone, 1977.

deanei (Lourenço de Oliveira, 1983) (p. 501) – species of *Wyeomyia* (*Phoniomyia*) [original combination: *Phoniomyia deanei*].

deaneorum Rosa-Freitas, 1989 (p. 535) – species of *Anopheles* (*Nyssorhynchus*).

debilis Dyar & Knab, 1914 (p. 58) – synonym of *Culex* (*Culex*) *apicinus* Philippi, 1865 [original combination: *Phalangomyia debilis*]; synonymy established before Knight & Stone, 1977.

deboeri Edwards, 1926a (p. 128) – species of *Aedes* (*Stegomyia*).

deccanus (Barraud, 1923b) (p. 217) – species of *Aedes* (*Phagomyia*) [original combination: *Finlaya deccana*].

decens Theobald, 1901d (p. 7) – species of *Culex* (*Culex*).
Synonyms: *masculus* Theobald, 1901c; *minutus* Theobald, 1905g; *nigrocostalis* Theobald, 1909b; *lividocostalis* Graham, 1910.

deceptor Dönitz, 1902 (p. 60) – synonym of *Anopheles* (*Cellia*) *tessellatus* Theobald, 1901a; synonymy established before Knight & Stone, 1977.

deceptor Dyar & Knab, 1909d (p. 257) – synonym of *Culex* (*Melanoconion*) *pilosus* (Dyar & Knab, 1906d); synonymy established before Knight & Stone, 1977.

declarator Dyar & Knab, 1906d (p. 211) – species of *Culex* (*Culex*).
Synonyms: *inquisitor* Dyar & Knab, 1906d; *proclamator* Dyar & Knab, 1906d; *jubilator* Dyar & Knab, 1907c; *revelator* Dyar & Knab, 1907c; *dictator* Dyar & Knab, 1909d; *vindicator* Dyar & Knab, 1909d; *forattinii* Corrêa & Ramalho, 1959a.

decorabilis (Leicester, 1908) (p. 239) – species of *Kimia* [original combination: *Topomyia decorabilis*].

decorator Dyar & Knab, 1906d (p. 218) – *nomen dubium*, *Culex*; status by Sirivanakarn, 1983.

decticus Howard, Dyar & Knab, 1917 (p. 737) – species of *Aedes* (*Ochlerotatus*).
Synonym: *pseudodiantaeus* Smith, 1952.

deemingi Service, 1970 (p. 647) – species of *Anopheles* (*Cellia*).

deficiens Arnell, 1976 (p. 24) – species of *Aedes* (*Ochlerotatus*).

degustator Dyar, 1921c (p. 39) – synonym of *Culex* (*Melanoconion*) *erraticus* (Dyar & Knab, 1906d); synonymy by King & Bradley, 1937.

deguzmanae (Mattingly, 1957a) (p. 53) – species of *Mimomyia* (*Ingramia*) [original combination: *Ficalbia* (*Ravenalites*) *deguzmanae*].

dejesusi Baisas & Feliciano, 1953 (p. 173) – species of *Topomyia* (*Topomyia*).

delae Baisas, 1935c (p. 68) – synonym of *Uranotaenia* (*Uranotaenia*) *clara* Dyar & Shannon, 1925a; synonymy by Belkin, 1953.

delfinadoae Sirivanakarn, 1973b (p. 118) – species of *Culex* (*Culiciomyia*).

delpilari Baisas & Ubaldo-Pagayon, 1953 (p. 91) – species of *Tripteroides* (*Tricholeptomyia*).

delpontei Duret, 1969a (p. 8) – species of *Culex* (*Melanoconion*).

delpontei Lane & Cerqueira, 1942 (p. 613) – synonym of *Wyeomyia rooti* (del Ponte, 1939); synonymy by Lane & Cerqueira, 1957.

delpontei Martínez & Prosen, 1955 (p. 613) – synonym of *Aedes* (*Howardina*) *pseudodominicii* Komp, 1936b; synonymy by Berlin, 1969a.

delta Anduze, 1948 (p. 18) – synonym of *Anopheles* (*Nyssorhynchus*) *aquasalis* Curry, 1932; synonymy by Lane, 1953.

delta Séguy, 1924 (p. 80) – synonym of *Aedes* (*Stegomyia*) *cretinus* Edwards, 1921d; synonymy by Monchadskii, 1951.

deltaorinoquensis Cova Garcia, Pulido F. & Amarista M., 1977 (p. 150) – synonym of *Anopheles* (*Nyssorhynchus*) *aquasalis* Curry, 1932; synonymy by Faran, 1980.

delys Howard, Dyar & Knab, 1915 (p. 317) – species of *Culex* (*Culex*).

demansis Strickland, 1911b (p. 202) – synonym of *Aedes* (*Dobrotworskyius*) *rubrithorax* (Macquart, 1850) [original combination: *Culicada demansis*]; synonymy established before Knight & Stone, 1977.

demeilloni Doucet, 1950 (p. 53) – species of *Culex* (*Culex*).

demeilloni Edwards, 1936 (p. 55) – species of *Aedes* (*Stegomyia*) [original combination: *Aedes* (*Stegomyia*) *deboeri* variety *demeilloni*]; specific status by Mattingly, 1952.

demeilloni Evans, 1933 (p. 266) – species of *Anopheles* (*Cellia*).

demeilloni Mattingly, 1970 (p. 21) – species of *Heizmannia* (*Heizmannia*).

demeilloni Peyton & Rattanarithikul, 1970 (p. 404) – species of *Uranotaenia* (*Pseudoficalbia*).

demissus Colless, 1965 (p. 296) – species of *Culex* (*Lophoceraomyia*).
Synonym: *fuscosiphonis* Bram & Rattanarithikul, 1967.

denbesteni Brug, 1925 (p. 345) – species of *Armigeres* (*Armigeres*).
Synonym: *spathulatus* Brug, 1939.

denderensis Wolfs, 1949 (p. 190) – species of *Aedes* (*Stegomyia*) [original combination: *Aedes* (*Stegomyia*) *apicoargenteus* variety *denderensis*]; specific status by Huang, 2004.

dendrophilus Edwards, 1921c (p. 74) – species of *Aedes* (*Stegomyia*).
Synonym: *trinidad* Gil Collado, 1936.

deniedmannii Ludlow, 1904b (p. 234) – synonym of *Aedes* (*Ochlerotatus*) *squamiger* (Coquillett, 1902c) [original combination: *Grabhamia deniedmannii*]; synonymy established before Knight & Stone, 1977.

dentatus Barraud, 1927b (p. 547) – species of *Armigeres* (*Leicesteria*).

dentatus (Theobald, 1904b) (p. 75) – species of *Aedes* (*Aedimorphus*) [original combination: *Culex dentatus*].
Synonym: *pallidopunctata* Theobald, 1911b.

denticulatus Delfinado & Hodges, 1968 (p. 362) – species of *Tripteroides* (*Tripteroides*).

depereti (Meunier, 1915) (p. 17) – fossil species of *Neoculicites* Evenhuis, 1994 [original combination: *Culicites depereti*].

derivator Dyar & Knab, 1906d (p. 216) – species of *Culex* (*Neoculex*).

dermajoensis Brug, 1931a (p. 250) – synonym of *Verrallina* (*Neomacleaya*) *incerta* (Edwards, 1922a) [original combination: *Aedes* (*Aedes*) *dermajoensis*]; synonymy by Reinert, 1974.

derooki Brug, 1932 (p. 75) – species of *Aedes* (*Patmarksia*).

derooki Soesilo & van Slooten, 1931a [1931b] (pp. 1256 [126]) – species of *Bironella* (*Neobironella*).

derricki Taylor, 1943 (p. 155) – synonym of *Anopheles* (*Cellia*) *annulipes* Walker, 1856a; synonymy by Lee & Woodhill, 1944.

desbansi Séguy, 1923 (p. 205) – synonym of *Aedes* (*Acartomyia*) *mariae* (Sergent & Sergent, 1903); synonymy established before Knight & Stone, 1977.

deserticola Kirkpatrick, 1925 (p. 378) – species of *Culex* (*Maillotia*).

deserticola Zavortink, 1970a (p. 1) – species of *Aedes* (*Jarnellius*).

desmotes (Giles, 1904a) (p. 367) – species of *Aedes* (*Stegomyia*) [original combination: *Stegomyia desmotes*].
Synonyms: *albitarsis* Ludlow, 1905a; *gracilis* Leicester, 1908; *albipes* Theobald, 1910a; *montana* Koidzumi, 1918.

detritus (Haliday, 1833) (p. 151) – species of *Aedes* (*Ochlerotatus*) [original combination: *Culex detritus*].
Synonyms: *salinus* Ficalbi, 1896; *terriei* Theobald, 1903a; *maculosa* Theobald, 1905f.

devemyi Hamon, 1955 (p. 672) – species of *Uranotaenia* (*Pseudoficalbia*).

diabolica (Lane & Forattini, 1952) (p. 254) – species of *Wyeomyia* (*Phoniomyia*) [original combination: *Phoniomyia diabolica*].

diaeretus Dyar, 1920i (p. 181) – synonym of *Coquillettidia* (*Coquillettidia*) *crassipes* (van der Wulp, 1881) [original combination: *Mansonia* (*Coquillettidia*) *crassipes*]; synonymy by Bohart, 1945.

diagonalis Brug, 1934 (p. 510) – species of *Uranotaenia* (*Pseudoficalbia*).

dialloi Hamon & Brengues, 1965 (p. 103) – species of *Aedes* (*Catageiomyia*).

diamphidius Peyton & Harbach, 1991 (p. 199) – species of *Culex* (*Melanoconion*).

diantaeus Howard, Dyar & Knab, 1913, 1917 (fig. 167, p. 758, respectively) – species of *Aedes* (*Ochlerotatus*).
Synonym: *serus* Martini, 1920a.

diazi Schick, 1970a (p. 89) – species of *Aedes* (*Protomacleaya*).

dibrugarhensis Barraud, 1927b (p. 539) – synonym of *Armigeres* (*Armigeres*) *durhami* Edwards, 1917 [original combination: *Armigeres* (*Armigeres*) *kuchingensis* form *dibrugarhensis*]; synonymy by Thurman, 1959.

dibrugarhensis Bhattacharyya, Prakash, Mohapatra & Mahanta, 2004 (p. 1) – species of *Uranotaenia* (*Pseudoficalbia*).

dicellaphora Howard, Dyar & Knab, 1913 [1915] (figs 43, 305 [166]) – synonym of *Runchomyia* (*Ctenogoeldia*) *magna* (Theobald, 1905f) [original combination: *Lesticocampa dicellaphora*]; synonymy established before Knight & Stone, 1977.

dictator Dyar & Knab, 1909d (p. 255) – synonym of *Culex* (*Culex*) *declarator* Dyar & Knab, 1906d; synonymy established before Knight & Stone, 1977.

didieri Neveu-Lemaire, 1906 (p. 257) – synonym of *Culex* (*Culex*) *quinquefasciatus* Say, 1823; synonymy established before Knight & Stone, 1977.

diengensis Brug, 1931c (p. 35) – species of *Culex* (*Culex*).

digitatum (Rondani, 1848) (p. 109) – species of *Trichoprosopon* [original combination: *Culex digitatus*].
Synonyms: *nivipes* Theobald, 1901c; *splendens* Lutz, 1904; *wilsoni* Ludlow, 1918; *subsplendens* Martini, 1931a.

digitatus (Edwards, 1914a) (p. 262) – species of *Armigeres* (*Leicesteria*) [original combination: *Leicesteria digitata*].

digoelensis Brug, 1932 (p. 81) – species of *Culex* (*Lophoceraomyia*).
Synonym: *caeruleus* King & Hoogstraal, 1947c.

digoelensis Brug, 1934 (p. 506) – species of *Tripteroides* (*Polylepidomyia*).

diluvialis Reinert, 1997 (in Reinert *et al.*, 1997) (p. 11) – species of *Anopheles* (*Anopheles*).

dimidiata Cerqueira, 1943 (p. 5) – species of *Psorophora* (*Grabhamia*).

diplolineata Schneider, 1913 (p. 37) – synonym of *Aedes* (*Ochlerotatus*) *communis* (de Geer, 1776) [original combination: *Culicada nemorosa* form *diplolineata*]; synonymy established before Knight & Stone, 1977.

diplophyllum Dyar, 1929b (p. 509) – species of *Culex* (*Culex*).

dipseticus Dyar & Knab, 1909a (race of *quinquefasciatus* Say, 1823) (p. 34) – synonym of *Culex* (*Culex*) *pipiens pallens* Coquillett, 1898; synonymy by Mattingly, 1951a.

diraphati Peyton & Klein, 1970 (p. 245) – species of *Uranotaenia* (*Uranotaenia*).

dirus Peyton & Harrison, 1979 (p. 41) – species of *Anopheles* (*Cellia*).

discolor (Coquillett, 1903a) (p. 256) – species of *Psorophora* (*Grabhamia*) [original combination: *Culex discolor*].

discors Harbach, 2014 (in Harbach & Culverwell, 2014) (p. 576) – species of *Topomyia* (*Miyagiella*).

discrepans (Edwards, 1922a) (p. 291) – species of *Heizmannia* (*Mattinglyia*) [original combination: *Haemagogus discrepans*].

discrucians (Walker, 1856a) (p. 430) – species of *Psorophora* (*Janthinosoma*) [original combination: *Culex discrucians*].

Synonym: *arribalzagae* Giles, 1902.

disjunctus Roubaud, 1957 (p. 3116) – synonym of *Culex* (*Culex*) *pipiens* Linnaeus, 1758 [original combination: *Culex pipiens disjunctus*]; synonymy by Stone *et al.*, 1959.

dispar Rattanarithikul & Harbach, 1991 (p. 176) – species of *Anopheles* (*Cellia*).

dispectus Bram, 1966 (p. 75) – species of *Culex* (*Culiciomyia*).

dissidens Taai & Harbach, 2015 (p. 253) – species of *Anopheles* (*Anopheles*).

dissimilierodes Dong, Zhou & Dong, 2002a (p. 516) – species of *Aedes* (*Kenknightia*).

dissimilis (Leicester, 1908) (p. 91) – species of *Aedes* (*Kenknightia*) [original combination: *Stegomyia dissimilis*].

dissimilis Theobald, 1901a (p. 376) – synonym of *Culex* (*Culex*) *duttoni* Theobald, 1901d; synonymy established before Knight & Stone, 1977.

distigma (Edwards, 1925) (p. 257) – species of *Tripteroides* (*Tripteroides*) [original combination: *Rachionotomyia distigma*].

distinctus (Newstead & Carter, 1911b) (p. 234) – species of *Anopheles* (*Cellia*) [original combination: *Pyretophorus distinctus*].

Synonym: *melanocosta* Newstead & Carter, 1911b.

distinguendus Dyar, 1928 (p. 305) – species of *Culex* (*Melanoconion*).

ditaenionota Prado, 1931b (p. 209) – species of *Uranotaenia* (*Uranotaenia*).

Synonym: *burkii* Lane, 1936b.

diurna Theobald, 1903f (p. 259) – synonym of *Verrallina* (*Verrallina*) *butleri* (Theobald, 1901c) [original combination: *Skusea diurna*]; synonymy established before Knight & Stone, 1977.

diversus Theobald, 1901c (p. 73) – synonym of *Aedes* (*Rusticoidus*) *rusticus* (Rossi, 1790) [original combination: *Culex diversus*]; synonymy established before Knight & Stone, 1977.

dives (Schiner, 1868) (p. 31) – species of *Mansonia* (*Mansonioides*) [original combination: *Culex dives*].

Synonyms: *annulipes* Walker, 1856b; *longipalpis* van der Wulp, 1881.

divisior Dyar & Knab, 1906d (p. 222) – synonym of *Culex* (*Anoedioporpa*) *conservator* Dyar & Knab, 1906d; synonymy by Lane, 1951.

djajasanensis Brug, 1926 (p. 805) – synonym of *Anopheles* (*Anopheles*) *annandalei* Prashad, 1918 [original combination: *Lophoscelomyia annandalei* variety *djajasanensis*]; synonymy by Puri, 1929a.

dmitryi Gornostaeva, 2005 (p. 485) – species of *Aedes* (*Aedes*).

dobodurus King & Hoogstraal, 1946b (p. 37) – species of *Aedes* (*Patmarksia*).

dobrotworskyi Marks, 1958 (p. 65) – species of *Aedes* (*Finlaya*).

doddi Taylor, 1919 (p. 831) – synonym of *Aedes* (*Macleaya*) *tremulus* (Theobald, 1903e) [original combination: *Mimeteomyia doddi*]; synonymy established before Knight & Stone, 1977.

dofleini (Guenther, 1913) (p. 204) – species of *Tripteroides* (*Rachionotomyia*) [original combination: *Ficalbia dofleini*].

dohenyi Hogue, 1975 (p. 357) – species of *Culex* (*Culex*).

doleschallii Giles, 1900a (p. 338) – synonym of *Culex* (*Culex*) *quinquefasciatus* Say, 1823 [*nomen novum* for *cingulatus* Doleschall, 1856]; synonymy established before Knight & Stone, 1977.

dolichocephalus (Leicester, 1908) (p. 95) – species of *Armigeres* (*Leicesteria*) [original combination: *Leicesteria dolichocephala*].

dolichophyllus Clastrier, 1970b (p. 858) – species of *Culex* (*Melanoconion*).

doliorum Edwards, 1912a (p. 263) – synonym of *Culex* (*Culex*) *pipiens* Linnaeus, 1758 [original combination: *Culex pipiens* variety *doliorum*]; synonymy established before Knight & Stone, 1977.

dolosus (Lynch Arribálzaga, 1891b) (p. 156) – species of *Culex* (*Culex*) [original combination: *Heteronycha dolosa*].

Synonym: *bonariensis* Brèthes, 1916.

domesticus Galvão & Damasceno, 1944 (p. 78) – synonym of *Anopheles* (*Nyssorhynchus*) *marajoara* Galvão & Damasceno, 1942b [original combination: *Anopheles* (*Nyssorhynchus*) *albitarsis domesticus*]; synonymy by Linthicum, 1988.

domesticus Germar, 1817 (p. 290) – synonym of *Culex* (*Culex*) *pipiens* Linnaeus, 1758; synonymy by Harbach, 1988.

domesticus Leicester, 1908 (p. 169) – synonym of *Culex* (*Oculeomyia*) *infula* Theobald, 1901a [original combination: *Taeniorhynchus domesticus*]; synonymy by Sirivanakarn, 1976.

domesticus (Theobald, 1901d) (p. xiv) – species of *Aedes* (*Aedimorphus*) [original combination: *Uranotaenia domestica*].

domicolus Edwards, 1916 (p. 363) – species of *Anopheles* (*Cellia*).

Synonym: *gingeroi* Corradetti & Archetti, 1947.

dominguezi Duret, 1971b (p. 83) – synonym of *Haemagogus* (*Haemagogus*) *anastasionis* Dyar, 1921k; synonymy by Arnell, 1973.

dominicanus Zavortink & Poinar, 2000 (p. 1231) – fossil species of *Anopheles* (*Nyssorhynchus*).

dominicii Rangel & Romero Sierra, 1907 (in Romero Sierra, 1907) (p. 48) – synonym of *Aedes* (*Howardina*) *sexlineata*

(Theobald, 1901a) [original combination: *Stegomyia dominicii*]; synonymy established before Knight & Stone, 1977.

donai da Cunha Ramos & Brunhes, 2004 (p. 254) – species of *Uranotaenia* (*Pseudoficalbia*).

donaldi Reid, 1962 (p. 17) – species of *Anopheles* (*Anopheles*).

doonii Wattal, Bhatia & Kalra, 1958 (p. 220) – species of *Aedes* (*Bruceharrisonius*) [original combination: *Aedes* (*Finlaya*) *aureostriatus* variety *doonii*]; specific status by Reinert, 2003.

dornarum Dyar & Shannon, 1924a (p. 46) – synonym of *Culex* (*Melanoconion*) *elevator* Dyar & Knab, 1906d; synonymy by Komp, 1935.

dorsalis (Meigen, 1830b) (p. 242) – species of *Aedes* (*Ochlerotatus*) [original combination: *Culex dorsalis*].
Synonyms: *curriei* Coquillett, 1901; *onondagensis* Felt, 1904; *lativittatus* Coquillett, 1906c; *quaylei* Dyar & Knab, 1906d; *broquetii* Theobald, 1913c; *grahami* Ludlow, 1920b.

dorseyi Knight, 1946 (p. 277) – species of *Aedes* (*Downsiomyia*).

dorso-vittatus Villeneuve, 1919 (p. 57) – synonym of *Aedes* (*Ochlerotatus*) *sticticus* (Meigen, 1838) [original combination: *Culex nemorosus* variety *dorso-vittatus*]; synonymy established before Knight & Stone, 1977.

douceti Adam & Hamon, 1959 (p. 525) – subspecies of *Eretmapodites wansoni* Edwards, 1941.

douceti Grjebine, 1953a (p. 459) – species of *Uranotaenia* (*Pseudoficalbia*).

douglasi Dobrotworsky, 1956 (p. 110) – species of *Culex* (*Neoculex*).

downsi Bohart & Ingram, 1946a (species) (p. 51) – subspecies of *Aedes* (*Stegomyia*) *flavopictus* Yamada, 1921; subspecific status by Tanaka *et al.*, 1979.

downsi Lane, 1945a (p. 140) – species of *Wyeomyia* (*Wyeomyia*).

dracaenae Edwards, 1916 (p. 362) – species of *Eretmapodites*.
Synonym: *ferox* Haddow, 1946.

draconarius Dyar, 1922g (p. 194) – synonym of *Aedes* (*Georgecraigius*) *fluviatilis* (Lutz, 1904); synonymy established before Knight & Stone, 1977.

draconis Ingram & de Meillon, 1927 (species) (p. 67) – subspecies of *Culex* (*Culex*) *vansomereni* Edwards, 1926a; subspecific status by Edwards, 1941.

drapetes Dyar & Knab, 1909d (p. 264) – synonym of *Wyeomyia* (*Wyeomyia*) *arthrostigma* (Lutz, 1905); synonymy by Lane & Cerqueira, 1942.

dravidicus Christophers, 1924c (p. 297) – species of *Anopheles* (*Cellia*) [original combination: *Anopheles maculatus* variety *dravidicus*]; specific status by Rattanarithikul & Green, 1987.

drummondi (Dobrotworsky, 1960b) (p. 241) – species of *Culiseta* (*Culicella*) [original combination: *Theobaldia drummondi*].

drymoecius Speiser, 1909 (p. 42) – synonym of *Coquillettidia* (*Coquillettidia*) *fuscopennata* (Theobald, 1903a) [original combination: *Culex drymoecius*]; synonymy established before Knight & Stone, 1977.

dthali Patton, 1905 (p. 627) – species of *Anopheles* (*Cellia*).
Synonym: *wardi* Leeson & Theodor, 1948.

dthalisimilis Corradetti, 1939a (p. 58) – synonym of *Anopheles* (*Cellia*) *rhodesiensis rupicolus* Lewis, 1937 [original combination: *Anopheles rhodesiensis* variety *dthalisimilis*]; synonymy by Edwards, 1941.

dualaensis Brunhes, Le Goff & Geoffroy, 1999 (p. 554) – species of *Anopheles* (*Cellia*).

dubia Theobald, 1910c (p. 133) – synonym of *Aedes* (*Stegomyia*) *metallicus* (Edwards, 1912c) [original combination: *Quasistegomyia dubia*]; synonymy established before Knight & Stone, 1977. **Note**: this nominal taxon is the senior secondary homonym of *dubia* Theobald, 1910c (p. 170), based on page priority.

dubia Theobald, 1910c (p. 170) – synonym of *Aedes* (*Stegomyia*) *africanus* (Theobald, 1901d) [original combination: *Stegomyia africana*]; synonymy established before Knight & Stone, 1977. **Note**: this nominal taxon is the junior secondary homonym of *dubia* Theobald, 1910c (p. 133), based on page priority.

dubitans Christophers, 1924a (p. 133) – *nomen nudum*, *Anopheles*.

dubitans Lane & Whitman, 1951 (p. 357) – species of *Culex* (*Microculex*).

dubitans Leicester, 1908 (p. 246) – species of *Topomyia* (*Topomyia*).

dubius Blanchard, 1905 (p. 205) – *nomen nudum*, *Anopheles*.

dudgeonii Theobald, 1907 (p. 113) – synonym of *Anopheles* (*Cellia*) *willmori* (James, 1903) [original combination: *Neocellia dudgeonii*]; synonymy established before Knight & Stone, 1977.

dufouri Hamon, 1953 (p. 40) – species of *Aedes* (*Ochlerotatus*).

dugesi Dyar & Knab, 1906a (p. 134) – synonym of *Culiseta* (*Culiseta*) *particeps* (Adams, 1903); synonymy by Stone, 1958b.

dulongensis Gong & Lu, 1995 (p. 479) – species of *Topomyia* (*Suaymyia*).

dumbletoni Belkin, 1962 (p. 244) – species of *Culex* (*Neoculex*).

dumonti Doucet, 1949b (p. 327) – species of *Uranotaenia* (*Uranotaenia*).

dundo da Cunha Ramos, 1993 (p. 424) – subspecies of *Uranotaenia* (*Uranotaenia*) *palmeirimi* de Meillon & Rebêlo, 1941.

dundo da Cunha Ramos & Ribeiro, 1992 (p. 33) – species of *Eretmapodites*.

dundo Ribeiro, 1991 (p. 51) – species of *Toxorhynchites* (*Toxorhynchites*).

dunhami Causey, 1945 (p. 231) – species of *Anopheles* (*Nyssorhynchus*).

dunni Dyar, 1918e (p. 123) – species of *Culex* (*Melanoconion*).
Synonyms: *ruffinis* Dyar & Shannon, 1924b; *exedrus* Root, 1927c.

dunni Evans, 1928 (p. 39) – subspecies of *Aedes* (*Pseudarmigeres*) *argenteoventralis* (Theobald, 1909b) [original combination: *Aedes* (*Armigeres*) *albomarginata* variety *dunni*]; subspecific status by White, 1975.
duplex Martini, 1926 (p. 203) – species of *Aedes* (*Ochlerotatus*).
duplicator Dyar & Knab, 1909d (p. 256) – species of *Culex* (*Culex*).
dupreei (Coquillett, 1904a) (p. 10) – species of *Aedes* (*Ochlerotatus*) [original combination: *Culex dupreei*].
durbanensis (Theobald, 1903a) (p. 246) – species of *Aedes* (*Aedimorphus*) [original combination: *Grabhamia durbanensis*].
Synonym: *ocellata* Theobald, 1910c.
dureni Edwards, 1938 (in Evans, 1938) (p. 123) – species of *Anopheles* (*Cellia*).
dureti Casal & García, 1968d (p. 123) – species of *Culex* (*Melanoconion*).
durhami Edwards, 1917 (p. 206) – species of *Armigeres* (*Armigeres*).
Synonyms: *dibrugarhensis* Barraud, 1927b; *shillongensis* Barraud, 1927b; *javanensis* Brug, 1939.
durhami Sirivanakarn, 1968 (p. 177) – species of *Culex* (*Lophoceraomyia*).
durhamii Theobald, 1901c (p. 350) – species of *Limatus*.
Synonyms: *curvirostris* Laveran, 1902c; *exhibitor* Shannon & del Ponte, 1928.
duttoni Theobald, 1901d (p. 5) – species of *Culex* (*Culex*).
Synonyms: *dissimilis* Theobald, 1901a; *hirsutipalpis* Theobald, 1901a; *anarmostus* Theobald, 1903a; *bifoliata* Theobald, 1905g; *minutus* Theobald, 1905g; *condylodesmus* Grünberg, 1905; *albovirgatus* Graham, 1910.
dux (Dyar & Shannon, 1925a) (p. 81) – species of *Verrallina* (*Verrallina*) [original combination: *Aedes* (*Aedes*) *dux*].
Synonym: *sigmoides* Barraud, 1928b.
dyari Belkin, Heinemann & Page, 1970 (p. 109) – species of *Mansonia* (*Mansonia*).
dyari Belkin & Hogue, 1959 (p. 428) – species of *Deinocerites*.
dyari Bohart & Farner, 1944 (p. 72) – species of *Tripteroides* (*Tripteroides*).
dyari Coquillett, 1902b (p. 192) – synonym of *Culiseta* (*Culicella*) *morsitans* (Theobald, 1901c) [original combination: *Culex dyari*]; synonymy by Wood *et al*., 1979.
dyari Lane & Cerqueira, 1942 (p. 580) – species of *Wyeomyia* (*Cruzmyia*).
dyari Petrocchi, 1923 (p. 92) – *nomen nudum*, *Psorophora*.
dybasi Bohart, 1957 (p. 53) – species of *Aedes* (*Stegomyia*).
dyi Baisas & Ubaldo-Pagayon, 1953 (p. 128) – species of *Tripteroides* (*Tripteroides*).
dyius Root, 1927c (p. 587) – species of *Culex* (*Melanoconion*).
dymodora Dyar & Knab, 1908 (p. 68) – synonym of *Wyeomyia* (*Wyeomyia*) *simmsi* (Dyar & Knab, 1908); synonymy by Lane, 1953.
dysanor Dyar, 1921e (p. 70) – synonym of *Aedes* (*Ochlerotatus*) *abserratus* (Felt & Young, 1904); synonymy established before Knight & Stone, 1977.
dysmathes Dyar & Ludlow, 1921a (p. 47) – synonym of *Culex* (*Melanoconion*) *conspirator* Dyar & Knab, 1906d; synonymy established before Knight & Stone, 1977.
dytes Martini, 1922 (p. 164) – synonym of *Aedes* (*Ochlerotatus*) *excrucians* (Walker, 1856a) [original combination: *Aedes excrucians* variety *dytes*]; synonymy established before Knight & Stone, 1977.
dzeta Séguy, 1924 (p. 98) – species of *Aedes* (*Ochlerotatus*).

Ee

eadithae Barraud, 1924d (p. 1004) – synonym of *Culex* (*Barraudius*) *modestus* Ficalbi, 1890b; synonymy established before Knight & Stone, 1977.
ealaensis Huang, 2004 (p. 36) – species of *Aedes* (*Stegomyia*).
earlei Vargas, 1943b (p. 9) – species of *Anopheles* (*Anopheles*).
eastor Dyar, 1920d (p. 71) – species of *Culex* (*Melanoconion*).
Synonym: *manaosensis* Evans, 1924.
eatoni (Edwards, 1916) (p. 358) – species of *Aedes* (subgenus uncertain) [original combination: *Ochlerotatus eatoni*].
ebogoensis Rickenbach & Ferrara, 1965 (p. 26) – species of *Aedes* (*Aedimorphus*).
echinata Grabham, 1906b (p. 311) – synonym of *Psorophora* (*Janthinosoma*) *ferox* (von Humboldt, 1819) [original combination: *Janthinosoma echinata*]; synonymy established before Knight & Stone, 1977.
echinus (Edwards, 1920a) (p. 133) – species of *Aedes* (*Dahliana*) [original combination: *Ochlerotatus* (*Finlaya*) *echinus*].
ecuadoriensis Berlin, 1969a (p. 115) – species of *Aedes* (*Howardina*).
edgari Stone & Rosen, 1952 (p. 425) – species of *Aedes* (*Ochlerotatus*).
eduardoi Casal & García, 1968e (p. 105) – species of *Culex* (*Culex*).
educator Dyar & Knab, 1906d (p. 217) – species of *Culex* (*Melanoconion*).
Synonyms: *apeteticus* Howard, Dyar & Knab, 1913; *aneles* Dyar & Ludlow, 1922; *keenani* Galindo & Mendez, 1961.
edwardsi (Barraud, 1923a) (p. 784) – species of *Aedes* (*Stegomyia*) [original combination: *Stegomyia edwardsi*].
edwardsi Barraud, 1923d (p. 507) – species of *Culex* (*Culex*).
edwardsi (Barraud, 1924d) (p. 999) – species of *Toxorhynchites* (*Toxorhynchites*) [original combination: *Megarhinus edwardsi*].
edwardsi Barraud, 1926 (p. 337) – species of *Uranotaenia* (*Uranotaenia*).
edwardsi (Barraud, 1929b) (p. 264) – species of *Tripteroides* (*Rachionotomyia*) [original combination: *Heizmannia edwardsi*].

edwardsi (Lane & Cerqueira, 1942) (p. 637) – species of *Wyeomyia* (*Phoniomyia*) [original combination: *Phoniomyia edwardsi*].

edwardsi Yamada, 1924 (p. 238) – synonym of *Anopheles* (*Anopheles*) *koreicus* Yamada & Watanabe, 1918; synonymy by Stone *et al*., 1959.

edwardsianus Lane & Cerqueira, 1942 (p. 528) – synonym of *Johnbelkinia longipes* (Fabricius, 1805) [original combination: *Trichoprosopon* (*Hyloconops*) *edwardsianus*]; synonymy by Zavortink, 1979.

egberti Dyar & Knab, 1907d (p. 214) – synonym of *Culex* (*Melanoconion*) *erraticus* (Dyar & Knab, 1906d); synonymy of King & Bradley, 1937.

egcymon Dyar, 1923b (p. 67) – species of *Culex* (*Melanoconion*).

eidsvoldensis Mackerras, 1927 (p. 295) – species of *Aedes* (*Ochlerotatus*) [original combination: *Aedes* (*Ochlerotatus*) *theobaldi* form *eidsvoldensis*]; specific status by Lee *et al*., 1984.

eiseni Coquillett, 1902b (p. 192) – species of *Anopheles* (*Anopheles*).
Synonym: *niveopalpis* Ludlow, 1919.

ejercitoi Baisas, 1935b (p. 486) – species of *Armigeres* (*Armigeres*).

ejercitoi Mendoza, 1947 (p. 171) – species of *Anopheles* (*Anopheles*).

eknomios Forattini & Sallum, 1992 (p. 265) – species of *Culex* (*Melanoconion*).

elchoensis (Taylor, 1929) (p. 276) – species of *Aedes* (*Macleaya*) [original combination: *Chaetocruiomyia elchoensis*].

eldridgei Adames & Galindo, 1973 (p. 56) – species of *Culex* (*Belkinomyia*).

eldridgei Reinert, 1973e (p. 14) – species of *Aedes* (*Bothaella*).

eleanorae Berlin, 1969a (p. 105) – species of *Aedes* (*Howardina*).

elegans Brug, 1934 (p. 507) – species of *Tripteroides* (*Tripteroides*).

elegans Ficalbi, 1890c (p. 95) – synonym of *Aedes* (*Stegomyia*) *aegypti* (Linnaeus, 1762) [original combination: *Culex elegans*]; synonymy established before Knight & Stone, 1977.

elegans (James, 1903) (in Theobald, 1903a) (p. 51) – species of *Anopheles* (*Cellia*) [original combination: *Myzomyia elegans*].

elegans (Taylor, 1914c) (p. 703) – species of *Mimomyia* (*Etorleptiomyia*) [original combination: *Dixomyia elegans*].

elephas Komp, 1936c (p. 328) – species of *Culex* (*Melanoconion*).

eleuthera Dyar, 1918a (p. 184) – synonym of *Culex* (*Culex*) *bahamensis* Dyar & Knab, 1906d; synonymy established before Knight & Stone, 1977.

elevator Dyar & Knab, 1906d (p. 217) – species of *Culex* (*Melanoconion*).
Synonyms: *dornarum* Dyar & Shannon, 1924a; *curryi* Dyar, 1926; *bonneti* Senevet, 1938; *vogelsangi* Anduze, 1949a.

elgonicus Edwards, 1941 (p. 328) – subspecies of *Culex* (*Culex*) *vansomereni* Edwards, 1926a.

ellinorae Edwards, 1941 (p. 212) – species of *Aedes* (*Neomelaniconion*).

ellinorae Ovazza, Hamon & Neri, 1956 (p. 167) – subspecies of *Culex* (*Oculeomyia*) *aurantapex* Edwards, 1914b [*nomen novum* for *abyssinicus* van Someren, 1945; original combination: *Culex* (*Culex*) *aurantapex* variety *ellinorae*]. **Note**: although originally designated a variety, *ellinorae* is a replacement name for a recognized subspecies and, hence, has been afforded subspecific status since Stone *et al*., 1959.
Synonym: *abyssinicus* van Someren, 1945.

elnora Paterson & Shannon, 1927 (p. 10) – subspecies of *Uranotaenia* (*Uranotaenia*) *pulcherrima* Lynch Arribálzaga, 1891b.

elocutilis Dyar & Knab, 1909d (p. 255) – synonym of *Culex* (*Culex*) *mollis* Dyar & Knab, 1906b; synonymy established before Knight & Stone, 1977.

eloisa Howard, Dyar & Knab, 1913, 1915 (figs 26, 287 and p. 121, respectively) – synonym of *Wyeomyia* (*Decamyia*) *pseudopecten* Dyar & Knab, 1906a; synonymy by Lane, 1953.

elongatus Rozeboom & Komp, 1950b (p. 147) – species of *Culex* (*Microculex*).

elsiae (Barraud, 1923e) (p. 406) – species of *Aedes* (*Collessius*) [original combination: *Finlaya elsiae*].
Synonym: *simulatus* Barraud, 1931b.

elutior Martini, 1930 (p. 145) – synonym of *Anopheles* (*Anopheles*) *martinius* Shingarev, 1926 [original combination: *Anopheles* (*Anopheles*) *elutus* variety *elutior*]; synonymy by White, 1978.

elutus Edwards, 1921d (p. 273) – synonym of *Anopheles* (*Anopheles*) *sacharovi* Favre, 1903; synonymy established before Knight & Stone, 1977.

embiensis (Huang, 1968a) (p. 24) – species of *Verrallina* (*Verrallina*) [original combination: *Aedes* (*Verrallina*) *embiensis*].

embuensis Edwards, 1930a (p. 295) – species of *Aedes* (*Hopkinsius*).

emilianus Komp, 1941 (p. 794) – synonym of *Anopheles* (*Nyssorhynchus*) *aquasalis* Curry, 1932; synonymy by Causey *et al*., 1945.

eminentia (Leicester, 1908) (p. 131) – species of *Culex* (*Lophoceraomyia*) [original combination: *Lophoceratomyia eminentia*].

engarensis Kanda & Oguma, 1978 (p. 46) – species of *Anopheles* (*Anopheles*).

enigmatica Peyton, 1977 (p. 133) – species of *Uranotaenia* (*Pseudoficalbia*).

ensiformis Bonne-Wepster & Bonne, 1920 (p. 176) – species of *Culex* (*Melanoconion*).

entebbiensis Evans, 1938 (p. 388) – synonym of *Anopheles* (*Cellia*) *cydippis* de Meillon, 1931a [original combination: *Anopheles* (*Myzomyia*) *squamosus* variety *entebbiensis*]; synonymy by de Meillon, 1947b.

eouzani Brunhes, Le Goff & Boussès, 2003 (p. 182) – species of *Anopheles* (*Cellia*).

eouzani Geoffroy, 1971a (p. 279) – species of *Culex* (*Culiciomyia*).

eouzani Rickenbach & Lombrici, 1974 (p. 218) – species of *Eretmapodites*.
epactius Dyar & Knab, 1908 (p. 53) – species of *Aedes* (*Georgecraigius*).
Synonyms: *perichares* Dyar, 1921b; *nielseni* O'Meara & Craig, 1970.
epanastasis Dyar, 1922g (p. 191) – species of *Culex* (*Melanoconion*).
Synonym: *pseudotaeniopus* Galindo & Blanton, 1954.
epidesmus (Theobald, 1910a) (p. 22) – species of *Culex* (*Oculeomyia*) [original combination: *Taeniorhynchus epidesmus*].
Synonyms: *ochracea* Theobald, 1905g; *luteoabdominalis* Theobald, 1910a.
epinolus Dyar & Knab, 1914 (p. 61) – synonym of *Aedes* (*Ochlerotatus*) *taeniorhynchus* (Wiedemann, 1821); synonymy established before Knight & Stone, 1977.
epiroticus Linton & Harbach, 2005 (in Linton *et al.*, 2005) (p. 332) – species of *Anopheles* (*Cellia*).
epirus Aiken, 1909 (p. 8) – *nomen dubium*, *Culex*; status by Sirivanakarn, 1983.
epitedeus (Knab, 1907b) (p. 120) – species of *Deinocerites* [original combination: *Dinomimetes epitedeus*].
epsilonn Séguy, 1924 (p. 97) – synonym of *Aedes* (*Ochlerotatus*) *caspius* (Pallas, 1771); synonymy by Monchadskii, 1951.
equinoxialis Floch & Abonnenc, 1945c (p. 3) – species of *Culex* (*Melanoconion*).
equinus Theobald, 1903g (p. 282) – species of *Haemagogus* (*Haemagogus*).
Synonyms: *affirmatus* Dyar & Knab, 1906b; *philosophicus* Dyar & Knab, 1906d.
equivocator Dyar & Knab, 1907c (p. 203) – synonym of *Culex* (*Culex*) *mollis* Dyar & Knab, 1906b; synonymy established before Knight & Stone, 1977.
erectus Iglisch, 1977 (p. 269) – synonym of *Culex* (*Culex*) *pipiens* Linnaeus, 1758 [original combination: *Culex pipiens erectus*]; synonymy by Harbach, 1988.
eremita Howard, Dyar & Knab, 1913, 1915 (figs 95, 331 and p. 261, respectively) – synonym of *Culex* (*Culex*) *habilitator* Dyar & Knab, 1906d; synonymy established before Knight & Stone, 1977.
erepens Gillies, 1958 (p. 9) – species of *Anopheles* (*Cellia*) [original combination: *Anopheles wellcomei erepens*]; specific status by Gillies & Coetzee, 1987.
erethyzonfer Galindo & Blanton, 1954 (p. 244) – species of *Culex* (*Micraedes*).
erikae Szadziewski & Szadziewska, 1985 (p. 515) – fossil species of *Culex* Linnaeus, 1758.
Synonym: *perkunas* Podėnas, 1999.
eritreae Lewis, 1942 (p. 165) – species of *Aedes* (*Aedimorphus*).
erlindae Baisas & Ubaldo-Pagayon, 1953 (p. 130) – species of *Tripteroides* (*Tripteroides*).
ernanii Duret, 1968f (p. 74) – species of *Culex* (*Melanoconion*).
ernsti Anduze, 1949a (p. 63) – species of *Culex* (*Melanoconion*).
errabunda Swellengrebel, 1925 (species of *Cellia* Theobald, 1902b) (p. 1913) – rejected name, suppression by International Commission on Zoological Nomenclature, 1969.
erraticus (Dyar & Knab, 1906d) (p. 224) – species of *Culex* (*Melanoconion*) [original combination: *Mochlostyrax erraticus*].
Synonyms: *leprincei* Dyar & Knab, 1907c; *egberti* Dyar & Knab, 1907d; *trachycampa* Dyar & Knab, 1909c; *moorei* Dyar, 1918c; *peribleptus* Dyar & Knab, 1918c; *pose* Dyar & Knab, 1918c; *degustator* Dyar, 1921c; *homoeopas* Dyar & Ludlow, 1921a; *borinqueni* Root, 1922; *tovari* Evans, 1924.
error Theobald, 1903a (p. 353) – synonym of *Anopheles* (*Cellia*) *subpictus* Grassi, 1899 [original combination: *Aldrichia error*]; synonymy established before Knight & Stone, 1977.
eruthrosops Theobald, 1910c (p. 299) – synonym of *Aedes* (*Aedimorphus*) *vexans* (Meigen, 1830b) [original combination: *Culicada eruthrosops*]; synonymy by Knight & Hull, 1953 (as synonym of *nocturnus* Theobald, 1903a).
erythraeus Corradetti, 1939c (p. 287) – species of *Anopheles* (*Cellia*).
erythrothorax Dyar, 1907 (p. 124) – species of *Culex* (*Culex*).
Synonyms: *federalis* Dyar, 1923d; *badgeri* Dyar, 1924e.
erythrurus (Edwards, 1941) (pp. 30, 361) – species of *Toxorhynchites* (*Afrorhynchus*) [original combination: *Megarhinus erythrurus*].
eschirasi Galliard, 1931b (p. 227) – subspecies of *Culex* (*Culex*) *pruina* Theobald, 1901d [original combination: *Culex pruina* variety *eschirasi*]; subspecific status by Harbach & Howard, 2007.
escodae Baisas & Ubaldo-Pagayon, 1953 (p. 167) – subspecies of *Tripteroides* (*Tripteroides*) *powelli* (Ludlow, 1909c).
escomeli Brèthes, 1920 (p. 41) – synonym of *Culex* (*Culex*) *apicinus* Philippi, 1865; synonymy established before Knight & Stone, 1977.
esmeraldasi (Levi-Castillo, 1955b) (p. 390) – species of *Wyeomyia* (*Hystatomyia*) [original combination: *Phoniomyia esmeraldasi*].
esoensis Yamada, 1921 (p. 77) – species of *Aedes* (*Aedes*).
Synonyms: *asanumai* Sasa, Kano & Takahasi, 1950; *pseudoesoensis* Yamaguti & Tamaboko, 1954.
espartana Dyar & Knab, 1906a (p. 140) – synonym of *Wyeomyia* (*Wyeomyia*) *arthrostigma* (Lutz, 1905); synonymy by Lane & Cerqueira, 1942.
espini (Martini, 1914) (p. 65) – species of *Isostomyia* [original combination: *Lesticocampa espini*].
Synonym: *shropshirei* Ludlow, 1919.
ethiopicus Edwards, 1912c (p. 30) – synonym of *Culex* (*Oculeomyia*) *bitaeniorhynchus* Giles, 1901a [original combination: *Culex agar* variety *ethiopicus*]; synonymy by Harbach, 1988.
ethiopicus Gillies & Coetzee, 1987 (p. 76) – species of *Anopheles* (*Cellia*).
ethiopiensis Huang, 2004 (p. 39) – species of *Aedes* (*Stegomyia*).
eucephalaeus Dyar, 1918e (p. 127) – species of *Aedes* (*Ochlerotatus*).

euclastus Theobald, 1903c (p. 8) – synonym of *Culex* (*Culex*) *invidiosus* Theobald, 1901d; synonymy established before Knight & Stone, 1977.

eucleptes Dyar, 1921i (p. 147) – synonym of *Aedes* (*Hulecoeteomyia*) *japonicus* (Theobald, 1901a); synonymy established before Knight & Stone, 1977.

euedes Howard, Dyar & Knab, 1913, 1917 (fig. 191, p. 714, respectively) – species of *Aedes* (*Ochlerotatus*).
Synonyms: *beklemishevi* Denisova, 1955; *grandilarva* Sazanova, 1955; *barri* Rueger, 1958.

euethes Dyar & Knab, 1909d (p. 263) – synonym of *Wyeomyia* (*Miamyia*) *hosautos* Dyar & Knab, 1907c; synonymy established before Knight & Stone, 1977.

euiris Dyar, 1922g (p. 193) – species of *Aedes* (*Ochlerotatus*).

eukrines Bram & Rattanarithikul, 1967 (p. 11) – species of *Culex* (*Lophoceraomyia*).

eumimetes Dyar & Knab, 1908 (p. 61) – synonym of *Culex* (*Culex*) *stigmatosoma* Dyar, 1907; synonymy by Strickman, 1988.

euplocamus Dyar & Knab, 1906d (p. 199) – species of *Aedes* (*Ochlerotatus*).

eurochrus Howard, Dyar & Knab, 1917 (p. 716) – synonym of *Aedes* (*Aedimorphus*) *vexans* (Meigen, 1830b); synonymy established before Knight & Stone, 1977.

europaeus da Cunha Ramos, Ribeiro & Harrison, 2003 (p. 7) – species of *Culex* (*Neoculex*).

evandroi da Costa Lima, 1937 (p. 5) – species of *Anopheles* (*Anopheles*).

evansae Antunes, 1942 (in Lane & Cerqueira, 1942) (p. 529) – species of *Trichoprosopon*.

evansae (Brèthes, 1926b) (p. 106) – species of *Anopheles* (*Nyssorhynchus*) [original combination: *Cellia evansi*]; emendation by Peyton *et al.*, 1983.
Synonyms: *metcalfi* Galvão & Lane, 1937b; *noroestensis* Galvão & Lane, 1937b; *ayrozai* Unti, 1940a; *clarki* Komp, 1943.

evansae (Edwards, 1936) (p. 53) – species of *Toxorhynchites* (*Toxorhynchites*) [original combination: *Megarhinus evansae*].

evansae Root, 1927c (p. 593) – species of *Culex* (*Melanoconion*).

exagitans Walker, 1856a (p. 430) – synonym of *Aedes* (*Stegomyia*) *aegypti* (Linnaeus, 1762) [original combination: *Culex exagitans*]; synonymy established before Knight & Stone, 1977.

exallos Rocha, Lourenço-de-Oliveira & Motta, 2012 (p. 928) – species of *Wyeomyia* (subgenus uncertain, determined by the authors).

excitans Walker, 1848 (p. 4) – synonym of *Aedes* (*Stegomyia*) *aegypti* (Linnaeus, 1762) [original combination: *Culex excitans*]; synonymy established before Knight & Stone, 1977.

excrucians (Walker, 1856a) (p. 429) – species of *Aedes* (*Ochlerotatus*) [original combination: *Culex excrucians*].
Synonyms: *siphonalis* Grossbeck, 1904a; *abfitchii* Felt, 1904; *sansoni* Dyar & Knab, 1909c; *surcoufi* Theobald, 1912; *dytes* Martini, 1922.

exedrus Root, 1927c (p. 580) – synonym of *Culex* (*Melanoconion*) *dunni* Dyar, 1918e; synonymy established before Knight & Stone, 1977.

exhibitor Shannon & del Ponte, 1928 (p. 92) – synonym of *Limatus durhamii* Theobald, 1901c; synonymy by Edwards, 1932a.

exilis Dyar, 1924f (p. 127) – synonym of *Culex* (*Culex*) *vagans* Wiedemann, 1828; synonymy established before Knight & Stone, 1977.

exnebulis Bonne-Wepster, 1948 (p. 308) – species of *Tripteroides* (*Rachisoura*).

explorator Marks, 1964a (p. 60) – species of *Aedes* (*Ochlerotatus*).

extricator Dyar & Knab, 1906d (p. 211) – synonym of *Culex* (*Culex*) *inflictus* Theobald, 1901c; synonymy established before Knight & Stone, 1977.

Ff

fabricii Lane & Cerqueira, 1942 (p. 687) – species of *Sabethes* (*Peytonulus*).

factor Dyar & Knab, 1906d (p. 212) – synonym of *Culex* (*Culex*) *nigripalpus* Theobald, 1901c; synonymy established before Knight & Stone, 1977.

faini Leleup, 1952 (p. 151) – species of *Anopheles* (*Cellia*).
Synonym: *vanthieli* Laarman, 1958.

fairchildi Galindo & Blanton, 1954 (p. 234) – species of *Culex* (*Melanoconion*).

fajardi (Lutz, 1904) (in Bourroul, 1904) (p. 64) – species of *Chagasia* [original combination: *Pyretophorus fajardi*].
Synonyms: *neivae* Cruz, 1906; *maculata* Peryassú, 1921b; *stigmopteryx* Martini, 1932a.

fajardoi Baisas & Feliciano, 1953 (p. 172) – species of *Zeugnomyia*.

falabreguesi Hamon, 1957 (p. 685) – species of *Aedes* (*Catageiomyia*).

falcipes Banks, 1906b (p. 1004) – species of *Uranotaenia* (*Uranotaenia*).

falco Kumm, Osorno-Mesa & Boshell-Manrique, 1946 (p. 14) – synonym of *Haemagogus* (*Haemagogus*) *janthinomys* Dyar, 1921g [original combination: *Haemagogus spegazzinii falco*]; synonymy by Arnell, 1973.

fallax Bonne-Wepster & Bonne, 1919 (p. 110) – synonym of *Wyeomyia* (*Miamyia*) *oblita* (Lutz, 1904); synonymy established before Knight & Stone, 1977.

fallax Roubaud, 1934 (p. 740) – synonym of *Anopheles* (*Anopheles*) *atroparvus* van Thiel, 1927 [original combination: *Anopheles maculipennis* variety *fallax*]; synonymy by Bates, 1940.

falsificator Dyar & Knab, 1909d (p. 257) – synonym of *Culex* (*Melanoconion*) *atratus* Theobald, 1901c; synonymy established before Knight & Stone, 1977.

farakoensis Hamon, 1955 (p. 674) – subspecies of *Culex* (*Culex*) *grahamii* Theobald, 1910c [original combination: *Culex grahami* [*sic*] variety *farakoensis*]; subspecific status by Harbach & Howard, 2007.

farauti Laveran, 1902b (p. 908) – species of *Anopheles* (*Cellia*).
Synonym: *moluccensis* Swellengrebel & Swellengrebel de Graaf, 1920a.

farquharsoni (Edwards, 1922c) (p. 500) – species of *Malaya* [original combination: *Harpagomyia farquharsoni*].

fasciatus Fabricius, 1805 (p. 36) – synonym of *Aedes* (*Stegomyia*) *aegypti* (Linnaeus, 1762) [original combination: *Culex fasciatus*]; synonymy established before Knight & Stone, 1977.

fasciatus Meigen, 1804 (p. 4) – synonym of *Aedes* (*Ochlerotatus*) *communis* (de Geer, 1776) [original combination: *Culex fasciatus*]; synonymy established before Knight & Stone, 1977.

fasciatus Müller, 1764 (p. 87) – synonym of *Culex* (*Culex*) *pipiens* Linnaeus, 1758; synonymy established before Knight & Stone, 1977.

fasciolata (Lynch Arribálzaga, 1891b) (p. 150) – species of *Coquillettidia* (*Rhynchotaenia*) [original combination: *Taeniorhynchus fasciolatus*].

fasciolatus (Lutz, 1904) (in Bourroul, 1904) (p. 70) – *nomen dubium*, *Culex* [original combination *Melanoconion fasciolatum*]; status by Sirivanakarn, 1983.

fascipalpis (Edwards, 1912c) (p. 19) – species of *Aedes* (*Diceromyia*) [original combination: *Ochlerotatus fascipalpis*].

fascipes (Coquillett, 1906a) (p. 182) – species of *Orthopodomyia* [original combination: *Mansonia fascipes*].
Synonyms: *longipalpis* Newstead & Thomas, 1910; *townsendi* da Costa Lima, 1935b; *bacigalupoi* Martinez & Prosen, 1958.

fasyi Baisas, 1938 (p. 215) – species of *Culex* (*Culex*).

fatigans Wiedemann, 1828 (p. 10) – synonym of *Culex* (*Culex*) *quinquefasciatus* Say, 1823; synonymy by Stone, 1957a (conclusive synonymy by Sirivanakarn & White, 1978).

fatuator Dyar & Shannon, 1924a (p. 47) – synonym of *Culex* (*Melanoconion*) *conspirator* Dyar & Knab, 1906d; synonymy by Komp, 1935.

fauna Dyar & Knab, 1919 (p. 2) – synonym of *Wyeomyia melanocephala* Dyar & Knab, 1906a; synonymy established before Knight & Stone, 1977.

faurani Duret, 1968f (p. 77) – species of *Culex* (*Melanoconion*).

fausti Vargas, 1943a (p. 66) – species of *Anopheles* (*Anopheles*).

favor Dyar & Núñez Tovar, 1928 (in Dyar, 1928) (p. 73) – synonym of *Wyeomyia* (*Dendromyia*) *jocosa* (Dyar & Knab, 1908) [original combination: *Dendromyia* (*Melanolepis*) *favor*]; synonymy by Lane, 1953.

febrifera Banks, 1914 (p. 405) – synonym of *Anopheles* (*Cellia*) *mangyanus* (Banks, 1906b) [original combination: *Myzomyia febrifera*]; synonymy by King, 1932b.

federalis Dyar, 1923d (p. 186) – synonym of *Culex* (*Culex*) *erythrothorax* Dyar, 1907; synonymy established before Knight & Stone, 1977.

feegradei Barraud, 1934 (p. 164) – species of *Aedes* (*Phagomyia*).

felicia (Dyar & Núñez Tovar, 1927b) (p. 3) – species of *Wyeomyia* (*Decamyia*) [original combination: *Dendromyia felicia*].

felicitatis Bonne-Wepster, 1948 (p. 307) – species of *Tripteroides* (*Rachisoura*).

femineus Edwards, 1926a (p. 107) – species of *Culex* (*Eumelanomyia*).

femorata (Edwards, 1936) (p. 54) – species of *Mimomyia* (*Mimomyia*) [original combination: *Ficalbia* (*Mimomyia*) *femorata*].

fengi Edwards, 1935a (p. 131) – species of *Aedes* (*Luius*).
Synonym: *loi* Lien, 1968b.

fergusoni (Taylor, 1914b) (p. 459) – species of *Culex* (*Neoculex*) [original combination: *Culicada fergusoni*].

ferinus Knight, 1947a (p. 316) – species of *Aedes* (*Mucidus*).

fernandezi Casal, García & Cavalieri, 1966 (p. 185) – species of *Culex* (*Culex*).

fernandezyepezi (Cova Garcia, Sutil Oramas & Pulido F., 1974b) (p. 23) – species of *Wyeomyia* (*Caenomyiella*) [original combination: *Sabethes fernández-yepezi* [*sic*]].

ferox Haddow, 1946 (p. 69) – synonym of *Eretmapodites dracaenae* Edwards, 1916; synonymy by Ovazza *et al.*, 1956.

ferox (von Humboldt, 1819) (p. 340) – species of *Psorophora* (*Janthinosoma*) [original combination: *Culex ferox*].
Synonyms: *posticatus* Wiedemann, 1821; *musicus* Say, 1829; *echinata* Grabham, 1906b; *sayi* Dyar & Knab, 1906d; *terminalis* Coquillett, 1906d; *vanhalli* Dyar & Knab, 1906a; *coquillettii* Theobald, 1907; *jamaicensis* Theobald, 1907; *sayi* Theobald, 1907; *pazosi* Pazos, 1908; *centrale* Brèthes, 1910.

ferox Wiedemann, 1828 (p. 1) – synonym of *Toxorhynchites* (*Lynchiella*) *theobaldi* (Dyar & Knab, 1906c) [original combination: *Culex ferox*]; synonymy established before Knight & Stone, 1977.

ferrarai Rickenbach & Eouzan, 1970 (p. 131) – species of *Eretmapodites*.

ferraroi Duret, 1971a (p. 25) – synonym of *Psorophora* (*Janthinosoma*) *varipes* (Coquillett, 1904a); synonymy by Belkin & Heinemann, 1975.

ferreri Duret, 1968f (p. 79) – species of *Culex* (*Melanoconion*).

ferruginata Martini, 1924 (p. 438) – synonym of *Culiseta* (*Culiseta*) *subochrea* (Edwards, 1921) [original combination: *Theobaldia annulata* variety *ferruginata*]; synonymy by Peus, 1930.

ferruginosus Wiedemann, 1828 (p. 12) – synonym of *Anopheles* (*Anopheles*) *atropos* Dyar & Knab, 1906b; synonymy by Belkin, 1968b.

ficalbii Noè, 1899 (p. 238) – synonym of *Culiseta* (*Culicella*) *fumipennis* (Stephens, 1825) [original combination: *Culex ficalbii*]; synonymy established before Knight & Stone, 1977.

fidelis Dyar, 1920i (p. 180) – synonym of *Culex* (*Eumelanomyia*) *brevipalpis* (Giles, 1902); synonymy established before Knight & Stone, 1977.

fiebrigi Edwards, 1922d (p. 77) – species of *Psorophora* (*Janthinosoma*).

fijiensis (Belkin, 1962) (p. 308) – species of *Coquillettidia* (*Coquillettidia*) [original combination: *Mansonia* (*Coquillettidia*) *fijiensis*].

fijiensis Marks, 1947a (p. 26) – species of *Aedes* (*Finlaya*).

filicis Ingram & de Meillon, 1927 (p. 58) – species of *Aedes* (*Catageiomyia*).

filipes (Walker, 1861) (p. 229) – species of *Tripteroides* (*Rachisoura*) [original combination: *Culex filipes*].
Synonym: *papua* Brug, 1934.

filipinae Manalang, 1930 (p. 258) – species of *Anopheles* (*Cellia*) [original combination: *Anopheles aconitus* variety *filipinae*]; specific status by Christophers & Puri, 1931a.

fimbriata King & Hoogstraal, 1947a (p. 592) – species of *Uranotaenia* (*Uranotaenia*).

fimbriatus Edwards, 1930a (p. 305) – species of *Armigeres* (*Armigeres*).

fimbriforceps Edwards, 1935a (p. 135) – species of *Culex* (*Eumelanomyia*).

fimbripes Edwards, 1924b (p. 390) – species of *Aedes* (*Geoskusea*).

finlayi Lane & Cerqueira, 1942 (p. 594) – species of *Wyeomyia* (*Spilonympha*).

finlayi Perez Vigueras, 1956 (p. 382) – synonym of *Culex* (*Culex*) *chidesteri* Dyar, 1921h; synonymy by Montchadsky & Garcia Avila, 1966.

fisheri Barraud, 1928a (p. 662) – synonym of *Aedes* (*Aedimorphus*) *syntheticus* Barraud, 1928b; synonymy established before Knight & Stone, 1977.

fisheri Dyar, 1917a (p. 19) – synonym of *Aedes* (*Ochlerotatus*) *ventrovittis* Dyar, 1916b; synonymy established before Knight & Stone, 1977.

fishi Zavortink, 1986 (p. 55) – species of *Wyeomyia* (*Zinzala*).

fitchii (Felt & Young, 1904) (p. 312) – species of *Aedes* (*Ochlerotatus*) [original combination: *Culex fitchii*].
Synonyms: *palustris* Dyar, 1916b; *pricei* Dyar, 1917a; *mimesis* Dyar, 1917c.

flabellata (Lane & Cerqueira, 1942) (p. 641) – species of *Wyeomyia* (*Phoniomyia*) [original combination: *Phoniomyia flabellata*].

flabellifer Komp, 1936c (p. 323) – species of *Culex* (*Melanoconion*).

flabelliger Bonne-Wepster, 1948 (p. 308) – species of *Tripteroides* (*Rachisoura*).

flava Ludlow, 1908a (p. 32) – synonym of *Anopheles* (*Cellia*) *kochi* Dönitz, 1901 [original combination: *Cellia flava*]; synonymy established before Knight & Stone, 1977. **Note:** this nominal taxon is the senior secondary homonym of *flava* Swellengrebel, 1917.

flava Swellengrebel, 1917 (p. 807) – synonym of *Anopheles* (*Cellia*) *vagus* Dönitz, 1902 [original combination: *Myzomyia flava*]; synonymy established before Knight & Stone, 1977. **Note:** this nominal taxon is the junior secondary homonym of *flava* Ludlow, 1908a.

flavens (King & Hoogstraal, 1946e) (p. 188) – species of *Mimomyia* (*Mimomyia*) [original combination: *Ficalbia* (*Mimomyia*) *flavens*].

flaveola (Coquillett, 1906a) (p. 182) – species of *Mansonia* (*Mansonia*) [original combination: *Taeniorhynchus flaveolus*].

flavescens (Müller, 1764) (p. 87) – species of *Aedes* (*Ochlerotatus*) [original combination: *Culex flavescens*]. **Note:** this nominal taxon is the senior primary homonym of *flavescens* Theobald, 1901a.
Synonyms: *lutescens* Fabricius, 1775; *variegatus* Schrank, 1781; *bipunctatus* Robineau-Desvoidy, 1827; *flavus* Motschulsky, 1859; *flavescens* Theobald, 1901a; *fletcheri* Coquillett, 1902c; *arcanus* Blanchard, 1905.

flavescens Swellengrebel, 1921 (p. 47) – synonym of *Anopheles* (*Cellia*) *vagus* Dönitz, 1902 [original combination: *Myzomyia ludlowi* [*sic*] variety *flavescens*]; synonymy by Reid, 1968.

flavescens Theobald, 1901a (p. 410) – synonym of *Aedes* (*Ochlerotatus*) *flavescens* (Müller, 1764) [original combination: *Culex flavescens*]; synonymy established before Knight & Stone, 1977. **Note:** this nominal taxon is the junior primary homonym of *flavescens* (Müller, 1764).

flaviceps Edwards, 1921c (p. 69) – synonym of *Anopheles* (*Cellia*) *turkhudi* Liston, 1901; synonymy in Evans, 1938.

flavicollis Edwards, 1928a (p. 269) – species of *Aedes* (*Diceromyia*).

flavicornis Barraud, 1924h (p. 45) – species of *Culex* (*Lophoceraomyia*).

flavicosta Barraud, 1927a (p. 531) – species of *Orthopodomyia*.

flavicosta Edwards, 1911a (p. 142) – species of *Anopheles* (*Cellia*).

flavicosta Walker, 1856a (p. 431) – synonym of *Aedes* (*Ochlerotatus*) *fulvus* (Wiedemann, 1828) [original combination: *Culex flavicosta*]; synonymy established before Knight & Stone, 1977.

flavidorsalis Luh & Lee, 1975 (p. 428) – synonym of *Aedes* (*Ochlerotatus*) *albineus* Séguy, 1923; synonymy by Danilov, 1979.

flavifacies Edwards, 1922d (p. 82) – species of *Wyeomyia* (*Antunesmyia*).

flavifrons (Skuse, 1889) (p. 1735) – species of *Aedes* (*Ochlerotatus*) [original combination: *Culex flavifrons*].
Synonyms: *vandema* Strickland, 1911b; *variegatans* Strickland, 1911b.

flavimargo Edwards, 1941 (p. 209) – species of *Aedes* (*Neomelaniconion*).

flavinervis Edwards, 1915a (p. 161) – *nomen nudum*, *Aedes* [original combination: *Banksinella luteolateralis* variety *flavinervis*].

flavipennis (Giles, 1904a) (p. 366) – species of *Aedes* (*Finlaya*) [original combination: *Finlaya flavipennis*].
Synonym: *aranetana* Banks, 1906b.

flavipes Macquart, 1838b (p. 35) – synonym of *Aedes* (*Ochlerotatus*) *albifasciatus* (Macquart, 1838b) [original combination: *Culex flavipes*]; synonymy established before Knight & Stone, 1977.

flavirostris (Ludlow, 1914a) (p. 30) – species of *Anopheles* (*Cellia*) [original combination: *Myzomyia flavirostris*].
Synonym: *merak* (*cohesia*) Mangkoewinoto, 1919.

flavirostris (Meigen, 1830b) (p. 242) – *nomen dubium*, *Culiseta* [original combination: *Culex flavirostris*]; status by Edwards, 1932a.

flavisetosus de Oliveira Castro, 1935 (p. 143) – species of *Limatus*.

flavithorax Barraud, 1927a (p. 529) – species of *Orthopodomyia*.

flavocincta (Edwards, 1936) (p. 54) – species of *Coquillettidia* (*Coquillettidia*) [original combination: *Taeniorhynchus* (*Coquillettidia*) *flavocinctus*].

flavopicta (Edwards, 1936) (p. 54) – species of *Mimomyia* (*Mimomyia*) [original combination: *Ficalbia* (*Mimomyia*) *flavopicta*].

flavopictus Yamada, 1921 (p. 52) – species of *Aedes* (*Stegomyia*).

flavus Gistl, 1831 (p. 247) – fossil species of *Culex* Linnaeus, 1758.

flavus (Leicester, 1908) (p. 101) – species of *Armigeres* (*Leicesteria*) [original combination: *Chaetomyia flava*].
Synonyms: *apicalis* Theobald, 1908b; *apicalis* Theobald, 1910a.

flavus Motschulsky, 1859 (p. 503) – synonym of *Aedes* (*Ochlerotatus*) *flavescens* (Müller, 1764) [original combination: *Culex flavus*]; synonymy by Natvig, 1948. **Note**: this nominal taxon is the senior secondary homonym of *flavus* Yamada, 1927.

flavus Ventrillon, 1904 (p. 550) – synonym of *Coquillettidia* (*Coquillettidia*) *grandidieri* (Blanchard, 1905) [original combination: *Culex flavus*]; synonymy established before Knight & Stone, 1977.

flavus Yamada, 1927 (p. 575) – *nomen nudum*, *Aedes* [original combination: *Aedes esoensis* variety *flavus*]. **Note**: this nominal taxon is the junior secondary homonym of *flavus* Motschulsky, 1859.

flerowi Portschinsky, 1910 (p. 605) – synonym of *Anopheles* (*Anopheles*) *hyrcanus* (Pallas, 1771); synonymy established before Knight & Stone, 1977.

fletcheri Coquillett, 1902c (p. 84) – synonym of *Aedes* (*Ochlerotatus*) *flavescens* (Müller, 1764) [original combination: *Culex fletcheri*]; synonymy established before Knight & Stone, 1977.

flindersi Taylor, 1914c (p. 686) – synonym of *Aedes* (*Ochlerotatus*) *stricklandi* (Edwards, 1912b) [original combination: *Grabhamia flindersi*]; synonymy established before Knight & Stone, 1977.

flochi Duret, 1969d (p. 119) – species of *Culex* (*Melanoconion*) (subgenus uncertain, determined by Pecor *et al.*, 1992).

florense Petrocchi, 1923 (p. 91) – *nomen nudum*, *Culex*.

florestan Dyar, 1925a (p. 20) – species of *Wyeomyia* (*Wyeomyia*).

floridanus Dyar & Knab, 1906b (p. 171) – synonym of *Culex* (*Melanoconion*) *pilosus* (Dyar & Knab, 1906d) [original combination: *Mochlostyrax floridanus*]; synonymy established before Knight & Stone, 1977.

floridense Dyar & Knab, 1906a (p. 135) – synonym of *Psorophora* (*Grabhamia*) *columbiae* (Dyar & Knab, 1906a); synonymy based on type locality being within the distribution of *columbiae* defined by Lanzaro *et al.* (2015).

floridensis Belkin, 1950 (p. 269) – species of *Tripteroides* (*Polylepidomyia*).

flui (Bonne-Wepster & Bonne, 1920) (p. 169) – species of *Wyeomyia* (subgenus uncertain, determined by Lourenço-de-Oliveira *et al.*, 1999) [original combination: *Prosopolepis flui*].
Synonym: *kerri* del Ponte & Cerqueira, 1938.

fluminensis Neiva, 1908 (in Peryassú, 1908) (p. 128) – synonym of *Toxorhynchites* (*Lynchiella*) *theobaldi* (Dyar & Knab, 1906c) [original combination: *Megarhinus fluminensis*]; synonymy by da Costa Lima, 1931b.

fluminensis Root, 1927d (p. 602) – species of *Anopheles* (*Anopheles*).
Synonym: *holmbergi* del Ponte & Heredia, 1945.

fluviatilis James, 1902 (p. 31) – species of *Anopheles* (*Cellia*).
Synonyms: *listonii* Liston, 1901; *leptomeres* Theobald, 1903a; *arabica* Christophers & Chand, 1915.

fluviatilis Leicester, 1908 (p. 111) – synonym of *Aedes* (*Hulecoeteomyia*) *saxicola* Edwards, 1922a [original combination: *Hulecoeteomyia fluviatilis*]; synonymy established before Knight & Stone, 1977. **Note**: this nominal taxon is the junior secondary homonym of *fluviatilis* (Lutz, 1904).

fluviatilis (Lutz, 1904) (in Bourroul, 1904) (p. 72) – species of *Aedes* (*Georgecraigius*) [original combination: *Culex fluviatilis*]. **Note**: this nominal taxon is the senior secondary homonym of *fluviatilis* Leicester, 1908.
Synonyms: *mediomaculata* Theobald, 1907; *tripunctata* Theobald, 1907; *lithoecetor* Dyar & Knab, 1907c; *draconarius* Dyar, 1922g.

fluviatilis (Theobald, 1903a) (p. 331) – species of *Shannoniana* [original combination: *Goeldia fluviatilis*].
Synonym: *longipalpis* Lutz, 1905.

foliaceus Lane, 1945b (p. 206) – species of *Culex* (*Culex*).

foliafer Komp & Rozeboom, 1951 (p. 121) – species of *Culex* (*Melanoconion*).

foliatus Brug, 1931c (p. 32) – species of *Armigeres* (*Armigeres*).

foliatus Brug, 1932 (p. 82) – species of *Culex* (*Eumelanomyia*) [original combination: *Culex* (*Culex*) *castrensis* variety *foliatus*]; specific status by Sirivanakarn, 1971a.
Synonyms: *chiyutoi* Baisas, 1935d; *shrivastavii* Wattal, Kalra & Krishnan, 1966; *chungkiangensis* Chang & Chang, 1974.

folicola Belkin, 1955 (p. 230) – species of *Tripteroides* (*Polylepidomyia*).

foliformis (King & Hoogstraal, 1947b) (p. 121) – species of *Verrallina* (*Verrallina*) [original combination: *Aedes* (*Aedes*) *foliformis*].

folquei de Mello, 1918 (p. 363) – synonym of *Anopheles* (*Cellia*) *stephensi* Liston, 1901; synonymy established before Knight & Stone, 1977.

fonsecai (Pinto, 1932) (p. 289) – species of *Mansonia* (*Mansonia*) [original combination: *Taeniorhynchus* (*Taeniorhynchus*) *fonsecai*].

fontenillei Brunhes & Hervy, 1995 (p. 46) – species of *Orthopodomyia*.

fontenillei (Le Goff, Boussès & Brunhes, 2007) (p. 196) – species of *Aedes* (*Neomelaniconion*) [original combination: *Neomelaniconion fontenillei*].

fontenillei Ribeiro, 2004 (p. 252) – species of *Toxorhynchites* (*Afrorhynchus*).
fontinalis Gillies & de Meillon, 1968 (p. 115) – species of *Anopheles* (*Cellia*).
forattinii Cerqueira, 1961 (p. 37) – species of *Sabethes* (*Sabethes*).
forattinii Clastrier, 1974 (p. 155) – species of *Wyeomyia* (*Cruzmyia*).
forattinii Corrêa & Ramalho, 1959a (p. 55) – synonym of *Culex* (*Culex*) *declarator* Dyar & Knab, 1906d; synonymy established before Knight & Stone, 1977.
forattinii Wilkerson & Sallum, 1999 (p. 345) – species of *Anopheles* (*Anopheles*).
forceps Cerqueira, 1939 (p. 83) – species of *Psorophora* (*Janthinosoma*).
forcipenis Lourenço-de-Oliveira & da Silva, 1985 (p. 321) – species of *Wyeomyia* (*Spilonympha*).
forcipulatus Edwards, 1936 (p. 53) – species of *Eretmapodites*.
formosae Hatori, 1901 (p. 275) – synonym of *Anopheles* (*Cellia*) *tessellatus* Theobald, 1901a; synonymy established before Knight & Stone, 1977.
formosaensis Koidzumi, 1917 (p. 354) – synonym of *Anopheles* (*Cellia*) *ludlowae* (Theobald, 1903a) [original combination: *Anopheles ludlowi* [*sic*] variety *formosaensis*]; synonymy established before Knight & Stone, 1977.
formosaensis Ogasawara, 1939a (p. 242) – synonym of *Toxorhynchites* (*Toxorhynchites*) *christophi auriflaus* (Edwards, 1921a) [original combination: *Megarhinus auriflaus* variety *formosaensis*]; synonymy by Lien, 1962.
formosaensis I Tsuzuki, 1902a (p. 288) – synonym of *Anopheles* (*Cellia*) *minimus* Theobald, 1901a; synonymy confirmed by Harbach *et al*., 2006.
formosaensis II Tsuzuki, 1902c (p. 28) – synonym of *Anopheles* (*Cellia*) *indefinitus* (Ludlow, 1904c); synonymy by Reid, 1968.
formosensis Yamada, 1921 (p. 67) – species of *Aedes* (*Hulecoeteomyia*).
formosus Ludlow, 1909a (species) (p. 22) – subspecies of *Anopheles* (*Anopheles*) *gigas* Giles, 1901b; subspecific status by Harbach & Howard, 2007.
formosus (Walker, 1848) (species) (p. 4) – subspecies of *Aedes* (*Stegomyia*) *aegypti* (Linnaeus, 1762) [original combination: *Culex formosus*]; subspecific status by Mattingly, 1957b.
fouchowensis Theobald, 1901c (p. 137) – synonym of *Culex* (*Culex*) *quinquefasciatus* Say, 1823; synonymy established before Knight & Stone, 1977.
fowleri Christophers, 1911 (p. 64) – synonym of *Anopheles* (*Cellia*) *pallidus* Theobald, 1901a [original combination: *Neocellia fowleri*]; synonymy established before Knight & Stone, 1977.
fowleri (de Charmoy, 1908) (p. 258) – species of *Aedes* (*Aedimorphus*) [original combination: *Culex fowleri*].
Synonym: *nigeriensis* Theobald, 1910c.
fragilis Leicester, 1908 (p. 199) – species of *Verrallina* (*Harbachius*).
Synonym: *malayi* Leicester, 1908.
fragilis Ludlow, 1903 (p. 142) – species of *Culex* (*Culiciomyia*).
Synonyms: *fuscus* Theobald, 1905h; *ceylonica* Theobald, 1907; *inornata* Theobald, 1907; *graminis* Leicester, 1908.
fragilis (Theobald, 1903f) (p. 257) – species of *Anopheles* (*Anopheles*) [original combination: *Stethomyia fragilis*].
Synonyms: *treacherii* Leicester, 1908; *stantoni* Puri, 1949.
franciscanus McCracken, 1904 (p. 12) – species of *Anopheles* (*Anopheles*).
Synonyms: *boydi* Vargas, 1939; *willardi* Vargas, 1941d.
franciscoi Mattingly, 1959 (p. 42) – species of *Aedes* (*Petermattinglyius*).
franciscoi Reid, 1962 (p. 27) – species of *Anopheles* (*Anopheles*).
franclemonti Belkin, 1962 (p. 256) – synonym of *Culex* (*Lophoceraomyia*) *atracus* Colless, 1960; synonymy by Sirivanakarn, 1968.
franclemonti Belkin, 1962 (p. 367) – species of *Aedes* (*Finlaya*).
fraseri (Edwards, 1912c) (p. 11) – species of *Aedes* (*Stegomyia*) [original combination: *Stegomyia fraseri*].
Synonym: *blacklocki* Evans, 1925a.
fraseri Edwards, 1912c (p. 39) – species of *Uranotaenia* (*Uranotaenia*) [original combination: *Uranotaenia bilineata* variety *fraseri*]; specific status by da Cunha Ramos, 1993.
Synonym: *obsoleta* Edwards, 1936.
fraseri (Edwards, 1914b) (p. 74) – species of *Culiseta* (*Theomyia*) [original combination: *Leptosomatomyia fraseri*].
fraseri (Edwards, 1922c) (p. 499) – species of *Malaya* [original combination: *Harpagomyia fraseri*].
fraseri (Theobald, 1911a) (p. 22) – species of *Coquillettidia* (*Coquillettidia*) [original combination: *Chrysoconops fraseri*].
Synonym: *atroapicalis* Gillett, 1946.
frater Robineau-Desvoidy, 1827 (p. 408) – synonym of *Aedes* (*Stegomyia*) *aegypti* (Linnaeus, 1762) [original combination: *Culex frater*]; synonymy established before Knight & Stone, 1977.
fratercula Dyar & Knab, 1906a (p. 139) – synonym of *Wyeomyia* (*Wyeomyia*) *sororcula* Dyar & Knab, 1906a; synonymy by Belkin, 1970a.
fraudatrix (Theobald, 1905f) (p. 94) – species of *Culex* (*Lophoceraomyia*) [original combination: *Lophoceratomyia fraudatrix*].
Synonym: *molestus* Wiedemann, 1828.
freeborni Aitken, 1939 (p. 192) – species of *Anopheles* (*Anopheles*) [original combination: *Anopheles maculipennis freeborni*]; specific status by King & Bradley, 1941.
freerae Banks, 1906b (p. 993) – synonym of *Anopheles* (*Cellia*) *philippinensis* Ludlow, 1902b [original combination: *Pyretophorus freerae*]; synonymy established before Knight & Stone, 1977.
freetownensis Evans, 1925c (p. 461) – species of *Anopheles* (*Cellia*) [original combination: *Anopheles marshallii* variety *freetownensis*]; specific status by Christophers & Puri, 1931b.

freetownensis Theobald, 1901c (p. 69) – synonym of *Culex* (*Culiciomyia*) *cinereus* Theobald, 1901c (syntype ♂) and *Culex* (*Culiciomyia*) *nebulosus* Theobald, 1901d (syntype ♀); synonymies established before Knight & Stone (1977) and Edwards (1941), respectively.

frenchii (Theobald, 1901c) (p. 66) – species of *Culiseta* (*Neotheobaldia*) [original combination: *Culex frenchii*].

freycinetiae Laird, 1957 (p. 343) – species of *Aedes* (*Finlaya*).

freyi Edwards, 1921d (p. 306) – synonym of *Aedes* (*Ochlerotatus*) *cyprius* Ludlow, 1920a; synonymy established before Knight & Stone, 1977.

freyi Meng, 1957 (p. 36) – species of *Anopheles* (*Anopheles*).

frickii Ludlow, 1906b (p. 132) – synonym of *Culex* (*Neoculex*) *territans* Walker, 1856a; synonymy established before Knight & Stone, 1977.

frontosa Theobald, 1903a (p. 319) – species of *Runchomyia* (*Runchomyia*).

fryeri (Theobald, 1912b) (p. 84) – species of *Aedes* (*Coetzeemyia*) [original combination: *Culicelsa fryeri*].
Synonym: *mombasaensis* Mattingly, 1963.

fulgens da Cunha Ramos & Brunhes, 2004 (p. 257) – species of *Uranotaenia* (*Pseudoficalbia*).

fulgens (Edwards, 1917) (p. 213) – species of *Aedes* (*Zavortinkius*) [original combination: *Ochlerotatus* (*Finlaya*) *fulgens*].

fuliginosus Giles, 1900a (p. 161) – synonym of *Anopheles* (*Cellia*) *annularis* van der Wulp, 1884; synonymy established before Knight & Stone, 1977.

fuliginosus Lee, 1946 (p. 248) – species of *Tripteroides* (*Rachisoura*).

fulleri (Ludlow, 1909b) (p. 97) – species of *Culex* (*Lophoceraomyia*) [original combination: *Oculiomyia* [*sic*] *fulleri*].

fulvithorax (Lutz, 1904) (in Bourroul, 1904) (p. 66) – species of *Aedes* (*Howardina*) [original combination: *Haemagogus fulvithorax*].
Synonyms: *palliatus* Coquillett, 1906e; *tachirensis* Anduze, 1947.

fulvus (Wiedemann, 1828) (p. 546) – species of *Aedes* (*Ochlerotatus*) [original combination: *Culex fulvus*].
Synonym: *flavicosta* Walker, 1956a.

fumidus Edwards, 1928a (p. 274) – species of *Aedes* (*Lorrainea*).

fumipennis (Stephens, 1825) (p. 453) – species of *Culiseta* (*Culicella*) [original combination: *Culex fumipennis*].
Synonyms: *ficalbii* Noè, 1899; *theobaldi* de Meijere, 1911; *setivalva* Montschadsky, 1936.

funerea (Leicester, 1908) (p. 252) – species of *Heizmannia* (*Heizmannia*) [original combination: *Wyeomyia funerea*].

funerea (Theobald, 1903a) (p. 292) – species of *Verrallina* (*Verrallina*) [original combination: *Skusea funerea*].
Synonym: *basalis* Taylor, 1912b.

funerearoides Dong, Zhou, Gong *et al.* [*sic*], 2004 (p. 187) – *nomen nudum*, *Heizmannia*.

funestus Giles, 1900b (p. 50) – species of *Anopheles* (*Cellia*).
Synonyms: *kumasii* Chalmers, 1900; *hebes* Dönitz, 1902; *subumbrosa* Theobald, 1903a; *umbrosa* Theobald, 1903a; *anisochloros* Theobald, 1903c; *bisignata* Grünberg, 1905; *imerinensis* Monier & Treillard, 1935.

funestus (Leicester, 1908) (p. 58) – species of *Toxorhynchites* (*Toxorhynchites*) [original combination: *Teromyia funestus*].

funiculus Dyar, 1920f (p. 141) – species of *Psorophora* (*Grabhamia*); elevated from synonymy with *confinnis* Lynch Arribálzaga, 1891b by Lanzaro *et al.*, 2015.
Synonym: *trigonophorus* Lutz, 1928.

fur Dyar & Knab, 1907a (p. 13) – synonym of *Culex* (*Melanoconion*) *spissipes* (Theobald, 1903a); synonymy established before Knight & Stone, 1977.

furcifer (Edwards, 1913b) (p. 48) – species of *Aedes* (*Diceromyia*) [original combination: *Ochlerotatus* (*Finlaya*) *furcifer*].
Synonyms: *nigra* Theobald, 1906b; *africana* Theobald, 1911c.

furfurea (Enderlein, 1923) (p. 26) – species of *Aedeomyia* (*Lepiothauma*) [original combination: *Lepiothauma furfurea*].

furlongi van Someren, 1954a (p. 107) – species of *Culex* (*Culiciomyia*).

furtivus Dyar, 1922h (p. 419) – *nomen nudum*, *Culex*.

furvus Edwards, 1928a (p. 274) – synonym of *Aedes* (*Lorrainea*) *amesii* (Ludlow, 1903); synonymy by Knight & Hull, 1952.

fusca Leicester, 1908 (p. 92) – synonym of *Aedes* (*Lorrainea*) *amesii* (Ludlow, 1903) [original combination: *Stegomyia fusca*]; synonymy established before Knight & Stone, 1977.

fusca (Leicester, 1908) (p. 102) – species of *Mimomyia* (*Ingramia*) [original combination: *Dasymyia fusca*].

fusca Leicester, 1908 (p. 227) – synonym of *Uranotaenia* (*Pseudoficalbia*) *bicolor* Leicester, 1908; synonymy established before Knight & Stone, 1977. **Note:** this nominal taxon is the junior primary homonym of *fusca* Theobald, 1907.

fusca Leicester, 1908 (p. 248) – synonym of *Tripteroides* (*Rachionotomyia*) *aranoides* (Theobald, 1901c) [original combination: *Skeiromyia fusca*]; synonymy established before Knight & Stone, 1977.

fusca Theobald, 1903a (p. 135) – synonym of *Armigeres* (*Armigeres*) *malayi* (Theobald, 1901c) [original combination: *Desvoidea fusca*]; synonymy established before Knight & Stone, 1977.

fusca Theobald, 1907 (p. 564) – species of *Uranotaenia* (*Pseudoficalbia*). **Note:** this nominal taxon is the senior primary homonym of *fusca* Leicester, 1908. (p. 227)
Synonym: *inornata* Theobald, 1908a.

fusca Theobald, 1909b (p. 17) – synonym of *Lutzia* (*Metalutzia*) *tigripes* de Grandpré & de Charmoy, 1901 [original combination: *Culex tigripes* variety *fusca*]; synonymy by Edwards, 1941. **Note:** if *Lutzia* is returned to subgeneric status in *Culex*, this nominal taxon will become the senior secondary homonym of *fusca* Theobald, 1909b (p. 20), based on page priority.

fusca Theobald, 1909b (p. 20) – synonym of *Culex* (*Eumelanomyia*) *horridus* Edwards, 1922d [original combination: *Protomelanoconion fusca*]; synonymy established before Knight & Stone, 1977. **Note**: if *Lutzia* is returned to subgeneric status in *Culex*, this nominal taxon will become the junior secondary homonym of *fusca* Theobald, 1909b (p. 17), based on page priority.

fusca Theobald, 1911a (p. 32) – synonym of *Ficalbia nigra* (Theobald, 1901d) [original combination: *Mimomyia fusca*]; synonymy established before Knight & Stone, 1977.

fuscana (Wiedemann, 1820) (p. 9) – species of *Lutzia* (*Metalutzia*) [original combination: *Culex fuscanus*].
Synonyms: *concolor* Robineau-Desvoidy, 1827; *luridus* Doleschall, 1857; *setulosus* Doleschall, 1857.

fuscatus Lane & Whitman, 1951 (p. 361) – subspecies of *Culex* (*Microculex*) *inimitabilis* Dyar & Knab, 1906d.

fuscicinctus King & Hoogstraal, 1946a (p. 145) – species of *Culex* (*Culiciomyia*).

fuscicolor van Someren, 1947b (p. 43) – species of *Anopheles* (*Anopheles*).

fuscifurcatus Edwards, 1934 (in Barraud, 1934) (p. 452) – synonym of *Culex* (*Culex*) *jacksoni* Edwards, 1934; synonymy by Sirivanakarn, 1976.

fuscinervis (Edwards, 1914b) (p. 73) – species of *Aedes* (*Neomelaniconion*) [original combination: *Banksinella fuscinervis*].

fuscipalpis Belkin, 1962 (p. 370) – species of *Aedes* (*Finlaya*).

fuscipes Edwards, 1922d (p. 76) – species of *Wyeomyia* (*Phoniomyia*).

fuscipleura Lee, 1946 (p. 246) – species of *Tripteroides* (*Rachisoura*).

fuscitarsis Barraud, 1924e (p. 1272) – synonym of *Culex* (*Culex*) *fuscocephala* Theobald, 1907; synonymy established before Knight & Stone, 1977.

fuscitarsis Belkin, 1962 (p. 371) – species of *Aedes* (*Finlaya*).

fuscivenosus Leeson, 1930 (p. 125) – species of *Anopheles* (*Cellia*) [original combination: *Anopheles funestus* variety *fuscivenosus*]; specific status by Gillies & de Meillon, 1968.

fusco Petrocchi, 1923 (p. 91) – *nomen nudum*, *Culex*.

fuscocephala Theobald, 1907 (p. 420) – species of *Culex* (*Culex*).
Synonyms: *minimus* Leicester, 1908; *uniformis* Leicester, 1908; *taytayensis* Banks, 1909; *luteola* Theobald, 1910c; *inelegans* Dyar, 1920i; *fuscitarsis* Barraud, 1924e.

fuscopalpalis Theobald, 1909b (p. 18) – synonym of *Aedes* (*Aedimorphus*) *cumminsii* (Theobald, 1903a) [original combination: *Culicada fuscopalpalis*]; synonymy established before Knight & Stone, 1977.

fuscopennata (Theobald, 1903a) (p. 265) – species of *Coquillettidia* (*Coquillettidia*) [original combination: *Taeniorhynchus fuscopennatus*].
Synonyms: *drymoecius* Speiser, 1909; *bakeri* Theobald, 1911a.

fuscopteron (Theobald, 1911d) (p. 239) – species of *Coquillettidia* (*Coquillettidia*) [original combination: *Chrysoconops fuscopteron*].

fuscosiphonis Bram & Rattanarithikul, 1967 (p. 14) – synonym of *Culex* (*Lophoceraomyia*) *demissus* Colless, 1965; synonymy by Sirivanakarn, 1977a.

fusculus Zetterstedt, 1850 (p. 3459) – synonym of *Aedes* (*Ochlerotatus*) *punctor* (Kirby, 1837) [original combination: *Culex fusculus*]; synonymy by Natvig, 1948.

fuscus Hutton, 1902 (p. 188) – species of *Opifex* (*Opifex*).

fuscus Osten Sacken, 1877 (p. 191) – synonym of *Aedes* (*Aedes*) *cinereus* Meigen, 1818; synonymy by Carpenter & LaCasse, 1955.

fuscus Taylor, 1914c (p. 699) – synonym of *Culex* (*Culex*) *quinquefasciatus* Say, 1823 [original combination: *Culicelsa fuscus*]; synonymy established before Knight & Stone, 1977. **Note**: this nominal taxon is the junior primary homonym of *fuscus* Theobald, 1909b (p. 20); the junior secondary homonym of *fuscus* Theobald, 1905h and *fuscus* Theobald, 1909b (p. 11).

fuscus Theobald, 1905h (p. 242) – synonym of *Culex* (*Culiciomyia*) *fragilis* Ludlow, 1903 [original combination: *Taeniorhynchus fuscus*]; synonymy established before Knight & Stone, 1977. **Note**: this nominal taxon is the senior secondary homonym of *fuscus* Theobald, 1909b (p. 11), *fuscus* Theobald, 1909b (p. 20) and *fuscus* Taylor, 1914c.

fuscus Theobald, 1909b (p. 11) – synonym of *Culex* (*Culiciomyia*) *nebulosus* Theobald, 1901d [original combination: *Pectinopalpus fuscus*]; synonymy established before Knight & Stone, 1977. **Note**: this nominal taxon is the junior secondary homonym of *fuscus* Theobald, 1905h and *fuscus* Taylor, 1914c; the senior secondary homonym of *fuscus* Theobald, 1909b (p. 20), based on page priority.

fuscus Theobald, 1909b (p. 20) – synonym of *Culex* (*Culex*) *thalassius* Theobald, 1903a [original combination: *Culicelsa accraensis* variety *fuscus*]; synonymy established before Knight & Stone, 1977. **Note**: this nominal taxon is the senior primary homonym of *fuscus* Taylor, 1914c; the junior secondary homonym of *fuscus* Theobald, 1905h and *fuscus* Theobald, 1909b (p. 11), the latter based on page priority.

futunae Belkin, 1962 (p. 455) – species of *Aedes* (*Stegomyia*).

Gg

gabaldoni Cova García, Pulido F., Escalante de Ugueto & Mora R., 1987b (p. 94) – species of *Uranotaenia* (*Uranotaenia*).

gabaldoni Vargas, 1941c (p. 389) – species of *Anopheles* (*Anopheles*).

gabonensis Rahola, Makanga & Paupy, 2014 (in Rahola *et al.*, 2014) (p. 630) – species of *Anopheles* (*Cellia*).

gabriel Schick, 1970a (p. 69) – species of *Aedes* (*Protomacleaya*).
gaffigani Reinert, 1990 (p. 24) – species of *Aedes* (*Kenknightia*).
gagnei Evenhuis, 1989 (in Evenhuis & Gon, 1989) (p. 209) – species of *Culex* (*Lophoceraomyia*) [*nomen novum* for *ornatus* Theobald, 1905f].
Synonym: *ornatus* Theobald, 1905f.
gahnicola Marks, 1947a (p. 43) – species of *Aedes* (*Finlaya*).
gairus Root, 1927c (p. 583) – species of *Culex* (*Microculex*).
galindoi Komp & Rozeboom, 1951 (p. 131) – species of *Culex* (*Melanoconion*).
galindoi Schick, 1970a (p. 77) – species of *Aedes* (*Protomacleaya*).
galliardi Edwards, 1941 (p. 266) – species of *Culex* (*Eumelanomyia*).
gallii Martini, 1920a (p. 110) – synonym of *Aedes* (*Ochlerotatus*) *pullatus* (Coquillett, 1904c); synonymy established before Knight & Stone, 1977.
galloisi Yamada, 1921 (p. 47) – species of *Aedes* (*Stegomyia*).
galloisioides Liu & Lu, 1984 (p. 76) – species of *Aedes* (*Stegomyia*).
galoa Dyar & Knab, 1906a (p. 140) – synonym of *Wyeomyia* (*Decamyia*) *pseudopecten* Dyar & Knab, 1906a; synonymy established before Knight & Stone, 1977.
galvaoi Causey, Deane & Deane, 1943 (p. 293) – species of *Anopheles* (*Nyssorhynchus*).
galvaoi (Correa & Ramalho, 1956) (p. 26) – species of *Wyeomyia* (*Phoniomyia*) [original combination: *Phoniomyia galvaoi*].
galvaoi Duret, 1968b (p. 60) – species of *Culex* (*Melanoconion*).
gambiae Giles, 1902 (p. 511) – species of *Anopheles* (*Cellia*).
Synonyms: *costalis* Giles, 1900b; *gracilis* Dönitz, 1902; *minor* Holstein, 1949; *typicus* Holstein, 1949.
gambiensis Theobald, 1903c (p. 5) – synonym of *Culex* (*Oculeomyia*) *annulioris* Theobald, 1901a [original combination: *Culex annulioris* variety *gambiensis*]; synonymy by Edwards, 1941.
gameti Bailly-Choumara, 1966b (p. 148) – species of *Culex* (*Culex*).
gamma Séguy, 1924 (p. 47) – species of *Culex* (*Neoculex*).
ganapathi Colless, 1958 (p. 474) – species of *Aedes* (*Downsiomyia*).
ganapathi Colless, 1965 (p. 294) – species of *Culex* (*Lophoceraomyia*).
gandaensis Huang, 2004 (p. 41) – species of *Aedes* (*Stegomyia*).
gandarai da Cunha Ramos, Capela & Ribeiro, 1995 (p. 98) – species of *Aedes* (*Polyleptiomyia*).
gani Bonne-Wepster, 1940 (p. 158) – species of *Aedes* (*Finlaya*).
garcesi Duret, 1968b (p. 63) – species of *Culex* (*Melanoconion*).
garciai Broche, 2000 (p. 30) – species of *Culex* (*Culex*).
garciai Leví-Castillo, 1953e (p. 102) – synonym of *Psorophora* (*Grabhamia*) *cingulata* (Fabricius, 1805) [original combination: *Aedes* (*Lanesia*) *garciai*]; synonymy by Stone, 1957a.
garciai Leví-Castillo, 1955a (p. 361) – synonym of *Haemagogus* (*Haemagogus*) *boshelli* Osorno-Mesa, 1944; synonymy by Arnell, 1973.
gardnerii (Ludlow, 1905a) (p. 99) – species of *Aedes* (*Stegomyia*) [original combination: *Stegomyia gardnerii*].
garioui Bailly-Choumara & Rickenbach, 1966 (p. 144) – species of *Culex* (*Eumelanomyia*).
garnhamellus Evans & Leeson, 1937 (p. 384) – synonym of *Anopheles* (*Cellia*) *rivulorum* Leeson, 1935 [original combination: *Anopheles rivulorum* variety *garnhamellus*]; synonymy by de Meillon, 1947b.
garnhami Edwards, 1930a (p. 292) – species of *Anopheles* (*Cellia*).
Synonym: *walshi* Evans & de Meillon, 1933.
garnhami van Someren, 1948 (p. 130) – species of *Uranotaenia* (*Pseudoficalbia*).
gateri Baisas, 1936 (p. 78) – synonym of *Anopheles* (*Anopheles*) *baezai* Gater, 1933; synonymy by Reid, 1950.
gaudeator Dyar & Knab, 1907c (p. 204) – species of *Culex* (*Microculex*) [original combination: *Culex jenningsi* variety *gaudeator*]; specific status by Lane, 1953.
gaudians Dyar & Núñez Tovar, 1927b (p. 5) – species of *Wyeomyia* (*Wyeomyia*).
gaufini Belkin, 1962 (p. 243) – species of *Culex* (*Neoculex*).
gausapata Dyar & Núñez Tovar, 1927b (p. 7) – species of *Wyeomyia* (*Wyeomyia*).
gebeleinensis Theobald, 1910c (p. 157) – synonym of *Aedes* (*Stegomyia*) *unilineatus* (Theobald, 1906b) [original combination: *Stegomyia gebeleinensis*]; synonymy established before Knight & Stone, 1977.
geberti Grjebine, 1953a (p. 465) – synonym of *Orthopodomyia vernoni* van Someren, 1949a; synonymy by Brunhes & Hervey, 1995.
gedanica Szadziewski & Gilka, 2011 (p. 766) – fossil species of *Culiseta* Felt, 1904.
gediensis Edwards, 1941 (p. 345) – subspecies of *Culex* (*Culex*) *weschei* Edwards, 1935a.
gelidus Theobald, 1901c (p. 20) – species of *Culex* (*Culex*).
Synonyms: *cuneatus* Theobald, 1901c; *bipunctata* Theobald, 1907.
geminus Colless, 1955b (p. 314) – species of *Culex* (*Oculeomyia*).
geminus Peus, 1970 (p. 22) – species of *Aedes* (*Aedes*).
geniculata da Cunha Ramos & Brunhes, 2004 (p. 361) – species of *Uranotaenia* (*Uranotaenia*).
geniculatus (Olivier, 1791) (p. 134) – species of *Aedes* (*Dahliana*) [original combination: *Culex geniculatus*].
Synonyms: *guttatus* Meigen, 1818; *lateralis* Meigen, 1818; *ornatus* Meigen, 1818; *guttatus* Curtis, 1835; *albo-punctatus* Rondani, 1872.
genumaculata Cruz, 1907b (p. 329) – synonym of *Psorophora* (*Psorophora*) *saeva* Dyar & Knab, 1906a; synonymy by Lane, 1951.
genurostris Leicester, 1908 (p. 258) – species of *Malaya*.
Synonym: *caeruleovittata* Ludlow, 1911b.
geoffroyi Reinert, 1999a (p. 31) – species of *Aedes* (*Zavortinkius*).

geometrica Theobald, 1901c (p. 247) – species of *Uranotaenia* (*Uranotaenia*).

geometricus Corrêa, 1944 (p. 408) – subspecies of *Anopheles* (*Anopheles*) *eiseni* Coquillett, 1902b; original subspecific status re-confirmed by Harbach & Howard, 2007.

georgianus King, 1939 (p. 462) – species of *Anopheles* (*Anopheles*) [original combination: *Anopheles crucians* variety *georgianus*]; specific status by King & Bradley, 1941.

geoskusea Amos, 1944 (p. 32) – species of *Aedes* (*Levua*).
Synonym: *suvae* Stone & Bohart, 1944.

gerbergi Belkin, 1977a (p. 329) – species of *Toxorhynchites* (*Lynchiella*).

gerdae Slooff, 1963 (p. 84) – species of *Uranotaenia* (*Uranotaenia*).

germaini Geoffroy, 1974 (p. 93) – species of *Culex* (*Eumelanomyia*).

germaini Rickenbach & Eouzan, 1970 (p. 133) – species of *Eretmapodites*.

gibbinsi Edwards, 1935a (p. 135) – species of *Aedes* (*Aedimorphus*).

gibbinsi Evans, 1935 (p. 469) – species of *Anopheles* (*Cellia*) [original combination: *Anopheles marshallii* variety *gibbinsi*]; specific status by Gillies & de Meillon, 1968.
Synonym: *kibena* Peters, 1955b.

gibbosa (Delfinado, 1967) (p. 20) – species of *Verrallina* (*Neomacleaya*) [original combination: *Aedes* (*Neomacleaya*) *gibbosus*].

gibbulus Delfinado, 1966a (p. 105) – species of *Culex* (*Lophoceraomyia*).

giblini (Taylor, 1914d) (p. 198) – species of *Coquillettidia* (*Coquillettidia*) [original combination: *Pseudotaeniorhynchus conopas* variety *giblini*]; specific status by ? Taylor, 1944, as *Taeniorhynchus* (*Coquillettidia*) *giblini*.

gigantea Brug, 1931c (p. 6) – species of *Uranotaenia* (*Pseudoficalbia*).

giganteus Ventrillon, 1906a (p. 100) – species of *Culex* (*Oculeomyia*).
Synonym: *neireti* Ventrillon, 1906a.

gigantulus (Dyar & Shannon, 1925a) (p. 67) – species of *Toxorhynchites* (*Toxorhynchites*) [original combination: *Megarhinus gigantulus*].

gigas Giles, 1901b (p. 196) – species of *Anopheles* (*Anopheles*).

gilcolladoi Sánchez-Covisa Villa, Rodríguez Rodríguez & Guillén Llera, 1985 (p. 441) – species of *Aedes* (*Dahliana*).

gilesi (Neiva, 1908) (in Peryassú, 1908) (p. 103) – species of *Anopheles* (*Lophopodomyia*) [original combination: *Myzorhynchella gilesi*].

gilesii Theobald, 1901a (p. 227) – synonym of *Toxorhynchites* (*Toxorhynchites*) *splendens* (Wiedemann, 1819) [original combination: *Megarhina gilesii*]; synonymy established before Knight & Stone, 1977.

gilletti van Someren, 1949b (p. 125) – species of *Eretmapodites*.

gilli (Barraud, 1924d) (p. 1000) – species of *Aedes* (*Himalaius*) [original combination: *Finlaya gilli*].

gilliesi Hamon & van Someren, 1961a (p. 724) – species of *Culex* (*Culiciomyia*).

gilliesi van Someren, 1962 (p. 24) – species of *Aedes* (*Albuginosus*).

gilroyi Service, 1960 (p. 87) – synonym of *Anopheles* (*Cellia*) *brohieri* Edwards, 1929b [original combination: *Anopheles hancocki* variety *gilroyi*]; synonymy established before Knight & Stone, 1977.

gingeroi Corradetti & Archetti, 1947 (p. 677) – synonym of *Anopheles* (*Cellia*) *domicolus* Edwards, 1916; synonymy established before Knight & Stone, 1977.

giveni Edwards, 1926a (p. 118) – species of *Armigeres* (*Armigeres*).

gladiator Dyar, 1921g (p. 108) – synonym of *Haemagogus* (*Haemagogus*) *argyromeris* Dyar & Ludlow, 1921b; synonymy established before Knight & Stone, 1977.

glaphyroptera (Schiner, 1864) (p. 628) – species of *Culiseta* (*Culiseta*) [original combination: *Culex glaphyropterus*].
Synonym: *zottae* Ungureanu, 1956.

glaucocephala Dyar & Knab, 1906a (p. 140) – synonym of *Wyeomyia* (*Wyeomyia*) *mitchellii* (Theobald, 1905d); synonymy established before Knight & Stone, 1977.

glaucodaemon (Dyar & Shannon, 1925b) (p. 39) – species of *Sabethes* (*Sabethoides*) [original combination: *Sabethoides glaucodaemon*].

globocoxitus Dobrotworsky, 1953 (p. 357) – species of *Culex* (*Culex*).

glyptosalpinx Harbach, Peyton & Harrison, 1984 (p. 186) – species of *Culex* (*Melanoconion*).

gnomatos Sallum, Hutchings, Leila & Ferreira, 1997 (p. 215) – species of *Culex* (*Melanoconion*).

gnophodes Theobald, 1903a (p. 163) – synonym of *Culex* (*Culex*) *sitiens* Wiedemann, 1828; synonymy established before Knight & Stone, 1977.

goeldii (Giles, 1904b) (p. 383) – *nomen dubium*, *Psorophora* [original combination: *Taeniorhynchus goeldii*]; status established before Knight & Stone, 1977.

goeldii Howard, Dyar & Knab, 1915 (p. 24) – synonym of *Sabethes* (*Sabethes*) *belisarioi* Neiva, 1908a; synonymy established before Knight & Stone, 1977.

goeldii Rozeboom & Gabaldon, 1941 (p. 89) – species of *Anopheles* (*Nyssorhynchus*).

gombakensis Mattingly, 1959 (p. 56) – species of *Aedes* (*Christophersiomyia*).

gomezdelatorrei Leví-Castillo, 1955d (p. 510) – species of *Anopheles* (*Lophopodomyia*).

gonguoensis Gong & Lu, 1986 (p. 1) – species of *Aedes* (*Jihlienius*).

gonimus Dyar & Knab, 1918b (p. 165) – synonym of *Aedes* (*Ochlerotatus*) *sticticus* (Meigen, 1838); synonymy by Natvig, 1948.

gonzalezrinconesi Cova Garcia, Pulido F. & Escalante de Ugueto, 1977a (p. 140) – species of *Anopheles* (*Kerteszia*).

gordoni Evans, 1924 (p. 369) – synonym of *Culex* (*Melanoconion*) *albinensis* Bonne-Wepster & Bonne, 1920; synonymy established before Knight & Stone, 1977.

gorgasi Duret, 1971a (p. 15) – species of *Sabethes* (*Peytonulus*).

gorgasi Dyar & Knab, 1907c (p. 198) – synonym of *Anopheles* (*Nyssorhynchus*) *albimanus* Wiedemann, 1820; synonymy by Rozeboom & Gabaldon, 1941.

gorgasi Galindo & Trapido, 1956 (p. 228) – synonym of *Haemagogus* (*Haemagogus*) *mesodentatus* Komp & Kumm, 1938 [original combination: *Haemagogus* (*Haemagogus*) *mesodentatus gorgasi*]; synonymy by Arnell, 1973.

gossi Bohart, 1957 (p. 68) – species of *Culex* (*Lophoceraomyia*).

goughii Theobald, 1911b (p. 268) – synonym of *Culex* (*Culex*) *quinquefasciatus* Say, 1823; synonymy by White, 1975 (status by lectotype designation).

gouldi Peyton & Klein, 1970 (p. 248) – species of *Uranotaenia* (*Pseudoficalbia*).

gouldi Reinert, 1972b (p. 357) – species of *Aedes* (*Aedimorphus*).

grabhami Berlin, 1969a (p. 42) – species of *Aedes* (*Howardina*).
Synonym: *aureostriata* Grabham, 1906a.

grabhamii Theobald, 1901a (p. 205) – species of *Anopheles* (*Anopheles*).

gracicornis Sirivanakarn, 1977a (p. 58) – species of *Culex* (*Lophoceraomyia*).

gracilelineatus Bonne-Wepster, 1937 (p. 93) – species of *Aedes* (subgenus uncertain).

gracilis Dönitz, 1902 (p. 76) – synonym of *Anopheles* (*Cellia*) *gambiae* Giles, 1902; synonymy established before Knight & Stone, 1977.

gracilis Leicester, 1908 (p. 81) – synonym of *Aedes* (*Stegomyia*) *desmotes* (Giles, 1904a) [original combination: *Stegomyia gracilis*]; synonymy established before Knight & Stone, 1977.

gracilis Leicester, 1908 (p. 232) – species of *Zeugnomyia*.

gracilis Leicester, 1908 (p. 244) – species of *Topomyia* (*Topomyia*).

gracilis Theobald, 1905f (p. 69) – species of *Bironella* (*Bironella*).
Synonyms: *walchi* Soesilo, 1932; *occulta* Bonne-Wepster, 1951a.

grahami Edwards, 1911b (p. 71) – species of *Eretmapodites*.

grahami Ludlow, 1920b (p. 154) – synonym of *Aedes* (*Ochlerotatus*) *dorsalis* (Meigen, 1830b); synonymy established before Knight & Stone, 1977.

grahamii (Theobald, 1909b) (p. 2) – species of *Aedes* (*Mucidus*) [original combination: *Mucidus grahamii*].
Synonym: *africanus* Theobald, 1901d.

grahamii Theobald, 1910c (p. 628) – species of *Culex* (*Culex*) [*nomen novum* for *pullatus* Graham, 1910].
Synonym: *pullatus* Graham, 1910.

graminis Leicester, 1908 (p. 158) – synonym of *Culex* (*Culiciomyia*) *fragilis* Ludlow, 1903; synonymy established before Knight & Stone, 1977.

grandidieri (Blanchard, 1905) (p. 627) – species of *Coquillettidia* (*Coquillettidia*) [*nomen novum* for *flavus* Ventrillon, 1904; original combination: *Culex grandidieri*].
Synonym: *flavus* Ventrillon, 1904.

grandilarva Sazanova, 1955 (p. 99) – synonym of *Aedes* (*Ochlerotatus*) *euedes* Howard, Dyar & Knab, 1913; synonymy by Wood, 1977.

grandiosus (Williston, 1900) (p. 224) – species of *Toxorhynchites* (*Lynchiella*) [original combination: *Megarrhina* [*sic*] *grandiosa*].
Synonym: *longipes* Theobald, 1901a.

grantii (Theobald, 1901a) (p. 306) – species of *Aedes* (*Stegomyia*) [original combination: *Stegomyia grantii*].

grassei da Cunha Ramos & Brunhes, 2004 (p. 366) – species of *Uranotaenia* (*Uranotaenia*).

grassei Doucet, 1951a (p. 69) – species of *Aedes* (*Paulianius*).

grassei Grjebine, 1953b (p. 203) – species of *Anopheles* (*Cellia*).

grata Banks, 1906a (p. 780) – synonym of *Toxorhynchites* (*Toxorhynchites*) *amboinensis* (Doleschall, 1857) [original combination: *Worcesteria grata*]; synonymy by Bohart, 1945.

gravelyi (Edwards, 1921c) (p. 73) – species of *Toxorhynchites* (*Toxorhynchites*) [original combination: *Megarhinus* (*Toxorhynchites*) *gravelyi*].

gravitator Dyar & Knab, 1906d (p. 218) – *nomen dubium*, *Culex*; status by Sirivanakarn, 1983.

grayii Theobald, 1901c (p. 269) – species of *Wyeomyia* (*Wyeomyia*).
Synonyms: *colsoni* Senevet & Quiévreux, 1941; *antillarum* Floch & Abonnenc, 1945a.

greeni Rattanarithikul & Harbach, 1991 (p. 170) – species of *Anopheles* (*Cellia*).

greenii (Theobald, 1903a) (p. 289) – species of *Aedes* (*Bruceharrisonius*) [original combination: *Howardina greenii*].
Synonym: *kanaranus* Barraud, 1924a.

greenii (Theobald, 1905h) (p. 247) – species of *Heizmannia* (*Heizmannia*) [original combination: *Wyeomyia greenii*].

gregoryi Ludlow, 1911b (p. 128) – synonym of *Topomyia* (*Topomyia*) *argyropalpis* Leicester, 1908 [original combination: *Kingia gregoryi*]; synonymy by Bohart, 1945.

greigi Barraud, 1923e (p. 406) – synonym of *Aedes* (*Hulecoeteomyia*) *saxicola* Edwards, 1922a [original combination: *Finlaya greigi*]; synonymy established before Knight & Stone, 1977.

grenadensis Edwards, 1916 (p. 363) – synonym of *Wyeomyia melanocephala* Dyar & Knab, 1906a; synonymy by Lane, 1951.

grenieri Doucet, 1951b (p. 110) – species of *Uranotaenia* (*Pseudoficalbia*).

grenieri Eouzan, 1969 (p. 163) – species of *Culex* (*Culiciomyia*).

grenieri Grjebine, 1964 (p. 38) – species of *Anopheles* (*Cellia*).

grenieri Hamon, 1954b (p. 836) – synonym of *Ficalbia uniformis* (Theobald, 1904b); synonymy by Mattingly & Hamon, 1955.

grenieri Hamon & van Someren, 1961b (p. 907) – species of *Eretmapodites*.

grenieri Hamon, Service, Adam & Taufflieb, 1961 (p. 376) – species of *Aedes* (*Catageiomyia*).

gressitti Bohart, 1957 (p. 23) – species of *Aedes* (*Finlaya*).

gressitti Sirivanakarn, 1968 (p. 153) – species of *Culex* (*Lophoceraomyia*).

griffithi (Thurman, 1954a) (p. 198) – species of *Aedes* (*Ayurakitia*) [original combination: *Ayurakitia griffithi*].

grisea Ludlow, 1907 (p. 130) – synonym of *Aedes* (*Ochlerotatus*) *nigromaculis* (Ludlow, 1906a) [original combination: *Grabhamia grisea*]; synonymy established before Knight & Stone, 1977.

grisescens Stephens, 1828 (p. 503) – synonym of *Anopheles* (*Anopheles*) *claviger* (Meigen, 1804); synonymy by Edwards, 1932a.

griveaudi Grjebine, 1961 (p. 939) – species of *Anopheles* (*Cellia*).

grjebinei (Brunhes, 1977b) (p. 153) – species of *Mimomyia* (*Ingramia*) [original combination: *Ficalbia* (*Ingramia*) *grjebinei*].

grjebinei da Cunha Ramos & Brunhes, 2004 (p. 300) – species of *Uranotaenia* (*Pseudoficalbia*).

grjebinei Hamon, Taufflieb & Maillot, 1957 (p. 695) – species of *Aedes* (*Pseudalbuginosus*).

grjebinei Ribeiro, 2004 (p. 251) – species of *Toxorhynchites* (*Afrorhynchus*).

grossbecki Dyar & Knab, 1906d (p. 201) – species of *Aedes* (*Ochlerotatus*).
Synonym: *sylvicola* Grossbeck, 1906.

guadeloupensis (Dyar & Knab, 1906c) (p. 254) – species of *Toxorhynchites* (*Lynchiella*) [original combination: *Megarhinus guadeloupensis*].
Synonyms: *guianensis* Bonne-Wepster & Bonne, 1920; *horei* Gordon & Evan, 1922; *tucumanus* Brèthes, 1926a; *arborealis* Shannon & del Ponte, 1928.

guamensis Farner & Bohart, 1944 (p. 117) – species of *Aedes* (*Stegomyia*).

guarani Shannon, 1928 (in Dyar, 1928) (p. 444) – species of *Anopheles* (*Nyssorhynchus*); current specific status by Nagaki *et al*., 2011.

guarao Anduze & Capdevielle, 1949 (p. 119) – species of *Anopheles* (*Anopheles*).

guarauno Anduze, 1948 (p. 18) – synonym of *Anopheles* (*Nyssorhynchus*) *aquasalis* Curry, 1932 [original combination: *Anopheles aquasalis* variety *guarauno*]; synonymy by Lane, 1953.

guarujaensis Ramos, 1942 (p. 61) – synonym of *Anopheles* (*Nyssorhynchus*) *aquasalis* Curry, 1932 [original combination: *Anopheles oswaldoi* variety *guarujaensis*]; synonymy by Lane, 1949.

guatemala Berlin, 1969a (p. 103) – species of *Aedes* (*Howardina*).

guatemala Dyar & Knab, 1906a (p. 139) – species of *Wyeomyia* (*Wyeomyia*).

guatemalensis de Leon, 1938 (p. 416) – subspecies of *Anopheles* (*Anopheles*) *parapunctipennis* Martini, 1932b [original combination: *Anopheles chiriquiensis* variety *guatemalensis*]; subspecific status by Wilkerson, 1990.

guayasi Leví-Castillo, 1953b (p. 15) – species of *Culex* (*Culex*).

guayasi Leví-Castillo, 1954a (p. 251) – species of *Limatus*.

gubernatoris (Giles, 1901a) (p. 607) – species of *Aedes* (*Phagomyia*) [original combination: *Culex gubernatoris*].
Synonym: *magna* Theobald, 1905e.

gudouensis Chang, Zhao, Hang & Chen, 1975 (p. 338) – synonym of *Culex* (*Lophoceraomyia*) *curtipalpis sumatranus* Brug, 1931a; synonymy by Meng & Chen, 1980.

guedesi da Silva Mattos & Xavier, 1991 (p. 193) – species of *Culex* (*Melanoconion*).

guerrero Berlin, 1969a (p. 102) – species of *Aedes* (*Howardina*).

guerreroi Cova García, Sutil & Pulido, 1971 (p. 203) – species of *Culex* (*Carrollia*).

guianensis Bonne-Wepster & Bonne, 1920 (p. 180) – synonym of *Toxorhynchites* (*Lynchiella*) *guadeloupensis* (Dyar & Knab, 1906c) [original combination: *Megarhinus guadeloupensis guianensis*]; synonymy by Lane, 1953.

guiarti Blanchard, 1905 (p. 629) – species of *Culex* (*Culex*) [*nomen novum* for *viridis* Theobald, 1903a].
Synonym: *viridis* Theobald, 1903a.

guizhouensis Chen & Zhao, 1985 (p. 265) – species of *Culex* (*Culex*).

gurneri van Someren, 1946c (p. 5) – subspecies of *Aedes* (*Pseudarmigeres*) *michaelikati* van Someren, 1946c.

gurneyi (Belkin, 1962) (p. 296) – species of *Mimomyia* (*Mimomyia*) [original combination: *Ficalbia* (*Mimomyia*) *gurneyi*].

gurneyi Stone & Bohart, 1944 (p. 217) – species of *Aedes* (*Stegomyia*).

gutierrezi Duret, 1971a (p. 17) – species of *Wyeomyia* (subgenus uncertain, determined by Motta & Lourenço-de-Oliveira, 1995).

guttatus Curtis, 1835 (p. 537) – synonym of *Aedes* (*Dahliana*) *geniculatus* (Olivier, 1791) [original combination: *Culex guttatus*]; synonymy established before Knight & Stone, 1977.

guttatus Meigen, 1818 (p. 5) – synonym of *Aedes* (*Dahliana*) *geniculatus* (Olivier, 1791) [original combination: *Culex guttatus*]; synonymy established before Knight & Stone, 1977.

guttulatus Harris, 1835 (p. 595) – *nomen nudum*, *Culex* [original combination: *Anopheles quadrimaculatus*? variety *guttulatus*].

gutzevichi Dubitsky & Deshevykh, 1978 (p. 177) – species of *Aedes* (*Ochlerotatus*).

guyanensis Clastrier, 1970a (p. 116) – species of *Culex* (*Aedinus*).

gyirongensis Ma, 1982 (p. 162) – subspecies of *Aedes* (*Ochlerotatus*) *lasaensis* Meng, 1962.

gymnothorax Harbach & Petersen, 1992 (p. 103) – species of *Sabethes* (*Sabethes*).

gynaecopus Dyar & Knab, 1908 (p. 66) – synonym of *Wyeomyia* (*Wyeomyia*) *hemisagnosta* Dyar & Knab, 1906d; synonymy established before Knight & Stone, 1977.

Hh

habanicus Dyar & Knab, 1906d (p. 198) – synonym of *Aedes* (*Ochlerotatus*) *tortilis* (Theobald, 1903g); synonymy established before Knight & Stone, 1977.

habibi Mulligan & Puri, 1936 (p. 68) – synonym of *Anopheles* (*Anopheles*) *claviger* (Meigen, 1804); synonymy by Glick, 1992.

habilitator Dyar & Knab, 1906d (p. 212) – species of *Culex* (*Culex*).
Synonym: *eremita* Howards, Dyar & Knab, 1913.

hackeri Edwards, 1921c (p. 70) – species of *Anopheles* (*Cellia*) [original combination: *Anopheles leucosphyrus* variety *hackeri*]; specific status by Bonne-Wepster & Swellengrebel, 1953.

hackeri Edwards, 1923b (p. 4) – species of *Culex* (*Eumelanomyia*).

haddowi da Cunha Ramos & Brunhes, 2004 (p. 116) – species of *Uranotaenia* (*Pseudoficalbia*).

haddowi Gillett, 1975 (p. 366) – *nomen nudum*, *Anopheles* (*Cellia*) *gambiae* species D [= *Anopheles* (*Cellia*) *bwambae* White, 1985].

haddowi van Someren, 1949b (p. 122) – species of *Eretmapodites*.

hadrognathus Harbach, 1995a (p. 88) – species of *Sabethes* (*Peytonulus*).

haematophagus Ficalbi, 1893 (p. 143) – synonym of *Culex* (*Culex*) *pipiens* Linnaeus, 1758; synonymy by Marshall & Staley, 1937.

haemorrhoidalis (Fabricius, 1787) (p. 364) – species of *Toxorhynchites* (*Lynchiella*) [original combination: *Culex haemorrhoidalis*].

hailarensis Xu & Luo, 1998 (p. 45) – species of *Anopheles* (*Anopheles*).

hainanensis Chen, 1977 (p. 95) – species of *Culex* (*Culiciomyia*).

hainanensis Takei, 1941 (pp. 417, 418) – synonym of *Anopheles* (*Cellia*) *philippinensis* Ludlow, 1902b [original combination: *Anopheles* (*Myzomyia*) *philippinensis* variety *hainanensis*]; synonymy by Reid, 1968.

haitiensis Dyar & Knab, 1906c (p. 253) – synonym of *Toxorhynchites* (*Lynchiella*) *portoricensis* (von Röder, 1885) [original combination: *Megarhinus haitiensis*]; synonymy established before Knight & Stone, 1977.

hakanssoni Knight & Hurlbut, 1949 (p. 23) – species of *Aedes* (*Stegomyia*).

hakusanensis Yamaguti & Tamaboko, 1954 (p. 414) – species of *Aedes* (*Ochlerotatus*).

halifaxii (Theobald, 1903a) (p. 231) – species of *Lutzia* (*Metalutzia*) [original combination: *Culex halifaxii*].
Synonyms: *multimaculosus* Leicester, 1908; *aureopunctis* Ludlow, 1910; *raptor* Edwards, 1922a.

halli James, 1910a (p. 33) – synonym of *Anopheles* (*Cellia*) *kochi* Dönitz, 1901 [original combination: *Christophersia halli*]; synonymy established before Knight & Stone, 1977.

halophylus Silva do Nascimento & Lourenço-de-Oliveira, 2002 (p. 801) – species of *Anopheles* (*Nyssorhynchus*).

hamadryadis Cooling, 1924 (p. 24) – *nomen nudum*, *Aedes* [original combination: *Aedes* (*Finlaya*) *priestleyi hamadryadis*]; status by Lee *et al.*, 1982; Ward, 1984.

hamistylus (Laffoon, 1946) (p. 232) – species of *Verrallina* (*Harbachius*) [original combination: *Aedes* (*Aedes*) *hamistylus*].

hamoni Adam, 1962 (p. 154) – species of *Anopheles* (*Cellia*).

hamoni Brunhes, Adam & Bailly-Choumara, 1967 (p. 48) – species of *Culex* (*Eumelanomyia*).

hamoni Grjebine, 1953a (p. 465) – species of *Uranotaenia* (*Uranotaenia*) [original combination: *Uranotaenia chorleyi* variety *hamoni*]; specific status by White, 1975.

hamoni Grjebine, 1972 (p. 308) – species of *Eretmapodites*.

hamoni Mattingly, 1963 (p. 166) – synonym of *Aedes* (*Aedimorphus*) *rickenbachi* Hamon & Adam, 1959b; synonymy by White, 1975.

hanabusai Yamada, 1925 (p. 471) – synonym of *Anopheles* (*Cellia*) *maculatus* Theobald, 1901a [original combination: *Myzomyia hanabusai*]; synonymy established before Knight & Stone, 1977.

hancocki Edwards, 1929b (p. 322) – species of *Anopheles* (*Cellia*) [original combination: *Anopheles theileri* variety *hancocki*]; specific status by Evans, 1934.

hancocki Edwards, 1930a (p. 294) – species of *Culex* (*Culex*).

hancocki van Someren, 1962 (p. 21) – species of *Aedes* (*Vansomerenis*).

hansfordi Huang, 1997 (p. 33) – species of *Aedes* (*Stegomyia*).

hapla Dyar & Knab, 1909d (p. 265) – synonym of *Wyeomyia* (*Wyeomyia*) *scotinomus* (Dyar & Knab, 1907c); synonymy established before Knight & Stone, 1977.

haplolineata Schneider, 1913 (p. 37) – synonym of *Aedes* (*Ochlerotatus*) *punctor* (Kirby, 1837) [original combination: *Culicada nemorosa* form *haplolineata*]; synonymy established before Knight & Stone, 1977.

harbachi Reinert, 1990 (p. 27) – species of *Aedes* (*Kenknightia*).

hardini Miyagi, Toma & Ramalingam, 1989 (p. 91) – species of *Topomyia* (*Topomyia*).

hargreavesi (Edwards, 1920a) (p. 130) – subspecies of *Aedes* (*Ochlerotatus*) *caspius* (Pallas, 1771) [original combination: *Ochlerotatus caspius* variety *hargreavesi*]; subspecific status by Harbach & Howard, 2007.

hargreavesi Evans, 1927 (p. 40) – species of *Anopheles* (*Cellia*) [original combination: *Anopheles marshalli* [*sic*] variety *hargreavesi*].

harinasutai Knight, 1978a (p. 106) – species of *Aedes* (*Downsiomyia*).

harleyi Peters, 1955a (p. 81) – species of *Culex* (*Culiciomyia*).

harpagophallus Wang & Feng, 1964 (p. 38) – synonym of *Culex* (*Lophoceraomyia*) *minor* (Leicester, 1908); synonymy by Meng & Chen, 1980.

harperi Evans, 1936 (p. 533) – species of *Anopheles* (*Cellia*).

harperi Knight, 1948b (p. 4) – species of *Aedes* (*Danielsia*).

harperi van Someren, 1949b (p. 123) – species of *Eretmapodites*.

harrisoni Harbach & Manguin, 2007 (in Harbach, Garros *et al.*, 2007) (p. 44) – species of *Anopheles* (*Cellia*).

harrisoni Muspratt, 1953 (p. 51) – species of *Aedes* (*Ochlerotatus*).

harrisoni Peyton, 1977 (p. 29) – species of *Uranotaenia* (*Pseudoficalbia*).

harrisoni Sirivanakarn, 1977c (p. 102) – species of *Culex* (*Culiciomyia*).

harrisonica (Reinert, 1974) (p. 46) – species of *Verrallina* (*Neomacleaya*) [original combination: *Aedes* (*Verrallina*) *harrisonicus*].

haruspicus Dyar & Knab, 1908 (p. 56) – synonym of *Psorophora* (*Grabhamia*) *insularia* (Dyar & Knab, 1906a) [original combination: *Aedes haruspicus*]; synonymy established before Knight & Stone, 1977.

harveyi (Barraud, 1923e) (p. 407) – species of *Aedes* (*Hulecoeteomyia*) [original combination: *Finlaya harveyi*].

hassardii Grabham, 1906a (p. 167) – synonym of *Culex* (*Phenacomyia*) *corniger* Theobald, 1903a; synonymy established before Knight & Stone, 1977.

hastatus Dyar, 1922f (p. 160) – species of *Aedes* (*Ochlerotatus*).

hatiensis Carter, 1910b (p. 275) – synonym of *Verrallina* (*Verrallina*) *butleri* (Theobald, 1901c) [original combination: *Stegomyia hatiensis*]; synonymy established before Knight & Stone, 1977.

hatorii Koidzumi, 1920 (p. 24) – synonym of *Anopheles* (*Cellia*) *ludlowae* (Theobald, 1903a); synonymy by King, 1932a.

hatorii Yamada, 1921 (p. 70) – species of *Aedes* (*Collessius*).

haworthi Edwards, 1923d (p. 499) – species of *Aedes* (*Albuginosus*).

hayashii Yamada, 1917 (p. 67) – species of *Culex* (*Eumelanomyia*).

haynei Dodge, 1947 (p. 118) – synonym of *Wyeomyia* (*Wyeomyia*) *smithii* (Coquillett, 1901); synonymy by Bradshaw & Lounibos, 1977.

haynei Komp & Curry, 1932 (p. 82) – synonym of *Culex* (*Melanoconion*) *spissipes* (Theobald, 1903a); synonymy by Komp, 1935; Takahashi, 1968.

hebes Barraud, 1931b (p. 609) – species of *Uranotaenia* (*Uranotaenia*).
Synonym: *nii* Lien, 1968a.

hebes Dönitz, 1902 (p. 84) – synonym of *Anopheles* (*Cellia*) *funestus* Giles, 1900b; synonymy established before Knight & Stone, 1977.

hebrardi da Cunha Ramos & Brunhes, 2004 (p. 376) – species of *Uranotaenia* (*Uranotaenia*).

hebrideus Edwards, 1926a (p. 102) – species of *Aedes* (*Stegomyia*) [original combination: *Aedes variegatus* variety *hebrideus*]; specific status by Belkin, 1962.

hectoris Giaquinto Mira, 1931 (p. 606) – species of *Anopheles* (*Anopheles*).
Synonym: *cricillium* Martini, 1932b.

hedys Root, 1927c (p. 584) – species of *Culex* (*Microculex*).

hegneri Causey, 1937a (p. 543) – synonym of *Aedes* (*Collessius*) *macfarlanei* (Edwards, 1914b); synonymy by Harrison *et al.*, 1991.

heiheensis Ma, 1981b (p. 66) – species of *Anopheles* (*Anopheles*).

heischi van Someren, 1951 (p. 1) – species of *Aedes* (*Stegomyia*).

heiseri Baisas, 1935c (p. 72) – species of *Uranotaenia* (*Uranotaenia*).

helenae Brunhes, Adam & Bailly-Choumara, 1967 (p. 50) – species of *Culex* (*Eumelanomyia*).

helenae Reinert, 1973e (p. 9) – species of *Aedes* (*Bothaella*).

helenae Ribeiro, 1992c (p. 191) – species of *Toxorhynchites* (*Afrorhynchus*).

hellenicus Peus, 1954 (p. 76) – synonym of *Anopheles* (*Cellia*) *superpictus* Grassi, 1899; synonymy established before Knight & Stone, 1977.

hemisagnosta Dyar & Knab, 1906d (p. 230) – species of *Wyeomyia* (*Wyeomyia*).
Synonyms: *baria* Dyar & Knab, 1908; *gynaecopus* Dyar & Knab, 1908.

hemisiris Dyar & Shannon, 1925c (p. 234) – synonym of *Wyeomyia melanocephala* Dyar & Knab, 1906a [original combination: *Prosopolepis hemisiris*]; synonymy established before Knight & Stone, 1977.

hemisurus Dyar & Knab, 1906d (p. 199) – synonym of *Aedes* (*Ochlerotatus*) *scapularis* (Rondani, 1848); synonymy by Arnell, 1976.

hemiteleus Dyar, 1924h (p. 179) – synonym of *Aedes* (*Aedes*) *cinereus* Meigen, 1818 [original combination: *Aedes cinereus hemiteleus* (as race)]; synonymy by Carpenter & LaCasse, 1955.

hendersoni Cockerell, 1918 (p. 199) – species of *Aedes* (*Protomacleaya*) [original combination: *Aedes triseriatus* variety *hendersoni*]; specific status by Breland, 1960.

henrardi Edwards, 1935c (p. 96) – species of *Uranotaenia* (*Pseudoficalbia*).

henrardi Wolfs, 1945a (p. 174) – synonym of *Anopheles* (*Christya*) *implexus* (Theobald, 1903b) [original combination: *Anopheles* (*Anopheles*) *implexus* variety *henrardi*]; synonymy by de Meillon, 1947b.

henriquei da Cunha Ramos, 1993 (p. 158) – species of *Uranotaenia* (*Pseudoficalbia*).

hensemaeon Dyar, 1920i (p. 178) – synonym of *Culex* (*Culex*) *quinquefasciatus* Say, 1823; synonymy established before Knight & Stone, 1977.

hensilli Farner, 1945 (p. 59) – species of *Aedes* (*Stegomyia*).

hepperi Casal & García, 1967c (p. 88) – species of *Culex* (*Phytotelmatomyia*).

heracleensis Callot, 1944a (p. 56) – synonym of *Aedes* (*Ochlerotatus*) *berlandi* Séguy, 1921; synonymy by Callot, 1944c.

hermanoi (Lane & Coutinho, 1940) (p. 593) – species of *Coquillettidia* (*Rhynchotaenia*) [original combination: *Mansonia* (*Rhynchotaenia*) *hermanoi*].

hermsi Barr & Guptavanij, 1989 (p. 353) – species of *Anopheles* (*Anopheles*).

hernandoi Baisas & Feliciano, 1953 (p. 177) – species of *Topomyia* (*Topomyia*).

herrerai Sutil Oramas, Pulido Florenzano & Amarista Meneses, 1987 (p. 81) – species of *Culex* (*Melanoconion*).

herrickii Theobald, 1906a (p. 241) – synonym of *Toxorhynchites* (*Lynchiella*) *rutilus septentrionalis* (Dyar & Knab, 1906c) [original combination: *Megarhinus herrickii*]; synonymy established before Knight & Stone, 1977.

hervyi Brunhes, Le Goff & Geoffroy, 1999 (p. 554) – species of *Anopheles* (*Cellia*).

hervyi da Cunha Ramos & Brunhes, 2004 (p. 262) – species of *Uranotaenia* (*Pseudoficalbia*).

hesitator Dyar & Knab, 1907c (p. 205) – synonym of *Culex* (*Melanoconion*) *pilosus* (Dyar & Knab, 1906d); synonymy by Lane, 1951.

hesperonotius Marks, 1959a (p. 131) – species of *Aedes* (*Ochlerotatus*).

heteropus Dyar, 1921j (p. 152) – species of *Aedes* (*Protomacleaya*).

heterospina Gong & Lu, 1986 (p. 183) – species of *Heizmannia* (*Heizmannia*).

hewitti (Edwards, 1914b) (p. 80) – species of *Culex* (*Lophoceraomyia*) [original combination: *Lophoceratomyia hewitti*].

hewitti Hearle, 1923 (p. 5) – synonym of *Aedes* (*Ochlerotatus*) *increpitus* Dyar, 1916b; synonymy established before Knight & Stone, 1977.

hexacis (Martini, 1931a) (p. 217) – species of *Toxorhynchites* (*Ankylorhynchus*) [original combination: *Megarrhinus* [*sic*] (*Ancylorhynchus* [*sic*]) *hexacis*].

hexodontus Dyar, 1916b (p. 83) – species of *Aedes* (*Ochlerotatus*).
Synonyms: *cyclocerculus* Dyar, 1920a; *leuconotips* Dyar, 1920a; *masamae* Dyar, 1920g; *labradorensis* Dyar & Shannon, 1925d.

hightoni van Someren, 1947c (p. 13) – species of *Eretmapodites*.

hildebrandi Evans, 1923c (p. 380) – synonym of *Culex* (*Aedinus*) *amazonensis* (Lutz, 1905); synonymy established before Knight & Stone, 1977.

hilli Cova Garcia, Pulido F. & Escalante de Ugueto, 1976 (p. 344) – synonym of *Anopheles* (*Kerteszia*) *rollai* Cova Garcia, Pulido F. & Escalante de Ugueto, 1977b; synonymy by Cova Garcia *et al.*, 1977b. **Note**: this nominal taxon is the junior primary homonym of *hilli* Woodhill & Lee, 1944a.

hilli Edwards, 1922d (p. 95) – species of *Culex* (*Lophoceraomyia*).
Synonym: *australis* Taylor, 1915.

hilli (Edwards, 1926a) (p. 111) – species of *Culiseta* (*Neotheobaldia*) [original combination: *Theobaldia hilli*].

hilli Taylor, 1914b (p. 456) – synonym of *Tripteroides* (*Rachisoura*) *sylvestris* (Theobald, 1910c) [original combination: *Stegomyia hilli*]; synonymy by Lee *et al.*, 1989c.

hilli Taylor, 1919 (p. 841) – synonym of *Aedes* (*Rhinoskusea*) *longirostris* (Leicester, 1908) [original combination: *Uranotaenia hilli*]; synonymy established before Knight & Stone, 1977.

hilli Woodhill & Lee, 1944a (p. 63) – species of *Anopheles* (*Cellia*) [original combination: *Anopheles amictus hilli*]; specific status by Colless, 1976. **Note**: this nominal taxon is the senior primary homonym of *hilli* Cova Garcia, Pulido F. & Escalante de Ugueto, 1976.

himalayana Giles, 1904b (p. 384) – synonym of *Aedes* (*Gilesius*) *pulchriventer* (Giles, 1901a) [original combination: *Howardina himalayana*]; synonymy established before Knight & Stone, 1977.

himalayensis Edwards, 1922a (p. 290) – species of *Heizmannia* (*Heizmannia*).

hinesorum Schmidt, 2001 (in Schmidt *et al.*, 2001) (p. 400) – species of *Anopheles* (*Cellia*).

hinglungensis Chu, 1957 (p. 159) – species of *Culex* (*Eumelanomyia*).
Synonym: *culionicus* Delfinado, 1966a.

hirsuta Boussès & Brunhes, 2013 (p. 403) – species of *Uranotaenia* (*Pseudoficalbia*).

hirsuta (Hill & Hill, 1946) (p. 39) – species of *Wyeomyia* (*Wyeomyia*) [original combination: *Phoniomyia hirsuta*].

hirsuteron Theobald, 1901c (p. 98) – synonym of *Aedes* (*Ochlerotatus*) *sticticus* (Meigen, 1838) [original combination: *Culex hirsuteron*]; synonymy established before Knight & Stone, 1977.

hirsutifemora Peters, 1964 (p. 21) – species of *Uranotaenia* (*Pseudoficalbia*).

hirsutipalpis Theobald, 1901a (p. 378) – synonym of *Culex* (*Culex*) *duttoni* Theobald, 1901d; synonymy established before Knight & Stone, 1977.

hirsutipleura Barraud, 1928b (p. 370) – synonym of *Verrallina* (*Neomacleaya*) *unca* (Theobald, 1901c) [original combination: *Aedes* (*Aedes*) *hirsutipleura*]; synonymy by Reinert, 1974.

hirsutus (Brunhes & Boussès, 2017) (in Brunhes *et al.*, 2017) (p. 352) – species of *Aedes* (*Paulianius*) [original combination: *Paulianius hirsutus*].

hirsutus (Theobald, 1901a) (p. 392) – species of *Aedes* (*Aedimorphus*) [original combination: *Culex hirsutum* [*sic*]].
Synonym: *transvaalensis* Theobald, 1903a.

hirtipalpis Sirivanakarn, 1977a (p. 152) – species of *Culex* (*Lophoceraomyia*).

hirtusa Gong, 1989a (p. 90) – species of *Topomyia* (*Topomyia*).

hisaoe Tsuchimoto, 1948 (p. 366) – synonym of *Anopheles* (*Anopheles*) *koreicus* Yamada & Watanabe, 1918 [original combination: *Anopheles koreicus* variety *hisaoe*]; synonymy by Kamimura, 1966 (cited by Tanaka *et al.*, 1979).

hispaniola (Theobald, 1903a) (species) (p. 49) – subspecies of *Anopheles* (*Cellia*) *cinereus* Theobald, 1901a [original combination: *Myzomyia hispaniola*]; subspecific status by Ribeiro *et al.*, 1980.
Synonyms: *myzomyifacies* Theobald, 1907; *italicus* Raffaele, 1928; *rifenus* Baeza Cuéllar, 1933.

hispida (Delfinado, 1967) (p. 21) – species of *Verrallina* (*Neomacleaya*) [original combination: *Aedes* (*Neomacleaya*) *hispidus*].

hispida (Theobald, 1910c) (p. 245) – species of *Mimomyia* (*Mimomyia*) [original combination: *Hispidimyia hispida*].
Synonyms: *palustris* Theobald, 1911a; *sunyaniensis* Edwards, 1935a.

hispidosus Skuse, 1889 (p. 1726) – synonym of *Aedes* (*Mucidus*) *alternans* (Westwood, 1835) [original combination: *Culex hispidosus*]; synonymy established before Knight & Stone, 1977.

hodgkini Marks, 1959a (p. 112) – species of *Aedes* (*Ochlerotatus*).

hodgkini Reid, 1962 (p. 20) – species of *Anopheles* (*Anopheles*).

hodgkini (Wharton, 1962) (p. 12) – species of *Coquillettidia* (*Coquillettidia*) [original combination: *Mansonia* (*Coquillettidia*) *hodgkini*].

hoffeinsorum (Szadziewski, 1998) (p. 235) – fossil species of *Aetheapnomyia* Harbach & Greenwalt, 2012 [original combination: *Aedes* (*Finlaya*) *hoffeinsorum*].

hoffmani Root, 1927a (p. 465) – species of *Limatus*.

hogsbackensis Huang, 2004 (p. 43) – species of *Aedes* (*Stegomyia*).

hoguei Belkin, 1962 (p. 460) – species of *Aedes* (*Stegomyia*).

hokkaidensis Tanaka, Mizusawa & Saugstad, 1979 (p. 286) – species of *Aedes* (*Ochlerotatus*) [original combination: *Aedes* (*Ochlerotatus*) *hexodontus hokkaidensis*]; specific status by Tanaka, 1999.

hollandi Taylor, 1934 (p. 229) – species of *Bironella* (*Brugella*).

hollandius King & Hoogstraal, 1946b (p. 38) – species of *Aedes* (*Patmarksia*).

hollingsheadi Belkin, 1962 (p. 372) – species of *Aedes* (*Finlaya*).

holmbergi del Ponte & Heredia, 1945 (p. 382) – synonym of *Anopheles* (*Anopheles*) *fluminensis* Root, 1927d; synonymy by Lane, 1953.

holmbergii Lynch Arribálzaga, 1891b (p. 142) – species of *Psorophora* (*Psorophora*).
Synonym: *agoggylia* Dyar, 1922e.

holocinctus Edwards, 1941 (p. 193) – species of *Aedes* (*Aedimorphus*) [original combination: *Aedes* (*Aedimorphus*) *cumminsi* [*sic*] *holocinctus*]; specific status by Ovazza *et al.*, 1956.

holoneus Dyar, 1921b (p. 35) – synonym of *Culex* (*Melanoconion*) *conspirator* Dyar & Knab, 1906d; synonymy by Komp, 1935.

homoeopas Dyar & Ludlow, 1921a (p. 46) – synonym of *Culex* (*Melanoconion*) *erraticus* (Dyar & Knab, 1906d); synonymy by King & Bradley, 1937.

homoeopus Dyar, 1922d (p. 92) – species of *Aedes* (*Protomacleaya*) [original combination: *Aedes terrens homoeopus*]; specific status by Schick, 1970a.

homothe Dyar & Knab, 1907c (p. 211) – synonym of *Wyeomyia* (*Wyeomyia*) *scotinomus* (Dyar & Knab, 1907c); synonymy by Lane, 1953.

homotina Dyar & Knab, 1906a (p. 141) – synonym of *Runchomyia* (*Ctenogoeldia*) *magna* (Theobald, 1905f) [original combination: *Phoniomyia homotina*]; synonymy established before Knight & Stone, 1977.

homunculus Komp, 1937b (p. 509) – species of *Anopheles* (*Kerteszia*).
Synonym: *anoplus* Komp, 1937b.

hongayi Galliard & Ngu, 1947 (p. 77) – species of *Uranotaenia* (*Pseudoficalbia*).

hoogstraali Baisas, 1947 (p. 121) – species of *Tripteroides* (*Tripteroides*).

hoogstraali Knight & Rozeboom, 1946 (p. 92) – species of *Aedes* (*Scutomyia*).

hopkinsi Edwards, 1932b (p. 561) – species of *Uranotaenia* (*Uranotaenia*).

hopkinsi Edwards, 1932b (p. 562) – species of *Culex* (*Culex*).

hopkinsi Edwards, 1936 (p. 49) – species of *Aedes* (*Catageiomyia*).

horei Gordon & Evans, 1922 (p. 98) – synonym of *Toxorhynchites* (*Lynchiella*) *guadeloupensis* (Dyar & Knab, 1906c) [original combination: *Megarhinus horei*]; synonymy by Lane, 1951.

horishensis Yamada, 1921 (p. 58) – synonym of *Aedes* (*Stegomyia*) *annandalei* (Theobald, 1910a); synonymy by Huang, 1977.

horotoi Taylor, 1972 (p. 317) – species of *Aedes* (*Finlaya*).

horrescens Edwards, 1935a (p. 128) – species of *Aedes* (*Stegomyia*) [original combination: *Aedes* (*Stegomyia*) *scutellaris* variety *horrescens*]; specific status prior to ? Stone & Farner, 1945.

horrida (Dyar & Knab, 1908) (p. 56) – species of *Psorophora* (*Janthinosoma*) [original combination: *Aedes horridus*].

horridus Edwards, 1922d (p. 88) – species of *Culex* (*Eumelanomyia*) [*nomen novum* for *fusca* Theobald, 1909b].
Synonym: *fusca* Theobald, 1909b.

hortator Dyar & Knab, 1907a (p. 12) – species of *Aedes* (*Sallumia*).

hortensis Ficalbi, 1889a (p. 27) – species of *Culex* (*Maillotia*).
Synonyms: *pilifera* Theobald, 1907; *lavieri* Larrousse, 1925.

hosautos Dyar & Knab, 1907c (as *hosautus*) (p. 211) – species of *Wyeomyia* (*Miamyia*).
Synonyms: *euethes* Dyar & Knab, 1909d; *symmachus* Dyar & Knab, 1909d;.

houghtoni Feng, 1941 (p. 243) – species of *Topomyia* (*Suaymyia*).

howardi Belkin & Hogue, 1959 (p. 441) – species of *Deinocerites*.

howardi Lane & Cerqueira, 1942 (p. 595) – species of *Wyeomyia* (*Spilonympha*).

howardii Coquillett, 1901 (p. 258) – species of *Psorophora* (*Psorophora*).
Synonyms: *virescens* Dyar & Knab, 1906a; *simplex* Martini, 1935.

huangae Meng, 1958 (p. 351) – species of *Culex* (*Culex*).

huangae Reinert, 1999a (p. 34) – species of *Aedes* (*Zavortinkius*).

hughi Lambert & Coetzee, 1982 (p. 329) – species of *Anopheles* (*Cellia*).

hui Bohart, 1957 (p. 25) – species of *Aedes* (*Finlaya*).

hui Lien, 1968d (p. 227) – synonym of *Culex* (*Lophoceraomyia*) *spiculosus* Bram & Rattanarithikul, 1967; synonymy by Sirivanakarn, 1977a.
humboldti (Lane & Cerqueira, 1942) (p. 520) – species of *Runchomyia* (*Runchomyia*) [original combination: *Trichoprosopon* (*Hyloconops*) *humboldti*].
humeralis Dyar & Knab, 1916 (p. 65) – species of *Mansonia* (*Mansonia*).
humeralis Edwards, 1922d (p. 93) – species of *Aedes* (*Macleaya*).
humilis Theobald, 1901c (p. 336) – *nomen dubium*, *Culex*; status by Sirivanakarn, 1983.
hungaricus Mihályi, 1955 (p. 343) – species of *Aedes* (*Ochlerotatus*).
hunteri (Strickland, 1916a) (p. 263) – species of *Anopheles* (*Anopheles*) [original combination: *Myzorhynchus hunteri*].
hurlbuti Belkin, 1962 (p. 266) – species of *Culex* (*Lophoceraomyia*).
hurlbuti Lien, 1967 (p. 179) – species of *Aedes* (*Bruceharrisonius*).
husaini Qutubuddin, 1947 (p. 117) – species of *Uranotaenia* (*Pseudoficalbia*).
hutchinsoni Barraud, 1924e (p. 1261) – species of *Culex* (*Culex*).
hybrida (Leicester, 1908) (p. 115) – species of *Mimomyia* (*Mimomyia*) [original combination: *Conopomyia hybrida*].
hybrida Taylor, 1916 (p. 568) – synonym of *Aedes* (*Dobrotworskyius*) *rubrithorax* (Macquart, 1850) [original combination: *Culicada hybrida*]; synonymy established before Knight & Stone, 1977.
hybridus Edwards, 1914c (p. 126) – species of *Armigeres* (*Armigeres*).
hybridus (Leicester, 1908) (p. 237) – species of *Tripteroides* (*Tripteroides*) [original combination: *Colonemyia hybrida*].
hyemalis Fitch, 1847 (p. 281) – synonym of *Anopheles* (*Anopheles*) *punctipennis* (Say, 1823) [original combination: *Culex hyemalis*]; synonymy established before Knight & Stone, 1977.
hylaius Harbach (herein) (p. 3) – species of *Aedes* (*Neomelaniconion*) [*nomen novum* for *sylvaticum* Le Goff, Boussès & Brunhes, 2007].
Synonym: *sylvaticum* Le Goff, Boussès & Brunhes, 2007.
hylephilus Dyar & Knab, 1917 (p. 38) – synonym of *Anopheles* (*Kerteszia*) *neivai* Howard, Dyar & Knab, 1913; synonymy by Komp, 1937b.
hyperleucum (Martini, 1931a) (p. 201) – *nomen dubium*, *Trichoprosopon* [original combination: *Goeldia leucopus* variety *hyperleuca*]; status by Zavortink, 1979.
hypocrindyna Dyar, 1918d (p. 115) – synonym of *Coquillettidia* (*Rhynchotaenia*) *juxtamansonia* (Chagas, 1907) [original combination: *Mansonia hypocrindyna*]; synonymy established before Knight & Stone, 1977.
hypoptes (Knab, 1907a) (p. 50) – species of *Toxorhynchites* (*Lynchiella*) [original combination: *Megarhinus hypoptes*].
hyrcanus (Pallas, 1771) (p. 475) – species of *Anopheles* (*Anopheles*) [original combination: *Culex hyrcanus*].
Synonyms: *pictus* Loew, 1845; *flerowi* Portschinsky, 1910; *mesopotamiae* Christophers & Chand, 1915; *marzinovski* Shingarev, 1926; *popovi* Shingarev, 1928; *mahmuti* Martini, 1930.

Ii

ibis Barraud, 1931b (p. 610) – species of *Aedes* (*Christophersiomyia*).
Synonym: *brayi* Knight, 1947b.
ichiromiyagii Toma & Higa, 2004 (p. 195) – species of *Ficalbia*.
idahoensis (Theobald, 1903a) (p. 250) – subspecies of *Aedes* (*Ochlerotatus*) *spencerii* (Theobald, 1901c) [original combination: *Grabhamia spencerii* variety *idahoensis*]; subspecific status by Myers, 1967.
idanus Schick, 1970a (p. 70) – species of *Aedes* (*Protomacleaya*).
identicus Dyar & Knab, 1907c (p. 207) – species of *Sabethes* (*Peytonulus*).
Synonym: *lutzianus* Lane & Cerqueira, 1942.
idiogenes Harbach, 1994 (p. 227) – species of *Sabethes* (*Sabethinus*).
idjenensis Brug, 1934 (p. 513) – species of *Aedes* (*Downsiomyia*) [original combination: *Aedes* (*Finlaya*) *niveus* variety *idjenensis*]; specific status by Knight, 1946.
idottus Dyar, 1920d (p. 77) – species of *Culex* (*Melanoconion*).
Synonym: *terepaima* Anduze, 1949a.
ignobilis Dyar & Knab, 1909b (p. 39) – synonym of *Culex* (*Melanoconion*) *pilosus* (Dyar & Knab, 1906d); synonymy established before Knight & Stone, 1977.
ignotus Harbach, 1995b (p. 586) – species of *Sabethes* (*Peytonulus*).
iguassuensis Barbosa, da Silva & Sallum, 2007 (p. 46) – species of *Mansonia* (*Mansonia*).
iguazu Shannon & del Ponte, 1928 (in Dyar, 1928) (p. 168) – synonym of *Aedes* (*Ochlerotatus*) *crinifer* (Theobald, 1903a); synonymy established before Knight & Stone, 1977.
ikelos Forattini & Sallum, 1995 (p. 132) – species of *Culex* (*Melanoconion*).
imerinensis Monier & Treillard, 1935 (p. 125) – synonym of *Anopheles* (*Cellia*) *funestus* Giles, 1900b [original combination: *Anopheles* (*Myzomyia*) *funestus* variety *imerinensis*]; synonymy by White, 1980.
imitata (Baisas, 1946d) (p. 32) – species of *Kimia* [original combination: *Topomyia imitatus* [*sic*]].
imitator (Leicester, 1908) (species) (p. 89) – subspecies of *Aedes* (*Stegomyia*) *gardnerii* (Ludlow, 1905a) [original combination: *Stegomyia imitator*]; subspecific status by Mattingly, 1965.
Synonyms: *argenteomaculata* Theobald, 1907; *minutissima* Theobald, 1910a; *christianus* Dyar, 1921i; *indosinensis* Borel, 1928.

imitator Leicester, 1908 (p. 201) – synonym of *Verrallina* (*Harbachius*) *indecorabilis* Leicester, 1908; synonymy by Stone *et al*., 1959.

imitator Theobald, 1903a (p. 175) – species of *Culex* (*Microculex*).
Synonyms: *daumasturus* Dyar & Knab, 1906d; *vector* Dyar & Knab, 1906d; *argenteoumbrosus* Theobald, 1907.

immaculata James, 1902 (p. 35) – synonym of *Anopheles* (*Cellia*) *vagus* Dönitz, 1902; synonymy established before Knight & Stone, 1977.

immaculatus Theobald, 1903a (p. 23) – synonym of *Anopheles* (*Cellia*) *vagus* Dönitz, 1902 [same type specimen as *immaculata* James, 1902]; synonymy established before Knight & Stone, 1977.

immisericors Walker, 1859 (p. 90) – synonym of *Toxorhynchites* (*Toxorhynchites*) *amboinensis* (Doleschall, 1857) [original combination: *Megarhina immisericors*]; synonymy established before Knight & Stone, 1977.

imparis Peyton & Harbach, 2000 (in Harbach & Peyton, 2000) (p. 130) – species of *Onirion*.

impatibilis (Walker, 1859) (p. 91) – species of *Aedes* (*Scutomyia*) [original combination: *Culex impatibilis*].

impatiens (Walker, 1848) (p. 5) – species of *Culiseta* (*Culiseta*) [original combination: *Culex impatiens*].
Synonyms: *pinguis* Walker, 1866; *absobrinus* Felt, 1904.

impellens Walker, 1859 (p. 91) – synonym of *Culex* (*Culex*) *sitiens* Wiedemann, 1828; synonymy established before Knight & Stone, 1977.

imperfectus Bonne-Wepster & Bonne, 1920 (p. 165) – synonym of *Sabethes* (*Sabethoides*) *chloropterus* (von Humboldt, 1819) [original combination: *Sabethoides imperfectus*]; synonymy by Lane, 1951.

imperfectus Corrêa & Ramos, 1943 (p. 246) – synonym of *Anopheles* (*Nyssorhynchus*) *albitarsis* Lynch Arribálzaga, 1878 [original combination: *Anopheles* (*Nyssorhynchus*) *albitarsis imperfectus*]; synonymy by Lane, 1953.

imperfectus Dobrotworsky, 1962 (p. 296) – species of *Aedes* (*Ochlerotatus*).

impiger (Walker, 1848) (p. 6) – species of *Aedes* (*Ochlerotatus*) [original combination: *Culex impiger*].
Synonyms: *nearcticus* Dyar, 1919f; *parvulus* Edwards, 1921d.

implacabilis Walker, 1848 (p. 7) – synonym of *Aedes* (*Ochlerotatus*) *punctor* (Kirby, 1837) [original combination: *Culex implacabilis*]; synonymy by Vockeroth, 1954.

implexus (Theobald, 1903b) (p. 34) – species of *Anopheles* (*Christya*) [original combination: *Christya implexa*].
Synonym: *henrardi* Wolfs, 1945a.

implicatus Senevet & Abonnenc, 1939a (p. 99) – synonym of *Culex* (*Melanoconion*) *saramaccensis* Bonne-Wepster & Bonne, 1920; synonymy by Sirivanakarn, 1983.

implicatus Vockeroth, 1954 (p. 110) – species of *Aedes* (*Ochlerotatus*).

imposter Sirivanakarn, 1977a (p. 80) – species of *Culex* (*Lophoceraomyia*).

impostor Schick, 1970a (p. 66) – species of *Aedes* (*Protomacleaya*).

imprimens (Walker, 1860) (p. 144) – species of *Aedes* (*Edwardsaedes*) [original combination: *Culex imprimens*].
Synonyms: *auratus* Leicester, 1908; *suknaensis* Theobald, 1910a; *brugi* Edwards, 1924b.

impudicus Ficalbi, 1890a (p. 81) – species of *Culex* (*Neoculex*).
Synonym: *sergentii* Theobald, 1903a.

impunctus Dönitz, 1902 (p. 67) – synonym of *Anopheles* (*Cellia*) *multicolor* Cambouliu, 1902; synonymy established before Knight & Stone, 1977.

inadmirabilis Dyar, 1928 (p. 297) – species of *Culex* (*Melanoconion*).
Synonym: *mistura* Komp & Rozeboom, 1951.

inaequalis (Grabham, 1907) (p. 25) – species of *Aedes* (*Howardina*) [original combination: *Howardina inaequalis*].

inatomii Kamimura & Wada, 1974 (p. 13) – species of *Culex* (*Barraudius*) [original combination: *Culex* (*Barraudius*) *modestus inatomii*]; specific status by Tanaka *et al*., 1979.

incana Dyar, 1922g (p. 189) – synonym of *Wyeomyia* (*Wyeomyia*) *scotinomus* (Dyar & Knab, 1907c); synonymy established before Knight & Stone, 1977.

incaudata Root, 1928 (in Dyar, 1928) (p. 54) – species of *Wyeomyia* (*Phoniomyia*).

incertus Edwards, 1922a, 1922b (pp. 264, 468, respectively) – synonym of *Verrallina* (*Neomacleaya*) *taeniata* (Leicester, 1908); original combination: *Aedes* (*Aedes*) *incertus*]; originally proposed as a replacement name for *taeniata* Leicester, *non* Wiedemann, 1828, when both were included in *Aedes* Meigen.

inchoatus Barraud, 1927b (p. 544) – species of *Armigeres* (*Leicesteria*).

incidens (Thomson, 1869) (p. 443) – species of *Culiseta* (*Culiseta*) [original combination: *Culex incidens*].

inclinata Thurman, 1959 (p. 41) – species of *Topomyia* (*Topomyia*).

incognita Galindo, Blanton & Peyton, 1954 (p. 150) – species of *Uranotaenia* (*Uranotaenia*).

incognitus Baisas, 1938 (p. 203) – species of *Culex* (*Culex*).

incognitus Brug, 1931b (p. 136) – species of *Anopheles* (*Cellia*).

incomptas Ramalingam & Pillai, 1972 (p. 451) – species of *Malaya*.

incomptus Arnell, 1976 (p. 20) – species of *Aedes* (*Ochlerotatus*).

incomptus Bram & Rattanarithikul, 1967 (p. 16) – species of *Culex* (*Lophoceraomyia*).

inconspicua (Lee, 1937) (p. 294) – species of *Culiseta* (*Culicella*) [original combination: *Theobaldia inconspicua*].

inconspicuosa Theobald, 1909b (p. 10) – synonym of *Culex* (*Eumelanomyia*) *albiventris* Edwards, 1922d [original combination: *Eumelanomyia inconspicuosa*]; synonymy established before Knight & Stone, 1977.

inconspicuosus (Theobald, 1908a) (p. 109) – species of *Culex* (*Eumelanomyia*) [original combination: *Aedes inconspicuosus*].
Synonym: *nyangae* Galliard, 1931b.

inconspicuus Grossbeck, 1904b (p. 332) – synonym of *Aedes* (*Ochlerotatus*) *trivittatus* (Coquillett, 1902b) [original combination: *Culex inconspicuus*]; synonymy established before Knight & Stone, 1977.

increpitus Dyar, 1916b (p. 87) – species of *Aedes* (*Ochlerotatus*).
Synonyms: *vittata* Theobald, 1903h; *mutatus* Dyar, 1919b; *hewitti* Hearle, 1923.

incriminator Dyar & Knab, 1909d (p. 257) – synonym of *Culex* (*Melanoconion*) *peccator* Dyar & Knab, 1909d; synonymy established before Knight & Stone, 1977.

inculus Colless, 1965 (p. 280) – species of *Culex* (*Lophoceraomyia*).

indecorabilis Leicester, 1908 (p. 200) – species of *Verrallina* (*Harbachius*).
Synonym: *imitator* Leicester, 1908.

indecorabilis (Theobald, 1903a) (p. 241) – *nomen dubium*, *Culex* [original combination: *Melanoconion indecorabilis*]; status by Sirivanakarn, 1983.

indefinitus (Ludlow, 1904c) (p. 299) – species of *Anopheles* (*Cellia*) [original combination: *Myzomyia rossii* variety *indefinita*]; specific status by Reid, 1966.
Synonyms: *formosaensis II* Tsuzuki, 1902c; *trimaculatus* Tsuzuki, 1907; *malayensis* Hacker, 1921.

indeterminatus Baisas & Ubaldo-Pagayon, 1953 (p. 134) – species of *Tripteroides* (*Tripteroides*).

indiana Edwards, 1930b (p. 541) – species of *Mansonia* (*Mansonioides*).

indica (Edwards, 1920a) (species) (p. 137) – subspecies of *Culiseta* (*Culiseta*) *alaskaensis* (Ludlow, 1906c) [original combination: *Theobaldia indica*]; subspecific status by Maslov, 1964.
Synonym: *wassilievi* Shingarev, 1927.

indica Theobald, 1901a (p. 183) – synonym of *Anopheles* (*Cellia*) *culicifacies* Giles, 1901b; synonymy established before Knight & Stone, 1977. **Note**: this nominal taxon is the senior secondary homonym of *indica* Theobald, 1907 (p. 111).

indica (Theobald, 1905f) (p. 115) – species of *Heizmannia* (*Heizmannia*) [original combination: *Phoniomyia indica*].
Synonym: *metallica* Leicester, 1908.

indica Theobald, 1907 (p. 111) – synonym of *Anopheles* (*Cellia*) *willmori* (James, 1903) [original combination: *Neocellia indica*]; synonymy by Christophers, 1931. **Note**: this nominal taxon is the junior secondary homonym of *indica* Theobald, 1901a.

indica (Theobald, 1907) (p. 238) – species of *Verrallina* (*Neomacleaya*) [original combination: *Neomacleaya indica*].
Synonym: *mediofasciata* Theobald, 1907.

indicus (Barraud, 1929a) (p. 1061) – species of *Tripteroides* (*Tripteroides*) [original combination: *Rachionotomyia powelli* variety *indica*]; specific status by Thurman, 1959.

indicus Evenhuis & Steffan, 1986 (p. 558) – species of *Toxorhynchites* (*Toxorhynchites*).

indiensis Theobald, 1901a (p. 145) – synonym of *Anopheles* (*Anopheles*) *nigerrimus* Giles, 1900a [original combination: *Anopheles sinensis indiensis*]; synonymy by Harrison *et al.*, 1973.

indiensis Theobald, 1903a (p. 99) – synonym of *Anopheles* (*Cellia*) *stephensi* Liston, 1901 [original combination: *Nyssorhynchus maculipalpis* variety *indiensis*]; synonymy established before Knight & Stone, 1977.

indoctum Dyar & Knab, 1906b (p. 161) – synonym of *Psorophora* (*Grabhamia*) *cingulata* (Fabricius, 1805) [original combination: *Janthinosoma indoctum*]; synonymy established before Knight & Stone, 1977.

indolescens Dyar & Knab, 1907a (p. 11) – synonym of *Aedes* (*Ochlerotatus*) *scapularis* (Rondani, 1848); synonymy by Arnell, 1976.

indonesiae Mattingly, 1958a (p. 53) – species of *Aedes* (*Cancraedes*).

indosinensis Borel, 1928 (p. 93) – synonym of *Aedes* (*Stegomyia*) *gardnerii imitator* Leicester, 1908 [original combination: *Stegomyia indosinensis*]; synonymy by Huang, 1977.

indubitans Dyar & Shannon, 1925b (p. 41) – species of *Mansonia* (*Mansonia*).
Synonym: *fonsecai* Pinto, 1932.

inducens Root, 1928 (in Dyar, 1928) (p. 307) – synonym of *Culex* (*Melanoconion*) *conspirator* Dyar & Knab, 1906d; synonymy by Komp, 1935.

inelegans Dyar, 1920i (p. 179) – synonym of *Culex* (*Culex*) *fuscocephala* Theobald, 1907; synonymy established before Knight & Stone, 1977.

inermis Colless, 1958 (p. 473) – species of *Aedes* (*Downsiomyia*).

inexorabilis Walker, 1848 (p. 4) – synonym of *Aedes* (*Stegomyia*) *aegypti* (Linnaeus, 1762) [original combination: *Culex inexorabilis*]; synonymy established before Knight & Stone, 1977.

inexpectatus Bonne-Wepster, 1948 (p. 318) – species of *Aedes* (*Ochlerotatus*).

infantulus Edwards, 1922a (p. 287) – species of *Culex* (*Lophoceraomyia*).
Synonym: *parainfantulus* Menon, 1944.

infinis (Dyar & Knab, 1906d) (p. 182) – species of *Psorophora* (*Grabhamia*) [original combination: *Janthinosoma infine*].

infirmatus Dyar & Knab, 1906d (p. 197) – species of *Aedes* (*Ochlerotatus*).

inflictus Theobald, 1901c (p. 115) – species of *Culex* (*Culex*).
Synonyms: *scholasticus* Theobald, 1901c; *extricator* Dyar & Knab, 1906d.

infoliatus Bonne-Wepster & Bonne, 1920 (p. 170) – species of *Culex* (*Carrollia*).

infula Theobald, 1901a (p. 370) – species of *Culex* (*Oculeomyia*).
Synonyms: *tenax* Theobald, 1901a; *ambiguus* Theobald, 1903a; *ocellata* Theobald, 1907; *sarawaki* Theobald, 1907; *taeniarostris* Theobald, 1907; *domesticus* Leicester, 1908; *afridii* Qutubuddin, 1956.

ingrami Edwards, 1916 (p. 360) – species of *Culex* (*Culex*).
ingrami Edwards, 1929b (p. 322) – synonym of *Anopheles* (*Cellia*) *rufipes broussesi* Edwards, 1929a [original combination: *Anopheles rufipes* variety *ingrami*]; synonymy established before Knight & Stone, 1977.
ingrami Edwards, 1930a (p. 296) – species of *Aedes* (*Hopkinsius*).
inhibitator Dyar & Knab, 1906d (p. 216) – species of *Culex* (*Melanoconion*).
Synonym: *investigator* Dyar & Knab, 1906d.
inimitabilis Dyar & Knab, 1906d (p. 221) – species of *Culex* (*Microculex*).
ininicola Fauran & Pajot, 1974 (p. 145) – species of *Wyeomyia* (subgenus uncertain, determined by Harbach & Peyton, 1991).
ininii Senevet & Abonnenc, 1938 (p. 494) – species of *Anopheles* (*Nyssorhynchus*).
innominata Stoker & Waktoedi Koesoemawinangoen, 1949 (p. 10) – ? synonym of *Anopheles* (*Anopheles*) *barbirostris* van der Wulp, 1884 [original combination: *Anopheles* (*Anopheles*) *barbirostris innominata* [*sic*]]; synonymy by Reid, 1968.
innominatus Evans, 1924 (p. 363) – synonym of *Culex* (*Melanoconion*) *bastagarius* Dyar & Knab, 1906b; synonymy by Komp, 1935.
innotata Dyar & Shannon, 1925a (p. 69) – synonym of *Uranotaenia* (*Uranotaenia*) *metatarsata* Edwards, 1914b; synonymy by Peyton, 1972.
innovator Evans, 1924 (p. 373) – species of *Culex* (*Melanoconion*).
innuitus Dyar & Knab, 1918b (p. 166) – synonym of *Aedes* (*Ochlerotatus*) *nigripes* (Zetterstedt, 1838); synonymy established before Knight & Stone, 1977.
inornata Strickland, 1911b (p. 201) – synonym of *Aedes* (*Ochlerotatus*) *camptorhynchus* (Thomson, 1869) [original combination: *Culicada inornata*]; synonymy established before Knight & Stone, 1977.
inornata Theobald, 1907 (p. 227) – synonym of *Culex* (*Culiciomyia*) *fragilis* Ludlow, 1903 [original combination: *Culiciomyia inornata*]; synonymy established before Knight & Stone, 1977. **Note**: this nominal taxon is the junior secondary homonym of *inornatus* (Theobald, 1905g).
inornata Theobald, 1908a (p. 108) – synonym of *Uranotaenia* (*Pseudoficalbia*) *fusca* Theobald, 1907 [original combination: *Ficalbia inornata*]; synonymy by Edwards, 1941.
inornata Theobald, 1910a (p. 28) – synonym of *Tripteroides* (*Rachionotomyia*) *aranoides* (Theobald, 1901c) [original combination: *Squamomyia inornata*]; synonymy established before Knight & Stone, 1977.
inornata (Williston, 1893) (p. 253) – species of *Culiseta* (*Culiseta*) [original combination: *Culex inornatus*].
Synonym: *magnipennis* Felt, 1904.
inornatus Newstead, 1907 (in Newstead *et al.*, 1907) (p. 12) – species of *Eretmapodites*.
inornatus (Theobald, 1905g) (p. 21) – species of *Culex* (subgenus uncertain, determined by Belkin, 1970b) [original combination: *Gnophodeomyia inornata*]. **Note**: this nominal taxon is the senior secondary homonym of *inornata* Theobald, 1907. **Also note**: the name of this species is incorrectly spelled as *Culex inornata* in the world catalogue of Knight & Stone, 1977 and the online catalogue of the Walter Reed Biosystematics Unit (http://www.mosquitocatalog.org/).
inornatus (Walker, 1865) (p. 102) – species of *Toxorhynchites* (*Toxorhynchites*) [original combination: *Megarhina inornata*].
inquinatus Edwards, 1922a (p. 267) – species of *Aedes* (*Phagomyia*).
inquisitor Dyar & Knab, 1906d (p. 211) – synonym of *Culex* (*Culex*) *declarator* Dyar & Knab, 1906d; synonymy established before Knight & Stone, 1977.
insatiabilis Bigot, 1859 (p. 118) – synonym of *Aedes* (*Stegomyia*) *aegypti* (Linnaeus, 1762) [original combination: *Culex insatiabilis*]; synonymy established before Knight & Stone, 1977.
insequens Marks, 1989 (in Lee *et al.*, 1989b) (p. 120) – species of *Culex* (*Lophoceraomyia*) [*nomen novum* for *cairnsensis* Taylor, 1919].
Synonym: *cairnsensis* Taylor, 1919.
insigniforceps Clastrier & Claustre, 1978a (p. 359) – species of *Culex* (*Carrollia*).
insignis (Carter, 1911) (p. 37) – species of *Culex* (*Eumelanomyia*) [original combination: *Culiciomyia* (?) *insignis*].
insolens Edwards, 1936 (p. 50) – species of *Aedes* (*Catageiomyia*).
insolitus (Coquillett, 1906e) (p. 62) – species of *Aedes* (*Protomacleaya*) [original combination: *Verrallina insolita*].
Synonym: *laternaria* Coquillett, 1906a.
insulaeflorum (Swellengrebel & Swellengrebel de Graaf, 1920a) (p. 2) – species of *Anopheles* (*Anopheles*) [original combination: *Stethomyia aitkenii* variety *insulaeflorum*]; specific status by Christophers, 1933.
insularis Sirivanakarn, 1968 (p. 166) – species of *Culex* (*Lophoceraomyia*).
insularius (Dyar & Knab, 1906a) (p. 135) – species of *Psorophora* (*Grabhamia*) [original combination: *Janthinosoma insularius*]. **Note**: *insularius* is a Latin masculine noun, meaning 'a keeper or porter of a detached house (insula)', in apposition to its original placement in *Janthinosoma* (neuter). It was incorrectly emended to *insularia* by Edwards (1932a) and subsequently retained that spelling in the catalogues of Stone *et al.* (1959), Knight & Stone (1977) and the Walter Reed Biosystematics Unit (http://www.mosquitocatalog.org/).
Synonym: *haruspicus* Dyar & Knab, 1908.
interfor Dyar, 1928 (p. 372) – species of *Culex* (*Culex*).
intermedia Barraud, 1929a (p. 1055) – species of *Mimomyia* (*Mimomyia*) [original combination: *Mimomyia chamberlaini* variety *intermedia*]; specific status by Mattingly, 1957a.
intermedia Rothwell, 1907 (p. 34) – synonym of *Anopheles* (*Cellia*) *stephensi* Liston, 1901 [original combination: *Neocellia intermedia*]; synonymy established before Knight & Stone, 1977. **Note**: this nominal taxon is the senior secondary homonym of *intermedium* Chagas, 1908 and *intermedius* Shingarev, 1928.

intermediatus Baisas & Ubaldo-Pagayon, 1953 (p. 135) – species of *Tripteroides* (*Tripteroides*).

intermedium Chagas, 1908 (in Peryassú, 1908) (p. 85) – synonym of *Anopheles* (*Anopheles*) *medialis* Harbach (herein) [original combination: *Cycloleppteron intermedium*]. **Note:** this nominal taxon is the junior secondary homonym of *intermedia* Rothwell, 1907 and the senior secondary homonym of *intermedius* Shingarev, 1928.

intermedius Danilov & Gornostaeva, 1987 (p. 612) – species of *Aedes* (*Ochlerotatus*).

intermedius Edwards, 1936 (p. 53) – species of *Eretmapodites*.

intermedius Lane & Whitman, 1951 (p. 348) – species of *Culex* (*Microculex*).

intermedius (Lutz, 1904) (in Bourroul, 1904) (p. 48) – species of *Sabethes* (*Sabethinus*) [original combination: *Sabethinus intermedius*].

intermedius Shingarev, 1928 (p. 49) – synonym of *Anopheles* (*Anopheles*) *plumbeus* Stephens, 1828; synonymy by Bates *et al.*, 1949. **Note:** this nominal taxon is the junior secondary homonym of *intermedia* Rothwell, 1907 and *intermedium* Chagas, 1908.

interrogator Dyar & Knab, 1906d (p. 209) – species of *Culex* (*Culex*).
Synonym: *reflector* Dyar & Knab, 1909d.

interruptus Puri, 1929a (p. 387) – species of *Anopheles* (*Anopheles*) [original combination: *Anopheles annandalei* variety *interruptus*]; specific status by Puri, 1949.

interruptus Reinert, 1999a (p. 47) – species of *Aedes* (*Zavortinkius*).

intonca Dyar & Knab, 1910 (p. 173) – species of *Wyeomyia* (*Hystatomyia*).

intonsus Galindo & Blanton, 1954 (p. 231) – species of *Culex* (*Melanoconion*).

intrincatus Brèthes, 1916 (p. 214) – species of *Culex* (*Melanoconion*).
Synonyms: *xivylis* Dyar, 1920d; *cenus* Root, 1927c.

introlatus Colless, 1957b (p. 137) – species of *Anopheles* (*Cellia*) [original combination: *Anopheles balabacensis introlatus*]; specific status by Hii *et al.*, 1988.

intrudens Dyar, 1919b (p. 23) – species of *Aedes* (*Ochlerotatus*).

inundatus Reinert, 1997 (in Reinert *et al.*, 1997) (p. 15) – species of *Anopheles* (*Anopheles*).

invenustus Theobald, 1901d (p. 9) – synonym of *Culex* (*Culiciomyia*) *nebulosus* Theobald, 1901d; synonymy established before Knight & Stone, 1977.

investigator Dyar & Knab, 1906d (p. 216) – synonym of *Culex* (*Melanoconion*) *inhibitator* Dyar & Knab, 1906d; synonymy established before Knight & Stone, 1977.

invidiosus Theobald, 1901d (p. 11) – species of *Culex* (*Culex*).
Synonyms: *euclastus* Theobald, 1903c; *chloroventer* Theobald, 1909b; *aquilus* Graham, 1910.

invocator Pazos, 1908 (p. 426) – species of *Culex* (*Melanoconion*).

iolambdis Dyar, 1918c (p. 106) – species of *Culex* (*Melanoconion*).

ioliota Dyar & Knab, 1913 (p. 77) – species of *Aedes* (*Howardina*).

iphis Barraud, 1924f (p. 1279) – species of *Culex* (*Eumelanomyia*).

iracunda Dyar & Knab, 1906a (p. 133) – synonym of *Psorophora* (*Psorophora*) *cilipes* (Fabricius, 1805); synonymy established before Knight & Stone, 1977.

iracunda (Walker, 1848) (p. 6) – species of *Coquillettidia* (*Coquillettidia*) [original combination: *Culex iracundus*].

irenicus Schmidt, 2003 (in Schmidt *et al.*, 2003) (p. 517) – species of *Anopheles* (*Cellia*).

irianensis Miyagi & Toma, 1997 (p. 134) – species of *Topomyia* (*Topomyia*).

iriartei Cova García, Pulido F., Escalante de Ugueto & Mora R., 1987a (p. 45) – species of *Uranotaenia* (*Uranotaenia*).

iridescens (Lutz, 1905) (p. 81) – species of *Culex* (*Carrollia*) [original combination: *Carrollia iridescens*].

iridicolor Dyar, 1921g (p. 106) – species of *Haemagogus* (*Haemagogus*).

iridipennis Dyar, 1922d (p. 92) – synonym of *Aedes* (*Lewnielsenius*) *muelleri* Dyar, 1920j; synonymy established before Knight & Stone, 1977.

iriomotensis (Tanaka & Mizusawa, 1973) (p. 633) – species of *Verrallina* (*Verrallina*) [original combination: *Aedes* (*Neomacleaya*) *iriomotensis*].

iris Knab, 1913 (p. 35) – synonym of *Toxorhynchites* (*Lynchiella*) *mariae* (Bourroul, 1904) [original combination: *Megarhinus iris*]; synonymy by Lane, 1944.

irritans (Theobald, 1901d) (p. 3) – species of *Aedes* (*Catageiomyia*) [original combination: *Stegomyia irritans*].
Synonyms: *senegalensis* Theobald, 1903d; *albotaeniatus* Theobald, 1909b; *confusus* Theobald, 1909b.

isabelae Duret, 1968f (p. 72) – species of *Culex* (*Melanoconion*).

ishigakiensis Bohart, 1956 (p. 32) – synonym of *Verrallina* (*Harbachius*) *nobukonis* (Yamada, 1932) [original combination: *Aedes* (*Aedes*) *ishigakiensis*]; synonymy by Reinert, 1974.

italicus Raffaele, 1928 (p. 11) – synonym of *Anopheles* (*Cellia*) *cinereus hispaniola* (Theobald, 1903a); synonymy by Aitken, 1953.

iwi Marks, 1955 (p. 11) – species of *Aedes* (*Phagomyia*).

iyengari Edwards, 1923b (p. 4) – species of *Aedes* (*Petermattinglyius*).

iyengari Mattingly & Rageau, 1958 (p. 244) – species of *Culex* (*Culex*).

Jj

jacksoni Edwards, 1934 (in Barraud, 1934) (p. 452) – species of *Culex* (*Culex*).
Synonyms: *fuscifurcatus* Edwards, 1934; *kangi* Lien, 1968d.

jacksoni Edwards, 1935a (p. 130) – species of *Uranotaenia* (*Pseudoficalbia*).
Synonym: *stonei* Bohart & Ingram, 1946a.

jacksoni Mattingly, 1949a (p. 9) – species of *Ficalbia*.

jacobi (Hill & Haydon, 1907) (p. 144) – *nomen dubium*, *Anopheles* [original combination: *Cellia jacobi*]; status by Gillies & de Meillon, 1968.

jacobinae Serafim & Davis, 1933 (p. 14) – species of *Aedes* (*Ochlerotatus*).

jacobsoni (Edwards, 1930b) (p. 543) – species of *Malaya* [original combination: *Harpagomyia jacobsoni*].

jalisco Berlin, 1974 (p. 273) – species of *Culex* (*Micraedes*).

jamaicensis Grabham, 1906b (p. 318) – synonym of *Culex* (*Melanoconion*) *pilosus* (Dyar & Knab, 1906d) [original combination: *Mochlostyrax jamaicensis*]; synonymy established before Knight & Stone, 1977.

jamaicensis (Theobald, 1901a) (p. 345) – species of *Psorophora* (*Grabhamia*) [original combination: *Culex jamaicensis*]. **Note**: this nominal taxon is the senior secondary homonym of *jamaicensis* Theobald, 1907.
Synonym: *walsinghamii* Theobald, 1907.

jamaicensis Theobald, 1905d (p. 11) – synonym of *Wyeomyia* (*Wyeomyia*) *mitchellii* (Theobald, 1905d) [original combination: *Dendromyia jamaicensis*]; synonymy established before Knight & Stone, 1977.

jamaicensis Theobald, 1907 (p. 157) – synonym of *Psorophora* (*Janthinosoma*) *ferox* (von Humboldt, 1819) [original combination: *Janthinosoma sayi* variety *jamaicensis*]; synonymy established before Knight & Stone, 1977. **Note**: this nominal taxon is the junior secondary homonym of *jamaicensis* (Theobald, 1901a).

jamesi (Edwards, 1914b) (p. 77) – species of *Aedes* (*Aedimorphus*) [original combination: *Ochlerotatus jamesi*].

jamesii Liston, 1901 (p. 441) – synonym of *Anopheles* (*Cellia*) *annularis* van der Wulp, 1884; synonymy established before Knight & Stone, 1977.

jamesii Theobald, 1901a (p. 134) – species of *Anopheles* (*Cellia*).

jamoti Hamon & Rickenbach, 1954 (p. 932) – species of *Aedes* (*Neomelaniconion*).

janconnae Wilkerson & Sallum, 2009 (in Motoki *et al.*, 2009) (p. 846) – species of *Anopheles* (*Nyssorhynchus*).

jani Rickenbach & Lombrici, 1975 (p. 42) – species of *Eretmapodites*.

janitor Theobald, 1903a (p. 183) – species of *Culex* (*Culex*).

janthinomys Dyar, 1921g (p. 112) – species of *Haemagogus* (*Haemagogus*).
Synonyms: *obscurescens* Martini 1931a; *falco* Kumm *et al.*, 1946; *petrocchiae* Martínez *et al.*, 1961.

japonicus (Theobald, 1901a) (p. 385) – species of *Aedes* (*Hulecoeteomyia*) [original combination: *Culex japonicus*].
Synonyms: *eucleptes* Dyar, 1921i; *tokushimaensis* Tanimura, 1952; *bisanensis* Suzuki, Tanimura, Miyagawa & Murata, 1953.

japonicus Yamada, 1918 (species) (p. 689) – subspecies of *Anopheles* (*Anopheles*) *lindesayi* Giles, 1900a; subspecific status by ? La Casse & Yamaguti, 1950.

javaensis Miyagi & Toma, 1995 (p. 179) – species of *Topomyia* (*Topomyia*).

javaensis Theobald, 1911d (p. 233) – synonym of *Toxorhynchites* (*Toxorhynchites*) *quasiferox* (Leicester, 1908); synonymy established before Knight & Stone, 1977.

javanensis Bonne-Wepster, 1934 (p. 274) – species of *Culex* (*Culiciomyia*).

javanensis Brug, 1939 (p. 100) – synonym of *Armigeres* (*Armigeres*) *durhami* Edwards, 1917 [original combination: *Armigeres* (*Armigeres*) *kuchingensis* variety *javanensis*]; synonymy by Ramalingam, 1987.

javanensis Swellengrebel & Swellengrebel de Graaf, 1920b (p. 91) – *nomen dubium*, *Anopheles*; status established before Knight & Stone, 1977.

jeansottei (Doucet, 1950) (p. 49) – species of *Mimomyia* (*Ingramia*) [original combination: *Ravenalites jeansottei*].

jebudensis Froud, 1944 (p. 73) – species of *Anopheles* (*Cellia*).

jefferyi Sirivanakarn, 1977b (p. 77) – species of *Culex* (*Eumelanomyia*).

jehafi Patton, 1905 (p. 630) – synonym of *Anopheles* (*Cellia*) *cinereus* Theobald, 1901a; synonymy established before Knight & Stone, 1977.

jenkinsi Knight, 1953 (p. 323) – synonym of *Culex* (*Maillotia*) *arbieeni* Salem, 1938; synonymy by Mattingly, 1955.

jenningsi Dyar & Knab, 1907c (p. 204) – species of *Culex* (*Microculex*).

jenseni (de Meijere, 1910) (p. 922) – species of *Culex* (*Lophoceraomyia*) [original combination: *Cyathomyia jenseni*].

jepsoni Bahr, 1912 (p. 18) – synonym of *Culex* (*Culex*) *annulirostris* Skuse, 1889; synonymy established before Knight & Stone, 1977. **Note**: this nominal taxon is the junior primary homonym of *jepsoni* Theobald, 1910b.

jepsoni Theobald, 1910b (p. 158) – synonym of *Culex* (*Culex*) *sitiens* Wiedemann, 1828; synonymy established before Knight & Stone, 1977. **Note**: this nominal taxon is the senior primary homonym of *jepsoni* Bahr, 1912.

jesoensis Tsuzuki, 1902b (p. 764) – synonym of *Anopheles* (*Anopheles*) *sinensis* Wiedemann, 1828; synonymy established before Knight & Stone, 1977.

jeyporensis Theobald, 1903a (p. 66) – synonym of *Anopheles* (*Cellia*) *jeyporiensis* James, 1902 [original combination: *Pyretophorus jeyporiensis*]; synonymy established before Knight & Stone, 1977. **Note**: this nominal taxon is the junior secondary homonym *jeyporiensis* James, 1902 (Harrison, 1980).

jeyporiensis James, 1902 (p. 32) – species of *Anopheles* (*Cellia*).
Synonyms: *vincenti* Laveran, 1901c; *jeyporensis* Theobald, 1903a; *candidiensis* Koidzumi, 1924; *tonkinensis* Toumanoff, 1931. **Note**: established usage of *jeyporiensis* James over the senior synonym *vincenti* Laveran is maintained pending application for suppression of the latter by the International Commission on Zoological

Nomenclature (Harbach, Garros *et al.*, 2007). **Also note:** this nominal taxon is the senior secondary homonym of *jeyporensis* Theobald, 1903a.

jinhongensis Dong, Dong & Zhou, 2003 (p. 204) – species of *Uranotaenia* (*Pseudoficalbia*).

jinjaensis Edwards, 1941 (p. 293) – subspecies of *Culex* (*Oculeomyia*) *aurantapex* Edwards, 1914b [original combination: *Culex* (*Culex*) *aurantapex* variety *jinjaensis*]; subspecific status by Harbach & Howard, 2007.

joanae Muspratt, 1955 (p. 185) – subspecies of *Culex* (*Culex*) *striatipes* Edwards, 1941.

jocasta Komp & Rozeboom, 1951 (p. 128) – species of *Culex* (*Melanoconion*).

jocosa (Dyar & Knab, 1908) (p. 64) – species of *Wyeomyia* (*Dendromyia*) [original combination: *Prosopolepis jocosa*].
Synonyms: *prolepidis* Dyar & Knab, 1919; *favor* Dyar & Núñez Tovar, 1928.

johni Cova Garcia, Pulido F. & Escalante de Ugueto, 1979 (p. 21) – species of *Culex* (*Neoculex*).

johnnyi Duret, 1968d (p. 17) – species of *Culex* (*Melanoconion*).

johnsoni Galindo & Mendez, 1961 (p. 1) – species of *Culex* (*Melanoconion*).

johnsoni (Laffoon, 1946) (p. 235) – species of *Verrallina* (*Neomacleaya*) [original combination: *Aedes* (*Aedes*) *johnsoni*].

johnstonii (Grabham, 1905) (p. 410) – species of *Psorophora* (*Janthinosoma*) [original combination: *Janthinosoma johnstonii*].
Synonyms: *coffini* Dyar & Knab, 1906a; *schwarzi* Dyar & Knab, 1906a.

johorensis (Reinert, 1974) (p. 58) – species of *Verrallina* (*Neomacleaya*) [original combination: *Aedes* (*Verrallina*) *johorensis*].

joloensis (Ludlow, 1904b) (p. 236) – species of *Armigeres* (*Armigeres*) [original combination: *Desvoidea fusca* variety *joloensis*]; specific status by Stone & Thurman, 1958.

jonistes Dyar, 1920d (p. 76) – synonym of *Culex* (*Melanoconion*) *ybarmis* Dyar, 1920d; synonymy by Komp, 1935.

jorgi Carpintero & Leguizamón, 2000 (p. 20) – species of *Aedes* (*Ochlerotatus*).

josephinae Marks, 1958 (p. 58) – species of *Aedes* (*Finlaya*).

josephineae Baisas, 1935d (p. 172) – species of *Culex* (*Lophoceraomyia*).

josiahae Huang, 1988a (p. 155) – species of *Aedes* (*Stegomyia*).

joucouri da Cunha Ramos & Brunhes, 2004 (p. 379) – species of *Uranotaenia* (*Uranotaenia*).

joyoni Brunhes, 1977a (p. 140) – species of *Orthopodomyia*.

jubifer Komp & Brown, 1935 (p. 254) – species of *Culex* (*Melanoconion*).

jubilator Dyar & Knab, 1907c (p. 201) – synonym of *Culex* (*Culex*) *declarator* Dyar & Knab, 1906d; synonymy established before Knight & Stone, 1977.

judaicus Edwards, 1926c (p. 127) – species of *Culex* (*Neoculex*) [original combination: *Culex apicalis* variety *judaicus*]; specific status by Mattingly, 1953b.

judithae Zavortink, 1969 (p. 28) – species of *Anopheles* (*Anopheles*).

jugorum Villeneuve, 1919 (p. 58) – synonym of *Aedes* (*Ochlerotatus*) *pullatus* (Coquillett, 1904c) [original combination: *Culex jugorum*]; synonymy established before Knight & Stone, 1977.

jugraensis (Leicester, 1908) (p. 77) – species of *Armigeres* (*Armigeres*) [original combination: *Desvoidya jugraensis*].

jugraensis (Leicester, 1908) (p. 109) – species of *Aedes* (*Hulecoeteomyia*) [original combination *Hulecocteomyia* [*sic*] *jugraensis*].

junlianensis Lei, 1996 (p. 1) – synonym of *Anopheles* (*Anopheles*) *pullus* Yamada, 1937; synonymy with *yatsushiroensis* Miyazaki (= *pullus*) by Ma *et al.*, 2000.

juppi Dumas, Atyame, Malcolm, Le Goff, Unal, Makoundou, Pasteur, Weill & Duron, 2016 (p. 807) – *nomen nudum, Culex*; status by Harbach (11 September 2017, http://mosquito-taxonomic-inventory.info/whats-new).

juppi McIntosh, 1973 (p. 262) – species of *Aedes* (*Ochlerotatus*).

juxtahirsuta Belkin, Heinemann & Page, 1970 (p. 220) – species of *Wyeomyia* (*Wyeomyia*).

juxtamansonia (Chagas, 1907) (p. 23) – species of *Coquillettidia* (*Rhynchotaenia*) [original combination: *Taeniorhynchus justa mansonia* [*sic*]].
Synonym: *hypocrindyna* Dyar, 1918d.

juxtapallidiceps Theobald, 1910c (p. 456) – synonym of *Culex* (*Lophoceraomyia*) *minutissimus* (Theobald, 1907) [original combination: *Melanoconion juxtapallidiceps*]; synonymy established before Knight & Stone, 1977.

Kk

kabaenensis Brug, 1939 (p. 108) – species of *Aedes* (*Geoskusea*).

kabwachensis Edwards, 1941 (p. 181) – subspecies of *Aedes* (*Aedimorphus*) *abnormalis* (Theobald, 1909b).

kaimosi (van Someren, 1946b) (p. 184) – species of *Toxorhynchites* (*Afrorhynchus*) [original combination: *Megarhinus kaimosi*].

kalabahensis Haga, 1925 (p. 43) – synonym of *Uranotaenia* (*Pseudoficalbia*) *bicolor* Leicester, 1908; synonymy by Peyton, 1977.

kalawara Stoker & Waktoedi Koesoemawinangoen, 1949 (p. 54) – subspecies of *Anopheles* (*Cellia*) *tessellatus* Theobald, 1901a [original combination: *Anopheles* (*Myzomyia*) *tessellatus* variety *kalawara*]; subspecific status by Harbach & Howard, 2007.

kana Tanaka, Mizusawa & Saugstad, 1979 (p. 249) – species of *Heizmannia* (*Heizmannia*).

kanaranus Barraud, 1924a (pp. 848, 850) – synonym of *Aedes* (*Bruceharrisonius*) *greenii* (Theobald, 1903a) [original combination: *Finlaya greenii* variety *kanaranus* [*sic*]]; synonymy by Harrison *et al.*, 1991.

kanarensis Edwards, 1934 (in Barraud, 1934) (p. 444) – species of *Aedes* (*Dendroskusea*).

kanayamensis Yamada, 1932 (p. 218) – synonym of *Culiseta* (*Culiseta*) *bergrothi* (Edwards, 1921b) [original combination: *Theobaldia kanayamensis*]; synonymy by Danilov, 1976.

kangi Lien, 1968d (p. 235) – synonym of *Culex* (*Culex*) *jacksoni* Edwards, 1934; synonymy by Sirivanakarn, 1976.

kanhsienensis Tung, 1955 (p. 309) – species of *Heizmannia* (*Heizmannia*).

kanyamwerima van Someren, 1951 (p. 4) – species of *Culex* (*Eumelanomyia*).

kappleri Bonne, 1923 (p. 122) – synonym of *Sabethes* (*Sabethes*) *amazonicus* Gordon & Evans, 1922; synonymy established before Knight & Stone, 1977.

kapretwae Edwards, 1941 (p. 162) – species of *Aedes* (*Albuginosus*).

karandalaensis (Wolfs, 1951) (p. 12) – species of *Coquillettidia* (*Coquillettidia*) [original combination: *Taeniorhynchus* (*Coquillettidia*) *karandalaensis*].

karatsuensis Mochizuki, 1913 (p. 28) – synonym of *Culex* (*Oculeomyia*) *bitaeniorhynchus* Giles, 1901a; synonymy established before Knight & Stone, 1977.

karooensis Muspratt, 1961 (p. 93) – species of *Aedes* (*Aedimorphus*) [original combination: *Aedes* (*Aedimorphus*) *eritreae karooensis*]; specific status by McIntosh, 1975.

kartalae Brunhes, 1977a (p. 140) – subspecies of *Culex* (*Culex*) *comorensis* Brunhes, 1977a.

karwari (Barraud, 1924a) (p. 865) – species of *Aedes* (*Kenknightia*) [original combination: *Finlaya dissimilis* variety *karwari*]; specific status by Reinert, 1990.

karwari (James, 1903) (in Theobald, 1903a) (p. 102) – species of *Anopheles* (*Cellia*) [original combination: *Nyssorhynchus karwari*].

Synonym: *nigrans* Stanton, 1912.

kasachstanicus Gutsevich, 1962 (p. 886) – species of *Aedes* (*Ochlerotatus*).

katherinensis Woodhill, 1949 (p. 141) – species of *Aedes* (*Stegomyia*) [original combination: *Aedes scutellaris katherinensis*]; specific status by Lee *et al.*, 1987.

katibasensis Miyagi & Toma, 2012 (in Miyagi *et al.*, 2012) (p. 98) – species of *Topomyia* (*Topomyia*).

kaviengensis Sirivanakarn, 1968 (p. 167) – species of *Culex* (*Lophoceraomyia*).

keefei King & Hoogstraal, 1946f (p. 311) – species of *Aedes* (subgenus uncertain).

keenani Galindo & Mendez, 1961 (p. 2) – synonym of *Culex* (*Melanoconion*) *educator* Dyar & Knab, 1906d; synonymy by Forattini & Sallum, 1993b.

kelabitensis Miyagi & Toma, 2010b (as *kelabitense*) (p. 354) – species of *Topomyia* (*Suaymyia*); emendation by Harbach (31 December 2010, http://mosquito-taxonomic-inventory.info/whats-new).

kelloggii Theobald, 1903i (p. 211) – synonym of *Culex* (*Culex*) *tarsalis* Coquillett, 1896; synonymy established before Knight & Stone, 1977.

kempi (Edwards, 1921c) (p. 72) – species of *Toxorhynchites* (*Toxorhynchites*) [original combination: *Megarhinus* (*Toxorhynchites*) *kempi*].

keniensis Evans, 1931a (p. 130) – species of *Anopheles* (*Cellia*) [original combination: *Anopheles marshalli* [*sic*] variety *keniensis*]; specific status by Evans, 1933.

keniensis van Someren, 1946a (p. 111) – species of *Aedes* (*Stegomyia*).

kennethi Muspratt, 1956 (p. 38) – species of *Aedes* (*Albuginosus*).

kenyae van Someren, 1946c (p. 2) – species of *Aedes* (*Stegomyia*) [original combination: *Aedes* (*Stegomyia*) *bambusae kenyae*]; specific status by Huang, 1997.

kermorganti Laveran, 1901a (p. 569) – synonym of *Aedes* (*Mucidus*) *alternans* (Westwood, 1835) [original combination: *Culex kermorganti*]; synonymy by Rageau & Hamon, 1957.

kerri del Ponte & Cerqueira, 1938 (p. 228) – synonym of *Wyeomyia flui* (Bonne-Wepster & Bonne, 1920); synonymy by Lourenço-de-Oliveira *et al.*, 1999.

kerri Duret, 1968a (p. 11) – synonym of *Culex* (*Melanoconion*) *phlabistus* Dyar, 1920d; synonymy by Sirivanakarn, 1983.

kesseli Belkin, 1962 (p. 200) – species of *Culex* (*Culex*).

kesseli Huang & Hitchcock, 1980 (p. 23) – species of *Aedes* (*Stegomyia*).

kesseli Ramalingam, 1987 (p. 56) – species of *Armigeres* (*Armigeres*).

keybergi Lips, 1960a (p. 49) – *nomen nudum*, *Anopheles*.

keyensis Buren, 1947 (p. 228) – synonym of *Aedes* (*Ochlerotatus*) *thelcter* Dyar, 1918e; synonymy by Thurman *et al.*, 1949.

khasiana Barraud, 1923e (p. 407) – synonym of *Aedes* (*Hulecoeteomyia*) *pallirostris* Edwards, 1922a [original combination: *Finlaya khasiana*]; synonymy by Tewari & Hiriyan, 1996.

khazani Edwards, 1922a (p. 265) – species of *Aedes* (*Phagomyia*).

khazani Edwards, 1922a (p. 286) – species of *Culex* (*Eumelanomyia*).

khmerus Klein, 1973 (p. 7) – synonym of *Verrallina* (*Neomacleaya*) *rara* (Delfinado, 1968) [original combination: *Aedes* (*Neomacleaya*) *khmerus*]; synonymy by Reinert, 1974.

kiangsiensis Tung, 1955 (p. 313) – species of *Aedes* (*Phagomyia*).

kiangsuensis Xu & Feng, 1975 (p. 86) – synonym of *Anopheles* (*Anopheles*) *lesteri* Baisas & Hu, 1936; synonymy by Yang *et al.*, 1983; Qu & Xhu, 2008.

kibena Peters, 1955b (p. 100) – synonym of *Anopheles* (*Cellia*) *gibbinsi* Evans, 1935; synonymy by Gillies & de Meillon, 1968.

kilara van Someren, 1951 (p. 5) – species of *Culex* (*Eumelanomyia*).

killertonis (Huang, 1968a) (p. 31) – species of *Verrallina* (*Verrallina*) [original combination: *Aedes* (*Verrallina*) *killertonis*].

kinabaluensis Ramalingam, 1972 (p. 460) – species of *Armigeres* (*Armigeres*).

kinabaluensis Sirivanakarn, 1976 (p. 91) – species of *Culex* (*Oculeomyia*).

kingi Christophers, 1923a (p. 1011) – species of *Anopheles* (*Cellia*).

kingi Lee, 1946 (p. 256) – species of *Tripteroides* (*Rachisoura*).

kingianus Edwards, 1922d (p. 86) – species of *Culex* (*Eumelanomyia*).

kingii Theobald, 1908c (p. 258) – synonym of *Aedes* (*Aedimorphus*) *ochraceus* (Theobald, 1901a) [original combination: *Mimeteculex kingii*]; synonymy established before Knight & Stone, 1977.

kingii (Theobald, 1913d) (species) (p. 601) – subspecies of *Culex* (*Culex*) *argenteopunctatus* (Ventrillon, 1905a) [original combination: *Heptaphlebomyia kingii*]; subspecific status by Edwards, 1941.

kinoshitai Koidzumi, 1917 (p. 135) – synonym of *Anopheles* (*Cellia*) *tessellatus* Theobald, 1901a; synonymy established before Knight & Stone, 1977.

kiriensis Klein & Sirivanakarn, 1970 (p. 583) – species of *Culex* (*Eumelanomyia*).

kiriromi (Klein, 1969) (p. 215) – species of *Mimomyia* (*Ingramia*) [original combination: *Ficalbia* (*Ravenalites*) *kiriromi*].

kirkpatricki Stackelberg, 1927 (p. 148) – synonym of *Culex* (*Lasiosiphon*) *adairi* Kirkpatrick, 1926 [*nomen novum* for *pluvialis* Kirkpatrick, 1925]; synonymy established before Knight & Stone, 1977.

kirkpatriki Edwards, 1926c (p. 123) – synonym of *Culex* (*Lasiosiphon*) *adairi* Kirkpatrick, 1926 [*nomen novum* for *pluvialis* Kirkpatrick, 1925]; synonymy established before Knight & Stone, 1977.

kishenehn Harbach &Greenwalt, 2012 (p. 26) – fossil species of *Culiseta* Felt, 1904.

kivuensis Edwards, 1941 (p. 132) – species of *Aedes* (*Stegomyia*) [original combination: *Aedes* (*Stegomyia*) *subargenteus* variety *kivuensis*]; specific status by Mattingly, 1953a.

kleini Reinert, 1973e (p. 18) – species of *Aedes* (*Bothaella*).

kleini Rueda, 2005 (p. 13) – species of *Anopheles* (*Anopheles*).

klossi (Edwards, 1921c) (p. 72) – species of *Toxorhynchites* (*Toxorhynchites*) [original combination: *Megarhinus* (*Toxorhynchites*) *klossi*].

klotsi Matheson, 1933 (p. 69) – synonym of *Aedes* (*Ochlerotatus*) *melanimon* Dyar, 1924e; synonymy by Carpenter & LaCasse, 1955.

knabi (Coquillett, 1906a) (p. 183) – species of *Aedes* (*Protomacleaya*) [original combination: *Culex knabi*].

knabi Lane & Cerqueira, 1942 (p. 600) – species of *Wyeomyia* (subgenus uncertain, determined by Motta & Lourenço-de- Oliveira, 1995).

knighti Baisas & Ubaldo-Pagayon, 1953 (p. 139) – species of *Tripteroides* (*Tripteroides*).

knighti Stone & Bohart, 1944 (p. 210) – species of *Aedes* (*Finlaya*).

kobayashii Nakata, 1956 (p. 135) – species of *Aedes* (*Bruceharrisonius*) [original combination: *Aedes* (*Finlaya*) *kobayashii*]; ? = *Aedes* (*Finlaya*) *alektorovi* Stackelberg, 1943 (Tanaka, 2002).

kochi Dönitz, 1901 (p. 36) – species of *Anopheles* (*Cellia*).
Synonyms: *ocellatus* Theobald, 1901a; *flava* Ludlow, 1908a; *halli* James, 1910a.

kochi (Dönitz, 1901) (p. 38) – species of *Aedes* (*Finlaya*) [original combination: *Culex kochi*].

kogievnikovi Shingarev, 1927 (p. 547) – synonym of *Culiseta* (*Culiseta*) *bergrothi* (Edwards, 1921b) [original combination: *Theobaldia kogievnikovi*]; synonymy by Stone *et al.*, 1959.

kohkutensis Mattingly, 1958a (p. 55) – species of *Aedes* (*Cancraedes*).

kokhani Vythilingam, Jeffery & Harbach, 2007 (in Vythilingam *et al.*, 2007) (p. 464) – species of *Anopheles* (*Cellia*).

kolambuganensis Baisas, 1932 (as *kulambuganensis*) (p. 252) – species of *Anopheles* (*Cellia*).

kolhapuriensis Sathe & Girhe, 2002 (p. 197) – *nomen nudum*, *Aedes*; unavailable name [published after 1999, fails to conform to Articles 16.4 and 72.3 of the *Code*, which require fixation of a name-bearing type and a statement naming the collection in which it is deposited (International Commission on Zoological Nomenclature, 1999)]; status by Harbach (22 November 2016, http://mosquito-taxonomic-inventory.info/whats-new).

koli Peyton & Klein, 1970 (p. 246) – species of *Uranotaenia* (*Pseudoficalbia*).

koliensis Owen, 1945 (p. 236) – species of *Anopheles* (*Cellia*).

kompi Edwards, 1930a (p. 288) – species of *Anopheles* (*Stethomyia*) [original combination: *Anopheles nimbus* variety *kompi*]; specific status by ? Edwards, 1932a.

kompi Valencia, 1973 (p. 81) – species of *Culex* (*Carrollia*).

kompi Vargas & Downs, 1950 (p. 167) – species of *Aedes* (*Protomacleaya*).

komponga (Klein, 1973) (p. 8) – species of *Verrallina* (*Neomacleaya*) [original combination: *Aedes* (*Neomacleaya*) *kompongus*].

konderi Galvão & Damasceno, 1942a (p. 115) – species of *Anopheles* (*Nyssorhynchus*).

koreicoides Sasa, Kano & Hayashi, 1950 (p. 627) – species of *Aedes* (subgenus uncertain).

koreicus (Edwards, 1917) (p. 212) – species of *Aedes* (*Hulecoeteomyia*) [original combination: *Ochlerotatus* (*Finlaya*) *koreicus*].

koreicus Yamada & Watanabe, 1918 (p. 206) – species of *Anopheles* (*Anopheles*).
Synonyms: *punctibasis* Edwards, 1921d; *edwardsi* Yamada, 1924; *hisaoe* Tsuchimoto, 1948.

kosiensis Coetzee, Segerman & Hunt, 1987 (p. 26) – species of *Anopheles* (*Cellia*).

kotiensis Barraud, 1934 (p. 161) – subspecies of *Aedes* (*Phagomyia*) *gubernatoris* (Giles, 1901a) [original combination:

Aedes (*Finlaya*) *gubernatoris* variety *kotiensis*]; subspecific status by Harbach & Howard, 2007.

koumbai Galliard, 1931c (p. 514) – synonym of *Culex* (*Eumelanomyia*) *rima* Theobald, 1901d [original combination: *Culex rima* variety *koumbai*]; synonymy by Edwards, 1941.

kounoupi Brullé, 1833 (p. 289) – synonym of *Aedes* (*Stegomyia*) *aegypti* (Linnaeus, 1762) [original combination: *Culex kounoupi*]; synonymy established before Knight & Stone, 1977.

kowiroensis Sirivanakarn, 1968 (p. 133) – species of *Culex* (*Lophoceraomyia*).

kraussi Grjebine, 1953a (p. 459) – species of *Uranotaenia* (*Pseudoficalbia*).

krishnai Sathe & Japtag, 2012 (p. 71) – *nomen nudum*, *Anopheles*; unavailable name [the name fails to conform to Articles 13.1 and 16.4 of the *Code* (International Commission on Zoological Nomenclature, 1999); status by Harbach (27 June 2012, http://mosquito-taxonomic-inventory.info/whats-new).

krombeini Huang, 1975 (p. 345) – species of *Aedes* (*Stegomyia*).

krymmontanus Alekseev, 1989 (p. 173) – species of *Aedes* (*Rusticoidus*).

kuchingensis Edwards, 1915c (p. 283) – species of *Armigeres* (*Armigeres*).

Synonym: *nongpohensis* Barraud, 1927b.

kuhnsi King & Hoogstraal, 1955 (p. 1) – species of *Culex* (*Lophoceraomyia*).

Synonym: *mercedesae* Baisas, 1974.

kukenan Anduze, 1942b (p. 558) – species of *Culex* (*Microculex*).

kumasii Chalmers, 1900 (p. 1262) – synonym of *Anopheles* (*Cellia*) *funestus* Giles, 1900b; synonymy established before Knight & Stone, 1977.

kumbae Chwatt, 1948 (p. 184) – subspecies of *Aedes* (*Bifidistylus*) *boneti* Gil Collado, 1936.

kummi Edwards, 1930a (p. 298) – species of *Aedes* (*Pseudarmigeres*).

kummi Edwards, 1939b (p. 121) – species of *Orthopodomyia*.

kummi Komp & Rozeboom, 1951 (p. 130) – species of *Culex* (*Melanoconion*).

kummi Lane & Cerqueira, 1942 (p. 582) – species of *Wyeomyia* (*Cruzmyia*).

kunmingensis Dong & Wang, 1985 (p. 117) – species of *Anopheles* (*Anopheles*).

kusaiensis Bohart, 1957 (p. 69) – species of *Culex* (*Lophoceraomyia*).

kweiyangensis Yao & Wu, 1944 (p. 1) – species of *Anopheles* (*Anopheles*).

kyondawensis Abraham, 1947 (p. 173) – species of *Anopheles* (*Baimaia*).

kyotoensis Yamaguti & LaCasse, 1952 (p. 1) – species of *Culex* (*Culiciomyia*).

Ll

labeculosus Coquillett, 1905 (p. 116) – synonym of *Aedes* (*Ochlerotatus*) *camptorhynchus* (Thomson, 1869) [original combination: *Culex labeculosus*]; synonymy established before Knight & Stone, 1977.

labesba Howard, Dyar & Knab, 1913 (fig. 282) – species of *Wyeomyia* (*Wyeomyia*).

labradorensis Dyar & Shannon, 1925d (p. 78) – synonym of *Aedes* (*Ochlerotatus*) *hexodontus* Dyar, 1916b; synonymy by Vockeroth, 1954.

labranchiae Falleroni, 1926 (p. 564) – species of *Anopheles* (*Anopheles*) [original combination: *Anopheles claviger* variety *labranchiae*]; specific status by Bates, 1940.

Synonyms: *pergusae* Missiroli, 1935; *sicaulti* Roubaud, 1935a.

lacani Grjebine, 1953a (p. 483) – species of *Anopheles* (*Cellia*).

lacani Rickenbach & Eouzan, 1970 (p. 135) – species of *Eretmapodites*.

lacertosus Komp & Rozeboom, 1951 (p. 123) – species of *Culex* (*Melanoconion*).

lachrimans Dyar & Knab, 1909d (p. 259) – synonym of *Culex* (*Culex*) *quinquefasciatus* Say, 1823 [*nomen novum* for *aikenii* Dyar & Knab, 1908]; synonymy established before Knight & Stone, 1977.

lactator Dyar & Knab, 1906d (p. 209) – species of *Culex* (*Phenacomyia*).

lacteus Knight, 1946 (p. 275) – species of *Aedes* (*Downsiomyia*).

lacustris (Edwards, 1935a) (p. 134) – species of *Mimomyia* (*Mimomyia*) [original combination: *Ficalbia* (*Mimomyia*) *lacustris*].

lacuum Edwards, 1922d (p. 97) – species of *Armigeres* (*Armigeres*).

laffooni Baisas & Ubaldo-Pagayon, 1953 (p. 168) – subspecies of *Tripteroides* (*Tripteroides*) *powelli* (Ludlow, 1909c).

laffooni Belkin, 1962 (p. 261) – species of *Culex* (*Lophoceraomyia*).

laffooni Knight & Rozeboom, 1946 (p. 94) – species of *Aedes* (*Scutomyia*).

laffosseae da Cunha Ramos & Brunhes, 2004 (p. 193) – species of *Uranotaenia* (*Pseudoficalbia*).

laguna Arnell & Nielsen, 1972 (p. 12) – species of *Aedes* (*Jarnellius*).

lagunensis Baisas, 1935c (p. 70) – synonym of *Uranotaenia* (*Pseudoficalbia*) *bicolor* Leicester, 1908; synonymy by Peyton, 1977.

lahillei Bachmann & Casal, 1962 (p. 267) – species of *Culex* (*Culex*).

lairdi Belkin, 1962 (p. 257) – species of *Culex* (*Lophoceraomyia*).

lakei Sirivanakarn, 1968 (p. 120) – species of *Culex* (*Lophoceraomyia*).

lamberti Ventrillon, 1904 (p. 552) – synonym of *Aedes* (*Stegomyia*) *aegypti* (Linnaeus, 1762); synonymy by Rodhain & Boutonnier, 1985.

lamborni Edwards, 1923a (p. 398) – species of *Aedes* (*Bifidistylus*).

lambrechti van Someren, 1971 (p. 21) – species of *Aedes* (*Skusea*).

lamellata (Bonne-Wepster & Bonne, 1920) (p. 168) – species of *Wyeomyia* (*Hystatomyia*) [original combination: *Hystatomyia lamellata*].

lamelliferus Bohart & Ingram, 1946b (p. 29) – species of *Aedes* (*Lorrainea*).

lamentator Dyar & Knab, 1906d (p. 219) – synonym of *Culex* (*Culex*) *secutor* Theobald, 1901c; synonymy established before Knight & Stone, 1977.

lampangensis Sirivanakarn, 1973b (p. 113) – species of *Culex* (*Culiciomyia*).

lampangensis Thurman, 1959 (p. 52) – species of *Hodgesia*.

lampropus (Howard, Dyar & Knab, 1913) (fig. 44) – species of *Trichoprosopon* [original combination: *Lesticocampa lampropus*].

laneanus Corrêa & Cerqueira, 1944 (p. 112) – species of *Anopheles* (*Kerteszia*).

lanei (Antunes, 1937) (p. 71) – species of *Trichoprosopon* [original combination: *Goeldia lanei*].

lanei Cerqueira, 1961 (p. 46) – species of *Sabethes* (*Sabethes*).

lanei Galvão & Franco do Amaral, 1938 (p. 8) – species of *Anopheles* (*Nyssorhynchus*).

lanei Martinez & Prosen, 1953a (p. 4) – species of *Uranotaenia* (*Uranotaenia*).

lanei Oliveira Coutinho & Forattini, 1962 (p. 210) – species of *Culex* (*Microculex*).

lanei Shannon & Cerqueira, 1943 (p. 135) – species of *Psorophora* (*Janthinosoma*).

langata van Someren, 1946c (p. 1) – species of *Aedes* (*Stegomyia*).

laniger (Wiedemann, 1820) (p. 9) – species of *Aedes* (*Mucidus*) [original combination: *Culex laniger*].

lankaensis Stone & Knight, 1958 (p. 69) – synonym of *Verrallina* (*Neomacleaya*) *ceylonica* (Edwards, 1917) [original combination: *Aedes* (*Aedes*) *lankaensis*]; originally proposed as a replacement name for *ceylonicus* Edwards, 1917, *non ceylonica* Theobald (1910c), when both were included in *Aedes* Meigen.

lanyuensis Lien, 1968a (p. 4) – species of *Orthopodomyia*.

laoagensis Knight, 1946 (p. 276) – species of *Aedes* (*Downsiomyia*).

laoensis Toma & Miyagi, 2003 (p. 169) – species of *Armigeres* (*Armigeres*).

laplantei (Hamon, Adam & Mouchet, 1955) (p. 862) – species of *Culex* (*Eumelanomyia*) [original combination: *Neoculex laplantei*].

lasaensis Meng, 1962 (p. 159) – species of *Aedes* (*Ochlerotatus*).

lasiopalpis Sirivanakarn, 1977a (p. 123) – species of *Culex* (*Lophoceraomyia*).

lassalli (Bonne-Wepster & Bonne, 1921) (p. 8) – species of *Wyeomyia* (*Phoniomyia*) [original combination: *Dyarina lassalli*].

latens Sallum & Peyton, 2005 (in Sallum *et al.*, 2005b) (p. 160) – species of *Anopheles* (*Cellia*).

lateralis Ludlow, 1905c (p. 385) – species of *Uranotaenia* (*Uranotaenia*) [original combination: *Uranotaenia caeruleocephala* variety *lateralis*]; elevated from synonymy with *atra* Theobald, 1905f by Stone, 1957b. Synonyms: *cancer* Leicester, 1908; *ceylonica* Theobald, 1910c; *propria* Taylor, 1914c; *cairnsensis* Taylor, 1919.

lateralis Meigen, 1818 (p. 5) – synonym of *Aedes* (*Dahliana*) *geniculatus* (Olivier, 1791) [original combination: *Culex lateralis*]; synonymy established before Knight & Stone, 1977. **Note**: this nominal taxon is the senior secondary homonym of *lateralis* Theobald, 1905f.

lateralis Petrocchi, 1927 (in Shannon & del Ponte, 1927) (p. 726) – species of *Wyeomyia* (*Nunezia*).

lateralis Theobald, 1905f (p. 110) – synonym of *Aedes* (*Leptosomatomyia*) *aurimargo* Edwards, 1922d; synonymy established before Knight & Stone, 1977. **Note**: this nominal taxon is the junior secondary homonym of *lateralis* Meigen, 1818.

laternaria Coquillett, 1906a (p. 184) – synonym of *Aedes* (*Protomacleaya*) *insolitus* (Coquillett, 1906a) [original combination: *Verrallina laternaria*]; synonymy by Schick, 1970a.

lateropunctata Theobald, 1907 (p. 458) – synonym of *Culex* (*Culex*) *mollis* Dyar & Knab, 1906b; synonymy established before Knight & Stone, 1977.

laticinctus Edwards, 1913a (p. 49) – species of *Culex* (*Culex*).

laticlasper Galindo & Blanton, 1954 (p. 246) – species of *Culex* (*Culex*).

latifoliatus Delfinado, 1966a (p. 131) – species of *Culex* (*Eumelanomyia*).

latipennis (Delfinado, 1967) (p. 26) – species of *Verrallina* (*Neomacleaya*) [original combination: *Aedes* (*Neomacleaya*) *latipennis*].

latispinus Gong & Ji, 1989 (p. 482) – species of *Tripteroides* (*Tripteroides*).

latisquama (Coquillett, 1906a) (p. 185) – species of *Culex* (*Tinolestes*) [original combination: *Tinolestes latisquama*].

latisquama (Edwards, 1927b) (p. 354) – species of *Tripteroides* (*Rachisoura*) [original combination: *Rachionotomyia latisquama*].

lativittatus Coquillett, 1906c (p. 109) – synonym of *Aedes* (*Ochlerotatus*) *dorsalis* (Meigen, 1830b) [original combination: *Culex lativittatus*]; synonymy by Wood *et al.*, 1979.

latus Dobrotworsky, 1956 (p. 108) – species of *Culex* (*Neoculex*).

laureli Baisas, 1935d (p. 176) – species of *Culex* (*Eumelanomyia*).

laurenti Newstead, 1907 (in Newstead *et al.*, 1907) (p. 24) – synonym of *Culex* (*Culex*) *antennatus* (Becker, 1903); synonymy by Hopkins, 1936.

lauriei (Carter, 1920) (p. 623) – species of *Aedes* (subgenus uncertain) [original combination: *Ochlerotatus laurei*]; emendation by Edwards, 1924b.

lavatae Stone & Bohart, 1944 (p. 220) – species of *Culex* (*Lophoceraomyia*).

lavieri Doucet, 1950 (p. 40) – species of *Uranotaenia* (*Pseudoficalbia*).

lavieri Larrousse, 1925 (p. 68) – synonym of *Culex* (*Maillotia*) *hortensis* Ficalbi, 1889a; synonymy by Callot, 1944b.

lawtoni Baisas, 1946b (p. 27) – species of *Zeugnomyia*.
lazarensis Felt & Young, 1904 (p. 312) – synonym of *Aedes* (*Ochlerotatus*) *communis* (de Geer, 1776) [original combination: *Culex lazarensis*]; synonymy established before Knight & Stone, 1977.
leberi Boreham, 1970 (p. 384) – species of *Mansonia* (*Mansonia*).
lebiedi da Cunha Ramos & Brunhes, 2004 (p. 381) – species of *Uranotaenia* (*Uranotaenia*).
ledgeri Huang, 1981 (p. 96) – species of *Aedes* (*Stegomyia*).
leei Berlin, 1969a (p. 110) – species of *Aedes* (*Howardina*).
leei King & Hoogstraal, 1955 (p. 4) – species of *Culex* (*Lophoceraomyia*).
leei Peters, 1959 (p. 141) – species of *Tripteroides* (*Rachisoura*).
leei Stone & Barreto, 1969 (p. 145) – species of *Galindomyia*.
leei (Wharton, 1947) (p. 60) – species of *Malaya* [original combination: *Harpagomyia leei*].
leesoni Edwards, 1932b (p. 562) – species of *Aedes* (*Aedimorphus*).
Synonym: *verna* Lewis, 1944.
leesoni Evans, 1931b (p. 546) – species of *Anopheles* (*Cellia*) [original combination: *Anopheles funestus leesoni*]; specific status by Evans & Leeson, 1935.
legoffi da Cunha Ramos & Brunhes, 2004 (p. 305) – species of *Uranotaenia* (*Pseudoficalbia*).
lehcharlesi Miyagi & Toma, 2008 (p. 163) – species of *Topomyia* (*Suaymyia*).
leiboensis Chu, 1981 (p. 273) – species of *Uranotaenia* (*Pseudoficalbia*).
leicesteri Edwards, 1913c (p. 239) – synonym of *Uranotaenia* (*Pseudoficalbia*) *bicolor* Leicester, 1908; synonymy established before Knight & Stone, 1977.
leicesteri (Edwards, 1917) (p. 222) – species of *Verrallina* (*Neomacleaya*) [original combination: *Aedes leicesteri*].
leicesteri Theobald, 1904c (p. 36) – species of *Toxorhynchites* (*Toxorhynchites*).
leilae (King & Hoogstraal, 1947b) (p. 121) – species of *Verrallina* (*Verrallina*) [original combination: *Aedes* (*Aedes*) *leilae*].
lemmonae Thurman, 1959 (p. 58) – synonym of *Orthopodomyia anopheloides* (Giles, 1903); synonymy established before Knight & Stone, 1977.
lemniscata Harbach & Greenwalt, 2012 (p. 31) – fossil species of *Culiseta* Felt, 1904.
lemuriae Ribeiro, 2004 (p. 254) – species of *Toxorhynchites* (*Afrorhynchus*).
leonardi Belkin, 1962 (p. 245) – species of *Culex* (*Neoculex*).
leonis Colless, 1958 (p. 476) – species of *Aedes* (*Downsiomyia*) [original combination: *Aedes* (*Finlaya*) *niveus leonis*]; specific status by Harrison *et al.*, 1991.
leontiniae Brèthes, 1910 (p. 471) – synonym of *Wyeomyia* (*Menolepis*) *leucostigma* Lutz, 1904 [original combination: *Limatus leontiniae*]; synonymy established before Knight & Stone, 1977.
lepchanus (Barraud, 1923b) (p. 217) – species of *Aedes* (*Danielsia*) [original combination: *Finlaya lepchana*].
lepidocoxitus Dong, Zhou & Dong, 1995b (p. 281) – species of *Armigeres* (*Leicesteria*).
lepidonotus (Edwards, 1920a) (p. 132) – species of *Aedes* (*Rusticoidus*) [original combination: *Ochlerotatus lepidonotus*].
lepidotus Zavortink, 1973 (p. 17) – species of *Anopheles* (*Kerteszia*).
lepidus Cerqueira & Paraense, 1945 (p. 12) – species of *Aedes* (*Ochlerotatus*).
lepostenis Dyar, 1923b (p. 70) – synonym of *Culex* (*Culex*) *mollis* Dyar & Knab, 1906b; synonymy by Komp, 1935.
leprincei Dyar & Knab, 1907c (p. 202) – synonym of *Culex* (*Melanoconion*) *erraticus* (Dyar & Knab, 1906d); synonymy by King & Bradley, 1937.
leptolabis Edwards, 1936 (p. 51) – species of *Aedes* (*Aedimorphus*).
leptomeres Theobald, 1903a (p. 38) – synonym of *Anopheles* (*Cellia*) *fluviatilis* James, 1902 [original combination: *Myzomyia leptomeres*]; synonymy established before Knight & Stone, 1977.
lerozeboomi Reinert, 1990 (p. 33) – species of *Aedes* (*Kenknightia*).
lesnei Séguy, 1920 (p. 328) – synonym of *Aedes* (*Ochlerotatus*) *sticticus* (Meigen, 1838) [original combination: *Ochlerotatus lesnei*]; synonymy established before Knight & Stone, 1977.
lesteri Baisas & Hu, 1936 (p. 229) – species of *Anopheles* (*Anopheles*) [original combination: *Anopheles hyrcanus* variety *lesteri*]; specific status by Reid, 1953.
Synonyms: *paraliae* Sandosham, 1959; *anthropophagus* Xu & Feng, 1975;.
letabensis Lambert & Coetzee, 1982 (p. 328) – species of *Anopheles* (*Cellia*).
letifer Sandosham, 1944 (p. 129) – species of *Anopheles* (*Anopheles*).
leucacanthus Loew, 1873 (p. 1) – synonym of *Aedes* (*Ochlerotatus*) *pulcritarsis* (Rondani, 1872) [original combination: *Culex leucacanthus*]; synonymy established before Knight & Stone, 1977.
leucarthrius (Speiser, 1909) (p. 43) – species of *Aedes* (*Aedimorphus*) [original combination: *Culex leucarthrius*].
leucocelaenus (Dyar & Shannon, 1924d) (p. 484) – species of *Haemagogus* (*Conopostegus*) [*nomen novum* for *leucomelas* Lutz, 1904; original combination: *Aedes leucocelaenus*].
Synonym: *leucomelas* Lutz, 1904.
leucocnemis Martini, 1931a (p. 214) – species of *Psorophora* (*Grabhamia*).
leucogrammus Loew, 1874 (p. 413) – synonym of *Culiseta* (*Allotheobaldia*) *longiareolata* (Macquart, 1838a) [original combination: *Culex leucogrammus*]; synonymy established before Knight & Stone, 1977.
leucomelas Lutz, 1904 (in Bourroul, 1904) (pp. 47, 66) – synonym of *Haemagogus* (*Conopostegus*) *leucocelaenus* (Dyar & Shannon, 1924d); synonymy by Dyar & Shannon, 1924d.

leucomelas (Meigen, 1804) (p. 3) – species of *Aedes* (*Ochlerotatus*) [original combination: *Culex leucomelas*].
Synonym: *salinellus* Edwards, 1921b.
leucomeres (Giles, 1904a) (p. 367) – species of *Aedes* (*Kenknightia*) [original combination: *Stegomyia leucomeres*].
leuconotips Dyar, 1920a (p. 24) – synonym of *Aedes* (*Ochlerotatus*) *hexodontus* Dyar, 1916b; synonymy established before Knight & Stone, 1977.
leucophoebus (Galindo, Carpenter & Trapido, 1953) (p. 535) – species of *Haemagogus* (*Conopostegus*) [original combination: *Aedes* (*Finlaya*) *leucophoebus*].
leucopisthepus Dyar & Knab, 1907c (p. 212) – synonym of *Wyeomyia* (*Wyeomyia*) *scotinomus* (Dyar & Knab, 1907c); synonymy established before Knight & Stone, 1977.
leucopleurus Rozeboom, 1946 (p. 588) – synonym of *Aedes* (*Bruceharrisonius*) *aureostriatus* (Doleschall, 1857); synonymy by Reinert, 1990.
leucopous Graham, 1909 (p. 88) – species of *Eretmapodites*.
leucoptera (Theobald, 1907) (p. 575) – species of *Uranotaenia* (*Uranotaenia*) [original combination: *Anisocheleomyia leucoptera*].
leucopus Dönitz, 1901 (p. 37) – synonym of *Anopheles* (*Cellia*) *annularis* van der Wulp, 1884; synonymy established before Knight & Stone, 1977.
leucopus (Dyar & Knab, 1906a) (p. 137) – species of *Johnbelkinia* [original combination: *Lesticocampa leucopus*].
leucopygos Eysell, 1903 (p. 299) – synonym of *Aedes* (*Aedes*) *cinereus* Meigen, 1818; synonymy established before Knight & Stone, 1977.
leucosphyrus Dönitz, 1901 (p. 37) – species of *Anopheles* (*Cellia*).
leucostigma Lutz, 1904 (in Bourroul, 1904) (pp. 48, 67) – species of *Wyeomyia* (*Menolepis*).
Synonym: *leontiniae* Brèthes, 1910.
leucotaeniatus (Komp, 1938) (p. 261) – species of *Haemagogus* (*Conopostegus*) [original combination: *Aedes leucotaeniatus*].
leucotarsis Lane, 1936b (p. 180) – subspecies of *Wyeomyia* (*Miamyia*) *hosautos* Dyar & Knab, 1907c [original combination: *Wyeomyia hosautos* variety *leucotarsis*]; subspecific status by Harbach & Howard, 2007.
leucotarsis Thurman, 1959 (p. 47) – species of *Topomyia* (*Suaymyia*).
leucotelus McCormack, 1919 (p. 29) – *nomen nudum*, *Culex*.
levicastilloi Grjebine, 1986 (p. 145) – species of *Mimomyia* (*Ingramia*).
levicastilloi Lane, 1945b (p. 209) – species of *Culex* (*Culex*).
Synonym: *tejerai* Cova García, 1962.
levicastilloi Levi Castillo, 1944 (p. 1) – subspecies of *Anopheles* (*Anopheles*) *pseudopunctipennis* Theobald, 1901c.
lewaldii Ludlow, 1904b (p. 233) – synonym of *Toxorhynchites* (*Toxorhynchites*) *amboinensis* (Doleschall, 1857) [original combination: *Megarhinus lewaldii*]; synonymy established before Knight & Stone, 1977.
lewelleni Starkey & Webb, 1946 (p. 179) – species of *Aedes* (*Finlaya*).
lewisi Ludlow, 1920a (p. 74) – species of *Anopheles* (*Anopheles*).
Synonym: *selengensis* Ludlow, 1920a.
lewisi Shannon, 1931 (p. 154) – synonym of *Anopheles* (*Stethomyia*) *thomasi* Shannon, 1933a [original combination: *Stethomyia lewisi*]; synonymy by Shannon, 1933a.
lewisi Ribeiro, 1991 (p. 53) – species of *Toxorhynchites* (*Toxorhynchites*).
liangshanensis Kang, Tan, Cao, Cheng, Yang & Huang, 1984 (p. 11) – species of *Anopheles* (*Anopheles*).
liberiensis Peters, 1955a (p. 85) – species of *Culex* (*Culiciomyia*).
ligator Dyar, 1924d (p. 123) – synonym of *Culex* (*Melanoconion*) *sursumptor* Dyar, 1924d; synonymy established before Knight & Stone, 1977.
lii Wu, 1936 (p. 46) – species of *Heizmannia* (*Heizmannia*).
lilii (Theobald, 1910c) (p. 160) – species of *Aedes* (*Stegomyia*) [original combination: *Stegomyia lilii*].
limacifer Komp, 1936c (p. 325) – species of *Culex* (*Melanoconion*).
limai da Fonseca & da Silva Ramos, 1940 (p. 384) – synonym of *Anopheles* (*Anopheles*) *mediopunctatus* Lutz, 1903; synonymy established before Knight & Stone, 1977. **Note**: this nominal taxon is the primary junior homonym of *limai* Galvão & Lane, 1937b.
limai Galvão & Lane, 1937b (p. 227) – synonym of *Anopheles* (*Nyssorhynchus*) *albitarsis* Lynch Arribálzaga, 1878 [original combination: *Anopheles* (*Nyssorhynchus*) *albitarsis* variety *limai*]; synonymy by Lane, 1953. **Note**: this nominal taxon is the primary senior homonym of *limai* da Fonseca & da Silva Ramos, 1940.
limai Lane & Cerqueira, 1942 (p. 571) – species of *Wyeomyia* (*Miamyia*).
limosus King, 1932a (p. 330) – species of *Anopheles* (*Cellia*) [original combination: *Anopheles vagus* variety *limosus*]; specific status by Ramalingam, 1974.
lindesayi Giles, 1900a (as *lindesaii*) (p. 166) – species of *Anopheles* (*Anopheles*).
Synonym: *maculata* Theobald, 1910a.
lindneri Martini, 1931b (p. 118) – synonym of *Haemagogus* (*Haemagogus*) *spegazzinii* Brèthes, 1912; synonymy with *uriartei* Shannon & del Ponte, 1928 by Lane, 1939.
lindsayi Thurman, 1959 (p. 42) – species of *Topomyia* (*Topomyia*).
lindtropi Shingarev, 1927 (p. 545) – synonym of *Aedes* (*Stegomyia*) *cretinus* Edwards, 1921d; synonymy established before Knight & Stone, 1977.
lineatis (Skuse, 1889) (p. 1747) – species of *Coquillettidia* (*Coquillettidia*) [original combination: *Culex lineatis*].
lineatis Taylor, 1913 (p. 57) – synonym of *Aedes* (*Neomelaniconion*) *lineatopennis* (Ludlow, 1905a) [original

combination: *Pseudohowardina linealis*]; synonymy established before Knight & Stone, 1977.

lineata Ludlow, 1908a (p. 50) – synonym of *Anopheles* (*Cellia*) *annularis* van der Wulp, 1884 [original combination: *Chagasia* (?) *lineata*]; synonymy established before Knight & Stone, 1977. **Note**: this nominal taxon is the junior secondary homonym of *lineata* Lutz, 1905.

lineata Lutz, 1905 (p. 202) – synonym of *Anopheles* (*Stethomyia*) *nimbus* (Theobald, 1902b) [original combination: *Rhynchomyia lineata*]; synonymy established before Knight & Stone, 1977. **Note**: this nominal taxon is the senior secondary homonym of *lineata* Ludlow, 1908a.

lineata (Taylor, 1914d) (p. 191) – species of *Verrallina* (*Verrallina*) [original combination: *Lepidotomyia lineatus* [*sic*]].
Synonym: *ornata* Theobald, 1905f.

lineata (von Humboldt, 1819) (p. 340) – species of *Psorophora* (*Psorophora*) [original combination: *Culex lineatus*].
Synonym: *blanchardi* Surcouf & Gonzalez-Rincones, 1911.

lineatopennis (Ludlow, 1905a) (p. 133) – species of *Aedes* (*Neomelaniconion*) [original combination: *Taeniorhynchus lineatopennis*].
Synonym: *linealis* Taylor, 1913.

lineatus (Theobald, 1912a) (p. 92) – species of *Culex* (*Afroculex*) [original combination: *Pseudohowardina lineata*].
Synonym: *pulchrithorax* Edwards, 1914b.

linesi Marks, 1964a (p. 63) – species of *Aedes* (*Ochlerotatus*).

lini Lien, 1968d (p. 225) – synonym of *Culex* (*Eumelanomyia*) *okinawae* Bohart, 1953; synonymy by Sirivanakarn, 1972.

lipovskyi Belkin, 1950 (p. 256) – species of *Tripteroides* (*Tripteroides*).

lishanensis Lien, 1968a (p. 6) – synonym of *Culiseta* (*Culiseta*) *niveitaeniata* (Theobald, 1907); synonymy by Dobrotworsky, 1971.

listeri de Meillon, 1931b (p. 482) – species of *Anopheles* (*Cellia*).

listoni Giles, 1901b (p. 197) – synonym of *Anopheles* (*Cellia*) *culicifacies* Giles, 1901b; synonymy established before Knight & Stone, 1977.

listonii Liston, 1901 (p. 365) – synonym of *Anopheles* (*Cellia*) *fluviatilis* James, 1902; synonymy established before Knight & Stone, 1977.

lithoecetor Dyar & Knab, 1907c (p. 201) – synonym of *Aedes* (*Georgecraigius*) *fluviatilis* (Lutz, 1904); synonymy by Zavortink, 1972.

litoralis Bohart, 1946 (p. 43) – species of *Culex* (*Culex*).

litoralis Halcrow, 1957 (p. 133) – synonym of *Anopheles* (*Cellia*) *merus* Dönitz, 1902 [original combination: *Anopheles gambiae litoralis*]; synonymy by ? Stone *et al.*, 1959.

litoralis King, 1932a (p. 311) – species of *Anopheles* (*Cellia*).

litorea (Shute, 1928) (p. 186) – species of *Culiseta* (*Culicella*) [original combination: *Culicella morsitans* variety *litorea*]; specific status by Marshall & Staley, 1933.

litoreus Colless, 1958 (p. 475) – species of *Aedes* (*Downsiomyia*).

littlechildi (Edwards, 1930b) (p. 544) – species of *Tripteroides* (*Tripteroides*) [original combination: *Rachionotomyia littlechildi*].

littlechildi Taylor, 1933 (p. 370) – species of *Aedes* (*Macleaya*).

littleri (Taylor, 1914c) (p. 702) – species of *Culiseta* (*Austrotheobaldia*) [original combination: *Chrysoconops littleri*].

littoralis Barraud, 1927c (p. 551) – synonym of *Aedes* (*Aedimorphus*) *trimaculatus* (Theobald, 1905f) [original combination: *Ochlerotatus littoralis*]; synonymy by Barraud, 1934.

litwakae Harbach, 1985 (p. 255) – species of *Culex* (*Culex*).

litwakae Reinert, 1990 (p. 39) – species of *Aedes* (*Kenknightia*).

lividocostalis Graham, 1910 (p. 269) – synonym of *Culex* (*Culex*) *decens* Theobald, 1901d; synonymy established before Knight & Stone, 1977.

lloreti Gil Collado, 1936 (pp. 313, 318) – species of *Anopheles* (*Cellia*).

lloydi Unti, 1941 (p. 10) – synonym of *Anopheles* (*Nyssorhynchus*) *strodei* Root, 1926 [original combination: *Anopheles* (*Nyssorhynchus*) *strodei* variety *lloydi*]; synonymy by Faran, 1980.

loashanensis Chu, 1980 (p. 238) – *lapsus* of *loshanensis* Wang & Feng, 1965.

locuples Robineau-Desvoidy, 1827 (p. 412) – synonym of *Sabethes* (*Sabethes*) *cyaneus* (Fabricius, 1805); synonymy established before Knight & Stone, 1977.

loewi Giebel, 1862 (p. 317) – synonym of *Toxorhynchites* (*Toxorhynchites*) *brevipalpis* Theobald, 1901a [original combination: *Culex loewi*]; name suppressed (International Commission on Zoological Nomenclature, 1982: Opinion 1213).

loi Lien, 1968b (p. 98) – synonym of *Aedes* (*Luius*) *fengi* (Edwards, 1935); synonymy by Lu & Ji, 1997.

lokojensis Service, 1959b (p. 244) – species of *Aedes* (*Catageiomyia*).

longefurcatus Becker, 1903 (p. 68) – synonym of *Culex* (*Culex*) *pipiens* Linnaeus, 1758; synonymy established before Knight & Stone, 1977.

longfieldae Edwards, 1928a (p. 283) – synonym of *Sabethes* (*Sabethes*) *amazonicus* Gordon & Evans, 1922; synonymy by Lane & Cerqueira, 1942.

longiareolata (Macquart, 1838a) (p. 38) – species of *Culiseta* (*Allotheobaldia*) [original combination: *Culex longiareolatus*].
Synonyms: *spathipalpis* Rondani, 1872; *leucogrammus* Loew, 1874; *serratipes* Becker, 1908; *marocanus* D'Anfreville, 1916.

longicornis Grjebine, 1986 (p. 149) – species of *Mimomyia* (*Ingramia*).

longicornis Sirivanakarn, 1976 (p. 81) – species of *Culex* (*Oculeomyia*).

longifilamentus Su & Zhang, 1988 (p. 11) – species of *Aedes* (*Ochlerotatus*).

longiforceps Edwards, 1929 (in Paine & Edwards, 1929) (p. 315) – species of *Aedes* (*Geoskusea*).
longifurcatus Theobald, 1910a (p. 19) – synonym of *Culex* (*Culiciomyia*) *viridiventer* Giles, 1901a; synonymy established before Knight & Stone, 1977.
longipalpatus Lee, 1946 (as *longipalpata*) (p. 249) – species of *Tripteroides* (*Rachisoura*).
longipalpis Dong, Zhou & Dong, 1997b (p. 197) – species of *Tripteroides* (*Rachionotomyia*).
longipalpis (Grünberg, 1905) (p. 383) – species of *Aedes* (*Zavortinkius*) [original combination: *Stegomyia longipalpis*].
longipalpis (Leicester, 1904) (in Theobald, 1904c) (p. 211) – species of *Armigeres* (*Leicesteria*) [original combination: *Leicesteria longipalpis*].
Synonym: *cingulata* Leicester, 1908.
longipalpis Lutz, 1905 (p. 127) – synonym of *Shannoniana fluviatilis* (Theobald, 1903a) [original combination: *Hyloconops longipalpis*]; synonymy by Lane & Cerqueira, 1942.
longipalpis Newstead & Thomas, 1910 (p. 145) – synonym of *Orthopodomyia fascipes* (Coquillett, 1906a) [original combination: *Mansonia longipalpis*]; synonymy established before Knight & Stone, 1977.
longipalpis Roth, 1945 (p. 13) – synonym of *Psorophora* (*Janthinosoma*) *longipalpus* Randolph & O'Neill, 1944; synonymy by Carpenter & LaCasse, 1955.
longipalpis (Theobald, 1903a) (p. 37) – species of *Anopheles* (*Cellia*) [original combination: *Myzomyia longipalpis*].
longipalpis van der Wulp, 1881 (p. 9) – synonym of *Mansonia* (*Mansonioides*) *dives* (Schiner, 1868) [original combination: *Culex longipalpis*]; synonymy by Stone, 1957a.
longipalpus Randolph & O'Neill, 1944 (in Division of Medical Entomology, Bureau of Laboratories, 1944) (p. 88) – species of *Psorophora* (*Janthinosoma*).
Synonym: *longipalpis* Roth, 1945.
longipes (Fabricius, 1805) (p. 34) – species of *Johnbelkinia* [original combination: *Culex longipes*].
Synonyms: *trichopus* Dyar, 1919a; *edwardsianus* Lane & Cerqueira, 1942.
longipes Theobald, 1901a (p. 241) – synonym of *Toxorhynchites* (*Lynchiella*) *grandiosus* (Williston, 1900) [original combination: *Megarhinus longipes*]; synonymy established before Knight & Stone, 1977.
longipes Theobald, 1901c (p. 68) – synonym of *Culex* (*Eumelanomyia*) *brevipalpis* (Giles, 1902); synonymy established before Knight & Stone, 1977.
longirostris Brug, 1928a (p. 278) – species of *Anopheles* (*Cellia*).
longirostris Leicester, 1908 (p. 217) – species of *Uranotaenia* (*Uranotaenia*).
longirostris (Leicester, 1908) (p. 228) – species of *Aedes* (*Rhinoskusea*) [original combination: *Ficalbia longirostris*].
Synonym: *hilli* Taylor, 1919.
longirostris Theobald, 1901c (p. 275) – species of *Wyeomyia* (*Phoniomyia*).
longiseta Edwards, 1936 (p. 50) – species of *Aedes* (*Aedimorphus*).
longisetosa Gong, 1994 (p. 210) – species of *Topomyia* (*Topomyia*).
longisiphonus Dong, Zhou & Dong, 2001 (p. 221) – species of *Tripteroides* (*Tripteroides*).
longisquamosa Theobald, 1905f (p. 103) – synonym of *Aedes* (*Ochlerotatus*) *caspius* (Pallas, 1771) [original combination: *Grabhamia longisquamosa*]; synonymy by ? Edwards, 1921d.
longitubus Cambournac, 1938 (p. 76) – synonym of *Aedes* (*Ochlerotatus*) *berlandi* Séguy, 1921; synonymy by Rioux & Arnold, 1955.
longitubus da Cunha Ramos & Brunhes, 2004 (p. 126) – species of *Uranotaenia* (*Pseudoficalbia*).
lopesi (Correa & Ramalho, 1956) (p. 37) – species of *Wyeomyia* (*Phoniomyia*) [original combination: *Phoniomyia lopesi*].
lopesi Sirivanakarn & Jakob, 1979 (p. 140) – species of *Culex* (*Melanoconion*).
lopezii Cova Garcia, Sutil Oramas & Pulido F., 1974a (p. 21) – species of *Wyeomyia* (*Hystatomyia*).
lophoventralis (Theobald, 1910a) (p. 13) – species of *Aedes* (*Phagomyia*) [original combination: *Pseudocarrollia lophoventralis*].
loquaculus Dyar & Knab, 1909d (p. 254) – synonym of *Culex* (*Phenacomyia*) *corniger* Theobald, 1903a [original combination: *Culex lactator* variety *loquaculus*]; synonymy established before Knight & Stone, 1977.
lorengaui Peters, 1963b (p. 67) – species of *Tripteroides* (*Tripteroides*).
loricatus Leicester, 1908 (p. 151) – synonym of *Culex* (*Culex*) *whitmorei* (Giles, 1904a); synonymy established before Knight & Stone, 1977.
lorraineae Berlin, 1969a (p. 80) – species of *Aedes* (*Howardina*).
loshanensis Wang & Feng, 1965 (p. 206) – synonym of *Uranotaenia* (*Pseudoficalbia*) *nivipleura* Leicester, 1908; synonymy by Chu, 1980 (as *loashanensis*, *lapsus*).
lottei Hamon & Brengues, 1965 (p. 101) – species of *Aedes* (*Catageiomyia*).
loturus Dyar, 1925g (p. 214) – synonym of *Culex* (*Melanoconion*) *zeteki* Dyar, 1918e; synonymy established before Knight & Stone, 1977.
lounibosi Gillies & Coetzee, 1987 (p. 73) – species of *Anopheles* (*Cellia*).
lousthei Boussès & Brunhes, 2013 (p. 406) – species of *Uranotaenia* (*Pseudoficalbia*).
lovettae Evans, 1934 (p. 564) – species of *Anopheles* (*Cellia*).
lowii Theobald, 1901c (p. 339) – species of *Uranotaenia* (*Uranotaenia*).
Synonyms: *continentalis* Dyar & Knab, 1906d; *minuta* Theobald, 1907; *monilis* Shannon & del Ponte, 1928.
lowisii (Theobald, 1910c) (p. 257) – species of *Aedes* (*Aedimorphus*) [original combination: *Reedomyia lowisii*].
Synonym: *mindoroensis* Knight & Hull, 1951a.

lucaris Colless, 1965 (p. 299) – species of *Culex* (*Lophoceraomyia*).
lucaris Macdonald & Mattingly, 1960 (p. 22) – species of *Udaya*.
luciae Senevet, Chabelard & Abonnenc, 1942 (p. 343) – synonym of *Wyeomyia chalcocephala* Dyar & Knab, 1906a; synonymy by Lane, 1951.
lucianus Muspratt, 1959 (p. 64) – species of *Aedes* (*Mucidus*).
luciensis Theobald, 1901a (p. 297) – synonym of *Aedes* (*Stegomyia*) *aegypti* (Linnaeus, 1762) [original combination: *Stegomyia fasciata* variety *luciensis*]; synonymy established before Knight & Stone, 1977.
lucifer (Howard, Dyar & Knab, 1913) (p. 23, fig. 164) – species of *Haemagogus* (*Haemagogus*) [original combination: *Stegoconops lucifer*].
lucifugus Komp, 1936c (p. 331) – species of *Culex* (*Melanoconion*).
lucyae van Someren, 1954a (p. 103) – species of *Uranotaenia* (*Pseudoficalbia*).
ludlowae (Blanchard, 1905) (species) (p. 630) – subspecies of *Aedes* (*Ochlerotatus*) *vigilax* (Skuse, 1889) [*nomen novum* for *annulifera* Ludlow, 1903; original combination: *Culex ludlowi* [*sic*]]; subspecific status by Knight & Hull, 1951b.
Synonym: *annulifera* Ludlow, 1903.
ludlowae Brunetti, 1920 (p. 173) – species of *Ficalbia* [*nomen novum* for *minima* Ludlow, 1907].
Synonym: *minima* Ludlow, 1907.
ludlowae Dyar & Shannon, 1925a (p. 68) – species of *Uranotaenia* (*Uranotaenia*).
ludlowae (Theobald, 1903a) (p. 42) – species of *Anopheles* (*Cellia*) [original combination: *Myzomyia ludlowii*].
Synonyms: *formosensis* Koidzumi, 1917; *hatorii* Koidzumi, 1920; *cabrerai* Cagampang-Ramos & Darsie, 1969.
luederwaldti Lane, 1936a (p. 6) – synonym of *Onirion personatum* (Lutz, 1904) [original combination: *Goeldia* (*Isogoeldia*) *luederwaldti*]; synonymy by Harbach & Peyton, 2000.
lugens Lutz, 1905 (p. 28) – *nomen dubium*, *Culex*; status by Sirivanakarn, 1983.
lugubris (Barraud, 1928b) (p. 372) – species of *Verrallina* (*Verrallina*) [original combination: *Aedes* (*Aedes*) *lugubris*].
lui Gong & Li, 1999 (p. 96) – species of *Heizmannia* (*Mattinglyia*).
lui Lien, 1968a (p. 9) – species of *Uranotaenia* (*Pseudoficalbia*).
lukisii Christophers, 1916b (p. 120) – synonym of *Anopheles* (*Anopheles*) *algeriensis* Theobald, 1903a; synonymy established before Knight & Stone, 1977.
luna Belkin, Heinemann & Page, 1970 (p. 213) – species of *Wyeomyia* (*Wyeomyia*).
lunata (Theobald, 1901c) (p. 279) – species of *Isostomyia* [original combination: *Wyeomyia lunata*].
lunda da Cunha Ramos, 1993 (p. 180) – species of *Uranotaenia* (*Pseudoficalbia*).
lungae Belkin & Schlosser, 1944 (p. 269) – species of *Anopheles* (*Cellia*).
lunulatus King & Hoogstraal, 1946c (p. 95) – species of *Aedes* (*Geoskusea*).
luridus Doleschall, 1857 (p. 384) – synonym of *Lutzia* (*Metalutzia*) *fuscana* Wiedemann, 1820 [original combination: *Culex luridus*]; synonymy established before Knight & Stone, 1977.
luridus McIntosh, 1971 (p. 327) – species of *Aedes* (*Neomelaniconion*).
lutea (Belkin, 1962) (p. 309) – species of *Coquillettidia* (*Coquillettidia*) [original combination: *Mansonia* (*Coquillettidia*) *lutea*].
luteifemur Edwards, 1926a (p. 112) – species of *Aedes* (*Ochlerotatus*).
luteoabdominalis Theobald, 1910a (p. 23) – synonym of *Culex* (*Oculeomyia*) *epidesmus* (Theobald, 1910a) [original combination: *Taeniorhynchus luteoabdominalis*]; synonymy established before Knight & Stone, 1977.
luteoannulatus Theobald, 1901c (p. 159) – synonym of *Culex* (*Culex*) *quinquefasciatus* Say, 1823 [original combination: *Culex fatigans luteoannulatus*]; synonymy established before Knight & Stone, 1977.
luteocephalus (Newstead, 1907) (in Newstead *et al.*, 1907) (p. 15) – species of *Aedes* (*Stegomyia*) [original combination: *Stegomyia luteocephala*].
luteola Edwards, 1934 (in Barraud, 1934) (p. 80) – species of *Uranotaenia* (*Pseudoficalbia*).
luteola Theobald, 1910c (p. 378) – synonym of *Culex* (*Culex*) *fuscocephala* Theobald, 1907; synonymy established before Knight & Stone, 1977.
luteolateralis (Theobald, 1901c) (p. 71) – species of *Aedes* (*Neomelaniconion*) [original combination: *Culex luteolateralis*].
luteopleurus (Theobald, 1903a) (p. 239) – species of *Culex* (*Anoedioporpa*) [original combination: *Melanoconion luteopleurus*].
luteostriatus Robinson, 1950 (p. 80) – species of *Aedes* (*Vansomerenis*).
luteoventralis Theobald, 1901a (p. 348) – species of *Wyeomyia* (*Dendromyia*).
Synonym: *quasiluteoventralis* Theobald, 1903a.
luteovittata Theobald, 1901c (p. 85) – synonym of *Aedes* (*Rusticoidus*) *rusticus* (Rossi, 1790) [original combination: *Culex nemorosus* variety *luteovittata* [*sic*]]; synonymy established before Knight & Stone, 1977.
lutescens Fabricius, 1775 (p. 800) – synonym of *Aedes* (*Ochlerotatus*) *flavescens* (Müller, 1764) [original combination: *Culex lutescens*]; synonymy established before Knight & Stone, 1977.
lutescens Leicester, 1908 (p. 222) – species of *Uranotaenia* (*Pseudoficalbia*).
lutescens (Theobald, 1901a) (p. 233) – species of *Toxorhynchites* (*Afrorhynchus*) [original combination: *Megarhinus lutescens*].
luteus (Ludlow, 1905a) (p. 96) – species of *Aedes* (*Finlaya*) [original combination: *Popea lutea*].
luteus Meigen, 1804 (p. 6) – synonym of *Culex* (*Culex*) *pipiens* Linnaeus, 1758; synonymy established before Knight & Stone, 1977.

lutzi Cruz, 1907a (p. 271) – synonym of *Anopheles* (*Anopheles*) *peryassui* Dyar & Knab, 1908 [original combination: *Manguinhosia lutzi*]; synonymy established before Knight & Stone, 1977.

lutzi (da Costa Lima, 1930a) (p. 75) – species of *Wyeomyia* (*Miamyia*) [original combination: *Miamyia* (*Miamyia*) *lutzi*].

lutzianus Lane & Cerqueira, 1942 (p. 679) – synonym of *Sabethes* (*Peytonulus*) *identicus* Dyar & Knab, 1907c; synonymy by Lane, 1951.

lutzii Cruz, 1901 (p. 426) – species of *Anopheles* (*Nyssorhynchus*). **Note**: this nominal taxon is the senior primary homonym of *lutzii* Theobald, 1901a (p. 177), (published 23 November 1901).

Synonym: *nigra* Theobald, 1907.

lutzii Theobald, 1901a (p. 177) – synonym of *Anopheles* (*Kerteszia*) *cruzii* Dyar & Knab, 1908; synonymy established before Knight & Stone, 1977. **Note**: this nominal taxon is the junior primary homonym of *lutzii* Cruz, 1901 (published 15 November 1901).

lutzii (Theobald, 1901a) (p. 257) – species of *Psorophora* (*Janthinosoma*) [original combination: *Janthinosoma lutzii*].

Synonym: *chaquensis* Paterson & Shannon, 1927.

lutzii Theobald, 1903a (p. 323) – *nomen dubium*, *Sabethes*; status by Belkin, 1968b.

luxodens Hall, Howard & Harbach, 1999 (p. 329) – species of *Sabethes* (*Peytonulus*).

luzonensis (Ludlow, 1905a) (p. 101) – species of *Mimomyia* (*Etorleptiomyia*) [original combination: *Oreillia luzonensis*].

Synonyms: *completiva* Leicester, 1908; *clavirostris* Stone & Bohart, 1944.

luzonensis Rozeboom, 1946 (p. 589) – species of *Aedes* (*Kenknightia*).

luzonensis Sirivanakarn, 1976 (p. 75) – species of *Culex* (*Oculeomyia*).

lygrus Root, 1927c (p. 579) – species of *Culex* (*Culex*).

lynchi Brèthes, 1916 (p. 204) – synonym of *Psorophora* (*Psorophora*) *ciliata* (Fabricius, 1794); synonymy established before Knight & Stone, 1977.

lynchi Dyar & Knab, 1906c (p. 244) – synonym of *Toxorhynchites* (*Lynchiella*) *haemorrhoidalis separatus* (Lynch Arribálzaga, 1891b) [original combination: *Megarhinus lynchi*]; synonymy by Lane, 1939.

lynchi (Shannon, 1931) (p. 152) – species of *Coquillettidia* (*Rhynchotaenia*) [original combination: *Mansonia lynchi*].

lynchii Brèthes, 1910 (p. 470) – synonym of *Aedes* (*Ochlerotatus*) *crinifer* (Theobald, 1903a) [original combination: *Culex lynchii*]; synonymy established before Knight & Stone, 1977.

Mm

mabinii Baisas & Ubaldo-Pagayon, 1953 (p. 60) – species of *Tripteroides* (*Tripteroides*).

macaensis Ribeiro, 1997 (p. 213) – species of *Toxorhynchites* (*Toxorhynchites*).

macaronensis Dyar & Nuñez Tovar, 1927a (p. 153) – synonym of *Culex* (*Melanoconion*) *conspirator* Dyar & Knab, 1906d; synonymy by Komp, 1935.

macarthuri Colless, 1956 (p. 67) – species of *Anopheles* (*Cellia*) [original combination: *Anopheles riparis macarthuri*]; specific status by Hii *et al.*, 1988.

maccrackenae Dyar & Knab, 1906a (p. 133) – synonym of *Culiseta* (*Culiseta*) *particeps* (Adams, 1903); synonymy by Stone, 1958b.

macdonaldi Colless, 1965 (p. 276) – species of *Culex* (*Lophoceraomyia*).

macdonaldi Mattingly, 1957c (p. 47) – species of *Heizmannia* (*Heizmannia*).

Synonym: *nivirostris* Lien, 1968c.

macdougalli Edwards, 1922a (p. 271) – species of *Aedes* (*Collessius*).

macedoniensis Cot & Hovasse, 1917 (p. 891) – synonym of *Anopheles* (*Cellia*) *superpictus* Grassi, 1899 [original combination: *Anopheles superpictus* variety *macedoniensis*]; synonymy established before Knight & Stone, 1977.

macfarlanei (Edwards, 1914b) (p. 78) – species of *Aedes* (*Collessius*) [original combination: *Ochlerotatus macfarlanei*].

Synonym: *hegneri* Causey, 1937a.

macfarlanei Edwards, 1914c (p. 127) – species of *Uranotaenia* (*Uranotaenia*).

Synonym: *zelena* Barraud, 1934.

macfiei Edwards, 1923a (p. 399) – species of *Culex* (*Culiciomyia*).

machadoi da Cunha Ramos, 1986 (p. 110) – species of *Uranotaenia* (*Uranotaenia*).

machadoi da Silva Mattos, da Silveira Guedes & Xavier, 1978 (p. 31) – species of *Culex* (*Phytotelmatomyia*).

machardyi Edwards, 1930b (p. 545) – species of *Anopheles* (*Cellia*).

macintoshi Marks, 1959a (p. 117) – species of *Aedes* (*Ochlerotatus*).

mackerrasi Taylor, 1927 (p. 68) – species of *Aedes* (*Patmarksia*).

macleayanus Mackerras, 1927 (p. 291) – synonym of *Aedes* (*Ochlerotatus*) *nigrithorax* (Macquart, 1847); synonymy established before Knight & Stone, 1977.

macleayi Skuse, 1889 (p. 1746) – synonym of *Culex* (*Culex*) *quinquefasciatus* Say, 1823; synonymy established before Knight & Stone, 1977.

macmahoni Evans, 1936 (species) (p. 538) – subspecies of *Anopheles* (*Cellia*) *sergentii* (Theobald, 1907); subspecific status by Mattingly & Knight, 1956.

Synonym: *barkhuusi* Giaquinto-Mira, 1950.

macmillani Marks, 1964b (p. 138) – species of *Aedes* (*Macleaya*).

macrodixoa (Dyar & Shannon, 1925a) (p. 79) – species of *Verrallina* (*Neomacleaya*) [original combination: *Aedes* (*Aedes*) *macrodixoa*].

macrophyllus Edwards & Gibbins, 1939 (subspecies of *vansomereni* Edwards, 1926a) (p. 31) – subspecies of

Culex (*Culex*) *toroensis* Edwards & Gibbins, 1939; subspecific status by Edwards, 1941 concomitant with elevation of *toroensis* to specific status.

macropus Blanchard, 1905 (p. 327) – synonym of *Culex* (*Eumelanomyia*) *brevipalpis* (Giles, 1902); synonymy established before Knight & Stone, 1977.

macrostylus Sirivanakarn & Ramalingam, 1976 (p. 210) – species of *Culex* (*Eumelanomyia*).

macrotus Dyar & Knab, 1907c (p. 212) – synonym of *Wyeomyia* (*Hystatomyia*) *circumcincta* Dyar & Knab, 1907c; synonymy established before Knight & Stone, 1977.

maculata Peryassú, 1921b (p. 141) – synonym of *Chagasia fajardi* (Lutz, 1904) [original combination: *Chagasia fajardoi* variety *maculata*]; synonymy established before Knight & Stone, 1977.

maculata Theobald, 1905g (p. 23) – synonym of *Aedes* (*Aedimorphus*) *taeniorhynchoides* (Christophers, 1911) [original combination: *Pecomyia maculata*]; synonymy established before Knight & Stone, 1977. **Note**: this nominal taxon is the junior secondary homonym of *maculatus* Meigen, 1804 and the senior secondary homonym of *maculata* Theobald, 1905h.

maculata Theobald, 1905h (p. 243) – synonym of *Aedes* (*Aedimorphus*) *pipersalatus* (Giles, 1902) [original combination: *Pseudograbhamia maculata*]; synonymy established before Knight & Stone, 1977. **Note**: this nominal taxon is the junior secondary homonym of *maculatus* Meigen, 1804 and *maculata* Theobald, 1905g.

maculata Theobald, 1910a (p. 1) – synonym of *Anopheles* (*Anopheles*) *lindesayi* Giles, 1900a [original combination: *Anopheles lindesayi* variety *maculata* [*sic*]]; synonymy established before Knight & Stone, 1977. **Note**: this nominal taxon is the junior primary homonym of *maculatus* Theobald, 1901a.

maculata Theobald, 1910a (p. 29) – synonym of *Orthopodomyia anopheloides* (Giles, 1903); synonymy established before Knight & Stone, 1977.

maculatus Meigen, 1804 (p. 4) – synonym of *Aedes* (*Rusticoidus*) *rusticus* (Rossi, 1790) [original combination: *Culex maculatus*]; synonymy established before Knight & Stone, 1977. **Note**: this nominal taxon is the senior secondary homonym of *maculata* Theobald, 1905g and *maculata* Theobald, 1905h.

maculatus Theobald, 1901a (p. 171) – species of *Anopheles* (*Cellia*). **Note**: this nominal taxon is the senior primary homonym of *maculata* Theobald, 1910a (p. 1).

Synonym: *hanabusai* Yamada, 1925.

maculatus von Humboldt, 1819 (p. 340) – *nomen dubium*, *Culex*; status established before Knight & Stone, 1977.

maculicosta Becker, 1903 (p. 69) – synonym of *Anopheles* (*Cellia*) *pharoensis* Theobald, 1901a; synonymy established before Knight & Stone, 1977.

maculicosta Edwards, 1936 (p. 52) – synonym of *Aedes* (*Neomelaniconion*) *carteri* Edwards, 1936 [original combination: *Aedes palpalis* variety *maculicosta*]; synonymy established before Knight & Stone, 1977.

maculicrures Theobald, 1901c (p. 34) – synonym of *Lutzia* (*Metalutzia*) *tigripes* de Grandpré & de Charmoy, 1901 [original combination: *Culex maculicrura*]; emendation by Theobald, 1901d; synonymy (as *maculicrura*) by Edwards, 1932a.

maculipalpis Giles, 1902 (p. 297) – species of *Anopheles* (*Cellia*).

maculipennis Meigen, 1818 (p. 11) – species of *Anopheles* (*Anopheles*).

Synonyms: *alexandraeschingarevi* Shingarev, 1928; *basilii* Falleroni, 1932; *typicus* Hackett & Missiroli, 1935.

maculipennis (Theobald, 1911a) (p. 27) – species of *Coquillettidia* (*Coquillettidia*) [original combination: *Chrysoconops maculipennis*].

maculipes (Theobald, 1903a) (p. 81) – species of *Anopheles* (*Anopheles*) [original combination: *Arribalzagia maculipes*].

maculipes Theobald, 1904b (p. 79) – synonym of *Culex* (*Oculeomyia*) *poicilipes* (Theobald, 1903a) [original combination: *Taeniorhynchus tenax* variety *maculipes*]; synonymy established before Knight & Stone, 1977.

maculipes Theobald, 1910c (p. 470) – synonym of *Orthopodomyia anopheloides* (Giles, 1903); synonymy established before Knight & Stone, 1977.

maculipesarabiensis Patton, 1905 (p. 635) – synonym of *Culex* (*Oculeomyia*) *bitaeniorhynchus* Giles, 1901a [original combination: *Taeniorhynchus tenax* variety *maculipes arabiensis*]; synonymy by White, 1980.

maculipleura Leicester, 1908 (p. 223) – species of *Uranotaenia* (*Pseudoficalbia*).

maculiventris Macquart, 1846a (p. 135) – synonym of *Aedes* (*Ochlerotatus*) *caspius* (Pallas, 1771) [original combination: *Culex maculiventris*]; synonymy established before Knight & Stone, 1977.

maculoabdominalis Theobald, 1913d (p. 597) – synonym of *Aedes* (*Stegomyia*) *apicoargenteus* (Theobald, 1909b) [original combination: *Kingia maculoabdominalis*]; synonymy by Edwards, 1941.

maculosa James & Liston, 1911 (p. 112) – synonym of *Anopheles* (*Cellia*) *willmori* (James, 1903) [original combination: *Neocellia willmori* variety *maculosa*]; synonymy established before Knight & Stone, 1977.

maculosa Theobald, 1905f (p. 105) – synonym of *Aedes* (*Ochlerotatus*) *detritus* (Haliday, 1833) [original combination: *Grabhamia maculosa*]; synonymy by Stone, 1957b.

madagascarensis da Cunha Ramos & Brunhes, 2004 (p. 267) – species of *Uranotaenia* (*Pseudoficalbia*).

madagascarensis Ribeiro, 2004 (p. 248) – species of *Toxorhynchites* (*Afrorhynchus*).

madagascarensis van Someren, 1949a (p. 7) – species of *Aedes* (*Paulianius*).

madagascarica Brunhes, Boussès & da Cunha Ramos, 2011 (p. 119) – species of *Aedeomyia* (*Aedeomyia*).

madagascarica da Cunha Ramos & Brunhes, 2004 (p. 383) – species of *Uranotaenia* (*Uranotaenia*).

madagascariensis Ventrillon, 1905b (p. 427) – synonym of *Culex* (*Oculeomyia*) *poicilipes* (Theobald, 1903a) [original combination: *Pseudoheptaphlebomyia madagascariensis*]; synonymy established before Knight & Stone, 1977.

madecassorum Rodhain & Boutonnier, 1984 (p. 228) – *nomen dubium*, *Orthopodomyia*; status by Brunhes & Hervy, 1995.

maderensis Mattingly, 1955 (p. 377) – subspecies of *Culex* (*Maillotia*) *hortensis* Ficalbi, 1889a.

madininensis Senevet, 1936 (p. 129) – species of *Culex* (*Melanoconion*).

madrensis Baisas, 1946a (p. 41) – species of *Orthopodomyia*.

maehleri Bohart, 1957 (p. 40) – species of *Aedes* (*Stegomyia*).

maffii Taylor & Tenorio, 1974 (p. 577) – species of *Aedes* (*Finlaya*).

magna Theobald, 1905e (p. 22) – synonym of *Aedes* (*Phagomyia*) *gubernatoris* (Giles, 1901a) [original combination: *Lepidotomyia magna*]; synonymy established before Knight & Stone, 1977.

magna (Theobald, 1905f) (p. 117) – species of *Runchomyia* (*Ctenogoeldia*) [original combination: *Phoniomyia magna*].
Synonyms: *homotina* Dyar & Knab, 1906a; *dicellaphora* Howard, Dyar & Knab, 1913.

magnesianus (Edwards, 1924b) (p. 361) – species of *Tripteroides* (*Tripteroides*) [original combination: *Rachionotomyia magnesiana*].

magnificus (Leicester, 1908) (p. 54) – species of *Toxorhynchites* (*Toxorhynchites*) [original combination: *Teromyia magnificus* [*sic*]].

magnipennis Felt, 1904 (p. 322) – synonym of *Culiseta* (*Culiseta*) *inornata* (Williston, 1893) [original combination: *Culex magnipennis*]; synonymy established before Knight & Stone, 1977.

magnus (Theobald, 1901c) (p. 344) – species of *Deinocerites* [original combination: *Brachiomyia magna*].
Synonym: *troglodytus* Dyar & Knab, 1909d.

magnus (Theobald, 1908b) (p. 293) – species of *Armigeres* (*Leicesteria*) [original combination: *Brevirhynchus magnus*].

mahaffyi van Someren, 1949b (p. 122) – species of *Eretmapodites*.

mahantai Bhattacharyya, Prakash, Mohapatra & Sarma, 2009 (p. 1) – species of *Armigeres* (*Armigeres*).

mahmuti Martini, 1930 (p. 149) – synonym of *Anopheles* (*Anopheles*) *hyrcanus* (Pallas, 1771) [original combination: *Anopheles* (*Anopheles*) *hyrcanus* variety *mahmuti*]; synonymy by Christophers, 1933.

maiae Edwards, 1917 (p. 208) – species of *Armigeres* (*Armigeres*).

maikalensis Natarajan, Rajavel & Jambulingam, 2017 (p. 251) – species of *Uranotaenia* (*Pseudoficalbia*).

majidi Young & Majid, 1928 (p. 469) – species of *Anopheles* (*Cellia*) [original combination: *Anopheles karwari* variety *majidi*]; specific status by Iyengar, 1929.

major Edwards, 1935a (p. 135) – synonym of *Culex* (*Oculeomyia*) *annulioris consimilis* Newstead, 1907 [original combination: *Culex annulioris* variety *major*]; synonymy by Hamon & Ovazza, 1956.

major Theobald, 1903a (p. 270) – synonym of *Mansonia* (*Mansonioides*) *africana* (Theobald, 1901d); synonymy established before Knight & Stone, 1977.

malariae Grassi, 1898 (p. 168) – synonym of *Aedes* (*Aedimorphus*) *vexans* (Meigen, 1830b) [original combination: *Culex malariae*]; synonymy established before Knight & Stone, 1977.

malariager Poinar, 2005 (p. 549) – fossil species of *Culex* Linnaeus, 1758.

malayensis Colless, 1962 (p. 314) – species of *Aedes* (*Stegomyia*) [original combination: *Aedes* (*Stegomyia*) *scutellaris malayensis*]; specific status by Huang, 1972.

malayensis Hacker, 1921 (p. 1) – synonym of *Anopheles* (*Cellia*) *indefinitus* (Ludlow, 1904c) [original combination: *Anopheles subpictus* variety *malayensis*]; synonymy by Reid, 1966.

malayensis Sirivanakarn, 1972 (p. 10) – species of *Culex* (*Eumelanomyia*).

malayi Delfinado & Hodges, 1968 (p. 368) – species of *Tripteroides* (*Tripteroides*).

malayi (Leicester, 1908) (p. 184) – species of *Culex* (*Eumelanomyia*) [original combination: *Aedes malayi*].
Synonyms: *nigrescens* Theobald, 1907; *aedes* Leicester, 1908.

malayi Leicester, 1908 (p. 198) – synonym of *Verrallina* (*Harbachius*) *fragilis* Leicester, 1908; synonymy established before Knight & Stone, 1977.

malayi Leicester, 1908 (p. 231) – species of *Hodgesia*.
Synonym: *ampyx* Dyar, 1920i.

malayi (Theobald, 1901c) (p. 258) – species of *Armigeres* (*Armigeres*) [original combination: *Uranotaenia malayi*].
Synonyms: *fusca* Theobald, 1903a; *russelli* Baisas, 1935b.

malaysiensis Ramalingam & Banu, 1987 (p. 119) – species of *Topomyia* (*Topomyia*).

malefactor Dyar & Knab, 1907c (p. 198) – species of *Anopheles* (*Anopheles*).

malfeyti (Newstead, 1907) (in Newstead *et al*., 1907) (p. 29) – species of *Ficalbia* [original combination: *Mimomyia malfeyti*].

maliensis Bailly-Choumara & Adam, 1959 (p. 3742) – species of *Anopheles* (*Cellia*).

malikuli Huang, 1973 (p. 225) – species of *Aedes* (*Stegomyia*).

mallochi Taylor, 1944 (p. 121) – species of *Aedes* (subgenus uncertain).

malvari Baisas & Ubaldo-Pagayon, 1953 (p. 141) – species of *Tripteroides* (*Tripteroides*).

mammilifer (Leicester, 1908) (p. 128) – species of *Culex* (*Lophoceraomyia*) [original combination: *Lophoceratomyia mammilifer*].
Synonym: *chiungchungensis* Hsu, 1963.

mamoedjoensis Mattingly, 1958a (p. 58) – species of *Aedes* (*Cancraedes*).

manaensis Floch & Fauran, 1955 (p. 3) – synonym of *Culex* (*Carrollia*) *antunesi* Lane & Whitman, 1943; synonymy by Valencia, 1973.

manakaraensis da Cunha Ramos & Brunhes, 2004 (p. 128) – species of *Uranotaenia* (*Pseudoficalbia*).
manalangi Baisas, 1935b (p. 492) – species of *Armigeres* (*Armigeres*).
manalangi Mendoza, 1940 (p. 368) – species of *Anopheles* (*Anopheles*).
manaosensis Evans, 1924 (p. 370) – synonym of *Culex* (*Melanoconion*) *eastor* Dyar, 1920d; synonymy by Rozeboom & Komp, 1950a.
manganus Baisas, 1946a (p. 35) – synonym of *Orthopodomyia anopheloides* (Giles, 1903); synonymy by Knight & Mattingly, 1950.
mangyanus (Banks, 1906b) (p. 991) – species of *Anopheles* (*Cellia*) [original combination: *Myzomyia mangyana*].
Synonym: *febrifera* Banks, 1914.
manhi (Harbach & Cook, 2010) (in Cook *et al.*, 2010) (p. 34) – species of *Aedes* (*Bothaella*) [original combination: *Bothaella manhi*].
manicatus (Edwards, 1921a) (p. 630) – species of *Toxorhynchites* (*Toxorhynchites*) [original combination: *Megarhinus manicatus*].
manopi Thurman, 1959 (p. 16) – species of *Toxorhynchites* (*Toxorhynchites*).
mansouri Qutubuddin, 1959 (p. 21) – species of *Aedes* (*Aedimorphus*).
manusensis Sirivanakarn, 1975 (p. 394) – species of *Culex* (*Eumelanomyia*).
maplei Knight & Hurlbut, 1949 (p. 31) – species of *Culex* (*Culiciomyia*) [original combination: *Culex* (*Culiciomyia*) *pullus maplei*]; specific status by Bohart, 1957.
mara Anduze, 1942a (p. 43) – synonym of *Toxorhynchites* (*Lynchiella*) *solstitialis* (Lutz, 1904) [original combination: *Mergarhinus* [*sic*] *mara*]; synonymy by Belkin *et al.*, 1965.
maracayensis Evans, 1923b (p. 102) – species of *Culex* (*Culex*).
Synonym: *aglischrus* Dyar, 1924d.
marajoara Galvão & Damasceno, 1942b (p. 424) – species of *Anopheles* (*Nyssorhynchus*).
Synonym: *domesticus* Galvão & Damasceno, 1944.
marajoensis Peryassú, 1923c (p. 76) – *nomen nudum*, *Culex*.
marceli (Mattingly, 1953c) (p. 315) – species of *Malaya* [original combination: *Harpagomyia marceli*].
marcelleae Adam & Hamon, 1959 (p. 524) – species of *Eretmapodites* [original combination: *Eretmapodites oedipodeios marcellei* [*sic*]]; specific status by Hamon, 1961.
marchettei Garcia, Jeffery & Rudnick, 1969 (p. 252) – species of *Culiseta* (*Climacura*).
margarsen (Dyar & Shannon, 1925a) (p. 80) – species of *Verrallina* (*Neomacleaya*) [original combination: *Aedes* (*Aedes*) *margarsen*].
margina Gong & Lu, 1995 (p. 476) – species of *Topomyia* (*Topomyia*).
marginalis Stephens, 1825 (p. 455) – synonym of *Culex* (*Culex*) *pipiens* Linnaeus, 1758; synonymy established before Knight & Stone, 1977.
mariae (Bourroul, 1904) (p. 3) – species of *Toxorhynchites* (*Lynchiella*) [original combination: *Megarhinus mariae*].
Synonym: *iris* Knab, 1913.
mariae (Sergent & Sergent, 1903) (p. 62) – species of *Aedes* (*Acartomyia*) [original combination: *Culex mariae*].
Synonym: *desbansi* Séguy, 1923.
marianae Bohart & Ingram, 1946b (p. 42) – subspecies of *Culex* (*Culex*) *annulirostris* Skuse, 1889.
marinkellei Berlin, 1969a (p. 107) – species of *Aedes* (*Howardina*).
marinus Theobald, 1901a (p. 396) – synonym of *Aedes* (*Ochlerotatus*) *vigilax* (Skuse, 1889) [original combination: *Culex marinus*]; synonymy established before Knight & Stone, 1977.
marksae Dobrotworsky, 1965 (p. 58) – species of *Tripteroides* (*Polylepidomyia*).
marksae Grjebine, 1986 (p. 154) – species of *Mimomyia* (*Ingramia*).
marksae King & Hoogstraal, 1955 (p. 6) – species of *Culex* (*Lophoceraomyia*).
marmorata (Philippi, 1865) (p. 597) – *nomen dubium*, ? *Psorophora* [original combination: *Culex marmoratus*]; status by Edwards, 1932a.
marocanus D'Anfreville, 1916 (p. 141) – synonym of *Culiseta* (*Allotheobaldia*) *longiareolata* (Macquart, 1838a) [original combination: *Culex annulatus* variety *marocanus*]; synonymy established before Knight & Stone, 1977.
maroniensis Bonne-Wepster & Bonne, 1920 (p. 175) – synonym of *Culex* (*Melanoconion*) *albinensis* Bonne-Wepster & Bonne, 1920; synonymy by Lane, 1951.
marquesensis Dyar, 1925e (p. 43) – synonym of *Mansonia* (*Mansonioides*) *uniformis* (Theobald, 1901c); synonymy established before Knight & Stone, 1977.
marquesensis Stone & Rosen, 1953 (p. 354) – species of *Culex* (*Culex*).
marshallensis Stone & Bohart, 1944 (p. 218) – species of *Aedes* (*Stegomyia*).
marshallii (Theobald, 1901a) (p. 310) – species of *Aedes* (*Albuginosus*) [original combination: *Stegomyia marshallii*].
marshallii (Theobald, 1903a) (p. 77) – species of *Anopheles* (*Cellia*) [original combination: *Pyretophorus marshallii*].
Synonyms: *pitchfordi* Giles, 1904c; *neireti* Blanchard, 1907; *pseudocostalis* Theobald, 1910c; *transvaalensis* Carter, 1910a.
marshallii Theobald, 1903a (p. 121) – synonym of *Toxorhynchites* (*Toxorhynchites*) *brevipalpis* Theobald, 1901a; synonymy established before Knight & Stone, 1977.
marteri Senevet & Prunnelle, 1927 (p. 529) – species of *Anopheles* (*Anopheles*).
Synonyms: *sogdianus* Keshishian, 1938; *conquensis* Torres Cañamares, 1946.
martiali Senevet & Abonnenc, 1939b (p. 276) – species of *Limatus*.

martinei (Doucet, 1951a) (p. 68) – species of *Mimomyia* (*Etorleptiomyia*) [original combination: *Ficalbia* (*Etorleptiomyia*) *martinei*].
martineti Senevet, 1937 (p. 364) – species of *Aedes* (*Ochlerotatus*).
martinezi Berlin, 1969a (p. 119) – species of *Aedes* (*Howardina*).
martinezi Casal & García, 1968c (p. 455) – species of *Culex* (*Melanoconion*).
martini Laveran, 1902a (p. 907) – synonym of *Anopheles* (*Anopheles*) *barbirostris* van der Wulp, 1884; synonymy by Reid, 1947.
martinii Lane, 1943 (p. 152) – synonym of *Uranotaenia* (*Uranotaenia*) *hystera* Dyar & Knab, 1913 [*nomen novum* for *bicolor* Martini, 1935]; synonymy of *bicolor* Martini by Galindo *et al.*, 1954.
martinii Medschid, 1930 (p. 364) – species of *Culex* (*Neoculex*).
martinii Statz, 1941 (p. 8) – *nomen nudum*, *Mansonia* (see *martinii* Statz, 1944).
martinii Statz, 1944 (p. 110) – fossil species of *Mansonia* Blanchard, 1901.
martinius Shingarev, 1926 (p. 47) – species of *Anopheles* (*Anopheles*).
Synonyms: *relictus* Shingarev, 1928; *elutior* Martini, 1930.
marzinovski Shingarev, 1926 (p. 48) – synonym of *Anopheles* (*Anopheles*) *hyrcanus* (Pallas, 1771) [original combination: *Anopheles* (*Anopheles*) *hyrcanus* variety *marzinovski*]; synonymy by Stone *et al.*, 1959.
masamae Dyar, 1920g (p. 166) – synonym of *Aedes* (*Ochlerotatus*) *hexodontus* Dyar, 1916b; synonymy by Wood, 1977.
mascarensis de Meillon, 1947a (p. 116) – species of *Anopheles* (*Cellia*).
mascarensis MacGregor, 1924 (p. 409) – species of *Aedes* (*Stegomyia*).
masculinus Mattingly, 1958a (p. 51) – species of *Aedes* (*Cancraedes*).
masculus Theobald, 1901c (p. 125) – synonym of *Culex* (*Culex*) *decens* Theobald, 1901d; synonymy established before Knight & Stone, 1977.
mashonaensis Theobald, 1901c (p. 259) – species of *Uranotaenia* (*Pseudoficalbia*).
masoalensis Fontenille & Brunhes, 1985 (p. 153) – species of *Aedes* (*Aedimorphus*).
masseguini Hamon, 1954a (p. 575) – synonym of *Anopheles* (*Cellia*) *brohieri* Edwards, 1929b [original combination: *Anopheles hancocki* variety *masseguini*]; synonymy established before Knight & Stone, 1977.
masseyi Edwards, 1923a (p. 397) – species of *Aedes* (*Stegomyia*).
mastersi Skuse, 1889 (p. 1757) – synonym of *Anopheles* (*Cellia*) *annulipes* Walker, 1856a; synonymy established before Knight & Stone, 1977.
mastigia Howard, Dyar & Knab, 1913 (fig. 90) – synonym of *Culex* (*Melanoconion*) *pilosus* (Dyar & Knab, 1906d); synonymy established before Knight & Stone, 1977.
mataea Dyar & Knab, 1908 (p. 70) – synonym of *Wyeomyia* (*Wyeomyia*) *celaenocephala* Dyar & Knab, 1906a; synonymy established before Knight & Stone, 1977.
mathesoni Anduze, 1942a (p. 47) – synonym of *Culex* (*Carrollia*) *urichii* (Coquillett, 1906e); synonymy by Valencia, 1973.
mathesoni Belkin, 1950 (p. 219) – species of *Tripteroides* (*Rachisoura*).
mathesoni Belkin & Heinemann, 1975 (p. 363) – species of *Psorophora* (*Janthinosoma*).
mathesoni Belkin & Hogue, 1959 (p. 426) – species of *Deinocerites*.
mathesoni Lane, 1943 (p. 153) – species of *Uranotaenia* (*Uranotaenia*).
mathesoni Middlekauff, 1944 (species) (p. 42) – subspecies of *Aedes* (*Ochlerotatus*) *canadensis* (Theobald, 1901c); subspecific status by Rings & Hill, 1948.
mathioti Fontenille & Brunhes, 1985 (p. 154) – species of *Aedes* (*Aedimorphus*).
mathisi Neveu-Lemaire, 1902b (p. 13) – synonym of *Aedes* (*Ochlerotatus*) *serratus* (Theobald, 1901c) [original combination: *Culex mathisi*]; synonymy established before Knight & Stone, 1977.
mattinglyi Baisas & Ubaldo-Pagayon, 1953 (p. 169) – subspecies of *Tripteroides* (*Tripteroides*) *powelli* (Ludlow, 1909c).
mattinglyi Grjebine, 1986 (p. 158) – species of *Mimomyia* (*Ingramia*).
mattinglyi Hamon & Rickenbach, 1954 (p. 930) – species of *Aedes* (*Aedimorphus*).
mattinglyi Hamon & van Someren, 1961a (p. 728) – species of *Eretmapodites*.
mattinglyi Knight, 1953 (p. 320) – species of *Culex* (*Culex*).
mattinglyi Lane, 1953 (p. 935) – species of *Wyeomyia* (*Cruzmyia*).
mattinglyi Qutubuddin, 1951 (p. 107) – species of *Uranotaenia* (*Pseudoficalbia*).
mattinglyi Thurman, 1959 (p. 70) – species of *Heizmannia* (*Heizmannia*).
mattinglyorum Huang, 1994 (p. 11) – species of *Aedes* (*Stegomyia*).
mattogrossensis Lutz & Neiva, 1911 (p. 297) – species of *Anopheles* (*Anopheles*).
Synonym: *amazonicus* Christophers, 1923b.
mauesensis Lane, 1945b (p. 208) – species of *Culex* (*Culex*).
mauritanicus Callot, 1940 (p. 362) – synonym of *Culex* (*Culex*) *simpsoni* Theobald, 1905g; synonymy by Harbach, 1988.
mauritianus de Grandpré & de Charmoy, 1901 (p. 8) – synonym of *Anopheles* (*Anopheles*) *coustani* Laveran, 1900; synonymy by Senevet, 1932.
maverlius Reinert, 1997 (in Reinert *et al.*, 1997) (p. 18) – species of *Anopheles* (*Anopheles*).
maxgermaini Huang, 1990 (p. 31) – species of *Aedes* (*Stegomyia*).
maxi Dyar, 1928 (p. 386) – species of *Culex* (*Culex*).
Synonym: *oswaldoi* Forattini, 1965.
maxima Leicester, 1908 (p. 221) – species of *Uranotaenia* (*Pseudoficalbia*).

maximalepido Dong, Zhou, Gong *et al.* [*sic*], 2004 (p. 187) – *nomen nudum*, *Heizmannia*.
maximus Edwards, 1922a (p. 292) – species of *Armigeres* (*Armigeres*).
maxinocca Dyar, 1920d (p. 71) – species of *Culex* (*Melanoconion*).
Synonym: *tosimus* Dyar, 1920d.
mayeri Edwards, 1912c (p. 40) – species of *Uranotaenia* (*Uranotaenia*).
mayottensis Brunhes, 1977b (p. 158) – species of *Uranotaenia* (*Uranotaenia*).
mayumbae Galliard, 1931b (p. 228) – synonym of *Culex* (*Oculeomyia*) *annulioris consimilis* Newstead, 1907 [original combination: *Culex bitaeniorhynchus* variety *mayumbae*]; synonymy by Edwards, 1941.
mccormicki (Belkin, 1962) (p. 420) – species of *Verrallina* (*Verrallina*) [original combination: *Aedes* (*Verrallina*) *mccormicki*].
mcdonaldi Belkin, 1962 (p. 396) – species of *Aedes* (*Ochlerotatus*).
mcdonaldi Belkin & Hogue, 1959 (p. 437) – species of *Deinocerites*.
mcgregori (Banks, 1909) (p. 548) – *nomen dubium*, *Orthopodomyia* [original combination: *Kerteszia mcgregori*]; status by Zavortink, 1968.
mcintoshi Huang, 1985 (p. 109) – species of *Aedes* (*Neomelaniconion*).
medialis Brug, 1932 (p. 79) – species of *Aedes* (*Huaedes*).
medialis Harbach (herein) (p. 3) – species of *Anopheles* (*Anopheles*) [*nomen novum* for *intermedium* Chagas, 1908].
Synonym: *intermedium* Chagas, 1908.
medioalbipes Lutz, 1904 (in Bourroul, 1904) (p. 68) – species of *Wyeomyia* (*Wyeomyia*).
Synonyms: *telestica* Dyar & Knab, 1906d; *abascanta* Dyar & Knab, 1908.
mediofasciata Theobald, 1907 (p. 544) – synonym of *Verrallina* (*Neomacleaya*) *indica* (Theobald, 1907) [original combination: *Skusea mediofasciata*]; synonymy established before Knight & Stone, 1977.
mediolineata Ludlow, 1907 (p. 129) – synonym of *Aedes* (*Ochlerotatus*) *melanimon* Dyar, 1924e [original combination: *Grabhamia mediolineata*]; synonymy by Wood, 1977. **Note**: this nominal taxon is the junior secondary homonym of *mediolineatus* (Theobald, 1901c).
mediolineata (Theobald, 1904b) (p. 71) – species of *Mimomyia* (*Etorleptiomyia*) [original combination: *Etorleptiomyia mediolineata*].
Synonym: *quadrimaculata* Newstead, 1907.
mediolineatus (Theobald, 1901c) (p. 113) – species of *Aedes* (*Aedimorphus*) [original combination: *Culex mediolineatus*]. **Note**: this nominal taxon is the senior secondary homonym of *mediolineata* Ludlow, 1907.
Synonym: *trilineatus* Theobald, 1901a.
mediomaculata Theobald, 1907 (p. 245) – synonym of *Aedes* (*Georgecraigius*) *fluviatilis* (Lutz, 1904) [original combination: *Danielsia mediomaculata*]; synonymy established before Knight & Stone, 1977.
mediopunctata Theobald, 1909b (species) (p. 18) – synonym of *Aedes* (*Aedimorphus*) *cumminsii mesostictus* Harbach (herein) [original combination: *Culicada mediopunctata*]; subspecific status as *Aedes cumminsii mediopunctatus* by Harbach & Howard, 2007. **Note**: this nominal taxon is the junior secondary homonym of *mediopunctatus* (Theobald, 1905h).
mediopunctatus (Lutz, 1903) (in Theobald, 1903a) (p. 60) – species of *Anopheles* (*Anopheles*) [original combination: *Cycloleppteron mediopunctatus*].
Synonyms: *rockefelleri* Peryassú, 1923b; *limai* da Fonseca & da Silva Ramos, 1940; *costalimai* Coutinho, 1944.
mediopunctatus (Theobald, 1905h) (p. 240) – species of *Aedes* (*Stegomyia*) [original combination: *Stegomyia mediopunctata*]. **Note**: this nominal taxon is the senior secondary homonym of *mediopunctata* Theobald, 1909b.
Synonym: *submediopunctata* Barraud, 1923a.
mediovittatus (Coquillett, 1906e) (p. 60) – species of *Aedes* (*Gymnometopa*) [original combination: *Stegomyia mediovittata*].
Synonym: *uncatus* Grabham, 1907.
medleri Knight & Laffoon, 1946 (p. 211) – species of *Aedes* (*Finlaya*).
mefouensis Ferrara, 1974 (p. 3) – species of *Aedes* (*Diceromyia*).
megafolius Chen & Dong, 1992 (p. 78) – species of *Culex* (*Eumelanomyia*).
megaloba Luh, Chao & Xu, 1974 (p. 458) – species of *Culiseta* (*Culiseta*).
megalodora Dyar & Knab, 1908 (p. 69) – synonym of *Wyeomyia* (*Wyeomyia*) *celaenocephala* Dyar & Knab, 1906a; synonymy established before Knight & Stone, 1977.
megaonychus Yang, Li & Chen, 1993 (p. 225) – species of *Culex* (*Culiciomyia*).
megapus Root, 1927c (p. 595) – synonym of *Culex* (*Melanoconion*) *alogistus* Dyar, 1918e; synonymy established before Knight & Stone, 1977.
meigenanus Dyar, 1921e (p. 72) – synonym of *Aedes* (*Ochlerotatus*) *punctor* (Kirby, 1837); synonymy established before Knight & Stone, 1977.
meirai Ribeiro, da Cunha Ramos, Capela & Pires, 1980 (p. 69) – subspecies of *Aedes* (*Ochlerotatus*) *caspius* (Pallas, 1771).
melanesiensis Belkin, 1955 (p. 233) – species of *Tripteroides* (*Polylepidomyia*).
melanesiensis Belkin, 1962 (p. 314) – species of *Mansonia* (*Mansonioides*).
melanimon Dyar, 1924e (p. 126) – species of *Aedes* (*Ochlerotatus*).
Synonyms: *mediolineata* Ludlow, 1907; *klotsi* Matheson, 1933.
melanocephala Dyar & Knab, 1906a (p. 140) – species of *Wyeomyia* (subgenus uncertain, determined by Motta & Lourenço-de-Oliveira 1995).
Synonyms: *agnostips* Dyar & Knab, 1907c; *canfieldi* Dyar & Knab, 1907c; *pandora* Dyar & Knab, 1909d;

grenadensis Edwards, 1916; *fauna* Dyar & Knab, 1919; *modalma* Dyar, 1922d; *hemisiris* Dyar & Shannon, 1925c; *typharum* Shannon & del Ponte, 1928; *melanoides* Root, 1928.

melanocosta Newstead & Carter, 1911b (p. 236) – synonym of *Anopheles* (*Cellia*) *distinctus* (Newstead & Carter, 1911b) [original combination: *Pyretophorus distinctus* variety *melanocosta*]; synonymy by Evans, 1934.

melanoides Root, 1928 (in Dyar, 1928) (p. 79) – synonym of *Wyeomyia melanocephala* Dyar & Knab, 1906a [original combination: *Dendromyia* (*Calladimyia*) *melanoides*]; synonymy by Lane & Cerqueira, 1942.

melanonymphe Dyar, 1924b (p. 100) – species of *Sabethes* (*Sabethinus*) [*nomen novum* for *albiprivatus* Theobald, 1907].
Synonym: *albiprivatus* Theobald, 1907.

melanoon Hackett, 1934 (p. 113) – species of *Anopheles* (*Anopheles*).
Synonym: *subalpinus* Hackett & Lewis, 1935.

melanophylum Dyar & Knab, 1907c (p. 200) – species of *Deinocerites* [original combination: *Deinocerites cancer* variety *melanophylum*]; specific status by Belkin & Hogue, 1959.
Synonym: *monospathus* Dyar & Knab, 1925.

melanopous Graham, 1909 (p. 158) – species of *Eretmapodites*.

melanopterus (Giles, 1904a) (p. 367) – species of *Aedes* (*Phagomyia*) [original combination: *Finlaya melanoptera*]. **Note**: the specific name is here interpreted as an adjectival form and corrected to agree in gender with the masculine *Aedes* Meigen.
Synonym: *palawanensis* Ludlow, 1914a.

melanopus Dyar, 1919d (p. 130) – species of *Wyeomyia* (*Wyeomyia*).
Synonym: *culebrae* Dyar, 1923b.

melanorhinus Giles, 1900a (p. 342) – synonym of *Culex* (*Culex*) *pipiens* Linnaeus, 1758 [*nomen novum* for *pallipes* Macquart, 1838a]; synonymy established before Knight & Stone, 1977.

melanota Cerqueira, 1943 (p. 4) – species of *Psorophora* (*Janthinosoma*).

melanotarsis Anonymous, 1943 (in Director General Medical Services, 1943) (p. 8) – *nomen nudum*, *Anopheles*.

melanura (Coquillett, 1902b) (p. 193) – species of *Culiseta* (*Climacura*) [original combination: *Culex melanurus*].

melas (Theobald, 1903a) (p. 76) – species of *Anopheles* (*Cellia*) [original combination: *Pyretophorus costalis* variety *melas*]; specific status by Ribbands, 1944a,b.

memorans (Bonne-Wepster, 1930) (p. 393) – species of *Coquillettidia* (*Coquillettidia*) [original combination: *Taeniorhynchus* (*Coquillettidia*) *memorans*].

mendacis (Daniels, 1908) (p. 266) – species of *Tripteroides* (*Tripteroides*) [original combination: *Colonemyia mendacis*].

mendiolai Baisas, 1935c (p. 71) – species of *Uranotaenia* (*Uranotaenia*).

mengi Chen, Wang & Zhao, 1989 (p. 224) – species of *Uranotaenia* (*Pseudoficalbia*).

mengi Dong, Wang & Lu, 1990 (p. 64) – species of *Topomyia* (*Suaymyia*).

menglaensis Dong, Zhou & Dong, 2002b (p. 155) – species of *Armigeres* (*Leicesteria*).

menglangensis Ma, 1981a (p. 277) – species of *Anopheles* (*Anopheles*).

menglianensis Lu & Gong, 1986 (p. 184) – species of *Heizmannia* (*Heizmannia*).

menglianeroides Dong, Dong & Zhou, 2003 (in Dong, Dong, Zhou *et al*., 2003) (p. 523) – species of *Heizmannia* (*Heizmannia*).

menoni Mattingly, 1958b (p. 77) – species of *Aedes* (*Paraedes*).

menui Clastrier, 1971 (p. 649) – synonym of *Culex* (*Anoedioporpa*) *damascenoi* Duret, 1969c; synonymy by Berlin & Belkin, 1980.

menytes Dyar, 1918e (p. 125) – synonym of *Culex* (*Melanoconion*) *spissipes* (Theobald, 1903a); synonymy established before Knight & Stone, 1977.

meprai Martínez & Prosen, 1953b (p. 1) – species of *Aedes* (*Ochlerotatus*).

merak (*cohesia*) Mangkoewinoto, 1919 (p. 57) – synonym of *Anopheles* (*Cellia*) *flavirostris* (Ludlow, 1914a) [original combination: *Myzomyia aconita* variety *merak* (*cohesia*)]; synonymy by Harrison, 1980.

meraukensis Venhuis, 1932 (p. 1040) – species of *Anopheles* (*Cellia*).

mercedesae Baisas, 1974 (p. 121) – synonym of *Culex* (*Lophoceraomyia*) *kuhnsi* King & Hoogstraal, 1955 [original combination: *Culex* (*Lophoceraomyia*) *uniformis mercedesae*]; synonymy by Sirivanakarn, 1977a.

mercurator Dyar, 1920a (p. 13) – species of *Aedes* (*Ochlerotatus*).
Synonyms: *albertae* Dyar, 1920e; *ater* Gutsevich, 1955.

meridionalis Dyar & Knab, 1906d (p. 195) – synonym of *Aedes* (*Ochlerotatus*) *serratus* (Theobald, 1901c); synonymy established before Knight & Stone, 1977.

meridionalis Leach, 1825 (p. 292) – synonym of *Culex* (*Culex*) *pipiens* Linnaeus, 1758; synonymy established before Knight & Stone, 1977.

merodaemon Dyar, 1921f (p. 100) – synonym of *Culex* (*Melanoconion*) *conspirator* Dyar & Knab, 1906d; synonymy established before Knight & Stone, 1977.

meronephada (Dyar & Shannon, 1925a) (p. 71) – species of *Aedes* (*Catatassomyia*) [original combination: *Catatassomyia meronephada*].
Synonym: *wainwrighti* Baisas, 1946c.

meroneus Dyar, 1925b (p. 22) – synonym of *Culex* (*Melanoconion*) *conspirator* Dyar & Knab, 1906d; synonymy established before Knight & Stone, 1977.

merus Dönitz, 1902 (p. 77) – species of *Anopheles* (*Cellia*).
Synonyms: *litoralis* Halcrow, 1957; *tangensis* Kuhlow, 1962.

mesodentatus Komp & Kumm, 1938 (p. 253) – species of *Haemagogus* (*Haemagogus*).
Synonyms: *gorgasi* Galindo & Trapido, 1956; *alticola* Galindo, Trapido & Boshell-Manrique, 1956.

mesodenticulatus Galindo & Mendez, 1961 (p. 2) – species of *Culex* (*Melanoconion*).

mesopotamiae Christophers & Chand, 1915 (p. 196) – synonym of *Anopheles* (*Anopheles*) *hyrcanus* (Pallas, 1771) [original combination: *Anopheles sinensis* variety *mesopotamiae*]; synonymy by Christophers, 1933.

mesostictus Harbach (herein) (p. 3) – subspecies of *Aedes* (*Aedimorphus*) *cumminsii* (Theobald, 1903a) [*nomen novum* for *mediopunctata* Theobald, 1909b]; formerly *Aedes cumminsii mediopunctatus* (Theobald, 1909b); subspecific status by Harbach & Howard, 2007.
Synonyms: *mediopunctata* Theobald, 1909b; *daruensis* Evans, 1925a.

messeae Falleroni, 1926 (p. 564) – species of *Anopheles* (*Anopheles*) [original combination: *Anopheles claviger* variety *messeae*]; specific status by Bates, 1940.

metaboles Theobald, 1902a (p. 374) – synonym of *Anopheles* (*Cellia*) *stephensi* Liston, 1901; synonymy established before Knight & Stone, 1977.

metalepticus Dyar, 1920c (p. 51) – synonym of *Aedes* (*Ochlerotatus*) *pullatus* (Coquillett, 1904c); synonymy established before Knight & Stone, 1977.

metallica (Leicester, 1908) (species) (p. 113) – subspecies of *Mimomyia* (*Mimomyia*) *chamberlaini* Ludlow, 1904c [original combination: *Conopomyia metallica*]; subspecific status by Rattanarithikul *et al*., 2006.

metallica Leicester, 1908 (p. 251) – synonym of *Heizmannia* (*Heizmannia*) *indica* (Theobald, 1905f) [original combination: *Wyeomyia metallica*]; synonymy by Mattingly, 1970.

metallica (Theobald, 1901d) (p. 8) – species of *Coquillettidia* (*Coquillettidia*) [original combination: *Culex metallicus*].
Synonyms: *nigrochaetae* Theobald, 1901c; *violaceus* Theobald, 1908c; *nigra* Theobald, 1910c.

metallicus (Edwards, 1912c) (p. 12) – species of *Aedes* (*Stegomyia*) [*nomen novum* for *dubia* Theobald, 1910c; original combination: *Stegomyia metallica*].
Synonym: *dubia* Theobald, 1910c.

metallicus Leicester, 1904 (in Theobald, 1904c) (p. 37) – species of *Toxorhynchites* (*Toxorhynchites*).

metatarsata Edwards, 1914b (p. 81) – species of *Uranotaenia* (*Uranotaenia*).
Synonym: *innotata* Dyar & Shannon, 1925a.

metcalfi Galvão & Lane, 1937b (p. 218) – synonym of *Anopheles* (*Nyssorhynchus*) *evansae* (Brèthes, 1926b) [original combination: *Anopheles* (*Nyssorhynchus*) *oswaldoi* variety *metcalfi*]; synonymy by Faran, 1981.

metempsytus Dyar, 1921k (as *metempsta*) (p. 154) – species of *Culex* (*Carrollia*).

methysticus Dyar & Knab, 1909d (p. 266) – synonym of *Limatus asulleptus* (Theobald, 1903a); synonymy established before Knight & Stone, 1977.

metoecopus Dyar, 1925d (p. 30) – species of *Aedes* (*Protomacleaya*).

mexicana (Bellardi, 1859) (p. 5) – species of *Psorophora* (*Janthinosoma*) [original combination: *Culex mexicanus*].

mexicanus Zavortink & Poinar, 2008 (p. 118) – fossil species of *Toxorhynchites* (*Toxorhynchites*).

miachaetessa Dyar & Shannon, 1925a (p. 78) – synonym of *Aedes* (*Cancraedes*) *curtipes* Edwards, 1915c; synonymy by Mattingly, 1958a.

miaolingensis Chen, 1982 (p. 312) – species of *Culex* (*Eumelanomyia*).

micans Leicester, 1908 (p. 206) – species of *Uranotaenia* (*Uranotaenia*).

michaeli de Meillon & Leeson, 1940 (p. 63) – synonym of *Anopheles* (*Cellia*) *schwetzi* Evans, 1934; synonymy by Gillies & de Meillon, 1968.

michaelikati van Someren, 1946c (p. 3) – species of *Aedes* (*Pseudarmigeres*).

mickevichae Huang, 1988b (p. 6) – species of *Aedes* (*Stegomyia*).

micolo Ribeiro, da Cunha Ramos, Capela & Pires, 1998 (p. 12) – species of *Culex* (*Eumelanomyia*).

microannulata Theobald, 1907 (p. 481) – synonym of *Culex* (*Culex*) *nigripalpus* Theobald, 1901c [original combination: *Trichopronomyia microannulata*]; synonymy established before Knight & Stone, 1977.

microannulata (Theobald, 1911a) (p. 26) – species of *Coquillettidia* (*Coquillettidia*) [original combination: *Chrysoconops microannulatus*].
Synonyms: *auripennis* Edwards, 1915b; *chubbi* Edwards, 1915b.

microannulatus Theobald, 1901a (p. 353) – synonym of *Culex* (*Culex*) *sitiens* Wiedemann, 1828; synonymy established before Knight & Stone, 1977.

microcala (Dyar, 1929c) (p. 61) – species of *Tripteroides* (*Tricholeptomyia*) [original combination: *Rachionotomyia microcala*].

microlepis (Edwards, 1927b) (p. 353) – species of *Tripteroides* (*Polylepidomyia*) [original combination: *Rachionotomyia microlepis*].

micromelas Edwards, 1934 (as *micomeles*) (p. 326) – species of *Uranotaenia* (*Pseudoficalbia*).

microphyllus Root, 1927c (p. 586) – species of *Culex* (*Microculex*).

micropterus (Giles, 1901a) (p. 609) – species of *Aedes* (*Dendroskusea*) [original combination: *Culex micropterus*].

microsquamosus Theobald, 1905 (in Grabham, 1905) (p. 407) – synonym of *Culex* (*Culex*) *nigripalpus* Theobald, 1901c; synonymy established before Knight & Stone, 1977.

microstictus Edwards, 1936 (p. 50) – species of *Aedes* (*Catageiomyia*).

microtaeniata Theobald, 1911d (p. 236) – *nomen dubium*, *Culex*; status by Colless, 1957a.

mijanae Brunhes, Adam & Bailly-Choumara, 1967 (p. 51) – species of *Culex* (*Eumelanomyia*).

mikiranus Edwards, 1922a (p. 269) – synonym of *Aedes* (*Danielsia*) *albotaeniatus* (Leicester, 1904); synonymy by Harrison *et al*., 1991.

mikrokopion Knight & Harrison, 1988 (p. 215) – species of *Aedes* (*Downsiomyia*).

milesi de Meillon & Evans, 1935 (p. 93) – synonym of *Anopheles* (*Cellia*) *wellcomei ugandae* Evans, 1934 [original combination: *Anopheles walravensi* variety *milesi*]; synonymy by Gillies & de Meillon, 1968.

millecampsi Lips, 1960a (p. 52) – species of *Anopheles* (*Cellia*) [original combination: *Anopheles dureni* variety *millecampsi*]; specific status by Brunhes *et al.*, 1997.

milleri Dyar, 1922g (p. 194) – species of *Aedes* (*Ochlerotatus*).
Synonyms: *araozi* Shannon & del Ponte, 1928; *oroecetor* Martini, 1931a.

millironi Belkin, 1962 (p. 242) – species of *Culex* (*Neoculex*).

milloti Doucet, 1949b (p. 331) – species of *Culex* (*Culiciomyia*).

milloti Doucet, 1951a (p. 66) – species of *Orthopodomyia*.

milloti Grjebine, 1986 (p. 163) – species of *Mimomyia* (*Ingramia*).

milloti Grjebine & Lacan, 1953 (in Grjebine, 1953a) (p. 495) – species of *Anopheles* (*Cellia*).

milnensis (King & Hoogstraal, 1947b) (p. 122) – species of *Verrallina* (*Verrallina*) [original combination: *Aedes* (*Aedes*) *milnensis*].

milnensis Lee, 1944b (p. 215) – species of *Armigeres* (*Armigeres*).

milni Taylor, 1914d (p. 196) – synonym of *Culex* (*Culex*) *sitiens* Wiedemann, 1828 [original combination: *Culicelsa annulirostris* variety *milni*]; synonymy established before Knight & Stone, 1977.

milsoni (Taylor, 1915) (p. 179) – species of *Aedes* (*Dobrotworskyius*) [original combination: *Culicada milsoni*].
Synonym: *milsoni* Taylor, 1916.

milsoni Taylor, 1916 (p. 566) – synonym of *Aedes* (*Dobrotworskyius*) *milsoni* (Taylor, 1915) [original combination: *Hulecoeteomyia milsoni*]; synonymy established before Knight & Stone, 1977.

milwardi Xavier & da Silva Mattos, 1972 (p. 569) – species of *Culex* (*Melanoconion*).

mimesis Dyar, 1917c (p. 116) – synonym of *Aedes* (*Ochlerotatus*) *fitchii* (Felt & Young, 1904); synonymy established before Knight & Stone, 1977.

mimeticus Noè, 1899 (p. 240) – species of *Culex* (*Culex*).
Synonym: *pseudomimeticus* Sergent, 1909.

mimomyiaformis (Newstead, 1907) (in Newstead *et al.*, 1907) (p. 34) – species of *Mimomyia* (*Mimomyia*) [original combination: *Boycia mimomyiaformis*].
Synonyms: *sudanensis* Theobald, 1907; *pincerna* Graham, 1910.

mimuloides Barraud, 1924c (p. 992) – species of *Culex* (*Culex*) [original combination: *Culex mimeticus* variety *mimuloides*]; specific status by F.W. Edwards (in Barraud, 1934).

mimulus Edwards, 1915c (p. 284) – species of *Culex* (*Culex*).
Synonyms: *mossmani* Taylor, 1915; *confusus* Baisas, 1938; *neomimulus* Lien, 1968d.

mindanaoensis Baisas, 1935d (p. 168) – *nomen dubium*, *Culex*; status by Sirivanakarn, 1977a.

mindoroensis Knight & Hull, 1951a (p. 199) – synonym of *Aedes* (*Aedimorphus*) *lowisii* (Theobald, 1910c); synonymy by Reinert, 1973b.

minima Ludlow, 1907 (p. 413) – synonym of *Ficalbia ludlowae* Brunetti, 1920 [original combination: *Ludlowia minima*]; synonymy established before Knight & Stone, 1977.

minima (Theobald, 1901c) (p. 262) – species of *Ficalbia* [original combination: *Uranotaenia minima*].
Synonym: *minuta* Theobald, 1908b.

minimus Leicester, 1908 (p. 160) – synonym of *Culex* (*Culex*) *fuscocephala* Theobald, 1907; synonymy established before Knight & Stone, 1977.

minimus Theobald, 1901a (p. 186) – species of *Anopheles* (*Cellia*).
Synonyms: *formosaensis I* Tsuzuki, 1902a; *christophersi* Theobald, 1902a; *cohaesa* Dönitz, 1903; *alboapicalis* Theobald, 1910c.

minimus (Theobald, 1905h) (p. 237) – species of *Toxorhynchites* (*Toxorhynchites*) [original combination: *Megarhinus minimus*].

minjensis Sirivanakarn, 1968 (p. 151) – species of *Culex* (*Lophoceraomyia*).

minnesotae Barr, 1957 (p. 163) – species of *Culiseta* (*Culicella*).

minor da Costa Lima, 1929 (p. 1101) – species of *Anopheles* (*Anopheles*). **Note:** this nominal taxon is the senior primary homonym of *minor* Holstein, 1949.

minor Dyar & Knab, 1906a (p. 138) – synonym of *Wyeomyia* (*Wyeomyia*) *bahama* Dyar & Knab, 1906a; synonymy established before Knight & Stone, 1977.

minor Holstein, 1949 (p. 156) – synonym of *Anopheles* (*Cellia*) *gambiae* Giles, 1902; synonymy established before Knight & Stone, 1977. **Note:** this nominal taxon is the junior primary homonym of *minor* da Costa Lima, 1929.

minor (Leicester, 1908) (p. 126) – species of *Culex* (*Lophoceraomyia*) [original combination: *Lophoceratomyia minor*]. **Note:** this nominal taxon is the senior secondary homonym of *minor* Theobald, 1908b.
Synonyms: *plantaginis* Barraud, 1924h; *nolledoi* Baisas, 1935d; *harpagophallus* Wang & Feng, 1964.

minor Leicester, 1908 (p. 238) – species of *Topomyia* (*Topomyia*).

minor Theobald, 1908b (p. 298) – synonym of *Culex* (*Culex*) *quinquefasciatus* Say, 1823; synonymy established before Knight & Stone, 1977. **Note:** this nominal taxon is the junior secondary homonym of *minor* (Leicester, 1908).

minuta Taylor, 1912a (p. 30) – synonym of *Aedes* (*Macleaya*) *tremulus* (Theobald, 1903e) [original combination: *Danielsia minuta*]; synonymy established before Knight & Stone, 1977. **Note:** this nominal taxon is the junior secondary homonym of *minutus* (Theobald, 1901a) and *minuta* Theobald, 1907 (p. 338).

minuta Theobald, 1907 (p. 338) – synonym of *Aedes* (*Aedimorphus*) *vexans* (Meigen, 1830b) [original combination: *Culicada minuta*]; synonymy established before Knight & Stone, 1977. **Note:** this nominal taxon is the junior secondary homonym of *minutus* (Theobald, 1901a) and the senior secondary homonym of *minuta* Taylor, 1912a.

minuta Theobald, 1907 (p. 559) – synonym of *Uranotaenia* (*Uranotaenia*) *lowii* Theobald, 1901c; synonymy established before Knight & Stone, 1977.

minuta Theobald, 1908b (p. 301) – synonym of *Ficalbia minima* (Theobald, 1901c) [original combination: *Mimomyia minuta*]; synonymy established before Knight & Stone, 1977.

minutissima Theobald, 1910a (p. 9) – synonym of *Aedes* (*Stegomyia*) *gardnerii imitator* (Leicester, 1908) [original combination: *Stegomyia minutissima*]; synonymy by Huang, 1977.

minutissimus (Theobald, 1907) (p. 235) – species of *Culex* (*Lophoceraomyia*) [original combination: *Culiciomyia minutissima*].
Synonyms: *juxtapallidiceps* Theobald, 1910c; *nigerrima* Theobald, 1910c.

minutus Eversmann, 1834 (p. 420) – *nomen nudum*, *Culex*.

minutus Macquart, 1834 (p. 33) – *nomen dubium*, *Anopheles*; status by Gillies & de Meillon, 1968.

minutus Poinar, Zavortink, Pike & Johnston, 2000 (p. 121) – fossil species of *Paleoculicis* Poinar, Zavortink, Pike & Johnston, 2000.

minutus (Theobald, 1901a) (p. 319) – species of *Aedes* (*Catageiomyia*) [original combination: *Stegomyia minuta*]. **Note:** this nominal taxon is the senior secondary homonym of *minuta* Taylor, 1912a and *minuta* Theobald, 1907.

minutus Theobald, 1903a (p. 91) – synonym of *Anopheles* (*Anopheles*) *nigerrimus* Giles, 1900a [original combination: *Myzorhynchus minutus*]; synonymy established before Knight & Stone, 1977.

minutus Theobald, 1905g (p. 30) – synonym of *Culex* (*Culex*) *decens* Theobald, 1901d; synonymy by Harbach, 1988.

miraculosus Bonne-Wepster, 1937 (p. 92) – species of *Culex* (*Culex*).

mirans Sallum & Peyton, 2005 (in Sallum *et al.*, 2005b) (p. 189) – species of *Anopheles* (*Cellia*).

mirificus Edwards, 1913b (p. 57) – species of *Culex* (*Culex*).

misionensis Duret, 1953, 1954a (pp. 75, 115, respectively) – species of *Culex* (*Melanoconion*).

missiroli del Vecchio, 1939 (p. 34) – synonym of *Anopheles* (*Anopheles*) *claviger* (Meigen, 1804) [original combination: *Anopheles claviger* (*bifurcatus*) variety *missiroli*]; synonymy by M. Bates (in Boyd, 1949).

mississippii Dyar, 1920d (p. 113) – synonym of *Aedes* (*Ochlerotatus*) *stimulans* (Walker, 1848) [original combination: *Aedes stimulans mississippii*]; synonymy established before Knight & Stone, 1977.

mistura Komp & Rozeboom, 1951 (p. 124) – synonym of *Culex* (*Melanoconion*) *inadmirabilis* Dyar, 1928; synonymy by Forattini & Sallum, 1993a.

mitchellae (Dyar, 1905b) (p. 74) – species of *Aedes* (*Ochlerotatus*) [original combination: *Culex mitchellae*].

mitchellii (Theobald, 1905d) (p. 37) – species of *Wyeomyia* (*Wyeomyia*) [original combination: *Dendromyia mitchellii*].
Synonyms: *jamaicensis* Theobald, 1905d; *glaucocephala* Dyar & Knab, 1906a; *violescens* Dyar & Knab, 1906a; *ochrura* Dyar & Knab, 1906d; *abia* Dyar & Knab, 1908; *antoinetta* Dyar & Knab, 1909d.

mixtus Edwards, 1936 (p. 49) – species of *Aedes* (*Catageiomyia*).

miyagii Mogi & Toma, 1999 (p. 473) – species of *Culex* (*Acallyntrum*).

miyagii (Toma & Mogi, 2003) (p. 25) – species of *Kimia* [original combination: *Topomyia* (*Suaymyia*) *miyagii*].

miyarai Tanaka, Mizusawa & Saugstad, 1979 (p. 390) – subspecies of *Aedes* (*Stegomyia*) *flavopictus* Yamada, 1921.

mjobergi (Edwards, 1926b) (p. 248) – species of *Aedes* (*Downsiomyia*) [original combination: *Armigeres mjobergi*].

moctezuma (Dyar & Knab, 1906c) (p. 251) – species of *Toxorhynchites* (*Lynchiella*) [original combination: *Megarhina moctezuma*].
Synonym: *trinidadensis* Dyar & Knab, 1906c.

modalma Dyar, 1922d (p. 97) – synonym of *Wyeomyia melanocephala* Dyar & Knab, 1906a; synonymy established before Knight & Stone, 1977.

modesta (King & Hoogstraal, 1946e) (p. 186) – species of *Mimomyia* (*Mimomyia*) [original combination: *Ficalbia* (*Mimomyia*) *modesta*].

modesta Leicester, 1908 (p. 218) – species of *Uranotaenia* (*Pseudoficalbia*).
Synonym: *tubanguii* Baisas, 1935c.

modesta Martini, 1935 (p. 29) – synonym of *Uranotaenia* (*Uranotaenia*) *pulcherrima* Lynch Arribálzaga, 1891b [original combination: *Uranotaenia pulcherrima* variety *modesta*]; synonymy by Lane, 1953.

modestus Ficalbi, 1890b (p. 93) – species of *Culex* (*Barraudius*).
Synonyms: *eadithae* Barraud, 1924d; *tanajcus* Stschelkanovzev, 1926; *nudipalpis* Shingarev, 1927.

moenensis Iyengar, 1955 (p. 30) – *nomen nudum*, *Aedes*.

moengoensis Bonne-Wepster & Bonne, 1923 (p. 7) – synonym of *Toxorhynchites* (*Lynchiella*) *theobaldi* (Dyar & Knab, 1906c) [original combination: *Megarhinus moengoensis*]; synonymy by Lane, 1939.

moerbista (Dyar & Knab, 1919) (p. 2) – species of *Wyeomyia* (subgenus uncertain, determined by Motta & Lourenço-de-Oliveira, 1995) [original combination: *Sabethinus moerbista*].
Synonym: *ocumarensis* Lutz, 1928.

moghulensis Christophers, 1924c (p. 296) – species of *Anopheles* (*Cellia*) [original combination: *Anopheles jeyporiensis* variety *moghulensis*]; specific status by Puri, 1928.

mogii Miyagi, Toma & Higa, 2004 (p. 11) – species of *Mimomyia* (*Ingramia*).

mogilasium (Dyar & Knab, 1907c) (species) (p. 206) – subspecies of *Trichoprosopon compressum* Lutz, 1905 [original combination: *Joblotia mogilasia*]; subspecific status by Harbach & Howard, 2007.

mohani Knight, 1969 (p. 382) – species of *Aedes* (*Downsiomyia*).

mohani Sirivanakarn, 1977d (p. 537) – species of *Culex* (*Eumelanomyia*).

mojuensis Duret & Damasceno, 1955 (p. 401) – synonym of *Culex* (*Tinolestes*) *breviculus* Senevet & Abonnenc, 1939a; synonymy by Belkin, 1968b.

molestus Forskål, 1775 (p. 85) – synonym of *Culex* (*Culex*) *pipiens* Linnaeus, 1758; synonymy by Harbach, Harrison & Gad, 1984. **Note**: this nominal taxon is the senior primary homonym of *molestus* Wiedemann, 1828.

molestus Kollar, 1832 (in Pohl & Kollar, 1832) (p. 18) – *nomen dubium*, *Culex*; status by Belkin, 1968b.

molestus Wiedemann, 1820 (p. 7) – synonym of *Psorophora* (*Psorophora*) *ciliata* (Fabricius, 1794) [original combination: *Culex molestus*]; synonymy established before Knight & Stone, 1977.

molestus Wiedemann, 1828 (p. 542) – synonym of *Culex* (*Lophoceraomyia*) *fraudatrix* (Theobald, 1905f); synonymy established before Knight & Stone, 1977. **Note**: this nominal taxon is the junior primary homonym of *molestus* Forskål, 1775.

mollis Dyar & Knab, 1906b (p. 171) – species of *Culex* (*Culex*) [original combination: *Culex carmodyae mollis*]; specific status by Dyar, 1918c.
Synonyms: *lateropunctata* Theobald, 1907; *equivocator* Dyar & Knab, 1907c; *elocutilis* Dyar & Knab, 1909d; *lepostenis* Dyar, 1923b; *tisseuli* Senevet, 1937.

moloiensis (Taylor, 1929) (p. 276) – species of *Aedes* (*Macleaya*) [original combination: *Chaetocruiomyia moloiensis*].

moluccensis Swellengrebel & Swellengrebel de Graaf, 1920a (p. 1) – synonym of *Anopheles* (*Cellia*) *farauti* Laveran, 1902b [original combination: *Nyssorhynchus annulipes* variety *moluccensis*]; synonymy by Knight & Farner, 1944.

mombasaensis Mattingly, 1963 (p. 165) – synonym of *Aedes* (*Coetzeemyia*) *fryeri* (Theobald, 1912b); synonymy by van Someren, 1972.

mombasaensis Theobald, 1901c (p. 36) – synonym of *Lutzia* (*Metalutzia*) *tigripes* de Grandpré & de Charmoy, 1901 [original combination: *Culex maculicrura* variety *mombasaensis*]; synonymy by Edwards, 1941.

monetifer (Dyar, 1920i) (p. 176) – species of *Tripteroides* (*Tripteroides*) [original combination: *Rachionotomyia monetifera*].

monetus Edwards, 1935a (p. 132) – species of *Aedes* (*Zavortinkius*).

mongiro van Someren, 1951 (p. 2) – species of *Culex* (*Culiciomyia*).

monilis Shannon & del Ponte, 1928 (p. 84) – synonym of *Uranotaenia* (*Uranotaenia*) *lowii* Theobald, 1901c; synonymy by Lane, 1943.

monocellatus Marks, 1948 (p. 14) – species of *Aedes* (subgenus uncertain).

monoleua Martini, 1931b (p. 116) – synonym of *Sabethes* (*Davismyia*) *petrocchiae* (Shannon & del Ponte, 1928) [original combination: *Miamyia monoleua*]; synonymy by Lane & Cerqueira, 1942.

monospathus Dyar & Knab, 1925 (in Dyar, 1925f) (p. 155) – synonym of *Deinocerites melanophylum* Dyar & Knab, 1907c; synonymy by Belkin & Hogue, 1959.

monotrichus Edwards, 1936 (p. 52) – species of *Aedes* (*Neomelaniconion*).

monstrosus Anonymous (in IMPE, 1987) (pp. 11, 33, 57) – *nomen nudum*, *Anopheles*; unavailable name [authorship cannot be determined from the content of the publication and is therefore deemed to be anonymous; a new name published after 1950 with anonymous authorship is not made available (International Commission on Zoological Nomenclature, 1999: *Code*, Articles 50.1 and 14, respectively)].

montana Ingram & de Meillon, 1927 (p. 80) – species of *Uranotaenia* (*Pseudoficalbia*).

montana Koidzumi, 1918 (p. 141) – synonym of *Aedes* (*Stegomyia*) *desmotes* (Giles, 1904a) [original combination: *Stegomyia montana*]; synonymy by Lien, 1962.

montanus Brug, 1939 (p. 104) – subspecies of *Aedes* (*Rampamyia*) *notoscriptus* (Skuse, 1889) [original combination: *Aedes* (*Finlaya*) *notoscriptus* variety *montana* [*sic*]]; subgeneric status by Stone *et al*., 1959.

montanus Stanton & Hacker, 1917 (p. 273) – species of *Anopheles* (*Anopheles*) [original combination: *Anopheles albotaeniatus* variety *montanus*]; specific status prior to Edwards, 1932a.

montcalmi Blanchard, 1905 (p. 307) – synonym of *Aedes* (*Aedimorphus*) *vexans* (Meigen, 1830b) [*nomen novum* for *sylvestris* Theobald, 1901a; original combination: *Culex montcalmi*]; synonymy established before Knight & Stone, 1977.

montchadskyi Dubitzky, 1968 (p. 218) – species of *Aedes* (*Ochlerotatus*).

montemor Corrêa, 1950 (p. 53) – synonym of *Anopheles* (*Kerteszia*) *cruzii* Dyar & Knab, 1908; synonymy by Barretto & Coutinho, 1951.

montforti Ventrillon, 1905a (p. 448) – synonym of *Culex* (*Culex*) *univittatus* Theobald, 1901c [original combination: *Heptaphlebomyia montforti*]; synonymy established before Knight & Stone, 1977.

monticola Belkin & McDonald, 1957 (p. 179) – species of *Aedes* (*Jarnellius*).

moorei Dyar, 1918c (p. 108) – synonym of *Culex* (*Melanoconion*) *erraticus* (Dyar & Knab, 1906d); synonymy by King & Bradley, 1937.

mooseri Vargas & Martínez Palacios, 1954 (p. 33) – synonym of *Culex* (*Culex*) *coronator* Dyar & Knab, 1906d [original combination: *Culex* (*Culex*) *coronator mooseri*]; synonymy by Bram, 1967b.

moralesi (Dyar & Knab, 1919) (p. 3) – species of *Shannoniana* [original combination: *Lesticocampa moralesi*].

moramangae da Cunha Ramos & Brunhes, 2004 (p. 393) – species of *Uranotaenia* (*Uranotaenia*).

moresbyensis Peters, 1963e (p. 204) – species of *Uranotaenia* (*Uranotaenia*).

morsitans (Theobald, 1901c) (p. 8) – species of *Culiseta* (*Culicella*) [original combination: *Culex morsitans*].

mortiauxi da Cunha Ramos & Ribeiro, 1990 (p. 139) – species of *Eretmapodites*.

mortiauxi Edwards, 1938 (in Evans, 1938) (p. 204) – species of *Anopheles* (*Cellia*).

mortificator Dyar & Knab, 1906d (p. 210) – synonym of *Culex* (*Culex*) *nigripalpus* Theobald, 1901c; synonymy established before Knight & Stone, 1977.

mosquito Robineau-Desvoidy, 1827 (p. 407) – synonym of *Aedes* (*Stegomyia*) *aegypti* (Linnaeus, 1762) [original combination: *Culex mosquito*]; synonymy established before Knight & Stone, 1977.

mossmani Taylor, 1915 (p. 181) – synonym of *Culex* (*Culex*) *mimulus* Edwards, 1915c; synonymy established before Knight & Stone, 1977.

moucheti Evans, 1923a (p. 89) – species of *Culex* (*Kitzmilleria*).

moucheti Evans, 1925b (p. 211) – species of *Anopheles* (*Cellia*) [original combination: *Anopheles marshalli* variety *moucheti*]; specific status by Evans, 1931a.

moucheti (Hamon & Adam, 1955) (p. 882) – species of *Malaya* [original combination: *Harpagomyia moucheti*].

moucheti Ravaonjanahary & Brunhes, 1977 (p. 213) – species of *Aedes* (*Skusea*). **Note**: the name of the second author is misspelled as 'Bruhnes' in the publication.

moufiedi Peyton, 1977 (p. 86) – species of *Uranotaenia* (*Pseudoficalbia*).

moultoni Edwards, 1914c (p. 125) – species of *Armigeres* (*Armigeres*).

moultoni Edwards, 1914c (p. 128) – species of *Uranotaenia* (*Pseudoficalbia*).
 Synonym: *brevirostris* Edwards, 1915c.

mousinhoi de Meillon & de Carvalho Pereira, 1940 (p. 80) – species of *Anopheles* (*Cellia*) [original combination: *Anopheles marshalli* [*sic*] variety *mousinhoi*]; specific status by Gillies & de Meillon, 1968.

mpusiensis Huang, 2004 (p. 47) – species of *Aedes* (*Stegomyia*).

mubiensis Luh & Shih, 1958 (as *mubienensis*) (p. 226) – species of *Aedes* (*Aedes*).

mucidus (Karsch, 1887) (p. 25) – species of *Aedes* (*Mucidus*) [original combination: *Culex mucidus*].

muehlensi Petrocchi, 1927 (in Shannon & del Ponte, 1927) (p. 727) – species of *Wyeomyia* (*Phoniomyia*).

muelleri Dyar, 1920j (p. 81) – species of *Aedes* (*Lewnielsenius*).
 Synonym: *iridipennis* Dyar, 1922d.

mulrennani Basham, 1948 (p. 1) – species of *Culex* (*Melanoconion*).

multicinctus Edwards, 1930a (p. 291) – species of *Anopheles* (*Cellia*).
 Synonym: *puselis* de Meillon, 1934.

multicolor Cambouliu, 1902 (p. 705) – species of *Anopheles* (*Cellia*).
 Synonyms: *impunctus* Dönitz, 1902; *chaudoyei* Theobald, 1903a; *nigrifasciatus* Theobald, 1907.

multifolium (King & Hoogstraal, 1947b) (p. 122) – species of *Verrallina* (*Verrallina*) [original combination: *Aedes* (*Aedes*) *multifolium*].

multimaculatus Gastiaburú, 1945 (in Aitken, 1945) (p. 327) – *nomen nudum*, *Anopheles*.

multimaculosus Leicester, 1908 (p. 155) – synonym of *Lutzia* (*Metalutzia*) *halifaxii* Theobald, 1903a [original combination: *Culex multimaculosus*]; synonymy established before Knight & Stone, 1977.

multiplex (Theobald, 1903a) (p. 293) – species of *Aedes* (*Pseudoskusea*) [original combination: *Skusea multiplex*].

multispinosus Bonne-Wepster & Bonne, 1920 (p. 177) – synonym of *Culex* (*Melanoconion*) *caudelli* (Dyar & Knab, 1906d); synonymy established before Knight & Stone, 1977.

mundulus Grünberg, 1905 (p. 388) – species of *Culex* (*Eumelanomyia*).

muroafcete Huang, 1997 (p. 54) – species of *Aedes* (*Stegomyia*).

murphyi Gillies & de Meillon, 1968 (p. 253) – species of *Anopheles* (*Cellia*).

murrelli Lien, 1968d (p. 243) – species of *Culex* (*Culex*).

muruae Sirivanakarn, 1968 (p. 130) – species of *Culex* (*Lophoceraomyia*).

mus Dyar, 1920i (p. 175) – synonym of *Tripteroides* (*Tricholeptomyia*) *nepenthicola* (Banks, 1909) [original combination: *Wyeomyia* (*Dodecamyia*) *mus*]; synonymy established before Knight & Stone, 1977.

musarum Edwards, 1932b (p. 562) – species of *Culex* (*Culex*).

musarum Edwards, 1936 (p. 54) – species of *Uranotaenia* (*Pseudoficalbia*) [original combination: *Uranotaenia ornata* variety *musarum*]; specific status by da Cunha Ramos, 1993.

musicus Leach, 1825 (p. 293) – synonym of *Aedes* (*Rusticoidus*) *rusticus* (Rossi, 1790) [original combination: *Culex musicus*]; synonymy established before Knight & Stone, 1977.

musicus Say, 1829 (p. 149) – synonym of *Psorophora* (*Janthinosoma*) *ferox* (von Humboldt, 1819) [original combination: *Culex musicus*]; synonymy established before Knight & Stone, 1977.

musivus Skuse, 1889 (p. 1754) – synonym of *Anopheles* (*Cellia*) *annulipes* Walker, 1856a; synonymy established before Knight & Stone, 1977.

muspratti Hamon & Lambrecht, 1959 (p. 582) – species of *Culex* (*Culiciomyia*).

mutator Dyar & Knab, 1906d (p. 216) – species of *Culex* (*Melanoconion*).

mutatus Dyar, 1919b (p. 24) – synonym of *Aedes* (*Ochlerotatus*) *increpitus* Dyar, 1916b; synonymy established before Knight & Stone, 1977.

mutica Enderlein, 1923 (p. 28) – synonym of *Aedes* (*Skusea*) *pembaensis* Theobald, 1901c [original combination: *Phagomyia mutica*]; synonymy established before Knight & Stone, 1977.

muticus Edwards, 1923b (p. 6) – synonym of *Culex* (*Culiciomyia*) *pullus* Theobald, 1905f; synonymy established before Knight & Stone, 1977.

mutilus Edwards, 1936 (p. 50) – species of *Aedes* (*Catageiomyia*).

mychonde Komp, 1928 (in Dyar, 1928) (p. 295) – synonym of *Culex* (*Melanoconion*) *taeniopus* Dyar & Knab, 1907b; synonymy by Sallum & Forattini, 1996.

mysorensis Sweet & Rao, 1937 (p. 673) – synonym of *Anopheles* (*Cellia*) *stephensi* Liston, 1901 [original combination: *Anopheles stephensi* variety *mysorensis*]; synonymy by Rutledge *et al.*, 1970.

mystes Dyar, 1924a (p. 92) – species of *Wyeomyia* (*Spilonympha*).

myzomyifacies Theobald, 1907 (p. 69) – synonym of *Anopheles* (*Cellia*) *cinereus hispaniola* (Theobald, 1903a) [original combination: *Pyretophorus myzomyifacies*]; synonymy established before Knight & Stone, 1977.

mzooi van Someren, 1962 (p. 19) – species of *Aedes* (*Zavortinkius*).

Nn

nagpori James & Liston, 1911 (p. 90) – synonym of *Anopheles* (*Cellia*) *annularis* van der Wulp, 1884 [original combination: *Nyssorhynchus fuliginosus* variety *nagpori*]; synonymy established before Knight & Stone, 1977.

nailoni King & Hoogstraal, 1946a (p. 143) – species of *Culex* (*Culiciomyia*).

nairobiensis (van Someren, 1946b) (p. 181) – species of *Toxorhynchites* (*Afrorhynchus*) [original combination: *Megarhinus nairobiensis*].

nakuruensis Mattingly, 1951b (p. 44) – species of *Culex* (*Culex*).

namibiensis Coetzee, 1984 (p. 2) – species of *Anopheles* (*Anopheles*).

nanseica Bohart & Ingram, 1946b (p. 56) – synonym of *Uranotaenia* (*Uranotaenia*) *annandalei* Barraud, 1926; synonymy by Bohart, 1953.

nanus Coquillett, 1903a (p. 256) – synonym of *Psorophora* (*Grabhamia*) *pygmaea* (Theobald, 1903a) [original combination: *Culex nanus*]; synonymy established before Knight & Stone, 1977.

natalensis Edwards, 1930a (p. 298) – species of *Aedes* (*Pseudarmigeres*).

natalensis (Hill & Haydon, 1907) (p. 152) – species of *Anopheles* (*Cellia*) [original combination: *Myzorhynchus natalensis*].

Synonym: *aureosquamiger* Theobald, 1907.

nataliae Belkin, 1945 (p. 315) – species of *Anopheles* (*Cellia*).

nataliae Gornostaeva, 2005 (p. 483) – species of *Aedes* (*Aedes*).

nataliae Lynch Arribálzaga, 1891b (p. 164) – species of *Uranotaenia* (*Uranotaenia*).

Synonyms: *rowlandii* Theobald, 1905g; *noctivaga* Neiva & Pinto, 1922c; *argenteopennis* Peryassú, 1923a; *capitis* Shannon & del Ponte, 1928.

natronius Edwards, 1932b (p. 562) – species of *Aedes* (*Aedimorphus*).

naudeanus Muspratt, 1961 (p. 97) – synonym of *Culex* (*Maillotia*) *salisburiensis* Theobald, 1901c [original combination: *Culex* (*Neoculex*) *salisburiensis naudeanus*]; synonymy by White, 1975.

navalis Edwards, 1926a (p. 121) – species of *Culex* (*Lophoceraomyia*).

nearcticus Dyar, 1919f (p. 32) – synonym of *Aedes* (*Ochlerotatus*) *impiger* (Walker, 1848); synonymy by Vockeroth, 1954.

neavei Theobald, 1906b (p. 76) – species of *Culex* (*Culex*).

nebulosus Arnell, 1973 (p. 35) – species of *Haemagogus* (*Haemagogus*).

nebulosus Theobald, 1901d (p. x) – species of *Culex* (*Culiciomyia*).

Synonyms: *freetownensis* Theobald, 1901c; *nigrochaetae* Theobald, 1901c (♀ only of type series); *invenustus* Theobald, 1901d; *fuscus* Theobald, 1909.

neghmei Mann, 1950 (p. 34) – subspecies of *Anopheles* (*Anopheles*) *pseudopunctipennis* Theobald, 1901c.

neglectus Lutz, 1904 (in Bourroul, 1904) (p. 14) – synonym of *Toxorhynchites* (*Ankylorhynchus*) *trichopygus* (Wiedemann, 1828) [original combination: *Ankylorhynchus neglectus*]; synonymy established before Knight & Stone, 1977.

neglectus Lutz, 1904 (in Bourroul, 1904) (p. 27) – species of *Culex* (*Microculex*).

negrensis Gordon & Evans, 1922 (p. 319) – species of *Wyeomyia* (subgenus uncertain, determined by Motta & Lourenço-de- Oliveira, 1995).

neireti Blanchard, 1907 (p. 188) – synonym of *Anopheles* (*Cellia*) *marshallii* (Theobald, 1903a) [original combination: *Myzomyia funesta* variety *neireti*]; synonymy established before Knight & Stone, 1977.

neireti Edwards, 1920b (p. 138) – species of *Uranotaenia* (*Uranotaenia*).

neireti Ventrillon, 1906a (p. 103) – synonym of *Culex* (*Oculeomyia*) *giganteus* Ventrillon, 1906a; synonymy established before Knight & Stone, 1977.

neivae Cruz, 1906 (p. 199) – synonym of *Chagasia fajardi* (Lutz, 1904); synonymy established before Knight & Stone, 1977.

neivai Howard, Dyar & Knab, 1913, 1917 (figs 8, 461; p. 986, respectively) – species of *Anopheles* (*Kerteszia*).

Synonym: *hylephilus* Dyar & Knab, 1917.

neivai Lane & Cerqueira, 1942 (p. 649) – synonym of *Wyeomyia* (*Phoniomyia*) *quasilongirostris* (Theobald, 1907) [original combination: *Phoniomyia neivai*]; synonymy by Correa & Ramalho, 1956.

neivai (Lane & Coutinho, 1940) (p. 592) – species of *Coquillettidia* (*Rhynchotaenia*) [original combination: *Mansonia* (*Rhynchotaenia*) *neivai*].

neivai Petrocchi, 1927 (in Shannon & del Ponte, 1927) (p. 725) – synonym of *Sabethes* (*Sabethes*) *albiprivus* Theobald, 1903a; synonymy established before Knight & Stone, 1977.

nematoides Dyar & Shannon, 1925a (p. 84) – synonym of *Culex* (*Neoculex*) *territans* Walker, 1856a; synonymy by Sirivanakarn, 1971b.

nemophilous Peyton & Ramalingam, 1988 (p. 276) – species of *Anopheles* (*Cellia*).

nemorosa (Gong, 1996) (p. 362) – species of *Kimia* [original combination: *Topomyia* (*Suaymyia*) *nemorosa*].

nemorosus Meigen, 1818 (p. 4) – synonym of *Aedes* (*Ochlerotatus*) *communis* (de Geer, 1776) [original combination: *Culex nemorosus*]; synonymy established before Knight & Stone, 1977.

neoafricanus Cornet, Valade & Dieng, 1978 (p. 228) – species of *Aedes* (*Stegomyia*).

neoapicalis Theobald, 1910c (p. 336) – synonym of *Psorophora* (*Grabhamia*) *cingulata* (Fabricius, 1805) [*nomen novum* for *apicalis* Theobald, 1903a]; synonymy established before Knight & Stone, 1977.

neobiannulata Theobald, 1910c (p. 255) – synonym of *Aedes* (*Catageiomyia*) *tarsalis* (Newstead, 1907) [original combination: *Reedomyia neobiannulata*]; synonymy by Edwards, 1941.

neogalloisi Chen & Chen, 2000 (p. 341) – species of *Aedes* (*Stegomyia*).

neogeorgianus Belkin, 1962 (p. 375) – species of *Aedes* (*Finlaya*).

neolitoralis Bram, 1967a (p. 236) – synonym of *Culex* (*Culex*) *alis* Theobald, 1903a; synonymy by Sirivanakarn, 1976.

neomacrodixoa (King & Hoogstraal, 1947b) (p. 124) – species of *Verrallina* (*Neomacleaya*) [original combination: *Aedes* (*Aedes*) *neomacrodixoa*].

neomaculipalpus Curry, 1931 (p. 643) – species of *Anopheles* (*Anopheles*).

neomimulus Lien, 1968d (p. 240) – synonym of *Culex* (*Culex*) *mimulus* Edwards, 1915c; synonymy by Sirivanakarn, 1976.

neopandani Bohart, 1957 (p. 36) – species of *Aedes* (*Stegomyia*).

neotaeniorhynchus Theobald, 1910c (p. 320) – synonym of *Culex* (*Culex*) *thalassius* Theobald, 1903a [original combination: *Culicelsa neotaeniorhynchus*]; synonymy established before Knight & Stone, 1977.

neotibialis King & Hoogstraal, 1947a (p. 590) – species of *Uranotaenia* (*Uranotaenia*).

neovishnui Lien, 1968d (p. 230) – synonym of *Culex* (*Culex*) *pseudovishnui* Colless, 1957a; synonymy by Sirivanakarn, 1976.

nepenthes (Theobald, 1912b) (p. 92) – species of *Uranotaenia* (*Pseudoficalbia*) [original combination: *Pseudoficalbia nepenthes*].

nepenthicola (Banks, 1909) (p. 550) – species of *Tripteroides* (*Tricholeptomyia*) [original combination: *Wyeomyia nepenthicola*].

Synonym: *mus* Dyar, 1920i.

nepenthicola Miyagi & Toma, 2007 (p. 167) – species of *Topomyia* (*Suaymyia*).

nepenthicola Steffan & Evenhuis, 1982 (p. 2) – species of *Toxorhynchites* (*Toxorhynchites*).

nepenthis (Dyar & Shannon, 1925a) (p. 66) – species of *Toxorhynchites* (*Toxorhynchites*) [original combination: *Megarhinus nepenthis*].

nepenthis (Edwards, 1915c) (p. 285) – species of *Tripteroides* (*Rachionotomyia*) [original combination: *Rachionotomyia nepenthis*].

nepenthisimilis Mattingly, 1981 (p. 41) – species of *Tripteroides* (*Rachionotomyia*).

nero (Doleschall, 1857) (p. 383) – *nomen dubium*, *Anopheles* [original combination: *Culex nero*]; status by Reid, 1953 (? = *nigerrimus* Giles, 1900a).

nevadensis Chapman & Barr, 1964 (p. 444) – species of *Aedes* (*Ochlerotatus*) [original combination: *Aedes communis nevadensis*]; specific status by Ellis & Brust, 1973.

ngong van Someren, 1950a (p. 7) – species of *Aedes* (*Albuginosus*).

niansimbii Gillett, 1975 (p. 366) – *nomen nudum*, *Anopheles* (*Cellia*) *gambiae* species D [= *Anopheles* (*Cellia*) *bwambae* White, 1985].

nicaensis Leach, 1825 (p. 292) – synonym of *Culiseta* (*Culiseta*) *annulata* (Schrank, 1776) [original combination: *Culex nicaensis*]; synonymy established before Knight & Stone, 1977.

nicaroensis Duret, 1967a (p. 77) – species of *Culex* (*Nicaromyia*).

nicceriensis Bonne-Wepster & Bonne, 1920 (p. 174) – species of *Culex* (*Melanoconion*).

nicksoni Miyagi & Toma, 2012 (in Miyagi *et al.*, 2012) (p. 109) – species of *Topomyia* (*Topomyia*).

nicoyae Adames & Hogue, 1970 (p. 9) – species of *Deinocerites*.

nielseni O'Meara & Craig, 1970 (in Carpenter, 1970) (p. 53) – *nomen nudum*, *Aedes*.

nielseni O'Meara & Craig, 1970 (p. 476) – synonym of *Aedes* (*Georgecraigius*) *epactius* Dyar & Knab, 1908; synonymy by Zavortink, 1972.

niger Giles, 1904b (p. 384) – synonym of *Aedes* (*Ochlerotatus*) *taeniorhynchus* (Wiedemann, 1821) [original combination: *Taeniorhynchus niger*]; synonymy established before Knight & Stone, 1977. **Note**: this nominal taxon is the senior secondary homonym of *niger* Theobald, 1913b.

niger Herrick, 1905 (p. 283) – *nomen nudum*, *Anopheles*; status as *lapsus calami* by Ward, 1979.

niger (Leicester, 1908) (p. 123) – species of *Culex* (*Lophoceraomyia*) [original combination: *Lophoceratomyia niger*].

Synonym: *atratulus* Edwards, 1922a.

niger Theobald, 1913b (p. 164) – synonym of *Aedes* (*Aedimorphus*) *vexans nocturnus* (Theobald, 1903a) [original combination: *Culex nocturnus niger*]; synonymy by Belkin, 1962. **Note**: this nominal taxon is the junior secondary homonym of *niger* Giles, 1904b.

nigeria Theobald, 1901a (p. 303) – synonym of *Aedes* (*Stegomyia*) *aegypti* (Linnaeus, 1762) [original combination: *Stegomyia nigeria*]; synonymy established before Knight & Stone, 1977.

nigeriae Edwards, 1930a (p. 300) – species of *Hodgesia*.

nigeriensis Evans, 1931a (p. 138) – subspecies of *Anopheles* (*Cellia*) *moucheti* Evans, 1925b [original combination: *Anopheles moucheti* variety *nigeriensis*]; subspecific status by Kengne *et al.*, 2007.

nigeriensis Ribeiro, 2005 (p. 227) – species of *Toxorhynchites* (*Afrorhynchus*).

nigeriensis Theobald, 1910c (p. 281) – synonym of *Aedes* (*Aedimorphus*) *fowleri* (de Charmoy, 1908) [original combination: *Grabhamia nigeriensis*]; synonymy by Edwards, 1941.

nigerrima Taylor, 1914d (p. 203) – synonym of *Uranotaenia* (*Pseudoficalbia*) *atra* Theobald, 1905f; synonymy by Stone, 1957b.

nigerrima Theobald, 1910c (p. 233) – synonym of *Culex* (*Lophoceraomyia*) *minutissimus* (Theobald, 1907) [original combination: *Culiciomyia nigerrima*]; synonymy established before Knight & Stone, 1977.

nigerrima Theobald, 1910c (species) (p. 450) – subspecies of *Mansonia* (*Mansonioides*) *africana* (Theobald, 1901d); subspecific status by Harbach & Howard, 2007.

nigerrimus Dong, Zhou, Gong *et al.* [*sic*], 2004 (as *nigerrmus*) (p. 187) – *nomen nudum*, *Toxorhynchites*.

nigerrimus Giles, 1900a (p. 162) – species of *Anopheles* (*Anopheles*).

Synonyms: *indiensis* Theobald, 1901a; *bentleyi* Bentley, 1902; *minutus* Theobald, 1903a; *williamsoni* Baisas & Hu, 1936; *venhuisi* Bonn-Wepster, 1951b.

nigerrimus (Theobald, 1913d) (p. 591) – species of *Aedes* (*Mucidus*) [original combination: *Mucidus nigerrimus*].

nigra Christophers, 1924a (p. 45) – synonym of *Anopheles* (*Cellia*) *nili* (Theobald, 1904b) [original combination: *Myzomyia funesta* variety *nigra*]; synonymy established before Knight & Stone, 1977.

nigra Leicester, 1908 (p. 245) – species of *Topomyia* (*Topomyia*).

nigra Ludlow, 1905c (p. 387) – synonym of *Aedes* (*Protomacleaya*) *triseriatus* (Say, 1823) [original combination: *Finlaya ? nigra*]; synonymy established before Knight & Stone, 1977. **Note**: this nominal taxon is the senior secondary homonym of *nigra* Theobald, 1906b and *nigra* Taylor, 1914c.

nigra Taylor, 1914c (p. 688) – synonym of *Aedes* (*Ochlerotatus*) *camptorhynchus* (Thomson, 1869) [original combination: *Culicada nigra*]; synonymy established before Knight & Stone, 1977. **Note**: this nominal taxon is the junior secondary homonym of *nigra* Ludlow, 1905c and *nigra* Theobald, 1906b.

nigra (Theobald, 1901d) (p. 14) – species of *Ficalbia* [original combination: *Aedes nigra*].

Synonyms: *africana* Newstead, 1907; *fusca* Theobald, 1911a.

nigra Theobald, 1906b (p. 80) – synonym of *Aedes* (*Diceromyia*) *furcifer* (Edwards, 1913b) [original combination: *Mansonia* (?) *nigra*]; synonymy established before Knight & Stone, 1977. **Note**: this nominal taxon is the junior secondary homonym of *nigra* Ludlow, 1905c and the senior secondary homonym of *nigra* Taylor, 1914c.

nigra Theobald, 1907 (p. 78) – synonym of *Anopheles* (*Nyssorhynchus*) *lutzii* Cruz, 1901 [original combination: *Myzorhynchella nigra*]; synonymy established before Knight & Stone, 1977.

nigra Theobald, 1910c (p. 434) – synonym of *Coquillettidia* (*Coquillettidia*) *metallica* (Theobald, 1901d) [original combination: *Chrysoconops nigra*]; synonymy established before Knight & Stone, 1977.

nigrans Stanton, 1912 (p. 7) – synonym of *Anopheles* (*Cellia*) *karwari* (James, 1903) [*nomen novum* in error for *karwari* James, 1903]; synonymy by Christophers, 1916a.

nigrescens (Edwards, 1929) (in Paine & Edwards, 1929) (p. 314) – species of *Aedes* (*Mucidus*) [original combination: *Pardomyia aurantia* variety *nigrescens*]; specific status by Belkin, 1962 (as *painei* Knight, 1948c). **Note**: *painei* Knight, 1948c was introduced as a replacement for *nigrescens* Edwards, 1929 when *nigrescens* Theobald, 1907 was classified in genus *Aedes* Meigen; however, the replacement name was no longer justified after *nigrescens* Theobald, 1907 was established as a synonym of *Culex* (*Eumelanomyia*) *malayi* (Leicester, 1908) by Sirivanakarn, 1972.

Synonym: *painei* Knight, 1948c.

nigrescens (Theobald, 1907) (p. 248) – *nomen dubium*, *Culex* [original combination: *Danielsia nigrescens*]; status by Sirivanakarn, 1983. **Note**: this nominal taxon is the senior secondary homonym of *nigrescens* Theobald, 1907 (p. 540), based on page priority.

nigrescens Theobald, 1907 (p. 540) – synonym of *Culex* (*Eumelanomyia*) *malayi* (Leicester, 1908) [original combination: *Aedes nigrescens*]; synonymy based on recognition of the single male of the type series as the holotype of *nigrescens* by Sirivanakarn, 1972; the female of the series is a specimen of *Culex* (*Eumelanomyia*) *castrensis* Edwards, 1922a (Townsend, 1990 incorrectly attributed the lectotype designation of *castrensis* by Sirivanakarn, 1972, to *nigrescens* Theobald, 1907). **Note**: this nominal taxon is the junior secondary homonym of *nigrescens* (Theobald, 1907) (p. 248), based on page priority.

nigricans (Coquillett, 1904c) (p. 166) – species of *Coquillettidia* (*Rhynchotaenia*) [original combination: *Taeniorhynchus nigricans*].

Synonym: *persephassa* Dyar & Knab, 1909d.

nigricephala Clastrier & Claustre, 1978b (p. 543) – species of *Wyeomyia* (subgenus uncertain, subgeneric placement not determined by the authors).

nigricephala da Cunha Ramos & Brunhes, 2004 (p. 199) – species of *Uranotaenia* (*Pseudoficalbia*).

nigricephala Leicester, 1908 (p. 149) – synonym of *Culex* (*Culex*) *sitiens* Wiedemann, 1828; synonymy established before Knight & Stone, 1977.

nigricephalus (Theobald, 1901d) (p. iv) – species of *Aedes* (*Aedimorphus*) [original combination: *Stegomyia nigricephala*].

Synonym: *paludosus* Graham, 1910.

nigriceps Buxton, 1927 (p. 307) – synonym of *Culex* (*Culex*) *atriceps* Edwards, 1926a [*lapsus* for *atriceps*].

nigricorpus (Theobald, 1901c) (p. 231) – *nomen dubium*, *Culex* [original combination: *Aedes nigricorpus*]; status by Sirivanakarn, 1983.

nigrifasciatus Theobald, 1907 (p. 65) – synonym of *Anopheles* (*Cellia*) *multicolor* Cambouliu, 1902 [original combination: *Pyretophorus nigrifasciatus*]; synonymy established before Knight & Stone, 1977.

nigrimacula Lane & Whitman, 1943 (p. 403) – species of *Culex* (subgenus uncertain, determined by Pecor *et al.*, 1992).

nigrinus (Eckstein, 1918) (p. 67) – species of *Aedes* (*Ochlerotatus*) [original combination: *Culicada nigrina*].

nigripalpus Theobald, 1901c (p. 322) – species of *Culex* (*Culex*).

Synonyms: *biocellatus* Theobald, 1903a; *palus* Theobald, 1903a; *similis* Theobald, 1903a; *microsquamosus* Theobald, 1905i; *carmodyae* Dyar & Knab, 1906d; *factor* Dyar & Knab, 1906d; *mortificator* Dyar & Knab, 1906d; *regulator* Dyar & Knab, 1906d; *microannulata* Theobald, 1907; *proximus* Dyar & Knab, 1909a; *caraibeus* Howard, Dyar & Knab, 1913; *prasinopleurus* Martini, 1914; *azuayus* Leví- Castillo, 1954b.

nigripes (Edwards, 1935b) (p. 580) – species of *Toxorhynchites* (*Toxorhynchites*) [original combination: *Megarhinus nigripes*].

nigripes Staeger, 1839 (p. 552) – synonym of *Anopheles* (*Anopheles*) *plumbeus* Stephens, 1828; synonymy established before Knight & Stone, 1977.

nigripes (Theobald, 1905i) (p. 199) – species of *Uranotaenia* (*Pseudoficalbia*) [original combination: *Ficalbia nigripes*].

nigripes (Zetterstedt, 1838) (p. 807) – species of *Aedes* (*Ochlerotatus*) [original combination: *Culex nigripes*].

Synonym: *innuitus* Dyar & Knab, 1918b.

nigripleura da Cunha Ramos & Brunhes, 2004 (p. 310) – species of *Uranotaenia* (*Pseudoficalbia*).

nigrirostris Enderlein, 1920 (p. 51) – synonym of *Culex* (*Culex*) *quinquefasciatus* Say, 1823 [original combination: *Culex fatigans* variety *nigrirostris*]; synonymy established before Knight & Stone, 1977.

nigritarsis (Chagas, 1907) (p. 12) – species of *Anopheles* (*Nyssorhynchus*) [original combination: *Myzorhynchella nigritarsis*].

nigritarsis Leicester, 1908 (p. 177) – synonym of *Orthopodomyia anopheloides* (Giles, 1903); synonymy by Barraud, 1934.

nigritarsis (Wolfs, 1958) (p. 297) – species of *Coquillettidia* (*Coquillettidia*) [original combination: *Taeniorhynchus* (*Coquillettidia*) *nigritarsis*].

nigrithorax (Macquart, 1847) (p. 25) – species of *Aedes* (*Ochlerotatus*) [original combination: *Culex nigrithorax*].

Synonym: *macleayanus* Mackerras, 1927.

nigrithorax (Theobald, 1910c) (p. 439) – species of *Coquillettidia* (*Coquillettidia*) [original combination: *Chrysoconops nigrithorax*].

nigritia Ludlow, 1910 (p. 194) – synonym of *Aedes* (*Stegomyia*) *albopictus* (Skuse, 1895) [original combination: *Stegomyia nigritia*]; synonymy established before Knight & Stone, 1977.

nigritubus Galindo, Carpenter & Trapido, 1951 (p. 89) – species of *Wyeomyia* (*Wyeomyia*).

nigritulus Zetterstedt, 1850 (p. 3459) – synonym of *Aedes* (*Aedes*) *cinereus* Meigen, 1818 [original combination: *Culex nigritulus*]; synonymy established before Knight & Stone, 1977.

nigrocanus Martini, 1927 (p. 390) – species of *Aedes* (*Ochlerotatus*).

nigrochaetae Theobald, 1901c (p. 60) – (♂ only of type series) synonym of *Coquillettidia* (*Coquillettidia*) *metallica* (Theobald, 1901d) [original combination: *Culex nigrochaetae*]; (♀ only of type series) synonym of *Culex* (*Culiciomyia*) *nebulosus* Theobald, 1901d; synonymies established before Knight & Stone, 1977.

nigrochracea (Bonne-Wepster, 1930) (p. 392) – species of *Coquillettidia* (*Coquillettidia*) [original combination: *Taeniorhynchus* (*Coquillettidia*) *nigrochraceus*].

nigrocostalis Theobald, 1909b (p. 16) – synonym of *Culex* (*Culex*) *decens* Theobald, 1901d; synonymy by Edwards, 1941.

nigromaculata Edwards, 1941 (p. 60) – species of *Uranotaenia* (*Pseudoficalbia*).

Synonym: *bimaculata* Theobald, 1909b.

nigromaculis (Ludlow, 1906a) (p. 83) – species of *Aedes* (*Ochlerotatus*) [original combination: *Grabhamia nigromaculis*].

Synonym: *grisea* Ludlow, 1907.

nigropterus (Le Goff, Boussès & Brunhes, 2007) (p. 194) – species of *Aedes* (*Neomelaniconion*) [original combination: *Neomelaniconion nigropterum*].

nigropunctatus Edwards, 1926a (p. 121) – species of *Culex* (*Culiciomyia*).

Synonym: *annulata* Theobald, 1907.

nigrorhynchus Brug, 1931c (p. 28) – species of *Aedes* (*Hulecoeteomyia*) [original combination: *Aedes* (*Finlaya*) *harveyi* variety *nigrorhynchus*]; specific status by Knight, 1968.

nigrosignata (Edwards, 1917) (p. 224) – species of *Coquillettidia* (*Coquillettidia*) [original combination: *Taeniorhynchus nigrosignatus*].

nigrostriatus (Barraud, 1927c) (p. 549) – species of *Aedes* (*Aedimorphus*) [original combination: *Aedimorphus nigrostriatus*].

nigrotarsis (Ludlow, 1908a) (p. 52) – species of *Verrallina* (*Neomacleaya*) [original combination: *Pseudoskusea nigrotarsis*].

nii Lien, 1968a (p. 10) – synonym of *Uranotaenia* (*Uranotaenia*) *hebes* Barraud, 1931b; synonymy by Peyton, 1972.

nikolskyi Shingarev, 1927 (as *nikolsky*) (p. 50) – synonym of *Coquillettidia* (*Coquillettidia*) *richiardii* (Ficalbi, 1889b) [original combination: *Taeniorhynchus nikolsky* [*sic*]]; synonymy by Natvig, 1948.

nilgiricus Christophers, 1924b (p. 13) – species of *Anopheles* (*Anopheles*) [original combination: *Anopheles lindesayi* variety *nilgiricus*]; specific status by Harrison *et al.*, 1991.

nilgiricus Edwards, 1916 (p. 358) – species of *Culex* (*Culex*).

nili (Theobald, 1904b) (p. 66) – species of *Anopheles* (*Cellia*) [original combination: *Myzomyia nili*].
Synonyms: *unicolor* Grünberg, 1905; *nigra* Christophers, 1924a.

nimbus (Theobald, 1902b) (p. 182) – species of *Anopheles* (*Stethomyia*) [original combination: *Stethomyia nimbus*].
Synonym: *lineata* Lutz, 1905.

nimpe Nguyen, Tran & Harbach, 2000 (in Nguyen *et al.*, 2000) (p. 190) – species of *Anopheles* (*Anopheles*).

ninagongoensis Edwards, 1928b (p. 324) – species of *Culex* (*Culex*).

ningheensis Lei, 1989b (p. 100) – species of *Aedes* (*Collessius*).

niphadopsis Dyar & Knab, 1918b (p. 166) – species of *Aedes* (*Ochlerotatus*).

nipponica La Casse & Yamaguti, 1948 (p. 264) – synonym of *Orthopodomyia anopheloides* (Giles, 1903); synonymy by Knight & Mattingly, 1950.

nipponica La Casse & Yamaguti, 1950 (p. 84) – species of *Culiseta* (*Culicella*).

nipponicus La Casse & Yamaguti, 1948 (p. 79) – species of *Aedes* (*Downsiomyia*) [original combination: *Aedes* (*Finlaya*) *niveus nipponicus*]; specific status by La Casse & Yamaguti, 1950.

nipponii (Theobald, 1907) (species) (p. 337) – subspecies of *Aedes* (*Aedimorphus*) *vexans* (Meigen, 1830b) [original combination: *Culicada nipponii*]; subspecific status by Bohart & Ingram, 1946b.

nippononiveus Sasa & Nakahashi, 1952 (p. 258) – species of *Aedes* (*Downsiomyia*).

nishikawai Tanaka, Mizusawa & Saugstad, 1979 (p. 356) – species of *Aedes* (*Downsiomyia*).

nissanensis Lee, 1946 (p. 239) – species of *Tripteroides* (*Tripteroides*).

nitens (Cerqueira, 1943) (p. 2) – species of *Coquillettidia* (*Rhynchotaenia*) [original combination: *Taeniorhynchus* (*Rhynchotaenia*) *nitens*].

nitidoventer (Giles, 1904a) (p. 368) – species of *Tripteroides* (*Tripteroides*) [original combination: *Uranotaenia nitidoventer*].
Synonym: *philippinensis* Giles, 1904a.

nitidus Harrison, Scanlon & Reid, 1973 (p. 266) – species of *Anopheles* (*Anopheles*).

nitidus Theobald, 1901c (p. 347) – species of *Sabethes* (*Sabethes*).

nivalis Edwards, 1926a (p. 112) – species of *Aedes* (*Ochlerotatus*).

nivea Leicester, 1908 (p. 211) – species of *Uranotaenia* (*Uranotaenia*).
Synonym: *triangulata* Ludlow, 1908a.

niveitaeniata (Theobald, 1907) (p. 272) – species of *Culiseta* (*Culiseta*) [original combination: *Pseudotheobaldia niveitaeniata*].
Synonyms: *sinensis* Meng & Wu, 1962; *lishanensis* Lien, 1968a.

niveocaputis Ludlow, 1911b (p. 130) – synonym of *Hodgesia quasisanguinae* Leicester, 1908; synonymy established before Knight & Stone, 1977.

niveoides Barraud, 1934 (p. 211) – species of *Aedes* (*Downsiomyia*).

niveopalpis Ludlow, 1919 (p. 166) – synonym of *Anopheles* (*Anopheles*) *eiseni* Coquillett, 1902b; synonymy established before Knight & Stone, 1977.

niveoscutellum Theobald, 1905g (p. 22) – synonym of *Aedes* (*Aedimorphus*) *pampangensis* (Ludlow, 1905a) [original combination: *Reedomyia niveoscutella* [*sic*]]; emendation by ? Barraud, 1934; synonymy by Reinert, 1973b.

niveoscutum Zavortink, 1972 (p. 49) – species of *Aedes* (*Protomacleaya*).

niveus de Meillon, 1943 (p. 94) – synonym of *Aedes* (*Paulianius*) *zethus* de Meillon & Lavoipierre, 1944; synonymy established before Knight & Stone, 1977. **Note**: this nominal taxon is the junior secondary homonym of *niveus* (von Eichwald, 1837) and *niveus* (Ludlow, 1903).

niveus (Ludlow, 1903) (in Theobald, 1903a) (p. 139) – species of *Aedes* (*Downsiomyia*) [original combination: *Stegomyia nivea*]. **Note**: this nominal taxon is the junior secondary homonym of *niveus* (von Eichwald, 1837) and the senior secondary homonym of *niveus* de Meillon, 1943; in the interests of stability, the prevailing usage of the junior homonym of *niveus* (Ludlow, 1903) is maintained based on criteria for reversal of precedence (International Commission on Zoological Nomenclature, 1999: *Code*, Article 23.9).

niveus (von Eichwald, 1837) (p. 183) – *nomen dubium*, *Aedes* [original combination: *Culex niveus*]; status by Edwards, 1932a, ? = *Aedes aegypti*. **Note**: this nominal taxon is the senior secondary homonym of *niveus* (Ludlow, 1903) and *niveus* de Meillon, 1943; in the interests of stability, the prevailing usage of *niveus* (Ludlow, 1903) is maintained based on criteria for reversal of precedence (International Commission on Zoological Nomenclature, 1999: *Code*, Article 23.9).

nivipes Theobald, 1901c (p. 285) – synonym of *Trichoprosopon digitatum* (Rondani, 1848); synonymy established before Knight & Stone, 1977.

nivipes (Theobald, 1903f) (p. 258) – species of *Anopheles* (*Cellia*) [original combination: *Nyssorhynchus nivipes*].

nivipes (Theobald, 1905a) (p. 53) – species of *Uranotaenia* (*Uranotaenia*) [original combination: *Anisocheleomyia nivipes*].
Synonym: *albofasciata* Taylor, 1920.

nivipleura Leicester, 1908 (p. 219) – species of *Uranotaenia* (*Pseudoficalbia*).
Synonym: *loshanensis* Wang & Feng, 1965.

nivipous Theobald, 1912a (p. 93) – species of *Uranotaenia* (*Pseudoficalbia*).
Synonym: *candidipes* Edwards, 1912c.

nivirostris Lien, 1968c (p. 139) – synonym of *Heizmannia* (*Heizmannia*) *macdonaldi* Mattingly, 1957c; synonymy by Mattingly, 1970.

nivitarsis Coquillett, 1904c (p. 168) – synonym of *Aedes* (*Ochlerotatus*) *canadensis* (Theobald, 1901c) [original combination: *Culex nivitarsis*]; synonymy established before Knight & Stone, 1977.

njombiensis Huang, 1997 (p. 56) – species of *Aedes* (*Stegomyia*).

njombiensis Peters, 1955b (p. 95) – species of *Anopheles* (*Cellia*).

nkolbissonensis Rickenbach & Hamon, 1966 (p. 1112) – species of *Orthopodomyia*.

nobukonis (Yamada, 1932) (p. 228) – species of *Verrallina* (*Harbachius*) [original combination: *Aedes nobukonis*].
Synonym: *ishigakiensis* Bohart, 1956.

nocticola Peyton, 1977 (p. 149) – species of *Uranotaenia* (*Pseudoficalbia*).

noctivaga Neiva & Pinto, 1922c (p. 374) – synonym of *Uranotaenia* (*Uranotaenia*) *nataliae* Lynch Arribálzaga, 1891b; synonymy established before Knight & Stone, 1977.

nocturnus (Theobald, 1903a) (species) (p. 159) – subspecies of *Aedes* (*Aedimorphus*) *vexans* (Meigen, 1830b) [original combination: *Culex nocturnus*]; subspecific status by Tsukamoto *et al.*, 1985.
Synonym: *niger* Theobald, 1913b.

nocturnus Theobald, 1913d (p. 593) – synonym of *Coquillettidia* (*Coquillettidia*) *cristata* (Theobald, 1904b) [original combination: *Chrysoconops nocturnus*]; synonymy established before Knight & Stone, 1977.

noei Mann, 1950 (p. 37) – subspecies of *Anopheles* (*Anopheles*) *pseudopunctipennis* Theobald, 1901c.

nolledoi Baisas, 1935d (p. 170) – synonym of *Culex* (*Lophoceraomyia*) *minor* (Leicester, 1908); synonymy by Colless, 1965.

nongpohensis Barraud, 1927b (p. 539) – synonym of *Armigeres* (*Armigeres*) *kuchingensis* Edwards, 1915c [original combination: *Armigeres* (*Armigeres*) *kuchingensis* form *nongpohensis*]; synonymy by Thurman, 1959.

noniae Reid, 1963 (p. 108) – species of *Anopheles* (*Anopheles*).

normanensis (Taylor, 1915) (p. 182) – species of *Aedes* (*Ochlerotatus*) [original combination: *Culex normanensis*].

noroestensis Galvão & Lane, 1937b (p. 220) – synonym of *Anopheles* (*Nyssorhynchus*) *evansae* (Brèthes, 1926b) [original combination: *Anopheles* (*Nyssorhynchus*) *oswaldoi* variety *noroestensis*]; synonymy by Faran, 1981.

notabilis (Delfinado, 1967) (p. 28) – species of *Verrallina* (*Neomacleaya*) [original combination: *Aedes* (*Neomacleaya*) *notabilis*].

notanandai Rattanarithikul & Green, 1987 (p. 259) – species of *Anopheles* (*Cellia*).

notleyi van Someren, 1949a (p. 3) – species of *Anopheles* (*Cellia*).

notoscriptus (Skuse, 1889) (p. 1738) – species of *Aedes* (*Rampamyia*) [original combination: *Culex notoscriptus*].

novaeguineae Evenhuis, 1989 (in Evenhuis & Gon, 1989) (p. 210) – *nomen dubium*, *Culex*; *nomen novum* for *ventralis* Walker, 1865, *non* Walker, 1859, both *nomina dubia*.

novaezealandiae Pillai, 1966 (p. 127) – species of *Culiseta* (*Climacura*).

novaguinensis Peters, 1963d (p. 141) – species of *Uranotaenia* (*Uranotaenia*).

novaguinensis Venhuis, 1933 (p. 203) – species of *Anopheles* (*Cellia*) [original combination: *Anopheles punctulatus* variety *novaguinensis*]; specific status by Woodhill & Lee, 1944b.
Synonym: *breinli* Taylor, 1943.

novalbitarsis King & Hoogstraal, 1946d (p. 146) – species of *Aedes* (*Patmarksia*) [*nomen novum* for *albitarsis* Taylor, 1914d].
Synonym: *albitarsis* Taylor, 1914d.

novalbopictus Barraud, 1931a (p. 224) – species of *Aedes* (*Stegomyia*).

novobscura Barraud, 1934 (p. 84) – species of *Uranotaenia* (*Pseudoficalbia*).

novochracea (Barraud, 1927d) (p. 558) – species of *Coquillettidia* (*Coquillettidia*) [original combination: *Taeniorhynchus* (*Coquillettidia*) *novochraceus*].

novohanoverae Peters, 1963b (p. 63) – species of *Tripteroides* (*Tripteroides*).

novoniveus Barraud, 1934 (p. 211) – species of *Aedes* (*Downsiomyia*).

novumbrosus Strickland, 1916b (p. 272) – synonym of *Anopheles* (*Anopheles*) *umbrosus* (Theobald, 1903a) [original combination: *Myzorhynchus novumbrosus*]; synonymy by Colless, 1948.

nowlini Evans, 1932 (p. 98) – synonym of *Anopheles* (*Anopheles*) *obscurus* (Grünberg, 1905) [original combination: *Anopheles obscurus* variety *nowlini*]; synonymy by White, 1980.

nubicolus (Laffoon, 1946) (p. 237) – species of *Verrallina* (*Neomacleaya*) [original combination: *Aedes* (*Aedes*) *nubicolus*].

nubilus (Theobald, 1903a) (p. 208) – species of *Aedes* (*Ochlerotatus*) [original combination: *Culex nubilus*].
Synonym: *polyagrus* Dyar, 1918b.

nudipalpis Shingarev, 1927 (p. 550) – synonym of *Culex* (*Barraudius*) *modestus* Ficalbi, 1890b; synonymy by Stackelberg, 1937.

nummatus Edwards, 1923b (p. 4) – species of *Aedes* (*Tewarius*).

nuneztovari Gabaldon, 1940 (p. 5) – species of *Anopheles* (*Nyssorhynchus*).

nursei Theobald, 1907 (p. 66) – synonym of *Anopheles* (*Cellia*) *superpictus* Grassi, 1899 [original combination: *Pyretophorus nursei*]; synonymy established before Knight & Stone, 1977.

nyangae Galliard, 1931b (p. 230) – synonym of *Culex* (*Eumelanomyia*) *inconspicuosus* (Theobald, 1908a); synonymy by White, 1980.

nyasae Edwards, 1930a (p. 296) – species of *Aedes* (*Hopkinsius*).

nyounae Hamon & Adam, 1959a (p. 531) – species of *Aedes* (*Catageiomyia*).

Oo

oakleyi Stone, 1939 (p. 163) – species of *Aedes* (*Aedimorphus*).

oblita (Lutz, 1904) (in Bourroul, 1904) (p. 68) – species of *Wyeomyia* (*Miamyia*) [original combination: *Dendromyia oblita*].
Synonyms: *fallax* Bonne-Wepster & Bonne, 1919; *pintoi* da Costa Lima, 1930b.

oblitus (Lynch Arribálzaga, 1891b) (p. 154) – *nomen dubium*, *Culex* [original combination: *Janthinosoma* ? *oblita* [*sic*]]; status by Belkin *et al*., 1968.

obscura Edwards, 1915c (p. 285) – species of *Uranotaenia* (*Pseudoficalbia*).
Synonyms: *papua* Brug, 1924; *philippinensis* Delfinado, 1966b.

obscura Tenorio, 1975 (p. 734) – species of *Bironella* (*Brugella*).

obscurescens Martini, 1931a (p. 212) – synonym of *Haemagogus* (*Haemagogus*) *janthinomys* Dyar, 1921g [original combination: *Haemagogus uriartei* variety *obscurescens*]; synonymy by Arnell, 1973.

obscurum Lane & Cerqueira, 1942 (p. 498) – species of *Trichoprosopon*.

obscurus Brug, 1934 (p. 504) – species of *Tripteroides* (*Rachisoura*).
Synonym: *subobscurus* Lee, 1946.

obscurus (Grünberg, 1905) (p. 380) – species of *Anopheles* (*Anopheles*) [original combination: *Myzorhynchus obscurus*].
Synonyms: *strachanii* Theobald, 1907; *nowlini* Evans, 1932.

obscurus (Leicester, 1908) (p. 194) – species of *Culex* (*Acalleomyia*) [original combination: *Acalleomyia obscurus* [*sic*]].

obscurus Meigen, 1830a (pl. 3, fig. 2) – synonym of *Aedes* (*Ochlerotatus*) *communis* (de Geer, 1776); synonymy established before Knight & Stone, 1977.

obsoleta Edwards, 1936 (p. 54) – synonym of *Uranotaenia* (*Uranotaenia*) *fraseri* Edwards, 1912c [original combination: *Uranotaenia bilineata* variety *obsoleta*]; synonymy by da Cunha Ramos, 1993.

obsoleta (Huang, 1968a) (p. 45) – species of *Verrallina* (*Verrallina*) [original combination: *Aedes* (*Verrallina*) *obsoletus*].

obturbans (Walker, 1859) (p. 91) – species of *Armigeres* (*Armigeres*) [original combination: *Culex obturbans*].

obturbator Dyar & Knab, 1907a (p. 9) – species of *Aedes* (*Ochlerotatus*).

occidentalis Brug, 1934 (p. 506) – synonym of *Tripteroides* (*Polylepidomyia*) *punctolateralis* (Theobald, 1903e) [original combination: *Tripteroides atripes* variety *occidentalis*]; synonymy by Lee, 1946.

occidentalis Dyar & Knab, 1906b (p. 159) – species of *Anopheles* (*Anopheles*).

occidentalis Evans, 1926 (p. 99) – synonym of *Aedes* (*Albuginosus*) *stokesi* Evans, 1929; synonymy established before Knight & Stone, 1977.

occidentalis (Skuse, 1889) (p. 1729) – species of *Aedes* (*Dobrotworskyius*) [original combination: *Culex occidentalis*].

occidentayunnana (Gong & Lu, 1991) (p. 55) – species of *Heizmannia* (*Mattinglyia*) [original combination: *Aedes* (*Sinoaedes*) *occidentayunnanus*].

occulta Bonne-Wepster, 1951a (p. 71) – synonym of *Bironella* (*Bironella*) *gracilis* Theobald, 1905f; synonymy by Stone, 1957b.

occulta Bonne-Wepster & Bonne, 1919 (p. 105) – species of *Wyeomyia* (subgenus uncertain, determined by Motta & Lourenço-de-Oliveira, 1995).

oceanicus Belkin, 1962 (p. 377) – species of *Aedes* (*Finlaya*).

ocellata Theobald, 1907 (p. 488) – synonym of *Culex* (*Oculeomyia*) *infula* Theobald, 1901a [original combination: *Taeniorhynchus tenax* variety *ocellata* [*sic*]]; synonymy by Sirivanakarn, 1976. **Note**: this nominal taxon is the junior secondary homonym of *ocellatus* Theobald, 1903a.

ocellata Theobald, 1910c (p. 284) – synonym of *Aedes* (*Aedimorphus*) *durbanensis* (Theobald, 1903a) [original combination: *Grabhamia ocellata*]; synonymy established before Knight & Stone, 1977. **Note**: this nominal taxon is listed as a synonym of *Aedes durbanensis angolae* Ribeiro & Ramos, 1974 in the online catalogue of the Walter Reed Biosystematics Unit (http://www.mosquitocatalog.org/), but it is not recognized here because no evidence of this synonymy could be found in the published literature.

ocellatus Theobald, 1901a (p. 174) – synonym of *Anopheles* (*Cellia*) *kochi* Dönitz, 1901; synonymy established before Knight & Stone, 1977.

ocellatus Theobald, 1903a (p. 222) – species of *Culex* (subgenus uncertain, determined by Pecor *et al*., 1992). **Note**: this nominal taxon is the senior secondary homonym of *ocellata* Theobald, 1907.
Synonym: *automartus* Root, 1927c.

ochracea (Theobald, 1903a) (p. 263) – species of *Coquillettidia* (*Coquillettidia*) [original combination: *Taeniorhynchus ochraceus*].
Synonyms: *chrysogona* Knab, 1909; *shakujiiensis* Ogasawara, 1939b.

ochracea Theobald, 1905g (p. 25) – synonym of *Culex* (*Oculeomyia*) *epidesmus* (Theobald, 1910a) [original combination: *Grabhamia ochracea*]; synonymy established before Knight & Stone, 1977.

ochraceus (Theobald, 1901c) (p. 103) – species of *Aedes* (*Aedimorphus*) [original combination: *Culex ochraceus*].
Synonym: *kingii* Theobald, 1908c.

ochripes (Macquart, 1850) (p. 315) – species of *Psorophora* (*Psorophora*) [original combination: *Culex ochripes*].

ochroptera (Peus, 1935) (p. 113) – species of *Culiseta* (*Culicella*) [original combination: *Theobaldia* (*Culicella*) *ochroptera*].

ochropus Dyar & Knab, 1907b (p. 100) – synonym of *Coquillettidia* (*Coquillettidia*) *perturbans* (Walker, 1856a) [original combination: *Culex ochropus*]; synonymy established before Knight & Stone, 1977.

ochrura Dyar & Knab, 1906d (p. 229) – synonym of *Wyeomyia* (*Wyeomyia*) *mitchellii* (Theobald, 1905d); synonymy established before Knight & Stone, 1977.

ocossa Dyar & Knab, 1919 (p. 6) – species of *Culex* (*Melanoconion*).

ocumarensis Lutz, 1928 (p. 34) – synonym of *Wyeomyia moerbista* (Dyar & Knab, 1919) [original combination: *Dendromyia ocumarensis*]; synonymy by Lane & Cerqueira, 1942.

oedipodeios Graham, 1909 (p. 86) – species of *Eretmapodites*.
Synonym: *stanleyi* Edwards, 1941.

oedipodius Edwards, 1912c (p. 48) – unjustified emendation of *oedipodeios* Graham, 1909; recorded by White, 1980.

oedipus Root, 1927c (p. 588) – species of *Culex* (*Melanoconion*).

oedjalikalah Nainggolan, 1939 (p. 163) – subspecies of *Anopheles* (*Anopheles*) *gigas* Giles, 1901b [original combination: *Anopheles gigas* variety *oedjalikalah*]; subspecific status by Harbach & Howard, 2007.

ohamai Ohama, 1947 (species) (p. 11) – subspecies of *Anopheles* (*Anopheles*) *saperoi* Bohart & Ingram, 1946a; subspecific status by Tanaka *et al*., 1975.

ohamai Tanaka, Mizusawa & Saugstad, 1975a (p. 27) – species of *Uranotaenia* (*Pseudoficalbia*).

oiketorakras Osorno-Mesa, 1947 (p. 431) – species of *Anopheles* (*Lophopodomyia*).

okinawae Bohart, 1953 (p. 187) – species of *Culex* (*Eumelanomyia*).
Synonym: *lini* Lien, 1968d.

okinawanus Bohart, 1946 (p. 39) – species of *Aedes* (*Bruceharrisonius*).

okinawensis Toma, Miyagi & Tanaka, 1990 (p. 344) – species of *Toxorhynchites* (*Toxorhynchites*).

okuensis Brunhes, Le Goff & Geoffroy, 1997 (p. 174) – species of *Anopheles* (*Anopheles*).

oligopistus Dyar, 1918b (p. 76) – species of *Aedes* (*Ochlerotatus*).

olimpioi Xavier, da Silva & da Silva Mattos, 1970a (p. 183) – species of *Culex* (*Melanoconion*).

omani Belkin, 1962 (p. 210) – species of *Culex* (*Culex*).

omissus (Edwards, 1914b) (p. 76) – species of *Armigeres* (*Leicesteria*) [original combination: *Leicesteria omissa*].

omorii Lien, 1968b (p. 105) – species of *Aedes* (*Downsiomyia*).

omorii Sakakibara, 1959 (p. 288) – species of *Anopheles* (*Anopheles*).

omurensis Yamada, 1921 (p. 73) – synonym of *Aedes* (*Aedimorphus*) *alboscutellatus* (Theobald, 1905f); synonymy established before Knight & Stone, 1977.

onderstepoortensis Theobald, 1911b (p. 265) – synonym of *Culex* (*Culex*) *theileri* Theobald, 1903a; synonymy established before Knight & Stone, 1977.

onidus Dyar & Knab, 1909d (p. 261) – synonym of *Wyeomyia* (*Decamyia*) *ulocoma* (Theobald, 1903a); synonymy established before Knight & Stone, 1977.

onondagensis Felt, 1904 (p. 304) – synonym of *Aedes* (*Ochlerotatus*) *dorsalis* (Meigen, 1830b) [original combination: *Culex onondagensis*]; synonymy by Wood *et al*., 1979.

opisthopus Komp, 1926 (p. 44) – synonym of *Culex* (*Melanoconion*) *taeniopus* Dyar & Knab, 1907b; synonymy by Sirivanakarn & Belkin, 1980.

opok Corbet & van Someren, 1962 (p. 73) – species of *Aedes* (*Stegomyia*).

orbitae Edwards, 1922a, 1922b (pp. 260, 466, respectively) – species of *Aedes* (*Aedimorphus*).
Synonym: *taeniata* Leicester, 1908.

orbostiensis Dobrotworsky, 1958a (p. 317) – species of *Culex* (*Lophoceraomyia*).

oreios Bangs & Harbach, 2014 (in Bangs *et al*., 2014, 2015) (p. 2) – species of *Anopheles* (*Cellia*).

oreophilus (Edwards, 1916) (p. 357) – species of *Aedes* (subgenus uncertain) [original combination: *Ochlerotatus oreophilus*].
Synonym: *bunanoki* Sasa & Ishimura, 1951.

oresbius Harbach & Rattanarithikul, 1988 (p. 69) – species of *Culex* (*Eumelanomyia*).

orfilai Duret, 1953, 1954a (pp. 74, 117, respectively) – species of *Culex* (*Melanoconion*).

orientalis Barraud, 1926 (p. 339) – species of *Uranotaenia* (*Uranotaenia*).

orientalis Edwards, 1921d (p. 338) – species of *Culex* (*Culex*).

orientalis (Swellengrebel & Swellengrebel de Graaf, 1920a) (p. 3) – subspecies of *Anopheles* (*Cellia*) *tessellatus* Theobald, 1901a [original combination: *Neomyzomyia punctulata* variety *orientalis*]; subspecific status by Harbach & Howard, 2007.

originator Gordon & Evans, 1922 (p. 323) – species of *Culex* (*Anoedioporpa*).

ornata Taylor, 1914d (p. 189) – synonym of *Tripteroides* (*Tripteroides*) *bimaculipes* (Theobald, 1905f) [original combination: *Stegomyia ornata*]; synonymy by Lee, 1946.

ornata Theobald, 1905f (p. 79) – synonym of *Verrallina* (*Verrallina*) *lineata* (Taylor, 1914d) [original combination: *Skusea funerea* variety *ornata*]; synonymy by King & Hoogstraal, 1947b.

ornata Theobald, 1909b (as *ornatus*) (p. 22) – species of *Uranotaenia* (*Pseudoficalbia*).

ornatothoracis Theobald, 1909b (p. 15) – species of *Culex* (*Culex*).

ornatus Meigen, 1818 (p. 5) – synonym of *Aedes* (*Dahliana*) *geniculatus* (Olivier, 1791) [original combination: *Culex ornatus*]; synonymy established before Knight & Stone, 1977.

ornatus Theobald, 1905f (p. 100) – synonym of *Culex* (*Lophoceraomyia*) *gagnei* Evenhuis, 1989 [original combination: *Melanoconion ornatus* [*sic*]]; synonymy by N.L. Evenhuis (in Evenhuis & Gon, 1989).

oroecetor Martini, 1931a (p. 204) – synonym of *Aedes* (*Ochlerotatus*) *milleri* Dyar, 1922g; synonymy by Lane, 1953.

orstom Brunhes, Adam & Bailly-Choumara, 1967 (p. 49) – species of *Culex* (*Eumelanomyia*).

orthodoxa Dyar, 1921h (p. 118) – species of *Uranotaenia* (*Uranotaenia*).
ortizi Vargas & Diaz Nájera, 1961 (p. 74) – species of *Sabethes* (*Sabethes*).
oryzalimnetes Wilkerson & Motoki, 2009 (in Motoki *et al.*, 2009) (p. 841) – species of *Anopheles* (*Nyssorhynchus*).
osakaensis Theobald, 1907 (p. 439) – synonym of *Culex* (*Culex*) *pipiens pallens* Coquillett, 1898; synonymy established before Knight & Stone, 1977.
osornoi Berlin, 1969a (p. 112) – species of *Aedes* (*Howardina*).
ostentatio (Leicester, 1908) (p. 193) – species of *Aedes* (*Paraedes*) [original combination: *Aioretomyia ostentatio*].
oswaldi Lutz, 1904 (in Bourroul, 1904) (pp. 47, 66) – synonym of *Aedes* (*Protomacleaya*) *terrens* (Walker, 1856a) [original combination: *Gualteria oswaldi*]; synonymy established before Knight & Stone, 1977.
oswaldoi Forattini, 1965 (p. 167) – synonym of *Culex* (*Culex*) *maxi* Dyar, 1928; synonymy by Harbach, Peyton & Jakob, 1984.
oswaldoi (Peryassú, 1922b) (p. 179) – species of *Anopheles* (*Nyssorhynchus*) [original combination: *Cellia oswaldoi*].
Synonym: *aquacaelestis* Curry, 1932.
otachati Klein & Sirivanakarn, 1970 (p. 585) – species of *Culex* (*Eumelanomyia*).
oteizai Perez Vigueras, 1956 (p. 443) – species of *Uranotaenia* (*Uranotaenia*).
ototomo da Cunha Ramos, 1993 (p. 261) – species of *Uranotaenia* (*Pseudoficalbia*).
ototomoensis Huang & Rueda, 2013 (p. 271) – species of *Orthopodomyia*.
otwayensis (Dobrotworsky, 1960b) (p. 246) – species of *Culiseta* (*Culicella*) [original combination: *Theobaldia otwayensis*].
ousqua Dyar, 1918c (p. 99) – species of *Culex* (*Culex*).
Synonym: *albertoi* Anduze, 1943a.
ovazzai Hamon & Adam, 1959b (p. 147) – species of *Aedes* (*Aedimorphus*).
ovengensis Awono-Ambene, Kengne, Simard, Antonio-Nkonkjio & Fontenille, 2004 (p. 563) – species of *Anopheles* (*Cellia*).
oweni Belkin, 1962 (p. 259) – species of *Culex* (*Lophoceraomyia*).

Pp

pachecoi Baisas, 1935d (p. 171) – synonym of *Culex* (*Lophoceraomyia*) *quadripalpis* (Edwards, 1914b); synonymy by Sirivanakarn, 1977a.
pachyurus Edwards, 1936 (p. 51) – species of *Aedes* (*Aedimorphus*).
pacificensis Hearle, 1927 (p. 101) – synonym of *Aedes* (*Ochlerotatus*) *cataphylla* Dyar, 1916b; synonymy by Curtis, 1967.
pacificus Edwards, 1916 (p. 360) – species of *Culex* (*Culex*).
paganus Evans, 1923b (p. 104) – synonym of *Culex* (*Anoedioporpa*) *conservator* Dyar & Knab, 1906d; synonymy by Berlin & Belkin, 1980.
pagei Ludlow, 1910 (in Theobald, 1910c) (p. 618) – synonym of *Coquillettidia* (*Coquillettidia*) *aureosquammata* (Ludlow, 1909c) [original combination: *Taeniorhynchus pagei*]; synonymy established before Knight & Stone, 1977.
pagei (Ludlow, 1911b) (p. 128) – species of *Aedes* (*Paraedes*) [original combination: *Danielsia pagei*].
pagetonotum Dyar & Knab, 1909d (p. 253) – synonym of *Aedes* (*Rusticoidus*) *provocans* (Walker, 1848); transferred synonymy by synonymy of senior synonym *trichurus* (Dyar) with *provocans* (Walker) by Wood, 1977.
pahangi (Delfinado, 1968) (p. 31) – species of *Verrallina* (*Harbachius*) [original combination: *Aedes* (*Neomacleaya*) *pahangi*].
painei Edwards, 1935a (p. 127) – species of *Uranotaenia* (*Pseudoficalbia*).
painei Knight, 1948c (p. 68) – synonym of *Aedes* (*Mucidus*) *nigrescens* (Edwards, 1929) [*nomen novum* for *nigrescens* (Edwards, 1929); original combination: *Aedes* (*Mucidus*) *aurantius painei*]. **Note:** *painei* Knight, 1948c was introduced as a replacement for *nigrescens* Edwards, 1929 when *nigrescens* Theobald, 1907 was classified in genus *Aedes* Meigen; however, the replacement name was no longer justified after *nigrescens* Theobald, 1907 was established as a synonym of *Culex* (*Eumelanomyia*) *malayi* (Leicester, 1908) by Sirivanakarn, 1972.
pairoji Sirivanakarn, 1977a (p. 63) – species of *Culex* (*Lophoceraomyia*).
pajoti da Cunha Ramos & Ribeiro, 1982 (p. 176) – species of *Culex* (*Culex*).
palaciosi Duret, 1968d (p. 15) – species of *Culex* (*Melanoconion*).
palauensis Bohart, 1957 (p. 51) – species of *Aedes* (*Stegomyia*).
palawanensis Ludlow, 1914a (p. 31) – synonym of *Aedes* (*Phagomyia*) *melanopterus* (Giles, 1904a) [original combination: *Popea palawanensis*]; synonymy established before Knight & Stone, 1977.
palawanicus Mattingly, 1958a (p. 56) – species of *Aedes* (*Cancraedes*).
palestinensis Theobald, 1903a (p. 71) – synonym of *Anopheles* (*Cellia*) *superpictus* Grassi, 1899 [original combination: *Pyretophorus palestinensis*]; synonymy established before Knight & Stone, 1977.
pallens Coquillett, 1898 (species) (p. 303) – subspecies of *Culex* (*Culex*) *pipiens* Linnaeus, 1758; subspecific status by Tanaka, 2004.
pallens (Ross, 1943) (p. 148) – subspecies of *Aedes* (*Ochlerotatus*) *fulvus* (Wiedemann, 1828).
pallescens Edwards, 1922d (p. 76) – species of *Psorophora* (*Psorophora*).
Synonym: *stigmatephora* Dyar, 1922e.

palliatus Coquillett, 1906e (p. 61) – synonym of *Aedes* (*Howardina*) *fulvithorax* (Lutz, 1904) [original combination: *Taeniorhynchus palliatus*]; synonymy established before Knight & Stone, 1977.

pallida Edwards, 1925 (p. 263) – species of *Mimomyia* (*Mimomyia*).

pallida Ludlow, 1905a (p. 129) – synonym of *Anopheles* (*Anopheles*) *aitkenii* James, 1903 [original combination: *Stethomyia pallida*]; synonymy by Bohart, 1945.

pallida Theobald, 1907 (p. 470) – synonym of *Aedes* (*Neomelaniconion*) *albothorax* (Theobald, 1907) [original combination: *Banksinella luteolateralis* variety *pallida*]; synonymy by Edwards, 1941.

pallidabdomen Dong, Zhou, Gong *et al.* [*sic*], 2004 (p. 187) – *nomen nudum*, *Armigeres*.

pallidiceps (Theobald, 1905f) (p. 101) – species of *Culex* (*Acallyntrum*) [original combination: *Melanoconion pallidiceps*].

pallidipleura da Cunha Ramos & Brunhes, 2004 (p. 273) – species of *Uranotaenia* (*Pseudoficalbia*).

pallidiventer (Lutz, 1905) (p. 125) – species of *Trichoprosopon* [original combination: *Hyloconops pallidiventer*].

pallidocephala Theobald, 1904b (p. 73) – synonym of *Culex* (*Culex*) *quinquefasciatus* Say, 1823; synonymy by ? Edwards, 1941.

pallidocephala Theobald, 1908c (p. 266) – species of *Uranotaenia* (*Uranotaenia*).

Synonyms: *caerulea* Theobald, 1908c; *similis* Theobald, 1908c; *abnormalis* Theobald, 1910c.

pallidohirta Grossbeck, 1905 (p. 359) – synonym of *Aedes* (*Aedes*) *cinereus* Meigen, 1818 [original combination: *Culex pallidohirta* [*sic*]]; synonymy established before Knight & Stone, 1977.

pallidopalpi Theobald, 1907 (p. 57) – synonym of *Anopheles* (*Cellia*) *smithii* Theobald, 1905b [original combination: *Feltinella pallidopalpi*]; synonymy established before Knight & Stone, 1977.

pallidopunctata Theobald, 1911b (p. 267) – synonym of *Aedes* (*Aedimorphus*) *dentatus* (Theobald, 1904b) [original combination: *Culex pallidopunctata* [*sic*]]; synonymy established before Knight & Stone, 1977.

pallidostriatus (Theobald, 1907) (p. 410) – species of *Aedes* (*Aedimorphus*) [original combination: *Culex pallidostriatus*].

Synonym: *parascelos* Theobald, 1910a.

pallidothoracis Theobald, 1909b (p. 17) – synonym of *Culex* (*Culex*) *pruina* Theobald, 1901d; synonymy established before Knight & Stone, 1977.

pallidothorax Dong, Dong & Wu, 2008 (p. 187) – species of *Tripteroides* (*Rachionotomyia*).

pallidothorax Theobald, 1905g (p. 32) – species of *Culex* (*Culiciomyia*).

Synonyms: *albopleura* Theobald, 1907; *annuloabdominalis* Theobald, 1910c.

pallidoventer Theobald, 1903a (p. 300) – species of *Uranotaenia* (*Uranotaenia*).

pallidoventer (Theobald, 1907) (p. 598) – species of *Wyeomyia* (*Phoniomyia*) [original combination: *Phoniomyia pallidoventer*].

pallidus Lee, 1946 (as *pallida*) (p. 255) – species of *Tripteroides* (*Rachisoura*).

pallidus Swellengrebel, 1919 (p. 8) – synonym of *Anopheles* (*Anopheles*) *barbumbrosus* Strickland & Chowdhury, 1927a [original combination: *Myzorhynchus barbirostris* variety *pallidus*]; synonymy established before Knight & Stone, 1977. **Note**: this nominal taxon is the junior secondary homonym of *pallidus* Theobald, 1901a.

pallidus Theobald, 1901a (p. 134) – species of *Anopheles* (*Cellia*) [original combination: *Anopheles fuliginosus* variety *pallida* [*sic*]]. **Note**: this nominal taxon is the senior secondary homonym of *pallidus* Swellengrebel, 1919.

Synonym: *fowleri* Christophers, 1911.

pallipes Macquart, 1838a (p. 37) – synonym of *Culex* (*Culex*) *pipiens* Linnaeus, 1758; synonymy established before Knight & Stone, 1977. **Note**: this nominal taxon is the junior primary homonym of *pallipes* Robineau-Desvoidy, 1827 and *pallipes* Waltl, 1835.

pallipes Robineau-Desvoidy, 1827 (p. 410) – *nomen dubium*, *Culex*; status by Belkin, 1968b. **Note**: this nominal taxon is the senior primary homonym of *pallipes* Waltl, 1835 and *pallipes* Macquart, 1838a.

pallipes Waltl, 1835 (p. 110) – synonym of *Culex* (*Culex*) *pipiens* Linnaeus, 1758; synonymy established before Knight & Stone, 1977. **Note**: this nominal taxon is the junior primary homonym of *pallipes* Robineau-Desvoidy, 1827 and the senior primary homonym of *pallipes* Macquart, 1838a.

pallirostris Edwards, 1922a (p. 270) – species of *Aedes* (*Hulecoeteomyia*).

Synonym: *khasiana* Barraud, 1923e.

pallithorax Dong, Zhou & Dong, 2004 (p. 123) – species of *Armigeres* (*Armigeres*).

palmarum Edwards, 1924b (p. 382) – species of *Aedes* (*Patmarksia*).

palmata (Lane & Cerqueira, 1942) (p. 648) – species of *Wyeomyia* (*Phoniomyia*) [original combination: *Phoniomyia palmata*].

palmatus (Rodenwaldt, 1926) (p. 794) – species of *Anopheles* (*Anopheles*) [original combination: *Stethomyia* [as *Stetomyia*] *aitkenii* variety *palmata*]; specific status by Lee & Woodhill, 1944.

palmeirimi de Meillon & Rebêlo, 1941 (p. 76) – species of *Uranotaenia* (*Uranotaenia*).

palmeni Edwards, 1921b (p. 52) – synonym of *Aedes* (*Ochlerotatus*) *communis* (de Geer, 1776) [original combination: *Ochlerotatus palmeni*]; synonymy by Edwards, 1921d.

palmi Baisas, 1938 (p. 207) – synonym of *Culex* (*Culex*) *annulirostris* Skuse, 1889; synonymy established before Knight & Stone, 1977.

palpalis (Newstead, 1907) (in Newstead *et al.*, 1907) (p. 31) – species of *Aedes* (*Neomelaniconion*) [original combination: *Neomelaniconion palpale*]. **Note**: the name of this species is incorrectly listed in its original form (neuter) in the online catalogue of the Walter Reed Biosystematics Unit (http://www.mosquitocatalog.org/).

palpalis Taylor, 1912a (p. 29) – species of *Culex* (*Culex*).

paltrinierii Shidrawi & Gillies, 1988 (p. 202) – species of *Anopheles* (*Cellia*).

paludis Taylor, 1913 (p. 56) – synonym of *Culex* (*Culex*) *sitiens* Wiedemann, 1828 [original combination: *Culicelsa paludis*]; synonymy established before Knight & Stone, 1977.

paludis Theobald, 1900b (p. 75) – species of *Anopheles* (*Anopheles*).

paludosa Galindo, Blanton & Peyton, 1954 (p. 162) – species of *Uranotaenia* (*Uranotaenia*).

paludosus Graham, 1910 (p. 270) – synonym of *Aedes* (*Aedimorphus*) *nigricephalus* (Theobald, 1901d) [original combination: *Myxosquamus paludosus*]; synonymy established before Knight & Stone, 1977.

palus Theobald, 1903a (p. 194) – synonym of *Culex* (*Culex*) *nigripalpus* Theobald, 1901c; synonymy established before Knight & Stone, 1977.

palustris Dyar, 1916b (p. 89) – synonym of *Aedes* (*Ochlerotatus*) *fitchii* (Felt & Young, 1904); synonymy established before Knight & Stone, 1977.

palustris Theobald, 1911a (p. 13) – synonym of *Mimomyia* (*Mimomyia*) *hispida* (Theobald, 1910c) [original combination: *Megaculex palustris*]; synonymy by Stone *et al.*, 1959.

pampanai Büttiker & Beales, 1959 (as *pampanae*) (p. 63) – species of *Anopheles* (*Cellia*); emendation by Stone *et al.*, 1959.

pampangensis Brunetti, 1920 (p. 114) – synonym of *Anopheles* (*Cellia*) *philippinensis* Ludlow, 1902b; *nomen novum* for *philippinensis* Ludlow, 1902b in error.

pampangensis (Ludlow, 1905a) (p. 94) – species of *Aedes* (*Aedimorphus*) [original combination: *Reedomyia pampangensis*].

Synonym: *niveoscutellum* Theobald, 1905g.

pampithes (Dyar & Núñez Tovar, 1928) (in Dyar, 1928) (p. 64) – species of *Wyeomyia* (subgenus uncertain, determined by Motta & Lourenço-de-Oliveira, 1995) [original combination: *Miamyia* (*Shropshirea*) *pampithes*].

panalectoris Giles, 1901a (p. 608) – synonym of *Armigeres* (*Armigeres*) *subalbatus* (Coquillett, 1898) [original combination: *Culex panalectoris*]; synonymy by Thurman, 1959.

panamena Dyar & Knab, 1907c (p. 210) – synonym of *Wyeomyia* (*Wyeomyia*) *arthrostigma* (Lutz, 1905); synonymy by Lane & Cerqueira, 1942.

panamensis Adames, 1971 (p. 61) – species of *Deinocerites*.

panarchys Dyar, 1921g (p. 104) – species of *Haemagogus* (*Haemagogus*).

panayensis (Ludlow, 1914c) (p. 159) – species of *Verrallina* (*Neomacleaya*) [original combination: *Aedes panayensis*].

panchgangee Sathe & Girhe, 2002 (p. 198) – *nomen nudum*, *Aedes*; unavailable name [published after 1999, fails to conform to Articles 16.4 and 72.3 of the *Code*, which require fixation of a name-bearing type and a statement naming the collection in which it is deposited (International Commission on Zoological Nomenclature, 1999); status by Harbach (22 November 2016, http://mosquito-taxonomic-inventory.info/whats-new)].

pandani Brunhes, 1969 (p. 175) – species of *Culex* (*Culiciomyia*).

pandani Stone, 1939 (p. 162) – species of *Aedes* (*Stegomyia*).

pandani (Theobald, 1912b) (p. 90) – species of *Uranotaenia* (*Pseudoficalbia*) [original combination: *Pseudoficalbia pandani*].

pandora Dyar & Knab, 1909d (p. 261) – synonym of *Wyeomyia melanocephala* Dyar & Knab, 1906a; synonymy established before Knight & Stone, 1977.

panocossa Dyar, 1923c (p. 120) – species of *Culex* (*Melanoconion*).

pantjarbatu Waktoedi Koesoemawinangoen, 1954 (p. 49) – subspecies of *Anopheles* (*Anopheles*) *gigas* Giles, 1901b [original combination: *Anopheles gigas* variety *pantjarbatu*]; subspecific status by Harbach & Howard, 2007.

pantoia Dyar & Knab, 1909d (p. 262) – synonym of *Wyeomyia* (*Decamyia*) *ulocoma* (Theobald, 1903a); synonymy established before Knight & Stone, 1977.

papago Zavortink, 1970c (p. 3) – species of *Aedes* (*Abraedes*).

papua Brug, 1924 (p. 441) – synonym of *Uranotaenia* (*Pseudoficalbia*) *obscura* Edwards, 1915c; synonymy by Peyton, 1972.

papua Brug, 1934 (p. 505) – synonym of *Tripteroides* (*Rachisoura*) *filipes* (Walker, 1861); synonymy by E.N. Marks (in Lee *et al.*, 1989c).

papuae (Swellengrebel & Swellengrebel de Graaf, 1920a) (p. 3) – species of *Bironella* (*Neobironella*) [original combination: *Stethomyia aitkenii* variety *papuae*]; specific status by de Rook & Soesilo, 1930.

Synonym: *brugi* Soesilo & van Slooten, 1931a, 1931b.

papuensis Dobrotworsky, 1957 (p. 185) – species of *Anopheles* (*Anopheles*).

papuensis de Rook, 1929 (pp. 29, 39, 45 [pp. 34, 48, 55 in English translation]) – *nomen nudum*, *Anopheles*; status by Lee *et al.*, 1988a.

papuensis Marks, 1960 (p. 91) – species of *Topomyia* (*Suaymyia*).

papuensis Peters, 1963a (p. 4) – species of *Armigeres* (*Armigeres*).

papuensis (Taylor, 1914d) (p. 193) – species of *Aedes* (*Patmarksia*) [original combination: *Leucomyia australiensis* variety *papuensis*]; specific status by Edwards, 1924b.

papuensis (Taylor, 1914d) (p. 200) – species of *Mansonia* (*Mansonioides*) [original combination: *Taeniorhynchus papuensis*].

papuensis (Taylor, 1914d) (p. 201) – species of *Culex* (*Culiciomyia*) [original combination: *Melanoconion papuensis*].

papuensis Zavortink, 1968 (p. 139) – species of *Orthopodomyia*.

par Newstead, 1907 (in Newstead *et al.*, 1907) (p. 25) – synonym of *Culex* (*Oculeomyia*) *poicilipes* (Theobald, 1903a); synonymy established before Knight & Stone, 1977.

paracrybda Komp, 1936c (p. 330) – species of *Culex* (*Melanoconion*).

paraculeatus Sirivanakarn, 1977a (p. 49) – species of *Culex* (*Lophoceraomyia*).

paradiantaeus Apfelbeck, 1929 (p. 290) – synonym of *Aedes* (*Ochlerotatus*) *sticticus* (Meigen, 1838); synonymy established before Knight & Stone, 1977.

paradissimilis Rozeboom, 1946 (p. 587) – species of *Aedes* (*Kenknightia*).

paradoxus Harbach, 2002 (in Harbach & Howard, 2002) (p. 363) – species of *Sabethes* (*Peytonulus*).

paraensis (Theobald, 1903a) (p. 316) – species of *Limatus* [original combination: *Dendromyia parensis*]; elevated from synonymy with *durhamii* Theobald, 1901c by Lane & Cerqueira, 1942.
Synonym: *cacophrades* Dyar & Knab, 1909d.

paraguayensis Strickland, 1911c (p. 268) – synonym of *Psorophora* (*Janthinosoma*) *albigenu* (Lutz, 1908) [original combination: *Janthinosoma paraguayensis*]; synonymy established before Knight & Stone, 1977.

parainfantulus Menon, 1944 (p. 389) – synonym of *Culex* (*Lophoceraomyia*) *infantulus* Edwards, 1922a [original combination: *Lophoceratomyia parainfantulus* [*sic*]]; synonymy by Mattingly, 1949b.

paraitepuyensis Anduze, 1941 (p. 824) – species of *Sabethes* (*Sabethes*).

paralateralis Peters, 1964 (p. 19) – species of *Uranotaenia* (*Uranotaenia*).

paraliae Sandosham, 1959 (p. 189) – synonym of *Anopheles* (*Anopheles*) *lesteri* Baisas & Hu, 1936 [original combination: *Anopheles* (*Anopheles*) *lesteri paraliae*]; synonymy by Taai *et al.*, 2013.

paramaxi Duret, 1968b (p. 57) – species of *Culex* (*Culex*).

paranensis (Brèthes, 1911) (p. 471) – species of *Isostomyia* [original combination: *Lynchiaria paranensis*].

parangensis Ludlow, 1909a (p. 24) – synonym of *Uranotaenia* (*Uranotaenia*) *argyrotarsis* Leicester, 1908 [original combination: *Pseudouranotaenia parangensis*]; synonymy established before Knight & Stone, 1977.

parangensis (Ludlow, 1914b) (p. 129) – species of *Anopheles* (*Cellia*) [original combination: *Myzomyia parangensis*].

paranovaguinensis Peters, 1963d (p. 135) – species of *Uranotaenia* (*Uranotaenia*).

paraplesia Dyar, 1922g (p. 192) – synonym of *Culex* (*Aedinus*) *amazonensis* (Lutz, 1905); synonymy established before Knight & Stone, 1977.

parapunctipennis Martini, 1932b (p. 101) – species of *Anopheles* (*Anopheles*).
Synonym: *guatemalensis* de Leon, 1938.

parascelos Theobald, 1910a (p. 18) – synonym of *Aedes* (*Aedimorphus*) *pallidostriatus* (Theobald, 1907) [original combination: *Culex parascelos*]; synonymy established before Knight & Stone, 1977.

parasimilis (King & Hoogstraal, 1947b) (p. 125) – species of *Verrallina* (*Verrallina*) [original combination: *Aedes* (*Aedes*) *parasimilis*].

parensis Gillies, 1962 (p. 82) – species of *Anopheles* (*Cellia*).

parenti (de Meillon & Lavoipierre, 1944) (p. 49) – species of *Mimomyia* (*Mimomyia*) [original combination: *Ficalbia* (*Mimomyia*) *parenti*].

parodites Dyar, 1928 (p. 244) – synonym of *Culiseta* (*Culicella*) *morsitans* (Theobald, 1901c) [original combination: *Culicella* (*Culicella*) *parodites*]; synonymy by Maslov, 1964.

particeps (Adams, 1903) (p. 26) – species of *Culiseta* (*Culiseta*) [original combination: *Culex particeps*].
Synonyms: *dugesi* Dyar & Knab, 1906a; *maccrackenae* Dyar & Knab, 1906a.

parvipluma Edwards, 1941 (p. 240) – species of *Eretmapodites* [original combination: *Eretmapodites oedipodius parvipluma*]; specific status by Hamon, 1961.

parvulus Edwards, 1921d (p. 314) – synonym of *Aedes* (*Ochlerotatus*) *impiger* (Walker, 1848); synonymy established before Knight & Stone, 1977.

parvus (Chagas, 1907) (p. 4) – species of *Anopheles* (*Nyssorhynchus*) [original combination: *Myzorhynchella parva*].

parvus Macquart, 1834 (p. 36) – synonym of *Aedes* (*Aedimorphus*) *vexans* (Meigen, 1830b) [original combination: *Culex parvus*]; synonymy established before Knight & Stone, 1977.

parvus Taylor, 1912a (p. 27) – synonym of *Culex* (*Culex*) *crinicauda* Edwards, 1921c; synonymy established before Knight & Stone, 1977.

pasadaemon Dyar, 1921f (p. 100) – synonym of *Culex* (*Melanoconion*) *conspirator* Dyar & Knab, 1906d; synonymy established before Knight & Stone, 1977.

patersoni Alvarado & Heredia, 1947 (p. 73) – subspecies of *Anopheles* (*Anopheles*) *pseudopunctipennis* Theobald, 1901c.

patersoni Shannon & del Ponte, 1928 (p. 73) – species of *Aedes* (*Ochlerotatus*).

patersoni Shannon & del Ponte, 1928 (in Dyar, 1928) (p. 268) – synonym of *Lutzia* (*Lutzia*) *bigoti* Bellardi, 1862; synonymy by Lane, 1951.

patientiae Floch & Fauran, 1955 (p. 1) – species of *Culex* (*Melanoconion*).

patriciae Mattingly, 1954a (p. 262) – species of *Aedes* (*Stegomyia*).

patriciae Peyton, 1977 (p. 103) – species of *Uranotaenia* (*Pseudoficalbia*).

pattoni Christophers, 1926 (p. 871) – species of *Anopheles* (*Cellia*).

paucartamboensis Porter, 2014 (p. 312) – species of *Wyeomyia* (*Nunezia*).

pauliani Doucet, 1949b (p. 325) – species of *Uranotaenia* (*Pseudoficalbia*).

pauliani (Doucet, 1951b) (p. 107) – species of *Toxorhynchites* (*Afrorhynchus*) [original combination: *Megarhinus pauliani*].

pauliani Grjebine, 1950 (p. 45) – species of *Eretmapodites*.

pauliani Grjebine, 1953a (p. 467) – species of *Aedeomyia* (*Aedeomyia*).

pauliani Grjebine, 1953a (p. 469) – species of *Anopheles* (*Cellia*).

paulistensis Galvão, Lane & Corrêa, 1937 (p. 37) – synonym of *Anopheles* (*Nyssorhynchus*) *darlingi* Root, 1926 [original combination: *Anopheles* (*Nyssorhynchus*) *darlingi* variety *paulistensis*]; synonymy by Lane, 1949.

paulli Paterson & Shannon, 1927 (pp. 4, 9) – species of *Psorophora* (*Grabhamia*).

paullusi Stone & Farner, 1945 (p. 155) – species of *Aedes* (*Stegomyia*).

pavlovsky Shingarev, 1928 (p. 52) – synonym of *Culex* (*Culex*) *torrentium* Martini, 1925; synonymy established before Knight & Stone, 1977.

pavlovskyi Casal & García, 1967a (p. 19) – species of *Culex* (*Melanoconion*).

pazosi Pazos, 1908 (p. 432) – synonym of *Psorophora* (*Janthinosoma*) *ferox* (von Humboldt, 1819) [original combination: *Aedes pazosi*]; synonymy established before Knight & Stone, 1977.

pearyi Dyar & Shannon, 1925d (p. 78) – synonym of *Aedes* (*Ochlerotatus*) *pullatus* (Coquillett, 1904c); synonymy by Vockeroth, 1954.

peccator Dyar & Knab, 1909d (p. 256) – species of *Culex* (*Melanoconion*).
Synonym: *incriminator* Dyar & Knab, 1909d.

pecori Reinert, 1990 (p. 52) – species of *Aedes* (*Kenknightia*).

pectinatus Arnell, 1976 (p. 74) – species of *Aedes* (*Ochlerotatus*).

pectinatus (Edwards, 1914a) (p. 263) – species of *Armigeres* (*Leicesteria*) [original combination: *Leicesteria pectinata*].

pecuniosus Edwards, 1922d (p. 94) – species of *Aedes* (*Molpemyia*).

pedicellus King & Hoogstraal, 1947c (p. 65) – species of *Culex* (*Neoculex*).

peditaeniatus (Leicester, 1908) (p. 31) – species of *Anopheles* (*Anopheles*) [original combination: *Myzorhynchus peditaeniatus*].

pedroi Sirivanakarn & Belkin, 1980 (p. 12) – species of *Culex* (*Melanoconion*).

pefflyi Stone, 1961a (p. 249) – subspecies of *Uranotaenia* (*Pseudoficalbia*) *unguiculata* Edwards, 1913a.

peipingensis Feng, 1938 (p. 520) – species of *Aedes* (subgenus uncertain).

pembaensis Theobald, 1901c (p. 235) – species of *Aedes* (*Skusea*).
Synonym: *mutica* Enderlein, 1923.

penafieli Sanchez, 1885 (p. 213, described), 1886 (p. 326, named) – synonym of *Culex* (*Culex*) *quinquefasciatus* Say, 1823; synonymy established before Knight & Stone, 1977.

penai Sirivanakarn, 1979 (p. 135) – species of *Culex* (*Melanoconion*).

pendleburyi (Edwards, 1930a) (p. 305) – species of *Toxorhynchites* (*Toxorhynchites*) [original combination: *Megarhinus pendleburyi*].

pendulus (Edwards, 1914a) (p. 261) – species of *Armigeres* (*Leicesteria*) [original combination: *Leicesteria pendula*].

penetrans (Robineau-Desvoidy, 1827) (p. 407) – *nomen dubium*, *Culiseta* [original combination: *Culex penetrans*]; status by Edwards, 1932a.

penghuensis Lien, 1968b (p. 96) – species of *Aedes* (*Cancraedes*).

penicillaris Rondani, 1872 (p. 31) – synonym of *Aedes* (*Ochlerotatus*) *caspius* (Pallas, 1771) [original combination: *Culex penicillaris*]; synonymy established before Knight & Stone, 1977.

penicillatus Edwards, 1941 (p. 234) – species of *Eretmapodites*.

pennai Antunes & Lane, 1938 (p. 607) – species of *Aedes* (*Ochlerotatus*).

perdita (Leicester, 1908) (p. 192) – *nomen dubium*, *Verrallina* [original combination: *Aioretomyia perdita*]; status by Reinert *et al.*, 2004.

perexiguus Theobald, 1903a (p. 199) – species of *Culex* (*Culex*).

pereyrai Duret, 1967a (p. 81) – species of *Culex* (*Melanoconion*).

perezi Shannon & del Ponte, 1928 (p. 56) – synonym of *Anopheles* (*Nyssorhynchus*) *triannulatus* (Neiva & Pinto, 1922b); synonymy by Faran, 1980.

perfidiosus Edwards, 1914b (p. 72) – species of *Culex* (*Culex*).

perfuscus Edwards, 1914b (p. 71) – species of *Culex* (*Culex*).

pergusae Missiroli, 1935 (p. 333) – synonym of *Anopheles* (*Anopheles*) *labranchiae* Falleroni, 1926 [original combination: *Anopheles maculipennis pergusae*]; synonymy by Bates, 1940.

peribleptus Dyar & Knab, 1918c (p. 181) – synonym of *Culex* (*Melanoconion*) *erraticus* (Dyar & Knab, 1906d); synonymy by King & Bradley, 1937.

perichares Dyar, 1921b (p. 36) – synonym of *Aedes* (*Georgecraigius*) *epactius* Dyar & Knab, 1908; synonymy by Zavortink, 1972.

peringueyi Edwards, 1924a (p. 162) – species of *Culex* (*Maillotia*).

periskelatus (Giles, 1902) (p. 371) – species of *Aedes* (*Dendroskusea*) [original combination: *Stegomyia periskelata*].
Synonym: *annulifemur* Edwards, 1914b.

perkinsi Marks, 1949 (p. 32) – species of *Aedes* (*Ochlerotatus*).

perkinsi Stone & Penn, 1948 (p. 110) – species of *Culex* (*Acallyntrum*).

perkunas Podėnas, 1999 (p. 113) – synonym of fossil species *Culex erikae* Szadziewski & Szadziewska, 1985 [original combination: *Aedes perkunas*]; synonymy by Szadziewski & Gilka, 2011.

perlisiensis Yong, Cheong, Chiang, Dhaliwal, Loong & Sarjan, 1983 (p. 613) – *nomen nudum*, *Anopheles* [original combination: *Anopheles balabacensis perlisiensis*]; status by Ward, 1984 (incorrectly listed as a synonym of *Anopheles* (*Cellia*) *cracens* Sallum & Peyton by Sallum *et al.*, 2005a).

permixtus Hsieh & Liao, 1956 (p. 123) – synonym of *Culex* (*Culex*) *annulus* Theobald, 1901a; synonymy by Chen, 1980.

pernotatus Farner & Bohart, 1944 (p. 118) – species of *Aedes* (*Stegomyia*).

perplexens (Edwards, 1932b) (p. 561) – species of *Mimomyia* (*Mimomyia*) [original combination: *Ficalbia* (*Mimomyia*) *perplexens*].

perplexens Ludlow, 1907 (p. 267) – species of *Anopheles* (*Anopheles*).

perplexus (Leicester, 1908) (p. 83) – species of *Aedes* (*Stegomyia*) [original combination: *Stegomyia perplexa*].

perplexus Leicester, 1908 (p. 150) – species of *Culex* (*Culex*).

perplexus Peters, 1963c (p. 91) – species of *Tripteroides* (*Polylepidomyia*).

perplexus Taylor, 1943 (p. 153) – synonym of *Anopheles* (*Cellia*) *annulipes* Walker, 1856a; synonymy by Lee & Woodhill, 1944.

perryi Belkin, 1962 (p. 268) – species of *Culex* (*Lophoceraomyia*).

perryi Belkin, 1962 (p. 336) – species of *Aedes* (*Geoskusea*).

persephassa Dyar & Knab, 1909d (p. 254) – synonym of *Coquillettidia* (*Rhynchotaenia*) *nigricans* (Coquillett, 1904c) [original combination: *Bancroftia persephassa*]; synonymy established before Knight & Stone, 1977.

persicus Edwards, 1921d (p. 280) – synonym of *Anopheles* (*Cellia*) *turkhudi* Liston, 1901 [original combination: *Anopheles* (*Myzomyia*) *turkhudi* variety *persicus*]; synonymy established before Knight & Stone, 1977.

persiensis Linton, Sedaghat & Harbach, 2003 (in Sedaghat *et al.*, 2003) (p. 533) – species of *Anopheles* (*Anopheles*).

persimilis Mattingly, 1970 (p. 30) – species of *Heizmannia* (*Heizmannia*).

persimilis Taylor, 1943 (p. 155) – synonym of *Anopheles* (*Cellia*) *annulipes* Walker, 1856a [original combination: *Anopheles* (*Myzomyia*) *perplexus* variety *persimilis*]; synonymy by Lee & Woodhill, 1944.

persistans Banks, 1906b (p. 996) – synonym of *Aedes* (*Stegomyia*) *aegypti* (Linnaeus, 1762) [original combination: *Stegomyia fasciata persistans*]; synonymy established before Knight & Stone, 1977.

personatum (Lutz, 1904) (in Bourroul, 1904) (p. 22) – species of *Onirion* [original combination: *Dendromyia personata*].

Synonym: *luederwaldti* Lane, 1936a.

perterrens (Walker, 1856a) (p. 431) – *nomen dubium*, *Psorophora* [original combination: *Culex perterrens*]; status by Stone, 1957a (see also Belkin *et al.*, 1971).

pertinans (Williston, 1896) (p. 271) – species of *Wyeomyia* (*Wyeomyia*) [original combination: *Aedes pertinans*].

pertinax Grabham, 1906b (p. 316) – species of *Aedes* (*Ochlerotatus*).

Synonym: *quasiserratus* Theobald, 1907.

perturbans (Walker, 1856a) (p. 428) – species of *Coquillettidia* (*Coquillettidia*) [original combination: *Culex perturbans*].

Synonyms: *testaceus* van der Wulp, 1867; *ochropus* Dyar & Knab, 1907b.

perturbans (Williston, 1896) (p. 271) – species of *Isostomyia* [original combination: *Aedes perturbans*].

peruvianus Tamayo, 1907 (in Tamayo & García, 1907) (p. 35) – synonym of *Anopheles* (*Anopheles*) *pseudopunctipennis* Theobald, 1901c; synonymy by Lane, 1953.

perventor Cerqueira & Costa, 1946 (p. 135) – species of *Aedes* (*Sallumia*).

pervigilans Bergroth, 1889 (p. 295) – species of *Culex* (*Culex*).

peryassui Dyar & Knab, 1908 (p. 53) – species of *Anopheles* (*Anopheles*) [*nomen novum* for *lutzi* Cruz, 1907a].

Synonyms: *lutzi* Cruz, 1907a; *alagoanii* Peryassú, 1925; *celidopus* Dyar & Shannon, 1925b.

pessoai (Barreto & Coutinho, 1944) (p. 69) – species of *Mansonia* (*Mansonia*) [original combination: *Taeniorhynchus* (*Taeniorhynchus*) *pessoai*].

pessoai Galvão & Lane, 1937a (p. 67) – synonym of *Anopheles* (*Nyssorhynchus*) *braziliensis* (Chagas, 1907); synonymy by Lane, 1953.

petersi Colless, 1960 (p. 388) – species of *Culex* (*Lophoceraomyia*).

petersoni Dyar, 1920b (p. 27) – synonym of *Culex* (*Culex*) *bahamensis* Dyar & Knab, 1906d; synonym established before Knight & Stone, 1977.

petragnani del Vecchio, 1939 (p. 30) – species of *Anopheles* (*Anopheles*) [original combination: *Anopheles claviger* (*bifurcatus*) variety *petragnani*]; specific status by Coluzzi, 1962.

Synonym: *portucaliensis* de Figueiredo, 1904.

petrifactellus Cockerell, 1916 (p. 489) – synonym of fossil species (Genus ?) *protolepis* Cockerell, 1916 [original combination: *Culex petrifactellus*]; synonymy by Edwards, 1923c.

petrocchiae Martínez, Carcavallo & Prosen, 1961 (p. 79) – synonym of *Haemagogus* (*Haemagogus*) *janthinomys* Dyar, 1921g [original combination: *Haemagogus* (*Stegoconops*) *capricornii petrocchiae*]; synonymy by Arnell, 1973.

petrocchiae (Shannon & del Ponte, 1928) (p. 94) – species of *Sabethes* (*Davismyia*) [original combination: *Miamyia petrocchiae*].

Synonym: *monoleua* Martini, 1931b.

petroelephantus (Wijesundara, 1951) (p. 176) – species of *Verrallina* (*Neomacleaya*) [original combination: *Aedes* (*Aedes*) *petroelephantus*].

pettigrewii Theobald, 1910a (p. 15) – synonym of *Culex* (*Culex*) *theileri* Theobald, 1903a; synonymy established before Knight & Stone, 1977.

peus Speiser, 1904 (p. 148) – rejected name, suppression by International Commission on Zoological Nomenclature, 1991; *nomen novum* for *affinis* Adams, 1903.

pexus Colless, 1958 (p. 469) – species of *Aedes* (*Downsiomyia*).

peytoni Bram & Rattanarithikul, 1967 (p. 7) – species of *Culex* (*Lophoceraomyia*).

peytoni Kulasekera, Harrison & Amerasinghe, 1989 (p. 303) – species of *Anopheles* (*Anopheles*).

peytoni Leguizamón & Carpintero, 2004 (p. 204) – species of *Orthopodomyia*.

peytoni Reinert, 1972a (p. 7) – species of *Aedes* (*Ayurakitia*).

phaecasiatus Marks, 1964a (p. 64) – species of *Aedes* (*Ochlerotatus*).

phaeonotus Arnell, 1976 (p. 56) – species of *Aedes* (*Ochlerotatus*).

phangngae Sirivanakarn, 1972 (p. 17) – species of *Culex* (*Eumelanomyia*).

pharoensis Theobald, 1901a (p. 169) – species of *Anopheles* (*Cellia*).

Synonyms: *albofimbriatus* Giles, 1902; *maculicosta* Becker, 1903; *albus* Ventrillon, 1904; *bozasi* Neveu-Lemaire, 1905a.

philipi Edwards, 1929b (p. 327) – species of *Culex* (*Culex*).

philippii Dyar, 1924g (p. 129) – synonym of *Aedes* (*Ochlerotatus*) *albifasciatus* (Macquart, 1838b) [*nomen novum* for *vittatus* Philippi, 1865]; synonymy established before Knight & Stone, 1977.

philippinensis Delfinado, 1966b (p. 36) – synonym of *Uranotaenia* (*Pseudoficalbia*) *obscura* Edwards, 1915c; synonymy by Peyton, 1972.

philippinensis (Delfinado, 1968) (p. 33) – species of *Verrallina* (*Neomacleaya*) [original combination: *Aedes* (*Neomacleaya*) *philippinensis*].

philippinensis Giles, 1904a (p. 368) – synonym of *Tripteroides* (*Tripteroides*) *nitidoventer* (Giles, 1904a) [original combination: *Runchomyia philippinensis*]; synonymy established before Knight & Stone, 1977.

philippinensis Ludlow, 1902b (p. 426) – species of *Anopheles* (*Cellia*).

Synonyms: *freerae* Banks, 1906b; *pampangensis* Brunetti, 1920; *hainanensis* Takei, 1941.

philippinensis Sirivanakarn, 1976 (p. 124) – species of *Culex* (*Culex*).

phillipi van Someren, 1949a (p. 7) – species of *Aedes* (*Zavortinkius*).

philonuxia Philip, 1931 (p. 188) – species of *Uranotaenia* (*Uranotaenia*).

philophone Dyar & Knab, 1907c (p. 209) – synonym of *Wyeomyia* (*Wyeomyia*) *celaenocephala* Dyar & Knab, 1906a [original combination: *Phoniomyia philophone*]; synonymy established before Knight & Stone, 1977.

philosophicus Dyar & Knab, 1906d (pp. 190, 195) – synonym of *Haemagogus* (*Haemagogus*) *equinus* Theobald, 1903g [original combination: *Aedes philosophicus*]; synonymy established before Knight & Stone, 1977.

phlabistus Dyar, 1920d (p. 63) – species of *Culex* (*Melanoconion*).

Synonym: *kerri* Duret, 1968a.

phlogistus Dyar, 1920d (p. 61) – species of *Culex* (*Melanoconion*).

phnoma (Klein, 1973) (p. 9) – species of *Verrallina* (*Neomacleaya*) [original combination: *Aedes* (*Neomacleaya*) *phnomus*].

phoeniciae Coluzzi & Sabatini, 1968 (p. 68) – species of *Aedes* (*Acartomyia*) [original combination: *Aedes* (*Ochlerotatus*) *mariae phoeniciae*]; specific status by Coluzzi *et al.*, 1970.

pholeocola (Linton & Harbach, 2013) (in Harbach *et al.*, 2013) (p. 162) – species of *Nyctomyia* Harbach, 2013 [original combination: *Nyx pholeocola*].

pholeter Bram & Rattanarithikul, 1967 (p. 13) – species of *Culex* (*Lophoceraomyia*).

pholidotus Zavortink, 1973 (p. 15) – species of *Anopheles* (*Kerteszia*).

phroso Howard, Dyar & Knab, 1915 (p. 149) – species of *Wyeomyia* (subgenus uncertain, determined by Motta & Lourenço-de-Oliveira, 1995).

Synonym: *proviolans* Dyar, 1919d.

phyllados Hutchings & Sallum, 2008 (p. 42) – species of *Culex* (*Melanoconion*).

phyllolabis Edwards, 1929b (p. 324) – species of *Aedes* (*Catageiomyia*).

phyllozoa (Dyar & Knab, 1907c) (p. 199) – species of *Orthopodomyia* [original combination: *Mansonia phyllozoa*].

phytophagus Ficalbi, 1890d (p. 126) – synonym of *Culex* (*Culex*) *pipiens* Linnaeus, 1758; synonymy established before Knight & Stone, 1977.

phytophagus Theobald, 1909b (p. 1) – species of *Toxorhynchites* (*Toxorhynchites*).

pia (Le Goff, Brengues & Robert, 2013) (p. 3) – species of *Aedes* (*Stegomyia*) [original combination: *Stegomyia pia*].

pictipennis (Philippi, 1865) (p. 596) – species of *Anopheles* (*Nyssorhynchus*) [original combination: *Culex pictipennis*].

Synonyms: *variegatus* Blanchard, 1852; *bigotii* Theobald, 1901a; *chilensis* Blanchard, 1905.

pictus Loew, 1845 (p. 4) – synonym of *Anopheles* (*Anopheles*) *hyrcanus* (Pallas, 1771); synonymy established before Knight & Stone, 1977.

pifanoi Anduze, 1949a (p. 60) – species of *Culex* (*Melanoconion*).

pifanoi Cova García, Pulido F. & Escalante de Ugueto, 1981 (p. 210) – species of *Uranotaenia* (*Uranotaenia*).

pilicauda Root, 1928 (in Dyar, 1928) (p. 55) – species of *Wyeomyia* (*Phoniomyia*).

pilifemoralis Wang & Feng, 1964 (p. 37) – species of *Culex* (*Lophoceraomyia*).

pilifera Theobald, 1907 (p. 274) – synonym of *Culex* (*Maillotia*) *hortensis* Ficalbi, 1889a [original combination: *Maillotia pilifera*]; synonymy established before Knight & Stone, 1977.

pilinotum Harrison & Scanlon, 1974 (p. 33) – species of *Anopheles* (*Anopheles*).

pilipes (Macquart, 1834) (p. 36) – species of *Psorophora* (*Psorophora*) [original combination: *Culex pilipes*].

pillaii Mattingly, 1958a (p. 40) – species of *Aedes* (*Rhinoskusea*).

pilosa Brug, 1931a (p. 245) – synonym of *Heizmannia* (*Heizmannia*) *scintillans* Ludlow, 1905a; synonymy by Mattingly, 1957c.

pilosa Brug, 1931c (p. 12) – species of *Topomyia* (*Topomyia*).

pilosa da Cunha Ramos & Brunhes, 2004 (p. 313) – species of *Uranotaenia* (*Pseudoficalbia*).

pilosa Duret, 1971a (p. 22) – species of *Psorophora* (*Janthinosoma*).

pilosus (Dyar & Knab, 1906d) (p. 224) – species of *Culex* (*Melanoconion*) [original combination: *Mochlostyrax pilosus*].
Synonyms: *floridanus* Dyar & Knab, 1906b; *cubensis* Dyar & Knab, 1906d; *jamaicensis* Grabham, 1906b; *agitator* Dyar & Knab, 1907b; *hesitator* Dyar & Knab, 1907c; *ignobilis* Dyar & Knab, 1909b; *deceptor* Dyar & Knab, 1909d; *reductor* Dyar & Knab, 1909d; *mastigia* Howard, Dyar & Knab, 1913; *curopinensis* Bonne-Wepster & Bonne, 1920; *colombiensis* Dyar, 1924i; *radiatus* Senevet & Abonnenc, 1939a.

pilosus Lee, 1946 (p. 255) – species of *Tripteroides* (*Rachisoura*).

pinarocampa Dyar & Knab, 1908 (p. 59) – species of *Culex* (*Culex*).

pincerna Graham, 1910 (p. 267) – synonym of *Mimomyia* (*Mimomyia*) *mimomyiaformis* (Newstead, 1907) [original combination: *Megaculex pincerna*]; synonymy established before Knight & Stone, 1977.

pingpaensis Chang, 1965 (p. 397) – species of *Aedes* (*Edwardsaedes*).
Synonym: *antuensis* Su, Wang & Li, 1978.

pinguis Walker, 1866 (p. 337) – synonym of *Culiseta* (*Culiseta*) *impatiens* (Walker, 1848) [original combination: *Culex pinguis*]; synonymy established before Knight & Stone, 1977.

pinjaurensis Barraud, 1932 (p. 353) – species of *Anopheles* (*Anopheles*) [original combination: *Anopheles aitkeni* variety *pinjaurensis*]; specific status by Christophers, 1933.

pintoi da Costa Lima, 1930b (p. 191) – synonym of *Wyeomyia* (*Miamyia*) *oblita* (Lutz, 1904) [original combination: *Miamyia* (*Miamyia*) *pintoi*]; synonymy by Lane & Cerqueira, 1942.

pionips Dyar, 1919b (p. 19) – species of *Aedes* (*Ochlerotatus*).

pipersalatus (Giles, 1902) (p. 372) – species of *Aedes* (*Aedimorphus*) [original combination: *Stegomyia pipersalata*].
Synonym: *maculata* Theobald, 1905h.

pipiens Linnaeus, 1758 (p. 602) – species of *Culex* (*Culex*).
Synonyms: *bifurcatus* Linnaeus, 1758; *fasciatus* Müller, 1764; *molestus* Forskål, 1775; *trifurcatus* Fabricius, 1794; *luteus* Meigen, 1804; *domesticus* Germar, 1817; *bicolor* Meigen, 1818; *rufus* Meigen, 1818; *marginalis* Stephens, 1825; *meridionalis* Leach, 1825; *calcitrans* Robineau-Desvoidy, 1827; *consobrinus* Robineau-Desvoidy, 1827; *thoracicus* Robineau-Desvoidy, 1827; *pallipes* Waltl, 1835; *unistriatus* Curtis, 1837; *pallipes* Macquart, 1838a; *rufinus* Bigot, 1888; *agilis* Bigot, 1889; *phytophagus* Ficalbi, 1890d; *haematophagus* Ficalbi, 1893; *melanorhinus* Giles, 1900a; *azoriensis* Theobald, 1903a; *varioannulatus* Theobald, 1903a; *longefurcatus* Becker, 1903; *quasimodestus* Theobald, 1905f; *osakaensis* Theobald, 1907; *comitatus* Dyar & Knab, 1909a; *dipseticus* Dyar & Knab, 1909a; *doliorum* Edwards, 1912a; *autogenicus* Roubaud, 1935b; *berbericus* Roubaud, 1935b; *sternopallidus* Roubaud, 1945; *sternopunctatus* Roubaud, 1945; *disjunctus* Roubaud, 1957; *calloti* Rioux & Pech, 1959; *erectus* Iglisch, 1977; *torridus* Iglisch, 1977.

pipkini (Bohart, 1957) (p. 65) – species of *Verrallina* (*Verrallina*) [original combination: *Aedes* (*Aedes*) *pipkini*].

pisces Lassmann, 1944 (p. 4) – synonym of *Psorophora* (*Janthinosoma*) *champerico* (Dyar & Knab, 1906a); synonymy by Lane, 1951.

pitchfordi Giles, 1904c (p. 34) – synonym of *Anopheles* (*Cellia*) *marshallii* (Theobald, 1903a) [original combination: *Pyretophorus pitchfordi*]; synonymy established before Knight & Stone, 1977.

pix Martini, 1935 (p. 55) – synonym of *Aedes* (*Ochlerotatus*) *taeniorhynchus* (Wiedemann, 1821); synonymy by Belkin, 1977b.

plagosus Marks, 1959b (p. 21) – species of *Aedes* (subgenus uncertain).

plantaginis Barraud, 1924h (p. 46) – synonym of *Culex* (*Lophoceraomyia*) *minor* (Leicester, 1908); synonymy by Sirivanakarn, 1977a.

platylepidus Knight & Hull, 1951a (p. 201) – species of *Aedes* (*Scutomyia*).

pleccau Koidzumi, 1924 (species) (p. 97) – subspecies of *Anopheles* (*Anopheles*) *lindesayi* Giles, 1900a; subspecific status by Puri, 1949.

plectoporpe Root, 1927c (p. 589) – species of *Culex* (*Melanoconion*).

plegepennis Theobald, 1907 (p. 375) – synonym of *Culex* (*Culex*) *whitmorei* (Giles, 1904a) [original combination: *Leucomyia plegepennis*]; synonymy established before Knight & Stone, 1977.

pleuristriatus Theobald, 1903a (p. 177) – species of *Culex* (*Microculex*).

plicatus Olivares, 1993 (in Angulo & Olivares, 1993) (p. 24) – species of *Culex* (*Culex*).

plioleucus Edwards, 1941 (p. 241) – species of *Eretmapodites*.

plumbeus Stephens, 1828 (p. 503) – species of *Anopheles* (*Anopheles*).
Synonyms: *nigripes* Staeger, 1839; *corsicanus* Edwards, 1928c; *intermedius* Shingarev, 1928.

plumiferus King & Hoogstraal, 1946f (p. 306) – species of *Aedes* (*Phagomyia*).

plumiger Bonne-Wepster, 1948 (p. 310) – species of *Tripteroides* (*Rachisoura*).

plumiger Dönitz, 1901 (p. 37) – synonym of *Anopheles* (*Anopheles*) *sinensis* Wiedemann, 1828; synonymy established before Knight & Stone, 1977.

plumosa (Theobald, 1901a) (p. 373) – species of *Mimomyia* (*Mimomyia*) [original combination: *Culex plumosus*].
Synonym: *albitarsis* Theobald, 1901c.

plumosus (Brug, 1931c) (p. 18) – species of *Tripteroides* (*Tripteroides*) [original combination: *Rachionotomyia plumosa*].

plutocraticus Dyar & Knab, 1907a (p. 11) – synonym of *Aedes* (*Ochlerotatus*) *tortilis* (Theobald, 1903g); synonymy established before Knight & Stone, 1977.

pluvialis Barraud, 1924f (p. 1281) – species of *Culex* (*Eumelanomyia*).

pluvialis Kirkpatrick, 1925 (p. 373) – synonym of *Culex* (*Lasiosiphon*) *adairi* Kirkpatrick, 1926; synonymy by Kirkpatrick, 1926.

podographicus Dyar & Knab, 1906b (p. 165) – species of *Aedes* (*Protomacleaya*).

pogonurus Edwards, 1936 (p. 52) – species of *Aedes* (*Neomelaniconion*).

poicilipes (Theobald, 1903a) (p. 236) – species of *Culex* (*Oculeomyia*) [original combination: *Lasioconops poicilipes*].

Synonyms: *quasigelidus* Theobald, 1903a; *taeniorhynchoides* Giles, 1904a; *maculipes* Theobald, 1904b; *madagascariensis* Ventrillon, 1905b; *par* Newstead, 1907; *punctipes* Theobald, 1907; *auritaenia* Enderlein, 1920.

poicilius (Theobald, 1903a) (p. 283) – species of *Aedes* (*Finlaya*) [original combination: *Finlaya poicilia*].

poliochros Dyar, 1919b (p. 35) – synonym of *Aedes* (*Ochlerotatus*) *provocans* (Walker, 1848); transferred synonymy by synonymy of senior synonym *trichurus* (Dyar) with *provocans* (Walker) by Wood, 1977.

pollicaris Reid, 1962 (p. 24) – species of *Anopheles* (*Anopheles*).

pollinctor (Graham, 1910) (p. 271) – species of *Aedes* (*Zavortinkius*) [original combination: *Stegomyia pollinctor*].

pollutus Torres Cañamares, 1945 (p. 242) – synonym of *Anopheles* (*Anopheles*) *claviger* (Meigen, 1804) [original combination: *Anopheles claviger* variety *pollutus*]; synonymy by M. Bates (in Boyd, 1949).

polyagrus Dyar, 1918b (p. 77) – synonym of *Aedes* (*Ochlerotatus*) *nubilus* (Theobald, 1903a); synonymy by Komp, 1949.

polynesiensis Marks, 1951 (p. 138) – species of *Aedes* (*Stegomyia*).

ponmeki Miyagi & Toma, 2001 (p. 269) – species of *Tripteroides* (*Rachionotomyia*).

popovi Shingarev, 1928 (p. 50) – synonym of *Anopheles* (*Anopheles*) *hyrcanus* (Pallas, 1771); synonymy established before Knight & Stone, 1977.

portesi Senevet & Abonnenc, 1941 (p. 41) – species of *Culex* (*Melanoconion*).

Synonym: *cayennensis* Floch & Abonnenc, 1945b.

portonovoensis Tewari & Hiriyan, 1992a (p. 123) – species of *Aedes* (*Rhinoskusea*).

portoricensis Ludlow, 1905c (p. 386) – synonym of *Aedes* (*Ochlerotatus*) *taeniorhynchus* (Wiedemann, 1821) [original combination: *Culex portoricensis*]; synonymy established before Knight & Stone, 1977.

portoricensis (von Röder, 1885) (p. 337) – species of *Toxorhynchites* (*Lynchiella*) [original combination: *Megarrhina portoricensis*].

Synonym: *haitiensis* Dyar & Knab, 1906c.

portoricensis Weinbren & Weinbren, 1970 (p. E-161) – *nomen nudum*, *Wyeomyia*.

portucaliensis de Figueiredo, 1904 (p. 223) – synonym of *Anopheles* (*Anopheles*) *petragnani* del Vecchio, 1939 [original combination: *Anopheles bifurcatus* variety *portucaliensis*]; synonymy by Ribeiro *et al*., 1988.

pose Dyar & Knab, 1918c (p. 182) – synonym of *Culex* (*Melanoconion*) *erraticus* (Dyar & Knab, 1906d); synonymy by King & Bradley, 1937.

posticatus Lutz & Neiva, 1913 (p. 139) – synonym of *Toxorhynchites* (*Lynchiella*) *theobaldi* (Dyar & Knab, 1906c) [original combination: *Megarrhinus posticatus*]; synonymy established before Knight & Stone, 1977.

posticatus Wiedemann, 1821 (p. 43) – synonym of *Psorophora* (*Janthinosoma*) *ferox* (von Humboldt, 1819) [original combination: *Culex posticatus*]; synonymy established before Knight & Stone, 1977.

postspiraculosus Dobrotworsky, 1961a (p. 261) – species of *Aedes* (*Pseudoskusea*).

postspiraculosus Lee, 1944b (p. 221) – species of *Culex* (*Neoculex*).

powderi Zavortink, 1970b (p. 17) – species of *Anopheles* (*Anopheles*).

powelli Lee, 1944a (p. 21) – species of *Anopheles* (*Anopheles*).

powelli (Ludlow, 1909c) (p. 235) – species of *Tripteroides* (*Tripteroides*) [original combination: *Uranotaenia powelli*].

poweri (Theobald, 1905g) (p. 19) – species of *Aedes* (*Stegomyia*) [original combination: *Stegomyia poweri*].

prachongae Rattanarithikul & Harrison, 2017 (in Harbach *et al*., 2017) (p. 66) – species of *Anopheles* (*Anopheles*).

praeteritus Séguy, 1923 (p. 207) – synonym of *Aedes* (*Ochlerotatus*) *berlandi* Séguy, 1921; synonymy by Rioux & Arnold, 1955.

prajimi Peyton & Rattanarithikul, 1970 (p. 406) – species of *Uranotaenia* (*Uranotaenia*).

prasinopleurus Martini, 1914 (p. 68) – synonym of *Culex* (*Culex*) *nigripalpus* Theobald, 1901c; synonymy established before Knight & Stone, 1977.

pretans Grossbeck, 1904b (p. 332) – synonym of *Aedes* (*Ochlerotatus*) *sticticus* (Meigen, 1838) [original combination: *Culex pretans*]; synonymy established before Knight & Stone, 1977.

pretoriensis Gough, 1910 (p. 117) – synonym of *Anopheles* (*Cellia*) *squamosus* Theobald, 1901a [original combination: *Cellia pretoriensis*]; synonymy established before Knight & Stone, 1977.

pretoriensis (Theobald, 1903a) (p. 99) – species of *Anopheles* (*Cellia*) [original combination: *Nyssorhynchus pretoriensis*].

Synonym: *tibani* Patton, 1905.

pricei Dyar, 1917a (p. 16) – synonym of *Aedes* (*Ochlerotatus*) *fitchii* (Felt & Young, 1904); synonymy established before Knight & Stone, 1977.

priestleyi Taylor, 1914c (p. 684) – synonym of *Aedes* (*Molpemyia*) *purpureus* (Theobald, 1910c) [original combination: *Calomyia priestleyi*]; synonymy by Taylor, 1944.

principensis da Cunha Ramos, 1993 (p. 264) – species of *Uranotaenia* (*Pseudoficalbia*).

prioekanensis (Brug, 1931a) (p. 249) – species of *Verrallina* (*Neomacleaya*) [original combination: *Aedes* (*Aedes*) *prioekanensis*].

pristinus Nagaki & Sallum, 2010 (in Nagaki *et al*., 2010) (p. 282) – species of *Anopheles* (*Nyssorhynchus*).

procax (Skuse, 1889) (p. 1742) – species of *Aedes* (*Ochlerotatus*) [original combination: *Culex procax*].

proclamator Dyar & Knab, 1906d (p. 211) – synonym of *Culex* (*Culex*) *declarator* Dyar & Knab, 1906d; synonymy established before Knight & Stone, 1977.

prodotes Dyar, 1917c (p. 118) – synonym of *Aedes* (*Ochlerotatus*) *cataphylla* Dyar, 1916b; synonymy established before Knight & Stone, 1977.

productus Edwards, 1941 (p. 241) – species of *Eretmapodites* [original combination: *Eretmapodites leucopous productus*]; specific status by Rickenbach & Lombrici, 1975.

productus Senevet & Abonnenc, 1939a (p. 107) – species of *Culex* (*Melanoconion*).

prolepidis Dyar & Knab, 1919 (p. 1) – synonym of *Wyeomyia* (*Dendromyia*) *jocosa* (Dyar & Knab, 1908); synonymy by Lane, 1953.

prolixus Dyar, 1922a (p. 2) – synonym of *Aedes* (*Ochlerotatus*) *communis* (de Geer, 1776); synonymy established before Knight & Stone, 1977.

prominens (Barraud, 1923c) (p. 228) – species of *Aedes* (*Phagomyia*) [original combination: *Finlaya prominens*].

propinqua Mattingly, 1970 (p. 32) – species of *Heizmannia* (*Heizmannia*).

propinqua Peyton, 1977 (p. 106) – species of *Uranotaenia* (*Pseudoficalbia*).

propinquus Colless, 1955b (p. 315) – species of *Culex* (*Culex*).

propria Taylor, 1914c (p. 704) – synonym of *Uranotaenia* (*Uranotaenia*) *lateralis* Ludlow, 1905c; synonymy established before Knight & Stone, 1977.

prosecutor Séguy, 1927 (p. 27) – species of *Culex* (*Culex*) [*nomen novum* for *pseudomimeticus* Séguy, 1925].
Synonym: *pseudomimeticus* Séguy, 1925.

protolepis Cockerell, 1916 (p. 488) – species of fossil Culicidae (genus uncertain) [original combination: *Culex protolepis*].
Synonym: *petrifactellus* Cockerell, 1916.

protorhinus Cockerell, 1916 (p. 488) – fossil species of *Culex* Linnaeus, 1758.

protuberans (Delfinado, 1967) (p. 29) – species of *Verrallina* (*Neomacleaya*) [original combination: *Aedes* (*Neomacleaya*) *protuberans*].

proviolans Dyar, 1919d (p. 117) – synonym of *Wyeomyia phroso* Howard, Dyar & Knab, 1915 [original combination: *Dinomyia proviolans*]; synonymy established before Knight & Stone, 1977.

provocans (Walker, 1848) (p. 7) – species of *Aedes* (*Rusticoidus*) [original combination: *Culex provocans*].
Synonyms: *trichurus* Dyar, 1904b; *cinereoborealis* Felt & Young, 1904; *pagetonotum* Dyar & Knab, 1909d; *poliochros* Dyar, 1919b.

proxima Mattingly, 1970 (p. 33) – species of *Heizmannia* (*Heizmannia*).

proximus Dyar & Knab, 1909a (p. 38) – synonym of *Culex* (*Culex*) *nigripalpus* Theobald, 1901c; synonymy established before Knight & Stone, 1977.

proximus (Edwards, 1915c) (p. 285) – species of *Tripteroides* (*Tripteroides*) [original combination: *Rachionotomyia proxima*].

pruina Theobald, 1901d (p. 8) – species of *Culex* (*Culex*).
Synonym: *pallidothoracis* Theobald, 1909b.

pruinosa Martini, 1935 (p. 47) – synonym of *Psorophora* (*Grabhamia*) *signipennis* (Coquillert, 1904c); synonymy by Díaz Nájera, 1965a, 1965b.

psatharus Dyar, 1920h (p. 173) – species of *Culex* (*Melanoconion*).

psectropus Edwards, 1930a (p. 300) – species of *Hodgesia*.

pseudalbolineatus Brug, 1939 (p. 103) – species of *Aedes* (*Scutomyia*).

pseudalbopictus (Borel, 1928) (p. 85) – species of *Aedes* (*Stegomyia*) [original combination: *Stegomyia pseudalbopicta*].

pseudes Dyar & Knab, 1909d (p. 260) – species of *Deinocerites*.

pseudoafricanus Chwatt, 1949 (p. 808) – species of *Aedes* (*Stegomyia*).

pseudoalbimanus da Cunha Ramos & Brunhes, 2004 (p. 132) – species of *Uranotaenia* (*Pseudoficalbia*).

pseudoalbipes Duret, 1971a (p. 20) – species of *Psorophora* (*Janthinosoma*).

pseudoandreanus Bailly-Choumara, 1966a (p. 660) – species of *Culex* (*Eumelanomyia*).

pseudoannulioris Theobald, 1909b (p. 15) – synonym of *Culex* (*Oculeomyia*) *annulioris* Theobald, 1901a; synonymy established before Knight & Stone, 1977.

pseudoauriceps Miyagi & Toma, 2010a (p. 35) – species of *Topomyia* (*Suaymyia*).

pseudobarbirostris Ludlow, 1902a (p. 129) – species of *Anopheles* (*Anopheles*).

pseudobarbus Baisas, 1946d (p. 46) – species of *Topomyia* (*Topomyia*).

pseudocinereus Theobald, 1901c (species) (p. 62) – subspecies of *Culex* (*Culiciomyia*) *nebulosus* Theobald, 1901d; subspecific status erroneously attributed by Stone *et al.* (1959) and Knight & Stone (1977) to Edwards, 1941, who clearly treated this taxon as a variety; these authors also listed Peters (1955c), who likewise treated the taxon as a variety. Despite this oversight, *pseudocinereus* is deemed to have subspecific rank because it was adopted (originally) as the valid name of a species before 1985, in accordance with the *Code*, Article 45.6.4.1 (International Commission on Zoological Nomenclature, 1999).

pseudoconfirmatus Giles, 1902 (pl. XIII) – *nomen nudum*, *Coquillettidia*? [name informally used to indicate resemblance of a '*Panoplites* sp.' to 'C. (*Ochlerotatus*) *confirmatus*'].

pseudoconopas (Theobald, 1910c) (p. 443) – species of *Coquillettidia* (*Coquillettidia*) [original combination: *Chrysoconops pseudoconopas*].

pseudocostalis Theobald, 1910c (p. 41) – synonym of *Anopheles* (*Cellia*) *marshallii* (Theobald, 1903a) [original combination: *Pyretophorus pseudocostalis*]; synonymy established before Knight & Stone, 1977.

pseudodiantaeus Smith, 1952 (p. 21) – synonym of *Aedes* (*Ochlerotatus*) *decticus* Howard, Dyar & Knab, 1917; synonymy by Carpenter & LaCasse, 1955.

pseudodiurna (Theobald, 1910a) (p. 32) – species of *Verrallina* (*Neomacleaya*) [original combination: *Skusea pseudodiurna*].

pseudodominicii Komp, 1936b (p. 75) – species of *Aedes* (*Howardina*).
Synonym: *delpontei* Martínez & Prosen, 1955.

pseudoesoensis Yamaguti & Tamaboko, 1954 (p. 418) – synonym of *Aedes* (*Aedes*) *esoensis* Yamada, 1921; synonymy by Tanaka *et al.*, 1975b.

pseudohenrardi Peters, 1955c (p. 117) – species of *Uranotaenia* (*Pseudoficalbia*).

pseudoinfula Theobald, 1911d (p. 237) – synonym of *Culex* (*Culex*) *annulus* Theobald, 1901a; synonymy by Colless, 1957a.

pseudojamesi Strickland & Chowdhury, 1927b (p. 241) – species of *Anopheles* (*Cellia*).
Synonym: *ramsayi* Covell, 1927.

pseudojanthinosoma Senevet & Abonnenc, 1946 (p. 139) – species of *Culex* (*Culex*).

pseudoleucotarsis Thurman, 1959 (p. 48) – species of *Topomyia* (*Suaymyia*).

pseudolongifurcatus Theobald, 1910c (p. 366) – synonym of *Culex* (*Culiciomyia*) *viridiventer* Giles, 1901a [*nomen novum* for *longifurcatus* Theobald, 1910a]; synonymy established before Knight & Stone, 1977.

pseudomaculipes (Chagas, 1908) (in Peryassú, 1908) (p. 108) – species of *Anopheles* (*Anopheles*) [original combination: *Arribalzagaia* [*sic*] *pseudomaculipes*].

pseudomaculipleura Peyton & Rattanarithikul, 1970 (p. 408) – species of *Uranotaenia* (*Pseudoficalbia*).

pseudomediofasciata (Theobald, 1910c) (p. 489) – species of *Verrallina* (*Neomacleaya*) [original combination: *Skusea pseudomediofasciata*].

pseudomelanoconia Theobald, 1907 (p. 416) – species of *Culex* (*Neoculex*).

pseudomelanota Barata & Cotrim, 1971 (p. 35) – species of *Psorophora* (*Janthinosoma*).

pseudomethysticus (Bonne-Wepster & Bonne, 1920) (p. 166) – species of *Limatus* [original combination: *Lemmamyia pseudomethysticus* [*sic*]].

pseudomimeticus Séguy, 1925 (p. 15) – synonym of *Culex* (*Culex*) *prosecutor* Séguy, 1927; synonymy by Séguy, 1927.

pseudomimeticus Sergent, 1909 (p. 445) – synonym of *Culex* (*Culex*) *mimeticus* Noè, 1899; synonymy by F.W. Edwards (in Barraud, 1934).

pseudonigeria (Theobald, 1910c) (p. 166) – species of *Aedes* (*Stegomyia*) [original combination: *Stegomyia pseudonigeria*].
Synonym: *wellmanii* Theobald, 1910c.

pseudoniveus (Theobald, 1905f) (p. 75) – species of *Aedes* (*Downsiomyia*) [original combination: *Stegomyia pseudonivea*].

pseudonormanensis Marks, 1949 (p. 26) – species of *Aedes* (*Ochlerotatus*).

pseudonummatus Reinert, 1973d (p. 29) – species of *Aedes* (*Tewarius*).

pseudopecten Dyar & Knab, 1906a (p. 139) – species of *Wyeomyia* (*Decamyia*).
Synonyms: *galoa* Dyar & Knab, 1906a; *cara* Dyar & Knab, 1909d; *eloisa* Howard, Dyar & Knab, 1913; *bicompressa* Lutz, 1928.

pseudopictus Grassi, 1899 (in Grassi *et al.*, 1899) (p. 102) – species of *Anopheles* (*Anopheles*).

pseudopruina van Someren, 1951 (p. 3) – species of *Culex* (*Culex*).

pseudopunctipennis Theobald, 1901c (p. 305) – species of *Anopheles* (*Anopheles*).
Synonyms: *peruvianus* Tamayo, 1907; *argentinus* Brèthes, 1912; *tucumanus* Lahille, 1912.

pseudornatus Colless, 1960 (p. 386) – species of *Culex* (*Lophoceraomyia*).

pseudorobusta Pajot & Fauran, 1975 (p. 125) – species of *Wyeomyia* (*Wyeomyia*).

pseudorubithoracis Sirivanakarn, 1968 (p. 178) – species of *Culex* (*Lophoceraomyia*).

pseudorusticus Callot, 1945 (p. 31) – synonym of *Aedes* (*Rusticoidus*) *refiki* (Medschid, 1928); synonymy by Callot *in litt.* (Stone *et al.*, 1959).

pseudoscutellaris (Theobald, 1910b) (p. 156) – species of *Aedes* (*Stegomyia*) [original combination: *Stegomyia pseudoscutellaris*].

pseudoshillitonis da Cunha Ramos & Brunhes, 2004 (p. 276) – species of *Uranotaenia* (*Pseudoficalbia*).

pseudosinensis Baisas, 1935a (p. 297) – species of *Anopheles* (*Anopheles*) [original combination: *Anopheles hyrcanus* variety *pseudosinensis*]; specific status by Reid, 1953.

pseudosinensis Colless, 1955b (p. 311) – species of *Culex* (*Oculeomyia*).

pseudosquamosa Newstead & Carter, 1911b (p. 236) – synonym of *Anopheles* (*Cellia*) *argenteolobatus* (Gough, 1910) [original combination: *Cellia pseudosquamosa*]; synonymy established before Knight & Stone, 1977.

pseudostenoetrus Theobald, 1910c (p. 343) – synonym of *Aedes* (*Aedimorphus*) *stenoetrus* (Theobald, 1907) [original combination: *Culex pseudostenoetrus*]; synonymy established before Knight & Stone, 1977.

pseudostigmaticus Dobrotworsky, 1957 (p. 184) – species of *Anopheles* (*Anopheles*).

pseudostigmatosoma Strickman, 1990 (p. 144) – species of *Culex* (*Culex*).

pseudostricklandi Natarajan, Rajavel & Jambulingam, 2018 (p. 181) – species of *Uranotaenia* (*Pseudoficalbia*).

pseudosundaicus Tyagi, Hiriyan, Tewari, Ayanar, Samuel, Arunachalam, Paramasivan, Krishnamoorthy, Dhananjeyan, Leo & Rajendran, 2009 (p. 50) – species of *Anopheles* (*Cellia*).

pseudotaeniatus (Giles, 1901a) (p. 607) – species of *Aedes* (*Collessius*) [original combination: *Culex pseudotaeniatus*].

pseudotaeniopus Galindo & Blanton, 1954 (p. 240) – synonym of *Culex* (*Melanoconion*) *epanastasis* Dyar, 1922g; synonymy by Sirivanakarn & Belkin, 1980.

pseudotarsalis van Someren, 1946a (p. 119) – species of *Aedes* (*Catageiomyia*).

pseudotibiamaculatus Galvão & Barretto, 1941 (p. 65) – species of *Anopheles* (*Lophopodomyia*).

pseudotitillans (Theobald, 1901c) (p. 178) – species of *Mansonia* (*Mansonia*) [original combination: *Panoplites pseudotitillans*].

pseudovarietas (Reinert, 1974) (p. 76) – species of *Verrallina* (*Neomacleaya*) [original combination: *Aedes* (*Verrallina*) *pseudovarietas*].

pseudovigilax Theobald, 1907 (p. 382) – synonym of *Aedes* (*Ochlerotatus*) *vigilax* (Skuse, 1889) [original combination: *Culicelsa pseudovigilax*]; synonymy established before Knight & Stone, 1977.

pseudovishnui Colless, 1957a (p. 88) – species of *Culex* (*Culex*).
Synonym: *neovishnui* Lien, 1968d.

pseudowillmori (Theobald, 1910c) (p. 65) – species of *Anopheles* (*Cellia*) [original combination: *Nyssorhynchus pseudowillmori*].

pseudumbrosus Haga, 1930 (pp. 366, 372, 377) – synonym of *Anopheles* (*Anopheles*) *barbumbrosus* Strickland & Chowdhury, 1927a; synonymy by Ward, 1979.

pubescens Edwards, 1925 (p. 264) – species of *Aedes* (*Aedimorphus*).

pujutensis Colless, 1948 (p. 101) – species of *Anopheles* (*Cellia*) [original combination: *Anopheles* (*Myzomyia*) *leucosphyrus pujutensis*]; specific status by Colless, 1956.

pulcherrima Lynch Arribálzaga, 1891b (p. 165) – species of *Uranotaenia* (*Uranotaenia*).
Synonyms: *urania* Shannon & del Ponte, 1928; *modesta* Martini, 1935.

pulcherrima Taylor, 1919 (p. 830) – synonym of *Aedes* (*Stegomyia*) *aegypti* (Linnaeus, 1762) [original combination: *Mimeteomyia pulcherrima*]; synonymy by Knight & Marks, 1952.

pulcherrimus Theobald, 1902a (p. 369) – species of *Anopheles* (*Cellia*).
Synonym: *atropotenae* Lindtrop, 1924.

pulchrithorax Edwards, 1914b (p. 73) – synonym of *Culex* (*Afroculex*) *lineatus* (Theobald, 1912a); synonymy by Danilov, 1989.

pulchrithorax Edwards, 1939a (p. 17) – species of *Aedes* (*Vansomerenis*).

pulchriventer (Giles, 1901a) (p. 608) – species of *Aedes* (*Gilesius*) [original combination: *Culex pulchriventer*].
Synonym: *himalayana* Giles, 1904b.

pulcripalpis (Rondani, 1872) (p. 31) – species of *Orthopodomyia* [original combination: *Culex pulcripalpis*].
Synonym: *albionensis* MacGregor, 1919.

pulcritarsis (Rondani, 1872) (p. 31) – species of *Aedes* (*Ochlerotatus*) [original combination: *Culex pulcritarsis*].
Synonyms: *leucacanthus* Loew, 1873; *simici* Baranoff, 1927.

pulidoi Cova Garcia & Sutil Oramas, 1974 (p. 29) – species of *Culex* (*Microculex*).

pullatus (Coquillett, 1904c) (p. 168) – species of *Aedes* (*Ochlerotatus*) [original combination: *Culex pullatus*].
Synonyms: *acrophilus* Dyar, 1917d; *jugorum* Villeneuve, 1919; *metalepticus* Dyar, 1920c; *gallii* Martini, 1920a; *pearyi* Dyar & Shannon, 1925d; *seguyi* Apfelbeck, 1929.

pullatus Graham, 1910 (p. 265) – synonym of *Culex* (*Culex*) *grahamii* Theobald, 1910c; synonymy established before Knight & Stone, 1977.

pullus Theobald, 1905f (p. 87) – species of *Culex* (*Culiciomyia*).
Synonym: *muticus* Edwards, 1923b.

pullus Yamada, 1937 (p. 238) – species of *Anopheles* (*Anopheles*).
Synonyms: *yatsushiroensis* Miyazaki, 1951; *xiaokuanus* Ma, 1981b; *junlianensis* Lei, 1996.

pulverulentus Edwards, 1922a (p. 273) – species of *Aedes* (*Indusius*).

punctatus Meigen, 1804 (p. 6) – synonym of *Aedes* (*Ochlerotatus*) *caspius* (Pallas, 1771) [original combination: *Culex punctatus*]; synonymy established before Knight & Stone, 1977.

punctibasis Edwards, 1921d (p. 274) – synonym of *Anopheles* (*Anopheles*) *koreicus* Yamada & Watanabe, 1918; synonymy established before Knight & Stone, 1977.

punctifemoris (Ludlow, 1921) (p. 690) – species of *Aedes* (*Aedimorphus*) [original combination: *Stegomyia punctifemore* [*sic*]].

punctimacula Dyar & Knab, 1906a (p. 136) – species of *Anopheles* (*Anopheles*).
Synonyms: *strigimacula* Dyar & Knab, 1906a; *venezuelae* Evans, 1922.

punctipennis (Say, 1823) (p. 9) – species of *Anopheles* (*Anopheles*) [original combination: *Culex punctipennis*].
Synonyms: *hyemalis* Fitch, 1847; *stonei* Vargas, 1941a.

punctipes Edwards, 1921c (p. 77) – species of *Aedes* (*Petermattinglyius*).

punctipes Theobald, 1907 (p. 316) – synonym of *Culex* (*Oculeomyia*) *poicilipes* (Theobald, 1903a) [original combination: *Aporoculex punctipes*]; synonymy established before Knight & Stone, 1977.

punctiscapularis Floch & Abonnenc, 1946b (p. 1) – species of *Culex* (subgenus uncertain, determined by Pecor *et al.*, 1992).

punctissimus Barraud, 1928b (p. 360) – synonym of *Aedes* (*Petermattinglyius*) *iyengari* Edwards, 1923b; synonymy established before Knight & Stone, 1977.

punctithorax Theobald, 1910c (p. 205) – synonym of *Aedes* (*Catageiomyia*) *punctothoracis* (Theobald, 1909b) [original combination: *Aedimorphus punctithorax*]; synonymy established before Knight & Stone, 1977.

punctocostalis (Theobald, 1909b) (p. 14) – species of *Aedes* (*Neomelaniconion*) [original combination: *Banksinella punctocostalis*].

punctodes Dyar, 1922a (p. 1) – species of *Aedes* (*Ochlerotatus*).

punctolateralis (Theobald, 1903e) (p. 156) – species of *Tripteroides* (*Polylepidomyia*) [original combination: *Stegomyia punctolateralis*].
Synonym: *occidentalis* Brug, 1934.

punctor (Kirby, 1837) (p. 309) – species of *Aedes* (*Ochlerotatus*) [original combination: *Culex punctor*].
Synonyms: *implacabilis* Walker, 1848; *fusculus* Zetterstedt, 1850; *auroides* Felt, 1905a; *wahlgreni* Theobald, 1910c; *alineata* Schneider, 1913; *haplolineata* Schneider, 1913; *meigenanus* Dyar, 1921e.

punctothoracis (Theobald, 1909b) (p. 4) – species of *Aedes* (*Catageiomyia*) [original combination: *Aedimorphus punctothoracis*].
Synonym: *punctithorax* Theobald, 1910c.

punctulatus Dönitz, 1901 (p. 36) – species of *Anopheles* (*Cellia*).

pungens Robineau-Desvoidy, 1827 (p. 407) – synonym of *Aedes* (*Rusticoidus*) *rusticus* (Rossi, 1790) [original combination: *Culex pungens*]; synonymy established before Knight & Stone, 1977.

pungens Wiedemann, 1828 (p. 9) – synonym of *Culex* (*Culex*) *quinquefasciatus* Say, 1823; synonymy established before Knight & Stone, 1977.

punjabensis James, 1911 (in James & Liston, 1911) (p. 72) – synonym of *Anopheles* (*Cellia*) *culicifacies* Giles, 1901b [original combination: *Myzomyia culicifacies* variety *punjabensis*]; synonymy established before Knight & Stone, 1977.

purii Barraud, 1931a (p. 226) – synonym of *Aedes* (*Stegomyia*) *craggi* (Barraud, 1923c); synonymy by Barraud, 1934.

purpuraceus Brug, 1932 (p. 77) – species of *Aedes* (*Ochlerotatus*).

purpurascens Edwards, 1922d (p. 77) – synonym of *Psorophora* (*Janthinosoma*) *cyanescens* (Coquillett, 1902a); synonymy established before Knight & Stone, 1977.

purpuratus (Edwards, 1921c) (p. 79) – species of *Tripteroides* (*Tripteroides*) [original combination: *Rachionotomyia purpurata*].

purpureifemur Marks, 1959a (p. 132) – species of *Aedes* (*Ochlerotatus*).

purpureipes Aitken, 1941 (p. 82) – species of *Aedes* (*Kompia*).

purpureus Peryassú, 1908 (p. 287) – synonym of *Sabethes* (*Sabethes*) *purpureus* (Theobald, 1907); synonymy established before Knight & Stone, 1977.

purpureus (Theobald, 1901a) (p. 230) – species of *Toxorhynchites* (*Ankylorhynchus*) [original combination: *Megarhinus purpureus*].

purpureus (Theobald, 1907) (p. 617) – species of *Sabethes* (*Sabethes*) [original combination: *Sabethoides purpureus*].
Synonyms: *purpureus* Peryassú, 1908; *remipusculus* Dyar, 1924b.

purpureus (Theobald, 1910c) (p. 479) – species of *Aedes* (*Molpemyia*) [original combination: *Molpemyia purpurea*].
Synonym: *priestleyi* Taylor, 1914c.

purpuriventris Edwards, 1926a (p. 113) – species of *Aedes* (*Ochlerotatus*).

pursati Laveran, 1902a (p. 907) – species of *Anopheles* (*Anopheles*).

puselis de Meillon, 1934 (p. 269) – synonym of *Anopheles* (*Cellia*) *multicinctus* Edwards, 1930a [original combination: *Anopheles natalensis* variety *puselis*]; synonymy by Evans, 1938.

pusillum Lutz, 1928 (p. 33) – synonym of *Wyeomyia* (*Prosopolepis*) *confusa* (Lutz, 1905) [original combination: *Trichoprosopon pusillum*]; synonymy by Lane & Cerqueira, 1942.

pusillus (da Costa Lima, 1931b) (p. 313) – species of *Toxorhynchites* (*Lynchiella*) [original combination: *Megarhinus* (*Megarhinus*) *pusillus*].

pusillus Macquart, 1850 (p. 313) – species of *Culex* (*Barraudius*).

putaneus Theobald, 1901a (p. 257) – *nomen nudum*, *Psorophora* [original combination: *Janthinosoma putaneus* (for specimen labelled *C. putaneus*)]. **Note**: the name may be a misspelling of *puteanus* (L. 'of a well').

putumayensis Matheson, 1934 (p. 121) – species of *Culex* (*Melanoconion*).
Synonym: *cavernicolus* Floch & Abonnenc, 1945c.

pygmaea Theobald, 1901c (p. 254) – species of *Uranotaenia* (*Uranotaenia*).

pygmaea (Theobald, 1903a) (p. 245) – species of *Psorophora* (*Grabhamia*) [original combination: *Grabhamia pygmaea*].
Synonyms: *nanus* Coquillett, 1903a; *antiguae* Giles, 1904b.

pygmaeus Curtis, 1831 (column 202) – *nomen nudum*, *Anopheles*.

pygmaeus Neveu-Lemaire, 1906 (p. 256) – synonym of *Culex* (*Culex*) *quinquefasciatus* Say, 1823; synonymy established before Knight & Stone, 1977.

pygmaeus Theobald, 1908b (p. 300) – synonym of *Coquillettidia* (*Coquillettidia*) *crassipes* (van der Wulp, 1881) [original combination: *Chrysoconops pygmaeus*]; synonymy by Bohart, 1945.

pylei Baisas, 1946a (p. 45) – species of *Uranotaenia* (*Pseudoficalbia*).

pyrenaicus Brolemann, 1919a (p. 427) – synonym of *Culex* (*Neoculex*) *territans* Walker, 1856a; synonymy established before Knight & Stone, 1977.

pyretophoroides Theobald, 1907 (p. 48) – synonym of *Anopheles* (*Cellia*) *ardensis* (Theobald, 1905g) [original combination: *Myzomyia pyretophoroides*]; synonymy established before Knight & Stone, 1977.

Qq

quadratimaculatus Macquart, 1834 (p. 34) – synonym of *Aedes* (*Rusticoidus*) *rusticus* (Rossi, 1790) [original combination: *Culex quadratimaculatus*]; synonymy established before Knight & Stone, 1977.

quadriannulatus (Theobald, 1911b) (p. 242) – species of *Anopheles* (*Cellia*) [original combination: *Pyretophorus quadriannulatus*].

quadricinctus (Barraud, 1923c) (p. 227) – subspecies of *Aedes* (*Stegomyia*) *annandalei* (Theobald, 1910a) [original combination: *Stegomyia annandalei* variety *quadricincta*]; subspecific status by Harbach & Howard, 2007. **Note**: the name of this species is incorrectly listed in its original form (feminine) in the catalogues of Knight & Stone, 1977 and the Walter Reed Biosystematics Unit (http://www.mosquitocatalog.org/).

quadrifoliatus Komp, 1936c (p. 322) – species of *Culex* (*Melanoconion*).

quadrifolium (Brug, 1934) (p. 512) – species of *Verrallina* (*Verrallina*) [original combination: *Aedes* (*Aedes*) *quadrifolium*].

quadrimaculata Edwards, 1929 (in Paine & Edwards, 1929) (p. 313) – species of *Uranotaenia* (*Pseudoficalbia*).

quadrimaculata Newstead, 1907 (in Newstead *et al.*, 1907) (p. 32) – synonym of *Mimomyia* (*Etorleptiomyia*) *mediolineata* (Theobald, 1904b) [original combination: *Anisocheleomyia quadrimaculata*]; synonymy established before Knight & Stone, 1977.

quadrimaculatus Say, 1824 (in Keating, 1824) (p. 356) – species of *Anopheles* (*Anopheles*).
  Synonym: *annulimanus* van der Wulp, 1867.

quadripalpis (Edwards, 1914b) (p. 80) – species of *Culex* (*Lophoceraomyia*) [original combination: *Lophoceratomyia quadripalpis*].
  Synonyms: *sylvestris* Leicester, 1908; *roubaudi* Borel, 1926; *pachecoi* Baisas, 1935d.

quadripunctis (Ludlow, 1910) (in Theobald, 1910c) (p. 608) – species of *Aedes* (*Mucidus*) [original combination: *Pardomyia quadripunctis*].

quadrispinata (King & Hoogstraal, 1947b) (p. 126) – species of *Verrallina* (*Verrallina*) [original combination: *Aedes* (*Aedes*) *quadrispinatus*].

quadrivittatus (Coquillett, 1902d) (p. 293) – species of *Aedes* (*Howardina*) [original combination: *Culex quadrivittatus*].

quartus Martini, 1920a (p. 128) – synonym of *Aedes* (*Ochlerotatus*) *annulipes* (Meigen, 1830b); synonymy established before Knight & Stone, 1977.

quasicyaneus Peryassú, 1922b (p. 179) – species of *Sabethes* (*Sabethes*).

quasiferinus Mattingly, 1961 (p. 32) – species of *Aedes* (*Mucidus*).

quasiferox (Leicester, 1908) (p. 51) – species of *Toxorhynchites* (*Toxorhynchites*) [original combination: *Teromyia quasiferox*].
  Synonym: *javaensis* Theobald, 1911d.

quasigelidus Theobald, 1903a (p. 181) – synonym of *Culex* (*Oculeomyia*) *poicilipes* (Theobald, 1903a); synonymy established before Knight & Stone, 1977.

quasiguiarti Theobald, 1910c (p. 374) – species of *Culex* (*Culex*).

quasihibridus Galindo & Blanton, 1954 (p. 232) – species of *Culex* (*Melanoconion*).

quasilinealis Theobald, 1907 (p. 415) – synonym of *Culex* (*Culex*) *quinquefasciatus* Say, 1823; synonymy established before Knight & Stone, 1977.

quasilongirostris (Theobald, 1907) (p. 598) – species of *Wyeomyia* (*Phoniomyia*) [original combination: *Phoniomyia quasilongirostris*].
  Synonym: *neivai* Lane & Cerqueira, 1942.

quasiluteoventralis Theobald, 1903a (p. 317) – synonym of *Wyeomyia* (*Dendromyia*) *luteoventralis* Theobald, 1901c [original combination: *Dendromyia quasiluteoventralis*]; synonymy by Belkin, 1968b.

quasimodesta Peyton, 1977 (p. 112) – species of *Uranotaenia* (*Pseudoficalbia*).

quasimodestus Theobald, 1905f (p. 88) – synonym of *Culex* (*Culex*) *pipiens* Linnaeus, 1758; synonymy by Stone, 1957b (as synonym of *molestus* Forskål).

quasinigritia Ludlow, 1911b (p. 129) – synonym of *Aedes* (*Stegomyia*) *albopictus* (Skuse, 1895); synonymy established before Knight & Stone, 1977.

quasioriginator Duret, 1972 (p. 3) – species of *Culex* (*Anoedioporpa*).

quasiornatus (Taylor, 1915) (p. 177) – species of *Tripteroides* (*Tripteroides*) [original combination: *Stegomyia quasiornata*].

quasipipiens Theobald, 1901c (p. 136) – synonym of *Culex* (*Culex*) *quinquefasciatus* Say, 1823; synonymy established before Knight & Stone, 1977.

quasirubithorax (Theobald, 1910c) (pp. 327, 348) – species of *Aedes* (subgenus uncertain) [original combination: *Culex quasirubithorax*].

quasirusticus Torres Cañamares, 1951 (p. 79) – species of *Aedes* (*Rusticoidus*).

quasisanguinae Leicester, 1908 (p. 230) – species of *Hodgesia*.
  Synonyms: *niveocaputis* Ludlow, 1911b; *triangulatus* Taylor, 1914d.

quasiscutellaris Farner & Bohart, 1944 (p. 120) – species of *Aedes* (*Stegomyia*).

quasisecutor Theobald, 1907 (p. 398) – synonym of *Culex* (*Culex*) *secutor* Theobald, 1901c; synonymy established before Knight & Stone, 1977.

quasiserratus Theobald, 1907 (p. 465) – synonym of *Aedes* (*Ochlerotatus*) *pertinax* Grabham, 1906b [original combination: *Protoculex quasiserratus*]; synonymy by Howard *et al.*, 1917.

quasiunivittatus (Theobald, 1901c) (p. 32) – species of *Aedes* (*Aedimorphus*) [original combination: *Culex quasiunivittatus*].

quaylei Dyar & Knab, 1906d (p. 202) – synonym of *Aedes* (*Ochlerotatus*) *dorsalis* (Meigen, 1830b); synonymy by Wood *et al.*, 1979.

queenslandensis Theobald, 1901a (p. 297) – synonym of *Aedes* (*Stegomyia*) *aegypti* (Linnaeus, 1762) [original combination: *Stegomyia fasciata* variety *queenslandensis*]; synonymy by Lee *et al.*, 1987.

queenslandis Strickland, 1911b (p. 179) – synonym of *Aedes* (*Dobrotworskyius*) *rubrithorax* (Macquart, 1850) [original combination: *Culicelsa queenslandis*]; synonymy by Lee *et al.*, 1982.

quettensis Mattingly, 1955 (p. 374) – species of *Culex* (*Maillotia*).

qui Dong, Dong & Zhou, 2003 (in Dong, Dong, Zhou *et al.*, 2003) (p. 519) – species of *Uranotaenia* (*Pseudoficalbia*).

quinquefasciatus Say, 1823 (p. 10) – species of *Culex* (*Culex*).

Synonyms: *aestuans* Wiedemann, 1828; *fatigans* Wiedemann, 1828; *pungens* Wiedemann, 1828; *acer* Walker, 1848; *cingulatus* Doleschall, 1856; *cubensis* Bigot, 1857; *anxifer* Bigot, 1859; *serotinus* Philippi, 1865; *autumnalis* Weyenbergh, 1882; *penafieli* Sanchez, 1885; *macleayi* Skuse, 1889; *doleschallii* Giles, 1900a; *skusii* Giles, 1900a; *albolineatus* Giles, 1901a; *fouchowensis* Theobald, 1901c; *luteoannulatus* Theobald, 1901c; *quasipipiens* Theobald, 1901c; *reesii* Theobald, 1901c; *sericeus* Theobald, 1901c; *trilineatus* Theobald, 1901c; *pallidocephala* Theobald, 1904b; *cartroni* Ventrillon, 1905b; *barbarus* Dyar & Knab, 1906d; *didieri* Neveu-Lemaire, 1906; *pygmaeus* Neveu-Lemaire, 1906; *zeltneri* Neveu-Lemaire, 1906; *christophersii* Theobald, 1907; *quasilinealis* Theobald, 1907; *stoehri* Theobald, 1907; *raymondii* Tamayo, 1907; *aikenii* Dyar & Knab, 1908; *minor* Theobald, 1908b; *lachrimans* Dyar & Knab, 1909d; *revocator* Dyar & Knab, 1909d; *goughii* Theobald, 1911b; *fuscus* Taylor, 1914c; *aseyehae* Dyar & Knab, 1915; *townsvillensis* Taylor, 1919; *hensemaeon* Dyar, 1920i; *nigrirostris* Enderlein, 1920.

quinquelineatus Edwards, 1922d (p. 93) – species of *Aedes* (*Rampamyia*).

quinquemaculata Bonne-Wepster, 1934 (p. 276) – species of *Uranotaenia* (*Pseudoficalbia*).

quinquepunctata Theobald, 1913d (p. 598) – synonym of *Aedes* (*Catageiomyia*) *argenteopunctata* (Theobald, 1901a) [original combination: *Aedimorphus quinquepunctata* [*sic*]]; synonymy established before Knight & Stone, 1977.

quinquevittatus Theobald, 1901d (p. ii) – species of *Eretmapodites*.

Synonyms: *condei* Ventrillon, 1905a; *austenii* Theobald, 1910c.

quintetti Brunhes, Adam & Bailly-Choumara, 1967 (p. 49) – species of *Culex* (*Eumelanomyia*).

quitensis Levi-Castillo, 1953a (p. 190) – species of *Culex* (*Culex*).

Rr

rabaiensis Lounibos, 1979 (p. 28) – *nomen nudum*, *Anopheles* [attributed to 'White, in press' but never published]; status by Ward, 1984.

rabanicolus Floch & Abonnenc, 1946a (p. 1) – species of *Culex* (*Melanoconion*).

rabelloi Forattini & Sallum, 1987a (p. 130) – species of *Culex* (*Melanoconion*).

rachoui Duret, 1968b (p. 58) – species of *Culex* (*Melanoconion*).

rachoui Galvão, 1952 (p. 69) – species of *Anopheles* (*Anopheles*).

rachoui Xavier, da Silva Mattos & da Silva, 1970b (p. 411) – species of *Uranotaenia* (*Uranotaenia*).

radama de Meillon, 1943 (p. 90) – species of *Anopheles* (*Cellia*).

radiatus Senevet & Abonnenc, 1939a (p. 120) – synonym of *Culex* (*Melanoconion*) *pilosus* (Dyar & Knab, 1906d); synonymy by Rozeboom & Komp, 1950a.

rageaui (Hamon & Rickenbach, 1955b) (p. 859) – subspecies of *Culex* (*Eumelanomyia*) *horridus* Edwards, 1922d [original combination: *Neoculex horridus* variety *rageaui*]; subspecific status by Harbach & Howard, 2007.

rageaui Mattingly & Adam, 1954 (p. 55) – species of *Anopheles* (*Cellia*).

raghavanii Rahman, Chowdhury & Kalra, 1969 (p. 28) – species of *Culex* (*Lophoceraomyia*).

rajah Tsukamoto, 1989 (p. 216) – species of *Culex* (*Culiciomyia*).

rajah Tsukamoto, 1989 (p. 221) – species of *Toxorhynchites* (*Toxorhynchites*).

rajaneeae Sirivanakarn, 1968 (p. 174) – species of *Culex* (*Lophoceraomyia*).

rajaonariveloi Brunhes & Hervy, 1995 (p. 75) – species of *Orthopodomyia*.

ramachandrai Reuben, 1967 (p. 234) – species of *Aedes* (*Dendroskusea*).

ramakrishnii Wattal & Kalra, 1965 (p. 309) – species of *Culex* (*Culiciomyia*).

ramalai Grjebine, 1986 (p. 169) – species of *Mimomyia* (*Ingramia*).

ramalingami Evenhuis & Steffan, 1986 (p. 566) – species of *Toxorhynchites* (*Toxorhynchites*).

ramalingami (Reinert, 1974) (p. 78) – species of *Verrallina* (*Harbachius*) [original combination: *Aedes* (*Verrallina*) *ramalingami*].

ramalingami Sirivanakarn, 1973b (p. 119) – species of *Culex* (*Culiciomyia*).

rami (Barraud, 1928b) (p. 370) – species of *Verrallina* (*Neomacleaya*) [original combination: *Aedes* (*Aedes*) *rami*].

ramirezi Vargas & Downs, 1950 (p. 164) – species of *Aedes* (*Aztecaedes*).

ramosa da Cunha Ramos, 1993 (p. 214) – species of *Uranotaenia* (*Pseudoficalbia*).

ramosi Unti, 1940b (p. 489) – synonym of *Anopheles* (*Nyssorhynchus*) *strodei* Root, 1926 [original combination: *Anopheles* (*Nyssorhynchus*) *strodei* variety *ramosi*]; synonymy by Faran, 1980; Sallum *et al.*, 2010.

rampae Harbach & Somboon, 2011 (in Somboon *et al.*, 2011) (p. 48) – species of *Anopheles* (*Cellia*).

rampae Peyton & Klein, 1970 (p. 243) – species of *Uranotaenia* (*Uranotaenia*).

ramsayi Covell, 1927 (p. 1019) – synonym of *Anopheles* (*Cellia*) *pseudojamesi* Strickland & Chowdhury, 1927b; synonymy by Nurul Huda & Harrison, 1985.

ranci Grjebine, 1953a (p. 488) – species of *Anopheles* (*Cellia*).

rangeli Gabaldon, Cova Garcia & Lopez, 1940 (p. 13) – species of *Anopheles* (*Nyssorhynchus*).

rangeli Surcouf & Gonzales-Rincones, 1911 (p. 251) – synonym of *Sabethes* (*Sabethoides*) *chloropterus* (von Humboldt, 1819) [original combination: *Sabethoides rangeli*]; synonymy established before Knight & Stone, 1977.

rapax (Dyar & Knab, 1906a) (p. 137) – species of *Runchomyia* (*Runchomyia*) [original combination: *Lesticocampa rapax*]; elevated from synonymy with *frontosa* Theobald, 1903a by Zavortink, 1979.

raptor Edwards, 1922a (p. 275) – synonym of *Lutzia* (*Metalutzia*) *halifaxii* Theobald, 1903a; synonymy by Bram, 1967a.

rara (Delfinado, 1968) (p. 35) – species of *Verrallina* (*Neomacleaya*) [original combination: *Aedes* (*Neomacleaya*) *rarus*].
Synonym: *khmerus* Klein, 1973.

raris (Leicester, 1908) (p. 56) – species of *Toxorhynchites* (*Toxorhynchites*) [original combination: *Teromyia raris*].

ratcliffei Marks, 1959a (p. 123) – species of *Aedes* (*Ochlerotatus*).

rausai Miyagi, 1980 (p. 178) – synonym of *Topomyia* (*Suaymyia*) *apsarae* Klein, 1977; synonymy by Miyagi, Toma & Cabrera, 1983.

rausseoi Cova García, Sutil O. & Pulido F., 1972 (p. 207) – species of *Culex* (*Carrollia*).

ravaonjanaharyi Brunhes & Hervy, 1995 (p. 78) – species of *Orthopodomyia*.

ravenalaphila Boussès & Brunhes, 2013 (p. 400) – species of *Uranotaenia* (*Pseudoficalbia*).

ravenalicola da Cunha Ramos & Brunhes, 2004 (p. 207) – species of *Uranotaenia* (*Pseudoficalbia*).

ravissei Rickenbach & Eouzan, 1970 (p. 134) – species of *Eretmapodites*.

raymondi del Ponte, Castro & Garcia, 1951 (p. 235) – species of *Aedes* (*Ochlerotatus*).

raymondii Tamayo, 1907 (in Tamayo & García, 1907) (p. 36) – synonym of *Culex* (*Culex*) *quinquefasciatus* Say, 1823; synonymy established before Knight & Stone, 1977.

reali Hamon & Adam, 1959a (p. 351) – species of *Aedes* (*Catageiomyia*).

recens Sallum & Peyton, 2005 (in Sallum *et al.*, 2005b) (p. 185) – species of *Anopheles* (*Cellia*).

recondita Edwards, 1922d (p. 91) – species of *Uranotaenia* (*Pseudoficalbia*).

rectirostris Theobald, 1910c (p. 214) – *nomen nudum*, *Toxorhynchites*.

reducens Lane & Whitman, 1951 (p. 354) – species of *Culex* (*Microculex*).

reductor Dyar & Knab, 1909d (p. 257) – synonym of *Culex* (*Melanoconion*) *pilosus* (Dyar & Knab, 1906d); synonymy established before Knight & Stone, 1977.

reesi (King & Hoogstraal, 1947b) (p. 127) – species of *Verrallina* (*Verrallina*) [original combination: *Aedes* (*Aedes*) *reesi*].

reesii Theobald, 1901c (p. 145) – synonym of *Culex* (*Culex*) *quinquefasciatus* Say, 1823; synonymy established before Knight & Stone, 1977.

reevesi Bohart, 1949 (p. 342) – synonym of *Culex* (*Neoculex*) *boharti* Brookman & Reeves, 1950; synonymy by Brookman & Reeves, 1950.

reevesi Wirth, 1948 (p. 230) – species of *Culex* (*Neoculex*).

refiki Medschid, 1928 (p. 310) – species of *Aedes* (*Rusticoidus*).
Synonyms: *stampari* Apfelbeck, 1929; *pseudorusticus* Callot, 1945.

reflector Dyar & Knab, 1909d (p. 256) – synonym of *Culex* (*Culex*) *interrogator* Dyar & Knab, 1906d; synonymy established before Knight & Stone, 1977.

refutans Alcock, 1913 (p. 161) – subspecies of *Anopheles* (*Anopheles*) *gigas* Giles, 1901b [original combination: *Anopheles gigas* variety *refutans*]; subspecific status by Harbach & Howard, 2007.

regale Peyton & Harbach, 2000 (in Harbach & Peyton, 2000) (p. 132) – species of *Onirion*.

regalis Dyar & Knab, 1906b (p. 167) – species of *Haemagogus* (*Haemagogus*).

reginae Edwards, 1922a (p. 272) – species of *Aedes* (*Dendroskusea*).

reginae Floch & Fauran, 1955 (p. 5) – species of *Culex* (*Microculex*).

regius Tennent, 1859 (p. 268) – synonym of *Toxorhynchites* (*Toxorhynchites*) *splendens* (Wiedemann, 1819) [original combination: *Culex regius*]; synonymy established before Knight & Stone, 1977.

regulator Dyar & Knab, 1906d (p. 213) – synonym of *Culex* (*Culex*) *nigripalpus* Theobald, 1901c; synonymy established before Knight & Stone, 1977.

reidi Colless, 1965 (p. 279) – species of *Culex* (*Lophoceraomyia*).

reidi Harrison, 1973 (p. 365) – species of *Anopheles* (*Anopheles*).

reidi Mattingly, 1957c (p. 38) – species of *Heizmannia* (*Heizmannia*).
Synonym: *cheni* Lien, 1968c.

reinerti Peyton, 1977 (p. 116) – species of *Uranotaenia* (*Pseudoficalbia*).

reinerti Rattanarithikul & Harrison, 1988 (p. 78) – species of *Aedes* (*Hulecoeteomyia*).

reiseni Basio, 1971 (p. 56) – species of *Tripteroides* (*Tripteroides*).

rejector Dyar & Knab, 1906d (p. 221) – species of *Culex* (*Microculex*).

relictus Shingarev, 1928 (p. 47) – synonym of *Anopheles* (*Anopheles*) *martinius* Shingarev, 1926; synonymy by White, 1978.

remipes Wiedemann, 1828 (p. 573) – synonym of *Sabethes* (*Sabethes*) *cyaneus* (Fabricius, 1805) [original combination: *Culex remipes*]; synonymy established before Knight & Stone, 1977.

remipusculus Dyar, 1924b (p. 100) – synonym of *Sabethes* (*Sabethes*) *purpureus* (Theobald, 1907) [*nomen novum* for *purpureus* Peryassú, 1908]; synonymy established before Knight & Stone, 1977.

rempeli Vockeroth, 1954 (p. 112) – species of *Aedes* (*Ochlerotatus*).

renatoi Lane & Ramalho, 1960 (p. 173) – species of *Culex* (*Phytotelmatomyia*).

rennellensis Taylor & Maffi, 1991 (p. 196) – species of *Anopheles* (*Cellia*).

reptans Meigen, 1804 (p. 3) – synonym of *Aedes* (*Ochlerotatus*) *cantans* (Meigen, 1818) [original combination: *Culex reptans*]; synonymy established before Knight & Stone, 1977.

restrictor Dyar & Knab, 1906d (p. 222) – species of *Culex* (*Anoedioporpa*).
Synonym: *consternator* Dyar & Knab, 1908.

restuans Theobald, 1901c (p. 142) – species of *Culex* (*Culex*).
Synonym: *brehmei* Knab, 1916.

retrosus Lane & Whitman, 1951 (p. 356) – subspecies of *Culex* (*Microculex*) *imitator* Theobald, 1903a.

reubenae Tewari & Hiriyan, 1992b (p. 164) – species of *Aedes* (*Tewarius*).

reunionensis Brunhes & Hervy, 1995 (p. 82) – species of *Orthopodomyia*.

revelator Dyar & Knab, 1907c (p. 202) – synonym of *Culex* (*Culex*) *declarator* Dyar & Knab, 1906d; synonymy established before Knight & Stone, 1977.

reversa (Lane & Cerqueira, 1942) (p. 522) – species of *Runchomyia* (*Runchomyia*) [original combination: *Trichoprosopon* (*Hyloconops*) *reversus* [*sic*]].

reversus Theobald, 1901c (p. 189) – synonym of *Mansonia* (*Mansonioides*) *uniformis* (Theobald, 1901c) [original combination: *Panoplites africanus* variety *reversus*]; synonymy established before Knight & Stone, 1977.

revocator Dyar & Knab, 1909d (p. 256) – synonym of *Culex* (*Culex*) *quinquefasciatus* Say, 1823; synonymy established before Knight & Stone, 1977.

reyi Baisas, 1935c (p. 74) – species of *Uranotaenia* (*Uranotaenia*).

rhecter Dyar, 1921d (p. 51) – synonym of *Aedes* (*Aedimorphus*) *dalzieli* (Theobald, 1910c); synonymy by Edwards, 1941.

rhodesiensis Theobald, 1901a (p. 184) – species of *Anopheles* (*Cellia*).

rhungkiangensis Chang & Chang, 1974 (p. 349) – species of *Aedes* (*Stegomyia*).

rhyacophilus da Costa Lima, 1933 (p. 404) – species of *Aedes* (*Ochlerotatus*).

ribeirensis Forattini & Sallum, 1985 (p. 171) – species of *Culex* (*Melanoconion*).

richardgarciai Jeffery, Oothuman & Rudnick, 1987 (p. 140) – species of *Culex* (*Eumelanomyia*).

richei Klein, 1970 (p. 504) – species of *Culex* (*Eumelanomyia*).

richeti Brunhes & Venard, 1966 (p. 75) – species of *Culex* (*Barraudius*).

richiardii (Ficalbi, 1889b) (p. 50) – species of *Coquillettidia* (*Coquillettidia*) [original combination: *Culex richiardii*].
Synonym: *nikolskyi* Shingarev, 1927.

richteri Ingram & de Meillon, 1927 (p. 72) – synonym of *Culex* (*Culex*) *simpsoni* Theobald, 1905g; synonymy established before Knight & Stone, 1977.

rickenbachi da Cunha Ramos, 1993 (p. 275) – species of *Uranotaenia* (*Pseudoficalbia*).

rickenbachi Ferrara & Eouzan, 1974 (p. 259) – species of *Eretmapodites*.

rickenbachi Hamon & Adam, 1959b (p. 151) – species of *Aedes* (*Aedimorphus*).
Synonym: *hamoni* Mattingly, 1963.

rickenbachi Ribeiro, 1991 (p. 54) – species of *Toxorhynchites* (*Toxorhynchites*).

rifenus Baeza Cuéllar, 1933 (p. 87) – synonym of *Anopheles* (*Cellia*) *cinereus hispaniola* (Theobald, 1903a) [original combination: *Myzomyia rifenus*]; synonymy by Romeo Viamonte, 1950.

rigidus Senevet & Abonnenc, 1939a (p. 68) – synonym of *Culex* (*Phenacomyia*) *corniger* Theobald, 1903a; synonymy by Lane, 1951.

rima Theobald, 1901d (p. xi) – species of *Culex* (*Eumelanomyia*).
Synonym: *koumbai* Galliard, 1931c.

rimandoi Basio, 1971 (p. 52) – *nomen dubium*, *Aedes*; status by Huang, 1979.

riojanus Duret, 1968e (p. 326) – species of *Culex* (*Culex*).

riparioides Su & Zhang, 1987 (p. 450) – species of *Aedes* (*Ochlerotatus*).

riparis King & Baisas, 1936 (p. 83) – species of *Anopheles* (*Cellia*) [original combination: *Anopheles* (*Myzomyia*) *leucosphyrus* variety *riparis*]; specific status by Colless, 1956.

riparius Dyar & Knab, 1907d (p. 213) – species of *Aedes* (*Ochlerotatus*).
Synonym: *semicantans* Martini, 1920b.

rivadeneirai Leví Castillo, 1945 (p. 33) – subspecies of *Anopheles* (*Anopheles*) *pseudopunctipennis* Theobald, 1901c.

riverai Duret, 1970 (p. 43) – species of *Uranotaenia* (*Uranotaenia*).

riverai Miyagi, Toma & Tsukamoto, 1983 (p. 6) – species of *Tripteroides* (*Tripteroides*).

riversi Bohart & Ingram, 1946a (p. 50) – species of *Aedes* (*Stegomyia*).

rivulorum Leeson, 1935 (p. 70) – species of *Anopheles* (*Cellia*) [original combination: *Anopheles funestus* variety *rivulorum*]; specific status by Evans & Garnham, 1936; de Meillon, 1947b.
Synonym: *garnhamellus* Evans & Leeson, 1937.

rizali (Banks, 1906b) (p. 999) – species of *Aedes* (*Hulecoeteomyia*) [original combination: *Culex rizali*].
Synonyms: *abadsantosi* Baisas, 1946c; *burgosi* Baisas, 1946c.

rizzoi (de Deus Palma & Galvão, 1969) (p. 3) – species of *Toxorhynchites* (*Lynchiella*) [original combination: *Lynchiella rizzoi*].

roai Belkin, 1962 (p. 353) – species of *Aedes* (subgenus uncertain).

roberti da Cunha Ramos & Brunhes, 2004 (p. 404) – species of *Uranotaenia* (*Uranotaenia*).

robertsi (Laffoon, 1946) (p. 230) – species of *Verrallina* (*Harbachius*) [original combination: *Aedes* (*Aedes*) *robertsi*].

robici Doucet, 1950 (p. 54) – synonym of *Culex* (*Maillotia*) *seyrigi* Edwards, 1941; synonymy established before Knight & Stone, 1977.

robinsoni Belkin, 1962 (p. 447) – species of *Aedes* (*Stegomyia*).

robusta Senevet & Abonnenc, 1939b (p. 253) – species of *Wyeomyia* (*Wyeomyia*).

rochei (Doucet, 1951b) (p. 109) – species of *Coquillettidia* (*Coquillettidia*) [original combination: *Taeniorhynchus* (*Coquillettidia*) *rochei*].

rockefelleri Peryassú, 1923b (p. 68) – synonym of *Anopheles* (*Anopheles*) *mediopunctatus* (Lutz, 1903) [original combination: *Cyclolepidopteron rockefelleri*]; synonymy established before Knight & Stone, 1977.

rodhaini (Brunhes & Boussès, 2017) (in Brunhes *et al*., 2017) (p. 364) – species of *Aedes* (*Paulianius*) [original combination: *Paulianius rodhaini*].

rodhaini Brunhes & Hervy, 1995 (p. 89) – species of *Orthopodomyia*.

rodhaini Leleup & Lips, 1950 (p. 303) – species of *Anopheles* (*Cellia*).

rodhaini Ribeiro, 1991 (p. 57) – species of *Toxorhynchites* (*Toxorhynchites*).

rollai Cova Garcia, Pulido F. & Escalante de Ugueto, 1977b (p. 241) – species of *Anopheles* (*Kerteszia*) [*nomen novum* for *hilli* Cova Garcia, Pulido F. & Escalante de Ugueto, 1976].
Synonym: *hilli* Cova Garcia, Pulido F. & Escalante de Ugueto, 1976.

rolonca Dyar & Knab, 1910 (p. 173) – synonym of *Wyeomyia* (*Wyeomyia*) *adelpha* Dyar & Knab, 1906a; synonymy by Stone, 1969.

roloncetta Dyar, 1919d (p. 131) – synonym of *Wyeomyia* (*Wyeomyia*) *simmsi* (Dyar & Knab, 1908); synonymy established before Knight & Stone, 1977.

romeroi Surcouf & Gonzalez-Rincones, 1912 (p. 295) – species of *Culex* (subgenus uncertain, subgenus *Culex* ?, Edwards, 1932a) [*nomen novum* for *nigritulus* Zetterstedt of Romero Sierra, 1907].

ronaldi de Charmoy, 1908 (p. 259) – synonym of *Culex* (*Culex*) *thalassius* Theobald, 1903a; synonymy by Edwards, 1941.

ronderosi de Linero, 1967 (p. 289) – species of *Culex* (*Melanoconion*).

rondoni (Neiva & Pinto, 1922a) (p. 322) – species of *Anopheles* (*Nyssorhynchus*) [original combination: *Cellia rondoni*].

rooti Brèthes, 1926b (p. 106) – synonym of *Anopheles* (*Nyssorhynchus*) *argyritarsis* Robineau-Desvoidy, 1827 [original combination: *Cellia rooti*]; synonymy by Lane & Cerqueira, 1957.

rooti (del Ponte, 1939) (p. 541) – species of *Wyeomyia* (subgenus uncertain, determined by Motta & Lourenço-de- Oliveira, 1995) [original combination: *Dendromyia rooti*].
Synonym: *delpontei* Lane & Cerqueira, 1942.

rooti Lane & Cerqueira, 1942 (p. 587) – synonym of *Wyeomyia* (*Antunesmyia*) *alani* Lane & Cerqueira, 1957; synonymy established before Knight & Stone, 1977.

rooti Rozeboom, 1935 (p. 251) – species of *Culex* (*Melanoconion*).

roperi Edwards, 1922a (p. 289) – synonym of *Uranotaenia* (*Uranotaenia*) *subnormalis* Martini, 1920b; synonymy established before Knight & Stone, 1977.

roperi Reid, 1950 (p. 298) – species of *Anopheles* (*Anopheles*).

rorotaensis Floch & Abonnenc, 1946a (p. 3) – species of *Culex* (*Melanoconion*).

rorotai Senevet, Chabelard & Abonnenc, 1942 (p. 336) – species of *Wyeomyia* (subgenus uncertain, determined by Motta & Lourenço-de-Oliveira, 1995).

roseni Belkin, 1962 (p. 203) – species of *Culex* (*Culex*).

roslihashimi Miyagi & Toma, 2005 (p. 275) – species of *Topomyia* (*Suaymyia*).

rossi Delfinado, 1966b (p. 36) – species of *Uranotaenia* (*Pseudoficalbia*).

rossicus Dolbeškin, Gorickaja & Mitrofanova, 1930 (p. 253) – species of *Aedes* (*Aedes*).
Synonym: *tarnogradskii* Martini, 1930.

rossii Giles, 1899 (p. 63) – synonym of *Anopheles* (*Cellia*) *subpictus* Grassi, 1899; synonymy established before Knight & Stone, 1977.

rossii Giles, 1899 (p. 64) – synonym of *Aedes* (*Stegomyia*) *aegypti* (Linnaeus, 1762) [original combination: *Culex rossii*]; synonymy established before Knight & Stone, 1977.

rostochiensis Martini, 1920b (p. 204) – synonym of *Aedes* (*Ochlerotatus*) *cataphylla* Dyar, 1916b; synonymy by Dahl & White, 1978.

rotanus Bohart & Ingram, 1946b (p. 39) – species of *Aedes* (*Stegomyia*).

rotoruae Belkin, 1968a (p. 64) – species of *Culex* (*Culex*).

rottensis Statz, 1944 (p. 109) – fossil species of ? *Anopheles* Meigen, 1818.

rotumae Belkin, 1962 (p. 472) – species of *Aedes* (*Stegomyia*).

rotumanus (Edwards, 1929b) (p. 337) – species of *Tripteroides* (*Polylepidomyia*) [original combination: *Rachionotomyia rotumana*].

roubaudi Borel, 1926 (p. 112) – synonym of *Culex* (*Lophoceraomyia*) *quadripalpis* (Edwards, 1914b) [original combination: *Lophoceratomyia roubaudi*]; synonymy established before Knight & Stone, 1977.

roubaudi (Doucet, 1950) (p. 51) – species of *Mimomyia* (*Ingramia*) [original combination: *Ravenalites roubaudi*].

roubaudi Grjebine, 1953a (p. 476) – species of *Anopheles* (*Cellia*).

roucouyana (Bonne-Wepster & Bonne, 1920) (p. 166) – species of *Wyeomyia* (subgenus uncertain, determined by Motta & Lourenço-de-Oliveira, 1995) [original combination: *Dendromyia roucouyana*].

rowlandii Theobald, 1905g (p. 34) – synonym of *Uranotaenia* (*Uranotaenia*) *nataliae* Lynch Arribálzaga, 1891b [original combination: *Pseudouranotaenia rowlandii*]; synonymy established before Knight & Stone, 1977.

roxasi Baisas & Ubaldo-Pagayon, 1953 (p. 103) – species of *Tripteroides* (*Tricholeptomyia*).

rozeboomi Baisas & Ubaldo-Pagayon, 1953 (p. 109) – species of *Tripteroides* (*Rachionotomyia*).

rozeboomi Causey, Deane & Deane, 1944 (p. 3) – species of *Chagasia*.

rozeboomi Vargas, 1941b (p. 394) – synonym of *Aedes* (*Ochlerotatus*) *bimaculatus* (Coquillett, 1902c); synonymy by Ross, 1943.

ruarinus Edwards, 1940 (p. 93) – species of *Anopheles* (*Cellia*).

rubensis Sasa & Takahashi, 1948 (p. 51) – species of *Culex* (*Neoculex*).

rubidus Robineau-Desvoidy, 1827 (p. 404) – synonym of *Psorophora* (*Psorophora*) *ciliata* (Fabricius, 1794) [original combination: *Culex rubidus*]; synonymy established before Knight & Stone, 1977.

rubiginosus Belkin, 1962 (p. 346) – species of *Aedes* (subgenus uncertain).

rubinotus Theobald, 1906b (p. 78) – species of *Culex* (*Eumelanomyia*).

rubithoracis (Leicester, 1908) (p. 119) – species of *Culex* (*Lophoceraomyia*) [original combination: *Lophoceratomyia rubithoracis*].

rubithoracis Leicester, 1908 (p. 243) – species of *Topomyia* (*Topomyia*).

rubrithorax (Macquart, 1850) (p. 313) – species of *Aedes* (*Dobrotworskyius*) [original combination: *Culex rubrithorax*].

Synonyms: *similis* Strickland, 1911a; *demansis* Strickland, 1911b; *queenslandis* Strickland, 1911b; *cumpstoni* Taylor, 1914c; *hybrida* Taylor, 1916.

ruffinis Dyar & Shannon, 1924b (p. 144) – synonym of *Culex* (*Melanoconion*) *dunni* Dyar, 1918e; synonymy by Komp, 1935.

rufinus Bigot, 1888 (p. 7) – synonym of *Culex* (*Culex*) *pipiens* Linnaeus, 1758; synonymy established before Knight & Stone, 1977.

rufipes (Gough, 1910) (p. 119) – species of *Anopheles* (*Cellia*) [original combination: *Nyssorhynchus pretoriensis* variety *rufipes*]; specific status by Edwards, 1912d.

Synonym: *watsoni* Edwards, 1911a.

rufus Gimmerthal, 1845 (p. 295) – synonym of *Aedes* (*Aedes*) *cinereus* Meigen, 1818; synonymy established before Knight & Stone, 1977.

rufus Meigen, 1818 (p. 7) – synonym of *Culex* (*Culex*) *pipiens* Linnaeus, 1758; synonymy established before Knight & Stone, 1977.

ruiliensis Dong, Zhou & Wang, 1997 (in Dong, Zhou, Wang & Wang, 1997) (p. 27) – species of *Heizmannia* (*Heizmannia*).

rupestris Dobrotworsky, 1959 (p. 136) – species of *Aedes* (*Dobrotworskyius*).

rupicolus Lewis, 1937 (p. 181) – subspecies of *Anopheles* (*Cellia*) *rhodesiensis* Theobald, 1901a.

Synonyms: *aegypti* Salem, 1938; *dthalisimilis* Corradetti, 1939a.

russelli Baisas, 1935b (p. 490) – synonym of *Armigeres* (*Armigeres*) *malayi* (Theobald, 1901c); synonymy by Bohart, 1945.

rusticus (Rossi, 1790) (p. 333) – species of *Aedes* (*Rusticoidus*) [original combination: *Culex rusticus*].

Synonyms: *maculatus* Meigen, 1804; *musicus* Leach, 1825; *pungens* Robineau-Desvoidy, 1827; *quadratimaculatus* Macquart, 1834; *diversus* Theobald, 1901c; *luteovittata* Theobald, 1901c.

ruthae Peters, 1958 (as *ruthi*) (p. 91) – species of *Culex* (*Culiciomyia*); emendation by Peters, 1959.

rutherfordi Edwards, 1922a (p. 288) – species of *Uranotaenia* (*Uranotaenia*).

rutilus (Coquillett, 1896) (p. 44) – species of *Toxorhynchites* (*Lynchiella*) [original combination: *Megarhinus rutila* [*sic*]].

Synonym: *herrickii* Theobald, 1906a.

ruwenzori Haddow & van Someren, 1950 (p. 281) – species of *Aedes* (*Stegomyia*).

ruwenzori (van Someren, 1948) (p. 128) – species of *Toxorhynchites* (*Afrorhynchus*) [original combination: *Megarhinus ruwenzori*].

ryukyensis Bohart, 1946 (p. 41) – species of *Culex* (*Culiciomyia*).

ryukyuana Tanaka, Mizusawa & Saugstad, 1979 (p. 465) – subspecies of *Uranotaenia* (*Pseudoficalbia*) *novobscura* Barraud, 1934.

ryukyuanus Tanaka, Mizusawa & Saugstad, 1979 (p. 190) – subspecies of *Culex* (*Eumelanomyia*) *hayashii* Yamada, 1917.

Ss

sabahensis Ramalingam & Ramakrishna, 1988 (p. 34) – species of *Topomyia* (*Topomyia*).

sabahensis (Reinert, 1974) (p. 82) – species of *Verrallina* (*Neomacleaya*) [original combination: *Aedes* (*Verrallina*) *sabahensis*].

sabethea Lane & Cerqueira, 1942 (p. 573) – species of *Wyeomyia* (*Miamyia*).

sacchettae Sirivanakarn & Jakob, 1982 (p. 192) – species of *Culex* (*Melanoconion*).

sacharovi Favre, 1903 (p. 189) – species of *Anopheles* (*Anopheles*).

Synonym: *elutus* Edwards, 1921d.

saeungae Taai & Harbach, 2015 (p. 256)– species of *Anopheles* (*Anopheles*).

saeva Dyar & Knab, 1906a (p. 133) – species of *Psorophora* (*Psorophora*).

Synonym: *genumaculata* Cruz, 1907b.

sagax (Skuse, 1889) (p. 1744) – species of *Aedes* (*Ochlerotatus*) [original combination: *Culex sagax*].

Synonym: *wilsoni* Taylor, 1919.

saheliensis Senevet & Andarelli, 1955 (p. 128) – *nomen nudum*, *Anopheles* [original combination: *Anopheles claviger petragnanii* [*sic*] race *saheliensis*].

saibaii Taylor, 1912b (p. 28) – synonym of *Culex* (*Culex*) *sitiens* Wiedemann, 1828; synonymy established before Knight & Stone, 1977.

saimedres Huang, 1988b (p. 10) – species of *Aedes* (*Stegomyia*).

saipanensis Stone, 1945b (p. 66) – species of *Aedes* (*Stegomyia*).

salauni Rickenbach, Ferrara & Eouzan, 1968 (p. 299) – species of *Eretmapodites*.

salbaii Maffi & Coluzzi, 1958 (p. 48) – species of *Anopheles* (*Cellia*).

salinarius Coquillett, 1904b (p. 73) – species of *Culex* (*Culex*).

salinellus Edwards, 1921b (p. 52) – synonym of *Aedes* (*Ochlerotatus*) *leucomelas* (Meigen, 1804) [original combination: *Ochlerotatus salinellus*]; synonymy established before Knight & Stone, 1977.

salinus Baisas, 1938 (p. 204) – synonym of *Culex* (*Culex*) *sitiens* Wiedemann, 1828; synonymy established before Knight & Stone, 1977.

salinus Ficalbi, 1896 (p. 29) – synonym of *Aedes* (*Ochlerotatus*) *detritus* (Haliday, 1833) [original combination: *Culex salinus*]; synonymy established before Knight & Stone, 1977.

salisburiensis Theobald, 1901c (p. 112) – species of *Culex* (*Maillotia*).
Synonyms: *bostocki* Theobald, 1905g; *amboannulatus* Theobald, 1913a; *capensis* de Meillon, 1935.

sallumae González & Reyes, 2017 (in González *et al.*, 2017) (p. 35) – species of *Aedes* (*Ochlerotatus*).

saltanensis Dyar, 1928 (p. 386) – species of *Culex* (*Culex*).
Synonym: *beauperthuyi* Anduze, 1943b.

salteno Petrocchi, 1923 (p. 91) – *nomen nudum*, *Culex*.

salus Theobald, 1908c (p. 256) – synonym of *Culex* (*Culex*) *sitiens* Wiedemann, 1828; synonymy established before Knight & Stone, 1977.

samarensis Ludlow, 1903 (p. 138) – synonym of *Aedes* (*Stegomyia*) *albopictus* (Skuse, 1895) [original combination: *Stegomyia scutellaris samarensis*]; synonymy established before Knight & Stone, 1977.

samarensis Rozeboom, 1951 (p. 502) – species of *Anopheles* (*Anopheles*).

samoaensis Stone, 1966 (p. 331) – species of *Coquillettidia* (*Coquillettidia*).

samoaensis (Theobald, 1914) (p. 36) – species of *Culex* (*Oculeomyia*) [original combination: *Pseudotaeniorhynchus samoaensis*].

samoanus (Grünberg, 1913) (p. 130) – species of *Aedes* (*Finlaya*) [original combination: *Finlaya samoana*].

sampaioi da Costa Lima, 1935b (as *sampaio*) (p. 176) – species of *Orthopodomyia*.

sampi Huang, 2004 (p. 49) – species of *Aedes* (*Stegomyia*).

sanctielii Senevet & Abonnenc, 1938 (as *sancti Elii*) (p. 501) – species of *Anopheles* (*Nyssorhynchus*).

sandrae Berlin, 1969b (p. 41) – species of *Culex* (*Micraedes*).

sandrae Zavortink, 1972 (p. 51) – species of *Aedes* (*Protomacleaya*).

sangengluoensis Wang, 1984 (p. 217) – species of *Culex* (*Lophoceraomyia*).

sangitee Sathe & Girhe, 2002 (p. 198) – *nomen nudum*, *Aedes*; unavailable name [published after 1999, fails to conform to Articles 16.4 and 72.3 of the *Code*, which require fixation of a name-bearing type and a statement naming the collection in which it is deposited (International Commission on Zoological Nomenclature, 1999); status by Harbach (22 November 2016, http://mosquito-taxonomic-inventory.info/whats-new)].

sangiti Girhe & Sathe, 2001 (p. 46) – *nomen nudum*, *Aedes*; unavailable name [published after 1999, fails to conform to Articles 16.4 and 72.3 of the *Code*, which require fixation of a name-bearing type and a statement naming the collection in which it is deposited (International Commission on Zoological Nomenclature, 1999); status by Harbach (22 November 2016, http://mosquito-taxonomic-inventory.info/whats-new)].

sanguinae Theobald, 1904a (p. 17) – species of *Hodgesia*.

sansoni Dyar & Knab, 1909c (p. 102) – synonym of *Aedes* (*Ochlerotatus*) *excrucians* (Walker, 1856a); synonymy established before Knight & Stone, 1977.

santamarinai Broche, 2000 (p. 27) – species of *Psorophora* (*Grabhamia*).

saperoi Bohart & Ingram, 1946a (p. 46) – species of *Anopheles* (*Anopheles*).

saperoi Knight, 1946 (p. 271) – species of *Aedes* (*Downsiomyia*).

sapiens Marks, 1964a (p. 66) – species of *Aedes* (*Ochlerotatus*).

sapphirina (Osten Sacken, 1868) (p. 47) – species of *Uranotaenia* (*Uranotaenia*) [original combination: *Aedes sapphirinus*].
Synonym: *coquilletti* Dyar & Knab, 1906d.

saramaccensis Bonne-Wepster & Bonne, 1920 (p. 172) – species of *Culex* (*Melanoconion*).

sarawaki Theobald, 1907 (p. 515) – synonym of *Culex* (*Oculeomyia*) *bitaeniorhynchus* Giles, 1901a [original combination: *Oculeomyia sarawaki*]; synonymy established before Knight & Stone, 1977.

sardinerae Fox, 1953 (p. 179) – species of *Culex* (*Melanoconion*).
Synonym: *bilobatus* Galindo & Blanton, 1954.

sasai Kano, Nitahara & Awaya, 1954 (p. 14) – species of *Culex* (*Culiciomyia*).
Synonym: *spiculothorax* Bram, 1967a.

sasai Tanaka, Mizusawa & Saugstad, 1975b (p. 41) – species of *Aedes* (*Aedes*).

satpuraensis Natarajan, Rajavel & Jambulingam, 2017 (p. 259) – species of *Uranotaenia* (*Pseudoficalbia*).

saungi Colless, 1955a (p. 334) – species of *Anopheles* (*Cellia*).

savoryi Bohart, 1957 (p. 31) – species of *Aedes* (*Tanakaius*).

sawadwongporni Rattanarithikul & Green, 1987 (p. 256) – species of *Anopheles* (*Cellia*).

sawyeri Causey, Deane, Deane & Sampaio, 1943 (p. 11) – species of *Anopheles* (*Nyssorhynchus*).

saxatilis Grossbeck, 1905 (p. 360) – synonym of *Culex* (*Neoculex*) *territans* Walker, 1856a; synonymy established before Knight & Stone, 1977.

saxicola Edwards, 1922a (p. 262) – species of *Aedes* (*Hulecoeteomyia*) [*nomen novum* for *fluviatilis* Leicester, 1908].

sayi Dyar & Knab, 1906d (p. 181) – synonym of *Psorophora* (*Janthinosoma*) *ferox* (von Humboldt, 1819) [*nomen novum* for *musicus* Say, 1829; original combination: *Janthinosoma sayi*]; synonymy established before Knight & Stone, 1977.

sayi Theobald, 1907 (p. 155) – synonym of *Psorophora* (*Janthinosoma*) *ferox* (von Humboldt, 1819) [*nomen novum* for *musicus* Say, 1829; original combination: *Janthinosoma sayi*]; synonymy established before Knight & Stone, 1977.

scanloni Bram, 1967a (p. 147) – species of *Culex* (*Culiciomyia*).

scanloni Mattingly, 1970 (p. 38) – species of *Heizmannia* (*Heizmannia*).

scanloni Reinert, 1970 (p. 17) – species of *Aedes* (*Petermattinglyius*).

scanloni Sallum & Peyton, 2005 (in Sallum *et al.*, 2005b) (p. 170) – species of *Anopheles* (*Cellia*).

scapularis (Rondani, 1848) (p. 109) – species of *Aedes* (*Ochlerotatus*) [original combination: *Culex scapularis*].

Synonyms: *confirmatus* Lynch Arribálzaga, 1891b; *hemisurus* Dyar & Knab, 1906d; *indolescens* Dyar & Knab, 1907a; *camposanus* Dyar, 1918e.

scatophagoides (Theobald, 1901a) (p. 277) – species of *Aedes* (*Mucidus*) [original combination: *Mucidus scatophagoides*].

schausi Dyar & Knab, 1908 (p. 63) – synonym of *Sabethes* (*Sabethes*) *belisarioi* Neiva, 1908a; synonymy established before Knight & Stone, 1977.

schedocyclia (Dyar & Knab, 1908) (p. 64) – species of *Shannoniana* [original combination: *Lesticocampa schedocyclia*].

scheuberi Carpintero & Leguizamón, 2004 (p. 502) – species of *Culex* (*Culex*).

schicki Berlin, 1969b (p. 39) – species of *Culex* (*Micraedes*).

schicki Zavortink, 1972 (p. 46) – species of *Aedes* (*Protomacleaya*).

schilfgaardei Sirivanakarn, 1968 (p. 163) – species of *Culex* (*Lophoceraomyia*).

schizopinax Dyar, 1929a (p. 1) – species of *Aedes* (*Ochlerotatus*).

schlosseri Belkin, 1962 (p. 382) – species of *Aedes* (*Finlaya*).

schnusei (Martini, 1931a) (p. 202) – species of *Sabethes* (*Sabethes*) [original combination: *Dendromyia schnusei*].

scholasticus Theobald, 1901c (p. 120) – synonym of *Culex* (*Culex*) *inflictus* Theobald, 1901c; synonymy established before Knight & Stone, 1977.

schoutedeni (Wolfs, 1948a) (p. 244) – species of *Coquillettidia* (*Coquillettidia*) [original combination: *Taeniorhynchus* (*Coquillettidia*) *schoutedeni*].

schroederi Schick, 1970a (p. 88) – species of *Aedes* (*Protomacleaya*).

schtakelbergi Shingarev, 1928 (p. 52) – species of *Aedes* (*Ochlerotatus*).

schueffneri Stanton, 1915 (p. 373) – species of *Anopheles* (*Cellia*).

schultzei Enderlein, 1931 (p. 123) – synonym of *Toxorhynchites* (*Toxorhynchites*) *brevipalpis conradti* Grünberg, 1907a; synonymy by Edwards, 1941.

schwarzi Dyar & Knab, 1906a (p. 135) – synonym of *Psorophora* (*Janthinosoma*) *johnstonii* (Grabham, 1905) [original combination: *Janthinosoma schwarzi*]; synonymy established before Knight & Stone, 1977.

schwetzi Edwards, 1926a (p. 127) – species of *Aedes* (*Stegomyia*).

schwetzi Edwards, 1929b (p. 326) – species of *Culex* (*Culex*).

schwetzi Evans, 1934 (p. 555) – species of *Anopheles* (*Cellia*).

Synonym: *michaeli* de Meillon & Leeson, 1940.

scimitar Branch & Seabrook, 1959 (p. 217) – species of *Culex* (*Culex*).

scintillans Ludlow, 1905a (p. 130) – species of *Heizmannia* (*Heizmannia*).

Synonym: *pilosa* Brug, 1931a.

scintillans Walker, 1848 (p. 1) – synonym of *Psorophora* (*Psorophora*) *cilipes* (Fabricius, 1805) [original combination: *Sabethes scintillans*]; synonymy established before Knight & Stone, 1977.

scotinomus (Dyar & Knab, 1907c) (p. 209) – species of *Wyeomyia* (*Wyeomyia*) [original combination; *Phoniomyia scotinomus*].

Synonyms: *homothe* Dyar & Knab, 1907c; *leucopisthepus* Dyar & Knab, 1907c; *abrachys* Dyar & Knab, 1909d; *chresta* Dyar & Knab, 1909d; *hapla* Dyar & Knab, 1909d; *incana* Dyar, 1922g; *camptocomma* Dyar, 1924d.

scottii Theobald, 1912b (p. 86) – species of *Culex* (*Culex*).

scutatus Rozeboom & Komp, 1948 (p. 396) – synonym of *Culex* (*Culex*) *abnormalis* Lane, 1936b; synonymy established before Knight & Stone, 1977.

scutellalbum Boshell-Manrique, 1939 (p. 309) – species of *Aedes* (*Ochlerotatus*).

scutellaris (Walker, 1859) (p. 77) – species of *Aedes* (*Stegomyia*) [original combination: *Culex scutellaris*].

Synonyms: *variegatus* Doleschall, 1858; *zonatipes* Walker, 1861.

scutipunctatus Lutz & Neiva, 1911 (p. 298) – synonym of *Psorophora* (*Grabhamia*) *confinnis* (Lynch Arribálzaga, 1891b) [original combination: *Culex scutipunctatus*]; synonymy established before Knight & Stone, 1977.

scutoscriptus Bohart & Ingram, 1946b (p. 27) – species of *Aedes* (*Stegomyia*).

scutostriata da Cunha Ramos & Brunhes, 2004 (p. 137) – species of *Uranotaenia* (*Pseudoficalbia*).

seampi Huang, 1974 (p. 137) – species of *Aedes* (*Stegomyia*).

seatoi Huang, 1969 (p. 234) – species of *Aedes* (*Stegomyia*).

sechani Brunhes & Boussès, 2009 (p. 21) – species of *Culex* (*Culex*).

seculata (Menon, 1950) (p. 139) – species of *Verrallina* (*Neomacleaya*) [original combination: *Aedes* (*Aedes*) *seculatus*].

secundus Bonne-Wepster & Bonne, 1920 (as *sucunda*) (p. 170) – species of *Culex* (*Carrollia*).

secutor Theobald, 1901c (p. 321) – species of *Culex* (*Culex*).

Synonyms: *lamentator* Dyar & Knab, 1906d; *quasisecutor* Theobald, 1907; *toweri* Dyar & Knab, 1907a.

sedaensis Lei, 1989a (p. 6) – species of *Aedes* (*Ochlerotatus*).

sedlacekae Sirivanakarn, 1968 (p. 180) – species of *Culex* (*Lophoceraomyia*).

segermanae Huang, 1997 (p. 57) – species of *Aedes* (*Stegomyia*).

seguini Laveran, 1901b (p. 992) – synonym of *Mansonia* (*Mansonioides*) *annulifera* (Theobald, 1901c) [original combination: *Panoplites seguini*]; synonymy established before Knight & Stone, 1977.

seguyi Apfelbeck, 1929 (p. 288) – synonym of *Aedes* (*Ochlerotatus*) *pullatus* (Coquillett, 1904c); synonymy established before Knight & Stone, 1977.

selai Klein & Sirivanakarn, 1970 (p. 589) – species of *Culex* (*Eumelanomyia*).

selangorensis Sirivanakarn, 1976 (p. 77) – species of *Culex* (*Oculeomyia*).

seldeslachtsi Wolfs, 1947c (p. 224) – species of *Culex* (*Culex*).

selengensis Ludlow, 1920a (p. 77) – synonym of *Anopheles* (*Anopheles*) *lewisi* Ludlow, 1920a; synonymy by Aitken, 1945.

sembeli Toma & Miyagi, 2002 (p. 257) – species of *Armigeres* (*Armigeres*).

semibrunneus Edwards, 1927a (p. 353) – species of *Culex* (*Culiciomyia*).

semicantans Martini, 1920b (p. 205) – synonym of *Aedes* (*Ochlerotatus*) *riparius* Dyar & Knab, 1907d; synonymy established before Knight & Stone, 1977.

semisimplicipes Edwards, 1914b (p. 76) – species of *Eretmapodites* [original combination: *Eretmapodites chrysogaster* variety *semisimplicipes*]; specific status by Edwards, 1941.

semlikiensis van Someren, 1950b (p. 40) – species of *Aedes* (*Aedimorphus*).

senegalensis Theobald, 1903d (p. i) – synonym of *Aedes* (*Catageiomyia*) *irritans* (Theobald, 1901d) [original combination: *Catageiomyia senegalensis*]; synonymy established before Knight & Stone, 1977.

seneveti Clastrier, 1970c (p. 464) – synonym of *Culex* (*Melanoconion*) *contei* Duret, 1968c; synonymy by Sirivanakarn, 1983.

seneveti Rioux, 1959 (p. 733) – synonym of *Anopheles* (*Cellia*) *rufipes broussesi* Edwards, 1929a [original combination: *Anopheles* (*Myzomyia*) *rufipes seneveti*]; synonymy established before Knight & Stone, 1977.

seniori Barraud, 1934 (p. 365) – species of *Culex* (*Lophoceraomyia*).

sentania (King & Hoogstraal, 1947b) (p. 128) – species of *Verrallina* (*Verrallina*) [original combination: *Aedes* (*Aedes*) *sentanius*].

senyavinensis Knight & Hurlbut, 1949 (p. 27) – species of *Aedes* (*Aedimorphus*).

seoulensis Yamada, 1921 (p. 61) – species of *Aedes* (*Hopkinsius*).

separatus (Leicester, 1908) (p. 36) – species of *Anopheles* (*Anopheles*) [original combination: *Myzorhynchus separatus*].

Synonym: *snijdersi* Swellengrebel, 1921.

separatus (Lynch Arribálzaga, 1891b) (species) (p. 133) – subspecies of *Toxorhynchites* (*Lynchiella*) *haemorrhoidalis* (Fabricius, 1787) [original combination: *Megarhina separata*]; subspecific status by Lane, 1951.

Synonym: *lynchi* Dyar & Knab, 1906c.

sepositus Leicester, 1908 (p. 152) – synonym of *Culex* (*Oculeomyia*) *sinensis* Theobald, 1903a; synonymy established before Knight & Stone, 1977.

septemguttata Theobald, 1907 (p. 499) – synonym of *Mansonia* (*Mansonioides*) *annulifera* (Theobald, 1901c) [original combination: *Mansonioides septemguttata*]; synonymy established before Knight & Stone, 1977.

septempunctata Theobald, 1905f (p. 107) – species of *Mansonia* (*Mansonioides*).

Synonym: *bonnewepsterae* van den Assem, 1958.

septemstriatus Dyar & Knab, 1907a (p. 10) – species of *Aedes* (*Howardina*). **Note**: incorrectly listed as *septemstriata* Dyar & Knab in the online catalogue of the Walter Reed Biosystematics Unit (http://www.mosquitocatalog.org/).

septentrionalis (Dyar & Knab, 1906c) (species) (p. 249) – subspecies of *Toxorhynchites* (*Lynchiella*) *rutilus* (Coquillett, 1896) [original combination: *Megarhinus septentrionalis*]; subspecific status by Jenkins & Carpenter, 1946.

Synonym: *herrickii* Theobald, 1906a.

septentrionalis Evans, 1934 (p. 558) – synonym of *Anopheles* (*Cellia*) *brohieri* Edwards, 1929b [original combination: *Anopheles theileri* variety *septentrionalis*]; synonymy by Adam *et al.*, 1956.

serafini (Szadziewski, 1998) (p. 240) – fossil species of *Aedes* (*Ochlerotatus*); transfer from subgenus *Finlaya* to subgenus *Ochlerotatus* (as genus) by Harbach (30 October 2012, http://mosquito-taxonomic-inventory.info/whats-new).

seretsei Abdulla-Khan, Coetzee & Hunt, 1998 (p. 248) – species of *Anopheles* (*Cellia*).

sergentii Theobald, 1903a (p. 218) – synonym of *Culex* (*Neoculex*) *impudicus* Ficalbi, 1890a; synonymy by Mattingly, 1955.

sergentii (Theobald, 1907) (p. 68) – species of *Anopheles* (*Cellia*) [original combination: *Pyretophorus sergentii*].

sergievi Danilov, Markovich & Proskuryakova, 1978 (p. 44) – species of *Aedes* (*Ochlerotatus*).

sericeus Theobald, 1901c (p. 147) – synonym of *Culex* (*Culex*) *quinquefasciatus* Say, 1823; synonymy established before Knight & Stone, 1977.

serotinus Philippi, 1865 (p. 595) – synonym of *Culex* (*Culex*) *quinquefasciatus* Say, 1823; synonymy established before Knight & Stone, 1977.

serrata (Lutz, 1905) (p. 287) – species of *Wyeomyia* (*Miamyia*) [original combination: *Dendromyia serrata*].

serratimarge Root, 1927c (p. 589) – species of *Culex* (*Melanoconion*).

serratipes Becker, 1908 (p. 78) – synonym of *Culiseta* (*Allotheobaldia*) *longiareolata* (Macquart, 1838a) [original combination: *Culex serratipes*]; synonymy established before Knight & Stone, 1977.

serratoria (Dyar & Nuñez Tovar, 1927b) (p. 7) – species of *Wyeomyia* (subgenus uncertain, determined by Motta & Lourenço-de-Oliveira, 1995) [original combination: *Sabethoides serratoria*].

serratus (Barraud, 1929a) (p. 1059) – species of *Tripteroides* (*Rachionotomyia*) [original combination: *Rachionotomyia aranoides* variety *serrata*]; specific status by Mattingly, 1981.

serratus (Theobald, 1901c) (p. 45) – species of *Aedes* (*Ochlerotatus*) [original combination: *Culex serratus*].

Synonyms: *mathisi* Neveu-Lemaire, 1902b; *meridionalis* Dyar & Knab, 1906d.

serus Martini, 1920a (p. 96) – synonym of *Aedes* (*Ochlerotatus*) *diantaeus* Howard, Dyar & Knab, 1913; synonymy established before Knight & Stone, 1977.

seticoxitus Luh & Li, 1981 (p. 188) – species of *Armigeres* (*Armigeres*).

setifer Delfinado, 1966a (p. 87) – species of *Armigeres* (*Armigeres*).

setivalva Monchadsky, 1936 (p. 211) – synonym of *Culiseta* (*Culicella*) *fumipennis* (Stephens, 1825) [original combination: *Theobaldia* (*Culicella*) *setivalva*]; synonymy by Danilov, 1984.

setosa King & Hoogstraal, 1947a (p. 585) – species of *Uranotaenia* (*Uranotaenia*).

setulosus Doleschall, 1857 (p. 384) – synonym of *Lutzia* (*Metalutzia*) *fuscana* Wiedemann, 1820 [original combination: *Culex setulosus*]; synonymy established before Knight & Stone, 1977.

sexaueri Belkin, 1953 (p. 359) – species of *Uranotaenia* (*Uranotaenia*).

sexlineatus (Theobald, 1901a) (p. 308) – species of *Aedes* (*Howardina*) [original combination: *Stegomyia sexlineata*].

Synonym: *dominicii* Rangel & Romero Sierra, 1907.

seychellensis Theobald, 1912b (p. 83) – synonym of *Aedes* (*Polyleptiomyia*) *albocephalus* (Theobald, 1903c) [original combination: *Reedomyia seychellensis*]; synonymy by Edwards, 1941.

seydeli Edwards, 1929b (p. 322) – species of *Anopheles* (*Cellia*) [original combination: *Anopheles theileri* variety *seydeli*]; specific status by de Meillon, 1947b.

seyrigi Edwards, 1941 (p. 256) – species of *Culex* (*Maillotia*).

Synonym: *robici* Doucet, 1950.

shakujiiensis Ogasawara, 1939b (p. 238) – synonym of *Coquillettidia* (*Coquillettidia*) *ochracea* (Theobald, 1903a) [original combination: *Culex shakujiiensis*]; synonymy by Kano & Hayashi, 1949.

shanahani Sirivanakarn, 1968 (p. 117) – species of *Culex* (*Lophoceraomyia*).

shannoni Cerqueira, 1961 (p. 40) – species of *Sabethes* (*Sabethes*).

shannoni Davis, 1931 (p. 345) – species of *Anopheles* (*Anopheles*).

shannoni (Lane & Antunes, 1937) (p. 227) – species of *Coquillettidia* (*Rhynchotaenia*) [original combination: *Mansonia* (*Rhynchotaenia*) *shannoni*].

shannoni Lane & Cerqueira, 1942 (p. 599) – species of *Wyeomyia* (subgenus uncertain, determined by Motta & Lourenço-de-Oliveira, 1995).

shannoni Vargas & Downs, 1950 (p. 161) – species of *Aedes* (*Ochlerotatus*).

shebbearei Barraud, 1924g (p. 19) – species of *Culex* (*Culiciomyia*).

shehzadae Qutubuddin, 1972 (p. 119) – species of *Aedes* (*Downsiomyia*).

sherki Knight, 1948a (p. 645) – species of *Aedes* (*Hulecoeteomyia*).

shillitonis Edwards, 1932b (p. 561) – species of *Uranotaenia* (*Pseudoficalbia*).

shillongensis Barraud, 1927b (p. 539) – synonym of *Armigeres* (*Armigeres*) *durhami* Edwards, 1917 [original combination: *Armigeres* (*Armigeres*) *kuchingensis* form *shillongensis*]; synonymy by Thurman, 1959.

shinonagai (Tanaka, Mizusawa & Saugstad, 1979) (p. 245) – species of *Lutzia* (*Insulalutzia*) [original combination: *Culex* (*Lutzia*) *shinonagai*].

shintienensis Tsai & Lien, 1950 (species) (p. 177) – subspecies of *Aedes* (*Hulecoeteomyia*) *japonicus* (Theobald, 1901a); subspecific status by Tanaka *et al.*, 1979.

shoae Hamon & Ovazza, 1954 (p. 416) – species of *Culex* (*Culex*).

shopei Forattini & Toda, 1966 (p. 507) – species of *Culex* (*Microculex*).

shortti (Barraud, 1923e) (p. 405) – species of *Aedes* (*Collessius*) [original combination: *Finlaya shortti*].

shrivastavii Wattal, Kalra & Krishnan, 1966 (p. 159) – synonym of *Culex* (*Eumelanomyia*) *foliatus* Brug, 1932; synonymy by Sirivanakarn, 1972.

shropshirei Ludlow, 1919 (p. 168) – synonym of *Isostomyia espini* (Martini, 1914) [original combination: *Trichoprosopon* (*Joblotia*) *shropshirei*]; synonymy established before Knight & Stone, 1977.

siamensis Barraud & Christophers, 1931 (p. 283) – synonym of *Culex* (*Culex*) *summorosus* Dyar, 1920i [original combination: *Culex tritaeniorhynchus* variety *siamensis*]; synonymy by Colless, 1957a.

siamensis Delfinado, 1968 (p. 36) – synonym of *Verrallina* (*Harbachius*) *yusafi* (Barraud, 1931b) [original combination: *Aedes* (*Neomacleaya*) *siamensis*]; synonymy by Reinert, 1974.

siamensis Zavortink, 1968 (p. 126) – species of *Orthopodomyia*.

siberiensis Ludlow, 1920b (p. 151) – synonym of *Culiseta* (*Culiseta*) *alaskaensis* (Ludlow, 1906c) [original combination: *Culex siberiensis*]; synonymy established before Knight & Stone, 1977.

sibiricus Danilov & Filippova, 1978 (p. 170) – species of *Aedes* (*Stegomyia*).

sicaulti Roubaud, 1935a (p. 107) – synonym of *Anopheles* (*Anopheles*) *labranchiae* Falleroni, 1926 [original combination: *Anopheles labranchiae* variety *sicaulti*, but written as *Anopheles labranchiae sicaulti*]; synonymy by de Zulueta *et al.*, 1983.

siculus Robineau-Desvoidy, 1827 (p. 406) – synonym of *Aedes* (*Ochlerotatus*) *caspius* (Pallas, 1771) [original combination: *Culex siculus*]; synonymy established before Knight & Stone, 1977.

sierraleonis Theobald, 1901c (p. 36) – synonym of *Lutzia* (*Metalutzia*) *tigripes* de Grandpré & de Charmoy, 1901 [original combination: *Culex maculicrura* variety *sierraleonis*]; synonymy by Edwards, 1941.

sierrensis (Ludlow, 1905b) (p. 231) – species of *Aedes* (*Jarnellius*) [original combination: *Taeniorhynchus sierrensis*].

sigmoides Barraud, 1928b (p. 373) – synonym of *Verrallina* (*Verrallina*) *dux* (Dyar & Shannon, 1925a) [original combination: *Aedes* (*Aedes*) *sigmoides*]; synonymy by Causey, 1937b.

signata da Cunha Ramos, 1993 (p. 219) – species of *Uranotaenia* (*Pseudoficalbia*).

signifera (Coquillett, 1896) (p. 43) – species of *Orthopodomyia* [original combination: *Culex signifer*].
Synonym: *californica* Bohart, 1950.

signipennis (Coquillett, 1904c) (p. 167) – species of *Psorophora* (*Grabhamia*) [original combination: *Taeniorhynchus signipennis*].
Synonym: *pruinosa* Martini, 1935

sikkimensis Giles, 1901a (p. 604) – synonym of *Toxorhynchites* (*Toxorhynchites*) *splendens* (Wiedemann, 1819) [original combination: *Megarhina sikkimensis*]; synonymy by Barraud, 1934.

silvai Duret, 1968f (p. 69) – species of *Culex* (*Melanoconion*).

silvestris Dobrotworsky, 1961b (p. 248) – species of *Aedes* (*Ochlerotatus*).
Synonym: *waterhousei* Dobrotworsky, 1960a.

silvestris Ingram & de Meillon, 1927 (p. 63) – species of *Eretmapodites*.

silvestris Ross, 1898 (p. 134) – *nomen nudum*, *Culex*.

silvestris (Shingarev, 1928) (p. 51) – *nomen dubium*, *Culiseta* [original combination: *Theobaldia silvestris*]; status by Dahl & White, 1978.

simanini Gutsevich, 1966 (p. 457) – species of *Aedes* (*Ochlerotatus*).

simici Baranoff, 1927 (p. 15) – synonym of *Aedes* (*Ochlerotatus*) *pulcritarsis* (Rondani, 1872); synonymy established before Knight & Stone, 1977.

simile Lane & Cerqueira, 1942 (p. 514) – species of *Trichoprosopon*.

similis (Leicester, 1908) (p. 235) – species of *Tripteroides* (*Tripteroides*) [original combination: *Colonemyia similis*].

similis Strickland, 1911a (p. 132) – synonym of *Aedes* (*Dobrotworskyius*) *rubrithorax* (Macquart, 1850) [original combination: *Culicelsa similis*]; synonymy established before Knight & Stone, 1977.

similis Strickland, 1917 (p. 611) – synonym of *Anopheles* (*Anopheles*) *similissimus* Strickland & Chowdhury, 1927a [original combination: *Myzorhynchus similis*]; synonymy established before Knight & Stone, 1977. **Note**: this nominal taxon is the junior secondary homonym of *similis* Theobald, 1901a.

similis Theobald, 1901a (p. 129) – synonym of *Anopheles* (*Anopheles*) *coustani* Laveran, 1900 [original combination: *Anopheles paludis* variety *similis*]; synonymy established before Knight & Stone, 1977. **Note**: this nominal taxon is the senior secondary homonym of *similis* Strickland, 1917.

similis Theobald, 1903a (p. 207) – synonym of *Culex* (*Culex*) *nigripalpus* Theobald, 1901c; synonymy established before Knight & Stone, 1977.

similis Theobald, 1908c (p. 257) – synonym of *Uranotaenia* (*Uranotaenia*) *pallidocephala* Theobald, 1908c; synonymy established before Knight & Stone, 1977.

similis (Theobald, 1910c) (p. 189) – species of *Verrallina* (*Verrallina*) [original combination: *Pseudoskusea similis*].

similissimus Strickland & Chowdhury, 1927a (p. 19) – species of *Anopheles* (*Anopheles*) [*nomen novum* for *similis* Strickland, 1917].
Synonym: *similis* Strickland, 1917.

simlensis Edwards, 1922a (p. 269) – species of *Aedes* (*Himalaius*).

simlensis (James, 1911) (in James & Liston, 1911) (species) (p. 66) – subspecies of *Anopheles* (*Anopheles*) *gigas* Giles, 1901b [original combination: *Patagiamyia simlensis*]; subspecific status by Harbach & Howard, 2007.

simmondsi Tenorio, 1977 (p. 331) – species of *Bironella* (*Bironella*).

simmsi (Dyar & Knab, 1908) (p. 65) – species of *Wyeomyia* (*Wyeomyia*) [original combination: *Phoniomyia simmsi*].
Synonyms: *dymodora* Dyar & Knab, 1908; *roloncetta* Dyar, 1919d.

simpla (King & Hoogstraal, 1947b) (p. 130) – species of *Verrallina* (*Verrallina*) [original combination: *Aedes* (*Aedes*) *simplus*].

simplex Baisas & Catipon, 1958 (p. 54) – *nomen nudum*, *Uranotaenia*.

simplex Brug, 1934 (p. 507) – species of *Tripteroides* (*Rachisoura*).

simplex Martini, 1935 (p. 36) – synonym of *Psorophora* (*Psorophora*) *howardii* Coquillett, 1901; synonymy by Lane, 1951.

simplex Taylor, 1914c (p. 698) – synonym of *Culex* (*Culex*) *annulirostris* Skuse, 1889 [original combination: *Culicelsa simplex*]; synonymy established before Knight & Stone, 1977. **Note**: this nominal taxon is the junior secondary homonym of *simplex* Theobald, 1903a (p. 337).

simplex (Theobald, 1903a) (p. 297) – species of *Aedes* (*Cancraedes*) [original combination: *Ficalbia simplex*].

simplex Theobald, 1903a (p. 337) – synonym of *Culex* (*Culex*) *univittatus* Theobald, 1901c [original combination: *Heptaphlebomyia simplex*]; synonymy established before Knight & Stone, 1977. **Note:** this nominal taxon is the senior secondary homonym of *simplex* Taylor, 1914c.

simplex Theobald, 1908b (p. 291) – synonym of *Verrallina* (*Neomacleaya*) *andamanensis* (Edwards, 1922a) [original combination: *Neomacleaya indica* variety *simplex*]; synonymy established before Knight & Stone, 1977.

simplicicornis Edwards, 1930a (p. 306) – species of *Culex* (*Eumelanomyia*).

simpliciforceps Edwards, 1941 (p. 279) – species of *Culex* (*Eumelanomyia*).

simpsoni (Theobald, 1905c) (p. 224) – species of *Aedes* (*Stegomyia*) [original combination: *Stegomyia simpsoni*].

simpsoni Theobald, 1905g (p. 28) – species of *Culex* (*Culex*).
Synonym: *richteri* Ingram & de Meillon, 1927.

simulans (Newstead & Carter, 1911b) (p. 240) – species of *Aedes* (*Elpeytonius*) [original combination: *Reedomyia simulans*].

simulator Dyar & Knab, 1906d (p. 218) – species of *Culex* (*Melanoconion*).
Synonym: *venezuelensis* Anduze, 1949a.

simulatus Baisas & Ubaldo-Pagayon, 1953 (p. 170) – species of *Tripteroides* (*Tripteroides*).

simulatus Barraud, 1931b (p. 611) – synonym of *Aedes* (*Collessius*) *elsiae* (Barraud, 1923a); synonymy by Barraud, 1934.

sinaiticus Kirkpatrick, 1925 (p. 383) – species of *Culex* (*Culex*).

sinensis Chow, 1950 (p. 283) – species of *Aedes* (*Downsiomyia*).

sinensis Meng & Wu, 1962 (p. 382) – synonym of *Culiseta* (*Culiseta*) *niveitaeniata* (Theobald, 1907) [original combination: *Theobaldia* (*Theobaldia*) *sinensis*]; synonymy by Maslov, 1967.

sinensis Theobald, 1903a (p. 180) – species of *Culex* (*Oculeomyia*).
Synonyms: *sepositus* Leicester, 1908; *tenax* Leicester, 1908; *tripunctatus* Mochizuki, 1913.

sinensis Wiedemann, 1828 (p. 547) – species of *Anopheles* (*Anopheles*).
Synonyms: *plumiger* Dönitz, 1901; *jesoensis* Tsuzuki, 1902b; *changfus* Ma, 1981b.

sineroides Yamada, 1924 (p. 233) – species of *Anopheles* (*Anopheles*).

singalesi Marlatt, 1903 (p. 122) – *nomen nudum*, *Culex*.

singalesi Theobald, 1901 (in Green, 1901) (p. 368) – *nomen nudum*, *Culex*.

singhbhumensis Natarajan & Rajavel, 2009 (p. 403) – species of *Culex* (*Lophoceraomyia*).

singuawaensis Sirivanakarn, 1969 (p. 119) – species of *Culex* (*Lophoceraomyia*) [*nomen novum* for *confusus* Sirivanakarn, 1968].
Synonym: *confusus* Sirivanakarn, 1968.

singularis (Leicester, 1908) (p. 188) – species of *Verrallina* (*Neomacleaya*) [original combination: *Aioretomyia singularis*].

sinkiangensis Hsiao, 1977 (p. 451) – species of *Aedes* (*Ochlerotatus*).

sintoni (Barraud, 1924b) (p. 967) – species of *Aedes* (subgenus uncertain) [original combination: *Finlaya sintoni*].

sintoni Puri, 1929b (p. 401) – species of *Anopheles* (*Anopheles*).

sintonoides Ho, 1938 (p. 279) – species of *Anopheles* (*Anopheles*).

siphanulatus Lourenço-de-Oliveira & da Silva, 1987 (p. 101) – species of *Culex* (*Microculex*).

siphonalis Grossbeck, 1904a (p. 332) – synonym of *Aedes* (*Ochlerotatus*) *excrucians* (Walker, 1856a) [original combination: *Culex siphonalis*]; synonymy established before Knight & Stone, 1977.

sirivanakarni (Duret, 1982) (p. 167) – species of *Onirion* [original combination: *Wyeomyia* (*Dendromyia*) *sirivanakarni*].

sitiens Wiedemann, 1828 (p. 224) – species of *Culex* (*Culex*).
Synonyms: *impellens* Walker, 1859; *bancroftii* Theobald, 1901a; *microannulatus* Theobald, 1901a; *gnophodes* Theobald, 1903a; *somaliensis* Neveu-Lemaire, 1906; *nigricephala* Leicester, 1908; *salus* Theobald, 1908c; *jepsoni* Theobald, 1910b; *saibaii* Taylor, 1912b; *paludis* Taylor, 1913; *annulata* Taylor, 1914c; *milni* Taylor, 1914d; *salinus* Baisas, 1938.

skusii Giles, 1900a (p. 292) – synonym of *Culex* (*Culex*) *quinquefasciatus* Say, 1823; synonymy established before Knight & Stone, 1977.

smaragdinus Reinert, 1997 (in Reinert *et al.*, 1997) (p. 28) – species of *Anopheles* (*Anopheles*).

smithburni van Someren, 1950b (p. 39) – species of *Aedes* (*Catageiomyia*).

smithii (Coquillett, 1901) (p. 260) – species of *Wyeomyia* (*Wyeomyia*) [original combination: *Aedes smithii*].
Synonym: *haynei* Dodge, 1947.

smithii Theobald, 1905b (p. 101) – species of *Anopheles* (*Cellia*).
Synonyms: *pallidopalpi* Theobald, 1907; *cavernicolus* Abonnenc, 1954.

smudges Dong, Zhou, Gong *et al.* [*sic*], 2004 (p. 187) – *nomen nudum*, *Uranotaenia*.

snijdersi Swellengrebel, 1921 (p. 114) – synonym of *Anopheles* (*Anopheles*) *separatus* (Leicester, 1908) [original combination: *Myzorhynchus sinensis* variety *snijdersi*]; synonymy established before Knight & Stone, 1977.

soalalaensis Grjebine, 1953a (p. 501) – *nomen dubium*, *Anopheles* [original combination: *Anopheles fuscicolor* variety *soalalaensis*]; status by Brunhes *et al.*, 1998a.

soaresi Lane & Cerqueira, 1942 (p. 499) – species of *Trichoprosopon*.

socialis Theobald, 1901c (p. 340) – species of *Uranotaenia* (*Uranotaenia*).

soesiloi Strickland & Chowdhury, 1931 (p. 12) – synonym of *Bironella* (*Neobironella*) *derooki* Soesilo & van Slooten, 1931a, 1931b [original combination: *Anopheles*

(*Bironella*) *soesiloi*]; synonymy by Evenhuis & Gon, 1989.

sogdianus Keshishian, 1938 (p. 888) – synonym of *Anopheles* (*Anopheles*) *marteri* Senevet & Prunnelle, 1927; synonymy by Ribeiro *et al.*, 1987.

sohni (Reinert, 1974) (p. 85) – species of *Verrallina* (*Neomacleaya*) [original combination: *Aedes* (*Verrallina*) *sohni*].

soleatus Edwards, 1924 (in Haworth, 1924) (p. 198) – species of *Aedes* (*Stegomyia*).

solitarius Bonne-Wepster, 1938 (p. 394) – species of *Culex* (*Culex*).

sollicitans (Walker, 1856a) (p. 427) – species of *Aedes* (*Ochlerotatus*) [original combination: *Culex sollicitans*].

solomonensis Cumpston, 1924 (p. 1402) – *nomen nudum*, *Anopheles*.

solomonis Belkin, 1953 (p. 350) – species of *Uranotaenia* (*Uranotaenia*).

solomonis Belkin, 1962 (p. 280) – species of *Hodgesia*.

solomonis (Belkin, 1962) (p. 292) – species of *Mimomyia* (*Etorleptiomyia*) [original combination: *Ficalbia* (*Etorleptiomyia*) *solomonis*].

solomonis Belkin, Knight & Rozeboom, 1945 (p. 247) – species of *Anopheles* (*Cellia*).

solomonis (Edwards, 1924b) (p. 363) – species of *Tripteroides* (*Polylepidomyia*) [original combination: *Rachionotomyia solomonis*].

solomonis Edwards, 1929 (in Paine & Edwards, 1929) (p. 316) – species of *Culex* (*Lophoceraomyia*) [original combination: *Culex* (*Lophoceraomyia*) *fraudatrix* variety *solomonis*]; specific status by Belkin, 1962.

solomonis Stone & Bohart, 1944 (p. 208) – species of *Aedes* (*Finlaya*).

solomonis (Wharton, 1947) (p. 63) – species of *Malaya* [original combination: *Harpagomyia solomonis*].

solstitialis (Lutz, 1904) (in Bourroul, 1904) (p. 10) – species of *Toxorhynchites* (*Lynchiella*) [original combination: *Megarhinus solstitialis*].

Synonyms: *chrysocephalus* Theobald, 1907; *mara* Anduze, 1942a.

somalicus Rivola & Holstein, 1957 (pp. 382, 387) – species of *Anopheles* (*Cellia*).

somaliensis Neveu-Lemaire, 1906 (p. 254) – synonym of *Culex* (*Culex*) *sitiens* Wiedemann, 1828; synonymy established before Knight & Stone, 1977.

somlooni Peyton & Klein, 1970 (p. 249) – species of *Uranotaenia* (*Uranotaenia*).

somerseti Taylor, 1912b (p. 28) – synonym of *Culex* (*Culex*) *annulirostris* Skuse, 1889; synonymy established before Knight & Stone, 1977.

soperi Antunes & Lane, 1937 (p. 21) – species of *Culex* (*Carrollia*).

soperi Lane & Cerqueira, 1942 (p. 682) – species of *Sabethes* (*Peytonulus*).

soperi Leví-Castillo, 1955c (p. 480) – species of *Haemagogus* (*Haemagogus*).

sororcula Dyar & Knab, 1906a (p. 139) – species of *Wyeomyia* (*Wyeomyia*).

Synonym: *fratercula* Dyar & Knab, 1906a.

sorsogonensis Bañez & Jueco, 1966 (p. 195) – species of *Aedes* (*Finlaya*).

spanius (Dyar & Knab, 1909d) (p. 259) – species of *Deinocerites* [original combination: *Dinanamesus spanius*].

spathifurca (Edwards, 1915c) (p. 284) – species of *Culex* (*Culiciomyia*) [original combination: *Culiciomyia spathifurca*].

Synonym: *stylifurcatus* Carter & Wijesundara, 1948.

spathipalpis Rondani, 1872 (p. 31) – synonym of *Culiseta* (*Allotheobaldia*) *longiareolata* (Macquart, 1838a) [original combination: *Culex spathipalpis*]; synonymy established before Knight & Stone, 1977.

spathulatus Brug, 1939 (p. 101) – synonym of *Armigeres* (*Armigeres*) *denbesteni* Brug, 1925; synonymy by Lee, 1944b.

spathulatus Forattini & Sallum, 1987b (p. 167) – species of *Culex* (*Melanoconion*).

spathulirostris Edwards, 1923b (p. 2) – species of *Topomyia* (*Suaymyia*).

speciosus (Skuse, 1889) (p. 1722) – species of *Toxorhynchites* (*Toxorhynchites*) [original combination: *Megarhina speciosa*].

spegazzinii Brèthes, 1912 (p. 39) – species of *Haemagogus* (*Haemagogus*).

Synonyms: *uriartei* Shannon & del Ponte, 1928; *lindneri* Martini, 1931b.

spencerii (Theobald, 1901c) (p. 99) – species of *Aedes* (*Ochlerotatus*) [original combination: *Culex spencerii*].

spermatheca (Wijesundara, 1951) (p. 175) – species of *Verrallina* (*Neomacleaya*) [original combination: *Aedes* (*Aedes*) *spermathecus*]. **Note**: the original spelling of the name as *spermathecus* appears to be the adjectival transformation of the Latin noun *spermatheca*; hence, the spelling is corrected here to agree in gender with the feminine *Verrallina*.

sphinx Howard, Dyar & Knab, 1913 (fig. 134) – species of *Culex* (*Culex*).

spiculosa Peyton & Rattanarithikul, 1970 (p. 409) – species of *Uranotaenia* (*Pseudoficalbia*).

spiculostylus Chen, 1989 (p. 86) – species of *Culex* (*Culiciomyia*).

spiculosus Bram & Rattanarithikul, 1967 (p. 3) – species of *Culex* (*Lophoceraomyia*).

Synonym: *hui* Lien, 1968d.

spiculothorax Bram, 1967a (p. 155) – synonym of *Culex* (*Culiciomyia*) *sasai* Kano, Nitahara & Awaya, 1954; synonymy by Phanitchakun *et al.*, 2017.

spilotus Marks, 1963b (p. 31) – species of *Aedes* (*Ochlerotatus*).

spinitubus da Cunha Ramos & Brunhes, 2004 (p. 142) – species of *Uranotaenia* (*Pseudoficalbia*).

spinophallus Dong, Zhou, Gong *et al.* [*sic*], 2004 (p. 187) – *nomen nudum*, *Uranotaenia*.

spinophallus Zhou, Zhu & Lu, 1999 (p. 107) – species of *Topomyia* (*Suaymyia*).

spinosa da Cunha Ramos & Brunhes, 2004 (p. 317) – species of *Uranotaenia* (*Pseudoficalbia*).

spinosa (Doucet, 1951a) (p. 77) – species of *Mimomyia* (*Ingramia*) [original combination: *Ravenalites spinosus*].
spinosipes Edwards, 1922d (p. 92) – species of *Aedes* (*Macleaya*).
spinosus Berlin, 1969a (p. 84) – species of *Aedes* (*Howardina*).
spinosus Lutz, 1905 (p. 26) – species of *Culex* (*Culex*).
spiraculata da Cunha Ramos & Brunhes, 2004 (p. 281) – species of *Uranotaenia* (*Pseudoficalbia*).
spissipes (Theobald, 1903a) (p. 242) – species of *Culex* (*Melanoconion*) [original combination: *Melanoconion spissipes*].
Synonyms: *fur* Dyar & Knab, 1907a; *menytes* Dyar, 1918e; *haynei* Komp & Curry, 1932; *alvarezi* Sutil Oramas, Pulido Florenzano & Amarista Meneses, 1987.
spixi Cerqueira, 1961 (p. 44) – species of *Sabethes* (*Sabethes*).
splendenroides Dong, Zhou, Gong *et al.* [*sic*], 2004 (p. 187) – *nomen nudum*, *Toxorhynchites*.
splendens (de Meijere, 1909) (p. 167) – species of *Malaya* [original combination: *Harpagomyia splendens*].
splendens Lee, 1946 (p. 233) – species of *Tripteroides* (*Tripteroides*).
splendens Lutz, 1904 (in Bourroul, 1904) (p. 68) – synonym of *Trichoprosopon digitatum* (Rondani, 1848); synonymy by Lane & Cerqueira, 1942.
splendens Theobald, 1903a (p. 304) – species of *Mimomyia* (*Mimomyia*).
splendens (Wiedemann, 1819) (p. 2) – species of *Toxorhynchites* (*Toxorhynchites*) [original combination: *Culex splendens*].
Synonyms: *regius* Tennent, 1859; *sikkimensis* Giles, 1901a; *gilesii* Theobald, 1901a.
splendens Williston, 1896 (p. 272) – species of *Haemagogus* (*Haemagogus*).
splendida Bonne-Wepster & Bonne, 1919 (p. 111) – species of *Wyeomyia* (*Phoniomyia*).
splendidus Koidzumi, 1920 (p. 23) – species of *Anopheles* (*Cellia*).
Synonym: *indiensis* Theobald, 1903a.
spoliata Edwards, 1923b (p. 8) – species of *Hodgesia*.
squamifemur Antunes, 1937 (p. 69) – species of *Anopheles* (*Lophopodomyia*).
squamiger (Coquillett, 1902c) (p. 85) – species of *Aedes* (*Ochlerotatus*) [original combination: *Culex squamiger*].
Synonym: *deniedmannii* Ludlow, 1904b.
squamipennis (Lynch Arribálzaga, 1878) (p. 151) – species of *Aedeomyia* (*Aedeomyia*) [original combination: *Aedes squamipennis*].
squamosus (Taylor, 1914c) (p. 691) – species of *Culex* (*Oculeomyia*) [original combination: *Culicada squamosa*].
Synonyms: *annulata* Theobald, 1905f; *annulirostris* Taylor, 1914c; *taylori* Edwards, 1921c.
squamosus Theobald, 1901a (p. 167) – species of *Anopheles* (*Cellia*).
Synonyms: *tananariviensis* Ventrillon, 1906c; *arnoldi* Stephens & Christophers, 1908; *pretoriensis* Gough, 1910.
srilankensis Peyton, 1974 (p. 222) – species of *Uranotaenia* (*Pseudoficalbia*).
srilankensis (Reinert, 1977) (p. 366) – species of *Verrallina* (*Harbachius*) [original combination: *Aedes* (*Verrallina*) *srilankensis*].
staminifera Lourenço-de-Oliveira, Motta & de Castro, 1992 (p. 115) – species of *Wyeomyia* (*Triamyia*).
stampari Apfelbeck, 1929 (p. 286) – synonym of *Aedes* (*Rusticoidus*) *refiki* Medschid, 1928; synonymy by Stone, 1963 (citing Peus, *in litt.*).
standfasti Peters, 1959 (p. 148) – species of *Tripteroides* (*Polylepidomyia*).
stanleyi Edwards, 1941 (p. 239) – synonym of *Eretmapodites oedipodeios* Graham, 1909 [original combination: *Eretmapodites oedipodeios stanleyi*]; synonymy by Hamon, 1961.
stanleyi Peters, 1963a (p. 1) – species of *Aedes* (subgenus uncertain).
stantoni Puri, 1949 (p. 483) – synonym of *Anopheles* (*Anopheles*) *fragilis* (Theobald, 1903f) [original combination: *Anopheles aitkeni stantoni*]; synonymy by Reid, 1965.
starckeae Stone & Knight, 1958 (p. 69) – species of *Culex* (*Oculeomyia*) [*nomen novum* for *basicinctus* Edwards, 1922d].
Synonym: *basicinctus* Edwards, 1922d.
steffani Sirivanakarn, 1968 (p. 112) – species of *Culex* (*Lophoceraomyia*).
stegomyina Stackelberg & Monchadskii, 1926 (in Montchadsky, 1926) (p. 151) – synonym of *Aedes* (*Ochlerotatus*) *pulcritarsis asiaticus* Edwards, 1926c [original combination: *Aedes* (*Ochlerotatus*) *pulchritarsis* [*sic*] variety *stegomyina*]; synonymy established before Knight & Stone, 1977.
stellata Belkin, Heinemann & Page, 1970 (p. 215) – species of *Wyeomyia* (*Wyeomyia*).
stellata Grjebine, 1986 (p. 192) – species of *Mimomyia* (*Ingramia*).
stellatus van Someren, 1947a (p. 29) – species of *Culex* (*Eumelanomyia*).
stenei Thompson, 1956 (p. 196) – species of *Aedes* (*Howardina*).
stenoetrus (Theobald, 1907) (p. 395) – species of *Aedes* (*Aedimorphus*) [original combination: *Culex stenoetrus*].
Synonym: *pseudostenoetrus* Theobald, 1910c.
stenolepis Dyar & Knab, 1908 (p. 60) – species of *Culex* (*Culex*).
stenoscutus (Edwards, 1912c) (p. 22) – species of *Aedes* (*Catageiomyia*) [*nomen novum* for *africanus* Theobald, 1909b; original combination: *Ochlerotatus minutus* variety *stenoscutus*]; specific status by White, 1975.
Synonyms: *africanus* Theobald, 1909b; *congolensis* Edwards, 1927a.
stephensi Liston, 1901 (p. 441) – species of *Anopheles* (*Cellia*).
Synonyms: *metaboles* Theobald, 1902a; *intermedia* Rothwell, 1907; *folquei* de Mello, 1918; *mysorensis* Sweet & Rao, 1937.

sternopallidus Roubaud, 1945 (p. 54) – synonym of *Culex* (*Culex*) *pipiens* Linnaeus, 1758 [original combination: *Culex autogenicus sternopallidus*]; synonymy with transfer of *molestus* Forskål, 1775 to synonymy with *pipiens* Linnaeus by Harbach, Harrison & Gad, 1984.

sternopunctatus Roubaud, 1945 (p. 54) – synonym of (*Culex*) *pipiens* Linnaeus, 1758 [original combination: *Culex autogenicus sternopunctatus*]; synonymy with transfer of *molestus* Forskål, 1775 to synonymy with *pipiens* Linnaeus by Harbach, Harrison & Gad, 1984.

stevensoni (Barraud, 1923b) (p. 219) – species of *Aedes* (*Phagomyia*) [original combination: *Finlaya stevensoni*].

sticticus (Meigen, 1838) (p. 1) – species of *Aedes* (*Ochlerotatus*) [original combination: *Culex sticticus*].
Synonyms: *concinnus* Stephens, 1846; *hirsuteron* Theobald, 1901c; *sylvae* Theobald, 1901c; *aestivalis* Dyar, 1904b; *pretans* Grossbeck, 1904b; *aldrichi* Dyar & Knab, 1908; *gonimus* Dyar & Knab, 1918b; *vinnipegensis* Dyar, 1919b; *dorso-vittatus* Villeneuve, 1919; *lesnei* Séguy, 1920; *paradiantaeus* Apfelbeck, 1929.

stigmatephora Dyar, 1922e (p. 116) – synonym of *Psorophora* (*Psorophora*) *pallescens* Edwards, 1922d; synonymy established before Knight & Stone, 1977.

stigmaticus Edwards, 1922d (p. 78) – species of *Aedes* (*Ochlerotatus*).

stigmaticus Skuse, 1889 (p. 1758) – species of *Anopheles* (*Anopheles*).

stigmatosoma Dyar, 1907 (p. 123) – species of *Culex* (*Culex*).
Synonym: *eumimetes* Dyar & Knab, 1908.

stigmopteryx Martini, 1932a (p. 276) – synonym of *Chagasia fajardi* (Lutz, 1904) [original combination: *Chagasia fajardi* variety *stigmopteryx*]; synonymy established before Knight & Stone, 1977.

stimulans (Walker, 1848) (p. 4) – species of *Aedes* (*Ochlerotatus*) [original combination: *Culex stimulans*].
Synonyms: *subcantans* Felt, 1905a; *classicus* Dyar, 1920e; *mississippii* Dyar, 1920e.

stoehri Theobald, 1907 (p. 419) – synonym of *Culex* (*Culex*) *quinquefasciatus* Say, 1823; synonymy established before Knight & Stone, 1977.

stokesi Evans, 1929 (p. 522) – species of *Aedes* (*Albuginosus*).
Synonym: *occidentalis* Evans, 1926.

stonei Belkin, 1950 (p. 228) – species of *Tripteroides* (*Rachisoura*).

stonei Bohart & Ingram, 1946a (p. 47) – synonym of *Uranotaenia* (*Pseudoficalbia*) *jacksoni* Edwards, 1935a; synonymy by Peyton, 1972.

stonei Knight & Laffoon, 1946 (p. 208) – species of *Aedes* (*Finlaya*).

stonei Lane & Whitman, 1943 (p. 405) – species of *Culex* (*Microculex*).

stonei Mattingly, 1957c (p. 51) – synonym of *Heizmannia* (*Heizmannia*) *complex* (Theobald, 1910a); synonymy by Mattingly, 1970.

stonei Vargas, 1941a (p. 179) – synonym of *Anopheles* (*Anopheles*) *punctipennis* (Say, 1823) [original combination: *Anopheles punctipennis* variety *stonei*]; synonymy by Freeborn, 1949.

stonei Vargas, 1956 (p. 15) – species of *Psorophora* (*Psorophora*).

stonei Vargas & Martínez Palacios, 1953 (p. 297) – species of *Wyeomyia* (*Wyeomyia*).

stoneorum Marks, 1977 (p. 33) – species of *Aedes* (*Macleaya*).

stookesi Colless, 1955a (p. 331) – species of *Anopheles* (*Cellia*).

strachanii Theobald, 1907 (p. 85) – synonym of *Anopheles* (*Anopheles*) *obscurus* (Grünberg, 1905) [original combination: *Myzorhynchus strachanii*]; synonymy established before Knight & Stone, 1977.

stramineus Dubitzky, 1970 (p. 408) – species of *Aedes* (*Ochlerotatus*).

strelitziae Muspratt, 1950 (p. 73) – species of *Aedes* (*Stegomyia*).

striatipes Edwards, 1941 (p. 311) – species of *Culex* (*Culex*).

striatus Sant'Ana & Sallum, 2016, 2017 (p. 137) – species of *Anopheles* (*Nyssorhynchus*).

stricklandi Barraud, 1926 (p. 345) – species of *Uranotaenia* (*Pseudoficalbia*).

stricklandi (Edwards, 1912b) (p. 523) – species of *Aedes* (*Ochlerotatus*) [original combination: *Ochlerotatus stricklandi*].
Synonyms: *australis* Strickland, 1911a; *flindersi* Taylor, 1914c.

stricklandi Reid, 1965 (p. 121) – species of *Anopheles* (*Anopheles*).

strigimacula Dyar & Knab, 1906a (p. 136) – synonym of *Anopheles* (*Anopheles*) *punctimacula* Dyar & Knab, 1906a; synonymy established before Knight & Stone, 1977.

striocrurus (Giles, 1904a) (p. 367) – *nomen dubium*, *Armigeres* [original combination: *Stegomyia striocrura*]; status by Edwards, 1932a [? = *magnus* (Theobald, 1908b)].

strodei Root, 1926 (p. 711) – species of *Anopheles* (*Nyssorhynchus*).
Synonyms: *ramosi* Unti, 1940b; *lloydi* Unti, 1941.

stunga (Klein, 1973) (p. 10) – species of *Verrallina* (*Harbachius*) [original combination: *Aedes* (*Neomacleaya*) *stungus*].

stylifurcatus Carter & Wijesundara, 1948 (p. 145) – synonym of *Culex* (*Culiciomyia*) *spathifurca* (Edwards, 1915c); synonymy by Mattingly, 1955 (in Iyengar & Menon, 1955).

suarezi Cova Garcia & Sutil O., 1976b (p. 360) – species of *Mansonia* (*Mansonia*).

subaequalis Edwards, 1941 (p. 276) – species of *Culex* (*Culiciomyia*).

subalbatus (Coquillett, 1898) (p. 302) – species of *Armigeres* (*Armigeres*) [original combination: *Culex subalbatus*].
Synonym: *panalectoris* Giles, 1901a.

subalbirostris Klein & Marks, 1960 (p. 115) – species of *Aedes* (*Ochlerotatus*).

subalbitarsis King & Hoogstraal, 1946d (p. 144) – species of *Aedes* (*Patmarksia*).

subalbopictus Barraud, 1931a (p. 225) – species of *Aedes* (*Stegomyia*).

subalpinus Hackett & Lewis, 1935 (p. 378) – synonym of *Anopheles* (*Anopheles*) *melanoon* Hackett, 1934 [original combination: *Anopheles maculipennis* variety *subalpinus*]; synonymy by Linton *et al.*, 2002.

subargenteus Edwards, 1925 (p. 262) – species of *Aedes* (*Stegomyia*).

subauridorsum Marks, 1948 (p. 28) – species of *Aedes* (subgenus uncertain).

subbasalis Dobrotworsky, 1962 (p. 298) – species of *Aedes* (*Dobrotworskyius*).

subcantans Felt, 1905a (pp. 448, 474) – synonym of *Aedes* (*Ochlerotatus*) *stimulans* (Walker, 1848) [original combination: *Culicada subcantans*]; synonymy established before Knight & Stone, 1977.

subcomplosa (del Ponte, 1939) (p. 541) – species of *Wyeomyia* (subgenus uncertain, determined by Motta & Lourenço-de-Oliveira, 1995) [original combination: *Dendromyia subcomplosa*].

subdentatus Edwards, 1936 (p. 51) – species of *Aedes* (*Aedimorphus*).

subdiversus Martini, 1926 (p. 204) – species of *Aedes* (*Rusticoidus*).

subfuscus Theobald, 1907 (p. 403) – synonym of *Culex* (*Phenacomyia*) *corniger* Theobald, 1903a; synonymy established before Knight & Stone, 1977.

submarginalis Sirivanakarn, 1973a (p. 212) – species of *Culex* (*Lophoceraomyia*).

submediopunctata Barraud, 1923a (p. 781) – synonym of *Aedes* (*Stegomyia*) *mediopunctatus* (Theobald, 1905h) [original combination: *Stegomyia mediopunctata* variety *submediopunctata*]; synonymy by Huang, 1977, as variety *submediopunctatus* (Barraud). **Note:** the name is incorrectly listed as *submediopunctatus* Barraud in the online catalogue of the Walter Reed Biosystematics Unit (http://www.mosquitocatalog.org/).

subniveus Edwards, 1922a (p. 269) – species of *Aedes* (*Downsiomyia*).

subnormalis Martini, 1920b (pl. II, fig. 6b) – species of *Uranotaenia* (*Uranotaenia*).

Synonym: *roperi* Edwards, 1922a.

subnudipennis (Edwards, 1927b) (p. 355) – species of *Tripteroides* (*Rachisoura*) [original combination: *Rachionotomyia subnudipennis*].

subobscura Lee, 1946 (p. 273) – synonym of *Tripteroides* (*Rachisoura*) *obscurus* Brug, 1934 [original combination: *Tripteroides* (*Mimeteomyia*) *subobscura*]; synonymy by E.N. Marks (in Lee *et al.*, 1989c).

subochrea (Edwards, 1921) (in Wesenberg-Lund, 1921) (p. 198) – species of *Culiseta* (*Culiseta*) [original combination: *Theobaldia annulata* variety *subochrea*]; specific status by Ribeiro *et al.*, 1977.

suborientalis Baisas, 1938 (p. 217) – *nomen dubium*, *Culex*; status by Delfinado, 1966a.

subpictus Grassi, 1899 (in Grassi *et al.*, 1899) (p. 101) – species of *Anopheles* (*Cellia*).

Synonyms: *rossii* Giles, 1899; *error* Theobald, 1903a.

subrima Edwards, 1941 (p. 262) – species of *Culex* (*Eumelanomyia*).

subsalisburiensis Hervé & Geoffroy, 1974 (p. 191) – species of *Culex* (*Maillotia*).

subsimilis (Barraud, 1927c) (p. 552) – species of *Udaya* [original combination: *Finlaya subsimilis*].

subsimplicipes Edwards, 1914b (p. 76) – species of *Eretmapodites* [original combination: *Eretmapodites chrysogaster* variety *subsimplicipes*]; specific status by Edwards, 1941.

subsplendens Martini, 1931a (p. 200) – synonym of *Trichoprosopon digitatum* (Rondani, 1848) [original combination: *Joblotia splendens* variety *subsplendens*]; synonymy by Lane & Cerqueira, 1942.

subtibioclada King & Hoogstraal, 1947a (p. 595) – species of *Uranotaenia* (*Uranotaenia*).

subtilis Pressat, 1905 (p. 48) – synonym of *Anopheles* (*Anopheles*) *tenebrosus* Dönitz, 1902; synonymy established before Knight & Stone, 1977.

subtilis Sergent & Sergent, 1905 (p. 673) – synonym of *Aedes* (*Ochlerotatus*) *caspius* (Pallas, 1771) [original combination: *Grabhamia subtilis*]; synonymy established before Knight & Stone, 1977.

subtrichurus Martini, 1927 (species) (p. 386) – subspecies of *Aedes* (*Rusticoidus*) *rusticus* (Rossi, 1790); subspecific status by Harbach & Howard, 2007.

subulifer Doleschall, 1857 (p. 382) – synonym of *Toxorhynchites* (*Toxorhynchites*) *amboinensis* (Doleschall, 1857) [original combination: *Culex subulifer*]; synonymy by Steffan, 1977.

subumbrosa Theobald, 1903a (p. 34) – synonym of *Anopheles* (*Cellia*) *funestus* Giles, 1900b [original combination: *Anopheles funestus* variety *subumbrosa* [*sic*]]; synonymy established before Knight & Stone, 1977.

subvexans Martini, 1922 (p. 159) – synonym of *Aedes* (*Ochlerotatus*) *cantans* (Meigen, 1818) [original combination: *Aedes cantans* variety *subvexans*]; synonymy by Natvig, 1948 (also see Harbach & Howard, 2007).

suchariti (Miyagi & Toma, 1989) (p. 16) – species of *Kimia* [original combination: *Topomyia* (*Suaymyia*) *suchariti*].

sudanensis Theobald, 1907 (p. 195) – synonym of *Mimomyia* (*Mimomyia*) *mimomyiaformis* (Newstead, 1907) [original combination: *Ludlowia sudanensis*]; synonymy established before Knight & Stone, 1977.

sudanensis (Theobald, 1908c) (p. 252) – species of *Aedes* (*Mucidus*) [original combination: *Mucidus sudanensis*]. **Note:** this nominal taxon is the senior secondary homonym of *sudanensis* Theobald, 1911c and *sudanensis* Theobald, 1913d.

sudanensis Theobald, 1911c (p. 154) – synonym of *Aedes* (*Aedimorphus*) *vexans* (Meigen, 1830b) [original combination: *Culex sudanensis*]; synonymy by Edwards, 1941 (via synonymy with *arabiensis* Patton, 1905). **Note:** this nominal taxon is the junior secondary homonym of *sudanensis* (Theobald, 1908c) and the senior secondary homonym of *sudanensis* Theobald, 1913d.

sudanensis Theobald, 1913d (p. 595) – synonym of *Aedes* (*Catageiomyia*) *tarsalis* (Newstead, 1907) [original combination: *Reedomyia sudanensis*]; synonymy established before Knight & Stone, 1977. **Note**: this nominal taxon is the junior secondary homonym of *sudanensis* (Theobald, 1908c) and *sudanensis* Theobald, 1911c; this nominal taxon is not listed in the online catalogue of the Walter Reed Biosystematics Unit (http://www.mosquitocatalog.org/).

sudanicus Edwards, 1941 (p. 345) – subspecies of *Culex* (*Culex*) *guiarti* Blanchard, 1905 [original combination: *Culex* (*Culex*) *guiarti* variety *sudanicus*]; subspecific status by Harbach & Howard, 2007.

suffusus Edwards, 1922a (p. 270) – species of *Aedes* (subgenus uncertain).

sugens Wiedemann, 1828 (p. 545) – synonym of *Aedes* (*Stegomyia*) *aegypti* (Linnaeus, 1762) [original combination: *Culex sugens*]; synonymy established before Knight & Stone, 1977.

suknaensis Theobald, 1910a (p. 21) – synonym of *Aedes* (*Edwardsaedes*) *imprimens* (Walker, 1859) [original combination: *Culicada suknaensis*]; synonymy established before Knight & Stone, 1977.

sulawesi Waktoedi Koesoemawinangoen, 1954 (p. 63) – species of *Anopheles* (*Cellia*) [original combination: *Anopheles leucosphyrus* variety *sulawesi*]; specific status by Reid, 1968.

sullivanae Baisas & Ubaldo-Pagayon, 1953 (p. 175) – species of *Tripteroides* (*Tripteroides*).

sumatranus Brug, 1931a (species) (p. 248) – subspecies of *Culex* (*Lophoceraomyia*) *curtipalpis* (Edwards, 1914c); subspecific status by Meng & Chen, 1980.

Synonym: *gudouensis* Chang, Zhao, Hang & Chen, 1975.

sumatranus (Brug, 1939) (p. 93) – species of *Toxorhynchites* (*Toxorhynchites*) [original combination: *Megarhinus sumatranus*].

sumatranus Swellengrebel & Rodenwaldt, 1932 (p. 78) – species of *Anopheles* (*Anopheles*) [original combination: *Anopheles gigas* variety *sumatrana* [*sic*]]; specific status by Harbach *et al.*, 2017. The original spelling of *sumatrana* (f.) is corrected to *sumatranus* to agree in gender with masculine *Anopheles*.

sumethi Peyton & Rattanarithikul, 1970 (p. 411) – species of *Uranotaenia* (*Pseudoficalbia*).

sumidero Schick, 1970a (p. 72) – species of *Aedes* (*Protomacleaya*).

summorosus Dyar, 1920i (p. 180) – species of *Culex* (*Culex*); removed from synonymy with *tritaeniorhynchus* Giles, 1901 by validation as species by Airi & Kaur, 2015.

Synonym: *siamensis* Barraud & Christophers, 1931.

sundaicus (Rodenwaldt, 1925) (p. 185) – species of *Anopheles* (*Cellia*) [original combination: *Myzomyia ludlowi* variety *sundaica*]; specific status by Christophers, 1933.

sunthorni Thurman, 1959 (p. 19) – species of *Toxorhynchites* (*Toxorhynchites*).

sunyaniensis Edwards, 1935a (p. 134) – synonym of *Mimomyia* (*Mimomyia*) *hispida* (Theobald, 1910c) [original combination: *Ficalbia* (*Mimomyia*) *hispida* variety *sunyaniensis*]; synonymy by Stone *et al.*, 1959.

sunyaniensis Edwards, 1941 (p. 263) – species of *Culex* (*Eumelanomyia*).

superbus (Dyar & Knab, 1906c) (species) (p. 255) – subspecies of *Toxorhynchites* (*Lynchiella*) *haemorrhoidalis* (Fabricius, 1787) [original combination: *Megarhinus superbus*]; subspecific status by Lane, 1951.

superpictus Grassi, 1899 (p. 560) – species of *Anopheles* (*Cellia*).

Synonyms: *palestinensis* Theobald, 1903a; *nursei* Theobald, 1907; *cardamatisi* Newstead & Carter, 1910; *vassilievii* Portschinsky, 1911; *macedoniensis* Cot & Hovasse, 1917; *berestnevi* Shingarev, 1926; *atheniensis* Cardamatis, 1931; *hellenicus* Peus, 1954.

surcoufi (Theobald, 1912) (in Surcouf, 1912) (p. 59) – species of *Aedes* (*Ochlerotatus*) [original combination: *Culicada surcoufi*].

sureilensis Barraud, 1934 (p. 231) – subspecies of *Aedes* (*Stegomyia*) *mediopunctatus* Theobald, 1905h [original combination: *Aedes* (*Stegomyia*) *mediopunctatus* variety *sureilensis*]; subspecific status by Harbach & Howard, 2007.

surinamensis Bruijning, 1959 (p. 135) – species of *Wyeomyia* (subgenus uncertain, determined by Motta & Lourenço-de-Oliveira, 1995).

surinamensis Dyar, 1918e (p. 121) – species of *Culex* (*Culex*).

sursumptor Dyar, 1924d (p. 123) – species of *Culex* (*Melanoconion*).

Synonym: *ligator* Dyar, 1924d.

surukumensis Anduze, 1941 (p. 833) – synonym of *Culex* (*Anoedioporpa*) *conservator* Dyar & Knab, 1906d; synonymy by Rozeboom & Komp, 1950a.

sutili Cova Garcia & Pulido F., 1974 (p. 27) – species of *Culex* (*Microculex*).

suvae Stone & Bohart, 1944 (p. 214) – synonym of *Aedes* (*Levua*) *geoskusea* Amos, 1944; synonymy by Stone & Bohart, 1944 (in a footnote).

svastii Thurman, 1959 (p. 43) – species of *Topomyia* (*Topomyia*).

swahilicus Gillies, 1964 (p. 67) – species of *Anopheles* (*Cellia*).

sylvae Theobald, 1901c (p. 96) – synonym of *Aedes* (*Ochlerotatus*) *sticticus* (Meigen, 1838) [original combination: *Culex nigripes* variety *sylvae*]; synonymy established before Knight & Stone, 1977.

sylvanensis (Dobrotworsky, 1960b) (p. 245) – species of *Culiseta* (*Culicella*) [original combination: *Theobaldia sylvanensis*].

sylvatica Lu, Dong & Wang, 1986 (p. 406) – species of *Topomyia* (*Topomyia*).

sylvaticum Le Goff, Boussès & Brunhes, 2007 (p. 195) – synonym of *Aedes* (*Neomelaniconion*) *hylaius* Harbach (herein) [original combination: *Neomelaniconion sylvaticum*]. **Note**: this nominal taxon is the junior secondary homonym of *sylvaticus* Meigen, 1818 and *sylvaticus* Brunhes, 1983.

sylvaticus Brunhes, 1983 (p. 316) – synonym of *Aedes* (*Paulianius*) *madagascarensis* van Someren, 1949a; synonymy by Brunhes *et al*., 2017. **Note**: this nominal taxon is the junior secondary homonym of *sylvaticus* Meigen, 1818 and the senior secondary homonym of *sylvaticus* Le Goff, Boussès & Brunhes, 2007.

sylvaticus Meigen, 1818 (p. 6) – synonym of *Aedes* (*Ochlerotatus*) *communis* (de Geer, 1776) [original combination: *Culex sylvaticus*]; synonymy established before Knight & Stone, 1977. **Note**: this nominal taxon is the senior secondary homonym of *sylvaticus* Brunhes, 1983 and *sylvaticus* Le Goff, Boussès & Brunhes, 2007.

sylvestris Leicester, 1908 (p. 125) – synonym of *Culex* (*Lophoceraomyia*) *quadripalpis* (Edwards, 1914b) [original combination: *Lophoceratomyia sylvestris*]; synonymy established before Knight & Stone, 1977.

sylvestris Theobald, 1901a (p. 406) – synonym of *Aedes* (*Aedimorphus*) *vexans* (Meigen, 1830b) [original combination: *Culex sylvestris*]; synonymy established before Knight & Stone, 1977. **Note**: this nominal taxon is the senior secondary homonym of *sylvestris* Theobald, 1910c (p. 196).

sylvestris Theobald, 1910c (p. 196) – synonym of *Aedes* (*Macleaya*) *spinosipes* Edwards, 1922d [original combination: *Chaetocruiomyia sylvestris*]; synonymy established before Knight & Stone, 1977. **Note**: this nominal taxon is the junior secondary homonym of *sylvestris* Theobald, 1901a.

sylvestris (Theobald, 1910c) (p. 208) – species of *Tripteroides* (*Rachisoura*) [original combination: *Rachisoura sylvestris*].
Synonym: *hilli* Taylor, 1914b.

sylvicola Grossbeck, 1906 (p. 129) – synonym of *Aedes* (*Ochlerotatus*) *grossbecki* Dyar & Knab, 1906d [original combination: *Culex sylvicola*]; synonymy established before Knight & Stone, 1977.

symbletos Sallum & Hutchings, 2003 (p. 620) – species of *Culex* (*Melanoconion*).

symesi Edwards, 1928a (p. 269) – species of *Anopheles* (*Anopheles*).

symmachus Dyar & Knab, 1909d (p. 262) – synonym of *Wyeomyia* (*Miamyia*) *hosautos* Dyar & Knab, 1907c; synonymy established before Knight & Stone, 1977.

synchytus Arnell, 1976 (p. 71) – species of *Aedes* (*Ochlerotatus*).

syntheta Dyar & Shannon, 1924c (species) (p. 189) – subspecies of *Uranotaenia* (*Pseudoficalbia*) *anhydor* Dyar, 1907; subspecific status by Belkin & McDonald, 1956.

syntheticus Barraud, 1928b (p. 375) – species of *Aedes* (*Aedimorphus*) [*nomen novum* for *fisheri* Barraud, 1928a].

szechwanensis Hsu, 1964 (p. 278) – synonym of *Tripteroides* (*Rachionotomyia*) *aranoides* (Theobald, 1901c); synonymy by Mattingly, 1980.

szemaonensis Wang & Feng, 1964 (p. 37) – species of *Culex* (*Lophoceraomyia*).

Tt

tabu Ramalingam & Belkin, 1965 (p. 1) – species of *Aedes* (*Stegomyia*).

tachirensis Anduze, 1947 (p. 355) – synonym of *Aedes* (*Howardina*) *fulvithorax* (Lutz, 1904); synonymy by Lane, 1951.

taeniarostris (Theobald, 1909b) (p. 13) – species of *Aedes* (*Neomelaniconion*) [original combination: *Banksinella taeniarostris*].
Synonym: *chrysothorax* Theobald, 1909b.

taeniarostris Theobald, 1907 (p. 299) – synonym of *Culex* (*Oculeomyia*) *infula* Theobald, 1901a [original combination: *Grabhamia taeniarostris*]; synonymy by Sirivanakarn, 1976.

taeniarostris (Theobald, 1911a) (p. 34) – species of *Malaya* [original combination: *Harpagomyia taeniarostris*].

taeniata Leicester, 1908 (p. 127) – synonym of *Culex* (*Lophoceraomyia*) *cinctellus* Edwards, 1922a [original combination: *Lophoceratomyia taeniata*]; synonymy established before Knight & Stone, 1977.

taeniata (Leicester, 1908) (p. 190) – species of *Verrallina* (*Neomacleaya*) [original combination: *Aioretomyia taeniata*].
Synonym: *incertus* Edwards, 1922a.

taeniatus Leicester, 1908 (p. 133) – synonym of *Aedes* (*Aedimorphus*) *orbitae* Edwards, 1922a [original combination: *Lepidotomyia taeniata*]; synonymy established before Knight & Stone, 1977. **Note**: this nominal taxon is the junior secondary homonym of *taeniatus* Wiedemann, 1828.

taeniatus Wiedemann, 1828 (p. 10) – synonym of *Aedes* (*Stegomyia*) *aegypti* (Linnaeus, 1762) [original combination: *Culex taeniatus*]; synonymy established before Knight & Stone, 1977. **Note**: this nominal taxon is the senior secondary homonym of *taeniatus* Leicester, 1908.

taeniopus Dyar & Knab, 1907b (p. 100) – species of *Culex* (*Melanoconion*).
Synonyms: *annulipes* Theobald, 1907; *opisthopus* Komp, 1926; *mychonde* Komp, 1928.

taeniorhynchoides (Christophers, 1911) (p. 68) – species of *Aedes* (*Aedimorphus*) [original combination: *Leslieomyia taeniorhynchoides*].
Synonym: *maculata* Theobald, 1905g.

taeniorhynchoides Giles, 1904a (p. 369) – synonym of *Culex* (*Oculeomyia*) *poicilipes* (Theobald, 1903a); synonymy established before Knight & Stone, 1977.

taeniorhynchus (Wiedemann, 1821) (p. 43) – species of *Aedes* (*Ochlerotatus*) [original combination: *Culex taeniorhynchus*].
Synonyms: *damnosus* Say, 1823; *niger* Giles, 1904b; *portoricensis* Ludlow, 1905c; *epinolus* Dyar & Knab, 1914; *pix* Martini, 1935.

tahoensis Dyar, 1916b (p. 82) – species of *Aedes* (*Ochlerotatus*).

taiwanensis Koidzumi, 1917 (p. 135) – synonym of *Anopheles* (*Cellia*) *tessellatus* Theobald, 1901a; synonymy established before Knight & Stone, 1977.

taiwanensis Lien, 1968c (p. 141) – species of *Heizmannia* (*Heizmannia*).

taiwanus Lien, 1968d (p. 223) – species of *Aedes* (*Bruceharrisonius*) [original combination: *Aedes* (*Finlaya*) *aureostriatus taiwanus*]; specific status by Reinert, 2003.

takasagoensis Morishita, 1946 (p. 65) – species of *Anopheles* (*Cellia*) [original combination: *Anopheles leucosphyrus* variety *takasagoensis*]; specific status by Peyton & Harrison, 1980.

tamsi Edwards, 1934 (p. 327) – species of *Culex* (*Culex*).

tanajcus Stschelkanavzev, 1926 (p. 133) – synonym of *Culex* (*Barraudius*) *modestus* Ficalbi, 1890b; synonymy by Montschadsky, 1936.

tanakai Miyagi & Toma, 2013 (p. 167) – species of *Uranotaenia* (*Pseudoficalbia*).

tananariviensis Ventrillon, 1906c (p. 198) – synonym of *Anopheles* (*Cellia*) *squamosus* Theobald, 1901a [original combination: *Cellia tananariviensis*]; synonymy established before Knight & Stone, 1977.

tangensis Kuhlow, 1962 (p. 448) – synonym of *Anopheles* (*Cellia*) *merus* Dönitz, 1902; synonymy established before Knight & Stone, 1977.

tanzaniae Capasso, 1991 (p. 173) – *nomen nudum*, fossil *Culex*.

tanzaniae da Cunha Ramos, 1993 (p. 287) – species of *Uranotaenia* (*Pseudoficalbia*).

tapena Dyar, 1919c (p. 150) – synonym of *Culex* (*Aedinus*) *amazonensis* (Lutz, 1905); synonymy established before Knight & Stone, 1977.

tapinops Brèthes, 1917 (p. 227) – synonym of *Aedes* (*Ochlerotatus*) *crinifer* (Theobald, 1903a) [original combination: *Culex tapinops*]; synonymy established before Knight & Stone, 1977.

tarnogradskii Martini, 1930 (p. 259) – synonym of *Aedes* (*Aedes*) *rossicus* Dolbeškin, Gorickaya & Mitrofanova, 1930; synonymy established before Knight & Stone, 1977.

tarsalis Coquillett, 1896 (p. 43) – species of *Culex* (*Culex*).
Synonyms: *willistoni* Giles, 1900a; *kelloggii* Theobald, 1903i.

tarsalis Delfinado & Hodges, 1968 (p. 372) – species of *Tripteroides* (*Tripteroides*).

tarsalis (Newstead, 1907) (in Newstead *et al.*, 1907) (p. 18) – species of *Aedes* (*Catageiomyia*) [original combination: *Duttonia tarsalis*].
Synonyms: *africana* Newstead, 1907; *biannulata* Theobald, 1907; *bipunctata* Theobald, 1909b; *uniannulata* Theobald, 1909b; *neobiannulata* Theobald, 1910c; *sudanensis* Theobald, 1913d.

tarsata Lane & Cerqueira, 1942 (p. 612) – species of *Wyeomyia* (*Exallomyia*).

tarsimaculata Goeldi, 1905 (p. 133) – synonym of *Anopheles* (*Nyssorhynchus*) *albimanus* Wiedemann, 1820 [*nomen novum* for *albipes* Theobald, 1901a, unjustified; intended as *Cellia* (*Anopheles*) *argyrotarsis* variety *tarsimaculata*]; synonymy by Rozeboom & Gabaldon, 1941.

tarsopus Dyar & Knab, 1908 (p. 62) – species of *Sabethes* (*Sabethes*).

tasmaniensis Dobrotworsky, 1966 (p. 130) – species of *Anopheles* (*Anopheles*).

tasmaniensis Strickland, 1911b (p. 181) – synonym of *Aedes* (*Halaedes*) *australis* (Erichson, 1842) [original combination: *Culicada tasmaniensis*]; synonymy by Mattingly & Marks, 1955. **Note**: this nominal taxon is the senior secondary homonym of *tasmaniensis* Taylor, 1914b.

tasmaniensis (Strickland, 1911b) (p. 249) – species of *Tripteroides* (*Polylepidomyia*) [original combination: *Stegomyia tasmaniensis*].
Synonym: *cephasi* Edwards, 1923b.

tasmaniensis Taylor, 1914b (p. 466) – synonym of *Aedes* (*Ochlerotatus*) *andersoni* Edwards, 1926a [original combination: *Menolepis* (?) *tasmaniensis*]; synonymy established before Knight & Stone, 1977. **Note**: this nominal taxon is the junior secondary homonym of *tasmaniensis* Strickland, 1911b.

tatoi Casal & García, 1971 (p. 631) – species of *Culex* (*Culex*).

tauffliebi Geoffroy & Hervé, 1976 (p. 151) – species of *Culex* (*Eumelanomyia*).

tauffliebi Rickenbach & Ferrara, 1965 (p. 24) – species of *Aedes* (*Aedimorphus*).

taurepana Anduze, 1941 (p. 825) – species of *Wyeomyia* (subgenus uncertain, determined by Motta & Lourenço-de-Oliveira, 1995).

taylori Edwards, 1921c (p. 78) – synonym of *Culex* (*Oculeomyia*) *squamosus* (Taylor, 1914c) [*nomen novum* for *annulirostris* Taylor, 1914c]; synonymy established before Knight & Stone, 1977.

taylori Edwards, 1936 (p. 55) – species of *Aedes* (*Diceromyia*).

taytayensis Banks, 1909 (p. 545) – synonym of *Culex* (*Culex*) *fuscocephala* Theobald, 1907; synonymy established before Knight & Stone, 1977.

tchekedii de Meillon & Leeson, 1940 (p. 64) – species of *Anopheles* (*Cellia*).

tecmarsis Dyar, 1918e (p. 124) – species of *Culex* (*Melanoconion*).

teesdalei van Someren, 1954a (p. 104) – species of *Aedes* (*Albuginosus*).

tehuantepec Schick, 1970a (p. 87) – species of *Aedes* (*Protomacleaya*).

tejerai Cova García, 1962 (p. 312) – synonym of *Culex* (*Culex*) *levicastilloi* Lane, 1945b; synonymy established before Knight & Stone, 1977.

telamali Saliternik & Theodor, 1942 (p. 433) – subspecies of *Anopheles* (*Cellia*) *turkhudi* Liston, 1901 [original combination: *Anopheles turkhudi* variety *telamali*]; subspecific status by Harbach & Howard, 2007.

telesilla de Meillon & Lavoipierre, 1945 (in de Meillon *et al.*, 1945) (p. 91) – species of *Culex* (*Culex*).

telestica Dyar & Knab, 1906d (p. 230) – synonym of *Wyeomyia* (*Wyeomyia*) *medioalbipes* Lutz, 1904; synonymy by Bruijning, 1959.
telmatophila Galindo, Blanton & Peyton, 1954 (p. 168) – species of *Uranotaenia* (*Uranotaenia*).
tenagius van Someren, 1954b (p. 122) – species of *Culex* (*Culex*).
tenax (de Meijere, 1910) (p. 928) – species of *Tripteroides* (*Rachionotomyia*) [original combination: *Ficalbia tenax*].
tenax Leicester, 1908 (p. 167) – synonym of *Culex* (*Culeomyia*) *sinensis* Theobald, 1903a [original combination: *Taeniorhynchus tenax*]; synonymy established before Knight & Stone, 1977. **Note**: this nominal taxon is the junior primary homonym of *tenax* Theobald, 1901c.
tenax Theobald, 1901c (p. 198) – synonym of *Culex* (*Culeomyia*) *infula* Theobald, 1901a [original combination: *Taeniorhynchus tenax*]; synonymy by Sirivanakarn, 1976. **Note**: this nominal taxon is the senior primary homonym of *tenax* Leicester, 1908.
tendeiroi da Cunha Ramos, Ribeiro & de Barros Machado, 1992 (p. 133) – species of *Eretmapodites*.
tenebrosus Dönitz, 1902 (p. 53) – species of *Anopheles* (*Anopheles*).
Synonym: *subtilis* Pressat, 1905.
tengchongensis Dong, Wang & Zhou, 2002 (p. 131) – species of *Heizmannia* (*Heizmannia*).
tenuipalpis Barraud, 1924f (p. 1278) – species of *Culex* (*Eumelanomyia*).
tenuipalpis (Edwards, 1924b) (p. 366) – species of *Coquillettidia* (*Austromansonia*) [original combination: *Taeniorhynchus* (*Coquillettidia*) *tenuipalpis*].
tenuis Edwards, 1922a (p. 290) – species of *Topomyia* (*Topomyia*).
terebor Dyar, 1920d (p. 56) – species of *Culex* (*Melanoconion*).
terepaima Anduze, 1949a (p. 65) – synonym of *Culex* (*Melanoconion*) *idottus* Dyar, 1920d; synonymy by Lane, 1953.
termi Thurman, 1955 (p. 18) – species of *Culex* (*Culiciomyia*).
terminalis Coquillett, 1906d (pp. 8, 17) – synonym of *Psorophora* (*Janthinosoma*) *ferox* (von Humboldt, 1819) [original combination: *Janthinosoma terminalis*]; synonymy established before Knight & Stone, 1977.
terrens (Walker, 1856a) (p. 429) – species of *Aedes* (*Protomacleaya*) [original combination: *Culex terrens*].
Synonyms: *oswaldi* Lutz, 1904; *apollo* Schick, 1970a.
terriei Theobald, 1903a (p. 193) – synonym of *Aedes* (*Ochlerotatus*) *detritus* (Haliday, 1833) [original combination: *Culex terriei*]; synonymy established before Knight & Stone, 1977.
territans Walker, 1856a (p. 428) – species of *Culex* (*Neoculex*).
Synonyms: *saxatilis* Grossbeck, 1905; *frickii* Ludlow, 1906b; *pyrenaicus* Brolemann, 1919a; *nematoides* Dyar & Shannon, 1925a.
terzii Edwards, 1941 (p. 312) – species of *Culex* (*Culex*).
tessellatus Theobald, 1901a (p. 175) – species of *Anopheles* (*Cellia*).
Synonyms: *formosae* Hatori, 1901; *deceptor* Dönitz, 1902; *thorntonii* Ludlow, 1904a; *ceylonica* Newstead & Carter, 1910; *kinoshitai* Koidzumi, 1917; *taiwanensis* Koidzumi, 1917.
tessmanni Enderlein, 1931 (p. 124) – synonym of *Toxorhynchites* (*Toxorhynchites*) *brevipalpis conradti* Grünberg, 1907a; synonymy by Edwards, 1941.
testacea Theobald, 1905f (p. 113) – species of *Uranotaenia* (*Uranotaenia*).
Synonym: *unilineata* Leicester, 1908.
testaceus van der Wulp, 1867 (p. 128) – synonym of *Coquillettidia* (*Coquillettidia*) *perturbans* (Walker, 1856a) [original combination: *Culex testaceus*]; synonymy established before Knight & Stone, 1977.
testei Senevet & Abonnenc, 1939b (p. 269) – species of *Wyeomyia* (*Dendromyia*).
tetraspathus Dyar & Knab, 1909d (p. 260) – synonym of *Deinocerites cancer* Theobald, 1901c; synonymy established before Knight & Stone, 1977.
texanum Dyar & Knab, 1906a (p. 135) – ? synonym of *Psorophora* (*Grabhamia*) *columbiae* (Dyar & Knab, 1906a); synonymy uncertain based on type locality in relation to the 'zone of overlap with hybridization' of *columbiae* and *tolteca* (Dyar & Knab, 1906a).
thailandensis Reinert, 1976a (p. 319) – species of *Aedes* (*Paraedes*).
thalassius Theobald, 1903a (p. 168) – species of *Culex* (*Culex*).
Synonyms: *ronaldi* de Charmoy, 1908; *accraensis* Theobald, 1909b; *fuscus* Theobald, 1909b; *neotaeniorhynchus* Theobald, 1910c.
thaxteri Dyar & Knab, 1919 (p. 8) – synonym of *Aedes* (*Howardina*) *busckii* (Coquillett, 1906e); synonymy established before Knight & Stone, 1977.
theileri Edwards, 1912d (p. 247) – species of *Anopheles* (*Cellia*) [*nomen novum* for *albipes* Theobald, 1911b].
Synonym: *albipes* Theobald, 1911b.
theileri Theobald, 1903a (p. 187) – species of *Culex* (*Culex*).
Synonyms: *creticus* Theobald, 1903a; *pettigrewii* Theobald, 1910a; *onderstepoortensis* Theobald, 1911b; *annulata* Theobald, 1913a; *alpha* Séguy, 1924.
thelcter Dyar, 1918e (p. 129) – species of *Aedes* (*Ochlerotatus*).
Synonym: *keyensis* Buren, 1947.
thelmae Mattingly, 1970 (p. 48) – species of *Heizmannia* (*Heizmannia*).
theobaldi Barraud, 1934 (p. 319) – species of *Armigeres* (*Armigeres*) [*nomen novum* for *apicalis* Theobald, 1910a].
Synonym: *apicalis* Theobald, 1910a.
theobaldi de Meijere, 1911 (p. 142) – synonym of *Culiseta* (*Culicella*) *fumipennis* (Stephens, 1825) [original combination: *Culicada theobaldi*]; synonymy established before Knight & Stone, 1977.
theobaldi (Dyar & Knab, 1906c) (p. 246) – species of *Toxorhynchites* (*Lynchiella*) [original combination: *Megarhinus theobaldi*].

Synonyms: *ferox* Wiedemann, 1828; *ambiguus* Dyar & Knab, 1906c; *wiedemanni* Dyar & Knab, 1906c; *fluminensis* Neiva, 1908; *posticatus* Lutz & Neiva, 1913; *moengoensis* Bonne-Wepster & Bonne, 1923.

theobaldi Giles, 1901b (p. 198) – species of *Anopheles* (*Cellia*).

theobaldi (Lane & Cerqueira, 1942) (p. 523) – species of *Runchomyia* (*Runchomyia*) [original combination: *Trichoprosopon* (*Hyloconops*) *theobaldi*].

theobaldi (Lane & Cerqueira, 1942) (p. 638) – species of *Wyeomyia* (*Phoniomyia*) [original combination: *Phoniomyia theobaldi*].

theobaldi (Lutz, 1904) (in Bourroul, 1904) (p. 70) – species of *Culex* (*Melanoconion*) [original combination: *Melanoconion theobaldi*].

Synonyms: *chrysonotum* Dyar & Knab, 1908; *aurilatus* Senevet & Abonnenc, 1939a.

theobaldi (Taylor, 1914a) (p. 751) – species of *Aedes* (*Ochlerotatus*) [original combination: *Grabhamia theobaldi*].

theresae Harbach, 2018 (p. 275) – species of *Sabethes* (*Sabethinus*).

thibaulti Dyar & Knab, 1910 (p. 174) – species of *Aedes* (*Ochlerotatus*).

thomasi Evans, 1924 (p. 372) – synonym of *Culex* (*Melanoconion*) *bastagarius* Dyar & Knab, 1906b; synonymy by Sirivanakarn, 1983.

thomasi Shannon, 1933a (p. 58) – species of *Anopheles* (*Stethomyia*) [*nomen novum* for *lewisi* Shannon, 1931].

Synonym: *lewisi* Shannon, 1931.

thomsoni (Theobald, 1905e) (p. 18) – species of *Aedes* (*Christophersiomyia*) [original combination: *Stegomyia thomsoni*].

thoracicus Robineau-Desvoidy, 1827 (p. 409) – synonym of *Culex* (*Culex*) *pipiens* Linnaeus, 1758; synonymy established before Knight & Stone, 1977.

thorntoni Dyar & Knab, 1907a (p. 10) – species of *Aedes* (*Protomacleaya*).

thorntonii Ludlow, 1904a (p. 69) – synonym of *Anopheles* (*Cellia*) *tessellatus* Theobald, 1901a [original combination: *Myzomyia thorntonii*]; synonymy established before Knight & Stone, 1977.

thriambus Dyar, 1921a (p. 33) – species of *Culex* (*Culex*).

Synonym: *affinis* Adams, 1903.

thurmanae Mattingly, 1958a (p. 55) – species of *Aedes* (*Cancraedes*).

thurmanorum Bram, 1967a (p. 159) – species of *Culex* (*Culiciomyia*).

tianpingensis Chen, 1981 (p. 426) – species of *Culex* (*Culex*).

tibani Patton, 1905 (p. 629) – synonym of *Anopheles* (*Cellia*) *pretoriensis* (Theobald, 1903a); synonymy established before Knight & Stone, 1977.

tibialis Robineau-Desvoidy, 1827 (p. 404) – synonym of *Psorophora* (*Psorophora*) *ciliata* (Fabricius, 1794) [original combination: *Culex tibialis*]; synonymy established before Knight & Stone, 1977.

tibialis Taylor, 1919 (p. 839) – species of *Uranotaenia* (*Uranotaenia*).

tibiamaculatus (Neiva, 1906) (p. 288) – species of *Anopheles* (*Anopheles*) [original combination: *Myzomyia tibiamaculata*].

tibioclada King & Hoogstraal, 1947a (p. 595) – species of *Uranotaenia* (*Uranotaenia*).

tigertti Scanlon & Peyton, 1967 (p. 19) – species of *Anopheles* (*Anopheles*).

tigripes (de Grandpré & de Charmoy, 1901) (p. 6) – species of *Lutzia* (*Metalutzia*) [original combination: *Culex tigripes*].

Synonyms: *maculicrures* Theobald, 1901d; *mombasaensis* Theobald, 1901c; *sierraleonis* Theobald, 1901c; *fusca* Theobald, 1909b; *bimaculata* Theobald, 1910c.

timendus Walker, 1874 (p. 25) – *nomen nudum*, *Culex*.

timorensis (Miyagi, Toma & Lien, 2004) (p. 107) – species of *Aedes* (*Geoskusea*) [original combination: *Ochlerotatus* (*Geoskusea*) *timorensis*].

tiptoni Grjebine, 1953a (p. 469) – species of *Aedes* (*Paulianius*).

tipuliformis Leicester, 1908 (p. 247) – species of *Topomyia* (*Topomyia*).

tipuliformis Theobald, 1901c (p. 325) – synonym of *Culex* (*Culex*) *vagans* Wiedemann, 1828; synonymy established before Knight & Stone, 1977.

tisseuli Senevet, 1937 (p. 375) – synonym of *Culex* (*Culex*) *mollis* Dyar & Knab, 1906b; synonymy established before Knight & Stone, 1977.

titillans (Walker, 1848) (p. 5) – species of *Mansonia* (*Mansonia*) [original combination: *Culex titillans*].

tityae Slooff, 1961 (p. 22) – species of *Tripteroides* (*Rachisoura*).

toffaletii Baisas & Ubaldo-Pagayon, 1953 (p. 182) – species of *Tripteroides* (*Tripteroides*).

togoensis Enderlein, 1923 (p. 28) – synonym of *Aedes* (*Stegomyia*) *apicoargenteus* (Theobald, 1909b [original combination: *Aniella togoensis*]; synonymy established before Knight & Stone, 1977.

togoi (Theobald, 1907) (p. 379) – species of *Aedes* (*Tanakaius*) [original combination: *Culicelsa togoi*].

tokushimaensis Tanimura, 1952 (p. 176) – synonym of *Aedes* (*Hulecoeteomyia*) *japonicus* (Theobald, 1901a) [original combination: *Aedes japonicus* variety *tokushimaensis*]; synonymy by Sasa & Tanimura, 1971.

toltecum (Dyar & Knab, 1906a) (p. 135) – species of *Psorophora* (*Grabhamia*) [original combination: *Janthinosoma toltecum*]; elevated from synonymy with *columbiae* (Dyar & Knab, 1906a) by Lanzaro *et al.*, 2015.

tongae Edwards, 1926a (p. 103) – species of *Aedes* (*Stegomyia*) [original combination: *Aedes variegatus* variety *tongae*]; specific status prior to ? Marks, 1954.

tonkinensis Galliard & Ngu, 1940 (p. 595) – synonym of *Anopheles* (*Anopheles*) *alongensis* Venhuis, 1940; synonymy by Senevet, 1947.

tonkinensis Galliard & Ngu, 1947 (p. 77) – species of *Aedes* (*Collessius*).

tonkinensis Toumanoff, 1931 (p. 576) – synonym of *Anopheles* (*Cellia*) *jeyporiensis* James, 1902 [original combination: *Anopheles* (*Myzomyia*) *aconitus* variety *tonkinensis*]; synonymy by Harrison, 1980.

tonkingi Gebert, 1948 (p. 96) – species of *Aedes* (*Mucidus*). Note: the title of the article by Gebert (1948) indicated that the author intended to place this species in genus *Aedes*, subgenus *Mucidus*, but it is referred as the 'new species *Mucidus tonkingi*' and described under the heading '**Mucidus tonkingi** sp. n.'.

tonnoiri (Edwards, 1925) (p. 258) – species of *Culiseta* (*Climacura*) [original combination: *Theobaldia tonnoiri*].

tonsus Edwards, 1924b (p. 390) – species of *Aedes* (*Geoskusea*).

tonsus Edwards, 1941 (p. 235) – species of *Eretmapodites*.

torakala Stoker & Waktoedi Koesoemawinangoen, 1949 (p. 58) – subspecies of *Anopheles* (*Cellia*) *ludlowae* (Theobald, 1903a) [original combination: *Anopheles* (*Myzomyia*) *ludlowi* variety *torakala*]; subspecific status by Harbach & Howard, 2007.

tormentor Dyar & Knab, 1906d (p. 191) – species of *Aedes* (*Ochlerotatus*).

toroensis Edwards & Gibbins, 1939 (p. 32) – species of *Culex* (*Culex*) [original combination: *Culex vansomereni toroensis*]; specific status by Edwards, 1941.

torokinae Belkin, 1950 (p. 236) – species of *Tripteroides* (*Rachisoura*).

torosa (Delfinado, 1967) (p. 30) – species of *Verrallina* (*Neomacleaya*) [original combination: *Aedes* (*Neomacleaya*) *torosus*].

torrentium Martini, 1925 (p. 336) – species of *Culex* (*Culex*).
Synonym: *pavlovsky* Shingarev, 1928.

torresiensis Schmidt, 2001 (in Schmidt *et al.*, 2001) (p. 400) – species of *Anopheles* (*Cellia*).

torridus Iglisch, 1977 (p. 269) – synonym of *Culex* (*Culex*) *pipiens* Linnaeus, 1758 [original combination: *Culex pipiens torridus*]; synonymy by Harbach, 1988.

tortilis (Theobald, 1903g) (p. 281) – species of *Aedes* (*Ochlerotatus*) [original combination: *Culex tortilis*].
Synonyms: *bracteatus* Coquillett, 1906a; *habanicus* Dyar & Knab, 1906d; *balteatus* Dyar & Knab, 1907a; *plutocraticus* Dyar & Knab, 1907a; *virginensis* Dyar, 1922c.

tosimus Dyar, 1920d (p. 72) – synonym of *Culex* (*Melanoconion*) *maxinocca* Dyar, 1920d; synonymy established before Knight & Stone, 1977.

totonaci Lassmann, 1951 (p. 285) – species of *Psorophora* (*Janthinosoma*).

tournieri Senevet & Abonnenc, 1939a (p. 105) – species of *Culex* (*Melanoconion*).

tovari Evans, 1922 (p. 218) – synonym of *Psorophora* (*Janthinosoma*) *cyanescens* (Coquillett, 1902a); synonymy established before Knight & Stone, 1977.

tovari Evans, 1924 (p. 367) – synonym of *Culex* (*Melanoconion*) *erraticus* (Dyar & Knab, 1906d); synonymy by King & Bradley, 1937.

toviiensis Klein, Rivière & Séchan, 1984 (p. 72) – species of *Culex* (*Culex*).

towadensis (Matsumara, 1916) (p. 445) – species of *Toxorhynchites* (*Toxorhynchites*) [original combination: *Megarhina towadensis*].

toweri Dyar & Knab, 1907a (p. 13) – synonym of *Culex* (*Culex*) *secutor* Theobald, 1901c; synonymy established before Knight & Stone, 1977.

townsendi da Costa Lima, 1935b (p. 178) – synonym of *Orthopodomyia fascipes* (Coquillett, 1906a); synonymy by Lane, 1951.

townsendi Stone, 1944 (p. 337) – subspecies of *Trichoprosopon digitatum* (Rondani, 1848) [original combination: *Trichoprosopon* (*Trichoprosopon*) *digitatus* variety *townsendi*]; subspecific status by Harbach & Howard, 2007.

townsvillensis Taylor, 1919 (p. 836) – synonym of *Culex* (*Culex*) *quinquefasciatus* Say, 1823 [*nomen novum* for *fuscus* Taylor, 1914c]; synonymy established before Knight & Stone, 1977.

toxopeusi Bonne-Wepster, 1948 (p. 315) – species of *Aedes* (subgenus uncertain).

toxorhynchus Macquart, 1838a (p. 39) – synonym of *Aedes* (*Stegomyia*) *aegypti* (Linnaeus, 1762) [original combination: *Culex toxorhynchus*]; synonymy by Belkin, 1968b.

tracei Senevet & Quiévreux, 1941 (p. 251) – synonym of *Aedes* (*Howardina*) *busckii* (Coquillett, 1906c); synonymy established before Knight & Stone, 1977.

trachycampa Dyar & Knab, 1909c (p. 101) – synonym of *Culex* (*Melanoconion*) *erraticus* (Dyar & Knab, 1906d); synonymy by King & Bradley, 1937.

tramazayguesi Duret, 1954b (p. 23) – species of *Culex* (*Allimanta*).

transvaalensis Carter, 1910a (p. 237) – synonym of *Anopheles* (*Cellia*) *marshallii* (Theobald, 1903a) [original combination: *Pyretophorus transvaalensis*]; synonymy by Evans, 1938.

transvaalensis Theobald, 1903a (p. 165) – synonym of *Aedes* (*Aedimorphus*) *hirsutus* (Theobald, 1901a) [original combination: *Culex transvaalensis*]; synonymy established before Knight & Stone, 1977.

trapidoi Galindo, Blanton & Peyton, 1954 (p. 170) – species of *Uranotaenia* (*Uranotaenia*).

traubi Colless, 1965 (p. 295) – species of *Culex* (*Lophoceraomyia*).

traubi Macdonald, 1960 (p. 132) – species of *Armigeres* (*Leicesteria*).

traversus Dyar, 1925g (p. 215) – synonym of *Aedes* (*Ochlerotatus*) *angustivittatus* Dyar & Knab, 1907a; synonymy established before Knight & Stone, 1977.

travestita (Brug, 1928b) (p. 921) – species of *Bironella* (*Brugella*) [original combination: *Anopheles* (*Bironella*) *travestitus*].

treacherii Leicester, 1908 (p. 19) – synonym of *Anopheles* (*Anopheles*) *fragilis* (Theobald, 1903f); synonymy by Reid, 1965.

tremulus (Theobald, 1903e) (p. 155) – species of *Aedes* (*Macleaya*) [original combination: *Macleaya tremula*].
Synonyms: *alboannulata* Taylor, 1912a; *minuta* Taylor, 1912a; *australis* Taylor, 1914b; *darwini* Taylor, 1914b; *doddi* Taylor, 1919.

treubi (de Meijere, 1910) (p. 918) – species of *Aedes* (*Alanstonea*) [original combination: *Scutomyia treubi*].

triangulata Ludlow, 1908a (p. 331) – synonym of *Uranotaenia* (*Uranotaenia*) *nivea* Leicester, 1908 [original combination: *Pseudo-uranotaenia triangulata*]; synonymy established before Knight & Stone, 1977.

triangulatus Taylor, 1914d (p. 204) – synonym of *Hodgesia quasisanguinae* Leicester, 1908; synonymy established before Knight & Stone, 1977.

triannulatus (Neiva & Pinto, 1922b) (p. 356) – species of *Anopheles* (*Nyssorhynchus*) [original combination: *Cellia triannulata*].
Synonyms: *cuyabensis* Neiva & Pinto, 1923; *bachmanni* Petrocchi, 1925; *davisi* Paterson & Shannon, 1927; *perezi* Shannon & del Ponte, 1928; *chagasi* Galvão, 1941a.

tricholabis Edwards, 1941 (p. 179) – species of *Aedes* (*Aedimorphus*).

trichopus Dyar, 1919a (p. 10) – synonym of *Johnbelkinia longipes* (Fabricius, 1805) [original combination: *Lesticocampa trichopus*]; synonymy by Lane, 1939.

trichopygus (Wiedemann, 1828) (p. 4) – species of *Toxorhynchites* (*Ankylorhynchus*) [original combination: *Culex trichopygus*].
Synonym: *neglectus* Lutz, 1904.

trichorostris (Theobald, 1909b) (p. 23) – species of *Malaya* [original combination: *Grahamia trichorostris*].

trichorryes Dyar & Knab, 1907c (p. 206) – synonym of *Trichoprosopon compressum* Lutz, 1905 [original combination: *Joblotia trichorryes*]; synonymy established before Knight & Stone, 1977.

trichurus Dyar, 1904a (p. 170) – synonym of *Aedes* (*Rusticoidus*) *provocans* (Walker, 1848) [original combination: *Culex trichurus*]; synonymy by Wood, 1977.

tricolor da Cunha Ramos & Brunhes, 2004 (p. 322) – species of *Uranotaenia* (*Pseudoficalbia*).

tricontus Delfinado, 1966a (p. 133) – synonym of *Culex* (*Eumelanomyia*) *cataractarum* Edwards, 1923b; synonymy by Sirivanakarn, 1972.

tricuspis Edwards, 1930a (p. 294) – species of *Culex* (*Culiciomyia*) [*nomen novum* for *trifidus* Edwards, 1926a].
Synonym: *trifidus* Edwards, 1926a.

tridentata da Cunha Ramos & Brunhes, 2004 (p. 287) – species of *Uranotaenia* (*Pseudoficalbia*).

tridentatus Cerqueira, 1961 (p. 48) – species of *Sabethes* (*Sabethoides*).

trifida Edwards, 1922a (p. 289) – species of *Topomyia* (*Topomyia*).

trifidus Dyar, 1921h (p. 115) – species of *Culex* (*Melanoconion*).

trifidus Edwards, 1926a (p. 108) – synonym of *Culex* (*Culiciomyia*) *tricuspis* Edwards, 1930a; synonymy established before Knight & Stone, 1977.

trifilatus Edwards, 1914b (p. 65) – species of *Culex* (*Culex*).

trifoliatus Edwards, 1914b (p. 70) – species of *Culex* (*Culex*).

trifurcata Clastrier, 1973 (p. 39) – species of *Wyeomyia* (*Dendromyia*).

trifurcatus Fabricius, 1794 (p. 401) – synonym of *Culex* (*Culex*) *pipiens* Linnaeus, 1758; synonymy established before Knight & Stone, 1977.

trigeminatus Clastrier, 1970c (p. 473) – species of *Culex* (*Melanoconion*).

trigonophorus Lutz, 1928 (p. 32) – synonym of *Psorophora* (*Grabhamia*) *funiculus* Dyar, 1920f [original combination: *Taeniorhynchus trigonophorus*]; synonymy based on proximity of the type locality to the Columbian distribution of *funiculus* established by Lanzaro *et al.* (2015).

trilineata Leicester, 1904 (in Theobald, 1904c) (p. 163) – synonym of *Aedes* (*Hulecoeteomyia*) *chrysolineatus* (Theobald, 1907); synonymy established before Knight & Stone, 1977.

trilineata Leicester, 1908 (p. 204) – species of *Uranotaenia* (*Uranotaenia*).

trilineatus Theobald, 1901c (p. 105) – synonym of *Aedes* (*Aedimorphus*) *mediolineatus* (Theobald, 1901c) [original combination: *Culex trilineatus*]; synonymy established before Knight & Stone, 1977.

trilineatus Theobald, 1901c (p. 159) – synonym of *Culex* (*Culex*) *quinquefasciatus* Say, 1823 [original combination: *Culex fatigans trilineatus*]; synonymy established before Knight & Stone, 1977.

trilobulatus Duret & Barreto, 1956 (p. 93) – species of *Culex* (*Melanoconion*).

trimaculata Tsuzuki, 1907 (p. 549) – synonym of *Anopheles* (*Cellia*) *indefinitus* (Ludlow, 1904c) [original combination: *Anopheles formosaensis II* variety *trimaculata* [*sic*]]; synonymy by Reid, 1968.

trimaculatus (Theobald, 1905f) (p. 86) – species of *Aedes* (*Aedimorphus*) [original combination: *Culex trimaculatus*].
Synonym: *littoralis* Barraud, 1927c.

trinidad Gil Collado, 1936 (pp. 315, 326) – synonym of *Aedes* (*Stegomyia*) *dendrophilus* Edwards, 1921c; synonymy by Mattingly, 1952.

trinidadensis Dyar & Knab, 1906c (p. 252) – synonym of *Toxorhynchites* (*Lynchiella*) *moctezuma* (Dyar & Knab, 1906c) [original combination: *Megarhinus trinidadensis*]; synonymy by Zavortink & Chaverri, 2009.

trinidadensis Theobald, 1901c (p. 277) – species of *Wyeomyia* (*Phoniomyia*).

trinkae Faran, 1979 (p. 27) – species of *Anopheles* (*Nyssorhynchus*).

tripartita (Bonne-Wepster & Bonne, 1921) (p. 7) – species of *Wyeomyia* (*Phoniomyia*) [original combination: *Dyarina tripartita*].

tripunctata Theobald, 1907 (p. 247) – synonym of *Aedes* (*Georgecraigius*) *fluviatilis* (Lutz, 1904) [original combination: *Danielsia tripunctata*]; synonymy established before Knight & Stone, 1977.

tripunctata (Theobald, 1908b) (p. 288) – species of *Heizmannia* (*Mattinglyia*) [original combination: *Stegomyia tripunctata*].

tripunctatus Mochizuki, 1913 (p. 24) – synonym of *Culex* (*Oculeomyia*) *sinensis* Theobald, 1903a; synonymy established before Knight & Stone, 1977.

triseriatus (Say, 1823) (p. 12) – species of *Aedes* (*Protomacleaya*) [original combination: *Culex triseriatus*].
Synonym: *nigra* Ludlow, 1905c.

trisetosus Fauran, 1961 (p. 1) – species of *Culex* (*Melanoconion*).

trisignatus Hoffmann, 1938 (p. 177) – synonym of *Anopheles* (*Nyssorhynchus*) *albimanus* Wiedemann, 1820 [original combination: *Anopheles albimanus* variety *trisignatus*]; synonymy by Ross & Roberts, 1943.

trispinata (King & Hoogstraal, 1947b) (p. 130) – species of *Verrallina* (*Verrallina*) [original combination: *Aedes* (*Aedes*) *trispinatus*].

tritaeniorhynchus Giles, 1901a (p. 606) – species of *Culex* (*Culex*).
Synonyms: *biroi* Theobald, 1905g; *summorosus* Dyar, 1920i; *siamensis* Barraud & Christophers, 1931.

trivittatus (Coquillett, 1902b) (p. 193) – species of *Aedes* (*Ochlerotatus*) [original combination: *Culex trivittatus*].
Synonym: *inconspicuus* Grossbeck, 1904b.

troglodytus Dyar & Knab, 1909d (p. 260) – synonym of *Deinocerites magnus* (Theobald, 1901c); synonymy by Belkin & Hogue, 1959.

tropicalis Cerqueira & Antunes, 1938 (p. 1) – species of *Haemagogus* (*Haemagogus*).

trujilloi Pulido F. & Sutil O., 1981 (p. 219) – synonym of *Wyeomyia* (*Nunezia*) *bicornis* (Root, 1928); synonymy by Navarro & Liria, 2007.

trukensis Bohart, 1957 (p. 63) – species of *Aedes* (*Aedimorphus*).

trychnus Root, 1927c (p. 585) – synonym of *Culex* (*Microculex*) *consolator* Dyar & Knab, 1906b; synonymy established before Knight & Stone, 1977.

tsaratananae Doucet, 1950 (p. 42) – species of *Uranotaenia* (*Pseudoficalbia*).

tsengi Lien, 1968d (p. 237) – species of *Culex* (*Culex*).

tsiliensis King & Hoogstraal, 1946f (p. 305) – species of *Aedes* (subgenus uncertain).

tubanguii Baisas, 1935c (p. 69) – synonym of *Uranotaenia* (*Pseudoficalbia*) *modesta* Leicester, 1908; synonymy by Peyton, 1972.

tubbutiensis Dobrotworsky, 1959 (p. 139) – species of *Aedes* (*Dobrotworskyius*).

tuberis Bohart, 1946 (p. 42) – species of *Culex* (*Lophoceraomyia*).

tucumanus Brèthes, 1926a (p. 318) – synonym of *Toxorhynchites* (*Lynchiella*) *guadeloupensis* (Dyar & Knab, 1906c) [original combination: *Megarhinus tucumanus*]; synonymy by Lane, 1951.

tucumanus Lahille, 1912 (p. 253) – synonym of *Anopheles* (*Anopheles*) *pseudopunctipennis* Theobald, 1901c; synonymy established before Knight & Stone, 1977.

tulagiensis Edwards, 1926a (p. 109) – species of *Aedes* (*Stegomyia*) [original combination: *Aedes* (*Stegomyia*) *edwardsi* variety *tulagiensis*]; specific status by Belkin, 1962.

tulliae (Taylor, 1929) (p. 275) – species of *Aedes* (*Macleaya*) [original combination: *Chaetocruiomyia tulliae*].

tumetarsalis Chen & Zhang, 1988 (p. 213) – species of *Topomyia* (*Suaymyia*).

turkestani Shingarev, 1926 (p. 47) – synonym of *Anopheles* (*Anopheles*) *claviger* (Meigen, 1804) [original combination: *Anopheles* (*Anopheles*) *algeriensis* variety *turkestani*]; synonymy by Edwards, 1932a.

turkhudi Liston, 1901 (p. 441) – species of *Anopheles* (*Cellia*).
Synonyms: *azriki* Patton, 1905; *flaviceps* Edwards, 1921c; *persicus* Edwards, 1921d; *amutis* de Burca, 1943.

turneri Marks, 1963b (p. 42) – species of *Aedes* (*Ochlerotatus*).

tutuilae Ramalingam & Belkin, 1965 (p. 3) – species of *Ades* (*Finlaya*).

tyagii Krishnamoorthy, Munirathinam, Dhananjeyan, Hiriyan, Mariappan, Samuel & Venkatesh, 2013 (p. 448) – species of *Toxorhynchites* (*Toxorhynchites*).

typharum Shannon & del Ponte, 1928 (p. 96) – synonym of *Wyeomyia melanocephala* Dyar & Knab, 1906a [original combination: *Dendromyia* (*Calladimyia* ?) *typharum*]; synonymy by Lane & Cerqueira, 1942.

typhlosomata Dyar & Knab, 1907c (p. 200) – species of *Uranotaenia* (*Uranotaenia*).

typicus Hackett & Missiroli, 1935 (p. 45) – synonym of *Anopheles* (*Anopheles*) *maculipennis* Meigen, 1818 [original combination: *Anopheles maculipennis* form *typicus*]; synonymy by Bates, 1940.

typicus Holstein, 1949 (p. 156) – synonym of *Anopheles* (*Cellia*) *gambiae* Giles, 1902 [original combination: *Anopheles gambiae typicus*]; synonymy established before Knight & Stone,1977.

Uu

ugandae da Cunha Ramos, 1993 (p. 290) – species of *Uranotaenia* (*Pseudoficalbia*).

ugandae Evans, 1934 (p. 553) – subspecies of *Anopheles* (*Cellia*) *wellcomei* Theobald, 1904b [original combination: *Anopheles distinctus* variety *ugandae*]; subspecific status by Gillies & de Meillon, 1968.
Synonym: *milesi* de Meillon & Evans, 1935.

ugandae van Someren, 1967 (p. 14) – subspecies of *Culex* (*Culex*) *shoae* Hamon & Ovazza, 1954.

uichancoi Baisas & Ubaldo-Pagayon, 1953 (p. 189) – species of *Tripteroides* (*Tripteroides*).

ulocoma (Theobald, 1903a) (p. 313) – species of *Wyeomyia* (*Decamyia*) [original combination: *Dendromyia ulocoma*].
Synonyms: *cacodela* Dyar & Knab, 1909d; *onidus* Dyar & Knab, 1909d; *pantoia* Dyar & Knab, 1909d.

ulopus (Dyar & Knab, 1906a) (p. 137) – species of *Johnbelkinia* [original combination: *Lesticocampa ulopus*].
Synonym: *culicivora* Dyar & Knab, 1907c.

umbripes Edwards, 1941 (p. 321) – species of *Culex* (*Culex*).

umbrosa Theobald, 1903a (p. 34) – synonym of *Anopheles* (*Cellia*) *funestus* Giles, 1900b [original combination: *Myzomyia funesta* variety *umbrosa*]; synonymy by Evans, 1938. **Note**: this nominal taxon and *umbrosus* (Theobald, 1903a) are secondary homonyms; if based on page priority it would be the senior secondary homonym.

umbrosus Brug, 1924 (p. 437) – synonym of *Verrallina* (*Verrallina*) *butleri* (Theobald, 1901c) [original combination:

Aedes (*Skusea*) *umbrosus*]; synonymy established before Knight & Stone, 1977.

umbrosus (Theobald, 1903a) (p. 87) – species of *Anopheles* (*Anopheles*) [original combination: *Myzorhynchus umbrosus*]. **Note**: this nominal taxon and *umbrosa* Theobald, 1903a are secondary homonyms; if based on page priority it would be the junior secondary homonym.
Synonym: *novumbrosus* Strickland, 1916b.

unca (Theobald, 190lc) (p. 53) – species of *Verrallina* (*Neomacleaya*) [original combination: *Culex uncus*].
Synonym: *hirsutipleura* Barraud, 1928b.

uncatus Grabham, 1907 (p. 25) – synonym of *Aedes* (*Gymnometopa*) *mediovittatus* (Coquillett, 1906e); synonymy established before Knight & Stone, 1977.

uncinatus Delfinado, 1966a (p. 134) – species of *Culex* (*Eumelanomyia*).

undosus (Coquillett, 1906a) (p. 186) – species of *Sabethes* (*Peytonulus*) [original combination: *Sabethoides undosus*].

undulata del Ponte & Cerqueira, 1938 (p. 233) – species of *Wyeomyia* (subgenus uncertain, determined by Motta & Lourenço-de-Oliveira, 1995).

unguiculata Edwards, 1913a (p. 51) – species of *Uranotaenia* (*Pseudoficalbia*).

ungujae White, 1975 (p. 316) – subspecies of *Anopheles* (*Cellia*) *wellcomei* Theobald, 1904b.

uniannulata Theobald, 1909b (p. 12) – synonym of *Aedes* (*Catageiomyia*) *tarsalis* (Newstead, 1907) [original combination: *Neopecomyia uniannulata*]; synonymy established before Knight & Stone, 1977.

unicinctus Edwards, 1922a (p. 266) – species of *Aedes* (*Jihlienius*).

unicolor Grünberg, 1905 (p. 379) – synonym of *Anopheles* (*Cellia*) *nili* (Theobald, 1904b) [original combination: *Myzomyia unicolor*]; synonymy established before Knight & Stone, 1977.

unicornis Root, 1928 (in Dyar, 1928) (p. 291) – species of *Culex* (*Melanoconion*).

unidentatus McIntosh, 1971 (p. 325) – species of *Aedes* (*Neomelaniconion*). **Note**: incorrectly listed as *unidentatum* McIntosh in the online catalogue of the Walter Reed Biosystematics Unit (http://www.mosquitocatalog.org/).

uniformis Leicester, 1908 (p. 136) – synonym of *Culex* (*Eumelanomyia*) *brevipalpis* (Giles, 1902) [original combination: *Melanoconion uniformis*]; synonymy established before Knight & Stone, 1977. **Note**: this nominal taxon is the junior secondary homonym of *uniformis* (Theobald, 1905h); the senior secondary homonym of *uniformis* Leicester, 1908 (p. 159), based on page priority, and *uniformis* Theobald, 1909b.

uniformis Leicester, 1908 (p. 159) – synonym of *Culex* (*Culex*) *fuscocephala* Theobald, 1907; synonymy established before Knight & Stone, 1977. **Note**: this nominal taxon is the junior secondary homonym of *uniformis* (Theobald, 1905h) and *uniformis* Leicester, 1908 (p. 136), based on page priority; the senior secondary homonym of *uniformis* Theobald, 1909b.

uniformis Strickland, 1911a (p. 131) – synonym of *Aedes* (*Ochlerotatus*) *vigilax* (Skuse, 1889) [original combination: *Culicelsa uniformis*]; synonymy established before Knight & Stone, 1977.

uniformis (Theobald, 1901c) (p. 180) – species of *Mansonia* (*Mansonioides*) [original combination: *Panoplites uniformis*].
Synonyms: *reversus* Theobald, 1901c; *australiensis* Giles, 1902; *marquesensis* Dyar, 1925e.

uniformis (Theobald, 1904b) (p. 80) – species of *Ficalbia* [original combination: *Mimomyia uniformis*].
Synonym: *grenieri* Hamon, 1954b.

uniformis (Theobald, 1905h) (p. 245) – species of *Culex* (*Lophoceraomyia*) [original combination: *Lophoceraomyia uniformis*]. **Note**: this nominal taxon is the senior secondary homonym of *uniformis* Leicester, 1908 (p. 136), *uniformis* Leicester, 1908 (p. 159) and *uniformis* Theobald, 1909b.

uniformis Theobald, 1909b (p. 8) – synonym of *Culex* (*Culiciomyia*) *cinereus* Theobald, 1901c [original combination: *Culiciomyia uniformis*]; synonymy by Knight & Stone, 1977. **Note**: this nominal taxon is the junior secondary homonym of *uniformis* (Theobald, 1905h), *uniformis* Leicester, 1908 (p. 136) and *uniformis* Leicester, 1908 (p. 159).

uniformis (Theobald, 1910a) (p. 33) – species of *Verrallina* (*Harbachius*) [original combination: *Skusea uniformis*].

unilineata Leicester, 1908 (p. 220) – synonym of *Uranotaenia* (*Uranotaenia*) *testacea* Theobald, 1905f; synonymy established before Knight & Stone, 1977.

unilineatus (Theobald, 1906b) (p. 70) – species of *Aedes* (*Stegomyia*) [original combination: *Quasistegomyia unilineata*].
Synonym: *gebeleinensis* Theobald, 1910c.

unimaculiala Leicester, 1908 (p. 210) – species of *Uranotaenia* (*Uranotaenia*).

unispinosa Thurman, 1959 (p. 44) – species of *Topomyia* (*Topomyia*).

unistriatus Curtis, 1837 (p. 231) – synonym of *Culex* (*Culex*) *pipiens* Linnaeus, 1758; synonymy by Stone *et al.*, 1959.

univittatus Theobald, 1901c (p. 29) – species of *Culex* (*Culex*).
Synonyms: *simplex* Theobald, 1903a; *montforti* Ventrillon, 1905a; *ataeniatus* Theobald, 1911b; *goughii* Theobald, 1911b.

upatensis Anduze & Hecht, 1943 (p. 185) – species of *Aedes* (*Ochlerotatus*).

upemba Lips, 1960b (p. 303) – *nomen dubium*, *Anopheles*; status by Brunhes *et al.*, 1997.

upolensis Marks, 1957b (p. 51) – species of *Aedes* (*Stegomyia*).

urania Shannon & del Ponte, 1928 (p. 83) – synonym of *Uranotaenia* (*Uranotaenia*) *pulcherrima* Lynch Arribálzaga, 1891b; synonymy by Lane, 1951.

uriartei Shannon & del Ponte, 1928 (p. 68) – synonym of *Haemagogus* (*Haemagogus*) *spegazzinii* Brèthes, 1912; synonymy by Martínez *et al.*, 1961.

urichii (Coquillett, 1906e) (p. 61) – species of *Culex* (*Carrollia*) [original combination: *Melanoconion urichii*].
Synonym: *mathesoni* Anduze, 1942a.

usambara Mattingly, 1953d (p. 294) – species of *Aedes* (*Stegomyia*).

usquatissimus Dyar, 1922b (p. 19) – species of *Culex* (*Culex*).

usquatus Dyar, 1918e (p. 122) – species of *Culex* (*Culex*).

Vv

vadakadiensis Doraisamy, 1963 (p. 42) – unavailable name [original combination: *Anopheles subpictus* variety *vadakadiensis*]; a name published after 1960 expressly as a 'variety' is deemed to be infrasubspecific and is not an available name (International Commission on Zoological Nomenclature, 1999: *Code*, Articles 15.2 and 45.5).

vagans Wiedemann, 1828 (p. 545) – species of *Culex* (*Culex*).
Synonyms: *tipuliformis* Theobald, 1901c; *virgatipes* Edwards, 1914c; *exilis* Dyar, 1924f.

vagus Dönitz, 1902 (p. 80) – species of *Anopheles* (*Cellia*).
Synonyms: *immaculata* James, 1902; *immaculatus* Theobald, 1903a; *flava* Swellengrebel, 1917; *flavescens* Swellengrebel, 1921; *albino* Stoker & Waktoedi Koesoemawinangoen, 1949.

valeryi Gornostaeva, 2005 (p. 488) – species of *Aedes* (*Aedes*).

vallistris (Barraud, 1928b) (p. 369) – species of *Verrallina* (*Neomacleaya*) [original combination: *Aedes* (*Aedes*) *vallistris*].

vanapa (Huang, 1968a) (p. 63) – species of *Verrallina* (*Verrallina*) [original combination: *Aedes* (*Verrallina*) *vanapus*].

vandema Strickland, 1911b (p. 202) – synonym of *Aedes* (*Ochlerotatus*) *flavifrons* (Skuse, 1889) [original combination: *Culicada vandema*]; synonymy established before Knight & Stone, 1977.

vanderwulpi Townson & Harbach, 2013 (in Townson *et al.*, 2013) (p. 183) – species of *Anopheles* (*Anopheles*).

vanduzeei Dyar & Knab, 1906a (p. 138) – species of *Wyeomyia* (*Wyeomyia*).
Synonyms: *argyrura* Dyar & Knab, 1908; *conchita* Dyar & Knab, 1909d.

vaneedeni Gillies & Coetzee, 1987 (p. 81) – species of *Anopheles* (*Cellia*).

vanemdeni Martini, 1931a (p. 208) – species of *Aedes* (*Howardina*).

vanhalli Dyar & Knab, 1906a (p. 134) – synonym of *Psorophora* (*Janthinosoma*) *ferox* (von Humboldt, 1819) [original combination: *Janthinosoma vanhalli*]; synonymy established before Knight & Stone, 1977.

vanhoofi Wanson & Lebied, 1945 (p. 118) – species of *Anopheles* (*Cellia*).

vanleeuweni (Edwards, 1927b) (p. 355) – species of *Tripteroides* (*Rachisoura*) [original combination: *Rachionotomyia vanleeuweni*].

vanoyei (Wolfs, 1948b) (p. 83) – species of *Coquillettidia* (*Coquillettidia*) [original combination: *Taeniorhynchus* (*Coquillettidia*) *vanoyei*].

vansomerenae Grjebine, 1986 (p. 199) – species of *Mimomyia* (*Ingramia*).

vansomerenae Mattingly, 1955 (in Mattingly & Brown, 1955) (p. 78) – subspecies of *Aedes* (*Ochlerotatus*) *vigilax* (Skuse, 1889).

vansomereni Edwards, 1926a (p. 129) – species of *Culex* (*Culex*).

vansomereni Hamon, 1962 (p. 253) – species of *Eretmapodites*.

vanthieli Laarman, 1958 (p. 18) – synonym of *Anopheles* (*Cellia*) *faini* Leleup, 1952; synonymy by Gillies & de Meillon, 1968.

vanus Colless, 1958 (p. 471) – species of *Aedes* (*Downsiomyia*).

vanus Walker, 1859 (p. 91) – species of *Anopheles* (*Anopheles*).

vapulans Dyar, 1920d (p. 69) – synonym of *Culex* (*Melanoconion*) *bastagarius* Dyar & Knab, 1906b; synonymy by Komp, 1935.

vargasi Gabaldon, Cova Garcia & Lopez, 1941b (p. 25) – species of *Anopheles* (*Lophopodomyia*).

vargasi Schick, 1970a (p. 73) – species of *Aedes* (*Protomacleaya*).

variabilis (Huang, 1968a) (p. 65) – species of *Verrallina* (*Verrallina*) [original combination: *Aedes* (*Verrallina*) *variabilis*].

variatus (Leicester, 1908) (p. 121) – species of *Culex* (*Lophoceraomyia*) [original combination: *Lophoceratomyia variata*].

variegata (Dobrotworsky, 1962) (p. 293) – species of *Coquillettidia* (*Coquillettidia*) [original combination: *Mansonia variegata*].

variegatans Strickland, 1911b (p. 204) – synonym of *Aedes* (*Ochlerotatus*) *flavifrons* (Skuse, 1889) [original combination: *Culicada vandema* variety *variegatans*]; synonymy established before Knight & Stone, 1977.

variegatus Blanchard, 1852 (p. 333) – synonym of *Anopheles* (*Nyssorhynchus*) *pictipennis* (Philippi, 1865) [original combination: *Culex variegatus*]; synonymy by Belkin, 1968b.

variegatus Doleschall, 1858 (p. 77) – synonym of *Aedes* (*Stegomyia*) *scutellaris* (Walker, 1859) [original combination: *Culex variegatus*]; synonymy established before Knight & Stone, 1977. **Note**: this nominal taxon is the junior primary homonym of *variegatus* Schrank, 1781.

variegatus Schrank, 1781 (p. 482) – synonym of *Aedes* (*Ochlerotatus*) *flavescens* (Müller, 1764) [original combination: *Culex variegatus*]; synonymy established before Knight & Stone, 1977. **Note**: this nominal taxon is the senior primary homonym of *variegatus* Doleschall, 1858.

variepictus King & Hoogstraal, 1946c (p. 98) – species of *Aedes* (*Huaedes*).

varietas (Leicester, 1908) (p. 185) – species of *Verrallina* (*Neomacleaya*) [original combination: *Aioretomyia varietas*].

varinervis Edwards, 1922d (p. 78) – species of *Psorophora* (*Grabhamia*).

varioannulatus Theobald, 1903a (p. 198) – synonym of *Culex* (*Culex*) *pipiens* Linnaeus, 1758; synonymy established before Knight & Stone, 1977.

varipalpus (Coquillett, 1902d) (p. 292) – species of *Aedes* (*Jarnellius*) [original combination: *Culex varipalpus*].
varipes (Coquillett, 1904a) (p. 10) – species of *Psorophora* (*Janthinosoma*) [original combination: *Conchyliastes varipes*].
Synonym: *ferraroi* Duret, 1971a.
varivestita Statz, 1941 (p. 11) – *nomen nudum*, *Mansonia*.
varivestita Statz, 1944 (p. 111) – fossil species of *Mansonia* Blanchard, 1901.
varovum Breeland, 1953 (pp. 2, 7) – *nomen nudum*, *Anopheles* [original combination: *Anopheles crucians varovum*].
varuae Belkin, 1962 (p. 478) – species of *Aedes* (*Stegomyia*).
varuna Iyengar, 1924 (p. 24) – species of *Anopheles* (*Cellia*).
vassilievii Portschinsky, 1911 (p. 58) – synonym of *Anopheles* (*Cellia*) *superpictus* Grassi, 1899 [original combination: *Anopheles superpictus* variety *vassilievii*]; synonymy established before Knight & Stone, 1977.
vattieri Geoffroy, 1971c (p. 265) – species of *Culex* (*Eumelanomyia*).
vaxus Dyar, 1920d (p. 73) – species of *Culex* (*Melanoconion*).
Synonym: *bibulus* Dyar, 1920d.
vectensis Edwards, 1923c (p. 151) – fossil species of *Culex* Linnaeus, 1758.
vector Dyar & Knab, 1906d (p. 220) – synonym of *Culex* (*Microculex*) *imitator* Theobald, 1903a; synonymy established before Knight & Stone, 1977.
veeniae McIntosh, 1975 (p. 263) – species of *Aedes* (*Catageiomyia*).
venezuelae Evans, 1922 (p. 214) – synonym of *Anopheles* (*Anopheles*) *punctimacula* Dyar & Knab, 1906a; synonymy established before Knight & Stone, 1977.
venezuelensis Anduze, 1949a (p. 64) – synonym of *Culex* (*Melanoconion*) *simulator* Dyar & Knab, 1906d; synonymy by Sirivanakarn, 1983.
venezuelensis (Theobald, 1912) (in Surcouf, 1912) (p. 61) – species of *Coquillettidia* (*Rhynchotaenia*) [original combination: *Pseudotaeniorhynchus venezuelensis*].
Synonym: *araozi* Shannon & del Ponte, 1928.
venhuisi Bonne-Wepster, 1951b (p. 284) – synonym of *Anopheles* (*Anopheles*) *nigerrimus* Giles, 1900a; synonymy by Reid, 1953.
ventralis (Walker, 1860) (p. 144) – *nomen dubium*, *Armigeres* [original combination: *Culex ventralis*]; status by Edwards, 1932a.
ventralis Walker, 1865 (p. 103) – *nomen dubium*, *Culex*; status established before Knight & Stone, 1977.
ventrilloni Edwards, 1920b (p. 135) – species of *Culex* (*Culex*).
Synonym: *albigenu* Enderlein, 1920.
ventrovittis Dyar, 1916b (p. 84) – species of *Aedes* (*Ochlerotatus*).
Synonyms: *fisheri* Dyar, 1917a; *cacothius* Dyar, 1923a.
venustipes (Skuse, 1889) (p. 1761) – species of *Aedeomyia* (*Aedeomyia*) [original combination: *Aedes venustipes*].
verna Lewis, 1944 (p. 27) – synonym of *Aedes* (*Aedimorphus*) *leesoni* Edwards, 1932b; synonymy by McIntosh, 1975.
vernoni van Someren, 1949a (p. 5) – species of *Orthopodomyia*.
Synonyms: *geberti* Grjebine, 1953a; *antanosyorum* Rodhain & Boutonnier, 1984.
vernus Gillies & de Meillon, 1968 (p. 89) – species of *Anopheles* (*Cellia*).
versabilis Sirivanakarn, 1968 (p. 126) – species of *Culex* (*Lophoceraomyia*).
versicolor (Barraud, 1924i) (p. 73) – species of *Aedes* (subgenus uncertain) [original combination: *Finlaya versicolor*].
versicolor (Edwards, 1913b) (p. 50) – species of *Coquillettidia* (*Coquillettidia*) [original combination: *Taeniorhynchus versicolor*].
veruslanei Vargas, 1979 (p. 87) – species of *Anopheles* (*Anopheles*).
verutus Harbach, 1987 (p. 216) – species of *Culex* (*Culex*).
vestitipennis Dyar & Knab, 1906a (p. 136) – species of *Anopheles* (*Anopheles*).
vexans (Meigen, 1830b) (p. 241) – species of *Aedes* (*Aedimorphus*) [original combination: *Culex vexans*].
Synonyms: *parvus* Macquart, 1834; *articulatus* Rondani, 1872; *malariae* Grassi, 1898; *sylvestris* Theobald, 1901a; *montcalmi* Blanchard, 1905; *minuta* Theobald, 1907; *eruthrosops* Theobald, 1910c; *sudanensis* Theobald, 1911c; *eurochrus* Howard, Dyar & Knab, 1917.
vexillatus Edwards, 1941 (p. 339) – subspecies of *Culex* (*Culex*) *invidiosus* Theobald, 1901d [original combination: *Culex* (*Culex*) *invidiosus* variety *vexillatus*]; subspecific status by Harbach & Howard, 2007.
vexillifer Komp, 1936c (p. 320) – species of *Culex* (*Melanoconion*).
vicarius Lien, 1968d (p. 218) – subspecies of *Aedes* (*Collessius*) *elsiae* (Barraud, 1923e).
vicinalis de Meillon & Lavoipierre, 1944 (p. 49) – subspecies of *Culex* (*Culex*) *invidiosus* Theobald, 1901d.
vicinus (Edwards, 1914c) (p. 128) – species of *Tripteroides* (*Tripteroides*) [original combination: *Rachionotomyia vicina*].
vicinus (Taylor, 1916) (p. 569) – species of *Culex* (*Culex*) [original combination: *Leucomyia vicina*].
Synonyms: *annulata* Taylor, 1914c; *basicinctus* Edwards, 1921c.
victoriensis (Dobrotworsky, 1954) (p. 73) – species of *Culiseta* (*Culicella*) [original combination: *Theobaldia victoriensis*].
victoriensis Taylor, 1914b (p. 460) – synonym of *Aedes* (*Ochlerotatus*) *camptorhynchus* (Thomson, 1869) [original combination: *Culicada victoriensis*]; synonymy established before Knight & Stone, 1977.
vidali Floch & Fauran, 1954 (p. 4) – species of *Culex* (*Melanoconion*).
vietnamensis Anonymous (in IMPE, 1987) (pp. 13, 33, 57) – *nomen nudum*, *Anopheles*; unavailable name [authorship cannot be determined from the content of the publication and is therefore deemed to be anonymous; a new name published after 1950 with anonymous

authorship is not made available (International Commission on Zoological Nomenclature, 1999: *Code*, Articles 50.1 and 14, respectively)]. See *vietnamensis* Nguyen, Tran & Nguyen, 1993.

vietnamensis Nguyen, Tran & Nguyen, 1993 (p. 215) – species of *Anopheles* (*Anopheles*).

vigilax (Skuse, 1889) (p. 1731) – species of *Aedes* (*Ochlerotatus*) [original combination: *Culex vigilax*].
Synonyms: *albirostris* Macquart, 1850; *marinus* Theobald, 1901a; *pseudovigilax* Theobald, 1907; *uniformis* Strickland, 1911a.

vijayae Ramalingam, 1975 (p. 185) – species of *Topomyia* (*Topomyia*).

villosus Robineau-Desvoidy, 1827 (p. 411) – synonym of *Anopheles* (*Anopheles*) *claviger* (Meigen, 1804); synonymy by Edwards, 1932a.

vimoli Thurman & Thurman, 1958 (p. 187) – species of *Armigeres* (*Leicesteria*).

vincenti Laveran, 1901c (p. 993) – synonym of *Anopheles* (*Cellia*) *jeyporiensis* James, 1902. **Note**: *Anopheles vincenti* is currently the senior name as a result of lectotype designation by Harbach, Garros *et al.* (2007). Established usage of *jeyporiensis* James over *vincenti* Laveran is maintained pending application for suppression of the latter by the International Commission on Zoological Nomenclature.

vinckei de Meillon, 1942 (p. 87) – species of *Anopheles* (*Cellia*).

vinckei Hamon, Holstein & Rivola, 1957 (p. 681) – species of *Culex* (*Eumelanomyia*).

vindicator Dyar & Knab, 1909d (p. 255) – synonym of *Culex* (*Culex*) *declarator* Dyar & Knab, 1906d; synonymy established before Knight & Stone, 1977.

vinnipegensis Dyar, 1919b (p. 34) – synonym of *Aedes* (*Ochlerotatus*) *sticticus* (Meigen, 1838); synonymy established before Knight & Stone, 1977.

vinsoni Mattingly, 1953a (p. 16) – species of *Aedes* (*Stegomyia*).

violaceus Theobald, 1908c (p. 262) – synonym of *Coquillettidia* (*Coquillettidia*) *metallica* (Theobald, 1901d) [original combination: *Taeniorhynchus violaceus*]; synonymy established before Knight & Stone, 1977.

violaceus (Wiedemann, 1820) (p. 7) – species of *Toxorhynchites* (*Lynchiella*) [original combination: *Culex violaceus*].

violescens Dyar & Knab, 1906a (p. 138) – synonym of *Wyeomyia* (*Wyeomyia*) *mitchellii* (Theobald, 1905d); synonymy established before Knight & Stone, 1977.

virescens Dyar & Knab, 1906a (p. 133) – synonym of *Psorophora* (*Psorophora*) *howardii* Coquillett, 1901; synonymy established before Knight & Stone, 1977.

virgatipes Edwards, 1914c (p. 126) – synonym of *Culex* (*Culex*) *vagans* Wiedemann, 1828; synonymy established before Knight & Stone, 1977.

virginensis Dyar, 1922c (p. 56) – synonym of *Aedes* (*Ochlerotatus*) *tortilis* (Theobald, 1903g); synonymy established before Knight & Stone, 1977.

virgultus Theobald, 1901c (p. 123) – *nomen dubium*, *Culex*; status by Belkin, 1968b.

viridibasis (Edwards, 1935a) (p. 133) – species of *Toxorhynchites* (*Afrorhynchus*) [original combination: *Megarhinus aeneus* variety *viridibasis*]; specific status by Edwards, 1941.

viridifrons Walker, 1848 (p. 3) – synonym of *Aedes* (*Stegomyia*) *aegypti* (Linnaeus, 1762) [original combination: *Culex viridifrons*]; synonymy established before Knight & Stone, 1977.

viridis Barraud, 1929b (p. 266) – species of *Heizmannia* (*Heizmannia*).

viridis Theobald, 1903a (p. 212) – synonym of *Culex* (*Culex*) *guiarti* Blanchard, 1905; synonymy established before Knight & Stone, 1977.

viridiventer Giles, 1901a (p. 609) – species of *Culex* (*Culiciomyia*).
Synonyms: *angulatus* Theobald, 1901c; *longifurcatus* Theobald, 1910a; *pseudolongifurcatus* Theobald, 1910c.

virilis Leicester, 1908 (p. 197) – species of *Verrallina* (*Neomacleaya*).

vishnui Theobald, 1901a (p. 355) – species of *Culex* (*Culex*).

vittata Theobald, 1903h (p. 313) – synonym of *Aedes* (*Ochlerotatus*) *increpitus* Dyar, 1916b [original combination: *Grabhamia vittata*]; synonymy established before Knight & Stone, 1977. **Note**: this nominal taxon is the junior secondary homonym of *vittatus* (Bigot, 1861) and *vittatus* Philippi, 1865.

vittatus (Bigot, 1861) (p. 227) – species of *Aedes* (*Fredwardsius*) [original combination: *Culex vittatus*]. **Note**: this nominal taxon is the senior primary homonym of *vittatus* Philippi, 1865 and the senior secondary homonym of *vittata* Theobald, 1903h.
Synonyms: *brumpti* Neveu-Lemaire, 1905b; *albopunctata* Theobald, 1907.

vittatus Philippi, 1865 (p. 596) – synonym of *Aedes* (*Ochlerotatus*) *albifasciatus* (Macquart, 1838b) [original combination: *Culex vittatus*]; synonymy established before Knight & Stone, 1977. **Note**: this nominal taxon is the junior primary homonym of *vittatus* (Bigot, 1861) and the senior secondary homonym of *vittata* Theobald, 1903h.

vittiger (Skuse, 1889) (p. 1728) – species of *Aedes* (*Ochlerotatus*) [original combination: *Culex vittiger*].

vogelsangi Anduze, 1949a (p. 61) – synonym of *Culex* (*Melanoconion*) *elevator* Dyar & Knab, 1906d; synonymy by Lane, 1953.

voltaensis (Danilov, 1982) (p. 26) – species of *Coquillettidia* (*Coquillettidia*) [original combination: *Mansonia* (*Coquillettidia*) *voltaensis*].

vomerifer Komp, 1932 (p. 79) – species of *Culex* (*Melanoconion*).

vonplesseni (Dyar & Knab, 1906a) (p. 137) – species of *Trichoprosopon* [original combination: *Lesticocampa vonplesseni*].

vorax Edwards, 1921d (p. 327) – species of *Lutzia* (*Metalutzia*).

vulgaris Hatori, 1901 (p. 274) – *nomen dubium*, *Anopheles*; status established before Knight & Stone, 1977.
vulgaris Linnaeus, 1792 (p. 363) – *nomen dubium*, *Culex*; status by Howard *et al.*, 1915 (mixture of a simuliid and perhaps an aedine species).

Ww

wadai Tanaka, Mizusawa & Saugstad, 1979 (p. 400) – species of *Aedes* (*Stegomyia*).
wahlbergi (Edwards, 1936) (p. 55) – species of *Coquillettidia* (*Coquillettidia*) [original combination: *Taeniorhynchus* (*Coquillettidia*) *wahlbergi*].
wahlgreni Theobald, 1910c (p. 396) – synonym of *Aedes* (*Ochlerotatus*) *punctor* (Kirby, 1837) [original combination: *Culex wahlgreni*]; synonymy established before Knight & Stone, 1977.
wainwrighti Baisas, 1946c (p. 34) – synonym of *Aedes* (*Catatassomyia*) *meronephada* (Dyar & Shannon, 1925a); synonymy by Reinert, 1973a.
w-albus (Theobald, 1905f) (p. 74) – species of *Aedes* (*Stegomyia*) [original combination: *Stegomyia w-alba*].
Synonym: *argenteomaculata* Theobald, 1907.
walchi Soesilo, 1932 (p. 1035) – synonym of *Bironella* (*Bironella*) *gracilis* Theobald, 1905f [original combination: *Anopheles* (*Bironella*) *walchi*]; synonymy established before Knight & Stone, 1977.
walcotti (Lane & Cerqueira, 1942) (p. 510) – species of *Runchomyia* (*Ctenogoeldia*) [original combination: *Trichoprosopon* (*Ctenogoeldia*) *walcotti*].
walkeri Theobald, 1901a (p. 199) – species of *Anopheles* (*Anopheles*).
walkeri (Theobald, 1901a) (p. 424) – species of *Aedes* (*Howardina*) [original combination: *Culex* (*Stegomyia* ?) *walkeri*].
wallacei Edwards, 1926a (p. 105) – species of *Aedes* (*Finlaya*).
walravensi Edwards, 1930a (p. 290) – species of *Anopheles* (*Cellia*).
walshi Evans & de Meillon, 1933 (p. 280) – synonym of *Anopheles* (*Cellia*) *garnhami* Edwards, 1930a [original combination: *Anopheles garnhami* variety *walshi*]; synonymy by Gillies & de Meillon, 1968.
walsinghamii Theobald, 1907 (p. 484) – synonym of *Psorophora* (*Grabhamia*) *jamaicensis* Theobald, 1901a [original combination: *Taeniorhynchus walsinghamii*]; synonymy established before Knight & Stone, 1977.
walukasi Belkin, 1962 (p. 264) – species of *Culex* (*Lophoceraomyia*).
wamanguae Sirivanakarn, 1968 (p. 135) – species of *Culex* (*Lophoceraomyia*).
wannonii Cova Garcia & Sutil O., 1976a (p. 354) – species of *Culex* (*Carrollia*).
wansoni Edwards, 1941 (p. 240) – species of *Eretmapodites*.
wansoni Wolfs, 1945b (p. 221) – species of *Culex* (*Eumelanomyia*).
wanxianensis Lei & Li, 1990 (p. 8) – species of *Orthopodomyia*.
wardangensis Brust, Ballard, Driver, Hartley, Galway & Curran, 1998 (p. 1241) – species of *Aedes* (*Halaedes*).
wardi Leeson & Theodor, 1948 (p. 222) – synonym of *Anopheles* (*Cellia*) *dthali* Patton, 1905 [original combination: *Anopheles* (*Myzomyia*) *dthali* variety *wardi*]; synonymy by Mattingly & Knight, 1956.
wardi Reinert, 1976b (p. 18) – species of *Aedes* (*Rhinoskusea*).
wardi Sirivanakarn, 1977a (p. 90) – species of *Culex* (*Lophoceraomyia*).
washinoi Lanzaro & Eldridge, 1992 (p. 93) – species of *Aedes* (*Ochlerotatus*).
wasselli Marks, 1947b (p. 9) – species of *Aedes* (subgenus uncertain).
wassilievi Shingarev, 1927 (p. 548) – synonym of *Culiseta* (*Culiseta*) *alaskaensis indica* (Edwards, 1920a) [original combination: *Theobaldia wassilievi*]; synonymy by Maslov, 1967.
watasei Yamada, 1921 (p. 64) – species of *Aedes* (*Phagomyia*).
waterhousei Dobrotworsky, 1960a (p. 57) – synonym of *Aedes* (*Ochlerotatus*) *silvestris* Dobrotworsky, 1961b; synonymy established before Knight & Stone, 1977. **Note**: this nominal taxon is the junior secondary homonym of *waterhousei* Theobald, 1905.
waterhousei Theobald, 1905 (in Waterhouse, 1905) (p. 674) – synonym of *Aedes* (*Ochlerotatus*) *cantans* (Meigen, 1818) [original combination: *Culex waterhousei*]; synonymy established before Knight & Stone, 1977. **Note**: this nominal taxon is the senior secondary homonym of *waterhousei* Dobrotworsky, 1960a.
watsoni Edwards, 1911a (p. 143) – synonym of *Anopheles* (*Cellia*) *rufipes* (Gough, 1910); synonymy established before Knight & Stone, 1977.
watsonii (Leicester, 1908) (p. 38) – species of *Anopheles* (*Cellia*) [original combination: *Pyretophorus watsonii*].
watteni Lien, 1968b (p. 103) – species of *Aedes* (*Downsiomyia*).
wattensis Taylor, 1929 (p. 275) – species of *Aedes* (*Macleaya*).
watti Edwards, 1920a (p. 135) – species of *Culex* (*Culex*).
wauensis Huang, 1968b (p. 170) – species of *Aedes* (*Huaedes*).
waverleyi (Grabham, 1907) (p. 25) – species of *Orthopodomyia* [original combination: *Mansonia waverleyi*].
weindorferi (Edwards, 1926a) (p. 111) – species of *Culiseta* (*Culicella*) [original combination: *Theobaldia weindorferi*].
wejchoochotei Taai & Harbach, 2015 (p. 259) – species of *Anopheles* (*Anopheles*).
wellcomei Theobald, 1904b (p. 64) – species of *Anopheles* (*Cellia*).
wellingtonianus Alcock, 1912 (p. 1) – species of *Anopheles* (*Anopheles*).
wellmanii (Theobald, 1905b) (p. 103) – species of *Aedes* (*Hopkinsius*) [original combination: *Danielsia wellmanii*].
wellmanii Theobald, 1910c (p. 163) – synonym of *Aedes* (*Stegomyia*) *pseudonigeria* (Theobald, 1910c); synonymy established before Knight & Stone, 1977.

wendyae Service, 1959a (p. 73) – species of *Aedes* (*Catageiomyia*).
wepsterae Komp & Rozeboom, 1951 (as *wepsteri*) (p. 127) – species of *Culex* (*Melanoconion*).
werneri Baisas & Ubaldo-Pagayon, 1953 (p. 111) – species of *Tripteroides* (*Tricholeptomyia*).
weschei Edwards, 1935a (p. 136) – species of *Culex* (*Culex*).
westralis Strickland, 1911a (p. 130) – synonym of *Aedes* (*Ochlerotatus*) *camptorhynchus* (Thomson, 1869) [original combination: *Culicelsa westralis*]; synonymy established before Knight & Stone, 1977.
whartoni Colless, 1965 (p. 275) – species of *Culex* (*Lophoceraomyia*).
whartoni Mattingly, 1965 (p. 65) – species of *Aedes* (*Petermattinglyius*).
whartoni Reid, 1963 (p. 106) – species of *Anopheles* (*Anopheles*).
whitei Barraud, 1923d (p. 508) – species of *Culex* (*Culex*).
whitmani Lane & Cerqueira, 1942 (p. 685) – species of *Sabethes* (*Peytonulus*).
whitmorei Dunn, 1918 (p. 128) – species of *Aedes* (*Howardina*).
whitmorei (Giles, 1904a) (p. 367) – species of *Culex* (*Culex*) [original combination: *Taeniorhynchus whitmorei*].
Synonyms: *argenteus* Ludlow, 1905a; *plegepennis* Theobald, 1907; *albus* Leicester, 1908; *loricatus* Leicester, 1908.
whittingtoni Belkin, 1962 (p. 204) – species of *Culex* (*Culex*).
wiedemanni Dyar & Knab, 1906c (p. 246) – synonym of *Toxorhynchites* (*Lynchiella*) *theobaldi* (Dyar & Knab, 1906c) [*nomen novum* for *ferox* Wiedemann, 1828 (in part); original combination: *Megarhinus wiedemanni*]; synonymy by Lane, 1953.
wigglesworthi Edwards, 1941 (p. 182) – species of *Aedes* (*Aedimorphus*).
wigglesworthi Edwards, 1941 (p. 264) – species of *Culex* (*Eumelanomyia*).
wilfredi Colless, 1965 (p. 297) – species of *Culex* (*Lophoceraomyia*).
wilkersoni Reinert, 1990 (p. 56) – species of *Aedes* (*Kenknightia*).
willardi Vargas, 1941d (p. 74) – synonym of *Anopheles* (*Anopheles*) *franciscanus* McCracken, 1904 [original combination: *Anopheles pseudopunctipennis* variety *willardi*]; synonymy by Russell *et al*., 1943.
willcocksii Theobald, 1907 (p. 296) – synonym of *Aedes* (*Ochlerotatus*) *caspius* (Pallas, 1771) [original combination: *Grabhamia willcocksii*]; synonymy established before Knight & Stone, 1977.
williamsoni Baisas & Hu, 1936 (p. 222) – synonym of *Anopheles* (*Anopheles*) *nigerrimus* Giles, 1900a [original combination: *Anopheles hyrcanus* variety *williamsoni*]; synonymy by Reid, 1953.
willistoni Giles, 1900a (p. 281) – synonym of *Culex* (*Culex*) *tarsalis* Coquillett, 1896; synonymy established before Knight & Stone, 1977.
willmori (James, 1903) (in Theobald, 1903a) (p. 100) – species of *Anopheles* (*Cellia*) [original combination: *Nyssorhynchus willmori*].
Synonyms: *dudgeonii* Theobald, 1907; *indica* Theobald, 1907; *maculosa* James & Liston, 1911.
wilsoni (Barreto & Coutinho, 1944) (p. 71) – species of *Mansonia* (*Mansonia*) [original combination: *Taeniorhynchus* (*Taeniorhynchus*) *wilsoni*].
wilsoni Evans, 1934 (p. 560) – species of *Anopheles* (*Cellia*).
wilsoni Lane & Whitman, 1943 (p. 393) – species of *Culex* (*Carrollia*).
wilsoni Ludlow, 1918 (p. 66) – synonym of *Trichoprosopon digitatum* (Rondani, 1848); synonymy established before Knight & Stone, 1977.
wilsoni Macdonald, 1958 (p. 121) – species of *Orthopodomyia*.
wilsoni Taylor, 1919 (p. 833) – synonym of *Aedes* (*Ochlerotatus*) *sagax* (Skuse, 1889) [original combination: *Culicada wilsoni*]; synonymy established before Knight & Stone, 1977.
winchesteri Cockerell, 1919 (p. 44) – fossil species of *Culex* Linnaeus, 1758.
winkleri Belkin, 1962 (p. 260) – species of *Culex* (*Lophoceraomyia*).
winter Dong, Wu & Mao, 2006 (p. 217) – species of *Topomyia* (*Topomyia*).
wolfsi Ribeiro, 2005 (p. 224) – species of *Toxorhynchites* (*Afrorhynchus*).
woodi Edwards, 1922d (p. 82) – species of *Aedes* (*Stegomyia*).
worontzowi Pessoa & Galvão, 1936 (p. 86) – species of *Culex* (*Microculex*).
wysockii Belkin, 1953 (p. 366) – species of *Uranotaenia* (*Uranotaenia*).

Xx

xanthogaster (Edwards, 1924b) (p. 366) – species of *Coquillettidia* (*Coquillettidia*) [original combination: *Taeniorhynchus* (*Coquillettidia*) *xanthogaster*].
xanthomelaena Edwards, 1925 (p. 259) – species of *Uranotaenia* (*Pseudoficalbia*).
xanthozona (van Someren, 1948) (p. 132) – species of *Mimomyia* (*Etorleptiomyia*) [original combination: *Ficalbia* (*Etorleptiomyia*) *xanthozona*].
xelajuensis de Leon, 1938 (p. 421) – species of *Anopheles* (*Anopheles*).
xenismus Harbach, 1995b (p. 583) – species of *Sabethes* (*Peytonulus*).
xenophobus Ronderos, 1965 (p. 249) – species of *Culex* (*Microculex*).
xhyphydes Harbach, 1994 (p. 222) – species of *Sabethes* (*Sabethinus*).
xiaokuanus Ma, 1981b (p. 67) – synonym of *Anopheles* (*Anopheles*) *pullus* Yamada, 1937; synonymy with *yatsushiroensis* Miyazaki, 1951 (= *pullus*) by Miao *et al*., 1988.

xivylis Dyar, 1920d (p. 78) – synonym of *Culex* (*Melanoconion*) ***intrincatus*** Brèthes, 1916; synonymy by Rozeboom & Komp, 1950a.
xui Dong, Zhou, Dong & Mao, 2007 (p. 37) – species of *Anopheles* (*Anopheles*).

Yy

yaeyamae Bohart, 1956 (species) (p. 29) – subspecies of *Toxorhynchites* (*Toxorhynchites*) *manicatus* (Edwards, 1921a); subspecific status by Tanaka *et al.*, 1979.
yaeyamaensis Somboon & Harbach, 2010 (in Somboon *et al.*, 2010) (p. 44) – species of *Anopheles* (*Cellia*).
yaeyamana Tanaka, Mizusawa & Saugstad, 1975a (p. 31) – species of *Uranotaenia* (*Pseudoficalbia*).
yaeyamensis Tanaka, Mizusawa & Saugstad, 1979 (p. 313) – subspecies of *Aedes* (*Hulecoeteomyia*) *japonicus* (Theobald, 1901a).
yaeyamensis Tanaka, Mizusawa & Saugstad, 1979 (p. 481) – subspecies of *Tripteroides* (*Tripteroides*) *bambusa* (Yamada, 1917).
yamadai (Ôuchi, 1939) (species) (p. 223) – subspecies of *Toxorhynchites* (*Toxorhynchites*) *manicatus* (Edwards, 1921a) [original combination: *Megarhinus yamadai*]; subspecific status by Tanaka *et al.*, 1979.
yamadai Sasa, Kano & Takahasi, 1950 (p. 635) – species of *Aedes* (*Aedes*).
yanbarensis Miyagi, 1976 (p. 202) – species of *Topomyia* (*Suaymyia*).
yanbarereoides Dong & Miyagi, 1995 (p. 343) – species of *Topomyia* (*Suaymyia*).
yangambiensis de Meillon & Lavoipierre, 1944 (p. 50) – species of *Aedes* (*Catageiomyia*).
yaoi Tung, 1955 (p. 311) – species of *Culex* (*Culiciomyia*).
yatsushiroensis Miyazaki, 1951 (p. 196) – synonym of *Anopheles* (*Anopheles*) ***pullus*** Yamada, 1937; synonymy by Wilkerson *et al.*, 2003.
ybarmis Dyar, 1920d (p. 57) – species of *Culex* (*Melanoconion*).
  Synonym: *jonistes* Dyar, 1920d.
yeageri Baisas, 1935d (p. 175) – species of *Culex* (*Eumelanomyia*).
yerburyi (Edwards, 1917) (p. 222) – species of *Verrallina* (*Neomacleaya*) [original combination: *Aedes yerburyi*].
yojoae Strickman, 1990 (p. 148) – species of *Culex* (*Culex*).
yongi Miyagi, Toma & Ramalingam, 1991 (p. 185) – species of *Topomyia* (*Topomyia*).
yovani van Someren, 1951 (p. 6) – species of *Uranotaenia* (*Pseudoficalbia*).
ypsipola Dyar, 1922d (p. 97) – species of *Wyeomyia* (*Dendromyia*).
yunnanensis Dong, Dong & Wu, 2004 (p. 566) – species of *Uranotaenia* (*Uranotaenia*).
yunnanensis Dong, Zhou & Dong, 1995a (p. 97) – species of *Armigeres* (*Armigeres*).
yunnanensis (Gaschen, 1934) (p. 331) – species of *Aedes* (*Hulecoeteomyia*) [original combination: *Finlaya yunnanensis*].
yunquensis Weinbren & Weinbren, 1970 (p. E-161) – *nomen nudum*, *Wyeomyia*.
yusafi (Barraud, 1931b) (p. 612) – species of *Verrallina* (*Harbachius*) [original combination: *Aedes* (*Aedes*) *yusafi*].
  Synonym: *siamensis* Delfinado, 1968.
yvonneae Edwards, 1941 (p. 176) – species of *Aedes* (*Catageiomyia*).

Zz

zairensis Ribeiro, 2005 (p. 226) – species of *Toxorhynchites* (*Afrorhynchus*).
zammitii (Theobald, 1903a) (p. 252) – species of *Aedes* (*Acartomyia*) [original combination: *Acartomyia zammitii*].
zavortinki Schick, 1970a (p. 45) – species of *Aedes* (*Protomacleaya*).
zelena Barraud, 1934 (p. 69) – synonym of *Uranotaenia* (*Uranotaenia*) ***macfarlanei*** Edwards, 1914c [original combination: *Uranotaenia campestris* variety *zelena*]; synonymy by Peyton, 1972.
zeltneri Neveu-Lemaire, 1906 (p. 251) – synonym of *Culex* (*Culex*) ***quinquefasciatus*** Say, 1823; synonymy by White, 1975.
zeteki Dyar, 1918e (as *zeteci*) (p. 122) – species of *Culex* (*Melanoconion*).
  Synonym: *loturus* Dyar, 1925g.
zethus de Meillon & Lavoipierre, 1944 (p. 57) – species of *Aedes* (*Paulianius*) [*nomen novum* for *niveus* de Meillon, 1943].
  Synonym: *niveus* de Meillon, 1943.
zhangi Gong, 1991 (p. 228) – species of *Topomyia* (*Topomyia*).
ziemanni Enderlein, 1923 (p. 27) – synonym of *Aedes* (*Stegomyia*) ***apicoargenteus*** (Theobald, 1909b) [original combination: *Aniella ziemanni*]; synonymy by Edwards, 1941.
ziemanni Grünberg, 1902 (p. 550) – species of *Anopheles* (*Anopheles*).
zinzala Zavortink, 1986 (p. 50) – species of *Wyeomyia* (*Zinzala*).
zombaensis Theobald, 1901c (p. 143) – species of *Culex* (*Culex*).
zonatipes Walker, 1961 (p. 229) – synonym of *Aedes* (*Stegomyia*) ***scutellaris*** (Walker, 1859) [original combination: *Culex zonatipes*]; synonymy established before Knight & Stone, 1977.
zoosophus Dyar & Knab, 1918b (p. 165) – species of *Aedes* (*Protomacleaya*).
  Synonym: *alleni* Turner, 1924.
zottae Ungureanu, 1956 (p. 4) – synonym of *Culiseta* (*Culiseta*) ***glaphyroptera*** (Schiner, 1864) [original combination: *Taeniorhynchus glaphyroptera* variety *zottae*]; synonymy established before Knight & Stone, 1977.

Section 2: Genus-group Names

Species that presumably evolved from a common ancestor are grouped together under genus-group names – genera and subgenera. For Linnaeus and his immediate successors, '*Culex*' was tantamount to 'mosquito' until the classification of species into other genera began in the 19th century. *Aedes* and *Anopheles* were introduced by Meigen (1818), *Psorophora* and *Sabethes* by Robineau-Desvoidy (1827), *Uranotaenia* by Lynch Arribálzaga (1891a, 1891b) and *Haemagogus* was established by Williston (1896).

The discovery at the turn of the 19th century that mosquitoes transmit the agents that cause malaria and yellow fever fuelled increased interest in these insects. In 1899, Fred. V. Theobald was commissioned to prepare a monograph on the mosquitoes of the world, which was published in five volumes between 1901 and 1910. Theobald (1910c) recognized 149 genera in a system of classification based principally on the distribution and shape of scales on adult mosquitoes that was neither practical nor natural. It is interesting to note that Theobald, as well as Howard, Dyar & Knab (1915, 1917), only recognized genera, no subgenera. Significant changes were made towards a much more conservative system of classification during the two decades following the publication of Theobald's *Monograph of the Culicidae*. The practice of recognizing fewer genera and many subgenera was firmly established by Edwards (1932a), who reduced 327 genus-group names to 30 genera and 59 subgenera, which provided the framework on which the 'traditional' classification of the family was built (Stone *et al.*, 1959; Knight & Stone, 1977). Increases in the number of valid genus-group names recognized since Edwards (1932a) are shown in the table at the bottom of the page. The table clearly shows that the practice of recognizing a greater number of subgenera has continued, with two times more subgenera than genera recognized since Edwards (1932a) – there has been a 2.73% increase in the number of subgenera compared with a 1.32% increase in the number of genera.

The list below contains 441 genus-group names: 209 currently valid names and 232 synonyms. In addition to the 41 extant genera and 162 extant subgenera noted in the table, the valid names include five genera of fossil mosquitoes and one *nomen dubium*.

There are two principal types of genus-group names: names derived from Greek and Latin and names formed from personal names. Both types are included in the list below, but the latter are treated in more detail in Section 6. Both types of names are, or are treated as, singular nouns in the nominative case (*Code*, Article 11.8).

While this tome was in its final stage of preparation for publication, Foster *et al.* (2017) proposed the elevation of ***Kerteszia*** Theobald, ***Lophopodomyia*** Antunes, ***Nyssorhynchus*** Blanchard and ***Stethomyia*** Theobald from subgeneric status in ***Anopheles*** Meigen to generic status based on phylogenetic analyses of mitochondrial protein coding genes. Although the approach and results of the study are laudable, the justification for changing the rank of the four groups is weak and untenable and neglects to consider the consequences of the proposal. For example, in speaking with the first author of the paper, it is obvious that the authors failed to realize that the spelling of adjectival names (currently masculine as names of *Anopheles*) would have to be changed to agree in gender with the feminine *Kerteszia*, *Lophopodomyia* and *Stethomyia*. As the groups are fully diagnosable and no less manageable as subgenera, their elevation to generic status is clearly unnecessary, and the assertion that the change of status adds stability to the classification of Anophelinae is unsubstantiated. For these reasons, and because much of the scientific community, as in the case of the multi-generic reclassification of *Aedes* Meigen by Reinert *et al.* (2009), is unlikely to accept the proposed generic rank of the four groups, their subgeneric status is retained herein.

Format

The list below includes all genus-group names applied to mosquitoes from Linnaeus (1758). The names are listed in alphabetical order. Valid names (senior synonyms), including

	Stone *et al.* (1959)	Knight & Stone (1977)	Knight (1978b)	Ward (1984); Gaffigan & Ward (1985)	Ward (1992)	June 2018, post-Wilkerson *et al.* (2015)
Genera	31	34	34	37	37	41
Subgenera	95	130	131	134	144	162

the sole *nomen dubium*, are in ***boldface italics*** and all synonymous names (junior synonyms) are in *italicized normal type*. Each name is followed by authorship, date of publication, page of citation, gender (indicated by f., feminine; m., masculine; n, neuter), taxonomic rank and source of taxonomic status (if applicable). Junior synonyms are also listed for the valid taxa, in chronological order. Literature citations are provided for synonymies recorded in the catalogue of Knight & Stone (1977) and later publications; however, in most cases where the source of a synonymy is not recorded in the catalogue, synonymous names are followed by 'synonymy established before Knight & Stone, 1977'.

Gender

It is important to know the gender of genus-group names to ensure that adjectival and participial species-group names agree in gender with the genus name with which they are combined. The requirement for gender agreement (*Code*, Article 32.2) was one of the reasons many mosquito workers, principally non-taxonomists, opposed the splitting of the 'traditional' broad concept of *Aedes* into many genera by Reinert *et al.* (2009), e.g. the change of the familiar masculine *Aedes albopictus* to the feminine *Stegomyia albopicta* (see Section 8). The gender of most genus-group names is easily identified by the terminal letter or letters of the Greek, Latin or latinized name. The common terminal letters of genus-group names in Culicidae are listed by gender in the following table.

Feminine	Masculine	Neuter
-a (Gr., from *-a* in *-ia*)	*-es* (Gr.)	*-ion* (Gr.)
-ella (L.)	*-ius* (L.)	*-on* (Gr., from *-on* in *ion*)
-ia (L., Gr.)	*-is* (L.)	*-um* (L.)
-iana (L.)	*-ulus* (L.)	
-ina (L.)	*-us* (L.)	
-is (L.)	*-x* (L.)	

Naturally there are exceptions to these typical endings. A number of names ending in *-a* are masculine rather than feminine, including ***Borichinda***, *Mukwaya*, ***Udaya*** and *Yamada* which are masculine patronymics formed from the surnames of men. In the case of names that are compound words formed from two or more components, the gender of the final component, whether a noun or suffix, determines the gender of the name (*Code*, Article 30.1), and the terminal letter or letters of such names may betray their true gender. ***Lasiosiphon***, for example, is a compound word formed by combining two Greek words, *lasios* (m., hairy, woolly) and *siphon* (m., pipe, bent tube), in reference to the setose siphon of larvae. The *-on* ending would suggest that the gender of the name is neuter when in fact it is masculine. Similarly, the gender of *Heteraspidion* and *Lepidokeneon* is masculine because *aspidion* of the former (meaning 'little shield') and *keneon* of the latter (meaning 'flank') are Greek masculine nouns. ***Janthinosoma*** and ***Lepiothauma*** would appear to be feminine when in fact they are neuter. Words ending in *-soma* (Gr. *soma*) are neuter and the Greek noun *thauma*, meaning wonder or marvel, is neuter. However, mistakes are easily made. Enderlein (1923) clearly considered *Lepiothauma* to be feminine when he named the new genus and new species, the type species, as *Lepiothauma furfurea*. *Culicada* and *Culicelsa*, proposed as genera by Felt (1904), are feminine but were used in combination with species names that are clearly masculine.

Names ending in *-is* can be feminine or masculine. As indicated by Sabrosky (1999), '*-is* names are a mixture of divergent usage and difficulties'. To avoid difficulties in determining the gender of some such names, it is sometimes helpful to refer to the binomials established by the author of the genus-group name. A check of original binomials reveals that *Melanolepis* and ***Menolepis*** were both correctly used as feminine names because they end in a Greek word transliterated into Latin without change (*Code*, Article 30.1.2) – names ending in *-lepis* (Gr. *lepis*) are feminine. However, original combinations do not always reveal the correct gender of the generic name. Lutz (1905) incorrectly introduced ***Prosopolepis*** as a masculine name in the combination *Prosopolepis confusus*. There is no doubt that ***Paleoculicis*** is masculine as the terminal component is derived from the Latin masculine word *Culex*, *-icis*, meaning mosquito.

Whereas *Mochlostyrax* ends in *-x* and is masculine (the terminal component of this compound word is the Greek masculine noun *styrax*, meaning 'a spike at the butt end of a spear'), *Nyx*, a single noun ending in *-x*, is feminine, named for the Greek goddess of the night. *Actinothrix*, another compound word ending in *-x*, is also feminine because the final component, *thrix*, is a Greek feminine noun. *Actinothrix* (Gr. *aktis*, *-inos*, meaning 'ray' or 'beam', and *thrix*, meaning 'hair') refers to the unusual development of larval setae 1–3-P ('setae like rays').

The gender of another ten genus-group names is not apparent because they end in letters that are not listed in the table. One of these names, *Xyele*, is a single noun and the others are compound words. *Xyele*, named in reference to the dagger-like gonostylar claw of the male genitalia, is derived from the Greek masculine noun *xyele*, meaning 'a kind of dagger'. *Chrysoconops*, *Cyanoconops*, *Hyloconops*, *Lasioconops* and *Stegoconops* are all masculine because the terminal component (*conops*) derives from the Greek masculine noun *konops*, meaning 'gnat' (note, however, that Article 30.1.4.3 of the *Code* mandates that compound genus-group names ending in *-ops* are always treated as masculine regardless of their derivation). *Empihals*, *Lepidoplatys* and *Longipalpifer* are masculine. *Empihals* is derived from the Greek masculine nouns *empis*, meaning 'mosquito' or 'gnat', and *hals*, meaning 'salt' or 'sea', in reference to larvae occurring in brackish water; *Lepidoplatys* is formed from *lepis*, *-idos*, *lepido-* (f., Gr. scale) and *platys* (m., Gr. broad, wide, level, flat); and *Longipalpifer* is compounded from three masculine Latin elements: *longus* (long; *longius*, longer), *palpo* (touch, stroke, feel), and *-fer*, *-a*, *-um*, suffix meaning 'bear, carry, have' (from *fero*, carry, bear). The final *-e* of *Diphalangarpe* might initially suggest that the word is masculine when in fact it is feminine. It consists of three Greek components: *di-* (two, double), *phalanga* (f., pole, roller) and *harpe* (f., sickle, hook), so named because the male genitalia bear 'two finger-shaped processes on top of the harpes [paraprocts]' (Dyar, 1919d).

Genus-group names in Culicidae

Abraedes Zavortink, 1970c (p. 2) – m., subgenus of *Aedes* Meigen, 1818; subgeneric status re-established by Wilkerson *et al.*, 2015.

Acallyntrum Stone & Penn, 1948 (p. 109) – n., subgenus of *Culex* Linnaeus, 1758.

Acartomyia Theobald, 1903a (f., genus) (p. 251) – subgenus of *Aedes* Meigen, 1818; subgeneric status by Wilkerson *et al.*, 2015.

Actinothrix Reinert, Harbach & Kitching, 2009 (f., subgenus of *Stegomyia* Theobald, 1901) (p. 781) – synonym of *Stegomyia* Theobald, 1901; synonymy by Wilkerson *et al.*, 2015.

Aedeomyia Theobald, 1901b(a,c) (pp. 235, 98, 218 respectively) – f., genus (subgenus) of tribe Aedeomyiini Theobald, 1901d.

Synonym: *Aedomyia* Giles, 1902.

Aedomyia Giles, 1902 (pp. 475, 478) – unjustified emendation of *Aedeomyia* Theobald, 1901b(a,c).

Aedes Meigen, 1818 (p. 13) – m., genus (subgenus) of tribe Aedini Neveu-Lemaire, 1902c.

Aedimorphus Theobald, 1903a (m., genus) (p. 290) – subgenus of *Aedes* Meigen, 1818; subgeneric status re-established by Wilkerson *et al.*, 2015.

Synonyms: *Ecculex* Felt, 1904; *Reedomyia* Ludlow, 1905a; *Lepidotomyia* Theobald, 1905f; *Pecomyia* Theobald, 1905g; *Pseudograbhamia* Theobald, 1905h; *Mimeteculex* Theobald, 1908c; *Geitonomyia* Leicester, 1908; *Bathosomyia* Theobald, 1909b; *Leslieomyia* Christophers, 1911.

Aedinus Lutz, 1904 (in Bourroul, 1904) (m., genus) (p. 43) – subgenus of *Culex* Linnaeus, 1758; subgeneric status by Edwards, 1932a, redefined by Belkin, 1968b.

Synonym: *Eubonnea* Dyar, 1919e.

Aetheapnomyia Harbach & Greenwalt, 2012 (p. 33) – f., genus of fossil Culicidae, subfamily Culicinae Meigen, 1818.

Afroculex Danilov, 1989 (p. 790) – m., subgenus of *Culex* Linnaeus, 1758.

Afrorhynchus Ribeiro, 1992a (p. 195) – m., subgenus of *Toxorhynchites* Theobald, 1901b.

Aglaonotus Reinert, Harbach & Kitching, 2009 (m., subgenus of *Petermattinglyius* Reinert, Harbach & Kitching, 2009) (p. 779) – synonym of *Petermattinglyius* Reinert, Harbach & Kitching, 2009; synonymy by Wilkerson *et al.*, 2015.

Aioretomyia Leicester, 1908 (f., genus) (p. 185) – synonym of *Neomacleaya* Theobald, 1907; synonymy established before Knight & Stone, 1977.

Alanstonea Mattingly, 1960 (p. 170) – f., subgenus of *Aedes* Meigen, 1818; subgeneric status re-established by Wilkerson *et al.*, 2015.

Albuginosus Reinert, 1987 (p. 308) – m., subgenus of *Aedes* Meigen, 1818; subgeneric status re-established by Wilkerson *et al.*, 2015.

Aldrichia Theobald, 1903a (f., genus) (p. 353) – synonym of *Cellia* Theobald, 1902b; synonymy established before Knight & Stone, 1977.

Aldrichinella Theobald, 1910c (f., genus) (pp. 3, 77) – synonym of *Cellia* Theobald, 1910c [*nomen novum* for *Aldrichia* Theobald, 1903a]; synonymy established before Knight & Stone, 1977.

Allimanta Casal & García, 1968a (p. 331) – f., subgenus of *Culex* Linnaeus, 1758.

Alloeomyia Reinert, Harbach & Kitching, 2008 (f., subgenus of *Collessius* Reinert, Harbach & Kitching, 2006) (p. 97) – synonym of *Collessius* Reinert, Harbach & Kitching, 2006; synonymy by Wilkerson *et al.*, 2015.

Allotheobaldia Brolemann, 1919b (p. 91) – f., subgenus of *Culiseta* Felt, 1904.

Andersonia Strickland, 1911b (f., genus) (pp. 250) – synonym of *Ochlerotatus* Lynch Arribálzaga, 1891a,b; synonymy established before Knight & Stone, 1977.

Aniella Enderlein, 1923 (f., genus) (p. 26) – synonym of *Stegomyia* Theobald, 1901; synonymy established before Knight & Stone, 1977.

Anisocheleomyia Theobald, 1905a (f., genus) (p. 52) – synonym of *Uranotaenia* Lynch Arribálzaga, 1891b; synonymy established before Knight & Stone, 1977.

Ankylorhynchus Lutz, 1904 (in Bourroul, 1904) (m., genus) (pp. 53, 65) – subgenus of *Toxorhynchites* Theobald, 1901b; subgeneric status by ? Dyar, 1928.

Anoedioporpa Dyar, 1923e (f., genus) (p. 190) – subgenus of *Culex* Linnaeus, 1758; subgeneric status by Belkin, 1968b.

Anopheles Meigen, 1818 (p. 10) – m., genus (subgenus) of subfamily Anophelinae Grassi, 1900.

Synonyms: *Cyclolepidopteron* Blanchard, 1901; *Cycloleppteron* Theobald, 1901b; *Myzorhynchus* Blanchard, 1902; *Rossia* Theobald, 1902b; *Lophomyia* Giles, 1904a; *Lophoscelomyia* Theobald, 1904c; *Arribalzagaia* Blanchard, 1905; *Nototricha* Coquillett, 1906d; *Coelodiazesis* Dyar & Knab, 1906d; *Neostethopheles* James, 1910b; *Patagiamyia* James, 1910b; *Cyclophorus* Eysell, 1912; *Proterorhynchus* Brèthes, 1912; *Memnemyia* Strickland, 1915a; *Shannonesia* da Fonseca & Ramos, 1940; *Shannoniella* da Fonseca & da Silva Ramos, 1940; *Ayrozamyia* Coutinho & Farias, 1942; *Russellia* Vargas, 1943a; *Maculipennia* Buonomini & Mariani, 1946 (not valid as no type was designated; not absolute tautonymy); *Missirolimyia* Vargas, 1950a; *Maculipennia* Buonomini & Mariani, 1953.

Antunesmyia Lane & Cerqueira, 1942 (p. 587) – f., subgenus of *Wyeomyia* Theobald, 1901b.

Aporoculex Theobald, 1907 (m., genus) (p. 316) – synonym of *Culex* Linnaeus, 1758; synonymy established before Knight & Stone, 1977.

Armigeres Theobald, 1901b (p. 235) – m., genus (subgenus) of tribe Aedini Neveu-Lemaire, 1902c.
Synonyms: *Desvoidya* Blanchard, 1901; *Blanchardiomyia* Brunetti, 1912; *Neosquamomyia* Taylor, 1914d.

Arribalzagaia Blanchard, 1905 (f., genus) (p. 634) – emendation of *Arribalzagia* Theobald, 1903a; synonym of *Anopheles* Meigen, 1818; synonymy established before Knight & Stone, 1977.

Arribalzagia Theobald, 1903a (f., genus) (pp. 13, 81) – synonym of *Anopheles* Meigen, 1818; synonymy established before Knight & Stone, 1977.

Arthuromyia Galvão, 1941b (f., subgenus of *Anopheles* Meigen, 1818) (p. 564) – synonym of *Lophopodomyia* Antunes, 1937; synonymy established before Knight & Stone, 1977.

Asebeomyia Aiken, 1911 (f., genus) (p. 193) – *nomen dubium*, Culicini; status by Sirivanakarn, 1983.

Austromansonia Belkin, 1968a (p. 76) – f., subgenus of *Coquillettidia* Dyar, 1905a.

Austrotheobaldia Dobrotworsky, 1954 (p. 68) – f., subgenus of *Culiseta* Felt, 1904.

Ayrozamyia Coutinho & Farias, 1942 (f., subgenus of *Anopheles* Meigen, 1818) (p. 343) – synonym of *Anopheles* Meigen, 1818; synonymy established before Knight & Stone, 1977.

Ayurakitia Thurman, 1954a (f., genus) (p. 198) – subgenus of *Aedes* Meigen, 1818; subgeneric status re-established by Wilkerson *et al.*, 2015.

Aztecaedes Zavortink, 1972 (p. 66) – m., subgenus of *Aedes* Meigen, 1818; subgeneric status re-established by Wilkerson *et al.*, 2015.

Baimaia Harbach, Rattanarithikul & Harrison, 2005 (p. 752) – f., subgenus of *Anopheles* Meigen, 1818.

Bancroftia Lutz, 1904 (in Bourroul, 1904) (f., genus) (pp. 40, 59) – synonym of *Orthopodomyia* Theobald, 1904c; synonymy established before Knight & Stone, 1977.

Banksinella Theobald, 1907 (f., genus) (p. 468) – synonym of *Neomelaniconion* Newstead, 1907; synonymy established before Knight & Stone, 1977.

Barraudius Edwards, 1921d (p. 332) – m., subgenus of *Culex* Linnaeus, 1758.

Bathosomyia Theobald, 1909b (f., genus) (p. 9) – synonym of *Aedimorphus* Theobald, 1903a; synonymy established before Knight & Stone, 1977.

Belkinius Reinert, 1982 (p. 102) – m., subgenus of *Aedes* Meigen, 1818; subgeneric status re-established by Wilkerson *et al.*, 2015.

Belkinomyia Adames & Galindo, 1973 (p. 55) – f., subgenus of *Culex* Linnaeus, 1758.

Bifidistylus Reinert, Harbach & Kitching, 2009 (m., genus) (p. 766) – subgenus of *Aedes* Meigen, 1818; subgeneric status by Wilkerson *et al.*, 2015.

Binotia Blanchard, 1904 (f., substitute name) (p. 478) – synonym of *Runchomyia* Theobald, 1903a; synonymy established before Knight & Stone, 1977.

Bironella Theobald, 1905f (p. 69) – f., genus (subgenus) of subfamily Anophelinae Grassi, 1900.

Blanchardiomyia Brunetti, 1912 (f., genus) (p. 440) – synonym of *Armigeres* Theobald, 1901b; synonymy established before Knight & Stone, 1977.

Bohartius Reinert, Harbach & Kitching, 2009 (m., subgenus of *Stegomyia* Theobald, 1901) (p. 781) – synonym of *Stegomyia* Theobald, 1901; synonymy by Wilkerson *et al.*, 2015.

Bolbodeomyia Theobald, 1910a (f., genus) (p. 31) – synonym of *Heizmannia* Ludlow, 1905a; synonymy established before Knight & Stone, 1977.

Borichinda Harbach & Rattanarithikul, 2007 (m., genus) (in Harbach, Rattanarithikul, Howard *et al.*, 2007) (p. 471) – subgenus of *Aedes* Meigen, 1818; subgeneric status by Wilkerson *et al.*, 2015.

Bothaella Reinert, 1973e (p. 2) – f., subgenus of *Aedes* Meigen, 1818; subgeneric status re-established by Wilkerson *et al.*, 2015.

Boycia Newstead, 1907 (in Newstead *et al.*, 1907) (f., genus) (p. 33) – synonym of *Mimomyia* Theobald, 1903a; synonymy established before Knight & Stone, 1977.

Brachiomyia Theobald, 1901c (f., genus) (p. 343) – synonym of *Deinocerites* Theobald, 1901b; synonymy established before Knight & Stone, 1977.

Brachiosoma Theobald, 1901b (n., genus) (p. 235) – synonym of *Deinocerites* Theobald, 1901b; synonymy established before Knight & Stone, 1977.

Brevirhynchus Theobald, 1908b (m., genus) (p. 293) – synonym of *Leicesteria* Theobald, 1904c; synonymy established before Knight & Stone, 1977.

Bruceharrisonius Reinert, 2003 (p. 309) – m., subgenus of *Aedes* Meigen, 1818; subgeneric status re-established by Wilkerson *et al.*, 2015.

Brugella Edwards, 1930a (p. 288) – f., subgenus of *Bironella* Theobald, 1905f.

Burmaculex Borkent & Grimaldi, 2004 (p. 883) – m., genus of fossil Culicidae, subfamily Burmaculicinae Borkent & Grimaldi, 2016.

Buvirilia Reinert, Harbach & Kitching, 2008 (f., subgenus of *Ochlerotatus* Lynch Arribálzaga, 1891a,b) (p. 104) – synonym of *Ochlerotatus* Lynch Arribálzaga, 1891a,b; synonymy by Wilkerson *et al.*, 2015.

Cacoculex Dyar, 1918c (m., subgenus of *Culex* Linnaeus, 1758) (p. 100) – synonym of *Culex* Linnaeus, 1758; synonymy established before Knight & Stone, 1977.

Cacomyia Coquillett, 1906d (f., genus) (pp. 16, 25) – synonym of *Haemagogus* Williston, 1896; synonymy established before Knight & Stone, 1977.

Caenocephalus Taylor, 1914c (m., genus) (p. 700) – synonym of *Pseudoskusea* Theobald, 1907; synonymy established before Knight & Stone, 1977.

Caenomyiella Harbach & Peyton, 1990 (p. 16) – f., subgenus of *Wyeomyia* Theobald, 1901b.

Calladimyia Dyar, 1919d (f., genus) (p. 137) – synonym of *Wyeomyia* Theobald, 1901b; synonymy by transfer of type species (*pandora* Dyar & Knab, 1909d, as synonym of *melanocephala* Dyar & Knab, 1906a) from subgenus *Dendromyia* Theobald, 1903a to subgenus uncertain by Motta & Lourenço-de-Oliveira, 1995.

Calomyia Taylor, 1914c (f., genus) (p. 684) – synonym of *Molpemyia* Theobald, 1910c; synonymy by transfer of type species (as synonym of *purpurea* Theobald, 1910c) from *Finlaya* Theobald, 1903a by Reinert, 1993.

Calvertia Ludlow, 1908b, 1909a (f., genus) (pp. 10, 22, respectively) – synonym of *Cellia* Theobald, 1902b; synonymy established before Knight & Stone, 1977.

Calvertina Ludlow, 1909c (f., genus) (p. 234) – synonym of *Cellia* Theobald, 1902b [*nomen novum* for *Calvertia* Ludlow, 1908b]; synonymy established before Knight & Stone, 1977.

Calviana Strand, 1928 (f., genus) (p. 48) – synonym of *Cellia* Theobald, 1902b [*nomen novum* for *Calvertina* Ludlow, 1909c]; synonymy established before Knight & Stone, 1977.

Cancraedes Edwards, 1929b (p. 342) – m., subgenus of *Aedes* Meigen, 1818; subgeneric status re-established by Wilkerson *et al.*, 2015.

Carrollella Lutz, 1921 (f., genus, as *Carollella*) (p. 163) – synonym of *Carrollia* Lutz, 1905; synonymy established before Knight & Stone, 1977.

Carrollia Lutz, 1905 (f., genus) (p. 81) – subgenus of *Culex* Linnaeus, 1758.
Synonym: *Carrollella* Lutz, 1921.

Catageiomyia Theobald, 1903d (f., genus) (p. i) – subgenus of *Aedes* Meigen, 1818; subgeneric status by Wilkerson *et al.*, 2015.
Synonyms: *Duttonia* Newstead, 1907; *Myxosquamus* Theobald, 1909b; *Neopecomyia* Theobald, 1909b; *Stenoscutus* Theobald, 1909b.

Catatassomyia Dyar & Shannon, 1925a (f., genus) (p. 71) – subgenus of *Aedes* Meigen, 1818; subgeneric status by Wilkerson *et al.*, 2015.

Cellia Theobald, 1902b, 1903a (f., genus) (pp. 181, 107, respectively) – subgenus of *Anopheles* Meigen, 1818; subgeneric status by Stone *et al.*, 1959.
Synonyms: *Grassia* Theobald, 1902b; *Howardia* Theobald, 1902b; *Myzomyia* Blanchard, 1902; *Pyretophorus* Blanchard, 1902; *Aldrichia* Theobald, 1903a; *Feltinella* Theobald, 1907; *Neocellia* Theobald, 1907; *Pseudomyzomyia* Theobald, 1907; *Calvertia* Ludlow, 1908b, 1909a; *Calvertina* Ludlow, 1909c; *Aldrichinella* Theobald, 1910c; *Neomyzomyia* Theobald, 1910c; *Christophersia* James, 1910a; *Nyssomyzomyia* James, 1910a, 1910b; *Dactylomyia* Newstead & Carter, 1910; *Calviana* Strand, 1928; *Paramyzomyia* Christophers & Barraud, 1931; *Palaeomyzomyia* Swellengrebel & Rodenwaldt, 1932; *Eomyzomyia* Evans, 1934.

Ceratocystia Dyar & Knab, 1906d (f., genus) (p. 183) – synonym of *Grabhamia* Theobald, 1903a; synonymy established before Knight & Stone, 1977.

Chaetocruiomyia Theobald, 1910c (f., genus) (pp. 113, 195) – synonym of *Macleaya* Theobald, 1903e; synonymy re-established by Wilkerson *et al.*, 2015.

Chaetomyia Leicester, 1908 (f., genus) (p. 100) – synonym of *Leicesteria* Theobald, 1904c; synonymy established before Knight & Stone, 1977.

Chagasia Cruz, 1906 (p. 199) – f., genus of subfamily Anophelinae Grassi, 1900.

Choeroporpa Dyar, 1918c (f., subgenus of *Culex* Linnaeus, 1758) (p. 103) – synonym of *Melanoconion* Theobald, 1903a; synonymy established before Knight & Stone, 1977.

Christophersia James, 1910a, 1910b (f., genus) (pp. 33, 103, respectively) – synonym of *Cellia* Theobald, 1902b; synonymy established before Knight & Stone, 1977.

Christophersiomyia Barraud, 1923a (f., genus) (p. 786) – subgenus of *Aedes* Meigen, 1818; subgeneric status re-established by Wilkerson *et al.*, 2015.

Christya Theobald, 1903b (f., genus) (p. 34) – subgenus of *Anopheles* Meigen, 1818; resurrected from synonymy with *Anopheles* Meigen, 1818 by Harbach & Kitching, 2015.

Chrysoconops Goeldi, 1905 (m., genus) (p. 114) – synonym of *Ochlerotatus* Lynch Arribálzaga, 1891a,b; synonymy re-established by Wilkerson *et al.*, 2015.

Cleobonnea Dyar, 1919d (f., genus) (p. 134) – synonym of *Wyeomyia* Theobald, 1901b (formerly a synonym of *Dendromyia* Theobald, 1903a; type species *occulta* Bonne-Wepster & Bonne, 1919 currently of uncertain subgeneric placement).

Climacura Howard, Dyar & Knab, 1915 (subgenus of *Culex* Linnaeus, 1758) (p. 452) – subgenus of *Culiseta* Felt, 1904; subgeneric status by Edwards, 1932a, as subgenus of *Theobaldia* Neveu-Lemaire, 1902a.

Coelodiazesis Dyar & Knab, 1906d (m., genus) (p. 177) – synonym of *Anopheles* Meigen, 1818; synonymy established before Knight & Stone, 1977.

Coetzeemyia Huang, Mathis & Wilkerson, 2010 (p. 14) – f., subgenus of *Aedes* Meigen, 1818; subgeneric status re-established by Wilkerson *et al.*, 2015.

Collessius Reinert, Harbach & Kitching, 2006 (m., genus) (p. 64) – subgenus of *Aedes* Meigen, 1818; subgeneric status by Wilkerson *et al.*, 2015.

Colonemyia Leicester, 1908 (f., genus) (p. 233) – synonym of *Tripteroides* Giles, 1904a; synonymy established before Knight & Stone, 1977.

Conchyliastes Theobald, 1901 (in Howard, 1901) (m., genus) (pp. 155, 235) – synonym of *Janthinosoma* Lynch Arribálzaga; synonymy established before Knight & Stone, 1977.

Conopomyia Leicester, 1908 (f., genus) (p. 113) – synonym of *Mimomyia* Theobald, 1903a; synonymy established before Knight & Stone, 1977.

Conopostegus Dyar, 1925f (m., subgenus of *Aedes* Meigen, 1818) (p. 141) – subgenus of *Haemagogus* Williston, 1896; transfer from subgeneric status in *Aedes* Meigen by Zavortink, 1972.

Coquillettidia Dyar, 1905a (p. 47) – f., genus (subgenus) of tribe Mansoniini Belkin, 1962.

Cornetius Huang, 2005 (p. 517) – m., subgenus of *Aedes* Meigen, 1818; subgeneric status re-established by Wilkerson *et al.*, 2015.

Cruzmyia Lane & Cerqueira, 1942 (p. 579) – f., subgenus of *Wyeomyia* Theobald, 1901b.

Ctenogoeldia Edwards, 1930a (f., subgenus of *Goeldia* Theobald, 1903a) (p. 301) – subgenus of *Runchomyia* Theobald, 1903a; transfer from subgeneric status in *Trichoprosopon* Theobald1901a,c by Zavortink, 1979.

Culex Linnaeus, 1758 (p. 602) – m., genus (subgenus) of tribe Culicini Meigen, 1818.

Synonyms: *Heteronycha* Lunch Arribálzaga, 1891a,b; *Heptaphlebomyia* Theobald, 1903a; *Lasioconops* Theobald, 1903c; *Pseudoheptaphlebomyia* Ventrillon, 1905b; *Trichopronomyia* Theobald, 1905f; *Aporoculex* Theobald, 1907; *Leucomyia* Theobald, 1907; *Theobaldiomyia* Brunetti, 1912; *Phalangomyia* Dyar & Knab, 1914; *Laiomyia* Izquierdo, 1916; *Transculicia* Dyar, 1918a; *Cacoculex* Dyar, 1918c.

Culicada Felt, 1904 (m., genus) (p. 391b) – synonym of *Ochlerotatus* Lynch Arribálzaga, 1891a,b; synonymy re-established by Wilkerson *et al.*, 2015.

Culicella Felt, 1904 (f., genus) (p. 391c) – subgenus of *Culiseta* Felt, 1904; subgeneric status by Edwards, 1932a, as subgenus of *Theobaldia* Neveu-Lemaire, 1902a.

Culicelsa Felt, 1904 (m., genus) (p. 391b) – synonym of *Ochlerotatus* Lynch Arribálzaga, 1891a,b; synonymy re-established by Wilkerson *et al.*, 2015.

Culiciomyia Theobald, 1907 (f., genus) (p. 227) – subgenus of *Culex* Linnaeus, 1758; subgeneric status by Borel, 1926.

Synonyms: *Trichorhynchus* Theobald, 1905h; *Neomelanoconion* Theobald, 1907; *Pectinopalpus* Theobald, 1909b; *Trichorhynchomyia* Brunetti, 1912.

Culicites Meunier, 1915 (m., genus) (p. 16) – synonym of *Neoculicites* Evenhuis, 1994; synonymy by Evenhuis, 1994.

Culiseta Felt, 1894 (p. 391c) – f., genus (subgenus) of tribe Culisetini Belkin, 1962.

Synonyms: *Theobaldia* Neveu-Lemaire, 1902a; *Theobaldinella* Blanchard, 1905; *Pseudotheobaldia* Theobald, 1907.

Cyanoconops Antunes, 1939 (p. 106) – m., *nomen nudum*, *Haemagogus*.

Cyathomyia de Meijere, 1910 (f., genus) (p. 921) – synonym of *Lophoceraomyia* Theobald, 1905h; synonymy established before Knight & Stone, 1977.

Cyclolepidopteron Blanchard, 1901 (n., genus) (p. 1046) – synonym of *Anopheles* Meigen, 1818; synonymy established before Knight & Stone, 1977.

Cycloleppteron Theobald, 1901b, 1901c (n., genus) (pp. 234, 312, respectively) – synonym of *Anopheles* Meigen, 1818; synonymy established before Knight & Stone, 1977.

Cyclophorus Eysell, 1912 (m., genus) (p. 422) – synonym of *Anopheles* Meigen, 1818; synonymy established before Knight & Stone, 1977.

Dactylomyia Newstead & Carter, 1910 (f., genus) (p. 377) – synonym of *Cellia* Theobald, 1902b; synonymy established before Knight & Stone, 1977.

Dahliana Reinert, Harbach & Kitching, 2006 (f., genus) (p. 66) – subgenus of *Aedes* Meigen, 1818; subgeneric status by Wilkerson *et al.*, 2015.

Danielsia Theobald, 1904c (f., genus) (p. 78) – subgenus of *Aedes* Meigen, 1818; subgeneric status by Wilkerson *et al.*, 2015.

Dasymyia Leicester, 1908 (f., genus) (p. 102) – synonym of *Ingramia* Edwards, 1912c; synonymy by White, 1974.

Davismyia Lane & Cerqueira, 1942 (f., subgenus of *Wyeomyia* Theobald, 1901b) (p. 582) – subgenus of *Sabethes* Robineau-Desvoidy, 1827; transfer from *Wyeomyia* Theobald, 1901b by Harbach & Peyton, 1991.

Decamyia Dyar, 1919d (f., genus) (p. 135) – subgenus of *Wyeomyia* Theobald, 1901b; subgeneric status by Harbach & Peyton, 1990.

Synonym: *Heliconiamyia* Dyar, 1919d.

Deinocerites Theobald, 1901a,b,c (pp. 98, 235, 215, respectively) – m., genus of tribe Culicini Meigen, 1818.

Synonyms: *Brachiosoma* Theobald, 1901b; *Brachiomyia* Theobald, 1901c; *Deinokerides* Giles, 1902; *Dinocerites* Blanchard, 1905; *Dinomimetes* Knab, 1907b; *Dinanamesus* Dyar & Knab, 1909d.

Deinokerides Giles, 1902 (m., genus) (p. 472) – synonym of *Deinocerites* Theobald, 1901a,b,c [as emendation?]; synonymy established before Knight & Stone, 1977.

Dendromyia Theobald, 1903a (f., genus) (p. 313) – subgenus of *Wyeomyia* Theobald, 1901b; subgeneric status by Edwards, 1932a, redefined by Motta & Lourenço-de-Oliveira, 2000.

Synonyms: *Shropshirea* Dyar, 1922d; *Melanolepis* Dyar, 1928.

Dendropaedium Dyar & Knab, 1918a (n., subgenus of *Anopheles* Meigen, 1818) (p. 141) – synonym of *Kerteszia* Theobald, 1905f; synonymy established before Knight & Stone, 1977.

Dendroskusea Edwards, 1929b (p. 341) – f., subgenus of *Aedes* Meigen, 1818; subgeneric status re-established by Wilkerson *et al.*, 2015.

Desvoidea Theobald, 1903a (f., genus) (p. 134) – variant spelling of *Desvoidya* Blanchard, 1901.

Desvoidya Blanchard, 1901 (f., genus) (p. 1046) – synonym of *Armigeres* Theobald, 1901b; synonymy established before Knight & Stone, 1977.

Diceromyia Theobald, 1911c (f., genus) (p. 151) – subgenus of *Aedes* Meigen, 1818; subgeneric status re-established by Wilkerson *et al.*, 2015.

Dinanamesus Dyar & Knab, 1909d (m., genus) (p. 259) – synonym of *Deinocerites* Theobald, 1901a,b,c; synonymy established before Knight & Stone, 1977.

Dinocerites Blanchard, 1905 (m., genus) (p. 413) – unjustified emendation of *Deinocerites* Theobald, 1901a,b,c.

Dinomimetes Knab, 1907b (m., genus) (p. 120) – synonym of *Deinocerites* Theobald, 1901a,b,c; synonymy established before Knight & Stone, 1977.

Dinomyia Dyar, 1919d (f., genus) (p. 117) – synonym of *Wyeomyia* Theobald, 1901b; synonymy by transfer of type species (*proviolans* Dyar, 1919d, as synonym of *phroso* Howard, Dyar & Knab, 1915) from subgenus *Dendromyia* Theobald, 1903a to subgenus uncertain by Motta & Lourenço-de-Oliveira, 1995.

Dinoporpa Dyar, 1928 (f., section of *Culex* subgenus *Mochlostyrax* Dyar & Knab, 1906d) (p. 286) – synonym of *Melanoconion* Theobald, 1903a; synonymy established before Knight & Stone, 1977.

Diphalangarpe Dyar, 1919d (f., genus) (p. 126) – synonym of *Wyeomyia* Theobald, 1901b; synonymy established before Knight & Stone, 1977.

Dixomyia Taylor, 1914c (f., genus) (p. 702) – synonym of *Etorleptiomyia* Theobald, 1904b; synonymy established before Knight & Stone, 1977.

Dobrotworskyius Reinert, Harbach & Kitching, 2006 (m., genus) (p. 68) – subgenus of *Aedes* Meigen, 1818; subgeneric status by Wilkerson *et al.*, 2015.

Dodecamyia Dyar, 1919d (f., genus) (p. 138) – subgenus of *Wyeomyia* Theobald, 1901b; subgeneric status by Heinemann & Belkin, 1978; Peyton *et al.*, 1983.

Downsiomyia Vargas, 1950c (p. 63) – f., subgenus of *Aedes* Meigen, 1818; subgeneric status re-established by Wilkerson *et al.*, 2015.

Dunnius Edwards, 1930a (m., subgenus of *Anopheles* Meigen, 1818) (p. 297) – synonym of *Pseudarmigeres* Stone & Knight, 1956; synonymy established before Knight & Stone, 1977.

Duttonia Newstead, 1907 (in Newstead *et al.*, 1907) (f., genus) (p. 17) – synonym of *Catageiomyia* Theobald, 1903d; synonymy by transfer of type species to *Catageiomyia* by Reinert *et al.*, 2009.

Dyarina Bonne-Wepster & Bonne, 1921 (f., genus) (p. 6) – synonym of *Phoniomyia* Theobald, 1903a; synonymy established before Knight & Stone, 1977.

Ecculex Felt, 1904 (m., genus) (p. 391c) – synonym of *Aedimorphus* Theobald, 1903a; synonymy established before Knight & Stone, 1977.

Edwardsaedes Belkin, 1962 (p. 408) – m., subgenus of *Aedes* Meigen, 1818; subgeneric status re-established by Wilkerson *et al.*, 2015.

Ekrinomyia Leicester, 1908 (f., genus) (p. 71) – synonym of *Mucidus* Theobald, 1901b; synonymy established before Knight & Stone, 1977.

Elpeytonius Reinert, Harbach & Kitching, 2009 (m., genus) (p. 768) – subgenus of *Aedes* Meigen, 1818; subgeneric status by Wilkerson *et al.*, 2015.

Empihals Reinert, Harbach & Kitching, 2008 (m., subgenus of *Ochlerotatus* Lynch Arribálzaga, 1891a,b) (p. 105) – synonym of *Ochlerotatus* Lynch Arribálzaga, 1891a,b; synonymy by Wilkerson *et al.*, 2015.

Eoaedes Harbach & Greenwalt, 2012 (p. 33) – m., genus of fossil Culicidae.

Eomyzomyia Evans, 1934 (f., group of *Anopheles* subgenus *Myzomyia* Blanchard, 1902) (p. 560) – synonym of *Cellia* Theobald, 1902b; synonymy established before Knight & Stone, 1977.

Eretmapodites Theobald, 1901d (p. ii) – m., genus of tribe Aedini Neveu-Lemaire, 1902c.

Eretmopodites Edwards, 1912c (p. 46) – m., unjustified emendation of *Eretmapodites* Theobald, 1901d.

Etorleptiomyia Theobald, 1904b (f., genus) (p. 71) – subgenus of *Mimomyia* Theobald, 1903a.

Synonyms: *O'Reillia* Ludlow, 1904a; *Dixomyia* Taylor, 1914c; *Luzonus* Stone & Bohart, 1944.

Eubonnea Dyar, 1919c (f., subgenus of *Culex* Linnaeus, 1758) (p. 150) – synonym of *Aedinus* Lutz, 1904; synonymy established before Knight & Stone, 1977.

Eumelanomyia Theobald, 1909b (f., genus) (p. 10) – subgenus of *Culex* Linnaeus, 1758; subgeneric status by Sirivanakarn, 1971a.

Synonyms: *Protomelanoconion* Theobald, 1909b; *Mochthogenes* Edwards, 1930a.

Eunicemyia Dyar & Shannon, 1924d (f., subgenus of *Prosopolepis* Lutz, 1905) (p. 481) – synonym of *Wyeomyia* Theobald, 1901b; synonymy by transfer of type species (*albosquamata* Bonne-Wepster & Bonne, 1919) from subgenus *Dendromyia* Theobald, 1903a to subgenus uncertain by Motta & Lourenço-de-Oliveira, 1995.

Exallomyia Harbach & Peyton, 1992a (p. 92) – f., subgenus of *Wyeomyia* Theobald, 1901b.

Feltianus Martini, 1930 (m., globus of subgenus *Ochlerotatus* Lynch Arribálzaga, 1891a,b) (p. 264) – synonym of *Rusticoidus* Shevchenko & Prudkina, 1973; synonymy established before Knight & Stone, 1977.

Feltidia Dyar, 1905a (f., genus) (p. 47) – synonym of *Grabhamia* Theobald, 1903a; synonymy established before Knight & Stone, 1977.

Feltinella Theobald, 1907 (f., genus) (p. 56) – synonym of *Cellia* Theobald, 1902b; synonymy established before Knight & Stone, 1977.

Ficalbia Theobald, 1903a (p. 296) – f., genus of tribe Ficalbiini Belkin, 1962.

Finlaya Theobald, 1903a (f., genus) (p. 281) – subgenus of *Aedes* Meigen, 1818; subgeneric status re-established by Wilkerson *et al.*, 2015.

Synonym: *Popea* Ludlow, 1905a.

Fredwardsius Reinert, 2000a (p. 1) – m., subgenus of *Aedes* Meigen, 1818; subgeneric status re-established by Wilkerson *et al.*, 2015.

Galindomyia Stone & Barreto, 1969 (p. 143) – f., genus of tribe Culicini Meigen, 1818.

Geitonomyia Leicester, 1908 (f., genus) (p. 134) – synonym of *Aedimorphus* Theobald, 1903a; synonymy established before Knight & Stone, 1977.

Georgecraigius Reinert, Harbach & Kitching, 2006 (m., genus) (p. 70) – subgenus of *Aedes* Meigen, 1818; subgeneric status by Wilkerson *et al.*, 2015.

Geoskusea Edwards, 1929b (p. 342) – f., subgenus of *Aedes* Meigen, 1818; subgeneric status re-established by Wilkerson *et al.*, 2015.

Gilesia Theobald, 1903a (f., genus) (p. 233) – synonym of *Ochlerotatus* Lynch Arribálzaga, 1891a,b; synonymy re-established by Wilkerson *et al.*, 2015.

Gilesius Reinert, Harbach & Kitching, 2006 (m., genus) (p. 73) – subgenus of *Aedes* Meigen, 1818; subgeneric status by Wilkerson *et al.*, 2015.

Gnophodeomyia Theobald, 1905g (f., genus) (p. 21) – synonym of *Melanoconion* Theobald, 1903a; synonymy established before Knight & Stone, 1977.

Goeldia Theobald, 1903a (f., genus) (p. 330) – synonym of *Shannoniana* Lane & Cerqueira, 1942; synonymy established before Knight & Stone, 1977.

Grabhamia Theobald, 1903a (f., genus) (p. 243) – subgenus of *Psorophora* Robineau-Desvoidy, 1827; subgeneric status by ? Dyar, 1928.

Synonyms: *Feltidia* Dyar, 1905a; *Ceratocystia* Dyar & Knab, 1906d; *Lanesia* Leví-Castillo, 1953e.

Grahamia Theobald, 1909b (f., genus) (p. 23) – synonym of *Malaya* Leicester, 1908; synonymy established before Knight & Stone, 1977.

Grassia Theobald, 1902b (f., genus) (p. 181) – synonym of *Cellia* Theobald, 1902b; synonymy established before Knight & Stone, 1977.

Gualteria Lutz, 1904 (in Bourroul, 1904) (f., genus) (pp. 47, 54) – synonym of *Protomacleaya* Theobald, 1907; synonymy by Zavortink, 1972.

Gymnometopa Coquillett, 1906 (f., genus) (p. 183) – subgenus of *Aedes* Meigen, 1818; subgeneric status re-established by Wilkerson *et al.*, 2015.

Haemagogus Williston, 1896 (p. 271) – m., genus (subgenus) of tribe Aedini Neveu-Lemaire, 1902c.

Synonyms: *Stegoconops* Lutz, 1905; *Cacomyia* Coquillett, 1906d; *Longipalpifer* Levi-Castillo, 1951.

Halaedes Belkin, 1962 (p. 328) – m., subgenus of *Aedes* Meigen, 1818; subgeneric status re-established by Wilkerson *et al.*, 2015.

Harbachius Reinert, 1999b (p. 18) – m., subgenus of *Verrallina* Theobald, 1903a.

Harpagomyia de Meijere, 1909 (f., genus) (p. 165) – synonym of *Malaya* Leicester, 1908; synonymy established before Knight & Stone, 1977.

Heizmannia Ludlow, 1905a (p. 130) – f., genus (subgenus) of tribe Aedini Neveu-Lemaire, 1902c.

Synonym: *Bolbodeomyia* Theobald, 1910a.

Helcoporpa Dyar, 1918e (f., subgenus of *Culex* Linnaeus, 1758) (p. 125) – synonym of *Melanoconion* Theobald, 1903a; synonymy established before Knight & Stone, 1977.

Heliconiamyia Dyar, 1919d (f., genus) (p. 123) – synonym of *Decamyia* Dyar, 1919d; synonymy established before Knight & Stone, 1977.

Heptaphlebomyia Theobald, 1903a (f., genus) (p. 336) – synonym of *Culex* Linnaeus, 1758; synonymy established before Knight & Stone, 1977.

Heteraspidion Reinert, Harbach & Kitching, 2009 (m., subgenus of *Stegomyia* Theobald, 1901) (p. 781) – synonym of *Stegomyia* Theobald, 1901; synonymy by Wilkerson *et al.*, 2015.

Heteronycha Lynch Arribálzaga, 1891a,b (f., genus) (pp. 373, 155, respectively) – synonym of *Culex* Linnaeus, 1758; synonymy established before Knight & Stone, 1977.

Himalaius Reinert, Harbach & Kitching, 2006 (m., genus) (p. 74) – subgenus of *Aedes* Meigen, 1818; subgeneric status by Wilkerson *et al.*, 2015.

Hispidimyia Theobald, 1910c (f., genus) (p. 245) – synonym of *Mimomyia* Theobald, 1903a; synonymy established before Knight & Stone, 1977.

Hodgesia Theobald, 1904a (p. 17) – f., genus of tribe Hodgesiini Belkin, 1962.

Hopkinsius Reinert, Harbach & Kitching, 2008 (m., genus) (p. 98) – subgenus of *Aedes* Meigen, 1818; subgeneric status by Wilkerson *et al.*, 2015.

Horsfallius Reinert, Harbach & Kitching, 2006 (m., subgenus of *Georgecraigius* Reinert, Harbach & Kitching, 2006) (p. 72) – synonym of *Georgecraigius* Reinert, Harbach & Kitching, 2006; synonymy by Wilkerson *et al.*, 2015.

Howardia Theobald, 1902b (f., genus) (p. 181) – synonym of *Cellia* Theobald, 1902b; synonymy established before Knight & Stone, 1977.

Howardina Theobald, 1903a (f., genus) (p. 287) – subgenus of *Aedes* Meigen, 1818; subgeneric status re-established by Wilkerson *et al.*, 2015.

Synonym: *Soperia* Komp, 1936b.

Huaedes Huang, 1968b (p. 169) – m., subgenus of *Aedes* Meigen, 1818; subgeneric status re-established by Wilkerson *et al.*, 2015.

Huangmyia Reinert, Harbach & Kitching, 2009 (f., subgenus of *Stegomyia* Theobald, 1901) (p. 782) – synonym of *Stegomyia* Theobald, 1901; synonymy by Wilkerson *et al.*, 2015.

Hulecoeteomyia Theobald, 1904c (f., genus) (p. 163) – subgenus of *Aedes* Meigen, 1818; subgeneric status by Wilkerson *et al.*, 2015.

Hyloconops Lutz, 1904 (in Bourroul, 1904) (m., genus) (pp. 49, 55) – synonym of *Trichoprosopon* Theobald, 1901a; synonymy established before Knight & Stone, 1977.

Hyparctius Martini, 1930 (m., globus of subgenus *Ochlerotatus* Lynch Arribálzaga, 1891a,b) (p. 264) – synonym of *Ochlerotatus* Lynch Arribálzaga, 1891a,b; synonymy established before Knight & Stone, 1977.

Hystatomyia Dyar, 1919d (f., genus) (p. 140) – subgenus of *Wyeomyia* Theobald, 1901b; resurrected from synonymy with *Dendromyia* Theobald, 1903a by Judd, 1998.

Ianthinosoma Lynch Arribálzaga, 1902 (in Goeldi, 1902) (p. 9) – n., unjustified emendation of *Janthinosoma* Lynch Arribálzaga, 1891a,b.

Indusius Edwards, 1934 (in Barraud, 1934) (pp. 268, 445) – m., subgenus of *Aedes* Meigen, 1818; subgeneric status re-established by Wilkerson *et al.*, 2015.

Ingramia Edwards, 1912c (f., genus) (p. 43) – subgenus of *Mimomyia* Theobald, 1903a; subgeneric status by White, 1974.

Synonyms: *Dasymyia* Leicester, 1908; *Ravenalites* Doucet, 1957.

Insulalutzia Tanaka, 2003 (p. 160) – f., subgenus of *Lutzia* Theobald, 1903a.

Isoaedes Reinert, 1979 (p. 144) – m., subgenus of *Aedes* Meigen, 1818; subgeneric status re-established by Wilkerson *et al.*, 2015.

Isogoeldia Edwards, 1930a (f., subgenus of *Goeldia* Theobald, 1903a) (p. 301) – synonym of *Isostomyia* Coquillett, 1906d; synonymy established before Knight & Stone, 1977.

Isostomyia Coquillett, 1906d (pp. 16, 24) – f., genus of tribe Sabethini Blanchard, 1905.
Synonym: *Isogoeldia* Edwards, 1930a.

Jamesia Christophers, 1906 (f., genus) (p. 12) – synonym of *Metalutzia* Tanaka, 2003; synonymy by transfer of the logotype, *concolor* Robineau-Desvoidy, 1827, as synonym of *fuscana* (Wiedemann), from *Lutzia* Theobald, 1903a to *Metalutzia* Tanaka, 2003.

Janicemyia Dyar & Shannon, 1924d (f., subgenus of *Prosopolepis* Lutz, 1905) (p. 481) – synonym of *Wyeomyia* Theobald, 1901b; synonymy by transfer of type species (*clasoleuca* Dyar & Knab, 1908) from subgenus *Dendromyia* Theobald, 1903a to *Wyeomyia* (subgenus uncertain) by Motta & Lourenço-de-Oliveira, 1995.

Janthinosoma Lynch Arribálzaga, 1891a,b (n., genus) (pp. 374, 153, respectively) – subgenus of *Psorophora* Robineau-Desvoidy, 1827; subgeneric status by ? Dyar, 1928.
Synonyms: *Conchyliastes* Theobald, 1901; *Ianthinosoma* Lynch Arribálzaga, 1902; *Lepidosia* Coquillett, 1906b.

Jarnellius Reinert, Harbach & Kitching, 2006 (m., genus) (p. 76) – subgenus of *Aedes* Meigen, 1818; subgeneric status by Wilkerson *et al*., 2015.

Jihlienius Reinert, Harbach & Kitching, 2006 (m., genus) (p. 79) – subgenus of *Aedes* Meigen, 1818; subgeneric status by Wilkerson *et al*., 2015.

Joblotia Blanchard, 1901 (f., genus) (p. 1046) – *nomen novum* for *Trichoprosopon* Theobald, 1901a (erroneously considered preoccupied by *Trichoprosopus* Macquart, 1843); synonymy established before Knight & Stone, 1977.

Johnbelkinia Zavortink, 1979 (p. 18) – f., genus of tribe Sabethini Blanchard, 1905.

Juppius Reinert, Harbach & Kitching, 2009 (m., subgenus of *Ochlerotatus* Lynch Arribálzaga, 1891a,b) (p. 770) – synonym of *Ochlerotatus* Lynch Arribálzaga, 1891a,b; synonymy by Wilkerson *et al*., 2015.

Kenknightia Reinert, 1990 (p. 4) – f., subgenus of *Aedes* Meigen, 1818; subgeneric status re-established by Wilkerson *et al*., 2015.

Kerteszia Theobald, 1905f (f., genus) (p. 66) – subgenus of *Anopheles* Meigen, 1818; subgeneric status by Komp, 1937b.
Synonym: *Dendropaedium* Dyar & Knab, 1918a.

Kimia Vu & Harbach, 2007 (in Harbach, Vu & Kitching, 2007) (p. 103) – f., genus of tribe Sabethini Blanchard, 1905.

Kingia Theobald, 1910c (f., genus) (p. 135) – synonym of *Stegomyia* Theobald, 1901; synonymy established before Knight & Stone, 1977.

Kitzmilleria Danilov, 1989 (p. 792) – f., subgenus of *Culex* Linnaeus, 1758.

Kompia Aitken, 1941 (p. 81) – f., subgenus of *Aedes* Meigen, 1818; subgeneric status re-established by Wilkerson *et al*., 2015.

Kummyia Vargas, 1950b (f., subgenus of *Aedes* Meigen, 1818) (p. 155) – synonym of *Ochlerotatus* Lynch Arribálzaga, 1891a,b; synonymy established before Knight & Stone, 1977.

Laiomyia Izquierdo, 1916 (f., genus) (p. 65) – synonym of *Culex* Linnaeus, 1758; synonymy established before Knight & Stone, 1977.

Lanesia Leví-Castillo, 1953e (f., subgenus of *Aedes* Meigen, 1818) (p. 102) – synonym of *Grabhamia* Theobald, 1903a; synonymy established before Knight & Stone, 1977.

Lasioconops Theobald, 1903c (m., genus) (p. viii) – synonym of *Culex* Linnaeus, 1758; synonymy established before Knight & Stone, 1977.

Lasiosiphon Kirkpatrick, 1925 (p. 372) – m., subgenus of *Culex* Linnaeus, 1758.

Laverania Theobald, 1902b (f., genus) (p. 183) – synonym of *Nyssorhynchus* Blanchard, 1902; synonymy established before Knight & Stone, 1977.

Leicesteria Theobald, 1904c (f., genus) (p. 211) – subgenus of *Armigeres* Theobald, 1901b; subgeneric status by Barraud, 1927b.
Synonyms: *Chaetomyia* Leicester, 1908; *Brevirhynchus* Theobald, 1909b; *Leicesteriomyia* Brunetti, 1912.

Leicesteriomyia Brunetti, 1912 (f., genus) (p. 452) – synonym of *Leicesteria* Theobald, 1904c; synonymy by Macdonald, 1960.

Lemmamyia Dyar, 1919d (f., genus) (p. 140) – synonym of *Limatus* Theobald, 1901a; synonymy established before Knight & Stone, 1977.

Lepidokeneon Reinert, Harbach & Kitching, 2009 (m., subgenus of *Ochlerotatus* Lynch Arribálzaga, 1891a,b) (p. 773) – synonym of *Ochlerotatus* Lynch Arribálzaga, 1891a,b; synonymy by Wilkerson *et al*., 2015.

Lepidoplatys Coquillett, 1906b (m., genus) (p. 314) – synonym of *Ochlerotatus* Lynch Arribálzaga, 1891a,b; synonymy established before Knight & Stone, 1977.

Lepidosia Coquillett, 1906b (f., genus) (p. 314) – synonym of *Janthinosoma* Lynch Arribálzaga, 1891a,b; synonymy established before Knight & Stone, 1977.

Lepidotomyia Theobald, 1905e (f., genus) (p. 22) – synonym of *Finlaya* Theobald, 1903a; synonymy established before Knight & Stone, 1977.

Lepidotomyia Theobald, 1905f (f., genus) (p. 80) – synonym of *Aedimorphus* Theobald, 1903a; synonymy established before Knight & Stone, 1977.

Lepiothauma Enderlein, 1923 (f., genus) (p. 25) – subgenus of *Aedeomyia* Theobald, 1901b.

Leptosomatomyia Theobald, 1905f (f., genus) (p. 110) – subgenus of *Aedes* Meigen, 1818; subgeneric status by Wilkerson *et al*., 2015.

Leslieomyia Christophers, 1911 (f., genus) (p. 68) – synonym of *Aedimorphus* Theobald, 1903a; synonymy established before Knight & Stone, 1977.

Lesticocampa Dyar & Knab, 1906d (f., genus) (p. 226) – synonym of *Runchomyia* Theobald, 1903a; synonymy established before Knight & Stone, 1977.

Leucomyia Theobald, 1907 (f., genus) (p. 372) – synonym of *Culex* Linnaeus, 1758; synonymy established before Knight & Stone, 1977.

Levua Stone & Bohart, 1944 (p. 214) – f., subgenus of *Aedes* Meigen, 1818; subgeneric status re-established by Wilkerson *et al.*, 2015.

Lewnielsenius Reinert, Harbach & Kitching, 2006 (m., subgenus of *Jarnellius* Reinert *et al.*, 2004) (p. 79) – subgenus of *Aedes* Meigen, 1818; change of status from genus of Reinert *et al.* (2009) by Wilkerson *et al.*, 2015.

Limamyia Lane & Cerqueira, 1942 (f., subgenus of *Trichoprosopon* Theobald, 1901a) (p. 500) – synonym of *Trichoprosopon* Theobald, 1901a; synonymy by Zavortink, 1979.

Limatus Theobald, 1901c (p. 349) – m., genus of tribe Sabethini Blanchard, 1905.

Synonyms: *Simondella* Laveran, 1902c; *Lemmamyia* Dyar, 1919d.

Longipalpifer Levi-Castillo, 1951 (m., subgenus; corrigendum of Masculongipalpi) (p. 10) – synonym of *Haemagogus* Williston, 1896; synonymy by Arnell, 1973.

Lophoceraomyia Theobald, 1905h (f., genus) (p. 245) – subgenus of *Culex* Linnaeus, 1758; subspecific status by Borel, 1926 (as *Lophoceratomyia*).

Synonyms: *Philodendromyia* Theobald, 1907; *Cyathomyia* de Meijere, 1910.

Lophoceratomyia Theobald, 1905f (p. 93) – f., *lapsus* of *Lophoceraomyia* Theobald, 1905h.

Lophomyia Giles, 1904a (f., genus) (p. 366) – synonym of *Anopheles* Meigen, 1818; synonymy established before Knight & Stone, 1977.

Lophopodomyia Antunes, 1937 (p. 69) – f., subgenus of *Anopheles* Meigen, 1818.

Synonym: *Arthuromyia* Galvão, 1941b.

Lophoscelomyia Theobald, 1904c (f., genus) (p. 12) – synonym of *Anopheles* Meigen, 1818; synonymy established before Knight & Stone, 1977.

Lorrainea Belkin, 1962 (p. 430) – f., subgenus of *Aedes* Meigen, 1818; subgeneric status re-established by Wilkerson *et al.*, 2015.

Ludlowia Theobald, 1907 (f., genus) (p. 193) – synonym of *Mimomyia* Theobald, 1903a; synonymy established before Knight & Stone, 1977.

Luius Reinert, Harbach & Kitching, 2008 (m., genus) (p. 102) – subgenus of *Aedes* Meigen, 1818; subgeneric status by Wilkerson *et al.*, 2015.

Lutzia Theobald, 1903a (p. 155) – f., genus (subgenus) of tribe Culicini Meigen, 1818.

Luzonus Stone & Bohart, 1944 (m., subgenus of *Aedes* Meigen, 1818) (p. 212) – synonym of *Etorleptiomyia* Theobald, 1904b; synonymy established before Knight & Stone, 1977.

Lynchiaria Brèthes, 1910 (f., genus) (p. 471) – synonym of *Runchomyia* Theobald, 1903a; synonymy established before Knight & Stone, 1977.

Lynchiella Lahille, 1904 (f., genus) (p. 82 [14]) – subgenus of *Toxorhynchites* Theobald, 1901b [*nomen novum* for *Megarhinus* Robineau-Desvoidy, 1827]; subgeneric status by Lane, 1953.

Synonyms: *Megarhina* Macquart, 1838b; *Megarhinus* Robineau-Desvoidy, 1927.

Macleaya Theobald, 1903e (f., genus) (p. 154) – subgenus of *Aedes* Meigen, 1818; subgeneric status re-established by Wilkerson *et al.*, 2015.

Maculipennia Buonomini & Mariani, 1953 (f., subgenus of *Anopheles* Meigen, 1818) (p. 178) – synonym of *Anopheles* Meigen [not valid, no type designated; not absolute tautonymy]; synonymy established before Knight & Stone, 1977.

Maillotia Theobald, 1907 (f., genus) (p. 274) – subgenus of *Culex* Linnaeus, 1758; subgeneric status by Sirivanakarn, 1971a.

Malaya Leicester, 1908 (p. 258) – f., genus of tribe Sabethini Blanchard, 1905.

Synonyms: *Grahamia* Theobald, 1909; *Harpagomyia* de Meijere, 1909.

Manguinhosia Cruz, 1907a (f., genus) (p. 271) – synonym of *Nyssorhynchus* Blanchard, 1902; synonymy established before Knight & Stone, 1977.

Mansonia Blanchard, 1901 (p. 1046) – f., genus (subgenus) of tribe Mansoniini Belkin, 1962 [*nomen novum* for *Panoplites* Theobald, 1900a].

Synonym: *Panoplites* Theobald, 1900a.

Mansonioides Theobald, 1907 (m., genus) (p. 498) – subgenus of *Mansonia* Blanchard, 1901; subgeneric status by ? Edwards, 1932a.

Maorigoeldia Edwards, 1930a (f., subgenus of *Rachionotomyia* Theobald, 1905h) (p. 302) – genus of tribe Sabethini Blanchard, 1905; generic status by Belkin, 1962.

Mattinglyia Lien, 1968c (f., genus) (p. 128) – subgenus of *Heizmannia* Ludlow, 1905a; subgeneric status by Mattingly, 1970.

Synonym: *Sinoaedes* Gong & Lu, 1991.

Megaculex Theobald, 1907 (m., genus) (p. 282) – synonym of *Mimomyia* Theobald, 1903a; synonymy established before Knight & Stone, 1977.

Megarhina Macquart, 1838b (f., genus) (p. 32) – unjustified emendation of *Megarhinus* Robineau-Desvoidy, 1827; preoccupied by *Megarhina* Le Peletier & Audinet-Serville, 1828; senior (but invalid) synonym of *Lynchiella* Lahille, 1904; synonymy by Evenhuis *et al.*, 2016.

Megarhinus Robineau-Desvoidy, 1827 (m., genus) (p. 412) – senior (but invalid) synonym of *Lynchiella* Lahille, 1904 [preoccupied by *Megarhinus* Rafinesque, 1820; see Evenhuis *et al.*, 2016]; synonymy established before Knight & Stone, 1977.

Megarrhina Skuse, 1889 (p. 1720) – f., variant spelling of *Megarhinus* Robineau-Desvoidy, 1827.

Megarrhinus Stackelberg, 1926 (p. 50) – m., incorrect spelling of *Megarhinus* Robineau-Desvoidy, 1827.

Melanoconion Theobald, 1903a (n., genus) (p. 238) – subgenus of *Culex* Linnaeus, 1758; subgeneric status as defined by Belkin, 1968b.

Synonyms: *Melanoconium* Blanchard, 1905; *Gnophodeomyia* Theobald, 1905g; *Mochlostyrax* Dyar &

Knab, 1906d; *Asebeomyia* Aiken, 1911; *Choeroporpa* Dyar, 1918c; *Helcoporpa* Dyar, 1918e; *Dinoporpa* Dyar, 1928; *Upsiloporpa* Komp & Curry, 1932.

Melanoconium Blanchard, 1905 (p. 238) – n., unjustified emendation of *Melanoconion* Theobald, 1903a.

Melanolepis Dyar, 1928 (f., subgenus of *Dendromyia* Theobald, 1903a) (pp. 69, 72) – synonym of *Dendromyia* Theobald, 1903a; synonymy established before Knight & Stone, 1977.

Memnemyia Strickland, 1915a (f., genus) (p. 204) – synonym of *Anopheles* Meigen, 1818; synonymy established before Knight & Stone, 1977.

Menolepis Lutz, 1905 (f., genus) (p. 269) – subgenus of *Wyeomyia* Theobald, 1901b; subgeneric status by Edwards, 1932a.

Metalutzia Tanaka, 2003 (p. 160) – f., subgenus of *Lutzia* Theobald, 1903a.

Synonym: *Jamesia* Christophers, 1906.

Miamyia Dyar, 1919d (f., genus) (p. 116) – subgenus of *Wyeomyia* Theobald, 1901b; resurrected from synonymy with *Dendromyia* Theobald by Motta *et al.*, 2007.

Micraedes Coquillett, 1906d (m., genus) (pp. 16, 24) – subgenus of *Culex* Linnaeus, 1758; subgeneric status by Edwards, 1932a; Belkin, 1968b.

Microculex Theobald, 1907 (m., genus) (p. 461) – subgenus of *Culex* Linnaeus 1758; subgeneric status by Edwards, 1932a.

Mimeteculex Theobald, 1908c (m., genus) (p. 258) – synonym of *Aedimorphus* Theobald, 1903a; synonymy established before Knight & Stone, 1977.

Mimeteomyia Theobald, 1910c (f., genus) (p. 210) – synonym of *Polylepidomyia* Theobald, 1905f; synonymy by transfer of type species, *apicotriangulatus* (Theobald, 1910c), as synonym of *atripes* (Skuse, 1889), from subgenus *Rachionotomyia* Theobald, 1905h to subgenus *Polylepidomyia* Theobald, 1905f by Mattingly, 1980.

Mimomyia Theobald, 1903a (p. 304) – f., genus (subgenus) of tribe Ficalbiini Belkin, 1962.

Synonyms: *Boycia* Newstead, 1907; *Ludlowia* Theobald, 1907; *Megaculex* Theobald, 1907; *Radioculex* Theobald, 1908b; *Conopomyia* Leicester, 1908; *Hispidimyia* Theobald, 1910c.

Missirolimyia Vargas, 1950a (f., subgenus of *Anopheles* Meigen, 1818) (p. 80) – synonym of *Anopheles* Meigen, 1818; synonymy established before Knight & Stone, 1977.

Miyagiella Harbach, 2014 (in Harbach & Culverwell, 2014) (p. 576) – f., subgenus of *Topomyia* Leicester, 1908.

Mochlostyrax Dyar & Knab, 1906d (m., genus) (p. 223) – synonym of *Melanoconion* Theobald, 1903a; synonymy by Belkin *et al.*, 1970.

Mochthogenes Edwards, 1930a (m., subgenus of *Culex* Linnaeus, 1758) (p. 305) – synonym of *Eumelanomyia* Theobald, 1909b; synonymy established before Knight & Stone, 1977.

Molpemyia Theobald, 1910c (f., genus) (p. 479) – subgenus of *Aedes* Meigen, 1818; subgeneric status re-established by Wilkerson *et al.*, 2015.

Synonym: *Calomyia* Taylor, 1914c.

Mucidus Theobald, 1901b (m., genus) (p. 235) – subgenus of *Aedes* Meigen, 1818; subgeneric status re-established by Wilkerson *et al.*, 2015.

Mukwaya Reinert, Harbach & Kitching, 2009 (m., subgenus of *Stegomyia* Theobald, 1901) (p. 782) – synonym of *Stegomyia* Theobald, 1901; synonymy by Wilkerson *et al.*, 2015.

Myxosquamus Theobald, 1909b (m., genus) (p. 7) – synonym of *Catageiomyia* Theobald, 1903d; synonymy by transfer of type species to *Catageiomyia* by Reinert *et al.*, 2009.

Myzomyia Blanchard, 1902 (f., genus) (p. 795) – synonym of *Cellia* Theobald, 1902b [*nomen novum* for *Grassia* Theobald, 1902b]; synonymy established before Knight & Stone, 1977.

Myzorhynchella Theobald, 1907 (f., genus) (pp. 23, 78) – synonym of *Nyssorhynchus* Blanchard, 1902; synonymy established before Knight & Stone, 1977.

Myzorhynchus Blanchard, 1902 (m., genus) (p. 795) – synonym of *Anopheles* Meigen, 1818 [*nomen novum* for *Rossia* Theobald, 1902b]; synonymy established before Knight & Stone, 1977.

Neobironella Tenorio, 1977 (p. 345) – f., subgenus of *Bironella* Theobald, 1905f.

Neocellia Theobald, 1907 (f., genus) (p. 111) – synonym of *Cellia* Theobald, 1902b; synonymy established before Knight & Stone, 1977.

Neoculex Dyar, 1905a (m., genus) (p. 48) – subgenus of *Culex* Linnaeus, 1758; subgeneric status by Edwards, 1932a.

Neoculicites Evenhuis, 1994 (p. 234) – m., genus of fossil Culicidae subfamily Culicinae Meigen, 1818 [*nomen novum* for *Culicites* Meunier, 1915].

Synonym: *Culicites* Meunier, 1915.

Neomacleaya Theobald, 1907 (f., genus) (p. 238) – subgenus of *Verrallina* Theobald, 1903a; transfer from subgeneric status in *Aedes* Meigen by Reinert, 1999b.

Synonym: *Aioretomyia* Leicester, 1908.

Neomelaniconion Newstead, 1907 (in Newstead *et al.*, 1907) (n., genus) (p. 31) – subgenus of *Aedes* Meigen, 1818; subgeneric status re-established by Wilkerson *et al.*, 2015.

Synonym: *Banksinella* Theobald, 1907.

Neomelanoconion Theobald, 1907 (n., genus) (p. 514) – synonym of *Culiciomyia* Theobald, 1907; synonymy by Stone, 1961b.

Neomyzomyia Theobald, 1910c (f., genus) (p. 29) – synonym of *Cellia* Theobald, 1902b; synonymy established before Knight & Stone, 1977.

Neopecomyia Theobald, 1909b (f., genus) (p. 12) – synonym of *Catageiomyia* Theobald, 1903d; synonymy by transfer of type species to *Catageiomyia* by Reinert *et al.*, 2009.

Neosquamomyia Taylor, 1914d (f., genus) (p. 186) – synonym of *Armigeres* Theobald, 1901b; synonymy established before Knight & Stone, 1977.

Neostethopheles James, 1910b (m., genus) (p. 98) – synonym of *Anopheles* Meigen, 1818; synonymy established before Knight & Stone, 1977.

Neotheobaldia Dobrotworsky, 1958b (p. 186) – f., subgenus of *Culiseta* Felt, 1904. **Note**: original description (Dobrotworsky, 1954) invalid until Dobrotworsky (1958) designated a type species.

Newsteadina Theobald, 1909a (f., genus) (p. 297) – synonym of *Orthopodomyia* Theobald, 1904c; synonymy established before Knight & Stone, 1977.

Nicaromyia González Broche & Rodríguez R., 2001 (p. 66) – f., subgenus of *Culex* Linnaeus, 1758.

Nothoskusea Dumbleton, 1962 (f., subgenus of *Aedes* Meigen, 1818) (p. 20) – subgenus of *Opifex* Hutton, 1902; transfer from *Aedes* Meigen, 1818 to *Opifex* Hutton, 1902 by Reinert *et al.*, 2004.

Nototricha Coquillett, 1906d (f., genus) (p. 13) – synonym of *Anopheles* Meigen, 1818; synonymy established before Knight & Stone, 1977.

Nunezia Dyar, 1928 (pp. 28, 50) – f., subgenus of *Wyeomyia* Theobald, 1901b.

Nyctomyia Harbach, 2013 (f., genus) (p. 199) – subgenus of *Aedes* Meigen, 1818 [*nomen novum* for *Nyx* Harbach & Linton, 2013]; subgeneric status by Wilkerson *et al.*, 2015.
Synonym: *Nyx* Harbach & Linton, 2013.

Nyssomyzomyia James, 1910a (f., genus, as *Nyssomzomyia*), 1910b (pp. 37, 101, respectively) – synonym of *Cellia* Theobald, 1902b; synonymy established before Knight & Stone, 1977.

Nyssorhynchus Blanchard, 1902 (m., genus) (p. 795) – subgenus of *Anopheles* Meigen, 1818 [*nomen novum* for *Laverania* Theobald, 1902b]; subgeneric status by Christophers, 1924a.
Synonyms: *Laverania* Theobald, 1902b; *Myzorhynchella* Theobald, 1907; *Manguinhosia* Cruz, 1907a.

Nyx Harbach & Linton, 2013 (in Harbach *et al.*, 2013) (f., genus) (p. 160) – synonym of *Nyctomyia* Harbach, 2013; synonymy established by Harbach, 2013.

Ochlerotatus Lynch Arribálzaga, 1891a,b (m., genus) (pp. 374, 143, respectively) – subgenus of *Aedes* Meigen, 1818; subgeneric status re-established by Wilkerson *et al.*, 2015.
Synonyms: *Pseudoculex* Dyar, 1905a; *Lepidoplatys* Coquillett, 1906b; *Pseudohowardina* Theobald, 1907; *Andersonia* Strickland, 1911b; *Feltianus* Martini, 1930.

Oculeomyia Theobald, 1907 (f., genus) (p. 515) – subgenus of *Culex* Linnaeus, 1758; resurrected from synonymy with *Culex* Linnaeus, 1758 by Tanaka, 2004.

Onirion Peyton & Harbach, 2000 (in Harbach & Peyton, 2000) (p. 118) – n., genus of Sabethini Blanchard, 1905.

Opifex Hutton, 1902 (p. 188) – m., genus (subgenus) of tribe Aedini Neveu-Lemaire, 1902c.

O'Reillia Ludlow, 1905a (f., genus) (p. 101) – synonym of *Etorleptiomyia* Theobald, 1904b; synonymy established before Knight & Stone, 1977.

Orthopodomyia Theobald, 1904c (p. 236) – f., genus of tribe Orthopodomyiini Belkin, Heinemann & Page, 1970.
Synonyms: *Bancroftia* Lutz, 1904; *Pneumaculex* Dyar, 1905a; *Newsteadina* Theobald, 1909a; *Thomasina* Newstead & Carter, 1911a.

Osornomyia Leví-Castillo, 1955a (f., subgenus of *Haemagogus* Williston, 1896) (p. 361) – *nomen nudum*, *Haemagogus*.

Palaeomyzomyia Swellengrebel & Rodenwaldt, 1932 (f., subgenus of *Anopheles* Meigen, 1818) (p. 114) – synonym of *Cellia* Theobald, 1902b; synonymy established before Knight & Stone, 1977.

Paleoculicis Poinar, Zavortink, Pike & Johnston, 2000 (p. 121) – m., genus of fossil Culicidae, subfamily Culicinae Meigen, 1818.

Panoplites Theobald, 1900a (m., genus) (p. 261) – synonym of *Mansonia* Blanchard, 1901; synonymy established before Knight & Stone, 1977.

Paraedes Edwards, 1934 (in Barraud, 1934) (m., genus) (p. 446) – subgenus of *Aedes* Meigen, 1818; subgeneric status re-established by Wilkerson *et al.*, 2015.

Paramyzomyia Christophers & Barraud, 1931 (f., group of subgenus *Myzomyia* Blanchard, 1902 of *Anopheles* Meigen, 1818) (p. 168) – synonym of *Cellia* Theobald, 1902b; synonymy established before Knight & Stone, 1977.

Pardomyia Theobald, 1907 (f., genus) (p. 280) – synonym of *Mucidus* Theobald, 1901b; synonymy re-established by Wilkerson *et al.*, 2015.

Patagiamyia James, 1910b (f., genus) (p. 98) – synonym of *Anopheles* Meigen, 1818; synonymy established before Knight & Stone, 1977.

Patmarksia Reinert, Harbach & Kitching, 2006 (f., genus) (p. 81) – subgenus of *Aedes* Meigen, 1818; subgeneric status by Wilkerson *et al.*, 2015.

Paulianius Brunhes & Boussès, 2017 (in Brunhes *et al.*, 2017) (m., genus) (p. 345) – subgenus of *Aedes* Meigen, 1818, ***stat. nov.***

Pecomyia Theobald, 1905g (f., genus) (p. 23) – synonym of *Aedimorphus* Theobald, 1903a; synonymy established before Knight & Stone, 1977.

Pectinopalpus Theobald, 1909b (m., genus) (p. 11) – synonym of *Culiciomyia* Theobald, 1907; synonymy established before Knight & Stone, 1977.

Pentemyia Dyar, 1919d (f., genus) (p. 122) – synonym of *Wyeomyia* Theobald, 1901b; synonymy established before Knight & Stone, 1977.

Petermattinglyius Reinert, Harbach & Kitching, 2009 (m., genus) (p. 778) – subgenus of *Aedes* Meigen, 1818; subgeneric status by Wilkerson *et al.*, 2015.

Peytonulus Harbach, 1991a (p. 1) – m., subgenus of *Sabethes* Robineau-Desvoidy, 1827.

Phagomyia Theobald, 1905e (f., genus) (p. 21) – subgenus of *Aedes* Meigen, 1818; subgeneric status by Wilkerson *et al.*, 2015.
Synonyms: *Lepidotomyia* Theobald, 1905e; *Pseudocarrollia* Theobald, 1910a.

Phalangomyia Dyar & Knab, 1914 (f., genus) (p. 58) – synonym of *Culex* Linnaeus, 1758; synonymy established before Knight & Stone, 1977.

Phenacomyia Harbach & Peyton, 1992b (p. 242) – f., subgenus of *Culex* Linnaeus, 1758.

Philodendromyia Theobald, 1907 (f., genus) (p. 623) – synonym of *Lophoceraomyia* Theobald, 1905h; synonymy established before Knight & Stone, 1977.

Pholeomyia Reinert, Harbach & Kitching, 2008 (f., subgenus of *Ochlerotatus* Lynch Arribálzaga, 1891a,b) (p. 162) – synonym of *Ochlerotatus* Lynch Arribálzaga, 1891a,b; synonymy by Wilkerson *et al*., 2015. **Note**: this name is preoccupied by *Pholeomyia* Bilimek, 1867.

Phoniomyia Theobald, 1903a (f., genus) (p. 311) – subgenus of *Wyeomyia* Theobald, 1901b; subgeneric status by Judd, 1998.

Synonym: *Dyarina* Bonne-Wepster & Bonne, 1921.

Phyllozomyia Dyar, 1924c (f., subgenus of *Wyeomyia* Theobald, 1901b) (p. 112) – synonym of *Wyeomyia* Theobald, 1901b; synonymy established before Knight & Stone, 1977.

Phytotelmatomyia Rossi & Harbach, 2008 (p. 2) – f., subgenus of *Culex* Linnaeus, 1758.

Pneumaculex Dyar, 1905a (m., genus) (p. 46) – synonym of *Orthopodomyia* Theobald, 1904c; synonymy established before Knight & Stone, 1977.

Polylepidomyia Theobald, 1905f (f., genus) (p. 118) – subgenus of *Tripteroides* Giles, 1904a; resurrected from synonymy with *Rachionotomyia* Theobald by Mattingly, 1980.

Synonym: *Mimeteomyia* Theobald, 1910c.

Polyleptiomyia Theobald, 1905e (f., genus) (p. 21) – subgenus of *Aedes* Meigen, 1818; subgeneric status by Wilkerson *et al*., 2015.

Popea Ludlow, 1905a (f., genus) (p. 95) – synonym of *Finlaya* Theobald, 1903a; synonymy established before Knight & Stone, 1977.

Promacleaya Surcouf & Gonzalez-Rincones, 1912 (p. 277) – f., unjustified emendation of *Protomacleaya* Theobald, 1907.

Prosopolepis Lutz, 1905 (m., genus) (p. 312) – subgenus of *Wyeomyia* Theobald, 1901b; resurrected from synonymy with *Dendromyia* Theobald by Lourenço-de-Oliveira *et al*., 1999.

Proterorhynchus Brèthes, 1912 (m., genus) (pp. 10, 14) – synonym of *Anopheles* Meigen, 1818; synonymy established before Knight & Stone, 1977.

Protoculex Felt, 1904 (m., genus) (p. 391d) – synonym of *Ochlerotatus* Lynch Arribálzaga, 1891a,b; synonymy re-established by Wilkerson *et al*., 2015.

Protomacleaya Theobald, 1907 (f., genus) (pp. 149, 253) – subgenus of *Aedes* Meigen, 1818; subgeneric status re-established by Wilkerson & Linton, 2015.

Synonym: *Gualteria* Lutz, 1904.

Protomelanoconion Theobald, 1909b (n., genus) (p. 20) – synonym of *Eumelanomyia* Theobald, 1909b; synonymy established before Knight & Stone, 1977.

Pseudalbuginosus Huang & Rueda, 2015 (p. 382) – m., subgenus of *Aedes* Meigen, 1818.

Pseudarmigeres Stone & Knight, 1956 (p. 34) – m., subgenus of *Aedes* Meigen, 1818; subgeneric status re-established by Wilkerson *et al*., 2015.

Synonym: *Dunnius* Edwards, 1930a.

Pseudocarrollia Theobald, 1910a (f., genus) (p. 12) – synonym of *Phagomyia* Theobald, 1905e; synonymy by transfer of type species *lophoventralis* Theobald, 1910a from *Finlaya* Theobald, 1903a by Reinert *et al*., 2006.

Pseudoculex Dyar, 1905a (m., genus) (p. 47) – synonym of *Ochlerotatus* Lynch Arribálzaga, 1891a,b; synonymy established before Knight & Stone, 1977.

Pseudoficalbia Theobald, 1912b (f., genus) (p. 89) – subgenus of *Uranotaenia* Lynch Arribálzaga, 1891a,b; resurrected from synonymy with *Uranotaenia* Lynch Arribálzaga by Peyton, 1972.

Pseudograbhamia Theobald, 1905h (f., genus) (p. 243) – synonym of *Aedimorphus* Theobald, 1903a; synonymy established before Knight & Stone, 1977.

Pseudograhamia Theobald, 1910a (f., genus) (p. 26) – synonym of *Topomyia* Leicester, 1908; synonymy established before Knight & Stone, 1977.

Pseudoheptaphlebomyia Ventrillon, 1905b (f., genus) (p. 427) – synonym of *Culex* Linnaeus, 1758; synonymy established before Knight & Stone, 1977.

Pseudohowardina Theobald, 1907 (f., genus) (p. 223) – synonym of *Ochlerotatus* Lynch Arribálzaga, 1891a,b; synonymy established before Knight & Stone, 1977.

Pseudomyzomyia Theobald, 1907 (f., genus) (errata et addenda for p. 74) – synonym of *Cellia* Theobald, 1902b; synonymy established before Knight & Stone, 1977.

Pseudoskusea Theobald, 1907 (f., genus) (p. 192) – subgenus of *Aedes* Meigen, 1818; subgeneric status re-established by Wilkerson & Linton, 2015.

Synonym: *Caenocephalus* Taylor, 1914c.

Pseudostegomyia Ludlow, 1908b (f., genus) (p. 10) – synonym of *Stegomyia* Theobald, 1901; synonymy with *Quasistegomyia* Theobald, 1906b by Theobald, 1910c.

Pseudotaeniorhynchus Theobald, 1911a (m., genus) (p. 19) – synonym of *Rhynchotaenia* Brèthes, 1910; synonymy established before Knight & Stone, 1977.

Pseudotheobaldia Theobald, 1907 (f., genus) (p. 271) – synonym of *Culiseta* Felt, 1904; synonymy established before Knight & Stone, 1977.

Pseudouranotaenia Theobald, 1905g (f., genus) (p. 33) – synonym of *Uranotaenia* Lynch Arribálzaga, 1891a,b; synonymy established before Knight & Stone, 1977.

Psorophora Robineau-Desvoidy, 1827 (p. 412) – f., genus (subgenus) of tribe Aedini Neveu-Lemaire, 1902c.

Pyretophorus Blanchard, 1902 (m., genus) (p. 795) – synonym of *Cellia* Theobald, 1902b [*nomen novum* for *Howardia* Theobald, 1902b]; synonymy established before Knight & Stone, 1977.

Quasistegomyia Theobald, 1906b (f., genus) (p. 69) – synonym of *Stegomyia* Theobald, 1901; synonymy established before Knight & Stone, 1977.

Rachionotomyia Theobald, 1905h (f., genus) (p. 248) – subgenus of *Tripteroides* Giles, 1904a; resurrected from synonymy with *Tripteroides* Giles, 1904a by Belkin, 1962, redefined by Mattingly, 1981.

Synonyms: *Skeiromyia* Leicester, 1908; *Squamomyia* Theobald, 1910a.

Rachisoura Theobald, 1910c (f., genus) (p. 207) – subgenus of *Tripteroides* Giles, 1904a; subgeneric status by Edwards, 1932a.

Radioculex Theobald, 1908b (m., genus) (p. 295) – synonym of *Mimomyia* Theobald, 1903a; synonymy established before Knight & Stone, 1977.

Rampamyia Reinert, Harbach & Kitching, 2006 (f., genus) (p. 83) – subgenus of *Aedes* Meigen, 1818; subgeneric status by Wilkerson *et al.*, 2015.

Ravenalites Doucet, 1957 (in Mattingly, 1957a) (m., subgenus of *Ficalbia* Theobald, 1903a) (pp. 2, 18) – synonym of *Ingramia* Edwards, 1912c; synonymy by White, 1974.

Reedomyia Ludlow, 1905a (f., genus) (p. 94) – synonym of *Aedimorphus* Theobald, 1903a; synonymy established before Knight & Stone, 1977.

Rhinoskusea Edwards, 1929b (p. 342) – f., subgenus of *Aedes* Meigen, 1818; subgeneric status re-established by Wilkerson *et al.*, 2015.

Rhynchomyia Lutz, 1904 (in Bourroul, 1904) (p. 55) – f., unjustified emendation of *Runchomyia* Theobald, 1903a.

Rhynchotaenia Brèthes, 1910 (f., genus) (p. 470) – subgenus of *Coquillettidia* Dyar, 1905a; subgeneric status by Dyar, 1928, as subgenus of *Mansonia* Blanchard, 1901.

Synonym: *Pseudotaeniorhynchus* Theobald, 1911a.

Rossia Theobald, 1902b (f., genus) (p. 183) – synonym of *Anopheles* Meigen, 1818; synonymy established before Knight & Stone, 1977.

Runchomyia Theobald, 1903a (p. 319) – f., genus (subgenus) of tribe Sabethini Blanchard, 1905; elevated from subgeneric status in *Trichoprosopon* Theobald, 1901a,c by Zavortink, 1979.

Synonyms: *Binotia* Blanchard, 1904; *Rhynchomyia* Lutz, 1904; *Lesticocampa* Dyar & Knab, 1906d.

Russellia Vargas, 1943a (f., subgenus of *Anopheles* Meigen, 1818) (pp. 60, 68) – synonym of *Anopheles* Meigen, 1818; synonymy established before Knight & Stone, 1977.

Rusticoidus Shevchenko & Prudkina, 1973 (p. 46) – m., subgenus of *Aedes* Meigen, 1818; subgeneric status re-established by Wilkerson & Linton, 2015.

Sabethes Robineau-Desvoidy, 1827 (p. 411) – m., genus (subgenus) of tribe Sabethini Blanchard, 1905.

Synonym: *Sabettus* Scudder, 1882.

Sabethinus Lutz, 1904 (in Bourroul, 1904) (m., genus) (pp. 48, 57) – subgenus of *Sabethes* Robineau-Desvoidy, 1827; subgeneric status by Lane & Cerqueira, 1942, redefined by Harbach, 1991a.

Synonym: *Sabettinus* Blanchard, 1905.

Sabethoides Theobald, 1903a (m., genus) (p. 328) – subgenus of *Sabethes* Robineau-Desvoidy, 1827; subgeneric status by Lane & Cerqueira, 1942.

Synonym: *Sabettoides* Blanchard, 1905.

Sabettinus Blanchard, 1905 (p. 634) – m., unjustified emendation of *Sabethinus* Lutz, 1904.

Sabettoides Blanchard, 1905 (p. 423) – m., unjustified emendation of *Sabethoides* Theobald, 1903a.

Sabettus Agassiz, 1846 (p. 329); Scudder, 1882 (p. 297) – m., unjustified emendation of *Sabethes* Robineau-Desvoidy, 1827.

Sallumia Reinert, Harbach & Kitching, 2008 (f., subgenus of *Ochlerotatus* Lynch Arribálzaga, 1891a,b) (p. 110) – subgenus of *Aedes* Meigen, 1818; generic status, Reinert *et al.*, 2009; subgeneric status by Wilkerson *et al.*, 2015.

Scutomyia Theobald, 1904c (f., genus) (p. 77) – subgenus of *Aedes* Meigen, 1818; subgeneric status re-established by Wilkerson *et al.*, 2015.

Shannonesia da Fonseca & Ramos, 1940 (f., subgenus of *Anopheles* Meigen, 1818) (p. 966) – synonym of *Anopheles* Meigen, 1818; synonymy established before Knight & Stone, 1977.

Shannoniana Lane & Cerqueira, 1942 (f., subgenus of *Trichoprosopon* Theobald, 1901a) (p. 502) – genus of tribe Sabethini Blanchard, 1905; elevated from subgeneric status in *Trichoprosopon* Theobald, 1901a by Zavortink, 1979.

Synonym: *Goeldia* Theobald, 1903a.

Shannoniella da Fonseca & da Silva Ramos, 1940 (f., subgenus of *Anopheles* Meigen, 1818) (p. 384) – synonym of *Anopheles* Meigen, 1818; synonymy established before Knight & Stone, 1977.

Shropshirea Dyar, 1922d (f., subgenus of *Wyeomyia* Theobald, 1901b) (p. 97) – synonym of *Dendromyia* Theobald, 1903a; synonymy established before Knight & Stone, 1977.

Simondella Laveran, 1902c (f., genus) (p. 1160) – synonym of *Limatus* Theobald, 1901c; synonymy established before Knight & Stone, 1977.

Sinoaedes Gong & Lu, 1991 (m., subgenus of *Aedes* Meigen, 1818) (p. 55) – synonym of *Mattinglyia* Lien, 1968c; synonymy by Reinert, 2000b.

Sirivanakarnius Tanaka, 2004 (p. 224) – m., subgenus of *Culex* Linnaeus, 1758.

Skeiromyia Leicester, 1908 (f., genus) (p. 248) – synonym of *Rachionotomyia* Theobald, 1905h; synonymy established before Knight & Stone, 1977.

Skusea Theobald, 1903a (f., genus) (p. 291) – subgenus of *Aedes* Meigen, 1818; subgeneric status re-established by Wilkerson *et al.*, 2015.

Soperia Komp, 1936b (f., subgenus of *Aedes* Meigen, 1818) (p. 71) – synonym of *Howardina* Theobald, 1903a; synonymy established before Knight & Stone, 1977.

Spilonympha Motta & Lourenço-de-Oliveira, 2005 (p. 839) – f., subgenus of *Wyeomyia* Theobald, 1901b.

Squamomyia Theobald, 1910a (f., genus) (p. 28) – synonym of *Rachionotomyia* Theobald, 1905h; synonymy established before Knight & Stone, 1977.

Stegoconops Lutz, 1905 (m., genus) (p. 83) – synonym of *Haemagogus* Williston, 1896; synonymy established before Knight & Stone, 1977.

Stegomyia Theobald, 1901 (in Howard, 1901) (f., genus) (p. 235) – subgenus of *Aedes* Meigen, 1818; subgeneric status re-established by Wilkerson *et al.*, 2015.

Stenoscutus Theobald, 1909b (m., genus) (p. 7) – synonym of *Catageiomyia* Theobald, 1903d; synonymy by transfer of type species to *Catageiomyia* by Reinert *et al.*, 2009.

Stethomyia Theobald, 1902b (f., genus) (p. 182) – subgenus of *Anopheles* Meigen, 1818; subgeneric status by Edwards, 1932a.

Suaymyia Thurman, 1959 (p. 44) – f., subgenus of *Topomyia* Leicester, 1908.

Taeniorhynchus Lynch Arribálzaga, 1891a,b (m., genus) (pp. 374, 143, respectively) – synonym of *Ochlerotatus* Lynch Arribálzaga, 1891a,b; synonymy established before Knight & Stone, 1977.

Tanakaius Reinert, Harbach & Kitching, 2004 (m., genus) (p. 349) – subgenus of *Aedes* Meigen, 1818; subgeneric status by Wilkerson *et al.*, 2015.

Technicomyia Dyar, 1925a (f., subgenus of *Wyeomyia* Theobald, 1901b) (p. 20) – synonym of *Wyeomyia* Theobald, 1901b; synonymy established before Knight & Stone, 1977.

Teromyia Leicester, 1908 (f., genus) (p. 49) – synonym of *Toxorhynchites* Theobald, 1901b; synonymy established before Knight & Stone, 1977.

Tewarius Reinert, 2006a (m., genus) (p. 640) – subgenus of *Aedes* Meigen, 1818; subgeneric status by Wilkerson *et al.*, 2015.

Thaiomyia Bram, 1966 (f., subgenus of *Culex* Linnaeus, 1758) (p. 73) – synonym of *Culiciomyia* Theobald, 1907; synonymy by Harrison, 1987.

Theobaldia Neveu-Lemaire, 1902a (f., genus) (p. 1331) – preoccupied senior synonym of *Culiseta* Felt, 1904; synonymy established before Knight & Stone, 1977.

Theobaldinella Blanchard, 1905 (f., genus) (p. 390) – synonym of *Culiseta* Felt, 1904; synonymy established before Knight & Stone, 1977.

Theobaldiomyia Brunetti, 1912 (f., genus) (p. 462) – synonym of *Culex* Linnaeus, 1758; synonymy established before Knight & Stone, 1977.

Theomyia Edwards, 1930a (f., subgenus of *Theobaldia* Neveu-Lemaire, 1902a) (p. 303) – subgenus of *Culiseta* Felt, 1904.

Thomasina Newstead & Carter, 1911a (f., genus) (p. 553) – synonym of *Orthopodomyia* Theobald, 1904c; synonymy established before Knight & Stone, 1977.

Tinolestes Coquillett, 1906d (m., genus) (pp. 16, 24) – subgenus of *Culex* Linnaeus, 1758.

Topomyia Leicester, 1908 (p. 238) – f., genus (subgenus) of tribe Sabethini Blanchard, 1905.
Synonym: *Pseudograhamia* Theobald, 1910a.

Toxorhynchites Theobald, 1901b (p. 234) – m., genus (subgenus) of tribe Toxorhynchitini Lahille, 1904.
Synonyms: *Megarhinus* Robineau-Desvoidy, 1827; *Worcesteria* Banks, 1906a; *Teromyia* Leicester, 1908.

Transculicia Dyar, 1918a (f., subgenus of *Culex* Linnaeus, 1758) (p. 84) – synonym of *Culex* Linnaeus, 1758; synonymy established before Knight & Stone, 1977.

Triamyia Dyar, 1919d (f., genus) (p. 120) – subgenus of *Wyeomyia* Theobald, 1901b; resurrected from synonymy with *Dendromyia* Theobald, 1901b by Motta *et al.*, 2007.

Tricholeptomyia Dyar & Shannon, 1925a (f., genus) (p. 72) – subgenus of *Tripteroides* Giles, 1904a; resurrected from synonymy with *Rachionotomyia* Theobald, 1905h by Mattingly, 1980.

Trichopronomyia Theobald, 1905f (f., genus) (p. 98) – synonym of *Culex* Linnaeus, 1758; synonymy established before Knight & Stone, 1977.

Trichoprosopon Theobald, 1901a, 1901c (pp. 98, 283, respectively) – n., genus of tribe Sabethini Blanchard, 1905.
Synonyms: *Joblotia* Blanchard, 1901; *Hyloconops* Lutz, 1904.

Trichorhynchomyia Brunetti, 1912 (f., genus) (p. 477) – synonym of *Culiciomyia* Theobald, 1907; synonymy established before Knight & Stone, 1977.

Trichorhynchus Theobald, 1905h (m., genus) (p. 241) – synonym of *Culiciomyia* Theobald, 1907; synonymy established before Knight & Stone, 1977.

Tripteroides Giles, 1904a (p. 369) – m., genus (subgenus) of tribe Sabethini Blanchard, 1905.
Synonym: *Colonemyia* Leicester, 1908.

Udaya Thurman, 1954b (m., subgenus of *Paraedes* Edwards, 1934) (p. 84) – genus of tribe Aedini Neveu-Lemaire, 1902c; generic status by Mattingly, 1958a.

Upsiloporpa Komp & Curry, 1932 (f., section of *Culex* Linnaeus, 1758) (p. 83) – synonym of *Melanoconion* Theobald, 1903a; synonymy established before Knight & Stone 1977.

Uranotaenia Lynch Arribálzaga, 1891a,b (pp. 375, 163, respectively) – f., genus (subgenus) of tribe Uranotaeniini Lahille, 1904.

Vansomerenis Reinert, Harbach & Kitching, 2006 (f., genus) (p. 85) – subgenus of *Aedes* Meigen, 1818; subgeneric status by Wilkerson *et al.*, 2015.

Verralina Lutz, 1904 (in Bourroul, 1904) (synoptic table, p. 45) – f., incorrect spelling of *Verrallina* Theobald, 1903a.

Verrallina Theobald, 1903a (p. 295) – f., genus (subgenus) of tribe Aedini Neveu-Lemaire, 1902c; elevated from subgeneric status in *Aedes* Meigen by Reinert, 1999b.

Vonplessenia Lane & Cerqueira, 1942 (f., subgenus of *Trichoprosopon* Theobald, 1901a,c) (p. 501) – synonym of *Trichoprosopon* Theobald, 1901a,c; synonymy by Zavortink, 1979.

Woodius Reinert, Harbach & Kitching, 2009 (m., subgenus of *Ochlerotatus* Lynch Arribálzaga, 1891a,b) (p. 775) – synonym of *Aedes* Meigen, 1818; synonymy by Wilkerson *et al.*, 2015.

Worcesteria Banks, 1906a (f., genus) (p. 779) – synonym of *Toxorhynchites* Theobald, 1901b; synonymy established before Knight & Stone, 1977.

Wyeomyia Theobald, 1901b, 1901a, 1901c (pp. 235, 98, 267, respectively) – f., genus (subgenus) of tribe Sabethini Blanchard, 1905.
Synonyms: *Diphalangarpe* Dyar, 1919d; *Pentemyia* Dyar, 1919d; *Phyllozomyia* Dyar, 1924c; *Technicomyia* Dyar, 1925a.

Xyele Reinert, Harbach & Kitching, 2009 (m., subgenus of *Stegomyia* Theobald, 1901) (p. 783) – synonym of *Stegomyia* Theobald, 1901; synonymy by Wilkerson *et al.*, 2015.

Yamada Reinert, Harbach & Kitching, 2008 (m., subgenus of *Hopkinsius* Reinert, Harbach & Kitching, 2008) (p. 101) – synonym of *Hopkinsius* Reinert, Harbach & Kitching, 2008; synonymy by Wilkerson *et al.*, 2015.

Zavortinkius Reinert, 1999a (p. 7) – m., subgenus of *Aedes* Meigen, 1818; subgeneric status re-established by Wilkerson *et al.*, 2015.

Zeugnomyia Leicester, 1908 (p. 231) – f., genus of tribe Aedini Neveu-Lemaire, 1902c.

Zinzala Zavortink, 1986 (p. 46) – f., subgenus of *Wyeomyia* Theobald, 1901a,b,c.

Zoromorphus Reinert, Harbach & Kitching, 2009 (m., subgenus of *Stegomyia* Theobald, 1901) (p. 783) – synonym of *Stegomyia* Theobald, 1901; synonymy by Wilkerson *et al.*, 2015.

Section 3: Family-group and Other Suprageneric Names

Family-group names are nouns in the nominative plural (*Code*, Article 11.7.1.1). They are based on the name of a genus, which automatically becomes the type genus of the family group (*Code*, Article 63). Under the Principle of Coordination (*Code*, Article 36), a name proposed for a family-group name at any rank simultaneously establishes it for all higher and lower ranks in the family group, each having the same type genus. Thus the family Culicidae, based on *Culex* Linnaeus, 1758 was established in 1818 by Meigen (as Culiciformes). Culiciformes Meigen, 1818 is the earliest family-group name that established the date of priority for Culicidae, not Billberg, 1820 (as Culicides). Meigen is therefore deemed to have simultaneously established the coordinate subfamily name Culicinae and the coordinate tribal name Culicini, and *Culex* is the type genus of all three.

It is apparent from the previous paragraph that the endings of family-group taxa are added to the stem of the type genus (Article 29.1). Article 29.2 requires the addition of -idae to the stem for family names, -inae for subfamily names, -ini for the names of tribes and -ina for the names of subtribes. Although a few subtribes have been recognized for mosquitoes in the past, none are recognized in contemporary culicid classification. It is interesting to note that a few family-group names are formed from the stems of generic names that are patronymics. These include **Ficalbiini**, **Hodgesiini**, Joblotina, Joblotinae, Lynchiellina, Lynchinellinae and **Mansoniini**.

Before the four endings became fully accepted and eventually required, some authors used endings for ranks other than the ranks indicated above. Theobald, for example, used the -ina ending for a number of subfamily names (Aedeomyina, etc.). Some authors, especially prior to the 20th century, used various other endings for family-level taxa. Such endings include -idea (e.g. Culicidea), -es (e.g. Culiciformes) and -idi (e.g. Culicidi). A number of authors at the start of the 20th century, notably Lahille and Lutz, used vernacular names for groupings of taxa that would be considered to be family-level groupings. Endings for such groups include -ae (e.g. Asiphonatae), -es (e.g. Micropsélaphes) and -i (e.g. Epialurgi).

The list below contains 112 names, including two family names (one valid), 41 subfamily names (three valid: two extant, one fossil; seven duplicated as corrected spellings), 21 tribal names (11 valid, one duplicated as a corrected spelling), seven subtribal names, 17 groups/groupings, five sections, three incorrect spellings, three errant spellings and five variant spellings. Nine names proposed for subfamilies are formed with the -ina ending and one name proposed as a tribe is formed with the -idi ending.

Format

Family-group names that are currently accepted as valid are in **boldface**. Disused, synonymic and other names are not in boldface. Names are followed by author and date, page of citation, taxonomic rank, and taxon equivalence or composition (i.e. subordinate taxa which comprise the taxon). Unlike species- and genus-group names, family-group names are not (and never should be) italicized.

Family-group names in Culicidae

Aedeinae Neveu-Lemaire, 1902c (pp. 201, 220) – subfamily comprising genera *Aedes*, *Aedeomyia*, *Haemagogus*, *Sabethes*, *Uranotaenia* and *Wyeomyia*; coordinate with Aedini (see).

Aedeomyiinae – corrected spelling of Aedeomyinae Neveu-Lemaire, 1902c (see).

Aedeomyiini Theobald, 1901b (coordinate tribal name of Aedeomyina Theobald, 1901b) (p. 235) – tribe of subfamily Culicinae Meigen, 1818.

Aedeomyina Theobald, 1901d (p. 235) – subfamily of true mosquitoes; corresponds to Aedeomyiini Theobald, 1901b.

Aedeomyinae Neveu-Lemaire, 1902c (p. 201) – subfamily of Culicidae Meigen, 1818 (not currently recognized); corresponds to Aedeomyiini Theobald, 1901b; [corrected spelling = Aedeomyiinae].

Aedina Lahille, 1904 (pp. 79, 84 [11, 16], tableau) – subtribe of Lahille's tribe 'Culicinae', comprising genera *Aedes*, *Aedeomyia*, *Haemagogus*, *Sabethes* and *Wyeomyia*.

Aedinae Lutz, 1904 (in Bourroul, 1904) (synoptic table, pp. 45, 53, 65) – subfamily comprising genera *Aedes*, *Aedinus*, *Aedeomyia*, *Deinocerites*, *Ficalbia*, *Mimomyia*, *Uranotaenia* and *Verrallina* (as *Verralina*).

Aedini Neveu-Lemaire, 1902c (coordinate tribal name of Aedeinae Neveu-Lemaire, 1902c) (p. 220) – tribe of subfamily Culicinae Meigen, 1818.

Aedomina Giles, 1902 (pp. 255, 256, 475) – subfamily comprising genera *Aedes*, *Aedeomyia* (as *Aedomyia*), *Haemagogus*, *Sabethes*, *Uranotaenia* and *Wyeomyia*.

Ankylorhynchae Lutz, 1904 (in Bourroul, 1904) (synoptic table, p. 53) – group of Lutz's Siphonatae; equivalent to Lutz's subfamily Megarhininae; corresponds to Toxorhynchitini Lahille, 1904.

Anopheleinae Eysell, 1905a, 1905b (pp. 24, 55, respectively) – subfamily (incorrect spelling) of Anophelidae Eysell, 1905a,b; corresponds to Anophelinae Grassi, 1900.

Anophelidae Eysell, 1905a, 1905b (pp. 24, 55, respectively) – family of Diptera; corresponds to Anophelinae Grassi, 1900.

Anophelina Theobald, 1901a (pp. 97, 115) – subfamily of true mosquitoes; corresponds to Anophelinae Grassi, 1900.

Anophelinae Grassi, 1900 (p. 359) – subfamily of Culicidae Meigen, 1818.

Anophelini Edwards, 1912c (p. 53) – tribe of Culicinae (= Culicidae Meigen, 1818 as currently defined); corresponds to Anophelinae Grassi, 1900.

Anophilini Pearse, 1936 (p. 53) – incorrect spelling of Anophelini Edwards, 1912c (see).

Asiphonatae Lutz, 1904 (in Bourroul, 1904) (synoptic table, p. 51) – group of Lutz's Euculicidae; corresponds to Anophelinae Grassi, 1900.

Branchiopodiformes Bouché, 1834 (p. 5) – a group for *Culex*, in a classification of insect larvae.

Burmaculicinae Borkent & Grimaldi, 2016 (p. 465) – subfamily of fossil Culicidae.

Campestrae Lynch Arribálzaga, 1891a (p. 367) – a grouping of culicid genera, including *Heteronycha*, *Janthinosoma* and *Ochlerotatus*.

Culicales Alcock, 1911 (pp. 241, 247) – section of Culicinae (= Culicidae Meigen, 1818 as currently defined) comprising Aedinae, Culicinae, Deinoceratinae (as Dinoceratinae), Heptaphlebomyinae and Uranotaeniinae as used by Theobald, 1907, 1910c.

Culices Griffith & Pidgeon, 1832 (p. 726) – family name (errant spelling) for mosquitoes; corresponds to Culicidae Meigen, 1818.

Culicida Perty, 1841 (p. 888) – family name (errant spelling) for mosquitoes; corresponds to Culicidae Meigen, 1818.

Culicidae Meigen, 1818 (coordinate family-group name of Culiciformes Meigen, 1818) (p. xxxiii) – family, the true mosquitoes, of Diptera.

Culicidea Bellardi, 1859 (p. 5) – family name (errant spelling) for mosquitoes; corresponds to Culicidae Meigen, 1818.

Culicides Billberg, 1820 (p. 122) – group name ('Natio') in Diptera for genera *Culex*, *Corethra*, *Chironomus*, *Ceratopogon*, *Cecdidomya*, *Tanypus*, *Psycoda* and *Cnips*.

Culicidi Lioy, 1863 (p. 202) – tribe comprising genera *Aedes*, *Anopheles*, *Culex* and *Phlaebotomus*.

Culiciformes Meigen, 1818 (p. xxxiii) – family-group name of Diptera; coordinate with Culicidae (see), Culicinae (see) and Culicini (see).

Culicinae Meigen, 1818 (coordinate family-group name of Culiciformes Meigen, 1818) (p. xxxiii) – subfamily of Culicidae Meigen, 1818.

Culicini Meigen, 1818 (coordinate family-group name of Culiciformes Meigen, 1818) (p. xxxiii) – tribe of Culicinae Meigen, 1818.

Culisetina Maslov, 1964 (p. 198) – subtribe of Culisetini Belkin, 1962 [attributed to Stone, 1957a] (not currently recognized).

Culisetini Belkin, 1962 (p. 282) – tribe of Culicinae Meigen, 1818.

Culisetini Stone, 1957a (p. 333) – *nomen nudum* (see Culisetini Belkin, 1962).

Deinoceritinae Mitchell, 1906 (p. 19) – subfamily of Culicidae Meigen, 1818 (not currently recognized); corresponds to *Deinocerites* Theobald, 1901a,b,c.

Dendromyiinae – corrected spelling of Dendromyinae Lutz, 1904 (see).

Dendromyinae Lutz, 1904 (in Bourroul, 1904) (synoptic table, pp. 47, 56, 67) – subfamily equivalent to Lutz's Micropalpae group within his Metanotrichae group; comprising genera *Dendromyia*, *Limatus*, *Phoniomyia*, *Sabethes*, *Sabethoides*, *Sabethinus* and *Wyeomyia*; [corrected spelling = Dendromyiinae].

Dinoceratinae Alcock, 1911 (p. 241) – incorrect spelling of Deinoceratinae Theobald, 1907 (= Deinoceritinae Mitchell, 1906).

Domesticae Lynch Arribálzaga, 1891a (p. 367) – a behavioural grouping of culicid genera, including *Culex* and *Taeniorhynchus*.

Epialurgi Alcock, 1911 (pp. 241, 247) – section of Culicinae (= Culicidae Meigen, 1818 as currently defined) for Anophelinae as used by Theobald, 1907, 1910c.

Euculicidae Lutz, 1904 (in Bourroul, 1904) (synoptic table, pp. 51, 63) – group name for true mosquitoes (i.e. Culicidae Meigen, 1818 as currently defined).

Ficalbiini Belkin, 1962 (p. 287) – tribe of Culicinae Meigen, 1818.

Haemagoginae Lutz, 1904 (in Bourroul, 1904) (synoptic table, pp. 47, 66) – subfamily of Lutz's Micropalpae group of mosquitoes; comprising genera *Gualteria* and *Haemagogus*.

Harpagomyiinae – corrected spelling of Harpagomyinae Ludlow, 1911b (see).

Harpagomyinae Ludlow, 1911b (pp. 127, 132) – subfamily of Culicidae Meigen, 1818; comprising genera *Harpagomyia* and *Hodgesia*; [corrected spelling = Harpagomyiinae].

Heptaphlebomyiinae – corrected spelling of Heptaphlebomyinae Lutz, 1904 (see).

Heptaphlebomyiini Handlirsch, 1924 (p. 973) – tribe of true mosquitoes; comprising genus *Heptaphlebomyia* Theobald, 1903a.

Heptaphlebomyina Theobald, 1903a (p. 336) – subfamily of true mosquitoes; corresponds to *Culex* Linnaeus, 1758.

Heptaphlebomyinae Lutz, 1904 (in Bourroul, 1904) (synoptic table) – subfamily of Lutz's Heteropalpae group of mosquitoes; comprising genus *Heptaphlebomyia* Theobald, 1903a; [corrected spelling = Heptaphlebomyiinae].

Heteropalpae Lutz, 1904 (in Bourroul, 1904) (synoptic table, p. 55) – group name in mosquitoes listed in synoptic table as (1) a group of Lutz's Metanotopsilae for subfamilies Culicinae and Heptaphlebomyinae and (2) a group of Metanotrichae (as Metanototrichae on page 55) for subfamily Hyloconopinae.

Hétéropsélaphes Lahille, 1904 (p. 79 [11], tableau) – collective vernacular name for the group of mosquitoes comprising Lahille's Culicina (Protoptères in part) and Toxorhynchina (Téléoptères in part).

Hodgesiini Belkin, 1962 (p. 278) – tribe of Culicinae Meigen, 1818.

Hyloconopinae Lutz 1904 (in Bourroul, 1904) (synoptic table, pp. 49, 55, 48) – subfamily equivalent to Lutz's Heteropalpae group of mosquitoes (in error as Hycolonopinae on p. 49); comprising genera *Goeldia*, *Hyloconops*, *Joblotia*, *Rhynchomyia* (incorrect spelling of *Runchomyia*) and *Trichoprosopon*.

Joblotiinae Grünberg, 1907b (pp. 63, 107) – incorrect spelling of Joblotinae Theobald, 1905e (see).

Joblotina Theobald, 1903a (p. 334) – subfamily of true mosquitoes; corresponds to *Trichoprosopon* Theobald, 1901a,c [however, Theobald also listed *Joblotia lunata* Theobald, 1901c (= *Isostomyia lunata*) as a member of the subfamily].

Joblotinae Theobald, 1905e (pp. 5, 33) – subfamily of Culicidae Meigen, 1818; corresponds to *Trichoprosopon* Theobald, 1901a,c.

Limatinae Theobald, 1907 (pp. 18, 627) – subfamily of Culicidae Meigen, 1818 (not currently recognized); corresponds to *Limatus* Theobald, 1901c.

Lynchiellina Lahille, 1904 (pp. 79, 81 [11, 13], tableau) – subtribe of Culicinae (= Culicidae Meigen, 1818 as currently defined); proposed by da Costa Lima *et al.* (1962) as a subtribe of Toxorhynchitini Lahille, 1904 (not currently recognized).

Lynchinellinae Peryassú, 1908 (p. 29) – variant spelling of Lynchiellina Lahille, 1904 in synonymy under Megarhininae.

Macropsélaphes Lahille, 1904 (p. 79 [11], tableau) – collective vernacular name for the group of mosquitoes comprising Lahille's Anophelina (Protoptères in part) and Lynchiellina (Téléoptères in part).

Mansoniini Belkin, 1962 (p. 299) – tribe of subfamily Culicinae Meigen, 1818.
Synonym: Mansonini Ronderos & Bachmann, 1963.

Mansonini Ronderos & Bachmann, 1963 (p. 45) – proposed tribal name anteceded by Mansoniini Belkin, 1962.

Megalorhini Alcock, 1911 (p. 241) – section of Culicinae (= Culicidae Meigen, 1818 as currently defined); corresponds to Toxorhynchitini Lahille, 1904.

Megalorrhini Brunetti, 1920 (p. 90) – variant spelling of Megarhininae Neveu-Lemaire, 1902c.

Megalorhinina Marshall, 1923 (footnote in Neveu-Lemaire, 1923) (p. 99) – variant spelling of Megalorhini Alcock, 1911.

Megarhinae Giles, 1902 (p. 265) – variant spelling (originally italicized) of Giles's subfamily Megarhinina and genus *Megarhina*; corresponds to Toxorhynchitini Lahille, 1904 and *Toxorhynchites* Theobald, 1901b, respectively.

Megarhini Brunetti, 1914 (pp. 32, 34) – section of Culicinae (= Culicidae Meigen, 1818 as currently defined); corresponds to Toxorhynchitini Lahille, 1904.

Megarhinina Theobald, 1901d (pp. 97, 215) – subfamily of true mosquitoes; corresponds to Toxorhynchitini Lahille, 1904.

Megarhininae Neveu-Lemaire, 1902c (pp. 201, 206) – subfamily of Culicidae Meigen, 1818 (not currently recognized); corresponds to Toxorhynchitini Lahille, 1904.

Megarhinini Edwards, 1932a (pp. 8, 29, 57) – tribe of Culicinae (= Culicidae Meigen, 1818 as currently defined); corresponds to Toxorhynchitini Lahille, 1904.

Megarinina Theobald, 1904c (p. 36) – subfamily (incorrect spelling) of true mosquitoes; corresponds to Toxorhynchitini Lahille, 1904.

Megarrhini Bertrand, 1954 (p. 200) – tribe comprising genus *Megarhinus* (= Toxorhynchitini Lahille, 1904).

Megarrhininae Bezzi, 1906 (p. 209) – subfamily of Culicinae (= Culicidae Meigen, 1818 as currently defined); corresponds to Toxorhynchitini Lahille, 1904.

Megharinae Gonzalez-Rincones & Guyon, 1953 (p. 169, 182, 199) – misspelling of Megarhininae Neveu-Lemaire, 1902c (as Megarhininae Theobald, 1907).

Metanotopsilae Lutz, 1904 (in Bourroul, 1904) (synoptic table, p. 53) – group of mosquitoes comprising Lutz's Heteropalpae (subfamilies Culicinae and Heptaphlebomyinae) and Micropalpae (subfamilies Aedinae and Haemagoginae).

Metanototricha Alcock, 1911 (pp. 241, 249) – section of Culicinae (= Culicidae Meigen, 1818 as currently defined) comprising Dendromyinae and Limatinae as used by Theobald, 1907, 1910c.

Metanototrichae Lutz, 1904 – see Metanotrichae Lutz, 1904.

Metanototrichina Marshall, 1923 (footnote in Neveu-Lemaire, 1923) (p. 99) – variant spelling of Metanototricha Alcock, 1911.

Metanotrichae Lutz, 1904 (in Bourroul, 1904) (synoptic table, p. 55, as Metanototrichae) – group of mosquitoes comprising Lutz's Heteropalpae (subfamily Hyloconopinae) and Micropalpae (subfamily Dendromyinae).

Micropalpae Lutz, 1904 (in Bourroul, 1904) (synoptic table, pp. 53, 56, as Isomicropalpae) – group name in mosquitoes listed in synoptic table as (1) a group of Lutz's Metanotopsilae for subfamilies Aedinae and Haemagoginae and (2) a group of Metanotrichae for subfamily Dendromyinae. Isomicropalpae (of Metanototrichae) is used for the two groups in keys on pages 53 and 56.

Micropsélaphes Lahille, 1904 (p. 79 [11], tableau) – collective vernacular name for the group of mosquitoes comprising Lahille's Aedina (Protoptères in part) and Uranotaenina (Téléoptères in part).

Opificinae Miller, 1922 (p. 115) – subfamily of Culicidae Meigen, 1818 (not currently recognized); corresponds to subgenus *Opifex* of *Opifex* Hutton, 1902.

Orthopodomyiini Belkin, Heinemann & Page, 1970 (p. 112) – tribe of subfamily Culicinae Meigen, 1818.

Orthorhynchae Lutz, 1904 (in Bourroul, 1904) (synoptic table) – group of Lutz's Euculicidae (= Culicidae Meigen, 1818 as currently defined) for all mosquitoes of subfamily Culicinae (his Siphonatae) except Toxorhynchitini (as Megarhininae).

Palustrae Lynch Arribálzaga, 1891a (p. 367) – a behavioural grouping of culicid genera, including *Aedes*, *Anopheles* and *Uranotaenia*.

Protoptères Lahille, 1904 (pp. 78, 79 [10,11], tableau) – collective vernacular name for mosquitoes that share more primitive characters of the wings, as opposed to the more specialized Téléoptères (see); comprising Lahille's Aedina, Anophelina and Culicina.

Psorophorinae Mitchell, 1906 (p. 21) – subfamily of Culicidae Meigen, 1818 (not currently recognized); corresponds to *Psorophora* Robineau-Desvoidy, 1827.

Sabethinae Dyar, 1906 (p. 191) – subfamily of Culicidae Meigen, 1818 (not currently recognized); corresponds to Sabethini Blanchard, 1905.

Sabethini Blanchard, 1905 (coordinate tribal name of Sabettinae Blanchard, 1905) (p. 420) – tribe of subfamily Culicinae Meigen, 1818.

Sabethinini Shannon, 1927 (p. 12) – misspelling (?) of tribe Sabethini.

Sabettinae Blanchard, 1905 (p. 420) – subfamily of Culicinae (= Culicidae Meigen, 1818 as currently defined) (not currently recognized; see Sabethini).

Silvicolae Lynch Arribálzaga, 1891a (p. 367) – a behavioural grouping of culicid genera, including *Megarhina*, *Psorophora* and *Sabethes*.

Siphonatae Lutz, 1904 (in Bourroul, 1904) (synoptic table, p. 53) – group name for all mosquitoes with siphon-bearing larvae; corresponds to Culicinae Meigen, 1818.

Stegomyiinae – corrected spelling of Stegomyinae Peryassú, 1908 (see).

Stegomyina Christophers, 1906 (pp. 16, 17) – subfamily (originally italicized) of Culicidae (including subfamily Corethrina = Corethrellidae) comprising genera *Desvoidea* (= *Desvoidya* Blanchard, 1901) and *Stegomyia*.

Stegomyinae Peryassú, 1908 (p. 364) – subfamily of Culicidae Meigen, 1818 (not currently recognized); comprising genera *Bancroftia*, *Carrollia*, *Gualteria*, *Haemagogus*, *Stegoconops* and *Stegomyia*; [corrected spelling = Stegomyiinae].

Téléoptères Lahille, 1904 (pp. 78, 79 [10, 11], tableau) – collective vernacular name for mosquitoes that share specialized characters of the wings, as opposed to the more primitive Protoptères (see); comprising Lahille's Lynchiellina, Toxorhynchina and Uranotaenina.

Toxorhynchina Lahille, 1904 (pp. 79, 84 [11, 16], tableau) – subtribe of Lahille's tribe Culicinae (not currently recognized); coordinate with Toxorhynchitinae (see) and Toxorhynchitini (see).

Toxorhynchitina da Costa Lima, Guitton & Ferreira, 1962 (p. 226) – subtribe of Toxorhynchitini Lahille, 1904 (not currently recognized).

Toxorhynchitinae Theobald, 1905e (pp. 5, 13) – subfamily of Culicidae Meigen, 1818 (not currently recognized); coordinate with Toxorhynchina (see); placed on the Official List of Family-Group Names in Zoology (International Commission on Zoological Nomenclature, 1959: Opinion 548), but that name is antedated by Toxorhynchina Lahille, 1904.

Toxorhynchitini Lahille, 1904 (coordinate family-group name of Toxorhynchina Lahille, 1904) (p. 79 [11]) – tribe of subfamily Culicinae Meigen, 1818.

Trichoprosopinae – corrected spelling of Trichoprosoponinae Coquillett, 1906d (but see Sabrosky, 1999: 313).

Trichoprosopini – corrected spelling of Trichoprosoponina Theobald, 1901a,c (see).

Trichoprosoponina Theobald, 1901a (p. 98) – subfamily of mosquitoes; corresponds to *Trichoprosopon* Theobald, 1901a,c; [corrected spelling = Trichoprosopini (but see Sabrosky, 1999: 313)].

Trichoprosoponinae Coquillett, 1906d (pp. 11, 26) – revised spelling of Trichoprosoponina Theobald, 1901a; corresponds to Sabethini Blanchard, 1905 (comprising genera *Dendromyia*, *Limatus*, *Phoniomyia*, *Sabethes*, *Sabethoides*, *Trichoprosopon* and *Wyeomyia*; [corrected spelling = Trichoprosopinae (but see Sabrosky, 1999: 313)].

Uranotaeniani Pearse, 1936 (p. 53) – tribe (incorrect spelling) of true mosquitoes; corresponds to Uranotaeniini Lahille, 1904.

Uranotaeniinae Coquillett, 1906d (pp. 11, 26) – subfamily of true mosquitoes; corresponds to *Uranotaenia* Lynch Arribálzaga, 1891a,b.

Uranotaeniini Lahille, 1904 (coordinate family-group name of Uranotaenina Lahille, 1904) (p. 79 [11]) – tribe of subfamily Culicinae Meigen, 1818.

Uranotaenina Lahille, 1904 (pp. 79, 88 [11, 16], tableau) – subtribe of Lahille's tribe Culicinae (= Culicidae Meigen, 1818 as currently defined); coordinate with Uranotaeniini (see).

Uranotaeninae Mitchell, 1907 (p. 297) – subfamily of true mosquitoes (not currently recognized); corresponds to Uranotaeniini Lahille, 1904.

Wyeomiinae Lahille, 1904 (p. 77 [9]) – 'les Wyeomiinae de Theobald', unavailable, not adopted.

Wyeomyiinae – corrected spelling of Wyeomyinae Leicester, 1908 (see).

Wyeomyinae Leicester, 1908 (p. 250) – subfamily of mosquitoes comprising genera *Dendromyia*, *Goeldia*, *Limatus*, *Malaya*, *Phoniomyia*, *Runchomyia*, *Sabethes*, *Sabethoides* and *Wyeomyia*; [corrected spelling = Wyeomyiinae].

Section 4: Species-group Names Formed from Personal Names

Of the 3689 extant species-group taxa, 3559 species and 130 subspecies (Section 1) that are currently recognized (fossil species, synonyms and *nomina dubia* are not included here), nearly 40% have names that are formed from personal names. These include the names of 1254 species and 33 subspecies. Most of the names commemorate men (1181). Only 102 are formed from the names of women and five honour a man and a woman.

The majority of genus-group patronymics are nouns in the genitive case, formed by adding an appropriate first or second declension Latin ending to the stem of the personal name. These endings, which denote commemoration or dedication, include:

- *-ae*, singular (*-arum*, plural) first declension (feminine) genitives;
- *-i*, singular (*-orum*, plural) second declension (masculine) genitives.

If the personal name is that of a woman, *-ae* is added to form a genitive singular noun; if the name is that of a man, *-i* is added to form a genitive singular noun. If the name is that of women, *-arum* is added to form a plural (feminine) genitive noun, and if it commemorates a man and a woman, *-orum* is added to form a plural (masculine) genitive noun. As indicated above, there are no plural first declension patronymics in Culicidae, i.e. none commemorates more than one woman. It should be noted that the *-ae* and *-i* endings are also used to form species-group epithets from geographical names (see Section 5). The endings of species-group names formed in the genitive case remain unchanged irrespective of the gender of the genus with which they are combined.

The stem of a noun is usually, but not always, determined by removing the genitive-case ending. The *-ae* ending can be added to all feminine proper names except those ending in *-a*, e.g. Amelia, to which only the *-e* is attached to form the species name ***ameliae*** Casal. Technically, the *-i* ending can also be attached to all proper nouns ending in any letter except *-a*, but in practice this is not the case. Fifty-five currently valid species-group names have been formed by affixing *-i* to the terminal *-a* of masculine names, e.g. ***cabrerai*** Miyagi, Toma & Rivera, dedicated to Dr Benjamin D. Cabrera. *Anopheles pampanai* Büttiker & Beales was originally incorrectly published as *pampanae* and later changed to its presently accepted form. A few other names dedicated to men and incorrectly formed but unchanged include ***andreae*** Doucet, ***messeae*** Falleroni and ***labranchiae*** Falleroni.

The formation of two nounal names, ***anastasionis*** Dyar and ***nobukonis*** (Yamada), is unusual. Patronyms ending in the Latin *-onis* are rare in nomenclature. These are genitive-case nouns formed from names ending in *-o*, in this case Anastasio and Nobuko.

All but 17 valid species-group patronymics in Culicidae are nouns in the genitive case – eight are nouns in apposition and nine are adjectives in the nominative singular. The nouns in apposition include the names of two modern men (***gabriel*** Schick and ***radama*** de Meillon), the probable name of a young girl (***eastor*** Dyar) and the names of five fictional or mythical characters (***apsarae*** Klein, ***oedipus*** Root, ***sphinx*** Howard, Dyar & Knab, ***varuna*** Iyengar and ***zethus*** de Meillon & Lavoipierre), all of which are treated as indeclinable and the ending unchanged if transferred to a genus of another gender. The adjectival names include ***adersianus*** Edwards (m.), ***andreanus*** Edwards (m.), ***andricus*** Root (m.), ***bancroftianus*** Edwards (m.), ***barberellus*** Evans (m.), ***bathana*** (Dyar) (f.), ***harrisonica*** (Reinert) (f.), ***laneanus*** Corrêa & Cerqueira (m.) and ***martinius*** Shingarev (m.). Unlike genitive and appositional names, the endings of adjectival names, when proposed, must agree in gender with the name of the genus with which they are combined, and must be changed if transferred to a genus of a different gender, e.g. *Chagasia bathana* was original described as *Anopheles bathanus*.

The five names that are dedicated to a man and a woman include ***deaneorum*** Rosa-Freitas, ***hinesorum*** Schmidt, ***mattinglyorum*** Huang, ***stoneorum*** Marks and ***thurmanorum*** Bram. Oddly, four names clearly intended to honour a man and his wife were formed by adding the masculine singular secondary declension ending (*-i*) to the surname: ***boneti*** Gil Collado, ***bonnei*** Dyar, ***gerbergi*** Belkin and ***harleyi*** Peters. Unfortunately, these names cannot be amended as there is no evidence of an inadvertent error in the published articles (*Code*, Article 32.5).

Format

The patronymics currently recognized for extant species-group taxa are listed below in alphabetical order. Each name is followed by authorship, date of publication, the higher taxon (genus or species) with which it is currently associated, the name of the individual or individuals for whom the taxon is dedicated (if indicated or inferable from information contained in the article) and a brief indication of the reason or justification for the dedication (if indicated or inferable). The names of honourees are printed as they appeared in the original publications, often incomplete, some with titles, many without titles and institutional affiliation. Little or no information is provided for most names proposed before the first half of the 20th century; consequently, many honourees will unfortunately remain obscure or unknown. Many of the early taxonomists dedicated taxa to collectors, whose names were generally indicated by italicized initials and surname enclosed in parentheses, a practice particularly attributable to British workers, e.g. P.J. Barraud and F.W. Edwards. No attempt has been made to contact relatives and surviving colleagues of honourees or to search published sources of potential information about them. In the case of anopheline names, reference is made to the magnificent works of Kitzmiller (1982) and Service (2010), who painstakingly collected information and provided accounts of the lives of individuals who have had mosquito species and subspecies named in their honour.

Valid Patronymic Species-group Names in Culicidae

abonnenci Clastrier, 1970c: *Culex* – Mr E. Abonnenc (entomologist).

acaci Baisas, 1946a: *Anopheles* – Capt. Alejandro B. Ac-ac (see Kitzmiller, 1982).

adairi Kirkpatrick, 1926: *Culex* – Mr E.W. Adair (contributions).

adamesi Sirivanakarn & Galindo, 1980: *Culex* – Dr Abdiel J. Adames (honorary).

adami Ferrara & Eouzan, 1974: *Eretmapodites* – Jean-Paul Adam (appreciation).

adami Geoffroy, 1971b: *Aedes* – J.-P. Adam (entomologist).

adami (Hamon & Mouchet, 1955): *Culex* – J.-P. Adam (collected holotype).

adersi (Edwards, 1917): *Aedes* – Dr W.M. Aders (collected type specimens).

adersianus Edwards, 1941: *Culex* – [Dr W.M.] Aders (collected type female).

aguilari Baisas & Feliciano, 1953: *Zeugnomyia* – Dr Eusebio Aguilar (in memoriam).

airosai Lane & Cerqueira, 1942: *Wyeomyia* – Dr A.L. Airosa Galvão (honorary).

airozai Lane, 1945b: *Culex* – Dr A.L. Ayroza Galvão (friend).

aitkeni Schick, 1970a: *Aedes* – T.H.G. Aitken (honorary).

aitkenii James, 1903: *Anopheles* – E.H. Aitken (collected type female) (see Kitzmiller, 1982).

alani Forattini, 1965: *Culex* – Dr Alan Stone (by Bram, 1967b).

alani Lane & Cerqueira, 1957: *Wyeomyia* – Dr Alan Stone (information).

albertoi Unti, 1941: *Anopheles* – Sr Alberto da Silva Ramos (entomologist) (see Kitzmiller, 1982).

alcasidi Huang, 1972: *Aedes* – Dr Godofredo L. Alcasid (appreciation).

alcocki Bonne-Wepster & Bonne, 1920: *Culex* – Alcock (see Belkin, 1968b).

alektorovi Stackelberg, 1943: *Aedes* – not mentioned, probably A.A. Alektorov (entomologist).

alexandrei Gornostaeva, 2005: *Aedes* – Alexander A. Shtakelberg (dipterologist).

aliciae Duret, 1953: *Culex* – mi hija (my daughter), Alicia Diana (see Duret, 1954a).

aliyusopi Miyagi & Toma, 2014: *Topomyia* – Haji Ali Bin Yusop (appreciation).

alkatirii Toma, Miyagi & Syafruddin, 1995: *Armigeres* – Jonus Alkatiri (contributions and support).

ameliae Casal, 1967: *Culex* – Amelia, mi esposa (my wife).

amesii (Ludlow, 1903): *Aedes* – Dr Roger P. Ames (provided specimens).

anastasionis Dyar, 1921k: *Haemagogus* – not mentioned, probably derived from Anastasio, perhaps the first name of the collector, A. Alfaro.

anchietai Corrêa & Ramalho, 1968: *Anopheles* – Padre José de Anchieta S. J. (in memoriam) (see Kitzmiller, 1982).

andersoni Edwards, 1914b: *Culex* – T.J. Anderson (collected type specimens).

andersoni Edwards, 1926a: *Aedes* – ? Dr Anderson (see *Andersonia* Strickland, 1911b).

andreae Doucet, 1962: *Uranotaenia* – Mme Andrée Doucet (? author's wife).

andreae Harbach, 2018: *Sabethes* – Andrea Lynn Harbach (author's daughter, in memoriam).

andreanus Edwards, 1927a: *Culex* – Major André (collected type specimens).

andrewsi Edwards, 1926a: *Aedes* – Dr C.W. Andrews (contributed specimens to type series).

andricus Root, 1927c: *Culex* – not mentioned, ? a servant of Cicero (Andricus, Andrici).

anduzei Cerqueira & Lane, 1944: *Culex* – Dr Pablo Anduze (honorary).

annandalei Barraud, 1926: *Uranotaenia* – Annandale (collected one female used in the description).

annandalei Prashad, 1918: *Anopheles* – Dr N. Annandale (regard and gratitude) (see Kitzmiller, 1982).

annandalei (Theobald, 1910a): *Aedes* – Annandale (collected holotype).

annettii (Theobald, 1901d): *Coquillettidia* – not mentioned, Dr Annett (collected type specimens) (see Theobald, 1901c).

antunesi Galvão & Franco do Amaral, 1940: *Anopheles* – Dr P.C. de Azevedo Antunes (see Kitzmiller, 1982).

antunesi Lane & Guimarães, 1937: *Wyeomyia* – Dr Paula C. Azevedo Antunes (friend).

antunesi Lane & Whitman, 1943: *Culex* – Dr P.C.A. Antunes (contributions).

apsarae Klein, 1977: *Topomyia* – not mentioned, ? female spirit of the clouds and waters in Hindu and Buddhist mythology.

arguellesi Baisas, 1935c: *Uranotaenia* – Dr Angel S. Argüelles (honorary).

arribalzagae (Theobald, 1903a): *Coquillettidia* – not mentioned, Arribálzaga, surname of the mother of Felix Lynch Arribálzaga (see Stone, 1957a, unjustified emendation to *arribalzagai*).

artemievi Gordeev, Zvantsov, Goriacheva, Sheĭkevich & Ezhov, 2005: *Anopheles* – M.M. Аргемьева [M.M. Artemiev] (see Service, 2010).

arthuri Unti, 1941: *Anopheles* – Dr Arthur Costa Filho (honorary) (see Kitzmiller, 1982).

aruni Sobti, 1968: *Anopheles* – not mentioned, Mr Arun (collected larvae) (see Kitzmiller, 1982).

ashworthi Edwards, 1921c: *Aedes* – Dr J.H. Ashworth (collected type specimens).

austenii (Theobald, 1905b): *Anopheles* – Mr Austen (provided specimens), Major E.E. Austen (see Kitzmiller, 1982).

azevedoi Ribeiro, 1969: *Anopheles* – Prof. J. Fraga de Azevedo (honorary) (see Kitzmiller, 1982).

azurini Basio, 1971: *Armigeres* – Dr Jesus C. Azurin (contributions).

azurini Toma, Miyagi & Cabrera, 1984: *Culex* – Dr Jesus C. Azurin (honorary).

baezai Gater, 1933: *Anopheles* – Dr J.I. Baeza (collected type specimens) (see Kitzmiller, 1982).

bahri (Edwards, 1914b): *Culex* – Dr P.H. Bahr (collected type specimens).

baileyi Edwards, 1929b: *Anopheles* – Lt.-Col. F.M. Bailey (collected holotype and a female paratype) (see Kitzmiller, 1982).

bailyi Barraud, 1929a: *Hodgesia* – J.D. Baily (collected holotype).

bailyi Barraud, 1934: *Culex* – [J.D.] Baily (collected type specimens).

baimaii Sallum & Peyton, 2005: *Anopheles* – Prof. Visut Baimai (contributions) (see Service, 2010).

baisasi Colless, 1957b: *Anopheles* – Mr F.E. Baisas (honorary) (see Kitzmiller, 1982).

baisasi Knight & Hull, 1951a: *Aedes* – Mr F.E. Baisas (contributions).

baisasi Sirivanakarn, 1972: *Culex* – Dr F.E. Baisas (contributions).

baisasi Stone & Thurman, 1958: *Armigeres* – Dr F.E. Baisas (contributions).

balfouri Theobald, 1904b: *Uranotaenia* – Dr Andrew Balfour (collected holotype).

baltae Porter, 2009: *Wyeomyia* – Mercedes Rosario Balta Leon (honorary).

bancrofti (Taylor, 1914b): *Verrallina* – Dr T.L. Bancroft (collected type specimens).

bancroftianus Edwards, 1921c: *Aedes* – Dr T.L. Bancroft (collected type specimens).

bancroftii Giles, 1902: *Anopheles* – not mentioned, Dr T.L. Bancroft (see Kitzmiller, 1982).

banksi Edwards, 1922a: *Aedes* – C.S. Banks (collected type specimens).

baolini Gong, 1989b: *Topomyia* – not mentioned, obviously Mr Lu Baolin.

barberellus Evans, 1932: *Anopheles* – Dr M.A. Barber (first bred specimens from larvae) (see Kitzmiller, 1982).

barberi Coquillett, 1903b: *Anopheles* – Mr H.S. Barber (collected type specimens) (see Kitzmiller, 1982).

barnardi Edwards, 1924a: *Aedes* – Mr K.H. Barnard (collected type specimens).

barnesi Belkin, 1953: *Uranotaenia* – Arthur W. Barnes, Jr (reared collections).

barraudi Baisas & Ubaldo-Pagayon, 1953: *Tripteroides* – P.J. Barraud (honorary).

barraudi Edwards, 1922a: *Culex* – Captain P.J. Barraud (collected type specimens).

barraudi (Edwards, 1934): *Aedes* – not mentioned, obviously P.J. Barraud.

barretoi Adames, 1971: *Deinocerites* – not mentioned, obviously Pablo Barreto.

basilewskyi Leleup, 1957: subspecies of *Anopheles garnhami* Edwards, 1930a – Monsieur P. Basilewsky (collected type specimens with the author) (see Kitzmiller, 1982).

batesi Lane & Cerqueira, 1942: *Sabethes* – H.W. Bates (naturalist).

batesi Rozeboom & Komp, 1948: *Culex* – Dr Marston Bates (honorary).

bathana (Dyar, 1928): *Chagasia* – not mentioned, Mr C.H. Bath, sanitarian (see Kitzmiller, 1982).

becki Belkin, 1962: *Aedes* – D. Elden Beck (contributions).

becki Belkin, 1962: *Culex* – D. Elden Beck (contributions).

bedfordi Edwards, 1936: *Aedes* – Bedford (collected type specimens).

behningi Martini, 1926: *Aedes* – Mr Behning (colleague).

bejaranoi Duret, 1953: *Culex* – Lieutenant Colonel Dr J.F.R. Bejarano (see Duret, 1954a).

bejaranoi Martínez, Carcavallo & Prosen, 1960: *Aedes* – Medical Doctor Colonel Don Juan Francisco Ricardo Bejarano (friend and consultant).

bekkui Mogi, 1977: *Aedes* – H. Bekku (collected type specimens).

beklemishevi Stegnii & Kabanova, 1976: *Anopheles* – V.N. Beklemishev (scientist and malariologist) (see Kitzmiller, 1982).

belenrae Rueda, 2005: *Anopheles* – Belen P. Rueda (author's wife) (see Service, 2010).

belisarioi Neiva, 1908a: *Sabethes* – Dr Belisario Penna (collected holotype).

belkini Adames, 1971: *Deinocerites* – not mentioned, obviously Dr John N. Belkin.

belkini Baisas & Ubaldo-Pagayon, 1953: *Tripteroides* – Professor John N. Belkin (contributions).

belkini Grjebine, 1979: *Uranotaenia* – Professor John N. Belkin (contributions).

belkini Stone & Penn, 1948: *Culex* – Dr John N. Belkin (collected holotype and several paratypes).

belleci (Le Goff, Boussès & Brunhes, 2007): *Aedes* – Christian Bellec (entomologist).

benarrochi Gabaldon, Cova-Garcia & Lopez, 1941a: *Anopheles* – Dr E.I. Benarroch (honorary) (see Kitzmiller, 1982).

benoiti Wolfs, 1964: *Uranotaenia* – Professor P. Benoit (collected type specimens).

bequaerti Dyar & Shannon, 1925b: *Culex* – J. Bequaert (collected type specimens).
bequaerti Wolfs, 1947a: *Aedes* – Dr Mich. Bequaert (collected type specimens).
bergerardi Pajot & Geoffroy, 1971: *Aedes* – Professor J. Bergerard (honorary).
berghei Vincke & Leleup, 1949: *Anopheles* – Prof. L. Van den Berghe (honorary) (see Kitzmiller, 1982).
bergi Belkin, 1962: *Culex* – Clifford O. Berg (contributions).
bergrothi (Edwards, 1921b): *Culiseta* – not mentioned, perhaps Ernst Evald Bergroth, Finnish physician and entomologist.
berlandi Séguy, 1921: *Aedes* – M.L. Berland (collected one female of type series).
berlini Schick, 1970a: *Aedes* – O.G.W. Berlin (honorary).
bernardi (Doucet, 1950): *Mimomyia* – not mentioned.
bertii Cova García & Rausseo, 1964: *Uranotaenia* – Ing. Dr Arturo Luis Berti (contributions).
bertrami Schick, 1970a: *Aedes* – D.S. Bertram (collected holotype).
bervoetsi D'Haenens, 1961: *Anopheles* – Docteur W. Bervoets (colleague) (see Kitzmiller, 1982).
bevisi (Edwards, 1915b): *Aedes* – L. Bevis (collected type specimens).
beytouti (Doucet, 1951b): *Mimomyia* – médecin-lieutenant Beytout (collaboration).
bhayungi Thurman & Thurman, 1958: *Armigeres* – Dr Luang Bhayung Vejjasastr (posthumous recognition for achievements in malaria control).
bicki Stone & Penn, 1947: *Culex* – George H. Bick (collected some paratypes).
bickleyi Forattini, 1965: *Culex* – Dr William E. Bickley (see Bram, 1967b).
bickleyi Thurman, 1959: *Toxorhynchites* – Dr William E. Bickley (appreciation and contributions).
bigoti (Bellardi, 1862): *Lutzia* – not mentioned, obviously Jacques Marie Frangile Bigot.
blacklocki Evans, 1925a: *Aedes* – Professor B. Blacklock (collected holotype and two paratypes).
boharti Knight & Rozeboom, 1946: *Aedes* – Dr Richard M. Bohart (honorary).
boharti Brookman & Reeves, 1950: *Culex* – not mentioned, obviously Dr Richard M. Bohart.
bolii Sirivanakarn, 1968: *Culex* – Mr Boli (contributions).
boneti Gil Collado, 1936: *Aedes* – 'Mr. (J. Gil-F.) Bonet and his wife, who unfortunately died shortly after returning [from the expedition]' [translated from Spanish]. **Note**: although clearly intended to honour a man and a woman, requiring the addition of *-orum* to the surname, the name should not be amended as there is no evidence of an inadvertent error in the published paper (International Commission on Zoological Nomenclature, 1999: *Code*, Article 32.5). See *bonnei* Dyar, 1921k, *gerbergi* Belkin, 1977a and *harleyi* Peters, 1955a.
bonneae Dyar & Knab, 1919: *Culex* – Mrs J. Bonne-Wepster (collected type specimens).
bonneae Edwards, 1930b: *Mansonia* – Mrs Bonne (obviously J. Bonne-Wepster), for earlier description as a form similar to *Mansonia annulipes* [= *dives* Schiner, 1868].
bonneae Mattingly, 1958a: *Aedes* – Dr J. Bonne-Wepster (contributions and appreciation).
bonneae Root, 1927b: *Chagasia* – Mrs J. Bonne-Wepster (collected holotype) (see Kitzmiller, 1982).
bonnei Dyar, 1921k: *Culex* – 'Dr. and Mrs. Bonne identified the species ... as *iridescens* Lutz.' **Note**: although clearly intended to honour a man and a woman, requiring the addition of *-orum* to the surname, the name should not be amended as there is no evidence of an inadvertent error in the published paper (International Commission on Zoological Nomenclature, 1999: *Code*, Article 32.5). See *boneti* Gil Collado, 1936, *gerbergi* Belkin, 1977a and *harleyi* Peters, 1955a.
bonnei (Lane & Cerqueira, 1942): *Wyeomyia* – J. Bonne (obviously J. Bonne-Wepster) (honorary).
bonneti Belkin, 1962: *Tripteroides* – David D. Bonnet (contributions).
boshelli Osorno-Mesa, 1944: *Haemagogus* – Dr Jorge Boshell-Manrique (friend/associate).
bosseri Grjebine, 1979: *Uranotaenia* – Jean Bosser (botanist; director ORSTOM).
bourrouli (Lutz, 1905): *Wyeomyia* – not mentioned, obviously Celestino Bourroul.
boussesi da Cunha Ramos & Brunhes, 2004: *Uranotaenia* – Philippe Boussès (friendly tribute).
bradleyi King, 1939: *Anopheles* – G.H. Bradley (contribuitons) (see Kitzmiller, 1982).
brami Forattini, Rabello & Lopes, 1967: *Culex* – Dr Ralph A. Bram (honorary).
breinli (Taylor, 1914d): *Armigeres* – Dr Breinl (bred type specimens from larvae).
brelandi Zavortink, 1972: *Aedes* – Ormond P. Breland (contributions).
brenguesi Brunhes & Ravaonjanahary, 1973: *Culex* – not mentioned, obviously Jacques Brengues.
brenguesi Rickenbach & Lombrici, 1975: *Eretmapodites* – Jacques Brengues (colleague).
brethesi Dyar, 1919e: *Culex* – not mentioned, obviously Juan Brèthes.
bricenoi Cova García, Pulido, Escalante de Ugueto, Amarista & Mora R., 1987c: *Uranotaenia* – Dr Leopoldo Briceño Iragorry (contributions).
briseis Dyar, 1925g: *Uranotaenia* – Briseis, in Greek Mythology, daughter of Briseus and princess of Lyrnessus.
britteni Marks & Hodgkin, 1958: *Aedes* – Mr E.J. Britten (contributions).
brohieri Edwards, 1929b: *Anopheles* – Dr E.L. Brohier [should be S.L. Brohier] (collected type specimens) (see Kitzmiller, 1982).
brottesi Rickenbach, 1967: *Eretmapodites* – Dr Henri Brottes (Director, Institute Pasteur, Cameroun).
broussesi Edwards, 1929a: subspecies of *Anopheles rufipes* (Gough, 1910) – Dr A. Brousses (collected type specimens) (see Kitzmiller, 1982).

browni Komp, 1936c: *Culex* – Mr C.G. Brown (collected holotype).
browni Mattingly, 1955: *Uranotaenia* – E.S. Brown (co-author of paper with Mattingly).
brucei (del Ponte & Cerqueira, 1938): *Onirion* – Dr D. Bruce Wilson (honorary).
brucei Service, 1960: *Anopheles* – Dr L.J. Bruce-Chwatt (honorary) (see Kitzmiller, 1982).
brumpti Doucet, 1951b: *Uranotaenia* – not mentioned, obviously Émile Brumpt.
brumpti Galliard, 1931a: *Culex* – not mentioned, obviously Émile Brumpt.
brumpti Hamon & Rickenbach, 1955a: *Anopheles* – E. Brumpt (honorary) (see Kitzmiller, 1982).
brunhesi Grjebine, 1979: *Uranotaenia* – Jacques Brunhes (honorary).
brunhesi Reinert, 1999a: *Aedes* – Professor Jacques Brunhes (contributions).
brunhesi Ribeiro, 2004: *Toxorhynchites* – Prof. Jacques Brunhes (contributions).
brygooi Brunhes, 1971: *Aedes* – not mentioned, obviously Dr Edouard Brygoo.
brygooi Grjebine, 1986: *Mimomyia* – Dr Edouard Brygoo (honorary).
bulkleyi Causey, 1937a: *Anopheles* – L.C. Bulkley (honorary) (see Kitzmiller, 1982).
burgeri Zavortink, 1972: *Aedes* – John F. Burger (original collector, specimens identified as *Aedes kompi*).
burnetti Belkin, 1962: *Aedes* – G.F. Burnett (contributions).
burnsi Basio & Reisen, 1971: *Aedes* – SMSGT [Senior Master Sergeant (US Air Force)] John P. Burns (contribution to mosquito survey).
busckii (Coquillett, 1906e): *Aedes* – Mr August Busck (collected type specimens).
bustamantei Galvão, 1955: *Anopheles* – Dr Fernando Machado de Bustamante (contributions) (see Kitzmiller, 1982).
butleri (Theobald, 1901c): *Verrallina* – Mr A.L. Butler (collected type specimens).
buxtoni Belkin, 1962: *Aedes* – P.A. Buxton (pioneer investigator, South Pacific).
buxtoni (Edwards, 1923b): *Coquillettidia* – P.A. Buxton (collected type specimens).
buxtoni Edwards, 1926a: *Culex* – P.A. Buxton (collected holotype).
buxtoni Service, 1958: *Anopheles* – Prof. P.A. Buxton (in memoriam) (see Kitzmiller, 1982).
cabrerai Miyagi, Toma & Rivera, 1983: *Topomyia* – Dr Benjamin D. Cabrera (contributions).
cachani (Doucet, 1950): *Uranotaenia* – [Pierre] Cachan (collected part of type series).
caillardi Rickenbach, Ferrara & Eouzan, 1968: *Eretmapodites* – M. le Médecin-Général Caillard (friendly tribute).
calabyi Marks, 1963a: *Aedes* – Mr J.H. Calaby (collected type specimens).
calderoni Wilkerson, 1991: *Anopheles* – Guillermo Calderon Felero (friend, collaborator and colleague) (see Service, 2010).
cambournaci Hamon & Gandara, 1955: *Culex* – M. le Professeur Cambourna (honorary).
cameroni de Meillon & Evans, 1935: *Anopheles* – Mr W.A.J. Cameron (honorary) (see Kitzmiller, 1982).
campilunati Carter & Wijesundara, 1948: *Culex* – not mentioned.
camposi Dyar, 1925d: *Culex* – Professor Francisco Campos R. (collected holotype).
capelai da Cunha Ramos, 1993: *Uranotaenia* – Prof. Doutor R.A. Capela (collected part of type series).
capelai Ribeiro, 1992b: *Toxorhynchites* – Prof. R.A. Capela (collected holotype).
carleti Brunhes & Ravaonjanahary, 1971: *Culex* – not mentioned.
carmenti (Edwards, 1924b): *Verrallina* – Dr A.G. Carment (collected most specimens of type series).
carnevalei Brunhes, Le Goff & Geoffroy, 1999: *Anopheles* – Pierre Carnevale (colleague and friend) (see Service, 2010).
caroni Adam, 1961: *Anopheles* – René Caron (friendly tribute) (see Kitzmiller, 1982).
carrilloi (Sutil O. & Pulido F., 1978): *Wyeomyia* – Ing. Salvador José Carrillo (in memoriam).
carteri Amerasinghe, 1993: *Heizmannia* – Henry F. Carter (contributions).
carteri Edwards, 1936: *Aedes* – Carter (recorded collection locality).
carteri Evans & de Meillon, 1933: *Anopheles* – not mentioned, Henry Francis Carter (see Kitzmiller, 1982).
cartroni (Ventrillon, 1906b): *Aedes* – Dr Cartron (provided first specimens).
casali Schick, 1970b: *Aedes* – Osvaldo Casal (honorary).
castelli Hamon, 1957: *Culex* – ? Technical Agent of Health (collected 'by the technical agent of Health R. CHATEAU to whom I dedicate it' [translated from French].
castroi Casal & García, 1967b: *Culex* – Sr Manuel P. Castro (friend).
castroi Lane & Cerqueira, 1942: *Trichoprosopon* – Dr G.M. de Oliveira Castro (honorary).
catesi (Lien, 1968c): *Heizmannia* – Dr M.D. Cates (entomologist).
caudelli (Dyar & Knab, 1906d): *Culex* – Mr A.N. Caudell (friend and co-worker).
cavalierii García & Casal, 1967: *Toxorhynchites* – Dr Fiorello Cavalieri (friend and colleague).
cerqueirai (Barreto & Coutinho, 1944): *Mansonia* – N. Cerqueira (specimen 'identified by N. Cerqueira as *Mansonia pseudotitillans*' [translated from Portuguese].
cerqueirai (Stone, 1944): *Runchomyia* – Dr Nelson Cerqueira (identified specimens as *Trichoprosopon rapax*).
cerqueirai Valencia, 1973: *Culex* – not mentioned, obviously Dr Nelson Cerqueira.
cesari del Ponte & Cerqueira, 1938: *Wyeomyia* – Dr Gastão Cesar (collected type specimens).
chagasi (da Costa Lima, 1935a): *Mansonia* – Prof. C. Chagas (collected holotype).

chaii Miyagi & Toma, 2012: *Topomyia* – Dr Paul Chai (contributions).

chamberlaini Ludlow, 1904c: *Mimomyia* – Capt. W.P. Chamberlain (collected holotype).

chamboni Cornet, 1968: *Aedes* – Dr L. Chambon (Director, Institute Pasteur, Dakar).

chandi Edwards, 1922a: *Heizmannia* – Surgeon Khazan Chand (reared larvae).

chaussieri Edwards, 1923a: *Aedes* – Dr Chaussier (collected holotype).

chauveti Brunhes & Rambelo, 1968: *Culex* – Guy Cauvet (colleague and friend).

cheesmanae Mattingly & Marks, 1955: *Culex* – Miss Evelyn Cheesman (collected part of type series).

chelli (Edwards, 1915b): *Aedes* – Dr G.R.H. Chell (collected type specimens).

chengi Lien, 1968c: *Heizmannia* – Mr Che I Cheng (field participation).

cheni Dong, Wang & Lu, 2003: *Culex* – apparently H.B. Chen (honorary).

cheni Lien, 1968a: *Tripteroides* – Mr Pao Shu Chen (illustrations and collections).

chidesteri Dyar, 1921h: *Culex* – W.S. Chidester (collected type specimens).

chitae Duret, 1967b: *Culex* – mi esposa (my wife).

chodukini Martini, 1929: *Anopheles* – Nikolai Chodukin (provided type specimens) (see Kitzmiller, 1982).

chorleyi Edwards, 1936: *Uranotaenia* – Mr T.W. Chorley (assistance).

chorleyi Edwards, 1941: *Culex* – not mentioned, obviously Mr T.W. Chorley.

christiani Colless, 1960: *Culex* – Mr S.H. Christian (collected type series).

christophersi Baisas & Ubaldo-Pagayon, 1953: *Tripteroides* – Sir S. Rickard Christophers (honorary).

christophersi Barraud, 1926: *Uranotaenia* – Christophers (obviously Sir S. Rickard Christophers) (collected type series).

christophersi Edwards, 1922a: *Aedes* – Major S.R. Christophers (collected most of type series).

christophi (Portschinsky, 1884): *Toxorhynchites* – not mentioned.

christyi (Newstead & Carter, 1911b): *Anopheles* – Dr C. Christy (collected holotype) (see Kitzmiller, 1982).

chungi Lien, 1968b: *Aedes* – Mr Chao Lin Chung (collected type series with the author).

civinskii Belkin, 1953: *Uranotaenia* – Stanley Civinski (collections; contributions).

claggi Bohart & Farner, 1944: *Tripteroides* – C.S. Clagg (collected holotype).

clarki Evans, 1924: *Culex* – Dr A. Aitken Clark (collected type specimens).

clarki (Galindo, Carpenter & Trapido, 1953): *Haemagogus* – Dr Herbert C. Clark (Director, Gorgas Memorial Laboratory).

clastrieri Casal & Garcia, 1968b: *Culex* – Dr [Jean] Clastrier (colleague).

clelandi (Taylor, 1914c): *Aedes* – not mentioned, obviously Dr J.B. Cleland (collected several species described in the paper).

clintoni Taylor, 1946: *Aedes* – K.J. Clinton (collected type specimens).

clowi Rozeboom & Knight, 1946: *Anopheles* – R.J. [Richard James] Clow (collected holotype) (see Kitzmiller, 1982).

cogilli Edwards, 1922d: *Aedes* – Dr H. Cogill (collected type specimens).

coheni Belkin, 1950: *Tripteroides* – Murray Cohen (collected holotype).

colledgei Marks, 1956: *Anopheles* – W.R. [William Robert] Colledge (honorary) (see Kitzmiller, 1982).

collessi Grjebine, 1986: *Mimomyia* – Dr Donald H. Colless (honorary).

collessi Lee, 1946: *Tripteroides* – D.H Colless (collected type specimens).

collessi Mattingly, 1958a: *Aedes* – Dr Donald Colless (contributions).

collessi Reid, 1963: *Anopheles* – Dr Donald H. Colless (appreciation) (see Kitzmiller, 1982).

collessi Sirivanakarn, 1968: *Culex* – Dr Donald H. Colless (contributions).

coluzzii Coetzee & Wilkerson, 2013: *Anopheles* – Professor Mario Coluzzi (honorary).

coluzzii Rioux, Guilvard & Pasteur, 1998: *Aedes* – Professeur Mario Coluzzi (contributions).

combesi Doucet, 1950: *Uranotaenia* – not mentioned, probably Prof. Rauol Combes (Director, Institut Scientifique de Madagascar).

connali Edwards, 1912c: *Uranotaenia* – Dr A.C. Connal (collected type specimens).

conradti Grünberg, 1907a: subspecies of *Toxorhynchites brevipalpis* Theobald, 1901a – not mentioned.

contei Duret, 1968c: *Culex* – Sr Juvenal O. Conte (friend and collaborator).

cooki Belkin, 1962: *Aedes* – Captain James Cook (discoverer of Niue Island, 1774).

cooki Root, 1937: *Uranotaenia* – Dr S.S. Cook (collected type specimens).

corbeti Hamon, 1962: *Eretmapodites* – P. Corbet (colleague).

cordellieri Huang, 1986a: *Aedes* – Dr Roger Cordellier (contributions).

corneti Huang, 1986b: *Aedes* – Dr Michel Cornet (contributions).

corrigani Dyar & Knab, 1907c: *Culex* – Mr John Corrigan (honorary).

costai da Fonseca & da Silva Ramos, 1940: *Anopheles* – Arthur Costa Filho (honorary) (see Kitzmiller, 1982).

cottlei Sirivanakarn, 1968: *Culex* – Robert Cottle (collected holotype with the author).

coulangesi Rodhain & Boutonnier, 1983b: *Aedes* – Dr Coulanges (Director, Institute Pasteur, Madagascar).

coursi Doucet, 1949a: subspecies of *Culex salisburiensis* Theobald, 1901c – not mentioned.

coustani Laveran, 1900: *Anopheles* – M. le Dr Coustan (sent specimens collected by Dr Rasamimanana) (see Kitzmiller, 1982).

covagarciai Forattini, 1965: *Culex* – Dr Pablo Cova Garcia (by Bram, 1967b).

covagarciai Sutil Oramas & Pulido F., 1974: *Wyeomyia* – Dr Pablo Cova García (contributions).

covelli Barraud, 1929b: *Heizmannia* – Major G. Covell (collected type specimens).

cozi Cornet, 1973: *Aedes* – Jean Coz (comrade).

craggi (Barraud, 1923c): *Aedes* – Major F.W. Cragg (honorary).

crawfordi Reid, 1953: *Anopheles* – Crawford, 1938 [Dr Robert Crawford] (collected type specimens) (see Kitzmiller, 1982).

cristovaoi Duret, 1968a: *Culex* – Sr Cristóvao da Silva Maciel (colleague).

crossi Lien, 1967: *Aedes* – Dr John H. Cross (honorary).

crowei Sirivanakarn, 1968: *Culex* – Lockland Crowe (honorary).

cruzii Dyar & Knab, 1908: *Anopheles* – Dr Oswaldo Cruz (honorary) (see Kitzmiller, 1982).

cuccioi (Belkin, 1962): *Verrallina* – J. Cuccio (dedication).

cumminsii (Theobald, 1903a): *Aedes* – Captain Cummins (provided holotype).

cuttsi van den Assem, 1959: *Tripteroides* – Reverend W. Cutts (missionary).

daggyi Stone & Bohart, 1944: *Aedes* – Dr R.H. Daggy (collected holotype; collected paratypes with K.L. Knight).

dahlae (Nielsen, 2009): *Aedes* – Christine Dahl (friend and colleague).

dalzieli (Theobald, 1910c): *Aedes* – [J.M.] Dalziel (collected holotype).

damascenoi Duret, 1969c: *Culex* – Sr Reinaldo G. Damasceno (friend; entomologist).

damasei Grjebine, 1979: *Uranotaenia* – Damase Randriamananten (assistant and friend).

danaraji Ramalingam, 1988: *Topomyia* – Prof. Tan Sri T.J. Danaraj (contributions and appreciation).

darlingi Root, 1926: *Anopheles* – Dr Samuel T. Darling (in memoriam) (see Kitzmiller, 1982).

daryi Schick, 1970a: *Aedes* – Mario Dary R. (honorary).

davidi Basio, 1971: *Aedes* – Dr Romeo David (Executive Officer, Bureau of Quarantine, Department of Health, Manila).

davisi Kumm, 1933: *Culex* – Dr Nelson C. Davis (suggested the study).

davisi Lane, 1943: *Uranotaenia* – Dr Nelson C. Davis (in memoriam).

davisi (Lane & Cerqueira, 1942): *Wyeomyia* – Dr N.C. Davis (in memoriam).

deanei (Lourenço de Oliveira, 1983): *Wyeomyia* – Prof. Leonidas M. Deane (contributions).

deaneorum Rosa-Freitas, 1989: *Anopheles* – Professors Leonidas W. [*sic*, *lapsus* for M.] Deane and Maria P. Deane (honorary) (see Service, 2010).

deboeri Edwards, 1926a: *Aedes* – not mentioned, obviously H.S. de Boer (mentioned elsewhere in the paper).

deemingi Service, 1970: *Anopheles* – Mr J.C. Deeming (collected holotype) (see Kitzmiller, 1982).

deguzmanae (Mattingly, 1957a): *Mimomyia* – Dr Milagros de Guzman ('at the request of Mr. Baisas').

dejesusi Baisas & Feliciano, 1953: *Topomyia* – Dr Vicente de Jesus (in memoriam).

delfinadoae Sirivanakarn, 1973b: *Culex* – Dr M.D. Delfinado (collected type specimens).

delpilari Baisas & Ubaldo-Pagayon, 1953: *Tripteroides* – General Gregorio del Pilar (honorary).

delpontei Duret, 1969a: *Culex* – Dr Eduardo F. Del Ponte (friend; medical entomologist).

demeilloni Doucet, 1950: *Culex* – not mentioned, undoubtedly Dr Botha de Meillon.

demeilloni Edwards, 1936: *Aedes* – mot mentioned, obviously Dr Botha de Meillon.

demeilloni Evans, 1933: *Anopheles* – Dr B. de Meillon (collaborator) (see Kitzmiller, 1982).

demeilloni Mattingly, 1970: *Heizmannia* – not mentioned, obviously Dr Botha de Meillon.

demeilloni Peyton & Rattanarithikul, 1970: *Uranotaenia* – Dr Botha de Meillon (contributions).

denbesteni Brug, 1925: *Armigeres* – Dr den Besten (collected type specimens).

derooki Brug, 1932: *Aedes* – Dr [H.] de Rook (collected and reared type specimens).

derooki Soesilo & van Slooten, 1931a, 1931b: *Bironella* – not mentioned, Dr H. de Rook (see Kitzmiller, 1982).

devemyi Hamon, 1955: *Uranotaenia* – docteur P. Devemy (collected holotype).

dialloi Hamon & Brengues, 1965: *Aedes* – M. Bellan Diallo (collected the paratype).

diazi Schick, 1970a: *Aedes* – Alfonso Diaz Najera (honorary).

diraphati Peyton & Klein, 1970: *Uranotaenia* – Mr Chaliou Diraphat (technician-collector).

dmitryi Gornostaeva, 2005: *Aedes* – Дмитрия Константиновича Львова [Dmitry Konstantinovich Lvov] (academician).

dobrotworskyi Marks, 1958: *Aedes* – Mr N.V. Dobrotworsky (contributions).

dofleini (Guenther, 1913): *Tripteroides* – Franz Doflein (Freiburg zoologist).

dohenyi Hogue, 1975: *Culex* – Mr Timothy M. Doheny (expedition sponsor).

donai da Cunha Ramos & Brunhes, 2004: *Uranotaenia* – Donadieu 'Dona' Randrianambinina (assistant).

donaldi Reid, 1962: *Anopheles* – Dr Donald Colless (contributions) (see Kitzmiller, 1982).

dorseyi Knight, 1946: *Aedes* – C.K. Dorsey (collected type specimens with H. Dybas).

douceti Adam & Hamon, 1959: subspecies of *Eretmapodites wansoni* Edwards, 1941 – J. Doucet (colleague).

douceti Grjebine, 1953a: *Uranotaenia* – not mentioned, probably Jean-Paul Doucet.

douglasi Dobrotworsky, 1956: *Culex* – Mr G.W. Douglas (collected some specimens examined).

downsi Bohart & Ingram, 1946a: subspecies of *Aedes flavopictus* Yamada, 1921 – Maj. W. G. Downs (contributed specimens).

downsi Lane, 1945a: *Wyeomyia* – Dr W.G. Downs (honorary; contributed specimens for study).
drummondi (Dobrotworsky, 1960b): *Culiseta* – Dr F.H. Drummond (discovered larval habitat).
dufouri Hamon, 1953: *Aedes* – Dr Dufour (Director, Départemental de la Santé de la Réunion).
dumbletoni Belkin, 1962: *Culex* – L.J. Dumbleton (contributions).
dumonti Doucet, 1949b: *Uranotaenia* – administrateur Dumont.
dunhami Causey, 1945: *Anopheles* – General George Dunham (honorary) (see Kitzmiller, 1982).
dunni Dyar, 1918e: *Culex* – L.H. Dunn (collected and bred type specimens).
dunni Evans, 1928: subspecies of *Aedes argenteoventralis* (Theobald, 1909b) – Mr L.H. Dunn (collected type specimens).
dupreei (Coquillett, 1904a): *Aedes* – Mr J.W. Dupree (provided a specimen of each sex).
dureni Edwards, 1938: *Anopheles* – Dr A. Duren (collected first three females; arranged additional collections) (see Kitzmiller, 1982).
dureti Casal & García, 1968d: *Culex* – Doctor José P. Duret (contributions).
durhami Edwards, 1917: *Armigeres* – Dr Durham (collected some type specimens).
durhami Sirivanakarn, 1968: *Culex* – Dr S. Durham (honorary).
durhamii Theobald, 1901c: *Limatus* – Dr Durham (collected type specimens).
duttoni Theobald, 1901d: *Culex* – not mentioned, probably Dr Joseph Everett Dutton.
dyari Belkin, Heinemann & Page, 1970: *Mansonia* – not mentioned, certainly Dr Harrison G. Dyar.
dyari Belkin & Hogue, 1959: *Deinocerites* – not mentioned, certainly Dr Harrison G. Dyar.
dyari Bohart & Farner, 1944: *Tripteroides* – not mentioned, obviously Dr Harrison G. Dyar.
dyari Lane & Cerqueira, 1942: *Wyeomyia* – Dr H.G. Dyar (in memoriam).
dybasi Bohart, 1957: *Aedes* – H.S. Dybas (collected holotype and most paratypes).
dyi Baisas & Ubaldo-Pagayon, 1953: *Tripteroides* – Lt. Col. Dr Francisco Dy (honorary).
earlei Vargas, 1943b: *Anopheles* – Sr Dr Walter C. Earle (honorary) (see Kitzmiller, 1982).
eastor Dyar, 1920d: *Culex* – ? baby girl's name, derivative of Easter.
eatoni (Edwards, 1916): *Aedes* – Rev. A.E. Eaton (collected holotype).
edgari Stone & Rosen, 1952: *Aedes* – S.A. Edgar ('one of the first to discover the species').
eduardoi Casal & García, 1968e: *Culex* – Especie dedicada a mi hijo [son, Eduardo Casal].
edwardsi (Barraud, 1923a): *Aedes* –Mr F.W. Edwards (examined specimens).
edwardsi Barraud, 1923d: *Culex* – not mentioned, certainly F.W. Edwards.
edwardsi (Barraud, 1924d): *Toxorhynchites* – not mentioned, certainly F.W. Edwards.
edwardsi Barraud, 1926: *Uranotaenia* – not mentioned, certainly F.W. Edwards.
edwardsi (Barraud, 1929b): *Tripteroides* – Mr F.W. Edwards (identification of specimens).
edwardsi (Lane & Cerqueira, 1942): *Wyeomyia* – Dr F.W. Edwards (taxonomic assistance).
eiseni Coquillett, 1902b: *Anopheles* – Prof. Gustav Eisen (provided type specimen) (see Kitzmiller, 1982).
ejercitoi Baisas, 1935b: *Armigeres* – Dr A. [Antonio] Ejercito (indebtedness).
ejercitoi Mendoza,1947: *Anopheles* – Dr Antonio Ejercito (tribute for role in anti-malaria control) (see Kitzmiller, 1982).
eldridgei Adames & Galindo, 1973: *Culex* – LTC Bruce F. Eldridge (collected type specimens).
eldridgei Reinert, 1973e: *Aedes* – Lieutenant Colonel Bruce F. Eldridge (appreciation).
eleanorae Berlin, 1969a: *Aedes* – Dedicated to my wife, Eleanor.
ellinorae Edwards, 1941: *Aedes* – Miss Ellinor C. MacDonald (collected type specimens).
ellinorae Ovazza, Hamon & Neri, 1956: subspecies of *Culex aurantapex* Edwards, 1914b – E.C.C. van Someren (honorary).
elnora Paterson & Shannon, 1927: subspecies of *Uranotaenia pulcherrima* Lynch Arribálzaga, 1891b – señora Elnora S. Shannon (collector; probably wife of the second author).
elsiae (Barraud, 1923e): *Aedes* – not mentioned.
eouzani Brunhes, Le Goff & Boussès, 2003: *Anopheles* – Jean-Pierre Eouzan (colleague) (see Service, 2010).
eouzani Geoffroy, 1971a: *Culex* – M. J.P. [Jean-Pierre] Eouzan (colleague).
eouzani Rickenbach & Lombrici, 1974: *Eretmapodites* – J.-P. [Jean-Pierre] Eouzan (colleague).
erlindae Baisas & Ubaldo-Pagayon, 1953: *Tripteroides* – named for the only child of the second author.
ernanii Duret, 1968f: *Culex* – Dr Ernani Ferreira (collaborator).
ernsti Anduze, 1949a: *Culex* – Dr A. Ernst (in memoriam).
escodae Baisas & Ubaldo-Pagayon, 1953: subspecies of *Tripteroides powelli* (Ludlow, 1909c) – Mrs Josefa Escoda (in memoriam; social worker).
espini (Martini, 1914): *Isostomyia* – Dr Espin (co-collector of another species, *prasinopleurus*, described by Martini).
evandroi da Costa Lima, 1937: *Anopheles* – Dr Evandro Chagas (collected holotype) (see Kitzmiller, 1982).
evansae Antunes, 1942: *Trichoprosopon* – Miss A.M. Evans (honorary).
evansae (Brèthes, 1926b): *Anopheles* – not mentioned, obviously A.M. Evans (see Kitzmiller, 1982).
evansae (Edwards, 1936): *Toxorhynchites* – Evans (undoubtedly A.M. Evans) (collected holotype).
evansae Root, 1927c: *Culex* – not mentioned, evidently A.M. Evans.
fabricii Lane & Cerqueira, 1942: *Sabethes* – in honor of the author of 'Systema Antliatorum', i.e. Johann Christian Fabricius.

faini Leleup, 1952: *Anopheles* – Docteur M.A. Fain (Prof. A.M.A.J. Fain) (collected part of type series) (see Kitzmiller, 1982).
fairchildi Galindo & Blanton, 1954: *Culex* – Dr G.B. Fairchild (colleague).
fajardi (Lutz, 1904): *Chagasia* – not mentioned (Francisco de Paula Fajardo) (see Kitzmiller, 1982).
fajardoi Baisas & Feliciano, 1953: *Zeugnomyia* – Dr Jacobo Fajardo (in memoriam).
falabreguesi Hamon, 1957: *Aedes* – F. Falabrègues (colleague).
farauti Laveran, 1902b: *Anopheles* – Dr Faraut (collected type specimens) (see Kitzmiller, 1982).
farquharsoni (Edwards, 1922c): *Malaya* – Mr C.L. Farquharson (collected type specimens).
fasyi Baisas, 1938: *Culex* – not mentioned.
faurani Duret, 1968f: *Culex* – Dr P. Fauran (collected type specimens with the author).
fausti Vargas, 1943a: *Anopheles* – Dr Ernest C. Faust (honorary) (see Kitzmiller, 1982).
feegradei Barraud, 1934: *Aedes* – Feegrade (collected type specimens).
fengi Edwards, 1935a: *Aedes* – Lan Chou Feng (collected type specimens).
fergusoni (Taylor, 1914b): *Culex* – Dr Ferguson (collected type specimens).
fernandezi Casal, García & Cavalieri, 1966: *Culex* – Dr Héctor I. Fernández (friend).
fernandezyepezi (Cova Garcia, Sutil Oramas & Pulido F., 1974b): *Wyeomyia* – Ingeniero Agrónomo Francisco Fernández Yépez (university professor).
ferrarai Rickenbach & Eouzan, 1970: *Eretmapodites* – Léo Ferrara (collaborator).
ferreri Duret, 1968f: *Culex* – Dr H. Ferrer (colleague).
fiebrigi Edwards, 1922d: *Psorophora* – Fiebrig (collected type specimens).
finlayi Lane & Cerqueira, 1942: *Wyeomyia* – Dr C. Finlay (in memoriam).
fishi Zavortink, 1986: *Wyeomyia* – Durland Fish (collected type specimens).
fitchii (Felt & Young, 1904): *Aedes* – not mentioned.
flochi Duret, 1969d: *Culex* – Dr [H.] Floch (honorary).
flui (Bonne-Wepster & Bonne, 1920): *Wyeomyia* – not mentioned.
fonsecai (Pinto, 1932): *Mansonia* – Dr O. da Fonseca (collected type specimens).
fontenillei Brunhes & Hervy, 1995: *Orthopodomyia* – Didier Fontenille (colleague).
fontenillei (Le Goff, Boussès & Brunhes, 2007): *Aedes* – Didier Fontenille (collector; contributions).
fontenillei Ribeiro, 2004: *Toxorhynchites* – Dr Didier Fontenille (collected holotype).
forattinii Cerqueira, 1961: *Sabethes* – Dr Oswaldo P. Forattini (honorary).
forattinii Clastrier, 1974: *Wyeomyia* – Prof. O.P. Forattini (honorary).
forattinii Wilkerson & Sallum, 1999: *Anopheles* – Oswaldo Paulo Forattini (collaborator; contributions) (see Service, 2010).
fowleri (de Charmoy, 1908): *Aedes* – Major P. Fowler (discovered/collected the species).
franciscoi Mattingly, 1959: *Aedes* – Mr Francisco Baisas (contributions).
franciscoi Reid, 1962: *Anopheles* – Francisco Baisas (contributions) (see Kitzmiller, 1982).
franclemonti Belkin, 1962: *Aedes* – John G. Franclemont (contributions).
fraseri (Edwards, 1912c): *Aedes* – Capt. A.D. Fraser (collected holotype).
fraseri Edwards, 1912c: *Uranotaenia* – Capt. A.D. Fraser (collected type specimens).
fraseri (Edwards, 1914b): *Culiseta* – Captain A.D. Fraser (collected holotype and one paratype).
fraseri (Edwards, 1922c): *Malaya* – Capt. A.D. Fraser (collected type specimens).
fraseri (Theobald, 1911a): *Coquillettidia* – not mentioned, probably A.D. Fraser.
freeborni Aitken, 1939: *Anopheles* – not mentioned, Prof. Stanley B. Freeborn (see Kitzmiller, 1982).
frenchii (Theobald, 1901c): *Culiseta* – Mr French (collected type specimens).
freyi Meng, 1957: *Anopheles* – FRT [*sic*] Frey (Dr Richard Frey; contributions) (see Kitzmiller, 1982).
fryeri (Theobald, 1912b): *Aedes* – J.C.F. Fryer (collected type specimens).
fulleri (Ludlow, 1909b): *Culex* – Major Fuller, Surgeon U.S. Army (collected type specimens).
furlongi van Someren, 1954a: *Culex* – Mr M. Furlong (collected type specimens).
gabaldoni Cova García, Pulido F., Escalante de Ugueto & Mora R., 1987b: *Uranotaenia* – doctor Arnoldo Gabaldon (contributions).
gabaldoni Vargas, 1941c: *Anopheles* – Sr Dr Arnoldo Gabaldón (malariologist; former colleague and friend) (see Kitzmiller, 1982).
gabriel Schick, 1970a: *Aedes* – Gabriel Mariaca (collected holotype and most paratypes).
gaffigani Reinert, 1990: *Aedes* – Mr Thomas V. Gaffigan (technical assistance).
gagnei Evenhuis, 1989: *Culex* – not mentioned, probably Raymond J. Gagne.
galindoi Komp & Rozeboom, 1951: *Culex* – Pedro Galindo (provided holotype collected with Harold Trapido).
galindoi Schick, 1970a: *Aedes* – Pedro Galindo (honorary).
galliardi Edwards, 1941: *Culex* – 'specimen figured by [Henri] Galliard ... was probably this species'.
galloisi Yamada, 1921: *Aedes* – Edme Gallois (first found one female).
galvaoi Causey, Deane & Deane, 1943: *Anopheles* – Dr A.L. Ayroza Galvão (honorary) (see Kitzmiller, 1982).
galvaoi (Correa & Ramalho, 1956): *Wyeomyia* – Prof. Dr Augusto Leopoldo Ayroza Galvão (honorary).
galvaoi Duret, 1968b: *Culex* – Archibaldo B. Galvao (honorary; entomologist).
gameti Bailly-Choumara, 1966b: *Culex* – Médecin-Colonel Gamet (honorary; founder of Pasteur Institute, Cameroon).

ganapathi Colless, 1958: *Aedes* – Mr A. Ganapathipillai (honorary).

ganapathi Colless, 1965: *Culex* – not mentioned, but Mr A. Ganapathipillai is acknowledged for contributing specimens to the study.

gandarai da Cunha Ramos, Capela & Ribeiro, 1995: *Aedes* – Dr Álvaro F. Gândara (medical entomologist).

gani Bonne-Wepster, 1940: *Aedes* – Dr Gan Houw Hwat (collected type specimens).

garcesi Duret, 1968b: *Culex* – Sr Gilberto Garcés (friend; entomologist).

garciai Broche, 2000: *Culex* – Dr Israel García Avila (contributions).

gardnerii (Ludlow, 1905a): *Aedes* – Dr Fletcher Gardner (provided specimens).

garioui Bailly-Choumara & Rickenbach, 1966: *Culex* – Médecin-Colonel Gariou (honorary).

garnhami Edwards, 1930a: *Anopheles* – 'named, at the request of Mr. Symes, after Dr. P. C. Garnham, who was the discoverer of *Anopheles symesi*, Edw., and who also collected one of the specimens of *A. garnhami*' (see Kitzmiller, 1982).

garnhami van Someren, 1948: *Uranotaenia* – Dr P.C.C. Garnham (collected type specimens).

gaufini Belkin, 1962: *Culex* – A.R. Gaufin (collected type specimens; contributions).

geoffroyi Reinert, 1999a: *Aedes* – Dr Bernard Geoffroy (contributions).

gerbergi Belkin, 1977a: *Toxorhynchites* – Dr Eugene Jordan Gerberg and Ms Jo Betty Vick Gerberg (friends; colleagues). **Note**: although clearly intended to honour a man and a woman, requiring the addition of *-orum* to the surname, the name should not be amended as there is no evidence of an inadvertent error in the published paper (International Commission on Zoological Nomenclature, 1999: *Code*, Article 32.5). See *bonnei* Dyar, 1921k, *boneti* Gil Collado, 1936 and *harleyi* Peters, 1955a.

gerdae Slooff, 1963: *Uranotaenia* – named after the author's wife.

germaini Geoffroy, 1974: *Culex* – M.M. Germain (colleague).

germaini Rickenbach & Eouzan, 1970: *Eretmapodites* – Max Germain (colleague).

gibbinsi Edwards, 1935a: *Aedes* – E.G. Gibbins (collected specimens with the author).

gibbinsi Evans, 1935: *Anopheles* – Mr E.G. Gibbins (collected type specimens) (see Kitzmiller, 1982).

giblini (Taylor, 1914d): *Coquillettidia* – Dr Giblin (collected type specimens).

gilcolladoi Sánchez-Covisa Villa, Rodríguez Rodríguez & Guillén Llera, 1985: *Aedes* – not mentioned, undoubtedly J. Gil Callado.

gilesi (Peryassú, 1908): *Anopheles* – not mentioned, probably LTC G.M.J. Giles (see Kitzmiller, 1982).

gilletti van Someren, 1949b: *Eretmapodites* – Mr Gillett, undoubtedly J.D. Gillett (honorary).

gilli (Barraud, 1924d): *Aedes* – not mentioned.

gilliesi Hamon & van Someren, 1961a: *Culex* – M.T. Gillies (colleague).

gilliesi van Someren, 1962: *Aedes* – M.T. Gillies (collected type specimens).

giveni Edwards, 1926a: *Armigeres* – Dr D.H.C. Given (collected type specimens).

goeldii Rozeboom & Gabaldon, 1941: *Anopheles* – not specifically stated, but obviously Emilio Goeldi (see Kitzmiller, 1982).

gomezdelatorrei Leví-Castillo, 1955d: *Anopheles* – Dr José Gómez de la Torre (honorary) (see Kitzmiller, 1982).

gonzalezrinconesi Cova Garcia, Pulido F. & Escalante de Ugueto, 1977a: *Anopheles* – Dr Rafael González Rincones (honorary) (see Kitzmiller, 1982).

gossi Bohart, 1957: *Culex* – R.J. Goss (collected holotype).

gouldi Peyton & Klein, 1970: *Uranotaenia* – Dr Douglas J. Gould (personal support).

gouldi Reinert, 1972b: *Aedes* – Dr D.J. Gould (support in providing specimens).

grabhami Berlin 1969a: *Aedes* – not mentioned, obviously M. Grabham (same species as *aureostriatus* Grabham, 1907).

grabhamii Theobald, 1901a: *Anopheles* – Dr Grabham (collected holotype) (see Kitzmiller, 1982).

grahami Edwards, 1911b: *Eretmapodites* – Dr W.M. Graham (collected holotype).

grahamii (Theobald, 1909b): *Aedes* – Dr [W.M.] Graham (collected holotype).

grahamii Theobald, 1910c: *Culex* – not mentioned, undoubtedly Dr W.M. Graham.

grandidieri Blanchard, 1905: *Culex* – not mentioned, French naturalist Alfred Grandidier?

grantii (Theobald, 1901a): *Aedes* – Mr Grant (collected holotype).

grassei da Cunha Ramos & Brunhes, 2004: *Uranotaenia* – Prof. P.P. Grassé (honorary).

grassei Doucet, 1951b: *Aedes* – not mentioned, certainly Pierre-Paul Grassé.

grassei Grjebine, 1953b: *Anopheles* – not mentioned, obviously Pierre-Paul Grassé (see Kitzmiller, 1982).

gravelyi (Edwards, 1921c): *Toxorhynchites* – F.H. Gravely (collected holotype).

grayii Theobald, 1901c: *Wyeomyia* – [Harold] St George Gray (collected type specimens).

greeni Rattanarithikul & Harbach, 1991: *Anopheles* – Dr Christopher A. Green (cytotaxonomy) (see Service, 2010).

greenii (Theobald, 1903a): *Aedes* – Green (collected holotype), probably Mr E.E. Green.

greenii (Theobald, 1905h): *Heizmannia* – Mr E.E. Green (collected type specimens).

grenieri Doucet, 1951b: *Uranotaenia* – not mentioned, certainly P. Grenier (Institut Pasteur, Paris).

grenieri Eouzan, 1969: *Culex* – P. Grenier (honorary).

grenieri Grjebine, 1964: *Anopheles* – P. Grenier (honorary) (see Kitzmiller, 1982).

grenieri Hamon & van Someren, 1961b: *Eretmapodites* – P. Grenier (honorary).

grenieri Hamon, Service, Adam & Taufflieb, 1961: *Aedes* – P. Grenier (honorary).
gressitti Bohart, 1957: *Aedes* – J.L. Gressitt (collected many of the specimens used in the study).
gressitti Sirivanakarn, 1968: *Culex* – Dr J. Linsley Gressitt (contributions).
griffithi Thurman, 1954a: *Ayurakitia* – Dr Melvin E. Griffith (honorary).
griveaudi Grjebine, 1961: *Anopheles* – M. Griveaud (collected holotype) [Dr Paul Griveaud] (see Kitzmiller, 1982).
grjebinei (Brunhes, 1977b): *Mimomyia* – A. Grjebine (collected allotype).
grjebinei da Cunha Ramos & Brunhes, 2004: *Uranotaenia* – Alexis Grjebine (honorary).
grjebinei Hamon, Taufflieb & Maillot, 1957: *Aedes* – A. Grjebine (colleague; contributions).
grjebinei Ribeiro, 2004: *Toxorhynchites* – Alexis Grjebine (collected paratypes).
grossbecki Dyar & Knab, 1906d: *Aedes* – not mentioned, obviously John A. Grossbeck.
guedesi da Silva Mattos & Xavier, 1991: *Culex* – Prof. Arino da Silveira Guedes (contributions).
guerreroi Cova García, Sutil & Pulido, 1971: *Culex* – Dr Lacenio Guerrero (honorary).
guiarti Blanchard, 1905: *Culex* – not mentioned, apparently [Dr] J. Guiart.
gurneri van Someren, 1946c: subspecies of *Aedes michaelikati* van Someren, 1946c – Gurner R. Cunningham van Someren (collected type specimens).
gurneyi (Belkin, 1962): *Mimomyia* – Ashley B. Gurney (collected holotype; contributions).
gurneyi Stone & Bohart, 1944: *Aedes* – A.B. Gurney (collected holotype).
gutierrezi Duret, 1971a: *Wyeomyia* – Dr José Gutierrez, PAHO (honorary).
gutzevichi Dubitzky & Deshevykh, 1978: *Aedes* – not mentioned, obviously A.V. Gutsevich (А.В. Гуцевич).
hackeri Edwards, 1921c: *Anopheles* – Dr H.P. Hacker (collected holotype) (see Kitzmiller, 1982).
hackeri Edwards, 1923b: *Culex* – Dr H.P. Hacker (collected type specimens).
haddowi da Cunha Ramos & Brunhes, 2004: *Uranotaenia* – A.J. Haddow (honorary).
haddowi van Someren, 1949b: *Eretmapodites* – Dr [A.J.] Haddow (honorary).
hakanssoni Knight & Hurlbut, 1949: *Aedes* – Capt. E.G. Hakansson (support for mosquito research).
halifaxii (Theobald, 1903a): *Lutzia* – not mentioned, probably F.J. Halifax, President, Municipal Commission, Singapore.
hamoni Adam, 1962: *Anopheles* – J. Hamon (colleague) (see Kitzmiller, 1982).
hamoni Brunhes, Adam & Bailly-Choumara, 1967: *Culex* – J. Hamon (colleague).
hamoni Grjebine, 1953a: *Uranotaenia* – [J.] Hamon (collected type specimens).
hamoni Grjebine, 1972: *Eretmapodites* – Jacques Hamon (colleague).
hancocki Edwards, 1929b: *Anopheles* – G.L.R. Hancock (collected holotype and allotype) (see Kitzmiller, 1982).
hancocki Edwards, 1930a: *Culex* – G.L.R. Hancock (collected type specimens).
hancocki van Someren, 1962: *Aedes* – Mr E.C. Hancock (collected holotype and most other specimens of the type series).
hansfordi Huang, 1997: *Aedes* – Dr Clifford F. Hansford (contributions).
harbachi Reinert, 1990: *Aedes* – Dr Ralph E. Harbach (contributions).
hardini Miyagi, Toma & Ramalingam, 1989: *Topomyia* – Dr Stalin Hardin (contributions; support).
hargreavesi (Edwards, 1920a): subspecies of *Aedes caspius* (Pallas, 1771) – E. Hargreaves (sent specimens from Italy).
hargreavesi Evans, 1927: *Anopheles* – Mr E. Hargreaves (collected type specimens) (see Kitzmiller, 1982).
harinasutai Knight, 1978a: *Aedes* – Dr Chamlong Harinasuta (honorary).
harleyi Peters, 1955a: *Culex* – Dr George Way Harley and Mrs Harley, Ganta Methodist Mission, Liberia (honorary). **Note**: although clearly intended to honour a man and a woman, requiring the addition of *-orum* to the surname, the name should not be amended as there is no evidence of an inadvertent error in the published paper (International Commission on Zoological Nomenclature, 1999: *Code*, Article 32.5). See *bonnei* Dyar, 1921k, *boneti* Gil Collado, 1936 and *gerbergi* Belkin, 1977a.
harperi Evans, 1936: *Anopheles* – Mr J.O. Harper (collected type specimens) (see Kitzmiller, 1982).
harperi Knight, 1948b: *Aedes* – Dr Paul A. Harper, U.S. Army Medical Corps (contributions).
harperi van Someren, 1949b: *Eretmapodites* – Mr Harper, probably Mr J.O. Harper (honorary).
harrisoni Harbach & Manguin, 2007: *Anopheles* – Dr Bruce A. Harrison (contributions) (see Service, 2010).
harrisoni Muspratt, 1953: *Aedes* – Mr A.D. Harrison (collected first known specimens of the species).
harrisoni Peyton, 1977: *Uranotaenia* – Dr Bruce A. Harrison (contributions).
harrisoni Sirivanakarn, 1977c: *Culex* – Major Bruce A. Harrison (honorary; also collected type specimens with K. Mongkolpanya).
harrisonica (Reinert, 1974): *Verrallina* – Bruce A. Harrison (contributions).
harveyi (Barraud, 1923e): *Aedes* – not mentioned.
hatorii Yamada, 1921: *Aedes* – Mr Hatori (presented specimen to the author).
haworthi Edwards, 1923d: *Aedes* – Dr W.E. Haworth (collected type specimens).
hayashii Yamada, 1917: *Culex* – if mentioned, requires translation of Japanese.
hebrardi da Cunha Ramos & Brunhes, 2004: *Uranotaenia* – Georges Hebrard, Institut de Recherche pour le Développement (honorary).
hectoris Giaquinto Mira, 1931: *Anopheles* – Doctor Prof. Héctor Giaquinto (author's father, in memoriam) (see Kitzmiller, 1982).

heischi van Someren, 1951: *Aedes* – Dr R.B. Heisch, parasitologist (honorary).
heiseri Baisas, 1935c: *Uranotaenia* – Dr Victor G. Heiser, Rockefeller Foundation (honorary).
helenae Brunhes, Adam & Bailly-Choumara, 1967: *Culex* – not mentioned.
helenae Reinert, 1973e: *Aedes* – my wife, Mary Helen (Mollie).
helenae Ribeiro, 1992c: *Toxorhynchites* – Doctor Helena da Cunha Ramos (contributions).
hendersoni Cockerell, 1918: *Aedes* – Professor Junius Henderson (collected type specimens with E. Schwabe).
henrardi Edwards, 1935c: *Uranotaenia* – Dr C. Henrard (collected holotype).
henriquei da Cunha Ramos, 1993: *Uranotaenia* – Professor Doutor Henrique Ribeiro (contributions; also the author's husband).
hensilli Farner, 1945: *Aedes* – George S. Hensill (collected type specimens).
hepperi Casal & García, 1967c: *Culex* – Sr Héctor C. Hepper (first collector of the species).
hermanoi (Lane & Coutinho, 1940): *Coquillettidia* – Hermano Ribeira da Silva (honorary).
hermsi Barr & Guptavanij, 1989: *Anopheles* – William B. Herms (contributions) (see Service, 2010).
hernandoi Baisas & Feliciano, 1953: *Topomyia* – Dr Eugenio Hernando, 3rd Filipino Director of Health (honorary).
herrerai Sutil Oramas, Pulido Florenzano & Amarista Meneses, 1987: *Culex* – Jesús Herrera (in memoriam; field work)
hervyi Brunhes, Le Goff & Geoffroy, 1999: *Anopheles* – not mentioned, obviously Mr John-Paul Hervy (colleague) (see Service, 2010).
hervyi da Cunha Ramos & Brunhes, 2004: *Uranotaenia* – J.-P. Hervy (colleague).
hewitti (Edwards, 1914b): *Culex* – J. Hewitt (collected type specimens).
hightoni van Someren, 1947c: *Eretmapodites* – Mr R.B. Highton (collected type specimens with Dr P.C.C. Garnham and Mr J.O. Harper).
hilli Edwards, 1922d: *Culex* – [Mr] G.F. Hill (collected type specimens).
hilli (Edwards, 1926a): *Culiseta* – G.F. Hill (collected type specimens).
hilli Woodhill & Lee, 1944a: *Anopheles* – not mentioned, probably Gerald Freer Hill (see Kitzmiller, 1982).
hinesorum Schmidt, 2001: *Anopheles* – Judith and Michael Hines (assistance) (see Service, 2010).
hodgkini Marks, 1959a: *Aedes* – Dr E.P. Hodgkin (collaborator of collector E.J. Britten).
hodgkini Reid, 1962: *Anopheles* – E.P. [Ernest P.] Hodgkin (inspiration) (see Kitzmiller, 1982).
hodgkini (Wharton, 1962): *Coquillettidia* – Dr E.P. Hodgkin (contributions).
hoffmani Root, 1927a: *Limatus* – Dr William A. Hoffman (collected holotype).
hoguei Belkin, 1962: *Aedes* – not mentioned, obviously Charles L. Hogue, who is acknowledged in the Preface for extensive assistance.
hollandi Taylor, 1934: *Bironella* – Dr E.A. Holland (collections) (see Kitzmiller, 1982).
hollingsheadi Belkin, 1962: *Aedes* – Charles Hollingshead (honorary).
holmbergii Lynch Arribálzaga, 1891b: *Psorophora* – E.L. Holmberg (collected type specimen).
hoogstraali Baisas, 1947: *Tripteroides* – Captain Harry Hoogstraal (for help with restoration of the Philippine Museum).
hoogstraali Knight & Rozeboom, 1946: *Aedes* – Mr Harry Hoogstraal (honorary).
hopkinsi Edwards, 1932b: *Culex* – [Mr] G.H.E. Hopkins (collected type specimens).
hopkinsi Edwards, 1932b: *Uranotaenia* – [Mr] G.H.E. Hopkins (collected type specimens).
hopkinsi Edwards, 1936: *Aedes* – [G.H.E.] Hopkins (collected type specimens).
horotoi Taylor, 1972: *Aedes* – E. Sosimo Horoto (collected type specimens; assistance).
houghtoni Feng, 1941: *Topomyia* – Dr H.S. Houghton (director, Peiping Union Medical College).
howardi Belkin & Hogue, 1959: *Deinocerites* – not mentioned, probably L.O. Howard.
howardi Lane & Cerqueira, 1942: *Wyeomyia* – Dr L.O. Howard (honorary).
howardii Coquillett, 1901: *Psorophora* – Dr L.O. Howard (contributions).
huangae Meng, 1958: *Culex* – Miss W.H. Huang (collected holotype).
huangae Reinert, 1999a: *Aedes* – Dr Yiau-Min Huang (contributions).
hughi Lambert & Coetzee, 1982: *Anopheles* – Professor Hugh Paterson (stimulus, interest, contributions) (see Service, 2010).
hui Bohart, 1957: *Aedes* – S.M.K. Hu (collected type specimens).
humboldti (Lane & Cerqueira, 1942): *Runchomyia* – A. Humboldt (in memoriam).
hunteri (Strickland, 1916a): *Anopheles* – Dr [Peter S.] Hunter (independent discovery of the species) (see Kitzmiller, 1982).
hurlbuti Belkin, 1962: *Culex* – Herbert S. Hurlbut (contributions).
hurlbuti Lien, 1967: *Aedes* – Dr Herbert S. Hurlbut (support and advice).
husaini Qutubuddin, 1947: *Uranotaenia* – Mian Afzal Husain (Indian entomologist).
hutchinsoni Barraud, 1924e: *Culex* – not mentioned.
ichiromiyagii Toma & Higa, 2004: *Ficalbia* – Prof. Emeritus Ichiro Miyagi (contributions).
inatomii Kamimura & Wada, 1974: *Culex* – Seiiti Inatomi (contributions).
ingrami Edwards, 1916: *Culex* – Dr A. Ingram (collected type specimens).
ingrami Edwards, 1930a: *Aedes* – Dr A. Ingram (collected holotype).
iriartei Cova García, Pulido F., Escalante de Ugueto & Mora R., 1987a: *Uranotaenia* – Dr David Iriarte (honorary).

isabelae Duret, 1968f: *Culex* – wife of the author.

iyengari Edwards, 1923b: *Aedes* – M.O.T. Iyengar (collected type specimens).

iyengari Mattingly & Rageau, 1958: *Culex* – Dr M.O.T. Iyengar (contributions; collected two paratypes).

jacksoni Edwards, 1934: *Culex* – Dr R.B. Jackson (collected type specimens).

jacksoni Edwards, 1935a: *Uranotaenia* – Dr R.B. Jackson (collected holotype).

jacksoni Mattingly, 1949a: *Ficalbia* – R.B. Jackson (collected type specimens).

jacobsoni (Edwards, 1930b): *Malaya* – E. Jacobson (collected type specimens).

jamesi (Edwards, 1914b): *Aedes* – Major S.P. James (collected holotype and two paratypes).

jamesii Theobald, 1901a: *Anopheles* – James (Captain S.P. James) (collected holotype) (see Kitzmiller, 1982).

jamoti Hamon & Rickenbach, 1954: *Aedes* – docteur Jamot (contributions).

janconnae Wilkerson & Sallum, 2009: *Anopheles* – Jan Conn (collaborator, friend, contributions) (see Service, 2010).

jani Rickenbach & Lombrici, 1975: *Eretmapodites* – docteur Charles Jan (honorary; virologist?).

jeansottei (Doucet, 1950): *Mimomyia* – not mentioned.

jefferyi Sirivanakarn, 1977b: *Culex* – John Jeffery (collected type specimens with Tee Kern Tho).

jenningsi Dyar & Knab, 1907c: *Culex* – Mr A.H. Jennings (honorary).

jenseni (de Meijere, 1910): *Culex* – Dr Jensen (collected type specimens).

joanae Muspratt, 1955: subspecies of *Culex striatipes* Edwards, 1941 – Mrs Joan Wright (assistance in field collections).

johni Cova Garcia, Pulido F. & Escalante de Ugueto, 1979: *Culex* – not mentioned.

johnnyi Duret, 1968d: *Culex* – Dr A.J. Walker, PAHO (honorary).

johnsoni Galindo & Mendez, 1961: *Culex* – Dr Carl M. Johnson, Director, Gorgas Memorial Laboratory (honorary).

johnsoni (Laffoon, 1946): *Verrallina* – D.R. Johnson (collected holotype).

johnstonii (Grabham, 1905): *Psorophora* – not mentioned.

jorgi Carpintero & Leguizamón, 2000: *Aedes* – Dr Miguel E. Jorg (appreciation and admiration).

josephinae Marks, 1958: *Aedes* – Dr M.J. [Mabel Josephine] Mackerras (collaboration and companionship; also collected paratypes with the author).

josephineae Baisas, 1935d: *Culex* – not mentioned.

josiahae Huang, 1988a: *Aedes* – Mrs Phoebe A.O. Josiah (contributions).

joucouri da Cunha Ramos & Brunhes, 2004: *Uranotaenia* – Joucour (name suggested by A. Grjebine).

joyoni Brunhes, 1977a: *Orthopodomyia* – not mentioned (? Prof. Louis Joyon, Université de Clermont).

judithae Zavortink, 1969: *Anopheles* – Judith A. Bergland ('special friend') (see Kitzmiller, 1982).

juppi McIntosh, 1973: *Aedes* – Peter G. Jupp (contributions).

kana Tanaka, Mizusawa & Saugstad, 1979: *Heizmannia* – Etymology not indicated by the authors. Kana is a feminine Japanese given name. The name is very popular on Amami Island, the type locality of the species. An associated meaning of the name is 'one's girlfriend or lover', and often occurs in the popular song 'Lover in the island' (Ichiro Miyagi, personal communication).

keefei King & Hoogstraal, 1946f: *Aedes* – Chaplain A.M. Keefe (collected holotype).

kempi (Edwards, 1921c): *Toxorhynchites* – S. Kemp (collected type specimens).

kennethi Muspratt, 1956: *Aedes* – Dr Kenneth C. Smithburn, Rockefeller Foundation (organized expedition).

kesseli Belkin, 1962: *Culex* – John F. Kessel (contributions).

kesseli Huang & Hitchcock, 1980: *Aedes* – Dr John F. Kessel (contributions).

kesseli Ramalingam, 1987: *Armigeres* – Dr John F. Kessel ('my former professor'; contributions).

khazani Edwards, 1922a: *Aedes* – Sub-Assistant Surgeon Khazan Chand (collected type specimens).

khazani Edwards, 1922a: *Culex* – Sub-Assistant Surgeon Khazan Chand (collected type specimens).

kingi Christophers, 1923a: *Anopheles* – Major H.H. King (collected type specimens) (see Kitzmiller, 1982).

kingi Lee, 1946: *Tripteroides* – Lieut.-Col. W.V. King (provided extensive collection of *Tripteroides*).

kingianus Edwards, 1922d: *Culex* – H.H. King (collected holotype and two paratypes).

kingii (Theobald, 1913d): subspecies of *Culex argenteopunctatus* (Ventrillon, 1905a) – not specifically mentioned, Mr Harold King (sent collections from Sudan).

kleini Reinert, 1973e: *Aedes* – J.-M. Klein (contributions; also collected type specimens).

kleini Rueda, 2005: *Anopheles* – Colonel Terry A. Klein (contributions) (see Service, 2010).

klossi (Edwards, 1921c): *Toxorhynchites* – C. Boden Kloss (collected holotype).

knabi (Coquillett, 1906a): *Aedes* – Mr Frederick Knab (collected type specimens).

knabi Lane & Cerqueira, 1942: *Wyeomyia* – F. Knab (in memoriam).

knighti Baisas & Ubaldo-Pagayon, 1953: *Tripteroides* – Dr K.L. [Kenneth Lee] Knight (contributions).

knighti Stone & Bohart, 1944: *Aedes* – K.L. [Kenneth Lee] Knight (collected type specimens).

kobayashii Nakata, 1956: *Aedes* – Dr Harujiro Kobayashi (honorary) [Harujiro Kobayashi (1960–1969) was a famous Medical Parasitologist and Professor at Kyoto Prefectural University of Medicine (Ischiro Miyagi, personal communication)].

kochi Dönitz, 1901: *Anopheles* – Herrn Geh. Rath Koch (mentioned in the Introduction), Prof. Heinrich Hermann Robert Koch (see Kitzmiller, 1982).

kochi (Dönitz, 1901): *Aedes* – Herrn Geh. Rath Koch (see previous species).

kokhani Vythilingam, Jeffery & Harbach, 2007: *Anopheles* – Dr Ng Kok Han (support of research activities) (see Service, 2010).
koli Peyton & Klein, 1970: *Uranotaenia* – Mr Kol Mongkolpanya (contributions to collections).
kompi Edwards, 1930a: *Anopheles* – Mr W.H.W. Komp (collected holotype) (see Kitzmiller, 1982).
kompi Valencia, 1973: *Culex* – W.H. Komp (collected type specimens).
kompi Vargas & Downs, 1950: *Aedes* – Sr W.H.W. Komp (honorary).
konderi Galvão & Damasceno, 1942a: *Anopheles* – Dr Valerio Regis Konder (honorary) (see Kitzmiller, 1982).
kraussi Grjebine, 1953a: *Uranotaenia* – not mentioned.
krombeini Huang, 1975: *Aedes* – Dr Karl V. Krombein (contributions).
kuhnsi King & Hoogstraal, 1955: *Culex* – Dwight M. Kuhns (honorary).
kummi Edwards, 1930a: *Aedes* – H.W. Kumm (collected type specimens).
kummi Edwards, 1939b: *Orthopodomyia* – Dr H.W. Kumm (collected type specimens).
kummi Komp & Rozeboom, 1951: *Culex* – Dr Henry W. Kumm (collected type specimens).
kummi Lane & Cerqueira, 1942: *Wyeomyia* – Dr H.W. Kumm (collected type specimens).
labranchiae Falleroni, 1926: *Anopheles* – A. [Antonio] Labrancha (honorary) (see Kitzmiller, 1982).
lacani Grjebine, 1953a: *Anopheles* – Dr [André M.M.] Lacan (collected specimens with the author) (see Kitzmiller, 1982).
lacani Rickenbach & Eouzan, 1970: *Eretmapodites* – Médecin-Général Lacan (honorary).
laffooni Baisas & Ubaldo-Pagayon, 1953: subspecies of *Tripteroides powelli* (Ludlow, 1909c) – Dr Jean Laffoon (honorary).
laffooni Belkin, 1962: *Culex* – J. Laffoon (contributions).
laffooni Knight & Rozeboom, 1946: *Aedes* – Jean Laffoon (collected type specimens).
laffosseae da Cunha Ramos & Brunhes, 2004: *Uranotaenia* – Josée Laffosse (participation in field work).
lahillei Bachmann & Casal, 1962: *Culex* – Fernando Lahille (in memoriam).
lairdi Belkin, 1962: *Culex* – Marshal Laird (contributions).
lakei Sirivanakarn, 1968: *Culex* – Harry Lake (honorary).
lamborni Edwards, 1923a: *Aedes* – Dr W.A. Lamborn (collected type specimen).
lambrechti van Someren, 1971: *Aedes* – Dr F.L. Lambrecht (collected type specimens).
laneanus Corrêa & Cerqueira, 1944: *Anopheles* – Dr John Lane (entomologist and distinguished friend) (see Kitzmiller, 1982).
lanei (Antunes, 1937): *Trichoprosopon* – John Lane (close friend).
lanei Cerqueira, 1961: *Sabethes* – Prof. John Lane (honorary).
lanei Galvão & Franco do Amaral, 1938: *Anopheles* – John Lane (friend and distinguished entomologist) (see Kitzmiller, 1982).
lanei Martinez & Prosen, 1953a: *Uranotaenia* – not mentioned, obviously Prof. John Lane.
lanei Oliveira Coutinho & Forattini, 1962: *Culex* – Prof. John Lane (honorary).
lanei Shannon & Cerqueira, 1943: *Psorophora* – Sr John Lane (honorary).
laplantei (Hamon, Adam & Mouchet, 1955): *Culex* – [André] Laplante (colleague).
lassalli (Bonne-Wepster & Bonne, 1921): *Wyeomyia* – not mentioned.
laureli Baisas, 1935d: *Culex* – not mentioned.
lauriei Carter, 1920: *Aedes* – Dr R. Douglas Laurie (collected type specimens).
lavatae Stone & Bohart, 1944: *Culex* – ? Philippine family name.
lavieri Doucet, 1950: *Uranotaenia* – ? Prof. G. Lavier.
lawtoni Baisas, 1946b: *Zeugnomyia* – Gen. Lawton (in memoriam).
leberi Boreham, 1970: *Mansonia* – Canal Zone Governor Walter P. Leber (honorary; support).
lebiedi da Cunha Ramos & Brunhes, 2004: *Uranotaenia* – [B.] Lebied (name suggested by A. Grjebine).
ledgeri Huang, 1981: *Aedes* – Dr John A. Ledger (contributions).
leei Berlin, 1969a: *Aedes* – Dr V.H. Lee (contributions).
leei King & Hoogstraal, 1955: *Culex* – Mr T.G. Lee (contributions).
leei Peters, 1959: *Tripteroides* – Mr D.J. Lee (honorary).
leei Stone & Barreto, 1969: *Galindomyia* – Dr Vernon H. Lee (collections of mosquitoes).
leei (Wharton, 1947): *Malaya* – Mr D.J. Lee (criticism and advice).
leesoni Edwards, 1932b: *Aedes* – H.S. Lesson (collected type specimen).
leesoni Evans, 1931b: *Anopheles* – Mr H.S. Leeson (contributions) (see Kitzmiller, 1982).
legoffi da Cunha Ramos & Brunhes, 2004: *Uranotaenia* – Gilbert Le Goff (colleague).
lehcharlesi Miyagi & Toma, 2008: *Topomyia* – Dr Moi Ung Charles Leh (support and encouragement).
leicesteri (Edwards, 1917): *Verrallina* – Dr G.F. Leicester (collected holotype).
leicesteri Theobald, 1904c: *Toxorhynchites* – Dr Leicester (provided specimens).
leilae (King & Hoogstraal, 1947b): *Verrallina* – not mentioned, woman's name.
leonardi Belkin, 1962: *Culex* – Justin W. Leonard (contributions).
lerozeboomi Reinert, 1990: *Aedes* – Dr L.E. Rozeboom (contributions).
lesteri Baisas & Hu, 1936: *Anopheles* – Henry Lester (honorary) (see Kitzmiller, 1982).
levicastilloi Grjebine, 1986: *Mimomyia* – José Roberto Levi-Castillo (contributions).

levicastilloi Lane, 1945b: *Culex* – Major Roberto Levi-Castillo (collected holotype).

levicastilloi Levi Castillo, 1944: subspecies of *Anopheles pseudopunctipennis* Theobald, 1901c – 'variety described by the author' [translated from Spanish] (see Kitzmiller, 1982).

lewelleni Starkey & Webb, 1946: *Aedes* – Pfc. William B. Lewellen (collected holotype; collected paratypes along with others).

lewisi Ludlow, 1920a: *Anopheles* – not mentioned (Colonel William F. Lewis) (see Kitzmiller, 1982).

lewisi Ribeiro, 1991: *Toxorhynchites* – Dr D.J. Lewis (collected holotype).

lii Wu, 1936: *Heizmannia* – Mr F.S. Li (help and advice).

limai Lane & Cerqueira, 1942: *Wyeomyia* – Prof. Dr A. da Costa Lima (honorary).

lindesayi Giles, 1900a: *Anopheles* – Captain Victor Lindesay (sent specimen to author) (see Kitzmiller, 1982).

lindsayi Thurman, 1959: *Topomyia* – Dr Dale R. Lindsay (contributions).

linesi Marks, 1964a: *Aedes* – E.W. Lines (collected holotype).

lipovskyi Belkin, 1950: *Tripteroides* – L.J. Lipovsky (first collected the species).

listeri de Meillon, 1931b: *Anopheles* – Sir Spencer Lister (honorary) (see Kitzmiller, 1982).

littlechildi (Edwards, 1930b): *Tripteroides* – H.O.C. Littlechild (collected type specimens).

littlechildi Taylor, 1933: *Aedes* – '[I] dedicate this species to its discoverer' (not named).

littleri (Taylor, 1914c): *Culiseta* – H.M. Littler (collected holotype).

litwakae Harbach, 1985: *Culex* – Ms Taina Litwak (mosquito illustrations).

litwakae Reinert, 1990: *Aedes* – Miss Taina R. Litwak (mosquito illustrations).

lloreti Gil Collado, 1936: *Anopheles* – Dr Lloret (Abelardo Lloret Peralt) (support) (see Kitzmiller, 1982).

lopesi (Correa & Ramalho, 1956): *Wyeomyia* – Dr Tito Lopes da Silva (friend).

lopesi Sirivanakarn & Jakob, 1979: *Culex* – Dr Oscar de Souza Lopes (collected type specimens).

lopezii Cova Garcia, Sutil Oramas & Pulido F., 1974a: *Wyeomyia* – Sr José Antonio López (in memoriam; contributions).

lorraineae Berlin, 1969a: *Aedes* – Mrs Lorraine Belkin (in memoriam).

lottei Hamon & Brengues, 1965: *Aedes* – Docteur Michel Lotte (honorary).

lounibosi Gillies & Coetzee, 1987: *Anopheles* – Dr L.P. Lounibos (collected type specimens with Dr R. Subra) (see Service, 2010).

lousthei Boussès & Brunhes, 2013: *Uranotaenia* – Mathieu Lousthe-Robert (type specimens collected by Vincent Robert and his son Mathieu Lousthe-Robert).

lovettae Evans, 1934: *Anopheles* – Mrs D. Bagster Wilson (née Lovett) (assisted her husband in survey work that resulted in collection of the holotype) (see Kitzmiller, 1982).

lowii Theobald, 1901c: *Uranotaenia* – Dr Low (collected and reared the type specimens).

lowisii (Theobald, 1910c): *Aedes* – Lowis (collected type specimens).

lucyae van Someren, 1954a: *Uranotaenia* – Lucy Ellinor Tunstall (author's mother).

ludlowae (Blanchard, 1905): subspecies of *Aedes vigilax* (Skuse, 1889) – not specifically mentioned, obviously Dr Clara S. Ludlow (replacement name for *annulifera* Ludlow).

ludlowae Brunetti, 1920: *Ficalbia* – C.S. Ludlow (replacement name for *minima* Ludlow).

ludlowae Dyar & Shannon, 1925a: *Uranotaenia* – Dr Clara S. Ludlow (specimens from Ludlow estate).

ludlowae (Theobald, 1903a): *Anopheles* – Miss Ludlow (collected type specimens) (see Kitzmiller, 1982).

lui Gong & Li, 1999: *Heizmannia* – if mentioned, requires translation of Chinese, but evidence suggests dedication to Mr Lu Baolin.

lui Lien, 1968a: *Uranotaenia* – Mr Liang Cheng Lu (collection and rearing of mosquitoes).

lutzi (da Costa Lima, 1930a): *Wyeomyia* – [Dr Aldolfo] Lutz (contributions).

lutzii Cruz, 1901: *Anopheles* – Dr [Adolfo] Lutz (honorary) (see Kitzmiller, 1982).

lutzii (Theobald, 1901a): *Psorophora* – Dr Lutz (provided specimens).

lynchi (Shannon, 1931): *Coquillettidia* – not mentioned, probably Lynch Arribálzaga.

mabinii Baisas & Ubaldo-Pagayon, 1953: *Tripteroides* – Apolinario Mabini (honorary).

macarthuri Colless, 1956: *Anopheles* – not mentioned (Dr John McArthur) (see Kitzmiller, 1982).

macdonaldi Colless, 1965: *Culex* – not mentioned, obviously W.W. Macdonald.

macdonaldi Mattingly, 1957c: *Heizmannia* – Mr W.W. Macdonald (provided specimens).

macdougalli Edwards, 1922a: *Aedes* – Major A.J. MacDougall (collected type specimens).

macfarlanei (Edwards, 1914b): *Aedes* – Dr H. Macfarlane (collected type specimens).

macfarlanei Edwards, 1914c: *Uranotaenia* – Dr H. Macfarlane (collected type specimens).

macfiei Edwards, 1923a: *Culex* – Dr Macfie (collected type specimens with Dr Ingram).

machadoi da Cunha Ramos, 1986: *Uranotaenia* – Dr A. de Barros Machado (collected type specimens).

machadoi da Silva Mattos, da Silveira Guedes & Xavier, 1978: *Culex* – Professor Angelo B.M. Machado (contributions).

machardyi Edwards, 1930b: *Anopheles* – J.W. McHardy [*sic*] (collected type specimens) (see Kitzmiller, 1982).

macintoshi Marks, 1959a: *Aedes* – Mr D.L. McIntosh (extensive collections).

mackerrasi Taylor, 1927: *Aedes* – Dr I.M. Mackerras (friend).

macmahoni Evans, 1936: subspecies of *Anopheles sergentii* (Theobald, 1907) – Mr J.P. McMahon (incorrectly spelled 'Macmahon' by the author) (study of the larvae) (see Kitzmiller, 1982).

macmillani Marks, 1964b: *Aedes* – B. McMillan (holotype female collected biting B. McMillan [*sic*]).
maehleri Bohart, 1957: *Aedes* – K.L. Maehler (collected type specimens).
maffii Taylor & Tenorio, 1974: *Aedes* – Mario Maffi (contributions).
mahaffyi van Someren, 1949b: *Eretmapodites* – Dr Mahaffy (honorary; facilities support).
mahantai Bhattacharyya, Prakash, Mohapatra & Sarma, 2009: *Armigeres* – J. Mahanta (interest and support for mosquito taxonomic studies in northeastern India).
maiae Edwards, 1917: *Armigeres* – 'dedicated to my wife'.
majidi Young & Majid, 1928: *Anopheles* – 'after the junior writer who first collected it' (see Kitzmiller, 1982).
malfeyti (Newstead, 1907): *Ficalbia* – Major Malfeyt (courtesy and assistance).
malikuli Huang, 1973: *Aedes* – Mr Vichai Malikul (mosquito illustrations).
mallochi Taylor, 1944: *Aedes* – J.R. Malloch (contributions).
malvari Baisas & Ubaldo-Pagayon, 1953: *Tripteroides* – General Miguel Malvar (in memoriam).
manalangi Baisas, 1935b: *Armigeres* – Dr C. Manalang (collected male and female of the type series).
manalangi Mendoza, 1940: *Anopheles* – Dr C. Manalang (author's entomology tutor) (see Kitzmiller, 1982).
manhi (Harbach & Cook, 2010): *Aedes* – Dr Nguyen Duc Manh (colleague, in memoriam).
manopi Thurman, 1959: *Toxorhynchites* – Mr Manop Rattanopradith (valuable assistance).
mansouri Qutubuddin, 1959: *Aedes* – Dr Mansour Ali Haseeb (friend and colleague).
maplei Knight & Hurlbut, 1949: *Culex* – Dr John D. Maple, II (in memoriam).
marceli (Mattingly, 1953c): *Malaya* – Monsieur Marcel Lips (provided collections including the type specimens; also co-author of the publication).
marcelleae Adam & Hamon, 1959: *Eretmapodites* – Mme Marcelle Hamon (co-author's wife?).
marchettei Garcia, Jeffery & Rudnick, 1969: *Culiseta* – Nyven J. Marchette (contributions).
mariae (Bourroul, 1904): *Toxorhynchites* – not mentioned.
mariae (Sergent & Sergent, 1903): *Aedes* – not mentioned.
marinkellei Berlin, 1969a: *Aedes* – Dr C.J. Marinkelle (contributions).
marksae Dobrotworsky, 1965: *Tripteroides* – Dr E.N. Marks (contributions).
marksae Grjebine, 1986: *Mimomyia* – Elisabeth [*sic*] N. Marks (contributions).
marksae King & Hoogstraal, 1955: *Culex* – Dr Elizabeth N. Marks (contributions).
marshallii (Theobald, 1901a): *Aedes* – G.A.K. Marshall (collected type specimens).
marshallii (Theobald, 1903a): *Anopheles* – G.A.K. Marshall (collected holotype) (see Kitzmiller, 1982).
marteri Senevet & Prunnelle, 1927: *Anopheles* – not mentioned (Senevet's wife Marie Louise Marter) (see Kitzmiller, 1982).
martiali Senevet & Abonnenc, 1939b: *Limatus* – Médecin-Colonel Martial (honorary).
martinei (Doucet, 1951a): *Mimomyia* – not mentioned.
martineti Senevet, 1937: *Aedes* – Mme Martinet (in memoriam).
martinezi Berlin, 1969a: *Aedes* – Dr A. Martinez (contributions).
martinezi Casal & García, 1968c: *Culex* – Sr Antonio Martínez (colleague).
martinii Medschid, 1930: *Culex* – E. Martini (contributions).
martinius Shingarev, 1926: *Anopheles* – not mentioned (Prof. Erich Martini) (see Kitzmiller, 1982).
masseyi Edwards, 1923a: *Aedes* – Dr A. Yale Massey (collected holotype).
mathesoni Belkin, 1950: *Tripteroides* – Prof. Robert Matheson (encouragement and direction).
mathesoni Belkin & Heinemann, 1975: *Psorophora* – not mentioned, certainly Prof. Robert Matheson.
mathesoni Belkin & Hogue, 1959: *Deinocerites* – not mentioned, certainly Prof. Robert Matheson.
mathesoni Lane, 1943: *Uranotaenia* – Prof. R. Matheson (honorary).
mathesoni Middlekauff, 1944: subspecies of *Aedes canadensis* (Theobald, 1901c) – Dr Robert Matheson (friend; former professor).
mathioti Fontenille & Brunhes, 1985: *Aedes* – M. Christian Mathiot (virologist).
mattinglyi Baisas & Ubaldo-Pagayon, 1953: subspecies of *Tripteroides powelli* (Ludlow, 1909c) – Dr P.F. Mattingly (consulted on mosquitoes).
mattinglyi Grjebine, 1986: *Mimomyia* – Peter F. Mattingly (friend).
mattinglyi Hamon & Rickenbach, 1954: *Aedes* – P.F. Mattingly (honorary).
mattinglyi Hamon & van Someren, 1961a: *Eretmapodites* – P.F. Mattingly (colleague).
mattinglyi Knight, 1953: *Culex* – P.F. Mattingly (contributions).
mattinglyi Lane, 1953: *Wyeomyia* – not mentioned, obviously Peter F. Mattingly.
mattinglyi Qutubuddin, 1951: *Uranotaenia* – Mr Mattingly (taxonomic assistance).
mattinglyi Thurman, 1959: *Heizmannia* – Mr Peter F. Mattingly (contributions).
mattinglyorum Huang, 1994: *Aedes* – Dr Peter Mattingly and his wife, Christine (hospitality; contributions).
maxgermaini Huang, 1990: *Aedes* – Dr Max Germain (contributions).
maxi Dyar, 1928: *Culex* – M[ax] Kisliuk, Jr (collected type specimens).
mayeri Edwards, 1912c: *Uranotaenia* – Dr T.F.G. Mayer (collected holotype).
mccormicki (Belkin, 1962): *Verrallina* – E.J. McCormick, Jr (member of Malaria Survey Unit).

mcdonaldi Belkin, 1962: *Aedes* – Willian A. McDonald (contributions).
mcdonaldi Belkin & Hogue, 1959: *Deinocerites* – W.A. McDonald (collected holotype).
mcintoshi Huang, 1985: *Aedes* – Dr Bruce M. McIntosh (contributions).
medleri Knight & Laffoon, 1946: *Aedes* – Dr J.T. Medler (collected type specimens).
meirai Ribeiro, da Cunha Ramos, Capela & Pires, 1980: subspecies of *Aedes caspius* (Pallas, 1771) – Sr Dr Manuel T.V. de Meira (developed scientific activities in public health).
mendiolai Baisas, 1935c: *Uranotaenia* – Dr J. Mendiola (honorary).
mengi Chen, Wang & Zhao, 1989: *Uranotaenia* – in Chinese, probably Meng Ching-Hua.
mengi Dong, Wang & Lu, 1990: *Topomyia* – in Chinese, probably Meng Ching-Hua.
menoni Mattingly, 1958b: *Aedes* – Mr M.A.U. Menon (contributions).
messeae Falleroni, 1926: *Anopheles* – A. [Alessandro di Federico] Messea (honorary) (see Kitzmiller, 1982).
michaelikati van Someren, 1946c: *Aedes* – Michael Ikati (assistant, reared specimens).
mickevichae Huang, 1988b: *Aedes* – Dr Mary F. Mickevich (contributions).
mijanae Brunhes, Adam & Bailly-Choumara, 1967: *Culex* – not mentioned, girl's name.
millecampsi Lips, 1960a: *Anopheles* – Monsieur [Albert] Millecamps (first to discover the species, with Doctor d'Ursel) (see Kitzmiller, 1982).
milleri Dyar, 1922g: *Aedes* – Dr F.A. Miller (collected holotype).
millironi Belkin, 1962: *Culex* – H.E. Milliron (collected type specimens).
milloti Doucet, 1949b: *Culex* – not mentioned, surely Prof. Jacques Millot.
milloti Doucet, 1951a: *Orthopodomyia* – not mentioned, surely Prof. Jacques Millot.
milloti Grjebine, 1986: *Mimomyia* – Professeur Jacques Millot (contributions).
milloti Grjebine & Lacan, 1953: *Anopheles* – not mentioned (Jacques Millot) (see Kitzmiller, 1982).
milwardi Xavier & da Silva Mattos, 1972: *Culex* – Dr Roberto Milward de Andrade (contributions).
mitchellae Dyar, 1905b: *Culex* – Miss Evelyn G. Mitchell (honorary).
mitchellii (Theobald, 1905d): *Wyeomyia* – not mentioned, probably Mrs Alfred Mitchell (defrayed the cost of the publication).
miyagii Mogi & Toma, 1999: *Culex* – Ichiro Miyagi (contributions).
miyagii (Toma & Mogi, 2003): *Kimia* – Dr Ichiro Miyagi (contributions).
miyarai Tanaka, Mizusawa & Saugstad, 1979: subspecies of *Aedes flavopictus* Yamada, 1921 – not mentioned, probably Dr Kobo Miyara noted in Acknowledgements.
mjobergi (Edwards, 1926b): *Aedes* – Dr E. Mjoberg (collected type specimens).
mogii Miyagi, Toma & Higa, 2004: *Mimomyia* – Dr Yotoyoshi Mogi (contributions).
mogilasium (Dyar & Knab, 1907c): subspecies of *Trichoprosopon compressum* Lutz, 1905 – not mentioned.
mohani Knight, 1969: *Aedes* – B.N. Mohan (collected type specimens).
mohani Sirivanakarn, 1977d: *Culex* – B.N. Mohan (collected type specimens).
montchadskyi Dubitzky, 1968: *Aedes* – in Russian, surely Alexander S. Montchadsky.
moralesi (Dyar & Knab, 1919): *Shannoniana* – Dr Morales (collected type specimens).
mortiauxi da Cunha Ramos & Ribeiro, 1990: *Eretmapodites* – Mr Mortiaux (collected a female).
mortiauxi Edwards, 1938: *Anopheles* – Mons. [Roger] Mortiaux (collected type specimens) (see Kitzmiller, 1982).
moucheti Evans, 1923a: *Culex* – Dr [René] Mouchet (collected type specimens).
moucheti Evans, 1925b: *Anopheles* – Dr [René] Mouchet (collected type specimens) (see Kitzmiller, 1982).
moucheti (Hamon & Adam, 1955): *Malaya* – J. [Jean] Mouchet (collector; colleague).
moucheti Ravaonjanahary & Brunhes, 1977: *Aedes* – Monsieur J. Mouchet (honorary).
moufiedi Peyton, 1977: *Uranotaenia* – Dr Moufied A. Moussa (support of mosquito systematics).
moultoni Edwards, 1914c: *Armigeres* – J.C. Moulton (collected type specimens).
moultoni Edwards, 1914c: *Uranotaenia* – J.C. Moulton (collected type specimens).
mousinhoi de Meillon & de Carvalho Pereira, 1940: *Anopheles* – [Joaquin Augusto] Mousinho de Albuquerque (honorary) (see Kitzmiller, 1982).
muehlensi Petrocchi, 1927: *Wyeomyia* – Dr Pedro Mühlens (honorary).
muelleri Dyar, 1920: *Aedes* – Juan Müller (collected type specimens).
mulrennani Basham, 1948: *Culex* – Mr John A. Mulrennan (honorary).
murphyi Gillies & de Meillon, 1968: *Anopheles* – D.H. Murphy (collected and reared type specimens) (see Kitzmiller, 1982).
murrelli Lien, 1968d: *Culex* – Lt K.D. Murrell (advice and encouragement).
muspratti Hamon & Lambrecht, 1959: *Culex* – J. Muspratt (colleague).
mzooi van Someren, 1962: *Aedes* – Mr Fabian Mzoo (interest in mosquitoes).
nailoni King & Hoogstraal, 1946a: *Culex* – Sergeant William T. Nailon (collected type specimens).
nataliae Belkin, 1945: *Anopheles* – not mentioned (Natalia Yantsin, author's first wife) (see Kitzmiller, 1982).
nataliae Gornostaeva, 2005: *Aedes* – Наталии Алексндровны Филипповой (Natalia Alexandrovna Filippova, acarologist and doctor of biological sciences).

nataliae Lynch Arribálzaga, 1891b: *Uranotaenia* – not mentioned, girl's name.

neavei Theobald, 1906b: *Culex* – Sheftield Neave, Esq. (collected type specimens).

neghmei Mann, 1950: subspecies of *Anopheles pseudopunctipennis* Theobald, 1901c – Prof. Dr Amado Neghme (parasitologist) (see Kitzmiller, 1982).

neireti Edwards, 1920b: *Uranotaenia* – [Dr] Neiret (probable collector).

neivai Howard, Dyar & Knab, 1913: *Anopheles* – Dr [Arthur] Neiva (help and suggestions) (see Kitzmiller, 1982).

neivai (Lane & Coutinho, 1940): *Coquillettidia* – not mentioned.

nicksoni Miyagi & Toma, 2012: *Topomyia* – Mr Nickson Joseph Robi (contributions).

nishikawai Tanaka, Mizusawa & Saugstad, 1979: *Aedes* – Dr Masaru Nishikawa (contributed largely to collections).

nobukonis (Yamada, 1932): *Verrallina* – Nobuko (author's daughter) (Ischiro Miyagi, personal communication).

noei Mann, 1950: subspecies of *Anopheles pseudopunctipennis* Theobald, 1901c – Dr Don Juan Noé (sanitarian) (see Kitzmiller, 1982).

noniae Reid, 1963: *Anopheles* – 'Named for my wife [Nonie]' (see Kitzmiller, 1982).

notanandai Rattanarithikul & Green, 1987: *Anopheles* – Dr Vimol Notananda (honorary) (see Service, 2010).

notleyi van Someren, 1949a: *Anopheles* – not mentioned, obviously Major F.B. Notley, per Acknowledgements (see Kitzmiller, 1982).

nuneztovari Gabaldon, 1940: *Anopheles* – Dr M. Núñez Tovar (in memoriam) (see Kitzmiller, 1982).

nyounae Hamon & Adam, 1959a: *Aedes* – not mentioned, probably a girl's name, Nyouna.

oakleyi Stone, 1939: *Aedes* – R.G. Oakley (reared specimens).

oedipus Root, 1927c: *Culex* – reputed solver of the riddle of the Sphinx.

ohamai Ohama, 1947: subspecies of *Anopheles saperoi* Bohart & Ingram, 1946a – named after the author, himself (see Kitzmiller, 1982).

ohamai Tanaka, Mizusawa & Saugstad, 1975a: *Uranotaenia* – not mentioned, but apparently 'Dr Shinken Ohama of Ishigaki City'.

olimpioi Xavier, da Silva & da Silva Mattos, 1970a: *Culex* – Dr Olympio da Silva Pinto (honorary).

omani Belkin, 1962: *Culex* – Paul W. Oman (support for investigations).

omorii Lien, 1968b: *Aedes* – Prof. Nanzaburo Omori (encouragement).

omorii Sakakibara, 1959: *Anopheles* – Prof. Dr N. [Nanzaburo] Omori (advice and criticism) (see Kitzmiller, 1982).

orfilai Duret, 1953: *Culex* – Dr Ricardo N. Orfila (honorary).

ortizi Vargas & Diaz Nájera, 1961: *Sabethes* – Dr Carlos Ortiz Mariotte (honorary).

osornoi Berlin, 1969a: *Aedes* – Dr E. Osorno-Mesa (contributions).

oswaldoi (Peryassú, 1922b): *Anopheles* – not mentioned (presumably Oswaldo Gonçalves Cruz) (see Kitzmiller, 1982).

oteizai Perez Vigueras, 1956: *Uranotaenia* – Dr Alberto Oteiza y Zetién (honorary).

ovazzai Hamon & Adam, 1959b: *Aedes* – M. ovazza (colleague).

oweni Belkin, 1962: *Culex* – William B. Owen (contributions).

pagei (Ludlow, 1911b): *Aedes* – Major Henry Page (collected type specimens).

painei Edwards, 1935a: *Uranotaenia* – R.W. Paine (collected type specimens).

pairoji Sirivanakarn, 1977a: *Culex* – Mr Pairoj Suwanakorn (honorary).

pajoti da Cunha Ramos & Ribeiro, 1982: *Culex* – Dr F.-X. Pajot (contributions).

palaciosi Duret, 1968d: *Culex* – Dr Silvio Palacios, malariologist (honorary).

palmeirimi de Meillon & Rebêlo, 1941: *Uranotaenia* – Dr Vasco Palmeirim (honorary).

paltrinierii Shidrawi & Gillies, 1988: *Anopheles* – Dr A.B. Paltrinieri (recognition for assistance) (see Service, 2010).

pampanai Büttiker & Beales, 1959: *Anopheles* – Dr E.J. [Emilio J.] Pampana (honorary) (see Kitzmiller, 1982).

parenti (de Meillon & Lavoipierre, 1944): *Mimomyia* – Dr Parent (collected type specimens).

patersoni Alvarado & Heredia, 1947: subspecies of *Anopheles pseudopunctipennis* Theobald, 1901c – Dr Guillermo Páterson (William Cleland Paterson, malariologist) (see Kitzmiller, 1982).

patersoni Shannon & del Ponte, 1928: *Aedes* – Dr Guillermo Paterson (William Cleland Paterson) (honorary).

patriciae Mattingly, 1954a: *Aedes* – Dr Elizabeth Marks (contributions).

patriciae Peyton, 1977: *Uranotaenia* – Patricia Ann Peyton (author's daughter).

pattoni Christophers, 1926: *Anopheles* – Major W.S. Patton (sent specimens) (see Kitzmiller, 1982).

pauliani Doucet, 1949b: *Uranotaenia* – not mentioned, surely Prof. Renaud Paulian (honorary).

pauliani (Doucet, 1951b): *Toxorhynchites* – not mentioned, surely Prof. Renaud Paulian (honorary).

pauliani Grjebine, 1950: *Eretmapodites* – R. [Renaud] Paulian, obvious from Introduction (honorary).

pauliani Grjebine, 1953a: *Aedeomyia* – R. [Renaud] Paulian (collected type specimens).

pauliani Grjebine, 1953a: *Anopheles* – MM. [Renaud] Paulian (collaborator) (see Kitzmiller, 1982).

paulli Paterson & Shannon, 1927: *Psorophora* – Señor Archibaldo Paull ('our kind innkeeper' [translated from Spanish]).

paullusi Stone & Farner, 1945: *Aedes* – J.H. Paullus (collected holotype and most paratypes).

pavlovskyi Casal & García, 1967a: *Culex* – Dr Eduardo Alejo Pavlovsky (honorary).
pecori Reinert, 1990: *Aedes* – Mr James E. Pecor (technical assistance).
pedroi Sirivanakarn & Belkin, 1980: *Culex* – Pedro Galindo (contributions).
pefflyi Stone, 1961a: subspecies of *Uranotaenia unguiculata* Edwards, 1913a – R.L. Peffly (collected type specimens).
penai Sirivanakarn, 1979: *Culex* – Luis E. Peña G. (collected type specimens).
pendleburyi (Edwards, 1930a): *Toxorhynchites* – H.M. Pendlebury (collected holotype).
pennai Antunes & Lane, 1938: *Aedes* – Dr Henrique de Azevedo Penna (honorary).
pereyrai Duret, 1967a: *Culex* – Sr Atilio Pereyra (friend and support).
peringueyi Edwards, 1924a: *Culex* – Dr L. Péringuey (sent specimens for identification).
perkinsi Marks, 1949: *Aedes* – Mr F.A. Perkins (co-discoverer of the species).
perkinsi Stone & Penn, 1948: *Culex* – Mr F.A. Perkins (lecturer in entomology).
perryi Belkin, 1962: *Aedes* – William J. Perry (contributions).
perryi Belkin, 1962: *Culex* – W.J. Perry (contributions).
peryassui Dyar & Knab, 1908: *Anopheles* – [Prof. Antonio Gonçalves] Peryassú (honorary) (see Kitzmiller, 1982).
pessoai (Barreto & Coutinho, 1944): *Taeniorhynchus* – Prof. Dr Samuel B. Pessoa (honorary).
petersi Colless, 1960: *Culex* – Dr W. Peters (honorary).
petragnani del Vecchio, 1939: *Anopheles* – S.E. [Prof.] Giovanni Petragnani (honorary) (see Kitzmiller, 1982).
petrocchiae (Shannon & del Ponte, 1928): *Sabethes* – Juana Petrocchi (in memoriam).
peytoni Bram & Rattanarithikul, 1967: *Culex* – SFC E.L. Peyton (honorary).
peytoni Kulasekera, Harrison & Amerasinghe, 1989: *Anopheles* – E.L. Peyton (friend, collector, contributions) (see Service, 2010).
peytoni Leguizamón & Carpintero, 2004: *Orthopodomyia* – Mr E.L. Peyton (in memoriam).
peytoni Reinert, 1972a: *Aedes* – Mr E.L. Peyton (contributions).
philipi Edwards, 1929b: *Culex* – C.B. Philip (collected type specimens).
phillipi van Someren, 1949a: *Aedes* – Captain Phillip Glasgow (collection of mosquitoes made with Major F.B. Notley).
pifanoi Anduze, 1949a: *Culex* – Prof. Dr Félix Pifano C. (honorary).
pifanoi Cova Garcia, Pulido F. & Escalante de Ugueto, 1981: *Uranotaenia* – Dr Félix Pifano (honorary).
pillaii Mattingly, 1958a: *Aedes* – Mr Ganapathipillai (preparation of specimens for the study).
pipkini (Bohart, 1957): *Verrallina* – A.C. Pipkin (studied the species during his filariasis study).
ponmeki Miyagi & Toma, 2001: *Tripteroides* – Dr Ponmek Dalaloy (approval and encouragement for the study).
portesi Senevet & Abonnenc, 1941: *Culex* –Professeur Portes (honorary).
powderi Zavortink, 1970b: *Anopheles* – William A. [*sic*] Powder (Anthony William Powder) (contributions) (see Kitzmiller, 1982).
powelli Lee, 1944a: *Anopheles* – O.W. Powell (collected type specimens) (see Kitzmiller, 1982).
powelli (Ludlow, 1909c): *Tripteroides* – Dr W.A. Powell (collected holotype).
poweri (Theobald, 1905g): *Aedes* – Dr Power (collected holotype).
prachongae Rattanarithikul & Harrison, 2017: *Anopheles* – Mrs Prachong Panthusiri (friend and colleague; mosquito illustrations).
prajimi Peyton & Rattanarithikul, 1970: *Uranotaenia* – Mr Prajim Boonyakanist (technician-collector).
pulidoi Cova Garcia & Sutil Oramas, 1974: *Culex* – Sr Juan Pulido F. (contributions).
pylei Baisas, 1946a: *Uranotaenia* – not mentioned.
qui Dong, Dong & Zhou, 2003: *Uranotaenia* – in Chinese, probably Qu Feng-yi.
quintetti Brunhes, Adam & Bailly-Choumara, 1967: *Culex* – not mentioned.
rabelloi Forattini & Sallum, 1987a: *Culex* – Dr Ernesto X. Rabello (contributions).
rachoui Duret, 1968b: *Culex* – Doctor René G. Rachou (in memoriam).
rachoui Galvão, 1952: *Anopheles* – Dr René G. Rachou (contributions) (see Kitzmiller, 1982).
rachoui Xavier, da Silva Mattos & da Silva, 1970b: *Uranotaenia* – Dr René G. Rachou (collaborator; in memoriam).
radama de Meillon, 1943: *Anopheles* – not mentioned (Radàma I, 1st king of Madagascar) (see Kitzmiller, 1982).
rageaui (Hamon & Rickenbach, 1955b): subspecies of *Culex horridus* Edwards, 1922d – J. Rageau (Jean Georges Ferdinand Rageaui) (colleague).
rageaui Mattingly & Adam, 1954: *Anopheles* – Monsieur J. Rageau (Jean Georges Ferdinand Rageaui) (contributions) (see Kitzmiller, 1982).
raghavanii Rahman, Chowdhury & Kalra, 1969: *Culex* – Dr N.G.S. Raghavan (honorary).
rajaneeae Sirivanakarn, 1968: *Culex* – Rajanee Sirivanakarn (author's wife; contributions).
rajaonariveloi Brunhes & Hervy, 1995: *Orthopodomyia* – E. Rajaonarivelo (honorary).
ramachandrai Reuben, 1967: *Aedes* – Dr T. Ramachandra Rao (contributions).
ramakrishnii Wattal & Kalra, 1965: *Culex* – Dr S.P. Ramakrishnan (interest in medical entomology).
ramalai Grjebine, 1986: *Mimomyia* – Ramala (Malagasy friend; collector).
ramalingami Evenhuis & Steffan, 1986: *Toxorhynchites* – Dr Shivaji Ramalingam (collections of *Toxorhynchites*).
ramalingami (Reinert, 1974): *Verrallina* – Dr Shivaji Ramalingam (contributions).

ramalingami Sirivanakarn, 1973b: *Culex* – Dr S. Ramalingam (provided specimens for study).
rami (Barraud, 1928b): *Verrallina* – not mentioned, probably Sobha Ram, collector of another species (*Verrallina lugubris*).
ramirezi Vargas & Downs, 1950: *Aedes* – Dr Eliseo Ramírez (in memoriam).
rampae Harbach & Somboon, 2011: *Anopheles* – Dr Rampa Rattanarithikul (contributions).
rampae Peyton & Klein, 1970: *Uranotaenia* – Mrs Rampa Rattanarithikul (service; assistance).
ranci Grjebine, 1953a: *Anopheles* – not mentioned (Dr André Ranc) (see Kitzmiller, 1982).
rangeli Gabaldon, Cova-Garcia & Lopez, 1940: *Anopheles* – Señor Rafael Rangel (honorary) (see Kitzmiller, 1982).
ratcliffei Marks, 1959a: *Aedes* – Mr F.N. Ratcliffe (encouragement; collections).
rausseoi Cova Garcia, Sutil O. & Pulido F., 1972: *Culex* – José Antonio Rausseo Aguilera (colleague).
ravaonjanaharyi Brunhes & Hervy, 1995: *Orthopodomyia* – Charles Ravaonjanahary (collected type specimens).
ravissei Rickenbach & Eouzan, 1970: *Eretmapodites* – Docteur Pierre Ravisse (honorary).
raymondi del Ponte, Castro & Garcia, 1951: *Aedes* – Raymond C. Shannon (collected larvae).
reali Hamon & Adam, 1959a: *Aedes* – P. Real (colleague).
reesi (King & Hoogstraal, 1947b): *Verrallina* – B.E. Rees (collected holotype).
reevesi Wirth, 1948: *Culex* – not mentioned, credited to Galindo (1943) in key, probably William C. Reeves.
refiki Medschid, 1928: *Aedes* – not mentioned.
reginae Edwards, 1922a: *Aedes* – 'I retain the specific name proposed by Col. James.'
reidi Colless, 1965: *Culex* – Dr J.A. Reid (mentioned in Acknowledgements).
reidi Harrison, 1973: *Anopheles* – Dr John A. Reid (contributions) (see Kitzmiller, 1982).
reidi Mattingly, 1957c: *Heizmannia* – J.A. Reid (mentioned as collector of paratypes).
reinerti Peyton, 1977: *Uranotaenia* – Dr John F. Reinert (contributions).
reinerti Rattanarithikul & Harrison, 1988: *Aedes* – Dr John F. Reinert (friend; contributions).
reiseni Basio, 1971: *Tripteroides* – Captain William K. Reisen (appreciation; collections).
rempeli Vockeroth, 1954: *Aedes* – Dr J.G. Rempel (contributions).
renatoi Lane & Ramalho, 1960: *Culex* – Dr Renato R. Corrêa (friend).
reubenae Tewari & Hiriyan, 1992b: *Aedes* – Dr Rachel Reuben (contributions).
reyi Baisas, 1935c: *Uranotaenia* – not mentioned.
richardgarciai Jeffery, Oothuman & Rudnick, 1987: *Culex* – Dr Richard Garcia (contributions).
richei Klein, 1970: *Culex* – Dr Armand Riche (honorary).
richeti Brunhes & Venard, 1966: *Culex* – P. Richet (honorary).
richiardii (Ficalbi, 1889b): *Coquillettidia* – Prof. Richiardi (honorary).
rickenbachi da Cunha Ramos, 1993: *Uranotaenia* – Dr A. [André] Rickenbach (collected holotype).
rickenbachi Ferrara & Eouzan, 1974: *Eretmapodites* – André Rickenbach (colleague).
rickenbachi Hamon & Adam, 1959b: *Aedes* – A. [André] Rickenbach (colleague).
rickenbachi Ribeiro, 1991: *Toxorhynchites* – Dr A. [André] Rickenbach (collected type specimens).
rivadeneirai Leví Castillo, 1945: subspecies of *Anopheles pseudopunctipennis* Theobald, 1901c – Lic. Jaime Rivadeneira D. (supplied material for study, facilitated the author's work with suggestions and moral assistance) (see Kitzmiller, 1982).
riverai Duret, 1970: *Uranotaenia* – Sr Hernado Rivera (Colombian friend).
riverai Miyagi, Toma & Tsukamoto, 1983: *Tripteroides* – Dr Delfin G. Rivera (contributions).
riversi Bohart & Ingram, 1946a: *Aedes* – Commodore T.M. Rivers (interest furthered research on mosquitoes).
rizali (Banks, 1906b): *Aedes* – Dr José Rizal y Mercado (contributions).
rizzoi (de Deus Palma & Galvão, 1969): *Toxorhynchites* – Dr José Ângelo Rizzo (contributions).
roai Belkin, 1962: *Aedes* – Vernon R. Roa (honorary).
roberti da Cunha Ramos & Brunhes, 2004: *Uranotaenia* – Vincent Robert (participated in the discovery of the species).
robertsi (Laffoon, 1946): *Verrallina* – H.R. Roberts (collected holotype).
robinsoni Belkin, 1962: *Aedes* – W.A. Robinson (contributions).
rochei (Doucet, 1951b): *Coquillettidia* – not mentioned.
rodhaini Brunhes & Boussès, 2017: *Paulianius* – François Rodhain (colleague; contributions).
rodhaini Brunhes & Hervy, 1995: *Orthopodomyia* – Docteur [François] Rodhain (contributions).
rodhaini Leleup & Lips, 1950: *Anopheles* – Docteur J. [Jerome] Rodhain (honorary) (see Kitzmiller, 1982).
rodhaini Ribeiro, 1991: *Toxorhynchites* – Dr F. [François] Rodhain (collected holotype).
rollai Cova Garcia, Pulido F. & Escalante de Ugueto, 1977b: *Anopheles* – Dr Rolla B. Hill (contributions; see Cova Garcia, Pulido F. & Escalante de Ugueto, 1976) (see Kitzmiller, 1982).
romeroi Surcouf & Gonzalez-Rincones, 1912: *Culex* – [Señor Bachiller Jesús Maria Romero Sierra], *nomen novum* for *nigritulus* Zetterstedt of Romero Sierra, 1907.
ronderosi de Linero, 1967: *Culex* – Doctor Ricardo Ronderos (honorary).
rondoni (Neiva & Pinto, 1922a): *Anopheles* – Sr General Cândido Mariano Rondon (honorary) (see Kitzmiller, 1982).
rooti (del Ponte, 1939): *Wyeomyia* – not mentioned, probably Dr F.M. Root.
rooti Rozeboom, 1935: *Culex* – Dr F.M. Root (in memoriam).

roperi Reid, 1950: *Anopheles* – R. [Richard Snelling] Roper (collections; '*A. umbrosus*, largest variety of Roper, 1914') (see Kitzmiller, 1982).
roseni Belkin, 1962: *Culex* – L. [Leon] Rosen (collected type specimens; contributions).
roslihashimi Miyagi & Toma, 2005: *Topomyia* – Dr Rosli Hashim (contributions).
rossi Delfinado, 1966b: *Uranotaenia* – Dr Edward S. Ross (outstanding field work).
roubaudi (Doucet, 1950): *Mimomyia* – not mentioned, probably Emile Roubaud.
roubaudi Grjebine, 1953a: *Anopheles* – not mentioned (Prof. Emile Roubaud) (see Kitzmiller, 1982).
roxasi Baisas & Ubaldo-Pagayon, 1953: *Tripteroides* – Professor Hilario Roxas (in memoriam).
rozeboomi Baisas & Ubaldo-Pagayon, 1953: *Tripteroides* – Professor L.E. Rozeboom (honorary).
rozeboomi Causey, Deane & Deane, 1944: *Chagasia* – not mentioned (Lloyd E. Rozeboom) (see Kitzmiller, 1982).
ruthae Peters, 1958: *Culex* – author's wife.
rutherfordi Edwards, 1922a: *Uranotaenia* – A. Rutherford (collected holotype).
sacchettae Sirivanakarn & Jakob, 1982: *Culex* – Dr Lia de Abreu Sacchetta (contributions).
sacharovi Favre, 1903: *Anopheles* – not mentioned, obviously Nikolai A. Sacharov (see Kitzmiller, 1982).
saeungae Taai & Harbach, 2015: *Anopheles* – Dr Atiporn Saeung (contributions).
salauni Rickenbach, Ferrara & Eouzan, 1968: *Eretmapodites* – Docteur Jean-Jacques Salaün, virologist (honorary).
sallumae González & Reyes, 2017: *Aedes* – Dra. Maria Anice Mureb Sallum (friend; contributions).
sampaioi da Costa Lima, 1935b: *Orthopodomyia* – Snr Paulo C. Sampaio (first to send specimens, two males, of the type series).
sandrae Berlin, 1969b: *Culex* – Miss Sandra J. Heinemann (contributions).
sandrae Zavortink, 1972: *Aedes* – Sandra J. Heinemann (contributions).
santamarinai Broche, 2000: *Psorophora* – Dr Alberto Santamarina (contributions).
saperoi Bohart & Ingram, 1946a: *Anopheles* – Capt. J.J. Sapero (contributions) (see Kitzmiller, 1982).
saperoi Knight, 1946: *Aedes* – Capt. J.J. Sapero (cooperation).
sasai Kano, Nitahara & Awaya, 1954: *Culex* – Dr Manabu Sasa (honorary).
sasai Tanaka, Mizusawa & Saugstad, 1975b: *Aedes* – Dr M. Sasa (mentioned in Acknowledgements for permission to examine types and other specimens).
savoryi Bohart, 1957: *Aedes* – Nathaniel Savory (colonized the Bonin Islands, 1830).
sawadwongporni Rattanarithikul & Green, 1987: *Anopheles* – Mr Phorn Sawadwongporn (honorary) (see Service, 2010).
sawyeri Causey, Deane, Deane & Sampaio, 1943: *Anopheles* – Dr Wilbur A. Sawyer (honorary) (see Kitzmiller, 1982).
scanloni Bram, 1967a: *Culex* – Lt. Col. John E. Scanlon (contributions).
scanloni Mattingly, 1970: *Heizmannia* – not mentioned, certainly John E. Scanlon.
scanloni Reinert, 1970: *Aedes* – LTC John E. Scanlon (contributions).
scanloni Sallum & Peyton, 2005: *Anopheles* – John E. Scanlon (in memoriam) (see Service, 2010).
scheuberi Carpintero & Leguizamón, 2004: *Culex* – Mr Luis Scheuber (friend; collaborator; rancher).
schicki Berlin, 1969b: *Culex* – Robert X. Schick (contributions).
schicki Zavortink, 1972: *Aedes* – Robert X. Schick (contributions).
schilfgaardei Sirivanakarn, 1968: *Culex* – R. Schilfgaarde (honorary).
schlosseri Belkin, 1962: *Aedes* – Ralph J. Schlosser (contributions).
schnusei (Martini, 1931a): *Sabethes* – not mentioned.
schoutedeni (Wolfs, 1948a): *Coquillettidia* – Dr Schouteden (honorary).
schroederi Schick, 1970a: *Aedes* – D. Schroeder (collected holotype).
schtakelbergi Shingarev, 1928: *Aedes* – not mentioned, surely A.A. Shtakel'berg.
schueffneri Stanton, 1915: *Anopheles* – Dr Schüffner (Wilhelm A.P. Schüffner) (collected type specimens) (see Kitzmiller, 1982).
schwetzi Edwards, 1926a: *Aedes* – Dr J. [Jacques] Schwetz (collected type specimens).
schwetzi Edwards, 1929b: *Culex* – Dr J. [Jacques] Schwetz (collected type specimens).
schwetzi Evans, 1934: *Anopheles* – Dr J. [Jacques] Schwetz (collected holotype) (see Kitzmiller, 1982).
scottii Theobald, 1912b: *Culex* – H. Scott (collected type specimens).
sechani Brunhes & Boussès, 2009: *Culex* – Yves Séchan, medical entomologist (honorary).
sedlacekae Sirivanakarn, 1968: *Culex* – Marie Sedlacek (assistance in field work).
segermanae Huang, 1997: *Aedes* – Mrs Joyce Segerma (contributions).
seldeslachtsi Wolfs, 1947c: *Culex* – Dr Seldeslachts, physician (honorary).
sembeli Toma & Miyagi, 2002: *Armigeres* – Dr Dante T. Sembel (support and encouragement).
seniori Barraud, 1934: *Culex* – [R.] Senior-White (collected holotype).
seretsei Abdulla-Khan, Coetzee & Hunt, 1998: *Anopheles* – Sir Seretse Khama (honorary) (see Service, 2010).
sergentii (Theobald, 1907): *Anopheles* – Dr E. [Edmond] Sergent (collected type specimens) (see Kitzmiller, 1982).
sergievi Danilov, Markovich & Proskuryakova, 1978: *Aedes* – П. Г. Сергиева (P.G. Sergiev) (in memoriam).
sexaueri Belkin, 1953: *Uranotaenia* – Henry F. Sexauer (contributed to mosquito survey).
seydeli Edwards, 1929b: *Anopheles* – Ch. [Charles H.V.] Seydel (collected holotype) (see Kitzmiller, 1982).

seyrigi Edwards, 1941: *Culex* – A. Seyrig (collected type specimens).
shanahani Sirivanakarn, 1968: *Culex* – P. Shanahan (collected holotype with the author).
shannoni Cerqueira, 1961: *Sabethes* – Raymond C. Shannon (in memoriam).
shannoni Davis, 1931: *Anopheles* – Mr R.C. [Raymond Corbett] Shannon (honorary) (see Kitzmiller, 1982).
shannoni (Lane & Antunes, 1937): *Coquillettidia* – Sr Raymond C. Shannon (friend).
shannoni Lane & Cerqueira, 1942: *Wyeomyia* – Sr R.C. Shannon (honorary).
shannoni Vargas & Downs, 1950: *Aedes* – Mr R.C. Shannon (in memoriam).
shebbearei Barraud, 1924g: *Culex* – Mr E.O. Shebbeare (kindness and interest).
shehzadae Qutubuddin, 1972: *Aedes* – 'named after my daughter Shebzad'.
sherki Knight, 1948a: *Aedes* – Sgt Herman D. Sherk (aided collection of mosquitoes).
shillitonis Edwards, 1932b: *Uranotaenia* – J.F. Shillito (collected type specimens).
shinonagai (Tanaka, Mizusawa & Saugstad, 1979): *Lutzia* – Shinonaga (Dr Satoshi Shinonaga) (collected holotype and most paratypes).
shopei Forattini & Toda, 1966: *Culex* – Dr Robert E. Shope (honorary).
shortti (Barraud, 1923e): *Aedes* – not mentioned.
silvai Duret, 1968f: *Culex* – Dr Tito López da Silva, malariologist (honorary).
simanini Gutsevich, 1966: *Aedes* – P.I. Simanin (collected type specimens).
simmondsi Tenorio, 1977: *Bironella* – Dr J.H. Simmonds (first to find the species) (see Kitzmiller, 1982).
simmsi (Dyar & Knab, 1908): *Wyeomyia* – Mr H. Simms, sanitary inspector (honorary).
simpsoni (Theobald, 1905c): *Aedes* – Mr Simpson (collected type specimens).
simpsoni Theobald, 1905g: *Culex* – not mentioned.
sintoni (Barraud, 1924b): *Aedes* – Major J.A. [John Alexander] Sinton (collected type specimens).
sintoni Puri, 1929b: *Anopheles* – Major J.A. [John Alexander] Sinton (collected type specimens) (see Kitzmiller, 1982).
sirivanakarni (Duret, 1982): *Onirion* – Dr Sunthorn Sirivanakam (collaboration and help).
smithburni van Someren, 1950b: *Aedes* – Dr K. Smithburn (honorary).
smithii (Coquillett, 1901): *Wyeomyia* – Prof. J.B. Smith (collected type specimens).
smithii Theobald, 1905b: *Anopheles* – Major [F.] Smith (collected type specimens) (see Kitzmiller, 1982).
soaresi Lane & Cerqueira, 1942: *Trichoprosopon* – Gabriel Soares de Souza (contributions).
sohni (Reinert, 1974): *Verrallina* – Young T. Sohn (mosquito illustrations).
sombooni Peyton & Klein, 1970: *Uranotaenia* – Mr Somboon Maneechai (technician-collector).
soperi Antunes & Lane, 1937: *Culex* – Dr Fred L. Soper (honorary).
soperi Lane & Cerqueira, 1942: *Sabethes* – Dr Fred L. Soper (honorary).
soperi Leví-Castillo, 1955c: *Haemagogus* – Dr Fred L. Soper (honorary).
spegazzinii Brèthes, 1912: *Haemagogus* – Dr D. Carlos Spegazzini (collected holotype).
spencerii (Theobald, 1901c): *Aedes* – W.I. Spencer (collected type specimens).
sphinx Howard, Dyar & Knab, 1913: *Culex* – a mythical creature with the head of a human and the body of a lion which guarded the entrance to the Greek city of Thebes and asked riddles of travellers to allow passage.
spixi Cerqueira, 1961: *Sabethes* – Johann Batist von Spix, naturalist (in memoriam).
standfasti Peters, 1959: *Tripteroides* – Mr H.A. Standfast (collecting and rearing mosquitoes).
stanleyi Peters, 1963a: *Aedes* – Mr Stanley Hinton Christian (collected type specimens).
starckeae Stone & Knight, 1958: *Culex* – Miss Helle Starcke (discovered the need for this replacement name).
steffani Sirivanakarn, 1968: *Culex* – not mentioned, obviously Dr Wallace A. Steffan (support of field work mentioned in Acknowledgements).
stenei Thompson, 1956: *Aedes* – Professor Andrew E. Stene, entomologist (honorary).
stephensi Liston, 1901: *Anopheles* – not mentioned (J.W.W. Stephens) (see Kitzmiller, 1982).
stevensoni (Barraud, 1923b): *Aedes* – not mentioned.
stokesi Evans, 1929: *Aedes* – Dr Adrian Stokes (in memoriam; died of Yellow Fever while working on the experimental transmission of the virus to laboratory animals).
stonei Belkin, 1950: *Tripteroides* – Dr Alan Stone (encouragement and assistance).
stonei Knight & Laffoon, 1946: *Aedes* – Dr Alan Stone (contributions).
stonei Lane & Whitman, 1943: *Culex* – not mentioned, obviously Dr Alan Stone, per Acknowledgements.
stonei Vargas, 1956: *Psorophora* – Dr Alan Stone (honorary).
stonei Vargas & Martínez Palacios, 1953: *Wyeomyia* – Dr Alan Stone (honorary).
stoneorum Marks, 1977: *Aedes* – Alan Stone and his wife, Louise (appreciation, kindness; contributions).
stookesi Colless, 1955a: *Anopheles* – not mentioned (Valentine A. Stookes) (see Kitzmiller, 1982).
stricklandi Barraud, 1926: *Uranotaenia* – C. [Cyril] Strickland (collected holotype).
stricklandi (Edwards, 1912b): *Aedes* – Mr E.H. Strickland (replacement name for *australis* Strickland).
stricklandi Reid, 1965: *Anopheles* – '*aitkeni* larval type (fig. 2) of Strickland (1915[b])'. (Cyril Strickland) (see Kitzmiller, 1982).
strodei Root, 1926: *Anopheles* – Dr George K. Strode (assistance in obtaining mosquitoes) (see Kitzmiller, 1982).

suarezi Cova García & Sutil Oramas, 1976b: *Mansonia* – Sr Miguel A. Suárez F. (contributions).
suchariti (Miyagi & Toma, 1989): *Kimia* – Dr Supat Sucharit (support and encouragement).
sullivanae Baisas & Ubaldo-Pagayon, 1953: *Tripteroides* – 'The name is a compliment of the collector [T/Sgt. J.W. Enke] to Miss Dorothy B. Sullivan'.
sumethi Peyton & Rattanarithikul, 1970: *Uranotaenia* – Mr Sumeth Chunchulcsherm (technician-collector).
sunthorni Thurman, 1959: *Toxorhynchites* – Dr Sunthorn Srihongse, malaria control officer (contributions).
surcoufi (Theobald, 1912): *Aedes* – not mentioned, obviously M.J. Surcouf (author of the publication in which Theobald described the species, i.e. Surcouf, 1912).
sutili Cova Garcia & Pulido F., 1974: *Culex* – Sr Ezequiel Sutil Oramas (contributions).
svastii Thurman, 1959: *Topomyia* – Dr Svasti Daengsvang (leadership and appreciation).
symesi Edwards, 1928a: *Anopheles* – C.B. [Cecil Bernard] Symes (collected type specimens) (see Kitzmiller, 1982).
tamsi Edwards, 1934: *Culex* – Mr Tams, collector of Diptera, mentioned in the Introduction.
tanakai Miyagi & Toma, 2013: *Uranotaenia* – Dr Kazuo Tanaka (contributions).
tatoi Casal & García, 1971: *Culex* – Tato Pavlovsky (honorary).
tauffliebi Geoffroy & Hervé, 1976: *Culex* – Monsieur R. [Roger] Taufflieb, medical entomologist (honorary).
tauffliebi Rickenbach & Ferrara, 1965: *Aedes* – R. [Roger] Taufflieb (friend and colleague).
taylori Edwards, 1936: *Aedes* – A.W. Taylor (collected type specimens).
tchekedii de Meillon & Leeson, 1940: *Anopheles* – not mentioned (Tchekedi Khama) (see Kitzmiller, 1982).
teesdalei van Someren, 1954a: *Aedes* – Mr C. Teesdale (collected type specimens).
tendeiroi da Cunha Ramos, Ribeiro & de Barros Machado, 1992: *Eretmapodites* – Professor J. Leal da Silva Tendeiro (contributions).
termi Thurman, 1955: *Culex* – Dr Term Vejarasthira, malaria control officer (honorary).
terzii Edwards, 1941: *Culex* – Mr A.J.E. Terzi (mosquito illustrations).
testei Senevet & Abonnenc, 1939b: *Wyeomyia* – Médecin-Colonel Teste (honorary).
theileri Edwards, 1912d: *Anopheles* – not mentioned (probably Sir Arnold Theiler) (see Kitzmiller, 1982).
theileri Theobald, 1903a: *Culex* – Dr Theiler (collected type specimens).
thelmae Mattingly, 1970: *Heizmannia* – Miss Thelma Ford (prepared illustrations for the article).
theobaldi Barraud, 1934: *Armigeres* – not mentioned, obviously Frederic V. Theobald.
theobaldi (Dyar & Knab, 1906c): *Toxorhynchites* – '[Frederic V.] Theobald's species [*ferox*] is obviously distinct and we propose for it the name *theobaldi*.'
theobaldi Giles, 1901b: *Anopheles* – not specifically mentioned, obviously Frederic V. Theobald (see Kitzmiller, 1982).
theobaldi (Lane & Cerqueira, 1942): *Runchomyia* – F.V. Theobald (in memoriam).
theobaldi (Lane & Cerqueira, 1942): *Wyeomyia* – F.V. Theobald (in memoriam).
theobaldi (Lutz, 1904): *Culex* – not specifically mentioned, undoubtedly Frederic V. Theobald.
theobaldi (Taylor, 1914a): *Aedes* – Frederic V. Theobald (replacement name for *flavifrons* Theobald).
theresae Harbach, 2018: *Sabethes* – Theresa M. Howard (colleague; mosquito illustrations).
thibaulti Dyar & Knab, 1910: *Aedes* – J.K. Thiboult, Jr (collected type specimens).
thomasi Shannon, 1933a: *Anopheles* – Dr H. [Harold] Wolderstan Thomas (honorary) (see Kitzmiller, 1982).
thomsoni (Theobald, 1905e): *Aedes* – not mentioned.
thorntoni Dyar & Knab, 1907a: *Aedes* – W.F. Thornton (collected type specimens).
thurmanae Mattingly, 1958a: *Aedes* – Mrs Ernestine B. Thurman (contributions).
thurmanorum Bram, 1967a: *Culex* – D.C. Thurman and E.B. Thurman (Mrs C. Schwartzwalder) (contributions).
tigertti Scanlon & Peyton, 1967: *Anopheles* – Colonel William A. [*lapsus* for D.] Tigertt (honorary; promoted mosquito studies) (see Kitzmiller, 1982).
tiptoni Grjebine, 1953a: *Aedes* – Tipton (?Vernon J. Tipton) (collaboration in collecting mosquitoes).
tityae Slooff, 1961: *Tripteroides* – Mrs F.M.J. Metselaar-van Sluys, apparently wife of Dr D. Metsellaar, who collected specimens of the species.
toffaletii Baisas & Ubaldo-Pagayon, 1953: *Tripteroides* – Lt. P.J. Toffaleti (collected holotype and all but one paratype).
togoi (Theobald, 1907): *Aedes* – not mentioned, perhaps Togo Heihachiro, Japanese admiral, lived 27 Jan 1848–30 May 1934.
tonkingi Gebert, 1948: *Aedes* – Dr H.G. Tonking (appreciation for use of laboratory).
tonnoiri (Edwards, 1925): *Culiseta* – A. Tonnoir (collected holotype).
tournieri Senevet & Abonnenc, 1939a: *Culex* – Médecin-Colonel Tournier (collected holotype).
townsendi Stone, 1944: subspecies of *Trichoprosopon digitatum* (Rondani, 1848) – C.H.T. Townsend (collected type specimens).
toxopeusi Bonne-Wepster, 1948: *Aedes* – 'I take pleasure in naming this ... after its collector' Dr L.J. Toxopeus (mentioned in the Introduction).
tramazayguesi Duret, 1954b: *Culex* – Coronel Médico doctor M.M. Tramazaygues (in memoriam).
trapidoi Galindo, Blanton & Peyton, 1954: *Uranotaenia* – not specifically mentioned, surely Harold Trapido.
traubi Colless, 1965: *Culex* – not specifically mentioned, surely R. [Robert] Traub.
traubi Macdonald, 1960: *Armigeres* – not mentioned, certainly R. [Robert] Traub.

treubi (de Meijere, 1910): *Aedes* – not mentioned, perhaps M.W.F. Treub, politician.

trinkae Faran, 1979: *Anopheles* – 'my wife, Trinka' (see Kitzmiller, 1982).

tsengi Lien, 1968d: *Culex* – Dr Po Tsun Tseng (support and encouragement).

turkhudi Liston, 1901: *Anopheles* – Dr [Dnanesh A.] Turkhud (illustrated mosquitoes) (see Kitzmiller, 1982).

turneri Marks, 1963b: *Aedes* – Dr Alfred Jefferis Turner, lepidopterist (first to collect the species, the allotype).

tyagii Krishnamoorthy *et al.*, 2013: *Toxorhynchites* – Dr B.K. Tyagi (contributions).

uichancoi Baisas & Ubaldo-Pagayon, 1953: *Tripteroides* – Dr Leopoldo B. Uichanco (honorary).

urichii (Coquillett, 1906e): *Culex* – Mr F.W. Urich (collected holotype).

valeryi Gornostaeva, 2005: *Aedes* – in memoriam, 'The species is named in honor of Valery Nikolayevich Danilov (1940–1987), who discovered this species, but did not have time to describe it' [translated from Russian].

vanderwulpi Townson & Harbach, 2013: *Anopheles* – in memoriam, Frederik Maurits van der Wulp (1818–1899).

vanduzeei Dyar & Knab, 1906a: *Wyeomyia* – J.B VanDuzee (*sic*) (collected type specimens).

vaneedeni Gillies & Coetzee, 1987: *Anopheles* – Gideon van Eden (contributions) (see Service, 2010).

vanemdeni Martini, 1931a: *Aedes* – 'It was a coincidence that the Dresden Museum (colleague van Emden) offered me his South American mosquitoes' [translated from German], surely Wolfgang van Emden.

vanhoofi Wanson & Lebied, 1945: *Anopheles* – Docteur L. [Lucien] Van Hoof (honorary) (see Kitzmiller, 1982).

vanleeuweni (Edwards, 1927b): *Tripteroides* – Dr W. Docters van Leeuwen (collected type specimens).

vanoyei (Wolfs, 1948b): *Coquillettidia* – Prof. Paul van Oye (honorary).

vansomerenae Grjebine, 1986: *Mimomyia* – Madame E.C.C. van Someren (contributions).

vansomerenae Mattingly, 1955: subspecies of *Aedes vigilax* (Skuse, 1889) – not mentioned, certainly Mrs E.C.C. van Someren.

vansomereni Edwards, 1926a: *Culex* – not specifically mentioned, undoubtedly Dr V.G.L. van Someren.

vansomereni Hamon, 1962: *Eretmapodites* – E.C.C. van Someren (colleague).

vargasi Gabaldon, Cova-Garcia & Lopez, 1941b: *Anopheles* – Dr José María Vargas and Dr Luís Vargas (honorary) (see Kitzmiller, 1982).

vargasi Schick, 1970a: *Aedes* – Luis Vargas (honorary).

varuna Iyengar, 1924: *Anopheles* – Varuna, God of Rain (Sanskrit) (see Kitzmiller, 1982).

vattieri Geoffroy, 1971c: *Culex* – Mme Vattier-Bernard, entomologist (honorary).

veeniae McIntosh, 1975: *Aedes* – 'This species is dedicated to my wife [Veenia]' (assistance and encouragement).

ventrilloni Edwards, 1920b: *Culex* – Dr [Edouard] Ventrillon (collected paratypes).

vernoni van Someren, 1949a: *Orthopodomyia* – not mentioned, obviously Major Vernon van Someren, per Acknowledgements.

veruslanei Vargas, 1979: *Anopheles* – [Clarence] John Lane (facilitated the study of material procured from Belize) (see Kitzmiller, 1982).

vijayae Ramalingam, 1975: *Topomyia* – 'This species is dedicated to my wife, Vijaya.'

vimoli Thurman & Thurman, 1958: *Armigeres* – Dr Vimol Notananda (friend and co-worker).

vinckei de Meillon, 1942: *Anopheles* – Dr [Ignace H.] Vincke (collected and reared type specimens) (see Kitzmiller, 1982).

vinckei Hamon, Holstein & Rivola, 1957: *Culex* – I.H. [Ignace H.] Vincke, malariologist (honorary).

vinsoni Mattingly, 1953a: *Aedes* – not specifically mentioned, obviously Monsieur Vinson (collected holotype), per Acknowledgements.

vishnui Theobald, 1901a: *Culex* – not mentioned, obviously attributed to Vishnu, one of the principal deities of Hinduism.

vonplesseni (Dyar & Knab, 1906a): *Trichoprosopon* – Baron von Plessen (collected type specimens).

wadai Tanaka, Mizusawa & Saugstad, 1979: *Aedes* – not mentioned, probably Dr Yoshitake Wada, who 'allowed us to describe a new species which ... [he] obtained prior to us'.

wahlbergi (Edwards, 1936): *Coquillettidia* – Wahlberg (collected type specimens).

walcotti (Lane & Cerqueira, 1942): *Runchomyia* – Dr A.M. Walcott (collections of mosquitoes).

walkeri Theobald, 1901a: *Anopheles* – E.M. [Prof. Edmund Muton] Walker (collected type specimens) (see Kitzmiller, 1982).

walkeri (Theobald, 1901a): *Aedes* – not mentioned, probably Francis Walker, from reference made to a specimen labelled '*C. fasciatus*, F.'

wallacei Edwards, 1926a: *Aedes* – Dr H.G. Wallace (collected holotype).

walravensi Edwards, 1930a: *Anopheles* – Dr [Paul A.F.] Walravens (collected type specimens) (see Kitzmiller, 1982).

walukasi Belkin, 1962: *Culex* – Leo Walukas, 20th Malaria Survey Unit, Guadalcanal (honorary).

wannonii Cova Garcia & Sutil O., 1976a: *Culex* – Luis Wannoni Lander, sanitary engineer (honorary).

wansoni Edwards, 1941: *Eretmapodites* – [M.] Wanson (collected type specimens).

wansoni Wolfs, 1945b: *Culex* – not mentioned.

wardi Reinert, 1976b: *Aedes* – Ronald A. Ward (support).

wardi Sirivanakarn, 1977a: *Culex* – Dr Ronald A. Ward (honorary).

washinoi Lanzaro & Eldridge, 1992: *Aedes* – Dr Robert K. Washino (contributions).

wasselli Marks, 1947b: *Aedes* – Lieut. J.L. Wassell (collected holotype).
watasei Yamada, 1921: *Aedes* – Prof. S. Watasé (kindness; encouragement).
watsonii (Leicester, 1908): *Anopheles* – Dr [Malcolm] Watson (collected holotype) (see Kitzmiller, 1982).
watteni Lien, 1968b: *Aedes* – Capt. R.H. Watten (encouragement; support).
watti Edwards, 1920a: *Culex* – Dr W.G. Watt (collected holotype and several paratypes).
weindorferi (Edwards, 1926a): *Culiseta* – not mentioned.
wejchoochotei Taai & Harbach, 2015: *Anopheles* – Prof. Wej Choochote (in memoriam; contributions).
wellcomei Theobald, 1904b: *Anopheles* – not clearly indicated whether named for the Wellcome Research Laboratories or its founder Sir Henry Wellcome (see Kitzmiller, 1982).
wellingtonianus Alcock, 1912: *Anopheles* – Dr A.R. [Arthur Robartes] Wellington (collected and reared type specimens) (see Kitzmiller, 1982).
wellmanii (Theobald, 1905b): *Aedes* – Dr Creighton Wellman (collected holotype).
wendyae Service, 1959a: *Aedes* – '... named *wendyae* after my wife, who first found them'.
wepsterae Komp & Rozeboom, 1951: *Culex* – Mrs J. Bonne-Wepster (collected holotype with her husband).
werneri Baisas & Ubaldo-Pagayon, 1953: *Tripteroides* – F.G. Werner (collected type specimens with H. Hoogstraal).
weschei Edwards, 1935a: *Culex* – not mentioned, probably Walter Wesché.
whartoni Colless, 1965: *Culex* – not specifically mentioned, obviously Dr R.H. Wharton, per Acknowledgements.
whartoni Mattingly, 1965: *Aedes* – Dr R.H. Wharton (contributions).
whartoni Reid, 1963: *Anopheles* – Dr R.H. [Ronald Harry] Wharton (found the species to be a vector of *Wuchereria bancrofti*) (see Kitzmiller, 1982).
whitei Barraud, 1923d: *Culex* – Mr R. Senior-White (collected a specimen that 'probably belongs' to this species).
whitmani Lane & Cerqueira, 1942: *Sabethes* – Dr L. Whitman (collections of mosquitoes).
whitmorei Dunn, 1918: *Aedes* – Major Eugene R. Whitmore (collected type specimens).
whitmorei (Giles, 1904a): *Culex* – First Lieut. and Assist.-Surgeon Eugene R. Whitmore (collections of mosquitoes).
whittingtoni Belkin, 1962: *Culex* – Forrest B. Whittington (contributions).
wigglesworthi Edwards, 1941: *Aedes* – [Dr Vincent Brian] Wigglesworth (collected holotype).
wigglesworthi Edwards, 1941: *Culex* – [Dr Vincent Brian] Wigglesworth (collected holotype and several paratypes).
wilfredi Colless, 1965: *Culex* – Mr Wilfred Chellapah (former assistant, per Acknowledgements).
wilkersoni Reinert, 1990: *Aedes* – Dr Richard C. Wilkerson (contributions).
willmori (James, 1903): *Anopheles* – Lieut. [William Southwick] Willmore (collected holotype) (see Kitzmiller, 1982).
wilsoni (Barreto & Coutinho, 1944): *Mansonia* – not mentioned, probably Dr D. Bruce Wilson, '... we decided to keep the specific name proposed by [Nelson Cerqueira]...' [translated from Portuguese].
wilsoni Evans, 1934: *Anopheles* – Dr D. [Donald] Bagster Wilson (collected type specimens with his wife) (see Kitzmiller, 1982).
wilsoni Lane & Whitman, 1943: *Culex* – Dr D. Bruce Wilson (honorary).
wilsoni Macdonald, 1958: *Orthopodomyia* – Dr T. Wilson (encouragement; participation).
winkleri Belkin, 1962: *Culex* – Eric C. Winkler, 20th Malaria Survey Unit (honorary).
wolfsi Ribeiro, 2005: *Toxorhynchites* – Dr J. Wolfs (in memoriam; collected holotype).
woodi Edwards, 1922d: *Aedes* – R.C. Wood (collected holotype).
worontzowi Pessoa & Galvão, 1936: *Culex* – Snr Cesar Worontzow (assistant).
wysockii Belkin, 1953: *Uranotaenia* – F.B. Wysocki (collections made on Guadalcanal).
xui Dong, Zhou, Dong & Mao, 2007: *Anopheles* – Prof. Xu Jin-jiang (contributions) (see Service, 2010).
yamadai (Ôuchi, 1939): subspecies of *Toxorhynchites* (*Toxorhynchites*) *manicatus* (Edwards, 1921a) – Dr Shinichiro Yamada (in memoriam).
yamadai Sasa, Kano & Takahasi, 1950: *Aedes* – S. Yamada (collected type specimens).
yaoi Tung, 1955: *Culex* – Dr K.F. Yao (interest; support).
yeageri Baisas, 1935d: *Culex* – not mentioned.
yerburyi (Edwards, 1917): *Verrallina* – Lt.-Col. Yerbury (collected holotype).
yongi Miyagi, Toma & Ramalingam, 1991: *Topomyia* – Dr Yong Hoi Sen (contributions).
yovani van Someren, 1951: *Uranotaenia* – Yovani Senkubuge (collected type specimens).
yusafi (Barraud, 1931b): *Verrallina* – Mohamed Yusaf (laboratory assistant).
yvonneae Edwards, 1941: *Aedes* – Madame Schwetz (collector's wife).
zammitii (Theobald, 1903a): *Aedes* – Dr Zammit (collected type specimens).
zavortinki Schick, 1970a: *Aedes* – Thomas J. Zavortink (honorary).
zeteki Dyar, 1918e: *Culex* – J. Zetek (collected holotype).
zethus de Meillon & Lavoipierre, 1944: *Aedes* – Greek noun (m., third declension), a son of Jupiter by Antiopa, and brother of Amphion.
zhangi Gong, 1991: *Topomyia* – Benhua Zhang, entomologist (honorary).
ziemanni Grünberg, 1902: *Anopheles* – Herrn Dr med. Ziemann (contributions) (see Kitzmiller, 1982).

Section 5: Species-group Names Formed from Geographical Names

Of the 3689 extant species-group taxa, 3559 species and 130 subspecies (Section 1), that are currently recognized (fossil species, synonyms and *nomina dubia* are not included here), 638 have names that are formed from geographical names. These include the names of 583 species and 55 subspecies. Most of these names are adjectives in the nominative singular. The majority are formed by adding *-ensis* (L. masculine and feminine suffix denoting place, locality, country) to produce a latinized adjective from the geographical name. None of the names ends in *-ense*, the neuter form of the Latin suffix. Other adjectival names end in one or other of the following Latin suffixes (listed alphabetically by masculine form):

-alis, (f. *-alis* pertaining to, having the nature or quality of.
-anus (f. *-ana*, n. *-anum*), belonging to, pertaining to.
-arium (f. *-arium*, n. *-eria*), denoting place where, place for.
-icus (f. *-ica*, n. *-icum*), belonging to, pertaining to.
-inus (f. *-ina*, n. *-inum*), of, pertaining to.
-us (f. *-a*, n. *-um*), of, belonging to, pertaining to (a species from the geographical place).

Many geographical species-group names are nouns. Some retain the spelling of the geographical name, i.e. without the addition of a suffix, and are treated as latinized nouns in the nominative singular standing in apposition to the generic name, e.g. ***Aedes buenaventura*** Schick. ***Aedes bambiotai*** Geoffroy is unique among geographical names, formed by uniting the names of two geographical localities (Bambio + Tai). Other names are nouns formed by the addition of genitive case endings:

-ae, L. ending for first declension genitive singular nouns (feminine).
-i, L. ending for second declension genitive singular nouns (masculine and neuter).
-is, L. ending for third declension genitive singular nouns (masculine, feminine and neuter); also a Gr. patronymic suffix.

Nounal names ending in the third genitive singular suffix might be confused for adjectival species-group names that also end in *-is*. The adjectival ending (*-is* L. m., f.; n., *-e*) conveys the meaning of 'with' or 'having the nature of'. This ending is not used for names formed from geographical (or personal) names, but is often used for names derived from Latin and Greek words, e.g. ***acrostichalis*** Edwards (from *acro-* + ancient Greek στίχος 'row', 'line' + *-al*, after German acrostichal, + *-is*; meaning 'with acrostichal setae'). Some Latin adjectives are used directly as species-group names, e.g. ***australis*** (Erichson): *australis*, L. southern.

The formation of two geographical species names, ***antillummagnorum*** Dyar and ***palmarum*** Edwards, is unusual. They bear the Latin genitive plural suffixes *-orum* (m.) and *-arum* (f.), respectively. These names are constructed as plural nouns because they connote groups of islands. The former is in reference to the Greater Antilles, which comprise a grouping of islands in the Caribbean Sea that includes Cuba, Hispaniola (containing Haiti and the Dominican Republic), Puerto Rico, Jamaica and the Cayman Islands. The latter pertains to the 16 islands that make up the Greater Palm group of islands located off the eastern coast of northern Queensland, Australia.

Format

The list below includes all currently valid names of species and subspecies derived from geographical localities. The names are listed in alphabetical order. Each name is followed by authorship; genus or species with which the species or subspecies is combined, respectively; the locality or localities from which the name is derived, preceded by 'not stated (inferential)' if the derivation was not explicitly indicated by the author or authors; formation and composition; and type locality, as originally disclosed and, if applicable, corrected to the corresponding present-day toponym (place name).

Valid geographical species-group names in Culicidae

abyssinicus Edwards, 1941: subspecies of *Culex andersoni* Edwards, 1914b – not stated (inferential). Abyssinia; *-icus*, L. belonging to, like, of, pertaining to. Type locality: Addis Ababa, [Shawa], Ethiopia. The Ethiopian Empire, also known as Abyssinia, was a kingdom that spanned a geographical area covered by northern Ethiopia.

abyssinicus Ribeiro, 1991: subspecies of *Toxorhynchites brevipalpis* Theobald, 1901a – not stated (inferential).

See *abyssinicus* Edwards, 1941. Type locality: Keffa Province, Ethiopia.

adenensis Edwards, 1941: subspecies of *Aedes hirsutus* (Theobald, 1901a) – not stated (inferential). Aden; *-ensis*, L. suffix denoting place, locality, country. Aden Colony, located in the south of present-day Yemen, was a British Crown colony from 1937 to 1963. Type locality: Jebel Jihaf [Jebel Jehaf Plateau], Aden (present-day Yemen).

aegypti (Linnaeus, 1762): *Aedes* – not stated (inferential). Egypt; *-i*, L. ending for second declension genitive singular noun (masculine and neuter). *Culex aegypti* of Hasselquist (1757), currently known as *Aedes caspius* (Pallas), is mistakenly but irreconcilably attributed to Linnaeus. Type locality (neotype): Kuala Lumpur, Selangor, Malaya (Mattingly *et al*., 1962; validated by the International Commission on Zoological Nomenclature, 1964).

aequatorianna Leví-Castillo, 1953b: *Uranotaenia* – not stated (inferential). *Eequatus*, L. Ecuatoriana, belonging to, pertaining to Ecuador (Spanish *ecuatoriano*, f. *ecuatoriana* = English Ecuadorian). The country of Ecuador straddles the equator on the western coast of South America. Type locality: Babhoyo, Los Rios, Ecuador.

aequatorianna Leví-Castillo, 1954a: *Wyeomyia* – not stated (inferential). See *aequatorianna* Leví-Castillo, 1953b. Type locality: Hacienda Pichilingue, Los Rios, Ecuador.

africana Neveu-Lemaire, 1906: *Aedeomyia* – not stated (inferential). *Africanus* (f. *–a*), L. African, belonging to, pertaining to Africa. Type locality: Doufile, Central Africa [Uganda].

africana (Theobald, 1901d): *Mansonia* – not stated (inferential). See *africana* Neveu-Lemaire, 1906. Type locality: Asaba, [Eastern Provinces, Nigeria].

africanus (Theobald, 1901d): *Aedes* – not stated (inferential). See *africana* Neveu-Lemaire, 1906. Type locality: Duke Town and Bonny [Eastern Provinces, Nigeria].

agastyai Tewari & Hiriyan, 1992b: *Aedes* – not stated (inferential). Agastya; *-i*, L. ending for second declension genitive singular noun (masculine and neuter). Type locality: Agastya Hills, Western Ghats, southern India.

agranensis Singh & Prakash, 2008: *Lutzia* – Agra; *-ensis*, L. suffix denoting place, locality, country. Named 'after Dayalbagh at Agra, semi-arid region of India, which includes the type locality'. Type locality: Haryana, Rajasthan, Madhya Pradesh, Uttar Pradesh, India.

agrihanensis Bohart, 1957: *Aedes* – not stated (inferential). Agrihan; *-ensis*, L. suffix denoting place, locality, country. Type locality: southwest village, Agrihan, North Mariana Island. Agrihan is an island (currently uninhabited) in the Northern Mariana Islands, a US commonwealth in the Pacific Ocean.

akkeshiensis Tanaka, 1998: *Aedes* – not stated (inferential). Akkeshi; *-ensis*, L. suffix denoting place, locality, country. '... this species is known only from Akkeshi-cho, eastern Hokkaido', Japan. Type locality: Bekambeushi, Akkeshi-cho, Hokkaido, Japan.

alaskaensis (Ludlow, 1906c): *Culiseta* – not stated (inferential). Alaska; *-ensis*, L. suffix denoting place, locality, country. 'Described from ... collections from Fort Egbert, Alaska.' Type locality: Fort Egbert, [Eagle], Alaska, USA.

albertianus Edwards, 1941: *Culex* – not stated (inferential). Albert; *-anus*, L. belonging to, pertaining to. Type locality: Blukwa, Lake Albert, [Orientale], Belgian Congo (present-day Democratic Republic of the Congo).

algeriensis Theobald, 1903a: *Anopheles* – not stated (inferential). Algeria; *-ensis*, L. suffix denoting place, locality, country. Type locality: Algeria. See Kitzmiller (1982).

alongensis Venhuis, 1940: *Anopheles* – not stated (inferential). Along; *-ensis*, L. suffix denoting place, locality, country. Type locality: Baie d'Along [Hạ Long Bay], Tonkin [Red River Delta Region], Indochina [Vietnam]. See Kitzmiller (1982). See *alongi* Galliard & Ngu, 1947, *hongayi* Galliard & Ngu, 1947 and *tonkinensis* Galliard & Ngu, 1947.

alongi Galliard & Ngu, 1947: *Aedes* – not stated (inferential). Along; *-i*, L. ending for second declension genitive singular noun (masculine and neuter). Type locality: Baie d'Along [Hạ Long Bay], Tonkin [Red River Delta Region], Indochina [Vietnam]. See *alongensis* Venhuis, 1940, *hongayi* Galliard & Ngu, 1947 and *tonkinensis* Galliard & Ngu, 1947.

alorensis Bonne-Wepster & Brug, 1932: *Aedes* – not stated (inferential). Alor; *-ensis*, L. suffix denoting place, locality, country. Alor Island, largest island in the Alor Archipelago, an outlying island of Indonesia. Type locality: Alor Island, Lesser Sunda Islands, Netherland India [Indonesia].

alorensis Sirivanakarn, 1977a: *Culex* – not stated (inferential). See *alorensis* Bonne-Wepster & Brug, 1932. Type locality: Alor [Island], Indonesia.

amamiensis Tanaka, Mizusawa & Saugstad, 1979: subspecies of *Aedes japonicus* (Theobald, 1901a) – not stated (inferential). Amami; *-ensis*, L. suffix denoting place, locality, country. Type locality: Mount Yuwan, Amami Ôshima, Ryukyu Archipelago, [Amami Guntô, Japan].

amaniensis van Someren & Hamon, 1964: *Culex* – not stated (inferential). Amani; *-ensis*, L. suffix denoting place, locality, country. Type locality: Amani, Tanganyika (present-day United Republic of Tanzania).

amazonensis (Lutz, 1905): *Culex* – not stated (inferential). Amazon; *-ensis*, L. suffix denoting place, locality, country. '... no Rio Amazonas, abaixo de Manáos' (on the Amazon River below Manaus). Type locality: Manaus, Amazonas, Brazil.

amazonensis (Theobald, 1901c): *Mansonia* – not stated (inferential). See *amazonensis* (Lutz, 1905). Type locality: Lower Amazon, [Gurupa, Para, Brazil].

amazonica Cerqueira, 1960b: *Psorophora* – not stated (inferential). Amazon; *-icus* (f. *-ica*), L. belonging to, like, of, pertaining to. Type locality: 'Igarapé do Tarumã (cachoeira) Manaus, Amazonas', [Brazil].

amazonica Leví-Castillo, 1954a: *Wyeomyia* – not stated (inferential). See *amazonica* Cerqueira, 1960b. Type locality: Tena, Napo-Pastaza Province, Ecuador.

amazonicus Gordon & Evans, 1922: *Sabethes* – not stated (inferential). See *amazonica* Cerqueira, 1960b. Type locality: Mecapa, Manaus, Amazonas, Brazil.

ambodimanga da Cunha Ramos & Brunhes, 2004: *Uranotaenia* – Ambodimanga, 'nom de la localité de l'holotype'. Treated as a latinized noun in the nominative singular standing in apposition to the generic name. Type locality: 'à proximité du village d'Ambodimanga (S.-P. de Moramanga)', Madagascar.

amboinensis (Doleschall, 1857): *Toxorhynchites* – not stated (inferential). Amboina, Ambon; *-ensis*, L. suffix denoting place, locality, country. Type locality: Ambon Island, part of the Maluku Islands of Indonesia.

ambreensis Rodhain & Boutonnier, 1983a: *Aedes* – not stated (inferential). Ambre; *-ensis*, L. suffix denoting place, locality, country. Type locality: Montagne d'Ambre (Ambre Mountain), Madagascar.

ambremontis (Brunhes & Boussès, 2017): *Aedes* – Ambre, Fr. amber; *montis*, L. mountain. 'Name in relation to the place of collection, the "mountain of Amber", located in the north of Madagascar' [translated from French]. Type locality: forêt de la montagne d'Ambre [Amber Mountain Forest], Antsiranana Province, Madagascar.

ambremontis Brunhes & Hervy, 1995: *Orthopodomyia* – not stated (inferential). See *ambremontis* (Brunhes & Boussès, 2017). Type locality: Montagne d'Ambre [Ambre Mountain], Province de Diégo-Suarez (Antsiranana), Madagascar.

amiensis Peters, 1963e: *Uranotaenia* – not stated (inferential). Ami; *-ensis*, L. suffix denoting place, locality, country. Type locality: Ami, 10 miles NW of Maprik (Maprik District), Sepik District (East Sepik Province), New Guinea (present-day Papua New Guinea).

amurensis Maslov, 1964: *Culiseta* – not stated (inferential). Amur; *-ensis*, L. suffix denoting place, locality, country. Type locality: Central Amur regions [Amur Oblast, a federal subject of Russia], Ussuri River basin and southern Primorye; northeastern China.

andamanensis Barraud, 1934: *Orthopodomyia* – not stated (inferential). Andaman; *-ensis*, L. suffix denoting place, locality, country. Type locality: Andaman Islands.

andamanensis (Edwards, 1922a): *Verrallina* – not stated (inferential). See *andamanensis* Barraud, 1934. Type locality: Andaman Islands.

andavakae Doucet, 1950: *Uranotaenia* – not stated (inferential). Andavaka; *-ae*, L. ending for first declension genitive singular noun (feminine). Type locality: Cap Andavaka, [Mont Tsaratanana, Sambirano], Madagascar.

andinum Leví-Castillo, 1953c: *Trichoprosopon* – not stated (inferential). *Andinus* (f. *-ina*; n. *-inum*), NL pertaining to the Andes Mountains. Type locality: Macuchi, Cotopaxi Province, Ecuador.

andinus Leví-Castillo, 1954a: *Limatus* – not stated (inferential). See *andinum* Leví-Castillo, 1953c. Type locality: Valencia, Los Ríos Province, Ecuador.

andinus Osorno-Mesa, 1944: *Haemagogus* – not stated (inferential). See *andinum* Leví-Castillo, 1953c. Type locality: Fusagasugá, Cundinamarca Department, Colombia.

anggiensis Bonne-Wepster, 1937: *Aedes* – not stated (inferential). Anggi; *-ensis*, L. suffix denoting place, locality, country. Type locality: Anggi lakes, altitude 2000 m, Western part of New Guinea (present-day West Papua Province, Indonesia).

angkoris Klein, 1977: *Topomyia* – not stated (inferential). Angkor; *-is*, L. ending for third declension genitive singular noun (masculine, feminine and neuter). Angkor was the capital city of the Khmer Empire; an extensive archaeological site in Siem Reap Province of Cambodia, located far north of the type locality in Kampot Province. Type locality: Mont Bokor, km 15 of the route Kampot-Bokor, Kampot Province, Cambodia.

angolae Ribeiro & Ramos, 1974: subspecies of *Aedes durbanensis* (Theobald, 1903a) – not stated (inferential). Angola; *-ae*, L. ending for first declension genitive singular noun (feminine). Type locality: Moçâmedes, Angola.

angolensis da Cunha Ramos, 1985: *Uranotaenia* – not stated (inferential). Angola; *-ensis*, L. suffix denoting place, locality, country. Type locality: Dundo, Angola.

angolensis da Cunha Ramos & Ribeiro, 1992: *Eretmapodites* – not stated (inferential). See *angolensis* da Cunha Ramos, 1985. Type locality: Dundo (Carrisso Park), north-eastern Angola.

angolensis Ribeiro, 1992c: *Toxorhynchites* – not stated (inferential). See *angolensis* da Cunha Ramos, 1985. Type locality: Dundo, Angola.

ankaratrensis Brunhes & Hervy, 1995: *Orthopodomyia* – not stated (inferential). Ankaratra; *-ensis*, L. suffix denoting place, locality, country. 'This species has been collected only in the old volcanic massif of Ankaratra, whose altitude is between 1,800 and 2,643 meters' [translated from French]. Type locality: forest station of Manjakatompo (altitude 1800 m), Ambatolampy Sub-prefecture, Tananarive (Antananarivo) Province, Madagascar.

antalahaensis da Cunha Ramos & Brunhes, 2004: *Uranotaenia* – not stated. Antalaha; *-ensis*, L. suffix denoting place, locality, country. 'Name in relation to that of the town closest to the type locality' [translated from French]. Type locality: Antalaha is a commune and district in north-east Madagascar, which includes the type locality: Masoala National Park, Onive River.

antillummagnorum Dyar, 1928: *Culex* – not stated (inferential). Greater Antilles: *antillum*, L. Antilles; *magnorum*, L. genitive masculine plural of *magnus*, great. Type locality: Antonio de Los Baños, Cuba.

antipodea Dobrotworsky, 1962: *Culiseta* – not stated (inferential). Antipode, pl. Antipodes: *anti-*, Gr. opposite; *podos*, Gr. foot. Refers to the feet of inhabitants that are against each other on opposite sides of the world; thus, the plural 'antipodes' is used to refer to Australia and New Zealand because they are on the other side of the earth from Britain. Type locality: Cann River, Victoria, [Australia].

antipodeus (Edwards, 1920a): *Aedes* – not stated. See *antipodea* Dobrotworsky, 1962. Type locality: Karikari Bay, New Zealand.

aobae Belkin, 1962: *Aedes* – not stated (inferential). Aoba; *-ae*, L. ending for first declension genitive singular noun (feminine). Type locality: Crater Lake, Aoba Island, New Hebrides.

apoensis Baisas & Ubaldo-Pagayon, 1953: *Tripteroides* – Apo; *-ensis*, L. suffix denoting place, locality, country. 'The specific name is merely a reminder of the highest Philippine peak, Mt. Apo.' Type locality: Taglawig, Tagum, Davao, Mindanao Island, [Philippines].

apoensis Bohart & Farner, 1944: *Armigeres* – not stated (inferential). See *apoensis* Baisas & Ubaldo-Pagayon, 1953. Type locality: Sibulan River, 7000–8000 feet, Mt Apo, [Davao], Mindanao, Philippines.

arabiensis Patton, 1905: *Anopheles* – not stated (inferential). Arabia; *-ensis*, L. suffix denoting place, locality, country. '... found from Sheik Othaman to D'thala.' Type locality: Aden Hinterland, the remote areas of the Aden Protectorate (part of present-day Yemen) lying inland from the port of Aden. See Kitzmiller (1982).

arabiensis (Patton, 1905): subspecies of *Aedes vexans* (Meigen, 1830b) – not stated (inferential). See *arabiensis* Patton, 1905. Type locality: near Ulub Camp and Crater, Aden. Crater, also Kraytar, a district of the Aden Governorate, Yemen.

arizonensis Bohart, 1949: *Culex* – not stated (inferential). Arizona; *-ensis*, L. suffix denoting place, locality, country. Type locality: 8 miles S.E. Prescott, Arizona, [USA].

asiaticus Edwards, 1926c: subspecies of *Aedes pulcritarsis* (Rondani, 1872) – not stated (inferential). Asia, Asiatic; *-us*, L., of, pertaining to; *asiaticus*, L. Asian. Type locality: [Drosh, Chitral, N.W. Frontier Provinces, Pakistan].

asiaticus Leicester, 1903: *Anopheles* – not stated (inferential). *asiaticus*, L. Asian. Type locality: [Ampang Jungle, near Kuala Lumpur, Selangor], Malaya [British Malaya, 18th century–1946] (type locality located in present-day Peninsular Malaysia, or West Malaysia, of Malaysia). See Kitzmiller (1982).

assamensis Bhattacharyya, Tewari, Prakash, Mohapatra & Mahanta, 2004: *Verrallina* – Assam; *-ensis*, L. suffix denoting place, locality, country. 'The species is named after the state of Assam, in northeastern India where the collection site is situated.' Type locality: Dihing Reserve Forest, Dibrugarh District, Assam, India.

assamensis (Theobald, 1908b): *Aedes* – not stated (inferential). See *assamensis* Bhattacharyya *et al.*, 2004. Type locality: Sylhet, Assam, [India].

astridianus de Meillon, 1942: *Culex* – not stated (inferential). Astrida; *-anus*, L. belonging to pertaining to. Type locality: Astrida, Ruanda Urundi, Belgian Congo (present-day Democratic Republic of the Congo).

atacamensis González & Sallum, 2010: *Anopheles* – Atacama; *-ensis*, L. suffix denoting place, locality, country. 'The name, *atacamensis*, is derived from the name of the geographical locality where the specimens were first collected, the Atacama Region of Northern Chile.' Type locality: Los Perales, Provincia de Copiapó, III Región de Atacama, Chile. See Service (2010).

atlantica (Edwards, 1932b): *Culiseta* – Atlantic; *-us* (f. *-a*), L. belonging to, pertaining to. '... Atlantic Culicine exists in the Azores'. Type locality: Serra Gorda, Ilha da Pica, Azores.

atlanticus Adames, 1971: *Deinocerites* – not stated (inferential). See *atlantica* (Edwards, 1932b). Distributed on the 'Atlantic coast from the Canal Zone in Panama to the state of Aragua and the island of Orchila in Venezuela.' Type locality: La Boca, Ocumare de la Costa, Aragua, Venezuela.

atlanticus Dyar & Knab, 1906: *Aedes* – not stated (inferential). See *atlantica* (Edwards, 1932b). '... new name for the Atlantic coast form'. Type locality: New Jersey and Sanford, Florida, [USA].

australicus Dobrotworsky & Drummond, 1953: *Culex* – not stated (inferential). Australia; *-icus*, L. belonging to, like, of, pertaining to. Type locality: Melbourne [Victoria, Australia].

australiensis (Theobald, 1910c): *Aedes* – not stated (inferential). Australia; *-ensis*, L. suffix denoting place, locality, country. Type locality: Stannary Hills, Queensland, [Australia].

australis (Erichson, 1842): *Aedes* – not stated (inferential). *australis*, L. southern. Type locality: Vandiemensland [Tasmania, Australia]. Vandiemensland (Gr.) [Van Diemen's Land] was the original name used by most Europeans for the island of Tasmania, now part of Australia.

auyantepuiensis Harbach & Navarro, 1996: *Anopheles* – Auyantepui; *-ensis*, L. suffix denoting place, locality, country. '... we are formally naming and describing the "Auyan-Tepui Mesa form" of Zavortink (1973) as a new species'. Type locality: Auyantepui, Canaima National Park, Bolivar State, south-eastern Venezuela. Auyantepui is one of the largest tepuis (table-top mountains, mesas) in the Guiana Highlands. See Service (2010).

azaniae Bailly-Choumara, 1960: *Anopheles* – Azania; *-ae*, L. ending for first declension genitive singular noun (feminine). '... proposed, with reference to Azania, the former name of North-East Africa' [translated from French]. Type locality: British Somaliland Protectorate (present-day north-western Somalia). See Kitzmiller (1982).

babahoyensis Levi-Castillo, 1953d: *Culex* – Babahoyo; *-ensis*, L. suffix denoting place, locality, country. 'The name of **babahoyensis** has been received because the Babahoyo River is the main river system of the region and one of the main tributaries of the hydrographic system of the Guayas River, the main waterway of the coast of Ecuador' [translated from Spanish]. Type locality: Hacienda Mora, Juan Montalvo, Los Ríos Province [capital city Babahoyo], Ecuador.

bahama Dyar & Knab, 1906a: *Wyeomyia* – not stated (inferential). Bahama, treated as a latinized noun in the nominative singular standing in apposition to the generic name. Type locality: Tarpon Bay, Bahamas.

bahamensis Berlin, 1969a: *Aedes* – not stated (inferential). Bahama; *-ensis*, L. suffix denoting place, locality, country. Type locality: Village of Hunter, Grand Bahama, Bahamas.

bahamensis Dyar & Knab, 1906d: *Culex* – not stated (inferential). See *bahamensis* Berlin, 1969a. '… collected by Dr. T. H. Coffin in the Bahamas'. Type locality: Bahamas.

bahiensis Duret, 1969b: *Culex* – not stated (inferential). Bahía; *-ensis*, L. suffix denoting place, locality, country. Type locality: Uruçuca, Bahía State, Brazil.

balabacensis Baisas, 1936: *Anopheles* – not stated (inferential). Balabac; *-ensis*, L. suffix denoting place, locality, country. Type locality: Balabac, Balabac Island, Philippines. See Kitzmiller (1982).

balerensis Mendoza, 1947: *Anopheles* – Baler; *-ensis*, L. suffix denoting place, locality, country. Named for 'the town of Baler where the mosquito was first identified'. Type locality: Baler, Quezon, [Luzon], Philippines. See Kitzmiller (1982).

bambiotai Geoffroy, 1987: *Aedes* – Bambio + Tai, '… the specific name is a noun in apposition derived by combining the two very distinct geographic localities'. Type locality: Adiopodoume, near Abidjan, Côte d'Ivoire.

bananea Wolfs, 1958: *Aedes* – not stated (inferential). Banana; *-us* (f. *-a*), L. belonging to, pertaining to. Type locality: Banana, Bas-Congo, Congo Belge (present-day Democratic Republic of the Congo). Banana is a small Atlantic seaport.

bancoi Geoffroy, 1987: *Aedes* – not stated (inferential). Banco; *-i*, L. ending for second declension genitive singular noun (masculine and neuter). Type locality: Banco Forest, near Abijan, Côte d'Ivoire.

bandoengensis Brug, 1939: *Culex* – not stated (inferential). Bandoeng; *-ensis*, L. suffix denoting place, locality, country. 'One specimen was obtained from a larva, found in a bamboo-stump near Bandoeng (Java)'. Type locality: Bandoeng, Java, [Indonesia].

banksensis Maffi & Tenorio, 1977: *Culex* – not stated (inferential). Banks; *-ensis*, L. suffix denoting place, locality, country. Type locality: Qeteqavit, Gaua Island, Banks Islands, New Hebrides.

bannaensis Gong & Lu, 1995: *Topomyia* – not stated (inferential). Banna; *-ensis*, L. suffix denoting place, locality, country. Xishuangbanna or Sibsongbanna, abbreviated Banna, is an autonomous prefecture in south-western Yunnan Province of China. The tropical rainforest of the Xishuangbanna Nature Reserve is located in the prefecture. Type locality: Mengla County (21° 30′ N, 101° 35′ E, 600 m), Yunnan Province, [China].

barianensis James, 1911: *Anopheles* – not stated (inferential). Barian; *-ensis*, L. suffix denoting place, locality, country. Type locality: Barian, in the Murree Hills, Punjab, [India]. See Kitzmiller (1982).

belemensis Duret & Damasceno, 1955: *Culex* – not stated (inferential). Belém; *-ensis*, L. suffix denoting place, locality, country. Type locality: region of Belém, Pará, Brazil.

bengalensis Barraud, 1934: *Culex* – not stated (inferential). Bengal; *-ensis*, L. suffix denoting place, locality, country. Politically, Bengal is divided between Bangladesh and West Bengal State of India (resolution passed in August 2016 to rename the state 'Bangal' in Hindi and 'Bengal' in English). Type locality (lectotype): Nongpoh, Assam, India (Colless, 1965).

bengalensis Puri, 1930: *Anopheles* – not stated (inferential). See *bengalensis* Barraud, 1934. Type locality: Marianbarie, Bengal Terai, [near Sukna, West Bengal], India. See Kitzmiller (1982).

bengalensis Rosenberg & Evenhuis, 1985: *Toxorhynchites* – not stated (inferential). See *bengalensis* Barraud, 1934. Type locality: Chaklapunji Tea Estate, Chunaraghat, Sylhet District, Bangladesh.

benguetensis King, 1931: subspecies of *Anopheles lindesayi* Giles, 1900a – not stated (inferential). Benguet; *-ensis*, L. suffix denoting place, locality, country. Type locality: Baguio, Benguet Subprovince, Luzon, Philippines. See Kitzmiller (1982).

biscaynensis Zavortink & O'Meara, 1999: *Culex* – Biscayne; *-ensis*, L. suffix denoting place, locality, country. 'The specific name *biscaynensis* refers to Biscayne Bay that lies just off shore from the known distribution of this species.' Type locality: Fairchild Tropical Garden or Matheson Hammock County Park, Dade County, Florida, USA.

biskraensis Brunhes, 1999: *Aedes* – Biskra; *-ensis*, L. suffix denoting place, locality, country. Named 'with reference to the region of Biskra (Algeria) where it was collected' [translated from French]. Type locality: Sériana/Biskra, Algeria.

bogotanus Arnell, 1976: *Aedes* – Bogota; *-anus*, L. belonging to, pertaining to. 'The name *bogotanus* is a manuscript name that W. H. W. Komp placed on a number of specimens of this species that he collected near Bogota in May and June 1942.' Type locality: Ogamora, near Soacha, Cundinamarca, Colombia.

bokorensis Klein & Sirivanakarn, 1970: *Culex* – not stated (inferential). Bokor; *-ensis*, L. suffix denoting place, locality, country. Type locality: Bokor Hill, Kampot, Cambodia.

bolensis Edwards, 1936: *Aedes* – not stated (inferential). Bole; *-ensis*, L. suffix denoting place, locality, country. Type locality: Bole, Gold Coast (present-day Ghana).

boliviensis (Theobald, 1905f): *Anopheles* – not stated (inferential). Bolivia; *-ensis*, L. suffix denoting place, locality, country. Type locality: Songo, Bolivia. See Kitzmiller (1982).

boninensis Bohart, 1957: *Culex* – not stated (inferential). Bonin; *-ensis*, L. suffix denoting place, locality, country. Type locality: Chichi Jima, Bonin Islands. The Bonin Islands, or Ogasawara Islands, comprise an archipelago of more than 30 islands located approximately 1000 km south of Tokyo, Japan.

borneensis McArthur, 1949: *Anopheles* – not stated (inferential). Borneo; *-ensis*, L. suffix denoting place, locality, country. Type locality: Kamnong Naudu, Tambunan Plain, North Borneo (type locality located in present-day Sabah, Malaysia). See Kitzmiller (1982).

bougainvillensis (Belkin, 1962): *Mimomyia* – not stated (inferential). Bougainville; *-ensis*, L. suffix denoting place, locality, country. Type locality: Bougainville [Island], Solomon Islands (Bougainville Island is now the main island of the Autonomous Region of Bougainville of Papua New Guinea).

bougainvillensis Marks, 1947a: *Aedes* – not stated (inferential). See *bougainvillensis* Belkin, 1962. Type locality: Bougainville Island, Solomon Islands (Bougainville Island is now the main island of the Autonomous Region of Bougainville of Papua New Guinea).

bougainvillensis Steffan, 1979: *Culex* – not stated (inferential). See *bougainvillensis* Belkin, 1962. Type locality: Bougainville Island, North Solomon Islands, Papua New Guinea (PNG).

braziliensis (Chagas, 1907): *Anopheles* – not stated (inferential). Brazil; *-ensis*, L. suffix denoting place, locality, country. Type locality: Restricted to Lassance, Minas Gerais, Brazil by Belkin *et al.* (1971). See Kitzmiller (1982).

braziliensis Gordon & Evans, 1922: *Aedes* – not stated (inferential). See *braziliensis* (Chagas, 1907). Type locality: Mecapa, Amazonas, Brazil.

breedensis Muspratt, 1953: *Aedes* – not stated (inferential). Breede; *-ensis*, L. suffix denoting place, locality, country. Type locality: Worcester, Breede River Valley, Cape Province, South Africa.

britanniae Strickland & Chowdhury, 1931: subspecies of *Anopheles aitkenii* James, 1903 – not stated (inferential). Britannia; *-ae*, L. ending for first declension genitive singular noun (feminine). Britannia (female personification of Britain), probably applied with reference to the Asian colonies of the British Empire. See Kitzmiller (1982).

buenaventura Schick, 1970a: *Aedes* – not stated (inferential). Buenaventura, treated as a latinized noun in the nominative singular standing in apposition to the generic name. Type locality: Rio Raposo Virus Field Station, about 4 km W of nearest town Buenaventura, Valle, Colombia.

bukavuensis Wolfs, 1947b: *Culex* – Bukavu; *-ensis*, L. suffix denoting place, locality, country. 'The name we give to this new species originates in the native name of the city of Costermansville, that is to say Bukavu' [translated from French]. Type locality: Kawa, in Costermansville, Congo belge (present-day Democratic Republic of the Congo).

burjaticus Kuchartshuk, 1973: *Aedes* – Burjat; *-icus*, L. belonging to, like, of, pertaining to. Burjat, i.e. Buryat (usual translation of Russian бурят). Type locality: Khuzir, Buryat Mongol ASSR, USSR (type locality located in the autonomous Republic of Buryatia of Russia). The Buryat Autonomous Soviet Socialist Republic was created with the name Buryat-Mongol Autonomous Soviet Socialist Republic in 1923. 'Mongol' was removed from the name of the republic in 1958. It declared sovereignty in 1992 with the name Republic of Buryatia but remained an autonomous republic within the Russian Federation.

burpengaryensis (Theobald, 1905g): *Aedes* – not stated (inferential). Burpengary; *-ensis*, L. suffix denoting place, locality, country. Type locality: Burpengary, South Queensland, Australia.

bwamba van Someren, 1950c: subspecies of *Aedes tricholabis* Edwards, 1941 – not stated (inferential). Bwamba, treated as a latinized noun in the nominative singular standing in apposition to the generic name. Type locality: Bwamba [County], Toro District (now in Bundibugyo District), [Western Region], Uganda. The former Toro District constitutes the Toro Sub-region, which is coterminous with Toro Kingdom in Western Uganda. The Bundibugyo District was part of the Toro Sub-region prior to 1967.

bwambae White, 1985: *Anopheles* – Bwamba; *-ae*, L. ending for first declension genitive singular noun (feminine). 'Bwamba County, from where this locally abundant mosquito species takes its name.' Type locality: Mongiro, Semliki Forest, Bwamba County, [Bundibugyo District, Western Region], Uganda. See Service (2010).

bwambanus Edwards, 1941: subspecies of *Culex andersoni* Edwards, 1914b – not stated (inferential). Bwamba; *-anus*, L. belonging to, pertaining to. Type locality: Bwamba Pass, Ruwenzori (Rwenzori Mountains), [Bundibugyo District, Western Region], Uganda. See *ruwenzori* Haddow & van Someren, 1950.

cacharanus (Barraud, 1923e): *Aedes* – not stated (inferential). Cachar; *-anus*, L. belonging to, pertaining to. Cachar is an administrative district in Assam State of India. Type locality: Haflong, [Cachar Hills], Assam, India.

cairnsensis (Taylor, 1919): *Culex* – not stated (inferential). Cairns; *-ensis*, L. suffix denoting place, locality, country. Type locality: Cairns, Queensland, Australia.

cairnsensis Taylor, 1919: *Hodgesia* – not stated (inferential). See *cairnsensis* (Taylor, 1919). Type locality: Cairns, Queensland, Australia.

calabarensis Edwards, 1941: *Culex* – not stated (inferential). Calabar; *-ensis*, L. suffix denoting place, locality, country. Type locality: Calabar, [Eastern Province], Nigeria.

calcariae Marks, 1957a: *Aedes* – not stated (inferential). Calcaria; *-ae*, L. ending for first declension genitive singular noun (feminine). Type locality: 'Calcaria', near Lake Eliza, about 7 m south of Robe, South Australia, Australia.

caledonicus (Edwards, 1922d): *Tripteroides* – not stated (inferential). Caledonia; *-icus*, L. belonging to, like, of, pertaining to. Type locality: Hauailou, New Caledonia (a French territory comprising dozens of islands in the South Pacific).

camaronis Ribeiro, 1991: *Toxorhynchites* – Cameroon; *-is*, L. ending for third declension genitive singular noun (masculine, feminine and neuter). 'The species is named for the country of origin of the type series.' Type locality: Ototomo, [Ngoumou commune, Méfou-et-Akono Department, Center Region], Cameroon.

cameronensis Edwards, 1929b: subspecies of *Anopheles lindesayi* Giles, 1900a – not stated (inferential). Cameron; *-ensis*, L. suffix denoting place, locality, country. Type locality: Cameron's Highlands [Cameron Highlands District], [near Tapah, Perak], peninsular Malaysia. See Kitzmiller (1982).

campana Schick, 1970a: *Aedes* – not stated (inferential). Campana, treated as a latinized noun in the nominative singular standing in apposition to the generic name. Type locality: Cerro [hill] Campana [bell], Panama.

canaanensis Lane & Whitman, 1943: *Culex* – not stated (inferential). Canaã; *-ensis*, L. suffix denoting place, locality, country. Type locality: Vale do Canaã, S. João de Petrópolis, Espírito Santo State, Brazil.

canadensis (Theobald, 1901c): *Aedes* – not stated (inferential). Canada; *-ensis*, L. suffix denoting place, locality, country. Type locality (lectotype): Front Creek, near Toronto, Ontario, Canada (Belkin, 1968b).

canorii Floch & Abonnenc, 1945d: *Anopheles* – Canori; *-i*, L. ending for second declension genitive singular noun (masculine and neuter). 'This is certainly a new species that we describe as *A.* (*S.*) *canorii*, of the place of origin: Saut Canori (Approuague)' [translated from French]. Type locality: Saut Canori, Haut Approuague, French Guiana [overseas region and department of France]. See Kitzmiller (1982).

capensis de Meillon, 1935: subspecies of *Culex salisburiensis* Theobald, 1901c – not stated (inferential). Cape; *-ensis*, L. suffix denoting place, locality, country. Type locality: Cape Province. Houtbay, Pinelands and Palmiet River, Cape Province, South Africa.

capensis Edwards, 1924a: *Aedes* – not stated (inferential). See *capensis* de Meillon, 1935. Type locality: Oudebosch, Caledon Division, Cape Province, South Africa (incorrectly listed as 'Oudebosch, Caledon Division, Ovamboland' by Knight & Stone, 1977).

capricornii Lutz, 1904: *Haemagogus* – not stated. Capricorn; *-ii*, L. disused inappropriate genitive singular noun case-ending (masculine). Tropic of Capricorn, as revealed by Lutz (1905, as *Stegoconops capricorni*): 'This sylvatic species predominates in the zone crossed by the Tropic of Capricorn, thus the generic and derived name' [translated from Portuguese]. Type locality (neotype): Horto Florestal, Serra da Cantareira, São Paulo State, Brazil (Cerqueira & Lane, 1945).

caribeanus Galindo & Blanton, 1954: *Culex* –not stated (inferential). *caribe*, Sp. of the Caribbean region; *-anus*, L. belonging to, pertaining to. Type locality: Mojinga Swamp, Canal Zone, Panama.

carioca Lane & Whitman, 1951: *Culex* – not stated. *carioca*, Portuguese 'from Rio'. Type locality: Distrito Federal [Rio de Janeiro], [Guanabara], Brazil. The capital of Brazil was transferred from Rio de Janeiro to Brasília on 21 April 1960.

carolinensis Bohart & Ingram, 1946b: *Culex* – not stated (inferential). Caroline; *-ensis*, L. suffix denoting place, locality, country. Type locality: Fala Island, Truk Atoll, Caroline Islands. The Caroline Islands are an archipelago in the western Pacific Ocean that comprise the republics of Palau and the Federated States of Micronesia.

caspius (Pallas, 1771): *Aedes* – not stated (inferential). *caspius*, L. adjective. 'Versus mare Caspium [Towards the Caspian Sea]'. Type locality: Caspian Sea.

castrensis Edwards, 1922a: *Culex* – not stated (inferential). *castrum*, L. fort, garrison, castle; *-ensis*, L. suffix denoting place, locality, country. Type locality: Castle Rock, [North Kanara, Bombay], India.

catharinensis (da Costa Lima, Guitton & Ferreira, 1962): *Toxorhynchites* – not stated (inferential). Catarina, Catharina; *-ensis*, L. suffix denoting place, locality, country. Type locality: Brusque, Santo Catarina State, Brazil.

cauchensis Floch & Abonnenc, 1945b: *Culex* – Caux; *-ensis*, L. suffix denoting place, locality, country. 'We consider it a new species which we call *C. cauchensis* from the name of its place of origin Caux' [translated from French]. Type locality: Caux, French Guiana [overseas region and department of France].

ceramensis Sirivanakarn & Kurihara, 1973: *Culex* – not stated (inferential). Ceram; *-ensis*, L. suffix denoting place, locality, country. Type locality: Piru, Ceram, Maluku Province, Indonesia.

ceylonensis (Theobald, 1905h): *Tripteroides* – not stated (inferential). Ceylon; *-ensis*, L. suffix denoting place, locality, country. Type locality: Peradeniya, Ceylon (present-day Sri Lanka). The Democratic Socialist Republic of Sri Lanka was known as Ceylon until 1972.

ceylonica (Edwards, 1917): *Verrallina* – not stated (inferential). Ceylon; *-icus* (f. *-ica*), L. belonging to, like, of, pertaining to. Type locality: Colombo, Ceylon (present-day Sri Lanka).

champerico (Dyar & Knab, 1906a): *Psorophora* – not stated (inferential). Champerico, treated as a latinized noun in the nominative singular standing in apposition to the generic name. Type locality: Champerico, Guatemala.

changuinolae Galindo & Blanton, 1954: *Culex* – not stated (inferential). Changuinola; *-ae*, L. ending for first declension genitive singular noun (feminine). Changuinola is a city in the Bocas del Toro Province of Panama. Type locality: vicinity of Almirante, Bocas del Toro Province, Panama.

chathamicus (Dumbleton, 1962): *Opifex* – not stated (inferential). Chatham; *-icus*, L. belonging to, like, of, pertaining to. Type locality: South of Point Weeding near Waitangi, Chatham Island.

chemulpoensis Yamada, 1921: *Aedes* – not stated (inferential). Chemulpo; *-ensis*, L. suffix denoting place, locality, country. Type locality: Jinsen (Chemulpo), [South] Korea.

cherrapunjiensis (Natarajan, Rajavel & Jambulingam, 2016): *Aedes* – Cherrapunji; *-ensis*, L. suffix denoting place, locality, country. 'This species is named for the Cherrapunji area in the Khasi forest of the eastern hill range in Meghalaya state of India where it was collected.' Type locality: Cherrapunji area, Khasi Forest, Meghalya State, India.

chiriquiensis Komp, 1936b: *Anopheles* – not stated (inferential). Chiriqui; *-ensis*, L. suffix denoting place, locality, country. Type locality: Chiriqui volcano, north-western Panama. See Kitzmiller (1982).

chocoensis Porter & Wolff E., 2004: *Wyeomyia* – Chocó; *-ensis*, L. suffix denoting place, locality, country. 'The name *chocoensis* is derived from Chocó, the name of an Indian tribe in northwestern Colombia and the political division, Department of Chocó.' Type locality: Ensenada de Utriam Chocó, Colombia.

churchillensis Ellis & Brust, 1973: *Aedes* – not stated (inferential). Churchill; *-ensis*, L. suffix denoting place, locality, country. Type locality: Churchill, Manitoba, Canada.

coatzacoalcos Dyar & Knab, 1906d: *Uranotaenia* – Coatzacoalcos, treated as a latinized noun in the nominative singular standing in apposition to the generic name. 'The larvae were collected by the junior author at Santa Lucrecia, Mexico, in the valley of the river Coatzacoalcos.' Type locality: Santa Lucrecia, [Veracruz], Mexico.

colombiana Lane, 1945a: *Wyeomyia* – not stated (inferential). Colombia; *-anus* (f. *-ana*), L. belonging to, pertaining to. Type locality: Vegagrande, Intendencia de Meta, Restrepo, Colombia.

colombianus Adames, 1971: *Deinocerites* – not stated (inferential). See *colombiana* Lane, 1945a. Type locality: La Punta, Turbo, Antioquia, Colombia.

columbiae (Dyar & Knab, 1906a): *Psorophora* – Columbia; *-ae*, L. ending for first declension genitive singular noun (feminine). Columbia is a historical name used for the United States of America. The type series consists of specimens from five US states. Type locality: Grassymead, Virginia, USA.

commevynensis Bonne-Wepster & Bonne, 1920: *Culex* – not stated. Commevyne; *-ensis*, L. suffix denoting place, locality, country. Commevyne is apparently an alternative or older spelling of Commewijne, a district of Suriname. Type locality: Alkmaar, Suriname.

comorensis Brunhes, 1977a: *Culex* – not stated (inferential). Comores; *-ensis*, L. suffix denoting place, locality, country. Comores is the French spelling of Comoros. Type locality: near M'Jamaoué, Anjouan Island, Comoros.

comorensis Brunhes, 1977a: *Orthopodomyia* – not stated (inferential). See *comorensis* Brunhes, 1977a: *Culex*. Type locality: near Dembeni, Mayotte Island (Grande-Terre), [Mayotte], Comoros.

comorensis Brunhes, Le Goff & Geoffroy, 1997: *Anopheles* – not stated (inferential). See *comorensis* Brunhes, 1977a: *Culex*. Type locality: Sada, Mayotte Island (Grande-Tere), [Myotte], Comoros. See Service (2010).

comorensis da Cunha Ramos & Brunhes, 2004: *Uranotaenia* – Comore; *-ensis*, L. suffix denoting place, locality, country. 'Name in relation to the place of origin of the type series' [translated from French]. Type locality: Massif of La Grille (volcano), Grande Comore Island, Comoros.

consonensis (Reinert, 1973c): *Verrallina* – not stated (inferential). Con Son; *-ensis*, L. suffix denoting place, locality, country. Type locality: Con Son, Con Son Province, Vietnam.

continentalis Dobrotworsky, 1960a: *Aedes* – not stated. Continental, from L. *continent-*, *continens*, continuous mass of land, mainland; *-alis*, L. pertaining to, having the nature or quality of. The name apparently refers to the presence of the species on the continent of Australia. Type locality: Carpendeit, Victoria, Australia.

coonorensis Mattingly, 1981: *Tripteroides* – not stated (inferential). Coonor; *-ensis*, L. suffix denoting place, locality, country. Type locality: Sim's Park, Coonor, Madras State, India.

coppenamensis Bonne-Wepster & Bonne, 1920: *Culex* – not stated. Coppename; *-ensis*, L. suffix denoting place, locality, country. Type locality: Kabelstation (destroyed), [Sarakreek Municipality, Brokopondo District], Suriname. Kabelstation is a historical populated place in Brokopondo District that was destroyed when the large Brokopondo Reservoir was built between 1961 and 1964. However, the species name is apparently based on Coppename, a river in the Sipaliwini District, forming part of the boundary between the Coronie and Saramacca Districts. See *nicceriensis* Bonne-Wepster & Bonne, 1920 and *saramaccensis* Bonne-Wepster & Bonne, 1920.

corentynensis Dyar, 1920d: *Culex* – not stated. Corentyne; *-ensis*, L. suffix denoting place, locality, country. Corentyne is apparently an alternative or older spelling of the Courantyne River in Suriname. 'Dr. and Mrs. Bonne attached the name *corentynensis* to the specimens [two males in the US National Museum], but did not include this among their described species.' Type locality: Suriname.

costaricensis Adames & Hogue, 1970: *Deinocerites* – not stated (inferential). Costa Rica; *-ensis*, L. suffix denoting place, locality, country. Type locality: Boca del Rio Barranca, Hacienda Bonilla, Puntarenas Province, Costa Rica.

cozumelensis Diaz Nájera, 1966: *Aedes* – not stated (inferential). Cozumel; *-ensis*, L. suffix denoting place, locality, country. Type locality: San Miguel de Cozumel, Quintana Roo, Mexico.

crockeri Colless, 1955a: subspecies of *Anopheles gigas* Giles, 1901b – not stated. Crocker; *-i*, L. ending for second declension genitive singular noun (masculine and neuter). Crocker Range of North Borneo. Type locality: Saung-Saung, Sunsuran Trace, North Borneo (Sinsuran, Sabah, Malaysia). See Kitzmiller (1982).

cucphuongensis Vu, Nguyen, Tran & Nguyen, 1991: *Anopheles* – not stated (inferential). Cuc-Phuong; *-ensis*, L. suffix denoting place, locality, country. Type locality: Cuc-Phuong [National Forest], Hoang-Long, Ha Nam Ninh, Vietnam. See Service (2010).

cunabulanus Edwards, 1924b: *Aedes* – not stated. *cunabula*, L. cradle; *-anus*, L. belonging to, pertaining to. Apparently so named in reference to the collection of the type specimens in the Cradle Valley of Tasmania. Type locality: Cradle Valley, Tasmania, Australia.

curiche Adames, 1971: *Deinocerites* – not stated (inferential). Curiche, treated as a latinized noun in the nominative singular standing in apposition to the generic name. Type locality: El Naranjo, Curiche, El Choco, Colombia.

cuyanus Duret, 1968e: *Culex* – not stated (inferential). Cuyo; *-anus*, L. belonging to, pertaining to. 'In a study tour that we did at the end of 1953 in Cuyo, in the company of Dr. Juan F. R. Bejarano, we obtained a good number of *Culex* (*Culex*)' [translated from Spanish]. Cuyo is the name given to the mountainous area of west-central Argentina. Type locality: Jáchal, San Juan Province, Argentina.

daciae Linton, Nicolescu & Harbach, 2004: *Anopheles* – 'The specific name *daciae* is taken from the Latin name (*Dacia*, *-iae*, feminine) for the country corresponding principally to modern Romania.' Type locality: Budeni, Giurgiu, Muntenia, Romania. See Service (2010).

dahuricus Danilov, 1987b: *Aedes* – not stated? Dauria, or Dahuria; *-icus*, L. belonging to. like, of, pertaining to. Dauria is present-day Chita Province in Eastern Siberia. Type locality: Kubukhai, Onan District, Chita Province, Russia.

daisetsuzanus Tanaka, Mizusawa & Saugstad, 1979: subspecies of *Aedes impiger* (Walker, 1848) – not stated (inferential). Daisetsu; *-anus*, L. belonging to, pertaining to. Type locality: Yokomambetsu, Mt Daisetsu, Hokkaido, Japan.

daitensis Miyagi & Toma, 1981: *Aedes* – not stated (inferential). Daito; *-ensis*, L. suffix denoting place, locality, country. Type locality: Zaisho, Minami Daito Island, Ryukyus [Shimajiri District, Okinawa Prefecture], Japan.

daliensis (Taylor, 1916): *Aedes* – not stated (inferential). Daly; *-ensis*, L. suffix denoting place, locality, country. Type locality: Daly River, Northern Territory, Australia.

danaubento Mochtar & Walandouw, 1934: subspecies of *Anopheles gigas* Giles, 1901b – not stated (inferential). Danau Bento; combination treated as a latinized noun in the nominative singular standing in apposition to the generic name. 'The described larva from the swamp at Danau Bento' [translated from Dutch]. Type locality: Danau Bento, Westkust [West Coast], Sumatra. See Kitzmiller (1982).

dancalicus Corradetti, 1939b: *Anopheles* – not stated (inferential). Danakil; *-icus*, L. belonging to, like, of, pertaining to. 'In the month of January 1939 I found in the Dobi (southern Danakil Valley [Depression]) a new species of the anopheline *Neocellia* group for which I propose the name of *Anopheles* (*Neocellia*) *dancalicus* n. sp.' [translated from Italian]. The Danakil Depression is a desert plains area lying in the north of the Afar Region of Ethiopia, near the border with Eritrea. Type locality: Dobi, Danakil Depression, Ethiopia. See Kitzmiller (1982).

dandakaranyensis Natarajan, Rajavel & Jambulingam, 2017: *Uranotaenia* – Dandakaranya; *-ensis*, L. suffix denoting place, locality, country. 'This species is named for the Dandakaranya Forest, spread across the Araku Hills in Andhra Pradesh and the Bonda Hills in Odisha, India, where it was collected.' Type locality: Araku-Maliguda-Bora section, Araku District, Andhra Pradesh, India.

daudi Coluzzi, 1958: *Anopheles* – not stated (inferential). Daud; *-i*, L. ending for second declension genitive singular noun (masculine and neuter). 'The capture was performed in the village of Gololei Daud site, along the road from Mogadishu to the port Villaggio Duca degli Abruzzi, a short distance from Uebi Shabelle River' [translated from Italian]. Type locality: Golelei Daud, near Uebi Shabelle River, Somalia. See Kitzmiller (1982).

deccanus (Barraud, 1923b): *Aedes* – not stated (inferential). Deccan; *-us*, L. of, pertaining to. 'These [type male and female] were found in the same places and situations as *F. magna* stated above', i.e. 'Belgaum, and other places in the Bornbay Deccan'. Type locality: Belgaum, Bombay, Deccan, India.

denderensis Wolfs, 1949: *Aedes* – not stated (inferential). Dendere; *-ensis*, L. suffix denoting place, locality, country. 'The larvae of the mosquito that I describe here were found in a tree-hole on the Dendere peninsula, in Costermansville, Belgian Congo.' Type locality: Dendere Peninsula, Costermansville, [Kivu], Belgian Congo (present-day Democratic Republic of the Congo).

dibrugarhensis Bhattacharyya, Prakash, Mohapatra & Mahanta, 2004: *Uranotaenia* – Dibrugarh; *-ensis*, L. suffix denoting place, locality, country. 'The species is named after the Dibrugarh District, Assam, India, where the type locality is located.' Type locality: Dihing reserve forest, Soraipung, Dibrugarh District, Assam, India.

diengensis Brug, 1931c: *Culex* – not stated (inferential). Dieng; *-ensis*, L. suffix denoting place, locality, country. Diënghochebene [Dieng plateau]. '3 ♀ from a moat at the edge of Lake Merdodo (Diënghochebene [Dieng Plateau]), 2nd VI. 1929. – 2 ♂ from the Telaga Varna (on the bank, Diënghochebene), 3rd VI. 1929' [translated from German]. The Dieng Plateau is a marshy plateau in central Java, Indonesia. Type locality:

Merdodo Lake and Telaga Warna, Dieng Plateau, Java, Indonesia.

digoelensis Brug, 1932: *Culex* – not stated (inferential). Digoel; *-ensis*, L. suffix denoting place, locality, country. Digoel [Digul] River. 'A male and a female of this species were caught by Dr. H. de Rook near Upper Digoel River (South New Guinea).' The Digul is a major river in the south of Papua Province, Indonesia. Type locality: Upper Digoel River, Papua Province, Indonesia.

digoelensis Brug, 1934: *Tripteroides* – not stated (inferential). See *digoelensis* Brug, 1932. 'In 1932 Dr. de Rook and the author spent half a day in collecting mosquito larvae from *Nepenthes* in Tanah Merah on Upper Digoel River (South New Guinea).' Type locality: Tanah erah, Upper Digoel River, Papua Province, Indonesia.

dobodurus King & Hoogstraal, 1946b: *Aedes* – not stated (inferential). Dobodura; *-us*, L. of, pertaining to. Type locality: Dobodura, New Guinea (present-day Papua New Guinea).

doonii Wattal, Bhatia & Kalra, 1958: *Aedes* – not stated. Doon; *-ii*, L. disused inappropriate genitive singular noun case-ending (masculine). Type locality: Kalsi, Dehra Dun [Dehradun] District, Uttarakhand State, India. Uttarakhand State is in the Doon Valley on the foothills of the Himalayas. Although not stated by the authors, the species is apparently named for the Doon Valley.

dthali Patton, 1905: *Anopheles* – not stated (inferential). D'thala; *-i*, L. ending for second declension genitive singular noun (masculine and neuter). 'This anopheles [*sic*] was first found breeding in a spring near D'thala.' D'thala is a village on the northern slopes of the Jehaf range in Yemen. Type locality: near D'thala, Hardeba, Sulek and Nobat, Aden Hinterland (present-day Yemen). See Kitzmiller (1982).

dualaensis Brunhes, Le Goff & Geoffroy, 1999: *Anopheles* – Duala; *-ensis*, L. suffix denoting place, locality, country. 'We propose to call this new species *An. dualaensis* after its geographic origin.' Type locality: Douala [Duala] (4° 03′ N, 9° 42′ E), [Littoral Region], Cameroon. Douala is the largest city in Cameroon, and the capital of Cameroon's Littoral Region. See Service (2010).

dulongensis Gong & Lu, 1995: *Topomyia* – not stated (inferential). Dulong; *-ensis*, L. suffix denoting place, locality, country. Type locality: Dulong River, Gongshan County, Yunnan Province, China.

dundo da Cunha Ramos, 1993: subspecies of *Uranotaenia palmeirimi* de Meillon & Rebêlo, 1941 – not stated (inferential). Dundo, treated as a latinized noun in the nominative singular standing in apposition to the generic name. 'All 25 extant specimens of this new subspecies are female adults harvested by a "New Jersey" light trap installed on a balcony of a Dundo home in Northeast Angola' [translated from Portuguese]. Dundo, a former mining town, is now a city and provincial capital of Lunda Norte of Angola. Type locality: Dundo, Angola.

dundo da Cunha Ramos & Ribeiro, 1992: *Eretmapodites* – not stated (inferential). See *dundo* da Cunha Ramos, 1993. Type locality: Luachimo River, Dundo (Carrisso Park), Angola.

dundo Ribeiro, 1991: *Toxorhynchites* – Dundo. See *dundo* da Cunha Ramos, 1993. 'The species is named for the area of origin of the type series, in Northeastern Angola.' Type locality: Dundo, Angola.

durbanensis (Theobald, 1903a): *Aedes* – not stated (inferential). Durban; *-ensis*, L. suffix denoting place, locality, country. Durban is a coastal city in KwaZulu-Natal Province of eastern South Africa. Type locality: Durban, [KwaZulu-Natal, South Africa].

ealaensis Huang, 2004: *Aedes* – not stated (inferential). Eala; *-ensis*, L. suffix denoting place, locality, country. Type locality: Eala (0° 01′ N, 18° 30′ E), Cuvette-Centrale, Democratic Republic of the Congo. The Cuvette Centrale (Central Basin) is a region of forests and wetlands.

ebogoensis Rickenbach & Ferrara, 1965: *Aedes* – not stated (inferential). Ebogo; *-ensis*, L. suffix denoting place, locality, country. 'It is described from four males caught in Ebogo (Department of Nyong and Sanaga)' [translated from French]. Type locality: Ebogo, Nyong and Sanaga Department (present-day Lekie Department), [Center Region], Cameroon.

ecuadoriensis Berlin, 1969a: *Aedes* – not stated (inferential). Ecuador; *-ensis*, L. suffix denoting place, locality, country. Type locality: Leon River, Saraguro, Azuay, Ecuador.

eidsvoldensis Mackerras, 1927: *Aedes* – not stated (inferential). Eidsvold; *-ensis*, L. suffix denoting place, locality, country. 'Holotype: The unique female, taken biting by day in forest country at Eidsvold'. Eidsvold is a small town and locality in the North Burnett Region of South Queensland, Australia. Type locality: Eidsvold, Queensland, Australia.

elchoensis (Taylor, 1929): *Aedes* – not stated (inferential). Elcho; *-ensis*, L. suffix denoting place, locality, country. Elcho Island lies off the coast of Arnhem Land, Northern Territory, Australia. Type locality: Elcho Island, [Northern Territory], Australia.

elgonicus Edwards, 1941: subspecies of *Culex vansomereni* Edwards, 1926a – not stated (inferential). Elgon; *-icus*, L. belonging to, like, of, pertaining to. Type locality: Mt Elgon, Buginyanya, Uganda. Mount Elgon is an extinct shield volcano on the border of Uganda and Kenya.

embiensis (Huang, 1968a): *Verrallina* – not stated (inferential). Embi; *-ensis*, L. suffix denoting place, locality, country. Type locality: Embi Lakes, Popondetta, [Northern Province], New Guinea (present-day Papua New Guinea). Embi Lakes are four separate freshwater lakes located to the south of Dobodura and Popondetta.

embuensis Edwards, 1930a: *Aedes* – not stated (inferential). Embu; *-ensis*, L. suffix denoting place, locality, country. Type locality: Embu, Kenya. Embu is a town and headquarters of Eastern Province located north-east of Nairobi.

engarensis Kanda & Oguma, 1978: *Anopheles* – not stated (inferential). Engaru; *-ensis*, L. suffix denoting place, locality, country. Engaru is 'a town in the northeast part of Hokkaido Island'. Type locality: Engaru, Hokkaido Island, Japan. See Kitzmiller (1982).

epiroticus Linton & Harbach, 2005: *Anopheles* – *epeirotikos*, Gr. continental. 'In ancient Greek, the adjective "epiroticus" was used to describe people originating from the northwestern district of Epiros in mainland Greece. However, in modern Greek, the name has became [*sic*] synonymous with "from the mainland", and thus the ... [species] ... is named *A. epiroticus* to reflect its primary distribution in Southeast Asia' (Linton *et al.*, 2005). Type locality: Village Văn Đức A, An Trach Commune, Gia Rai District, Bac Liêu Province, Vietnam. See Service (2010).

eritreae Lewis, 1942: *Aedes* – not stated (inferential). Eritrea; *-ae*, L. ending for first declension genitive singular noun (feminine). Type locality: Acchele Guzai, Debre Axum, 20.3 km from Saganeiti along Dessie road, Eritrea (officially the State of Eritrea).

esmeraldasi (Levi-Castillo, 1955b): *Wyeomyia* – Esmeraldas; *-i*, L. ending for second declension genitive singular noun (masculine and neuter). '... because of the fact that it comes from Esmeraldas province it was named **Phoniomyia esmeraldasi** n. sp.' Type locality: Changuaral Island, Ancon de Sardinas Bay, Esmeraldas Province, Ecuador.

esoensis Yamada, 1921: *Aedes* – not stated. Eso; *-ensis*, L. suffix denoting place, locality, country. Type locality: Kanayama, [Hokkaido], Japan. Kanayama is a town in the Hokkaido region of Japan. Hokkaido, formerly known as Ezo (蝦夷, also spelled Yezo, Yeso or Yesso), is Japan's second largest island. Eso appears to be another transliteration of the Japanese; hence, *esoensis*.

ethiopicus Gillies & Coetzee, 1987: *Anopheles* – not mentioned (inferential). Ethiopia; *-icus*, L. belonging to, like, of, pertaining to. Type locality: Gambela, Illubabor Province, Ethiopia. Gambela is a city and separate woreda (district) in Ethiopia, and the capital of the Gambela Region, one of the nine ethnic divisions (kililoch) of the country. See Service (2010).

ethiopiensis Huang, 2004: *Aedes* – not mentioned (inferential). Ethiopia; *-ensis*, L. suffix denoting place, locality, country. Type locality: Ethiopia.

europaeus da Cunha Ramos, Ribeiro & Harrison, 2003: *Culex* – *europaeus*, L. European. 'The new species is named for Europe, its area of occurrence.' Type locality: Avô, Beira-Baixa Province, Portugal.

farakoensis Hamon, 1955: subspecies of *Culex grahamii* Theobald, 1910c – not stated (inferential). Farako; *-ensis*, L. suffix denoting place, locality, country. Type locality: Farako Waterfalls, Sikasso Circle and Subdivision, French Sudan (present-day Mali). Sikasso is the southern-most region of Mali.

fijiensis (Belkin, 1962): *Coquillettidia* – not stated (inferential). Fiji; *-ensis*, L. suffix denoting place, locality, country. Type locality: Koronivia, Viti Levu, Fiji.

fijiensis Marks, 1947a: *Aedes* – not stated (inferential). See *fijiensis* (Belkin, 1962). Type locality: Samabula, [Viti Levu], Fiji.

filipinae Manalang, 1930: *Anopheles* – not stated (inferential). Filipino, Sp. a native of the Philippine Islands; *-ae*, L. ending for first declension genitive singular noun (feminine). Type locality: Buluton stream, Tungcong Manga, Bulacan, Philippines. See Kitzmiller (1982).

floridensis Belkin, 1950: *Tripteroides* – not stated (inferential). Florida; *-ensis*, L. suffix denoting place, locality, country. Type locality: Halavo, Florida Islands, Solomon Islands. The Florida Islands, also known as the Nggela Islands, are a small island group in the Central Province of Solomon Islands nation.

fluminensis Root, 1927d: *Anopheles* – not stated (inferential). Fluminense; *-ensis*, L. suffix denoting place, locality, country. 'This locality ... was in the "baixada fluminense," that is, the coastal lowlands of the state of Rio de Janeiro.' Type locality: Itaperuna, Rio de Janeiro State, Brazil. See Kitzmiller (1982).

formosensis Yamada, 1921: *Aedes* – not stated (inferential). Formosa; *-ensis*, L. suffix denoting place, locality, country. Type locality: Kakubanzan, Formosa (present-day Taiwan, officially the Republic of China).

franciscanus McCracken, 1904: *Anopheles* – not stated (inferential). Francisco; *-anus*, L. belonging to, pertaining to. Type locality: Felt Lake, near Stanford University, [Santa Clara County], California, USA. McCracken collected the specimens 'in the vicinity of Stanford University ... situated near the base of the San Francisco peninsula'. See Kitzmiller (1982).

freetownensis Evans, 1925c: *Anopheles* – not stated (inferential). Freetown; *-ensis*, L. suffix denoting place, locality, country. Type locality: Kissy Bridge, near Freetown, Sierra Leone. See Kitzmiller (1982).

futunae Belkin, 1962: *Aedes* – not stated (inferential). Futuna; *-ae*, L. ending for first declension genitive singular noun (feminine). Type locality: Mu'a, Alofi Island, Horne Islands. Alofi Island belongs to the French overseas collectivity of Wallis and Futuna, a pair of volcanic islands that comprise the Horne Islands, also called Futuna Islands.

gabonensis Rahola, Makanga & Paupy, 2014: *Anopheles* – Gabon; *-ensis*, L. suffix denoting place, locality, country. 'The species name *gabonensis* refers to Gabon, the country where the species was discovered.' Type locality: La Lékédi Park, Bakoumba, Gabon.

gambiae Giles, 1902: *Anopheles* – not stated (inferential). Gambia; *-ae*, L. ending for first declension genitive singular noun (feminine). Type locality: Gambia Valley, West Africa (Gambia). The Gambia River originates in the Republic of Guinea, flows north-westwards through southern Senegal and westwards through The Gambia into the Atlantic Ocean. See Kitzmiller (1982).

gandaensis Huang, 2004: *Aedes* – not stated (inferential). Ganda; *-ensis*, L. suffix denoting place, locality, country. Type locality: Ganda, Kenya.

gediensis Edwards, 1941: subspecies of *Culex weschei* Edwards, 1935a – not stated (inferential). Gede; *-ensis*, L. suffix denoting place, locality, country. Type locality: Gede (near Mombasa), Kenya.

georgianus King, 1939: *Anopheles* – not stated (inferential). Georgia; *-anus*, L. belonging to, pertaining to. Type locality: near Quitman, Brooks County, Georgia, USA. The state of Georgia, established in 1733, was the last of the original Thirteen Colonies. See Kitzmiller (1982).

gombakensis Mattingly, 1959: *Aedes* – Gombak; *-ensis*, L. suffix denoting place, locality, country. 'This species is named from Ulu Gombak where it was collected.' Type locality: Ulu Gombak, Malaysia.

gonguoensis Gong & Lu, 1986: *Aedes* – not stated (inferential). Gonguo; *-ensis*, L. suffix denoting place, locality, country. Type locality: Gonguo, Yunnan Province, China.

guadeloupensis (Dyar & Knab, 1906c): *Toxorhynchites* – not stated (inferential). Guadeloupe; *-ensis*, L. suffix denoting place, locality, country. Type locality: Guadeloupe, West Indies. Guadeloupe, an archipelago of over a dozen islands, is an insular region of France located in the Leeward Islands, part of the Lesser Antilles in the Caribbean.

guamensis Farner & Bohart, 1944: *Aedes* – not stated (inferential). Guam; *-ensis*, L. suffix denoting place, locality, country. Type locality: Mata, Guam, [Mariana Islands]. Guam is a territory of the USA, in the Western Pacific.

guatemala Berlin, 1969a: *Aedes* – not stated (inferential). Guatemala, treated as a latinized noun in the nominative singular standing in apposition to the generic name. Type locality: Yepocapa, Chimeltenango, Guatemala.

guatemala Dyar & Knab, 1906a: *Wyeomyia* – not stated (inferential). See *guatemala* Berlin, 1969a. Type locality: Trece Aguas, Alta Verapaz, Guatemala.

guatemalensis de Leon, 1938: subspecies of *Anopheles parapunctipennis* Martini, 1932b – not stated (inferential). Guatemala; *-ensis*, L. suffix denoting place, locality, country. Type locality: Cumbre del Aire, Totonicapán, Gualemala. See Kitzmiller (1982).

guayasi Leví-Castillo, 1953b: *Culex* – not stated. Guayas; *-i*, L. ending for second declension genitive singular noun (masculine and neuter). Probably in reference to the Guayas River, which flows through the type locality. Type locality: Babahoyo, Los Ríos Province, Ecuador. See *guayasi* Leví-Castillo, 1954a.

guayasi Leví-Castillo, 1954a: *Limatus* – not stated (inferential). See *guayasi* Leví-Castillo, 1953b. Type locality: El Empalme, Guayas Province, Ecuador.

guerrero Berlin, 1969a: *Aedes* – not stated (inferential). Guerrero, treated as a latinized noun in the nominative singular standing in apposition to the generic name. Type locality: 21 miles south of Chilpancingo, Guerrero, Mexico.

guizhouensis Chen & Zhao, 1985: *Culex* – not stated (inferential). Guizhou; *-ensis*, L. suffix denoting place, locality, country. Type locality: Guiyang, Guizhou Province, China.

guyanensis Clastrier, 1970a: *Culex* – not stated (inferential). Guyane; *-ensis*, L. suffix denoting place, locality, country. Type locality: Forêt de Cabassou, près Cayenne (Forest of Cabassou, near Cayenne), French Guiana [overseas region and department of France].

gyirongensis Ma, 1982: subspecies of *Aedes lasaensis* Meng, 1962 – not stated (inferential). Gyirong; *-ensis*, L. suffix denoting place, locality, country. Type locality: Gyirong, Xizang, China. The Xizang Autonomous Region, Tibet Autonomous Region, called Tibet or Xizang for short, is a province-level autonomous region of the People's Republic of China.

hailarensis Xu & Luo, 1998: *Anopheles* – not stated (inferential). Hailar; *-ensis*, L. suffix denoting place, locality, country. Type locality: Hailaer, Nei Mongol Autonomous Region, China. Hailar is an urban district in northeastern Inner Mongolia (officially the Nei Mongol Autonomous Region or Inner Mongolia Autonomous Region), located in the north of China and containing most of China's border with Mongolia. See Service (2010).

hainanensis Chen, 1977: *Culex* – not stated (inferential). Hainan; *-ensis*, L. suffix denoting place, locality, country. Type locality: Diaolaoshan, Hainan, China. Hainan is an island province and southernmost point of China.

hakusanensis Yamaguti & Tamaboko, 1954: *Aedes* – not stated (inferential). Hakusan; *-ensis*, L. suffix denoting place, locality, country. Type locality: Mt Hakusan, Ishikawa Prefecture, [Honshu], Japan. Mount Hakusan, or Mount Haku, a potentially active volcano, is one of Japan's Three Sacred Mountains.

hebrideus Edwards, 1926a: *Aedes* – not stated (inferential). Hebrides; *-us*, L. of, pertaining to. Type locality: Hog Harbour, Espiritu Santo, New Hebrides (Vanuatu). New Hebrides was the Scottish colonial name for the island group in the South Pacific that is now the nation of Vanuatu.

heiheensis Ma, 1981b: *Anopheles* – not stated. Heihe; *-ensis*, L. suffix denoting place, locality, country. Type locality: Aihui, Heilungjiang Province, China. Aihui is an administrative district and the seat of the prefecture-level city of Heihe, Heilongjiang Province, People's Republic of China. See Service (2010).

himalayensis Edwards, 1922a: *Heizmannia* – not stated (inferential). Himalaya; *-ensis*, L. suffix denoting place, locality, country. Type locality: Darjeeling, [West Bengal], India. Darjeeling is a town in the Himalayan foothills in West Bengal State of India.

hinglungensis Chu, 1957: *Culex* – not stated (inferential). Hing-lung; *-ensis*, L. suffix denoting place, locality, country. Type locality: Hing-lung, Hainan Island, China.

hispaniola (Theobald, 1903a): subspecies of *Anopheles cinereus* Theobald, 1901a – not stated (inferential). *Hispaniola*, L. (Spanish: Española) little Spain. Type locality: Spain and Santa Cruz, Tenerife, Canary Islands. See Kitzmiller (1982).

hogsbackensis Huang, 2004: *Aedes* – not stated (inferential). Hogsback; *-ensis*, L. suffix denoting place, locality, country. Type locality: Hogsback, [Eastern] Cape Province, South Africa. Hogsback is a town in the Amathole Mountains.

hokkaidensis Tanaka, Mizusawa & Saugstad, 1979: *Aedes* – not stated (inferential). Hokkaido; *-ensis*, L. suffix denoting place, locality, country. Type locality: Ônuma-kôen, Hokkaido, Japan. Hokkaido is the northernmost of Japan's main islands.

hollandius King & Hoogstraal, 1946b: *Aedes* – not stated (inferential). Hollandia; *-us*, L. of, pertaining to. Type locality: Mt Dafonsero, Cyclops Range, Hollandia area, Netherlands New Guinea (present-day West Papua Province, Indonesia).

hongayi Galliard & Ngu, 1947: *Uranotaenia* – not stated. Hon Gay; *-i*, L. ending for second declension genitive singular noun (masculine and neuter). Type locality: Baie d'Along [Hạ Long Bay], Tonkin [Red River Delta Region], Indochina [Vietnam]. Hon Gay (Hồng Gai) is a town and port of Ha Long Bay. See *alongensis* Venhuis, 1940, *alongi* Galliard & Ngu, 1947 and *tonkinensis* Galliard & Ngu, 1947.

hungaricus Mihályi, 1955: *Aedes* – not stated (inferential). Hungary; *-icus*, L. belonging to. like, of, pertaining to. Type locality: Budafok, Háros-sziget, Hungary. Háros-sziget (Háros Island) is an island in the Danube River at Budapest.

hyrcanus (Pallas, 1771): *Anopheles* – not stated. Hyrcania; *-us*, L., of, pertaining to. Type locality: [near Caspian Sea]. Hyrcania is a historical region (country or province) encompassing land southeast of the Caspian Sea in modern-day Iran. See Kitzmiller (1982).

idahoensis (Theobald, 1903a): subspecies of *Aedes spencerii* (Theobald, 1901c) – not stated (inferential). Idaho; *-ensis*, L. suffix denoting place, locality, country. Type locality: Market Lake, Idaho, USA.

idjenensis Brug, 1934: *Aedes* – not stated. Idjen; *-ensis*, L. suffix denoting place, locality, country. Type locality: Ongop Ongop, Idjen, Java, Indonesia. The Ijen volcano complex is a group of composite volcanoes in the Banyuwangi Regency of East Java. The crater of Ijen volcano has a lake 1 km wide, 200 m deep.

iguassuensis Barbosa, da Silva & Sallum, 2007: *Mansonia* – Iguassu; *-ensis*, L. suffix denoting place, locality, country. 'The name iguassuensis [*sic*] is derived from Iguau [*sic*], the type locality of the new species.' Type locality: Curitiba (25° 25′ 04″ S 49° 14′ 30″ W), Paraná, Brazil. The Iguassu Falls, Iguazu Falls, Iguazú Falls or Iguaçu Falls are waterfalls (the largest in the world) of the Iguazu River on the border of the Misiones Province of Argentina and the Paraná State of Brazil.

indiana Edwards, 1930b: *Mansonia* – not stated. *indianus* (f. *indiana*), L. Indian. Type locality: Bandjaran, South Prenanger, Java, Indonesia. Preanger, Priangan or Parahyangan is the mountainous region in West Java Province. Edwards apparently based the name on females collected in India.

indica (Edwards, 1920a): subspecies of *Culiseta alaskaensis* (Ludlow, 1906c) – not stated (inferential). India; *-icus* (f. *-ica*), L. belonging to, like, of, pertaining to. Type locality: Bakloh, [Chamba District, Himachal Pradesh State], India. Bakloh is a cantonment town, a hill station, 4584 feet (1397 m) above sea level.

indica (Theobald, 1907): *Verrallina* – not stated (inferential). *indicus* (f. *indica*), L. Indian. Type locality: Dehra Dun (Dehradun), [Uttarakhand State], India. Dehradun is the capital of the Indian state of Uttarakhand, near the Himalayan foothills. **Note**: *indica* (Theobald, 1905f), a species of *Heizmannia* described from Singapore, is probably derived from Latin *indicum*, meaning indigo.

indicus (Barraud, 1929a): *Tripteroides* – not stated (inferential). See *indica* (Theobald, 1907). Type locality: Sukna, [Darjeeling District, West] Bengal, India.

indonesiae Mattingly, 1958a: *Aedes* – not stated (inferential). Indonesia; *-ae*, L. ending for first declension genitive singular noun (feminine). Type locality: Island in Bay of Batavia (Jakarta Bay), Java, Indonesia. Batavia was the name of the capital city of the Dutch East Indies and corresponds to the present-day Jakarta.

ininii Senevet & Abonnenc, 1938: *Anopheles* – not stated (inferential). Inini; *-i*, L. ending for second declension genitive singular noun (masculine and neuter). Type locality: Mines at Saint Elie, French Guiana [overseas region and department of France]. Saint Elie is a gold mine located in the former arrondissement of Inini, which was an inland territory of French Guiana, administered separately between July 1930 and March 1946, after which all of French Guiana became a department of France. See Kitzmiller (1982).

irianensis Miyagi & Toma, 1997: *Topomyia* – not stated (inferential). Irian; *-ensis*, L. suffix denoting place, locality, country. Type locality: Arso, Irian Jaya (present-day Western New Guinea), Indonesia. Western New Guinea is the Indonesian part of the island of New Guinea, which was known previously as Irian Jaya (1973–2001) and Papua (2002–2003).

iriomotensis (Tanaka & Mizusawa, 1973): *Verrallina* – not stated (inferential). Iriomote; *-ensis*, L. suffix denoting place, locality, country. Type locality: Iriomote Island, Yaeyama Guntô, [Okinawa Prefecture], Ryukyu Islands, Japan. Iriomote Island (Iriomote-jima) is the largest of the Yaeyama Islands and the second largest in the Okinawa Prefecture.

jacobinae Serafim & Davis, 1933: *Aedes* – not stated (inferential). Jacobina; *-ae*, L. ending for first declension genitive singular noun (feminine). Type locality: Jacobina, Bahia, Brazil. Jacobina is a city in the Centro Norte Baiano mesoregion of Bahia State.

jalisco Berlin, 1974: *Culex* – not stated (inferential). Jalisco, treated as a latinized noun in the nominative singular standing in apposition to the generic name. Type locality: near El Tuito, 16 km from Boca de

Tomatlan (on highway from Puerto Vallarta to Manzanillo), Jalisco, Mexico.

jamaicensis (Theobald, 1901a): *Psorophora* – not stated (inferential). Jamaica; *-ensis*, L. suffix denoting place, locality, country. Type locality: Spanish Road, Kingston, Jamaica.

japonicus (Theobald, 1901a): *Aedes* – not stated (inferential). *japonicus*, NL. of or relating to Japan. Type locality: Tokyo, [Honshu], Japan.

japonicus Yamada, 1918: subspecies of *Anopheles lindesayi* Giles, 1900a – not stated (inferential). See *japonicus* (Theobald, 1901a). Type locality: [Kanayama], Hokkaido [and Mount Myogi], Japan. See Kitzmiller (1982).

javaensis Miyagi & Toma, 1995: *Topomyia* – not stated (inferential). Java; *-ensis*, L. suffix denoting place, locality, country. Type locality: Mount Cibodas (1000 m), Bogor Botanical Garden, West Java, Indonesia. West Java is a province on the island of Java, extending south and east of Jakarta.

javanensis Bonne-Wepster, 1934: *Culex* – not stated (inferential). See *javaensis* Miyagi & Toma, 1995. Type locality: Tjisarua (Buitenzorg), [West] Java, Indonesia. Tjisarua is a town in West Java Province. Buitenzorg (Dutch, present-day Bogor) is a city located approximately 60 km south of Jakarta.

jebudensis Froud, 1944: *Anopheles* – not stated (inferential). Jebu; *-ensis*, L. suffix denoting place, locality, country. Type locality: Ijebu-ode, Southern Nigeria. Jebu (also known as Ijebu or Geebu) was a Yoruba kingdom in pre-colonial Nigeria. Ijebu-Ode is a town in Ogun State, south-western Nigeria. See Kitzmiller (1982).

jeyporiensis James, 1902: *Anopheles* – not stated (inferential). Jeypor; *-ensis*, L. suffix denoting place, locality, country. Type locality: Nagpur, Jeypur State (Nagpur is located in present-day Maharashtra State), India. Jeypur (Jaipur) State was a princely state of India from 1128 to 1948, located in the Central Provinces of British India that covered parts of present-day Madhya Pradesh, Chhattisgarh and Maharashtra states. See Kitzmiller (1982).

jinhongensis Dong, Dong & Zhou, 2003: *Uranotaenia* – not stated (inferential). Jinhong; *-ensis*, L. suffix denoting place, locality, country. Type locality: Jinhong County, Yunnan Province, China. Jinhong (Jinghong) is a city in the far south of Yunnan Province.

jinjaensis Edwards, 1941: subspecies of *Culex aurantapex* Edwards, 1914b – not stated (inferential). Jinja; *-ensis*, L. suffix denoting place, locality, country. Type locality: Jinja, Uganda. Jinja is in Jinja District, located on the north shore of Lake Victoria.

johorensis (Reinert, 1974): *Verrallina* – not stated (inferential). Johore; *-ensis*, L. suffix denoting place, locality, country. Type locality: Pontian Besar, Johore (Johor), Malaysia. Johor is a state in southern peninsular Malaysia linked to Singapore by causeways.

joloensis (Ludlow, 1904b): *Armigeres* – not stated (inferential). Jolo; *-ensis*, L. suffix denoting place, locality, country. Type locality: Jolo, Jolo Island, [Sulu Province], Philippines. Jolo is a volcanic island in the south-western Philippines and is the primary island of Sulu Province, the capital of which is the same as the name of the island.

judaicus Edwards, 1926c: *Culex* – not stated (inferential). Judea, Judæa; *-icus*, L. belonging to, like, of, pertaining to. Type locality: Palestine (Israel and State of Palestine). Palestine was a geopolitical region carved out of Ottoman Southern Syria after World War I and existed under British administration from 1920 until 1948. 'Judea' was the name used by English-speakers until the Jordanian occupation of the area in 1948. The name Judea is a Greek and Roman adaptation of 'Judah', which originally encompassed the territory of the Israelite tribe of that name, and later the ancient Kingdom of Judah, which encompassed a landlocked area on the western side of the Dead Sea, including the southern part of the present-day West Bank. Jordan called the area *ad-difa'a al-gharbiya*, which in English is translated as the 'West Bank'.

jugraensis (Leicester, 1908): *Armigeres* – not stated (inferential). Jugra; *-ensis*, L. suffix denoting place, locality, country. 'The larvae found in a bamboo in Ampang Jungle and water collected in a fallen leaf in the jungle at Jugra'. Type locality: Ampang Jungle, Kuala Lumpur, [Selangor], Malaysia. Jugra is a mukim (subdivision), historical town and former royal town in Kuala Langat District of Selangor State.

jugraensis (Leicester, 1908): *Aedes* – not stated (inferential). See *jugraensis* (Leicester, 1908): *Armigeres*. 'Described from a series bred from larvae taken from water collected in fallen leaves in the jungle at Jugra.' Type locality: Jugra, [Selangor], Malaysia.

kabaenensis Brug, 1939: *Aedes* – not stated (inferential). Kabaena; *-ensis*, L. suffix denoting place, locality, country. Type locality: Kabaena, [Bombana Regency, Southeast Sulawesi Province], Indonesia. Kabaena is an island in the Flores Sea off the coast of Sulawesi.

kabwachensis Edwards, 1941: subspecies of *Aedes abnormalis* (Theobald, 1909b) – not stated (inferential). Kabwach; *-ensis*, L. suffix denoting place, locality, country. Type locality: Kabwach Forest, Kisii, Kavirondo, Kenya.

kaimosi (van Someren, 1946b): *Toxorhynchites* – not stated (inferential). Kaimosi, treated as a latinized noun in the nominative singular standing in apposition to the generic name. Type locality: Kaimosi Forest, North Kavirondo [Vihiga County], Kenya. Kavirondo is the former name of the region surrounding Kavirondo Gulf (present-day Winam Gulf), an extension of Lake Victoria into western Kenya.

kalawara Stoker & Waktoedi Koesoemawinangoen, 1949: subspecies of *Anopheles tessellatus* Theobald, 1901a – not stated. Kalawara, treated as a latinized noun in the nominative singular standing in apposition to the generic name. Type locality: [Kalawara, Sulawesi Utara (North Sulawesi Province), Sulawesi], Indonesia. See Kitzmiller (1982).

kanarensis Edwards, 1934: *Aedes* – not stated (inferential). Kanara; *-ensis*, L. suffix denoting place, locality, country. Type locality: Yellapur, North Kanara (Uttara Kannada District), [Karnataka State], India. Yellapur is a town in Uttara Kannada District (North Kanara) located in the Western Ghats section of Karnataka State. The Kanara (also known as Canara, Karavali and Coastal Karnataka) region comprises three coastal districts, namely Dakshina Kannada (South Kanara), Udupi District and Uttara Kannada (North Kanara).

kanhsienensis Tung, 1955: *Heizmannia* – not stated (inferential). Kanhsien; *-ensis*, L. suffix denoting place, locality, country. Type locality: Kanhsien, Kiangsi (Jiangsu) Province, China. Jiangsu is a coastal province north of Shanghai.

kanyamwerima van Someren, 1951: *Culex* – not stated (inferential). Kanyamwerima, treated as a latinized noun in the nominative singular standing in apposition to the generic name. Type locality: Kanyamwerima, Bwamba County, Toro District (now in Bundibugyo District), [Western Region], Uganda. See *bwamba* van Someren, 1950c.

kapretwae Edwards, 1941: *Aedes* – not stated (inferential). Kapretwa; *-ae*, L. ending for first declension genitive singular noun (feminine). Type locality: Mount Elgon, 8000 ft (2438 m), forest zone above Kapretwa, Kenya.

karandalaensis (Wolfs, 1951): *Coquillettidia* – not stated (inferential). Karandala; *-ensis*, L. suffix denoting place, locality, country. Type locality: Karandala marsh, 300 meters from the Ruzizi River, near the Costermansville road in Shangugu, Kivu, [Rwanda]. Shangugu (present-day Cyangugu) is a city and capital of the Rusizi District in Western Province, Rwanda. The city lies at the southern end of Lake Kivu, and is contiguous with Bukavu, Democratic Republic of the Congo, but separated from it by the Ruzizi River. Bukavu (formerly Costermansville) is a city in eastern Democratic Republic of the Congo (formerly Belgium Congo), lying at the extreme south-western edge of Lake Kivu, west of Cyangugu in Rwanda.

karooensis Muspratt, 1961: *Aedes* – not stated. Karoo; *-ensis*, L. suffix denoting place, locality, country. Type locality: Near Noupoort, 22 miles north of Middleburg, Cape Province (Northern Cape Province), South Africa. The Karoo is a semi-desert natural region. Noupoort is a town in the eastern Karoo region of South Africa.

kartalae Brunhes, 1977a: subspecies of *Culex comorensis* Brunhes, 1977a – not stated (inferential). Kartala; *-ae*, L. ending for first declension genitive singular noun (feminine). Type locality: Kartala volcano (Mount Karthala), 1000 m, Grande Comore Island, Comoros (officially the Union of the Comoros). Mount Karthala is an active volcano and the highest point of the Comoros.

karwari (Barraud, 1924a): *Aedes* – not stated (inferential). Karwar; *-i*, L. ending for second declension genitive singular noun (masculine and neuter). Type locality: Karwar, North Kanara (Uttara Kannada District), [Maharashtra State], India. The Kanara (also known as Canara, Karavali and Coastal Karnataka) is a region of Karnataka State that encompasses the coastal districts of Dakshina Kannada (South Canara), Udupi and Uttara Kannada (North Canara). See *karwari* (James, 1903).

karwari (James, 1903): *Anopheles* – not stated (inferential). See *karwari* (Barraud, 1924a). Type locality: 'Karwar, Bombay Presidency, Goa', present-day Karwar, Uttara Kannada District, Maharashtra State, India. Karwar lies 15 km south of Goa. The Bombay Presidency was a former province of British India, which at its greatest extent comprised the present-day state of Gujarat, the western two-thirds of Maharashtra State, including the regions of Konkan, Desh and Kandesh, north-western Karnataka State of India, and also included Sindh Province of Pakistan and the British territory of Aden in Yemen. See Kitzmiller (1982).

kasachstanicus Gutsevich, 1962: *Aedes* – not stated (inferential). Kasachstan; *-icus*, L. belonging to, like, of, pertaining to. Type locality: 70 km north of Alma-Ata (Almaty Oblast), Ili River valley, Kazakhstan. Almaty Oblast (Region), formerly known as Alma-Ata, is south-eastern Kazakhstan. The Ili River flows from the Ili Kazakh Autonomous Prefecture of the Xinjiang Uighur Autonomous Region of China to the Almaty Region (Province) in Kazakhstan.

katherinensis Woodhill, 1949: *Aedes* – not stated (inferential). Katherine; *-ensis*, L. suffix denoting place, locality, country. Type locality: Katherine, Northern Territory, Australia.

katibasensis Miyagi & Toma, 2012: *Topomyia* – Katibas; *-ensis*, L. suffix denoting place, locality, country. 'The species name *katibasensis* refers to the name of the river, Sg. Katibas that flows through the LEWS [Lanjak-Entemau Wildlife Sanctuary], where this new species was found.' Type locality: Headquarters (01° 38.777′ N, 112° 167.709′ E) of Lanjak-Entemau Wildlife Sanctuary, Katibas (state constituency), Sarawak, Malaysia.

kaviengensis Sirivanakarn, 1968: *Culex* – not stated (inferential). Kavieng; *-ensis*, L. suffix denoting place, locality, country. Type locality: Kavieng, New Ireland, Bismark Archipelago, [Papua New Guinea]. The Bismarck Archipelago is a group of islands off the north-eastern coast of New Guinea and is part of the Islands Region of Papua New Guinea, which includes the islands that comprise New Ireland Province. Kavieng, the capital of New Ireland Province, is on New Ireland, the largest island of the province.

kelabitensis Miyagi & Toma, 2010b: *Topomyia* – corrected spelling. The authors clearly intended to name the species for the Kelabit people: 'The species name *kelabitense* refers to the name of the Kelabit, one of the smallest tribes in Sarawak, Malaysia', but unwittingly made it a place name that denotes the Kelabit Highlands, the mountainous interior region of Sarawak where the Kelabit people live. As the genus *Topomyia* is feminine, the original ending *-ense*

(L. neuter) has been changed to *-ensis* (L. feminine and masculine). Type locality: Arur Dalan (1000–1200 m above sea level), Bario, Sarawak, Malaysia.

keniensis Evans, 1931a: *Anopheles* – not stated. Kenya, Kenia, Kenie; *-ensis*, L. suffix denoting place, locality, country. The Republic of Kenya is named after Mount Kenya. The origin of the name is not clear, but it may be linked to the Kikuyu, Embu and Kamba words Kirinyaga, Kirenyaa and Kiinyaa, which mean 'God's resting place'. However, according to Kitzmiller (1982), the name of the mountain is derived from 'Kenie' or 'Kenye', which are modifications of a Masai word meaning 'mist surrounding the mountain'. Type locality: Taveta, Kenya. See Kitzmiller (1982).

keniensis van Someren, 1946a: *Aedes* – not stated. See *keniensis* Evans, 1931a. Type locality: City Part, Nairobi, Kenya.

kenyae van Someren, 1946c: *Aedes* – not stated (inferential). Kenya; *-ae*, L. ending for first declension genitive singular noun (feminine). Type locality: Kaimosi, North Kavirondo [Vihiga County], Kenya. Kavirondo is the former name of the region surrounding Kavirondo Gulf (present-day Winam Gulf), an extension of Lake Victoria into western Kenya. See *kaimosi* (van Someren, 1946b).

kiangsiensis Tung, 1955: *Aedes* – not stated (inferential). Kiangsi; *-ensis*, L. suffix denoting place, locality, country. Type locality: Kanhsien, Kiangsi (Jiangsu) Province, China. See *kanhsienensis* Tung, 1955.

kilara van Someren, 1951: *Culex* – not stated (inferential). Kilara, treated as a latinized noun in the nominative singular standing in apposition to the generic name. Type locality: Kilara, Bwamba County, Toro District (now in Bundibugyo District), [Western Region], Uganda. See *bwamba* van Someren, 1950c.

killertonis (Huang, 1968a): *Verrallina* – not stated (inferential). Killerton; *-is*, L. ending for third declension genitive singular noun (masculine, feminine and neuter). Type locality: Cape Killerton, [Oro Province], SE New Guinea (present-day Papua New Guinea). Cape Killerton is a point located on the north coast of Papua New Guinea.

kinabaluensis Ramalingam, 1972: *Armigeres* – not stated (inferential). Kinabalu; *-ensis*, L. suffix denoting place, locality, country. Type locality: Mount Kinabalu, at 7,500 ft (2286 m), Sabah, Malaysia.

kinabaluensis Sirivanakarn, 1976: *Culex* – not stated (inferential). See *kinabaluensis* Ramalingam, 1972. Type locality: Mount Kinabalu, at 950–1,075 m, Sabah, Malaysia.

kiriensis Klein & Sirivanakarn, 1970: *Culex* – not stated. Kiri; *-ensis*, L. suffix denoting place, locality, country. The name appears to be derived from the Khmer word គិរី (*kiri* 'mountain' from Sanskrit *giri*) in reference to paratypes collected in the 'Kirirom Hills, Kampong Speu', the mountains of Kirirom National Park located mostly in Phnom Sruoch District of Kampong Speu Province in south-western Cambodia. Type locality: Bokor Hill, Kampot Province, Cambodia. See *kriromi* (Klein, 1969).

kiriromi (Klein, 1969): *Mimomyia* – not stated (inferential). Kirirom; *-i*, L. ending for second declension genitive singular noun (masculine and neuter). Type locality: Kirirom Hills, O Tachat (altitude 300–400 m), Kompong Speu Province, Cambodia. See *kiriensis* Klein & Sirivanakarn, 1970.

kivuensis Edwards, 1941: *Aedes* – not stated (inferential). Kivu; *-ensis*, L. suffix denoting place, locality, country. Type locality: Kibati lava plain, North Kivu, Belgium Congo (present-day Democratic Republic of the Congo). North Kivu is a province bordering Lake Kivu, one of the African Great Lakes, located on the border between the Democratic Republic of the Congo and Rwanda.

kohkutensis Mattingly, 1958a: *Aedes* – not stated (inferential). Koh Kut; *-ensis*, L. suffix denoting place, locality, country. Type locality: Koh Kut (Ko Kut) Island, east Gulf of Siam (present-day Gulf of Thailand), Thailand. Ko Kut (also transliterated as Koh Kut, Ko Kood or Koh Kood) is located in the south-eastern part of Thailand near the maritime border with Cambodia.

kolambuganensis Baisas, 1932: *Anopheles* – not stated (inferential). Kolambugan (originally misspelled as kulambugan); *-ensis*, L. suffix denoting place, locality, country. Type locality: Kolambugan, [Lanao del Norte Province], Mindanao, Philippines. The name 'Kolambugan' is derived from *kolambog*, the elephant apple tree which grew abundantly in the area. See Kitzmiller (1982).

koliensis Owen, 1945: *Anopheles* – not stated (inferential). Koli; *-ensis*, L. suffix denoting place, locality, country. Type locality: Koli area, Guadalcanal, Solomon Islands. Guadalcanal is the principal island in Guadalcanal Province of the nation of Solomon Islands. The Koli area surrounds Koli Point at the mouth of the Nalimbu River on the mid-northern side of Guadalcanal Island. See Kitzmiller (1982).

komponga (Klein, 1973): *Verrallina* – not stated (inferential). Kampong; *-us*, *-a*, *-um*, L. nominative singular adjectival termination. Type locality: Kirirom Hills, Kompong Speu, Cambodia.

koreicus (Edwards, 1917): *Aedes* – not stated (inferential). Korea; *-icus*, L. belonging to, like, of, pertaining to. Type locality: Korea.

koreicus Yamada & Watanabe, 1918: *Anopheles* – not stated (inferential). See *koreicus* (Edwards, 1917). Type locality: Pyeoncheing (Pyeongchang) [County, Gangwon Province, South] Korea. See Kitzmiller (1982).

kosiensis Coetzee, Segerman & Hunt, 1987: *Anopheles* – not stated (inferential). Kosi; *-ensis*, L. suffix denoting place, locality, country. Type locality: Kosi Bay, northern Zululand [District Municipality], [KwaZulu-Natal Province], South Africa. Kosi Bay is a series of four interlinked lakes in the Maputaland area of KwaZulu-Natal. Zululand is one of the 11 district municipalities of the province. See Service (2010).

kotiensis Barraud, 1934: subspecies of *Aedes gubernatoris* (Giles, 1901a) – not stated (inferential). Koti; *-ensis*, L. suffix denoting place, locality, country. Type locality: Koti (Bhoj Koti village), near Kalka (city), Western Himalayas [Panchkula District, Haryana State], India.

kowiroensis Sirivanakarn, 1968: *Culex* – not stated (inferential). Kowiro; *-ensis*, L. suffix denoting place, locality, country. Type locality: Kowiro, inland of Wewak, Sepik (East Sepik Province), NE New Guinea (present-day Papua New Guinea). Wewak is the capital of the East Sepik Province, located on the northern coast of the island of New Guinea.

krymmontanus Alekseev, 1989: *Aedes* – not stated (inferential). Krym (refers to Crimea); *montanus*, L. pertaining to mountains. Type locality: southern slope of Crimean Mountains (300–600 m above sea level), Crimea, USSR (present-day Russia). The Crimean Mountains lie on the south-eastern coast of the Crimean Peninsula (known simply as Crimea), located on the northern coast of the Black Sea. The Ukrainian territory of Crimea was annexed by the Russian Federation on 18 March 2014.

kuchingensis Edwards, 1915c: *Armigeres* – not stated (inferential). Kuching; *-ensis*, L. suffix denoting place, locality, country. Type locality: Kuching Reservoir, Sarawak, Malaysia.

kukenan Anduze, 1942b: *Culex* – not stated. Kukenán, treated as a latinized noun in the nominative singular standing in apposition to the generic name. Type locality: La Gran Sabana, Surukum, [Bolívar State], Venezuela. Kukenán is a tepui in Guayana Region, Venezuela. La Gran Sabana (The Great Plain), also known as the Guianan savanna, is located inside Canaima National Park in the south of Bolívar State. Kukenán, also known as Matawi, is a tepui (table-top mountain or mesa) located in the national park.

kumbae Chwatt, 1948: subspecies of *Aedes boneti* Gil Collado, 1936 – not stated (inferential). Kumba; *-ae*, L. ending for first declension genitive singular noun (feminine). '... collected from a rock-pool in the Mambonjeze stream, near to the town of Kumba'. Type locality: Mambonjeze stream, Kumba, [Southwest Region], British Cameroons (Cameroon). British Cameroons was a British Mandate territory in British West Africa. The territory now comprises parts of Nigeria and Cameroon. Kumba is located north of Duala in the coastal region (Southwest Region) of Cameroon.

kunmingensis Dong & Wang, 1985: *Anopheles* – not stated (inferential). Kunming; *-ensis*, L. suffix denoting place, locality, country. Type locality: Long-Tan Ba, Kunming [Prefecture], Yunnan Province, China. See Service (2010).

kusaiensis Bohart, 1957: *Culex* – not stated (inferential). Kusaie; *-ensis*, L. suffix denoting place, locality, country. Type locality: Weye Cave, Kusaie (Kosrae) Island, Eastern Caroline Islands (present-day Federated States of Micronesia). The Caroline Islands are a widely scattered group of small islands north of New Guinea. Politically they are divided between the Federated States of Micronesia in the eastern part of the group and Palau at the extreme western end. Kosrae, formerly known as Kusaie, is an island in the Federated States of Micronesia.

kweiyangensis Yao & Wu, 1944: *Anopheles* – not stated (inferential). Kweiyang; *-ensis*, L. suffix denoting place, locality, country. Type locality: Kweiyang (Guiyang) Municipality, Kweichow (present-day Guizhou) Province, China. Guiyang, also spelled Kweiyang, is a city and capital of Guizhou Province. See Kitzmiller (1982).

kyondawensis Abraham, 1947: *Anopheles* – not stated (inferential). Kyondaw; *-ensis*, L. suffix denoting place, locality, country. Type locality: Kyondaw (Kyontaw), [Kawkareik District, Kayin State], Lower Burma (Myanmar). See Kitzmiller (1982).

kyotoensis Yamaguti & LaCasse, 1952: *Culex* – not stated. Kyoto; *-ensis*, L. suffix denoting place, locality, country. The authors did not provide locality data for the species but mentioned it being the species they previously identified as *Culex ryukyensis* Bohart, 1958 (La Casse & Yamaguti, 1948, 1950), which they described from specimens collected in Honshu and Kyushu, the main and most southern islands of Japan, respectively. Type locality: [Kyoto, Honshu], Japan. Kyoto, once the capital of Japan, is a city on the island of Honshu.

laguna Arnell & Nielsen, 1972: *Aedes* – not stated (inferential). Laguna, treated as a latinized noun in the nominative singular standing in apposition to the generic name. Type locality: La Laguna meadow, Sierra Laguna, 20 miles east-northeast of Todas Santos, Baja California Sur, Mexico.

lampangensis Sirivanakarn, 1973b: *Culex* – not stated (inferential). Lampang; *-ensis*, L. suffix denoting place, locality, country. Type locality: Ban Pha Daeng, Lampang Province, Thailand.

lampangensis Thurman, 1959: *Hodgesia* – not stated (inferential). See *lampangensis* Sirivanakarn, 1973b. Type locality: Ngao, Lampang Province, Thailand.

langata van Someren, 1946c: *Aedes* – not stated (inferential). Langata; treated as a latinized noun in the nominative singular standing in apposition to the generic name. Type locality: Langata Forest, Nairobi, Kenya.

lanyuensis Lien, 1968a: *Orthopodomyia* – not stated (inferential). Lanyu; *-ensis*, L. suffix denoting place, locality, country. Type locality: south of Hongtou, Lanyu (Orchid Island), Taitung Hsien (County), Taiwan (officially the Republic of China). Orchid Island lies off the southeastern coast of Taiwan Island. It is governed as Lanyu Township of Taitung County, Taiwan.

laoagensis Knight, 1946: *Aedes* – not stated (inferential). Laoag; *-ensis*, L. suffix denoting place, locality, country. Type locality: Laoag (City of), Ilocos Norte Province, Luzon, Philippines.

laoensis Toma & Miyagi, 2003: *Armigeres* – not stated (inferential). Laos; *-ensis*, L. suffix denoting place, locality, country. Type locality: Thapachon (Ban Thapachôn), [Boualapha District], Khammouane Province, Laos.

lasaensis Meng, 1962: *Aedes* – not stated (inferential). Lasa; *-ensis*, L. suffix denoting place, locality, country. Type locality: vicinity of Lasa (Lhasa), [Tibet Autonomous Region, China]. Lhasa is the capital of the Tibet Autonomous Region. It resides on the north bank of the Lhasa River in a valley of the Himalayan Mountains.

leiboensis Chu, 1981: *Uranotaenia* – not stated (inferential). Leibo; *-ensis*, L. suffix denoting place, locality, country. Type locality: Leibo County, Sichuan Province, China.

lemuriae Ribeiro, 2004: *Toxorhynchites* – Lemuria; *-ae*, L. ending for first declension genitive singular noun (feminine). 'The new specific name is derived from the word Lemuria, taken in the sense of the island of Madagascar, home of lemurs' [translated from French]. Type locality: Loharano, Manakara [District], Fianarantsoa Province (former province), [Vatovavy-Fitovinany Region], Madagascar. Manakara is the capital of the Vatovavy-Fitovinany Region.

letabensis Lambert & Coetzee, 1982: *Anopheles* – not stated. Letaba; *-ensis*, L. suffix denoting place, locality, country. Type locality: Grey Stones (farm), [Letaba District], Transvaal Province (Limpopo Province), South Africa. Transvaal Province was a province of South Africa from 1910 until the end of apartheid in 1994 when it was subdivided under a new constitution. The northern region of Transvaal Province was formally changed to Limpopo Province. Letaba is a forested, lush, mountainous area of the province and Grey Stones is a farm. See Service (2010).

liangshanensis Kang, Tan, Cao, Cheng, Yang & Huang, 1984: *Anopheles* – not stated. Liangshan; *-ensis*, L. suffix denoting place, locality, country. Type locality: Zhaojiao County, [Liangshan Yi], Sichuan Province, China. Liangshan, officially the Liangshan Yi Autonomous Prefecture, occupies much of the southern extremity of Sichuan Province. See Service (2010).

liberiensis Peters, 1955a: *Culex* – not stated (inferential). Liberia; *-ensis*, L. suffix denoting place, locality, country. Type locality: Kpain, [Nimba County], Liberia.

lokojensis Service, 1959b: *Aedes* – not stated (inferential). Lokoja; *-ensis*, L. suffix denoting place, locality, country. Type locality: Lokoja, [Kogi State], Nigeria.

lorengaui Peters, 1963b: *Tripteroides* – not stated (inferential). Lorengau; *-i*, L. ending for second declension genitive singular noun (masculine and neuter). Type locality: Lorengau River, Manus Island, [Papua] New Guinea. Manus Island is part of Manus Province in northern Papua New Guinea and is the largest of the Admiralty Islands.

lucianus Muspratt, 1959: *Aedes* – not stated (inferential). Lucia; *-anus*, L. belonging to, pertaining to. Type locality: M'canduzi, western side of Lake St. Lucia, Zululand, Natal (present day KwaZulu-Natal Province, South Africa). Lake St Lucia is an estuarine lake system in northern KwaZulu-Natal Province.

lunda da Cunha Ramos, 1993: *Uranotaenia* – not stated. Lunda; treated as a latinized noun in the nominative singular standing in apposition to the generic name. Type locality: Dundo, [Lunda Norte Province], Angola. Dundo is a former mining town, now a city and the capital of Lunda Norte Province.

lungae Belkin & Schlosser, 1944: *Anopheles* – not stated. Lunga; *-ae*, L. ending for first declension genitive singular noun (feminine). Type locality: Tassafaronga Swamp, [Lunga District], Guadalcanal Island, [Guadalcanal Province, Solomon Islands]. Solomon Islands is a nation of hundreds of islands in the South Pacific, and Guadalcanal Province is one of the largest islands. Lunga (Lungga) Point is a promontory on the northern coast of Guadalcanal, an area that includes the mouth of the Lunga River and Lunga village. See Kitzmiller (1982).

luzonensis (Ludlow, 1905a): *Mimomyia* – not stated (inferential). Luzon; *-ensis*, L. suffix denoting place, locality, country. Type locality: Bayambang [Province], Pangasinan, Luzon, Philippines.

luzonensis Rozeboom, 1946: *Aedes* – not stated (inferential). See *luzonensis* (Ludlow, 1905a). Type locality: Subic Bay, [Zambales Province], Luzon, Philippines.

luzonensis Sirivanakarn, 1976: *Culex* – not stated (inferential). See *luzonensis* (Ludlow, 1905a). Type locality: Imugan, Dalton, N. Vizcaya (Nueva Vizcaya Province), Luzon, Philippines.

macaensis Ribeiro, 1997: *Toxorhynchites* – Macau; *-ensis*, L. suffix denoting place, locality, country. 'The new species is named for Macau, the area of origin of the type series.' Type locality: Ilha de Coloane, Parque Siac Pai Van, Macau, [China]. Coloane, officially Freguesia de São Francisco Xavier, is one of the two main islands of Macau. Macau was a Portuguese territory until 1999 and is now an autonomous region on the south coast of China.

madagascarensis da Cunha Ramos & Brunhes, 2004: *Uranotaenia* – Madagascar; *-ensis*, L. suffix denoting place, locality, country. 'Name related to the Malagasy endemism of this species' [translated from French]. Type locality: Ambodimanga, [Andapa District, Sava Region], Madagascar. Ambodimanga is a commune in northern Madagascar.

madagascarensis Ribeiro, 2004: *Toxorhynchites* – see *madagascarensis* da Cunha Ramos & Brunhes, 2004. 'The specific epithet is derived from Madagascar, the country of origin of the type series' [translated from French]. Type locality: Fort Dauphin (Port d'Ehola), [Tôlanaro District, Anosy Region], Madagascar. Fort-Dauphin, present-day Port d'Ehola, was the first French settlement on the southeast coast of Madagascar.

madagascarensis van Someren, 1949a: *Aedes* – not stated (inferential). See *madagascarensis* da Cunha Ramos & Brunhes, 2004. Type locality: Sakaramy, near Diego Suarez (Antsiranana since 1975, currently Diana Region in part), Madagascar. Antsiranana is a port city in northern Madagascar.

madagascarica Brunhes, Boussès & da Cunha Ramos, 2011: *Aedeomyia* – Madagascar; *-icus* (f. *-ica*), L. belonging to, like, of, pertaining to. 'Name in relation to the Malagasy endemicity of the species' [translated from French]. Type locality: Ivoloina forest station (Parc Ivoloina), Toamasina Province, Madagascar. Parc Ivoloina, a regional centre for conservation and education, consists of a 4 ha zoological park within a 282 ha forestry station.

madagascarica da Cunha Ramos & Brunhes, 2004: *Uranotaenia* – see *madagascarica* Brunhes *et al.*, 2011. 'Name in relation to the endemicity of the species' [translated from French]. Type locality: Station of Mandena, Fort-Dauphin (Port d'Ehola), Tôlanaro District, [Anosy Region], Madagascar. See *madagascarensis* Ribeiro, 2004.

maderensis Mattingly, 1955: subspecies of *Culex hortensis* Ficalbi, 1889a – not stated (inferential). Madère; *-ensis*, L. suffix denoting place, locality, country. Type locality: Monte (Freguesia do Monte), altitude 450–900 m, Madeira (Madère). Madeira is an autonomous region of Portugal, an archipelago comprising four islands off the north-west coast of Africa. Monte (English: mountain) is a civil parish (Freguesia do Monte) in the municipality and a suburb of Funchal, the capital city on the main island.

madininensis Senevet, 1936: *Culex* – Madinina; *-ensis*, L. suffix denoting place, locality, country. '... a distinct species which I propose to name *madininensis* in honor of the ancient name of Martinique (Madinina)'. Type locality: village of Trinité, Martinique, [Lesser Antilles]. Martinique is an insular region of France located in the Lesser Antilles in the eastern Caribbean Sea.

madrensis Baisas, 1946a: *Orthopodomyia* – not stated (inferential). Madre; *-ensis*, L. suffix denoting place, locality, country. Type locality: Llavac, Sierra Madre Mountains, Laguna Province, Luzon, Philippines.

maikalensis Natarajan, Rajavel & Jambulingam, 2017: *Uranotaenia* – Maikal; *-ensis*, L. suffix denoting place, locality, country. The species is based on specimens collected in villages 'located in the part of the eastern hill range of India known as Maikal Hills and Satpura Hills'. Type locality: Lapha (22° 20.882′ N, 82° 13.106′ E), Korba District, Chattisgarh State, India.

malayensis Colless, 1962: *Aedes* – not stated. Malaya; *-ensis*, L. suffix denoting place, locality, country. Type locality: Pulau Hanatu, Keppel Harbour, Singapore. The species name is apparently based on paratype specimens collected in peninsular Malaysia (Kuantan and Pulau Jarak).

malayensis Sirivanakarn, 1972: *Culex* – not stated (inferential). See *malayensis* Colless, 1962. Type locality: Rantau Panjang, Kelantan State, peninsular Malaysia. Rantau Panjang is a daerah (administrative division) and also a parliamentary constituency in Pasir Mas District of Kelantan State.

malayi Delfinado & Hodges, 1968: *Tripteroides* – not stated (inferential). Malaya; *-i*, L. ending for second declension genitive singular noun (masculine and neuter). Type locality: 20 mile Pahang Road, Kuanta, Gunong Tabu, Trengganu, peninsular Malaysia. Terengganu, formerly spelled Trengganu or Tringganu, is a sultanate and constitutive state of federal Malaysia.

malayi (Leicester, 1908): *Culex* – not stated (inferential). See *malayi* Delfinado & Hodges, 1968. Type locality: pond in Kuala Lumpur, [Selangor State], Malaysia.

malayi Leicester, 1908: *Hodgesia* – not stated (inferential). See *malayi* Delfinado & Hodges, 1968. Type locality: pools in jungle near Kuala Lumpur, [Selangor State], Malaysia.

malayi (Theobald, 1901c): *Armigeres* – not stated (inferential). See *malayi* Delfinado & Hodges, 1968. Type locality: Klang Jungle, Selangor State, Malaysia.

malaysiensis Ramalingam & Banu, 1987: *Topomyia* – not stated (inferential). Malaysia; *-ensis*, L. suffix denoting place, locality, country. Type locality: Tambunan, Sabah, Malaysia. Tambunan is a valley district in the Interior Division of Sabah State.

maliensis Bailly-Choumara & Adam, 1959: *Anopheles* – Mali; *-ensis*, L. suffix denoting place, locality, country. 'We propose the name of *A. maliensis* in reference to the typical locality: Mali' [translated from French]. Type locality: Mali, Fouta-Kjalon, Haute-Guinée (Upper Guinea). Mali is a town and prefecture (Préfecture De Mali) located in the Labé Region of the Republic of Guinea. See Kitzmiller (1982).

mamoedjoensis Mattingly, 1958a: *Aedes* – not stated (inferential). Mamoedjoe; *-ensis*, L. suffix denoting place, locality, country. Type locality: Mamoedjoe (Mamudju), Celebes (Sulawesi) [present-day Mamudju, Karema, Mamuju Sub-district, Mamuju Regency, West Sulawesi Province, Indonesia].

manakaraensis da Cunha Ramos & Brunhes, 2004: *Uranotaenia* – Manakara; *-ensis*, L. suffix denoting place, locality, country. 'Name in relation to the typical locality of the species' [translated from French]. Type locality: coastal forest about 10 km north of Manakara, Fianarantsoa Province [Vatovavy-Fitovinany Region], Madagascar. Fianarantsoa Province is a former province of Madagascar, which, along with five other provinces, was abolished in 2007 in favour of the creation of smaller administrative regions.

manusensis Sirivanakarn, 1975: *Culex* – not stated (inferential). Manus; *-ensis*, L. suffix denoting place, locality, country. Type locality: Lorengau Village, Manus District, Manus Island, Papua New Guinea. Manus Island is part of Manus Province in northern

Papua New Guinea and is the largest of the Admiralty Islands.

maracayensis Evans, 1923b: *Culex* – not stated (inferential). Maracay; *-ensis*, L. suffix denoting place, locality, country. Type locality: Maracay, [Aragua State], Venezuela.

marajoara Galvão & Damasceno, 1942b: *Anopheles* – not stated (inferential). Marajó; *-arium*, *-arium*, *-aria*, L. denoting place where, place for. The name appears to have been latinized incorrectly, and should have been *maragoarium*, which the authors may have assumed would be neuter and not in gender agreement with the masculine *Anopheles*. Type locality: Ilha de Marajó, Pará, Brazil. Marajó is a coastal island, slightly larger than Switzerland, located at the mouth of the Amazon River. See Kitzmiller (1982).

marianae Bohart & Ingram, 1946b: subspecies of *Culex annulirostris* Skuse, 1889 – not stated (inferential). Mariana; *-ae*, L. ending for first declension genitive singular noun (feminine). Type locality: Cheran Konoa, Saipan, Mariana Islands. The Mariana Islands are a crescent-shaped archipelago comprising 15 volcanic mountains in the western North Pacific Ocean. The Northern Mariana Islands, including the island of Saipan, make up a Commonwealth of the USA.

marquesensis Stone & Rosen, 1953: *Culex* – not stated (inferential). Marquesas; *-ensis*, L. suffix denoting place, locality, country. Type locality: Atuona Bay, Hivaoa (Hiva Oa), Marquesas Islands. The Marquesas Islands are a group of volcanic islands in French Polynesia, an overseas territory of France in the southern Pacific Ocean. Atuona Bay is a small inlet on the southern coast of Hiva Oa, the second largest island.

marshallensis Stone & Bohart, 1944: *Aedes* – not stated (inferential). Marshall; *-ensis*, L. suffix denoting place, locality, country. Type locality: Airok Island, Ailinglaplap Atoll, Marshall Islands. The Marshall Islands, officially the Republic of the Marshall Islands, is a country comprising a sprawling chain of volcanic islands and coral atolls near the equator in the central Pacific Ocean. Ailinglaplap (or Ailinglapalap) is a coral atoll of 56 islands that form a legislative district of the Ralik Chain in the country.

mascarensis de Meillon, 1947a: *Anopheles* – not stated. Mascarene; *-ensis*, L. suffix denoting place, locality, country. Type locality: Diego Suarez area (Antsiranana area), [Diana Region], Madagascar. The city of Antsiranana was known as Diego Suarez prior to 1975. The derivation of the name *mascarensis* is nomenclaturally incorrect as Madagascar is not on the Mascarene Plateau, a submarine plateau in the Indian Ocean located north and east of Madagascar that bears the Mascarene Islands. However, Madagascar was generically included under the term 'Marcarenes' in some early biological literature. See Kitzmiller (1982).

mascarensis MacGregor, 1924: *Aedes* – not stated. See *mascarensis* de Meillon, 1947a. Type locality: Mauritius. Mauritius, officially the Republic of Mauritius, is an island nation in the Indian Ocean about 2000 km south-east of Africa. The country includes the islands of Mauritius, Rodrigues, Agaléga, St Brandon and two disputed territories. The islands of Mauritius and Rodrigues form part of the Mascarene Islands (or Mascarenhas Archipelago), along with nearby Réunion.

mashonaensis Theobald, 1901c: *Uranotaenia* – not stated (inferential). Mashona; *-ensis*, L. suffix denoting place, locality, country. Type locality: Salisbury (present-day Harare), Mashonaland, [Zimbabwe]. Harare, officially called Salisbury until 1982, is the capital and most populous city of Zimbabwe. Mashonaland, home of the Shona people, is a region in northern Zimbabwe which is currently divided into three provinces, Mashonaland West, Mashonaland Central and Mashonaland East (including Harare). Mashona is another name for the Shona people; consequently, the name *mashonaensis* is incorrectly formed.

masoalensis Fontenille & Brunhes, 1985: *Aedes* – not stated (inferential). Masoala; *-ensis*, L. suffix denoting place, locality, country. Type locality: Onive Valley, altitude 200 m, Masoala Peninsula, Antalaha District, Diego Suarez Province (current Sava Region in part), Madagascar. Diego Suarez Province became Antsiranana Province in 1975 and was abolished in 2007 in favour of the creation of smaller administrative regions.

mattogrossensis Lutz & Neiva, 1911: *Anopheles* – not stated (inferential). Matto (Mato) Grosso; *-ensis*, L. suffix denoting place, locality, country. Type locality: Manicoré (Mandicoré) Lagoon, Mato Grosso State, Brazil. See Kitzmiller (1982).

mauesensis Lane, 1945b: *Culex* – not stated (inferential). Maués; *-ensis*, L. suffix denoting place, locality, country. Type locality: Maués, Amazonas State, Brazil.

mayottensis Brunhes, 1977b: *Uranotaenia* – not stated (inferential). Mayotte; *-ensis*, L. suffix denoting place, locality, country. Type locality: Vahibé, route Passamainti-Combani, [Mamoudzou Prefecture], Mayotte Department, Grande-Terre (Maore) Island, Mayotte. Mayotte is an insular department and region of France officially named Département de Mayotte. It consists of a main island, Grande-Terre, a smaller island, Petite-Terre (or Pamanzi) and several smaller islets. The territory, also known as Maore, the native name of the main island, is geographically part of the Comoro Islands.

mefouensis Ferrara, 1974: *Aedes* – not stated (inferential). Méfou, *-ensis*, L. suffix denoting place, locality, country. Type locality: Zoatoupsi, Mbankomo District (present-day Bikok District), [Center Region], Méfou Department (present-day Méfou-et-Akono Department), Cameroon. Zoatoupsi is a village located in the Bikok District and Méfou-et-Akono Department in the Center Region of Cameroon. The former Méfou Department was divided into the Méfou-et-Afamba and Méfou-et-Akono departments in 1995.

melanesiensis Belkin, 1955: *Tripteroides* – not stated. Melanesia; *-ensis*, L. suffix denoting place, locality, country. Type locality: Segond Channel area, [Sanma Province], Espiritu Santo, New Hebrides (Vanuatu). New Hebrides was the colonial name for the island group in the South Pacific that is now the nation of Vanuatu. Espiritu Santo is the largest island in Vanuatu. The Segond Channel is located between southern Espiritu Santo and Aore Island.

melanesiensis Belkin, 1962: *Mansonia* – not stated. See *melanesiensis* Belkin, 1955. Type locality: Doma (Ndoma) Cove area, Guadalcanal, Solomon Islands. Guadalcanal is the principal island in Guadalcanal Province of the nation of Solomon Islands.

menglaensis Dong, Zhou & Dong, 2002b: *Armigeres* – not stated (inferential). Mengla; *-ensis*, L. suffix denoting place, locality, country. Type locality: Mengla County, [Yunnan Province], China. Mengla County is a county under the jurisdiction of the Xishuangbanna Dai Autonomous Prefecture in Yunnan Province.

menglangensis Ma, 1981a: *Anopheles* – not stated (inferential). Menglang; *-ensis*, L. suffix denoting place, locality, country. Type locality: Menglang, [Lancang Lahu Autonomous County], Yunnan Province, China. Lancang Lahu Autonomous County is an autonomous county under the jurisdiction of Pu'er City in south-western Yunnan Province. See Service (2010).

menglianensis Lu & Gong, 1986: *Heizmannia* – not stated (inferential). Menglian; *-ensis*, L. suffix denoting place, locality, country. Type locality: Menglian [Dai], Yunnan Province, China. Menglian Dai is part of the Menglian Dai, Lahu and Va Autonomous County (referred to as Menglian County), an autonomous county under the jurisdiction of the Pu'er Prefecture of Yunnan Province.

meraukensis Venhuis, 1932: *Anopheles* – not stated (inferential). Meraude; *-ensis*, L. suffix denoting place, locality, country. Type locality: Meraude, southeast coast of Dutch New Guinea (Papua Province, Indonesia). Merauke is a city and regency in Papua Province, Indonesia. The West Papua region of Indonesia was known as Dutch New Guinea until 1949 when it was an overseas territory of the Netherlands. The territory included what are now Papua and West Papua Provinces of Indonesia, which were previously administered under the name Irian Jaya. See Kitzmiller (1982).

mexicana (Bellardi, 1859): *Psorophora* – not stated (inferential). Mexico; *-anus* (f. *-ana*), L. belonging to, pertaining to. Type locality: Mexico. Interestingly, the species was described from an unspecified number of females collected by the French entomologist Auguste Sallé (1820–1896), who specialized in Coleoptera and conducted expeditions in the US southern states, West Indies, Central America (especially Mexico) and Venezuela.

miaolingensis Chen, 1982: *Culex* – not mentioned? Miaoling; *-ensis*, L. suffix denoting place, locality, country. Type locality: Guiyang, Guizhou Province, China. Guiyang, the capital city of Guizhou Province, lies in the central part of the Miaoling mountain range. The species is apparently named for the Miaoling Mountains.

micolo Ribeiro, da Cunha Ramos, Capela & Pires, 1998: *Culex* – not stated (inferential). Micóló, treated as a latinized noun in the nominative singular standing in apposition to the generic name. Type locality: Micóló settlement, [Lobata District], São Tomé Island, [São Tomé and Príncipe].

milnensis (King & Hoogstraal, 1947b): *Verrallina* – not stated (inferential). Milne; *-ensis*, L. suffix denoting place, locality, country. Type locality: Milne Bay, [Milne Bay Province], [Papua] New Guinea.

milnensis Lee, 1944b: *Armigeres* – not stated (inferential). See *milnensis* (King & Hoogstraal, 1947b). Type locality: Milne Bay, [Milne Bay Province], Papua, New Guinea (Papua New Guinea).

milsoni (Taylor, 1915): *Aedes* – not stated (inferential). Milson; *-i*, L. ending for second declension genitive singular noun (masculine and neuter). Type locality: Milson Island, [Hawkesbury River], New South Wales, Australia.

minjensis Sirivanakarn, 1968: *Culex* – not stated (inferential). Minj; *-ensis*, L. suffix denoting place, locality, country. Type locality: Tengen stream, Minj, Western Highlands Province, NE New Guinea (present-day Papua New Guinea).

minnesotae Barr, 1957: *Culiseta* – not stated (inferential). Minnesota; *-ae*, L. ending for first declension genitive singular noun (feminine). Type locality: Olcott Park, Virginia, St Louis Country, Minnesota, USA.

misionensis Duret, 1953: *Culex* – not stated (inferential). Misiones; *-ensis*, L. suffix denoting place, locality, country. Type locality: Aristóbula del Valle, Misiones Province, Argentina.

moctezuma (Dyar & Knab, 1906c): *Toxorhynchites* – not stated. Moctezuma, treated as a latinized noun in the nominative singular standing in apposition to the generic name. Type locality: Rio Aranjuez, near Puntarenas, [Puntarenas Province], Costa Rica. Since the authors did not mention the source of the species name, it is uncertain whether *moctezuma* is derived from a geographical or a personal name. Although Montezuma is a town in Puntarenas Province where the type locality is located, the name could relate to an Aztec emperor, bearing in mind that scholars use the term Moctezuma for more than one Aztec ruler.

moghulensis Christophers, 1924c: *Anopheles* – Moghul; *-ensis*, L. suffix denoting place, locality, country. 'The distribution of the variety is therefore reminiscent of the ancient Moghul Empire extending as it does over the north-western and central parts of the Peninsula, and on this account I have given to it the name of var. *moghulensis*.' Type locality: Quetta, [Balochistan Province, Pakistan]. See Kitzmiller (1982).

moloiensis (Taylor, 1929): *Aedes* – not stated (inferential). Molloy; *-ensis*, L. suffix denoting place, locality, country. Type locality: Mount Molloy, Queensland, Australia.

mongiro van Someren, 1951: *Culex* – not stated (inferential). Mongiro, treated as a latinized noun in the nominative singular standing in apposition to the generic name. Type locality: Mongiro, Bwamba County, Toro District (now in Bundibugyo District), [Western Region], Uganda. See *bwamba* van Someren, 1950c.

moramangae da Cunha Ramos & Brunhes, 2004: *Uranotaenia* – Moramanga; *-ae*, L. ending for first declension genitive singular noun (feminine). 'Name in relation to the locality of Moramanga (Province of Tamatave / Taomasina), where the type series was collected' [translated from French]. Type locality: Moramanga, on the road to Beparasy, Moramanga District, [Alaotra-Mangoro Region], Madagascar.

moresbyensis Peters, 1963e: *Uranotaenia* – not stated (inferential). Moresby; *-ensis*, L. suffix denoting place, locality, country. Type locality: Port Moresby, Central District, [Papua] New Guinea. Port Moresby, also known as Pom City or simply Moresby, located on the shores of the Gulf of Papua, is the capital of Papua New Guinea.

mpusiensis Huang, 2004: *Aedes* – not stated (inferential). Mpuse; *-ensis*, L. suffix denoting place, locality, country. Type locality: Mont Mpuse (Mount Mouse, i.e. Mount Kahuzi), Democratic Republic of the Congo. Mount Kahuzi is an extinct volcano. The extremely rare Mount Kahuzi Climbing Mouse (*Dendromus kahuziensis*) has only ever been found on its slopes.

mubiensis Luh & Shih, 1958: *Aedes* – not stated (inferential). Mubi; *-ensis*, L. suffix denoting place, locality, country. Type locality: Mubi, Guangxi Autonomous Region, China. The authors unintentionally duplicated the *-en* of *-ensis* thus rendering the name *mubienensis*. Ward (1992) corrected the spelling to *mubiensis* but incorrectly listed the type locality as Mubien, the spelling of which has gone uncorrected until now.

muruae Sirivanakarn, 1968: *Culex* – not stated (inferential). Murua; *-ae*, L. ending for first declension genitive singular noun (feminine). Type locality: Murua, Gulf Province, [Papua] New Guinea.

nairobiensis (van Someren, 1946b): *Toxorhynchites* – not stated (inferential). Nairobi; *-ensis*, L. suffix denoting place, locality, country. Type locality: City Park, Nairobi, Kenya.

nakuruensis Mattingly, 1951b: *Culex* – not stated (inferential). Nakuru; *-ensis*, L. suffix denoting place, locality, country. Type locality: Nakuru, [Nakuru County], Kenya.

namibiensis Coetzee, 1984: *Anopheles* – not stated (inferential). Namibia; *-ensis*, L. suffix denoting place, locality, country. Type locality: Kavango (Kavango East Region), Mahongo (Mahango Game Reserve, Bwabwata National Park), Namibia. Kavango was one of the 13 Regions of Namibia until it was split into the Kavango East and Kavango West Regions in 2013. See Service (2010).

natalensis Edwards, 1930a: *Aedes* – not stated (inferential). Natal; *-ensis*, L. suffix denoting place, locality, country. Type locality: Umbilo, Durban, Natal (KwaZulu-Natal Province), South Africa. Natal was a province of South Africa from 1910 until 1994. It included the Bantustan (territory) of KwaZulu, which was reincorporated into the territory of Natal and became KwaZulu-Natal Province.

natalensis (Hill & Haydon, 1907): *Anopheles* – not stated (inferential). See *natalensis* Edwards, 1930a. Type locality: Colony of Natal (KwaZulu-Natal Province), South Africa. See Kitzmiller (1982).

nevadensis Chapman & Barr, 1964: *Aedes* – not stated (inferential). Nevada; *-ensis*, L. suffix denoting place, locality, country. Type locality: Lamoille Canyon, Elko County, Ruby Mountains, eastern Nevada, USA. Knight & Stone (1977) incorrectly placed the type locality in the state of Utah. The Ruby Mountain range is primarily located within Elko County of Nevada with a small extension into White Pine County, Nevada.

ngong van Someren, 1950a: *Aedes* – not stated (inferential). Ngong, treated as a latinized noun in the nominative singular standing in apposition to the generic name. Type locality: Langata Forest, Ngong, Nairobi, Kenya. Ngong is a town near the Ngong Hills located in the south-west of Nairobi.

nicaroensis Duret, 1967a: *Culex* – not stated (inferential). Nicaro; *-ensis*, L. suffix denoting place, locality, country. Type locality: Nicaro, Mayarí Municipality, Oriente Province (Holguín Province), Cuba. Oriente Province was split up in 1976 and is now represented by five provinces. Nicaro, Mayarí Municipality is located in present-day Holguín Province.

nicceriensis Bonne-Wepster & Bonne, 1920: *Culex* – not stated. Niccerie; *-ensis*, L. suffix denoting place, locality, country. Type locality: Kabelstation (destroyed), [Sarakreek Municipality, Brokopondo District], Suriname. Kabelstation is a historical populated place in Brokopondo District that was destroyed when the large Brokopondo Reservoir was built between 1961 and 1964. However, the species name is seemingly based on Niccerie, apparently an alternative or older spelling of Nickerie, a district (and a river) in the north-west of Suriname. See *coppenamensis* Bonne-Wepster & Bonne, 1920 and *saramaccensis* Bonne-Wepster & Bonne, 1920.

nicoyae Adames & Hogue, 1970: *Deinocerites* – not stated. Nicoya; *-ae*, L. ending for first declension genitive singular noun (feminine). Type locality: Estero El Mero, Boca del Rio Barranca (mouth of the Barranca River), Puntarenas Province, Costa Rica. Puntarenas Province covers most of the Pacific coast of Costa Rica. The northern area of the province encompasses the southern part of the Nicoya Peninsula.

nigeriae Edwards, 1930a: *Hodgesia* – not stated (inferential). Nigeria; *-ae*, L. ending for first declension genitive singular noun (feminine). Type locality: Ebute Metta, near Lagos, [Lagos State], Nigeria.

Ebute Metta is today a neighbourhood of Lagos Mainland.

nigeriensis Evans, 1931a: subspecies of *Anopheles moucheti* Evans, 1925b – not stated (inferential). Nigeria; *-ensis*, L. suffix denoting place, locality, country. Type locality: Yaba, [Lagos State], Nigeria. Yaba is today a suburb located in Lagos Mainland. See Kitzmiller (1982).

nigeriensis Ribeiro, 2005: *Toxorhynchites* – 'The new species is named after the country of origin of the holotype.' See *nigeriensis* Evans, 1931a. Type locality: Shaki, [Oyo State], Nigeria. Shaki is a town situated in the northern part of Oyo State in western Nigeria.

nilgiricus Christophers, 1924b: *Anopheles* – not stated. Nilgiri; *-icus*, L. belonging to, like, of, pertaining to. Type locality: [Nilgiri Hills, Nilgiris District, Tamil Nadu State], India. The Nilgiri Hills are part of the Western Ghats mountain chain in western Tamil Nadu, Karnataka and Kerala States (formerly parts of Madras State), in southern India. Madras State was renamed Tamil Nadu State in 1969. See Kitzmiller (1982).

nilgiricus Edwards, 1916: *Culex* – not stated (inferential). See *nilgiricus* Christophers, 1924b. Type locality: Utakamand (Udhagamandalam), Nilgiri Hills, [Nilgiris District], Madras (present-day Tamil Nadu State), India. Udhagamandalam (also Ootacamund, abbreviated as Udhagai and Ooty) is a town (hill station), municipality and capital of the Nilgiris District located in the Nilgiri Hills.

nili (Theobald, 1904b): *Anopheles* – not stated (inferential). *Nilus*, (Gr. *Neilos*), the Nile River; *-i*, L. ending for second declension genitive singular noun (masculine and neuter). 'Habitat – Jebel Aket-Aga on the White Nile'. Type locality: Jebel Akmet-Aga (Jebel Ahmed Agha), [Upper Nile State], [South] Sudan. Jebel Ahmed Agha, also known as Bibban, Jabal Ahmad Agha and Jabal Aḥmad Āghā, is a hill in Upper Nile State of present-day South Sudan. See Kitzmiller (1982).

ninagongoensis Edwards, 1928b: *Culex* – not stated (inferential). Ninagongo; *-ensis*, L. suffix denoting place, locality, country. Type locality: Mount Ninagongo (Nyirangongo), Kivu Volcanoes, [North Kivu Province], Belgian Congo (present-day Democratic Republic of the Congo). Mount Nyiragongo is an active volcano located in the Virunga Mountains (Virunga National Park) north of Lake Kivu near the border with Rwanda.

ningheensis Lei, 1989b: *Aedes* – not stated (inferential). Ninghe; *-ensis*, L. suffix denoting place, locality, country. Type locality: Wushan County, Sichuan Province (Chongqing Municipality), China. Wushan County is a county located in Chongqing Municipality, which was created in 1997, succeeding the sub-provincial city administration that was part of Sichuan Province. The species is described from larvae collected in stone pits or pools in the Dangyang River, a branch of the Daninghe River. The species should have been named for the latter river, which means 'Large Ninghe River', but there is also a 'Small Nighe River' that is a branch of the Large Ninghe (Bin Chen, personal communication).

nipponica La Casse & Yamaguti, 1950: *Culiseta* – not stated. See *nipponicus* La Casse & Yamaguti, 1948. Type locality: Lake Akan, Hokkaido Island, Japan. Lake Akan is located in Akan National Park in Kushiro on Hokkaido Island.

nipponicus La Casse & Yamaguti, 1948: *Aedes* – not stated. Nippon; *-icus* (f. *-ica*), L. belonging to, like, of, pertaining to. Type locality: [Maizuru, Wakayama, Yodo Nara], Honshu Island and [Sasabo, Fukuoka, Kogoshima], Kyushu Island, Japan. Nippon (and Nihon) is a Japanese name for Japan. Both names literally mean 'the sun's origin', and are often translated as the Land of the Rising Sun.

nipponii (Theobald, 1907): subspecies of *Aedes vexans* (Meigen, 1830b) – not stated. Nippon; *-ii*, L. disused inappropriate genitive singular noun case-ending (masculine). See *nipponicus* La Casse & Yamaguti, 1948. Type locality: Karnizana, [Saitama Prefecture, Honshu Island] Japan.

nissanensis Lee, 1946: *Tripteroides* – not stated (inferential). Nissan; *-ensis*, L. suffix denoting place, locality, country. Type locality: Nissan Island, Green Islands, Bismarck Archipelago, [Papua New Guinea]. The Bismarck Archipelago is a group of islands off the north-eastern coast of New Guinea and is part of the Islands Region of Papua New Guinea. Nissan Island is the largest of the Green Islands within the Autonomous Region of Bougainville (previously known as North Solomons Province) in the north-eastern part of the country.

njombiensis Huang, 1997: *Aedes* – not stated (inferential). Njombe; *-ensis*, L. suffix denoting place, locality, country. Type locality: Njombe, [Njombe Region], Tanganyika (present-day United Republic of Tanzania). Njombe Region is one of the 30 administrative regions of Tanzania. It was established in March 2012, when it was split off from Iringa Region. Tanganyika was a sovereign state that gained independence from the UK as the Commonwealth realm of Tanganyika before becoming a republic within the Commonwealth of Nations. Tanganyika officially joined with the People's Republic of Zanzibar and Pemba in 1964 to form the United Republic of Tanganyika and Zanzibar. The new state changed its name to the United Republic of Tanzania within a year. Njombe is a town in and capital of Njombe Region in southern Tanzania.

njombiensis Peters, 1955b: *Anopheles* – not stated (inferential). See *njombiensis* Huang, 1997. Type locality: Njombe, Tanganyika (present-day United Republic of Tanzania). See Kitzmiller (1982).

nkolbissonensis Rickenbach & Hamon, 1966: *Orthopodomyia* – not stated (inferential). Nkolbisson; *-ensis*, L. suffix denoting place, locality, country. Type locality: Agricultural Research Centre, Nkolbisson, 8 km from Yaoundé, southern Cameroon.

normanensis (Taylor, 1915): *Aedes* – not stated (inferential). Norman; *-ensis*, L. suffix denoting place, locality, country. Type locality: Normanton, Queensland, Australia. Normanton is a small cattle town and locality in the Shire of Carpentaria (local government area) in Queensland.

novaezealandiae Pillai, 1966: *Culiseta* – not stated (inferential). Nova Zealandia; *-ae*, L. ending for first declension genitive singular noun (feminine). Type locality: Tahakopa, South Otago, [South Island], New Zealand. New Zealand was called Staten Landt by the Dutch explorer Abel Tasman who sighted it in 1642 and assumed it was connected to a landmass of the same name at the southern tip of South America. The Dutch renamed it Nova Zeelandia in 1645 after the Dutch Province of Zeeland. The name was subsequently anglicized to New Zealand by the British explorer James Cook. Tahakopa is a small settlement in the southeastern area of the South Island.

novaguinensis Peters, 1963d: *Uranotaenia* – not stated (inferential). *Nova* (from Latin *novus*) *Guinea*, L.; *-ensis*, L. suffix denoting place, locality, country. Type locality: Ami, 10 miles northwest of Maprik, Sepik District (East Sepik Province), New Guinea (present-day Papua New Guinea).

novaguinensis Venhuis, 1933: *Anopheles* – not stated (inferential). See *novaguinensis* Peters, 1963d. Type locality: Kakajoe and Konggeap on Merauke River (Maro River), [Merauke Regency], Western New Guinea (Papua Province, Indonesia). See Kitzmiller (1982).

novohanoverae Peters, 1963b: *Tripteroides* – not stated (inferential). *Novo* (from Latin *novus*) Hanover; *-ae*, L. ending for first declension genitive singular noun (feminine). Type locality: Noipuas, New Hanover Island, [New Ireland Province], New Guinea (present-day Papua New Guinea). The island of New Hanover is a large volcanic island in New Ireland Province (part of the Bismarck Archipelago), the most north-eastern province of Papua New Guinea.

occidentalis Dyar & Knab, 1906b: *Anopheles* – not stated (inferential). *occidens*, L. the west; *-alis*, L. pertaining to, having the nature or quality of (western, occidental, situated in the Occident). The name implies the western part of North America. Type locality: Stanford University, California, USA. See Kitzmiller (1982).

occidentalis (Skuse, 1889): *Aedes* – not stated (inferential). See *occidentalis* Dyar & Knab, 1906b. The name implies western Australia. Type locality: King George Sound, Western Australia State, Australia. King George Sound is located on the south coast of Western Australia. It was originally named King George the Third's Sound and was referred to as King George's Sound from 1826 until about 1934 when it subsequently gradually became known as 'King George Sound'.

occidentayunnana (Gong & Lu, 1991): *Heizmannia* – not stated (inferential). *occident-*, *occidens*, L. direction of the setting sun, west (occidental, related to or situated in the west, western); Yunnan; *-us* (f. *-a*), L. of, pertaining to. Type locality: Longchuan County, Yunnan Province, China. Longchuan County borders Myanmar in the western Dehong Prefecture of Yunnan Province.

okinawae Bohart, 1953: *Culex* – not stated (inferential). Okinawa; *-ae*, L. ending for first declension genitive singular noun (feminine). Type locality: East Taira, Okinawa Island, Okinawa Prefecture, Ryukyu Islands, Japan. Okinawa is a Japanese prefecture comprising more than 150 islands in the East China Sea. Okinawa Island is the largest of the Okinawa Islands.

okinawanus Bohart, 1946: *Aedes* – not stated (inferential). Okinawa; *-anus*, L. belonging to, pertaining to. Type locality: Okuma, Okinawa Island, Okinawa Prefecture, Ryukyu Islands, Japan. See *okinawae* Bohart, 1953.

okinawensis Toma, Miyagi & Tanaka, 1990: *Toxorhynchites* – not stated (inferential). Okinawa; *-ensis*, L. suffix denoting place, locality, country. Type locality: Ada, Okinawajima (Okinawa Island), Okinawa Prefecture, Ryukyu Islands, Japan.

okuensis Brunhes, Le Goff & Geoffroy, 1997: *Anopheles* – Oku; *-ensis*, L. suffix denoting place, locality, country. 'We have given this mosquito the name *okuensis* because it seems to be located, at least at present, on the southern part of the Adamaoua range (Mount Oku region)' [translated from French]. Type locality: Sabga, [North-West Province], Cameroon. See Service (2010).

orbostiensis Dobrotworsky, 1958a: *Culex* – not stated (inferential). Orbost; *-ensis*, L. suffix denoting place, locality, country. Type locality: Orbost, [Shire of East Gippsland], Victoria, Australia.

orientalis Barraud, 1926: *Uranotaenia* – not stated (inferential). *oriens*, L. east; *-alis*, L. pertaining to, having the nature or quality of (eastern, oriental, situated in the Orient). Type locality: Golaghat, [Golaghat District], Assam State, India.

orientalis Edwards, 1921d: *Culex* – not stated (inferential). See *orientalis* Barraud, 1926. Type locality: Tokyo, [Honshu Island], Japan.

orientalis (Swellengrebel & Swellengrebel de Graaf, 1920a): subspecies of *Anopheles tessellatus* Theobald, 1901a – not stated (inferential). See *orientalis* Barraud, 1926. Type locality: North Celebes (North Sulawesi Province, Sulawesi), Moluccas (Spice Islands; Indonesian Maluku) and eastern Java, [Indonesia]. See Kitzmiller (1982).

otachati Klein & Sirivanakarn, 1970: *Culex* – not stated (inferential). O Tachat; *-i*, L. ending for second declension genitive singular noun (masculine and neuter). Type locality: Kirirom Hills, O Tachat, Kompong Speu Province, Cambodia.

ototomo da Cunha Ramos, 1993: *Uranotaenia* – not stated (inferential). Ototomo, treated as a latinized noun in the nominative singular standing in apposition to the generic name. Type locality: Ototomo, [Ngoumou commune, Méfou-et-Akono Department, Center Region], Cameroon.

ototomoensis Huang & Rueda, 2013: *Orthopodomyia* – not stated (inferential). Ototomo; *-ensis*, L. suffix denoting place, locality, country. Type locality: Ototomo, [Ngoumou commune, Méfou-et-Akono Department, Center Region], Cameroon.

otwayensis (Dobrotworsky, 1960b): *Culiseta* – not stated (inferential). Otway; *-ensis*, L. suffix denoting place, locality, country. Type locality: Cape Horn, Otway [Ranges], Victoria, Australia.

ovengensis Awono-Ambene, Kengne, Simard, Antonio-Nkonkjio & Fontenille, 2004: *Anopheles* – Oveng; *-ensis*, L. suffix denoting place, locality, country. '... we propose to elevate the formerly described *An. nili* "Oveng form" to specific rank, and we name it *Anopheles ovengensis*, from its type locality.' Type locality: Oveng, near Campo, [South Province], Cameroon. See Service (2010).

pahangi (Delfinado, 1968): *Verrallina* – not stated (inferential). Pahang; *-i*, L. ending for second declension genitive singular noun (masculine and neuter). Type locality: Kuala Terengan, Pahang State, peninsular Malaysia. Kuala Terengganu, often abbreviated as K.T., is the administrative capital, royal capital and the main economic centre of Terengganu, Malaysia. Kuala Terengganu is also the administrative centre for the eponymous district of Kuala Terengganu. Terengganu, formerly spelled Trengganu or Tringganu, is a sultanate and constitutive state of federal Malaysia.

palauensis Bohart, 1957: *Aedes* – not stated (inferential). Palau; *-ensis*, L. suffix denoting place, locality, country. Type locality: Ngergoi Island (Carp Island, also known as Garakayo and Ngercheu Island), Palau. Carp Island is located within Peleliu State of the Republic of Palau. The island is approximately 1 km^2 and densely forested.

palawanicus Mattingly, 1958a: *Aedes* – not stated (inferential). Palawan; *-icus*, L. belonging to, like, of, pertaining to. Type locality: Puerto Princesa, Palawan Island, Philippines. Puerto Princesa is a coastal city on Palawan Island in the western Philippines.

palmarum Edwards, 1924b: *Aedes* – not stated (inferential). *palmarum*, L. (genitive feminine plural of *palma*, L. palm), feminine noun in apposition to the masculine generic name *Aedes*. Type locality: [Greater] Palm Island, [Aboriginal Shire of Palm Island], Queensland, Australia. The Aboriginal Shire of Palm Island is a special local government area of Queensland State. Sixteen islands make up the Greater Palm group.

pampangensis (Ludlow, 1905a): *Aedes* – not stated (inferential). Pampanga; *-ensis*, L. suffix denoting place, locality, country. Type locality (neotype): Sison, Pangasinan, Luzon Island, Philippines (Reinert, 1973c). Ludlow based the description of *pampangensis* on specimens collected at Angeles in Pampanga Province. Angeles is today a highly urbanized city.

panamensis Adames, 1971: *Deinocerites* – not stated (inferential). Panama; *-ensis*, L. suffix denoting place, locality, country. Type locality: Portobelo, [Portobelo District], Colón Province, Panama.

panayensis (Ludlow, 1914c): *Verrallina* – not stated (inferential). Panay; *-ensis*, L. suffix denoting place, locality, country. Type locality: Iloilo [City, officially the City of Iloilo], Panay [Island], Philippines.

pantjarbatu Waktoedi Koesoemawinangoen, 1954: subspecies of *Anopheles gigas* Giles, 1901b – not stated. Pantjarbatu, treated as a latinized noun in the nominative singular standing in apposition to the generic name. Type locality: [Pancur Batu], Sumatra. The name is a variant spelling of Pancurbatu, a compound derived from Pancur Batu, a sub-district of Deli Serdang Regency in North Sumatra Province, Indonesia. See Kitzmiller (1982).

papuae (Swellengrebel & Swellengrebel de Graaf, 1920a): *Bironella* – not stated (inferential). Papua; *-ae*, L. ending for first declension genitive singular noun (feminine). Type locality: West New Guinea (Western New Guinea, Indonesia). Western New Guinea, or West Papua (region), formerly known as Irian Jaya, is the Indonesian part of the island of New Guinea. The eastern part of the island belongs to the nation of Papua New Guinea. The Papuan people are indigenous to New Guinea and the Solomon Islands. See Kitzmiller (1982).

papuensis Dobrotworsky, 1957: *Anopheles* – not stated (inferential). Papua; *-ensis*, L. suffix denoting place, locality, country. Type locality: Minj, [Papua] New Guinea. See Kitzmiller (1982).

papuensis Marks, 1960: *Topomyia* – not stated (inferential). See *papuensis* Dobrotworsky, 1957. Type locality: Koipa, 4–5 miles from Saiho on the Popondetta Road, Northern District [Oro (Northern) Province], Papua [New Guinea]. Koipa is a populated place in Oro Province (capital Popondetta) of Papua New Guinea.

papuensis Peters, 1963a: *Armigeres* – not stated (inferential). See *papuensis* Dobrotworsky, 1957. Type locality: Maprik (Maprik District), Sepik District (East Sepik Province), New Guinea (present-day Papua New Guinea).

papuensis (Taylor, 1914d): *Aedes* – not stated (inferential). See *papuensis* Dobrotworsky, 1957. Type locality: Milne Bay, [Milne Bay Province], Papua [New Guinea].

papuensis (Taylor, 1914d): *Culex* – not stated (inferential). See *papuensis* Dobrotworsky, 1957. Type locality: Lakekamu Gold Field (Lakekamu Goldfield mining district), [Lakekamu River, Gulf Province, Papua Region], Papua [New Guinea].

papuensis (Taylor, 1914d): *Mansonia* – not stated (inferential). See *papuensis* Dobrotworsky, 1957. Type locality: Lakekamu Gold Field (Lakekamu Goldfield mining district), [Lakekamu River, Gulf Province, Papua Region, Papua New Guinea].

papuensis Zavortink, 1968: *Orthopodomyia* – not stated (inferential). See *papuensis* Dobrotworsky, 1957. Type locality: Finschhafen, [Finschhafen District, Morobe Province], Territory of New Guinea (present-day Papua New Guinea).

paraensis (Theobald, 1903a): *Limatus* – not stated (inferential). Pará; *-ensis*, L. suffix denoting place, locality, country. Type locality: Pará [State], Brazil. Pará is the northern Brazilian state through which the lower Amazon River flows to the sea.

paraitepuyensis Anduze, 1941: *Sabethes* – not stated (inferential). Paraitepuy; *-ensis*, L. suffix denoting place, locality, country. Type locality: Foot of Paraitepuy (Paraitepui), Bolívar State, Venezuela. Tepuis (table-top mountains, mesas) populate the Gran Sabana region in the Guiana Highlands in south-eastern Venezuela.

paranensis (Brèthes, 1911): *Isostomyia* – not stated (inferential). Paraná; *-ensis*, L. suffix denoting place, locality, country. Type locality: Islas del Paraná, [Argentina]. The Islas del Paraná consist of several islands in the delta of the Paraná River (Paraná Delta). Most of the delta is in the jurisdiction of Entre Ríos Province, with parts in the north of Buenos Aires Province.

parangensis (Ludlow, 1914b): *Anopheles* – not stated (inferential). Parang; *-ensis*, L. suffix denoting place, locality, country. Type locality: Parang, [Maguindanao Province], Mindanao, Philippines. Parang is a municipality in Maguindanao Province, located in the Autonomous Region in Muslim Mindanao (ARMM). The ARMM is located in the Mindanao island group that consists of five predominantly Muslim provinces. It is the only region that has its own government, seated in Cotabato City, although this self-governing city is outside of its jurisdiction. See Kitzmiller (1982).

parensis Gillies, 1962: *Anopheles* – not stated (inferential). Pare; *-ensis*, L. suffix denoting place, locality, country. Type locality: Kihurio, South Pare District, Tanganyika (present-day United Republic of Tanzania). The Pare District is named for the Pare people, members of an ethnic group indigenous to the Pare Mountains, part of the Kilimanjaro Region (administrative region) of northern Tanzania. See Kitzmiller (1982).

patientiae Floch & Fauran, 1955: *Culex* – not stated. *patientia*, L. patience; *-ae*, L. ending for first declension genitive singular noun (feminine). Type locality: Patience (Haute-Mana), [Saül, Saint], French Guiana [overseas region and department of France]. Modern French *patience*, from Latin *patientia*.

paucartamboensis Porter, 2014: *Wyeomyia* – Paucartambo; *-ensis*, L. suffix denoting place, locality, country. 'The name *paucartamboensis* is derived from Paucartambo Province, one of 13 provinces of the Cisco Region of Peru.' Type locality: San Pedro, Kosñipata District, Paucartambo Province, Cusco Region, Peru. The Kosñipata District is one of six districts of the province.

peipingensis Feng, 1938: *Aedes* – not stated (inferential). Peiping; *-ensis*, L. suffix denoting place, locality, country. Type locality: Peiping (Beijing), China. Peiping (or Beiping), meaning 'Northern Peace', is a former name of Beijing, which means 'Northern Capital'. The city was called Beiping from 1368 to 1403 and from 1928 to 1949, when the Chinese capital was at Nanjing in Jiangsu Province. In 1403 and again in 1949, the name of the city was changed from Beiping to Beijing. Beijing and Tianjin Municipalities were carved out of Hebei (also Hopeh) Province.

pembaensis Theobald, 1901c: *Aedes* – not stated (inferential). Pemba; *-ensis*, L. suffix denoting place, locality, country. Type locality: Pemba Island, [Zanzibar, Tanzania], East Africa. Zanzibar is a semi-autonomous part of Tanzania in East Africa comprising the Zanzibar Archipelago in the Indian Ocean, which consists of many small islands and the two large islands of Pemba and Unguja (the main island, referred to informally as Zanzibar).

penghuensis Lien, 1968b: *Aedes* – not stated (inferential). Penghu; *-ensis*, L. suffix denoting place, locality, country. Type locality: Kunpei, Huhsi [Township], Penghu Hsien (County), Taiwan (officially the Republic of China). Penghu County of the Republic of China, known as Taiwan, comprises the Penghu or Pescadores Islands, an archipelago of 90 islands and islets in the Taiwan Strait. The largest island is also named Penghu.

persiensis Linton, Sedaghat & Harbach, 2003: *Anopheles* – Persia; *-ensis*, L. suffix denoting place, locality, country. 'The specific name, *persiensis*, is a Latinized geographical name meaning "from Persia". The final vowel of Persia is omitted for euphony and smoothness of pronunciation.' Type locality: Amol, [Amol County], Mazandaran Province, Iran (commonly known as Persia before 1955). Mazandaran Province is located along the southern coast of the Caspian Sea. See Service (2010).

phangngae Sirivanakarn, 1972: *Culex* – not stated (inferential). Phangnga; *-ae*, L. ending for first declension genitive singular noun (feminine). Type locality: Nam Tai, Phangnga (Phang Nga) Province, Thailand.

pharoensis Theobald, 1901a: *Anopheles* – not stated (inferential). Pharaoh, Pharao or Pharoah (*pharo-*); *-ensis*, L. suffix denoting place, locality, country. Type locality: Cairo, [Egypt]. The name is intended to mean 'from the Land of Pharaoh'. See Kitzmiller (1982).

philippinensis (Delfinado, 1968): *Verrallina* – not stated (inferential). Philippine; *-ensis*, L. suffix denoting place, locality, country. Type locality: Mahaplag, [Leyte Province], Leyte Island, Philippines.

philippinensis Ludlow, 1902b: *Anopheles* – not stated (inferential). See *philippinensis* (Delfinado, 1968). Type locality: San Jose, Abra Province, Luzon Island, Philippines. See Kitzmiller (1982).

philippinensis Sirivanakarn, 1976: *Culex* – not stated (inferential). See *philippinensis* (Delfinado, 1968). Type locality: Tay-Tay, Rizal Province, Luzon Island, Philippines.

phoeniciae Coluzzi & Sabatini, 1968: *Aedes* – not stated. Phoenicia; *-ae*, L. ending for first declension genitive singular noun (feminine). 'For the third subspecies [now a species], referred to as the "Middle East" form, there appears to be no available name in the literature; it is therefore considered appropriate to propose the new name *Aedes mariae phoeniciae*' [translated from Italian]. Type locality: Beirut, Lebanon. Phoenicia was an ancient civilization that occupied lands along the eastern coast of the Mediterranean Sea that correspond to present-day Lebanon, Israel, Gaza, Syria and south-western Turkey.

pingpaensis Chang, 1965: *Aedes* – not stated (inferential). Pingpa; *-ensis*, L. suffix denoting place, locality, country. Type locality: Pingpa (Pingba) District, Kweichow (present-day Guizhou) Province, China. Pingba is a district of Guizhou Province. Guizhou, located in the south-western part of China, was formerly anglicized as Kweichow.

pinjaurensis Barraud, 1932: *Anopheles* – not stated (inferential). Pinjaur; *-ensis*, L. suffix denoting place, locality, country. Type locality: Pinjaur (Pinjore), near Kalka, Patiala State (in part now Haryana State), India. Pinjore is a town in the Panchkula District of Haryana State. Patiala State was a self-governing princely state outside British India. When the British left India in 1947, Patiala acceded to the new Union of India. Patiala State corresponds to present-day Punjab and Haryana States. See Kitzmiller (1982).

polynesiensis Marks, 1951: *Aedes* – Polynesia; *-ensis*, L. suffix denoting place, locality, country. 'The widespread pest species and filaria-carrier of Polynesia therefore requires a new name, and I propose to call it *Aëdes* (*Stegomyia*) *polynesiensis* sp. nov.' Type locality: Taveuni Island, Fiji. The country of Fiji is an archipelago of more than 300 islands in the South Pacific. Taveuni is the fifth-largest island in Fiji.

portonovoensis Tewari & Hiriyan, 1992a: *Aedes* – Porto Novo; *-ensis*, L. suffix denoting place, locality, country. 'This species is named after a well known, small, coastal town Porto Novo situated nearby the type locality.' Type locality: Pichavaram mangrove swamp, South Arcot District, Tamil Nadu State, India.

portoricensis (von Röder, 1885): *Toxorhynchites* – not stated (inferential). Portorico; *-ensis*, L. suffix denoting place, locality, country. 'This collection from Portorico' [translated from German]. Type locality: Porto Rico. Puerto Rico is an island in the Caribbean Sea and unincorporated territory of the USA.

pretoriensis (Theobald, 1903a): *Anopheles* – not stated (inferential). Pretoria; *-ensis*, L. suffix denoting place, locality, country. Type locality: Pretoria, [Gauteng Province], South Africa. Gauteng Province was part of Transvaal, a province of South Africa from 1910 until the end of apartheid in 1994 when it was subdivided under a new constitution. See Kitzmiller (1982).

principensis da Cunha Ramos, 1993: *Uranotaenia* – not stated (inferential). Príncipe; *-ensis*, L. suffix denoting place, locality, country. Type locality: Bela Vista, Príncipe Island, São Tomé and Príncipe.

prioekanensis (Brug, 1931a): *Verrallina* – not stated (inferential). Prioekan; *-ensis*, L. suffix denoting place, locality, country. Type locality: virgin forest near Air Priokan village, near Benkoelen (Bengkulu), [Bengkulu Province], Sumatra, [Indonesia].

pujutensis Colless, 1948: *Anopheles* – not stated (inferential). Pujut; *-ensis*, L. suffix denoting place, locality, country. Type locality: Pujut, Sarawak [State], North-west Borneo, [Malaysia]. Pujut is today a state constituency that has been represented in the Sarawak State Legislative Assembly since 2006. See Kitzmiller (1982).

pursati Laveran, 1902a: *Anopheles* – not stated (inferential). Pursat; *-i*, L. ending for second declension genitive singular noun (masculine and neuter). Type locality: west of Pursat, [Pursat Province], Cambodia. Laveran states that the mosquito was collected 'in the mountainous and wooded area, west of Pursat' [translated from French]. See Kitzmiller (1982).

putumayensis Matheson, 1934: *Culex* – not stated (inferential). Putumayo; *-ensis*, L. suffix denoting place, locality, country. Type locality: Amazon River (restricted to Santo Antonio do Ica, Amazonas State, Brazil by Belkin *et al*., 1971). Although seemingly ironical, the author obviously chose to name the species for paratypes collected along the Putumayo River of Peru rather than the locality of the holotype because *amazonensis* was already in use for another species of *Culex*, i.e. *Cx*. (*Aedinus*) *amazonensis* (Lutz, 1905).

quettensis Mattingly, 1955: *Culex* – not stated (inferential). Quetta; *-ensis*, L. suffix denoting place, locality, country. Type locality: Quetta, Béloutschistan (Balochistan) Province, [Pakistan]. Quetta is the provincial capital of Balochistan Province.

quitensis Levi-Castillo, 1953a: *Culex* – not stated (inferential). Quito; *-ensis*, L. suffix denoting place, locality, country. Type locality: Quito, [Pichincha Province], Ecuador. Quito is the capital city of Ecuador and Pichincha Province.

reginae Floch & Fauran, 1955: *Culex* – not stated (inferential). Régina; *-ae*, L. ending for first declension genitive singular noun (feminine). Type locality: Régina (Approuague River), [Cayenne Arrondissement], French Guiana [overseas region and department of France]. Régina is a commune lying on the Approuague River.

rennellensis Taylor & Maffi, 1991: *Anopheles* – not stated (inferential). Rennell; *-ensis*, L. suffix denoting place, locality, country. Type locality (lectotype): Matangi, Rennell Island, [Solomon Islands] (Harbach & Ward, 1994). Rennell Island, locally known as Mugaba, is the main island of two inhabited islands that make up the Rennell and Bellona Province in the sovereign country of Solomon Islands. See Kitzmiller (1982).

reunionensis Brunhes & Hervy, 1995: *Orthopodomyia* – not stated (inferential). Réunion; *-ensis*, L. suffix denoting place, locality, country. Type locality: Réunion Island, 'near the volcanic flow that extends from the Piton de la Fournaise and reaches the sea at the place known as "Grand Brûlé" (commune of St-Philippe)' [translated from French]. Réunion Island is a French department in the Indian Ocean. Piton de la Fournaise (Peak of the Furnace) is a shield volcano on the eastern side of the island.

rhodesiensis Theobald, 1901a: *Anopheles* – not stated (inferential). Rhodesia; *-ensis*, L. suffix denoting place, locality, country. Type locality: [Harare], Mashonaland, [Zimbabwe]. Mashonaland (land of the Shona people) is a region in northern Zimbabwe that is divided into three provinces, Mashonaland West, Mashonaland Central and Mashonaland East. Additionally, Harare (officially called Salisbury until 1982), is itself a province that lies entirely in Mashonaland. See Kitzmiller (1982).

rhungkiangensis Chang & Chang, 1974: *Aedes* – not stated (inferential). Rhungkiang; *-ensis*, L. suffix denoting place, locality, country. Type locality: Rhungkiang (Rongjiang) [County], Kweichow (present-day Guizhou) Province, China. Rongjiang County, under the administration of the Qiandongnan Miao and Dong Autonomous Prefecture, is located in the south-eastern region of Guizhou Province. The province, formerly anglicized as Kweichow, is located in south-western China.

ribeirensis Forattini & Sallum, 1985: *Culex* – not stated (inferential). Ribeira; *-ensis*, L. suffix denoting place, locality, country. Type locality: Experimental Station of Ribeira Valley (Vale do Rio Ribeira), Pariquera-Açú County, São Paulo State, Brazil.

riojanus Duret, 1968e: *Culex* – Not stated (inferential). Rioja; *-anus*, L. belonging to, pertaining to. Type locality: Chamical, [Chamical Department], La Rioja Province, Argentina. Chamical is a small city and the seat of government of the Chamical Department in the south of La Rioja Province.

rorotaensis Floch & Abonnenc, 1946a: *Culex* – Rorota; *-anus*, L. belonging to, pertaining to. 'We regard this Culex [*sic*] as belonging to a species not described and call it *C. rorotaensis* for the name of the place of capture (Rorota)' [translated from French]. Type locality: Rorota, [Cayenne Arrondissement], French Guiana [overseas region and department of France].

rorotai Senevet, Chabelard & Abonnenc, 1942: *Wyeomyia* – not stated (inferential). Rorota; *-i*, L. ending for second declension genitive singular noun (masculine and neuter). '*W. rorotai* was found on the heights of [Mount] Rorota (near Cayenne, French Guiana on 6-7-39, in the leaves of Bromeliaceae' [translated from French]. Type locality: [Mount] Rorota, near Cayenne, [Cayenne Arrondissement], French Guiana [overseas region and department of France].

rossicus Dolbeškin, Gorickaja & Mitrofanova, 1930: *Aedes* – not stated (inferential). *rossicus*, Mod. L. Russian; geographical indicator meaning 'from Russia'. Type locality: Dnepropetrovsk (present-day Dnipro), Perm and Orenburg, U.S.S.R. Dnipro is a city in present-day Ukraine; Perm is a city and the administrative centre of Perm Krai and Orenburg is the administrative centre of Orenburg Oblast, both in present-day Russia.

rotanus Bohart & Ingram, 1946b: *Aedes* – not stated (inferential). Rota; *-anus*, L. belonging to, pertaining to. Type locality: north side Rota Island, Marianas (Mariana Islands). The Mariana Islands (or simply the Marianas) consist of 15 mostly dormant volcanic mountains of the Marianas Archipelago that lies south-southeast of Japan. They are politically divided into two jurisdictions of the USA: the Commonwealth of the Northern Mariana Islands and the southern territory of Guam. Rota is the southernmost island of the Commonwealth of the Northern Mariana Islands.

rotoruae Belkin, 1968a: *Culex* – not stated (inferential). Rotorua; *-ae*, L. ending for first declension genitive singular noun (feminine). Type locality: Kuirau Hot Springs Reserve, Rotorua, [Rotorua District], South Auckland (inappropriate), [North Island], New Zealand. South Auckland is an imprecisely defined area of Auckland. Prior to the 1970s, South Auckland encompassed a broad region immediately south of Auckland, which does not extend as far south as Rotorua.

rotumae Belkin, 1962: *Aedes* – not stated (inferential). Rotuma; *-ae*, L. ending for first declension genitive singular noun (feminine). Type locality: Oinafa village, Rotuma Island, Rotuma [Fijian dependency]. Rotuma comprises Rotuma Island and nearby islets. See *rotumanus* (Edwards, 1929b).

rotumanus (Edwards, 1929b): *Tripteroides* – not stated (inferential). Rotuma; *-anus*, L. belonging to, pertaining to. Type locality: Rotuma Island, Fiji [Fijian dependency]. Fiji is divided administratively into four divisions, which in turn are subdivided into 14 provinces. Rotuma (Rotuma Island and nearby islets) is self-governing and separate from the four divisions.

ruiliensis Dong, Zhou & Wang, 1997: *Heizmannia* – not stated (inferential). Ruili; *-ensis*, L. suffix denoting place, locality, country. Type locality: Ruili County, [Dehong Prefecture], Yunnan Province, China. Ruili is a county-level city on the border with Myanmar.

ruwenzori Haddow & van Someren, 1950: *Aedes* – not stated (inferential). Ruwenzori, treated as a latinized noun in the nominative singular standing in apposition to the generic name. Type locality: Ruwenzori (Rwenzori Mountains), [Western Uganda Region], Uganda. The Rwenzori Mountains were previously called the Ruwenzori Range (Mountains of the Moon). The spelling was changed circa 1980 to be more akin to the local name 'Rwenjura', meaning 'the rain maker'.

ruwenzori (van Someren, 1948): *Toxorhynchites* – not stated (inferential). See *ruwenzori* Haddow & van Someren, 1950. Type locality: Bwamba Pass, Ruwenzori (Rwenzori Mountains), [Bundibugyo District, Western Uganda Region], Uganda.

ryukyensis Bohart, 1946: *Culex* – not stated (inferential). Ryukyu; *-ensis*, L. suffix denoting place, locality, country. Type locality: Chizuka, Okinawa (Island),

[Okinawa Prefecture, Japan]. The Okinawa Prefecture encompasses hundreds of the Ryukyu Islands of Japan. Okinawa Island is the largest of the Ryukyu Islands.

ryukyuana Tanaka, Mizusawa & Saugstad, 1979: subspecies of *Uranotaenia novobscura* Barraud, 1934 – not stated (inferential). Ryukyu; *-anus* (f. *-ana*), L. belonging to, pertaining to. Type locality: Okinawa Island, Ryukyu Archipelago (Okinawa Prefecture), Japan. See *ryukyensis* Bohart, 1946.

ryukyuanus Tanaka, Mizusawa & Saugstad, 1979: subspecies of *Culex hayashii* Yamada, 1917 – not stated (inferential). See *ryukyuana* Tanaka, Mizusawa & Saugstad, 1979. Type locality: near Shirahama, Iriomote Island, Ryukyu Archipelago (Okinawa Prefecture), Japan. Iriomote Island is the largest of the Yaeyama Islands and the second largest island in Okinawa Prefecture.

sabahensis Ramalingam & Ramakrishna, 1988: *Topomyia* – not stated (inferential). Sabah; *-ensis*, L. suffix denoting place, locality, country. Type locality: Double Hill Forest Reserve, Semporna, [Tawau Division], Sabah State, Malaysia.

sabahensis (Reinert, 1974): *Verrallina* – not stated (inferential). See *sabahensis* Ramalingam & Ramakrishna, 1988. Type locality: Kg. Bingul (Kampung Bingul), Beaufort, [Interior Division], Sabah State, Malaysia.

saipanensis Stone, 1945b: *Aedes* – not stated (inferential). Saipan; *-ensis*, L. suffix denoting place, locality, country. Type locality: Saipan, Mariana Islands. Saipan is the largest of the Northern Mariana Islands, a US commonwealth in the Western Pacific.

salbaii Maffi & Coluzzi, 1958: *Anopheles* – not stated (inferential). Salbai; *-i*, L. ending for second declension genitive singular noun (masculine and neuter). 'The *tuc Salbai* is a right tributary of the upper course of the Juba River, and flows in this one 15 km downstream from the town of Lugh Ferrandi' [translated from Italian]. Type locality: Tuc Salbai (intermittent stream), Upper Juba Region (present-day Gebo Region), Somalia. The Gedo Region, formerly part of the historic Upper Juba Region, is an administrative region in southern Somalia. See Kitzmiller (1982).

salisburiensis Theobald, 1901c: *Culex* – not stated (inferential). Salisbury; *-ensis*, L. suffix denoting place, locality, country. Type locality: Salisbury (present-day Harare), Mashonaland, [Zimbabwe]. Harare was officially called Salisbury until 1982. Mashonaland is a region in northern Zimbabwe which is currently divided into three provinces, Mashonaland West, Mashonaland Central and Mashonaland East (including Harare). See *mashonaensis* Theobald, 1901c and *rhodesiensis* Theobald, 1901a.

saltanensis Dyar, 1928: *Culex* – not stated (inferential). Salta; *-ensis*, L. suffix denoting place, locality, country. Type locality: Campo Santa (Santo) de Salta [Province], Argentina. Campo Santo is a locality some 50 km from the city of Salta in the General Güemes Department of Salta Province.

samarensis Rozeboom, 1951: *Anopheles* – Samar; *-ensis*, L. suffix denoting place, locality, country. 'Because of the insular nature of the distribution [Samar Island] of this mosquito, which has several close relatives in the Malay Peninsula and neighboring regions, a name based upon the type locality would appear to be appropriate.' Type locality: Osmena, [Eastern Samar Province], Samar Island, Philippines. Samar Island is divided into three provinces: Samar Province (the western region of the island), Northern Samar Province and Eastern Samar Province. See Kitzmiller (1982).

samoaensis Stone, 1966: *Coquillettidia* – not stated (inferential). Samoa; *-ensis*, L. suffix denoting place, locality, country. Type locality: Tavalogi Ridge, Tau [Ta'u] Island, Manua [Manu'a Islands] Group, Eastern [American] Samoa. Ta'u is the largest island in the Manu'a Islands Group, which consists of three main islands, Ta'u, Ofu and Olosega. Ta'u Island is part of American Samoa, an unincorporated territory of the United States. See *samoaensis* (Theobald, 1914).

samoaensis (Theobald, 1914): *Culex* – not stated (inferential). See *samoaensis* Stone, 1966. Type locality: Apia, [Upolu Island], Samoa. The Independent State of Samoa, commonly known as Samoa, is a unitary parliamentary democracy with eleven administrative divisions. Apia, the capital and the largest city of Samoa, is located on the central north coast of Upolu island.

samoanus (Grünberg, 1913): *Aedes* – not stated (inferential). Samoa; *-anus*, L. belonging to, pertaining to. Type locality: Apia, [Upolu Island], Samoa. See *samoaensis* (Theobald, 1914).

sanctielii Senevet & Abonnenc, 1938: *Anopheles* – not stated (inferential). Saint-Élie; *sancti* (genitive singular of L. *sanctus*, venerable, august, divine, blessed, holy, saintly, sacred) + *eli-* and *-i*, L. ending for second declension genitive singular noun (masculine and neuter). Type locality: Mines de Saint-Elie, Sinnamary River, French Guiana. Saint-Élie is a commune and former capital of the inland Inini territory that was administered separately from 1930 until 1946 when all of French Guiana became a department of France. See Kitzmiller (1982).

sangengluoensis Wang, 1984: *Culex* – not stated (inferential). Sangengluo; *-ensis*, L. suffix denoting place, locality, country. Type locality: Sangengluo Commune, Wanning, Hainan [Province], China. Wanning is a county-level land area ('city') in the south-east of Hainan Province. Wanning City lies within this area.

saramaccensis Bonne-Wepster & Bonne, 1920: *Culex* – not stated. Saramacca; *-ensis*, L. suffix denoting place, locality, country. Type locality: Kabelstation (destroyed), Surinam River, [Brokopondo District], Suriname. Kabelstation is a historical populated place in Brokopondo District that was destroyed when the large Brokopondo Reservoir was built between 1961 and 1964. However, the species name is apparently based on Saramacca, which is a district and a river in the north of Suriname. See *coppenamensis* Bonne-Wepster & Bonne, 1920 and *nicceriensis* Bonne-Wepster & Bonne, 1920.

sardinerae Fox, 1953: *Culex* – not stated (inferential). Sardinera; *-ae*, L. ending for first declension genitive singular noun (feminine). Type locality: Finca Sardinera (Sardinera Estate), Dorado, Puerto Rico. Puerto Rico is a Caribbean island and unincorporated US territory. Dorado is a town and municipality on the northern coast of the island.

satpuraensis Natarajan, Rajavel & Jambulingam, 2017: *Uranotaenia* – Satpura; *-ensis*, L. suffix denoting place, locality, country. 'This species is named for the Satpura hill range of Eastern Ghats in Chattisgarh State of India where it was collected.' Type locality: Lapha, Korba District, Chattisgarh State, India.

saungi Colless, 1955a: *Anopheles* – not stated (inferential). Saung; *-i*, L. ending for second declension genitive singular noun (masculine and neuter). Type locality: Saung-Saung, Sunsuran Trace (footpath between Jesselton and Sunsuran), [Crocker Range], North Borneo (Sabah, Malaysia). See Kitzmiller (1982).

sedaensis Lei, 1989a: *Aedes* – not stated (inferential). Seda; *-ensis*, L. suffix denoting place, locality, country. Type locality: Seda and Songpan Counties, Sichuan Province, China. Seda, or 'Sertar', is a remote county under administration of the Garzê Tibetan Autonomous Prefecture and Songpan is a county under the administration of the Ngawa Tibetan and Qiang Autonomous Prefecture, both in the north-western region of Sichuan Province.

selai Klein & Sirivanakarn, 1970: *Culex* – not stated (inferential). Sela; *-i*, L. ending for second declension genitive singular noun (masculine and neuter). Type locality: Kompong Sela, Koh-Kong Province, Cambodia.

selangorensis Sirivanakarn, 1976: *Culex* – not stated (inferential). Selangor; *-ensis*, L. suffix denoting place, locality, country. Type locality: Bukit (Bt.) Kutu, Selangor State, (peninsular) Malaysia. Bukit Kutu is a mountain, height 1053 m.

semlikiensis van Someren, 1950b: *Aedes* – not stated (inferential). Semiliki; *-ensis*, L. suffix denoting place, locality, country. Type locality: Semliki River, Bwamba [County], Toro District (now in Bundibugyo District), [Western Region], Uganda. See *bwamba* van Someren, 1950c.

sentania (King & Hoogstraal, 1947b): *Verrallina* – not stated (inferential). Sentani; *-ius* (f. *-ia*), L. of, belonging to, pertaining to. Type locality: Poee Village, south side of Lake Sentani, Hollandia area (now Jayapura City area), Netherlands New Guinea (present-day Papua Province, Indonesia). Lake Sentani is a tropical, shallow, low-altitude open lake located at the northeast extremity of the Jayapura Regency in Papua Province of Indonesia.

senyavinensis Knight & Hurlbut, 1949: *Aedes* – Senyavin; *-ensis*, L. suffix denoting place, locality, country. 'This species is named after the Senyavin Islands, which is the eastern island group in the Carolines that includes the island of Ponape.' Type locality: Rekisau, Not District, Ponape (Pohnpei) Island, Caroline Island Group, [Federated States of Micronesia]. Pohnpei is an island of the Senyavin Islands, which are part of the larger Caroline Islands group. It belongs to Pohnpei State, one of the four states in the Federated States of Micronesia.

seoulensis Yamada, 1921: *Aedes* – not stated (inferential). Seoul; *-ensis*, L. suffix denoting place, locality, country. Type locality: Keijo (Seoul), [South] Korea. During the Japanese invasion and colonization of Korea from 1910 to 1945, Seoul was named Keijo (Japanese) or Gyeongseong (Korean). Seoul, the capital of South Korea, is a huge metropolis where the species may no longer exist.

shoae Hamon & Ovazza, 1954: *Culex* – Shoa; *-ae*, L. ending for first declension genitive singular noun (feminine). 'During surveys carried out in the Shoa ... one of us harvested *Culex* larvae ... whose breeding provided ♂ and ♀ adults for which we propose the name of *Culex shoæ*.' Type locality: Shoa (Shewa Province), Ethiopia. Shewa, formerly romanized as Shoa, is part of the mountainous plateau in the central area of Ethiopia, which includes the centrally located Ethiopian capital Addis Ababa.

siamensis Zavortink, 1968: *Orthopodomyia* – not stated (inferential). Siam; *-ensis*, L. suffix denoting place, locality, country. Type locality: Muang (Mueang) Trang, [Trang Province], Thailand. Siam is an exonym formerly used as the name of Thailand. Mueang Trang (previously Mueang Thap Thiang) is the capital district of Trang Province, which borders the Andaman Sea in the southern peninsular area of Thailand.

sibiricus (Danilov & Filippova, 1978): *Aedes* – not stated (inferential). Siberia (Ger. Sibirien); *-icus*, L. belonging to, like, of, pertaining to. Type locality: Kedrovaya Pad Nature Reserve, Maritime Territory (Primorsky Krai Province), USSR (present-day Russia). Kedrovaya Pad is a nature reserve in Khasansky District in the southern portion of Primorsky Krai, informally known as Primorye. Primorsky in Russian means 'maritime'; thus, it is translated in English as the Maritime Province or Maritime Territory.

sierrensis (Ludlow, 1905b): *Aedes* – not stated (inferential). Sierra; *-ensis*, L. suffix denoting place, locality, country. Type locality: Three Rivers (?), Sierra Nevada Mountains, California State, U.S.A. The author was uncertain whether the specimens sent to her were from Three Rivers; hence, the question mark. Three Rivers is an unincorporated community in Tulare County of California, located in the foothills of the Sierra Nevada.

simlensis Edwards, 1922a: *Aedes* – not stated (inferential). Simla; *-ensis*, L. suffix denoting place, locality, country. Type locality: Simla (Shimla), [Himachal Pradesh State], India. Shimla, also known as Simla, a community of the former East Punjab Province, is today the capital of Himachal Pradesh State located in the Himalayan foothills.

simlensis (James, 1911): subspecies of *Anopheles gigas* Giles, 1901b – not stated (inferential). See *simlensis* Edwards, 1922a. Type locality: Mahasu, near Simla (Shimla), [Himachal Pradesh State], India. Mahasu village is located in Kotkhai Tehsil of Shimla District in Himachal Pradesh State. Kotkhai is the sub-district headquarters of Mahasu village. Mahasu village is also a gram panchayat (local government at the village and district level). See Kitzmiller (1982).

sinaiticus Kirkpatrick, 1925: *Culex* – not stated (inferential). Sinaitic; *-us*, L. (Gr. *-os*) of, pertaining to. Type locality (syntypes): Kossaima, El Moweilleh, Ain Kadeis, Ain Gedeirat and Ain Musa, Sinai [Peninsula], Egypt. This oddly formed species name appears to be derived from the adjective Sinaitic, meaning relating to the Sinai Peninsula (or Mount Sinai). The name, therefore, is an adjective derived from an adjective.

sinensis Chow, 1950: *Aedes* – not stated (inferential). *Sina* (combining form *sino-*), L. China (the country); *-ensis*, L. suffix denoting place, locality, country. Type locality: Ta-keng, Taichung Hsien (County), Taiwan (officially the Republic of China). Taichung County was a county in central Taiwan, established in 1945 following the handover of Taiwan from Japan to the Republic of China. On 25 December 2010, Taichung County and Taichung City merged to form a single Taichung special municipality.

sinensis Theobald, 1903a: *Culex* – not stated (inferential). See *sinensis* Chow, 1950. Type locality: Shaohyling (Shaoxing), [Keqiao District, Zhejiang Province], China. Keqiao District, formerly Shaoxing County, is a district of the city of Shaoxing in Zhejiang Province.

sinensis Wiedemann, 1828: *Anopheles* – not stated (inferential). See *sinensis* Chow, 1950. Type locality: [Guangzhou], China. Guangzhou (formerly officially named Canton) is the capital of Guangdong Province (formerly officially named Kwangtung), or Canton Province. See Kitzmiller (1982).

singhbhumensis Natarajan & Rajavel, 2009: *Culex* – Singhbhum; *-ensis*, L. suffix denoting place, locality, country. 'The species is named after these hills [Singhbhum hills], known for their extensive forest cover and rich flora and fauna.' Type locality: Mamalapusi, Keonjhar District, Orissa (Odisha State), India. Odisha (formerly Orissa) is an eastern state on the Bay of Bengal.

singuawaensis Sirivanakarn, 1969: *Culex* – not stated (inferential). Singuawa; *-ensis*, L. suffix denoting place, locality, country. Type locality: Singuawa River [Bumbu River?] area, Lae [City], Morobe [Province], NE New Guinea (Papua New Guinea). Morobe is a province on the northern coast of Papua New Guinea. Lae is the provincial capital.

sinkiangensis Hsiao, 1977: *Aedes* – not stated. Sinkiang; *-ensis*, L. suffix denoting place, locality, country. Type locality: Halakule pasture, Akesu Prefecture, Xinjiang (Xinjiang Uyghur Autonomous Region), China. Sinkiang (also Shinjang) appears to be a transliteration from the Uyghur (or Uighur) language, equivalent to Xinjiang.

solomonis Belkin, 1953: *Uranotaenia* – not stated (inferential). Solomon; *-is*, L. ending for third declension genitive singular noun (masculine, feminine and neuter). Type locality: Lunga [District?], Guadalcanal Island, [Guadalcanal Province], Solomon Islands. Lunga (Lungga) is a river (Lunga River, Lunga Valley) and a promontory (Lunga Point) on the northern coast of Guadalcanal, an area the includes the mouth of the river and Lunga village. See *lungae* Belkin & Schlosser, 1944.

solomonis Belkin, 1962: *Hodgesia* – not stated (inferential). See *solomonis* Belkin, 1953. Type locality: West Poha swamp, Guadalcanal Island, [Guadalcanal Province], Solomon Islands. Poha is a stream (Poha River) on the northern coast of Guadalcanal.

solomonis (Belkin, 1962): *Mimomyia* – not stated (inferential). See *solomonis* Belkin, 1953. Type locality: Chacon swamp, Lunga area, Guadalcanal Island, [Guadalcanal Province], Solomon Islands.

solomonis Belkin, Knight & Rozeboom, 1945: *Anopheles* – not stated (inferential). See *solomonis* Belkin, 1953. Type locality: Poha River tributary, 3 miles south of the coast, Guadalcanal Island, [Guadalcanal Province], Solomon Islands. See Kitzmiller (1982).

solomonis (Edwards, 1924b): *Tripteroides* – not stated (inferential). See *solomonis* Belkin, 1953. Type locality: Tulagi, [Central Province], Solomon Islands. Tulagi (or Tulaghi) is a town on a small island of the same name located off the southern coast of Nggela Sule Island, one of two main islands of the Nggela Islands (also known as the Florida Islands), which comprise a small island group in the Central Province of the state of Solomon Islands.

solomonis Edwards, 1929: *Culex* – not stated (inferential). See *solomonis* Belkin, 1953. Type locality: Manaba, Malaita Island, [Malaita Province], Solomon Islands. Malaita is the largest island of Malaita Province, Solomon Islands. The smaller South Malaita Island lies at the southern tip of Malaita Island.

solomonis Stone & Bohart, 1944: *Aedes* – not stated (inferential). See *solomonis* Belkin, 1953. Type locality: Guadalcanal, Solomon Islands.

solomonis (Wharton, 1947): *Malaya* – Not stated (inferential). See *solomonis* Belkin, 1953. Type locality: Bougainville Island, [Papua New Guinea]. Bougainville Island is the main island of the Autonomous Region of Bougainville of Papua New Guinea. This region is also known as Bougainville Province or the North Solomon Islands.

somalicus Rivola & Holstein, 1957: *Anopheles* – not stated (inferential). Somalia; *-icus*, L. belonging to, like, of, pertaining to. Type locality: Villabruzzi (Villaggio Abruzzi), Audegle (Awdhegele) and Uebi Scebeli (Shebeli) River, Somalia. Shebeli River, also spelled Shebelle, Wabē Shebelē (Amharic), Uebi Scebeli (Italian) and Webi Shabeellex (Somali), begins in Ethiopia, flows south-east into Somalia, turns sharply south-westwards near Mogadishu, and follows the coast of the Indian Ocean. See Kitzmiller (1982).

sorsogonensis Bañez & Jueco, 1966: *Aedes* – not stated (inferential). Sorsogon; *-ensis*, L. suffix denoting place, locality, country. Type locality: Sorsogon Province, [Luzon Island], Philippines. Sorsogon is the southernmost province in Luzon.

srilankensis Peyton, 1974: *Uranotaenia* – not stated (inferential). Sri Lanka; *-icus*, L. belonging to, like, of, pertaining to. Type locality: Wilpatti (Wilpttu National Park), Hunuwilagama, Anuradhapura District, North Central Province, Sri Lanka. Wilpttu National Park is an important protected area in Sri Lanka. The park is entered at Hunuwilagama.

srilankensis (Reinert, 1977): *Verrallina* – not stated (inferential). See *srilankensis* Peyton, 1974. Type locality: Yakkala (30 km northeast of Colombo), Western Province, Sri Lanka. Yakkala is a township located on the Colombo-Kandy Highway 30 km from Colombo.

stunga (Klein, 1973) *Verrallina* – not stated (inferential). Stung; *-us* (f. *-a*), L. (Gr. *-os*) of, pertaining to. Type locality: Stung (Stueng) Chral (village), Compong Sela (Kampong Seila District), [Preah Sihanouk Province], Cambodia. Kampong Seila District, located in south-western Cambodia, formerly belonged to Koh Kong Province. The district was incorporated into Preah Sihanouk Province in 2009 by Royal Decree.

sudanensis (Theobald, 1908c): *Aedes* – not stated (inferential). Sudan; *-ensis*, L. suffix denoting place, locality, country. Type locality: Upper White Nile River, South Sudan; also at Sangikia?, Kodok (Kothok), South Sudan, and Khartoum, Sudan (also known as North Sudan and officially the Republic of the Sudan).

sudanicus Edwards, 1941: subspecies of *Culex guiarti* Blanchard, 1905 – not stated (inferential). Sudan; *-icus*, L. belonging to, like, of, pertaining to. Type locality: Bole, [Northern Region], Gold Coast (present-day Ghana). Bole is a town and the capital of Bole District in the Northern Region of north Ghana. Why Edwards named the subspecies after Sudan is a mystery, perhaps an unintended error due to mental confusion.

sulawesi Waktoedi Koesoemawinangoen, 1954: *Anopheles* – not stated (inferential). Sulawesi, treated as a latinized noun in the nominative singular standing in apposition to the generic name. Type locality: Sulawesi, [Indonesia]. As Hii *et al.* (1988) cogently explained, the author did not intend to formally describe a new taxon when he listed morphological characters for the recognition of 'Sulawesi' as a variety of *Anopheles leucosphyrus* Dönitz. He probably intended to refer to a *Terra typica* variety or form but used the Indonesian 'Sulawesi' in preference to 'Celebes', which was used by previous authors. That aside, the morphological characters provided for distinguishing 'Sulawesi' from other forms satisfy the requirements for formal application of the name. See Kitzmiller (1982).

sumatranus Brug, 1931a: subspecies of *Culex curtipalpis* (Edwards, 1914c) – Sumatra; *-anus*, L. belonging to, pertaining to. Type locality: Dermajoe, near Benkoelen (Bengkulu], [Bengkulu Province], Sumatra, [Indonesia]. Bengkulu is the capital and largest city of Bengkulu Province, historically known as Bencoolen or British Bencoolen, located on the south-west coast of Sumatra.

sumatranus (Brug, 1939): *Toxorhynchites* – not stated (inferential). See *sumatranus* Brug, 1931a. Type locality: Benkoelen (Bengkulu), [Bengkulu Province], Sumatra, [Indonesia].

sumatranus Swellengrebel & Rodenwaldt, 1932: *Anopheles* – not stated (inferential). See *sumatranus* Brug, 1931a. Type locality: Karoo-Hochebene and Kotaradja, Sumatra [Indonesia]. The original spelling of *sumatrana* (f.) is corrected to *sumatranus* to agree in gender with masculine *Anopheles*. See Kitzmiller (1982).

sumidero Schick, 1970a: *Aedes* – not stated (inferential). Sumidero, treated as a latinized noun in the nominative singular standing in apposition to the generic name. Type locality: Sumidero [Canyon], Chiapas State, Mexico. Sumidero is a narrow and deep canyon surrounded by a national park located 24 km north of Tuxtla Gutiérrez, the capital of Chiapas State.

sundaicus (Rodenwaldt, 1925): *Anopheles* – not stated (inferential). Sunda; *-icus*, L. belonging to, like, of, pertaining to. Type locality (neotype): Panda Beach, near Lundu, Sarawak, Malaysia (Linton *et al.*, 2001). The taxon was originally described from specimens collected on the Sunda Islands, Java, Sumatra and Borneo, principally the Lesser Sunda Islands. See Kitzmiller (1982).

sunyaniensis Edwards, 1941: *Culex* – not stated (inferential). Sunyani; *-ensis*, L. suffix denoting place, locality, country. Type locality: Sunyani, [Brong-Ahafo Region], Gold Coast (present-day Ghana). Sunyani is a city and the capital of the Brong-Ahafo Region and the Sunyani Municipal District of Ghana.

sureilensis Barraud, 1934: subspecies of *Aedes mediopunctatus* Theobald, 1905h – not stated (inferential). Sureil; *-ensis*, L. suffix denoting place, locality, country. Type locality: Sureil, Darjeeling District, [West Bengal State], India. The Darjeeling District is the northernmost district of the state of West Bengal located in the foothills of the Himalayas in eastern India.

surinamensis Bruijning, 1959: *Wyeomyia* – not stated (inferential). Suriname; *-ensis*, L. suffix denoting place, locality, country. Type locality: Ornamibo, [Wanica District], Suriname. Wanica District is located in the northeast of Suriname.

surinamensis Dyar, 1918e: *Culex* – not stated (inferential). See *surinamensis* Bruijning, 1959. Type locality: Marowyne (Maroni or Marowijne) and Lawa Rivers, Suriname. The Maroni River forms the border between Suriname and French Guiana and the Lawa River forms part of the international boundary between the two countries.

sylvanensis (Dobrotworsky, 1960b): *Culiseta* – not stated (inferential). Sylvan; *-ensis*, L. suffix denoting place, locality, country. Type locality: Sylvan, Victoria State, Australia.

szemaonensis Wang & Feng, 1964: *Culex* – not stated (inferential). Szemao; *-ensis*, L. suffix denoting place, locality, country. Type locality: Szemao (Simao District), Yunnan Province, China. Simao is a district under the jurisdiction of the Pu'er Prefecture of Yunnan Province. Szemao appears to be another transliteration of the Chinese; however, the addition of an 'n' to the ending of the place name was not necessary as *szemaoensis* would have been properly formed and euphonic.

tahoensis Dyar, 1916b: *Aedes* – not stated (inferential). Tahoe; *-ensis*, L. suffix denoting place, locality, country. Type locality: Fallen Leaf [Lake], [and] Lake Tahoe, [El Dorado County], California, U.S.A. Fallen Leaf is an unincorporated community located at the south end of Fallen Leaf Lake, but the community does not appear to be the type locality *per se* as Dyar made it clear that specimens were collected in the 'Fallen Leaf Lake region' and areas near Lake Tahoe.

taiwanensis Lien, 1968c: *Heizmannia* – not stated (inferential). Taiwan; *-ensis*, L. suffix denoting place, locality, country. Type locality: Kuantzulin (Kuan Tzu Ling), Paiho Township, Tainan Hsien (Tainan County), Taiwan (officially the Republic of China). Kuan Tzu Ling (also Kuan-tzu Ling or Kuan-Tzu-Ling) is a mountain.

taiwanus Lien, 1968d: *Aedes* – not stated (inferential). Taiwan; *-us*, L., of, pertaining to. Type locality: Mashihshih, Taiwu [Township], Pingtung Hsien (Pingtung County), Taiwan (officially the Republic of China).

takasagoensis Morishita, 1946: *Anopheles* – not stated. Takasago; *-ensis*, L. suffix denoting place, locality, country. Type locality: Dairi, Karenko Prefecture (defunct) and Kanshirei, Tainan Prefecture (defunct), Formosa (present-day Taiwan, officially the Republic of China). Takasago (State of Takasago) is the name Japan called Taiwan during the Edo period (or Tokugawa period) between 1603 and 1868. Karenko and Tainan Prefectures were administrative divisions of Taiwan during the Japanese rule (1895 and 1945). The former consisted of modern-day Hualien County and the latter consisted of present-day Tainan City, Chiayi City, Chiayi County and Yunlin County. See Kitzmiller (1982).

tanzaniae da Cunha Ramos, 1993: *Uranotaenia* – not stated (inferential). Tanzania; *-ae*, L. ending for first declension genitive singular noun (feminine). Type locality: Amani forest (Amani Nature Reserve), [Tanga Region], Tanzania. The Amani Nature Reserve is a protected area located within the Muheza and Korogwe Districts of the Tanga Region of Tanzania.

tasmaniensis Dobrotworsky, 1966: *Anopheles* – not stated (inferential). Tasmania; *-ensis*, L. suffix denoting place, locality, country. Type locality: Birralee, Tasmania, [Australia]. See Kitzmiller (1982).

tasmaniensis (Strickland, 1911b): *Tripteroides* – not stated (inferential). See *tasmaniensis* Dobrotworsky, 1966. Type locality: Tasmania, [Australia].

tehuantepec Schick, 1970a: *Aedes* – not stated (inferential). Tehuantepec, treated as a latinized noun in the nominative singular standing in apposition to the generic name. Type locality: Tehuantepec, Oaxaca State, Mexico. Tehuantepec city and municipality are part of the Tehuantepec District in the west of the Istmo Region (Istmo de Tehuantepec). It is not certain whether the species name is derived from the name of the city, district or region, but it seems unlikely that the species was collected in the city.

telamali Saliternik & Theodor, 1942: subspecies of *Anopheles turkhudi* Liston, 1901 – not stated (inferential). Tel Amal; *-i*, L. ending for second declension genitive singular noun (masculine and neuter). Type locality: Tel Amal (present-day Nir David), Plain of Esdraelon, Palestine (present-day Israel). Nir David is a kibbutz located to the south of the Plain of Esdraelon, also called Valley of Jezreel or Jezrell Valley, a large fertile plain and inland valley south of the Lower Galilee region of Israel. See Kitzmiller (1982).

tengchongensis Dong, Wang & Zhou, 2002: *Heizmannia* – not stated (inferential). Tengchong; *-ensis*, L. suffix denoting place, locality, country. Type locality: Gudong (Town), Tengchong County (City), Yunnan Province, China. Gudong is a town in Tengchong, a county-level city of Baoshan City (prefecture-level city) in western Yunnan Province. Tengchong city is named after the town of Tengchong, which serves as its political centre.

thailandensis Reinert, 1976a: *Aedes* – not stated (inferential). Thailand; *-ensis*, L. suffix denoting place, locality, country. Type locality: Ban Bu Phram, [Tambon Bu Phram, Amphoe Na Di], Prachin Buri Province, Thailand. Transliterations: Ban (village), Tambon (township or sub district), Amphoe (district), Province (Changwat).

tianpingensis Chen, 1981: *Culex* – not stated (inferential). Tianping; *-ensis*, L. suffix denoting place, locality, country. Type locality: Tian Ping Shan (Tianping Mountain), Guangxi Zhuang Autonomous Region, South China, China. Tianping Mountain, also called Baiyun Mountain or Ci Mountain, is in Guangxi, officially the Guangxi Zhuang Autonomous Region, which is located in South China, a geographical and cultural region that covers the southernmost part of China.

timorensis (Miyagi, Toma & Lien, 2004): *Aedes* – not stated (inferential). Timor; *-ensis*, L. suffix denoting place, locality, country. Type locality: Sulamu (Subdistrict), Kupang (District), Timor [East Nusa Tenggara Province, Indonesia]. The island of Timor is divided between the sovereign states of East Timor (eastern part of the island) and Indonesia (western part). In addition to the western part of Timor, East Nusa Tenggara Province encompasses the islands of Adonara, Alor, Flores, Komodo, Lembata, Palue, Rote, Sabu, Solor and Sumba.

tongae Edwards, 1926a: *Aedes* – not stated (inferential). Tonga; *-ae*, L. ending for first declension genitive singular noun (feminine).Type locality: Haapai (Ha'apai), Tonga Islands (Kingdom of Tonga). Ha'apai is a group of 62 islands, islets, reefs and shoals with an area of 109.30 km^2 in the central part of the Kingdom of Tonga.

tonkinensis Galliard & Ngu, 1947: *Aedes* – not stated (inferential). Tonga; *-ensis*, L. suffix denoting place, locality, country. Type locality: Baie d'Along [Hạ Long Bay], Tonkin [Red River Delta Region], Indochina [Vietnam]. See *alongensis* Venhuis, 1940, *alongi* Galliard & Ngu, 1947 and *hongayi* Galliard & Ngu, 1947.

torakala Stoker & Waktoedi Koesoemawinangoen, 1949: subspecies of *Anopheles ludlowae* (Theobald, 1903a) – not stated. Torakala, treated as a latinized noun in the nominative singular standing in apposition to the generic name. Type locality: [Torakala, Lamuru District, Bone Regency, South Sulawesi Province, Indonesia]. See Kitzmiller (1982).

toroensis Edwards & Gibbins, 1939: *Culex* – not stated. Toro; *-ensis*, L. suffix denoting place, locality, country. Type locality: Kisomoro, [Kisomoro Subcountry, Kabarole District, Western Region], Uganda. At the time of publication, Kisomoro was located in the former Toro District, which constitutes the Toro Sub-region and is coterminous with the Toro Kingdom in Western Uganda.

torokinae Belkin, 1950: *Tripteroides* – not stated (inferential). Torokina; *-ae*, L. ending for first declension genitive singular noun (feminine). Type locality: Torokina, Empress Augusta Bay, Bougainville Island, Solomon Islands (now part of Papua New Guinea). Torokina is a coastal village in the Autonomous Region of Bougainville, eastern Papua New Guinea. Bougainville Island is the main island of the Region, which is also known as Bougainville Province or the North Solomon Islands. Torokina and Empress Augusta Bay are on the western side of Bougainville Island.

torresiensis Schmidt, 2001: *Anopheles* – Torres; *-ensis*, L. suffix denoting place, locality, country. 'The name is based on the known distribution of this species, encompassing both the Torresian zoogeographic province of northern Australia and southern [Papua] New Guinea'. Type locality: edge of Jabiru, Northern Territory, Australia. See Service (2010).

toviiensis Klein, Rivière & Séchan, 1984: *Culex* – not stated (inferential). Tovii; *-ensis*, L. suffix denoting place, locality, country. Type locality: Tovii Plateau, [Nuku Hiva Island], Marquesas Islands, French Polynesia. The Tovii Plateau is the only flat agricultural area on Nuku Hiva, the largest of the Marquesas Islands in French Polynesia, an overseas territory of France in the Pacific Ocean.

towadensis (Matsumara, 1916): *Toxorhynchites* – not stated (inferential). Towad; *-ensis*, L. suffix denoting place, locality, country. Type locality: Towad (Towada), [Aomori Prefecture], Honshu, Japan. Towada is a city located in central Aomori Prefecture in the Tohoku region of northern Japan.

trinidadensis Theobald, 1901c: *Wyeomyia* – not stated (inferential). Trinidad; *-ensis*, L. suffix denoting place, locality, country. Type locality: Trinidad, [Trinidad and Tobago]. Trinidad is the larger of the two major islands which make up the island nation of Trinidad and Tobago.

trukensis Bohart, 1957: *Aedes* – not stated (inferential). Truk; *-ensis*, L. suffix denoting place, locality, country. Type locality: Ton (Tol), Truk (Chuuk State), Caroline Islands (Federated State of Micronesia). The Caroline Islands are a widely scattered archipelago of small islands in the north-western Pacific Ocean that are politically divided between the eastern Federated States of Micronesia (FSM) and western Palau. Tol is the largest island in the Faichuk group of islands of Chuuk State (also known as Truk), one of the four states of the FSM.

tsaratananae Doucet, 1950: *Uranotaenia* – not stated (inferential). Tsaratanana; *-ae*, L. ending for first declension genitive singular noun (feminine). Type locality: Mont Tsaratanana (Tsaratanana Massif), Sambirano (River), Madagascar. The Tsaratanana Massif is a mountain range in northern Madagascar that includes Maromokotro Mountain, the highest point on the island. Sambirano is a river that originates in the Tsaratanana Massif, but the name Sambirano is also used to designate the whole plain where the river flows towards the Indian Ocean.

tsiliensis King & Hoogstraal, 1946f: *Aedes* – not stated (inferential). Tsili; *-ensis*, L. suffix denoting place, locality, country. Type locality: Tsili Tsili, Watut River [Valley], [Morobe Province], [Papua] New Guinea. Tsili Tsili is a village in the Watut Valley of Morobe Province, Papua New Guinea.

tubbutiensis Dobrotworsky, 1959: *Aedes* – not stated (inferential). Tubbut; *-ensis*, L. suffix denoting place, locality, country. Type locality: Tubbut, East Gippsland, Victoria State, Australia. East Gippsland is the eastern region of Gippsland, an economic rural region of Victoria State.

tulagiensis Edwards, 1926a: *Aedes* – not stated (inferential). Tulagi; *-ensis*, L. suffix denoting place, locality, country. Type locality: Tulagi Hospital, [Tulagi], Santa Cruz Island (Nendo Island), Solomon Islands. Tulagi was the capital of the British Solomon Islands Protectorate from 1896 to 1942. Nendo (also known as Santa Cruz, Ndeni, Nitendi or Ndende Island) is the largest of the Santa Cruz Islands, located in the Temotu Province of the Solomon Islands, a sovereign country consisting of six major islands and over 900 smaller islands lying to the east of Papua New Guinea and northwest of Vanuatu.

tulliae (Taylor, 1929): *Aedes* – not stated (inferential). Tully; *-ae*, L. ending for first declension genitive singular noun (feminine). Type locality: Tully, Queensland State, Australia. Tully is a town and locality in the Cassowary Coast Region, a local government area in Far North Queensland (or Tropical North Queensland) located in the northernmost part of the state of Queensland.

tutuilae Ramalingam & Belkin, 1965: *Ades* – not stated (inferential). Tutuila; *-ae*, L. ending for first declension genitive singular noun (feminine). Type locality: Aoloau

(A'oloau), Tutuila Island, American Samoa. A'oloau (also known as A'oloaufou) is a village in the west of Tutuila Island, the largest and main island of American Samoa. American Samoa is a territory of the USA consisting of seven South Pacific islands and atolls in the archipelago of Samoan Islands.

ugandae da Cunha Ramos, 1993: *Uranotaenia* – not stated (inferential). Uganda; *-ae*, L. ending for first declension genitive singular noun (feminine). Type locality: Kalinzu Forest [Reserve], Uganda. The Kalinzu Forest Reserve is located in the Bushenyi District of the Western Region in the southwest of Uganda.

ugandae Evans, 1934: subspecies of *Anopheles wellcomei* Theobald, 1904b – not stated (inferential). See *ugandae* da Cunha Ramos, 1993. Type locality: [Buganda] King's Lake Area, Kampala, Uganda. King's Lake belongs to the current Buganda King. Buganda is a subnational kingdom of the Ganda people, the largest of the traditional kingdoms in present-day Uganda. See Kitzmiller (1982).

ugandae van Someren, 1967: subspecies of *Culex shoae* Hamon & Ovazza, 1954 – not stated (inferential). See *ugandae* da Cunha Ramos, 1993. Type locality: Ngogwe, Uganda. Ngogwe is a municipality in the Buikwe District of the Central Region of Uganda.

ungujae White, 1975: subspecies of *Anopheles wellcomei* Theobald, 1904b – Unguja; *-ae*, L. ending for first declension genitive singular noun (feminine). 'The epithet derives from Unguja, the Swahili term for Zanzibar.' Type locality: Kibonde, Nzungu, Zanzibar (Unguja), [Tanzania]. Zanzibar is a Tanzanian archipelago off the coast of East Africa. The main island, Unguja (also referred to as Zanzibar Island or simply Zanzibar), is the largest island of the Zanzibar Archipelago. See Kitzmiller (1982).

upatensis Anduze & Hecht, 1943: *Aedes* – not stated (inferential). Upata; *-ensis*, L. suffix denoting place, locality, country. Type locality: Región de Upata (Upata area), Distrito Piar (Piar Municipality), Bolívar State, Venezuela. Upata is a city in the Guayana Region (administrative region) at the north of Bolívar State. It is the capital of Piar Municipality.

upolensis Marks, 1957b: *Aedes* – not stated (inferential). Upola; *-ensis*, L. suffix denoting place, locality, country. Type locality: Afiamalu, Upolu Island, Samoa Islands (Samoa). Upolu is an island of the Independent State of Samoa, commonly known as Samoa (known as Western Samoa until 1997). Samoa comprises the westernmost group of the Samoan Islands in Polynesia.

usambara Mattingly, 1953d: *Aedes* – not stated. Usambara, treated as a latinized noun in the nominative singular standing in apposition to the generic name. Type locality: Amani Bustani (Garden), Tanganyika (present-day United Republic of Tanzania). The species is apparently named for the Usambara Mountains of north-eastern Tanzania. The Amani Nature Reserve, including the Amani Botanical Garden, is a protected area located within the Muheza and Korogwe Districts of the Tanga Region in Tanzania. The Amani Nature Reserve was established in 1997 in a forested area in the East Usambara Mountains.

vanapa (Huang, 1968a): *Verrallina* – not stated (inferential). Vanapa; *-us* (f. *-a*), L. belonging to, pertaining to. Type locality: Vanapa River, Brown River Road, SE New Guinea (Papua New Guinea). The species was originally described as *Aedes vanapus*.

venezuelensis (Theobald, 1912): *Coquillettidia* – not stated (inferential). Venezuela; *-ensis*, L. suffix denoting place, locality, country. Type locality: Caño de la Viuda (Caño La Viuda), [Monagas State], Venezuela. Caño La Viuda is a hydrographic feature, a watercourse.

victoriensis (Dobrotworsky, 1954): *Culiseta* – not stated (inferential). Victoria; *-ensis*, L. suffix denoting place, locality, country. Type locality: Ringwood, Victoria, Australia. Ringwood is a suburb of Melbourne.

vidali Floch & Fauran, 1954: *Culex* – not stated (inferential). Vidal; *-i*, L. ending for second declension genitive singular noun (masculine and neuter). '... captured in a burrow in the undergrowth near Moulin-de-Vidal in the Isle of Cayenne' [translated from French]. Type locality: Moulin-de-Vidal, Ile de Cayenne (Île de Cayenne), French Guiana [overseas region and department of France]. The Île de Cayenne (island of Cayenne or peninsula of Cayenne) is a landscape unit that groups together the three French communes of Cayenne, Matoury and Remire-Montjoly.

vietnamensis Nguyen, Tran & Nguyen, 1993: *Anopheles* – not stated (inferential). Vietnam; *-ensis*, L. suffix denoting place, locality, country. Type locality: Muong Va commune, Song Ma District (Sông Mã District), Son La Province (Sơn La Province), Vietnam. See Service (2010).

voltaensis (Danilov, 1982): *Coquillettidia* – not stated (inferential). Volta; *-ensis*, L. suffix denoting place, locality, country. Type locality: Boromo, [Boromo Department, Balé Province], Upper Volta (present-day Burkina Faso).

wamanguae Sirivanakarn, 1968: *Culex* – not stated (inferential). Wamangu; *-ae*, L. ending for first declension genitive singular noun (feminine). Type locality: Wamangu, inland of Wewak, Sepik (East Sepik Province), NE New Guinea (present-day Papua New Guinea). See *amiensis* Peters, 1963e, *kowiroensis* Sirivanakarn, 1968, *novaguinensis* Peters, 1963d and *papuensis* Peters, 1963a.

wanxianensis Lei & Li, 1990: *Orthopodomyia* – not stated (inferential). Wanxian; *-ensis*, L. suffix denoting place, locality, country. Type locality: Wanxian County (present-day Wanzhou District), Sichuan Province (formerly), China. Wanzhou District, formerly known as Wanxian or Wan County, is a city governed as a district of Chongqing Municipality, bordered by Sichuan Province to the north-west and Hubei Province to the south-east. Wanzhou was part of Sichuan Province prior to the formation of Chongqing Municipality.

wardangensis Brust, Ballard, Driver, Hartley, Galway & Curran, 1998: *Aedes* – not stated (inferential).

Waradang; *-ensis*, L. suffix denoting place, locality, country. Type locality: north-west coast of Wardang Island, South Australia State, Australia. Wardang Island is located in the Spencer Gulf close to the western coast of the Yorke Peninsula of the Australian state of South Australia. Wardang Island is also known as Wauraltee Island.

wattensis Taylor, 1929: *Aedes* – not stated (inferential). Watten; *-ensis*, L. suffix denoting place, locality, country. Type locality: Watten, Queensland State, Australia.

wauensis Huang, 1968b: *Aedes* – not stated (inferential). Wau; *-ensis*, L. suffix denoting place, locality, country. Type locality: Wau, [Morobe Province], NE New Guinea (present-day Papua New Guinea).

waverleyi (Grabham, 1907): *Orthopodomyia* – not stated (inferential). Waverley; *-i*, L. ending for second declension genitive singular noun (masculine and neuter). 'The larvae were collected from thick coffee-like water found in hollow mango trees at Waverley Estate, Constant Spring, Jamaica.' Type locality: Waverley Estate (?Maverly Estate), Constant Spring, [Saint Andrew Parish, Surrey County], Jamaica. Constant Spring is a residential community in the north of Kingston, Jamaica.

xelajuensis de Leon, 1938: *Anopheles* – Xelajú; *-ensis*, L. suffix denoting place, locality, country. 'Finally, I describe in the present work a new species, the *Anopheles Xelajuensis* [*sic*], designating the site of its discovery in the city of Quezaltenango (N. ind. Xelajú)' [translated from Spanish]. Type locality: Cerro Quemado (Almolonga volcano), near Xelajú, Quezaltenango, [Quetzaltenango Department], Guatemala. The Almolonga volcano, usually called Cerro Quemado (meaning Burned Mountain) is located near the town of Almolonga, just south of Quetzaltenango. Xelajú (or Xela) is the Mayan name of Quetzaltenango, the second largest city of Guatemala. See Kitzmiller (1982).

yaeyamae Bohart, 1956: subspecies of *Toxorhynchites manicatus* (Edwards, 1921a) – not stated. Yaeyama; *-ae*, L. ending for first declension genitive singular noun (feminine). Type locality: east fork of Nakara River, Iriomote Island (Iriomote-jima), Yaeyama Guntô, [Okinawa Prefecture], Ryukyu Islands, Japan. See *iriomotensis* (Tanaka & Mizusawa, 1973) and *ryukyuanus* Tanaka, Mizusawa & Saugstad, 1979.

yaeyamaensis Somboon & Harbach, 2010: *Anopheles* – Yaeyama: *-ensis*, L. suffix denoting place, locality, country. 'The specific name, *yaeyamaensis*, is a Latinised geographical name (yaeyama and the Latin suffix *-ensis*) denoting the type locality of the species in the Yaeyama Islands.' Type locality: Nishihama stream, Nosoko, Ishigaki City, Ishigaki Island, [Okinawa Prefecture], Ryukyu Archipelago, Japan. See *yaeyamae* Bohart, 1956.

yaeyamana Tanaka, Mizusawa & Saugstad, 1975a: *Uranotaenia* – not stated (inferential). Yaeyama; *-anus* (f. *-ana*), L. belonging to, pertaining to. Type locality: Funaura, Iriomote Island (Iriomote-jima), Yaeyama Guntô, [Okinawa Prefecture], Ryukyu Islands, Japan. See *yaeyamae* Bohart, 1956 and *yaeyamaensis* Somboon & Harbach, 2010.

yaeyamensis Tanaka, Mizusawa & Saugstad, 1979: subspecies of *Aedes japonicus* (Theobald, 1901a) – not stated (inferential). See *yaeyamaensis* Somboon & Harbach, 2010. Type locality: near Yonehara, Ishigaki Island, [Yaeyama Guntô (Okinawa Prefecture)], Ryukyu Archipelago, Japan. Ishigaki Island, also known as Ishigaki-jima, lies west of Okinawa Island and is the second-largest island of the Yaeyama Island group.

yaeyamensis Tanaka, Mizusawa & Saugstad, 1979: subspecies of *Tripteroides bambusa* (Yamada, 1917) – not stated (inferential). See *yaeyamaensis* Somboon & Harbach, 2010 and *yaeyamensis* Tanaka *et al.*, 1979. Type locality: Mount Banna, Ishigaki Island, [Yaeyama Guntô (Okinawa Prefecture)], Ryukyu Archipelago, Japan. 'Mount Banna' is a prominence (200 m) loosely described as a mountain within the confines of Banna Park (Banna kouen).

yanbarensis Miyagi, 1976: *Topomyia* – Yanbaru; *-ensis*, L. suffix denoting place, locality, country. 'The northern sylvan region of Okinawa-honto is usually called "yanbaru" by the local people in the island.' Type locality: Yona, Kunigami District, northern Okinawa-honto (Okinawa Island), [Okinawa Prefecture], Ryukyu Islands, Japan. Yona is a community of Kunigami, a village in the Kunigami District of Okinawa Prefecture.

yangambiensis de Meillon & Lavoipierre, 1944: *Aedes* – not stated (inferential). Yangambi; *-ensis*, L. suffix denoting place, locality, country. Type locality: Yangambi, [Isangi Territory, Tshopo Province], Belgian Congo (present-day Democratic Republic of the Congo). Yangambi is the largest community in the territory of Isangi, which is located west of Kisangani, the capital city of Tshopo Province, and mostly to the south of the Congo River.

yojoae Strickman, 1990: *Culex* – Yojoa; *-ae*, L. ending for first declension genitive singular noun (feminine). '*Culex yojoae* is named for Lake Yojoa, a large natural lake adjacent to the type locality.' Type locality: La Joya (?La Guama), Cartés Department, Honduras. The author states that the type locality lies 'within a kilometer of Lake Yojoa'. La Joya, meaning 'the jewel', may refer to the region that includes the Joya Grande Zoo at Santa Cruz de Yojoa, located approximately 30 km from Lake Yojoa.

yunnanensis Dong, Dong & Wu, 2004: *Uranotaenia* – not stated (inferential). Yunnan; *-ensis*, L. suffix denoting place, locality, country. Type locality: Jindong [Yi Autonomous] County, Yunnan Province, China. Jingdong Yi Autonomous County is an autonomous county under the jurisdiction of Pu'er Prefecture of Yunnan Province.

yunnanensis Dong, Zhou & Dong, 1995a: *Armigeres* – not stated (inferential). See *yunnanensis* Dong, Dong & Wu, 2004. Type locality: Jinhong (Jinghong),

[Xishuangbanna Dai Autonomous Prefecture], Yunnan Province, China. Jinghong is a city in and the seat of the Xishuangbanna Dai Autonomous Prefecture, which borders Myanmar and Laos in the far south of Yunnan Province.

yunnanensis (Gaschen, 1934): *Aedes* – not stated (inferential). See *yunnanensis* Dong, Dong & Wu, 2004. 'The mosquito described below was collected as a larva on August 23, 1934, during a survey on Si-Chang Mountain, west of Yunnanfou' [translated from French]. Type locality: Si-Chang Mountain (Cang Mountain), west of Yunnanfou (Kunming), Yunnan Province, China. Chang or Cangshan Mountain is a mountain range in the west of Yunnan Province. Yunnanfou (Yunnan-Fu) was the name of present-day Kunming city until the 1920s. Kunming is today the governmental centre of Yunnan Province.

zairensis Ribeiro, 2005: *Toxorhynchites* – Zaire; *-ensis*, L. suffix denoting place, locality, country. 'The new species is named after the country of origin of the holotype.' Type locality: Mapanda Fuma, Bakali-Inzia [streams], Kwanga, [Bandundu Province], Congo (Democratic Republic of the Congo). Zaire, officially the Republic of Zaire, was the name of the Democratic Republic of the Congo between 1971 and 1997. Kwanga is a place in the region of Bandundu.

zombaensis Theobald, 1901c: *Culex* – not stated (inferential). Zomba; *-ensis*, L. suffix denoting place, locality, country. Type locality: Zomba, British Central Africa (present-day Malawi). British Central Africa was a British protectorate proclaimed in 1889 (renamed Nyasaland in 1907) that occupied the same area as present-day Malawi. Zomba is a city in southern Malawi.

Section 6: Genus-group Names Formed from Personal Names

Of the 203 genus-group names that are currently recognized as valid extant taxa (Section 2), excluding the sole *nomen dubium* (*Asebeomyia* Aiken, 1911), 76 (listed below) are formed from personal names. Most of these commemorative names are formed by adding the suffixes *-ius* (masculine) or *-ia* (feminine) to produce latinized nominative singular nouns. However, traditional latinization is relaxed in many cases for the practical purpose of preserving the terminal letters of the name of the honouree; hence, for example, *Alanstonea* instead of the less attractive *Alanstonia*. Other genus-group names formed in this manner include (see) *Baimaia*, *Cellia*, *Christya*, *Ficalbia*, *Finlaya*, *Lorrainea*, *Macleaya* and *Skusea*. All of these names are feminine in gender, even though only *Lorrainea* is named for a woman.

Thirty-one of the 76 genus-group eponyms are formed by addition of the feminine *-ia* (including the nine from which the '*i*' is omitted). Not surprisingly, 28 commemorate men and only three commemorate women (*Kimia*, *Patmarksia* and *Sallumia*). As would be expected, all 21 eponyms that end in the masculine *-ius* are named in honour of men. In fact, only eight of the 76 eponymic generic-group names commemorate women (*Coetzeemyia*, *Dahliana*, *Kimia*, *Lorrainea*, *Patmarksia*, *Rampamyia*, *Sallumia* and *Vansomerenis*), clearly reflecting the male-dominated history of mosquito taxonomy.

Genus-group eponyms of family Culicidae may be classified as either conventional or unconventional names. Conventional genus-group eponyms are those whereby personal names are latinized by the addition of an appropriate Latin suffix, whereas unconventional eponymic names are formed by combining personal names with other words to form compound genus-group names. A recommendation contained in the third edition of the *International Code of Zoological Nomenclature* (International Commission on Zoological Nomenclature, 1985) emphatically states that 'The use of personal names in the formation of compound genus-group names is objectionable.' Well, so much for recommendations. The practice of forming compound names for mosquito genera and subgenera began long before and continued after the recommendation was published. It is interesting to note that the recommendation seems to have been excluded from the fourth edition of the *Code* (International Commission on Zoological Nomenclature. 1999).

Twelve unconventional genus-group honorary names are currently recognized as valid names of extant taxa. These include *Antunesmyia*, *Belkinomyia*, *Borichinda*, *Christophersiomyia*, *Coetzeemyia*, *Cruzmyia*, *Davismyia*, *Downsiomyia*, *Edwardsaedes*, *Galindomyia*, *Huaedes* and *Rampamyia*. *Borichinda* is not a compound name, but it is unconventional because it is the English transliteration of the Thai honouree's name which has not been latinized by the addition of a Latin suffix – the English spelling is treated as a latinized masculine noun in the nominative singular.

Nine currently valid unconventional honorary names (see above) and at least 13 synonymous genus-group eponyms terminate in *myia* (Gr., f., fly). The practice of forming such unorthodox compound names, i.e. combining words of different languages, appears to have been introduced by Ludlow (1905a) with *Reedomyia*, followed by Christophers (1911) (*Leslieomyia*) and Brunetti (1912) (*Blanchardiomyia*, *Leicesteriomyia*, *Theobaldiomyia*), all of which are currently regarded as synonyms of other names. Two currently valid subgeneric names terminate in *aedes* (Gr., m., distasteful, disagreeable, odious, unpleasant), i.e. *Edwardsaedes* and *Huaedes*.

Suffixes other than *-ius* and *-ia* have been used to form genus-group eponyms. These include the diminutive Latin suffixes *-ella* (feminine) and *-ulus* (masculine). It is rather amusing that mosquitoes, being small animals, might be classified as members of a little person's group, e.g. the latinized diminutive patronym *Bothaella*. If *Bothaella* happened to be an adjective instead of a noun (genus-group names must be or be treated as nouns in the nominative singular), the suffix would impart the meaning of 'rather' or 'somewhat', but as a noun it can only mean 'little' Botha, named in honour of the esteemed dipterist Botha de Meillon. Other currently valid diminutive genus-group eponyms include (see) *Bironella*, *Brugella*, *Lynchiella*, *Miyagiella* and *Peytonulus*.

Two genus-group names are formed differently than all other currently valid genus-group names of the family Culicidae. These are *Dahliana* and *Vansomerenis*. The formation of *Dahliana* is unorthodox as the ending is a Latin adjectival suffix. The feminine adjectival ending *-ana* (masculine *-anus*), with an '*i*' added for euphony and smoothness, means 'pertaining to' or 'belonging to'. The *-is* ending of *Vansomerenis* is a Greek patronymic suffix denoting descendant of, which could be masculine or feminine but is treated as feminine as stated by the authors.

Each genus-group honorary name in the list below is followed by the name and authorship of the genus or tribe to which it is subordinate, and mention of the individual for whom the name is dedicated, usually by quoting the etymology provided by the author or authors who established the nominal taxon. In some cases, the name of the honouree is deduced from direct or indirect evidence, usually but not always contained in the original publication. Whether a name was originally proposed for a subgenus or genus that is currently recognized as a genus or subgenus, respectively, and its gender, are also indicated. No attempt has been made to include biographical information about the persons who are honoured by the names.

Valid genus-group patronymics in Culicidae

Alanstonea Mattingly, 1960: subgenus of *Aedes* Meigen, 1818 – 'This subgenus [feminine] is named after Dr. Alan Stone of the U.S. National Museum in recognition of his many services to mosquito studies and, in particular, of his contribution to the *Synoptic Catalog of the Mosquitoes of the World* ...'.

Antunesmyia Lane & Cerqueira, 1942: subgenus of *Wyeomyia* Theobald, 1901b – 'This subgenus [feminine] has already been characterized in the key. We dedicate it to our friend DR. P. C. A. ANTUNES, from the Institute of Hygiene of São Paulo' [translated from Portuguese].

Ayurakitia Thurman, 1954a: subgenus of *Aedes* Meigen, 1818 – originally proposed as a genus: 'The genus [feminine] is named in honor of Dr. Luang Ayurakit Kosol, director of the Division of Malaria and Filariasis Control of the Department of Public Health of the Ministry of Health of Thailand.'

Baimaia Harbach, Rattanarithikul & Harrison, 2005: subgenus of *Anopheles* Meigen, 1818 – '*Baimaia* [feminine] is a patronymic honouring Prof. Visut Baimai of Mahidol University, Bangkok, for his many important contributions to our knowledge of the cytogenetics and systematics of *Anopheles* mosquitoes in southeastern Asia.' See Service (2010).

Barraudius Edwards, 1921d: subgenus of *Culex* Linnaeus, 1758 – not stated, but obviously in honour of 'Capt. P. J. Barraud', who is mentioned many times in the publication for collecting specimens and providing information. The name is masculine.

Belkinius Reinert, 1982: subgenus of *Aedes* Meigen, 1818 – 'This new subgenus [masculine] is dedicated to the late Dr. John N. Belkin for his unique and extraordinary contributions to the systematics of the family Culicidae.'

Belkinomyia Adames & Galindo, 1973: subgenus of *Culex* Linnaeus, 1758 – 'We take pleasure in naming this taxon [feminine] after Dr. John N. Belkin, whose numerous and brilliant contributions to the morphology and systematics of the Culicidae of the World have led to a better understanding of this important group of insects.'

Bironella Theobald, 1905f: genus (subgenus) of subfamily Anophelinae Grassi, 1900 – named for Lajos Bíró, a Hungarian zoologist who deposited extensive collections made in New Guinea in the Hungarian National Museum where the mosquitoes were studied by Theobald. The name is feminine. See Kitzmiller (1982).

Borichinda Harbach & Rattanarithikul, 2007: subgenus of *Aedes* Meigen, 1818 – originally proposed as a genus (masculine): Borichinda is the name of the cave where the type species of the genus was discovered. The cave bears the surname of Mr Thanom Borichinda, the 11th chief of Chomthong District (1927–1929) in Chiangmai Province, Thailand, where the cave is located.

Bothaella Reinert, 1973e: subgenus of *Aedes* Meigen, 1818 – 'This subgenus [feminine] is dedicated to Dr. Botha de Meillon, Principal Investigator, Southeast Asia Mosquito Project, Smithsonian Institution, in acknowledgement of his many contributions to the study of the world Culicidae and his guidance and inspiration.'

Bruceharrisonius Reinert, 2003: subgenus of *Aedes* Meigen, 1818 – 'It is a pleasure to name the new subgenus *Bruceharrisonius* [masculine] in honor of Dr. Bruce A. Harrison, North Carolina Department of Environment and Natural Resources, Winston-Salem, North Carolina, in recognition of his outstanding contributions to the biosystematics of the Culicidae, especially to the genus *Anopheles* Meigen.'

Brugella Edwards, 1930a: subgenus of *Bironella* Theobald, 1905f – attribution not mentioned, but undoubtedly dedicated to Dr Steffen Lambert Brug, Instituut voor Tropische Hygiene, Amsterdam. The name is feminine. See Kitzmiller (1982).

Carrollia Lutz, 1905: subgenus of *Culex* Linnaeus, 1758 – originally proposed as a genus (feminine). Attribution not mentioned: in the absence of direct evidence, it would appear that this taxon may have been named for Major James Carroll (1854–1907), a US Army physician who worked with Walter Reed, Jesse William Lazear and Aristides Agramonte as members of the Yellow Fever Commission in Cuba.

Cellia Theobald, 1902b, 1903a: subgenus of *Anopheles* Meigen, 1818 – originally proposed as a genus (feminine): Blanchard (1905) stated the genus is 'dedicated to Professor A. [Angelo] Celli of the University of Rome [translated from French]'. See Kitzmiller (1982).

Chagasia Cruz, 1906: genus of subfamily Anophelinae Grassi, 1900 – 'The genus [feminine] described was dedicated to Dr. Carlos Chagas ...' [translated from Portuguese]. See Kitzmiller (1982).

Christophersiomyia Barraud, 1923a: subgenus of *Aedes* Meigen, 1818 – originally proposed as a genus (feminine). Attribution not mentioned, but obviously in honour of Sir S. Rickard Christophers: 'I am indebted to Lieut.-Colonel Christophers for a very large amount of assistance and advice in the completion of this work.'

Christya Theobald, 1903b: subgenus of *Anopheles* Meigen, 1818 – originally proposed as a genus (feminine): 'It was first taken by Dr. [Cuthbert] Christy, after whom the genus is named.' See Kitzmiller (1982).

Coetzeemyia Huang, Mathis & Wilkerson, 2010: subgenus of *Aedes* Meigen, 1818 – 'The subgeneric name, *Coetzeemyia*, honors Prof. (Dr.) Maureen Coetzee (feminine, formed from her surname and *myia*, the Greek noun for fly) of the Vector Control Reference Unit, National Institute for Communicable Diseases (NICD), Johannesburg, South Africa.'

Collessius Reinert, Harbach & Kitching, 2006: subgenus of *Aedes* Meigen, 1818 – originally proposed as a genus (masculine): '*Collessius* is named in honour of Dr Donald H. Colless in recognition of his contributions to the taxonomy of Oriental mosquitoes.'

Coquillettidia Dyar, 1905a: genus (subgenus) of tribe Mansoniini Belkin, 1962 – attribution not mentioned, but obviously in honour of Daniel William Coquillett. The name is feminine.

Cornetius Huang, 2005: subgenus of *Aedes* Meigen, 1818 – 'The subgeneric name, *Cornetius* (gender, masculine), honors Dr. Michel Cornet, Medical Entomologist, Services Scientifiques Centraux, Office de la Recherche Scientifique et Technique Outre-Mer (ORSTOM), and Chief, Medical Entomology Laboratory, Institute Pasteur de Dakar, Senegal.'

Cruzmyia Lane & Cerqueira, 1942: subgenus of *Wyeomyia* Theobald, 1901b – 'This subgenus [feminine]... is dedicated to the memory of Dr. Oswaldo Cruz.'

Dahliana Reinert, Harbach & Kitching, 2006: subgenus of *Aedes* Meigen, 1818 – originally proposed as a genus (feminine): '*Dahliana* is named in honour of Dr Christine Dahl in recognition of her many important contributions to the taxonomy and bionomics of Palaearctic mosquitoes.'

Danielsia Theobald, 1904c: subgenus of *Aedes* Meigen, 1818 – originally proposed as a genus (feminine); attribution not mentioned, but although there is no mention of Dr Charles Wilberforce Daniels in the publication, the taxon was undoubtedly dedicated to him. *Danielsia* is one of six new genera that Theobald based on specimens collected near Kuala Lumpur by Dr G.F. Leicester. Dr Daniels was director of the Institute for Medical Research in Kuala Lumpur from 1903 to 1905.

Davismyia Lane & Cerqueira, 1942: subgenus of *Sabethes* Robineau-Desvoidy, 1827 – originally proposed as a subgenus of *Wyeomyia* Theobald, 1901b: 'The new subgenus [feminine] is dedicated to the memory of Dr. Nelson C. Davis, who greatly contributed to the knowledge of Neotropical Culicidae' [translated from Portuguese].

Dobrotworskyius Reinert, Harbach & Kitching, 2006: subgenus of *Aedes* Meigen, 1818 – originally proposed as a genus (masculine): '*Dobrotworskyius* is named in honour of Dr Nikolai V. Dobrotworsky in recognition of his important contributions to the taxonomy of Australian mosquitoes, especially of Victoria.'

Downsiomyia Vargas, 1950c: subgenus of *Aedes* Meigen, 1818 – 'The subgenus [feminine] is named in honor of Dr. Wilbur G. Downs' [translated from Spanish].

Edwardsaedes Belkin, 1962: subgenus of *Aedes* Meigen, 1818 – attribution not mentioned, but obviously in honour of Frederick Wallace Edwards.

Elpeytonius Reinert, Harbach & Kitching, 2009: subgenus of *Aedes* Meigen, 1818 – originally proposed as a genus (masculine): '*Elpeytonius* is named in honour of Mr E. L. Peyton in recognition of his important contributions to the taxonomy and biology of Culicidae ...'.

Ficalbia Theobald, 1903a: genus of tribe Ficalbiini Belkin, 1962 – attribution not mentioned, but obviously in honour of the Italian zoologist Eugenio Ficalbi. The name is feminine.

Finlaya Theobald, 1903a: subgenus of *Aedes* Meigen, 1818 – originally proposed as a genus (feminine): 'I have named this genus after Dr. [Carlos Juan] Finlay, the originator of the yellow-fever mosquito theory.'

Fredwardsius Reinert, 2000a: subgenus of *Aedes* Meigen, 1818 – 'I take pleasure in naming the new subgenus, *Fredwardsius* [masculine], for Frederick W. Edwards, who during his approximately 30 years of work at the British Museum (Natural History) (now The Natural History Museum) published numerous outstanding articles on the systematics of the Culicidae.'

Galindomyia Stone & Barreto, 1969: genus of tribe Culicini Meigen, 1818 – 'Dr [Pedro] Galindo ... assisted us extensively in preliminary taxonomic studies. We take pleasure in dedicating the genus [feminine] to him ...'.

Georgecraigius Reinert, Harbach & Kitching, 2006: subgenus of *Aedes* Meigen, 1818 – originally proposed as a genus (masculine): '*Georgecraigius* is named in honour of Dr George Brownlee Craig, Jr in recognition of his pioneering contributions to the genetics of Nearctic mosquitoes and descriptions of the eggs of aedine species of this region.'

Gilesius Reinert, Harbach & Kitching, 2006: subgenus of *Aedes* Meigen, 1818 – originally proposed as a genus (masculine): '*Gilesius* is named in honour of Dr (Lt-Col) George Michael James Giles in recognition of his contributions to the taxonomy of mosquitoes, primarily of India and surrounding areas.'

Grabhamia Theobald, 1903a: subgenus of *Psorophora* Robineau-Desvoidy, 1827 – originally proposed as a genus (feminine); attribution not mentioned, but obviously in recognition of Dr Michael Grabham for his contributions and information on mosquitoes while serving as surgeon to the Lying-in Hospital and General Penitentiary in Kingston, Jamaica.

Harbachius Reinert, 1999b: subgenus of *Verrallina* Theobald, 1903a – 'It is a pleasure to name the new subgenus *Harbachius* [masculine] in honor of Dr. Ralph E. Harbach, The Natural History Museum (NHM), London, United Kingdom, in recognition of his many outstanding contributions to the biosystematics and morphology of the Culicidae.'

Heizmannia Ludlow, 1905a: genus (subgenus) of tribe Aedini Neveu-Lemaire, 1902c – 'It is named after Col. C. L. Heizmann, Asst. Surgeon-General, U. S. A., whose continued interest in and effective support of this research, extending over several years, have [*sic*] been invaluable in making possible such success as has been attained.' The name is feminine.

Hodgesia Theobald, 1904a: genus of tribe Hodgesiini Belkin, 1962 – 'The genus [feminine] is named after Dr. Aubrey Hodges, who discovered the species [*Hodgesia sanguinae*].'

Hopkinsius Reinert, Harbach & Kitching, 2008: subgenus of *Aedes* Meigen, 1818 – originally proposed as a genus (masculine): '*Hopkinsius* is named in honour of Mr G. H. E. Hopkins in recognition of his contributions to the taxonomy of the larval stages of Afrotropical mosquitoes.'

Howardina Theobald, 1903a: subgenus of *Aedes* Meigen, 1818 – originally proposed as a genus (feminine); attribution not mentioned, but there is no doubt that the taxon was dedicated to the American entomologist Dr Leland Ossian Howard, who was employed by the U.S. Department of Agriculture and chief of the Bureau of Entomology.

Huaedes Huang, 1968b: subgenus of *Aedes* Meigen, 1818 – 'This subgenus [masculine] is named in honor of Dr Stephen M. K. Hu, head of the Entomology Department, U. S. Naval Medical Research Unit No. 2 in Taipei, Taiwan, Republic of China in 1959 …'.

Ingramia Edwards, 1912c: subgenus of *Mimomyia* Theobald, 1903a – originally proposed as a genus (feminine): 'It is with much pleasure that I name the genus after Dr. A. [Alexander] Ingram, who has done such excellent work in breeding many African Culicidae, and is the first to have reared any member of this genus from the larva.'

Jarnellius Reinert, Harbach & Kitching, 2006: subgenus of *Aedes* Meigen, 1818 – originally proposed as a genus (masculine): '*Jarnellius* is named in honour of Dr J. Hal Arnell in recognition of his important work on species of this genus and other New World aedines.'

Jihlienius Reinert, Harbach & Kitching, 2006: subgenus of *Aedes* Meigen, 1818 – originally proposed as a genus (masculine): '*Jihlienius* is named in honour of Dr Jih Ching Lien in recognition of his important contributions to the taxonomy of the mosquitoes of Taiwan.'

Johnbelkinia Zavortink, 1979: genus of tribe Sabethini Blanchard, 1905 – 'This genus [feminine] is dedicated to John N. Belkin, one of the world's leading culicidologists, who happens also to be my mentor and friend.'

Kenknightia Reinert, 1990: subgenus of *Aedes* Meigen, 1818 – 'It is a pleasure to name this new subgenus, *Kenknightia* [feminine], in honor of Dr. Kenneth L. Knight, in recognition of his many outstanding contributions to the biosystematics and morphology of the Culicidae.'

Kerteszia Theobald, 1905f: subgenus of *Anopheles* Meigen, 1818 – originally proposed as a genus (feminine): 'The material so beautifully mounted and collected by M. Biró has been graciously placed at my disposed [*sic*] by Dr. Kertész to whom I owe many thanks for the honour he has paid me.' Dr Kálmán Kertész, founder of the Diptera collection of the Hungarian National Museum, is regarded as the father of Hungarian Dipterology.

Kimia Vu & Harbach, 2007: genus of tribe Sabethini Blanchard, 1905 – 'This genus [feminine] is dedicated to Tran Thi Bach Kim, medical biologist and wife of the second author of this paper, Dr Vu Due Huong, who recognized *Kimia* as a new genus more that [*sic*] 25 years ago.'

Kitzmilleria Danilov, 1989: subgenus of *Culex* Linnaeus, 1758 – 'The subgenus [feminine] is named for J. B. [James Blaine] Kitzmiller, U.S.A., in recognition of his great contributions to the study of the genetics of mosquitoes' [translated from Russian].

Kompia Aitken, 1941: subgenus of *Aedes* Meigen, 1818 – 'The writer takes great pleasure in naming this subgenus [feminine] in honor of Mr. W. H. W. Komp, Senior Medical Entomologist, U. S. Public Health Service, who has done much to further our knowledge of mosquitoes, and who confirmed the writer's opinion concerning the status of this mosquito [*Aedes* (*Kompia*) *purpureipes*].'

Leicesteria Theobald, 1904c: subgenus of *Armigeres* Theobald, 1901b – originally proposed as a genus (feminine): 'I have proposed six new genera, one named after Dr. Leicester.' George Frederick Leicester, Assistant to the Director of the Institute of Medical Research, Kuala Lumpur, collected the specimens on which Theobald based the six new genera. See *Danielsia* Theobald, 1904c.

Lewnielsenius Reinert, Harbach & Kitching, 2006: subgenus of *Aedes* Meigen, 1818 – originally proposed as a subgenus of *Jarnellius* Reinert *et al*., 2004: 'Subgenus *Lewnielsenius* [masculine] is named in honour of Dr Lewis T. Nielsen in recognition of his many contributions to the taxonomy, distribution and bionomics of North American mosquitoes, especially aedine species.'

Lorrainea Belkin, 1962: subgenus of *Aedes* Meigen, 1818 – 'This subgenus [feminine] is dedicated to my wife, Lorraine.'

Luius Reinert, Harbach & Kitching, 2008: subgenus of *Aedes* Meigen, 1818 – originally proposed as a genus (masculine): '*Luius* is named in honour of Dr Lu Baolin (Luh Paoling) in recognition of his numerous contributions to the taxonomy of Chinese Culicidae.'

Lutzia Theobald, 1903a: genus (subgenus) of tribe Culicini Meigen, 1818 – 'Dr. [Adolfo] Lutz originally suggested to me that this very marked *Culicine* [*sic*] [*Lutzia bigoti*] should be placed in a new genus, consequently I have named the genus after this great Brazilian dipterist …'. The name is feminine.

Lynchiella Lahille, 1904: subgenus of *Toxorhynchites* Theobald, 1901b – originally proposed as a genus (feminine): 'I will therefore replace this name [*Megarhina*] with that of *Lynchiella* in honour of the learned and late Argentine dipterologist: Félix Lynch Arribálzaga' [translated from French].

Macleaya Theobald, 1903e: subgenus of *Aedes* Meigen, 1818 – originally proposed as a genus (masculine); attribution not mentioned, but it is assumed that the taxon was established in recognition of the pre-eminent entomologist William Sharp Macleay, who donated the vast Macleay family Natural History Collection to the museum which bears the name of the Macleay family, the Macleay Museum. The name is feminine.

Maillotia Theobald, 1907: subgenus of *Culex* Linnaeus, 1758 – originally proposed as a genus (feminine). Theobald stated that 'The genus … is quite distinct, its peculiar characters were noticed by Dr. Edmond Serpent, who suggested the name under which it is described.' Snow (2001) erroneously listed the etymology of the name as 'French, maillot = swaddling-clothes; -ia, suffix commonly used in generic names but normally reserved for those derived from personal names.' Theobald, who was skilled in Latin and Greek, would only have attached the *-ia* suffix to a personal name. The name suggested by Dr Serpent, who collected the type species in Algeria, is almost certainly attributable to François Clément Maillot (1804–1894), a French army officer who was posted to Algeria in 1832 and in 1834 became the first in the French Colonial Empire to use quinine to treat French soldiers suffering from malaria. Near the end of his life, Maillot was lionized as a hero of French science.

Mansonia Blanchard, 1901: genus (subgenus) of tribe Mansoniini Belkin, 1962 – 'I therefore propose: … The genus *Mansonia* [feminine], in honor of [Sir] Patrick Manson …' [translated from French].

Mattinglyia Lien, 1968c: subgenus of *Heizmannia* Ludlow, 1905a – originally proposed as a genus (feminine): 'This genus is named in honor of Dr. P. F. Mattingly of the British Museum (Natural History) in recognition of his excellent contributions to the study of Culicidae.'

Miyagiella Harbach, 2014: subgenus of *Topomyia* Leicester, 1908 – '*Miyagiella* [feminine], a Latinized diminutive patronymic, is established here to honour Professor Dr Ichiro Miyagi for his many contributions to the taxonomy of Oriental mosquitoes, especially those of genus *Topomyia*.'

Nunezia Dyar, 1928: subgenus of *Wyeomyia* Theobald, 1901b – 'I propose NUNEZIA, in honor of Dr. M. Núñez Tovar, my esteemed collaborator.' The name is feminine.

Patmarksia Reinert, Harbach & Kitching, 2006: subgenus of *Aedes* Meigen, 1818 – originally proposed as a genus (feminine): '*Patmarksia* is named in honour of Dr Elizabeth Nesta Marks in recognition of her pioneering and outstanding contributions to the taxonomy of Australasian mosquitoes.'

Paulianius Brunhes & Boussès, 2017: subgenus of *Aedes* Meigen, 1818 – originally proposed as a genus (masculine): 'In homage to Renaud Paulian (1913–2003) who has done so much for the identification of the Malagasy biodiversity in general and the entomofauna in particular.'

Petermattinglyius Reinert, Harbach & Kitching, 2009: subgenus of *Aedes* Meigen, 1818 – originally proposed as a genus (masculine): '*Petermattinglyius* is named in honour of Dr Peter Frederick Mattingly in recognition of his many important contributions to the taxonomy of family Culicidae.'

Peytonulus Harbach, 1991a: subgenus of *Sabethes* Robineau-Desvoidy, 1827 – 'The subgeneric name [masculine] is a Latinized diminutive patronym to recognize E.L. Peyton for his special knowledge of sabethine mosquitoes and his many contributions to the field of mosquito systematics.'

Rampamyia Reinert, Harbach & Kitching, 2006: subgenus of *Aedes* Meigen, 1818 – originally proposed as a genus (feminine): '*Rampamyia* is named in honour of Dr Rampa Rattanarithikul in recognition of her outstanding contributions to the taxonomy of Oriental mosquitoes.'

Sallumia Reinert, Harbach & Kitching, 2008: subgenus of *Aedes* Meigen, 1818 – originally proposed as a subgenus (feminine) of *Ochlerotatus* Lynch Arribálzaga, 1891a,b: '*Sallumia* is named in honour of Dr Maria Anice Mureb Sallum in recognition of her many contributions to the taxonomy of Neotropical and Oriental mosquitoes.'

Shannoniana Lane & Cerqueira, 1942: genus of tribe Sabethini Blanchard, 1905 – originally proposed as a subgenus of *Trichoprosopon* Theobald, 1901a: 'We will dedicate it to Mr. R. C. Shannon, to whom we owe valuable assistance in the preparation of this paper.' The name is feminine.

Sirivanakarnius Tanaka, 2004: subgenus of *Culex* Linnaeus, 1758 – 'I believe the peculiar adult characteristics … are sufficient to establish a distinct subgenus, and dedicate its name to Dr. Sunthorn Sirivanakarn.' The name is masculine.

Skusea Theobald, 1903a: subgenus of *Aedes* Meigen, 1818 – originally proposed as a genus (feminine); attribution not mentioned, but obviously in honour of the Australian entomologist Frederick A.A. Skuse. The name is feminine.

Tanakaius Reinert, Harbach & Kitching, 2004: subgenus of *Aedes* Meigen, 1818. – originally proposed as a genus: '*Tanakaius* is masculine and named in honour of Dr Kazuo Tanaka in recognition of his outstanding contributions to the taxonomy of the mosquitoes of Japan and adjacent areas.'

Tewarius Reinert, 2006a: subgenus of *Aedes* Meigen, 1818 – originally proposed as a genus: '*Tewarius* is masculine and named in honor of Satish C. Tewari, Centre for Research in Medical Entomology, Tamil Nadu, India.'

Udaya Thurman, 1954b: genus of tribe Aedini Neveu-Lemaire, 1902c – originally proposed as a subgenus of *Paraedes* Edwards, 1934: 'This subgenus is named in honor Dr. Udaya Sandhinand, Chief Malaria Control Officer, Northern Thailand, a valued co-worker and loyal friend.' The name is masculine (see Reinert & Harbach, 2005).

Vansomerenis Reinert, Harbach & Kitching, 2006: subgenus of *Aedes* Meigen, 1818 – originally proposed as a genus (feminine): '*Vansomerenis* is named in honour of Mrs E. C. C. van Someren in recognition of her many important contributions to the taxonomy of Afrotropical mosquitoes, especially those in tribe Aedini.'

Verrallina Theobald, 1903a: genus (subgenus) of tribe Aedini Neveu-Lemaire, 1902c – attribution not mentioned, but most certainly named in honour of the influential British dipterist George Henry Verrall. The name is feminine.

Zavortinkius Reinert, 1999a: subgenus of *Aedes* Meigen, 1818 – 'It is a pleasure to name the new subgenus *Zavortinkius* [masculine] in honor of Dr. Thomas J. Zavortink, Department of Biology, University of San Francisco, San Francisco, California, in recognition of his important contributions to the systematics of the Culicidae.'

Section 7: Some Interesting Species- and Genus-group Names

The practical application of Latin and Greek in the formation of scientific names has not been without grammatical problems. Some names are hybrids compounded from two different languages, a reprehensible practice from a purely grammatical point of view. In some cases, in addition to forming impure names, authors have created epithets that are not euphonious, mnemonically attractive, or conceptually or descriptively appropriate. Some names are fanciful, frivolous or otherwise make little or no sense, which makes one wonder what the authors were thinking at the time such names were coined.

In general, it has not been the practice of authors to reveal the formation or derivation of species-group names, which in the majority of cases must be determined by inferring whether a name is: (i) adopted directly from Latin, Greek or another language, with or without grammatical modification for gender agreement; (ii) constructed by uniting two or more words, with or without addition of a prefix and/or suffix; or (iii) invented, without identifiable root or stem words. For this reason, elucidating the formation and derivation of many species-group names can be a gruelling and time-consuming task for anyone who is not a scholar of classical Latin and Greek. With that in mind, the derivations of some names in the lists below may not be completely accurate, but most of them are surely correct. The ambiguity of the derivations provided for ***epitedeus*** (Knab), ***labesba*** Howard, Dyar & Knab and ***ocossa*** Dyar & Knab in the list of species names (see below) exemplifies the difficulty and uncertainly of determining the source and construction of certain names.

Unlike the names listed in Sections 4, 5 and 6 that are derived from the proper names of particular real or fictional persons and geographical places, all other species-group names are derived from a variety of other sources. The sources of these names, and some genus-group names as well, fall into the following broad categories.

Acronyms and institutions

Names based on acronyms are basically arbitrary combinations of letters that form nouns, all of which in Culicidae are masculine, either: (i) used directly as a noun in apposition to a masculine genus name (e.g. ***Anopheles nimpe*** Nguyen, Tran & Harbach; National Institute of Malariology, Parasitology and Entomology); (ii) formed by adding the singular second declension (masculine) genitive ending (e.g. ***Aedes meprai*** Martínez & Prosen; Misión de Estudios de Patoiogía Regional Argentina); or (iii) by adding the common masculine adjectival suffix (e.g. ***Anopheles maverlius*** Reinert; Medical and Veterinary Entomology Research Laboratory).

Cultures and peoples

The naming of species in recognition of cultures and peoples naturally depends on geographical location; hence, a particular name is an indicator of the area of the world in which the taxon resides. The majority of such names are masculine, a few are feminine, formed in a similar manner to names derived from acronyms and institutions. This can be illustrated by a few examples: ***amharicus*** Hunt, Wilkerson & Coetzee (masculine adjective; Amharic, Semitic language of Ethiopia); ***guarao*** Anduze & Capdevielle (noun in apposition; guarao, indigenous tribe of Venezuela); and ***nyasae*** Edwards (feminine first declension genitive singular noun; Nyasa people of south-eastern Africa).

At this point it may be appropriate to consider pronunciation. It is safe to assume that the majority of authors do not consider pronunciation when proposing names, which of course, although Latin or latinized, will normally be pronounced as English words. The culicid nomenclature abounds with names that are distorted, humorous or even derogatory when enunciated. Certainly Banks (1906b) was not aware when he proposed *mangyana* as a species of *Myzomyia* that it would be transferred to *Anopheles* and changed to ***mangyanus***, the pronunciation of which might refer to a species with a scabby or scaly opening at the end of the alimentary canal (mangy anus) – no disrespect intended for the indigenous Philippine people for whom the name is dedicated.

Habitats and plants

Habitats and plants form a single category because plants and plant ecotopes harbour the immature and adult life stages of mosquitoes, respectively. A number of species are named for the types of plants with phytotelm habitats where the larvae of the new species were discovered. The names of these species are nouns in the genitive case, e.g. ***asteliae*** Belkin (larvae found in the axils of a species of *Astelia*) and ***musarum*** Edwards (larvae found in the axils of a species of wild banana of the genus *Musa*). The names of species derived from other types of habitats may be nouns but are more often adjectives, e.g. ***natronius*** Edwards (larvae found in alkaline pools) and ***diluvialis*** Reinert (larvae occur in flooded swamps). The formation of two adjectival habitat-based names, ***barrinus*** Bram and ***conchobius*** Edwards, are particularly apropos and pleasing. The former is a Latin adjective meaning 'of elephants', coined in reference to the collection of larvae from elephant footprints, and the latter is a compound word formed from two Greek words, which in combination loosely mean 'life in snail shells', so named for larvae found in snail shells.

Morphological traits

Names based on anatomical features of larval and adult mosquitoes, more often adults, are more numerous than other types of species-group names in Culicidae. The authors of most of these names did not provide etymologies, but it is fairly easy to determine the construction and derivation of most descriptive names. Some are single words and some are compound words, the latter less numerous but more interesting in terms of deciphering their derivation, construction and meaning. The euphonious and aesthetic ***glaucodaemon*** (Dyar & Shannon) is a good example, the two components of the compound word engendering the meaning 'bluish-green demon', a species of *Sabethes*, the *Morpho* butterflies of the mosquito world.

Environment and collection sites

Names based on environmental conditions where specimens of assumed new species were first collected are fairly common in Culicidae. There are no objections to the creation of names from such sources, but their derivations, unlike patronymics, are generally more reflective of an imaginative mind. Three examples, among many, are typical of imaginative nomenclaturists: ***cubiculi*** Marks, L. *cubiculi*, bedroom, in reference to specimens collected in a bedroom; ***oiketorakras*** Osorno-Mesa, from the Greek *oiketor* (inhabitant) and *akras* (of height) in reference to specimens collected above 2700 m; and ***monticola*** Belkin & McDonald (mountain dweller).

Perceptions and behavioural attributes

This is an interesting class of names which generally originate from actual observations or presumed notions of bionomical attributes and outcomes. The specific name ***funestus*** (L. causing death), for example, illustrates both plausible and false perceptions: ***Anopheles funestus*** Giles, a vector of malarial parasites, may be a harbinger of death whereas the epithet is entirely inappropriate in the case of ***Toxorhynchites funestus*** (Leicester), a non-blood feeder (the name is especially inappropriate since the species was described from a single male). While it is not a stretch of the imagination to understand the connotation of names such as ***Wyeomyia gaudians*** Dyar & Núñez Tovar (L., delightful) and ***Uranotaenia modestus*** Ficalbi (L. moderate, gentle, unassuming, chaste), it is difficult to fathom the choice of names like ***Culex hedys*** Root (Gr. sweet, pleasant).

Animals

Few nominal taxa have been given the names of types of animals. Five are known in Culicidae: ***Culex*** Linnaeus (L. gnat, midge, mosquito), ***iwi*** Marks (the word for mosquito in the language of indigenous people of western islands of the Torres Strait), *mosquito* Robineau-Desvoidy (self-explanatory, a synonym of ***Aedes aegypti***), ***pleccau*** Koidzumi (English transliteration of the Taiwanese aboriginal word for mosquito), and both ***Zinzala*** and ***zinzala*** Zavortink (L. a kind of gnat). A number of synonymous genus-group names, *Chrysoconops* Goeldi (=*Ochlerotatus* Lynch Arribálzaga), *Cyanoconops* Antunes (*nomen nudum*, *Haemagogus*), *Hyloconops* Lutz (=*Trichoprosopon* Theobald), *Lasioconops* Theobald (=*Culex* Linnaeus) and *Stegoconops* Lutz (=*Haemagogus* Williston), are compound words with the terminal component derived from *konops* (Gr. gnat). And, of course, it is obvious that mosquito nomenclators have favoured the use of *myia* (Gr. fly) in the formation of many genus-group names (126, with only 55 currently recognized as valid in the 'traditional' classification of the Culicidae).

Relationships

Names based on perceived relationships are relatively few in number. Examples include ***adelpha*** Dyar & Knab (*adelphos*, Gr. brother, twin, near kinsman), ***propinqua*** Mattingly (*propinquus*, L. near, neighbouring) and ***sororcula*** Dyar & Knab (L. *sororcula*, little sister).

Distinction and uniqueness

Taxonomists are sometimes impelled to let the scientific community know they are certain that the species they discovered is indeed new by choosing a name which indicates that this is the case. Such names are more common than one might expect. Here are just a few examples: ***alienus*** Colless (L. foreign, strange, not belonging, not related), ***dissimilis*** (Leicester) (L. unlike, dissimilar), ***distinctus*** (Newstead & Carter) (L. different, separate), ***eukrines*** Bram & Rattanarithikul (Gr. clear, distinct, well separated), ***Exallomyia*** Harbach & Peyton (Gr. *exallos*, quite different), ***incognitus*** Baisas (L. unknown, unidentified), ***singularis*** (Leicester) (L. alone, solitary).

Whimsical, unconventional and ludicrous names

One wonders what authors may have had in mind when proposing certain names. Truly, some are farcical or otherwise not befitting of mosquitoes. Three names serve as examples: ***jocosa*** (Dyar & Knab) (*jocosus*, L. facetious), ***pacificus*** Edwards (L. peace-making, peaceful) and ***summorosus*** Dyar (L. somewhat peevish). Harrison Dyar must be awarded for proposing the most ludicrous name, ***psatharus***, a transliteration of the ancient Greek ψαθαρός (*psatharos*), meaning 'all I know is that I know nothing'. ***Aedes varuae***, named for the brigantine 'Varua', a sailing vessel with two masts, shows that even the eminent John Belkin had a whimsical side. ***Topomyia yanbareroides*** Dong & Miyagi is a nonsensical name formed by combining *-oides* (L. meaning 'like', 'resembling', 'having the form of') with the stem of the specific name ***yanbarensis***, which is named for a forested region known as 'yanbaru' – the species cannot resemble a forested region.

Abraedes Zavortink is almost a contradiction, derived from *habr* (Gr. pretty, delicate, dainty) and *aedes* (Gr. distasteful, disagreeable, odious, unpleasant). At least the name is appropriately formed of two Greek words, unlike ***Miamyia*** Dyar, which is a combination of Latin and Greek words (*mia*, L. my; *myia*, Gr. fly), not to mention that the author effectively claimed personal possession of that group of mosquitoes. ***Maorigoeldia*** Edwards is a rather disagreeable imitative (false) patronymic, which is the hybridization of Maori (people of New Zealand) and *Goeldia* (named after Dr Emilio Augusto Goeldi; currently a synonym of ***Shannoniana*** Lane & Cerqueira). ***Suaymyia*** Thurman, unorthodox as it may seem, being formed from the transliterated Thai word meaning pretty (สวย, transliterated 'suay') and the Greek *myia*, is one of the most attractive genus-group names in Culicidae.

Below is a short list of some currently valid species-group names and their derivations, followed by an even shorter list of some currently valid genus-group names and their formations. The names were more or less randomly selected to provide further examples of the categories listed above. The meanings of the constructions are for the most part left for personal interpretation.

Some Interesting Species-group Names

aboriginis Dyar, 1917b: *Aedes* – *aborigineus*, L. ancestral, native, original.

achaetae (Leicester, 1908): *Heizmannia* – 'There are no chaetae on the metanotum'; *a-*, Gr. not, without; *chaite*, Gr. (f.) long hair, mane.

adelpha Dyar & Knab, 1906a: *Wyeomyia* – *adelphos*, Gr. brother, twin, near kinsman.

agrestis (Barraud, 1931): *Verrallina* – *agrestis*, L. of the land, rural, wild.

ainshamsi Gad, Harbach & Harrison, 2006: *Anopheles* – Ain Shams University, Cairo.

alienus Colless, 1957a: *Culex* – *alienus*, L. foreign, strange, not related.

alocasicola Marks, 1947a: *Aedes* – *alocasia*, NL. (f.) apparently derived from *colocasia*, L. (m.) (Gr. *kolokasia*) an arum; *-cola*, L. (m., f.) dweller, inhabitant. The name is a noun in apposition meaning *Alocasia* inhabitant. Marks merely indicates that the holotype was 'reared from larvae taken from axils of cunjevoi'. In Australia, a *cunjevoi* is a species of *Alocasia*.

alticola Bonne-Wepster, 1948: *Aedes*; ***alticola*** Peters, 1963d: subspecies of *Uranotaenia novaguinensis* Peters, 1963d – *altus*, L. (m.) high; *-cola*, L. (m., f.) dweller, inhabitant. Bonne-Wepster did not provide explicit information needed to explain the meaning of the name whereas Peters indicated that 'The highland subspecies *alticola* has been taken ... at altitudes of 5000 feet and above'.

altivallis Bonne-Wepster, 1948: *Tripteroides* – *altus*, L. high; *valles* (*vallis*), L. vale, valley, hollow.

ambaricus Hunt, Wilkerson & Coetzee, 2013: *Anopheles* – 'This species is named after Amharic, the official Semitic language of Ethiopia and is derived as a Latinized adjective.'

ananae Knight & Laffoon, 1946: *Aedes* – named from the Spanish word (anana) for pineapple. The holotype male was reared from a larva collected from pineapple axil.

aningae Motta & Lourenço-de-Oliveira, 2005: *Wyeomyia* – 'The name "Aningae" is in allusion to the habitat of the immatures of the species, the aninga plant *Montrichardia arborescens* (L.) Schott (Araceae).'

apoci Marsh, 1933: *Anopheles* – acronym: Anglo-Persian Oil Company.

arawak Berlin, 1969b: *Culex* – Arawak, a group of indigenous peoples of South America and historically of the Caribbean.

arboricolus Galindo & Mendez, 1961: *Culex*; ***arboricolus*** Knight & Rozeboom, 1946: *Aedes*; ***arboricolus*** Zavortink, 1970b: *Anopheles* – *arbor*, *-is*, L. (f.) tree; *-colus*, NL. (m.), dwelling in, living among.

ardensis (Theobald, 1905g): *Anopheles* – not mentioned. ? *ardens*, L. burning, glowing, fiery, hot; *-is*, L. with, having the nature of. The name would seem to be derived from a geographical location, but Theobald made no mention of a place or source of the name. As noted by Kitzmiller (1982), evidence suggests that the derivation given here is probably correct, although it is possible that Theobald may have intended but incorrectly formed the genitive *ardentis*, meaning 'of the burning one'.

arenivaga Marks, 1968: *Culiseta* – '*C. arenivaga* is named from its habitat "a wanderer in sandy places"'. *arens*, L. dry, parched; *vagus*, L. wandering, roaming.

ascidiicola de Meijere, 1910: *Uranotaenia* – *askidion*, Gr. diminutive of *askos* (m.) bag, bladder, sac; *-cola*, L. (m., f.) dweller, inhabitant. Larvae are found 'in the cups of *Nepenthes*' [translated from German].

asteliae Belkin, 1968a: *Culex* – larvae collected in axils of the broad-leaf astelia *Collospermum hastatum*.

avianus de Meillon, 1943: *Culex* – not mentioned. ? *avium*, L. byway, desert, wilderness; *-anus*, L. belonging to, pertaining to.

axillicolus Steffan, 1979: *Culex* – *axilla*, L. (f.) armpit, crotch; *-colus*, NL. (m.), dwelling in, living among. 'Holotype ... reared from taro axil ...'. 'The immatures ... are found primarily in cultivated or wild taro'.

aztecus Hoffmann, 1935: *Anopheles* – Aztec (not mentioned). Aztec people/Aztec Empire. The Aztecs were a Native American people who lived in central Mexico.

balteatus Macdonald, 1960: *Armigeres* – not mentioned. ? *balteus*, L. girdle, belt, border, edge; *-atus*, L. provided with, having the nature of, pertaining to.

bamborum Rozeboom & Komp, 1948: *Culex* – not mentioned. Probably derived from the Malay *bambu* and the Latin genitive plural suffix *-orum* (m.), meaning 'of bamboo'. Holotype 'from uncut bamboo internodes perforated by holes bored by insects'.

bambusa (Yamada, 1917): *Tripteroides* – not mentioned. *Bambusa*, a genus of clumping bamboos, probable larval habitat. *bambusa*, NL. a kind of grass.

bambusae Edwards, 1935a: *Aedes* – *Bambusa*, genus of bamboos. '... bamboo-forest ... some specimens captured, many others reared from larvae found in bamboo-stems'.

bambusaihole Dong, Zhou & Dong, 1997a: *Topomyia* – not specifically mentioned; apparently a corrupt combination of *bambusa*, NL. a type of grass (*Bambusa*, a genus of bamboos) and *hol* (ME., Fr., OE.), hole, perforation. '... collected from bamboo holes'.

bambusicola da Cunha Ramos & Brunhes, 2004: *Uranotaenia*; ***bambusicola*** (Lutz & Neiva, 1913): *Toxorhynchites* – *bambusa*, NL. (f.) a kind of grass (bamboo); *-cola*, L. (m., f.) dweller, inhabitant. These species are named for the development of the immature stages in collections of water in bamboo habitats.

bambusicolus Knight & Rozeboom, 1946: *Aedes*; ***bambusicolus*** Komp, 1937b: *Anopheles* – *bambusa*, NL. (f.) a kind of grass (bamboo); *-colus*, NL. (m.), dwelling in, living among. These species are named for larvae found in collections of water in bamboo.

baresi Cerqueira, 1960a: *Haemagogus* – named for the Baré Indian tribe of south-western Venezuela and north-western Brazil.

barrinus Bram, 1967a: *Culex* – obviously named (not mentioned) for the collection of larvae from elephant footprints; *barrinus*, L. of elephants.

bihaicolus Dyar & Nuñez Tovar, 1928: *Culex* – *bihai*, species of *Heliconia*; *-colus*, NL. (m.), dwelling in, living among. 'Bred from larvae found in the flower-bracts of Heliconia [*sic*] [*Heliconia bihai*]'. *Heliconia bihai* of the family Heliconiaceae is a tall (≥ 1.5 m) erect plant found in northern South America and the West Indies.

bihamatus Edwards, 1926a: *Culex* – *bi-*, L. two, double; *hamatus*, L. hooked. Apparently named for 'two rather stout and blunt spines' borne on 'a small lobe' of the gonocoxite.

bromeliae (Theobald, 1911a): *Aedes* – apparently based on larvae found in pineapples (Bromeliaceae – type genus *Bromelia*).

cacozelus Marks, 1963b: *Aedes* – *caco-* < Gr. *kakos*; L. *zelus* (Gr. *zelos*), emulation, rivalry, ardour.

caracula Dyar & Núñez Tovar, 1928: *Wyeomyia* – *kara*, Gr. head; *-culus*, L. diminutive suffix, small, little.

carcinicola da Cunha Ramos & Brunhes, 2004: *Uranotaenia* – *karkinos*, Gr. (m.) crab; *-cola*, L. (m., f.) dweller, inhabitant. 'Name related to larval development in crab holes' [translated from French]. An adjectival name meaning 'living among crabs'.

carcinoxenus de Oliveira Castro, 1932: *Culex* – *carcino-* < Gr. *karkinos*, crab; *xenos*, Gr. stranger, guest.

castaneus Sirivanakarn, 1973a: *Culex* – *castanea*, L. chestnut; *castaneus*, of the colour of chestnuts.

cataractarum Edwards, 1923b: *Culex* – 'on rocks under waterfall', *cataract* < Gr. *kataraktes*, waterfall.

cavernicola Mattingly, 1954b: *Uranotaenia*; ***cavernicola*** (Rattanarithikul & Harbach, 2007): *Aedes* – *caverna*, L (f.) hollow, cave, vault; *-cola*, L. (m., f.) dweller, inhabitant. These are cave-inhabiting species.

cedecei Stone & Hair, 1968: *Culex* – acronym: National Communicable Disease Center.

charmion Dyar, 1928: *Wyeomyia* – *charma*, Gr. source of joy, delight; *-ion*, L. act, process, result of, state of, having the nature of.

chaguanco Casal, García & Fernández, 1968: *Culex* – Gustavo Rossi contends that the name is a derogatory epithet for the original settlers of Salta and Jujuy Provinces of northern Argentina, the Chiriguanos.

chionodes Belkin, 1962: *Aedes* – *chiono-* < Gr. *chion*, *-os* >, snow; *-odes*, Gr. *eides*, like. The specific name refers to the snow-white scaling of the thorax.

clivis Lanzaro & Eldridge, 1992: *Aedes* – the authors erroneously stated that 'The species name is descriptive, taken from the Latin adjective *clivis*, which means sloping or inclining, in this case referring to the mountainous habitat in which this species occurs.' The Latin adjective meaning the 'sloping side of a hill' is actually *clivus* (m.). The word '*clivis*' is a noun, which in music is a neume denoting two descending notes or tones, with the first higher in pitch than the second. The word is also the dative plural and the ablative plural of *clivus*.

colocasiae Edwards, 1928a: *Uranotaenia* – named for larvae collected from taro axils (*Colocasia*).

conchobius Edwards, 1941: subspecies of *Eretmapodites silvestris* Ingram & de Meillon, 1927 – *concha*, L. (Gr. *konche*; *conchos*) snail, shell; *bios*, Gr. life. 'The specimens from Shimba Hills and Tiwi (near Mombasa) were reared from larvae found in snail-shells in forest'.

creole Anduze, 1949b: *Culex* – Creole people of Venezuela?

cubiculi Marks, 1989: *Culex* – *cubiculi*, the genitive of the Latin noun for bedroom, refers to G.F. Hill's (1917) statement that 'The type and co-type specimens were taken before sunrise in a bedroom'.

cyprius Ludlow, 1920b: *Aedes* – *cyprius*, *a*, *um*, L. Cyprian, copper; f. as substantive *Venus*.

deserticola Kirkpatrick, 1925: *Culex*; ***deserticola*** Zavortink, 1970a: *Aedes* – *desertum*, L. (n.) desert, wasteland; *-cola*, L. (m., f.) dweller, inhabitant. These species are inhabitants of desert environments.

diluvialis Reinert, 1997: *Anopheles* – *diluvium*, L. deluge, flood; *diluvialis*, of a flood. 'The species name *diluvialis* is of Latin origin and refers to the habitat of this species, i.e., intermittently flooded swamps.'

domicolus Edwards, 1916: *Anopheles* – *domus*, L. (f.) house, home (house dweller); *-colus*, NL. (m.), dwelling in, living among. This species is described from 'six females taken in bungalow'.

dracaenae Edwards, 1916: *Eretmapodites* – 'A series was bred by Mr. Bacot from pupae found in water collected at the bases of the leaves of the "Cooked Hat" plant (*Dracaena* sp.)'.

draconis Ingram & de Meillon, 1927: subspecies of *Culex vansomereni* Edwards, 1926a – *draco*, *-onis*, L. a kind of snake, dragon.

dravidicus Christophers, 1924c: *Anopheles* – Dravidians, Sanskrit name for early inhabitants of southern India.

dzeta Séguy, 1924: *Aedes* – Zeta (ζῆτα, Ζζ), the sixth letter of the Greek alphabet.

epitedeus (Knab, 1907b): *Deinocerites* – Knab gave no clue as to the etymology of this name. It would seem to be derived from the ancient Greek word ἐπιτήδειος (*epitedeios*), which is derived from ἐπιτηδές (*epitedes*) meaning 'enough' and has variable meaning when applied to inanimate objects or people. When applied to people it means 'close friends' or more broadly 'friendly'. The latter meaning could also apply to non-human animals in the sense of pets. It is doubtful that it would be applicable to bloodsucking insects, although they certainly become 'close friends' with their hosts (Christian Kammerer, personal communication).

equinoxialis Floch & Abonnenc, 1945c: *Culex* – etymology not mentioned. *equinoxium*, ML., alteration of *aequinoctium*, L. from *aequi-*, *equi-* + *noct-*, *nox* night; *-alis*, L. pertaining to, having the nature or quality of. Equinox, relating to a state or the time of equal day and night.

eschirasi Galliard, 1931b: subspecies of *Culex pruina* Theobald, 1901 – 'In the plain of the Eschiras' [translated from French]. 'Habitat: stagnant water in the bed of a stream near the mission of the Holy Cross of the Eschiras' [translated from French] The Eschira (Echira, Eshira or Sira) are a Bantu population of central Africa living mainly in the coastal regions of Gabon.

euiris Dyar, 1922g: *Aedes* – *eu-*, Gr. *eu*, good, well, true, beautiful; *iris*, L., rainbow.

eukrines Bram & Rattanarithikul, 1967: *Culex* – 'The specific name is derived from the Greek adjective meaning clear or distinct.' *eukrines*, Gr. clear, distinct, well separated.

folicola Belkin, 1955: *Tripteroides* – *folium*, L. (n.) leaf; *-cola*, L. (m., f.) dweller, inhabitant. Larvae are found 'in wild banana axils containing very small quantities of water'.

freycinetiae Laird, 1957: *Aedes* – 'Four larval collections … were made from a forest climbing plant belonging to the family Pandanaceae, *Freycinetia milnei* Seem.'

funestus Giles, 1900b: *Anopheles*; ***funestus*** (Leicester, 1908): *Toxorhynchites* – *funestus*, L. causing death, mournful. Ironic in the case of *Toxorhynchites funestus*.

gahnicola Marks, 1947a: *Aedes* – *Gahnia*, genus of sedges, family Cyperaceae; *-cola*, L. (m., f.) dweller, inhabitant. Marks provided the derivation '**gahnicola**—one living in sword grass, **Gahnia** sp.'

gaudians Dyar & Núñez Tovar, 1927b: *Wyeomyia* – *gaudium*, L. joy, gladness, delight, a source of delight; *-ans*, L. present participle, having the quality of, belonging to.

gibbulus Delfinado, 1966a: *Culex* – *gibbus*, L. humped, humpbacked, protuberant, bent; *-ulus*, L. diminutive suffix, small, little.

gigas Giles, 1901b: *Anopheles* – *gigas*, Gr. giant.

glaucodaemon (Dyar & Shannon, 1925b): *Sabethes* – *glaucus*, L. (Gr. *glaukos*), bluish-green or grey, sea-coloured; *daemon*, L. genius, lar, guardian spirit; from Ancient Greek δαίμων, which refers to the daemons of ancient Greek religion and mythology and of later Hellenistic religion and philosophy.

glyptosalpinx Harbach, Peyton & Harrison, 1984: *Culex* – '… named for the salient character of the pupal trumpet.' *glyphis* (verbal stem *gtyph-*, combining stem *gtypt-*), notch or groove (of an arrow); *salpinx*, trumpet, tube; literally translated 'notched trumpet'. The name stands in apposition to the generic name of *Culex*.

gorgasi Duret, 1971a: *Sabethes* – 'La especie lleva el nombre del Institute Gorgas.'

guarani Shannon, 1928: *Anopheles* – 'Named for the Guaraní, a tribe of Indians.' The Guaraní are a group of culturally related indigenous peoples of South America.

guarao Anduze & Capdevielle, 1949: *Anopheles* – 'The name *guarao* is that of the indigenous tribe that inhabits the Federal Territory Delta Amacuro and the [music] pipes of the neighboring States Sucre and Monagas.'

gubernatoris (Giles, 1901a): *Aedes* – *gubernator*, L. pilot, director; *-is*, L. with, having the nature of. The name *gubernatoris* implies 'of a governor/government', specifically in reference to the 'Allahabad Government House Garden' in Uttar Pradesh State of India where the species was discovered.

gymnothorax Harbach & Petersen, 1992: *Sabethes* – *gymnos*, Gr. bare, naked; *thorax*, Gr. breastplate, chest. Named for the absence of scales on the lower proepisternum.

habilitator Dyar & Knab, 1906d: *Culex* – *habilitas*, L. aptitude, suitability; *-tor*, L. signifying agent, doer.

hadrognathus Harbach, 1995a: *Sabethes* – *hadros*, Gr. well developed, bulky, stout, strong, great; *gnathos*, Gr. jaw. Named for the enormous, protruding maxillae of the larva.

hedys Root, 1927c: *Culex* – *hedys*, Gr. sweet, pleasant. An ironic name for a mosquito. The species was described from a single male reared from larvae collected 'from

Bromeliads in the mountains near Angra dos Reis', Rio de Janeiro, Brazil. The name could refer the 'pleasant' provenance of the species.

hispida (Delfinado, 1967): *Verrallina*; ***hispida*** (Theobald, 1910c): *Mimomyia* – *hispidus*, L. hairy, bristly, rough.

hortator Dyar & Knab, 1907a: *Aedes* – *hortator*, L. inciter, encourager.

hortensis Ficalbi, 1889a: *Culex* – *hortensis*, L. (adjective) third declension, garden (attributive).

hosautos Dyar & Knab, 1907c: *Wyeomyia* – *hosautos*, Gr. in like manner, likewise. Application of the name makes no sense unless it refers to the preceding new species described by the authors that were collected in bamboo woods at the same locality.

idanus Schick, 1970a: *Aedes* – *idanos*, Gr. fair, comely.

impudicus Ficalbi, 1890a: *Culex* – *impudicus*, L. lewd, unchaste, unashamed.

inermis Colless, 1958: *Aedes* – *inermis*, L. unarmed, without weapons. An ironic name for a mosquito.

ininicola Fauran & Pajot, 1974: *Wyeomyia* – Inini, a river and previously an inland territory of French Guiana; *-cola*, L. (m., f.) dweller, inhabitant. 'The name *ininicola* recalls that this species was found in the experimental station of the Institut Pasteur of Guyana, located on the Inini River' [translated from French].

inornata (Williston, 1893): *Culiseta*; ***inornatus*** Newstead, 1907: *Eretmapodites*; ***inornatus*** (Theobald, 1905g): *Culex*; ***inornatus*** (Walker, 1865): *Toxorhynchites* – *in-*, L. not, without; *ornatus* (f. *ornata*), L. adorned, decorated.

iracunda (Walker, 1848): *Coquillettidia* – *iracundus*, L. inclined to anger, wrathful.

irenicus Schmidt, 2003: *Anopheles* – 'The species name is derived from the Greek *eir ēnikos*, aiming at peace. Evidence to date indicates that this species does not bite humans but lives in peaceful coexistence.'

iwi Marks, 1955: *Aedes* – 'The specific name *iwi* is the word for mosquito in the language of Badu Island and other western islands of Torres Strait.'

jocosa (Dyar & Knab, 1908): *Wyeomyia* – *jocosus*, L. facetious.

labesba Howard, Dyar & Knab, 1913: *Wyeomyia* – derivation not provided by the authors (Howard *et al.*, 1915). It is not a patronymic or a Latin word, and *labes* (L. a fall, subsidence, stroke of misfortune, first slip) and the terminal *-ba* make no sense. Howard *et al.* (1915) note that 'This species is closely allied to *Wyeomyia ablabes*, but differs slightly both as adult and larva.' Thus, it appears that *labesba* may be an anagram of *ablabes*.

lacustris (Edwards, 1935a): *Mimomyia* – *lacustris*, L. lake (attributive). Probably named in reference to the type specimens being collected on or near Lake Victoria at Entebbe, Uganda.

lacuum Edwards, 1922d: *Armigeres* – genitive plural of *lacus*, L. a lake, pond, basin, reservoir.

lepchanus (Barraud, 1923b): *Aedes* – Lepcha people (Lepchas of Darjeeling, India).

letifer Sandosham, 1944: *Anopheles* – *letifer*, L. causing death.

leucarthrius (Speiser, 1909): *Aedes* – *leuco-* (*leukos*, Gr. white); *arthro-* (*arthron*, Gr. joint); *-ius* (Gr. *-ios*), L. of, pertaining to. The name undoubtedly refers to the distinctive white apical area of the hindtibia.

lilii (Theobald, 1910c): *Aedes* – named for larvae collected in water at the base of lily leaves.

limosus King, 1932a: *Anopheles* – *limosus*, L. miry, muddy, marshy, slimy. Probably named in reference to larvae found in a 'small, muddy unshaded pool without vegetation'.

madecassorum Rodhain & Boutonnier, 1984: *nomen dubium*, *Orthopodomyia* – named after the old name of the inhabitants of Madagascar.

mangyanus (Banks, 1906b): *Anopheles* – 'The species is named from a tribe of semi-civilized peoples inhabiting the interior of Mindoro and known as the Mangyanes.' Mangyan is the generic name for the eight indigenous groups found on the island of Mindoro, each with its own tribal name, language and customs.

manicatus (Edwards, 1921a): *Toxorhynchites* – *manicatus*, L. having long sleeves. Probably so named in reference to the 'broad creamy-white rings' of the tarsi.

masculinus Mattingly, 1958a: *Aedes* – *masculinus*, L. manly, male. 'This is so named because it is the species on which Edwards (1928a:274), followed by Knight & Hull, 1953:478), based his description of the male (but not the holotype female) of *Aë. curtipes*', which is distinct from *Aedes curtipes* Edwards, 1915c.

maverlius Reinert, 1997: *Anopheles* – acronym: Medical and Veterinary Entomology Research Laboratory of the United States Department of Agriculture.

meprai Martínez & Prosen, 1953b: *Aedes* – acronym: Misión de Estudios de Patoiogía Regional Argentina.

merus Dönitz, 1902: *Anopheles* – *merus*, L. pure, unadulterated, genuine.

micans Leicester, 1908: *Uranotaenia* – *micans*, L. shining, twinkling.

modesta (King & Hoogstraal, 1946e): *Mimomyia*; ***modesta*** Leicester, 1908: *Uranotaenia*; ***modestus*** Ficalbi, 1890b: *Culex* – *modestus* (f. *modesta*), L. moderate, gentle, unassuming, chaste.

montana Brug, 1939: *Aedes*; ***montana*** Ingram & de Meillon, 1927: *Uranotaenia*; ***montanus*** Stanton & Hacker, 1917: *Anopheles* – *montanus* (f. *montana*), L. pertaining to mountains.

monticola Belkin & McDonald, 1957: *Aedes* – *mons*, *montis*, L. mountain; *-cola*, L. dweller, inhabitant. This is a mountain species collected at 5000 ft (1520 m) and above.

mucidus (Karsch, 1887): *Aedes* – *mucidus*, L. snivelling, mouldy, musty.

muroafcete Huang, 1997: *Aedes* – acronym: Musee Royale de 1'Afrique Centrale at Tervuren, Belgium.

musarum Edwards, 1932b: *Culex*; ***musarum*** Edwards, 1936: *Uranotaenia* – *Musa*, NL. a genus of the banana family (Musaceae); *-orum*, L. suffix denoting belonging to, place where, place for, pertaining to. Larvae collected in leaf-axils of wild banana.

natronius Edwards, 1932b: *Aedes* – *natrium*, L.; *natrum*, *natro-*, Ar. sodium; natron, English; *-ius*, L. of, belonging to, pertaining to. 'Larvae in alkaline pools.'

navalis Edwards, 1926a: *Culex* – *navalis*, L. of ships, naval, nautical. Refers to adults 'reared from pitcher-plants'.

nemorosa (Gong, 1996): *Kimia* – *nemorosus* (f. *nemorosa*), L. full of groves, thickly leaved, full of foliage. Apparently named for larvae collected in a bamboo grove.

nepenthicola (Banks, 1909): *Tripteroides*; ***nepenthicola*** Miyagi & Toma, 2007: *Topomyia*; ***nepenthicola*** Steffan & Evenhuis, 1982: *Toxorhynchites* – *Nepenthes*, genus of pitcher plants, family Nepenthaceae; *-cola*, L. (m., f.) dweller, inhabitant. The immature stages of these species develop in pitcher plants.

nimbus (Theobald, 1902b): *Anopheles* – *nimbus*, L. raincloud. *Stethomyia* are dark mosquitoes, with no pale scaling on the maxillary palpi, wings or legs; thus, the connotation with a "dark-coloured" rain-cloud.

nimpe Nguyen, Tran & Harbach, 2000: *Anopheles* – acronym: National Institute of Malariology, Parasitology and 'Entomology, Hanoi.

niphadopsis Dyar & Knab, 1918b: *Aedes* – *niphados*, Gr. snowflake; *opsis*, Gr. sight, appearance. Probably so named in reference to the rather extensive white scaling of the body and legs.

nitens (Cerqueira, 1943): *Coquillettidia* – *nitens* (from *niteo*), L. shining, glittering, bright.

nocticola Peyton, 1977: *Uranotaenia* – *nocticola*, L. (m.) night-lover. 'Holotype male ... collected as a larva from a pool inside a cave'. Not indicated but the species is probably named for the absence of light inside the cave.

nubicola (Laffoon, 1946): *Aedes* – *nubes*, *-is*, L. (f.) cloud; *-cola*, L. dweller, inhabitant. Although not stated, Laffoon apparently named the species for its discovery at an altitude of '7,000-8,000 feet' (2130–2440 m).

nyasae Edwards, 1930a: *Aedes* – Nyasa people of south-eastern Africa.

obturbans (Walker, 1859): *Armigeres* – *obturbans*, L. disturbing.

obturbator Dyar & Knab, 1907a: *Aedes* – *obturbo*, L. to disturb, confuse, distract, harass; *-tor*, L. suffix signifying one who, actor, agent. This is an appropriate name for an annoying mosquito.

ocossa Dyar & Knab, 1919: *Culex* – three possibilities: (1) *o-*, L. to, toward, against, opposite, upon; *cossus* (f. *cossa*), L. a kind of larva under the bark of trees. Indirect evidence suggests the species may have been named with reference to larvae found in a hollow tree trunk. (2) *oeco-*, *oikos*, Gr., house; *-so*, common ending of Greek verbs, latinized to *ocossa*, connoting adults entering houses. (3) *oc-*, L. toward, to, upon; *ossa*, Gr. rumour. Perhaps in reference to uncertainty (based on partial evidence) about the validity of the species expressed by the authors – 'but on account of the doubt, propose the name *ocossa*, as above, so as not to leave our species possibly nameless'. See *panoccosa* Dyar, 1923c.

oiketorakras Osorno-Mesa, 1947: *Anopheles* – 'The Greek word *oiketorakras* from *oiketor* (= inhabitant) and *akras* (= of height) has been chosen according to the peculiar habitat of this species' [translated from Spanish]. The type locality is in the Cordillera Central east of Bogotá at an altitude of 2700–2800 m. See Kitzmiller (1982).

opok Corbet & van Someren, 1962: *Aedes* – collected in *Terminalia velutina* woodland, tress locally known as 'Opok'.

orstom Brunhes, Adam & Bailly-Choumara, 1967: *Culex* – acronym: Office de la Recherche Scientifique & Technique Outre-Mer, Paris.

pacificus Edwards, 1916: *Culex* – *pacificus*, L., peacemaking, peaceful. Connotation not apparent, but some specimens were captured 'at the French Hospital'.

palmata (Lane & Cerqueira, 1942): *Wyeomyia*; ***palmatus*** (Rodenwaldt, 1926): *Anopheles* – *palmatus* (f. *palmata*), L. like the palm of the hand or a palm leaf. Refers to the form of larval setae. See Kitzmiller (1982).

pandani Brunhes, 1969: *Culex*; ***pandani*** Stone, 1939: *Aedes*; ***pandani*** (Theobald, 1912b): *Uranotaenia* – *pandanus*, NL. genus of monocotyledons; *-i*, L. belong to, pertaining to. First two species in reference to larvae collected from *Pandanus* leaf axils; last species in reference to adults 'Caught in morning flying amongst dead leaves of *Pandanus hornei* on ground under a large group of those trees'.

panocossa Dyar, 1923c: *Culex* – *pan-* (*pas*), Gr. all, the whole, every, very; *ocossa* (see *ocossa* Dyar & Knab, 1919). Probably in reference to the overall similarity to *ocossa*.

papago Zavortink, 1970c: *Aedes* – papago, an archaic term for Tohono O'odham Native Americans in southern Arizona.

perexiguus Theobald, 1903a: *Culex* – *perexiguus*, L. very small.

periskelatus (Giles, 1902): *Aedes* – *peri*, Gr. very, exceedingly; *skelatus*, Gr. dried up, withered, mummified. Probably named with reference to the condition of the specimen at hand (type male).

phnoma (Klein, 1973): *Verrallina* – the author did not mention the source of the name. *Phnom* is Khmer for 'hill'; however, the type locality is in Kandal Province, which completely surrounds the capital Phnom Penh (literally translated as Penh's hill).

pholeocola (Linton & Harbach, 2013): *Nyctomyia* – *pholeos*, Gr. (m.) hole, cave or den; *-cola*, L. dweller, inhabitant. This is a cavernicolous species.

pia (Le Goff, Brengues & Robert, 2013): *Aedes* – 'The term "pia" means "fresh" or "new" in the Shimaore language, the main language spoken in Mayotte. In addition, Pia is the first name of a stepdaughter of V. Robert.'

pinarocampa Dyar & Knab, 1908: *Culex* – *pinaros*, Gr. dirty; *campa*, L. (Gr. *kampe*), caterpillar.

pipiens Linnaeus, 1758: *Culex* – *pipiens*, L. (adjective: m., f., n.), chirping, peeping, piping, twittering (from *pipio*, L. chirp, peep, pipe, twitter).

pleccau Koidzumi, 1924: subspecies of *Anopheles lindesayi* Giles, 1900a – the name *pleccau* (alternatively *pleccaw*) is the English transliteration of the Taiwanese aboriginal word for mosquito. See Kitzmiller (1982).

plagosus Marks, 1959b: *Aedes* – the author stated that 'The specific name means "full of stripes"'; however, this appears to be incorrect as the meaning of *plagosus* given in Latin dictionaries is 'fond of beating, flogging or striking'. The word is related to the noun *plaga* (pl. *plagae*), which has many meanings: (1) a blow, stroke, wound, stripe, welt, scar; (2) shock; (3) cut, thrust; and in zoology it is used to mean 'a stripe of colour'. The formation of *plagosus* (*sensu* Marks) by combining *plaga* and the suffix *-osus*, L. having the nature or quality of (usually in fullness or abundance), is incongruous with the classical Latin meaning of *plagosus*.

pogonurus Edwards, 1936: *Aedes* – *pogon*, Gr. beard; *oura*, Gr. tail. Latinized *pogonourus*, 'bearded tail', probably refers to the tufts of setae on the gonocoxite and gonostylus of the male genitalia.

propinqua Mattingly, 1970: *Heizmannia*; ***propinqua*** Peyton, 1977: *Uranotaenia*; ***propinquus*** Colless, 1955b: *Culex* – *propinquus* (f. *propinqua*), L. near, neighbouring. The species are presumably named in reference to being very similar to a related species.

psatharus Dyar, 1920h: *Culex* – *psatharos*, transliteration of ancient Gr. ψαθαρός, meaning 'all I know is that I know nothing'.

rabanicolus Floch & Abonnenc, 1946a: *Culex* – Raban, a place in French Guiana; *-colus*, NL. (m.), dwelling in, living among. 'We consider our black *Culex* as belonging to a new species which we have named after the place of capture (way of Raban)' [translated from French].

rajah Tsukamoto, 1989: *Culex*; ***rajah*** Tsukamoto, 1989: *Toxorhynchites* – named for larvae collected from water in pitchers of *Nepenthes rajah* Hooker f.

ramosa da Cunha Ramos, 1993: *Uranotaenia* – *ramosus* (f. *ramosa*), L. full of branches. 'The new name emphasizes the very particular morphology of the upper caudal seta (2-X) with numerous branches' [translated from Portuguese].

ravenalaphila Boussès & Brunhes, 2013: *Uranotaenia* – *philia*, Gr. friendly love, affection, fondness (*philos*, beloved, dear). 'Name in connection with the development of larvae in the trunk and cut hollows of *Ravenala madagascariensis*' [translated from French].

ravenalicola da Cunha Ramos & Brunhes, 2004: *Uranotaenia* – *Ravenala*; *-cola*, L. dweller, inhabitant. See *ravenalaphila* Boussès & Brunhes, 2013. The species is named for 'the place of larval development ... in the axils of the leaves of *Ravenala* that grow in the swamp of Ambila' [translated from French]. *Ravenala* is a genus of flowering plants with a single species in Madagascar, *Ravenala madagascariensis*, a member of the monocotyledonous family Strelitziaceae.

rivulorum Leeson, 1935: *Anopheles* – named after larval and adult habitats: 'Adults along streams, in crevices and cavities in the banks. Larvae in slowly moving streams near banks and among boulders'. Genitive plural of *rivulus* (L. small brook, rill, stream, rivulet), diminutive of *rivus* (L. stream, brook).

roucouyana (Bonne-Wepster & Bonne, 1920): *Wyeomyia* – roucou, roucouyer; *-anus* (f. *-ana*), L. having the nature, quality, state or condition of. The roucouyer, also referred to as roucou or annatto, is a small tropical American tree (*Bixa orellana*), the seeds of which were originally used by Native Americans to make red body paint and lipstick. The name was probably inspired by the violaceous reflections of scaling on the antepronota and scutum.

ruarinus Edwards, 1940: *Anopheles* – ruar- + *-inus*, L. pertaining to. 'Larvae were collected from rock pools on the top of a flat rock kopje which is termed a "ruari" by the natives.'

rupicolus Lewis, 1937: *Anopheles* – *rupes*, L. (m.), rock; *-colus*, NL. (m.), dwelling in, living among. 'Larvae have been collected ... many hundreds in a deep cleft in the rock.'

saeva Dyar & Knab, 1906a: *Psorophora* – *saevus*, L. raging, wild, violent, fierce.

sagax (Skuse, 1889): *Aedes* – *sagax*, L. wise.

saimedres Huang, 1988b: *Aedes* – acronym: South African Institute for Medical Research.

salinarius Coquillett, 1904b: *Culex* – *salinarius*, L. a salter, of or belonging to salt-works.

sampi Huang, 2004: *Aedes* – acronym: not specifically stated, but obviously derived from Systematics of Aedes Mosquitoes Project, a cooperative venture between the Smithsonian Institution and the US Army Medical Research and Development Command.

saxicola Edwards, 1922a: *Aedes* – *saxum*, *-i*, L. rock; *-cola*, L. dweller, inhabitant. This is a replacement name for *Hulecoeteomyia fluviatilis* Leicester, 1908. Leicester stated that 'The larvae are quite distinct and found in rock pools'.

schedocyclia (Dyar & Knab, 1908): *Shannoniana* – *schedon*, Gr. near, almost; *kyklos*, Gr. (L. *cyclus*) circle, ring. Possibly named in reference to a 'large silver patch on the vertex'.

seampi Huang, 1974: *Aedes* – acronym: Southeast Asia Mosquito Project, Smithsonian Institution, supported by the US Army Medical Research and Development Command.

seatoi Huang, 1969: *Aedes* – acronym: Southeast Asia Treaty Organization, Medical Research Laboratory, Bangkok.

smaragdinus Reinert, 1997: *Anopheles* – name derived from Latin for the emerald green colour of the live fourth-instar larva and early pupa: *smaragdus*, L. (Gr. *smaragdos*), a precious stone of green colour, emerald; *smaragdinus*, emerald-green.

solstitialis (Lutz, 1904): *Toxorhynchites* – *solstitium*, L. longest day of the year, summer solstice; *-alis*, L. pertaining to, having the nature or quality of.

sororcula Dyar & Knab, 1906a: *Wyeomyia* – *soror*, L. sister; *sororcula* (dim.), little sister.

spanius (Dyar & Knab, 1909d): *Deinocerites* – *spanios*, Gr. (latinized) rare, scarce.

spoliata Edwards, 1923b: *Hodgesia* – *spoliator*, L. plunderer; *spolio*, *-atus*, strip, rob, plunder.

stenoetrus (Theobald, 1907): *Aedes* – *stenos*, *steno-*, Gr. narrow; *-tros*, Gr. tool, instrument, means. It seems likely that the species was named for its 'markedly narrowed abdomen'.

stenolepis Dyar & Knab, 1908: *Culex* – *stenos*, *steno-*, Gr. narrow; *lepis*, Gr. scale.

stenoscutus (Edwards, 1912c): *Aedes* – *stenos*, *steno-*, Gr. narrow; *scutum*, L. shield.

sticticus (Meigen, 1838): *Aedes* – *sticto-* (Gr. *stiktos*), punctured, spotted, dappled; *-icus* (Gr. *-ikos*), belonging to, like, pertaining to.

stramineus Dubitzky, 1970: *Aedes* – *stramineus*, L. of straw (*stramen*, L. straw).

strelitziae Muspratt, 1950: *Aedes* – this species is 'a prolific breeder in the axils of the so-called wild banana, *Strelitzia nicolai*'.

suffusus Edwards, 1922a: *Aedes* – *suffusus*, L. suffused; having been tinged, imbued, stained or coloured. 'Pale scales of mesonotum showing indications of the same arrangement as in *A. oreophilus*, but the markings suffused and very indefinite.'

summorosus Dyar, 1920i: *Culex* – *summorosus*, L. somewhat peevish. What was Dyar thinking when he proposed this name?

swahilicus Gillies, 1964: *Anopheles* – Swahili; *-icus*, L. belonging to, like, of, pertaining to. The species is named for the Swahili people of the Swahili Coast, which refers to a coastal area of eastern Africa that mainly includes littoral Kenya, Tanzania and northern Mozambique. See Kitzmiller (1982).

tabu Ramalingam & Belkin, 1965: *Aedes* – tabu, treated as a latinized noun in the nominative singular standing in apposition to the generic name. Type locality: Eua Island, [Tongatapu Group], Tonga. The authors indicated that the species occurs on islands of the 'Tongatabu Group' (Tongatapu Group). The Tongatapu Group, also spelled Tongataboo Group, is the southernmost island cluster of the kingdom of Tonga. Tapu (also tabu or kapu) is a Polynesian traditional concept denoting something holy or sacred – something that is tapu is considered inviolable or sacrosanct; it involves rules and prohibitions, the meaning from which the English word 'taboo' is derived.

taurepana Anduze, 1941: *Wyeomyia* – '*Taurepan* is the name of the indigenous tribe that inhabits the region' [translated from Spanish], i.e. the headwaters of the Surukum River, Bolívar State, Venezuela.

tecmarsis Dyar, 1918e: *Culex* – *tecmar*, latinization of Gr. *tekmar*, goal, end, mark, fixed sign; *-is*, L. adjectival suffix meaning with, having, nature of.

tenebrosus Dönitz, 1902: *Anopheles* – *tenebrosus*, L. dark, gloomy.

terrens (Walker, 1856a): *Aedes* – *terrens*, L. alarming, scaring, etc. (present active participle of *terreo*, verb meaning frighten, terrify, alarm).

testacea Theobald, 1905f: *Uranotaenia* – *testaceus*, L. with a shell, of earthenware (entomological definition: bearing a test or hard covering, brownish-yellow). 'Thorax bright testaceous with azure blue prothoracic lobes.'

thalassius Theobald, 1903a: *Culex* – *thalassios*, Gr. of the sea (from *thalassa*, sea). 'The larvae were mostly taken in a drain of tidal water and others from a pool in a mangrove swamp, others from a canoe on the foreshore, and some from a pool of tidal water that had soaked through sand into a drain.'

tigripes (de Grandpré & de Charmoy, 1901): *Lutzia* – *tigris*, L. tiger; *pes*, L. foot. The species is apparently named for the tiger-like appearance of the spotted femora and tibiae.

toltecum (Dyar & Knab, 1906a): *Psorophora* – apparently named for the archaeological Toltec culture that dominated a state centred in Tula, Hidalgo, Mexico.

torosa (Delfinado, 1967): *Verrallina* – *torosus* (f. *torosa*), L. muscular, bulging, lusty.

torrentium Martini, 1925: *Culex* – *torrentium*, L. genitive masculine plural of *torrens*, swift stream.

tortilis (Theobald, 1903g): *Aedes* – *tortilis*, L. twisted, winding.

totonaci Lassmann, 1951: *Psorophora* – Totonacas (Totonac people) antedated the conquest of Mexico by the Spaniards. 'The entire region around Cerro Guzman [type locality] is covered with Indian ruins which antedate the conquest of Mexico by the Spaniards and was populated by a tribe known as the Totonacas, and following the custom of Dyar and Knab who named *Psorophora toltecum* [as *Janthinosoma toltecum*] after an Indian tribe I will christen another representative of this genus with the name of another Indian tribe.'

travestita (Brug, 1928b): *Bironella* – species of Dutch/German *Transvestit*, from Latin *trans-* + *vestire* to clothe.

tremulus (Theobald, 1903e): *Aedes* – *tremulus*, L. shaking, quivering.

tuberis Bohart, 1946: *Culex* – *tuber -eris*, L. swelling, lump.

ulocoma (Theobald, 1903a): *Wyeomyia* – *ulo-* (from *oulos*), Gr. woolly, curly; *kome*, Gr. hair of the head, mane.

unicornis Root, 1928: *Culex* – *unus*, *uni-*, L. one; *cornis* (m, f; neuter *corne*), L. horned, third declension singular of *cornus*, horn.

varuae Belkin, 1962: *Aedes* – 'This species is named for the brigantine "Varua" [boat] of the Robinson-Peabody Museum expedition to the South Pacific, 1956.' Varua, designed by Starling Burgess in 1940, was gradually changed over to the brigantine after sailing from Gloucester in 1945. The brigantine was a two-masted sailing vessel.

venustipes (Skuse, 1889): *Aedeomyia* – *venu*, ablative singular of *venus*, L. goddess of love, charm, beauty, desire;

stipes, L. stock, stem, trunk. Probably in reference to the 'Proboscis ... somewhat spotted with white scales, with a ring of white in the middle and another rather broader one immediately before the terminal lobes'.

versicolor (Barraud, 1924i): *Aedes*; ***versicolor*** (Edwards, 1913b): *Coquillettidia* – *versicolor*, L. of various colours, variegated.

vorax Edwards, 1921d: *Lutzia* – *vorax*, L. gluttonous, greedy.

winter Dong, Wu & Mao, 2006: *Topomyia* – 'The species was named as "winter" because the type specimens were collected in the "winter season"' [translated from Chinese].

xanthogaster (Edwards, 1924b): *Coquillettidia* – *xanthos*, Gr. yellow; *gaster*, Gr. stomach, belly.

xanthomelaena Edwards, 1925: *Uranotaenia* – *xanthos*, Gr. yellow; *melaina*, Gr. black, dark.

xanthozona (van Someren, 1948): *Mimomyia* – *xanthos*, Gr. yellow; *zone* Gr. (L. *zona*), belt, girdle.

xenophobus Ronderos, 1965: *Culex* – *xenos*, Gr. stranger, guest; *phobe*, Gr. (L. *phoba*), corymb (flower cluster), curl, tuft.

yanbareroides Dong & Miyagi, 1995: *Topomyia* – so named because the species resembles *Topomyia yanbarensis* Miyagi, 1976. The ending *-oides* (from Gr. *-eides*) meaning like, resembling, having the form of, is nonsensical as *yanbarensis* is named for the northern forested region of Okinawa Island known as 'yanbaru' by the local people. The species cannot resemble a 'forested region'.

zinzala Zavortink, 1986: *Wyeomyia* – *zinzala*, L. (f.), a type of gnat.

zoosophus Dyar & Knab, 1918b: *Aedes* – English *zoon*, *zoo-*, *zo-* (*zoon*, NL., from Gr. *zôion*, animal, beast); *zophos*, Gr. (L. *sophus*), darkness, blackness, murky.

Some Interesting Genus-group Names

Abraedes Zavortink, 1970c: subgenus of *Aedes* Meigen – *habr*, Gr. delicate; *aedes*, Gr. distasteful, disagreeable, odious, unpleasant. Etymology is not mentioned; however, in a personal communication the author stated 'The "abr" of *Abraedes* is just the Greek "abr" (or "habr") meaning pretty, delicate, or dainty.'

Aedes Meigen, 1818: genus of tribe Aedini Meigen – *aedes*, Gr. distasteful, disagreeable, odious, unpleasant.

Allotheobaldia Brolemann, 1919b: subgenus of *Culiseta* Felt – imitative (false) patronymic. *allo-* (Gr. *allos*), other: the other *Theobaldia* (named after Frederic V. Theobald; preoccupied senior synonym of *Culiseta* Felt, 1904).

Ctenogoeldia Edwards, 1930a: subgenus of *Runchomyia* Theobald – imitative (false) patronymic. *cteno-* (Gr. *kteis*, *ktenos*), comb; *Goeldia* (named after Dr Emilio Augusto Goeldi; currently a synonym of *Shannoniana* Lane & Cerqueira, 1942).

Culex Linnaeus, 1758: genus of Culicini Meigen – *culex*, L. gnat, midge, mosquito.

Deinocerites Theobald, 1901a,b,c: genus of Culicini Meigen – *deinos*, Gr. terrible, fearful; *keros*, Gr. horn; *-ites*, Gr. having the nature of, like.

Dendroskusea Edwards, 1929b: subgenus of *Aedes* Meigen – imitative (false) patronymic. *dendro-* (Gr. *dendron*), tree; *Skusea*.

Eretmapodites Theobald, 1901d: genus of Aedini Meigen – *eretmon*, Gr. oar; *podo-*, *podi-*, *poda* (Gr. *pous*, *podos*), foot; *-ites*, Gr. having the nature of, like.

Geoskusea Edwards, 1929b: subgenus of *Aedes* Meigen – imitative (false) patronymic. *geo-* (*ge*, Gr. earth); *Skusea*.

Halaedes Belkin, 1962: subgenus of *Aedes* Meigen – *hali-* (Gr. *hals*, *halos*), sea salt; *aedes*, Gr. distasteful, disagreeable, odious, unpleasant. '*A. australis* [type species] breeds only in salt or brackish rock pools at or slightly above high-tide mark'.

Hulecoeteomyia Theobald, 1904c: subgenus of *Aedes* Meigen – Theobald stated that six new genera are described from specimens collected by Dr Leicester, and 'Most are jungle-living species'. Other than this, Theobald, as usual, gave no information about the etymology of his new names. Assuming that *Hulecoeteomyia* is in reference to the jungle, the name appears to be derived from the following: *hyle*, Gr. (f.) wood; *hylikos* Gr. of wood - implying 'of the woods'; *eteos*, Gr. true, real, genuine; *myia*, Gr. fly. Putting these together generates Hylikoeteomyia. When transliterated from Greek, *hyliko-* becomes *huleco-*; hence, *huleco-* + *eteo-* + *myia* becomes *Hulecoeteomyia*, which loosely translated means 'true forest fly'.

Indusius Edwards, 1934: subgenus of *Aedes* Meigen – named for the Indus Valley, Indus River, Pakistan.

Insulalutzia Tanaka, 2003: subgenus of *Lutzia* Theobald – imitative (false) patronymic. *insula*, L. island; *Lutzia*.

Janthinosoma Lynch Arribálzaga, 1891a,b: subgenus of *Psorophora* Robineau-Desvoidy – *ianyhinos*, Gr. violet-coloured; *soma*, Gr. body.

Lepiothauma Enderlein, 1923: subgenus of *Aedeomyia* Theobald – *lepis*, Gr. scale; *thauma*, Gr. wonder, marvel.

Levua Stone & Bohart, 1944: subgenus of *Aedes* Meigen – not stated. Levu; *-us* (f. *-a*), L. of, pertaining to. Type locality: Suva, [Viti Levu Island], Fiji. Viti Levu is the largest of the islands that comprise the South Pacific nation of Fiji. The island is home of the capital city of Suva, the type locality of the orthotype of the subgenus.

Limatus Theobald, 1901c: genus of Sabethini Blanchard – *limatus*, L. polished, finished, having been polished.

Lophoceraomyia Theobald, 1905h: subgenus of *Culex* Linnaeus – *lopho-* (Gr. *lophos*), mane, crest, comb, tuft; *keros*, Gr. horn; *-ites*, Gr. having the nature of, like; *myia*, Gr. fly.

Lophopodomyia Antunes, 1937: subgenus of *Anopheles* Meigen – *lopho-* (Gr. *lophos*), mane, crest, comb, tuft; *podo-*, *podi-*, *poda* (Gr. *pous*, *podos*), foot; *myia*, Gr. fly.

Malaya Leicester, 1908: genus of Sabethini Blanchard – not stated, but obviously in recognition of British Malaya, the area of the Malay Peninsula and the island of Singapore that was under British control between the 18th and 20th centuries.

Mansonioides Theobald, 1907: subgenus of *Mansonia* Blanchard – imitative (false) patronymic. 'Closely related to *Mansonia*'. *Mansonia* + *-oides*, like, resembling, having the form of (Gr. *-eides*), like.

Maorigoeldia Edwards, 1930a: genus of Sabethini Blanchard – imitative (false) patronymic. Māori (people of New Zealand) + *Goeldia* (named after Dr Emilio Augusto Goeldi; currently a synonym of *Shannoniana* Lane &Cerqueira, 1942).

Menolepis Lutz, 1905: subgenus of *Wyeomyia* Theobald – *meno-* (Gr. *menos*), force, courage, spirit, strength, strong; *lepis*, Gr. scale.

Metalutzia Tanaka, 2003: subgenus of *Lutzia* Theobald – imitative (false) patronymic. *meta-* (Gr. *meta*), between, among, near; *Lutzia*.

Miamyia Dyar, 1919d: subgenus of *Wyeomyia* Theobald – *mia*, L. my; *myia*, Gr. fly.

Molpemyia Theobald, 1910c: subgenus of *Aedes* Meigen – *molpe*, Gr. song; *myia*, Gr. fly.

Mucidus Theobald, 1901b: subgenus of *Aedes* Meigen – *mucidus* (L. from *muceo*, be musty; *-idus* L., state or condition of), mouldy, musty.

Neobironella Tenorio, 1977: subgenus of *Bironella* Theobald – imitative (false) patronymic. *neo-* (Gr. *neos*), new, young, recent; *Bironella*.

Neomacleaya Theobald, 1907: subgenus of *Verrallina* Theobald – imitative (false) patronymic. *neo-* (Gr. *neos*), new, young, recent; *Macleaya*.

Neotheobaldia Dobrotworsky, 1958b: subgenus of *Culiseta* Felt – imitative (false) patronymic. *neo-* (Gr. *neos*), new, young, recent; *Theobaldia* (named after Frederic V. Theobald; preoccupied senior synonym of *Culiseta* Felt, 1904).

Nothoskusea Dumbleton, 1962: subgenus of *Opifex* Hutton – imitative (false) patronymic. *notho-* (Gr. *nothos*), spurious, bastard; *Skusea*.

Nyssorhynchus Blanchard, 1902: subgenus of *Anopheles* Meigen – *nysso*, Gr. prick, spur, pierce, puncture; *rhynchos*, Gr. nose, snout, muzzle.

Ochlerotatus Lynch Arribálzaga, 1891a,b: subgenus of *Aedes* Meigen – *ochleros*, Gr. troublesome; *-tatos* Gr. superlative, of the highest degree.

Orthopodomyia Theobald, 1904c: genus of Orthopodomyiini Belkin, Heinemann & Page – *ortho-* (Gr. *orthos*), straight, correct, normal, right, direct; *podo-*, *podi-*, *poda* (Gr. *pous*, *podos*), foot; *myia*, Gr. fly.

Paraedes Edwards, 1934: subgenus of *Aedes* Meigen – *para*, Gr. beside, near, by; *aedes*, Gr. distasteful, disagreeable, odious, unpleasant.

Protomacleaya Theobald, 1907: subgenus of *Aedes* Meigen – imitative (false) patronymic. *protos*, Gr. first; *Macleaya*.

Pseudoficalbia Theobald, 1912b: subgenus of *Uranotaenia* Lynch Arribálzaga – imitative (false) patronymic. *pseudo-* (*pseudos*, Gr. lie; *pseudes*, false); *Ficalbia*.

Pseudoskusea Theobald, 1907: subgenus of *Aedes* Meigen – imitative (false) patronymic. *pseudo-* (*pseudos*, Gr. lie; *pseudes*, false); *Skusea*.

Psorophora Robineau-Desvoidy, 1827: genus of Aedini Meigen – *psoro-* (Gr. *psora*), itch, scurvy, scab, mange; *phoras*, Gr. fruitful, bearing.

Rhinoskusea Edwards, 1929b: subgenus of *Aedes* Meigen – imitative (false) patronymic. *rhino-* (*rhis*, *rhinos*, Gr. nose, snout, beak, bill); *Skusea*.

Stethomyia Theobald, 1902b: subgenus of *Anopheles* Meigen – *stethos*, Gr. breast, chest; *myia*, Gr. fly.

Suaymyia Thurman, 1959: subgenus of *Topomyia* Leicester – *suay*, Thai (transliteration) beautiful; *myia*, Gr. fly.

Theomyia Edwards, 1930a: subgenus of *Culiseta* Felt – *theos*, Gr. god, spirit; *myia*, Gr. fly.

Tinolestes Coquillett, 1906d: subgenus of *Culex* Linnaeus – *tino-* (Gr. *teino*), to stretch, to extend; *lestes*, Gr. robber, plunderer, pirate.

Toxorhynchites Theobald, 1901b: genus of Toxorhynchitini Lahille – *toxo-* (Gr. *toxon*), bow; *rhynchos*, Gr. nose, snout, muzzle; *-ites*, Gr. having the nature of, like.

Trichoprosopon Theobald, 1901a,c: genus of Sabethini Blanchard – *trich-* (Gr. *thrix*, *trichos*), hair; *prosopon*, Gr. face, front.

Tripteroides Giles, 1904a: genus of Sabethini Blanchard – *tri-*, L. (Gr. *treis*, *trion*) three; *ptero-*, *pteron*, Gr. feather, wing; *-eides*, Gr. (L. *-oides*) like, resembling, having the form of.

Uranotaenia Lynch Arribálzaga, 1891a,b: genus of Uranotaeniini Lynch Arribálzaga – *urano-* (Gr. *ouranos*), heaven, sky; *taenia*, L. (Gr. *tainia*), ribbon.

Wyeomyia Theobald, 1901b,a,c: genus of Sabethini Blanchard – wye (?), English, a Y-shaped part or object; *myia*, Gr. fly. Theobald (1901c) states: 'These "wood mosquitoes" have a peculiar habit of, so to speak, staying in the air with the last pair of legs thrown over the back, and, with their thin body and long legs, looking very much like spiders.' This seems to conjure up an image of a hovering or dangling Y-shaped object, but why 'wye' and its intended meaning is anyone's guess.

Zeugnomyia Leicester, 1908: genus of Aedini Meigen – '... it links on the Uranotaenias [*sic*] to Colonemyia [*sic*] and Skeiromyia [*sic*] and through them to the Wyeomyia [*sic*]; hence its name.' *zygos*, Gr. yoke, *zeugnymi*, to connect, join; *myia*, Gr. fly.

Section 8: Phylogenetic Classification of Composite Genus *Aedes*

The traditional classification of tribe Aedini (Knight & Stone, 1977) included nine genera, the largest of which was *Aedes* Meigen. The first revision of aedine mosquitoes was conducted by Edwards (1932a), who recognized an *Aedes* group of seven genera – *Aedes* (the largest with 29 subgenera, many of which had previously been described as genera), *Armigeres*, *Eretmapodites*, *Haemagogus*, *Heizmannia*, *Opifex* and *Psorophora*. Aedini was defined by Belkin (1962) to consist of the *Aedes* group of Edwards and genera *Ayurakitia*, *Udaya* and *Zeugnomyia*. *Ayurakitia* was later reduced to subgeneric status in *Aedes* by Mattingly (1971); hence, the nine genera recognized by Knight & Stone (1977). Belkin recognized that some subgenera of *Aedes* were heterogeneous groups and noted that the genus would likely have to be revised into smaller more natural groups of species. Before the end of the 20th century, the genus included 56 subgenera.

Radical changes to the traditional classification of the genus *Aedes* were introduced in a series of studies which began with the restoration of *Verrallina* and *Ayurakitia* to generic status (Reinert, 1999b, 2000c, respectively) and were followed by the landmark study of Reinert (2000d), who split the composite genus *Aedes* into the genera *Aedes* and *Ochlerotatus* (originally proposed as a genus), with 22 of the 43 recognized subgenera placed in the former and 21 placed in the latter. This was followed by a series of morphology-based phylogenetic studies (Reinert *et al.*, 2004, 2006, 2008, 2009), which resulted in the traditional classification of the genus being divided into 73 genera and 39 subgenera.

The nomenclatural changes of Reinert (2000d) and Reinert *et al.* (2004) provoked a heated debate and much controversy. Given the major changes to the classification of *Aedes* and the implications for those working in applied taxonomy, this was perhaps not surprising. Some critics were opposed to nomenclatural changes based on an exemplar approach and direct translation of the results of cladistic analyses into a reclassification of the genus (though one wonders why there has never been any objection to the elevation of *Verrallina* and *Ayurakitia* to generic status by Reinert, 1999b, 2000c). However, despite the criticism and controversy, Reinert *et al.* went on to publish the results of three further detailed studies, which appear to have been largely ignored by mosquito workers, due at least in part to the earlier criticism. A consequence of the heated debate was a deep impasse between researchers and applied taxonomists who supported the reclassification advanced by Reinert *et al.* (2009) and those who ignored or summarily rejected it.

The ranking of taxa as genera or subgenera is subjective, but essentially involves the criteria of monophyly and equivalent rank in a phylogenetic analysis. In the analyses of Reinert *et al.*, taxa that were widely accepted as genera (i.e. *Armigeres*, *Eretmapodites*, *Haemagogus*, *Heizmannia*, *Opifex*, *Psorophora*, *Udaya* and *Zeugnomyia*) were taken as the reference genera for equivalence, i.e. monophyletic clades (groups) of equivalent rank to those taxa were afforded generic rank and monophyletic groups within those clades were afforded subgeneric rank. It is noted that intuitive interpretation is not without merit as the application of explicit methodology confirmed the monophyly of most subgenera and species groups that were diagnosed by unique combinations of characters. The principal problem, then, is not in recognizing monophyletic groups, but in deciding which taxonomic ranks should be assigned to such taxa. Bearing in mind that acceptance of broad genus-group concepts for convenience is phylogenetically insupportable, in cases where a taxon is found to be paraphyletic or polyphyletic it should be reclassified to ensure that taxonomic ranking reflects monophyly. On that basis, and because most taxonomists do not accept paraphyletic or polyphyletic classifications, Reinert *et al.* raised many of the subgenera and species groups of the genus *Aedes* to generic level. It should be noted that many of the taxa recognized as subgenera in the traditional classification of *Aedes* were originally described as genera.

Wilkerson *et al.* (2015) reanalysed the morphological data of Reinert *et al.* (2009), using equal weighting rather than the implied weighting used by Reinert *et al.*, to 'gauge the stability' of the generic groupings of the latter authors. Disregarding the weighting of characters and treating them all as equal produces impoverished results that do not reflect reality (Williams *et al.*, 2010); consequently, the result was the retention of all the genera as terminal taxa within an otherwise largely collapsed tree. The point to be made here is that although the deeper relationships were lost, all of the generic-level taxa of Reinert *et al.* survived, thus supporting the integrity (monophyly) of all of these groups. A further point to be made is that no large morphological data set analysed in any study conducted to date has provided strong support for deeper relationships, e.g. the phylogenetic studies of Anophelinae (Sallum *et al.*, 2000; Harbach & Kitching, 2005, 2016; Collucci & Sallum (2007)), Culicini (Harbach *et al.*, 2012, 2017) and Sabethini (Judd, 1996; Harbach, Vu & Kitching, 2007; Motta *et al.*, 2007)).

Differences between phylogenetic trees and classifications derived from them occur when taxonomists choose to recognize paraphyletic and/or polyphyletic taxa to preserve traditional nomenclature. The issue is how to deal with such groups. One way is to eliminate them, which usually requires new names and changes to categorical ranks in a classification; a second way is simply to leave the classification in an unnatural state. Reinert *et al.* (2004) chose the first way and recognized all of the former subgenera of *Aedes* that were 'weighting independent', i.e. those that were common to the results of both equal and implied weighting analyses, as genera to reflect monophyly and rank equivalence with *Armigeres*, *Eretmapodites*, *Haemagogus*, *Heizmannia*, *Opifex*, *Psorophora*, *Udaya* and *Zeugnomyia*. Wilkerson *et al.* (2015) effectively chose the second way and recognized all the genera of Reinert *et al.* (2009) as subgenera within a very large and highly polyphyletic genus *Aedes* to preserve a backbone traditional classification for convenience.

In the phylogeny of Wilkerson *et al.* (2015: fig. 3), *Aedes* (presumably subgenus *Aedes* despite boldface type indicating it is the genus *Aedes*) is sister to *Paraedes* + *Verrallina*. *Paraedes* is afforded subgeneric rank whereas *Verrallina* is retained as a genus; thus, the authors, being traditional systematists, interpret *Verrallina* as a group morphologically very different from *Paraedes*, and this dissimilarity is the basis for recognizing it as a genus. This clade alone (there are 13 problematic clades within a polytomy that contains all the genera of Reinert *et al.* except *Psorophora*) demonstrates the inconsistent application of the 'equivalent rank' criterion that they claimed to use (in part?) to recognize groups as genera or subgenera. Reinert *et al.* (2004) faced an identical problem with a clade comprising *Aedes* (*Alanstonea*) and *Eretmapodites*. Two options were available: either elevate *Ae.* (*Alanstonea*) to genus rank or treat it as a subgenus of *Eretmapodites*. They chose the first option because the support for the *Ae.* (*Alanstonea*) + *Eretmapodites* clade was very weak and the supporting characters were all homoplastic. Species of *Alanstonea* would never be recognized as members of the genus *Eretmapodites* based on other morphological attributes. As has been pointed out many times, and is clearly stated or otherwise evident in major taxonomic treatments, the majority of subgeneric taxa in *Aedes* are diagnosed by combinations of characters (polythetically diagnosed), most of which are homoplastic, i.e. they can occur in many different taxa. Wilkerson *et al.* seem to have ignored or conveniently forgotten that the traditional classification of *Aedes* is likewise based on unique combinations of homoplastic characters. They stated that 'most of the genera [of Reinert *et al.*] cannot be identified except by combinations of character states, i.e., they do not have simple, unreversed diagnostic character states', which conceals the fact that this is also the case when the genera are reduced to subgenera.

The application of explicit methods of phylogenetic analysis often reveals weaknesses in traditional classifications, yet the explicit methodology may also often confirm the monophyly of taxonomic groups based intuitively on unique combinations of characters. Consequently, the main issue is not in recognizing monophyletic groups per se, but deciding what taxonomic ranks should be assigned to such groups. For phylogenetic systematists, the acceptance of broad polyphyletic genus groups for convenience is not an option, and where groups are found to be paraphyletic or polyphyletic, it is justifiable to remove them from the classification and give equal ranks to their constituent monophyletic groups.

The way in which Wilkerson *et al.* translated their phylogeny into a classification is problematic. Despite the various criteria discussed by the authors, their classification seems to have been based on a determination to retain the traditional polyphyletic concept of *Aedes*. If their huge polytomy is taken to be genus *Aedes*, then some branches in that clade should not be afforded generic rank while others are recognized as subgenera. For greater consistency in ranking taxa within a large genus *Aedes*, the seven traditional genera included in the polytomy, i.e. *Armigeres*, *Eretmapodites*, *Haemagogus*, *Heizmannia*, *Opifex*, *Udaya* and *Zeugnomyia*, should be ranked as subgenera. Because their generic status is retained, a reasonable alternative course of action would have been to recognize all the generic-level taxa as genera, thus preserving the genera and subgenera of Reinert *et al.* (2009).

Taxonomists rank organisms into different categories based principally on the degree of apparent similarities and dissimilarities – the greater the degree of similarity, the more closely two taxa are related. Despite the weak support for deeper relationships within the phylogeny of Reinert *et al.* (2009), the combinations of homoplastic morphological characters on the branches indicate broad degrees of similarity and dissimilarity, which in many cases parallel the intuitive assessments of relationships hypothesized in taxonomic publications. Based on this and the commentary above, and also because (i) some mosquito workers prefer the classification of Reinert *et al.* (2009), e.g. Batovska *et al.* (2016), Natarajan *et al.* (2016) and Brunhes *et al.* (2017), and (ii) molecular studies are finding support for elements of their classification, e.g. Soghigian *et al.*, 2017), the species-group and genus-group names that comprise their multi-generic reclassification of the 'traditional' composite *Aedes* are compiled in this section. The two classifications are compared in Appendix 2 to aid the conversion of one to the other.

Format

Entries in this section are formatted in the same way as the entries in Sections 1 and 2, and include the same elements of information. The section is divided into two subsections. The first includes species-group names and the second includes genus-group names. The names in each section are listed in alphabetical order. Valid names are in ***boldface italics*** and all synonymous names and *nomina nuda* are in *italicized normal type*. The endings of Latin and latinized adjectival and participle-based species-group names that are currently recognized as valid have been changed to agree in gender with the genus in which they are classified, and the authorship-date of those names included in a genus other than the original one are enclosed in parentheses to indicate that they have been transferred. Synonymous species-group names, however, are spelled exactly as originally spelled, or should have been spelled for gender agreement, with the generic name with which they were originally combined.

Species-group names

Aa

abadsantosi Baisas, 1946c (p. 25) – synonym of *Hulecoeteomyia rizali* (Banks, 1906b) [original combination: *Aedes* (*Finlaya*) *abadsantosi*]; synonymy by Knight, 1968.

abfitchii Felt, 1904 (p. 381) – synonym of *Ochlerotatus excrucians* (Walker, 1856a) [original combination: *Culex abfitchii*]; synonymy established before Knight & Stone, 1977.

abnormalis (Theobald, 1909b) (p. 10) – species of *Aedimorphus* [original combination: *Bathosomyia abnormalis*].

aboriginis (Dyar, 1917b) (p. 99) – species of *Ochlerotatus* (subgenus uncertain, determined by Reinert *et al*., 2008) [original combination: *Aedes aboriginis*].

abserratus (Felt & Young, 1904) (p. 312) – species of *Ochlerotatus* (subgenus uncertain, determined by Reinert *et al*., 2008) [original combination: *Culex abserratus*].
Synonyms: *centrotus* Howard, Dyar & Knab, 1917; *dysanor* Dyar, 1921e.

acrophilus Dyar, 1917d (p. 127) – synonym of *Ochlerotatus pullatus* (Coquillett, 1904c) [original combination: *Aedes acrophilus*]; synonymy established before Knight & Stone, 1977.

aculeatus (Theobald, 1903a) (p. 233) – species of *Ochlerotatus* (*Gilesia*) [original combination: *Gilesia aculeata*].

adami (Geoffroy, 1971b) (p. 273) – species of *Catageiomyia* [original combination: *Aedes* (*Aedimorphus*) *adami*].

adenensis (Edwards, 1941) (p. 198) – subspecies of *Aedimorphus hirsutus* (Theobald, 1901a) [original combination: *Aedes* (*Aedimorphus*) *hirsutus* variety *adenensis*]; subspecific status by White, 1975.

adersi (Edwards, 1917) (p. 214) – species of *Diceromyia* [original combination: *Ochlerotatus* (*Diceromyia*) *adersi*].

aegypti (Linnaeus, 1762) (p. 470) – species of *Stegomyia* (*Stegomyia*) [original combination: *Culex aegypti*].
Synonyms: *argenteus* Poiret, 1787; *fasciatus* Fabricius, 1805; *calopus* Meigen, 1818; *frater* Robineau-Desvoidy, 1827; *mosquito* Robineau-Desvoidy, 1827; *sugens* Wiedemann, 1828; *taeniatus* Wiedemann, 1828; *kounoupi* Brullé, 1833; *toxorhynchus* Macquart, 1838a; *annulitarsis* Macquart, 1846a; *excitans* Walker, 1848; *inexorabilis* Walker, 1848; *viridifrons* Walker, 1848; *exagitans* Walker, 1856a; *insatiabilis* Bigot, 1859; *bancrofti* Skuse, 1889; *elegans* Ficalbi, 1890c; *rossii* Giles, 1899; *luciensis* Theobald, 1901a; *nigeria* Theobald, 1901a; *queenslandensis* Theobald, 1901a; *lamberti* Ventrillon, 1904; *canariensis* Pittaluga, 1905; *persistans* Banks, 1906b; *albopalposus* Becker, 1908; *anguste-alatus* Becker, 1908; *alboannulis* Ludlow, 1911b; *pulcherrima* Taylor, 1919; *atritarsis* Edwards, 1920a.

aenigmaticus (Cerqueira & Costa, 1946) (p. 133) – species of *Ochlerotatus* (*Protoculex*) [original combination: *Aedes* (*Ochlerotatus*) *aenigmaticus*].

aerarius (McIntosh, 1975) (p. 281) – species of *Aedimorphus* [original combination: *Aedes* (*Aedimorphus*) *aerarius*].

aestivalis Dyar, 1904b (p. 245) – synonym of *Ochlerotatus sticticus* (Meigen, 1838) [original combination: *Culex aestivalis*]; synonymy established before Knight & Stone, 1977.

affinis Theobald, 1909b (p. 25) – synonym of *Pseudarmigeres argenteoventralis* (Theobald, 1909b) [original combination: *Dendromyia affinis*]; synonymy established before Knight & Stone, 1977.

africana Newstead, 1907 (in Newstead *et al*., 1907) (p. 20) – synonym of *Catageiomyia tarsalis* (Newstead, 1907) [original combination: *Duttonia africana*]; synonymy established before Knight & Stone, 1977.

africana Theobald, 1901d (p. 3) – species of *Stegomyia* (subgenus uncertain, determined by Reinert *et al*., 2009).

africana Theobald, 1911c (p. 151) – synonym of *Diceromyia furcifer* (Edwards, 1913b); synonymy established before Knight & Stone, 1977.

africanus Neveu-Lemaire, 1906 (p. 271) – synonym of *Ochlerotatus caspius* (Pallas, 1771) [original combination: *Taeniorhynchus africanus*]; synonymy established before Knight & Stone, 1977.

africanus Theobald, 1901d (p. 2) – synonym of *Mucidus* (*Mucidus*) *grahamii* Theobald, 1909b; synonymy established before Knight & Stone, 1977.

africanus (Theobald, 1909b) (p. 8) – species of *Aedimorphus* (Edwards, 1912c) [original combination: *Stenoscutus africanus*].
Synonym: *stenoscutus* Edwards, 1912c.

agastyai (Tewari & Hiriyan, 1992b) (p. 154) – species of *Tewarius* [original combination: *Aedes* (*Diceromyia*) *agastyai*].

agrihanensis (Bohart, 1957) (p. 39) – species of *Stegomyia* (*Bohartius*) [original combination: *Aedes* (*Stegomyia*) *agrihanensis*].

aitkeni (Schick, 1970a) (p. 61) – species of '*Ochlerotatus* (*Protomacleaya*)' *sensu auctorum* [original combination: *Aedes* (*Finlaya*) *aitkeni*].

akkeshiensis (Tanaka, 1998) (p. 215) – species of *Ochlerotatus* (subgenus uncertain, determined by Reinert *et al*., 2008) [original combination: *Aedes* (*Ochlerotatus*) *akkeshiensis*].

albertae Dyar, 1920e (p. 115) – synonym of *Ochlerotatus mercurator* (Dyar, 1920a) [original combination: *Aedes stimulans albertae*]; synonymy by Wood, 1977.

albertii Theobald, 1912 (in Schouteden, 1912) (p. 78) – synonym of *Stegomyia apicoargentea* Theobald, 1909b [original combination: *Kingia albertii*]; synonymy established before Knight & Stone, 1977.

albescens (Edwards, 1921d) (p. 308) – species of *Ochlerotatus* (*Rusticoidus*) [original combination: *Aedes* (*Ochlerotatus*) *albescens*].

albicosta (Edwards, 1913b) (p. 47) – species of *Neomelaniconion* [original combination: *Banksinella luteolateralis* variety *albicosta*]; specific status by ? Edwards, 1941.

albifasciatus (Macquart, 1838b) (p. 35) – species of *Ochlerotatus* (subgenus uncertain, determined by Reinert *et al.*, 2008) [original combination: *Culex albifasciatus*].

Synonyms: *flavipes* Macquart, 1838b; *annuliferus* Blanchard, 1852; *vittatus* Philippi, 1865; *colonarius* Dyar, 1924g; *philippii* Dyar, 1924g.

albilabris (Edwards, 1925) (p. 258) – species of *Rampamyia* [original combination: *Aedes* (*Finlaya*) *albilabris*].

albineus (Séguy, 1923) (p. 206) – species of *Ochlerotatus* (subgenus uncertain, determined by Reinert *et al.*, 2008) [original combination: *Aedes albineus*].

Synonym: *flavidorsalis* Luh & Lee, 1975.

albipes Theobald, 1910a (p. 11) – synonym of *Stegomyia* (*Xyele*) *desmotes* Giles, 1904a; synonymy established before Knight & Stone, 1977.

albiradius Le Goff, Boussès & Brunhes, 2007 (p. 197) – species of *Neomelaniconion*.

albitarsis Ludlow, 1905a (p. 131) – synonym of *Stegomyia* (*Xyele*) *desmotes* Giles, 1904a [original combination: *Anisocheleomyia albitarsis*]; synonymy established before Knight & Stone, 1977.

albitarsis Taylor, 1914d (p. 194) – synonym of *Patmarksia novalbitarsis* (King & Hoogstraal, 1946d) [original combination: *Leucomyia albitarsis*]; synonymy established before Knight & Stone, 1977.

alboannulata Taylor, 1912a (p. 31) – synonym of *Macleaya* (*Macleaya*) *tremula* Theobald, 1903e [original combination: *Danielsia alboannulata*]; synonymy established before Knight & Stone, 1977.

alboannulatus (Macquart, 1850) (p. 314) – species of *Dobrotworskyius* [original combination: *Culex albo-annulatus*].

alboannulatus Theobald, 1905b (p. 154) – synonym of *Elpeytonius apicoannulatus* (Edwards, 1912c) [original combination: *Aedimorphus alboannulatus*]; synonymy established before Knight & Stone, 1977.

alboannulis Ludlow, 1911b (p. 132) – synonym of *Stegomyia* (*Stegomyia*) *aegypti* (Linnaeus, 1762) [original combination: *Duttonia alboannulis*]; synonymy established before Knight & Stone, 1977.

alboapicus (Schick, 1970a) (p. 50) – species of '*Ochlerotatus* (*Protomacleaya*)' *sensu auctorum* [original combination: *Aedes* (*Finlaya*) *alboapicus*].

albocephala (Theobald, 1903c) (p. 4) – species of *Polyleptiomyia* (original combination: *Stegomyia albocephala*].

Synonym: *seychellensis* Theobald, 1912b.

albocinctus (Barraud, 1924d) (p. 1002) – species of *Hopkinsius* (*Yamada*) [original combination: *Finlaya albocincta*].

albodorsalis (Fontenille & Brunhes, 1985) (p. 151) – species of *Aedimorphus* [original combination: *Aedes* (*Aedimorphus*) *albodorsalis*].

albolateralis (Theobald, 1908b) (p. 289) – species of *Downsiomyia* [original combination: *Stegomyia albolateralis*].

albolineata Theobald, 1904c (p. 77) – species of *Scutomyia*.

albomarginatus (Newstead, 1907) (in Newstead *et al.*, 1907) (p. 16) – species of *Pseudarmigeres* [original combination: *Stegomyia albomarginata*].

albonivea (Barraud, 1934) (p. 210) – species of *Downsiomyia* [original combination: *Aedes* (*Finlaya*) *alboniveus*].

albonotata (Coquillett, 1906a) (p. 183) – species of *Howardina* [original combination: *Gymnometopa albonotata*].

albopalposus Becker, 1908 (p. 80) – synonym of *Stegomyia* (*Stegomyia*) *aegypti* (Linnaeus, 1762) [original combination: *Culex albopalposus*]; synonymy established before Knight & Stone, 1977.

albopicta (Skuse, 1895) (p. 20) – species of *Stegomyia* (subgenus uncertain, determined by Reinert *et al.*, 2009) [original combination: *Culex albopictus*].

Synonyms: *samarensis* Ludlow, 1903; *nigritia* Ludlow, 1910; *quasinigritia* Ludlow, 1911b.

albopunctata Theobald, 1907 (p. 262) – synonym of *Fredwardsius vittatus* (Bigot, 1861) [original combination: *Reedomyia albopunctata*]; synonymy established before Knight & Stone, 1977.

albo-punctatus Rondani, 1872 (p. 31) – synonym of *Dahliana geniculata* (Olivier, 1791) [original combination: *Culex albo-punctatus*]; synonymy established before Knight & Stone, 1977.

alboscutellatus (Theobald, 1905f) (p. 80) – species of *Aedimorphus* [original combination: *Lepidotomyia alboscutellata*].

Synonyms: *argentinotus* Banks, 1909; *omurensis* Yamada, 1921.

albotaeniata Leicester, 1904 (in Theobald, 1904c) (p. 111) – species of *Danielsia*.

Synonym: *mikiranus* Edwards, 1922a.

albotaeniatus Theobald, 1909b (p. 5) – synonym of *Catageiomyia irritans* (Theobald, 1901d) [original combination: *Aedimorphus albotaeniatus*]; synonymy established before Knight & Stone, 1977.

albothorax (Theobald, 1907) (p. 470) – species of *Neomelaniconion* [original combination: *Banksinella luteolateralis* variety *albothorax*]; specific status by ? Edwards, 1941.

alboventralis (Theobald, 1910c) (p. 251) – species of *Aedimorphus* [original combination: *Protomacleaya alboventralis*].

albus LaCasse & Yamaguti, 1950 (p. 136) – *nomen nudum*, *Aedes* [original combination: *Aedes* (*Ochlerotatus*) *dorsalis albus*].

alcasidi (Huang, 1972) (p. 37) – species of *Stegomyia* (subgenus uncertain, determined by Reinert *et al.*,

2009) [original combination: *Aedes (Stegomyia) alcasidi*].

aldrichi Dyar & Knab, 1908 (p. 57) – synonym of *Ochlerotatus sticticus* (Meigen, 1838) [original combination: *Aedes aldrichi*]; synonymy established before Knight & Stone, 1977.

alektorovi (Stackelberg, 1943) (p. 311) – species of *Bruceharrisonius* [original combination: *Aedes (Finlaya) alektorovi*]; see *kobayashii* (Nakata, 1956).

alexandrei Gornostaeva, 2005 (p. 485) – species of *Aedes*.

alineata Schneider, 1913 (p. 37) – synonym of *Ochlerotatus punctor* (Kirby, 1837) [original combination: *Culicada nemorosa* form *alineata*]; synonymy established before Knight & Stone, 1977.

alius (Lien, 1968d) (p. 221) – species of *Gilesius* [original combination: *Aedes (Finlaya) pulchriventer alius*]; specific status by Meng, 1981.

alleni Turner, 1924 (p. 84) – synonym of '*Ochlerotatus (Protomacleaya)*' *zoosophus* (Dyar & Knab, 1918b) [original combination: *Aedes alleni*]; synonymy by Gjullin, 1946.

allotecnon (Kumm, Komp & Ruiz, 1940) (p. 417) – species of *Howardina* [original combination: *Aedes allotecnon*].

alocasicola (Marks, 1947a) (p. 35) – species of *Finlaya* [original combination: *Aedes (Finlaya) alocasicola*].

alongi (Galliard & Ngu, 1947) (as *halongi*) (p. 77) – species of *Bothaella* [original combination: *Aedes (Finlaya) halongi*]; emendation by Galliard & Ngu, 1950.

aloponotum (Dyar, 1917b) (p. 98) – species of *Ochlerotatus* (subgenus uncertain, determined by Reinert *et al.*, 2008) [original combination: *Aedes aloponotum*].

alorensis (Bonne-Wepster & Brug, 1932) (p. 92) – species of *Stegomyia* [original combination: *Aedes (Stegomyia) variegatus* variety *alorensis*]; specific status by Stone & Farner, 1945.

alternans (Westwood, 1835) (p. 681) – species of *Mucidus (Mucidus)* [original combination: *Culex alternans*].

Synonyms: *commovens* Walker, 1856a; *hispidosus* Skuse, 1889; *kermorganti* Laveran, 1901a.

alticola (Bonne-Wepster, 1948) (p. 313) – species of '*Ochlerotatus (Finlaya)*' *sensu auctorum* [original combination: *Aedes (Finlaya) alticola*].

altiusculus Dyar, 1917b (p. 100) – synonym of *Ochlerotatus communis* (de Geer, 1776) [original combination: *Culex altiusculus*]; synonymy established before Knight & Stone, 1977.

amabilis (Schick, 1970a) (p. 68) – species of '*Ochlerotatus (Protomacleaya)*' *sensu auctorum* [original combination: *Aedes (Finlaya) amabilis*].

amalthea (de Meillon & Lavoipierre, 1944) (p. 52) – species of *Stegomyia* (subgenus uncertain, determined by Reinert *et al.*, 2009) [original combination: *Aedes (Stegomyia) amaltheus*].

amamiensis (Tanaka, Mizusawa & Saugstad, 1979) (p. 312) – subspecies of *Hulecoeteomyia japonica* (Theobald, 1901a) [original combination: *Aedes (Finlaya) japonicus amamiensis*].

ambreensis (Rodhain & Boutonnier, 1983a) (p. 826) – species of *Ochlerotatus* (subgenus uncertain, determined by Reinert *et al.*, 2008) [original combination: *Aedes (Ochlerotatus) ambreensis*].

ambremontis Brunhes & Boussès, 2017 (in Brunhes *et al.*, 2017) (p. 356) – species of *Paulianius*.

amesii (Ludlow, 1903) (p. 139) – species of *Lorrainea* [original combination: *Stegomyia amesii*].

Synonyms: *fusca* Leicester, 1908; *furvus* Edwards, 1928a.

ananae (Knight & Laffoon, 1946) (p. 218) – species of *Finlaya* [original combination: *Aedes (Finlaya) ananae*].

andersoni (Edwards, 1926a) (p. 112) – species of *Ochlerotatus* (subgenus uncertain, determined by Reinert *et al.*, 2008) [original combination: *Aedes (Ochlerotatus) andersoni*].

Synonyms: *tasmaniensis* Strickland, 1911b; *tasmaniensis* Taylor, 1941b.

andrewsi (Edwards, 1926a) (p. 103) – species of *Stegomyia* (subgenus uncertain, determined by Reinert *et al.*, 2009) [original combination: *Aedes variegatus* variety *andrewsi*]; specific status by Stone & Farner, 1945.

anggiensis (Bonne-Wepster, 1937) (p. 97) – species of *Patmarksia* [original combination: *Aedes (Finlaya) anggiensis*].

angolae (Ribeiro & Ramos, 1974) (p. 94) – subspecies of *Aedimorphus durbanensis* (Theobald, 1903a) [original combination: *Aedes durbanensis angolae*].

angusta (Edwards, 1935a) (p. 135) – species of *Stegomyia* (subgenus uncertain, determined by Reinert *et al.*, 2009) [original combination: *Aedes (Stegomyia) angustus*].

anguste-alatus Becker, 1908 (p. 79) – synonym of *Stegomyia (Stegomyia) aegypti* (Linnaeus, 1762) [original combination: *Culex anguste-alatus*]; synonymy established before Knight & Stone, 1977.

angustivittatus (Dyar & Knab, 1907a) (p. 9) – species of *Ochlerotatus (Ochlerotatus)* [original combination: *Aedes angustivittatus*].

Synonyms: *argentescens* Dyar & Knab, 1908; *cuneatus* Dyar & Knab, 1908; *traversus* Dyar, 1925g.

annandalei Theobald, 1910a (p. 10) – species of *Stegomyia (Heteraspidion)*.

Synonym: *horishensis* Yamada, 1921.

annulifemur Edwards, 1914b (p. 77) – synonym of *Dendroskusea periskelata* (Giles, 1902) [original combination: *Ochlerotatus annulifemur*]; synonymy established before Knight & Stone, 1977.

annulifera Ludlow, 1903 (p. 141) – synonym of *Ochlerotatus (Empihals) vigilax ludlowae* (Blanchard, 1905) [original combination: *Culex annulifera* [*sic*]]. **Note:** this nominal taxon is the junior primary homonym of *annuliferus* Blanchard, 1852.

annuliferus Blanchard, 1852 (p. 333) – synonym of *Ochlerotatus albifasciatus* (Macquart, 1838b) [original combination: *Culex annuliferus*]; synonymy by Belkin, 1968b. **Note:** this nominal taxon is the senior primary homonym of *annulifera* Ludlow, 1903.

annulipes (Meigen, 1830b) (p. 241) – species of *Ochlerotatus* (subgenus uncertain, determined by Reinert *et al.*,

2008) [original combination: *Culex annulipes*]. **Note:** this nominal taxon is the senior secondary homonym of *annulipes* Taylor, 1914c.

Synonym: *quartus* Martini, 1920a.

annulipes Taylor, 1914c (p. 693) – synonym of *Ochlerotatus camptorhynchus* (Thomson, 1869) [original combination: *Culicada annulipes*]; synonymy established before Knight & Stone, 1977. **Note:** this nominal taxon is the junior secondary homonym of *annulipes* (Meigen, 1830b).

annulirostris (Theobald, 1905h) (p. 239) – species of *Christophersiomyia* [original combination: *Stegomyia annulirostris*].

annulitarsis Macquart, 1846a (p. 136) – synonym of *Stegomyia* (*Stegomyia*) *aegypti* (Linnaeus, 1762) [original combination: *Culex annulitarsis*]; synonymy by Mattingly & Bruce-Chwatt, 1954.

antipodeus Edwards, 1920a (p. 132) – species of *Ochlerotatus* (subgenus uncertain, determined by Reinert *et al.*, 2008).

antuensis Su, Wang & Li, 1978 (p. 439) – synonym of *Edwardsaedes pingpaensis* (Chang, 1965) [original combination: *Aedes* (*Edwardsaedes*) *antuensis*]; synonymy by Danilov, 1985.

aobae (Belkin, 1962) (p. 452) – species of *Stegomyia* (subgenus uncertain, determined by Reinert *et al.*, 2009) [original combination: *Aedes* (*Stegomyia*) *aobae*].

apicoannulatus (Edwards, 1912c) (p. 18) – species of *Elpeytonius* [*nomen novum* for *alboannulatus* Theobald, 1905b; original combination: *Ochlerotatus apicoannulatus*].

Synonym: *alboannulatus* Theobald, 1905b.

apicoargentea Theobald, 1909b (p. 3) – species of *Stegomyia* (subgenus uncertain, determined by Reinert *et al.*, 2009).

Synonyms: *albertii* Theobald, 1912; *maculoabdominalis* Theobald, 1913d; *togoensis* Enderlein, 1923; *ziemanni* Enderlein, 1923.

apollo Schick, 1970a (p. 46) – synonym of '*Ochlerotatus* (*Protomacleaya*)' *terrens* (Walker, 1856a) [original combination: *Aedes* (*Finlaya*) *apollo*]; synonymy by Schick, 1970b.

arabica Giles, 1906 (p. 130) – synonym of *Ochlerotatus caspius* (Pallas, 1771) [original combination: *Mansonia arabica*]; synonymy established before Knight & Stone, 1977. **Note:** this nominal taxon is the senior secondary homonym of *arabicus* Becker, 1910.

arabicus Becker, 1910 (p. 140) – synonym of *Ochlerotatus caspius* (Pallas, 1771) [original combination: *Culex arabicus*]; synonymy established before Knight & Stone, 1977. **Note:** this nominal taxon is the junior secondary homonym of *arabica* Giles, 1906.

arabiensis (Patton, 1905) (species) (p. 633) – subspecies of *Aedimorphus vexans* (Meigen, 1830b) [original combination: *Culex arabiensis*]; subspecific status by White, 1975.

Synonym: *davidsoni* Ribeiro, da Cunha Ramos, Pires & Capela, 1979.

aranetana Banks, 1906b (p. 1001) – synonym of *Finlaya flavipennis* Giles, 1904a; synonymy established before Knight & Stone, 1977.

araozi Shannon & del Ponte, 1928 (p. 74) – synonym of *Ochlerotatus milleri* (Dyar, 1922g) [original combination: *Aedes* (*Taeniorhynchus*) *araozi*]; synonymy established before Knight & Stone, 1977.

arborealis (Bonne-Wepster & Bonne, 1920) (p. 178) – species of *Howardina* [original combination: *Aedes arborealis*].

arboricolus (Knight & Rozeboom, 1946) (p. 88) – species of *Scutomyia* [original combination: *Aedes* (*Stegomyia*) *arboricolus*].

arcanus Blanchard, 1905 (p. 303) – synonym of *Ochlerotatus flavescens* (Müller, 1764) [*nomen novum* for *flavescens* Theobald, 1901a; original combination: *Culex arcanus*]; synonymy established before Knight & Stone, 1977.

argenteitarsis (Brug, 1932) (p. 76) – species of *Patmarksia* [original combination: *Aedes* (*Finlaya*) *argenteitarsis*].

argenteomaculata Theobald, 1907 (p. 184) – ? synonym of *Stegomyia gardnerii imitator* Leicester, 1908; synonymy by Huang, 1977.

argenteopunctata (Theobald, 1901a) (p. 316) – species of *Catageiomyia* [original combination: *Stegomyia argenteopunctata*].

Synonym: *quinquepunctata* Theobald, 1913d.

argenteoscutellatus (Carter & Wijesundara, 1948) (p. 139) – species of *Aedimorphus* [original combination: *Aedes* (*Aedimorphus*) *argenteoscutellatus*].

argenteoventralis (Theobald, 1909b) (p. 24) – species of *Pseudarmigeres* [original combination: *Dendromyia argenteoventralis*].

Synonym: *affinis* Theobald, 1909b.

argentescens Dyar & Knab, 1908 (p. 55) – synonym of *Ochlerotatus* (*Ochlerotatus*) *angustivittatus* (Dyar & Knab, 1907a) [original combination: *Aedes argentescens*]; synonymy established before Knight & Stone, 1977.

argenteus Poiret, 1787 (p. 245) – synonym of *Stegomyia* (*Stegomyia*) *aegypti* (Linnaeus, 1762) [original combination: *Culex argenteus*]; synonymy established before Knight & Stone, 1977.

argentinotus Banks, 1909 (p. 547) – synonym of *Aedimorphus alboscutellatus* (Theobald, 1905f) [original combination: *Culex argentinotus*]; synonymy established before Knight & Stone, 1977.

argyrites (Dyar & Nuñez Tovar, 1927a) (p. 190) – species of *Howardina* [original combination: *Aedes* (*Howardina*) *argyrites*].

argyronotum (Belkin, 1962) (p. 354) – species of *Patmarksia* [original combination: *Aedes* (*Finlaya*) *argyronotum*].

argyrothorax (Bonne-Wepster & Bonne, 1920) (p. 179) – species of '*Ochlerotatus* (*Protomacleaya*)' *sensu auctorum* [original combination: *Aedes argyrothorax*].

articulatus Rondani, 1872 (p. 30) – synonym of *Aedimorphus vexans* (Meigen, 1830b) [original combination: *Culex articulatus*]; synonymy established before Knight & Stone, 1977.

asanumai Sasa, Kano & Takahasi, 1950 (p. 637) – synonym of *Aedes esoensis* Yamada, 1921; synonymy by Tanaka *et al.*, 1975b.

ashworthi (Edwards, 1921c) (p. 75) – species of *Halaedes* [original combination: *Aedes* (*Ochlerotatus*) *ashworthi*].

asiaticus (Edwards, 1926c) (p. 73) – subspecies of *Ochlerotatus pulcritarsis* (Rondani, 1872) [original combination: *Aedes* (*Ochlerotatus*) *pulchritarsis* [*sic*] variety *asiaticus*]; subspecific status by Monchadskii, 1951.

Synonym: *stegomyina* Stackelberg & Montchadsky, 1926.

assamensis (Theobald, 1908b) (p. 290) – species of *Phagomyia* [original combination: *Stegomyia assamensis*].

atactavittatus (Arnell, 1976) (p. 26) – species of *Ochlerotatus* (*Ochlerotatus*) [original combination: *Aedes* (*Ochlerotatus*) *atactavittatus*].

ater Gutsevich, 1955 (p. 321) – synonym of *Ochlerotatus mercurator* (Dyar, 1920a) [original combination: *Aedes* (*Ochlerotatus*) *riparius ater*]; synonymy by Danilov, 1974.

atlanticus (Dyar & Knab, 1906d) (p. 198) – species of *Ochlerotatus* (*Protoculex*) [original combination: *Aedes atlanticus*].

atritarsis Edwards, 1920a (p. 129) – synonym of *Stegomyia* (*Stegomyia*) *aegypti* (Linnaeus, 1762) [original combination: *Stegomyia fasciata* variety *atritarsis*]; synonymy established before Knight & Stone, 1977.

atropalpus (Coquillett, 1902d) (p. 292) – species of *Georgecraigius* (*Georgecraigius*) [original combination: *Culex atropalpus*].

aurantius (Theobald, 1907) (p. 280) – species of *Mucidus* (*Pardomyia*) [original combination: *Pardomyia aurantia*].

Synonym: *aureostriata* Leicester, 1908.

auratus (Grabham, 1906b) (p. 313) – species of *Ochlerotatus* (*Ochlerotatus*) [original combination: *Aedes auratus*].

auratus Leicester, 1908 (p. 153) – synonym of *Edwardsaedes imprimens* (Walker, 1860) [original combination: *Culex auratus*]; synonymy established before Knight & Stone, 1977.

aureolineata (Berlin, 1969a) (p. 25) – species of *Howardina* [original combination: *Aedes* (*Howardina*) *aureolineatus*].

aureostriata Grabham, 1906a (p. 171) – synonym of *Howardina grabhami* (Berlin 1969a); synonymy by Berlin, 1969a.

aureostriata Leicester, 1908 (p. 71) – synonym of *Mucidus* (*Pardomyia*) *aurantius* (Theobald, 1907) [original combination: *Ekrinomyia aureostriata*]; synonymy established before Knight & Stone, 1977.

aureostriatus (Doleschall, 1857) (p. 385) – species of *Bruceharrisonius* [original combination: *Culex aureostriatus*].

Synonym: *leucopleurus* Rozeboom, 1946.

aureum (Gutsevich, 1955) (species) (p. 320) – subspecies of *Neomelaniconion lineatopenne* (Ludlow, 1905a) [original combination: *Aedes* (*Aedes*?) *aureus*]; subspecific status by Danilov, 1977.

auridorsum (Edwards, 1922d) (p. 93) – species of *Molpemyia* [original combination: *Aedes* (*Finlaya*) *auridorsum*].

aurifer (Coquillett, 1903a) (p. 255) – species of *Ochlerotatus* (subgenus uncertain, determined by Reinert *et al.*, 2008) [original combination: *Culex aurifer*].

aurimargo (Edwards, 1922d) (p. 94) – species of *Leptosomatomyia* [original combination: *Aedes* (? *Skusea*) *aurimargo*].

Synonym: *lateralis* Theobald, 1905f.

aurites Theobald, 1907 (p. 216) – species of *Howardina*.

aurivittata (Cerqueira, 1943) (p. 7) – species of *Howardina* [original combination: *Aedes* (*Howardina*) *aurivittatus*].

auroides Felt, 1905a (p. 449) – synonym of *Ochlerotatus punctor* (Kirby, 1837) [original combination: *Culicelsa auroides*]; synonymy established before Knight & Stone, 1977.

auronitens (Edwards, 1922a) (p. 268) – species of '*Ochlerotatus* (*Finlaya*)' *sensu auctorum* [original combination: *Aedes* (*Finlaya*) *auronitens*].

aurostriata Banks, 1906b (p. 995) – synonym of *Belkinius aurotaeniatus* (Edwards, 1922a) [original combination: *Stegomyia aureostriata*]; synonymy by Edwards, 1922b.

aurotaeniatus (Edwards, 1922a) (p. 256) – species of *Belkinius* [*nomen novum* for *aureostriata* Banks, 1906b; original combination: *Aedes* (*Stegomyia*) *aurotaeniatus*].

aurovenatum (Worth, 1960) (p. 312) – species of *Neomelaniconion* [original combination: *Aedes* (*Neomelaniconion*) *aurovenatus*].

australiensis (Theobald, 1910c) (p. 313) – species of '*Ochlerotatus* (*Finlaya*)' *sensu auctorum* [original combination: *Leucomyia australiensis*].

australis (Erichson, 1842) (p. 270) – species of *Halaedes* [original combination: *Culex australis*].

Synonyms: *crucians* Walker, 1856a; *tasmaniensis* Strickland, 1911b; *concolor* Taylor, 1914c.

australis Strickland, 1911a (p. 133) – synonym of *Ochlerotatus* (*Lepidokeneon*) *stricklandi* Edwards, 1912b [original combination: *Grabhamia australis*]; synonymy established before Knight & Stone, 1977.

australis Taylor, 1914b (p. 457) – synonym of *Macleaya* (*Macleaya*) *tremula* Theobald, 1903e [original combination: *Aedimorphus australis*]; synonymy established before Knight & Stone, 1977.

avistylus (Brug, 1939) (p. 107) – species of *Finlaya* [original combination: *Aedes* (*Finlaya*) *flavipennis* variety *avistyla* [*sic*]]; specific status by Stone & Bohart, 1944.

axitiosa (Kulasekera, Knight & Harbach, 1990) (p. 26) – species of *Downsiomyia* [original combination: *Aedes* (*Finlaya*) *axitiosus*].

Bb

bahamensis (Berlin, 1969a) (p. 61) – species of *Howardina* [original combination: *Aedes* (*Howardina*) *bahamensis*].

baisasi (Knight & Hull, 1951a) (p. 197) – species of *Geoskusea* [original combination: *Aedes* (*Geoskusea*) *baisasi*].

balteatus Dyar & Knab, 1907a (p. 9) – synonym of *Ochlerotatus* (*Ochlerotatus*) *tortilis* (Theobald, 1903g) [original combination: *Aedes balteatus*]; synonymy established before Knight & Stone, 1977.

bambiotai (Geoffroy, 1987) (p. 101) – species of *Aedimorphus* [original combination: *Aedes* (*Aedimorphus*) *bambiotai*].

bambusae (Edwards, 1935a) (p. 134) – species of *Stegomyia* (subgenus uncertain, determined by Reinert *et al.*, 2009) [original combination: *Aedes* (*Stegomyia*) *bambusae*].

bambusicolus (Knight & Rozeboom, 1946) (p. 94) – species of *Scutomyia* [original combination: *Aedes* (*Stegomyia*) *bambusicolus*].

bananea (Wolfs, 1958) (p. 298) – species of *Diceromyia* [original combination: *Aedes* (*Diceromyia*) *bananea*].

bancoi (Geoffroy, 1987) (p. 103) – species of *Aedimorphus* [original combination: *Aedes* (*Aedimorphus*) *bancoi*].

bancrofti Skuse, 1889 (p. 1740) – synonym of *Stegomyia* (*Stegomyia*) *aegypti* (Linnaeus, 1762) [original combination: *Culex bancrofti*]; synonymy established before Knight & Stone, 1977.

bancroftianus (Edwards, 1921c) (p. 74) – species of *Ochlerotatus* (*Pseudoskusea*) [original combination: *Aedes* (*Ochlerotatus*) *bancroftianus*].

banksi (Edwards, 1922a) (p. 270) – species of *Collessius* (*Alloeomyia*) [original combination: *Aedes* (*Finlaya*) *banksi*].

barnardi (Edwards, 1924a) (p. 161) – species of *Hopkinsius* (*Hopkinsius*) [original combination: *Aedes* (*Finlaya*) *barnardi*].

barraudi Edwards, 1934 (in Barraud, 1934) (p. 447) – species of *Paraedes*.

barri Rueger, 1958 (p. 34) – synonym of *Ochlerotatus euedes* (Howard, Dyar & Knab, 1913) [original combination: *Aedes* (*Ochlerotatus*) *barri*]; synonymy by Wood, 1977.

becki (Belkin, 1962) (p. 334) – species of *Geoskusea* [original combination: *Aedes* (*Geoskusea*) *becki*].

bedfordi (Edwards, 1936) (p. 50) – species of *Catageiomyia* [original combination: *Aedes* (*Aedimorphus*) *bedfordi*].

behningi (Martini, 1926) (p. 200) – species of *Ochlerotatus* (subgenus uncertain, determined by Reinert *et al.*, 2008) [original combination: *Aedes behningi*].

bejaranoi (Martínez, Carcavallo & Prosen, 1960) (p. 26) – species of *Ochlerotatus* (subgenus uncertain, determined by Reinert *et al.*, 2008) [original combination: *Aedes* (*Ochlerotatus*) *bejaranoi*].

bekkui (Mogi, 1977) (p. 129) – species of *Edwardsaedes* [original combination: *Aedes* (*Edwardsaedes*) *bekkui*].

beklemishevi Denisova, 1955 (p. 58) – synonym of *Ochlerotatus euedes* (Howard, Dyar & Knab, 1913) [original combination: *Aedes* (*Ochlerotatus*) *beklemishevi*]; synonymy by Wood, 1977.

belleci Le Goff, Boussès & Brunhes, 2007 (p. 191) – species of *Neomelaniconion*.

bequaerti (Wolfs, 1947a) (p. 40) – species of *Neomelaniconion* [original combination: *Aedes* (*Banksinella*) *bequaerti*].

bergerardi (Pajot & Geoffroy, 1971) (p. 269) – species of *Neomelaniconion* [original combination: *Aedes* (*Neomelaniconion*) *bergerardi*].

berlandi (Séguy, 1921) (p. 192) – species of *Ochlerotatus* (subgenus uncertain, determined by Reinert *et al.*, 2008) [original combination: *Aedes* (*Ochlerotatus*) *berlandi*].

Synonyms: *praeteritus* Séguy, 1923; *longitubus* Cambournac, 1938; *heracleensis* Callot, 1944a.

berlini (Schick, 1970a) (p. 48) – species of '*Ochlerotatus* (*Protomacleaya*)' *sensu auctorum* [original combination: *Aedes* (*Finlaya*) *berlini*].

bertrami (Schick, 1970a) (p. 38) – species of '*Ochlerotatus* (*Protomacleaya*)' *sensu auctorum* [original combination: *Aedes* (*Finlaya*) *bertrami*].

bevisi (Edwards, 1915b) (p. 275) – species of *Aedimorphus* [original combination: *Ochlerotatus bevisi*].

biannulata Theobald, 1907 (p. 263) – synonym of *Catageiomyia tarsalis* (Newstead, 1907) [original combination: *Reedomyia biannulata*]; synonymy established before Knight & Stone, 1977.

bicristatus (Thurman & Winkler, 1950) (p. 239) – species of *Ochlerotatus* (*Rusticoidus*) [original combination: *Aedes* (*Ochlerotatus*) *bicristatus*].

bimaculatus (Coquillett, 1902c) (p. 84) – species of *Ochlerotatus* (*Chrysoconops*) [original combination: *Culex bimaculatus*].

Synonym: *rozeboomi* Vargas, 1941b.

biocellatus (Taylor, 1914b) (p. 463) – species of '*Ochlerotatus* (*Finlaya*)' *sensu auctorum* [original combination: *Culex biocellatus*].

bipunctata Theobald, 1909b (p. 5) – synonym of *Catageiomyia tarsalis* (Newstead, 1907) [original combination: *Reedomyia bipunctata*]; synonymy established before Knight & Stone, 1977.

bipunctatus Robineau-Desvoidy, 1827 (p. 405) – synonym of *Ochlerotatus flavescens* (Müller, 1764) [original combination: *Culex bipunctatus*]; synonymy established before Knight & Stone, 1977.

bisanensis Suzuki, Tanimura, Miyagawa & Murata, 1953 (p. 9) – synonym of *Hulecoeteomyia japonica* (Theobald, 1901a) [original combination: *Aedes japonicus* variety *bisanensis*]; synonymy by Sasa & Kamimura, 1971.

biskraensis (Brunhes, 1999) (p. 25) – species of *Ochlerotatus* (subgenus uncertain, determined by Reinert *et al.*, 2008) [original combination: *Aedes* (*Ochlerotatus*) *biskraensis*].

biunguiculata Harbach, 2014 (in Harbach & Taai, 2014) (p. 428) – species of *Nyctomyia*.

blacklocki (Evans, 1925a) (p. 121) – species of *Stegomyia* (subgenus uncertain, determined by Reinert *et al.*, 2009) [original combination: *Aedes* (*Stegomyia*) *blacklocki*].

bogotanus (Arnell, 1976) (p. 22) – species of *Ochlerotatus* (*Ochlerotatus*) [original combination: *Aedes* (*Ochlerotatus*) *bogotanus*].

boharti (Knight & Rozeboom, 1946) (p. 90) – species of *Scutomyia* [original combination: *Aedes* (*Stegomyia*) *boharti*].
bolense (Edwards, 1936) (p. 52) – species of *Neomelaniconion* [original combination: *Aedes bolensis*].
boneti (Gil Collado, 1936) (p. 323) – species of *Bifidistylus* [original combination: *Aedes* (*Aedimorphus*) *boneti*].
bonneae (Mattingly, 1958a) (p. 32) – species of *Paraedes* [original combination: *Aedes* (*Paraedes*) *bonneae*].
borealis Ludlow, 1911a (p. 178) – synonym of *Ochlerotatus communis* (de Geer, 1776) [original combination: *Culex borealis*]; synonymy established before Knight & Stone, 1977.
bougainvillensis (Marks, 1947a) (p. 19) – species of *Finlaya* [original combination: *Aedes* (*Finlaya*) *bougainvillensis*].
bracteatus Coquillett, 1906a (p. 184) – synonym of *Ochlerotatus* (*Ochlerotatus*) *tortilis* (Theobald, 1903g) [original combination: *Culex bracteatus*]; synonymy established before Knight & Stone, 1977.
brayi Knight, 1947b (p. 73) – synonym of *Christophersiomyia ibis* (Barraud, 1931b) [original combination: *Aedes* (*Christophersiomyia*) *brayi*]; synonymy by Abercrombie, 1977.
braziliensis (Gordon & Evans, 1922) (p. 329) – species of '*Ochlerotatus* (*Protomacleaya*)' *sensu auctorum* [original combination: *Aedes* (*Finlaya*) *oswaldi* [*sic*] variety *braziliensis*]; specific status by Belkin, 1968b.
breedensis (Muspratt, 1953) (p. 55) – species of *Ochlerotatus* (subgenus uncertain, determined by Reinert *et al*., 2008) [original combination: *Aedes* (*Ochlerotatus*) *breedensis*].
brelandi (Zavortink, 1972) (p. 25) – species of '*Ochlerotatus* (*Protomacleaya*)' *sensu auctorum* [original combination: *Aedes* (*Protomacleaya*) *brelandi*].
brevis (Berlin, 1969a) (p. 82) – species of *Howardina* [original combination: *Aedes* (*Howardina*) *brevis*].
brevitibia (Edwards, 1914c) (p. 125) – species of *Alanstonea* [original combination: *Armigeres brevitibia*].
brevivittata (Berlin, 1969a) (p. 116) – species of *Howardina* [original combination: *Aedes* (*Howardina*) *brevivittatus*].
britteni (Marks & Hodgkin, 1958) (p. 33) – species of '*Ochlerotatus* (*Finlaya*)' *sensu auctorum* [original combination: *Aedes* (*Finlaya*) *britteni*].
bromeliae Theobald, 1911a (p. 10) – species of *Stegomyia* (*Mukwaya*).
broquetii Theobald, 1913c (p. 179) – synonym of *Ochlerotatus dorsalis* (Meigen, 1830b) [original combination: *Grabhamia broquetii*]; synonymy established before Knight & Stone, 1977.
brownscutuma (Dong, Zhou & Dong, 1999) (p. 133) – species of *Bothaella* [original combination: *Aedes* (*Bothaella*) *brownscutumus*).
brugi Edwards, 1924b (p. 371) – synonym of *Edwardsaedes imprimens* (Walker, 1859) [original combination: *Aedes* (*Banksinella*) *brugi*]; synonymy by Mattingly, 1961.
brumpti Neveu-Lemaire, 1905b (p. 9) – synonym of *Fredwardsius vittatus* (Bigot, 1861) [original combination: *Stegomyia brumpti*]; synonymy established before Knight & Stone, 1977.
brunhesi (Reinert, 1999a) (p. 55) – species of *Zavortinkius* [original combination: *Aedes* (*Zavortinkius*) *brunhesi*].
brygooi (Brunhes, 1971) (p. 335) – species of *Zavortinkius* [original combination: *Aedes* (*Finlaya*) *brygooi*].
buenaventura (Schick, 1970a) (p. 52) – species of '*Ochlerotatus* (*Protomacleaya*)' *sensu auctorum* [original combination: *Aedes* (*Finlaya*) *buenaventura*].
bunanoki Sasa & Ishimura, 1951 (p. 103) – synonym of '*Ochlerotatus* (*Finlaya*)' *oreophilus* Edwards, 1916 [original combination: *Aedes* (*Finlaya*) *bunanoki*]; synonymy by Kamimura, 1968.
burgeri (Zavortink, 1972) (p. 42) – species of '*Ochlerotatus* (*Protomacleaya*)' *sensu auctorum* [original combination: *Aedes* (*Protomacleaya*) *burgeri*].
burgosi Baisas, 1946c (p. 27) – synonym of *Hulecoeteomyia rizali* (Banks, 1906b) [original combination: *Aedes* (*Finlaya*) *burgosi*]; synonymy by Knight, 1968.
burjaticus (Kuchartshuk, 1973) (p. 98) – species of *Ochlerotatus* (subgenus uncertain, determined by Reinert *et al*., 2008) [original combination: *Aedes* (*Ochlerotatus*) *burjaticus*].
burnetti (Belkin, 1962) (p. 363) – species of *Finlaya* [original combination: *Aedes* (*Finlaya*) *burnetti*].
burnsi (Basio & Reisen, 1971) (p. 58) – species of *Stegomyia* (subgenus uncertain, determined by Reinert *et al*., 2009) [original combination: *Aedes* (*Stegomyia*) *burnsi*].
burpengaryensis (Theobald, 1905g) (p. 26) – species of *Ochlerotatus* (subgenus uncertain, determined by Reinert *et al*., 2008) [original combination: *Culicada burpengaryensis*].
busckii (Coquillett, 1906e) (p. 60) – species of *Howardina* [original combination: *Stegomyia busckii*].
Synonyms: *thaxteri* Dyar & Knab, 1919; *tracei* Senevet & Quiévreux, 1941.
buxtoni (Belkin, 1962) (p. 356) – species of *Patmarksia* [original combination: *Aedes* (*Finlaya*) *buxtoni*].
bwamba (van Someren, 1950c) (p. 67) – subspecies of *Aedimorphus tricholabis* (Edwards, 1941) [original combination: *Aedes* (*Aedimorphus*) *tricholabis bwamba*].

Cc

caballus (Theobald, 1912a) (p. 93) – species of *Ochlerotatus* (*Juppius*) [original combination: *Grabhamia caballa*].
cacharana (Barraud, 1923e) (p. 406) – species of *Phagomyia* [original combination: *Finlaya cacharana*].
cacothius Dyar, 1923a (p. 44) – synonym of *Ochlerotatus ventrovittis* (Dyar, 1916b) [original combination: *Aedes cacothius*]; synonymy established before Knight & Stone, 1977.
cacozelus (Marks, 1963b) (p. 45) – species of *Ochlerotatus* (subgenus uncertain, determined by Reinert *et al*.,

2008) [original combination: *Aedes* (*Ochlerotatus*) *cacozelus*].

caecus (Theobald, 1901a) (p. 413) – species of *Aedimorphus* [original combination: *Culex caecus*].

calabyi (Marks, 1963a) (p. 204) – species of *Macleaya* (*Chaetocruiomyia*) [original combination: *Aedes* (*Chaetocruiomyia*) *calabyi*].

calcariae (Marks, 1957a) (p. 74) – species of *Ochlerotatus* (*Pholeomyia*) [original combination: *Aedes* (*Ochlerotatus*) *calcariae*].

calceata (Edwards, 1924) (in Haworth, 1924) (p. 197) – species of *Stegomyia* (subgenus uncertain, determined by Reinert *et al*., 2009) [original combination: *Aedes* (*Stegomyia*) *calceatus*].

caliginosus (Graham, 1910) (p. 268) – species of *Aedimorphus* [original combination: *Culex caliginosus*].

callithotrys Dyar, 1920a (p. 16) – synonym of *Ochlerotatus campestris* (Dyar & Knab, 1907d) [original combination: *Aedes callithotrys*]; synonymy established before Knight & Stone, 1977.

calopus Meigen, 1818 (p. 3) – synonym of *Stegomyia* (*Stegomyia*) *aegypti* (Linnaeus, 1762) [original combination: *Culex calopus*]; synonymy established before Knight & Stone, 1977.

calumnior (Belkin, Heinemann & Page, 1970) (p. 169) – species of *Ochlerotatus* (subgenus uncertain, determined by Reinert *et al*., 2008) [original combination: *Aedes* (*Ochlerotatus*) *calumnior*].

campana (Schick, 1970a) (p. 79) – species of '*Ochlerotatus* (*Protomacleaya*)' *sensu auctorum* [original combination: *Aedes* (*Finlaya*) *campana*].

campestris (Dyar & Knab, 1907d) (p. 213) – species of *Ochlerotatus* (subgenus uncertain, determined by Reinert *et al*., 2008) [original combination: *Aedes campestris*].

camposanus Dyar, 1918e (p. 128) – synonym of *Ochlerotatus* (*Ochlerotatus*) *scapularis* (Rondani, 1848) [original combination: *Aedes* (*Ochlerotatus*) *camposanus*]; synonymy by Arnell, 1976.

camptorhynchus (Thomson, 1869) (p. 443) – species of *Ochlerotatus* (subgenus uncertain, determined by Reinert *et al*., 2008) [original combination: *Culex camptorhynchus*].
Synonyms: *labeculosus* Coquillett, 1905; *westralis* Strickland, 1911a; *inornata* Strickland, 1911b; *victoriensis* Taylor, 1914b; *annulipes* Taylor, 1914c; *nigra* Taylor, 1914c.

canadensis (Theobald, 1901c) (p. 3) – species of *Ochlerotatus* (*Culicada*) [original combination: *Culex canadensis*].
Synonym: *nivitarsis* Coquillett, 1904c.

canariensis Pittaluga, 1905 (p. 432) – synonym of *Stegomyia* (*Stegomyia*) *aegypti* (Linnaeus, 1762) [original combination: *Stegomyia calopus* variety *canariensis*]; synonymy established before Knight & Stone, 1977.

cancricomes (Edwards, 1922a) (p. 272) – species of *Cancraedes* [original combination: *Aedes* (*Skusea*) *cancricomes*].

candidoscutellum (Marks, 1947b) (p. 1) – species of '*Ochlerotatus* (*Finlaya*)' *sensu auctorum* [original combination: *Aedes* (*Finlaya*) *candidoscutellum*].

cantans (Meigen, 1818) (p. 6) – species of *Ochlerotatus* (subgenus uncertain, determined by Reinert *et al*., 2008) [original combination: *Culex cantans*].
Synonyms: *reptans* Meigen, 1804; *waterhousei* Theobald, 1905; *subvexans* Martini, 1922.

cantator (Coquillett, 1903a) (p. 255) – species of *Ochlerotatus* (subgenus uncertain, determined by Reinert *et al*., 2008) [original combination: *Culex cantator*].

capensis (Edwards, 1924a) (p. 162) – species of *Albuginosus* [original combination: *Aedes* (*Ecculex*) *capensis*].

carteri (Edwards, 1936) (p. 52) – species of *Neomelaniconion* [original combination: *Aedes palpalis* variety *carteri*]; specific status by Haddow *et al*., 1951.
Synonym: *maculicosta* Edwards, 1936.

cartroni (Ventrillon, 1906b) (p. 143) – species of *Skusea* [original combination: *Stegomyia cartroni*].

casali (Schick, 1970b) (p. 21) – species of '*Ochlerotatus* (*Protomacleaya*)' *sensu auctorum* [original combination: *Aedes* (*Finlaya*) *casali*].

caspius (Pallas, 1771) (p. 475) – species of *Ochlerotatus* (subgenus uncertain, determined by Reinert *et al*., 2008) [original combination: *Culex caspius*].
Synonyms: *punctatus* Meigen, 1804; *siculus* Robineau-Desvoidy, 1827; *maculiventris* Macquart, 1846b; *penicillaris* Rondani, 1872; ?*longisquamosa* Theobald, 1905f; *subtilis* Sergent & Sergent, 1905; *arabica* Giles, 1906; *africanus* Neveu-Lemaire, 1906; *willcocksii* Theobald, 1907; *arabicus* Becker, 1910; *albineus* Séguy, 1923; *epsilonn* Séguy, 1924.

cataphylla (Dyar, 1916b) (p. 86) – species of *Ochlerotatus* (subgenus uncertain, determined by Reinert *et al*., 2008) [original combination: *Aedes cataphylla*].
Synonyms: *prodotes* Dyar, 1917c; *rostochiensis* Martini, 1920b; *pacificensis* Hearle, 1927.

cavaticus (Reinert, 1979) (p. 148) – species of *Isoaedes* [original combination: *Aedes* (*Isoaedes*) *cavaticus*].

cavernicola Rattanarithikul & Harbach, 2007 (in Harbach, Rattanarithikul, Howard *et al*., 2007) (p. 475) – species of *Borichinda*.

celebica (Mattingly, 1959) (p. 32) – species of *Lorrainea* [original combination: *Aedes* (*Skusea*) *celebicus*].

centropunctatus (Theobald, 1913d) (p. 599) – species of *Aedimorphus* [original combination: *Culicelsa centropunctata*].

centrotus Howard, Dyar & Knab, 1917 (p. 747) – synonym of *Ochlerotatus abserratus* (Felt & Young, 1904) [original combination: *Aedes centrotus*]; synonymy established before Knight & Stone, 1977.

ceylonica Theobald, 1910c (p. 391) – synonym of *Hulecoeteomyia chrysolineata* (Theobald, 1907) [original combination: *Culex japonicus* variety *ceylonica* [*sic*]]; synonymy by Knight, 1948a.

chamboni (Cornet, 1968) (p. 398) – species of *Catageiomyia* [original combination: *Aedes* (*Aedimorphus*) *chamboni*].

chaussieri (Edwards, 1923a) (p. 397) – species of *Stegomyia* (subgenus uncertain, determined by Reinert *et al.*, 2009) [original combination: *Aedes* (*Stegomyia*) *chaussieri*].

chelli Edwards, 1915b (p. 275) – species of *Ochlerotatus* (*Juppius*).

chemulpoensis (Yamada, 1921) (p. 54) – species of *Stegomyia* (subgenus uncertain, determined by Reinert *et al.*, 2009) [original combination: *Aedes chemulpoensis*].

cherrapunjiensis Natarajan, Rajavel & Jambulingam, 2016 (p. 330) – species of *Hulecoeteomyia*.

chionodes (Belkin, 1962) (p. 434) – species of *Christophersiomyia* [original combination: *Aedes* (*Christophersiomyia*) *chionodes*].

chionotum (Zavortink, 1972) (p. 47) – species of '*Ochlerotatus* (*Protomacleaya*)' *sensu auctorum* [original combination: *Aedes* (*Protomacleaya*) *chionotum*].

christianus Dyar, 1921i (p. 148) – synonym of *Stegomyia gardnerii imitator* Leicester, 1908 [original combination: *Aedes* (*Stegomyia*) *christianus*]; synonymy by Mattingly, 1965.

christophersi (Edwards, 1922a) (p. 267) – species of *Bruceharrisonius* [original combination: *Aedes* (*Finlaya*) *christophersi*].

chrysogaster Taylor, 1927 (p. 67) – subspecies of *Mucidus* (*Pardomyia*) *aurantius* (Theobald, 1907).

chrysolineata (Theobald, 1907) (p. 218) – species of *Hulecoeteomyia* [original combination: *Howardina chrysolineata*].
Synonyms: *trilineata* Leicester, 1904; *ceylonica* Theobald, 1910c.

chrysoscuta (Theobald, 1910c) (p. 228) – species of *Paraedes* [original combination: *Pseudohowardina chrysoscuta*]; current specific status by Reinert, 1981.

chrysothorax Theobald, 1909b (p. 13) – synonym of *Neomelaniconion taeniarostre* (Theobald, 1909b) [original combination: *Banksinella chrysothorax*]; synonymy by Edwards, 1941.

chungi (Lien, 1968b) (p. 100) – species of *Jihlienius* [original combination: *Aedes* (*Finlaya*) *chungi*].

churchillensis (Ellis & Brust, 1973) (p. 922) – species of *Ochlerotatus* (subgenus uncertain, determined by Reinert *et al.*, 2008) [original combination: *Aedes* (*Ochlerotatus*) *churchillensis*].

ciliaris (Linnaeus, 1767) (p. 1002) – *nomen dubium*, *Aedes* [original combination: *Culex ciliaris*]; status established before Knight & Stone, 1977.

cinereoborealis Felt & Young, 1904 (p. 312) – synonym of *Ochlerotatus* (*Rusticoidus*) *provocans* (Walker, 1848) [original combination: *Culex cinereoborealis*]; transferred synonymy by synonymy of senior synonym *trichurus* (Dyar) with *provocans* (Walker) by Wood, 1977.

cinereus Meigen, 1818 (p. 13) – species of *Aedes*.
Synonyms: *rufus* Gimmerthal, 1845; *nigritulus* Zetterstedt, 1850; *fuscus* Osten Sacken, 1877; *leucopygos* Eysell, 1903; *pallidohirta* Grossbeck, 1905; *hemiteleus* Dyar, 1924h.

circumluteolum (Theobald, 1908a) (p. 107) – species of *Neomelaniconion* [original combination: *Banksiella* [*sic*] *luteolateralis* variety *circumluteola*]; specific status by Edwards, 1941.

classicus Dyar, 1920e (p. 113) – synonym of *Ochlerotatus stimulans* (Walker, 1848) [original combination: *Aedes stimulans classicus*]; synonymy established before Knight & Stone, 1977.

clelandi (Taylor, 1914c) (p. 690) – species of *Ochlerotatus* (*Pholeomyia*) [original combination: *Culicada clelandi*].

clintoni (Taylor, 1946) (p. 211) – species of *Patmarksia* [original combination: *Aedes* (*Finlaya*) *clintoni*].

clivis (Lanzaro & Eldridge, 1992) (p. 89) – species of *Ochlerotatus* (subgenus uncertain, determined by Reinert *et al.*, 2008) [original combination: *Aedes* (*Ochlerotatus*) *clivis*].

cogilli (Edwards, 1922d) (p. 92) – species of *Phagomyia* [original combination: *Aedes* (*Finlaya*) *cogilli*].

collessi (Mattingly, 1958a) (p. 28) – species of *Paraedes* [original combination: *Aedes* (*Paraedes*) *collessi*].

colonarius Dyar, 1924g (p. 130) – synonym of *Ochlerotatus albifasciatus* (Macquart, 1838b) [original combination: *Aedes colonarius*]; synonymy established before Knight & Stone, 1977.

coluzzii (Rioux, Guilvard & Pasteur, 1998) (p. 357) – species of *Ochlerotatus* (subgenus uncertain, determined by Reinert *et al.*, 2008) [original combination: *Aedes* (*Ochlerotatus*) *coluzzii*].

comitatus (Arnell, 1976) (p. 66) – species of *Ochlerotatus* (*Ochlerotatus*) [original combination: *Aedes* (*Ochlerotatus*) *comitatus*].

commovens Walker, 1856a (p. 432) – synonym of *Mucidus* (*Mucidus*) *alternans* (Westwood, 1835); synonymy established before Knight & Stone, 1977.

communis (de Geer, 1776) (p. 316) – species of *Ochlerotatus* (subgenus uncertain, determined by Reinert *et al.*, 2008) [original combination: *Culex communis*].
Synonyms: *fasciatus* Meigen, 1804; *nemorosus* Meigen, 1818; *sylvaticus* Meigen, 1818; *obscurus* Meigen, 1830a; *lazarensis* Felt & Young, 1904; *borealis* Ludlow, 1911a; *diplolineata* Schneider, 1913; *altiusculus* Dyar, 1917b; *palmeni* Edwards, 1921b; *prolixus* Dyar, 1922a.

concinnus Stephens, 1846 (p. 19) – synonym of *Ochlerotatus sticticus* (Meigen, 1838) [original combination: *Culex concinnus*]; synonymy established before Knight & Stone, 1977.

concolor Taylor, 1914c (p. 700) – synonym of *Halaedes australis* (Erichson, 1842) [original combination: *Caenocephalus concolor*]; synonymy by Mattingly & Marks, 1955.

condolescens (Dyar & Knab, 1907a) (p. 11) – species of *Ochlerotatus* (*Ochlerotatus*) [original combination: *Aedes condolescens*].

confirmatus Lynch Arribálzaga, 1891b (p. 146) – synonym of *Ochlerotatus* (*Ochlerotatus*) *scapularis* (Rondani, 1848); synonymy established before Knight & Stone, 1977.

confusus Theobald, 1909b (p. 10) – synonym of *Catageiomyia irritans* (Theobald, 1901d) [original combination: *Myxosquamus confusus*]; synonymy established before Knight & Stone, 1977.

congolensis Edwards, 1927a (p. 351) – synonym of *Aedimorphus stenoscutus* (Edwards, 1912c) [original combination: *Aedes* (*Aedimorphus*) *congolensis*]; synonymy by White, 1975.

contigua (Edwards, 1936) (p. 55) – species of *Stegomyia* (subgenus uncertain, determined by Reinert *et al.*, 2009) [original combination: *Aedes* (*Stegomyia*) *contiguus*].

continentalis (Dobrotworsky, 1960a) (p. 71) – species of *Ochlerotatus* (subgenus uncertain, determined by Reinert *et al.*, 2008) [original combination: *Aedes continentalis*].

cooki (Belkin, 1962) (p. 454) – species of *Stegomyia* (subgenus uncertain, determined by Reinert *et al.*, 2009) [original combination: *Aedes* (*Stegomyia*) *cooki*].

cordellieri (Huang, 1986a) (p. 636) – species of *Diceromyia* [original combination: *Aedes* (*Diceromyia*) *cordellieri*].

corneti (Huang, 1986b) (p. 765) – species of *Stegomyia* (subgenus uncertain, determined by Reinert *et al.*, 2009) [original combination: *Aedes* (*Stegomyia*) *corneti*].

coulangesi (Rodhain & Boutonnier, 1983b) (p. 332) – species of *Paulianius* [original combination: *Aedes* (*Diceromyia*) *coulangesi*].

cozi (Cornet, 1973) (p. 175) – species of *Cornetius* [original combination: *Aedes* (*Stegomyia*) *cozi*].

cozumelensis (Diaz Nájera, 1966) (p. 334) – species of *Howardina* [original combination: *Aedes* (*Howardina*) *cozumelensis*].

craggi Barraud, 1923c (p. 227) – species of *Stegomyia* (*Heteraspidion*).
Synonym: *purii* Barraud, 1931d.

crassiforceps (Edwards, 1927a) (p. 352) – species of *Neomelaniconion* [original combination: *Aedes* (*Banksinella*) *crassiforceps*].

cretina (Edwards, 1921d) (p. 325) – species of *Stegomyia* (subgenus uncertain, determined by Reinert *et al.*, 2009) [original combination: *Aedes* (*Stegomyia*) *cretinus*].
Synonyms: *delta* Séguy, 1924; *lindtropi* Shingarev, 1927.

crinifer (Theobald, 1903a) (p. 209) – species of *Ochlerotatus* (*Ochlerotatus*) [original combination: *Culex crinifer*].
Synonyms: *lynchii* Brèthes, 1910; *tapinops* Brèthes, 1917; *iguazu* Shannon & del Ponte, 1928.

crocea (Knight & Laffoon, 1946) (p. 213) – species of *Finlaya* [original combination: *Aedes* (*Finlaya*) *croceus*].

crossi (Lien, 1967) (p. 177) – species of '*Ochlerotatus* (*Finlaya*)' *sensu auctorum* [original combination: *Aedes* (*Finlaya*) *crossi*].

crucians Walker, 1856a (p. 432) – synonym of *Halaedes australis* (Erichson, 1842) [original combination: *Culex crucians*]; synonymy by Mattingly & Marks, 1955.

culiciformis (Theobald, 1905f) (p. 77) – species of *Ochlerotatus* (*Pseudoskusea*) [original combination: *Skusea culiciformis*].

culicinus (Edwards, 1922a) (p. 271) – species of *Aedimorphus* [original combination: *Aedes* (*Ecculex*) *culicinus*].

cumminsii (Theobald, 1903a) (p. 214) – species of *Aedimorphus* [original combination: *Culex cumminsii*].
Synonyms: *fuscopalpalis* Theobald, 1909; *daruensis* Evans, 1925a.

cumpstoni Taylor, 1914c (p. 692) – synonym of *Dobrotworskyius rubrithorax* (Macquart, 1850) [original combination: *Culicada cumpstoni*]; synonymy established before Knight & Stone, 1977.

cunabulanus (Edwards, 1924b) (p. 378) – species of *Ochlerotatus* (subgenus uncertain, determined by Reinert *et al.*, 2008) [original combination: *Aedes* (*Ochlerotatus*) *cunabulanus*].

cuneatus Dyar & Knab, 1908 (p. 54) – synonym of *Ochlerotatus* (*Ochlerotatus*) *angustivittatus* (Dyar & Knab, 1907a) [original combination: *Aedes cuneatus*]; synonymy established before Knight & Stone, 1977.

curriei Coquillett, 1901 (p. 259) – synonym of *Ochlerotatus dorsalis* (Meigen, 1830b) [original combination: *Culex curriei*]; synonymy by Wood *et al.*, 1979.

curtipes (Edwards, 1915c) (p. 283) – species of *Cancraedes* [original combination: *Aedes curtipes*].
Synonym: *miachaetessa* Dyar & Shannon, 1925a.

cyclocerculus Dyar, 1920a (p. 23) – synonym of *Ochlerotatus hexodontus* (Dyar, 1916b) [original combination: *Aedes cyclocerculus*]; synonymy established before Knight & Stone, 1977.

cyprioides (Danilov & Stupin, 1982) (p. 824) – species of *Ochlerotatus* (subgenus uncertain, determined by Reinert *et al.*, 2008) [original combination: *Aedes* (*Ochlerotatus*) *cyprioides*].

cyprius (Ludlow, 1920b) (p. 158) – species of *Ochlerotatus* (subgenus uncertain, determined by Reinert *et al.*, 2008) [original combination: *Aedes cyprius*].
Synonym: *freyi* Edwards, 1921d.

Dd

daggyi (Stone & Bohart, 1944) (p. 215) – species of *Geoskusea* [original combination: *Aedes* (*Geoskusea*) *daggyi*].

dahlae Nielsen, 2009 (p. 123) – species of *Ochlerotatus* (subgenus uncertain, determined by Reinert *et al.*, 2009).

dahuricus Danilov, 1987b (p. 35) – species of *Aedes*.

daisetsuzanus (Tanaka, Mizusawa & Saugstad, 1979) (p. 271) – subspecies of *Ochlerotatus impiger* (Walker, 1848) [original combination: *Aedes* (*Ochlerotatus*) *daisetsuzanus*].

daitensis (Miyagi & Toma, 1981) (p. 428) – species of *Stegomyia* (subgenus uncertain, determined by Reinert *et al.*, 2009) [original combination: *Aedes* (*Stegomyia*) *daitensis*].

daliensis (Taylor, 1916) (p. 565) – species of '*Ochlerotatus*' *sensu auctorum* [original combination: *Stegomyia daliensis*].

dalzieli (Theobald, 1910c) (p. 234) – species of *Aedimorphus* [original combination: *Culiciomyia dalzieli*].
Synonym: *rhecter* Dyar, 1921d.

damnosus Say, 1823 (p. 11) – synonym of *Ochlerotatus* (*Culicelsa*) *taeniorhynchus* (Wiedemann, 1821) [original combination: *Culex damnosus*]; synonymy established before Knight & Stone, 1977.

daruensis Evans, 1925a (p. 123) – synonym of *Aedimorphus cumminsii mediopunctatus* (Theobald, 1909) [original combination: *Aedes* (*Aedimorphus*) *cumminsi* [*sic*] variety *daruensis*]; synonymy by Edwards, 1941.

darwini Taylor, 1914b (p. 458) – synonym of *Macleaya* (*Macleaya*) *tremula* Theobald, 1903e [original combination: *Aedimorphus australis* variety *darwini*]; synonymy established before Knight & Stone, 1977.

daryi (Schick, 1970a) (p. 80) – species of '*Ochlerotatus* (*Protomacleaya*)' *sensu auctorum* [original combination: *Aedes* (*Finlaya*) *daryi*].

dasyorrhus (King & Hoogstraal, 1946c) (p. 100) – species of *Lorrainea* [original combination: *Aedes* (*Skusea*) *dasyorrhus*].

davidi (Basio, 1971) (p. 51) – species of *Aedimorphus* [original combination: *Aedes* (*Aedimorphus*) *davidi*].

deboeri (Edwards, 1926a) (p. 128) – species of *Stegomyia* (subgenus uncertain, determined by Reinert *et al.*, 2009) [original combination: *Aedes* (*Stegomyia*) *deboeri*].

deccana (Barraud, 1923b) (p. 217) – species of *Phagomyia* [original combination: *Finlaya deccana*].

decticus (Howard, Dyar & Knab, 1917) (p. 737) – species of *Ochlerotatus* (subgenus uncertain, determined by Reinert *et al.*, 2008) [original combination: *Aedes decticus*].
Synonym: *pseudodiantaeus* Smith, 1952.

deficiens (Arnell, 1976) (p. 24) – species of *Ochlerotatus* (*Ochlerotatus*) [original combination: *Aedes* (*Ochlerotatus*) *deficiens*].

delpontei Martínez & Prosen, 1955 (p. 613) – synonym of *Howardina pseudodominicii* (Komp, 1936a) [original combination: *Aedes* (*Howardina*) *delpontei*]; synonymy by Berlin, 1969a.

delta Séguy, 1924 (p. 80) – synonym of *Stegomyia cretina* (Edwards, 1921d) [original combination: *Aedes* (*Stegomyia*) *delta*]; synonymy by Monchadskii, 1951.

demansis Strickland, 1911b (p. 202) – synonym of *Dobrotworskyius rubrithorax* (Macquart, 1850) [original combination: *Culicada demansis*]; synonymy established before Knight & Stone, 1977.

demeilloni (Edwards, 1936) (p. 55) – species of *Stegomyia* (subgenus uncertain, determined by Reinert *et al.*, 2009) [original combination: *Aedes* (*Stegomyia*) *deboeri* variety *demeilloni*]; specific status by Mattingly, 1952.

denderensis (Wolfs, 1949) (p. 190) – species of *Stegomyia* (subgenus uncertain, determined by Reinert *et al.*, 2009) [original combination: *Aedes* (*Stegomyia*) *apicoargenteus* variety *denderensis*]; specific status by Huang, 2004.

dendrophila (Edwards, 1921c) (p. 74) – species of *Stegomyia* (subgenus uncertain, determined by Reinert *et al.*, 2009) [original combination: *Aedes* (*Stegomyia*) *dendrophila* [*sic*]].
Synonym: *trinidad* Gil Collado, 1936.

deniedmannii Ludlow, 1904b (p. 234) – synonym of *Ochlerotatus squamiger* (Coquillett, 1902c) [original combination: *Grabhamia deniedmannii*]; synonymy established before Knight & Stone, 1977.

dentatus (Theobald, 1904b) (p. 75) – species of *Aedimorphus* [original combination: *Culex dentatus*].
Synonym: *pallidopunctata* Theobald, 1911b.

derooki (Brug, 1932) (p. 75) – species of *Patmarksia* [original combination: *Aedes* (*Finlaya*) *derooki*].

desbansi Séguy, 1923 (p. 205) – synonym of *Acartomyia mariae* (Sergent & Sergent, 1903) [original combination: *Aedes desbansi*]; synonymy established before Knight & Stone, 1977.

deserticola (Zavortink, 1970a) (p. 1) – species of *Jarnellius* [original combination: *Aedes* (*Ochlerotatus*) *deserticola*].

desmotes Giles, 1904a (p. 367) – species of *Stegomyia* (*Xyele*).
Synonyms: *albitarsis* Ludlow, 1905a; *gracilis* Leicester, 1908; *albipes* Theobald, 1910a; *montana* Koidzumi, 1918.

detritus (Haliday, 1833) (p. 151) – species of *Ochlerotatus* (subgenus uncertain, determined by Reinert *et al.*, 2008) [original combination: *Culex detritus*].
Synonyms: *salinus* Ficalbi, 1896; *terriei* Theobald, 1903a; *maculosa* Theobald, 1905f.

dialloi (Hamon & Brengues, 1965) (p. 103) – species of *Catageiomyia* [original combination: *Aedes dialloi*].

diantaeus (Howard, Dyar & Knab, 1913, 1917) (fig. 167, p. 758, respectively) – species of *Ochlerotatus* (*Woodius*) [original combination: *Aedes diantaeus*].
Synonym: *serus* Martini, 1920a.

diazi (Schick, 1970a) (p. 89) – species of '*Ochlerotatus* (*Protomacleaya*)' *sensu auctorum* [original combination: *Aedes* (*Finlaya*) *diazi*].

diplolineata Schneider, 1913 (p. 37) – synonym of *Ochlerotatus communis* (de Geer, 1776) [original combination: *Culicada nemorosa* form *diplolineata*]; synonymy established before Knight & Stone, 1977.

dissimilierodes (Dong, Zhou & Dong, 2002a) (p. 516) – species of *Kenknightia* [original combination: *Aedes* (*Kenknightia*) *dissimilierodes*].

dissimilis (Leicester, 1908) (p. 91) – species of *Kenknightia* [original combination: *Stegomyia dissimilis*].

diversus Theobald, 1901c (p. 73) – synonym of *Ochlerotatus* (*Rusticoidus*) *rusticus* (Rossi, 1790) [original combination: *Culex diversus*]; synonymy established before Knight & Stone, 1977.

dmitryi Gornostaeva, 2005 (p. 485) – species of *Aedes*.

dobodura (King & Hoogstraal, 1946b) (p. 37) – species of *Patmarksia* [original combination: *Aedes* (*Finlaya*) *dobodurus*].

dobrotworskyi (Marks, 1958) (p. 65) – species of *Finlaya* [original combination: *Aedes* (*Finlaya*) *dobrotworskyi*].

doddi Taylor, 1919 (p. 831) – synonym of *Macleaya* (*Macleaya*) *tremula* Theobald, 1903e [original combination: *Mimeteomyia doddi*]; synonymy established before Knight & Stone, 1977.

domesticus (Theobald, 1901d) (p. xiv) – species of *Aedimorphus* [original combination: *Uranotaenia domestica*].

dominicii Rangel & Romero Sierra, 1907 (in Romero Sierra, 1907) (p. 48) – synonym of *Howardina sexlineata* (Theobald, 1901a) [original combination: *Stegomyia dominicii*]; synonymy established before Knight & Stone, 1977.

doonii (Wattal, Bhatia & Kalra, 1958) (p. 220) – species of *Bruceharrisonius* [original combination: *Aedes* (*Finlaya*) *aureostriatus* variety *doonii*]; specific status by Reinert, 2003.

dorsalis (Meigen, 1830b) (p. 242) – species of *Ochlerotatus* (subgenus uncertain, determined by Reinert *et al.*, 2008) [original combination: *Culex dorsalis*].
Synonyms: *curriei* Coquillett, 1901; *onondagensis* Felt, 1904; *lativittatus* Coquillett, 1906c; *quaylei* Dyar & Knab, 1906d; *broquetii* Theobald, 1913c; *grahami* Ludlow, 1920b.

dorseyi (Knight, 1946) (p. 277) – species of *Downsiomyia* [original combination: *Aedes* (*Finlaya*) *dorseyi*].

dorso-vittatus Villeneuve, 1919 (p. 57) – synonym of *Ochlerotatus sticticus* (Meigen, 1838) [original combination: *Culex nemorosus* variety *dorso-vittatus*]; synonymy established before Knight & Stone, 1977.

downsi (Bohart & Ingram, 1946a) (species) (p. 51) – subspecies of *Stegomyia flavopicta* (Yamada, 1921) [original combination: *Aedes* (*Stegomyia*) *downsi*]; subspecific status by Tanaka *et al.*, 1979.

draconarius Dyar, 1922g (p. 194) – synonym of *Georgecraigius* (*Horsfallius*) *fluviatilis* (Lutz, 1904) [original combination: *Aedes draconarius*]; synonymy established before Knight & Stone, 1977.

dubia Theobald, 1910c (p. 133) – synonym of *Stegomyia metallica* (Edwards, 1912) [original combination: *Quasistegomyia dubia*]; synonymy established before Knight & Stone, 1977. **Note**: this nominal taxon is the senior secondary homonym of *dubia* Theobald, 1910c (p. 170), based on page priority.

dubia Theobald, 1910c (p. 170) – synonym of *Stegomyia africana* Theobald, 1901d; synonymy established before Knight & Stone, 1977. **Note**: this nominal taxon is the junior secondary homonym of *dubia* Theobald, 1910c (p. 133), based on page priority.

dufouri (Hamon, 1953) (p. 40) – species of *Ochlerotatus* (subgenus uncertain, following removal from *Levua* to *Aedes* subgenus *Ochlerotatus* by Huang *et al.*, 2010) [original combination: *Aedes* (*Ochlerotatus*) *dufouri*].

dunni (Evans, 1928) (p. 39) – subspecies of *Pseudarmigeres argenteoventralis* (Theobald, 1909b) [original combination: *Aedes* (*Armigeres*) *albomarginata* variety *dunni*]; subspecific status by White, 1975.

duplex (Martini, 1926) (p. 203) – species of *Ochlerotatus* (subgenus uncertain, determined by Reinert *et al.*, 2008) [original combination: *Aedes duplex*].

dupreei (Coquillett, 1904a) (p. 10) – species of *Ochlerotatus* (*Protoculex*) [original combination: *Culex dupreei*].

durbanensis (Theobald, 1903a) (p. 246) – species of *Aedimorphus* [original combination: *Grabhamia durbanensis*].
Synonym: *ocellata* Theobald, 1910c.

dybasi (Bohart, 1957) (p. 53) – species of *Stegomyia* (subgenus uncertain, determined by Reinert *et al.*, 2009) [original combination: *Aedes* (*Stegomyia*) *dybasi*].

dysanor Dyar, 1921e (p. 70) – synonym of *Ochlerotatus abserratus* (Felt & Young, 1904) [original combination: *Aedes* (*Ochlerotatus*) *dysanor*]; synonymy established before Knight & Stone, 1977.

dytes Martini, 1922 (p. 164) – synonym of *Ochlerotatus excrucians* (Walker, 1856a) [original combination: *Aedes excrucians* variety *dytes*]; synonymy established before Knight & Stone, 1977.

dzeta (Séguy, 1924) (p. 98) – species of *Ochlerotatus* (subgenus uncertain, determined by Reinert *et al.*, 2008) [original combination: *Aedes* (*Ochlerotatus*) *dzeta*]

Ee

ealaensis (Huang, 2004) (p. 36) – species of *Stegomyia* (subgenus uncertain, determined by Reinert *et al.*, 2009) [original combination: *Aedes* (*Stegomyia*) *ealaensis*].

eatoni Edwards, 1916 (p. 358) – species of '*Ochlerotatus* (*Finlaya*)' *sensu auctorum* [original combination: *Ochlerotatus eatoni*].

ebogoensis (Rickenbach & Ferrara, 1965) (p. 26) – species of *Aedimorphus* [original combination: *Aedes* (*Aedimorphus*) *ebogoensis*].

echinus (Edwards, 1920a) (p. 133) – species of *Dahliana* [original combination: *Ochlerotatus* (*Finlaya*) *echinus*].

ecuadoriensis (Berlin, 1969a) (p. 115) – species of *Howardina* [original combination: *Aedes* (*Howardina*) *ecuadoriensis*].

edgari (Stone & Rosen, 1952) (p. 425) – species of *Ochlerotatus* (*Buvirilia*) [original combination: *Aedes* (*Ochlerotatus*) *edgari*].

edwardsi Barraud, 1923a (p. 784) – species of *Stegomyia* (*Actinothrix*).

eidsvoldensis (Mackerras, 1927) (p. 295) – species of *Ochlerotatus* (subgenus uncertain, determined by Reinert *et al.*, 2008) [original combination: *Aedes* (*Ochlerotatus*) *theobaldi* form *eidsvoldensis*]; specific status by Lee *et al.*, 1984.

elchoensis (Taylor, 1929) (p. 276) – species of *Macleaya* (*Chaetocruiomyia*) [original combination: *Chaetocruiomyia elchoensis*].

eldridgei (Reinert, 1973e) (p. 14) – species of *Bothaella* [original combination: *Aedes* (*Bothaella*) *eldridgei*].

eleanorae (Berlin, 1969a) (p. 105) – species of *Howardina* [original combination: *Aedes* (*Howardina*) *eleanorae*].

elegans Ficalbi, 1890a (p. 95) – synonym of *Stegomyia* (*Stegomyia*) *aegypti* (Linnaeus, 1762) [original combination: *Culex elegans*]; synonymy established before Knight & Stone, 1977.

ellinorae (Edwards, 1941) (p. 212) – species of *Neomelaniconion* [original combination: *Aedes* (*Banksinella*) *ellinorae*].

elsiae (Barraud, 1923e) (p. 406) – species of *Collessius* (*Collessius*) [original combination: *Finlaya elsiae*].
Synonym: *simulatus* Barraud, 1931b.

embuensis (Edwards, 1930a) (p. 295) – species of *Hopkinsius* (*Hopkinsius*) [original combination: *Aedes* (*Finlaya*) *embuensis*].

epactius (Dyar & Knab, 1908) (p. 53) – species of *Georgecraigius* (*Georgecraigius*) [original combination: *Aedes epactius*].
Synonyms: *perichares* Dyar, 1921b; *nielseni* O'Meara & Craig, 1970.

epinolus Dyar & Knab, 1914 (p. 61) – synonym of *Ochlerotatus* (*Culicelsa*) *taeniorhynchus* (Wiedemann, 1821) [original combination: *Aedes epinolus*]; synonymy established before Knight & Stone, 1977.

epsilonn Séguy, 1924 (p. 97) – synonym of *Ochlerotatus caspius* (Pallas, 1771) [original combination: *Aedes* (*Ochlerotatus*) *epsilonn*]; synonymy by Monchadskii, 1951.

eritreae (Lewis, 1942) (p. 165) – species of *Aedimorphus* [original combination: *Aedes* (*Aedimorphus*) *eritreae*].

eruthrosops Theobald, 1910c (p. 299) – synonym of *Aedimorphus vexans* (Meigen, 1830b) [original combination: *Culicada eruthrosops*]; synonymy by Knight & Hull, 1953 (as synonym of *nocturnus* Theobald, 1903a).

esoensis Yamada, 1921 (p. 77) – species of *Aedes*.
Synonyms: *asanumai* Sasa, Kano & Takahasi, 1950; *pseudoesoensis* Yamaguti & Tamaboko, 1954.

ethiopiensis (Huang, 2004) (p. 39) – species of *Stegomyia* (subgenus uncertain, determined by Reinert *et al.*, 2009) [original combination: *Aedes* (*Stegomyia*) *ethiopiensis*].

eucephalaeus (Dyar, 1918e) (p. 127) – species of *Ochlerotatus* (*Protoculex*) [original combination: *Aedes* (*Ochlerotatus*) *eucephalaeus*].

eucleptes Dyar, 1921i (p. 147) – synonym of *Hulecoeteomyia japonica* (Theobald, 1901a) [original combination: *Aedes* (*Finlaya*) *eucleptes*]; synonymy established before Knight & Stone, 1977.

euedes (Howard, Dyar & Knab, 1913, 1917) (fig. 191, p. 714, respectively) – species of *Ochlerotatus* (subgenus uncertain, determined by Reinert *et al.*, 2008) [original combination: *Aedes euedes*].
Synonyms: *beklemishevi* Denisova, 1955; *grandilarva* Sazanova, 1955; *barri* Rueger, 1958.

euiris (Dyar, 1922g) (p. 193) – species of *Ochlerotatus* (subgenus uncertain, determined by Reinert *et al.*, 2008) [original combination: *Aedes euiris*].

euplocamus (Dyar & Knab, 1906d) (p. 199) – species of *Ochlerotatus* (*Ochlerotatus*) [original combination: *Aedes euplocamus*].

eurochrus Howard, Dyar & Knab, 1917 (p. 716) – synonym of *Aedimorphus vexans* (Meigen, 1830b) [original combination: *Aedes eurochrus*]; synonymy established before Knight & Stone, 1977.

exagitans Walker, 1856a (p. 430) – synonym of *Stegomyia* (*Stegomyia*) *aegypti* (Linnaeus, 1762) [original combination: *Culex exagitans*]; synonymy established before Knight & Stone, 1977.

excitans Walker, 1848 (p. 4) – synonym of *Stegomyia* (*Stegomyia*) *aegypti* (Linnaeus, 1762) [original combination: *Culex excitans*]; synonymy established before Knight & Stone, 1977.

excrucians (Walker, 1856a) (p. 429) – species of *Ochlerotatus* (subgenus uncertain, determined by Reinert *et al.*, 2008) [original combination: *Culex excrucians*].
Synonyms: *siphonalis* Grossbeck, 1904a; *abfitchii* Felt, 1904; *sansoni* Dyar & Knab, 1909c; *surcoufi* Theobald, 1912; *dytes* Martini, 1922.

explorator (Marks, 1964a) (p. 60) – species of *Ochlerotatus* (subgenus uncertain, determined by Reinert *et al.*, 2008) [original combination: *Aedes* (*Ochlerotatus*) *explorator*].

Ff

falabreguesi (Hamon, 1957) (p. 685) – species of *Catageiomyia* [original combination: *Aedes* (*Aedimorphus*) *falabreguesi*].

fasciatus Fabricius, 1805 (p. 36) – synonym of *Stegomyia* (*Stegomyia*) *aegypti* (Linnaeus, 1762) [original combination: *Culex fasciatus*]; synonymy established before Knight & Stone, 1977.

fasciatus Meigen, 1804 (p. 4) – synonym of *Ochlerotatus communis* (de Geer, 1776) [original combination: *Culex fasciatus*]; synonymy established before Knight & Stone, 1977.

fascipalpis (Edwards, 1912c) (p. 19) – species of *Diceromyia* [original combination: *Ochlerotatus fascipalpis*].

feegradei (Barraud, 1934) (p. 164) – species of *Phagomyia* [original combination: *Aedes* (*Finlaya*) *feegradei*].

fengi (Edwards, 1935a) (p. 131) – species of *Luius* [original combination: *Aedes* (*Finlaya*) *fengi*].
Synonym: *loi* Lien, 1968b.

ferinus (Knight, 1947a) (p. 316) – species of *Mucidus* (*Mucidus*) [original combination: *Aedes* (*Mucidus*) *ferinus*].

fijiensis (Marks, 1947a) (p. 26) – species of *Finlaya* [original combination: *Aedes* (*Finlaya*) *fijiensis*].

filicis (Ingram & de Meillon, 1927) (p. 58) – species of *Catageiomyia* [original combination: *Aedes* (*Aedimorphus*) *filicis*].

fimbripes (Edwards, 1924b) (p. 390) – species of *Geoskusea* [original combination: *Aedes* (*Skusea*) *fimbripes*].

fisheri (Barraud, 1928a) (p. 662) – species of *Aedimorphus* [original combination: *Aedes* (*Aedimorphus*) *fisheri*].
Synonym: *syntheticus* Barraud, 1928b.

fisheri Dyar, 1917a (p. 19) – synonym of *Ochlerotatus ventrovittis* (Dyar, 1916b) [original combination: *Aedes fisheri*]; synonymy established before Knight & Stone, 1977.

fitchii (Felt & Young, 1904) (p. 312) – species of *Ochlerotatus* (subgenus uncertain, determined by Reinert *et al.*, 2008) [original combination: *Culex fitchii*].
Synonyms: *palustris* Dyar, 1916b; *pricei* Dyar, 1917a; *mimesis* Dyar, 1917c.

flavescens (Müller, 1764) (p. 87) – species of *Ochlerotatus* (subgenus uncertain, determined by Reinert *et al.*, 2008) [original combination: *Culex flavescens*].
Synonyms: *lutescens* Fabricius, 1775; *variegatus* Schrank, 1781; *bipunctatus* Robineau-Desvoidy, 1827; *flavus* Motschulsky, 1859; *flavescens* Theobald, 1901a; *fletcheri* Coquillett, 1902c; *arcanus* Blanchard, 1905. **Note**: this nominal taxon is the senior primary homonym of *flavescens* Theobald, 1901a.

flavescens Theobald, 1901a (p. 410) – synonym of *Ochlerotatus flavescens* (Müller, 1764) [original combination: *Culex flavescens*]; synonymy established before Knight & Stone, 1977. **Note**: this nominal taxon is the junior primary homonym of *flavescens* (Müller, 1764).

flavicollis (Edwards, 1928a) (p. 269) – species of *Diceromyia* [original combination: *Aedes* (*Aedimorphus*) *flavicollis*].

flavicosta Walker, 1856a (p. 431) – synonym of *Ochlerotatus* (*Chrysoconops*) *fulvus* (Wiedemann, 1828) [original combination: *Culex flavicosta*]; synonymy established before Knight & Stone, 1977.

flavidorsalis Luh & Lee, 1975 (p. 428) – synonym of *Ochlerotatus albineus* (Séguy, 1923) [original combination: *Aedes* (*Ochlerotatus*) *flavidorsalis*]; synonymy by Danilov, 1979.

flavifrons (Skuse, 1889) (p. 1735) – species of *Ochlerotatus* (*Pholeomyia*) [original combination: *Culex flavifrons*].
Synonyms: *vandema* Strickland, 1911b; *variegatans* Strickland, 1911b.

flavimargo (Edwards, 1941) (p. 209) – species of *Neomelaniconion* [original combination: *Aedes* (*Banksinella*) *flavimargo*].

flavinervis Edwards, 1915a (p. 161) – *nomen nudum*, *Neomelaniconion* [original combination: *Banksinella luteolateralis* variety *flavinervis*].

flavipennis Giles, 1904a (p. 366) – species of *Finlaya*.
Synonym: *aranetana* Banks, 1906b.

flavipes Macquart, 1838b (p. 35) – synonym of *Ochlerotatus albifasciatus* (Macquart, 1838b) [original combination: *Culex flavipes*]; synonymy established before Knight & Stone, 1977.

flavopicta (Yamada, 1921) (p. 52) – species of *Stegomyia* (subgenus uncertain, determined by Reinert *et al.*, 2009) [original combination: *Aedes flavopictus*].

flavus Motschulsky, 1859 (p. 503) – synonym of *Ochlerotatus flavescens* (Müller, 1764) [original combination: *Culex flavus*]; synonymy by Natvig, 1948.

flavus Yamada, 1927 (p. 575) – *nomen nudum*, *Aedes* [original combination: *Aedes esoensis* variety *flavus*].

fletcheri Coquillett, 1902c (p. 84) – synonym of *Ochlerotatus flavescens* (Müller, 1764) [original combination: *Culex fletcheri*]; synonymy established before Knight & Stone, 1977.

flindersi Taylor, 1914c (p. 686) – synonym of *Ochlerotatus* (*Lepidokeneon*) *stricklandi* Edwards, 1912b [original combination: *Grabhamia flindersi*]; synonymy established before Knight & Stone, 1977.

fluviatilis Leicester, 1908 (p. 111) – species of *Hulecoeteomyia*.
Synonyms: *saxicola* Edwards, 1922a; *greigi* Barraud, 1923a.

fluviatilis (Lutz, 1904) (in Bourroul, 1904) (p. 72) – species of *Georgecraigius* (*Horsfallius*) [original combination: *Culex fluviatilis*].
Synonyms: *mediomaculata* Theobald, 1907; *tripunctata* Theobald, 1907; *lithoecetor* Dyar & Knab, 1907c; *draconarius* Dyar, 1922g.

fontenillei Le Goff, Boussès & Brunhes, 2007 (p. 196) – species of *Neomelaniconion*.

formosensis (Yamada, 1921) (p. 67) – species of *Hulecoeteomyia* [original combination: *Aedes formosensis*].

formosus (Walker, 1848) (species) (p. 4) – subspecies of *Stegomyia* (*Stegomyia*) *aegypti* (Linnaeus, 1762) [original combination: *Culex formosus*]; subspecific status by Mattingly, 1957b.

fowleri (de Charmoy, 1908) (p. 258) – species of *Aedimorphus* [original combination: *Culex fowleri*].
Synonym: *nigeriensis* Theobald, 1910c.

franciscoi (Mattingly, 1959) (p. 42) – species of *Petermattinglyius* [original combination: *Aedes* (*Diceromyia*) *franciscoi*].

franclemonti (Belkin, 1962) (p. 367) – species of *Finlaya* [original combination: *Aedes* (*Finlaya*) *franclemonti*].

fraseri Edwards, 1912c (p. 11) – species of *Stegomyia* (subgenus uncertain, determined by Reinert *et al.*, 2009).
Synonym: *blacklocki* Evans, 1925a.

frater Robineau-Desvoidy, 1827 (p. 408) – synonym of *Stegomyia* (*Stegomyia*) *aegypti* (Linnaeus, 1762) [original combination: *Culex frater*]; synonymy established before Knight & Stone, 1977.

freycinetiae (Laird, 1957) (p. 343) – species of *Finlaya* [original combination: *Aedes* (*Finlaya*) *freycinetiae*].

freyi Edwards, 1921d (p. 306) – synonym of *Ochlerotatus cyprius* (Ludlow, 1920a) [original combination: *Aedes* (*Ochlerotatus*) *freyi*]; synonymy established before Knight & Stone, 1977.

fryeri (Theobald, 1912b) (p. 84) – species of '*Ochlerotatus* (*Coetzeemyia*)' *sensu auctorum* [original combination: *Culicelsa fryeri*].
Synonym: *mombasaensis* Mattingly, 1963.

fulgens (Edwards, 1917) (p. 213) – species of *Zavortinkius* [original combination: *Ochlerotatus* (*Finlaya*) *fulgens*].

fulvithorax (Lutz, 1904) (in Bourroul, 1904) (p. 66) – species of *Howardina* [original combination: *Haemagogus fulvithorax*].
Synonyms: *palliatus* Coquillett, 1906e; *tachirensis* Anduze, 1947.

fulvus (Wiedemann, 1828) (p. 546) – species of *Ochlerotatus* (*Chrysoconops*) [original combination: *Culex fulvus*].
Synonym: *flavicosta* Walker, 1956a.

fumida (Edwards, 1928a) (p. 274) – species of *Lorrainea* [original combination: *Aedes* (*Skusea*) *fumidus*].

furcifer (Edwards, 1913b) (p. 48) – species of *Diceromyia* [original combination: *Ochlerotatus* (*Finlaya*) *furcifer*].
Synonyms: *nigra* Theobald, 1906b; *africana* Theobald, 1911c.

furvus Edwards, 1928a (p. 274) – synonym of *Lorrainea amesii* (Ludlow, 1903) [original combination: *Aedes*

(*Skusea*) *furvus*]; synonymy by Knight & Hull, 1952.

fusca Leicester, 1908 (p. 92) – synonym of *Lorrainea amesii* (Ludlow, 1903) [original combination: *Stegomyia fusca*]; synonymy established before Knight & Stone, 1977.

fuscinerve (Edwards, 1914b) (p. 73) – species of *Neomelaniconion* [original combination: *Banksinella fuscinervis*].

fuscipalpis (Belkin, 1962) (p. 370) – species of *Finlaya* [original combination: *Aedes* (*Finlaya*) *fuscipalpis*].

fuscitarsis (Belkin, 1962) (p. 371) – species of *Finlaya* [original combination: *Aedes* (*Finlaya*) *fuscitarsis*].

fuscopalpalis Theobald, 1909b (p. 18) – synonym of *Aedimorphus cumminsii* (Theobald, 1903a) [original combination: *Culicada fuscopalpalis*]; synonymy established before Knight & Stone, 1977.

fusculus Zetterstedt, 1850 (p. 3459) – synonym of *Ochlerotatus punctor* (Kirby, 1837) [original combination: *Culex fusculus*]; synonymy by Natvig, 1948.

fuscus Osten Sacken, 1877 (p. 191) – synonym of *Aedes cinereus* Meigen, 1818; synonymy by Carpenter & LaCasse, 1955.

futunae (Belkin, 1962) (p. 455) – species of *Stegomyia* (*Zoromorphus*) [original combination: *Aedes* (*Stegomyia*) *futunae*].

Gg

gabriel (Schick, 1970a) (p. 69) – species of '*Ochlerotatus* (*Protomacleaya*)' *sensu auctorum* [original combination: *Aedes* (*Finlaya*) *gabriel*].

gaffigani (Reinert, 1990) (p. 24) – species of *Kenknightia* [original combination: *Aedes* (*Kenknightia*) *gaffigani*].

gahnicola (Marks, 1947a) (p. 43) – species of *Finlaya* [original combination: *Aedes* (*Finlaya*) *gahnicola*].

galindoi (Schick, 1970a) (p. 77) – species of '*Ochlerotatus* (*Protomacleaya*)' *sensu auctorum* [original combination: *Aedes* (*Finlaya*) *galindoi*].

gallii Martini, 1920a (p. 110) – synonym of *Ochlerotatus pullatus* (Coquillett, 1904c) [original combination: *Aedes gallii*]; synonymy established before Knight & Stone, 1977.

galloisi (Yamada, 1921) (p. 47) – species of *Stegomyia* (subgenus uncertain, determined by Reinert *et al.*, 2009) [original combination: *Aedes galloisi*].

galloisioides (Liu & Lu, 1984) (p. 76) – species of *Stegomyia* (subgenus uncertain, determined by Reinert *et al.*, 2009) [original combination: *Aedes* (*Stegomyia*) *galloisioides*].

ganapathi (Colless, 1958) (p. 474) – species of *Downsiomyia* [original combination: *Aedes* (*Finlaya*) *ganapathi*].

gandaensis (Huang, 2004) (p. 41) – species of *Stegomyia* (*Mukwaya*) [original combination: *Aedes* (*Stegomyia*) *gandaensis*].

gandarai (da Cunha Ramos, Capela & Ribeiro, 1995) (p. 98) – species of *Polyleptiomyia* [original combination: *Aedes* (*Aedimorphus*) *gandarai*].

gani (Bonne-Wepster, 1940) (p. 158) – species of *Finlaya* [original combination: *Aedes* (*Finlaya*) *gani*].

gardnerii Ludlow, 1905a (p. 99) – species of *Stegomyia* (subgenus uncertain, determined by Reinert *et al*, 2009).

gebeleinensis Theobald, 1910c (p. 157) – synonym of *Stegomyia unilineata* (Theobald, 1906b); synonymy established before Knight & Stone, 1977.

geminus Peus, 1970 (p. 22) – species of *Aedes*.

geniculata (Olivier, 1791) (p. 134) – species of *Dahliana* [original combination: *Culex geniculatus*].

Synonyms: *guttatus* Meigen, 1818; *lateralis* Meigen, 1818; *ornatus* Meigen, 1818; *guttatus* Curtis, 1835; *albo-punctatus* Rondani, 1872.

geoffroyi (Reinert, 1999a) (p. 31) – species of *Zavortinkius* [original combination: *Aedes* (*Zavortinkius*) *geoffroyi*].

geoskusea (Amos, 1944) (p. 32) – species of *Levua* [original combination: *Aedes geoskusea*].

Synonym: *suvae* Stone & Bohart, 1944.

gibbinsi (Edwards, 1935a) (p. 135) – species of *Aedimorphus* [original combination: *Aedes* (*Aedimorphus*) *gibbinsi*].

gilcolladoi (Sánchez-Covisa Villa, Rodríguez Rodríguez & Guillén Llera, 1985) (p. 441) – species of *Dahliana* [original combination: *Aedes* (*Finlaya*) *gilcolladoi*].

gilli (Barraud, 1924d) (p. 1000) – species of *Himalaius* [original combination: *Finlaya gilli*].

gilliesi (van Someren, 1962) (p. 24) – species of *Albuginosus* [original combination: *Aedes* (*Aedimorphus*) *gilliesi*].

gombakensis (Mattingly, 1959) (p. 56) – species of *Christophersiomyia* [original combination: *Aedes* (*Christophersiomyia*) *gombakensis*].

gonguoensis (Gong & Lu, 1986) (p. 1) – species of *Jihlienius* [original combination: *Aedes* (*Finlaya*) *gonguoensis*].

gonimus Dyar & Knab, 1918b (p. 165) – synonym of *Ochlerotatus sticticus* (Meigen, 1838) [original combination: *Aedes gonimus*]; synonymy by Natvig, 1948.

gouldi (Reinert, 1972b) (p. 357) – species of *Aedimorphus* [original combination: *Aedes* (*Aedimorphus*) *gouldi*].

grabhami (Berlin, 1969a) (p. 42) – species of *Howardina* [original combination: *Aedes* (*Howardina*) *grabhami*].

Synonym: *aureostriata* Grabham, 1906a.

gracilelineatus (Bonne-Wepster, 1937) (p. 93) – species of '*Ochlerotatus* (*Finlaya*)'*sensu auctorum* [original combination: *Aedes* (*Finlaya*) *gracilelineatus*].

gracilis Leicester, 1908 (p. 81) – synonym of *Stegomyia* (*Xyele*) *desmotes* Giles, 1904a; synonymy established before Knight & Stone, 1977.

grahami Ludlow, 1920b (p. 154) – synonym of *Ochlerotatus dorsalis* (Meigen, 1830b) [original combination: *Aedes grahami*]; synonymy established before Knight & Stone, 1977.

grahamii Theobald, 1909b (p. 2) – species of *Mucidus* (*Mucidus*).

Synonym: *africanus* Theobald, 1901d.

grandilarva Sazanova, 1955 (p. 99) – synonym of *Ochlerotatus euedes* (Howard, Dyar & Knab, 1913) [original combination: *Aedes* (*Ochlerotatus*) *grandilarva*]; synonymy by Wood, 1977.

grantii Theobald, 1901a (p. 306) – species of *Stegomyia* (subgenus uncertain, determined by Reinert *et al.*, 2009).

grassei (Doucet, 1951a) (p. 69) – species of *Paulianius* [original combination: *Aedes* (*Diceromyia*) *grassei*].

greenii (Theobald, 1903a) (p. 289) – species of *Bruceharrisonius* [original combination: *Howardina greenii*].
Synonym: *kanaranus* Barraud, 1924a.

greigi Barraud, 1923e (p. 406) – synonym of *Hulecoeteomyia fluviatilis* Leicester, 1908 [original combination: *Finlaya greigi*]; synonymy established before Knight & Stone, 1977.

grenieri (Hamon, Service, Adam & Taufflieb, 1961) (p. 376) – species of *Catageiomyia* [original combination: *Aedes grenieri*].

gressitti (Bohart, 1957) (p. 23) – species of *Finlaya* [original combination: *Aedes* (*Finlaya*) *gressitti*].

griffithi Thurman, 1954a (p. 198) – species of *Ayurakitia*.

grisea Ludlow, 1907 (p. 130) – synonym of *Ochlerotatus* (*Culicelsa*) *nigromaculis* (Ludlow, 1906a) [original combination: *Grabhamia grisea*]; synonymy established before Knight & Stone, 1977.

grjebinei Hamon, Taufflieb & Maillot, 1957 (p. 695) – species of '*Aedes* (*Pseudalbuginosus*)' *sensu auctorum* [original combination: *Aedes* (*Aedimorphus*) *grjebinei*].

grossbecki (Dyar & Knab, 1906d) (p. 201) – species of *Ochlerotatus* (subgenus uncertain, determined by Reinert *et al.*, 2008) [original combination: *Aedes grossbecki*].
Synonym: *sylvicola* Grossbeck, 1906.

guamensis (Farner & Bohart, 1944) (p. 117) – species of *Stegomyia* (subgenus uncertain, determined by Reinert *et al.*, 2009) [original combination: *Aedes* (*Stegomyia*) *guamensis*].

guatemala (Berlin, 1969a) (p. 103) – species of *Howardina* [original combination: *Aedes* (*Howardina*) *guatemala*].

gubernatoris (Giles, 1901a) (p. 607) – species of *Phagomyia* [original combination: *Culex gubernatoris*].
Synonym: *magna* Theobald, 1905e.

guerrero (Berlin, 1969a) (p. 102) – species of *Howardina* [original combination: *Aedes* (*Howardina*) *guerrero*].

gurneri (van Someren, 1946c) (p. 5) – subspecies of *Pseudarmigeres michaelikati* (van Someren, 1946c) [original combination: *Aedes* (*Dunnius*) *michaelikati gurneri*].

gurneyi (Stone & Bohart, 1944) (p. 217) – species of *Stegomyia* (subgenus uncertain, determined by Reinert *et al.*, 2009) [original combination: *Aedes* (*Stegomyia*) *gurneyi*].

guttatus Curtis, 1835 (p. 537) – synonym of *Dahliana geniculata* (Olivier, 1791) [original combination: *Culex guttatus*]; synonymy established before Knight & Stone, 1977.

guttatus Meigen, 1818 (p. 5) – synonym of *Dahliana geniculata* (Olivier, 1791) [original combination: *Culex guttatus*]; synonymy established before Knight & Stone, 1977.

gutzevichi (Dubitsky & Deshevykh, 1978) (p. 177) – species of *Ochlerotatus* (subgenus uncertain, determined by Reinert *et al.*, 2008) [original combination: *Aedes* (*Ochlerotatus*) *gutzevichi*].

gyirongensis (Ma, 1982) (p. 162) – subspecies of *Ochlerotatus lasaensis* (Meng, 1962) [original combination: *Aedes* (*Ochlerotatus*) *lasaensis gyirongensis*].

Hh

habanicus Dyar & Knab, 1906d (p. 198) – synonym of *Ochlerotatus* (*Ochlerotatus*) *tortilis* (Theobald, 1903g) [original designation: *Aedes habanicus*]; synonymy established before Knight & Stone, 1977.

hakanssoni (Knight & Hurlbut, 1949) (p. 23) – species of *Stegomyia* (subgenus uncertain, determined by Reinert *et al.*, 2009) [original combination: *Aedes* (*Stegomyia*) *hakanssoni*].

hakusanensis (Yamaguti & Tamaboko, 1954) (p. 414) – species of *Ochlerotatus* (subgenus uncertain, determined by Reinert *et al.*, 2008) [original combination: *Aedes* (*Ochlerotatus*) *hakusanensis*].

hamadryadis Cooling, 1924 (p. 24) – *nomen nudum*, *Molpemyia* [original combination: *Aedes* (*Finlaya*) *priestleyi hamadryadis*]; status by Lee *et al.*, 1982; Ward, 1984.

hamoni Mattingly, 1963 (p. 166) – synonym of *Aedimorphus rickenbachi* (Hamon & Adam, 1959b) [original combination: *Aedes* (*Aedimorphus*) *hamoni*]; synonymy by White, 1975.

hancocki (van Someren, 1962) (p. 21) – species of *Vansomerenis* [original combination: *Aedes* (*Finlaya*) *hancocki*].

hansfordi (Huang, 1997) (p. 33) – species of *Stegomyia* (subgenus uncertain, determined by Reinert *et al.*, 2009) [original combination: *Aedes* (*Stegomyia*) *hansfordi*].

haplolineata Schneider, 1913 (p. 37) – synonym of *Ochlerotatus punctor* (Kirby, 1837) [original combination: *Culicada nemorosa* form *haplolineata*]; synonymy established before Knight & Stone, 1977.

harbachi (Reinert, 1990) (p. 27) – species of *Kenknightia* [original combination: *Aedes* (*Kenknightia*) *harbachi*].

hargreavesi Edwards, 1920a (p. 130) – subspecies of *Ochlerotatus caspius* (Pallas, 1771) [original combination: *Ochlerotatus caspius* variety *hargreavesi*]; subspecific status by Harbach & Howard, 2007.

harinasutai (Knight, 1978a) (p. 106) – species of *Downsiomyia* [original combination: *Aedes* (*Finlaya*) *harinasutai*].

harperi (Knight, 1948b) (p. 4) – species of *Danielsia* [original combination: *Aedes* (*Finlaya*) *harperi*].

harrisoni (Muspratt, 1953) (p. 51) – species of *Ochlerotatus* (subgenus uncertain, determined by Reinert *et al.*, 2008) [original combination: *Aedes* (*Ochlerotatus*) *harrisoni*].

harveyi (Barraud, 1923a) (p. 407) – species of *Hulecoeteomyia* [original combination: *Finlaya harveyi*].

hastatus (Dyar, 1922f) (p. 160) – species of *Ochlerotatus* (*Protoculex*) [original combination: *Aedes* (*Ochlerotatus*) *hastatus*].

hatorii (Yamada, 1921) (p. 70) – species of *Collessius* (*Collessius*) [original combination: *Aedes hatorii*].

haworthi (Edwards, 1923d) (p. 499) – species of *Albuginosus* [original combination: *Aedes* (*Aedimorphus*) *haworthi*].

hebridea (Edwards, 1926a) (p. 102) – species of *Stegomyia* (subgenus uncertain, determined by Reinert *et al.*, 2009) [original combination: *Aedes variegatus* variety *hebrideus*]; specific status by Belkin, 1962.

hegneri Causey, 1937a (p. 543) – synonym of *Collessius* (*Collessius*) *macfarlanei* (Edwards, 1914b) [original combination: *Aedes* (*Finlaya*) *hegneri*]; synonymy by Harrison *et al.*, 1991.

heischi (van Someren, 1951) (p. 1) – species of *Stegomyia* (subgenus uncertain, determined by Reinert *et al.*, 2009) [original combination: *Aedes* (*Stegomyia*) *heischi*].

helenae (Reinert, 1973e) (p. 9) – species of *Bothaella* [original combination: *Aedes* (*Bothaella*) *helenae*].

hemisurus Dyar & Knab, 1906d (p. 199) – synonym of *Ochlerotatus* (*Ochlerotatus*) *scapularis* (Rondani, 1848) [original combination: *Aedes hemisurus*]; synonymy by Arnell, 1976.

hemiteleus Dyar, 1924h (p. 179) – synonym of *Aedes cinereus* Meigen, 1818 [original combination: *Aedes cinereus hemiteleus* (as race)]; synonymy by Carpenter & LaCasse, 1955.

hendersoni (Cockerell, 1918) (p. 199) – species of '*Ochlerotatus* (*Protomacleaya*)' *sensu auctorum* [original combination: *Aedes triseriatus* variety *hendersoni*]; specific status by Breland, 1960.

hensilli (Farner, 1945) (p. 59) – species of *Stegomyia* (subgenus uncertain, determined by Reinert *et al.*, 2009) [original combination: *Aedes* (*Stegomyia*) *hensilli*].

heracleensis Callot, 1944a (p. 56) – synonym of *Ochlerotatus berlandi* (Séguy, 1921) [original combination: *Aedes* (*Finlaya*) *heracleensis*]; synonymy by Callot, 1944c.

hesperonotius (Marks, 1959a) (p. 131) – species of *Ochlerotatus* (subgenus uncertain, determined by Reinert *et al.*, 2008) [original combination: *Aedes* (*Ochlerotatus*) *hesperonotius*].

heteropus (Dyar, 1921j) (p. 152) – species of '*Ochlerotatus* (*Protomacleaya*)' *sensu auctorum* [original combination: *Aedes* (*Finlaya*) *heteropus*].

hewitti Hearle, 1923 (p. 5) – synonym of *Ochlerotatus increpitus* (Dyar, 1916b) [original combination: *Aedes hewitti*]; synonymy established before Knight & Stone, 1977.

hexodontus (Dyar, 1916b) (p. 83) – species of *Ochlerotatus* (subgenus uncertain, determined by Reinert *et al.*, 2008) [original combination: *Aedes hexodontus*].

Synonyms: *cyclocerculus* Dyar, 1920a; *leuconotips* Dyar, 1920a; *masamae* Dyar, 1920g; *labradorensis* Dyar & Shannon, 1925d.

hilli Taylor, 1919 (p. 841) – synonym of *Rhinoskusea longirostris* (Leicester, 1908) [original combination: *Uranotaenia hilli*]; synonymy established before Knight & Stone, 1977.

himalayana Giles, 1904b (p. 384) – synonym of *Gilesius pulchriventer* (Giles, 1901a) [original combination: *Howardina himalayana*]; synonymy established before Knight & Stone, 1977.

hirsuteron Theobald, 1901c (p. 98) – synonym of *Ochlerotatus sticticus* (Meigen, 1838) [original combination: *Culex hirsuteron*]; synonymy established before Knight & Stone, 1977.

hirsutus Brunhes & Boussès, 2017 (in Brunhes *et al.*, 2017) (p. 352) – species of *Paulianius*.

hirsutus (Theobald, 1901a) (p. 392) – species of *Aedimorphus* [original combination: *Culex hirsutum* [*sic*]].

Synonym: *transvaalensis* Theobald, 1903a.

hispidosus Skuse, 1889 (p. 1726) – synonym of *Mucidus* (*Mucidus*) *alternans* (Westwood, 1835) [original combination: *Culex hispidosus*]; synonymy established before Knight & Stone, 1977.

hodgkini (Marks, 1959a) (p. 112) – species of *Ochlerotatus* (subgenus uncertain, determined by Reinert *et al.*, 2008) [original combination: *Aedes* (*Ochlerotatus*) *hodgkini*].

hogsbackensis (Huang, 2004) (p. 43) – species of *Stegomyia* (subgenus uncertain, determined by Reinert *et al.*, 2009) [original combination: *Aedes* (*Stegomyia*) *hogsbackensis*].

hoguei (Belkin, 1962) (p. 460) – species of *Stegomyia* (subgenus uncertain, determined by Reinert *et al.*, 2009) [original combination: *Aedes* (*Stegomyia*) *hoguei*].

hokkaidensis (Tanaka, Mizusawa & Saugstad, 1979) (p. 286) – species of *Ochlerotatus* (subgenus uncertain, determined by Reinert *et al.*, 2008) [original combination: *Aedes* (*Ochlerotatus*) *hexodontus hokkaidensis*]; specific status by Tanaka, 1999.

hollandia (King & Hoogstraal, 1946b) (p. 38) – species of *Patmarksia* [original combination: *Aedes* (*Finlaya*) *hollandius*].

hollingsheadi (Belkin, 1962) (p. 372) – species of *Finlaya* [original combination: *Aedes* (*Finlaya*) *hollingsheadi*].

holocinctus (Edwards, 1941) (p. 193) – species of *Aedimorphus* [original combination: *Aedes* (*Aedimorphus*) *cumminsi* [*sic*] *holocinctus*]; specific status by Ovazza *et al.*, 1956.

homoeopus (Dyar, 1922d) (p. 92) – species of '*Ochlerotatus* (*Protomacleaya*)' *sensu auctorum* [original combination: *Aedes terrens homoeopus*]; specific status by Schick, 1970a.

hoogstraali (Knight & Rozeboom, 1946) (p. 92) – species of *Scutomyia* [original combination: *Aedes* (*Stegomyia*) *hoogstraali*].

hopkinsi (Edwards, 1936) (p. 49) – species of *Catageiomyia* [original combination: *Aedes hopkinsi*].

horishensis Yamada, 1921 (p. 58) – synonym of *Stegomyia* (*Heteraspidion*) *annandalei* Theobald, 1910a [original combination: *Aedes horishensis*]; synonymy by Huang, 1977.

horotoi (Taylor, 1972) (p. 317) – species of *Finlaya* [original combination: *Aedes* (*Finlaya*) *horotoi*].

horrescens (Edwards, 1935a) (p. 128) – species of *Stegomyia* (subgenus uncertain, determined by Reinert *et al.*,

2009) [original combination: *Aedes* (*Stegomyia*) *scutellaris* variety *horrescens*]; specific status prior to ? Stone & Farner, 1945.

hortator (Dyar & Knab, 1907a) (p. 12) – species of *Sallumia* [original combination: *Aedes hortator*].

huangae (Reinert, 1999a) (p. 34) – species of *Zavortinkius* [original combination: *Aedes* (*Zavortinkius*) *huangae*].

hui (Bohart, 1957) (p. 25) – species of *Finlaya* [original combination: *Aedes* (*Finlaya*) *hui*].

humeralis (Edwards, 1922d) (p. 93) – species of *Macleaya* (*Chaetocruiomyia*) [original combination: *Aedes* (*Chaetocruiomyia*) *humeralis*].

hungaricus (Mihályi, 1955) (p. 343) – species of *Ochlerotatus* (subgenus uncertain, determined by Reinert *et al.*, 2008) [original combination: *Aedes* (*Ochlerotatus*) *hungaricus*].

hurlbuti (Lien, 1967) (p. 179) – species of *Bruceharrisonius* [original combination: *Aedes* (*Finlaya*) *hurlbuti*].

hybrida Taylor, 1916 (p. 568) – synonym of *Dobrotworskyius rubrithorax* (Macquart, 1850) [original combination: *Culicada hybrida*]; synonymy established before Knight & Stone, 1977.

Ii

ibis (Barraud, 1931b) (p. 610) – species of *Christophersiomyia* [original combination: *Aedes* (*Christophersiomyia* ?) *ibis*].
Synonym: *brayi* Knight, 1947b.

idahoensis (Theobald, 1903a) (p. 250) – subspecies of *Ochlerotatus spencerii* (Theobald, 1901c) [original combination: *Grabhamia spencerii* variety *idahoensis*]; subspecific status by Myers, 1967.

idanus (Schick, 1970a) (p. 70) – species of '*Ochlerotatus* (*Protomacleaya*)' *sensu auctorum* [original combination: *Aedes* (*Finlaya*) *idanus*].

idjenensis (Brug, 1934) (p. 513) – species of *Downsiomyia* [original combination: *Aedes* (*Finlaya*) *niveus* variety *idjenensis*]; specific status by Knight, 1946.

iguazu Shannon & del Ponte, 1928 (in Dyar, 1928) (p. 168) – synonym of *Ochlerotatus* (*Ochlerotatus*) *crinifer* (Theobald, 1903a) [original combination: *Aedes* (*Ochlerotatus*) *iguazu*]; synonymy established before Knight & Stone, 1977.

imitator Leicester, 1908 (species) (p. 89) – subspecies of *Stegomyia gardnerii* Ludlow, 1905; subspecific status by Mattingly, 1965.
Synonyms: *argenteomaculata* Theobald, 1907; *minutissima* Theobald, 1910a; *christianus* Dyar, 1921i; *indosinensis* Borel, 1928.

impatibilis (Walker, 1859) (p. 91) – species of *Scutomyia* [original combination: *Culex impatibilis*].

imperfectus (Dobrotworsky, 1962) (p. 296) – species of *Ochlerotatus* (subgenus uncertain, determined by Reinert *et al.*, 2008) [original combination: *Aedes* (*Ochlerotatus*) *imperfectus*].

impiger (Walker, 1848) (p. 6) – species of *Ochlerotatus* (subgenus uncertain, determined by Reinert *et al.*, 2008) [original combination: *Culex impiger*].
Synonyms: *nearcticus* Dyar, 1919f; *parvulus* Edwards, 1921d.

implacabilis Walker, 1848 (p. 7) – synonym of *Ochlerotatus punctor* (Kirby, 1837) [original combination: *Culex implacabilis*]; synonymy by Vockeroth, 1954.

implicatus (Vockeroth, 1954) (p. 110) – species of *Ochlerotatus* (subgenus uncertain, determined by Reinert *et al.*, 2008) [original combination: *Aedes* (*Ochlerotatus*) *implicatus*].

impostor (Schick, 1970a) (p. 66) – species of '*Ochlerotatus* (*Protomacleaya*)' *sensu auctorum* [original combination: *Aedes* (*Finlaya*) *impostor*].

imprimens (Walker, 1860) (p. 144) – species of *Edwardsaedes* [original combination: *Culex imprimens*].
Synonyms: *auratus* Leicester, 1908; *suknaensis* Theobald, 1910a; *brugi* Edwards, 1924b.

inaequalis Grabham, 1907 (p. 25) – species of *Howardina*.

incomptus (Arnell, 1976) (p. 20) – species of *Ochlerotatus* (*Ochlerotatus*) [original combination: *Aedes* (*Ochlerotatus*) *incomptus*].

inconspicuus Grossbeck, 1904b (p. 332) – synonym of *Ochlerotatus* (*Ochlerotatus*) *trivittatus* (Coquillett, 1902b) [original combination: *Culex inconspicuus*]; synonymy established before Knight & Stone, 1977.

increpitus (Dyar, 1916b) (p. 87) – species of *Ochlerotatus* (subgenus uncertain, determined by Reinert *et al.*, 2008) [original combination: *Aedes increpitus*].
Synonyms: *vittata* Theobald, 1903h; *mutatus* Dyar, 1919b; *hewitti* Hearle, 1923.

indolescens Dyar & Knab, 1907a (p. 11) – synonym of *Ochlerotatus* (*Ochlerotatus*) *scapularis* (Rondani, 1848) [original combination: *Aedes indolescens*]; synonymy by Arnell, 1976.

indonesiae (Mattingly, 1958a) (p. 53) – species of *Cancraedes* [original combination: *Aedes* (*Cancraedes*) *indonesiae*].

indosinensis Borel, 1928 (p. 93) – synonym of *Stegomyia gardnerii imitator* Leicester, 1908; synonymy by Huang, 1977.

inermis (Colless, 1958) (p. 473) – species of *Downsiomyia* [original combination: *Aedes* (*Finlaya*) *inermis*].

inexorabilis Walker, 1848 (p. 4) – synonym of *Stegomyia* (*Stegomyia*) *aegypti* (Linnaeus, 1762) [original combination: *Culex inexorabilis*]; synonymy established before Knight & Stone, 1977.

inexpectatus (Bonne-Wepster, 1948) (p. 318) – species of *Ochlerotatus* (subgenus uncertain, determined by Reinert *et al.*, 2008) [original combination: *Aedes* (*Aedimorphus*) *inexpectatus*].

infirmatus (Dyar & Knab, 1906d) (p. 197) – species of *Ochlerotatus* (*Ochlerotatus*) [original combination: *Aedes infirmatus*].

ingrami (Edwards, 1930a) (p. 296) – species of *Hopkinsius* (*Hopkinsius*) [original combination: *Aedes* (*Finlaya*) *ingrami*].

innuitus Dyar & Knab, 1918b (p. 166) – synonym of *Ochlerotatus nigripes* (Zetterstedt, 1838) [original combination: *Aedes innuitus*]; synonymy established before Knight & Stone, 1977.

inornata Strickland, 1911b (p. 201) – synonym of *Ochlerotatus camptorhynchus* (Thomson, 1869) [original combination: *Culicada inornata*]; synonymy established before Knight & Stone, 1977.

inquinata (Edwards, 1922a) (p. 267) – species of *Phagomyia* [original combination: *Aedes* (*Finlaya*) *inquinatus*].

insatiabilis Bigot, 1859 (p. 118) – synonym of *Stegomyia* (*Stegomyia*) *aegypti* (Linnaeus, 1762) [original combination: *Culex insatiabilis*]; synonymy established before Knight & Stone, 1977.

insolens (Edwards, 1936) (p. 50) – species of *Catageiomyia* [original combination: *Aedes insolens*].

insolitus (Coquillett, 1906e) (p. 62) – species of '*Ochlerotatus* (*Protomacleaya*)' *sensu auctorum* [original combination: *Verrallina insolita*].
Synonym: *laternaria* Coquillett, 1906a.

intermedius (Danilov & Gornostaeva, 1987) (p. 612) – species of *Ochlerotatus* (subgenus uncertain, determined by Reinert *et al.*, 2008) [original combination: *Aedes* (*Ochlerotatus*) *intermedius*].

interruptus (Reinert, 1999a) (p. 47) – species of *Zavortinkius* [original combination: *Aedes* (*Zavortinkius*) *interruptus*].

intrudens (Dyar, 1919b) (p. 23) – species of *Ochlerotatus* (*Woodius*) [original combination: *Aedes intrudens*].

ioliota (Dyar & Knab, 1913) (p. 77) – species of *Howardina* [original combination: *Aedes ioliota*].

iridipennis Dyar, 1922d (p. 92) – synonym of *Lewnielsenius muelleri* (Dyar, 1920j) [original combination: *Aedes iridipennis*]; synonymy established before Knight & Stone, 1977.

irritans (Theobald, 1901d) (p. 3) – species of *Catageiomyia* [original combination: *Stegomyia irritans*].
Synonyms: *senegalensis* Theobald, 1903d; *albotaeniatus* Theobald, 1909b; *confusus* Theobald, 1909b.

iwi (Marks, 1955) (p. 11) – species of *Phagomyia* [original combination: *Aedes* (*Finlaya*) *iwi*].

iyengari (Edwards, 1923b) (p. 4) – species of *Petermattinglyius* (*Petermattinglyius*) [original combination: *Aedes* (*Skusea*) *iyengari*].

Jj

jacobinae (Serafim & Davis, 1933) (p. 14) – species of *Ochlerotatus* (subgenus uncertain, determined by Reinert *et al.*, 2008) [original combination: *Aedes* (*Taeniorhynchus*) *jacobinae*].

jamesi (Edwards, 1914b) (p. 77) – species of *Aedimorphus* [original combination: *Ochlerotatus jamesi*].

jamoti (Hamon & Rickenbach, 1954) (p. 932) – species of *Neomelaniconion* [original combination: *Aedes* (*Banksinella*) *jamoti*].

japonica (Theobald, 1901a) (p. 385) – species of *Hulecoeteomyia* [original combination: *Culex japonicus*].
Synonyms: *eucleptes* Dyar, 1921i; *tokushimaensis* Tanimura, 1952; *bisanensis* Suzuki, Tanimura, Miyagawa & Murata, 1953.

jorgi (Carpintero & Leguizamón, 2000) (p. 20) – species of *Ochlerotatus* (*Chrysoconops*) [original combination: *Aedes* (*Ochlerotatus*) *jorgi*].

josephinae (Marks, 1958) (p. 58) – species of *Finlaya* [original combination: *Aedes* (*Finlaya*) *josephinae*].

josiahae (Huang, 1988a) (p. 155) – species of *Stegomyia* (*Mukwaya*) [original combination: *Aedes* (*Stegomyia*) *josiahae*].

jugorum Villeneuve, 1919 (p. 58) – synonym of *Ochlerotatus pullatus* (Coquillett, 1904c) [original combination: *Culex jugorum*]; synonymy established before Knight & Stone, 1977.

jugraensis Leicester, 1908 (p. 109) – species of *Hulecoeteomyia* [as *Hulecocteomyia*].

juppi (McIntosh, 1973) (p. 262) – species of *Ochlerotatus* (*Juppius*) [original combination: *Aedes juppi*].

Kk

kabaenensis (Brug, 1939) (p. 108) – species of *Geoskusea* [original combination: *Aedes* (*Skusea*) *kabaenensis*].

kabwachensis (Edwards, 1941) (p. 181) – subspecies of *Aedimorphus abnormalis* (Theobald, 1909) [original combination: *Aedes* (*Aedimorphus*) *abnormalis kabwachensis*].

kanaranus Barraud, 1924a (pp. 848, 850) – synonym of *Bruceharrisonius greenii* (Theobald, 1903a) [original combination: *Finlaya greenii* variety *kanaranus* [*sic*]]; synonymy by Harrison *et al.*, 1991.

kanarensis (Edwards, 1934) (in Barraud, 1934) (p. 444) – species of *Dendroskusea* [original combination: *Aedes* (*Cancraedes* ?) *kanarensis*].

kapretwae (Edwards, 1941) (p. 162) – species of *Albuginosus* [original combination: *Aedes* (*Aedimorphus*) *kapretwae*].

karooensis (Muspratt, 1961) (p. 93) – species of *Aedimorphus* [original combination: *Aedes* (*Aedimorphus*) *eritreae karooensis*]; specific status by McIntosh, 1975.

karwari (Barraud, 1924a) (p. 865) – species of *Kenknightia* [original combination: *Finlaya dissimilis* variety *karwari*]; specific status by Reinert, 1990.

kasachstanicus (Gutsevich, 1962) (p. 886) – species of *Ochlerotatus* (subgenus uncertain, determined by Reinert *et al.*, 2008) [original combination: *Aedes* (*Ochlerotatus* ?) *kasachstanicus*].

katherinensis (Woodhill, 1949) (p. 141) – species of *Stegomyia* (subgenus uncertain, determined by Reinert *et al.*, 2009) [original combination: *Aedes scutellaris katherinensis*]; specific status by Lee *et al.*, 1987.

keefei (King & Hoogstraal, 1946f) (p. 311) – species of '*Ochlerotatus* (*Finlaya*)' *sensu auctorum* [original combination: *Aedes* (*Finlaya*) *keefei*].

keniensis (van Someren, 1946a) (p. 111) – species of *Stegomyia* (subgenus uncertain, determined by Reinert *et al.*, 2009) [original combination: *Aedes* (*Stegomyia*) *keniensis*].

kennethi (Muspratt, 1956) (p. 38) – species of *Albuginosus* [original combination: *Aedes* (*Aedimorphus*) *kennethi*].

kenyae (van Someren, 1946c) (p. 2) – species of *Stegomyia* (subgenus uncertain, determined by Reinert *et al.*, 2009) [original combination: *Aedes* (*Stegomyia*) *bambusae kenyae*]; specific status by Huang, 1997.

kermorganti Laveran, 1901a (p. 569) – synonym of *Mucidus* (*Mucidus*) *alternans* (Westwood, 1835) [original combination: *Culex kermorganti*]; synonymy by Rageau & Hamon, 1957.

kesseli (Huang & Hitchcock, 1980) (p. 23) – species of *Stegomyia* (subgenus uncertain, determined by Reinert *et al.*, 2009) [original combination: *Aedes* (*Stegomyia*) *kesseli*].

keyensis Buren, 1947 (p. 228) – synonym of *Ochlerotatus* (*Ochlerotatus*) *thelcter* (Dyar, 1918e) [original combination: *Aedes* (*Ochlerotatus*) *keyensis*]; synonymy by Thurman *et al.*, 1949.

khasiana Barraud, 1923e (p. 407) – synonym of *Hulecoeteomyia pallirostris* (Edwards, 1922a) [original combination: *Finlaya khasiana*]; synonymy by Tewari & Hiriyan, 1996.

khazani (Edwards, 1922a) (p. 265) – species of *Phagomyia* [original combination: *Aedes* (*Finlaya*) *khazani*].

kiangsiensis (Tung, 1955) (p. 313) – species of *Phagomyia* [original combination: *Aedes* (*Finlaya*) *kiangsiensis*].

kingii Theobald, 1908c (p. 258) – synonym of *Aedimorphus ochraceus* (Theobald, 1901a) [original combination: *Mimeteculex kingii*]; synonymy established before Knight & Stone, 1977.

kivuensis (Edwards, 1941) (p. 132) – species of *Stegomyia* (*Mukwaya*) [original combination: *Aedes* (*Stegomyia*) *subargenteus* variety *kivuensis*]; specific status by Mattingly, 1953a.

kleini (Reinert, 1973e) (p. 18) – species of *Bothaella* [original combination: *Aedes* (*Bothaella*) *kleini*].

klotsi Matheson, 1933 (p. 69) – synonym of *Ochlerotatus melanimon* (Dyar, 1924e) [original combination: *Aedes klotsi*]; synonymy by Carpenter & LaCasse, 1955.

knabi (Coquillett, 1906a) (p. 183) – species of '*Ochlerotatus* (*Protomacleaya*)' *sensu auctorum* [original combination: *Culex knabi*].

knighti (Stone & Bohart, 1944) (p. 210) – species of *Finlaya* [original combination: *Aedes* (*Finlaya*) *knighti*].

kobayashii (Nakata, 1956) (p. 135) – species of *Bruceharrisonius* [original combination: *Aedes* (*Finlaya*) *kobayashii*]; ? = *Bruceharrisonius alektorovi* Stackelber, 1943 (Tanaka, 2002).

kochi (Dönitz, 1901) (p. 38) – species of *Finlaya* [original combination: *Culex kochi*].

kohkutensis (Mattingly, 1958a) (p. 55) – species of *Cancraedes* [original combination: *Aedes* (*Cancraedes*) *kohkutensis*].

kolhapuriensis Sathe & Girhe, 2002 (p. 197) – *nomen nudum*, *Aedes*; unavailable name [published after 1999, fails to conform to Articles 16.4 and 72.3, which require fixation of a name-bearing type and a statement naming the collection in which it is deposited (International Commission on Zoological Nomenclature, 1999)].

kompi (Vargas & Downs, 1950) (p. 167) – species of '*Ochlerotatus* (*Protomacleaya*)' *sensu auctorum* [original combination: *Aedes* (*Gualteria*) *kompi*].

koreica (Edwards, 1917) (p. 212) – species of *Hulecoeteomyia* [original combination: *Ochlerotatus* (*Finlaya*) *koreicus*].

koreicoides (Sasa, Kano & Hayashi, 1950) (p. 627) – species of '*Ochlerotatus* (*Finlaya*)' *sensu auctorum* [original combination: *Aedes koreicoides*].

kotiensis (Barraud, 1934) (p. 161) – subspecies of *Phagomyia gubernatoris* (Giles, 1901a) [original combination: *Aedes* (*Finlaya*) *gubernatoris* variety *kotiensis*]; subspecific status by Harbach & Howard, 2007.

kounoupi Brullé, 1833 (p. 289) – synonym of *Stegomyia* (*Stegomyia*) *aegypti* (Linnaeus, 1762) [original combination: *Culex kounoupi*]; synonymy established before Knight & Stone, 1977.

krombeini (Huang, 1975) (p. 345) – species of *Stegomyia* (subgenus uncertain, determined by Reinert *et al.*, 2009) [original combination: *Aedes* (*Stegomyia*) *krombeini*].

krymmontanus (Alekseev, 1989) (p. 173) – species of *Ochlerotatus* (*Rusticoidus*) [original combination: *Aedes* (*Rusticoidus*) *krymmontanus*].

kumbae (Chwatt, 1948) (p. 184) – subspecies of *Bifidistylus boneti* (Gil Collado, 1936) [original combination: *Aedes* (*Aedimorphus*) *boneti kumbae*].

kummi (Edwards, 1930a) (p. 298) – species of *Pseudarmigeres* [original combination: *Aedes* (*Dunnius*) *kummi*].

Ll

labeculosus Coquillett, 1905 (p. 116) – synonym of *Ochlerotatus camptorhynchus* (Thomson, 1869) [original combination: *Culex labeculosus*]; synonymy established before Knight & Stone, 1977.

labradorensis Dyar & Shannon, 1925d (p. 78) – synonym of *Ochlerotatus hexodontus* (Dyar, 1916b) [original combination: *Aedes labradorensis*]; synonymy by Vockeroth, 1954.

lactea (Knight, 1946) (p. 275) – species of *Downsiomyia* [original combination: *Aedes* (*Finlaya*) *lacteus*].

laffooni (Knight & Rozeboom, 1946) (p. 94) – species of *Scutomyia* [original combination: *Aedes* (*Stegomyia*) *laffooni*].

laguna (Arnell & Nielsen, 1972) (p. 12) – species of *Jarnellius* [original combination: *Aedes* (*Ochlerotatus*) *laguna*].

lamberti Ventrillon, 1904 (p. 552) – synonym of *Stegomyia* (*Stegomyia*) *aegypti* (Linnaeus, 1762); synonymy by Rodhain & Boutonnier, 1985.

lamborni (Edwards, 1923a) (p. 398) – species of *Bifidistylus* [original combination: *Aedes* (*Aedimorphus*) *lamborni*].

lambrechti (van Someren, 1971) (p. 21) – species of *Skusea* [original combination: *Aedes* (*Skusea*) *lambrechti*].

lamellifera (Bohart & Ingram, 1946b) (p. 29) – species of *Lorrainea* [original combination: *Aedes* (*Skusea*) *lamelliferus*].

langata (van Someren, 1946c) (p. 1) – species of *Stegomyia* (subgenus uncertain, determined by Reinert *et al*., 2009) [original combination: *Aedes* (*Stegomyia*) *langata*].

laniger (Wiedemann, 1820) (p. 9) – species of *Mucidus* (*Mucidus*) [original combination: *Culex laniger*].

laoagensis (Knight, 1946) (p. 276) – species of *Downsiomyia* [original combination: *Aedes* (*Finlaya*) *laoagensis*].

lasaensis (Meng, 1962) (p. 159) – species of *Ochlerotatus* (subgenus uncertain, determined by Reinert *et al*., 2008) [original combination: *Aedes* (*Ochlerotatus*) *lasaensis*].

lateralis Meigen, 1818 (p. 5) – synonym of *Dahliana geniculata* (Olivier, 1791) [original combination: *Culex lateralis*]; synonymy established before Knight & Stone, 1977.

lateralis Theobald, 1905f (p. 110) – synonym of *Leptosomatomyia aurimargo* (Edwards, 1922d); synonymy established before Knight & Stone, 1977.

laternaria Coquillett, 1906a (p. 184) – synonym of '*Ochlerotatus* (*Protomacleaya*)' *insolitus* (Coquillett, 1906a) [original combination: *Verrallina laternaria*]; synonymy by Schick, 1970a.

lativittatus Coquillett, 1906c (p. 109) – synonym of *Ochlerotatus dorsalis* (Meigen, 1830b) [original combination: *Culex lativittatus*]; synonymy by Wood *et al*., 1979.

lauriei Carter, 1920 (as *laurei*) (p. 623) – species of '*Ochlerotatus* (*Finlaya*)' *sensu auctorum*; emendation by Edwards, 1924b.

lazarensis Felt & Young, 1904 (p. 312) – synonym of *Ochlerotatus communis* (de Geer, 1776) [original combination: *Culex lazarensis*]; synonymy established before Knight & Stone, 1977.

ledgeri (Huang, 1981) (p. 96) – species of *Stegomyia* (subgenus uncertain, determined by Reinert *et al*., 2009) [original combination: *Aedes* (*Stegomyia*) *ledgeri*].

leei (Berlin, 1969a) (p. 110) – species of *Howardina* [original combination: *Aedes* (*Howardina*) *leei*].

leesoni (Edwards, 1932b) (p. 562) – species of *Aedimorphus* [original combination: *Aedes* (*Aedimorphus*) *leesoni*].
Synonym: *verna* Lewis, 1944.

leonis (Colless, 1958) (p. 476) – species of *Downsiomyia* [original combination: *Aedes* (*Finlaya*) *niveus leonis*]; specific status by Harrison *et al*., 1991.

lepchana (Barraud, 1923b) (p. 217) – species of *Danielsia* [original combination: *Finlaya lepchana*].

lepidonotus Edwards, 1920a (p. 132) – species of *Ochlerotatus* (*Rusticoidus*).

lepidus (Cerqueira & Paraense, 1945) (p. 12) – species of *Ochlerotatus* (subgenus uncertain, determined by Reinert *et al*., 2008) [original combination: *Aedes* (*Ochlerotatus*) *lepidus*].

leptolabis (Edwards, 1936) (p. 51) – species of *Aedimorphus* [original combination: *Aedes leptolabis*].

lerozeboomi (Reinert, 1990) (p. 33) – species of *Kenknightia* [original combination: *Aedes* (*Kenknightia*) *lerozeboomi*].

lesnei Séguy, 1920 (p. 328) – synonym of *Ochlerotatus sticticus* (Meigen, 1838); synonymy established before Knight & Stone, 1977.

leucacanthus Loew, 1873 (p. 1) – synonym of *Ochlerotatus pulcritarsis* (Rondani, 1872) [original combination: *Culex leucacanthus*]; synonymy established before Knight & Stone, 1977.

leucarthrius (Speiser, 1909) (p. 43) – species of *Aedimorphus* [original combination: *Culex leucarthrius*].

leucomelas (Meigen, 1804) (p. 3) – species of *Ochlerotatus* (subgenus uncertain, determined by Reinert *et al*., 2008) [original combination: *Culex leucomelas*].
Synonym: *salinellus* Edwards, 1921b.

leucomeres (Giles, 1904a) (p. 367) – species of *Kenknightia* [original combination: *Stegomyia leucomeres*].

leuconotips Dyar, 1920a (p. 24) – synonym of *Ochlerotatus hexodontus* (Dyar, 1916b) [original combination: *Aedes leuconotips*]; synonymy established before Knight & Stone, 1977.

leucopleurus Rozeboom, 1946 (p. 588) – synonym of *Bruceharrisonius aureostriatus* (Doleschall, 1857) [original combination: *Aedes* (*Finlaya*) *leucopleurus*]; synonymy by Reinert, 1990.

leucopygos Eysell, 1903 (p. 299) – synonym of *Aedes cinereus* Meigen, 1818; synonymy established before Knight & Stone, 1977.

lewelleni (Starkey & Webb, 1946) (p. 179) – species of *Finlaya* [original combination: *Aedes* (*Finlaya*) *lewelleni*].

lilii Theobald, 1910c (p. 160) – species of *Stegomyia* (*Mukwaya*).

lindtropi Shingarev, 1927 (p. 545) – synonym of *Stegomyia cretina* (Edwards, 1921d) [original combination: *Aedes* (*Stegomyia*) *lindtropi*]; synonymy established before Knight & Stone, 1977.

linealis Taylor, 1913 (p. 57) – synonym of *Neomelaniconion lineatopenne* (Ludlow, 1905a) [original combination: *Pseudohowardina linealis*]; synonymy established before Knight & Stone, 1977.

lineatopenne (Ludlow, 1905a) (p. 133) – species of *Neomelaniconion* [original combination: *Taeniorhynchus lineatopennis*].
Synonym: *linealis* Taylor, 1913.

linesi (Marks, 1964a) (p. 63) – species of *Ochlerotatus* (subgenus uncertain, determined by Reinert *et al*., 2008) [original combination: *Aedes* (*Ochlerotatus*) *linesi*].

lithoecetor Dyar & Knab, 1907c (p. 201) – synonym of *Georgecraigius* (*Horsfallius*) *fluviatilis* (Lutz, 1904) [original combination: *Aedes lithoecetor*]; synonymy by Zavortink, 1972.

litorea (Colless, 1958) (p. 475) – species of *Downsiomyia* [original combination: *Aedes* (*Finlaya*) *litoreus*].

littlechildi (Taylor, 1933) (p. 370) – species of *Macleaya* (*Macleaya*) [original combination: *Aedes* (*Finlaya*) *littlechildi*].

littoralis Barraud, 1927c (p. 551) – synonym of *Aedimorphus trimaculatus* (Theobald, 1905f); synonymy by Barraud, 1934.

litwakae (Reinert, 1990) (p. 39) – species of *Kenknightia* [original combination: *Aedes* (*Kenknightia*) *litwakae*].

loi Lien, 1968b (p. 98) – synonym of *Luius fengi* (Edwards, 1935) [original combination: *Aedes* (*Finlaya*) *loi*]; synonymy Lu & Ji, 1997.

lokojensis (Service, 1959b) (p. 244) – species of *Catageiomyia* [original combination: *Aedes* (*Aedimorphus*) *lokojensis*].

longifilamentus (Su & Zhang, 1988) (p. 11) – species of *Ochlerotatus* (subgenus uncertain, determined by Reinert *et al.*, 2008) [original combination: *Aedes* (*Ochlerotatus*) *longifilamentus*].

longiforceps (Edwards, 1929) (in Paine & Edwards, 1929) (p. 315) – species of *Geoskusea* [original combination: *Aedes* (*Skusea*) *longiforceps*].

longipalpis (Grünberg, 1905) (p. 383) – species of *Zavortinkius* [original combination: *Stegomyia longipalpis*].

longirostris (Leicester, 1908) (p. 228) – species of *Rhinoskusea* [original combination: *Ficalbia longirostris*].
Synonym: *hilli* Taylor, 1919.

longiseta (Edwards, 1936) (p. 50) – species of *Aedimorphus* [original combination: *Aedes longiseta*].

longisquamosa Theobald, 1905f (p. 103) – synonym of *Ochlerotatus caspius* (Pallas, 1771) [original combination: *Grabhamia longisquamosa*]; synonymy by ? Edwards, 1921d.

longitubus Cambournac, 1938 (p. 76) – synonym of *Ochlerotatus berlandi* (Séguy, 1921) [original combination: *Aedes* (*Ochlerotatus*) *longitubus*]; synonymy by Rioux & Arnold, 1955.

lophoventralis (Theobald, 1910a) (p. 13) – species of *Phagomyia* [original combination: *Pseudocarrollia lophoventralis*].

lorraineae (Berlin, 1969a) (p. 80) – species of *Howardina* [original combination: *Aedes* (*Howardina*) *lorraineae*].

lottei (Hamon & Brengues, 1965) (p. 101) – species of *Catageiomyia* [original combination: *Aedes lottei*].

lowisii (Theobald, 1910c) (p. 257) – species of *Aedimorphus* [original combination: *Reedomyia lowisii*].
Synonym: *mindoroensis* Knight & Hull, 1951a.

lucianus (Muspratt, 1959) (p. 64) – species of *Mucidus* (*Mucidus*) [original combination: *Aedes* (*Mucidus*) *lucianus*].

luciensis Theobald, 1901a (p. 297) – synonym of *Stegomyia* (*Stegomyia*) *aegypti* (Linnaeus, 1762) [original combination: *Stegomyia fasciata* variety *luciensis*]; synonymy established before Knight & Stone, 1977.

ludlowae (Blanchard, 1905) (species) (p. 630) – subspecies of *Ochlerotatus* (*Empihals*) *vigilax* (Skuse, 1889) [*nomen novum* for *annulifera* Ludlow, 1903; original combination: *Culex ludlowi* [*sic*]]; subspecific status by Knight & Hull, 1951b.
Synonym: *annulifera* Ludlow, 1903.

lunulata (King & Hoogstraal, 1946c) (p. 95) – species of *Geoskusea* [original combination: *Aedes* (*Pseudoskusea*) *lunulatus*].

luridum (McIntosh, 1971) (p. 327) – species of *Neomelaniconion* [original combination: *Aedes luridus*].

lutea (Ludlow, 1905a) (p. 96) – species of *Finlaya* [original combination: *Popea lutea*].

luteifemur (Edwards, 1926a) (p. 112) – species of *Ochlerotatus* (subgenus uncertain, determined by Reinert *et al.*, 2008) [original combination: *Aedes* (*Ochlerotatus*) *luteifemur*].

luteocephala Newstead, 1907 (in Newstead *et al.*, 1907) (p. 15) – species of *Stegomyia* (subgenus uncertain, determined by Reinert *et al.*, 2009).

luteolaterale (Theobald, 1901c) (p. 71) – species of *Neomelaniconion* [original combination: *Culex luteolateralis*].

luteostriata (Robinson, 1950) (p. 80) – species of *Vansomerenis* [original combination: *Aedes* (*Finlaya*) *luteostriatus*].

luteovittata Theobald, 1901c (p. 85) – synonym of *Ochlerotatus* (*Rusticoidus*) *rusticus* (Rossi, 1790) [original combination: *Culex nemorosus* variety *luteovittata* [*sic*]]; synonymy established before Knight & Stone, 1977.

lutescens Fabricius, 1775 (p. 800) – synonym of *Ochlerotatus flavescens* (Müller, 1764) [original combination: *Culex lutescens*]; synonymy established before Knight & Stone, 1977.

luzonensis (Rozeboom, 1946) (p. 589) – species of *Kenknightia* [original combination: *Aedes* (*Finlaya*) *luzonensis*].

lynchii Brèthes, 1910 (p. 470) – synonym of *Ochlerotatus* (*Ochlerotatus*) *crinifer* (Theobald, 1903a) [original combination: *Culex lynchii*]; synonymy established before Knight & Stone, 1977.

Mm

macdougalli (Edwards, 1922a) (p. 271) – species of *Collessius* (*Collessius*) [original combination: *Aedes* (*Finlaya*) *macdougalli*].

macfarlanei (Edwards, 1914b) (p. 78) – species of *Collessius* (*Collessius*) [original combination: *Ochlerotatus macfarlanei*].
Synonym: *hegneri* Causey, 1937a.

macintoshi (Marks, 1959a) (p. 117) – species of *Ochlerotatus* (subgenus uncertain, determined by Reinert *et al.*, 2008) [original combination: *Aedes* (*Ochlerotatus*) *macintoshi*].

mackerrasi (Taylor, 1927) (p. 68) – species of *Patmarksia* [original combination: *Aedes* (*Finlaya*) *mackerrasi*].

macleayanus Mackerras, 1927 (p. 291) – synonym of *Ochlerotatus nigrithorax* (Macquart, 1847) [original combination: *Aedes* (*Ochlerotatus*) *macleayanus*]; synonymy established before Knight & Stone, 1977.

macmillani (Marks, 1964b) (p. 138) – species of *Macleaya* (*Chaetocruiomyia*) [original combination: *Aedes* (*Chaetocruiomyia*) *macmillani*].

maculata Theobald, 1905g (p. 23) – synonym of *Aedimorphus taeniorhynchoides* (Christophers, 1911) [original combination: *Pecomyia maculata*]; synonymy established before Knight & Stone, 1977. **Note**: this nominal taxon is the senior secondary homonym of *maculata* Theobald, 1905h.

maculata Theobald, 1905h (p. 243) – synonym of *Aedimorphus pipersalatus* (Giles, 1902) [original combination: *Pseudograbhamia maculata*]; synonymy established before Knight & Stone, 1977.

Note: this nominal taxon is the junior secondary homonym of *maculata* Theobald, 1905g.

maculatus Meigen, 1804 (p. 4) – synonym of *Ochlerotatus* (*Rusticoidus*) *rusticus* (Rossi, 1790) [original combination: *Culex maculatus*]; synonymy established before Knight & Stone, 1977.

maculicosta Edwards, 1936 (p. 52) – synonym of *Neomelaniconion carteri* (Edwards, 1936) [original combination: *Aedes palpalis* variety *maculicosta*]; synonymy established before Knight & Stone, 1977.

maculiventris Macquart, 1846a (p. 135) – synonym of *Ochlerotatus caspius* (Pallas, 1771) [original combination: *Culex maculiventris*]; synonymy established before Knight & Stone, 1977.

maculoabdominalis Theobald, 1913d (p. 597) – synonym of *Stegomyia apicoargentea* Theobald, 1909b [original combination: *Kingia maculoabdominalis*]; synonymy by Edwards, 1941.

maculosa Theobald, 1905f (p. 105) – synonym of *Ochlerotatus detritus* (Haliday, 1833) [original combination: *Grabhamia maculosa*]; synonymy by Stone, 1957b.

madagascarensis (van Someren, 1949a) (p. 7) – species of *Paulianius* [original combination: *Aedes* (*Finlaya*) *madagascarensis*].
Synonym: *sylvaticus* Brunhes, 1983.

maehleri (Bohart, 1957) (p. 40) – species of *Stegomyia* (subgenus uncertain, determined by Reinert *et al.*, 2009) [original combination: *Aedes* (*Stegomyia*) *maehleri*].

maffii (Taylor & Tenorio, 1974) (p. 577) – species of *Finlaya* [original combination: *Aedes* (*Finlaya*) *maffii*].

magna Theobald, 1905e (p. 22) – synonym of *Phagomyia gubernatoris* (Giles, 1901a) [original combination: *Lepidotomyia magna*]; synonymy established before Knight & Stone, 1977.

malariae Grassi, 1898 (p. 168) – synonym of *Aedimorphus vexans* (Meigen, 1830b) [original combination: *Culex malariae*]; synonymy established before Knight & Stone, 1977.

malayensis (Colless, 1962) (p. 314) – species of *Stegomyia* (subgenus uncertain, determined by Reinert *et al.*, 2009) [original combination: *Aedes* (*Stegomyia*) *scutellaris malayensis*]; specific status by Huang, 1972.

malikuli (Huang, 1973) (p. 225) – species of *Stegomyia* (*Huangmyia*) [original combination: *Aedes* (*Stegomyia*) *malikuli*].

mallochi (Taylor, 1944) (p. 121) – species of '*Ochlerotatus* (*Finlaya*)' *sensu auctorum* [original combination: *Aedes* (*Finlaya*) *mallochi*].

mamoedjoensis (Mattingly, 1958a) (p. 58) – species of *Cancraedes* [original combination: *Aedes* (*Cancraedes*) *mamoedjoensis*].

manhi Harbach & Cook, 2010 (in Cook *et al.*, 2010) (p. 34) – species of *Bothaella*.

mansouri (Qutubuddin, 1959) (p. 21) – species of *Aedimorphus* [original combination: *Aedes* (*Aedimorphus*) *mansouri*].

mariae (Sergent & Sergent, 1903) (p. 62) – species of *Acartomyia* [original combination: *Culex mariae*].
Synonym: *desbansi* Séguy, 1923.

marinkellei (Berlin, 1969a) (p. 107) – species of *Howardina* [original combination: *Aedes* (*Howardina*) *marinkellei*].

marinus Theobald, 1901a (p. 396) – synonym of *Ochlerotatus* (*Empihals*) *vigilax* (Skuse, 1889) [original combination: *Culex marinus*]; synonymy established before Knight & Stone, 1977.

marshallensis (Stone & Bohart, 1944) (p. 218) – species of *Stegomyia* (subgenus uncertain, determined by Reinert *et al.*, 2009) [original combination: *Aedes* (*Stegomyia*) *marshallensis*].

marshallii (Theobald, 1901a) (p. 310) – species of *Albuginosus* [original combination: *Stegomyia marshallii*].

martineti (Senevet, 1937) (p. 364) – species of *Ochlerotatus* (subgenus uncertain, determined by Reinert *et al.*, 2008) [original combination: *Aedes* (*Ochlerotatus*) *martineti*].

martinezi (Berlin, 1969a) (p. 119) – species of *Howardina* [original combination: *Aedes* (*Howardina*) *martinezi*].

masamae Dyar, 1920g (p. 166) – synonym of *Ochlerotatus hexodontus* (Dyar, 1916b) [original combination: *Aedes* (*Heteronycha*) *masamae*]; synonymy by Wood, 1977.

mascarensis (MacGregor, 1924) (p. 409) – species of *Stegomyia* (*Stegomyia*) [original combination: *Aedes mascarensis*].

masculinus (Mattingly, 1958a) (p. 51) – species of *Cancraedes* [original combination: *Aedes* (*Cancraedes*) *masculinus*].

masoalensis (Fontenille & Brunhes, 1985) (p. 153) – species of *Aedimorphus* [original combination: *Aedes* (*Aedimorphus*) *masoalensis*].

masseyi (Edwards, 1923a) (p. 397) – species of *Stegomyia* (subgenus uncertain, determined by Reinert *et al.*, 2009) [original combination: *Aedes* (*Stegomyia*) *masseyi*].

mathesoni (Middlekauff, 1944) (species) (p. 42) – subspecies of *Ochlerotatus* (*Culicada*) *canadensis* (Theobald, 1901c) [original combination: *Aedes* (*Ochlerotatus*) *mathesoni*]; subspecific status by Rings & Hill, 1948.

mathioti (Fontenille & Brunhes, 1985) (p. 154) – species of *Aedimorphus* [original combination: *Aedes* (*Aedimorphus*) *mathioti*].

mathisi Neveu-Lemaire, 1902b (p. 13) – synonym of *Ochlerotatus* (*Protoculex*) *serratus* (Theobald, 1901c) [original combination: *Culex mathisi*]; synonymy established before Knight & Stone, 1977.

mattinglyi (Hamon & Rickenbach, 1954) (p. 930) – species of *Aedimorphus* [original combination: *Aedes* (*Aedimorphus*) *mattinglyi*].

mattinglyorum (Huang, 1994) (p. 11) – species of *Stegomyia* (subgenus uncertain, determined by Reinert *et al.*, 2009) [original combination: *Aedes* (*Stegomyia*) *mattinglyorum*].

maxgermaini (Huang, 1990) (p. 31) – species of *Stegomyia* (subgenus uncertain, determined by Reinert *et al.*, 2009) [original combination: *Aedes* (*Stegomyia*) *maxgermaini*].

mcdonaldi (Belkin, 1962) (p. 396) – species of *Ochlerotatus* (*Gilesia*) [original combination: *Aedes* (*Ochlerotatus*) *mcdonaldi*].

mcintoshi (Huang, 1985) (p. 109) – species of *Neomelaniconion* [original combination: *Aedes* (*Neomelaniconion*) *mcintoshi*].

medialis (Brug, 1932) (p. 79) – species of *Huaedes* [original combination: *Aedes* (*Leptosomatomyia*) *medialis*].

mediolineata Ludlow, 1907 (p. 129) – synonym of *Ochlerotatus melanimon* (Dyar, 1924e) [original combination: *Grabhamia mediolineata*]; synonymy by Wood, 1977.

mediolineatus (Theobald, 1901c) (p. 113) – species of *Aedimorphus* [original combination: *Culex mediolineatus*].
Synonym: *trilineatus* Theobald, 1901a.

mediomaculata Theobald, 1907 (p. 245) – synonym of *Georgecraigius* (*Horsfallius*) *fluviatilis* (Lutz, 1904) [original combination: *Danielsia mediomaculata*]; synonymy established before Knight & Stone, 1977.

mediopunctata Theobald, 1905h (p. 240) – species of *Stegomyia* (*Huangmyia*).
Synonym: *submediopunctata* Barraud, 1923a.

mediopunctatus (Theobald, 1909b) (species) (p. 18) – subspecies of *Aedimorphus cumminsii* (Theobald, 1903a) [original combination: *Culicada mediopunctata*]; subspecific status by Harbach & Howard, 2007.

mediovittata (Coquillett, 1906e) (p. 60) – species of *Gymnometopa* [original combination: *Stegomyia mediovittata*].
Synonym: *uncatus* Grabham, 1907.

medleri (Knight & Laffoon, 1946) (p. 211) – species of *Finlaya* [original combination: *Aedes* (*Finlaya*) *medleri*].

mefouensis (Ferrara, 1974) (p. 3) – species of *Diceromyia* [original combination: *Aedes* (*Diceromyia*) *mefouensis*].

meigenanus Dyar, 1921e (p. 72) – synonym of *Ochlerotatus punctor* (Kirby, 1837) [original combination: *Aedes* (*Ochlerotatus*) *meigenanus*]; synonymy established before Knight & Stone, 1977.

meirai (Ribeiro, da Cunha Ramos, Capela & Pires, 1980) (p. 69) – subspecies of *Ochlerotatus caspius* (Pallas, 1771) [original combination: *Aedes* (*Ochlerotatus*) *caspius meirai*].

melanimon (Dyar, 1924e) (p. 126) – species of *Ochlerotatus* (subgenus uncertain, determined by Reinert *et al.*, 2008) [original combination: *Aedes melanimon*].
Synonyms: *mediolineata* Ludlow, 1907; *klotsi* Matheson, 1933.

melanoptera (Giles, 1904a) (p. 367) – species of *Phagomyia* [original combination: *Finlaya melanoptera*].
Synonym: *palawanensis* Ludlow, 1914a.

menoni (Mattingly, 1958b) (p. 77) – species of *Paraedes* [original combination: *Aedes* (*Paraedes*) *menoni*].

meprai (Martínez & Prosen, 1953b) (p. 1) – species of *Ochlerotatus* (*Ochlerotatus*) [original combination: *Aedes* (*Ochlerotatus*) *meprai*].

mercurator (Dyar, 1920a) (p. 13) – species of *Ochlerotatus* (subgenus uncertain, determined by Reinert *et al.*, 2008) [original combination: *Aedes mercurator*].
Synonyms: *albertae* Dyar, 1920e; *ater* Gutsevich, 1955.

meridionalis Dyar & Knab, 1906d (p. 195) – synonym of *Ochlerotatus* (*Protoculex*) *serratus* (Theobald, 1901c) [original combination: *Aedes meridionalis*]; synonymy established before Knight & Stone, 1977.

meronephada Dyar & Shannon, 1925a (p. 71) – species of *Catatassomyia*.
Synonym: *wainwrighti* Baisas, 1946c.

metalepticus Dyar, 1920c (p. 51) – synonym of *Ochlerotatus pullatus* (Coquillett, 1904c) [original combination: *Aedes metalepticus*]; synonymy established before Knight & Stone, 1977.

metallica Edwards, 1912c (p. 12) – species of *Stegomyia* (subgenus uncertain, determined by Reinert *et al.*, 2009) [*nomen novum* for *dubia* Theobald, 1910c (p. 133)].
Synonym: *dubia* Theobald, 1910c (p. 133).

metoecopus (Dyar, 1925d) (p. 30) – species of '*Ochlerotatus* (*Protomacleaya*)' *sensu auctorum* [original combination: *Aedes metoecopus*].

miachaetessa Dyar & Shannon, 1925a (p. 78) – synonym of *Cancraedes curtipes* (Edwards, 1915c) [original combination: *Aedes* (*Skusea*) *miachaetessa* [*sic*]]; synonymy by Mattingly, 1958a.

michaelikati (van Someren, 1946c) (p. 3) – species of *Pseudarmigeres* [original combination: *Aedes* (*Dunnius*) *michaelikati*].

mickevichae (Huang, 1988b) (p. 6) – species of *Stegomyia* (subgenus uncertain, determined by Reinert *et al.*, 2009) [original combination: *Aedes* (*Stegomyia*) *mickevichae*].

microptera (Giles, 1901a) (p. 609) – species of *Dendroskusea* [original combination: *Culex micropterus*].

microsticta (Edwards, 1936) (p. 50) – species of *Catageiomyia* [original combination: *Aedes microstictus*].

mikiranus Edwards, 1922a (p. 269) – synonym of *Danielsia albotaeniata* Leicester, 1904 [original combination: *Aedes* (*Finlaya*) *mikiranus*]; synonymy by Harrison *et al.*, 1991.

mikrokopion (Knight & Harrison, 1988) (p. 215) – species of *Downsiomyia* [original combination: *Aedes* (*Finlaya*) *mikrokopion*].

milleri (Dyar, 1922g) (p. 194) – species of *Ochlerotatus* (subgenus uncertain, determined by Reinert *et al.*, 2008) [original combination: *Aedes milleri*].
Synonyms: *araozi* Shannon & del Ponte, 1928; *oroecetor* Martini, 1931a.

milsoni (Taylor, 1915) (p. 179) – species of *Dobrotworskyius* [original combination: *Culicada milsoni*].
Synonym: *milsoni* Taylor, 1916.

milsoni Taylor, 1916 (p. 566) – synonym of *Dobrotworskyius milsoni* (Taylor, 1915) [original combination: *Hulecoeteomyia milsoni*]; synonymy established before Knight & Stone, 1977.

mimesis Dyar, 1917c (p. 116) – synonym of *Ochlerotatus fitchii* (Felt & Young, 1904) [original combination: *Aedes mimesis*]; synonymy established before Knight & Stone, 1977.

mindoroensis Knight & Hull, 1951a (p. 199) – synonym of *Aedimorphus lowisii* (Theobald, 1910c) [original combination: *Aedes* (*Aedimorphus*) *mindoroensis*]; synonymy by Reinert, 1973b.

minuta Taylor, 1912a (p. 30) – synonym of *Macleaya* (*Macleaya*) *tremula* Theobald, 1903e [original combination: *Danielsia minuta*]; synonymy established before Knight & Stone, 1977.

minuta (Theobald, 1901a) (p. 319) – species of *Catageiomyia* [original combination: *Stegomyia minuta*].

minuta Theobald, 1907 (p. 338) – synonym of *Aedimorphus vexans* (Meigen, 1830b) [original combination: *Culicada minuta*]; synonymy established before Knight & Stone, 1977.

minutissima Theobald, 1910a (p. 9) – synonym of *Stegomyia gardnerii imitator* Leicester, 1908; synonymy by Huang, 1977.

mississippii Dyar, 1920e (p. 113) – synonym of *Ochlerotatus stimulans* (Walker, 1848) [original combination: *Aedes stimulans mississippii*]; synonymy established before Knight & Stone, 1977.

mitchellae (Dyar, 1905b) (p. 74) – species of *Ochlerotatus* (*Culicelsa*) [original combination: *Culex mitchellae*].

mixta (Edwards, 1936) (p. 49) – species of *Catageiomyia* [original combination: *Aedes mixtus*].

miyarai (Tanaka, Mizusawa & Saugstad, 1979) (p. 390) – subspecies of *Stegomyia flavopicta* (Yamada, 1921) [original combination: *Aedes* (*Stegomyia*) *flavopictus miyarai*].

mjobergi (Edwards, 1926b) (p. 248) – species of *Downsiomyia* [original combination: *Armigeres mjobergi*].

moenensis Iyengar, 1955 (p. 30) – *nomen nudum*, *Aedes*.

mohani (Knight, 1969) (p. 382) – species of *Downsiomyia* [original combination: *Aedes* (*Finlaya*) *mohani*].

moloiensis (Taylor, 1929) (p. 276) – species of *Macleaya* (*Chaetocruiomyia*) [original combination: *Chaetocruiomyia moloiensis*].

mombasaensis Mattingly, 1963 (p. 165) – synonym of '*Ochlerotatus* (*Coetzeemyia*)' *fryeri* (Theobald, 1912b) [original combination: *Aedes* (*Ochlerotatus*) *mombasaensis*]; synonymy by van Someren, 1972.

monetus (Edwards, 1935a) (p. 132) – species of *Zavortinkius* [original combination: *Aedes* (*Finlaya*) *monetus*].

monocellatus (Marks, 1948) (p. 14) – species of '*Ochlerotatus* (*Finlaya*)' *sensu auctorum* [original combination: *Aedes* (*Finlaya*) *monocellatus*].

monotrichus (Edwards, 1936) (p. 52) – species of *Neomelaniconion* [original combination: *Aedes monotrichus*].

montana (Brug, 1939) (p. 104) – subspecies of *Rampamyia notoscripta* (Skuse, 1889) [original combination: *Aedes* (*Finlaya*) *notoscriptus* variety *montana*]; subspecific status by Stone *et al*., 1959.

montana Koidzumi, 1918 (p. 141) – synonym of *Stegomyia* (*Xyele*) *desmotes* Giles, 1904a; synonymy by Lien, 1962.

montcalmi Blanchard, 1905 (p. 307) – synonym of *Aedimorphus vexans* (Meigen, 1830b) [*nomen novum* for *sylvestris* Theobald, 1901a; original combination: *Culex montcalmi*]; synonymy established before Knight & Stone, 1977.

montchadskyi (Dubitzky, 1968) (p. 218) – species of *Ochlerotatus* (subgenus uncertain, determined by Reinert *et al*., 2008) [original combination: *Aedes* (*Ochlerotatus*) *montchadskyi*].

monticola (Belkin & McDonald, 1957) (p. 179) – species of *Jarnellius* [original combination: *Aedes* (*Ochlerotatus*) *monticola*].

mosquito Robineau-Desvoidy, 1827 (p. 407) – synonym of *Stegomyia* (*Stegomyia*) *aegypti* (Linnaeus, 1762) [original combination: *Culex mosquito*]; synonymy established before Knight & Stone, 1977.

moucheti (Ravaonjanahary & Brunhes, 1977) (p. 213) – species of *Skusea* [original combination: *Aedes* (*Skusea*) *moucheti*]. **Note**: the name of the second author is misspelled as 'Bruhnes' in the publication.

mpusiensis (Huang, 2004) (p. 47) – species of *Stegomyia* (subgenus uncertain, determined by Reinert *et al*., 2009) [original combination: *Aedes* (*Stegomyia*) *mpusiensis*].

mubiensis Luh & Shih, 1958 (as *mubienensis*) (p. 226) – species of *Aedes*.

mucidus (Karsch, 1887) (p. 25) – species of *Mucidus* (*Mucidus*) [original combination: *Culex mucidus*].

muelleri (Dyar, 1920j) (p. 81) – species of *Lewnielsenius* [original combination: *Aedes* (*Heteronycha*) *muelleri*].

Synonym: *iridipennis* Dyar, 1922d.

multiplex (Theobald, 1903a) (p. 293) – species of *Ochlerotatus* (*Pseudoskusea*) [original combination: *Skusea multiplex*].

muroafcete (Huang, 1997) (p. 54) – species of *Stegomyia* (subgenus uncertain, determined by Reinert *et al*., 2009) [original combination: *Aedes* (*Stegomyia*) *muroafcete*].

musicus Leach, 1825 (p. 293) – synonym of *Ochlerotatus* (*Rusticoidus*) *rusticus* (Rossi, 1790) [original combination: *Culex musicus*]; synonymy established before Knight & Stone, 1977.

mutatus Dyar, 1919b (p. 24) – synonym of *Ochlerotatus increpitus* (Dyar, 1916b) [original combination: *Aedes mutatus*]; synonymy established before Knight & Stone, 1977.

mutica Enderlein, 1923 (p. 28) – synonym of *Skusea pembaensis* (Theobald, 1901c) [original combination: *Phagomyia mutica*]; synonymy established before Knight & Stone, 1977.

mutila (Edwards, 1936) (p. 50) – species of *Catageiomyia* [original combination: *Aedes mutilus*].

mzooi (van Someren, 1962) (p. 19) – species of *Zavortinkius* [original combination: *Aedes* (*Finlaya*) *mzooi*].

Nn

natalensis (Edwards, 1930a) (p. 298) – species of *Pseudarmigeres* [original combination: *Aedes* (*Dunnius*) *natalensis*].

nataliae Gornostaeva, 2005 (p. 483) – species of *Aedes*.

natronius (Edwards, 1932b) (p. 562) – species of *Aedimorphus* [original combination: *Aedes* (*Aedimorphus*) *natronius*].
nearcticus Dyar, 1919f (p. 32) – synonym of *Ochlerotatus impiger* (Walker, 1848) [original combination: *Aedes* (*Ochlerotatus*) *nearcticus*]; synonymy by Vockeroth, 1954.
nemorosus Meigen, 1818 (p. 4) – synonym of *Ochlerotatus communis* (de Geer, 1776) [original combination: *Culex nemorosus*]; synonymy established before Knight & Stone, 1977.
neoafricana (Cornet, Valade & Dieng, 1978) (p. 228) – species of *Stegomyia* (subgenus uncertain, determined by Reinert *et al.*, 2009) [original combination: *Aedes* (*Stegomyia*) *neoafricanus*].
neobiannulata Theobald, 1910c (p. 255) – synonym of *Catageiomyia tarsalis* (Newstead, 1907) [original combination: *Reedomyia neobiannulata*]; synonymy by Edwards, 1941.
neogalloisi (Chen & Chen, 2000) (p. 341) – species of *Stegomyia* (subgenus uncertain, determined by Reinert *et al.*, 2009) [original combination: *Aedes* (*Stegomyia*) *neogalloisi*].
neogeorgiana (Belkin, 1962) (p. 375) – species of *Finlaya* [original combination: *Aedes* (*Finlaya*) *neogeorgianus*].
neopandani (Bohart, 1957) (p. 36) – species of *Stegomyia* (*Bohartius*) [original combination: *Aedes* (*Stegomyia*) *neopandani*].
nevadensis (Chapman & Barr, 1964) (p. 444) – species of *Ochlerotatus* (subgenus uncertain, determined by Reinert *et al.*, 2008) [original combination: *Aedes communis nevadensis*]; specific status by Ellis & Brust, 1973.
ngong (van Someren, 1950a) (p. 7) – species of *Albuginosus* [original combination: *Aedes* (*Aedimorphus*) *ngong*].
nielseni O'Meara & Craig, 1970 (in Carpenter, 1970) (p. 53) – *nomen nudum*, *Georgecraigius* [original combination: *Aedes* (*Finlaya*) *atropalpus nielseni*].
nielseni O'Meara & Craig, 1970 (p. 476) – synonym of *Georgecraigius* (*Georgecraigius*) *epactius* (Dyar & Knab, 1908) [original combination: *Aedes atropalpus nielseni*]; synonymy by Zavortink, 1972.
niger Giles, 1904b (p. 384) – synonym of *Ochlerotatus* (*Culicelsa*) *taeniorhynchus* (Wiedemann, 1821) [original combination: *Taeniorhynchus niger*]; synonymy established before Knight & Stone, 1977.
niger Theobald, 1913b (p. 164) – synonym of *Aedimorphus vexans nocturnus* (Theobald, 1903a) [original combination: *Culex nocturnus niger*]; synonymy by Belkin, 1962.
nigeria Theobald, 1901a (p. 303) – synonym of *Stegomyia* (*Stegomyia*) *aegypti* (Linnaeus, 1762); synonymy established before Knight & Stone, 1977.
nigeriensis Theobald, 1910c (p. 281) – synonym of *Aedimorphus fowleri* (de Charmoy, 1908) [original combination: *Grabhamia nigeriensis*]; synonymy by Edwards, 1941.
nigerrimus Theobald, 1913d (p. 591) – species of *Mucidus* (*Mucidus*).
nigra Ludlow, 1905c (p. 387) – synonym of '*Ochlerotatus* (*Protomacleaya*)' *triseriatus* (Say, 1823) [original combination: *Finlaya ? nigra*]; synonymy established before Knight & Stone, 1977. **Note**: this nominal taxon is the senior secondary homonym of *nigra* Taylor, 1914c.
nigra Taylor, 1914c (p. 688) – synonym of *Ochlerotatus camptorhynchus* (Thomson, 1869) [original combination: *Culicada nigra*]; synonymy established before Knight & Stone, 1977. **Note**: this nominal taxon is the junior secondary homonym of *nigra* Ludlow, 1905c.
nigra Theobald, 1906b (p. 80) – synonym of *Diceromyia furcifer* (Edwards, 1913b) [original combination: *Mansonia* (?) *nigra*]; synonymy established before Knight & Stone, 1977.
nigrescens (Edwards, 1929) (in Paine & Edwards, 1929) (p. 314) – species of *Mucidus* (*Pardomyia*) [original combination: *Pardomyia aurantia* variety *nigrescens*]; specific status by Belkin, 1962 (as *painei* Knight, 1948c). **Note:** *painei* Knight, 1948c was introduced as a replacement for *nigrescens* Edwards, 1929, *non* Theobald, 1907, when classified as members of genus *Aedes* Meigen. However, *nigrescens* Theobald, 1907 is a synonym of *Culex* (*Eumelanomyia*) *malayi* (Leicester, 1908) (see comment under *nigrescens* Theobald, 1907); hence, *nigrescens* Edwards, 1929 is not a homonym of *nigrescens* Theobald, 1907 and *painei* Knight, 1948c is not a justified replacement name for the former.
Synonym: *painei* Knight, 1948c.
nigricephalus (Theobald, 1901d) (p. iv) – species of *Aedimorphus* [original combination: *Stegomyia nigricephala*].
Synonym: *paludosus* Graham, 1910.
nigrinus (Eckstein, 1918) (p. 67) – species of *Ochlerotatus* (subgenus uncertain, determined by Reinert *et al.*, 2008) [original combination: *Culicada nigrina*].
nigripes (Zetterstedt, 1838) (p. 807) – species of *Ochlerotatus* (subgenus uncertain, determined by Reinert *et al.*, 2008) [original combination: *Culex nigripes*].
Synonym: *innuitus* Dyar & Knab, 1918b.
nigrithorax (Macquart, 1847) (p. 25) – species of *Ochlerotatus* (subgenus uncertain, determined by Reinert *et al.*, 2008) [original combination: *Culex nigrithorax*].
Synonym: *macleayanus* Mackerras, 1927.
nigritia Ludlow, 1910 (p. 194) – synonym of *Stegomyia albopicta* (Skuse, 1895); synonymy established before Knight & Stone, 1977.
nigritulus Zetterstedt, 1850 (p. 3459) – synonym of *Aedes cinereus* Meigen, 1818 [original combination: *Culex nigritulus*]; synonymy established before Knight & Stone, 1977.
nigrocanus (Martini, 1927) (p. 390) – species of *Ochlerotatus* (subgenus uncertain, determined by Reinert *et al.*, 2008) [original combination: *Aedes nigrocanus*].
nigromaculis (Ludlow, 1906a) (p. 83) – species of *Ochlerotatus* (*Culicelsa*) [original combination: *Grabhamia nigromaculis*].
Synonym: *grisea* Ludlow, 1907.
nigropterum Le Goff, Boussès & Brunhes, 2007 (p. 194) – species of *Neomelaniconion*.

nigrorhynchus (Brug, 1931c) (p. 28) – species of *Hulecoeteomyia* [original combination: *Aedes* (*Finlaya*) *harveyi* variety *nigrorhynchus*]; specific status by Knight, 1968.

nigrostriatus Barraud, 1927c (p. 549) – species of *Aedimorphus*.

ningheensis (Lei, 1989b) (p. 100) – species of *Collessius* (*Collessius*) [original combination: *Aedes* (*Finlaya*) *ningheensis*].

niphadopsis (Dyar & Knab, 1918b) (p. 166) – species of *Ochlerotatus* (subgenus uncertain, determined by Reinert *et al*., 2008) [original combination: *Aedes niphadopsis*].

nipponica (La Casse & Yamaguti, 1948) (p. 79) – species of *Downsiomyia* [original combination: *Aedes* (*Finlaya*) *niveus nipponicus*]; specific status by La Casse & Yamaguti, 1950.

nipponii (Theobald, 1907) (species) (p. 337) – subspecies of *Aedimorphus vexans* (Meigen, 1830b) [original combination: *Culicada nipponii*]; subspecific status by Bohart & Ingram, 1946b.

nippononivea (Sasa & Nakahashi, 1952) (p. 258) – species of *Downsiomyia* [original combination: *Aedes* (*Finlaya*) *nippononiveus*].

nishikawai (Tanaka, Mizusawa & Saugstad, 1979) (p. 356) – species of *Downsiomyia* [original combination: *Aedes* (*Finlaya*) *nishikawai*].

nivalis (Edwards, 1926a) (p. 112) – species of *Ochlerotatus* (subgenus uncertain, determined by Reinert *et al*., 2008) [original combination: *Aedes* (*Ochlerotatus*) *nivalis*].

nivea (Ludlow, 1903) (in Theobald, 1903a) (p. 139) – species of *Downsiomyia* [original combination: *Stegomyia nivea*].

niveoides (Barraud, 1934) (p. 211) – species of *Downsiomyia* [original combination: *Aedes* (*Finlaya*) *niveoides*].

niveoscutellum Theobald, 1905g (p. 22) – synonym of *Aedimorphus pampangensis* (Ludlow, 1905a) [original combination: *Reedomyia niveoscutella* [*sic*]]; emendation by ? Barraud, 1934; synonymy by Reinert, 1973b.

niveoscutum (Zavortink, 1972) (p. 49) – species of '*Ochlerotatus* (*Protomacleaya*)' *sensu auctorum* [original combination: *Aedes* (*Protomacleaya*) *niveoscutum*].

niveus (de Meillon, 1943) (p. 94) – species of *Paulianios* [original combination: *Aedes* (*Diceromyia*) *niveus*].
Synonym: *zethus* de Meillon & Lavoipierre, 1944.

niveus (von Eichwald, 1837) (p. 183) – *nomen dubium*, *Stegomyia* [original combination: *Culex niveus*]; Edwards, 1932a, ? = *Stegomyia aegypti* (as *Aedes aegypti*).

nivitarsis Coquillett, 1904c (p. 168) – synonym of *Ochlerotatus* (*Culicada*) *canadensis* (Theobald, 1901c) [original combination: *Culex nivitarsis*]; synonymy established before Knight & Stone, 1977.

njombiensis (Huang, 1997) (p. 56) – species of *Stegomyia* (subgenus uncertain, determined by Reinert *et al*., 2009) [original combination: *Aedes* (*Stegomyia*) *njombiensis*].

nocturnus (Theobald, 1903a) (species) (p. 159) – subspecies of *Aedimorphus vexans* (Meigen, 1830b) [original combination: *Culex nocturnus*]; subspecific status by Tsukamoto *et al*., 1985.
Synonym: *niger* Theobald, 1913b.

normanensis (Taylor, 1915) (p. 182) – species of *Ochlerotatus* (subgenus uncertain, determined by Reinert *et al*., 2008) [original combination: *Culex normanensis*].

notoscripta (Skuse, 1889) (p. 1738) – species of *Rampamyia* [original combination: *Culex notoscriptus*].

novalbitarsis (King & Hoogstraal, 1946d) (p. 146) – species of *Patmarksia* [*nomen novum* for *albitarsis* Taylor, 1914d; original combination: *Aedes* (*Finlaya*) *novalbitarsis*].
Synonym: *albitarsis* Taylor, 1914d.

novalbopicta (Barraud, 1931a) (p. 224) – species of *Stegomyia* (subgenus uncertain, determined by Reinert *et al*., 2009) [original combination: *Aedes* (*Stegomyia*) *novalbopictus*].

novonivea (Barraud, 1934) (p. 211) – species of *Downsiomyia* [original combination: *Aedes* (*Finlaya*) *novoniveus*].

nubilus (Theobald, 1903a) (p. 208) – species of *Ochlerotatus* (*Protoculex*) [original combination: *Culex nubilus*].
Synonym: *polyagrus* Dyar, 1918b.

nummatus (Edwards, 1923b) (p. 4) – species of *Tewarius* [original combination: *Aedes* (*Aedimorphus*) *nummatus*].

nyasae (Edwards, 1930a) (p. 296) – species of *Hopkinsius* (*Hopkinsius*) [original combination: *Aedes* (*Finlaya*) *nyasae*].

nyounae (Hamon & Adam, 1959a) (p. 531) – species of *Catageiomyia* [original combination: *Aedes* (*Aedimorphus*) *nyounae*].

Oo

oakleyi (Stone, 1939) (p. 163) – species of *Aedimorphus* [original combination: *Aedes* (*Aedimorphus*) *oakleyi*].

obscurus Meigen, 1830a (pl. 3, fig. 2) – synonym of *Ochlerotatus communis* (de Geer, 1776) [original combination: *Aedes obscurus*]; synonymy established before Knight & Stone, 1977.

obturbator (Dyar & Knab, 1907a) (p. 9) – species of *Ochlerotatus* (*Ochlerotatus*) [original combination: *Aedes obturbator*].

occidentalis Evans, 1926 (p. 99) – synonym of *Albuginosus stokesi* (Evans, 1929) [original combination: *Aedes* (*Aedimorphus*) *occidentalis*]; synonymy established before Knight & Stone, 1977.

occidentalis (Skuse, 1889) (p. 1729) – species of *Dobrotworskyius* [original combination: *Culex occidentalis*].

oceanica (Belkin, 1962) (p. 377) – species of *Finlaya* [original combination: *Aedes* (*Finlaya*) *oceanicus*].

ocellata Theobald, 1910c (p. 284) – synonym of *Aedimorphus durbanensis* (Theobald, 1903a) [original combination: *Grabhamia ocellata*]; synonymy established before

Knight & Stone, 1977. **Note**: this nominal taxon is listed as a synonym of *durbanensis angolae* Ribeiro & Ramos, 1974 in the online catalogue of the Walter Reed Biosystematics Unit (http://www.mosquitocatalog.org/), but it is not accepted here because no record of the synonymy could be found in the published literature.

ochraceus (Theobald, 1901c) (p. 103) – species of *Aedimorphus* [original combination: *Culex ochraceus*].
Synonym: *kingii* Theobald, 1908c.

okinawanus (Bohart, 1946) (p. 39) – species of *Bruceharrisonius* [original combination: *Aedes* (*Finlaya*) *okinawanus*].

oligopistus (Dyar, 1918b) (p. 76) – species of *Ochlerotatus* (*Protoculex*) [original combination: *Aedes oligopistus*].

omorii (Lien, 1968b) (p. 105) – species of *Downsiomyia* [original combination: *Aedes* (*Finlaya*) *omorii*].

omurensis Yamada, 1921 (p. 73) – synonym of *Aedimorphus alboscutellatus* (Theobald, 1905f) [original combination: *Aedes omurensis*]; synonymy established before Knight & Stone, 1977.

onondagensis Felt, 1904 (p. 304) – synonym of *Ochlerotatus dorsalis* (Meigen, 1830b) [original combination: *Culex onondagensis*]; synonymy by Wood *et al*., 1979.

opok (Corbet & van Someren, 1962) (p. 73) – species of *Stegomyia* (subgenus uncertain, determined by Reinert *et al*., 2009) [original combination: *Aedes* (*Stegomyia*) *opok*].

orbitae Edwards, 1922a, 1922b (pp. 260, 466, respectively) – synonym of *Aedimorphus taeniatus* (Leicester, 1908) [original combination: *Aedes* (*Ecculex*) *orbitae*]; originally proposed as a replacement name for *taeniata* Leicester, *non taeniatus* Wiedemann, 1828, when both were included in *Aedes* Meigen.

oreophilus Edwards, 1916 (p. 357) – species of '*Ochlerotatus* (*Finlaya*)' *sensu auctorum*.
Synonym: *bunanoki* Sasa & Ishimura, 1951.

ornatus Meigen, 1818 (p. 5) – synonym of *Dahliana geniculata* (Olivier, 1791) [original combination: *Culex ornatus*]; synonymy established before Knight & Stone, 1977.

oroecetor Martini, 1931a (p. 204) – synonym of *Ochlerotatus milleri* (Dyar, 1922g) [original combination: *Aedes oroecetor*]; synonymy by Lane, 1953.

osornoi (Berlin, 1969a) (p. 112) – species of *Howardina* [original combination: *Aedes* (*Howardina*) *osornoi*].

ostentatio (Leicester, 1908) (p. 193) – species of *Paraedes* [original combination: *Aioretomyia ostentatio*].

oswaldi Lutz, 1904 (in Bourroul, 1904) (pp. 47, 66) – synonym of '*Ochlerotatus* (*Protomacleaya*)' *terrens* (Walker, 1856a) [original combination: *Gualteria oswaldi*]; synonymy established before Knight & Stone, 1977.

ovazzai (Hamon & Adam, 1959b) (p. 147) – species of *Aedimorphus* [original combination: *Aedes* (*Aedimorphus*) *ovazzai*].

Pp

pachyurus (Edwards, 1936) (p. 51) – species of *Aedimorphus* [original combination: *Aedes pachyurus*].

pacificensis Hearle, 1927 (p. 101) – synonym of *Ochlerotatus cataphylla* (Dyar, 1916b) [original combination: *Aedes pacificensis*]; synonymy by Curtis, 1967.

pagei (Ludlow, 1911b) (p. 128) – species of *Paraedes* [original combination: *Danielsia pagei*].

pagetonotum Dyar & Knab, 1909d (p. 253) – synonym of *Ochlerotatus* (*Rusticoidus*) *provocans* (Walker, 1848) [original combination: *Aedes pagetonotum*]; transferred synonymy by synonymy of senior synonym *trichurus* (Dyar) with *provocans* (Walker) by Wood, 1977.

painei Knight, 1948c (p. 68) – synonym of *Mucidus* (*Pardomyia*) *nigrescens* (Edwards, 1929) [original combination: *Aedes* (*Mucidus*) *aurantius painei*]; originally proposed as a replacement name for *nigrescens* Edwards, 1929, *non* Theobald, 1907, when both were classified as nominal taxa of genus *Aedes* Meigen; however, *nigrescens* Theobald, 1907 is a synonym of *Culex* (*Eumelanomyia*) *malayi* (Leicester, 1908) (see comment under *nigrescens* Theobald, 1907) and, consequently, *painei* Knight, 1948c is not a justified replacement name for *nigrescens* Edwards, 1929.

palauensis (Bohart, 1957) (p. 51) – species of *Stegomyia* (subgenus uncertain, determined by Reinert *et al*., 2009) [original combination: *Aedes* (*Stegomyia*) *palauensis*].

palawanensis Ludlow, 1914a (p. 31) – synonym of *Phagomyia melanoptera* (Giles, 1904a) [original combination: *Popea palawanensis*]; synonymy established before Knight & Stone, 1977.

palawanicus (Mattingly, 1958a) (p. 56) – species of *Cancraedes* [original combination: *Aedes* (*Cancraedes*) *palawanicus*].

pallens (Ross, 1943) (p. 148) – subspecies of *Ochlerotatus* (*Chrysoconops*) *fulvus* (Wiedemann, 1828) [original combination: *Aedes* (*Ochlerotatus*) *fulvus pallens*].

palliatus Coquillett, 1906e (p. 61) – synonym of *Howardina fulvithorax* (Lutz, 1904) [original combination: *Taeniorhynchus palliatus*]; synonymy established before Knight & Stone, 1977.

pallida Theobald, 1907 (p. 470) – synonym of *Neomelaniconion albothorax* (Theobald, 1907) [original combination: *Banksinella luteolateralis* variety *pallida*]; synonymy by Edwards, 1941.

pallidohirta Grossbeck, 1905 (p. 359) – synonym of *Aedes cinereus* Meigen, 1818 [original combination: *Culex pallidohirta* [*sic*]]; synonymy established before Knight & Stone, 1977.

pallidopunctata Theobald, 1911b (p. 267) – synonym of *Aedimorphus dentatus* (Theobald, 1904b) [original combination: *Culex pallidopunctata* [*sic*]]; synonymy established before Knight & Stone, 1977.

pallidostriatus (Theobald, 1907) (p. 410) – species of *Aedimorphus* [original combination: *Culex pallidostriatus*].
Synonym: *parascelos* Theobald, 1910a.

pallirostris (Edwards, 1922a) (p. 270) – species of *Hulecoeteomyia* [original combination: *Aedes* (*Finlaya*) *pallirostris*].
Synonym: *khasiana* Barraud, 1923e.

palmarum (Edwards, 1924b) (p. 382) – species of *Patmarksia* [original combination: *Aedes* (*Finlaya*) *palmarum*].

palmeni Edwards, 1921b (p. 52) – synonym of *Ochlerotatus communis* (de Geer, 1776); synonymy by Edwards, 1921d.

palpale Newstead, 1907 (in Newstead *et al*., 1907) (p. 31) – species of *Neomelaniconion*.

paludosus Graham, 1910 (p. 270) – synonym of *Aedimorphus nigricephalus* (Theobald, 1901d) [original combination: *Myxosquamus paludosus*]; synonymy established before Knight & Stone, 1977.

palustris Dyar, 1916b (p. 89) – synonym of *Ochlerotatus fitchii* (Felt & Young, 1904) [original combination: *Aedes palustris*]; synonymy established before Knight & Stone, 1977.

pampangensis (Ludlow, 1905a) (p. 94) – species of *Aedimorphus* [original combination: *Reedomyia pampangensis*].
Synonym: *niveoscutellum* Theobald, 1905g.

panchgangee Sathe & Girhe, 2002 (p. 198) – *nomen nudum*, *Ochlerotatus*; unavailable name [published after 1999, fails to conform to Articles 16.4 and 72.3 of the *Code*, which require fixation of a name-bearing type and a statement naming the collection in which it is deposited (International Commission on Zoological Nomenclature, 1999)].

pandani (Stone, 1939) (p. 162) – species of *Stegomyia* (*Bohartius*) [original combination: *Aedes* (*Stegomyia*) *pandani*].

papago (Zavortink, 1970c) (p. 3) – species of *Abraedes* [original combination: *Aedes* (*Abraedes*) *papago*].

papuensis (Taylor, 1914d) (p. 193) – species of *Patmarksia* [original combination: *Leucomyia australiensis* variety *papuensis*]; specific status by Edwards, 1924b.

paradiantaeus Apfelbeck, 1929 (p. 290) – synonym of *Ochlerotatus sticticus* (Meigen, 1838) [original combination: *Aedes* (*Ochlerotatus*) *paradiantaeus*]; synonymy established before Knight & Stone, 1977.

paradissimilis (Rozeboom, 1946) (p. 587) – species of *Kenknightia* [original combination: *Aedes* (*Finlaya*) *paradissimilis*].

parascelos Theobald, 1910a (p. 18) – synonym of *Aedimorphus pallidostriatus* (Theobald, 1907) [original combination: *Culex parascelos*]; synonymy established before Knight & Stone, 1977.

parvulus Edwards, 1921d (p. 314) – synonym of *Ochlerotatus impiger* (Walker, 1848) [original combination: *Aedes* (*Ochlerotatus*) *parvulus*]; synonymy established before Knight & Stone, 1977.

parvus Macquart, 1834 (p. 36) – synonym of *Aedimorphus vexans* (Meigen, 1830b) [original combination: *Culex parvus*]; synonymy established before Knight & Stone, 1977.

patersoni (Shannon & del Ponte, 1928) (p. 73) – species of *Ochlerotatus* (*Ochlerotatus*) [original combination: *Aedes* (*Ochlerotatus*) *patersoni*].

patriciae (Mattingly, 1954a) (p. 262) – species of *Stegomyia* (subgenus uncertain, determined by Reinert *et al*., 2009) [original combination: *Aedes* (*Stegomyia*) *patriciae*].

paullusi (Stone & Farner, 1945) (p. 155) – species of *Stegomyia* (subgenus uncertain, determined by Reinert *et al*., 2009) [original combination: *Aedes* (*Stegomyia*) *paullusi*].

pearyi Dyar & Shannon, 1925d (p. 78) – synonym of *Ochlerotatus pullatus* (Coquillett, 1904c) [original combination: *Aedes pearyi*]; synonymy by Vockeroth, 1954.

pecori (Reinert, 1990) (p. 52) – species of *Kenknightia* [original combination: *Aedes* (*Kenknightia*) *pecori*].

pectinatus (Arnell, 1976) (p. 74) – species of *Ochlerotatus* (*Ochlerotatus*) [original combination: *Aedes* (*Ochlerotatus*) *pectinatus*].

pecuniosa (Edwards, 1922d) (p. 94) – species of *Molpemyia* [original combination: *Aedes* (*Finlaya*) *pecuniosus*].

peipingensis (Feng, 1938) (p. 520) – species of '*Ochlerotatus* (*Finlaya*)' *sensu auctorum* [original combination: *Aedes* (*Finlaya*) *peipingensis*].

pembaensis (Theobald, 1901c) (p. 235) – species of *Skusea* [original combination: *Aedes pembaensis*].
Synonym: *mutica* Enderlein, 1923.

penghuensis (Lien, 1968b) (p. 96) – species of *Cancraedes* [original combination: *Aedes* (*Cancraedes*) *penghuensis*].

penicillaris Rondani, 1872 (p. 31) – synonym of *Ochlerotatus caspius* (Pallas, 1771) [original combination: *Culex penicillaris*]; synonymy established before Knight & Stone, 1977.

pennai (Antunes & Lane, 1938) (p. 607) – species of *Ochlerotatus* (*Chrysoconops*) [original combination: *Aedes* (*Ochlerotatus*) *pennai*].

perichares Dyar, 1921b (p. 36) – synonym of *Georgecraigius* (*Georgecraigius*) *epactius* (Dyar & Knab, 1908) [original combination: *Aedes* (*Culicelsa* [as *Culiselsa*]) *perichares*]; synonymy by Zavortink, 1972.

periskelata (Giles, 1902) (p. 371) – species of *Dendroskusea* [original combination: *Stegomyia periskelata*].
Synonym: *annulifemur* Edwards, 1914b.

perkinsi (Marks, 1949) (p. 32) – species of *Ochlerotatus* (subgenus uncertain, determined by Reinert *et al*., 2008) [original combination: *Aedes* (*Ochlerotatus*) *perkinsi*].

pernotata (Farner & Bohart, 1944) (p. 118) – species of *Stegomyia* (subgenus uncertain, determined by Reinert *et al*., 2009) [original combination: *Aedes* (*Stegomyia*) *pernotatus*].

perplexa Leicester, 1908 (p. 83) – species of *Stegomyia* (*Huangmyia*).

perryi (Belkin, 1962) (p. 336) – species of *Geoskusea* [original combination: *Aedes* (*Geoskusea*) *perryi*].

persistans Banks, 1906b (p. 996) – synonym of *Stegomyia* (*Stegomyia*) *aegypti* (Linnaeus, 1762) [original combination: *Stegomyia fasciata persistans*]; synonymy established before Knight & Stone, 1977.

pertinax (Grabham, 1906b) (p. 316) – species of *Ochlerotatus* (*Protoculex*) [original combination: *Aedes pertinax*].
Synonym: *quasiserratus* Theobald, 1907.

perventor (Cerqueira & Costa, 1946) (p. 135) – species of *Sallumia* [original combination: *Aedes* (*Ochlerotatus*) *perventor*].

peus Speiser, 1904 (p. 148) – rejected name, suppression by International Commission on Zoological Nomenclature, 1991; *nomen novum* for *affinis* Adams, 1903.

pexa (Colless, 1958) (p. 469) – species of *Downsiomyia* [original combination: *Aedes* (*Finlaya*) *pexus*].

peytoni (Reinert, 1972a) (p. 7) – species of *Ayurakitia* [original combination: *Aedes* (*Ayurakitia*) *peytoni*].

phaecasiatus (Marks, 1964a) (p. 64) – species of *Ochlerotatus* (subgenus uncertain, determined by Reinert *et al.*, 2008) [original combination: *Aedes* (*Ochlerotatus*) *phaecasiatus*].

phaeonotus (Arnell, 1976) (p. 56) – species of *Ochlerotatus* (*Ochlerotatus*) [original combination: *Aedes* (*Ochlerotatus*) *phaeonotus*].

philippii Dyar, 1924g (p. 129) – synonym of *Ochlerotatus albifasciatus* (Macquart, 1838b) [*nomen novum* for *vittatus* Philippi, 1865; original combination: *Aedes philippii*]; synonymy established before Knight & Stone, 1977.

phillipi (van Someren, 1949a) (p. 7) – species of *Zavortinkius* [original combination: *Aedes* (*Finlaya*) *phillipi*].

phoeniciae (Coluzzi & Sabatini, 1968) (p. 68) – species of *Acartomyia* [original combination: *Aedes* (*Ochlerotatus*) *mariae phoeniciae*]; specific status by Coluzzi *et al.*, 1970.

pholeocola (Linton & Harbach, 2013) (in Harbach *et al.*, 2013) (p. 162) – species of *Nyctomyia* Harbach, 2013 [original combination: *Nyx pholeocola*].

phyllolabis (Edwards, 1929b) (p. 324) – species of *Catageiomyia* [original combination: *Aedes* (*Aedimorphus*) *phyllolabis*].

pia Le Goff, Brengues & Robert, 2013 (p. 3) – species of *Stegomyia* (*Stegomyia*).

pillaii (Mattingly, 1958a) (p. 40) – species of *Rhinoskusea* [original combination: *Aedes* (*Rhinoskusea*) *pillaii*].

pingpaensis (Chang, 1965) (p. 397) – species of *Edwardsaedes* [original combination: *Aedes* (*Edwardsaedes*) *pingpaensis*].
Synonym: *antuensis* Su, Wang & Li, 1978.

pionips (Dyar, 1919b) (p. 19) – species of *Ochlerotatus* (subgenus uncertain, determined by Reinert *et al.*, 2008) [original combination: *Aedes pionips*].

pipersalatus (Giles, 1902) (p. 372) – species of *Aedimorphus* [original combination: *Stegomyia pipersalata*].
Synonym: *maculata* Theobald, 1905h.

pix Martini, 1935 (p. 55) – synonym of *Ochlerotatus* (*Culicelsa*) *taeniorhynchus* (Wiedemann, 1821) [original combination: *Aedes pix*]; synonymy by Belkin, 1977b.

plagosus (Marks, 1959b) (p. 21) – species of '*Ochlerotatus* (*Finlaya*)' *sensu auctorum* [original combination: *Aedes* (*Finlaya*) *plagosus*].

platylepida (Knight & Hull, 1951a) (p. 201) – species of *Scutomyia* [original combination: *Aedes* (?) *platylepidus*].

plumifera (King & Hoogstraal, 1946f) (p. 306) – species of *Phagomyia* [original combination: *Aedes* (*Finlaya*) *plumiferus*].

plutocraticus Dyar & Knab, 1907a (p. 11) – synonym of *Ochlerotatus* (*Ochlerotatus*) *tortilis* (Theobald, 1903g) [original combination: *Aedes plutocraticus*]; synonymy established before Knight & Stone, 1977.

podographicus (Dyar & Knab, 1906b) (p. 165) – species of '*Ochlerotatus* (*Protomacleaya*)' *sensu auctorum* [original combination: *Aedes podographicus*].

pogonurum (Edwards, 1936) (p. 52) – species of *Neomelaniconion* [original combination: *Aedes pogonurus*].

poicilia Theobald, 1903a (p. 283) – species of *Finlaya*.

poliochros Dyar, 1919b (p. 35) – synonym of *Ochlerotatus* (*Rusticoidus*) *provocans* (Walker, 1848) [original combination: *Aedes poliochros*]; transferred synonymy by synonymy of senior synonym *trichurus* (Dyar) with *provocans* (Walker) by Wood, 1977.

pollinctor (Graham, 1910) (p. 271) – species of *Zavortinkius* [original combination: *Stegomyia pollinctor*].

polyagrus Dyar, 1918b (p. 77) – synonym of *Ochlerotatus* (*Protoculex*) *nubilus* (Theobald, 1903a) [original combination: *Aedes polyagrus*]; synonymy by Komp, 1949.

polynesiensis (Marks, 1951) (p. 138) – species of *Stegomyia* (subgenus uncertain, determined by Reinert *et al.*, 2009) [original combination: *Aedes polynesiensis*].

portonovoensis (Tewari & Hiriyan, 1992a) (p. 123) – species of *Rhinoskusea* [original combination: *Aedes* (*Rhinoskusea*) *portonovoensis*].

portoricensis Ludlow, 1905c (p. 386) – synonym of *Ochlerotatus* (*Culicelsa*) *taeniorhynchus* (Wiedemann, 1821) [original combination: *Culex portoricensis*]; synonymy established before Knight & Stone, 1977.

postspiraculosus (Dobrotworsky, 1961a) (p. 261) – species of *Ochlerotatus* (*Pseudoskusea*) [original combination: *Aedes postspiraculosus*].

poweri Theobald, 1905g (p. 19) – species of *Stegomyia* (subgenus uncertain, determined by Reinert *et al.*, 2009).

praeteritus Séguy, 1923 (p. 207) – synonym of *Ochlerotatus berlandi* (Séguy, 1921) [original combination: *Aedes praeteritus*]; synonymy by Rioux & Arnold, 1955.

pretans Grossbeck, 1904b (p. 332) – synonym of *Ochlerotatus sticticus* (Meigen, 1838) [original combination: *Culex pretans*]; synonymy established before Knight & Stone, 1977.

pricei Dyar, 1917a (p. 16) – synonym of *Ochlerotatus fitchii* (Felt & Young, 1904) [original combination: *Aedes palustris* variety *pricei*]; synonymy established before Knight & Stone, 1977.

priestleyi Taylor, 1914c (p. 684) – synonym of *Molpemyia purpurea* Theobald, 1910c [original combination: *Calomyia priestleyi*]; synonymy by Taylor, 1944.

procax (Skuse, 1889) (p. 1742) – species of *Ochlerotatus* (subgenus uncertain, determined by Reinert *et al.*, 2008) [original combination: *Culex procax*].

prodotes Dyar, 1917c (p. 118) – synonym of *Ochlerotatus cataphylla* (Dyar, 1916b) [original combination: *Aedes prodotes*]; synonymy established before Knight & Stone, 1977.

prolixus Dyar, 1922a (p. 2) – synonym of *Ochlerotatus communis* (de Geer, 1776) [original combination: *Aedes prolixus*]; synonymy established before Knight & Stone, 1977.

prominens (Barraud, 1923c) (p. 228) – species of *Phagomyia* [original combination: *Finlaya prominens*].

provocans (Walker, 1848) (p. 7) – species of *Ochlerotatus* (*Rusticoidus*) [original combination: *Culex provocans*].
Synonyms: *trichurus* Dyar, 1904b; *cinereoborealis* Felt & Young, 1904; *pagetonotum* Dyar & Knab, 1909d; *poliochros* Dyar, 1919b.

pseudalbolineata (Brug, 1939) (p. 103) – species of *Scutomyia* [original combination: *Aedes* (*Stegomyia*) *pseudalbolineatus*].

pseudalbopicta Borel, 1928 (p. 85) – species of *Stegomyia* (subgenus uncertain, determined by Reinert *et al.*, 2009).

pseudoafricana (Chwatt, 1949) (p. 808) – species of *Stegomyia* (subgenus uncertain, determined by Reinert *et al.*, 2009) [original combination: *Aedes* (*Stegomyia*) *pseudoafricanus*].

pseudodiantaeus Smith, 1952 (p. 21) – synonym of *Ochlerotatus decticus* (Howard, Dyar & Knab, 1917) [original combination: *Aedes* (*Ochlerotatus*) *pseudodiantaeus*]; synonymy by Carpenter & LaCasse, 1955.

pseudodominicii (Komp, 1936a) (p. 75) – species of *Howardina* [original combination: *Aedes* (*Soperia*) *pseudodominicii*].
Synonym: *delpontei* Martínez & Prosen, 1955.

pseudoesoensis Yamaguti & Tamaboko, 1954 (p. 418) – synonym of *Aedes esoensis* Yamada, 1921; synonymy by Tanaka *et al.*, 1975b.

pseudonigeria Theobald, 1910c (p. 166) – species of *Stegomyia* (subgenus uncertain, determined by Reinert *et al.*, 2009).
Synonym: *wellmanii* Theobald, 1910c.

pseudonivea (Theobald, 1905f) (p. 75) – species of *Downsiomyia* [original combination: *Stegomyia pseudonivea*].

pseudonormanensis (Marks, 1949) (p. 26) – species of *Ochlerotatus* (subgenus uncertain, determined by Reinert *et al.*, 2008) [original combination: *Aedes* (*Ochlerotatus*) *pseudonormanensis*].

pseudonummatus (Reinert, 1973d) (p. 29) – species of *Tewarius* [original combination: *Aedes* (*Diceromyia*) *pseudonummatus*].

pseudorusticus Callot, 1945 (p. 31) – synonym of *Ochlerotatus* (*Rusticoidus*) *refiki* (Medschid, 1928) [original combination: *Aedes* (*Ochlerotatus*) *pseudorusticus*]; synonymy by Callot *in litt.* (Stone *et al.*, 1959).

pseudoscutellaris Theobald, 1910b (p. 156) – species of *Stegomyia* (subgenus uncertain, determined by Reinert *et al.*, 2009).

pseudostenoetrus Theobald, 1910c (p. 343) – synonym of *Aedimorphus stenoetrus* (Theobald, 1907) [original combination: *Culex pseudostenoetrus*]; synonymy established before Knight & Stone, 1977.

pseudotaeniatus (Giles, 1901a) (p. 607) – species of *Collessius* (*Alloeomyia*) [original combination: *Culex pseudotaeniatus*].

pseudotarsalis (van Someren, 1946a) (p. 19) – species of *Catageiomyia* [original combination: *Aedes* (*Aedimorphus*) *pseudotarsalis*].

pseudovigilax Theobald, 1907 (p. 382) – synonym of *Ochlerotatus* (*Empihals*) *vigilax* (Skuse, 1889) [original combination: *Culicelsa pseudovigilax*]; synonymy established before Knight & Stone, 1977.

pubescens (Edwards, 1925) (p. 264) – species of *Aedimorphus* [original combination: *Aedes* (*Aedimorphus*) *pubescens*].

pulcherrima Taylor, 1919 (p. 830) – synonym of *Stegomyia* (*Stegomyia*) *aegypti* (Linnaeus, 1762) [original combination: *Mimeteomyia pulcherrima*]; synonymy by Knight & Marks, 1952.

pulchrithorax (Edwards, 1939a) (p. 17) – species of *Vansomerenis* [original combination: *Aedes* (*Finlaya*) *pulchrithorax*].

pulchriventer (Giles, 1901a) (p. 608) – species of *Gilesius* [original combination: *Culex pulchriventer*].
Synonym: *himalayana* Giles, 1904b.

pulcritarsis (Rondani, 1872) (p. 31) – species of *Ochlerotatus* (subgenus uncertain, determined by Reinert *et al.*, 2008) [original combination: *Culex pulcritarsis*].
Synonyms: *leucacanthus* Loew, 1873; *simici* Baranoff, 1927.

pullatus (Coquillett, 1904c) (p. 168) – species of *Ochlerotatus* (subgenus uncertain, determined by Reinert *et al.*, 2008) [original combination: *Culex pullatus*].
Synonyms: *acrophilus* Dyar, 1917d; *jugorum* Villeneuve, 1919; *metalepticus* Dyar, 1920c; *gallii* Martini, 1920a; *pearyi* Dyar & Shannon, 1925d; *seguyi* Apfelbeck, 1929.

pulverulentus (Edwards, 1922a) (p. 273) – species of *Indusius* [original combination: *Aedes pulverulentus*].

punctatus Meigen, 1804 (p. 6) – synonym of *Ochlerotatus caspius* (Pallas, 1771) [original combination: *Culex punctatus*]; synonymy established before Knight & Stone, 1977.

punctifemoris (Ludlow, 1921) (p. 690) – species of *Aedimorphus* [original combination: *Stegomyia punctifemore* [*sic*]].

punctipes (Edwards, 1921c) (p. 77) – species of *Petermattinglyius* (*Petermattinglyius*) [original combination: *Aedes* (*Skusea*) *punctipes*].

punctissimus Barraud, 1928b (p. 360) – synonym of *Petermattinglyius* (*Petermattinglyius*) *iyengari* (Edwards, 1923b) [original combination: *Aedes* (*Skusea*) *punctissimus*]; synonymy established before Knight & Stone, 1977.

punctithorax Theobald, 1910c (p. 205) – synonym of *Catageiomyia punctothoracis* (Theobald, 1909b) [original combination: *Aedimorphus punctithorax*]; synonymy established before Knight & Stone, 1977.

punctocostale (Theobald, 1909b) (p. 14) – species of *Neomelaniconion* [original combination: *Banksinella punctocostalis*].

punctodes (Dyar, 1922a) (p. 1) – species of *Ochlerotatus* (subgenus uncertain, determined by Reinert *et al.*, 2008) [original combination: *Aedes punctodes*].

punctor (Kirby, 1837) (p. 309) – species of *Ochlerotatus* (subgenus uncertain, determined by Reinert *et al.*, 2008) [original combination: *Culex punctor*].
Synonyms: *implacabilis* Walker, 1848; *fusculus* Zetterstedt, 1850; *auroides* Felt, 1905a; *wahlgreni* Theobald, 1910c; *alineata* Schneider, 1913; *haplolineata* Schneider, 1913; *meigenanus* Dyar, 1921e.

punctothoracis (Theobald, 1909b) (p. 4) – species of *Catageiomyia* [original combination: *Aedimorphus punctothoracis*].
Synonym: *punctithorax* Theobald, 1910c.

pungens Robineau-Desvoidy, 1827 (p. 407) – synonym of *Ochlerotatus* (*Rusticoidus*) *rusticus* (Rossi, 1790) [original combination: *Culex pungens*]; synonymy established before Knight & Stone, 1977.

purii Barraud, 1931a (p. 226) – synonym of *Stegomyia* (*Heteraspidion*) *craggi* Barraud, 1923c [original combination: *Aedes* (*Stegomyia*) *purii*]; synonymy by Barraud, 1934.

purpuraceus (Brug, 1932) (p. 77) – species of *Ochlerotatus* (*Gilesia*) [original combination: *Aedes* (*Ochlerotatus*) *purpuraceus*].

purpurea Theobald, 1910c (p. 479) – species of *Molpemyia*.
Synonym: *priestleyi* Taylor, 1914c.

purpureifemur (Marks, 1959a) (p. 132) – species of *Ochlerotatus* (subgenus uncertain, determined by Reinert *et al.*, 2008) [original combination: *Aedes* (*Ochlerotatus*) *purpureifemur*].

purpureipes (Aitken, 1941) (p. 82) – species of *Kompia* [original combination: *Aedes* (*Kompia*) *purpureipes*].

purpuriventris (Edwards, 1926a) (p. 113) – species of *Ochlerotatus* (*Pholeomyia*) [original combination: *Aedes* (*Ochlerotatus*) *purpuriventris*].

Qq

quadratimaculatus Macquart, 1834 (p. 34) – synonym of *Ochlerotatus* (*Rusticoidus*) *rusticus* (Rossi, 1790) [original combination: *Culex quadratimaculatus*]; synonymy established before Knight & Stone, 1977.

quadricincta Barraud, 1923c (p. 227) – subspecies of *Stegomyia* (*Heteraspidion*) *annandalei* Theobald, 1910a [original combination: *Stegomyia annandalei* variety *quadricincta*]; subspecific status by Harbach & Howard, 2007.

quadripunctis (Ludlow, 1910) (in Theobald, 1910c) (p. 608) – species of *Mucidus* (*Pardomyia*) [original combination: *Pardomyia quadripunctis*].

quadrivittata (Coquillett, 1902d) (p. 293) – species of *Howardina* [original combination: *Culex quadrivittatus*].

quartus Martini, 1920a (p. 128) – synonym of *Ochlerotatus annulipes* (Meigen, 1830b) [original combination: *Aedes quartus*]; synonymy established before Knight & Stone, 1977.

quasiferinus (Mattingly, 1961) (p. 32) – species of *Mucidus* (*Mucidus*) [original combination: *Aedes* (*Mucidus*) *quasiferinus*].

quasinigritia Ludlow, 1911b (p. 129) – synonym of *Stegomyia albopicta* (Skuse, 1895); synonymy established before Knight & Stone, 1977.

quasirubithorax (Theobald, 1910c) (pp. 327, 348) – species of '*Ochlerotatus* (*Finlaya*)' *sensu auctorum* [original combination: *Culex quasirubithorax*].

quasirusticus (Torres Cañamares, 1951) (p. 79) – species of *Ochlerotatus* (*Rusticoidus*) [original combination: *Aedes* (*Ochlerotatus*) *quasirusticus*].

quasiscutellaris (Farner & Bohart, 1944) (p. 120) – species of *Stegomyia* (subgenus uncertain, determined by Reinert *et al.*, 2009) [original combination: *Aedes* (*Stegomyia*) *quasiscutellaris*].

quasiserratus Theobald, 1907 (p. 465) – synonym of *Ochlerotatus* (*Protoculex*) *pertinax* (Grabham, 1906b) [original combination: *Protoculex quasiserratus*]; synonymy by Howard *et al.*, 1917.

quasiunivittatus (Theobald, 1901c) (p. 32) – species of *Aedimorphus* [original combination: *Culex quasiunivittatus*].

quaylei Dyar & Knab, 1906d (p. 202) – synonym of *Ochlerotatus dorsalis* (Meigen, 1830b) [original combination: *Aedes quaylei*]; synonymy by Wood *et al.*, 1979.

queenslandensis Theobald, 1901a (p. 297) – synonym of *Stegomyia* (*Stegomyia*) *aegypti* (Linnaeus, 1762) [original combination: *Stegomyia fasciata* variety *queenslandensis*]; synonymy by Lee *et al.*, 1987.

queenslandis Strickland, 1911b (p. 179) – synonym of *Dobrotworskyius rubrithorax* (Macquart, 1850) [original combination: *Culicelsa queenslandis*]; synonymy by Lee *et al.*, 1982.

quinquelineata (Edwards, 1922d) (p. 93) – species of *Rampamyia* [original combination: *Aedes* (*Finlaya*) *quinquelineatus*].

quinquepunctata Theobald, 1913d (p. 598) – synonym of *Catageiomyia argenteopunctata* (Theobald, 1901a) [original combination: *Aedimorphus quinquepunctata* [*sic*]]; synonymy established before Knight & Stone, 1977.

Rr

ramachandrai (Reuben, 1967) (p. 234) – species of *Dendroskusea* [original combination: *Aedes* (*Diceromyia*) *ramachandrai*].

ramirezi (Vargas & Downs, 1950) (p. 164) – species of *Aztecaedes* [original combination: *Aedes* (*Gualteria*) *ramirezi*].

ratcliffei (Marks, 1959a) (p. 123) – species of *Ochlerotatus* (subgenus uncertain, determined by Reinert *et al.*,

2008) [original combination: *Aedes* (*Ochlerotatus*) *ratcliffei*].

raymondi (del Ponte, Castro & Garcia, 1951) (p. 235) – species of *Ochlerotatus* (*Ochlerotatus*) [original combination: *Aedes* (*Ochlerotatus*) *raymondi*].

reali (Hamon & Adam, 1959a) (p. 351) – species of *Catageiomyia* [original combination: *Aedes* (*Aedimorphus*) *reali*].

refiki (Medschid, 1928) (p. 310) – species of *Ochlerotatus* (*Rusticoidus*) [original combination: *Aedes refiki*].
Synonyms: *stampari* Apfelbeck, 1929; *pseudorusticus* Callot, 1945.

reginae (Edwards, 1922a) (p. 272) – species of *Dendroskusea* [original combination: *Aedes* (*Skusea*) *reginae*].

reinerti (Rattanarithikul & Harrison, 1988) (p. 78) – species of *Hulecoeteomyia* [original combination: *Aedes* (*Finlaya*) *reinerti*].

rempeli (Vockeroth, 1954) (p. 112) – species of *Ochlerotatus* (subgenus uncertain, determined by Reinert *et al.*, 2008) [original combination: *Aedes* (*Ochlerotatus*) *rempeli*].

reptans Meigen, 1804 (p. 3) – synonym of *Ochlerotatus cantans* (Meigen, 1818) [original combination: *Culex reptans*]; synonymy established before Knight & Stone, 1977.

reubenae (Tewari & Hiriyan, 1992b) (p. 164) – species of *Tewarius* [original combination: *Aedes* (*Diceromyia*) *reubenae*].

rhecter Dyar, 1921d (p. 51) – synonym of *Aedimorphus dalzieli* (Theobald, 1910c) [original combination: *Aedes* (*Ecculex*) *rhecter*]; synonymy by Edwards, 1941.

rhungkiangensis (Chang & Chang, 1974) (p. 349) – species of *Stegomyia* (subgenus uncertain, determined by Reinert *et al.*, 2009) [original combination: *Aedes* (*Stegomyia*) *rhungkiangensis*].

rhyacophilus (da Costa Lima, 1933) (p. 404) – species of *Ochlerotatus* (*Ochlerotatus*) [original combination: *Aedes* (*Ochlerotatus*) *rhyacophilus*].

rickenbachi (Hamon & Adam, 1959b) (p. 151) – species of *Aedimorphus* [original combination: *Aedes* (*Aedimorphus*) *rickenbachi*].
Synonym: *hamoni* Mattingly, 1963.

rimandoi (Basio, 1971) (p. 52) – *nomen dubium*, *Stegomyia*; status by Huang, 1979 [original combination: *Aedes* (*Stegomyia*) *rimandoi*].

riparioides (Su & Zhang, 1987) (p. 450) – species of *Ochlerotatus* (subgenus uncertain, determined by Reinert *et al.*, 2008) [original combination: *Aedes* (*Ochlerotatus*) *riparioides*].

riparius (Dyar & Knab, 1907d) (p. 213) – species of *Ochlerotatus* (subgenus uncertain, determined by Reinert *et al.*, 2008) [original combination: *Aedes riparius*].
Synonym: *semicantans* Martini, 1920b.

riversi (Bohart & Ingram, 1946a) (p. 50) – species of *Stegomyia* (subgenus uncertain, determined by Reinert *et al.*, 2009) [original combination: *Aedes* (*Stegomyia*) *riversi*].

rizali (Banks, 1906b) (p. 999) – species of *Hulecoeteomyia* [original combination: *Culex rizali*].
Synonyms: *abadsantosi* Baisas, 1946c; *burgosi* Baisas, 1946c.

roai (Belkin, 1962) (p. 353) – species of '*Ochlerotatus* (*Finlaya*)' *sensu auctorum* [original combination: *Aedes* (*Finlaya*) *roai*].

robinsoni (Belkin, 1962) (p. 447) – species of *Stegomyia* (*Actinothrix*) [original combination: *Aedes* (*Stegomyia*) *robinsoni*].

rodhaini Brunhes & Boussès, 2017 (in Brunhes *et al.*, 2017) (p. 364) – species of *Paulianius*.

rossicus Dolbeškin, Gorickaja & Mitrofanova, 1930 (p. 253) – species of *Aedes*.
Synonym: *tarnogradskii* Martini, 1930.

rossii Giles, 1899 (p. 64) – synonym of *Stegomyia* (*Stegomyia*) *aegypti* (Linnaeus, 1762) [original combination: *Culex rossii*]; synonymy established before Knight & Stone, 1977.

rostochiensis Martini, 1920b (p. 204) – synonym of *Ochlerotatus cataphylla* (Dyar, 1916b) [original combination: *Aedes* (*Ochlerotatus*) *rostochiensis*]; synonymy by Dahl & White, 1978.

rotana (Bohart & Ingram, 1946b) (p. 39) – species of *Stegomyia* (*Bohartius*) [original combination: *Aedes* (*Stegomyia*) *rotanus*].

rotumae (Belkin, 1962) (p. 472) – species of *Stegomyia* (subgenus uncertain, determined by Reinert *et al.*, 2009) [original combination: *Aedes* (*Stegomyia*) *rotumae*].

rozeboomi Vargas, 1941b (p. 394) – synonym of *Ochlerotatus* (*Chrysoconops*) *bimaculatus* (Coquillett, 1902c) [original combination: *Aedes* (*Ochlerotatus*) *rozeboomi*]; synonymy by Ross, 1943.

rubiginosus (Belkin, 1962) (p. 346) – species of '*Ochlerotatus* (*Finlaya*)' *sensu auctorum* [original combination: *Aedes* (*Finlaya*) *rubiginosus*].

rubrithorax (Macquart, 1850) (p. 313) – species of *Dobrotworskyius* [original combination: *Culex rubrithorax*].
Synonyms: *similis* Strickland, 1911a; *demansis* Strickland, 1911b; *queenslandis* Strickland, 1911b; *cumpstoni* Taylor, 1914c; *hybrida* Taylor, 1916.

rufus Gimmerthal, 1845 (p. 295) – synonym of *Aedes cinereus* Meigen, 1818; synonymy established before Knight & Stone, 1977.

rupestris (Dobrotworsky, 1959) (p. 136) – species of *Dobrotworskyius* [original combination: *Aedes rupestris*].

rusticus (Rossi, 1790) (p. 333) – species of *Ochlerotatus* (*Rusticoidus*) [original combination: *Culex rusticus*].
Synonyms: *maculatus* Meigen, 1804; *musicus* Leach, 1825; *pungens* Robineau-Desvoidy, 1827; *quadratimaculatus* Macquart, 1834; *diversus* Theobald, 1901c; *luteovittata* Theobald, 1901c.

ruwenzori (Haddow & van Someren, 1950) (p. 281) – species of *Stegomyia* (subgenus uncertain, determined by Reinert *et al.*, 2009) [original combination: *Aedes* (*Stegomyia*) *ruwenzori*].

Ss

sagax (Skuse, 1889) (p. 1744) – species of *Ochlerotatus* (subgenus uncertain, determined by Reinert *et al.*, 2008) [original combination: *Culex sagax*].
Synonym: *wilsoni* Taylor, 1919.

saimedres (Huang, 1988b) (p. 10) – species of *Stegomyia* (subgenus uncertain, determined by Reinert *et al.*, 2009) [original combination: *Aedes* (*Stegomyia*) *saimedres*].

saipanensis (Stone, 1945b) (p. 66) – species of *Stegomyia* (*Bohartius*) [original combination: *Aedes* (*Stegomyia*) *saipanensis*].

salinellus Edwards, 1921b (p. 52) – synonym of *Ochlerotatus leucomelas* (Meigen, 1804); synonymy established before Knight & Stone, 1977.

salinus Ficalbi, 1896 (p. 29) – synonym of *Ochlerotatus detritus* (Haliday, 1833) [original combination: *Culex salinus*]; synonymy established before Knight & Stone, 1977.

sallumae (González & Reyes, 2017) (in González *et al.*, 2017) (p. 35) – species of *Ochlerotatus* (subgenus uncertain) [original combination: *Aedes* (*Ochlerotatus*) *sallumae*].

samarensis Ludlow, 1903 (p. 138) – synonym of *Stegomyia albopicta* (Skuse, 1895) [original combination: *Stegomyia scutellaris samarensis*]; synonymy established before Knight & Stone, 1977.

samoana Grünberg, 1913 (p. 130) – species of *Finlaya*.

sampi (Huang, 2004) (p. 49) – species of *Stegomyia* (*Mukwaya*) [original combination: *Aedes* (*Stegomyia*) *sampi*].

sandrae (Zavortink, 1972) (p. 51) – species of '*Ochlerotatus* (*Protomacleaya*)' *sensu auctorum* [original combination: *Aedes* (*Protomacleaya*) *sandrae*].

sangitee Sathe & Girhe, 2002 (p. 198) – *nomen nudum*, *Ochlerotatus*; unavailable name [published after 1999, fails to conform to Articles 16.4 and 72.3 of the *Code*, which require fixation of a name-bearing type and a statement naming the collection in which it is deposited (International Commission on Zoological Nomenclature, 1999)].

sangiti Girhe & Sathe, 2001 (p. 46) – *nomen nudum*, *Stegomyia*; unavailable name [published after 1999, fails to conform to Articles 16.4 and 72.3 of the *Code*, which require fixation of a name-bearing type and a statement naming the collection in which it is deposited (International Commission on Zoological Nomenclature, 1999)].

sansoni Dyar & Knab, 1909c (p. 102) – synonym of *Ochlerotatus excrucians* (Walker, 1856a) [original combination: *Aedes sansoni*]; synonymy established before Knight & Stone, 1977.

saperoi (Knight, 1946) (p. 271) – species of *Downsiomyia* [original combination: *Aedes* (*Finlaya*) *saperoi*].

sapiens (Marks, 1964a) (p. 66) – species of *Ochlerotatus* (subgenus uncertain, determined by Reinert *et al.*, 2008) [original combination: *Aedes* (*Ochlerotatus*) *sapiens*].

sasai Tanaka, Mizusawa & Saugstad, 1975b (p. 41) – species of *Aedes*.

savoryi (Bohart, 1957) (p. 31) – species of *Tanakaius* [original combination: *Aedes* (*Finlaya*) *savoryi*].

saxicola (Edwards, 1922a) (p. 262) – synonym of *Hulecoeteomyia fluviatilis* Leicester, 1908 [original combination: *Aedes saxicola*]; originally proposed as a replacement name for *fluviatilis* Leicester, 1908, *non* Lutz, 1904, when both were included in *Aedes* Meigen (before Reinert, 2000d and Reinert *et al.*, 2006).

scanloni (Reinert, 1970) (p. 17) – species of *Petermattinglyius* (*Petermattinglyius*) [original combination: *Aedes* (*Diceromyia*) *scanloni*].

scapularis (Rondani, 1848) (p. 109) – species of *Ochlerotatus* (*Ochlerotatus*) [original combination: *Culex scapularis*].
Synonyms: *camposanus* Dyar, 1918e; *confirmatus* Lynch Arribálzaga, 1891b; *hemisurus* Dyar & Knab, 1906d; *indolescens* Dyar & Knab, 1907a.

scatophagoides Theobald, 1901a (p. 277) – species of *Mucidus* (*Mucidus*).

schicki (Zavortink, 1972) (p. 46) – species of '*Ochlerotatus* (*Protomacleaya*)' *sensu auctorum* [original combination: *Aedes* (*Protomacleaya*) *schicki*].

schizopinax (Dyar, 1929a) (p. 1) – species of *Ochlerotatus* (subgenus uncertain, determined by Reinert *et al.*, 2008) [original combination: *Aedes schizopinax*].

schlosseri (Belkin, 1962) (p. 382) – species of *Finlaya* [original combination: *Aedes* (*Finlaya*) *schlosseri*].

schroederi (Schick, 1970a) (p. 88) – species of '*Ochlerotatus* (*Protomacleaya*)' *sensu auctorum* [original combination: *Aedes* (*Finlaya*) *schroederi*].

schtakelbergi (Shingarev, 1928) (p. 52) – species of *Ochlerotatus* (subgenus uncertain, determined by Reinert *et al.*, 2008) [original combination: *Aedes* (*Ochlerotatus*) *schtakelbergi*].

schwetzi (Edwards, 1926a) (p. 127) – species of *Stegomyia* (subgenus uncertain, determined by Reinert *et al.*, 2009) [original combination: *Aedes* (*Stegomyia*) *schwetzi*].

scutellalbum (Boshell-Manrique, 1939) (p. 309) – species of *Ochlerotatus* (subgenus uncertain, determined by Reinert *et al.*, 2008) [original combination: *Aedes* (*Finlaya*) *scutellalbum*].

scutellaris (Walker, 1859) (p. 77) – species of *Stegomyia* (subgenus uncertain, determined by Reinert *et al.*, 2009) [original combination: *Culex scutellaris*].
Synonyms: *variegatus* Doleschall, 1858; *zonatipes* Walker, 1861.

scutoscripta (Bohart & Ingram, 1946b) (p. 27) – species of *Stegomyia* (subgenus uncertain, determined by Reinert *et al.*, 2009) [original combination: *Aedes* (*Stegomyia*) *scutoscriptus*].

seampi (Huang, 1974) (p. 137) – species of *Stegomyia* (*Actinothrix*) [original combination: *Aedes* (*Stegomyia*) *seampi*].

seatoi (Huang, 1969) (p. 234) – species of *Stegomyia* (subgenus uncertain, determined by Reinert *et al.*, 2009) [original combination: *Aedes* (*Stegomyia*) *seatoi*].

sedaensis (Lei, 1989a) (p. 6) – species of *Ochlerotatus* (subgenus uncertain, determined by Reinert *et al.*, 2008) [original combination: *Aedes* (*Ochlerotatus*) *sedaensis*].

segermanae (Huang, 1997) (p. 57) – species of *Stegomyia* (subgenus uncertain, determined by Reinert *et al.*, 2009) [original combination: *Aedes* (*Stegomyia*) *segermanae*].

seguyi Apfelbeck, 1929 (p. 288) – synonym of *Ochlerotatus pullatus* (Coquillett, 1904c) [original combination: *Aedes* (*Ochlerotatus*) *seguyi*]; synonymy established before Knight & Stone, 1977.

semicantans Martini, 1920b (p. 205) – synonym of *Ochlerotatus riparius* (Dyar & Knab, 1907d) [original combination: *Aedes* (*Ochlerotatus*) *semicantans*]; synonymy established before Knight & Stone, 1977.

semlikiensis (van Someren, 1950b) (p. 40) – species of *Aedimorphus* [original combination: *Aedes* (*Aedimorphus*) *semlikiensis*].

senegalensis Theobald, 1903d (p. i) – synonym of *Catageiomyia irritans* (Theobald, 1901d); synonymy established before Knight & Stone, 1977.

senyavinensis (Knight & Hurlbut, 1949) (p. 27) – species of *Aedimorphus* [original combination: *Aedes* (*Aedimorphus*) *senyavinensis*].

seoulensis (Yamada, 1921) (p. 61) – species of *Hopkinsius* (*Yamada*) [original combination: *Aedes seoulensis*].

septemstriata (Dyar & Knab, 1907a) (p. 10) – species of *Howardina* [original combination: *Aedes septemstriatus*].

serafini (Szadziewski, 1998) (p. 240) – fossil species of *Ochlerotatus* Lynch Arribálzaga, 1891a,b [original combination: *Aedes* (*Finlaya*) *serafini*].

sergievi (Danilov, Markovich & Proskuryakova, 1978) (p. 44) – species of *Ochlerotatus* (subgenus uncertain, determined by Reinert *et al.*, 2008) [original combination: *Aedes* (*Ochlerotatus*) *sergievi*].

serratus (Theobald, 1901c) (p. 45) – species of *Ochlerotatus* (*Protoculex*) [original combination: *Culex serratus*].
Synonyms: *mathisi* Neveu-Lemaire, 1902b; *meridionalis* Dyar & Knab, 1906d.

serus Martini, 1920a (p. 96) – synonym of *Ochlerotatus* (*Woodius*) *diantaeus* (Howard, Dyar & Knab, 1913) [original combination: *Aedes serus*]; synonymy established before Knight & Stone, 1977.

sexlineata (Theobald, 1901a) (p. 308) – species of *Howardina* [original combination: *Stegomyia sexlineata*].
Synonym: *dominicii* Rangel & Romero Sierra, 1907.

seychellensis Theobald, 1912b (p. 83) – synonym of *Polyleptiomyia albocephala* (Theobald, 1903c) [original combination: *Reedomyia seychellensis*]; synonymy by Edwards, 1941.

shannoni (Vargas & Downs, 1950) (p. 161) – species of *Ochlerotatus* (subgenus uncertain, determined by Reinert *et al.*, 2008) [original combination: *Aedes* (*Ochlerotatus*) *shannoni*].

shehzadae (Qutubuddin, 1972) (p. 119) – species of *Downsiomyia* [original combination: *Aedes* (*Finlaya*) *shehzadae*].

sherki (Knight, 1948a) (p. 645) – species of *Hulecoeteomyia* [original combination: *Aedes* (*Finlaya*) *sherki*].

shintienensis (Tsai & Lien, 1950) (species) (p. 177) – subspecies of *Hulecoeteomyia japonica* (Theobald, 1901a) [original combination: *Aedes* (*Finlaya*) *shintienensis*]; subspecific status by Tanaka *et al.*, 1979.

shortti (Barraud, 1923e) (p. 405) – species of *Collessius* (*Collessius*) [original combination: *Finlaya shortti*].

sibirica (Danilov & Filippova, 1978) (p. 170) – species of *Stegomyia* (subgenus uncertain, determined by Reinert *et al.*, 2009) [original combination: *Aedes* (*Stegomyia*) *sibiricus*].

siculus Robineau-Desvoidy, 1827 (p. 406) – synonym of *Ochlerotatus caspius* (Pallas, 1771) [original combination: *Culex siculus*]; synonymy established before Knight & Stone, 1977.

sierrensis (Ludlow, 1905b) (p. 231) – species of *Jarnellius* [original combination: *Taeniorhynchus sierrensis*].

silvestris (Dobrotworsky, 1961b) (p. 248) – species of *Ochlerotatus* (subgenus uncertain, determined by Reinert *et al.*, 2008) [original combination: *Aedes silvestris*].
Synonym: *waterhousei* Dobrotworsky, 1960a.

simanini (Gutsevich, 1966) (p. 457) – species of *Ochlerotatus* (subgenus uncertain, determined by Reinert *et al.*, 2008) [original combination: *Aedes* (*Ochlerotatus*) *simanini*].

simici Baranoff, 1927 (p. 15) – synonym of *Ochlerotatus pulcritarsis* (Rondani, 1872) [original combination: *Aedes simici*]; synonymy established before Knight & Stone, 1977.

similis Strickland, 1911a (p. 132) – synonym of *Dobrotworskyius rubrithorax* (Macquart, 1850) [original combination: *Culicelsa similis*]; synonymy established before Knight & Stone, 1977.

simlensis (Edwards, 1922a) (p. 269) – species of *Himalaius* [original combination: *Aedes* (*Finlaya*) *simlensis*].

simplex (Theobald, 1903a) (p. 297) – species of *Cancraedes* [original combination: *Ficalbia simplex*].

simpsoni Theobald, 1905c (p. 224) – species of *Stegomyia* (*Mukwaya*).

simulans (Newstead & Carter, 1911b) (p. 240) – species of *Elpeytonius* [original combination: *Reedomyia simulans*].

simulatus Barraud, 1931b (p. 611) – synonym of *Collessius* (*Collessius*) *elsiae* (Barraud, 1923a) [original combination: *Aedes* (*Finlaya*) *simulatus*]; synonymy by Barraud, 1934.

sinensis (Chow, 1950) (p. 283) – species of *Downsiomyia* [original combination: *Aedes* (*Finlaya*) *sinensis*].

sinkiangensis (Hsiao, 1977) (p. 451) – species of *Ochlerotatus* (subgenus uncertain, determined by Reinert *et al.*, 2008) [original combination: *Aedes* (*Ochlerotatus*) *sinkiangensis*].

sintoni (Barraud, 1924b) (p. 967) – species of '*Ochlerotatus* (*Finlaya*)' *sensu auctorum* [original combination: *Finlaya sintoni*].

siphonalis Grossbeck, 1904a (p. 332) – synonym of *Ochlerotatus excrucians* (Walker, 1856a) [original combination: *Culex siphonalis*]; synonymy established before Knight & Stone, 1977.

smithburni (van Someren, 1950b) (p. 39) – species of *Catageiomyia* [original combination: *Aedes (Aedimorphus) smithburni*].
soleata (Edwards, 1924) (in Haworth, 1924) (p. 198) – species of *Stegomyia* (subgenus uncertain, determined by Reinert *et al.*, 2009) [original combination: *Aedes (Stegomyia) soleatus*].
sollicitans (Walker, 1856a) (p. 427) – species of *Ochlerotatus (Culicelsa)* [original combination: *Culex sollicitans*].
solomonis (Stone & Bohart, 1944) (p. 208) – species of *Finlaya* [original combination: *Aedes (Finlaya) solomonis*].
sorsogonensis (Bañez & Jueco, 1966) (p. 195) – species of *Finlaya* [original combination: *Aedes (Finlaya) sorsogonensis*].
spencerii (Theobald, 1901c) (p. 99) – species of *Ochlerotatus* (subgenus uncertain, determined by Reinert *et al.*, 2008) [original combination: *Culex spencerii*].
spilotus (Marks, 1963b) (p. 31) – species of *Ochlerotatus (Lepidokeneon)* [original combination: *Aedes (Ochlerotatus) spilotus*].
spinosa (Berlin, 1969a) (p. 84) – species of *Howardina* [original combination: *Aedes (Howardina) spinosus*].
spinosipes (Edwards, 1922d) (p. 92) – species of *Macleaya (Chaetocruiomyia)* [original combination: *Aedes (Chaetocruiomyia) spinosipes*].
squamiger (Coquillett, 1902c) (p. 85) – species of *Ochlerotatus* (subgenus uncertain, determined by Reinert *et al.*, 2008) [original combination: *Culex squamiger*].
Synonym: *deniedmannii* Ludlow, 1904b.
stampari Apfelbeck, 1929 (p. 286) – synonym of *Ochlerotatus (Rusticoidus) refiki* (Medschid, 1928) [original combination: *Aedes (Ochlerotatus) stampari*]; synonymy by Stone, 1963 (citing Peus, *in litt.*).
stanleyi (Peters, 1963a) (p. 1) – species of '*Ochlerotatus (Finlaya)*' *sensu auctorum* [original combination: *Aedes (Finlaya) stanleyi*].
stegomyina Stackelberg & Monchadskii, 1926 (in Montchadsky, 1926) (p. 151) – synonym of *Ochlerotatus pulcritarsis asiaticus* (Edwards, 1926c) [original combination: *Aedes (Ochlerotatus) pulchritarsis* [*sic*] variety *stegomyina*]; synonymy established before Knight & Stone, 1977.
stenei (Thompson, 1956) (p. 196) – species of *Howardina* [original combination: *Aedes (Howardina) stenei*].
stenoetrus (Theobald, 1907) (p. 395) – species of *Aedimorphus* [original combination: *Culex stenoetrus*].
Synonym: *pseudostenoetrus* Theobald, 1910c.
stenoscutus (Edwards, 1912c) (p. 22) – synonym of *Aedimorphus africanus* (Theobald, 1909b) [*nomen novum* for *africanus* (Theobald, 1909b); original combination: *Ochlerotatus minutus* variety *stenoscutus*]; synonymy with transfer of *africanus* (Newstead, 1907), as synonym of *tarsalis* Newstead, 1907, from *Aedimorphus* to *Catageiomyia* by Reinert *et al.* (2009).
stevensoni (Barraud, 1923b) (p. 219) – species of *Phagomyia* [original combination: *Finlaya stevensoni*].
sticticus (Meigen, 1838) (p. 1) – species of *Ochlerotatus* (subgenus uncertain, determined by Reinert *et al.*, 2008) [original combination: *Culex sticticus*].
Synonyms: *concinnus* Stephens, 1846; *hirsuteron* Theobald, 1901c; *sylvae* Theobald, 1901c; *aestivalis* Dyar, 1904b; *pretans* Grossbeck, 1904b; *aldrichi* Dyar & Knab, 1908; *gonimus* Dyar & Knab, 1918b; *vinnipegensis* Dyar, 1919b; *dorso-vittatus* Villeneuve, 1919; *lesnei* Séguy, 1920; *paradiantaeus* Apfelbeck, 1929.
stigmaticus (Edwards, 1922d) (p. 78) – species of *Ochlerotatus (Chrysoconops)* [original combination: *Aedes (Ochlerotatus) stigmaticus*].
stimulans (Walker, 1848) (p. 4) – species of *Ochlerotatus* (subgenus uncertain, determined by Reinert *et al.*, 2008) [original combination: *Culex stimulans*].
Synonyms: *subcantans* Felt, 1905a; *classicus* Dyar, 1920e; *mississippii* Dyar, 1920e.
stokesi (Evans, 1929) (p. 522) – species of *Albuginosus* [original combination: *Aedes (Aedimorphus) stokesi*].
Synonym: *occidentalis* Evans, 1926.
stonei (Knight & Laffoon, 1946) (p. 208) – species of *Finlaya* [original combination: *Aedes (Finlaya) stonei*].
stoneorum (Marks, 1977) (p. 33) – species of *Macleaya (Macleaya)* [original combination: *Aedes (Macleaya) stoneorum*].
stramineus (Dubitzky, 1970) (p. 408) – species of *Ochlerotatus* (subgenus uncertain, determined by Reinert *et al.*, 2008) [original combination: *Aedes (Ochlerotatus) stramineus*].
strelitziae (Muspratt, 1950) (p. 73) – species of *Stegomyia (Mukwaya)* [original combination: *Aedes (Stegomyia) strelitziae*].
stricklandi Edwards, 1912b (p. 523) – species of *Ochlerotatus (Lepidokeneon)*.
Synonyms: *australis* Strickland, 1911a; *flindersi* Taylor, 1914c.
subalbirostris (Klein & Marks, 1960) (p. 115) – species of *Ochlerotatus* (subgenus uncertain, determined by Reinert *et al.*, 2008) [original combination: *Aedes (Ochlerotatus) subalbirostris*].
subalbitarsis (King & Hoogstraal, 1946d) (p. 144) – species of *Patmarksia* [original combination: *Aedes (Finlaya) subalbitarsis*].
subalbopicta (Barraud, 1931a) (p. 225) – species of *Stegomyia* (subgenus uncertain, determined by Reinert *et al.*, 2009) [original combination: *Aedes (Stegomyia) subalbopictus*].
subargentea (Edwards, 1925) (p. 262) – species of *Stegomyia (Mukwaya)* [original combination: *Aedes (Stegomyia) subargenteus*].
subauridorsum (Marks, 1948) (p. 28) – species of '*Ochlerotatus (Finlaya)*' *sensu auctorum* [original combination: *Aedes (Finlaya) subauridorsum*].
subbasalis (Dobrotworsky, 1962) (p. 298) – species of *Dobrotworskyius* [original combination: *Aedes (Finlaya) subbasalis*].
subcantans Felt, 1905a (pp. 448, 474) – synonym of *Ochlerotatus stimulans* (Walker, 1848) [original combination:

Culicada subcantans]; synonymy established before Knight & Stone, 1977.

subdentatus (Edwards, 1936) (p. 51) – species of *Aedimorphus* [original combination: *Aedes subdentatus*].

subdiversus (Martini, 1926) (p. 204) – species of *Ochlerotatus* (*Rusticoidus*) [original combination: *Aedes subdiversus*].

submediopunctata Barraud, 1923a (p. 781) – synonym of *Stegomyia* (*Huangmyia*) *mediopunctata* Theobald, 1905h [original combination: *Stegomyia mediopunctata* variety *submediopunctata*]; synonymy by Huang, 1977.

subnivea (Edwards, 1922a) (p. 269) – species of *Downsiomyia* [original combination: *Aedes* (*Finlaya*) *subniveus*].

subtilis Sergent & Sergent, 1905 (p. 673) – synonym of *Ochlerotatus caspius* (Pallas, 1771) [original combination: *Grabhamia subtilis*]; synonymy established before Knight & Stone, 1977.

subtrichurus (Martini, 1927) (species) (p. 386) – subspecies of *Ochlerotatus* (*Rusticoidus*) *rusticus* (Rossi, 1790) [original combination: *Aedes subtrichurus*]; subspecific status by Harbach & Howard, 2007.

subvexans Martini, 1922 (p. 159) – synonym of *Ochlerotatus cantans* (Meigen, 1818) [original combination: *Aedes cantans* variety *subvexans*]; synonymy by Natvig, 1948 (also see Harbach & Howard, 2007).

sudanensis Theobald, 1908c (p. 252) – species of *Mucidus* (*Mucidus*).

sudanensis Theobald, 1911c (p. 154) – synonym of *Aedimorphus vexans* (Meigen, 1830b) [original combination: *Culex sudanensis*]; synonymy by Edwards, 1941 (via synonymy with *arabiensis* Patton, 1905).

sudanensis Theobald, 1913d (p. 595) – synonym of *Catageiomyia tarsalis* (Newstead, 1907) [original combination: *Reedomyia sudanensis*]; synonymy established before Knight & Stone, 1977.

suffusus (Edwards, 1922a) (p. 270) – species of '*Ochlerotatus* (*Finlaya*)' *sensu auctorum* [original combination: *Aedes* (*Finlaya*) *suffusus*].

sugens Wiedemann, 1828 (p. 545) – synonym of *Stegomyia* (*Stegomyia*) *aegypti* (Linnaeus, 1762) [original combination: *Culex sugens*]; synonymy established before Knight & Stone, 1977.

suknaensis Theobald, 1910a (p. 21) – synonym of *Edwardsaedes imprimens* (Walker, 1859) [original combination: *Culicada suknaensis*]; synonymy established before Knight & Stone, 1977.

sumidero (Schick, 1970a) (p. 72) – species of '*Ochlerotatus* (*Protomacleaya*)' *sensu auctorum* [original combination: *Aedes* (*Finlaya*) *sumidero*].

surcoufi (Theobald, 1912) (in Surcouf, 1912) (p. 59) – species of *Ochlerotatus* (subgenus uncertain, determined by Reinert *et al.*, 2008) [original combination: *Culicada surcoufi*].

sureilensis (Barraud, 1934) (p. 231) – subspecies of *Stegomyia* (*Huangmyia*) *mediopunctata* Theobald, 1905h [original combination: *Aedes* (*Stegomyia*) *mediopunctatus* variety *sureilensis*]; subspecific status by Harbach & Howard, 2007.

suvae Stone & Bohart, 1944 (p. 214) – synonym of *Levua geoskusea* (Amos, 1944) [original combination: *Aedes* (*Levua*) *suvae*]; synonymy by Stone & Bohart, 1944 (footnote).

sylvae Theobald, 1901c (p. 96) – synonym of *Ochlerotatus sticticus* (Meigen, 1838) [original combination: *Culex nigripes* variety *sylvae*]; synonymy established before Knight & Stone, 1977.

sylvaticum Le Goff, Boussès & Brunhes, 2007 (p. 195) – species of *Neomelaniconion*.

sylvaticus Brunhes, 1983 (p. 316) – synonym of *Paulianius madagascarensis* (van Someren, 1949a); synonymy by Brunhes *et al.*, 2017.

sylvaticus Meigen, 1818 (p. 6) – synonym of *Ochlerotatus communis* (de Geer, 1776) [original combination: *Culex sylvaticus*]; synonymy established before Knight & Stone, 1977.

sylvestris Theobald, 1901a (p. 406) – synonym of *Aedimorphus vexans* (Meigen, 1830b) [original combination: *Culex sylvestris*]; synonymy established before Knight & Stone, 1977.

sylvestris Theobald, 1910c (p. 196) – synonym of *Macleaya* (*Chaetocruiomyia*) *spinosipes* (Edwards, 1922d) [original combination: *Chaetocruiomyia sylvestris*]; synonymy established before Knight & Stone, 1977.

sylvicola Grossbeck, 1906 (p. 129) – synonym of *Ochlerotatus grossbecki* (Dyar & Knab, 1906d) [original combination: *Culex sylvicola*]; synonymy established before Knight & Stone, 1977.

synchytus (Arnell, 1976) (p. 71) – species of *Ochlerotatus* (*Ochlerotatus*) [original combination: *Aedes* (*Ochlerotatus*) *synchytus*].

syntheticus Barraud, 1928b (p. 375) – synonym of *Aedimorphus fisheri* (Barraud, 1928a) [original combination: *Aedes* (*Aedimorphus*) *syntheticus*]; originally proposed as a replacement name for *fisheri* Barraud, 1928a, *non fisheri* Dyar, 1917a, when both were included in *Aedes* Meigen.

Tt

tabu (Ramalingam & Belkin, 1965) (p. 1) – species of *Stegomyia* (subgenus uncertain, determined by Reinert *et al.*, 2009) [original combination: *Aedes* (*Stegomyia*) *tabu*].

tachirensis Anduze, 1947 (p. 355) – synonym of *Howardina fulvithorax* (Lutz, 1904) [original combination: *Aedes* (*Howardina*) *tachirensis*]; synonymy by Lane, 1951.

taeniarostre (Theobald, 1909b) (p. 13) – species of *Neomelaniconion* [original combination: *Banksinella taeniarostris*].

Synonym: *chrysothorax* Theobald, 1909b.

taeniatus (Leicester, 1908) (p. 133) – species of *Aedimorphus* [original combination: *Lepidotomyia taeniata*].

Synonym: *orbitae* Edwards, 1922a.

taeniatus Wiedemann, 1828 (p. 10) – synonym of *Stegomyia* (*Stegomyia*) *aegypti* (Linnaeus, 1762) [original combination: *Culex taeniatus*]; synonymy established before Knight & Stone, 1977.

taeniorhynchoides (Christophers, 1911) (p. 68) – species of *Aedimorphus* [original combination: *Leslieomyia taeniorhynchoides*].
Synonym: *maculata* Theobald, 1905g.

taeniorhynchus (Wiedemann, 1821) (p. 43) – species of *Ochlerotatus* (*Culicelsa*) [original combination: *Culex taeniorhynchus*].
Synonyms: *damnosus* Say, 1823; *niger* Giles, 1904b; *portoricensis* Ludlow, 1905c; *epinolus* Dyar & Knab, 1914; *pix* Martini, 1935.

tahoensis (Dyar, 1916b) (p. 82) – species of *Ochlerotatus* (subgenus uncertain, determined by Reinert *et al.*, 2008) [original combination: *Aedes tahoensis*].

taiwanus (Lien, 1968d) (p. 223) – species of *Bruceharrisonius* [original combination: *Aedes* (*Finlaya*) *aureostriatus taiwanus*]; specific status by Reinert, 2003.

tapinops Brèthes, 1917 (p. 227) – synonym of *Ochlerotatus* (*Ochlerotatus*) *crinifer* (Theobald, 1903a) [original combination: *Culex tapinops*]; synonymy established before Knight & Stone, 1977.

tarnogradskii Martini, 1930 (p. 259) – synonym of *Aedes rossicus* Dolbeškin, Gorickaja & Mitrofanova, 1930; synonymy established before Knight & Stone, 1977.

tarsalis (Newstead, 1907) (in Newstead *et al.*, 1907) (p. 18) – species of *Catageiomyia* [original combination: *Duttonia tarsalis*].
Synonyms: *africana* Newstead, 1907; *biannulata* Theobald, 1907; *bipunctata* Theobald, 1909b; *uniannulata* Theobald, 1909b; *neobiannulata* Theobald, 1910c; *sudanensis* Theobald, 1913d.

tasmaniensis Strickland, 1911b (p. 181) – synonym of *Halaedes australis* (Erichson, 1842) [original combination: *Culicada tasmaniensis*]; synonymy by Mattingly & Marks, 1955.

tasmaniensis Taylor, 1914b (p. 466) – synonym of *Ochlerotatus andersoni* (Edwards, 1926a) [original combination: *Menolepis* (?) *tasmaniensis*]; synonymy established before Knight & Stone, 1977.

tauffliebi (Rickenbach & Ferrara, 1965) (p. 24) – species of *Aedimorphus* [original combination: *Aedes* (*Aedimorphus*) *tauffliebi*].

taylori (Edwards, 1936) (p. 55) – species of *Diceromyia* [original combination: *Aedes* (*Diceromyia*) *taylori*].

teesdalei (van Someren, 1954a) (p. 104) – species of *Albuginosus* [original combination: *Aedes* (*Aedimorphus*) *teesdalei*].

tehuantepec (Schick, 1970a) (p. 87) – species of '*Ochlerotatus* (*Protomacleaya*)' *sensu auctorum* [original combination: *Aedes* (*Finlaya*) *tehuantepec*].

terrens (Walker, 1856a) (p. 429) – species of '*Ochlerotatus* (*Protomacleaya*)' *sensu auctorum* [original combination: *Culex terrens*].
Synonyms: *oswaldi* Lutz, 1904; *apollo* Schick, 1970a.

terriei Theobald, 1903a (p. 193) – synonym of *Ochlerotatus detritus* (Haliday, 1833) [original combination: *Culex terriei*]; synonymy established before Knight & Stone, 1977.

thailandensis (Reinert, 1976a) (p. 319) – species of *Paraedes* [original combination: *Aedes* (*Paraedes*) *thailandensis*].

thaxteri Dyar & Knab, 1919 (p. 8) – synonym of *Howardina busckii* (Coquillett, 1906e) [original combination: *Aedes thaxteri*]; synonymy established before Knight & Stone, 1977.

thelcter (Dyar, 1918e) (p. 129) – species of *Ochlerotatus* (*Ochlerotatus*) [original combination: *Aedes* (*Taeniorhynchus?*) *thelcter*].
Synonym: *keyensis* Buren, 1947.

theobaldi (Taylor, 1914a) (p. 751) – species of *Ochlerotatus* (subgenus uncertain, determined by Reinert *et al.*, 2008) [original combination: *Grabhamia theobaldi*].

thibaulti (Dyar & Knab, 1910) (p. 174) – species of *Ochlerotatus* (subgenus uncertain, determined by Reinert *et al.*, 2008) [original combination: *Aedes thibaulti*].

thomsoni (Theobald, 1905e) (p. 18) – species of *Christophersiomyia* [original combination: *Stegomyia thomsoni*].

thorntoni (Dyar & Knab, 1907a) (p. 10) – species of '*Ochlerotatus* (*Protomacleaya*)' *sensu auctorum* [original combination: *Aedes thorntoni*].

thurmanae (Mattingly, 1958a) (p. 55) – species of *Cancraedes* [original combination: *Aedes* (*Cancraedes*) *thurmanae*].

timorensis (Miyagi, Toma & Lien, 2004) (p. 107) – species of *Geoskusea* [original combination: *Ochlerotatus* (*Geoskusea*) *timorensis*].

tiptoni (Grjebine, 1953a) (p. 469) – species of *Paulianius* [original combination: *Aedes tiptoni*].

togoensis Enderlein, 1923 (p. 28) – synonym of *Stegomyia apicoargentea* Theobald, 1909b [original combination: *Aniella togoensis*]; synonymy established before Knight & Stone, 1977.

togoi (Theobald, 1907) (p. 379) – species of *Tanakaius* [original combination: *Culicelsa togoi*].

tokushimaensis Tanimura, 1952 (p. 176) – synonym of *Hulecoeteomyia japonica* (Theobald, 1901a) [original combination: *Aedes japonicus* variety *tokushimaensis*]; synonymy by Sasa & Tanimura, 1971.

tongae (Edwards, 1926a) (p. 103) – species of *Stegomyia* (subgenus uncertain, determined by Reinert *et al.*, 2009) [original combination: *Aedes variegatus* variety *tongae*]; specific status prior to ? Marks, 1954.

tonkinensis (Galliard & Ngu, 1947) (p. 77) – species of *Collessius* (*Alloeomyia*) [original combination: *Aedes* (*Finlaya*) *tonkinensis*].

tonkingi Gebert, 1948 (p. 96) – species of *Mucidus* (*Mucidus*). **Note**: the title of the article by Gebert (1948) indicated that the author intended to place this species in genus *Aedes*, subgenus *Mucidus*, but it is referred to as the 'new species *Mucidus tonkingi*' and described under the heading '**Mucidus tonkingi** sp. n.'.

tonsa (Edwards, 1924b) (p. 390) – species of *Geoskusea* [original combination: *Aedes* (*Skusea*) *tonsus*].

tormentor (Dyar & Knab, 1906d) (p. 191) – species of *Ochlerotatus* (*Protoculex*) [original combination: *Aedes tormentor*].

tortilis (Theobald, 1903g) (p. 281) – species of *Ochlerotatus* (*Ochlerotatus*) [original combination: *Culex tortilis*].
Synonyms: *bracteatus* Coquillett, 1906a; *habanicus* Dyar & Knab, 1906d; *balteatus* Dyar & Knab, 1907a; *plutocraticus* Dyar & Knab, 1907a; *virginensis* Dyar, 1922c.

toxopeusi (Bonne-Wepster, 1948) (p. 315) – species of '*Ochlerotatus* (*Finlaya*)' *sensu auctorum* [original combination: *Aedes* (*Finlaya*) *toxopeusi*].

toxorhynchus Macquart, 1838a (p. 39) – synonym of *Stegomyia* (*Stegomyia*) *aegypti* (Linnaeus, 1762) [original combination: *Culex toxorhynchus*]; synonymy by Belkin, 1968b.

tracei Senevet & Quiévreux, 1941 (p. 251) – synonym of *Howardina busckii* (Coquillett, 1906e) [original combination: *Aedes tracei*]; synonymy established before Knight & Stone, 1977.

transvaalensis Theobald, 1903a (p. 165) – synonym of *Aedimorphus hirsutus* (Theobald, 1901a) [original combination: *Culex transvaalensis*]; synonymy established before Knight & Stone, 1977.

traversus Dyar, 1925g (p. 215) – synonym of *Ochlerotatus* (*Ochlerotatus*) *angustivittatus* (Dyar & Knab, 1907a) [original combination: *Aedes traversus*]; synonymy established before Knight & Stone, 1977.

tremula Theobald, 1903e (p. 155) – species of *Macleaya* (*Macleaya*).
Synonyms: *alboannulata* Taylor, 1912a; *minuta* Taylor, 1912a; *australis* Taylor, 1914b; *darwini* Taylor, 1914b; *doddi* Taylor, 1919.

treubi (de Meijere, 1910) (p. 918) – species of *Alanstonea* [original combination: *Scutomyia treubi*].

tricholabis (Edwards, 1941) (p. 179) – species of *Aedimorphus* [original combination: *Aedes* (*Aedimorphus*) *tricholabis*].

trichurus Dyar, 1904a (p. 170) – synonym of *Ochlerotatus* (*Rusticoidus*) *provocans* (Walker, 1848) [original combination: *Culex trichurus*]; synonymy by Wood, 1977.

trilineata Leicester, 1904 (in Theobald, 1904c) (p. 163) – synonym of *Hulecoeteomyia chrysolineata* (Theobald, 1907); synonymy established before Knight & Stone, 1977.

trilineatus Theobald, 1901c (p. 105) – synonym of *Aedimorphus mediolineatus* (Theobald, 1901c) [original combination: *Culex trilineatus*]; synonymy established before Knight & Stone, 1977.

trimaculatus (Theobald, 1905f) (p. 86) – species of *Aedimorphus* [original combination: *Culex trimaculatus*].
Synonym: *littoralis* Barraud, 1927c.

trinidad Gil Collado, 1936 (pp. 315, 326) – synonym of *Stegomyia dendrophila* (Edwards, 1921c) [original combination: *Aedes* (*Stegomyia*) *trinidad*]; synonymy by Mattingly, 1952.

tripunctata Theobald, 1907 (p. 247) – synonym of *Georgecraigius* (*Horsfallius*) *fluviatilis* (Lutz, 1904) [original combination: *Danielsia tripunctata*]; synonymy established before Knight & Stone, 1977.

triseriatus (Say, 1823) (p. 12) – species of '*Ochlerotatus* (*Protomacleaya*)' *sensu auctorum* [original combination: *Culex triseriatus*].
Synonym: *nigra* Ludlow, 1905c.

trivittatus (Coquillett, 1902b) (p. 193) – species of *Ochlerotatus* (*Ochlerotatus*) [original combination: *Culex trivittatus*].
Synonym: *inconspicuus* Grossbeck, 1904b.

trukensis (Bohart, 1957) (p. 63) – species of *Aedimorphus* [original combination: *Aedes* (*Aedimorphus*) *trukensis*].

tsiliensis (King & Hoogstraal, 1946f) (p. 305) – species of '*Ochlerotatus* (*Finlaya*)' *sensu auctorum* [original combination: *Aedes* (*Finlaya*) *tsiliensis*].

tubbutiensis (Dobrotworsky, 1959) (p. 139) – species of *Dobrotworskyius* [original combination: *Aedes tubbutiensis*].

tulagiensis (Edwards, 1926a) (p. 109) – species of *Stegomyia* (*Actinothrix*) [original combination: *Aedes* (*Stegomyia*) *edwardsi* variety *tulagiensis*]; specific status by Belkin, 1962.

tulliae (Taylor, 1929) (p. 275) – species of *Macleaya* (*Chaetocruiomyia*) [original combination: *Chaetocruiomyia tulliae*].

turneri (Marks, 1963b) (p. 42) – species of *Ochlerotatus* (*Lepidokeneon*) [original combination: *Aedes* (*Ochlerotatus*) *turneri*].

tutuilae (Ramalingam & Belkin, 1965) (p. 3) – species of *Finlaya* [original combination: *Aedes* (*Finlaya*) *tutuilae*].

Uu

uncatus Grabham, 1907 (p. 25) – synonym of *Gymnometopa mediovittata* (Coquillett, 1906d) [original combination: *Aedes uncatus*]; synonymy established before Knight & Stone, 1977.

uniannulata Theobald, 1909b (p. 12) – synonym of *Catageiomyia tarsalis* (Newstead, 1907) [original combination: *Neopecomyia uniannulata*]; synonymy established before Knight & Stone, 1977.

unicinctus (Edwards, 1922a) (p. 266) – species of *Jihlienius* [original combination: *Aedes* (*Finlaya*) *unicinctus*].

unidentatum (McIntosh, 1971) (p. 325) – species of *Neomelaniconion* [original combination: *Aedes unidentatus*].

uniformis Strickland, 1911a (p. 131) – synonym of *Ochlerotatus* (*Empihals*) *vigilax* (Skuse, 1889) [original combination: *Culicelsa uniformis*]; synonymy established before Knight & Stone, 1977.

unilineata (Theobald, 1906b) (p. 70) – species of *Stegomyia* (subgenus uncertain, determined by Reinert *et al.*, 2009) [original combination: *Quasistegomyia unilineata*].
Synonym: *gebeleinensis* Theobald, 1910c.

upatensis (Anduze & Hecht, 1943) (p. 185) – species of *Ochlerotatus* (subgenus uncertain, determined by

Reinert *et al.*, 2008) [original combination: *Aedes (Finlaya) upatensis*].

upolensis (Marks, 1957b) (p. 51) – species of *Stegomyia* (subgenus uncertain, determined by Reinert *et al.*, 2009) [original combination: *Aedes (Stegomyia) upolensis*].

usambara (Mattingly, 1953d) (p. 294) – species of *Stegomyia* (subgenus uncertain, determined by Reinert *et al.*, 2009) [original combination: *Aedes (Stegomyia) usambara*].

Vv

valeryi Gornostaeva, 2005 (p. 488) – species of *Aedes*.

vana (Colless, 1958) (p. 471) – species of *Downsiomyia* [original combination: *Aedes (Finlaya) vanus*].

vandema Strickland, 1911b (p. 202) – synonym of *Ochlerotatus (Pholeomyia) flavifrons* (Skuse, 1889) [original combination: *Culicada vandema*]; synonymy established before Knight & Stone, 1977.

vanemdeni (Martini, 1931a) (p. 208) – species of *Howardina* [original combination: *Aedes van Emdeni* [*sic*]].

vansomerenae (Mattingly, 1955) (in Mattingly & Brown, 1955) (p. 78) – subspecies of *Ochlerotatus (Empihals) vigilax* (Skuse, 1889) [original combination: *Aedes (Ochlerotatus) vigilax vansomerenae*].

vargasi (Schick, 1970a) (p. 73) – species of '*Ochlerotatus (Protomacleaya)*' *sensu auctorum* [original combination: *Aedes (Finlaya) vargasi*].

variegatans Strickland, 1911b (p. 204) – synonym of *Ochlerotatus (Pholeomyia) flavifrons* (Skuse, 1889) [original combination: *Culicada vandema* variety *variegatans*]; synonymy established before Knight & Stone, 1977.

variegatus Doleschall, 1858 (p. 77) – synonym of *Stegomyia scutellaris* (Walker, 1859) [original combination: *Culex variegatus*]; synonymy established before Knight & Stone, 1977.

variegatus Schrank, 1781 (p. 482) – synonym of *Ochlerotatus flavescens* (Müller, 1764) [original combination: *Culex variegatus*]; synonymy established before Knight & Stone, 1977.

variepictus (King & Hoogstraal, 1946c) (p. 98) – species of *Huaedes* [original combination: *Aedes (Leptosomatomyia) variepictus*].

varipalpus (Coquillett, 1902d) (p. 292) – species of *Jarnellius* [original combination: *Culex varipalpus*].

varuae (Belkin, 1962) (p. 478) – species of *Stegomyia* (subgenus uncertain, determined by Reinert *et al.*, 2009) [original combination: *Aedes (Stegomyia) varuae*].

veeniae (McIntosh, 1975) (p. 263) – species of *Catageiomyia* [original combination: *Aedes veeniae*].

ventrovittis (Dyar, 1916b) (p. 84) – species of *Ochlerotatus* (subgenus uncertain, determined by Reinert *et al.*, 2008) [original combination: *Aedes ventrovittis*].
Synonyms: *fisheri* Dyar, 1917a; *cacothius* Dyar, 1923a.

verna Lewis, 1944 (p. 27) – synonym of *Aedimorphus leesoni* (Edwards, 1932b) [original combination: *Aedes (Aedimorphus) verna*]; synonymy by McIntosh, 1975.

versicolor (Barraud, 1924i) (p. 73) – species of '*Ochlerotatus (Finlaya)*' *sensu auctorum* [original combination: *Finlaya versicolor*].

vexans (Meigen, 1830b) (p. 241) – species of *Aedimorphus* [original combination: *Culex vexans*].
Synonyms: *parvus* Macquart, 1834; *articulatus* Rondani, 1872; *malariae* Grassi, 1898; *sylvestris* Theobald, 1901a; *montcalmi* Blanchard, 1905; *minuta* Theobald, 1907; *eruthrosops* Theobald, 1910c; *sudanensis* Theobald, 1911c; *eurochrus* Howard, Dyar & Knab, 1917.

vicarius (Lien, 1968d) (p. 218) – subspecies of *Collessius (Collessius) elsiae* (Barraud, 1923e) [original combination: *Aedes (Finlaya) elsiae vicarius*].

victoriensis Taylor, 1914b (p. 460) – synonym of *Ochlerotatus camptorhynchus* (Thomson, 1869) [original combination: *Culicada victoriensis*]; synonymy established before Knight & Stone, 1977.

vigilax (Skuse, 1889) (p. 1731) – species of *Ochlerotatus (Empihals)* [original combination: *Culex vigilax*].
Synonyms: *albirostris* Macquart, 1850; *marinus* Theobald, 1901a; *pseudovigilax* Theobald, 1907; *uniformis* Strickland, 1911a.

vinnipegensis Dyar, 1919b (p. 34) – synonym of *Ochlerotatus sticticus* (Meigen, 1838) [original combination: *Aedes vinnipegensis*]; synonymy established before Knight & Stone, 1977.

vinsoni (Mattingly, 1953a) (p. 16) – species of *Stegomyia* (subgenus uncertain, determined by Reinert *et al.*, 2009) [original combination: *Aedes (Stegomyia) vinsoni*].

virginensis Dyar, 1922c (p. 56) – synonym of *Ochlerotatus (Ochlerotatus) tortilis* (Theobald, 1903g) [original combination: *Aedes tortilis virginensis*]; synonymy established before Knight & Stone, 1977.

viridifrons Walker, 1848 (p. 3) – synonym of *Stegomyia (Stegomyia) aegypti* (Linnaeus, 1762) [original combination: *Culex viridifrons*]; synonymy established before Knight & Stone, 1977.

vittata Theobald, 1903h (p. 313) – synonym of *Ochlerotatus increpitus* (Dyar, 1916b) [original combination: *Grabhamia vittata*]; synonymy established before Knight & Stone, 1977.

vittatus (Bigot, 1861) (p. 227) – species of *Fredwardsius* [original combination: *Culex vittatus*].
Synonyms: *brumpti* Neveu-Lemaire, 1905b; *albopunctata* Theobald, 1907.

vittatus Philippi, 1865 (p. 596) – synonym of *Ochlerotatus albifasciatus* (Macquart, 1838b) [original combination: *Culex vittatus*]; synonymy established before Knight & Stone, 1977.

vittiger (Skuse, 1889) (p. 1728) – species of *Ochlerotatus* (subgenus uncertain, determined by Reinert *et al.*, 2008) [original combination: *Culex vittiger*].

Ww

wadai (Tanaka, Mizusawa & Saugstad, 1979) (p. 400) – species of *Stegomyia* (subgenus uncertain, determined by Reinert *et al*., 2009) [original combination: *Aedes* (*Stegomyia*) *wadai*].

wahlgreni Theobald, 1910c (p. 396) – synonym of *Ochlerotatus punctor* (Kirby, 1837) [original combination: *Culex wahlgreni*]; synonymy established before Knight & Stone, 1977.

wainwrighti Baisas, 1946c (p. 34) – synonym of *Catatassomyia meronephada* Dyar & Shannon, 1925a [original combination: *Aedes* (*Aedimorphus*) *wainwrighti*]; synonymy by Reinert, 1973a.

w-alba Theobald, 1905f (p. 74) – species of *Stegomyia* (subgenus uncertain, determined by Reinert *et al*., 2009).
Synonym: *argenteomaculata* Theobald, 1907.

walkeri (Theobald, 1901a) (p. 424) – species of *Howardina* [original combination: *Culex* (*Stegomyia* ?) *walkeri*].

wallacei (Edwards, 1926a) (p. 105) – species of *Finlaya* [original combination: *Aedes* (*Finlaya*) *wallacei*].

wardangensis (Brust, Ballard, Driver, Hartley, Galway & Curran, 1998) (p. 1241) – species of *Halaedes* [original combination: *Aedes* (*Halaedes*) *wardangensis*].

wardi (Reinert, 1976b) (p. 18) – species of *Rhinoskusea* [original combination: *Aedes* (*Rhinoskusea*) *wardi*].

washinoi (Lanzaro & Eldridge, 1992) (p. 93) – species of *Ochlerotatus* (subgenus uncertain, determined by Reinert *et al*., 2008) [original combination: *Aedes* (*Ochlerotatus*) *washinoi*].

wasselli (Marks, 1947b) (p. 9) – species of '*Ochlerotatus* (*Finlaya*)' *sensu auctorum* [original combination: *Aedes* (*Finlaya*) *wasselli*].

watasei (Yamada, 1921) (p. 64) – species of *Phagomyia* [original combination: *Aedes watasei*].

waterhousei Dobrotworsky, 1960a (p. 57) – synonym of *Ochlerotatus silvestris* (Dobrotworsky, 1961b) [original combination: *Aedes waterhousei*]; synonymy established before Knight & Stone, 1977. **Note**: this nominal taxon is the junior secondary homonym of *waterhousei* Theobald, 1905.

waterhousei Theobald, 1905 (in Waterhouse, 1905) (p. 674) – synonym of *Ochlerotatus cantans* (Meigen, 1818) [original combination: *Culex waterhousei*]; synonymy established before Knight & Stone, 1977. **Note**: this nominal taxon is the senior secondary homonym of *waterhousei* Dobrotworsky, 1960a.

watteni (Lien, 1968b) (p. 103) – species of *Downsiomyia* [original combination: *Aedes* (*Finlaya*) *watteni*].

wattensis (Taylor, 1929) (p. 275) – species of *Macleaya* (*Chaetocruiomyia*) [original combination: *Aedes* (*Chaetocruiomyia*) *wattensis*].

wauensis (Huang, 1968b) (p. 170) – species of *Huaedes* [original combination: *Aedes* (*Huaedes*) *wauensis*].

wellmanii (Theobald, 1905b) (p. 103) – species of *Hopkinsius* (*Hopkinsius*) [original combination: *Danielsia wellmanii*].

wellmanii Theobald, 1910c (p. 163) – synonym of *Stegomyia pseudonigeria* Theobald, 1910c; synonymy established before Knight & Stone, 1977.

wendyae (Service, 1959a) (p. 73) – species of *Catageiomyia* [original combination: *Aedes* (*Aedimorphus*) *wendyae*].

westralis Strickland, 1911a (p. 130) – synonym of *Ochlerotatus camptorhynchus* (Thomson, 1869) [original combination: *Culicelsa westralis*]; synonymy established before Knight & Stone, 1977.

whartoni (Mattingly, 1965) (p. 65) – species of *Petermattinglyius* (*Aglaonotus*) [original combination: *Aedes* (*Diceromyia*) *whartoni*].

whitmorei (Dunn, 1918) (p. 128) – species of *Howardina* [original combination: *Aedes whitmorei*].

wigglesworthi (Edwards, 1941) (p. 182) – species of *Aedimorphus* [original combination: *Aedes* (*Aedimorphus*) *wigglesworthi*].

wilkersoni (Reinert, 1990) (p. 56) – species of *Kenknightia* [original combination: *Aedes* (*Kenknightia*) *wilkersoni*].

willcocksii Theobald, 1907 (p. 296) – synonym of *Ochlerotatus caspius* (Pallas, 1771) [original combination: *Grabhamia willcocksii*]; synonymy established before Knight & Stone, 1977.

wilsoni Taylor, 1919 (p. 833) – synonym of *Ochlerotatus sagax* (Skuse, 1889) [original combination: *Culicada wilsoni*]; synonymy established before Knight & Stone, 1977.

woodi (Edwards, 1922d) (p. 82) – species of *Stegomyia* (*Mukwaya*) [original combination: *Aedes* (*Stegomyia*) *woodi*].

Yy

yaeyamensis (Tanaka, Mizusawa & Saugstad, 1979) (p. 313) – subspecies of *Hulecoeteomyia japonica* (Theobald, 1901a) [original combination: *Aedes* (*Finlaya*) *yaeyamensis*].

yamadai Sasa, Kano & Takahasi, 1950 (p. 635) – species of *Aedes*.

yangambiensis (de Meillon & Lavoipierre, 1944) (p. 50) – species of *Catageiomyia* [original combination: *Aedes* (*Aedimorphus*) *yangambiensis*].

yunnanensis (Gaschen, 1934) (p. 331) – species of *Hulecoeteomyia* [original combination: *Finlaya yunnanensis*].

yvonneae (Edwards, 1941) (p. 176) – species of *Catageiomyia* [original combination: *Aedes* (*Aedimorphus*) *yvonneae*].

Zz

zammitii Theobald, 1903a (p. 252) – species of *Acartomyia*.

zavortinki (Schick, 1970a) (p. 45) – species of '*Ochlerotatus* (*Protomacleaya*)' *sensu auctorum* [original combination: *Aedes* (*Finlaya*) *zavortinki*].

zethus de Meillon & Lavoipierre, 1944 (p. 57) – synonym of *Paulianius niveus* (de Meillon, 1943) [*nomen novum* for *niveus* de Meillon, 1943; original combination:

Aedes (*Diceromyia*) *zethus*]; synonymy established before Knight & Stone, 1977.
ziemanni Enderlein, 1923 (p. 27) – synonym of *Stegomyia apicoargentea* Theobald, 1909 [original combination: *Aniella ziemanni*]; synonymy by Edwards, 1941.
zonatipes Walker, 1961 (p. 229) – synonym of *Stegomyia scutellaris* (Walker, 1859) [original combination: *Culex zonatipes*]; synonymy established before Knight & Stone, 1977.
zoosophus (Dyar & Knab, 1918b) (p. 165) – species of '*Ochlerotatus* (*Protomacleaya*)' *sensu auctorum* [original combination: *Aedes zoosophus*].
Synonym: *alleni* Turner, 1924.

Genus-group Names

Abraedes Zavortink, 1970c (subgenus of *Aedes* Meigen, 1818) (p. 2) – genus of tribe Aedini Neveu-Lemaire, 1902c; generic status by Reinert *et al.*, 2004.
Acartomyia Theobald, 1903a (p. 251) – genus of tribe Aedini Neveu-Lemaire, 1902c; resurrected from synonymy with *Ochlerotatus* Lynch Arribálzaga by Reinert *et al.*, 2009.
Actinothrix Reinert, Harbach & Kitching, 2009 (p. 781) – subgenus of *Stegomyia* Theobald, 1901.
Aedes Meigen, 1818 (p. 13) – genus of tribe Aedini Neveu-Lemaire, 1902c.
Aedimorphus Theobald, 1903a (p. 290) – genus of tribe Aedini Neveu-Lemaire, 1902c.
Synonyms: *Ecculex* Felt, 1904; *Reedomyia* Ludlow, 1905a; *Lepidotomyia* Theobald, 1905f; *Pecomyia* Theobald, 1905g; *Pseudograbhamia* Theobald, 1905h; *Mimeteculex* Theobald, 1908c; *Geitonomyia* Leicester, 1908; *Bathosomyia* Theobald, 1909b; *Leslieomyia* Christophers, 1911.
Aglaonotus Reinert, Harbach & Kitching, 2009 (p. 779) – subgenus of *Petermattinglyius* Reinert, Harbach & Kitching, 2009.
Alanstonea Mattingly, 1960 (subgenus of *Aedes* Meigen, 1818) (p. 170) – genus of tribe Aedini Neveu-Lemaire, 1902c; generic status by Reinert *et al.*, 2004.
Albuginosus Reinert, 1987 (subgenus of *Aedes* Meigen, 1818) (p. 308) – genus of tribe Aedini Neveu-Lemaire, 1902c; generic status by Reinert *et al.*, 2004.
Alloeomyia Reinert, Harbach & Kitching, 2008 (p. 97) – subgenus of *Collessius* Reinert, Harbach & Kitching, 2006.
Andersonia Strickland, 1911b (genus) (p. 250) – synonym of *Ochlerotatus* Lynch Arribálzaga, 1891a,b; synonymy established before Knight & Stone, 1977.
Aniella Enderlein, 1923 (genus) (p. 26) – synonym of *Stegomyia* Theobald, 1901; synonymy established before Knight & Stone, 1977.
Ayurakitia Thurman, 1954a (p. 198) – genus of tribe Aedini Neveu-Lemaire, 1902c; elevated from subgeneric status in *Aedes* Meigen by Reinert, 2000c.
Aztecaedes Zavortink, 1972 (subgenus of *Aedes* Meigen, 1818) (p. 66) – genus of tribe Aedini Neveu-Lemaire, 1902c; generic status by Reinert *et al.*, 2004.
Banksinella Theobald, 1907 (genus) (p. 468) – synonym of *Neomelaniconion* Newstead, 1907; synonymy established before Knight & Stone, 1977.
Bathosomyia Theobald, 1909b (genus) (p. 9) – synonym of *Aedimorphus* Theobald, 1903a; synonymy established before Knight & Stone, 1977.
Belkinius Reinert, 1982 (subgenus of *Aedes* Meigen, 1818) (p. 102) – genus of tribe Aedini Neveu-Lemaire, 1902c; generic status by Reinert *et al.*, 2004.
Bifidistylus Reinert, Harbach & Kitching, 2009 (p. 766) – genus of tribe Aedini Neveu-Lemaire, 1902c.
Bohartius Reinert, Harbach & Kitching, 2009 (p. 781) – subgenus of *Stegomyia* Theobald, 1901.
Borichinda Harbach & Rattanarithikul, 2007 (in Harbach, Rattanarithikul, Howard *et al.*, 2007) (p. 471) – genus of tribe Aedini Neveu-Lemaire, 1902c.
Bothaella Reinert, 1973e (subgenus of *Aedes* Meigen, 1818) (p. 2) – genus of tribe Aedini Neveu-Lemaire, 1902c; generic status by Reinert *et al.*, 2004.
Bruceharrisonius Reinert, 2003 (subgenus of *Aedes* Meigen, 1818) (p. 309) – genus of tribe Aedini Neveu-Lemaire, 1902c; generic status by Reinert *et al.*, 2006.
Buvirilia Reinert, Harbach & Kitching, 2008 (p. 104) – subgenus of *Ochlerotatus* Lynch Arribálzaga, 1891a,b.
Caenocephalus Taylor, 1914c (genus) (p. 700) – synonym of *Pseudoskusea* Theobald, 1907; synonymy established before Knight &Stone, 1977.
Calomyia Taylor, 1914c (genus) (p. 684) – synonym of *Molpemyia* Theobald, 1910c; synonymy by transfer of type species (*priestleyi* Taylor, 1914c, as synonym of *purpurea* Theobald, 1910c) from *Finlaya* Theobald, 1903a by Reinert, 1993.
Cancraedes Edwards, 1929b (subgenus of *Aedes* Meigen, 1818) (p. 342) – genus of tribe Aedini Neveu-Lemaire, 1902c; generic status by Reinert *et al.*, 2009.
Catageiomyia Theobald, 1903d (p. i) – genus of tribe Aedini Neveu-Lemaire, 1902c; resurrected from synonymy with *Aedimorphus* Theobald by Reinert *et al.*, 2009.
Synonyms: *Duttonia* Newstead, 1907; *Myxosquamus* Theobald, 1909b; *Neopecomyia* Theobald, 1909b; *Stenoscutus* Theobald, 1909b.
Catatassomyia Dyar & Shannon, 1925a (p. 71) – genus of tribe Aedini Neveu-Lemaire, 1902c; resurrected from synonymy with *Diceromyia* Theobald by Reinert *et al.*, 2009.
Chaetocruiomyia Theobald, 1910c (genus) (pp. 113, 195) – subgenus of *Macleaya* Theobald, 1903e; transfer from subgeneric status in *Aedes* Meigen by Reinert *et al.*, 2006.
Christophersiomyia Barraud, 1923a (p. 786) – genus of tribe Aedini Neveu-Lemaire, 1902c; elevated from subgeneric status in *Aedes* Meigen by Reinert *et al.*, 2004.
Chrysoconops Goeldi, 1905 (genus) (p. 114) – subgenus of *Ochlerotatus* Lynch Arribálzaga, 1891a,b; resurrected

from synonymy with *Ochlerotatus* Lynch Arribálzaga by Reinert *et al.*, 2008.

'Coetzeemyia' Huang, Mathis & Wilkerson, 2010 (subgenus of *Aedes* Meigen, 1818) (p. 14) – subgenus of *'Ochlerotatus'* Lynch Arribálzaga, 1891a,b *sensu auctorum*; transfer from *Aedes* Meigen by Harbach (6 October 2010, http://mosquito-taxonomic-inventory.info/whats-new).

Collessius Reinert, Harbach & Kitching, 2006 (p. 64) – genus (subgenus) of tribe Aedini Neveu-Lemaire, 1902c.

Cornetius Huang, 2005 (subgenus of *Aedes* Meigen, 1818) (p. 517) – genus of tribe Aedini Neveu-Lemaire, 1902c; generic status by Reinert *et al.*, 2009.

Culicada Felt, 1904 (genus) (p. 391b) – subgenus of *Ochlerotatus* Lynch Arribálzaga, 1891a,b; resurrected from synonymy with *Ochlerotatus* Lynch Arribálzaga by Reinert *et al.*, 2008.

Culicelsa Felt, 1904 (genus) (p. 391b) – subgenus of *Ochlerotatus* Lynch Arribálzaga, 1891a,b; resurrected from synonymy with *Ochlerotatus* Lynch Arribálzaga by Reinert *et al.*, 2008.

Synonym: *Taeniorhynchus* Lynch Arribálzaga, 1891a,b.

Dahliana Reinert, Harbach & Kitching, 2006 (p. 66) – genus of tribe Aedini Neveu-Lemaire, 1902c.

Danielsia Theobald, 1904c (p. 78) – genus of tribe Aedini Neveu-Lemaire, 1902c; resurrected from synonymy with *Finlaya* Theobald by Reinert *et al.*, 2008.

Dendroskusea Edwards, 1929b (subgenus of *Aedes* Meigen, 1818) (p. 341) – genus of tribe Aedini Neveu-Lemaire, 1902c; resurrected from synonymy with *Diceromyia* Theobald by Reinert *et al.*, 2009.

Diceromyia Theobald, 1911c (p. 151) – genus of tribe Aedini Neveu-Lemaire, 1902c; elevated from subgeneric status in *Aedes* Meigen by Reinert *et al.*, 2004.

Dobrotworskyius Reinert, Harbach & Kitching, 2006 (p. 68) – genus of tribe Aedini Neveu-Lemaire, 1902c.

Downsiomyia Vargas, 1950c (subgenus of *Aedes* Meigen, 1818) (p. 63) – genus of tribe Aedini Neveu-Lemaire, 1902c; resurrected from synonymy with *Finlaya* Theobald by Reinert *et al.*, 2004.

Duttonia Newstead, 1907 (in Newstead *et al.*, 1907) (genus) (p. 17) – synonym of *Catageiomyia* Theobald, 1903d; synonymy by transfer of type species to *Catageiomyia* by Reinert *et al.*, 2009.

Ecculex Felt, 1904 (genus) (p. 391c) – synonym of *Aedimorphus* Theobald, 1903a; synonymy established before Knight & Stone, 1977.

Edwardsaedes Belkin, 1962 (subgenus of *Aedes* Meigen, 1818) (p. 408) – genus of tribe Aedini Neveu-Lemaire, 1902c; generic status by Reinert *et al.*, 2004.

Ekrinomyia Leicester, 1908 (genus) (p. 71) – synonym of *Pardomyia* Theobald, 1907; synonymy by transfer of type species, *aureostriata* Leicester, 1908, as synonym of *aurantius* (Theobald, 1907), from *Mucidus* Theobald, 1901b to *Mucidus* subgenus *Pardomyia* by Reinert, 2006b.

Elpeytonius Reinert, Harbach & Kitching, 2009 (p. 768) – genus of tribe Aedini Neveu-Lemaire, 1902c.

Empihals Reinert, Harbach & Kitching, 2008 (p. 105) – subgenus of *Ochlerotatus* Lynch Arribálzaga, 1891a,b.

Feltianus Martini, 1930 (globus of *Aedes* subgenus *Ochlerotatus* Lynch Arribálzaga, 1891a,b) (p. 264) – synonym of *Rusticoidus* Shevchenko & Prudkina, 1973; synonymy established before Knight & Stone, 1977.

Finlaya Theobald, 1903a (p. 281) – genus of tribe Aedini Neveu-Lemaire, 1902c; elevated from subgeneric status in *Aedes* Meigen by Reinert *et al.*, 2004.

Synonym: *Popea* Ludlow, 1905a.

'Finlaya' *sensu auctorum* – subgenus of *'Ochlerotatus'* Lynch Arribálzaga, 1891a,b *sensu auctorum*; repository introduced by Reinert *et al.* (2004) for species previously placed in *Ochlerotatus* subgenus *Finlaya* Theobald that could not be assigned with confidence to any newly proposed genera.

Fredwardsius Reinert, 2000a (subgenus of *Aedes* Meigen, 1818) (p. 1) – genus of tribe Aedini Neveu-Lemaire, 1902c; generic status by Reinert *et al.*, 2004.

Geitonomyia Leicester, 1908 (genus) (p. 134) – synonym of *Aedimorphus* Theobald, 1903a; synonymy established before Knight & Stone, 1977.

Georgecraigius Reinert, Harbach & Kitching, 2006 (p. 70) – genus (subgenus) of tribe Aedini Neveu-Lemaire, 1902c.

Geoskusea Edwards, 1929b (subgenus of *Aedes* Meigen, 1818) (p. 342) – genus of tribe Aedini Neveu-Lemaire, 1902c; generic status by Reinert *et al.*, 2004.

Gilesia Theobald, 1903a (genus) (p. 233) – subgenus of *Ochlerotatus* Lynch Arribálzaga, 1891a,b; resurrected from synonymy with *Ochlerotatus* Lynch Arribálzaga by Reinert *et al.*, 2008.

Gilesius Reinert, Harbach & Kitching, 2006 (p. 73) – genus of tribe Aedini Neveu-Lemaire, 1902c.

Gualteria Lutz, 1904 (in Bourroul, 1904) (genus) (pp. 47, 54) – synonym of *'Protomacleaya'* Theobald, 1907; synonymy by Zavortink, 1972.

Gymnometopa Coquillett, 1906a (p. 183) – genus of tribe Aedini Neveu-Lemaire, 1902c; elevated from subgeneric status in *Aedes* Meigen by Reinert *et al.*, 2004.

Halaedes Belkin, 1962 (subgenus of *Aedes* Meigen, 1818) (p. 328) – genus of tribe Aedini Neveu-Lemaire, 1902c; generic status by Reinert *et al.*, 2004.

Heteraspidion Reinert, Harbach & Kitching, 2009 (p. 781) – subgenus of *Stegomyia* Theobald, 1901.

Himalaius Reinert, Harbach & Kitching, 2006 (p. 74) – genus of tribe Aedini Neveu-Lemaire, 1902c.

Hopkinsius Reinert, Harbach & Kitching, 2008 (p. 98) – genus (subgenus) of tribe Aedini Neveu-Lemaire, 1902c.

Horsfallius Reinert, Harbach & Kitching, 2006 (p. 72) – subgenus of *Georgecraigius* Reinert, Harbach & Kitching, 2006.

Howardina Theobald, 1903a (p. 287) – genus of tribe Aedini Neveu-Lemaire, 1902c; elevated from subgeneric status in *Aedes* Meigen by Reinert *et al.*, 2004.

Synonym: *Soperia* Komp, 1936a.

Huaedes Huang, 1968b (subgenus of *Aedes* Meigen, 1818) (p. 169) – genus of tribe Aedini Neveu-Lemaire, 1902c; generic status by Reinert *et al.*, 2004.

Huangmyia Reinert, Harbach & Kitching, 2009 (p. 782) – subgenus of *Stegomyia* Theobald, 1901.

Hulecoeteomyia Theobald, 1904c (p. 163) – genus of tribe Aedini Neveu-Lemaire, 1902c; resurrected from synonymy with *Finlaya* Theobald by Reinert *et al.*, 2004.

Hyparctius Martini, 1930 (globus of *Aedes* subgenus *Ochlerotatus* Lynch Arribálzaga, 1891a,b) (p. 264) – synonym of *Ochlerotatus* Lynch Arribálzaga, 1891a,b; synonymy established before Knight & Stone, 1977.

Indusius Edwards, 1934 (in Barraud, 1934) (subgenus of *Aedes* Meigen, 1818) (pp. 268, 445) – genus of tribe Aedini Neveu-Lemaire, 1902c; generic status by Reinert *et al.*, 2004.

Isoaedes Reinert, 1979 (subgenus of *Aedes* Meigen, 1818) (p. 144) – genus of tribe Aedini Neveu-Lemaire, 1902c; generic status by Reinert *et al.*, 2004.

Jarnellius Reinert, Harbach & Kitching, 2006 (p. 76) – genus of tribe Aedini Neveu-Lemaire, 1902c.

Jihlienius Reinert, Harbach & Kitching, 2006 (p. 79) – genus of tribe Aedini Neveu-Lemaire, 1902c.

Juppius Reinert, Harbach & Kitching, 2009 (p. 770) – subgenus of *Ochlerotatus* Lynch Arribálzaga, 1891a,b.

Kenknightia Reinert, 1990 (subgenus of *Aedes* Meigen, 1818) (p. 4) – genus of tribe Aedini Neveu-Lemaire, 1902c; generic status by Reinert *et al.*, 2004.

Kingia Theobald, 1910c (genus) (p. 135) – synonym of *Stegomyia* Theobald, 1901; synonymy established before Knight & Stone, 1977.

Kompia Aitken, 1941 (subgenus of *Aedes* Meigen, 1818) (p. 81) – genus of tribe Aedini Neveu-Lemaire, 1902c; generic status by Reinert *et al.*, 2004.

Kummyia Vargas, 1950b (subgenus of *Aedes* Meigen, 1818) (p. 155) – synonym of *Protoculex* Felt, 1904; synonymy by transfer of type species (*serratus* Theobald, 1901c) from *Ochlerotatus* Lynch Arribálzaga, 1891a,b by Reinert *et al.*, 2008.

Lepidokeneon Reinert, Harbach & Kitching, 2009 (p. 773) – subgenus of *Ochlerotatus* Lynch Arribálzaga, 1891a,b.

Lepidoplatys Coquillett, 1906b (genus) (p. 314) – synonym of *Ochlerotatus* Lynch Arribálzaga, 1891a,b; synonymy established before Knight & Stone, 1977.

Lepidotomyia Theobald, 1905e (genus) (p. 22) – synonym of *Phagomyia* Theobald, 1905e; synonymy by transfer of type species (*magna* Theobald, 1905e, as synonym of *gubernatoris* Giles, 1901a) from *Finlaya* Theobald, 1903a by Reinert *et al.*, 2006.

Lepidotomyia Theobald, 1905f (genus) (p. 80) – synonym of *Aedimorphus* Theobald, 1903a; synonymy established before Knight & Stone, 1977.

Leptosomatomyia Theobald, 1905f (p. 110) – genus of tribe Aedini Neveu-Lemaire, 1902c; elevated from subgeneric status in *Aedes* Meigen by Reinert *et al.*, 2004.

Leslieomyia Christophers, 1911 (genus) (p. 68) – synonym of *Aedimorphus* Theobald, 1903a; synonymy established before Knight & Stone, 1977.

Levua Stone & Bohart, 1944 (subgenus of *Aedes* Meigen, 1818) (p. 214) – genus of tribe Aedini Neveu-Lemaire, 1902c; generic status by Reinert *et al.*, 2004.

Lewnielsenius Reinert, Harbach & Kitching, 2006 (subgenus of *Jarnellius* Reinert *et al.*, 2004) (p. 79) – genus of tribe Aedini Neveu-Lemaire, 1902c; generic status by Reinert *et al.*, 2009.

Lorrainea Belkin, 1962 (subgenus of *Aedes* Meigen, 1818) (p. 430) – genus of tribe Aedini Neveu-Lemaire, 1902c; generic status by Reinert *et al.*, 2004.

Luius Reinert, Harbach & Kitching, 2008 (p. 102) – genus of tribe Aedini Neveu-Lemaire, 1902c.

Macleaya Theobald, 1903e (p. 154) – genus (subgenus) of tribe Aedini Neveu-Lemaire, 1902c; elevated from subgeneric status in '*Ochlerotatus*' *sensu auctorum* by Reinert *et al.*, 2006.

Mimeteculex Theobald, 1908c (genus) (p. 258) – synonym of *Aedimorphus* Theobald, 1903a; synonymy established before Knight & Stone, 1977.

Molpemyia Theobald, 1910c (p. 479) – genus of tribe Aedini Neveu-Lemaire, 1902c; elevated from subgeneric status in *Aedes* Meigen by Reinert *et al.*, 2006.
Synonym: *Calomyia* Taylor, 1914c.

Mucidus Theobald, 1901b (p. 235) – genus (subgenus) of tribe Aedini Neveu-Lemaire, 1902c.

Mukwaya Reinert, Harbach & Kitching, 2009 (p. 782) – subgenus of *Stegomyia* Theobald, 1901.

Myxosquamus Theobald, 1909b (genus) (p. 7) – synonym of *Catageiomyia* Theobald, 1903d; synonymy by transfer of type species to *Catageiomyia* by Reinert *et al.*, 2009.

Neomelaniconion Newstead, 1907 (in Newstead *et al.*, 1907) (p. 31) – genus of tribe Aedini Neveu-Lemaire, 1902c.
Synonym: *Banksinella* Theobald, 1907.

Neopecomyia Theobald, 1909b (genus) (p. 12) – synonym of *Catageiomyia* Theobald, 1903d; synonymy by transfer of type species to *Catageiomyia* by Reinert *et al.*, 2009.

Nyctomyia Harbach, 2013 (p. 199) – genus of Aedini Neveu-Lemaire, 1902c [*nomen novum* for *Nyx* Harbach & Linton, 2013].
Synonym: *Nyx* Harbach & Linton, 2013.

Nyx Harbach & Linton, 2013 (in Harbach *et al.*, 2013) (genus) (p. 160) – synonym of *Nyctomyia* Harbach, 2013; synonymy established by Harbach, 2013.

Ochlerotatus Lynch Arribálzaga, 1891a,b (pp. 374, 143, respectively) – genus (subgenus) of tribe Aedini Neveu-Lemaire, 1902c.
Synonyms: *Pseudoculex* Dyar, 1905a; *Lepidoplatys* Coquillett, 1906b; *Pseudohowardina* Theobald, 1907; *Andersonia* Strickland, 1911b; *Feltianus* Martini, 1930.

'*Ochlerotatus*' *sensu auctorum* – subgenus of '*Ochlerotatus*' Lynch Arribálzaga, 1891a,b *sensu auctorum*; repository introduced by Reinert *et al.* (2004) for species previously placed in *Ochlerotatus* Lynch Arribálzaga that could not be assigned with confidence to any newly proposed genera.

Paraedes Edwards, 1934 (in Barraud, 1934) (p. 446) – genus of tribe Aedini Neveu-Lemaire, 1902c; elevated from subgeneric status in *Aedes* Meigen by Reinert *et al.*, 2004.

Pardomyia Theobald, 1907 (genus) (p. 280) – subgenus of *Mucidus* Theobald, 1901b; resurrected from synonymy with *Mucidus* Theobald by Reinert, 2006b.
Synonym: *Ekrinomyia* Leicester, 1908.

Patmarksia Reinert, Harbach & Kitching, 2006 (p. 81) – genus of tribe Aedini Neveu-Lemaire, 1902c.

Paulianius Brunhes & Boussès, 2017 (in Brunhes *et al.*, 2017) (p. 345) – genus of tribe Aedini Neveu-Lemaire, 1902c.

Pecomyia Theobald, 1905g (genus) (p. 23) – synonym of *Aedimorphus* Theobald, 1903a; synonymy established before Knight & Stone, 1977.

Petermattinglyius Reinert, Harbach & Kitching, 2009 (p. 778) – genus (subgenus) of tribe Aedini Neveu-Lemaire, 1902c.

Phagomyia Theobald, 1905e (p. 21) – genus of tribe Aedini Neveu-Lemaire, 1902c; resurrected from synonymy with *Finlaya* Theobald by Reinert *et al.*, 2006.
Synonyms: *Lepidotomyia* Theobald, 1905e; *Pseudocarrollia* Theobald, 1910a.

Pholeomyia Reinert, Harbach & Kitching, 2008 (p. 162) – subgenus of *Ochlerotatus* Lynch Arribálzaga, 1891a,b). **Note**: this name is preoccupied by *Pholeomyia* Bilimek, 1867. Should the subgenus be retained as valid, *Reinertomyia* should be considered as a replacement name.

Polyleptiomyia Theobald, 1905e (p. 21) – genus of tribe Aedini Neveu-Lemaire, 1902c; resurrected from synonymy with *Aedimorphus* Theobald by Reinert *et al.*, 2009.

Popea Ludlow, 1905a (genus) (p. 95) – synonym of *Finlaya* Theobald, 1903a; synonymy established before Knight & Stone, 1977.

Promacleaya Surcouf & Gonzalez-Rincones, 1912 (p. 277) – unjustified emendation of *Protomacleaya* Theobald, 1907.

Protoculex Felt, 1904 (genus) (p. 391d) – subgenus of *Ochlerotatus* Lynch Arribálzaga, 1891a,b; resurrected from synonymy with *Ochlerotatus* Lynch Arribálzaga by Reinert *et al.*, 2008.

'*Protomacleaya*' Theobald, 1907 (genus) (pp. 149, 253) – subgenus of '*Ochlerotatus*' Lynch Arribálzaga, 1891a,b *sensu auctorum*; repository introduced by Reinert *et al.* (2004) for species previously placed in *Ochlerotatus* subgenus *Protomacleaya* Theobald that could not be assigned with confidence to any newly proposed genera.
Synonym: *Gualteria* Lutz, 1904.

'*Pseudalbuginosus*' Huang & Rueda, 2015 (subgenus of *Aedes* Meigen, 1818) (p. 382) – subgenus of '*Aedes*' Meigen, 1818 *sensu auctorum*; transfer from *Aedes* Meigen by Harbach (24 August 2015, http://mosquito-taxonomic-inventory.info/simpletaxonomy/term/8743).

Pseudarmigeres Stone & Knight, 1956 (subgenus of *Aedes* Meigen, 1818) (p. 34) – genus of tribe Aedini Neveu-Lemaire, 1902c; generic status by Reinert *et al.*, 2004.
Synonym: *Dunnius* Edwards, 1930a.

Pseudocarrollia Theobald, 1910a (genus) (p. 12) – synonym of *Phagomyia* Theobald, 1905e; synonymy by transfer of type species (*lophoventralis* Theobald, 1910a) from *Finlaya* Theobald, 1903a by Reinert *et al.*, 2006.

Pseudoculex Dyar, 1905a (genus) (p. 47) – synonym of *Ochlerotatus* Lynch Arribálzaga, 1891a,b; synonymy established before Knight & Stone, 1977.

Pseudograbhamia Theobald, 1905h (genus) (p. 243) – synonym of *Aedimorphus* Theobald, 1903a; synonymy established before Knight & Stone, 1977.

Pseudohowardina Theobald, 1907 (genus) (p. 223) – synonym of *Ochlerotatus* Lynch Arribálzaga, 1891a,b; synonymy established before Knight & Stone, 1977.

Pseudoskusea Theobald, 1907 (genus) (p. 192) – subgenus of *Ochlerotatus* Lynch Arribálzaga, 1891a,b; subgeneric status by Reinert *et al.*, 2008.
Synonym: *Caenocephalus* Taylor, 1914c.

Pseudostegomyia Ludlow, 1908b (genus) (p. 10) – synonym of *Stegomyia* Theobald, 1901; synonymy with *Quasistegomyia* Theobald, 1906b by Theobald, 1910c.

Quasistegomyia Theobald, 1906b (genus) (p. 69) – synonym of *Stegomyia* Theobald, 1901; synonymy established before Knight & Stone, 1977.

Rampamyia Reinert, Harbach & Kitching, 2006 (p. 83) – genus of tribe Aedini Neveu-Lemaire, 1902c.

Reedomyia Ludlow, 1905a (genus) (p. 94) – synonym of *Aedimorphus* Theobald, 1903a; synonymy established before Knight & Stone, 1977.

Rhinoskusea Edwards, 1929b (subgenus of *Aedes* Meigen, 1818) (p. 342) – genus of tribe Aedini Neveu-Lemaire, 1902c; elevated from subgeneric status in *Aedes* Meigen by Reinert *et al.*, 2004.

Rusticoidus Shevchenko & Prudkina, 1973 (subgenus of *Aedes* Meigen, 1818) (p. 46) – subgenus of *Ochlerotatus* Lynch Arribálzaga, 1891a,b; transfer from subgeneric status in *Aedes* Meigen by Reinert, 2000d.

Sallumia Reinert, Harbach & Kitching, 2008 (subgenus of *Ochlerotatus* Lynch Arribálzaga, 1891a,b) (p. 110) – genus of tribe Aedini Neveu-Lemaire, 1902c; generic status by Reinert *et al.*, 2009.

Scutomyia Theobald, 1904c (p. 77) – genus of tribe Aedini Neveu-Lemaire, 1902c; elevated from subgeneric status in *Aedes* Meigen by Reinert *et al.*, 2004.

Skusea Theobald, 1903a (p. 291) – genus of tribe Aedini Neveu-Lemaire, 1902c; elevated from subgeneric status in *Aedes* Meigen by Reinert *et al.*, 2004.

Soperia Komp, 1936a (subgenus of *Aedes* Meigen, 1818) (p. 71) – synonym of *Howardina* Theobald, 1903a; synonymy established before Knight & Stone, 1977.

Stegomyia Theobald, 1901 (in Howard, 1901) (p. 235) – genus (subgenus) of tribe Aedini Neveu-Lemaire, 1902c.

Stenoscutus Theobald, 1909b (genus) (p. 7) – synonym of *Catageiomyia* Theobald, 1903d; synonymy by transfer of type species to *Catageiomyia* by Reinert *et al.*, 2009.

Taeniorhynchus Lynch Arribálzaga, 1891a,b (pp. 374, 143, respectively) – synonym of *Culicelsa* Felt, 1904; synonymy established before Knight & Stone, 1977.

Tanakaius Reinert, Harbach & Kitching, 2004 (p. 349) – genus of tribe Aedini Neveu-Lemaire, 1902c.

Tewarius Reinert, 2006a (p. 640) – genus of tribe Aedini Neveu-Lemaire, 1902c.

Vansomerenis Reinert, Harbach & Kitching, 2006 (p. 85) – genus of tribe Aedini Neveu-Lemaire, 1902c.

Woodius Reinert, Harbach & Kitching, 2009 (p. 775) – subgenus of *Ochlerotatus* Lynch Arribálzaga, 1891a,b.

Xyele Reinert, Harbach & Kitching, 2009 (p. 783) – subgenus of *Stegomyia* Theobald, 1901.

Yamada Reinert, Harbach & Kitching, 2008 (p. 101) – subgenus of *Hopkinsius* Reinert, Harbach & Kitching, 2008.

Zavortinkius Reinert, 1999a (subgenus of *Aedes* Meigen, 1818) (p. 7) – genus of tribe Aedini Neveu-Lemaire, 1902c; elevated from subgeneric status in *Aedes* Meigen by Reinert *et al.*, 2004.

Zoromorphus Reinert, Harbach & Kitching, 2009 (p. 783) – subgenus of *Stegomyia* Theobald, 1901.

Section 9: Changed Spellings of Species-group Names

Inspection of Section 1 reveals that 894 currently recognized extant species are included in genera other than the ones in which they were originally placed. When the composite *Aedes* is divided into numerous smaller genera (Section 8), many of which were removed from synonymy with *Aedes*, the number of species that have been transferred from the original to another genus is 1470. Surprisingly, the spelling of only 196 of the larger number of species has had to be changed to agree in gender with the genus to which they have been transferred.

Obviously a large proportion of species-group names have retained their original spellings when transferred to another genus. The reason is that many adjectival and participial names have retained their spelling because they were transferred to a genus of the same gender as the one from which they were removed. The spellings of many other names never change, including names taken from Latin and Greek nouns, and those derived from personal names and other proper names, e.g. the scientific names of plants and the names of geographical entities (see Sections 1, 4, 5 and 7). Many names have Latin endings that are both masculine and feminine (e.g. *-ensis*, see Section 5) that do not change when they are transferred from a masculine to a feminine genus, and vice versa.

Prior to the 21st century, changes to the spelling of adjectival and participial species-group names as a consequence of being transferred to a genus of a different gender were accepted as a linguistic and taxonomic requirement without concern or complaint. No concerns were raised, for example, when numerous changes to the spelling of species names occurred when Reinert (1999b) raised *Verrallina* (feminine) to generic status from subgeneric status in *Aedes* (masculine). For a number of reasons, further changes to the classification of composite *Aedes* during the first decade of the 21st century (Reinert, 2000d; Reinert *et al.*, 2004, 2006, 2008, 2009) were a major cause of grievance among mosquito workers, especially operational entomologists. Although a satisfactory resolution to objections concerning the reclassification of *Aedes* was established by Wilkerson *et al.* (2015), one of the concerns, 'spelling changes in hundreds of species names' (Savage, 2005), needs to be put in perspective. This is an over-exaggeration that has no basis in fact.

As noted above, among the 3559 species that are currently recognized as extant members of the family Culicidae, only a fraction (196, 0.55%) have been transferred to genera that has resulted in a change of spelling. Among the species that comprise the traditional classification of the family (*sensu* Wilkerson *et al.*, 2015), 129 names have been changed to agree in gender with the genus in which they are currently placed. When the reclassification of the composite *Aedes* is factored in, the spelling of 169 names have changed. (Note: some names are changed in one classification but not the other, and vice versa, the differences resulting in a total of 196 changed spellings.) These changes, of course, are for the entire family. Considering only species that comprise the traditional broad concept of *Aedes*, 31 are spelled differently than originally proposed. When the composite *Aedes* is reclassified into numerous smaller genera (*sensu* Reinert *et al.*, 2009), the spellings of 74 species (not hundreds) are changed. The number of changes between the traditional and multigeneric classifications appears to be 43, but this is not correct. Eight of the changed spellings are common to both classifications; therefore, the reclassification of *Aedes* only resulted in an additional 35 changed spellings.

Currently recognized extant species that have had the spelling of their names changed to agree in gender with the genus in which they are now placed are listed in Appendix 3. The original formation and generic placement of each name is contrasted with its generic assignment and formation in the current traditional classification (Section 1) and the classification that includes the reclassification of composite *Aedes* (Section 8).

Some Thoughts about Mosquito Taxonomy and Nomenclature

It is an undeniable fact that DNA sequencing has overtaken the study of mosquito biodiversity to the exclusion of morphological study. However, it is also obvious that molecular methods are not being used to conduct revisionary taxonomic studies of group taxa with the aim of resolving nomenclatural problems, making formal taxonomic changes, recognizing and naming new species, and refining the classification of mosquitoes. Although it is frequently argued that molecular data are needed to corroborate proposed changes to classifications based on phylogenetic analyses of morphological data, the presumed superiority of DNA sequence data for resolving relationships is fraught with inconsistent results and a failure of 'molecular taxonomists' to translate their findings into revised supraspecific classifications based on unambiguous natural affinities. Until DNA sequence data can be shown to consistently produce unequivocal phylogenies, they should not be viewed as a panacea for resolving deeper relationships within the Culicidae.

It is in the nature of humans to discover and name things. Museums house millions of specimens that have been discovered and named. Names are given to species to acknowledge their existence as separate from other species that are given different names. But naming species is not without problems – some end up with more than one name. The application of two or more names to a species presumed to be distinct from other species requires careful assessment because the names may actually apply to two or more morphologically very similar or cryptic species. The process of resolving the identities of such species is analogous to the identification of human remains by forensic anthropologists, which is sometimes an impossible task based on the condition of material at hand. The use of molecular methodology has the potential to determine whether the names of nominal species represent the same or different species, or species new to science. However, molecular techniques are not a substitute for morphological study, because they cannot be used by themselves to resolve formal taxonomic problems. The reason for this is that faunal classifications are based on morphology and the type specimen concept, which means that molecular techniques must be integrated with classical morphological analysis to determine the correct application of available names for species concepts previously based solely on anatomical characters. This is a problem, because molecular taxonomists are largely unfamiliar with the morphology-based taxonomy. They have limited knowledge of mosquito anatomy and morphological terminology, lack insights into the limitations of available identification keys, published illustrations and descriptions, and are unaware of the morphological diversity of mosquitoes. This predicates a perilous situation for the future of mosquito taxonomy. Unfortunately, the current generation seems to be of a different mind-set than the older, classical practitioners. The molecular methods are more attractive for a number of reasons, but ultimately it is easier and takes much less time to do DNA studies than to do morphological studies. Sadly, the whole-organism approach, the study of morphological diversity and anatomical distinctions, has lost its appeal in favour of molecular diversity and nucleotide distinctions. To overcome this bias, it would be prudent for young taxonomists to consult and collaborate with taxonomic experts. Unfortunately, there is a critical shortage of professional mosquito taxonomists, and in a few years there will be none left to work with the new generation of molecular taxonomists.

If the curtain is about to close on classical mosquito taxonomy, members of the taxonomic community should give some thought to an important question: what will become of mosquito nomenclature? The only thing which seems to be certain is that molecular taxonomy is forging ahead without appropriately interfacing with classical morphological analysis, and without resolving taxonomic problems and naming new species as was done in the heyday of taxonomic revisionary studies. Taxonomy has always focused upon species discovery, description, naming and classification. Surely numerous mosquito species await discovery and naming, but what is their fate, and the fate of named species, in the face of environmental alteration and destruction? Many undiscovered species, and even some known and named species, may have already vanished, particularly those adapted to thrive in specialized habitats and specific localities. When a species becomes extinct, taxonomic names will only apply to a few dead specimens that remain in museum collections – fossils of interest to future mosquito palaeontologists.

Mosquito taxonomy is a work in progress. Assuming that new-age taxonomy will bring about changes to the existing classification, involving the transfer of names into and out of synonymy and reassignments of species among genera, the requirement for changing the spellings of names for gender agreement is likely to be met with consternation, as happened when John Reinert (1999b) was informed that the spelling of many adjectival species names had to be changed when he elevated *Verrallina* (feminine) to generic status from subgeneric status in *Aedes* (masculine). Name changes, especially those involving medically important species, always cause confusion and difficulties for non-taxonomists and taxonomists alike.

It is not an easy chore to determine whether a given species-group name is an adjective or a noun in apposition, the former liable to a change of spelling (*Code*, Articles 31.2 and 34.2) whereas the latter is immutable. Unlike scientists of the 18th, 19th and early part of the 20th centuries who were educated and proficient in the use of Latin and Greek, most contemporary taxonomists have little practical knowledge of those languages. In view of the troublesome requirement for gender agreement, and in the interests of nomenclatural stability, should mosquito taxonomists disregard the gender agreement provisions of the *Code*? If the answer is yes, two options are available: (i) adopt the original spelling of all adjectival species-group names that have been changed; or (ii) accept the current spellings of all such names and treat them as immutable so that they will never again be altered. The principal advantage of ignoring gender agreement lies in the fact that stability and universality of usage fosters easy, clear and exact information retrieval and communication. It is worth noting that the requirement for gender agreement has been ignored by lepidopterists, who treat generic names as genderless and retain the original spellings of specific names (Ian Kitching, personal communication).

Literature Cited

References are listed in alphabetical order by author, alphabetized by the last name of the sole or first author of each publication. Articles by single authors are followed by articles with two authors, and those in turn are followed by articles with three or more authors. For multiple articles by the same author, or two or more authors listed in the same order, entries are in chronological order, from earliest to most recent. In the case of multiple articles by the same author or authors published in the same year, lower case letters are added to the date in alphabetical sequence to identify each when cited (in cases where the full authorship is not necessary, publications are identified in the text by the surname of the first author followed by *et al.*). For multi-author publications with the same first author, and also where one or more of the following authors are the same, the articles are listed in alphabetical order by the surname of the first co-author that is different than the authorship of the preceding reference, regardless of the number of authors. In a few places in the text, the first two or three authors of multi-author articles are cited so that those particular articles can be located in the list.

Authors' names and the wording of articles are spelled exactly as they were published, including compound names, diacritical marks, special symbols, improper italicization and the misspelling of mosquitoes (as 'mosquitos'). However, genus and species names are italicized in titles even if they were not italicized in the published article. Likewise, titles published in all capital letters are reproduced here in lower case letters, except for proper names that must begin with a capital letter, e.g. genus names and German nouns. The titles of books are italicized and the names of journals are written in full and italicized. Volume numbers of periodicals are in Arabic numerals even if denoted as Roman numerals. Issues that comprise individual volumes, if applicable or known from available copies of published articles, are enclosed in parentheses after volume numbers.

In addition to the exceptions mentioned above, the symbol v, which came into use to represent the Latin consonantal u, and was used in the titles of publications by Linnaeus and other authors of the 17th and 18th centuries, has been replaced with the letter u in applicable references. Also, the symbol æ, a ligature of the vowels a and e, has been replaced with ae, e.g. Linnæus and *Systema Naturæ* are spelled as Linnaeus and *Systema Naturae*, respectively.

In many cases, publications listed here and in the world catalogue of Knight & Stone (1977) and its supplements that have the same authorship and date do not have the same letter appended to the date, or no letter at all, e.g. Marks, 1958 and Reinert, 1972b correspond to Marks, 1958a and Reinert, 1972e in the catalogue of Knight & Stone, respectively. This is due to the many non-taxonomic articles included in the catalogue and its supplements that are not included herein.

The list below includes 2480 references, the majority of which contain the originally proposed names of mosquito taxa. The goal was to see the original sources of all names that comprise the culicid nomenclature, which was accomplished except for the following two names: *Anopheles peruvianus* Tamayo, 1907 (but see Tamayo & García, 1907) and *Culex raymondii* Tamayo, 1907 (but see Tamayo & García, 1907).

Abdulla-Khan, R., Coetzee, M. & Hunt, R.H. 1998. Description of *Anopheles* (*Cellia*) *seretsei* sp. nov. from Kasane, Botswana. *Journal of the American Mosquito Control Association* 14(3): 248–252.

Abercrombie, J. 1977. Medical entomology studies – IX. The subgenus *Christophersiomyia* of the genus *Aedes* (Diptera: Culicidae). *Contributions of the American Entomological Institute* 14(2): 1–72.

Abonnenc, E. 1954. Sur un anophele cavernicole de la Guinee: *Anopheles cavernicolus* n.sp. (Diptera – Culicidae). *Bulletins et Mémoires de l'École Préparatoire de Médecine et de Pharmacie de Dakar, Parasitologie* 2: 288–290.

Abonnenc, E. 1956. Speologica africana. Sur un anophèle cavernicole de la Guinée: *Anopheles cavernicolus* n. sp. (Diptera–Culicidae). *Bulletin de l'Institut Français d'Afrique Noire* 18(3): 802–805.

Abraham, O.K. 1947. *Anopheles* (*Anopheles*) *kyondawensis* n. sp., from Lower Burma. *Medical Journal of Malaya* 1: 173–176.

Adam, J.-P. 1961. *Anopheles caroni* n. sp. (*Diptera, Culicidae*) anophèle cavernicole nouveau de la République du Congo. *Bulletin de la Société de Pathologie exotique* 54(4): 714–717.

Adam, J.-P. 1962. Un anophèle cavernicole nouveau de la République du Congo (Brazzaville): *Anopheles* (*Neomyzomyia*) *hamoni* n. sp. (*Diptera-Culicidae*). *Bulletin de la Société de Pathologie exotique* 55(1): 153–165.

Adam, J.P. & Hamon, J. 1959. Description de deux nouvelles sous-espèces d'*Eretmapodites œdipodius* Graham, capturées en Afrique Occidentale. *Bulletin de la Société de Pathologie exotique* (for 1958) 51(4): 524–527.

Adam, J.-P., Hamon, J. Rickenbach, A. & Lips, M. 1956. I. – Description du male et du pharynx de la femelle d'*Anopheles brohieri* Edwards 1929 et *A. hancocki* var. *masseguini* Hamon 1954. II. – Étude des affinités existant entre *Anopheles hancocki*, *A. hancocki* var. *masseguini*, *A. brohieri*, *A. theileri*, *A. seydeli*. *Bulletin de la Société de Pathologie exotique* 49(4): 747–758.

Adames, A.J. 1971. Mosquito studies (Diptera, Culicidae) XXIV. A revision of the crabhole mosquitoes of the genus *Deinocerites*. *Contributions of the American Entomological Institute* 7(2): 1–154.

Adames, A.J. & Galindo, P. 1973. Mosquito studies (Diptera, Culicidae) XXX. A new subgenus and species of *Culex* from Colombia. *Contributions of the American Entomological Institute* 9(3): 55–61.

Adames, A.J. & Hogue, C.L. 1970. Mosquito studies (Diptera, Culicidae) XVII. Two new species of *Deinocerites* from Costa Rica. *Contributions of the American Entomological Institute* (for 1969) 5(1): 9–20.

Adams, C.F. 1903. Dipterological contributions. *Kansas University Science Bulletin* 2(2): 21–47.

Agassiz, L. 1846. *Nomenclatoris zoologici index universalis, continens nomina systematica classium, ordinum, familiarum et generum animalium omnium, tam viventium quam fossilium, secundum ordinem alphabeticum unicum disposita, adjectis homonymiis plantarum, nec non variis adnotationibus et emendationibus*. Jent et Gassmann, Soloduri [= Solothurn, Switzerland]. viii + 393 pp.

Agramonte, A. 1900. Los mosquitos maláricos. *El Progreso Médico, Habana* 10(12): 455–465.

Aiken, J. 1909. Notes on the mosquitoes of British Guiana. *British Guiana Medical Annual* 1908: 1–25, 4 pls.

Aiken, J. 1911. A synoptical view of the mosquitoes of British Guiana. *Timehri* 1(2): 187–204.

Aiken, J. & Rowland, E.D. 1906. Preliminary notes on the mosquitoes of British Guiana. *British Guiana Medical Annual* 1905: 13–38, 8 pls.

Airi, M. & Kaur, S. 2015. Confirmation of *Culex (Culex) tritaeniorhynchus summorosus* (Diptera; Culicidae) as a separate species. *Journal of Vector Borne Diseases* 52(3): 219–223.

Aitken, T.H.G. 1939. The *Anopheles maculipennis* complex of western America (Diptera, Culicidae). *Pan-Pacific Entomologist* 15(4): 191–192.

Aitken, T.H.G. 1941. A new American subgenus and species of *Aëdes* (Diptera, Culicidae). *Pan-Pacific Entomologist* 17(2): 81–84.

Aitken, T.H.G. 1945. Studies on the anopheline complex of western America. *University of California Publications in Entomology* 7(11): 273–364.

Aitken, T.H.G. 1953. The anopheline fauna of Sardinia. *American Journal of Hygiene Monographic Series* 20: 303–352, i–vi.

Alcock, A. 1911. Remarks on the classification of the Culicidae, with particular reference to the constitution of the genus *Anopheles*. *Annals and Magazine of Natural History* 8(44): 240–250.

Alcock, A. 1912. Description of a new species of *Anopheles* from the Malay Peninsula. *Journal of the London School of Tropical Medicine* 2(1): 1–2.

Alcock, A. 1913. Synopsis of the anopheline mosquitos of Africa and of the Oriental Region. *Journal of the London School of Tropical Medicine* 2(3): 153–166.

Alekseev, E.V. 1989. Bloodsucking mosquito *Aedes* (*Rusticoidus*) *krymmontanus* Alekseev, sp. n., a relic of the entomofauna of the Crimea [in Russian]. *Parazitologiia* 23(2): 173–178.

Alvarado, C.A. & Heredia, R.L. 1947. Observaciones sobre una nueva variedad del *Anopheles* (*A.*) *pseudopunctipennis* Theobald 1901, encontrada en la provincia de Tucuman. (Nota previa). *Anales del Instituto de Medicina Regional Universidad Tucuman* 2(1): 73–78.

Amerasinghe, F.P. 1993. A new species of *Heizmannia* from Sri Lanka (Diptera: Culicidae). *Mosquito Systematics* 25(2): 110–115.

Amos, D.W. 1944. *Mosquito Control Training Manual*. Suva, Fiji. 43 pp.

Anduze, P.J. 1941. Primer informe sobre entomologia medica del Estado Bolivar (Venezuela). Serie 1. La fauna culicidiana. Descripción de tres especies nuevas (*Diptera,Culicidae*). *Revista de Sanidad y Asistencia Social* 6: 812–836.

Anduze, P.J. 1942a. Fauna culicidiana de Venzuela. Descripción de dos especies nuevas (*Diptera: Culicidae*). *Boletín de Entomologia Venezolana* 1(2): 43–49.

Anduze, P.J. 1942b. II. La fauna culicidiana de Venezuela. Descripcion de una especie nueva (Diptera: Culicidae). *Revista de Sanidad y Asistencia Social* 7(4): 557–560.

Anduze, P.J. 1943a. Estudios de entomologia medica en el estado Merida –Venezuela. *La fauna culicidiana. – Descripción del Culex (Culex) albertoi* n. sp. *Boletín de Entomologia Venezolana* 2(4): 189–196.

Anduze, P.J. 1943b. *Culex* (*Culex*) *beauperthuyi* sp. nov. (Diptera: Culicidae). *Revista de Sanidad y Asistencia Social* 8(3): 459–461.

Anduze, P.J. 1947. Contribucion al estudio de los vectores de la fiebre amarilla en Venezuela. *Boletin de la Academia de Ciencias Físicas, Matemáticas y Naturales* 10(30): 331–373.

Anduze, P.J. 1948. Notas entomologicas I. Dos variedades nuevas de *A. aquasalis*. *Boletín médico* 1(1): 17–19.

Anduze, P.J. 1949a. Fauna culicina de Venezuela. Descripcion de cinco nuevas especies. *Boletín de Entomologia Venezolana* (for 1948) 7(1–2): 60–66.

Anduze, P.J. 1949b. Fauna culicina de Venezuela. Descripcion de una especie nueva. *Boletín de Entomologia Venezolana* (for 1948) 7(3–4): 149–150.

Anduze, P.J. & Capdevielle, A. 1949. *Anopheles* (*Anopheles*) *guarao* sp. nov. *Boletín de Entomologia Venezolana* 8(3–4): 119–124.

Anduze, P.J. & Hecht, O. 1943. *Aedes* (*Finlaya*) *upatensis* sp. nov. *Boletín de Entomologia Venezolana* 2(4): 185–188.

Angulo, A.O. & Olivares, T.S. 1993. Catalogo de los culicidos de Chile (Diptera Culicidae), y dos especies nuevas de *Culex* (*Culex*) Linnaeus. *Boletin Sociedad Biologica Concepción, Chile* 64: 21–26.

Antonio-Nkondjio, C., Ndo, C., Kengne, P., Mukwaya, L., Awono-Ambene, P., Fontenille, D. & Simard, F. 2008. Population structure of the malaria vector *Anopheles moucheti* in the equatorial forest region of Africa. *Malaria Journal* 7: 120.

Antunes, P.C.A. 1937. A new *Anopheles* and a new *Goeldia* from Colombia (Dipt. Culic.). *Bulletin of Entomological Research* 28(1): 69–73.

Antunes, P.C.A. 1939. Nota sobre o genero *Haemagogus* Williston — (Diptera, Culicidae). *Revista Paulista de Medicina (São Paulo)* 14: 106.

Antunes, P.C.A. & Lane, J. 1937. Um novo *Culex, Culex (Carrollia) soperi*, encontrado em São Paulo (*Diptera, Culicidae*). *Revista de Biologia e Hygiene* 8(1): 21–23.

Antunes, P.C.A. & Lane, J. 1938. Um novo *Aedes, Aedes (Ochlerotatus) pennai*, encontrado em São Paulo. *Revista do Museu Paulista* 23: 605–614, 3 pls.

Apfelbeck, V. 1929. Neue Culiciden aus Bosnien und Dalmatien. *Konowia: Zeitschrift für systematische Insektenkunde* 8(3): 286–292.

Arnell, J.H. 1973. Mosquito studies (Diptera, Culicidae) XXXII. A revision of the genus *Haemagogus*. *Contributions of the American Entomological Institute* 10(2): ii, 1–174.

Arnell, J.H. 1976. Mosquito studies (Diptera, Culicidae) XXXIII. A revision of the Scapularis Group of *Aedes (Ochlerotatus)*. *Contributions of the American Entomological Institute* 13(3): 1–144.

Arnell, J.H. & Nielsen, L.T. 1972. Mosquito studies (Diptera, Culicidae) XXVII. The Varipalpus Group of *Aedes (Ochlerotatus)*. *Contributions of the American Entomological Institute* 8(2): 1–48.

Awono-Ambene, H.P., Kengne, P., Simard, F., Antonio-Nkondjio, C. & Fontenille, D. 2004. Description and bionomics of *Anopheles (Cellia) ovengensis* (Diptera: Culicidae), a new malaria vector species of the *Anopheles nili* group from South Cameroon. *Journal of Medical Entomology* 41(4): 561–568.

Bachmann, A.O. & Casal, O.H. 1962. *Culex (Culex) lahillei* una nueva especie de la Argentina (Diptera: Culicidae). *Physis* 23(65): 267–270.

Baeza Cuéllar, M. 1933. *Estudio médico de los culícidos hematófagos*. Talleres Gráficos Herrera, Madrid. 158 pp.

Bahr, P.H. 1912. Filariasis and elephantiasis in Fiji: being a report to the London School of Tropical Medicine. Dysentry [*sic*] in Fiji during the year 1910. [Series] *Journal of the London School of Tropical Medicine, Supplement* 1: viii, 1–192, 5 pls, map.

Bailly-Choumara, H. 1960. Une espèce nouvelle d'anophèle du golfe d'Aden, *Anopheles (Myzomyia) azaniae* (Diptera, Culicidae), notes morphologiques, systématiques et écologiques. *Bulletin de la Société de Pathologie exotique* 53(3): 531–542.

Bailly-Choumara, H. 1966a. Contribution a l'étude des *Culex (Neoculex) (Diptera: Culicidae)* de la région Éthiopienne. Description de *Culex (Neoculex) pseudoandreanus* sp. n. *Bulletin de la Société de Pathologie exotique* (for 1965) 58(4): 660–664.

Bailly-Choumara, H. 1966b. *Culex (Culex) gameti (Diptera, Culicidae)*, moustique nouveau du Cameroun, description du male. *Bulletin de la Société de Pathologie exotique* 59(1): 148–150.

Bailly-Choumara, H. & Adam, J.-P. 1959. *Anopheles maliensis* n. sp. (Diptera Culicidae), une espèce nouvelle d'anophélinés du Fouta-Djalon (République de Guinée). *Compte Rendu de l'Académie des Sciences* 248: 3742–3743.

Bailly-Choumara, H. & Rickenbach, A. 1966. Contribution a l'étude des *Culex (Diptera, Culicidae)* de la région Éthiopienne. Description de *Culex (Neoculex) garioui* sp. n., moustique nouveau du Cameroun. *Bulletin de la Société de Pathologie exotique* 59(1): 144–148.

Baisas, F.E. 1932. The morphology of some Philippine *Anopheles*. *Transactions of the Eighth Congress of the Far East Association of Tropical Medicine* 1930: 252–257.

Baisas, F.E. 1935a. Notes on Philippine mosquitoes, V. The pupal characters of anophelines under the Myzorhynchus Series and group Neocellia, with further comments on the larvae and adults of *sinensis*. *Monthly Bulletin of the Bureau of Health (Manila)* 15(9): 291–339.

Baisas, F.E. 1935b. Notes on Philippine mosquitoes, I. The *Armigeres* group. *Philippine Journal of Science* 56(4): 485–497, 4 pls.

Baisas, F.E. 1935c. Notes on Philippine mosquitoes, II. *Uranotaenia* group. *Philippine Journal of Science* 57(1): 63–80, 4 pls.

Baisas, F.E. 1935d. Notes on Philippine mosquitoes, III. Genus *Culex*: groups *Lophoceratomyia*, *Mochthogenes*, and *Neoculex*. *Philippine Journal of Science* 57(2): 167–179, 4 pls.

Baisas, F.E. 1936. Notes on Philippine mosquitoes, IV. The pupal and certain adult characters of some rare species of *Anopheles*. *Philippine Journal of Science* 59(1): 65–84, 15 pls.

Baisas, F.E. 1938. Notes on Philippine mosquitoes, VII. A–*Culex (Culex)* with banded proboscis and tarsi. B–*Anopheles*: the pupae of three rare species; the *leucosphyrus*-subgroup. *Monthly Bulletin of the Bureau of Health (Manila)* 18(5): 175–232, 18 pls.

Baisas, F.E. 1946a. Notes on Philippine mosquitoes, VIII. Species found in the jungles of Llavac. *Monthly Bulletin of the Bureau of Health (Manila)* 22(1): 27–49.

Baisas, F.E. 1946b. Notes on Philippine mosquitoes, IX. A new species of *Zeugnomyia*. *Monthly Bulletin of the Bureau of Health (Manila)* 22(2): 27–35.

Baisas, F.E. 1946c. Notes on Philippine mosquitoes, X. Some species of *Aedes (Finlaya)* and *(Aedimorphus)*. *Monthly Bulletin of the Bureau of Health (Manila)* 22(3): 21–37.

Baisas, F.E. 1946d. Notes on Philippine mosquitoes—XII. *Topomyia*. *Monthly Bulletin of the Bureau of Health (Manila)* 22(4): 31–47.

Baisas, F.E. 1947. Notes on Philippine mosquitoes, XI. A new species of *Tripteroides*. *Fieldiana, Zoology* 31(15): 121–124.

Baisas, F.E. 1974. The mosquito fauna of Subic Bay Naval Reservation. Republic of the Philippines. Headquarters, First Medical Service Wing (PACAF), APO San Francisco 96274, Technical Report No. 72–2. 179 pp., 79 figs.

Baisas, F.E. & Catipon, R. 1958. Notes on Philippine mosquitoes, XX. Daytime observations in houses of two barrios in Laguna Province. *Philippine Journal of Science* 87(1): 47–63.

Baisas, F.E. & Feliciano, P. 1953. Notes on Philippine mosquitoes, XIII. Four new species of *Zeugnomyia* and *Topomyia*. *Fieldiana, Zoology* 33(3): 161–179.

Baisas, F.E. & Hu, S.M.K. 1936. *Anopheles hyrcanus* var. *sinensis* of the Philippines and certain parts of China, with some comments on *Anopheles hyrcanus* var. *nigerrimus* of the Philippines. *Monthly Bulletin of the Bureau of Health (Manila)* 16(6): 205–242, 6 pls.

Baisas, F.E. & Ubaldo-Pagayon, A. 1953. Notes on Philippine mosquitoes, XVI. Genus *Tripteroides*. Institute of Science and Technology, Manila (1952), Monograph. 198 pp., 23 pls.

Baker, F.C. 1936. A new species of *Orthopodomyia, O. alba* sp. n. (Diptera, Culicidae). *Proceedings of the Entomological Society of Washington* 38(1): 1–7.

Bañez, L.F.L. & Jueco, N.L. 1966. Adult and larva of *Aedes* (*Finlaya*) *sorsogonensis* Bañez, 1963. Comparison with *Aedes* (*Finlaya*) *ananae* Knight and Laffoon, 1946 and *Aedes* (*Finlaya*) *poecilus* [*sic*] (Theobald), 1903. *Acta Medica Philippina* 2(4): 194–200.

Bangs, M.J., Taai, K., Howard, T.M., Cook, S. & Harbach, R.E. 2014, 2015. The mosquito *Anopheles* (*Cellia*) *oreios* sp. n., formerly species 6 of the Australasian *Anopheles farauti* complex, and a critical review of its biology and relation to disease. *Medical and Veterinary Entomology* (2014) doi: 10.1111/mve.12092; (2015) 29(1): 68–81.

Banks, C.S. 1906a. A new genus and species of Culicidae. *Philippine Journal of Science* 1(7): 779–785.

Banks, C.S. 1906b. A list of Philippine Culicidae with descriptions of some new species. *Philippine Journal of Science* 1(9): 977–1005.

Banks, C.S. 1909. Four new Culicidae from the Philippines. *Philippine Journal of Science* 4(6): 545–551.

Banks, C.S. 1914. A new Philippine malaria mosquito. *Philippine Journal of Science, Section D, General Biology, Ethnology, and Anthropology* 9(4): 405–407.

Baranoff, N. 1927. Eine neue *Aëdes*-Art aus Süd-Serbien (Dipt., Culic.). *Neue Beiträge zur Systematischen Insektenkunde* 4(1/2): 15–16.

Barata, J.M.S. & Cotrim, M.D. 1971. Uma nova espécie de *Psorophora* (Diptera, Culicidae). *Revista Brasileira de Entomologia* 15(6): 35–40.

Barbosa, A.A., da Silva, M.A.N. & Sallum, M.A.M. 2007. *Mansonia* (*Mansonia*) *iguassuensis* sp. nov. (Diptera: Culicidae) from Brasil. *Zootaxa* 1527: 45–52.

Barr, A.R. 1957. A new species of *Culiseta* (Diptera: Culicidae) from North America. *Proceedings of the Entomological Society of Washington* 59(4): 163–167.

Barr, A.R. & Guptavanij, P. 1989. *Anopheles hermsi* N. SP., an unrecognized American species of the *Anopheles maculipennis* group (Diptera: Culicidae). *Mosquito Systematics* (for 1988) 20(3): 352–356.

Barraud, P.J. 1923a. A revision of the culicine mosquitoes of India. Part I. The genera *Stegomyia*, Theo. and *Christophersiomyia* gen. n. adult stage. *Indian Journal of Medical Research* 10(3): 772–788, 17 pls.

Barraud, P.J. 1923b. A revision of the culicine mosquitoes of India. Part III. Notes on certain Indian species of the genus *Finlaya*, Theo. and descriptions of new species. *Indian Journal of Medical Research* 11(1): 214–219.

Barraud, P.J. 1923c. A revision of the culicine mosquitoes of India. Part V. Further notes on the genera *Stegomyia*, Theo. and *Finlaya*, Theo. with descriptions of new species. *Indian Journal of Medical Research* 11(1): 224–228, 3 pls.

Barraud, P.J. 1923d. Two new species of *Culex* (Diptera, Culicidae) from Assam. *Indian Journal of Medical Research* 11(2): 507–509.

Barraud, P.J. 1923e. Some new culicine mosquitos found in India, and a note on *Finlaya assamensis* (Theo.). *Bulletin of Entomological Research* 13(4): 405–408.

Barraud, P.J. 1924a. A revision of the culicine mosquitoes of India. Part VIII. Further descriptions of Indian species of *Finlaya* Theo. *Indian Journal of Medical Research* 11(3): 845–865, 10 pls.

Barraud, P.J. 1924b. A new mosquito from Kashmir. *Indian Journal of Medical Research* 11(3): 967–968.

Barraud, P.J. 1924c. A revision of the culicine mosquitoes of India. Part XI. Some Indian species of *Culex* L. *Indian Journal of Medical Research* 11(4): 979–998, 4 pls.

Barraud, P.J. 1924d. Four new mosquitoes from the western Himalayas. *Indian Journal of Medical Research* 11(4): 999–1006.

Barraud, P.J. 1924e. A revision of the culicine mosquitoes of India. Part XII. Further descriptions of Indian species of *Culex* L. including two new species. *Indian Journal of Medical Research* 11(4): 1259–1274, 3 pls.

Barraud, P.J. 1924f. A revision of the culicine mosquitoes of India. Part XIII. Further descriptions of Indian species of *Culex* L. including three new species. *Indian Journal of Medical Research* 11(4): 1275–1282, 2 pls.

Barraud, P.J. 1924g. A revision of the culicine mosquitoes of India. Part XIV. The Indian species of the subgenus *Culiciomyia* (Theo.) Edw., including one new species. *Indian Journal of Medical Research* 12(1): 15–22, 1 pl.

Barraud, P.J. 1924h. A revision of the culicine mosquitoes of India. XV. The Indian species of the subgenus *Lophoceratomyia* (Theo.) Edw., including two new species. *Indian Journal of Medical Research* 12(1): 39–46, 2 pls.

Barraud, P.J. 1924i. A new mosquito from Kashmir and the North-west Frontier Province. *Indian Journal of Medical Research* 12(1): 73–74.

Barraud, P.J. 1926. A revision of the culicine mosquitoes of India. Part XVIII. The Indian species of *Uranotaenia* and *Harpagomyia*, with descriptions of five new species. *Indian Journal of Medical Research* 14(2): 331–350, 1 pl.

Barraud, P.J. 1927a. A revision of the culicine mosquitoes of India. Part XIX. The Indian species of *Aëdomyia* [*sic*] and *Orthopodomyia* with descriptions of two new species. *Indian Journal of Medical Research* 14(3): 523–532, 1 pl.

Barraud, P.J. 1927b. A revision of the culicine mosquitoes of India. XX. The Indian species of *Armigeres* (including *Leicesteria*), with descriptions of two new species. *Indian Journal of Medical Research* 14(3): 533–548, 2 pls.

Barraud, P.J. 1927c. A revision of the culicine mosquitoes of India. Part XXI. Descriptions of new species of *Aedimorphus* and *Finlaya*, and notes on *Stegomyia albolineata* (Theo.). *Indian Journal of Medical Research* 14(3): 549–554.

Barraud, P.J. 1927d. A revision of the culicine mosquitoes of India. Part XXII. The Indian species of the genus *Taeniorhynchus* (including *Mansonioides*), with a description of one new species. *Indian Journal of Medical Research* 14(3): 555–563, 2 pls.

Barraud, P.J. 1928a. A revision of the culicine mosquitoes of India. Part XXIII. The genus *Aedes* (sens. lat.) and the classification of the subgenera. Descriptions of the Indian species of *Aedes* (*Aedimorphus*), *Aedes* (*Ochlerotatus*), and *Aedes* (*Banksinella*), with notes on *Aedes* (*Stegomyia*) *variegatus*. *Indian Journal of Medical Research* 15(3): 653–669, 4 pls.

Barraud, P.J. 1928b. A revision of the culicine mosquitoes of India. Part XXIV. The Indian species of the subgenera *Skusea* and *Aedes*, with descriptions of eight new species, and remarks on a new method for identifying the females of the subgenus *Aedes*. *Indian Journal of Medical Research* 16(2): 357–375, 8 pls.

Barraud, P.J. 1929a. A revision of the culicine mosquitoes of India. Part XXV. The genera *Mucidus*, *Mimomyia*, *Ficalbia*, *Rachionotomyia*, and *Hodgesia*. *Indian Journal of Medical Research* 16(4): 1052–1063, 1 pl.

Barraud, P.J. 1929b. A revision of the culicine mosquitoes of India. Part XXVI. The genera *Heizmannia*, *Haemagogus*, *Topomyia*, and *Megarhinus*. *Indian Journal of Medical Research* 17(1): 259–280, 2 pls.

Barraud, P.J. 1931a. Notes on some Indian mosquitoes of the subgenus *Stegomyia*, with descriptions of new species. *Indian Journal of Medical Research* 19(1): 221–227, 1 pl.

Barraud, P.J. 1931b. Descriptions of eight new species of Indian culicine mosquitoes. *Indian Journal of Medical Research* 19(2): 609–616, 1 pl.

Barraud, P.J. 1932. *Anopheles aitkeni* James var. n. *pinjaurensis*. *Records of the Malaria Survey of India* 3(2): 353–355.

Barraud, P.J. 1934. *The fauna of British India, including Ceylon and Burma. Diptera. Volume V. Family Culicidae. Tribes Megarhinini and Culicini.* Taylor and Francis, London. xxviii, 463 pp., 8 pls.

Barraud, P.J. & Christophers, S.R. 1931. On a collection of anopheline and culicine mosquitoes from Siam. *Record of the Malaria Survey of India* 2(2): 269–285.

Barreto, M.P. & Coutinho, J.O. 1944. Sôbre o gênero *Taeniorhynchus* Arribálzaga, 1891, com a descrição de três novas espécies do subgênero *Taeniorhynchus* (*Diptera, Culicidae*). *Arquivos de Higiene e Saúde Pública* 9(21): 53–85.

Barreto, M.P. & Coutinho, J.O. 1951. Sobre o *Anopheles* (*Kerteszia*) *cruzii* Dyar e Knab e o *Anopheles* (*Kerteszia*) *laneanus* Correa e Cerqueira (*Diptera*, Culicidae). *Arquivos de Higiene e Saúde Pública* (for 1950) 15: 177–180.

Basham, E.H. 1948. *Culex* (*Melanoconion*) *mulrennani*, a new species from Florida (Diptera: Culicidae). *Annals of the Entomological Society of America* 41(1): 1–7.

Basio, R.G. 1971. On Philippine mosquitoes III. Some new species from Mt. Makiling and its vicinity in Luzon (Diptera: Culicidae). *Philippine Entomologist* 2(1): 51–57.

Basio, R.G. & Reisen, W.K. 1971. On some mosquitoes of Guam, Marianas Islands (Diptera: Culicidae). *Philippine Entomologist* 2(1): 57–61.

Bates, M. 1940. The nomenclature and taxonomic status of the mosquitoes of the *Anopheles maculipennis* complex. *Annals of the Entomological Society of America* 33(2): 343–356.

Bates, M., Beklemishev, W.N. & la Face, L. 1949. Anophelines of the Palearctic Region, pp. 419–442. *In* Boyd, M.F. (ed.), *Malariology. A comprehensive survey of all aspects of this group of diseases from a global standpoint*. Volume 1. W.B. Saunders Company, Philadelphia and London.

Batovska, J., Blacket, M.J., Brown, K. & Lynch, S.E. 2016. Molecular identification of mosquitoes (Diptera: Culicidae) in southeastern Australia. *Ecology and Evolution* 6(9): 3301–3011.

Bauhin, C. 1623. Pinax Theatri Botanici Caspari Bauhini Basileens. Archiatri & Professoris Ordin. sive Index in Theophrasti, Dioscoridis[,] Plinii et Botanicorum qui a Seculo scripserunt : Opera: Plantarum circiter sex millium ab ipsis exhibitarum nomina cum earundem Synonymiis & differentiis : Methodice secundum earum & genera & species proponens : Opus XL. annorum Hactenus non editum summopere expetitum & ad auctores intelligendos plurimum faciens. Sumptibus & typis Ludovici Regis. [xxiv], 522, [23] pp.

Becker, T. 1903. Ägyptische Dipteren. *Mitteilungen aus dem zoologischen Museum in Berlin* 2(3): 67–195, 4 pls.

Becker, T. 1908. Dipteren der Kanarischen Inseln. *Mitteilungen aus dem zoologischen Museum in Berlin* 4(1): 1–180, 4 pls.

Becker, T. 1910. Dipteren aus Südarabien und von der Insel Sokótra. *Denkschriften der Akademie der Wissenschaften* 71: 131–160.

Belkin, J.N. 1945. *Anopheles nataliae*, a new species from Guadalcanal. *Journal of Parasitology* 31(5): 315–318.

Belkin, J.N. 1950. Mosquitoes of the genus *Tripteroides* in the Solomon Islands. *Proceedings of the United States National Museum* 100(3262): 201–274.

Belkin, J.N. 1953. Mosquitoes of the genus *Uranotaenia* in the Solomon Islands (Diptera: Culicidae). *Pacific Science* 7(3): 312–391.

Belkin, J.N. 1955. The *Tripteroides caledonica* complex of mosquitoes in Melanesia (Diptera: Culicidae). *Pacific Science* 9(2): 221–246.

Belkin, J.N. 1962. *The mosquitoes of the South Pacific (Diptera, Culicidae)*. Volumes 1 and 2. University of California Press, Berkeley and Los Angeles. xii, 608 pp., 412 figs.

Belkin, J.N. 1968a. Mosquito studies (Diptera, Culicidae) VII. The Culicidae of New Zealand. *Contributions of the American Entomological Institute* 3(1): 1–182.

Belkin, J.N. 1968b. Mosquito studies (Diptera, Culicidae). IX. The type specimens of New World mosquitoes in European museums. *Contributions of the American Entomological Institute* 3(4): 1–69.

Belkin, J.N. 1969. *Culex* (*Melanoconion*) *Annulipes* invalid. *Mosquito Systematics Newsletter* 1(3): 68–70.

Belkin, J.N. 1970a. Corrected type localities for *Wyeomyia abia* D. & K., 1908 and *W. fratercula* D. & K., 1906. *Mosquito Systematics Newsletter* 2(2): 57–58.

Belkin, J.N. 1970b. *Culex* (*Melanoconion*) *aikenii* (A. & R., 1906) a *nomen dubium*; *ocossa* D. & K., 1919 and *panocossa* Dyar, 1923 both valid. *Mosquito Systematics Newsletter* 2(2): 59–60.

Belkin, J.N. 1977a. *Toxorhynchites* (*Lyn.*) *gerbergi*, a new species from the southern Lesser Antilles. *Mosquito Systematics* 9(3): 329–332.

Belkin, J.N. 1977b. *Aedes* (*Ochlerotatus*) *pix* Martini 1935 a synonym of *Aedes (0.) taeniorhynchus* (Wiedemann 1821). *Mosquito Systematics* 9(4): 535.

Belkin, J.N. & Heinemann, S.J. 1975. *Psorophora (Janthinosoma) mathesoni*, sp. nov. for "*varipes*" of the southeastern U.S.A. *Mosquito Systematics* 7(4): 363–366.

Belkin, J.N. & Hogue, C.L. 1959. A review of the crabhole mosquitoes of the genus *Deinocerites* (Diptera, Culicidae). *University of California Publications in Entomology* 14(6): 411–458.

Belkin, J.N. & McDonald, W.A. 1956. A population of *Uranotaenia anhydor* from Death Valley, with descriptions of all stages and discussion of the complex (Diptera, Culicidae). *Annals of the Entomological Society of America* 49(2): 105–132.

Belkin, J.N. & McDonald, W.A. 1957. A new species of *Aedes* (*Ochlerotatus*) from tree holes in southern Arizona and a discussion of the *varipalpus* complex (Diptera: Culicidae). *Annals of the Entomological Society of America* 50(2): 179–191.

Belkin, J.N. & Schlosser, R.J. 1944. A new species of *Anopheles* from the Solomon Islands. *Journal of the Washington Academy of Sciences* 34(8): 268–273.

Belkin, J.N., Heinemann, S.J. & Page, W.A. 1970. The Culicidae of Jamaica (Mosquito Studies. XXI). *Contributions of the American Entomological Institute* 6(1): 1–458.

Belkin, J.N., Knight, K.L. & Rozeboom, L.E. 1945. Anopheline mosquitoes of the Solomon Islands and New Hebrides. *Journal of Parasitology* 31(4): 241–265.

Belkin, J.N., Schick, R.X. & Heinemann, S.J. 1965. Mosquito studies (Diptera: Culicidae) V. Mosquitoes originally described from Middle America. *Contributions of the American Entomological Institute* 1(5): 1–95.

Belkin, J.N., Schick, R.X. & Heinemann, S.J. 1968. Mosquito studies (Diptera, Culicidae) XI. Mosquitoes originally described from Argentina, Bolivia, Chile, Paraguay, Peru, and Uruguay. *Contributions of the American Entomological Institute* 4(1): 9–29.

Belkin, J.N., Schick, R.X. & Heinemann, S.J. 1971. Mosquito studies (Diptera, Culicidae) XXV. Mosquitoes originally described from Brazil. *Contributions of the American Entomological Institute* 7(5): 1–64.

Bellardi, L. 1859. *Saggio di ditterologia messicana. Parte I.* Torino. 80 pp.

Bellardi, L. 1862. *Saggio di ditterologia messicana. Appendice.* Torino. 99 pp., 2 pls, index [2 pp.].

Benoit, P.L.G. 1964. Diptera – Culicidae. Observations biologiques et écologiques, pp. 297–320. Mission de Zoologie médicale au Maniema (Congo, Léopoldville) (P.L.G. Benoit, 1959): Résultats scientifiques. Musee Royal de l'Afrique centrale – Tervuren, Belgique. *Annales — Serie in-8° — Sciences Zoologiques* — n° 132.

Bentley, C.A. 1902. *Anopheles* mosquitos in Tezpur,Assam. *Indian Medical Gazette* 37(1): 15–16.

Bergroth, E. 1889. Zwei neue Diptera. *Wiener entomologische Zeitung* 8(9): 295–298.

Berlin, O.G.W. 1969a. Mosquito studies (Diptera, Culicidae) XII. A revision of the Neotropical subgenus *Howardina* of *Aedes*. *Contributions of the American Entomological Institute* 4(2): 1–190.

Berlin, O.G.W. 1969b. Mosquito studies (Diptera, Culicidae) XVIII. The subgenus *Micraedes* of *Culex*. *Contributions of the American Entomological Institute* 5(1): 21–63.

Berlin, O.G.W. 1974. A new species of bromeliad breeding *Culex* (*Micraedes*) from Mexico (Diptera, Culicidae). *Mosquito Systematics* 6(4): 273–278.

Berlin, O.G.W. & Belkin, J.N. 1980. Mosquito studies (Diptera, Culicidae) XXXVI. Subgenera *Aedinus*, *Tinolestes* and *Anoedioporpa* of *Culex*. *Contributions of the American Entomological Institute* 17(2): ii, 1–104.

Bertrand, H. 1954. Les insectes aquatiques d'Europe (Genres: Alrves, Nymphes, Imagos). Volume II. – *Trichoptères, Lépidoptères, Diptères, Hyménoptères. Encyclopédie Entomologique. Série A.– Travaux généraux* 31: 1–547.

Bezzi, M. 1906. Ditteri Eritrei raccolti dal Dott. Andreini e dal Prof. Tellini. *Bollettino della Società entomologica italiana* (for 1905) 37(4): 195–304.

Bhattacharyya, D.R., Prakash, A., Mohapatra, P.K. & Mahanta, J. 2004. *Uranotaenia dibrugarhensis*, a new species in subgenus *Pseudoficalbia* from Assam, India. *Journal of the American Mosquito Control Association* 20(1): 1–5.

Bhattacharyya, D.R., Prakash, A., Mohapatra, P.K. & Sarma, D.K. 2009. *Armigeres* (*Armigeres*) *mahantai*, a new mosquito species from India. *Journal of the American Mosquito Control Association* 25(1): 1–5.

Bhattacharyya, D.R., Tewari, S.C., Prakash, A., Mohapatra, P.K. & Mahanta, J. 2004. *Verrallina* (*Neomacleaya*) *assamensis*, a new species from Assam, Indian. *Journal of the American Mosquito Control Association* 20(2): 115–120.

Bianchi, U. 1968. Sulla tassonomia di *Anopheles labranchiae* ed *Anopheles atroparvus*. *Rivista di Parassitologia* 29(3): 221–226.

Bick, G.H. 1951. The ecology of the mosquito larvae of New Guinea. *Pacific Science* 5(4): 392–431.

Bigot, J. 1857. Dipteros, pp. 328–349. *In*: de la Sagra, R., *Historia fisica, politica y natural de la isla de Cuba*. Segunda parte. Historia Natural. Tomo VII. Crustaceos, Aragnides é Insectos. A. Bertrand, Paris.

Bigot, J. 1859. Diptères de Madagascar. Première partie. *Annales de la Société Entomologique de France* 7: 115–135, 3 pls.

Bigot, J. 1861. Trois diptères nouveaux de la Corse. *Annales de la Société Entomologique de France* 1: 227–229.

Bigot, J.-M.-F. 1888. *Énumération des diptères recueillis en Tunisie dans la Mission de 1884 par M. Valéry Mayet, member de la Mission de l'Exploration Scientifique de la Tunisie, et Description des Espèces Nouvelles, par J.-M.-F. Bigot.* [Series] *Exploration Scientifique de la Tunisie*. Imprimerie nationale, Paris. 11 pp.

Bigot, J. 1889. Bulletin entomologique. – M. J.-M.-F. Bigot envoie, de Quincy, une note sur une nouvelle espècie du genre *Culex*. *Annales de la Société Entomologique de France* 9: cxxii.

Bilimek, D. 1867. Fauna der Grotte Cacahuamilpa in Mexiko. *Verhandlungen der Kaiserlich-Königlichen Zoologisch-Botanischen Gesellschaft in Wien* 17: 901–908.

Billberg, G.J. 1820. *Enumeratio insectorum in Museo Gust. Joh. Billberg*. Typis Gadelianis, [Holmiae]. 138 pp.

Blanchard, E. 1852. Orden IX. Dipteros, pp. 327–468. *In*: Gay, C. (ed.), *Historia fisica y politica de Chile. Zoologia*. Volume 7. Paris.

Blanchard, R. 1901. Observations sur quelques moustiques. *Comptes Rendus Hebdomadaires des Séances et Mémoires de la Société de Biologie* 53: 1045–1046.

Blanchard, R. 1902. Nouvelle note sur les moustiques. *Comptes Rendus Hebdomadaires des Séances et Mémoires de la Société de Biologie* 54: 793–795.

Blanchard, R. 1904. In: Notes et informations. *Archives de Parasitologie* 8(3): 478.
Blanchard, R. 1905. *Les moustiques, histoire naturelle et médicale*. F.R. de Rudeval, Imprimeur-Éditeur, Paris. xiii, 673 pp.
Blanchard, R. 1907. Le paludisme a Madagascar. *Archives de Parasitologie* 11(2): 185–214.
Bohart, R.M. 1945. *A synopsis of the Philippine mosquitoes*. Navmed 580. Bureau of Medicine and Surgery, Navy Department, Washington, DC. 88 pp., 10 pls.
Bohart, R.M. 1946. New species of mosquitoes from the Marianas and Okinawa (Diptera, Culicidae). *Proceedings of the Biological Society of Washington* 59: 39–46.
Bohart, R.M. 1949. The subgenus *Neoculex* in America north of Mexico (Diptera, Culicidae). *Annals of the Entomological Society of America* (for 1948) 41(3): 330–345.
Bohart, R.M. 1950. A new species of *Orthopodomyia* from California (Diptera, Culicidae). *Annals of the Entomological Society of America* 43(3): 399–404.
Bohart, R.M. 1953. A new species of *Culex* and notes on other species of mosquitoes from Okinawa (Diptera, Culicidae). *Proceedings of the Entomological Society of Washington* 55(4): 183–188.
Bohart, R.M. 1956. New species of mosquitoes from the southern Ryukyu Islands. *Bulletin of the Brooklyn Entomological Society* 51(2): 29–34.
Bohart, R.M. 1957. Insects of Micronesia, Diptera: Culicidae. *Insects of Micronesia* (for 1956) 12(1): 1–85.
Bohart, R.M. & Farner, D.S. 1944. New culicine mosquitoes from the Philippine Islands (Diptera, Culicidae). *Proceedings of the Biological Society of Washington* 57: 69–74.
Bohart, R.M. & Ingram, R.L. 1946a. Four new species of mosquitoes from Okinawa (Diptera: Culicidae). *Journal of the Washington Academy of Sciences* 36(2): 46–52.
Bohart, R.M. & Ingram, R.L. 1946b. *Mosquitoes of Okinawa and islands in the central Pacific*. Navmed 1055. Bureau of Medicine and Surgery, Navy Department, Washington, D.C. 110 pp.
Bonne, C. 1923. A new *Sabethes* from Surinam (*Diptera, Culicidae*). *Insecutor Inscitiae Menstruus* 11(7–9): 122–123.
Bonne-Wepster, J. 1930. The genus *Taeniorhynchus* Arribalzaga in the Dutch East Indies. Part II. Subgenus *Coquillettidia* Dyar. *Mededeelingen van den Dienst Volksgezondheid in Nederlandsch-Indië* 19(3): 387–399.
Bonne-Wepster, J. 1934. New mosquitos (Dipt.) from the Netherlands Indies. *Stylops* 3(12): 272–276.
Bonne-Wepster, J. 1937. Notes on mosquitoes from the Netherlands Indies. A new *Culex* and two new *Finlaya*'s from New-Guinea. *Mededeelingen van den Dienst Volksgezondheid in Nederlandsch-Indië* 26(1–2): 92–98.
Bonne-Wepster, J. 1938. Notes on mosquitoes from the Netherlands Indies. A new *Culex* from New Guinea: *Culex* (*Culex*) *solitarius* nov. sp. Description of male terminalia of *Culex* (*C*) *miraculosus* B.-W. *Mededeelingen van den Dienst Volksgezondheid in Ned.-Indië* 27(3): 394–396.
Bonne-Wepster, J. 1940. Notes on mosquitoes from the Netherlands Indies: A new *Finlaya* from New-Guinea. *Mededeelingen van den Dienst der Volksgezondheid in Nederlandsch-Indië* 29(3–4): 158–159, 2 pls.
Bonne-Wepster, J. 1948. Results of the third Archbold Expedition 1938–1939. *Diptera Culicidae*. Notes on the mosquitoes collected by the Neth. Indian – American expedition to Central and North New Guinea. *Treubia* 19(2): 305–322.
Bonne-Wepster, J. 1951a. Notes on Oriental and Australasian mosquitoes of the Malay Archipelago. *Documenta Neerlandica et Indonesica de Morbis Tropicis* 3(1): 67–74.
Bonne-Wepster, J. 1951b. *Anopheles venhuisi n. sp. Documenta Neerlandica et Indonesica de Morbis Tropicis* 3: 284.
Bonne-Wepster, J. & Bonne, C. 1919. Four new South American mosquitoes (*Diptera, Culicidae*). *Insecutor Inscitiae Menstruus* 7(7–9): 105–113.
Bonne-Wepster, J. & Bonne, C. 1920. Diagnoses of new mosquitoes from Surinam, with a note on synonymy *(Diptera, Culicidae)*. *Insecutor Inscitiae Menstruus* (for 1919) 7(10–12): 165–180.
Bonne-Wepster, J. & Bonne, C. 1921. Notes on South American mosquitoes in the British Museum (*Diptera, Culicidae*). *Insecutor Inscitiae Menstruus* 9(1–3): 1–26.
Bonne-Wepster, J. & Bonne, C. 1923. A new *Megarhinus* from Surinam (*Diptera, Culicidae*). *Insecutor Inscitiae Menstruus* 11(1–3): 7–9.
Bonne-Wepster, J. & Brug, S.L. 1932. The subgenus *Stegomyia* in Netherland India. *Geneeskundig Tijdschrift voor Nederlandsch-Indië* 72: 39–119.
Bonne-Wepster, J. & Swellengrebel, N.H. 1953. *The anopheline mosquitoes of the Indo-Australian Region*. J.H. de Bussy, Amsterdam. 504 pp.
Boreham, M.M. 1970. *Mansonia leberi,* a new mosquito from the Panama Canal Zone, with notes on its biology (Diptera: Culicidae). *Journal of Medical Entomology* 7(3): 383–390.
Borel, E. 1926. Les moustiques de la Cochinchine et du Sud-Annam. *Archives des Instituts Pasteur d'Indochine* 3–4: 75–121, 19 pls.
Borel, E. 1928. Les moustiques de la Cochinchine et du Sud-Annam (*Suite*). *Archives des Instituts Pasteur d'Indochine* 7: 75–106, 8 pls.
Borkent, A. & Grimaldi, D.A. 2004. The earliest fossil mosquito (Diptera: Culicidae), in mid-Cretaceous amber. *Annals of the Entomological Society of America* 97(5): 882–888.
Borkent, A. & Grimaldi, D.A. 2016. The Cretaceous fossil *Burmaculex antiquus* confirmed as the earliest known lineage of mosquito (Diptera: Culicidae). *Zootaxa* 4079(4): 457–466.
Boshell-Manrique, J. 1939. A new species of *Aedes* from Colombia (Dipt. Culicidae). *Revista de Entomologia* 10(2): 309–312.
Bouché, P.F. 1834. *Naturgeschichte der Insekten, besonders in Hinsicht ihrer ersten Zustände als Larven und Puppen*. Erste Lieferung. Berlin. v, 216 pp., 10 pls.
Bourroul, C. 1904. Mosquitos do Brasil. These de Doutoramento, Faculdade de Medicina da Bahia. Officina Typographica de João Baptista de Oliveira Costa, Bahia. [Consisting of Prefacio (pp. i–viii), Novas especies (pp. l–6, 30–32), Quadro das especies encontradas na Bahia (4 pp.) and Proposições (18 pp.) by Bourroul; remainder and bulk of work by Adolpho Lutz. Continuous pagination used. See Belkin, 1968b: 50, 62 for details].

Boussès, P. & Brunhes, J. 2013. Trois nouveaux *Uranotaenia* Lynch Arribálzaga de Madagascar (Diptera, Culicidae). *Bulletin de la Société Entomologique de France* 118(3): 399–410.

Boyd, M.F. (ed.) 1949. *Malariology. A comprehensive survey of all aspects of this group of diseases from a global standpoint.* Volume 1. W.B. Saunders Company, Philadelphia and London. xxi, 787 pp.

Bradshaw, W.E. & Lounibos, L.P. 1977. Evolution of dormancy and its photoperiodic control in pitcher-plant mosquitoes. *Evolution* 31(3): 546–567.

Brahmachari Bahadur, U.N. 1912. On some new anophelines of Calcutta and on the seasonal prevalence and variations of anopheline *fuliginosus* of Calcutta. *Indian Medical Gazette* 47(5): 186–187.

Bram, R.A. 1966. *Culex (Thaiomyia) dispectus*, a new subgenus and species from Thailand (Diptera: Culicidae). *Proceedings of the Entomological Society of Washington* 68(2): 73–77.

Bram, R.A. 1967a. Contributions to the mosquito fauna of Southeast Asia. II. The genus *Culex* in Thailand (Diptera: Culicidae). *Contributions of the American Entomological Institute* 2(1): 1–296.

Bram, R.A. 1967b. Classification of *Culex* subgenus *Culex* in the New World (Diptera: Culicidae). *Proceedings of the United States National Museum* 120(3557): 1–122.

Bram, R.A. & Rattanarithikul, M. 1967. Six new species of the *Culex (Lophoceraomyia) mammilifer* group from Thailand (Diptera: Culicidae). *Proceedings of the Entomological Society of Washington* 69(1): 1–17.

Branch, N. & Seabrook, E.L. 1959. *Culex (Culex) scimitar*, a new species of mosquito from the Bahama Islands (Diptera, Culicidae). *Proceedings of the Entomological Society of Washington* 61(5): 216–218.

Breeland, S.G. 1953. A morphological investigation of an anopheline of the *crucians* complex reared from atypical ova (Diptera, Culicidae). Faculty Publications and Research, together with Abstracts of Doctoral Dissertations and Master's Theses, North Carolina State University, Raleigh. vi, 69 pp.

Breland, O.P. 1960. Restoration of the name, *Aedes hendersoni* Cockerell, and its elevation to full specific rank (Diptera: Culicidae). *Annals of the Entomological Society of America* 53(5): 600–606.

Brèthes, J. 1910. Dípteros nuevos ó poco conocidos de Sud-América. *Anales del Museo Nacional de Buenos Aires* 13: 469–484.

Brèthes, J. 1912. Los mosquitos de la República Argentina. *Boletín de Instituto Entomológia Patagonia Vegetal* 1: 1–48.

Brèthes, J. 1916. Algunas notas sobre mosquitos argentinos su relación con las enfermedades palúdicas, etc. y descripción de tres especies nuevas. *Anales del Museo Nacional de Buenos Aires* 28: 193–218.

Brèthes, J. 1917. Description d'une nouvelle espèce de moustique de Buenos Aires. *Physis* 3: 226–229.

Brèthes, J. 1920. Description d'un nouveau moustique du Pérou. *Revista Chilena de Historia Natural* 24: 41–43.

Brèthes, J. 1926a. Contribución para el conocimiento de los mosquitos argentinos. Descripción de un nuevo *Megarhinus*: *Megarhinus tucumanus*. *Boletin del Instituto de Clínica Quirúrgica (Buenos Aires)* 2: 318–321.

Brèthes, J. 1926b. Description provisoire de deux especes nouvelles d'Anophelinae argentins. *Prensa Medica Argentina* 13(3): 106–107.

Broche, R.G. 2000. Dos nuevas especies de mosquitos de los géneros *Psorophora (Grabhamia)* y *Culex (Culex)* (Diptera, Culicidae), de Cuba. *Revista Brasileira de Entomologia* 44(1/2): 27–33.

Brolemann, H.-W. 1919a. Sur quelques *Culex* des Pyrénées et description d'une espèce nouvelle I. – Campagne 1917. *Annales de la Société Entomologique de France* 87(3–4): 425–440.

Brolemann, H.-W. 1919b. Sur quelques *Culex* des Pyrénées II. Campagne 1918. *Annales de la Société Entomologique de France* 88(1–2): 65–103.

Brookman, B. & Reeves, W.C. 1950. A new name for a California mosquito (Diptera, Culicidae). *Pan-Pacific Entomologist* 26(4): 159–160.

Brug, S.L. 1924. Notes on Dutch-East-Indian mosquitos. *Bulletin of Entomological Research* 14(4): 433–442.

Brug, S.L. 1925. A new *Armigeres* (Diptera, Culicidae) from Ceram (Moluccas). *Bulletin of Entomological Research* 15(4): 345–347.

Brug, S.L. 1926. *Lophoscelomya* [*sic*] *annandalei* var. *djajasanensis* nov. var. (*Diptera, Culicidae*). *Bulletin de la Société de Pathologie exotique* 19(9): 805–806.

Brug, S.L. 1928a. *Anopheles (Neomyzomyia) longirostris* n. sp., (Diptera, Culicidae). *Geneeskundig Tijdschrift voor Nederlandsch-Indië* 68: 278–279, 1 pl.

Brug, S.L. 1928b. Aanteekeningen omtrent Muskieten (IV). *Geneeskundig Tijdschrift voor Nederlandsch-Indië* 68: 921–927, 3 pls.

Brug, S.L. 1931a. New Culicidae from Sumatra. *Tijdschrift voor Entomologie* 74: 245–250.

Brug, S.L. 1931b.*Anopheles incognitus* n. sp. *Geneeskundig Tijdschrift voor Nederlandsch-Indië* 71(2): 136–137.

Brug, S.L. 1931c. Culiciden der Deutschen Limnologischen Sunda-Expedition. *Archiv für Hydrobiologie Suppl.-Bd 9, Tropische Binnengewässer* 2: 1–42.

Brug, S.L. 1932. Notes on Dutch East Indian mosquitos. *Bulletin of Entomological Research* 23(1): 73–83.

Brug, S.L. 1934. Notes on Dutch East Indian mosquitos. *Bulletin of Entomological Research* 25(4): 501–519.

Brug, S.L. 1938. Waarnemingen bij anophelinen. *Geneeskundig Tijdschrift voor Nederlandsch-Indië* 78: 520–523.

Brug, S.L. 1939. Notes on Dutch East-Indian mosquitoes. *Tijdschrift voor Entomologie* 82: 91–113.

Bruijning, C.F.A. 1959. Notes on *Wyeomyia* mosquitoes of Suriname, with a description of *Wyeomyia surinamensis* sp.n. *Studies on the Fauna of Suriname and other Guyanas* 3(8): 99–146.

Brullé, A. 1833. *Expédition scientifique de Morée. Section des Sciences Physiques. Tome III. – 1.re Partie. Zoologie. Deuxième Section. – Des animaux articulés.* F.G. Levrault, Paris. 400 pp.

Brunetti, E. 1912. Annotated catalogue of Oriental Culicidae – Supplement. *Records of the Indian Museum* 4(10): 403–517.

Brunetti, E. 1914. Critical review of "general" in Culicidae. *Records of the Indian Museum* 10(1): 15–73.

Brunetti, E. 1920. Catalogue of Oriental and South Asiatic Nemocera. *Records of the Indian Museum* 17: preface, 300 pp.

Brunhes, J. 1969. Culicidés de Madagascar: IV. Description des imagos, de la nymphe et de la larve de *Culex (Culiciomyia) pandani* sp. n. *Cahiers ORSTOM Série Entomologie médicale et Parasitologie* 7(3): 175–180.
Brunhes, J. 1971. Culicidés de Madagascar. V. Quelques *Aedes* (sous-genre *Finlaya*) de Madagascar. *Cahiers ORSTOM Série Entomologie médicale et Parasitologie* 9(4): 335–349.
Brunhes, J. 1977a. Les moustiques de l'archipel des Comores I. – Inventaire, répartition et description de quatre espèces ou sous-espèces nouvelles. *Cahiers ORSTOM Série Entomologie médicale et Parasitologie* 15(2): 131–152.
Brunhes, J. 1977b. Les moustiques de l'archipel des Comores. II. – Description de quatre espèces nouvelles ou peu connues. Répartition des membres du sous-genre *Skusea* dans l'océan Indien accidental. Affinités de la faune culicidienne des Comores. *Cahiers ORSTOM Série Entomologie médicale et Parasitologie* 15(2): 153–170.
Brunhes, J. 1983. Culicidés de Madagascar IX. Compléments à la description d'*Aedes (Diceromyia) grassei* Doucet, 1951 et description d'*Aedes (Diceromyia) sylvaticus* n.sp. (Diptères, Nématocères). *Cahiers ORSTOM Série Entomologie médicale et Parasitologie* (for 1982) 20(4): 313–318.
Brunhes, J. 1999. Culicidae du Maghreb. Description d'*Aedes (Ochlerotatus) biskraensis* n. sp. d'Algérie (Diptera, Nematocera). *Bulletin de la Société Entomologique de France* 104(1): 25–30.
Brunhes, J. & Boussès, P. 2009. Description de *Culex* (*Culex*) *sechani* n. sp. de Tahiti (îles de la Société (Diptera, Culicidae). *Bulletin de la Société Entomologique de France* 114(1): 21–28.
Brunhes, J. & Hervy, J.-P. 1995. *Faune de Madagascar. 85. Insectes diptères Culicidae Culicinae genre* Orthopodomyia *de la sous-région malgache et de la région afrotropicale.* Muséum nationale d'Histoire naturelle, Paris. 119 pp.
Brunhes, J. & Rambelo, J. 1968. Contribution a l'étude des culicides de Madagascar. Description des adultes, nymphe et larve de *Culex (Neoculex) chauveti* sp. n.*Cahiers ORSTOM Entomologie médicale* 6(1): 113–118.
Brunhes, J. & Ravaonjanahary, C. 1971. Culicides de Madagascar: VI. Description de *Culex (Culex) carleti* sp. n. *Cahiers ORSTOM Série Entomologie médicale et Parasitologie* 9(2): 177–182.
Brunhes, J. & Ravaonjanahary, C. 1973. Culicides de Madagascar. VII. Description des imagos, de la nymphe et de la larve de *Culex (Eumelanomyia) brenguesi* sp. n. *Cahiers ORSTOM Série Entomologie médicale et Parasitologie* 11(3): 169–173.
Brunhes, J. & Razafindrasolo, E. 1975. Culicidés de Madagascar VIII. Description des imagos, de la nymphe et de la larve d'*Uranotaenia (Uranotaenia) anopheloides* sp. n. *Cahiers ORSTOM Série Entomologie médicale et Parasitologie* 13(4): 223–227.
Brunhes, J. & Venard, P. 1966. Contribution a l'étude des culicidés de la région éthiopienne. Description de *Culex (Barraudius) richeti* n. sp. Notes complémentaires sur *Culex (B) pusillus* Macquart et *Culex (B.) modestus* Ficalbi. *Cahiers ORSTOM Série Entomologie médicale et Parasitologie* 4(7): 75–84.
Brunhes, J., Adam, J.P. & Bailly-Choumara, H. 1967. Contribution a l'etude des *Culex* de la region Ethiopienne appartenant au sous-genre *Mochthogenes* (Diptères, Culicidés), avec description des males de cinq nouvelles especes. *Cahiers ORSTOM Série Entomologie médicale* 5(1): 43–52.
Brunhes, J., Boussès, P. & da Cunha Ramos, H. 2011. Les *Aedeomyia* Theobald, 1901, des régions afro-tropicale et malgache (Diptera, Culicidae). *Bulletin de la Société Entomologique de France* 116(1): 99–128.
Brunhes, J., Boussès, P., Tantely, M.L. & Kengne, P. 2017. Un nouveau genre de Culicidae (Diptera), *Paulianius* n. gen., avec la description de trois nouvelles espèces malgaches. *Annales de la Société Entomologique de France* 53(5): 344–373.
Brunhes, J., Le Goff, G. & Bousses, P. 2003. Anophèles afrotropicaux. – V. Description du mâle et des stades pré-imaginaux d'*An. deemingi* Service, 1970 et description d'*An. eouzani* n. sp. (Diptera: Culicidae). *Annales de la Société Entomologique de France* 39(2): 179–185.
Brunhes, J., Le Goff, G. & Geoffroy, B. 1997. Anophèles afro-tropicaux. I. – Descriptions d'espèces nouvelles et changements de statuts taxonomiques (Diptera: Culicidae). *Annales de la Société Entomologique de France* 33(2): 173–183.
Brunhes, J., Le Goff, G. & Geoffroy, B. 1998a. Anophèles afro-tropicaux. II. – Mises au point sur les espèces de la sous-région malgache (Diptera, Culicidae). *Bulletin de la Société Entomologique de France* 103(2): 139–152.
Brunhes, J., Le Goff, G. & Geoffroy, B. 1999. Afro-tropical anopheline mosquitoes. III. Description of three new species: *Anopheles carnevalei* sp. nov., *An. hervyi* sp. nov., and *An. dualaensis* sp. nov., and resurrection of *An. rageaui* Mattingly and Adam. *Journal of the American Mosquito Control Association* 15(4): 552–558.
Brunhes, J., Le Goff, G., Manga, L. & Geoffroy, B. 1998b. Anophèles afrotropicaux.– IV. Mises au point sur le complexe *Anopheles moucheti*, réhabilitation d'*An. garnhami basilewskyi* (Diptera: Culicidae). *Annales de la Société Entomologique de France* 34(4): 397–405.
Brust, R.A., Ballard, J.W.O., Driver, F., Hartley, D.M., Galway, N.J. & Curran, J. 1998. Molecular systematics, morphological analysis, and hybrid crossing identify a third taxon, *Aedes* (*Halaedes*) *wardangensis* sp. nov., of the *Aedes* (*Halaedes*) *australis* species-group (Diptera: Culicidae). *Canadian Journal of Zoology* 76(7): 1236–1246.
Buonomini, G. & Mariani, M. 1946. Qual'è il valore sistematico del nome *maculipennis* Meig.? *Rivista di Malariologia* 25(3–4): 101–107.
Buonomini, G. & Mariani, M. 1953. World Anophelines belonging to the subgenus *Maculipennia* Buonomini e Mariani, 1946. *Rivista di Malariologia* 32: 173–188.
Buren, W.F. 1947. A new *Aedes* from the Florida Keys (Diptera, Culicidae). *Proceedings of the Entomological Society of Washington* 49(8): 228–229.
Büttiker, W. & Beales, P. 1959. *Anopheles pampanae,* a new species of mosquito from Cambodia. *Acta Tropica* 16(1): 63–69.
Buxton, P.A. 1927. Sur les moustiques de Tahiti et du Groupe des Iles de la Société. *Bulletin de la Société des Études Océaniennes (Polynésie orientale)* 2(21): 306–308.
Callot, J. 1940. Sur quelques moustiques du Maroc. *Archives de l'Institut Pasteur du Maroc* 2(2): 361–365.
Callot, J. 1944a. Sur un nouveau moustique arboricole: *Aedes* (*Finlaya*) *heracleensis* sp. nov. *Bulletin de la Société de Pathologie exotique* 37(1–2): 56–59.

Callot, J. 1944b. Culicides des eaux acides d'Auvergne. II. A propos de *Culex lavieri* Larrousse. *Comptes rendus des séances de la Société de physique biologique de France* 18(80): 196–197.

Callot, J. 1944c. A propos de *Aedes (Finlaya) heracleensis* Callot 1944. *Annales de Parasitologie Humaine et Comparée* 20(1–2): 93–94.

Callot, J. 1945. Diptères nématocères piqueurs d'Auvergne. *Revue des Sciences naturelles d'Auvergne* 11: 28–36.

Cambouliu, M. 1902. Contribution à l'étude des *Anopheles* de l'isthme de Suez. *Compte Rendu de l'Académie des Sciences* 135(17): 704–706.

Cambournac, F.J. 1938. *Aëdes (Ochlerotatus) longitubus*, a new species from Portugal. (Diptera, Culicidae). *Proceedings of the Royal Entomological Society of London, Series B: Taxonomy* 7(3): 74–80.

Cambournac, F.J.C., Petrarca, V. & Coluzzi, M. 1982. *Anopheles arabiensis* in the Cape Verde Archipelago. *Parassitologia* 24(2,3): 265–267.

Capasso, L. 1991. Fossil mosquitoes and the spread of infectious diseases in man's ancestors. *Journal of Paleopathology* 3(3): 171–201.

Cardamatis, J.P. 1931. Les espèces de moustiques en Grèce et tout particulièrement d'Athènes. *Bulletin de la Société de Pathologie exotique* 24: 122–131.

Carpenter, S.J. 1970. Review of recent literature on mosquitoes of North America. Supplement I. *California Vector News* 17(6): 39–66.

Carpenter, S.J. & LaCasse, W.J. 1955. *Mosquitoes of North America (north of Mexico)*. University of California Press, Berkeley, Los Angeles, London. viii, 360 pp., 127 pls.

Carpintero, D.J. & Leguizamón, M.N. 2000. Una neuva especie de mosquito en la la [*sic*] Provincia de Buenos Aires *Aëdes (Ochlerotatus) jorgi*. Nueva especies. *Mariposas del Mundo* 5: 20–23.

Carpintero, D.J. & Leguizamón, M.N. 2004. Description of a new *Culex* (*Culex*) species (Diptera: Culicidae) from La Pampa Province, Argentina, and a preliminary list of the mosquitoes of the province. *Studia dipterologica* 11(2): 501–503.

Carter, H.F. 1910a. A new anopheline from South Africa. *Entomologist* 43(568): 237–238.

Carter, H.F. 1910b. Report on a collection of Culicidae from Cochin China. *Entomologist* 43(569): 274–276.

Carter, H.F. 1911. A new mosquito from Uganda. *Bulletin of Entomological Research* 2(1): 37–38.

Carter, H.F. 1920. Descriptions of the adult, larval, and pupal stages of a new mosquito from Lord Howe Island, S. Pacific. *Proceedings of the general meetings for scientific business of the Zoological Society of London* 1920: 623–628.

Carter, H.F. & Wijesundara, D.P. 1948. Notes on some Ceylon culicine mosquitoes. *Ceylon Journal of Science, Section B: Zoology and Geology* 23(3): 135–151.

Casal, O.H. 1967. *Culex (Culex) ameliae,* una nueva especie de la Republic Argentina *(Diptera, Culicidae)*. *Neotropica* 13(40): 29–30.

Casal, O.H. & García, M. 1966. Una nueva especie de *Wyeomyia (Dendromyia)* de la Republica Argentina *(Diptera, Culicidae)*. *Physis* 26(71): 155–162.

Casal, O.H. & Garcia, M. 1967a. *Culex (Melanoconion) pavlovskyi*, nueva especie de la Republica Argentina (Diptera, Culicidae). *Neotropica* 13(40): 19–20.

Casal, O.H. & García, M. 1967b. *Culex (Culex) castroi,* nueva especie de la República Argentina *(Diptera, Culicidae)*. *Physis* 26(73): 451–457.

Casal, O.H. & Garcia, M. 1967c. *Culex (Culex) hepperi*, nueva especie del Delta Bonaerense del Rio Parana *(Diptera, Culicidae)*. *Physis* 27(74): 87–94.

Casal, O.H. & García, M. 1968a. *Culex* (*Allimanta*), nuevo subgenero para *Culex* (*Culex*) *tramazayguesi* Duret, 1954 (Diptera, Culicidae). *Physis* 27(75): 329–335.

Casal, O.H. & Garcia, M. 1968b. *Culex (Eubonnea) clastrieri*, una nueva especie del Brasil (Diptera, Culicidae). *Neotropica* 14(45): 119–120.

Casal, O.H. & García, M. 1968c. *Culex (Melanoconion) martinezi,* una nueva especie de la Republica Argentina *(Diptera, Culicidae)*. *Physis* 27(75): 455–459.

Casal, O.H. & García, M. 1968d. *Culex (Melanoconion) dureti*, una nueva especie Argentino-Paraguaya (*Diptera, Culicidae)*. *Physis* 28(76): 123–125.

Casal, O.H. & García, M. 1968e. *Culex* (*Culex*) *eduardoi*, una nueva especie de la Republica Argentina (Diptera, Culicidae). *Revista de la Sociedad Entomólogica Argentina* 30(1–4): 105–111.

Casal, O.H. & García, M. 1971. *Culex (Culex) tatoi*, una nueva especie de la Republica Argentina *(Diptera, Culicidae)*. *Physis* 30(81): 631–635.

Casal, O.H., García, M. & Cavalieri, F. 1966. *Culex (Culex) fernandezi,* una nueva especie bromelicola de la *Republica Argentina (Diptera, Culicidae)*. *Physis* 26(71): 185–192.

Casal, O.H., García, M. & Fernández, H.I. 1968. El subgenero *Culex (Aedinus)* Bourroul, 1904, nuevo para la entomofauna Argentina, con la descripcion de una nueva especie *(Diptera, Culicidae)*. *Physis* 28(76): 217–218.

Castro, M. & Bressanello, M. 1952. Revisión de las especies de "*Taeniorhynchus* (*Rhynchotaenia*)" (Dipt. Cul.). *Revista Brasileira de Biologia* 12(3): 229–246.

Causey, O.R. 1937a. New anopheline and culicine mosquitoes from Siam. *Journal of Parasitology* 23(5): 543–545.

Causey, O.R. 1937b. Some anopheline and culicine mosquitoes of Siam with remarks on malaria control in Bangkok. *American Journal of Hygiene* 25(2): 400–420.

Causey, O.R. 1945. Description of *Anopheles* (*Nyssorhynchus*) *dunhami,* a new species from the Upper Amazon Basin. *Journal of the National Malaria Society* 4(3): 231–234.

Causey, O.R., Deane, L.M. & Deane, M.P. 1943. Descrição de um novo anofelino da parte alta do Vale do Amazonas "*Anopheles* (*Nyssorhynchus*) *galvãoi*" n. sp. *Revista Paulista de Medicina* 23: 293–296.

Causey, O.R., Deane, L.M. & Deane, M.P. 1944. An illustrated key to the eggs of thirty species of Brazilian anophelines, with several new descriptions. *American Journal of Hygiene* 39(1): 1–7.

Causey, O.R., Deane, L.M. & Deane, M.P. 1945. *Anopheles aquasalis* vs. *Anopheles tarsimaculatus* as the name for the brackish water anopheline of Central and South America and the Caribbean Islands. *Journal of the National Malaria Survey* 4(3): 243–250.

Causey, O.R., Deane, L.M., Deane, M.P. & Sampaio, M.M. 1943. *Anopheles (Nyssorhynchus) sawyeri,* a new anopheline mosquito from Ceara, Brazil. *Annals of the Entomological Society of America* 36(1): 11–20.

Cerqueira, N. 1939. Sobre *Psorophora lutzi* Theobald, *Psorophora albipes* Theobald e *Psorophora forceps* n. sp. (Dipt. Culicidae). *Revista de Entomologia* 10(1): 78–85.

Cerqueira, N.L. 1943. Algumas espécies novas da Bolívia, e referência a três espécies do *Haemagogus* (Diptera, Culicidae). *Memórias do Instituto Oswaldo Cruz* 39(1): 1–14, 4 pls.

Cerqueira, N.L. 1960a. Uma nova espécie de *Haemagogus* do Amazonas (Diptera Culicidae). *Boletim de Museu Paraense Emilio Goeldi* 25: 1–5, 2 pls.

Cerqueira, N.L. 1960b. Sôbre uma n. sp. de *Janthinosoma* da Série Lutzi e a descrição do macho de *Psorophora (J.) circumflava* Cerqueira, 1943 (Diptera Culicidae). *Boletim de Museu Paraense Emilio Goeldi* 26: 1–7, 3 pls.

Cerqueira, N.L. 1961. Cinco novos sabetinos da Amazônia (Diptera Culicidae). *Revista Brasileira de Entomologia* 10: 37–52.

Cerqueira, N. & Antunes, P.C.A. 1938. *Haemagogus tropicalis,* a new species from Pará, Brazil (Diptera, Culicidae). *Proceedings of the Entomological Society of Washington* 40(1): 1–9.

Cerqueira, N.L. & Costa, A.F. 1946. Duas novas espécies de *Aëdes (Ochlerotatus) (Diptera Culicidae)*. *Livro de homenagem a R. F. d'Almeida* 11: 133–140.

Cerqueira, N.L. & Lane, J. 1944. Sôbre algumas espécies de *Culex* (*Carrollia*) (Diptera, Culicidae). *Revista de Entomologia* 15(1–2): 216–220.

Cerqueira, N.L. & Lane, J. 1945. Note on *Haemagogus capricornii* Lutz, 1904 (Diptera, Culicidae). *Proceedings of the Entomological Society of Washington* 47(9): 279–288.

Cerqueira, N.L. & Paraense, W.L. 1945. Uma nova espécie de *Aedes,* transmissora de *Plasmodium gallinaceum* (Diptera Culicidae). *Memórias do Instituto Oswaldo Cruz* 42(1): 11–15.

Chagas, C. 1907. *Novas especies de culicidos brazileiros*. Besnard Freres, Rio de Janeiro. 28 pp.

Chalmers, A.J. 1900. Uncomplicated aestivo-autumnal fever in Europeans in the Gold Coast Colony,West Africa. *Lancet* 2(4027): 1262–1264.

Chang, H. 1965. Description of a mosquito *Aëdes* (*Edwardsaedes*) *pingpaensis,* sp. nov. [in Chinese]. *Acta Parasitologica Sinica* 2(4): 397–402.

Chang, H.-T., Zhao, S.X., Hang, B.-X. & Chen, S.-F. 1975. A new species of *Culex* [in Chinese]. *Acta Entomologica Sinica* 18(3): 338–340.

Chang, P.-X. & Chang, S.-M. 1974. Two new species of Culicini from Kweichow [in Chinese]. *Acta Entomologica Sinica* 17(3): 347–352.

Chapman, H.C. & Barr, A.R. 1964. *Aedes communis nevadensis,* a new subspecies of mosquito from western North America (Diptera: Culicidae). *Mosquito News* 24(4): 439–447.

Chen, H.-B. 1977. A new species of *Culex* (Diptera: Culicidae) [in Chinese]. *Acta Entomologica Sinica* 20(1): 95–98.

Chen, H.-B. 1980. Study on the *Culex vishnui* subgroup in China with discussions on the taxonomic status of *C. neovishnui* Lien, *C. permixtus* Hsieh & Liao and *C. cheni* Ho [in Chinese]. *Acta Entomologica Sinica* 23(4): 434–440.

Chen, H.-B. 1981. A new species of *Culex* (Diptera: Culicidae) [in Chinese]. *Acta Entomologica Sinica* 24(4): 426–429.

Chen, H.-B. 1982. A new species of *Culex* from Guizhou, *Culex* (*Eumelanomyia*) *miaolingensis,* sp. nov. [in Chinese]. *Acta Zootaxonomica Sinica* 7(3): 312–316.

Chen, H.-B. 1989. A new species of *Culex* (*Culiciomyia*) from Yunnan (Diptera: Culicidae) [in Chinese]. *Acta Zootaxonomica Sinica* 14(1): 86–90.

Chen, H.-L. & Chen, H.-B. 2000. Description of a new species of the *Aedes* from Henan, China (Diptera: Culicidae) [in Chinese]. *Acta Zootaxonomica Sinica* 25(3): 341–344.

Chen, H.-B. & Dong, X.-S. 1992. A new species of *Culex* (*Eumelanomyia*) from China (Diptera: Culicidae) [in Chinese]. *Acta Zootaxonomica Sinica* 17(1): 78–83.

Chen, H.B. & Zhang, P.-X. 1988. *Topomyia* (*Suaymyia*) *tumetarsalis,* a new species from China (Diptera: Culicidae) [in Chinese]. *Acta Entomologica Sinica* 31(2): 213–218.

Chen, H. & Zhao, H. 1985. A new species of the genus *Culex* from China (Diptera: Culicidae) [in Chinese]. *Entomotaxonomia* 7(4): 265–269.

Chen, H.-B., Wang, S.-H. & Zhao, H. 1989. A new species of *Uranotaenia* breeding in crab holes from Guizhou Province of China (Diptera: Culicidae) [in Chinese]. *Acta Zootaxonomica Sinica* 14(2): 224–228.

Chow, C.Y. 1950. Collection of culicine mosquitoes (Diptera, Culicidae) in Taiwan (Formosa), China, with description of a new species. *Quarterly Journal of the Taiwan Museum* 3(4): 281–287.

Chowdhury, K.L. 1929. A new variety of protanopheline *A. barbirostris* van der Wulp, var. *ahomi*, found in Upper Assam. *Indian Journal of Medical Research* 16(4): 986–988.

Christophers, S.R. 1906. On the importance of larval characters in the classification of mosquitoes. *Scientific memoirs by officers of the medical and sanitary departments of the government of India* 25: 1–18 pp., 3 pls.

Christophers, S.R. 1911. [Notes on mosquitoes.] II.–A new anopheline. *Paludism* 2: 64–68, 1 pl.

Christophers, S.R. 1916a. A revision of the nomenclature of Anophelini. *Indian Journal of Medical Research* 3(3): 454–488.

Christophers, S.R. 1916b. A new anopheline with unspotted wings from Mesopotamia (*Anopheles lukisii*). *Indian Journal of Medical Research* 4(1): 120–122.

Christophers, S.R. 1923a. A new East-African anopheline (*A. kingi*) related to *A. natalensis*, Hill and Hayden, and *A. watsonii*, Leicester. *Indian Journal of Medical Research* 10(4): 1008–1019, 2 pls.

Christophers, S.R. 1923b. An Anopheles [*sic*] of the *Myzorhynchus* group (*Anopheles amazonicus* sp.n.) from South America. *Annals of Tropical Medicine and Parasitology* 17(1): 71–78, 1 pl.

Christophers, S.R. 1924a. Provisional list and reference catalogue of the Anophelini. *Indian Medical Research Memoirs* 3: 1–105.

Christophers, S.R. 1924b. Some Himalayan and peninsular varieties of Indian species of *Anopheles*. *Indian Journal of Medical Research* 12(1): 11–13.

Christophers, S.R. 1924c. Some further varieties of Indian species of *Anopheles* with notes on the species *A. pallidus* Theobald and *A. philippinensis* Ludlow. *Indian Journal of Medical Research* 12(2): 295–301.

Christophers, S.R. 1926. *A.* (*Myzomyia*) *pattoni*, a new Anopheles [*sic*] from Shan-tung, North China; with notes on some other species of Anopheles [*sic*] from the same locality. *Indian Journal of Medical Research* 13(4): 871 –877, 1 pl.

Christophers, S.R. 1931. Studies on the anopheline fauna of India. (Parts I–IV). *Records of the Malaria Survey of India* 11(2): 305–332.

Christophers, S.R. 1933. *The fauna of British India, including Ceylon and Burma. Diptera. Vol. IV. Family Culicidae. Tribe Anophelini*. Taylor and Francis, London. vi [iii], 371 pp., 3 pls.

Christophers, S.R. & Barraud, P.J. 1931. The eggs of Indian *Anopheles*, with descriptions of the hitherto undescribed eggs of a number of species. *Records of the Malaria Survey of India* 2(1): 161–192, 5 pls.

Christophers, S.R. & Chand, K. 1915. Notes on some anophelines from Arabia and Mesopotamia. *Indian Journal of Medical Research* 3(1): 180–200, 2 pls.

Christophers, S.R. & Puri, I.M. 1931a. Species and varieties of the *funestus* series of *Anopheles*. *Records of the Malaria Survey of India* 2(3): 481–493.

Christophers, S.R. & Puri, I.M. 1931b. Notes on some anopheline mosquitoes collected in Sierra Leone including differentiation of *Anopheles dthali* Patton (Mediterranean) as a distinct species from *Anopheles rhodesiensis* Theo. (Ethiopian). *Indian Journal of Medical Research* 18(4): 1133–1166, 3 pls.

Chu, F.-I. 1957. Collection of megarhine and culicine mosquitoes from Hainan Island, South China, with description of a new species [in Chinese]. *Acta Zoologica Sinica* 9(2): 145–164, 3 pls.

Chu, F.-I. 1980. The synonymy, distribution and classification of Chinese *Uranotaenia* (Diptera: Culicidae) [in Chinese]. *Entomotaxonomia* 2(3): 237–245.

Chu, F.-I. 1981. A new species of Chinese *Uranotaenia* (Diptera: Culicidae) [in Chinese]. *Entomotaxonomia* 3(4): 273–275.

Chwatt, L.J. 1948. A new *Aëdes* from the Cameroons, *A.* (*Aëdimorphus*) *boneti* s.sp. *kumbae* s.sp.nov. *Annals of Tropical Medicine and Parasitology* 42(2): 184–189.

Chwatt, L.J. 1949. *Aedes* (*Stegomyia*) *pseudoafricanus* sp. nov.: a new species of *Aedes* from the coast of Nigeria (British West Africa). *Nature* 163(4151): 808–809.

Clastrier, J. 1970a. *Culex* (*Eubonnea*) *guyanensis n.sp. (Diptera, Culicidae)* nouveau moustique de la Guyane française. *Annales de Parasitologie* 45(1): 115–118.

Clastrier, J. 1970b. *Culex (Melanoconion) dolichophyllus* n. sp. et *Culex (Microculex) stonei* Lane et Whitman, 1943 *(Diptera, Culicidae)* en Guyane française. *Annales de Parasitologie Humaine et Comparée* 45(5): 857–861.

Clastrier, J. 1970c. Quatre nouveau *Melanoconion (Diptera, Culicidae)* de la Guyane française. *Annales de Parasitologie* 45(4): 463–476.

Clastrier, J. 1971. Sur deux *Melanoconion* et un nouvel *Aedinus* de la Guyane française [*Dipt. Culicidae*]. *Annales de la Société Entomologique de France* 7(3): 629–654.

Clastrier, J. 1973. *Wyeomyia* (*Dendromyia*) *trifurcata* n. sp. nouveau moustique de la Guyane française (Diptera, Culicidae). *Nouvelle Revue d'Entomologie* 3(1): 39–47.

Clastrier, J. 1974. *Wyeomyia (Dendromyia) bourrouli* (Lutz, 1905) et *Wyeomyia (Cruzmyia) forattinii* n.sp. en Guyane française *(Diptera, Culicidae)*. *Cahiers ORSTOM Série Entomologie médicale et Parasitologie* 12(3): 153–163.

Clastrier, J. & Claustre, J. 1978a. *Culex (Carrollia) insigniforceps* n. sp. de la Guyane française (Diptera, Culicidae). *Nouvelle Revue d'Entomologie* 8(3): 359–362.

Clastrier, J. & Claustre, J. 1978b. *Limatus martiali* Senevet et Abonnenc, 1939 et *Wyeomyia nigricephala* n. sp. (Diptera, Culicidae) de la Guyane française. *Annales de Parasitologie* 53(5): 539–546.

Cockerell, T.D.A. 1916. British fossil insects. *Proceedings of the United States National Museum* 49(2119): 469–499, 6 pls.

Cockerell, T.D.A. 1918. The mosquitoes of Colorado. *Journal of Economic Entomology* 11(2): 195–200.

Cockerell, T.D.A. 1919. The oldest mosquitoes. *Nature* 103(2577): 43–46.

Coetzee, M. 1984. A new species of *Anopheles* (*Anopheles*) from Namibia (Diptera: Culicidae). *Systematic Entomology* 9(1): 1–8.

Coetzee, M. 1995. *Anopheles crypticus*, new species from South Africa is distinguished from *Anopheles coustani* (Diptera: Culicidae). *Mosquito Systematics* (for 1994) 26(3): 125–131.

Coetzee, M., Hunt, R.H., Wilkerson, R., della Torre, A., Coulibaly, M.B. & Besansky, N.J. 2013. *Anopheles coluzzii* and *Anopheles amharicus*, new members of the *Anopheles gambiae* complex. *Zootaxa* 3619(3): 246–274.

Coetzee, M., Segerman, J. & Hunt, R.H. 1987. Description of a new species *Anopheles* (*Cellia*) *kosiensis* (Diptera: Culicidae) from Zululand, South Africa. *Systematic Entomology* 12(1): 23–28.

Cogill, H. 1903. The *Anopheles* of Karwar (North Kanara). *Journal of the Bombay Natural History Society* 15(2): 327–336, 1 pl.

Colless, D.H. 1948. The anopheline mosquitoes of north-west Borneo. *Proceedings of the Linnean Society of New South Wales* 73(3–4): 71–119.

Colless, D.H. 1955a. New anopheline mosquitoes from North Borneo. *Sarawak Museum Journal* 6(5): 331–342.

Colless, D.H. 1955b. Notes on the culicine mosquitoes of Singapore. I.–Three new species of *Culex* (Diptera, Culicidae) and a redescription of *Culex hutchinsoni* Barraud. *Annals of Tropical Medicine and Parasitology* 49(3): 311–319.

Colless, D.H. 1956. The *Anopheles leucosphyrus* group. *Transactions of the Royal Entomological Society of London* 108(3): 37–116.
Colless, D.H. 1957a. Notes on the culicine mosquitoes of Singapore II.–The *Culex vishnui* group (Diptera, Culicidae), with descriptions of two new species. *Annals of Tropical Medicine and Parasitology* 51(1): 87–101.
Colless, D.H. 1957b. Further notes on the systematics of the *Anopheles leucosphyrus* group (Diptera: Culicidae). *Proceedings of the Royal Entomological Society of London Series B Taxonomy* 26(7–8): 131–139.
Colless, D.H. 1958. Notes on the culicine mosquitoes of Singapore IV.–The *Aëdes niveus* subgroup (Diptera, Culicidae): introduction and description of five new species and of one new subspecies. *Annals of Tropical Medicine and Parasitology* 52(4): 468–483.
Colless, D.H. 1960. Some species of *Culex* (*Lophoceraomyia*) from New Guinea and adjacent islands, with descriptions of four new species and notes on the male of *Culex fraudatrix* Theobald (Diptera: Culicidae). *Proceedings of the Linnean Society of New South Wales* (for 1959) 84(3): 382–390.
Colless, D.H. 1962. Notes on the taxonomy of the *Aëdes scutellaris* group, and new records of *A. paullusi* and *A. albopictus* (Diptera: Culicidae). *Proceedings of the Linnean Society of New South Wales* 87(3): 312–315.
Colless, D.H. 1965. The genus *Culex*, subgenus *Lophoceraomyia* in Malaya (Diptera: Culicidae). *Journal of Medical Entomology* 2(3): 261–307.
Colless, D.H. 1976. *Anopheles amictus* Edwards and the subspecies concept. *Mosquito Systematics* 8(3): 221–222.
Collucci, E. & Sallum, M.A.M. 2007. Cladistic analysis of the subgenus *Anopheles* (*Anopheles*) Meigen (Diptera: Culicidae) based on morphological characters. *Memórias do Instituto Oswaldo Cruz* 102(3): 277–291.
Coluzzi, M. 1958. Descrizione di una nouva specie rinvenuta in Somalia. *Revista di Malariologia* 37(1–3): 69–71.
Coluzzi, M. 1962. Le forme di *Anopheles claviger* Meigen indicate con i nomi *missirolii* [*sic*] e *petragnanii* [*sic*], sono due specie riproduttivamente isolate. *Accademia Nazionale dei Lincei. Rendiconti della Classe di Scienze fisiche, matematiche e naturali* 32(6): 1025–1030.
Coluzzi, M. & Sabatini, A. 1968. Divergenze morfologiche e barriere di sterilita' nel complesso *Aedes mariae* (*Diptera, Culicidae*). *Rivista de Parassitologia* 29(1): 49–70, 5 pls.
Coluzzi, M., Gironi, A.M. & Muir, D.A. 1970. Ulteriori esperimenti d'incrocio tra le forme del complesso *mariae* del genere *Aedes*. *Parassitologia* 12(2–3): 119–123, 2 pls.
Cook, S., Lien, N.G., McAlister, E. & Harbach, R.E. 2010. *Bothaella manhi*, a new species of tribe Aedini (Diptera: Culicidae) from the Cuc Phuong National Park of Vietnam based on morphology and DNA sequence. *Zootaxa* 2661: 33–46.
Cooling, L.E. 1924. The larval stages and biology of the commoner species of Australian mosquitoes, with the biology of *Aëdes pecuniosus* Edwards. *Service publications / Department of Health, Australia, Tropical Division* 8: 1–40.
Coquillett, D.W. 1896. New Culicidae from North America. *Canadian Entomologist* 28(1): 43–44.
Coquillett, D.W. 1898. Report on a collection of Japanese Diptera, presented to the U. S. National Museum by the Imperial University of Tokyo. *Proceedings of the United Stated National Museum* 21(1146): 301–340.
Coquillett, D.W. 1901. Three new species of Culicidae. *Canadian Entomologist* 33(9): 258–260.
Coquillett, D.W. 1902a. New orthorrhaphous Diptera from Mexico and Texas. *Journal of the New York Entomological Society* 10(3): 136–141.
Coquillett, D.W. 1902b. New forms of Culicidae from North America. *Journal of the New York Entomological Society* 10(3): 191–194.
Coquillett, D.W. 1902c.New Diptera from North America. *Proceedings of the United States National Museum* 25(1280): 83–126.
Coquillett, D.W. 1902d. Three new species of *Culex*. *Canadian Entomologist* 34(11): 292–293.
Coquillett, D.W. 1903a. Four new species of *Culex*. *Canadian Entomologist* 35(9): 255–257.
Coquillett, D.W. 1903b. A new *Anopheles* with unspotted wings. *Canadian Entomologist* 35(11): 310.
Coquillett, D.W. 1904a. Several new Diptera from North America. *Canadian Entomologist* 36(1): 10–12.
Coquillett, D.W. 1904b. Notes on *Culex nigritulus*. *Entomological News and Proceedings of the Entomological Section of the Academy of Natural Sciences of Philadelphia* 15(2): 73–74.
Coquillett, D.W. 1904c. New North American Diptera. *Proceedings of the Entomological Society of Washington* 6(3): 166–192.
Coquillett, D.W. 1905. A new *Culex* from Australia. *Entomological News and Proceedings of the Entomological Section of the Academy of Natural Sciences of Philadelphia* 16(4): 116.
Coquillett, D.W. 1906a. New Culicidae from the West Indies and Central America. *Proceedings of the Entomological Society of Washington* (for 1905) 7(4): 182–186.
Coquillett, D.W. 1906b. On the breaking-up of the old genus *Culex*. *Science* 23(582): 312–314.
Coquillett, D.W. 1906c. A new *Culex* near *curriei*. *Entomological News and Proceedings of the Entomological Section of the Academy of Natural Sciences of Philadelphia* 17(4): 109.
Coquillett, D.W. 1906d. A classification of the mosquitoes of North and Middle America. *U. S. Department of Agriculture, Bureau of Entomology, Technical Series* 11: 1–31.
Coquillett, D.W. 1906e. Five new Culicidae from the West Indies. *Canadian Entomologist* 38(2): 60–62.
Corbet, P.S. & van Someren, E.C.C. 1962. *Aëdes (Stegomyia) opok* sp. nov., a new species of mosquito from Uganda. *Annals of Tropical Medicine and Parasitology* 56(1): 73–77.
Cornet, M. 1968. *Aedes (Aedimorphus) chamboni* n. sp., une nouvelle espèce de *Culicidae* du Sénégal. *Bulletin de la Société de Pathologie exotique* (for 1967) 60(4): 398–402.
Cornet, M. 1973. *Aedes (Stegomyia) cozi* n. sp., une nouvelle espèce de Culicidae au Sénégal. *Cahiers ORSTOM Série Entomologie médicale et Parasitologie* 11(3): 175–180.
Cornet, M., Valade, M. & Dieng, P.Y. 1978. *Aedes (Stegomyia) neoafricanus* une nouvelle espèce de moustique capturée au Sénégal Oriental *(Diptera: Culicidae)*. *Cahiers ORSTOM Série Entomologie médicale et Parasitologie* 16(3): 227–230.

Corradetti, A. 1939a. Studi morfologici sulla specie anofelica precedentemente identificata come *A. dthali* nel semien e sua classificazione come nuova varietà di *A. rhodesiensis.Rivista di Parassitologia* 3(1): 57–63.

Corradetti, A. 1939b. Una nuova specie di *Anopheles* Rinvenuta in Dancalia: *Anopheles (Neocellia) dancalicus* n. sp. Nota preliminare. *Rivista di Parassitologia* 3(3): 277–278.

Corradetti, A. 1939c. Su alcune larve di *Anopheles* rinvenutea Ghinda (Eritrea) e descrizione di una nuova larva: *Anopheles (Myzomyia) erythraeus* n.sp. *Rivista di Parassitologia* 3(4): 287–292.

Corradetti, A. & Archetti, I. 1947. Gli anofelini raccolti dalla Missione Biologica Sagan-Omo. *Rendiconti Instituto Superiore di Sanità* 10: 676–679.

Corrêa, R.R. 1944. Deserição de uma nova subespécie do subgénero "*Anopheles*" (Diptera, Culicidae). *Anals Paulistas de Medicina e Cirurgia* 47(5): 407–408.

Corrêa, R.R. 1950. Descrição de *Anopheles* (*Kerteszia*) *montemor* nova espécie de anofelino do Brasil. (Diptera, Culicidae). *Arquivos de Higiene e Saúde Pública* 14(39–42): 53–55, 1 pl.

Corrêa, R.R. & Cerqueira, F.M.C. 1944. Descrição de *Anopheles (Kerteszia) laneanus*, nova espécie de anofelino de Campos do Jordão *(Diptera, Culicidae)*. *Arquivos de Higiene e Saúde Pública* 9: 111–117.

Correa, R.R. & Ramalho, G.R. 1956. Revisão de *Phoniomyia* Theobald, 1903 (Diptera, Culicidae, Sabethini). *Folia Clinica et Biologica* 25(1/6): 1–176.

Corrêa, R.R. & Ramalho, G.R. 1959a. Deserição de uma nova espécie de *Culex (Diptera, Culicidae)*. *Revista Brasileira de Malariologia e Doenças Tropicais* 11(1): 55–57.

Corrêa, R.R. & Ramalho, G.R. 1959b. *Culex* (*Culex*) *deanei*, nova espécie do género *Culex* Linnaeus, 1758. (*Diptera, Culicidae*). *Revista do Instituto de Medicina Tropical de São Paulo* 1(2): 141–143.

Corrêa, R.R. & Ramalho, G.R. 1968. *Anopheles* (*Arribalzagia*) *anchietai* nova espécie de anofelino de São Paula, Brasil *(Diptera Culicidae)*. *Revista Brasileira de Malariologia e Doenças Tropicais* 20(8–9): 115–132.

Corrêa, R.R. & Ramos, A.S. 1943. Descrição de uma nova sub-espécie de anofelino do sub-género *Nyssorhynchus* Blanchard, 1902 (Diptera, Culicidae). *Revista Paulista de Medicina* 22: 246–248.

Cot & Hovasse. 1917. Quelques remarques sur les Anophélines de Macédoine. *Bulletin de la Société de Pathologie exotique* 10(10): 890–896.

Coutinho, J.O. 1944. *Anopheles (Shannoniezia) costalimai* Fonseca e Ramos, 1940, novo nome e a redescrição da espécie *(Diptera-Culicidae)*. *Memórias do Instituto Oswaldo Cruz* (for 1943) 39(3): 425–434.

Coutinho, J.O. & Farias, G.S. 1942. *Anopheles* (*Ayrozamyia*) *tibiamaculatus* (Neiva, 1906) – descrição do macho e criação de novo sub-genero (Diptera, Culicidae). *Anais da Academia Brasileira de Ciencias* 14(4): 343–347, 1 pl.

Cova García, P. 1962. Hallazgo del insecto *Culex* (*Culex*) *tejerai*, sp. n. en la cumbre del Choroni, Estado Aragua. *Revista Venezolana de Sanidad y Asistencia Social* 27(2): 312–316.

Cova Garcia, P. & Pulido F., J. 1974. Nueva especie de *Culex* (*Microculex*) (Diptera, Culicidae): *Culex* (*Microculex*) *sutili*. *Boletin de la Direccion de Malariologia y Saneamiento Ambiental* 14(1–2): 27–28.

Cova García, P. & Rausseo, J.A. 1964. *Uranotaenia bertii*, n. sp. *Revista Venezolana de Sanidad y Asistencia Social* 29(2): 221–229.

Cova Garcia, P. & Sutil Oramas, E. 1974. *Culex* (*Microculex*) *pulidoi* (Diptera, Culicidae). *Boletin de la Direccion de Malariologia y Saneamiento Ambiental* 14(1–2): 29–30.

Cova Garcia, P. & Sutil O., E. 1976a. *Culex (Carrollia) wannonii* (Diptera, Culicidae), una nueva especie de Venezuela. *Boletin de la Direccion de Malariologia y Saneamiento Ambiental* 16(4): 354–359.

Cova Garcia, P. & Sutil Oramas, E. 1976b. Nueva especie para Venezuela *Mansonia* (*Mansonia*) *suarezi* (Diptera, Culicidae). *Boletin de la Direccion de Malariologia y Saneamiento Ambiental* 16(4): 360–362.

Cova Garcia, P., Pulido F., J. & Amarista M., J.R. 1977. *Anopheles* (*Nyssorhynchus*) *deltaorinoquensis* n. sp. (Diptera, Culicidae) de Venezuela. *Boletin de la Direccion de Malariologia y Saneamiento Ambiental* 17(2): 150–161.

Cova Garcia, P., Pulido F., J. & Escalante de Ugueto, C. 1976. *Anopheles (Kerteszia) hilli* (Diptera, Culicidae), una nueva especie de Venezuela. *Boletin de la Direccion de Malariologia y Saneamiento Ambiental* 16(4): 344–353.

Cova Garcia, P., Pulido F., J. & Escalante de Ugueto, C. 1977a. *Anopheles (Kerteszia) gonzalezrinconesi* n. sp. (Diptera, Culicidae) de Venezuela. *Boletin de la Direccion de Malariologia y Saneamiento Ambiental* 17(2): 140–149.

Cova Garcia, P., Pulido F., J. & Escalante de Ugueto, C. 1977b. *Anopheles* (*Kerteszia*) *rollai*, nomen novum *Anopheles* (*Kerteszia*) *hilli* Cova García, 1976 (non *Anopheles hilli* Woodhill & Lee 1944). *Boletin de la Direccion de Malariologia y Saneamiento Ambiental* 17(3): 241.

Cova Garcia, P., Pulido F., J. & Escalante de Ugueto, C. 1979. *Culex (Neoculex) johni* (Diptera: Culicidae) nueva especie de Venezuela. *Boletin de la Direccion de Malariologia y Saneamiento Ambiental* 19(1): 21–28.

Cova Garcia, P., Pulido F., J. & Escalante de Ugueto, C. 1981. *Uranotaenia pifanoi* (Diptera, Culicidae) nueva especie de Venezuela. *Boletin de la Direccion de Malariologia y Saneamiento Ambiental* 21(3–4): 210–218.

Cova García, P., Pulido F., J., Escalante de Ugueto, C. & Mora F., J.D. 1987a. *Uranotaenia iriartei* (Diptera, Culicidae), nueva especie de Venezuela. *Boletin de la Direccion de Malariologia y Saneamiento Ambiental* 27(1–4): 45–52.

Cova García, P., Pulido F., J., Escalante de Ugueto, C. & Mora R., J.D. 1987b. *Uranotaenia gabaldoni* (Diptera, Culicidae), nueva especie de Venezuela. *Boletin de la Direccion de Malariologia y Saneamiento Ambiental* 27(1–4): 94–104.

Cova García, P., Pulido, J., Escalante de Ugueto, C., Amarista, J. & Mora, J.D. 1987c. *Uranotaenia briceñoi* (Diptera, Culicidae), nueva especie de Venezuela. *Boletin de la Direccion de Malariologia y Saneamiento Ambiental* 27(1–4): 65–73.

Cova García, P., Sutil, E. & Pulido, J. 1971. Descripcion de nueva especie de *Culex*, del subgenero *Carrollia* (Diptera, Culicidae). *Boletín de la Sociedad Venezolana de Ciencias Naturales* 29: 203–210.

Cova Garcia, P., Sutil O., E. & Pulido F., J. 1972. Neuva especie de *Culex* (*carrollia* [*sic*]) *rausseoi.Boletin de la Direccion de Malariologia y Saneamiento Ambiental* 12: 207–209.

Cova Garcia, P., Sutil Oramas, E. & Pulido F., J. 1974a. *Wyeomyia lopezii* n. sp. (Diptera, Culicidae). *Boletin de la Direccion de Malariologia y Saneamiento Ambiental* 14(1–2): 21–22.

Cova Garcia, P., Sutil Oramas, E. & Pulido F., J. 1974b. *Sabethes fernández-yepezi* n. sp. *Boletin de la Direccion de Malariologia y Saneamiento Ambiental* 14(1–2): 23–24.

Covell, G. 1927. A new species of *Anopheles* from eastern India, *A.* (*Myzomyia*) *ramsayi*; with a new description of *A.* (*Myzomyia*) *jamesii* Theobald. *Indian Journal of Medical Research* 14(4): 1019–1025, 1 pl.

Cruz, O.G. 1901. Contribuição para o estudo dos culicidios do Rio de Janeiro. *Brazil-Medico* 15(43): 423–426.

Cruz, O.G. 1906. Um novo genero da sub-familia «Anophelina». *Brazil-Medico* 20(20): 199–200.

Cruz, O.G. 1907a. Um novo genero brazileiro da sub-familia "Anophelinae". *Brazil-Medico* 21(28): 271–273.

Cruz, G. 1907b. Uma nova especie do genero *Psorophora*. *Brazil-Medico* 21(16): 329–330.

Cumpston, J.H.L. 1924. Disease distribution in the Pacific Basin.*Proceedings of the Pan-Pacific Science Congress, Australia, 1923.* 2: 1400–1407 [Conference proceedings].

Curry, D.P. 1928. A new anopheline mosquito, *Anopheles* (*Chagasia*) *bathanus* Dyar, discovered in the Canal Zone. *American Journal of Tropical Medicine* 8(3): 243–248.

Curry, D.P. 1931. *Anopheles* (*Anopheles*) *neomaculipalpus*. A new species of the *Arribalzagaia* group of *Anopheles* from Panama. *American Journal of Hygiene* 13(2): 643–647.

Curry, D.P. 1932. Some observations on the *Nyssorhynchus* group of the *Anopheles* (*Culicidae*) of Panama. *American Journal of Hygiene* 15(2): 566–572.

Curtis, J. 1831. *A guide to an arrangement of British insects; being a catalogue of all the named species hitherto discovered in Great Britain and Ireland* (for 1829). Printed for the author by Richard Taylor, London. vi, 129 pp. (256 columns).

Curtis, J. 1835. *Culex guttatus*. The white-spotted gnat or mosquito. Fascicle 537. *British entomology; being illustrations and descriptions of the genera of insects found in Great Britain and Ireland: containing coloured figures from nature of the most rare and beautiful species, and in many instances of the plants upon which they are found.* Volume VIII. Diptera. Omaloptera. London, printed for the author, 1823–1840.

Curtis, J. 1837. *A guide to an arrangement of British insects; being a catalogue of all the named species hitherto discovered in Great Britain and Ireland*. Second edition. London. vi, 294 pp.

Curtis, L.C. 1967. The mosquitoes of British Columbia. *Occasional Papers of the British Columbia Provincial Museum* 15: v +1–90.

da Costa Lima, A. 1929. Sobre um novo *Anopheles* do Brasil. *Brasil-Medico* 43(2): 1100–1101.

da Costa Lima, A. 1930a. Sobre especies do genero *Miamyia*, subgenero *Miamyia* (Diptera: Culicidae). *Memórias do Instituto Oswaldo Cruz* 24(2): 73–78, 3 pls.

da Costa Lima, A. 1930b. Sobre especies do genero *Miamyia*, subgenero *Miamyia* (2ª. Nota) (Diptera: Culicidae). *Memórias do Instituto Oswaldo Cruz* 24(3): 187–194, 3 pls.

da Costa Lima, A. 1931a. Nota sobre sabethineos do grupo *Joblotia* (Diptera: Culicidae). *Memórias do Instituto Oswaldo Cruz* 25(1): 65–71, 9 pls.

da Costa Lima, A. 1931b. Sobre as especies de *Megarhinus* do Brasil (Diptera: Culicidae). *Memórias do Instituto Oswaldo Cruz* 25(4): 307–315, 2 pls.

da Costa Lima, A. 1933. Sobre um novo *Aëdes* (*Ochlerotatus*) do Brasil. *Memórias do Instituto Oswaldo Cruz* 27(4): 403–406, 1 pl.

da Costa Lima, A. 1935a. Especies de *Taeniorhynchus* (*Taeniorhynchus*) (Diptera: Culicidae). *Memórias do Instituto Oswaldo Cruz* 30(3): 453–469, 10 pls.

da Costa Lima, A. 1935b. Sobre as especies de *Orthopodomyia* encontradasno Brasil (Diptera: Culicidae). *Revista Medico-Cirurgica do Brasil* 43(6): 175–179, 2 pls.

da Costa Lima, A. 1937. Um novo *Anopheles* da Baixada Fluminense (*Diptera: Culicidae*). *Revista Medico-Cirurgica do Brasil* 45(1): 5–7.

da Costa Lima, A., Guitton, N. & Ferreira, O. 1962. Comentários relativos às espécies da tribo Toxorhynchitini (Megarhinini) com a desçricão de uma espécie nova de *Lynchiella* (Diptera, Culicidae). *Memórias do Instituto Oswaldo Cruz* 60(2): 225–251, 4 pls.

da Cunha Ramos, H. 1985. Research on the mosquitoes of Angola (*Diptera: Culicidae*) XVII – Description of *Uranotaenia (Uranotaenia) angolensis* sp. nov. *Boletim da Sociedade Portuguesa de Entomologia* 1(Suplemento l): 73–82.

da Cunha Ramos, H. 1986. Research on the mosquitoes of Angola (Diptera: Culicidae). XVIII - Description of *Uranotaenia* (*Uranotaenia*) *machadoi* sp. nov. *Mosquito Systematics* 18(1): 110–117.

da Cunha Ramos, H. 1993. *Revisão do género* Uranotaenia *(Diptera, Culicidae) na região afrotropical.* Ministério do Planeamento e da Administração do Território, Secretaria de Estado da Ciêcia e Technologia, Instituto de Investigação Científica Tropical, Lisboa, Portugal. 486 pp.

da Cunha Ramos, H. & Brunhes, J. 2004. *Insecta Diptera Culicidae* Uranotaenia. Faune de Madagascar 91. Institut de recherche pour le développement (Montpellier); Centre de cooperation internationale en recherché agronomique pour le développement (Paris); Muséum national d'histoire naturelle (Paris). 463 pp.

da Cunha Ramos, H. & Ribeiro, H. 1982. Research on the mosquitoes of Angola (Diptera: Culicidae) XII - Description of *Culex (Culex) pajoti* sp. nov. *Mosquito Systematics* (for 1981) 13(2): 176–190.

da Cunha Ramos, H. & Ribeiro, H. 1990. Research on the mosquitoes of Angola (Diptera, Culicidae). XX - *Eretmapodites mortiauxi*, a new species of the *leucopus* group. *Boletim da Sociedade Portuguesa de Entomologia* 4(115): 138–144.

da Cunha Ramos, H. & Ribeiro, H. 1992. Research on the mosquitoes of Angola *(Diptera, Culicidae)* XXI - Description of *Eretmapodites angolensis* sp. nov. and *Eretmapodites dundo* sp. nov., of the «*oedipodeios* group». *Garcia de Orta, Série de Zoologia* (for 1990) 17(1–2): 31–35.

da Cunha Ramos, H., Capela, R.A. & Ribeiro, H. 1995. Description of *Aedes* (*Aedimorphus*) *gandarai*, a new species from the island of São Tomé, with keys to all known species of the Tarsalis Group (Diptera: Culicidae). *Mosquito Systematics* (for 1994) 26(3): 97–104.

da Cunha Ramos, H., Ribeiro, H. & de Barros Machado, A. 1992. Research on the mosquitoes of Angola (*Diptera, Culicidae*). XIX - *Eretmapodites tendeiroi*, a new species of the «*plioleucus* group». *Garcia de Orta, Série de Zoologia* (for 1989) 16(1–2): 133–136.

da Cunha Ramos, H., Ribeiro, H. & Harrison, B.A. 2003. A new European mosquito species: *Culex (Neoculex) europaeus* (Diptera: Culicidae). *European Mosquito Bulletin [Journal of the European Mosquito Control Association]* 15: 6–11.

da Fonseca, F. & da Silva Ramos, A. 1940. Novo subgenero e novas especies de anofelinas neotropicas (*Diptera. Culicidae*). (Nota previa). *Memorias do Instituto Butantan* (for 1939) 13: 383–387, 2 pls.

da Fonseca, F. & Ramos, A.S. 1940. *Shannonesia* nov. nom. (Dipt. Culicidae). *Revista de Entomologia* 11: 966.

Dahl, C. & White, G.B. 1978. Culicidae, pp. 390–395. *In*: Illies, J. (ed.), *Limnofauna Europaea. A checklist of the animals inhabiting European inland waters, with accounts of their distribution and ecology (except protozoa)*. Second edition. Gustav Fischer Verlag, Stuttgart - New York.

D'Anfreville, L. 1916. Les moustiques de Salé, Maroc. *Bulletin de la Société de Pathologie exotique* 9(3): 140–142.

Daniels, C.W. 1908. Notes on the mosquitoes on the river and coast districts of the eastern side of the Peninsula. *Studies from the Institute for Medical Research, Federated Malay States* 3(3): 262–266.

Danilov, V.N. 1974. On the restoration of the name *Aedes (O.) mercurator* Dyar for a mosquito known in the USSR as *Aedes riparius ater* Gutsevich (Diptera, Culicidae) [in Russian]. *Parazitologiia* 8(4): 322–328.

Danilov, V.N. 1976. Revision of some Holarctic species and subspecies of the genus *Culiseta* Felt. I. *Culiseta* (*Culiseta*) *kanayamensis* Yamada as a synonym *of C.* (*C.*) *bergrothi* Edwards [in Russian]. *Parazitologiia* 10(2): 185–187.

Danilov, V.N. 1977. On the synonymy of species names of *Aedes* mosquitoes (subgenera *Finlaya* and *Neomelaniconion*) in the Far East fauna [in Russian]. *Parasitologiia* 11(2): 181–184.

Danilov, V.N. 1979. Nomenclature, synonymy, differential diagnosis and distribution of *Aedes (Ochlerotatus) albineus* Séguy (Diptera, Culicidae) [in Russian]. *Vestnik Zoologii* 13(1): 29–35.

Danilov, V.N. 1982. Mosquito fauna (Diptera: Culicidae) of the Ethiopian zoogeographical region 2. A new species of *Mansonia* (*Coquillettidia*) and a key for adults of the genus *Mansonia* [in Russian]. *Meditsinskaya Parazitologiya i Parazitarnye Bolezni* 51(4): 26–30.

Danilov, V.N. 1984. *Culiseta* (*Culicella*) *setivalva* Maslov as a synonym of *C.* (*C.*) *fumipennis* Stephens [in Russian]. *Parazitologiia* 18(4): 313–317.

Danilov, V.N. 1985. Mosquitoes of the subgenus *Edwardsaedes* of the Palearctic with description of the larva of *Aedes* (*Edw.*) *bekkui* [in Russian]. *Parazitologiia* 19(5): 378–381.

Danilov, V.N. 1987a. Mosquitoes of the genus *Toxorhynchites* in the fauna of the USSR and related taxa in East and South-east Asia (Culicidae) [in Russian]. *Parazitologiia* 21(2): 151–155.

Danilov, V.N. 1987b. Mosquitoes of the subgenus *Aedes* (Diptera, Culicidae) of the USSR fauna. II. *Aedes dahuricus* sp. n. [in Russian]. *Vestnik Zoologii* 1987(4): 35–41.

Danilov, V.N. 1989. Two new Afrotropical subgenera of the mosquito genus *Culex* (Diptera, Culicidae) [in Russian]. *Entomologischeskoe Obozrenie* 68(4): 790–797.

Danilov, V.N. & Filippova, V.V. 1978. A new species of mosquito, *Aedes* (*Stegomyia*) *sibiricus* sp. n. (Culicidae) [in Russian]. *Parazitologiia* 12(2): 170–176.

Danilov, V.N. & Gornostaeva, R.M. 1987. A new species of mosquito, *Aedes* (*Ochlerotatus*) *intermedius* sp. n. (Diptera, Culicidae) [in Russian]. *Parazitologiia* 21(5): 612–619.

Danilov, V.N. & Stupin, V.I. 1982. A new species of the mosquito genus *Aedes* Mg. (Diptera, Culicidae) from southern Siberia and Priamuria [in Russian]. *Entomologischeskoe Obozrenie* 61(4): 824–829.

Danilov, V.N., Markovich, N.Ya. & Proskuryakova, A.M. 1978. A new species of mosquito, *Aedes* (*Ochlerotatus*) *sergievi* sp. n. [in Russian]. *Parazitologiya* 12(1): 44–49.

Darsie, R.F., Jr. & Cagampang-Ramos, A. 1969. *Manual of malaria entomology. A handbook of laboratory and field training exercises*. Second edition. Malaria Eradication Training Center, Manila. vi, 142 pp.

Darsie, R.F., Jr. & Cagampang-Ramos, A. 1977. A study of *Anopheles ludlowae* Theobald siblings, leading to new synonymy (Diptera, Culicidae). *Mosquito Systematics* 9(2): 123–127.

da Silva Mattos, S. & Xavier, S.H. 1991. *Culex (Melanoconion) guedesi* uma nova espécie do Brasil (Diptera, Culicidae). *Revista Brasileira de Biologia* 51(1): 193–196.

da Silva Mattos, S., de Silveira Guedes, A. & Xavier, S.H. 1978. Uma espécie nova de *Culex (Culex)* que se cria em umbelífera (Diptera, Culicidae). *Revista Brasileira de Biologia* 38(1): 31–35.

Davis, N.C. 1931. A new anopheline mosquito from Pará, Brazil. *American Journal of Hygiene* 13(1): 345–348, 2 pl.

de Burca, B. 1943. A new species of *Anopheles* from East Africa. *Journal of the Malaria Institute of India* 5(2): 233–234.

de Charmoy, d'E. 1908. On three new species of *Culex* collected during the anti-malarial campaign in Mauritius in 1908. *Annals of Tropical Medicine and Parasitology* 2(3): 257–264.

de Deus Palma, J. & Bello Galvão, A. 1969. *Lynchiella rizzoi* sp. n., toxorrinquitíneo bromelícola do Estado de Goiás (Diptera, Culicidae). *Revista Goiana de Medicina* 15(1,2): 3–13.

de Figueiredo, A.C. 1904. Da existencia do «*Anopheles bifurcatus*» em Portugal. *A Medicina Contemporanea* 22: 184–187, 223–225.

de Fourcroy, A.F. 1785. *Entomologia Parisiensis; sive catalogus Insectorum quoe in agro Parisiensi reperiuntur; secundum methodum Geoffrœanam in sectiones, genera & species distributus: Cui addita sunt nomina trivalia & fere trecentoe novœ species*. Volumes 1 and 2. Parisiis [Paris]. vii, 544 pp.

de Geer, C. 1776. *Memoires pour servir a l'histoire des insectes*. Tome sixieme. De l'imprimerie de Pierre Hesselberg, Stockholm. viii, 522 pp., 30 pls.

de Grandpré, A.D. & de Charmoy, D. d'E. 1901. *Les moustiques: Anatomie et biologie. Contribution à l'étude des culicidées et principalement des genres* Culex *et* Anopheles *de leur rôle dans la propagation de la malaria et de la filariose et des moyens de s'en préserver*. (for 1900) Ile Maurice, Port Louis. iv, 59 pp., 5pls.

de Leon, J.R. 1938. El anophelismo de altura en Guatemala. *Boletin de la Direccion General de Sanidad Publica (Guatemala)* 9: 411–424.

Delfinado, M.D. 1966a. The culicine mosquitoes of the Philippines, tribe Culicini (Diptera, Culicidae). *Memoirs of the American Entomological Institute* 7: 1–252.

Delfinado, M.D. 1966b. Two undescribed species of *Uranotaenia* from the Philippines (Diptera: Culicidae). *Journal of Medical Entomology* 3(1): 36–37.

Delfinado, M.D. 1967. Contributions to the mosquito fauna of Southeast Asia. I. The genus *Aedes*, subgenus *Neomacleaya* Theobald in Thailand. *Contributions of the American Entomological Institute* 1(8): 1–56.

Delfinado, M.D. 1968. Contributions to the mosquito fauna of Southeast Asia. III. The genus *Aedes*, subgenus *Neomacleaya* Theobald in Southeast Asia. *Contributions of the American Entomological Institute* 2(4): 1–74.

Delfinado, M.D. & Hodges, E.R. 1968. Three new species of the genus *Tripteroides*, subgenus *Tripteroides* Giles (Diptera: Culicidae). *Proceedings of the Entomological Society of Washington* 70(4): 361–375.

de Linero, M.Q. 1967. Nueva especie de "*Culex* (*Melanoconion*)". *Revista Brasileira de Biologia* 27(3): 289–291.

del Ponte, E. 1939. Identificación de «Sabethini» (Dip. Culicidae) por medio de tarjetas perforadas. *Physis* 17: 535–541.

del Ponte, E. & Cerqueira, N. 1938. Alguns sabethineos do Brasil (Diptera, Culicidae). *Revista de Entomologia* 8(3–4): 225–237.

del Ponte, E. & Heredia, R.L. 1945. *Anopheles* (*Anopheles*) *holmbergi* n. sp. de Misiones (Argentina) *(Diptera Culicidae). Revista de la Sociedad Entomólogica Argentina* 12: 382–386.

del Ponte, E., Castro, M.P. & Garcia, M. 1951. Clave para las especies de *Psorophora* y *Aëdes* de la Argentina y comarcas vecinas. Diagnosis de *Aëdes* (*O.*) *raymondi* n. sp. (Diptera, Culicidæ). *Anales de la Sociedad Cientifíca Agentina* 151(5): 228–243.

del Vecchio, G. 1939. Sulle varietà di *A. claviger* (*bifurcatus*). Nota I. *Rivista di Parassitologia* 3(1): 27–37, 1 pl.

de Meijere, J.C.H. 1909. Drei myrmecophile Dipteren aus Java. *Tijdschrift voor Entomologie* 52: 165–174, 1 pl.

de Meijere, J.C.H. 1910. *Nepenthes*-Tiere. I. Systematik. *Annales du Jardin Botanique de Buitenzorg, Supplément* 3(2): 917–940, 4 pls.

de Meijere, J.C.H. 1911. Zur Kenntnis niederländischer Culiciden. *Tijdschrift voor Entomologie* 54: 137–157, 3 pls.

de Meillon, B. 1931a. Illustrated keys to the full-grown larvae and adults of South African anopheline mosquitos. *Publications of the South African Institute for Medical Research* 4(28): 275–375.

de Meillon, B. 1931b. A new South African anopheline. *Journal of the Medical Association of South Africa* 5: 482–483.

de Meillon, B. 1934. Entomological studies, studies on insects of medical importance in South Africa. *Publications of the South African Institute for Medical Research* 6(33): 251–307.

de Meillon, B. 1935. Entomological studies. Studies on insects of medical importance in South Africa — Part II. *Publications of the South African Institute for Medical Research* 6(35): 323–365.

de Meillon, B. 1942. New Nematocera from the Ethiopian Region. *Journal of the Entomological Society of Southern Africa* 5: 87–98.

de Meillon, B. 1943. New records, and new species of Nematocera (Diptera) from the Ethiopian region. *Journal of the Entomological Society of Southern Africa* 6: 90–113.

de Meillon, B. 1947a. New records and species of biting insects from the Ethiopian region. II. *Journal of the Entomological Society of Southern Africa* 10(1): 110–124.

de Meillon, B. 1947b. The Anophelini of the Ethiopian geographical region. *Publications of the South African Institute for Medical Research* 10(49): 1–272.

de Meillon, B. & de Carvalho Pereira, M. 1940. Notes on some anophelines (Dipt. Culicidae) from Portuguese East Africa. *Moçambique Documentário Trimestral* 23: 69–83, 1 photo, 7 pls, map.

de Meillon, B. & Evans, A.M. 1935. Two new anophelines from South Africa. *Annals of Tropical Medicine and Parasitology* 29(1): 91–94.

de Meillon, B. & Lavoipierre, M. 1944. New records and species of biting insects from the Ethiopian region. *Journal of the Entomological Society of Southern Africa* 7: 38–67.

de Meillon, B. & Leeson, H.S. 1940. Notes on Ethiopian Anophelini. *Bulletin of Entomological Research* 31(1): 61–67.

de Meillon, B. & Rebêlo, A. 1941. Culicini (Diptera, Nematocera) from the Colony of Mocambique. *Moçambique Documentário trimestral* 27: 69–77.

de Meillon, B. & van Eeden, G. 1976. *Anopheles (Cellia) deaconi* n. sp., from South Africa (Diptera: Culicidae). *Mosquito Systematics* 8(4): 335–342.

de Meillon, B., Parent, M. & Black, L.O'C. 1945. Descriptions of new larvae and pupae of Ethiopian Culicini. *Bulletin of Entomological Research* 36(1): 85–101.

de Mello, F. 1918. *Anopheles folquei* n. sp. (Nouvelle espècie anophéline trouvée dans notre province de Pragana). *Anais Scientíficos da Faculdade de Medicina do Pôrto* 4: 363–371, post-scriptum, 1 pl.

Denisova, Z.M. 1955. New type of *Aëdes* (*Ochlerotatus*) [in Russian]. *Meditsinskaia Parazitologiia i Parazitarnye Bolezni* 24: 58–61.

de Oliveira Castro, G.M. 1932. Estudo sobre uma especie de *Culex*, que se cria em buracos de guayamú. (Dipt., Culicidae). *Revista de Entomologia* 2(1): 97–105.

de Oliveira Castro, G.M. 1935. Estudo sobre uma especie de "*Limatus*" que se cria em buracos de pau (Dipt., Culicidae). *Revista do Departamento Nacional da Produção Animal* 2: 143–155.

de Rook, H. 1929. *Malaria en Anophelinen aan de Boven-Digoel*. Dutch mimeographed report. [English translation: Malaria and *Anopheles* on the Upper Digul, with 63 numbered pp., 2 pp. references and 7 double-page tables].

de Rook, H. & Soesilo, R. 1930. *Anopheles* (*Bionella*) *papuae*. *Geneeskundig Tijdschrift voor Nederlandsch-Indië* 70: 475–481.

de Zulueta, J., Ramsdale, C., Cianchi, R., Bullini, L. & Coluzzi, M. 1983. Observations on the taxonomic status of *Anopheles sicaulti*. *Parassitologia* 23(1): 73–92.

D'Haenens, G. 1961. Description d'un Anophele [*sic*] nouveau du Congo (Kwango) *A. moucheti bervoetsi* n. subsp. *Bulletin et Annales de la Société Royale d'Entomologie de Belgique* 97(7–8): 188–200.

Díaz Nájera, A. 1965a. Mosquitos mexicanos. Claves para identificar hembras de *Mansonia* y *Psorophora* (*Diptera: Culcidae*). *Salud Publica de Mexica* 7(4): 519–525.

Díaz Nájera, A. 1965b. Claves para identificar especies mexicanas de *Mansonia* y *Psorophora* (Diptera: Culcidae). *Revista del Instituto de Salubridad y Enfermedades Tropicales* 25(2): 127–137.

Díaz Nájera, A. 1966. *Aedes* de Mexico. Descripcion de una nueva especie del subgenero *Howardina* (Diptera: Culicidae). *Revista de Investigacion en Salud Pública* 26(4): 331–344.

Director General Medical Services, Land Forces Headquarters. 1943. Notes on mosquito-borne diseases in Australasia. Australian Military Forces, by authority: Modern Printing Co. Pty. Ltd, Melbourne. Pages numbered 3–30.

Division of Medical Entomology, Bureau of Laboratories. 1944. The mosquitoes of Texas. Texas State Health Department. Austin, Texas. 100 pp. [Authorship of an included new species attributed to N.M. and K. O'Neill, Preface, p. 3].

Dobrotworsky, N.V. 1953. The *Culex pipiens* group in south-eastern Australia. I. *Proceedings of the Linnean Society of New South Wales* 77(5–6): 357–360.

Dobrotworsky, N.V. 1954. The genus *Theobaldia* (Diptera, Culicidae) in Victoria. *Proceedings of the Linnean Society of New South Wales* 79(3–4): 65–78.

Dobrotworsky, N.V. 1956. Notes on Australian mosquitoes (Diptera, Culicidae). I. Some species of the subgenus *Neoculex*. *Proceedings of the Linnean Society of New South Wales* 81(2): 105–114.

Dobrotworsky, N.V. 1957. Notes on Australian mosquitoes (Diptera, Culicidae). II. Notes on *Anopheles stigmaticus* Skuse and description of new species of *Anopheles* from Australia and New Guinea. *Proceedings of the Linnean Society of New South Wales* 82(2): 180–188.

Dobrotworsky, N.V. 1958a. Notes on Australian mosquitoes (Diptera, Culicidae). III. The subgenus *Lophoceraomyia* in Victoria. *Proceedings of the Linnean Society of New South Wales* (for 1957) 82(3): 317–321.

Dobrotworsky, N.V. 1958b. Designation of the type species of the subgenus *Neotheobaldia* Dobrotworsky (genus *Theobaldia* Neveu-Lemaire, 1902) (Diptera, Culicidae). *Proceedings of the Entomological Society of Washington* 60(4): 186.

Dobrotworsky, N.V. 1959. Notes on Australian mosquitoes (*Diptera, Culicidae*). IV. *Aëdes alboannulatus* complex in Victoria. *Proceedings of the Linnean Society of New South Wales* 84(1): 131–145.

Dobrotworsky, N.V. 1960a. The subgenus *Ochlerotatus* in the Australian Region (Diptera: Culicidae). III. Review of the Victorian species of *perkinsi* and *cunabulanus* sections with descriptions of two new species. *Proceedings of the Linnean Society of New South Wales* 85(1): 53–74.

Dobrotworsky, N.V. 1960b. The genus *Theobaldia* (Diptera, Culicidae) in Victoria. II. *Proceedings of the Linnean Society of New South Wales* 85(2): 240–247.

Dobrotworsky, N.V. 1961a. Notes on Australian mosquitoes (Diptera, Culicidae). V. Subgenus *Pseudoskusea* in Victoria. *Proceedings of the Linnean Society of New South Wales* (for 1960) 85(3): 257–267.

Dobrotworsky, N.V. 1961b. A new name for *Aedes waterhousei* Dobrotworsky (Diptera: Culicidae). *Proceedings of the Entomological Society of Washington* (for 1960) 62(4): 248.

Dobrotworsky, N.V. 1962. Notes on Australian mosquitoes (Diptera, Culicidae). VI. Five new Victorian species and a description of the larva of *Aëdes milsoni* (Taylor). *Proceedings of the Linnean Society of New South Wales* 87(3): 291–302.

Dobrotworsky, N.V. 1965. *The mosquitoes of Victoria* (*Diptera, Culicidae*). Melbourne University Press, Melbourne; Cambridge University Press, London, New York. [vii], 237 pp., 3 pls.

Dobrotworsky, N.V. 1966. Mosquitoes of Tasmania and Bass Straits Islands. *Proceedings of the Linnean Society of New South Wales* 91(2): 121–146.

Dobrotworsky, N.V. 1971. Contributions to the mosquito fauna of Southeast Asia. X. The genus *Culiseta* Felt in Southeast Asia. *Contributions of the American Entomological Institute* 7(3): 38–61.

Dobrotworsky, N.V. & Drummond, F.H. 1953. The *Culex pipiens* group in south-eastern Australia. II. *Proceedings of the Linnean Society of New South Wales* 78(3–4): 131–146.

Dodge, H.R. 1947. A new species of *Wyeomyia* from the pitcher plant (Diptera, Culicidae). *Proceedings of the Entomological Society of Washington* 49(5): 117–122.

Dolbeškin, B., Gorickaja, V. & Mitrofanova, J. 1930. Description of a new species of the genus *Aedes* (n. sp.) from eastern Europe [in Russian]. *Magazine of Parasitology Zoological Museum of Sciences of the USSR* 1: 253–260, 2 pls.

Doleschall, C.L. 1856. Eerste bijdrage tot de kennis der dipterologische fauna van Nederlandsch Indië. *Natuurkundig Tijdschrift voor Nederlandsch Indië* 10: 403–414.

Doleschall, C.L. 1857. Tweede bijdrage tot de kennis der dipterologische fauna van Nederlandsch Indië. *Natuurkundig Tijdschrift voor Nederlandsch Indië* 14: 377–418, 11 pls.

Doleschall, C.L. 1858. Derde bijdrage tot de kennis der dipteren fauna van Nederlandsch Indië. *Natuurkunde Tijdschrift voor Nederlandsch-Indië* 17: 73–128.

Dong, X. & Miyagi, I. 1995. A new species description of subgenus *Suaymyia* of genus *Topomyia* (Diptera: Culicidae) [in Chinese]. *Zoological Research* 16(4): 343–347.

Dong, X. & Wang, X. 1985. A new species of genus *Anopheles* (Diptera: Culicidae) [in Chinese]. *Zoological Research* 6(2): 117–122.

Dong, X.-S., Dong, L.-M. & Wu, C. 2004. A new species of the genus *Uranotaenia* from China (Diptera, Culicidae) [in Chinese]. *Acta Zootaxonomica Sinica* 29(3): 566–568.

Dong, X.-S., Dong, L.-M. & Wu, C. 2008. A new species of the genus *Tripteroides* (Diptera: Culicidae) from China [in Chinese]. *Acta Zootaxonomica Sinica* 33(1): 187–190.

Dong, X.-S., Dong, L.-M. & Zhou, H.-N. 2003. A new species of genus *Uranotaenia* (Diptera: Culicidae) [in Chinese]. *Entomotaxonomia* 25(3): 204–208.

Dong, X.-S., Dong, L.-Y.-M., Zhou, H.-N., Wang, X.-Z. & Lu, Y.-R. 2003. Three new species of mosquitoes from Yunnan, China (Diptera: Culicidae) [in Chinese]. *Acta Entomologica Sinica* 46(4): 519–532.

Dong, X.-S., Wang, X.-Z. & Lu, B.-L. 1990. A new species of *Topomyia* from Yunnan Province, China (Diptera: Culicidae) [in Chinese]. *Acta Zootaxonomica Sinica* 15(1): 64–66.

Dong, X.-S., Wang, X.-Z. & Lu, B.-L. 1995. A new species of *Topomyia* from Yunnan Province, China (Diptera: Culicidae) [in Chinese]. *Acta Zootaxonomica Sinica* 20(4): 384–486.

Dong, X.-S., Wang, X.-Z. & Zhou, H.-N. 2002. A new species of genus *Heizmannia* (Diptera: Culicidae) [in Chinese]. *Entomotaxonomia* 24(2): 131–133.

Dong, X.-S., Wu, C. & Mao, X.-H. 2006. A new species of the genus *Topomyia* (Diptera: Culicidae) [in Chinese]. *Entomotaxonomia* 28(3): 217–221.

Dong, X., Zhou, H. & Dong, L. 1995a. Studies on genus *Armigeres* of Yunnan with description of a new species (Diptera: Culicidae) [in Chinese]. *Zoological Research* 16(2): 95–104.

Dong, X., Zhou, H., & Dong, L. 1995b. A new species of the genus *Armigeres* (Diptera: Culicidae) [in Chinese]. *Entomotaxonomia* 17(4): 281–286.

Dong, X., Zhou, H. & Dong, L. 1997a. Study on *Topomyia* of Yunnan with one new species and one new record (Diptera: Culicidae) [in Chinese]. *Zoological Research* 18(3): 247–251.

Dong, X.-S., Zhou, H.N. & Dong, L.M. 1997b. A new species of subgenus *Rachionotomyia* of the genus *Tripteroides* (Diptera: Culicidae) [in Chinese]. *Acta Zootaxonomica Sinica* 22(2): 197–200.

Dong, X., Zhou, H., & Dong, L. 1999. A new species and a new Chinese record of the subgenus *Bothaella* of genus *Aedes* (Diptera: Culicidae) [in Chinese]. *Entomotaxonomia* 21(2): 133–137.

Dong, X.-S., Zhou, H.-N. & Dong, L.M. 2001. A new species of *Tripteroides* (s.str.) from Yunnan China (Diptera: Culicidae) [in Chinese]. *Acta Zootaxonomica Sinica* 26(2): 221–224.

Dong, X.-S., Zhou, H.-N. & Dong, L.-M. 2002a. One new species and two new records of the subgenus *Kenknightia* of the genus *Aedes* in Yunnan (Diptera: Culicidae) [in Chinese]. *Acta Entomologica Sinica* 45(4): 516 –521.

Dong, X., Zhou, H. & Dong, L. 2002b. A new species of *Armigeres* from Yunnan, China (Diptera: Culicidae) [in Chinese]. *Acta Parasitologica et Medica Entomologica* 9(3): 155–159.

Dong, X.-S., Zhou, H.N. & Dong, L.M. 2004. A new species of the genus *Armigeres* (Diptera: Culicidae) [in Chinese]. *Entomotaxonomia* 26(2): 123–125.

Dong, X.-S., Zhou, H.-N., Dong, L.-M. & Mao, X.-H. 2007. A new species of genus *Anopheles* (Diptera: Culicidae) [in Chinese]. *Entomotaxonomia* 29(1): 37–43.

Dong, X., Zhou, H., Gong, Z., *et al.* [*sic*]. 2004. Investigation of mosquito species in Yunnan Province with some new species [in Chinese]. *Chinese Journal of Vector Biology and Control* 15(3): 186–191.

Dong, X., Zhou, H., Wang, P. & Wang, X. 1997. A new species of the subgenus *Heizmannia* (Diptera: Culicidae) [in Chinese]. *Zoological Research* 18(1): 27–31.

Dönitz, W. 1901. Nachrichten aus dem Berliner Entomologischen Verein. *Insektenbörse* 18(5): 36–38.

Dönitz, W. 1902. Beiträge zur Kenntniss der *Anopheles*. *Zeitschrift für Hygiene und Infektionskrankheiten* 41: 15–88, 2 pls.

Dönitz, W. 1903. Beiträge zur Kenntniss der *Anopheles*. II. Mittheilung. *Zeitschrift für Hygiene und Infektionskrankheiten* 43: 215–238.

Doraisamy, L.V. 1963. A subspecific variation in the larva of *Anopheles subpictus*. *Bulletin of Entomology* 4: 41–42.

Doucet, J. 1949a. Étude des Culicidae (Diptera) du lac Alaotra. *Mémoires de l'Institut Scientifique de Madagascar Série A* 3(2): 121–145.

Doucet, J. 1949b. Recherches sur les Culicides de Madagascar. *Mémoires de l'Institut Scientifique de Madagascar* 3(3): 325–332.

Doucet, J. 1950. Les culicines de Madagascar [Dipt.]. *Mémoires de l'Institut Scientifique de Madagascar Série A* 4(1): 39–65.

Doucet, J. 1951a. Les moustiques de la region de Perinet. *Mémoires de l'Institut Scientifique de Madagascar Série A* 6(1): 63–82.

Doucet, J. 1951b. Etude des culicides de la region de Vangaindrano (Diptera). *Mémoires de l'Institut Scientifique de Madagascar Série A* 6(1): 83–114.

Doucet, J. 1962. Moustiques forestiers de la République de Côte-d'Ivoire. III. – Notes biologiques et systématiques sur *Uranotaenia andreae* sp. n. et *Culex (Neoculex) vinckei* Hamon, Holstein et Rivola, 1957. *Bulletin de la Société de Pathologie exotique* (for 1961) 54(5): 1156–1164.

Dubitzky, A.M. 1968. A new species *Aedes montchadskyi* sp. n. from the basin of the Ili River [in Russian]. *Parazitologiia* 2(3): 218–224.

Dubitzky, A.M. 1970. A new species of mosquito *Aedes stramineus* sp. n. (Diptera, Culicidae) [in Russian]. *Parazitologiia* 4(5): 408–413.

Dubitzky, A.M. & Deshevykh, N.D. 1978. A new species of mosquito *Aedes* (*Ochlerotatus*) *gutzevichi* sp.n. from central Asia (Culicidae) [in Russian]. *Parazitologiia* 12(2) 177–182. 9.

Dumas, E., Atyame, C.M., Malcolm, C.A., Le Goff, G., Unal, S., Makoundou, P., Pasteur, N., Weill, M. & Duron, O. 2016. Molecular data reveal a cryptic species within the *Culex pipiens* mosquito complex. *Insect Molecular Biology* 25(6): 800–809.

Dumbleton, L.J. 1962. A new species and new sub-genus of *Aedes* (Diptera: Culicidae) from New Zealand. *New Zealand Journal of Science* 5(1): 17–27.

Dunn, L.H. 1918. A new mosquito (*Aedes whitmorei*) from Colombia. *Proceedings of the Entomological Society of Washington* 20(6): 128–130.

Duret, J.P. 1953. Las especies argentinas de *Culex (Melanoconion) (Diptera-Culicidae)*. *Revista de la Sociedad Entomólogica Argentina* 16(3): 67–76.

Duret, J.P. 1954a. Las especies argentinas de *Culex (Melanoconion) (Diptera-Culicidae)*. *Revista de la Sociedad Entomológica Argentina* (for 1953) 16(4): 99–121.

Duret, J.P. 1954b. Nuevo mosquito argentino (Diptera, Culicidae). *Neotropica* 1(2): 23–26.

Duret, J.P. 1967a. Dos especies nuevas de *Culex* (*Melanoconion*) (Diptera - Culicidae). *Neotropica* 13(41): 77–84.

Duret, J.P. 1967b. *Culex (Culex) chitae* una nueva especie de Colombia (Diptera, Culicidae). *Neotropica* 13(42): 113–116.

Duret, J.P. 1968a. *Culex (Melanoconion) kerri* y *C. (M.) cristovaoi* dos nuevas especies del Brasil. (Diptera, Culicidae). *Neotropica* 14(43): 11–16.

Duret, J.P. 1968b. Cuatro especies nuevas de *Culex* (Diptera, Culicidae). *Neotropica* 14(44): 57–64.

Duret, J.P. 1968c. *Culex (Melanoconion) contei* una nueva especie del Brasil y Panama (Diptera, Culicidae). *Neotropica* 14(45): 121–124.

Duret, J.P. 1968d. Dos especies nuevas de *Culex* del Brasil *(Diptera, Culicidae)*. *Physis* 28(76): 15–19.

Duret, J.P. 1968e. Dos especies nuevas de *Culex (Culex)* de la Argentina *(Diptera, Culicidae)*. *Physis* 27(75): 323–328.

Duret, J.P. 1968f. Cinco especies nuevas de *Culex* (*Melanoconion*) (Diptera, Culicidae). *Revista de la Sociedad Entomológica Argentina* 30(1–4): 69–81.

Duret, J.P. 1969a. Contribucion al conocimiento de los *Culex* del Paraguay (Diptera- Culicidae). *Revista de la Sociedad Entomólogica Argentina* 31(1–4): 3–13.

Duret, J.P. 1969b. *Culex (Melanoconion) bahiense* una especie nueva del Brasil (Diptera, Culicidae). *Neotropica* 15(46): 40–42.

Duret, J.P. 1969c. *Culex (Anoedioporpa) damascenoi* una nueva especie del Brasil (Diptera - Culicidae). *Neotropica* 15(48): 143–145.

Duret, J.P. 1969d. Estudio sobre *Culex* (*Melanoconion*) *ocellatus* y sus especies afines *(Diptera. Culicidae)*. *Physis* 29(78): 113–129.

Duret, J.P. 1970. *Uranotaenia riverai* una nueva especie de Colombia *(Diptera, Culicidae)*. *Revista de la Sociedad entomólogica Argentina* 32(1–4): 43–46.

Duret, J.P. 1971a. Cinco especies nuevas de culicidos neotropicales (Diptera-Culicidae). *Neotropica* 17(52): 15–28.

Duret, J.P. 1971b. Una nueva especie de *Haemagogus* de Nicaragua (Diptera-Culicidae). *Neotropica* 17(53): 83–86.

Duret, J.P. 1972. *Culex (Anoedioporpa) quasioriginator*, una nueva especie del Brasil. *Neotropica* 18(55): 1–5.

Duret, J.P. 1982. *Wyeomyia (Dendromyia) sirivanakarni* una especie nueva de Panama (Diptera, Cuicidae). *Revista de la Sociedad Entomólogica Argentina* 41(1–4): 167–170.

Duret, J.P. & Barreto, M.P. 1956. Notas sôbre culícidas do Estado de São Paulo, Brasil, com descrições de três novas espécies de *Culex* (*Diptera, Culicidae*). *Revista Brasileira de Entomologia* 5: 81–99.

Duret, J.P. & Damasceno, R.G. 1955. Notas sobre *Culex* (*Tinolestes*) de Pará, Brasil (Diptera Culicidae). *Revista ecuatoriana de entomologia y parasitologia* 2(3–4): 393–409, 3 pls.

Dyar, H.G. 1904a. The larva of *Culex punctor* Kirby, with notes on an allied form. *Journal of the New York Entomological Society* 12(3): 169–171, 1 pl.

Dyar, H.G. 1904b. Brief notes on mosquito larvae. *Journal of the New York Entomological Society* 12(3): 172–174, 243–246.

Dyar, H.G. 1905a. Remarks on genitalic genera in the Culicidae. *Proceedings of the Entomological Society of Washington* 7(1): 42–49.

Dyar, H.G. 1905b. A new mosquito. *Journal of the New York Entomological Society* 13(2): 74.

Dyar, H.G. 1906. On the classification of the Culicidae. *Proceedings of the Entomological Society of Washington* 7(4): 188–191.

Dyar, H.G. 1907. Report on the mosquitoes of the coast region of California, with descriptions of new species. *Proceedings of the United States National Museum* 32(1516): 121–129.

Dyar, H.G. 1916a. Mosquitoes at San Diego, California. *Insecutor Inscitiae Menstruus* 4(4–6): 46–51.

Dyar, H.G. 1916b. New *Aedes* from the mountains of California *(Diptera, Culicidae)*. *Insecutor Inscitiae Menstruus* 4(7–9): 80–90.

Dyar, H.G. 1917a. The mosquitoes of the mountains of California (*Diptera, Culicidae*). *Insecutor Inscitiae Menstruus* 5(1–3): 11–21.

Dyar, H.G. 1917b. The mosquitoes of the Pacific Northwest *(Diptera, Culicidae)*. *Insecutor Inscitiae Menstruus* 5(7–9): 97–102, 1 pl.

Dyar, H.G. 1917c. Notes on the *Aedes* of Montana (*Diptera, Culicidae*). *Insecutor Inscitiae Menstruus* 5(7–9): 104–121.

Dyar, H.G. 1917d. A new *Aedes* from the Rocky Mountain region *(Diptera, Culicidae)* *Insecutor Inscitiae Menstruus* 5(7–9): 127–128.

Dyar, H.G. 1918a. A second note on the species of *Culex* of the Bahamas (*Diptera, Culicidae*). *Insecutor Inscitiae Menstruus* (for 1917) 5(10–12): 183–187.

Dyar, H.G. 1918b. The male genitalia of *Aedes* as indicative of natural affinities *(Diptera, Culicidae)*. *Insecutor Inscitiae Menstruus* 6(4–6): 71–86.

Dyar, H.G. 1918c. A revision of the American species of *Culex* on the male genitalia *(Diptera, Culicidae)*. *Insecutor Inscitiae Menstruus* 6(4–6): 86–111, 2 pls.

Dyar, H.G. 1918d. A note on the American species of *Mansonia* (*Diptera, Culicidae*). *Insecutor Inscitiae Menstruus* 6(4–6): 112–115.

Dyar, H.G. 1918e. New American mosquitoes (*Diptera, Culicidae*). *Insecutor Inscitiae Menstruus* 6(7–9): 120–129.

Dyar, H.G. 1919a. A note on *Lesticocampa*, and a new species (*Diptera, Culicidae*) [*sic*]. *Insecutor Inscitiae Menstruus* 7(1–3): 9–11.

Dyar, H.G. 1919b. Westward extension of the Canadian mosquito fauna (*Diptera, Culicidae*). *Insecutor Inscitiae Menstruus* 7(1–3): 11–39.

Dyar, H.G. 1919c. A new subgenus of *Culex* Linn (*Diptera, Culicidae*). *Insecutor Inscitiae Menstruus* 7(7–9): 150.

Dyar, H.G. 1919d. A revision of the American Sabethini of the *Sabethes* group by the male genitalia (*Diptera, Culicidae*). *Insecutor Inscitiae Menstruus* 7(7–9): 114–142, 1 pl.

Dyar, H.G. 1919e. A note on Argentine mosquitoes (*Diptera, Culicidae*). *Insecutor Inscitiae Menstruus* 7(4–6): 85–89.

Dyar, H.G. 1919f. The mosquitoes collected by the Canadian Arctic Expedition, 1913-18. (Diptera, Culicidae.), pp. 31–33. *In*: Report of the Canadian Arctic Expedition, 1913-18. Volume III: Insects. Part C: Diptera. J. de Labroquerie Taché, Ottawa.

Dyar, H.G. 1920a. The mosquitoes of British Columbia and Yukon Territory, Canada (*Diptera, Culicidae*). *Insecutor Inscitiae Menstruus* 8(1–3): 1–27.

Dyar, H.G. 1920b. A second *Culex* of the subgenus *Transculicia* Dyar (*Diptera, Culicidae*). *Insecutor Inscitiae Menstruus* 8(1–3): 27–29.

Dyar, H.G. 1920c. Notes on European mosquitoes (*Diptera, Culicidae*). *Insecutor Inscitiae Menstruus* 8(4–6): 51–54.

Dyar, H.G. 1920d. The species of *Choeroporpa*, a subgenus of *Culex* (*Diptera, Culicidae*). *Insecutor Inscitiae Menstruus* 8(4–6): 54–81.

Dyar, H.G. 1920e. The American *Aedes* of the *stimulans* group (*Diptera, Culicidae*). *Insecutor Inscitiae Menstruus* 8(7–9): 106–120.

Dyar, H.G. 1920f. The *Grabhamia* group of *Psorophora* (*Diptera, Culicidae*). *Insecutor Inscitiae Menstruus* 8(7–9): 140–141.

Dyar, H.G. 1920g. The *Aedes* of the mountains of California and Oregon (*Diptera, Culicidae*). *Insecutor Inscitiae Menstruus* 8(10–12): 165–173.

Dyar, H.G. 1920h. A new *Culex* from Panama (*Diptera, Culicidae*). *Insecutor Inscitiae Menstruus* 8(10–12): 173–174.

Dyar, H.G. 1920i. A collection of mosquitoes from the Philippine Islands (*Diptera, Culicidae*). *Insecutor Inscitiae Menstruus* 8(10–12): 175–186.

Dyar, H.G. 1920j. A new mosquito from Mexico (*Diptera, Culicidae*). *Insecutor Inscitiae Menstruus* 8(4–6): 81–82.

Dyar, H.G. 1921a. Ring-legged *Culex* in Texas (*Diptera, Culicidae*). *Insecutor Inscitiae Menstruus* 9(1–3): 32–34.

Dyar, H.G. 1921b. Three new mosquitoes from Costa Rica (*Diptera, Culicidae*). *Insecutor Inscitiae Menstruus* 9(1–3): 34–36.

Dyar, H.G. 1921c. Notes on the North American species of *Choeroporpa* (*Diptera, Culicidae*). *Insecutor Inscitiae Menstruus* 9(1–3): 37–39.

Dyar, H.G. 1921d. A new mosquito from East Africa (*Diptera, Culicidae*). *Insecutor Inscitiae Menstruus* 9(4–6): 51–52.

Dyar, H.G. 1921e. The American *Aedes* of the *punctor* group (*Diptera, Culicidae*). *Insecutor Inscitiae Menstruus* 9(4–6): 69–80, 1 pl.

Dyar, H.G. 1921f. Two new *Culex* from Costa Rica (*Diptera, Culicidae*). *Insecutor Inscitiae Menstruus* 9(4–6): 100.

Dyar, H.G. 1921g. The genus *Haemagogus* Williston (*Diptera, Culicidae*). *Insecutor Inscitiae Menstruus* 9(7–9): 101–114.

Dyar, H.G. 1921h. Illustrations of certain mosquitoes (*Diptera, Culicidae*). *Insecutor Inscitiae Menstruus* 9(7–9): 114–118, 1 pl.

Dyar, H.G. 1921i. Two new mosquitoes from China (*Diptera, Culicidae*). *Insecutor Inscitiae Menstruus* 9(7–9): 147–148.

Dyar, H.G. 1921j. The species of *Finlaya* allied to *terrens* Walker (*Diptera, Culicidae*). *Insecutor Inscitiae Menstruus* 9(10–12): 151–153.

Dyar, H.G. 1921k. New mosquitoes from Costa Rica (*Diptera, Culicidae*). *Insecutor Inscitiae Menstruus* 9(10–12): 154–155.

Dyar, H.G. 1922a. New mosquitoes from Alaska (*Diptera, Culicidae*). *Insecutor Inscitiae Menstruus* 10(1–3): 1–3.

Dyar, H.G. 1922b. Note on the male genitalia of *Culex coronator* and allied forms (*Diptera, Culicidae*). *Insecutor Inscitiae Menstruus* 10(1–3): 18–19, 1 pl.

Dyar, H.G. 1922c. The American *Aedes* of the *scapularis* group (*Diptera, Culicidae*). *Insecutor Inscitiae Menstruus* 10(4–6): 51–60.

Dyar, H.G. 1922d. Mosquito notes (*Diptera, Culicidae*). *Insecutor Inscitiae Menstruus* 10(4–6): 92–99.

Dyar, H.G. 1922e. The species of *Psorophora* of the *ciliata* group (*Diptera, Culicidae*). *Insecutor Inscitiae Menstruus* 10(7–9): 113–117, 1 pl.

Dyar, H.G. 1922f. The American *Aedes* of the *serratus* group (*Diptera, Culicidae*). *Insecutor Inscitiae Menstruus* 10(7–9): 157–166.

Dyar, H.G. 1922g. Notes on tropical American mosquitoes (*Diptera, Culicidae*). *Insecutor Inscitiae Menstruus* 10(10–12): 188–196, 1 pl.

Dyar, H.G. 1922h. [List of specimens identified by H. G. Dyar, pp. 418–422], in A. Alfaro, Museo Nacional, Memorias Instrucción Publicación República de Costa Rica 1921, pp. 416–423.

Dyar, H.G. 1923a. The mosquitoes of the Yellowstone National Park (*Diptera, Culicidae*). *Insecutor Inscitiae Menstruus* 11(1–3): 36–46.

Dyar, H.G. 1923b. Mosquito notes (*Diptera, Culicidae*). *Insecutor Inscitiae Menstruus* 11(4–6): 64–72, 3 pls.

Dyar, H.G. 1923c. Notes on American *Culex* (*Diptera, Culicidae*). *Insecutor Inscitiae Menstruus* 11(7–9): 118–121.

Dyar, H.G. 1923d. A new *Culex* from Mexico (*Diptera, Culicidae*). *Insecutor Inscitiae Menstruus* 11(10–12): 186–187.

Dyar, H.G. 1923e. On some of the American subgenera of *Culex* (*Diptera, Culicidae*). *Insecutor Inscitiae Menstruus* 11(10–12): 187–190, 2 pls.

Dyar, H.G. 1924a. A new sabethid from Brazil (*Diptera, Culicidae*). *Insecutor Inscitiae Menstruus* 12(7–9): 92.

Dyar, H.G. 1924b. A note on *Sabethes* Robineau-Desvoidy (*Diptera, Culicidae*). *Insecutor Inscitiae Menstruus* 12(7–9): 97–100.

Dyar, H.G. 1924c. *Phoniomyia* and *Dendromyia* Theobald (*Diptera, Culicidae*). *Insecutor Inscitiae Menstruus* 12: 107–113.

Dyar, H.G. 1924d. Some new mosquitoes from Colombia (*Diptera, Culicidae*). *Insecutor Inscitiae Menstruus* 12: 119–124.

Dyar, H.G. 1924e. Two new mosquitoes from California (*Diptera, Culicidae*). *Insecutor Inscitiae Menstruus* 12: 125–127.

Dyar, H.G. 1924f. A new mosquito from Siberia (*Diptera, Culicidae*). *Insecutor Inscitiae Menstruus* 12: 127–128.

Dyar, H.G. 1924g. Mosquitoes from Chile *(Diptera, Culicidae)*. *Insecutor Inscitiae Menstruus* 12: 128–131.

Dyar, H.G. 1924h. The American forms of *Aedes cinereus* Meigen (*Diptera, Culicidae*). *Insecutor Inscitiae Menstruus* 12: 179–180.

Dyar, H.G. 1924i. Some new mosquitoes from Colombia—II (*Diptera, Culicidae*). *Insecutor Inscitiae Menstruus* 12: 183–186.

Dyar, H.G. 1925a. A new sabethid from Panama *(Diptera, Culicidae)*. *Insecutor Inscitiae Menstruus* 13: 20–21.

Dyar, H.G. 1925b. Some new mosquitoes from Colombia—III (*Diptera, Culicidae*). *Insecutor Inscitiae Menstruus* 13: 21–24.

Dyar, H.G. 1925c. Note on bromelicolous *Anopheles (Diptera, Culicidae)*. *Insecutor Inscitiae Menstruus* 13: 25–27.

Dyar, H.G. 1925d. Some mosquitoes from Ecuador *(Diptera, Culicidae)*. *Insecutor Inscitiae Menstruus* 13: 27–31.

Dyar, H.G. 1925e. The subgenera of *Mansonia* Blanchard *(Diptera, Culicidae)*. *Insecutor Inscitiae Menstruus* 13(1–3): 40–43, 1 pl.

Dyar, H.G. 1925f. The mosquitoes of Panama (*Diptera, Culicidae*). *Insecutor Inscitiae Menstruus* 13(7–9): 101–195.

Dyar, H.G. 1925g. Some mosquitoes from Venezuela (*Diptera, Culicidae*). *Insecutor Inscitiae Menstruus* 13: 213–216.

Dyar, H.G. 1926. Notes on Panama mosquitoes (*Diptera, Culicidae*). *Insecutor Inscitiae Menstruus* 14: 111–114.

Dyar, H.G. 1928. *The mosquitoes of the Americas*. Publication no. 387. Carnegie Institution of Washington, Washington, D.C. 616 pp.

Dyar, H.G. 1929a. A new species of mosquito from Montana, with annotated list of the species known from the state. *Proceedings of the United States National Museum* 75(2794): 1–8.

Dyar, H.G. 1929b. Remarks on the subgenus *Phalangomyia* of *Culex* Linn. (Diptera, Culicidae). *American Journal of Hygiene* 9(2): 509–511.

Dyar, H.G. 1929c. A new mosquito from the Philippine Islands. *Proceedings of the Entomological Society of Washington* 31(3): 61–62.

Dyar, H.G. & Knab, F. 1906a. Diagnoses of new species of mosquitoes. *Proceedings of the Biological Society of Washington* 19: 133–141.

Dyar, H.G. & Knab, F. 1906b. Notes on some American mosquitoes with descriptions of new species. *Proceedings of the Biological Society of Washington* 19: 159–172.

Dyar, H.G. & Knab, F. 1906c. The species of mosquitoes in the genus *Megarhinus*. *Smithsonian Miscellaneous Collections* 48(3): 241–258.

Dyar, H.G. & Knab, F. 1906d. The larvae of Culicidae classified as independent organisms. *Journal of the New York Entomological Society* 14(4): 169–230, 13 pls.

Dyar, H.G. & Knab, F. 1907a. Descriptions of some American mosquitoes. *Journal of the New York Entomological Society* 15(1): 9–13.

Dyar, H.G. & Knab, F. 1907b. New American mosquitoes. *Journal of the New York Entomological Society* 15(2): 100–101.

Dyar, H.G. & Knab, F. 1907c. Descriptions of new mosquitoes from the Panama Canal Zone. *Journal of the New York Entomological Society* 15(4): 197–212.

Dyar, H.G. & Knab, F. 1907d. Descriptions of three new North American mosquitoes. *Journal of the New York Entomological Society* 15(4): 213–214.

Dyar, H.G. & Knab, F. 1908. Descriptions of some new mosquitoes from tropical America. *Proceedings of the United States National Museum* 35(1632): 53–70.

Dyar, H.G. & Knab, F. 1909a. On the identity of *Culex pipiens* Linnaeus. *Proceedings of the Entomological Society of Washington* 11(1): 30–39, 3 pls.

Dyar, H.G. & Knab, F. 1909b. Description of a new mosquito from Cuba. [Diptera, Culicidae]. *Proceedings of the Entomological Society of Washington* 11(1): 39.

Dyar, H.G. & Knab, F. 1909c. Mosquito comment. *Canadian Entomologist* 41(3): 101–102.

Dyar, H.G. & Knab, F. 1909d. Descriptions of some new species and a new genus of American mosquitoes. *Smithsonian Miscellaneous Collections* 52(1822): 253–266.

Dyar, H.G. & Knab, F. 1910. Description of three new American mosquitoes. *Proceedings of the Entomological Society of Washington* (for 1909) 11(1): 173–174.

Dyar, H.G. & Knab, F. 1913. Three new Neotropical mosquitoes. *Insecutor Inscitiae Menstruus* 1(6): 76–78.

Dyar, H.G. & Knab, F. 1914. New mosquitoes from Peru (*Diptera, Culicidae*). *Insecutor Inscitiae Menstruus* 2(4): 58–62.

Dyar, H.G. & Knab, F. 1915. Notes on the species of *Culex* of the Bahamas. *Insecutor Inscitiae Menstruus* 3(8–10): 112–115.

Dyar, H.G. & Knab, F. 1916. Eggs and oviposition in certain species of *Mansonia (Diptera; Culicidae)*. *Insecutor Inscitiae Menstruus* 4(4–6): 61–68.

Dyar, H.G. & Knab, F. 1917. Bromelicolous *Anopheles* (*Diptera, Culicidae*). *Insecutor Inscitiae Menstruus* 5(1–3): 38–40.

Dyar, H.G. & Knab, F. 1918a. Bromelicolous *Anopheles*—a correction *(Diptera, Culicidae)*. *Insecutor Inscitiae Menstruus* 6(7–9): 140–141.

Dyar, H.G. & Knab, F. 1918b. New American mosquitoes (*Diptera, Culicidae*). *Insecutor Inscitiae Menstruus* (for 1917) 5(10–12): 165–169.

Dyar, H.G. & Knab, F. 1918c. The genus *Culex* in the United States (*Diptera, Culicidae*). *Insecutor Inscitiae Menstruus* (for 1917) 5(10–12): 170–183.

Dyar, H.G. & Knab, F. 1919. New species of tropical American mosquitoes *(Diptera, Culicidae)*. *Insecutor Inscitiae Menstruus* 7(1–3): 1–9.

Dyar, H.G. & Ludlow, C.S. 1921a. Two new American mosquitoes *(Diptera, Culicidae)*. *Insecutor Inscitiae Menstruus* 9(1–3): 46–50.
Dyar, H.G. & Ludlow, C.S. 1921b. A note on two Panama mosquitos (*Diptera, Culicidae*). *Military Surgeon* 48: 677–680.
Dyar, H.G. & Ludlow, C.S. 1922. Notes on American mosquitoes (*Diptera, Culicidae*). *Military Surgeon* 50: 61–64.
Dyar, H.G. & Nuñez Tovar, M. 1927a. Notes on biting flies from Venezuela (*Diptera, Culicidae, Psychodidae*). *Insecutor Inscitiae Menstruus* (for 1926) 14: 152–155, 190–191.
Dyar, H.G. & Núñez Tovar, M. 1927b. *Notas sobre nuevos dipteros hematofagos de Venezuela (Culicidae)*. Maracay, Venezuela. 8 pp.
Dyar, H.G. & Núñez Tovar, M. 1928. Descriptions of new species of mosquitoes from Venezuela. *American Journal of Hygiene* 8(1): 89–92.
Dyar, H.G. & Shannon, R.C. 1924a. New *Culex* from Panama (*Diptera, Culicidae*). *Insecutor Inscitiae Menstruus* 12(4–6): 46–48.
Dyar, H.G. & Shannon, R.C. 1924b. Another new *Culex* from Panama *(Diptera, Culicidae)*. *Insecutor Inscitiae Menstruus* 12: 144.
Dyar, H.G. & Shannon, R.C. 1924c. The American species of *Uranotaenia (Diptera, Culicidae)*. *Insecutor Inscitiae Menstruus* 12(10–12): 187–192.
Dyar, H.G. & Shannon, R.C. 1924d. The subfamilies, tribes, and genera of American Culicidae. *Journal of the Washington Academy of Sciences* 14(20): 472–486.
Dyar, H.G. & Shannon, R.C. 1925a. The types of Philippine mosquitos described by Ludlow and other notes on the fauna (*Diptera, Culicidae*). *Insecutor Inscitiae Menstruus* 13(4–6): 66–89.
Dyar, H.G. & Shannon, R.C. 1925b. New mosquitoes from Brazil (Diptera, Culicidae). *Journal of the Washington Academy of Sciences* 15(2): 39–41.
Dyar, H.G. & Shannon, R.C. 1925c. A new sabethid mosquito from Panama (Diptera, Culicidae). *Journal of the Washington Academy of Sciences* 15(11): 234.
Dyar, H.G. & Shannon, R.C. 1925d. The mosquitoes of Peary's North Pole expedition of 1908. (Diptera, Culicidae). *Journal of the Washington Academy of Sciences* 15(4): 77–78.
Eckstein, F. 1918. Zur Systematik der einheimischen Stechmücken. *Zentralblatt fur Bakteriologie, Abt. 1, Originale* 82(2): 57–68.
Edwards, F.W. 1911a. Some new West African species of *Anopheles* (*sensu lato*), with notes on nomenclature. *Bulletin of Entomological Research* 2(2): 141–143.
Edwards, F.W. 1911b. On the culicid genus *Eretmapodites*, Theobald. *Annals and Magazine of Natural History* 8(43): 67–73.
Edwards, F.W. 1912a. Notes on the British mosquitos (Culicinae). *Entomologist* 45(592): 260–264.
Edwards, F.W. 1912b. A key to the Australasian species of *Ochlerotatus* (Culicidae). *Annals and Magazine of Natural History* 9(53): 521–527.
Edwards, F.W. 1912c. A synopsis of the species of African Culicidae, other than *Anopheles*. *Bulletin of Entomological Research* 3(1): 1–53.
Edwards, F.W. 1912d. A key for determining the African species of *Anopheles* (*sensu lato*). *Bulletin of Entomological Research* 3(3): 241–250.
Edwards, F.W. 1913a. Tipulidae and Culicidae from the Lake of Tiberias and Damascus. *Journal and Proceedings of the Asiatic Society of Bengal* 9: 47–51.
Edwards, F.W. 1913b. Further notes on African Culicidae. *Bulletin of Entomological Research* 4(1): 47–59.
Edwards, F.W. 1913c. New synonymy in Oriental Culicidae. *Bulletin of Entomological Research* 4(3): 221–242.
Edwards, F.W. 1914a. On the Oriental culicid genus *Leicesteria*, Theobald. *Bulletin of Entomological Research* 4(4): 255–263.
Edwards, F.W. 1914b. New species of Culicidae in the British Museum, with notes on the genitalia of some African *Culex*. *Bulletin of Entomological Research* 5(1): 63–81.
Edwards, F.W. 1914c. New Culicidae from Borneo and Hong Kong. *Bulletin of Entomological Research* 5(2): 125–128.
Edwards, F.W. 1915a. An annotated list of mosquitos occurring at Durban, Natal. *Annals of the Durban Museum* 1(2): 160–166.
Edwards, F.W. 1915b. New and little-known East African Culicidae. *Bulletin of Entomological Research* 5(4): 273–281.
Edwards, F.W. 1915c. Diagnoses of new Bornean Culicidae. *Bulletin of Entomological Research* 5(4): 283–285.
Edwards, F.W. 1916. Eight new mosquitos in the British Museum collection. *Bulletin of Entomological Research* 6(4): 357–364.
Edwards, F.W. 1917. Notes on Culicidae, with descriptions of new species. *Bulletin of Entomological Research* 7(3): 201–229.
Edwards, F.W. 1920a. Mosquito notes. *Bulletin of Entomological Research* 10(2): 129–137.
Edwards, F.W. 1920b. Notes on the mosquitos of Madagascar, Mauritius and Réunion. *Bulletin of Entomological Research* 11(2): 133–138.
Edwards, F.W. 1921a. H. Sauter's Formosan collections: Culicidae. *Annals and Magazine of Natural History* 8(48): 629–632.
Edwards, F.W. 1921b. A synonymic list of the mosquitoes hitherto recorded from Sweden, with keys for determining the genera and species. *Entomologisk Tidskrift* 42(1): 46–52.
Edwards, F.W. 1921c. Mosquito notes.–II. *Bulletin of Entomological Research* 12(1): 69–80.
Edwards, F.W. 1921d. A revision of the mosquitos of the Palaearctic Region. *Bulletin of Entomological Research* 12(3): 263–351.
Edwards, F.W. 1922a. A synopsis of adult Oriental culicine (including megarhinine and sabethine) mosquitoes. Part I. *Indian Journal of Medical Research* 10(1): 249–293, 3 pls.
Edwards, F.W. 1922b. A synopsis of adult Oriental culicine (including megarhinine and sabethine) mosquitoes. Part II. *Indian Journal of Medical Research* 10(2): 430–475, 3 pls.
Edwards, F.W. 1922c. A revision of the genus *Harpagomyia* de Meij. (Diptera, Culicidae). *Transactions of the Royal Entomological Society of London* 1921 (3–4): 496–503, 1 pl.
Edwards, F.W. 1922d. Mosquito notes.—III. *Bulletin of Entomological Research* 13(1): 75–102.

Edwards, F.W. 1923a. Four new African mosquitos. *Bulletin of Entomological Research* 13(4): 397–399.
Edwards, F.W. 1923b. Mosquito notes.—IV. *Bulletin of Entomological Research* 14(1): 1–9.
Edwards, F.W. 1923c. Oligocene mosquitoes in the British Museum; with a summary of our present knowledge concerning fossil Culicidae. *Quarterly Journal of the Geological Society of London* 79(2): 139–155, 1 pl.
Edwards, F.W. 1923d. Mosquitoes reared by Dr. W. E. Haworth from coconut palms in East Africa. *Transactions of the Royal Society of Tropical Medicine and Hygiene* 16(8): 498–501.
Edwards, F.W. 1924a. Some mosquitoes from Ovamboland, S. W. Africa, and from the Cape Province. *Annals of the South African Museum* 19(1): 159–163.
Edwards, F.W. 1924b. A synopsis of the adult mosquitos of the Australasian Region. *Bulletin of Entomological Research* 14(4): 351–401.
Edwards, F.W. 1925. Mosquito notes.–V. *Bulletin of Entomological Research* 15(3): 257–270.
Edwards, F.W. 1926a. Mosquito notes.–VI. *Bulletin of Entomological Research* 17(2): 101–131.
Edwards, F.W. 1926b. Diptera Nematocera from the mountains of Borneo. *Sarawak Museum Journal* 3(8–11): 243–278, 2 pls.
Edwards, F.W. 1926c. Una revisione delle zanzare delle regioni paleartiche. *Rivista di Malariologia* 5(3–6): 1–152.
Edwards, F.W. 1927a. Four new mosquitoes from the Belgian Congo. *Revue de Zoologie Africaine* 15(3): 351–355.
Edwards, F.W. 1927b. New mosquitoes of the genus *Rachionotomyia* from New Guinea. *Nova Guinea: Résultats des Expéditions Scientifiques Néerlandaises à la Nouvelle Guinée* 15(3): 352–356.
Edwards, F.W. 1928a. Mosquito notes.–VII. *Bulletin of Entomological Research* 18(3): 267–284.
Edwards, F.W. 1928b. A new mosquito from the Kivu volcanoes, Belgian Congo. *Revue de zoologie et de botanique Africaines* 16(3): 324–325.
Edwards, F.W. 1928c. The nematocerous Diptera of Corsica. *Encyclopédie Entomologique Série B Mémoires et Notes* 4(4): 157–189.
Edwards, F.W. 1929a. A new *Anopheles* from the central Sahara. *Archives de l'Institut Pasteur d'Algérie* 7(1): 82–87.
Edwards, F.W. 1929b. Mosquito notes.–VIII. *Bulletin of Entomological Research* 20(3): 321–343.
Edwards, F.W. 1930a. Mosquito notes.–IX. *Bulletin of Entomological Research* 21(3): 287–306.
Edwards, F.W. 1930b. Mosquito notes.–X. *Bulletin of Entomological Research* 21(4): 541–545.
Edwards, F.W. 1932a. *Genera Insectorum. Diptera, Fam. Culicidae*. Fascicle 194. M.P. Wytsman, Bruxelles. 258 pp., 5 pls.
Edwards, F.W. 1932b. Mosquito notes.–XI. *Bulletin of Entomological Research* 23(4): 559–562.
Edwards, F.W.1934. The Percy Sladen and Godman Trusts expedition to the islands in the Gulf of Guinea, October 1932–March 1933.—II. Diptera Nematocera. *Annals and Magazine of Natural History* 14(81): 321–336.
Edwards, F.W. 1935a. Mosquito notes.—XII. *Bulletin of Entomological Research* 26(2): 127–136.
Edwards, F.W. 1935b. Results of the Oxford University expedition to Borneo, 1932. Two new species of *Megarhinus* (Diptera, Culicidae). *Annals and Magazine of Natural History* 15(89): 579–581.
Edwards, F.W. 1935c. A new *Uranotaenia* from the Belgian Congo (Diptera, Culicidae). *Revue de zoologie et de botanique Africaines* 27(1): 96.
Edwards, F.W. 1936. New African culicine mosquitoes (Diptera, Culicidae). *Proceedings of the Royal Entomological Society of London, Series B: Taxonomy* 5(3): 49–55.
Edwards, F.W. 1939a. A new East African *Aëdes* (Dipt., Culicidae). *Proceedings of the Royal Entomological Society of London, Series B: Taxonomy* 8(2): 17.
Edwards, F.W. 1939b. A new species of *Orthopodomyia* (Diptera, Culicidae). *Proceedings of the Royal Entomological Society of London, Series B: Taxonomy* 8(6): 121–123.
Edwards, F.W. 1940. A new plain-winged *Anopheles* from Rhodesia. *Annals of Tropical Medicine and Parasitology* 34(2): 93–96.
Edwards, F.W. 1941. *Mosquitoes of the Ethiopian Region III. – Culicine adults and pupae*. Printed by order of the Trustees, British Museum (Natural History), London. viii, 499 pp.
Edwards, F.W. & Gibbins, E.G. 1939. *Mosquitoes. Ruwenzori Expedition 1934-5*. Volume 1, no. 3, pp. 29–33, 1 pl. British Museum (Natural History), London.
Ellis, R.A. & Brust, R.A. 1973. Sibling species delimitation in the *Aedes communis* (Degeer [*sic*]) aggregate (Diptera: Culicidae). *Canadian Journal of Zoology* 51(9): 915–959.
Enderlein, G. 1920. Die Culiciden–Fauna Madagascars. *Wiener Entomologische Zeitung* 38(1–3): 47–52.
Enderlein, G. 1923. Zur Kenntnis afrikanischer und orientalischer Culiciden. *Wiener Entomologische Zeitung* 40(1–4): 25–29.
Enderlein, G. 1931. Zwei neue *Toxohynchites* aus dem Kongogebiete. *Zoologischer Anzeiger* 94(3/4): 123–125.
Eouzan, J.-P. 1969. Description du mâle de *Culex (Culiciomyia) grenieri* sp. n. (Diptera, Culicidae). *Cahiers ORSTOM Série Entomologie médicale et Parasitologie* 7(3): 163–166.
Erichson, W.F. 1842. Beitrag zur Insecten-Fauna von Vandiemensland, mit besonderer Berücksichtigung der geographischen verbreitung der Insecten, vom Herausgeber. *Archiv für Naturgeschichte* 8(1): 83–287, 2 pls.
Evans, A.M. 1922. Notes on Culicidae in Venezuela, with descriptions of new species. Part II. *Annals of Tropical Medicine and Parasitology* 16(2): 213–222, 1 pl.
Evans, A.M. 1923a. A new species and a new variety of *Culex* from the Belgian Congo. *Annals of Tropical Medicine and Parasitology* 17(1): 89–91.
Evans, A.M. 1923b. Notes on Culicidae in Venezuela, with descriptions of new species. Part III. *Annals of Tropical Medicine and Parasitology* 17(1): 101–111.
Evans, A.M. 1923c. Note on *Aëdinus amazonensis*, Lutz. *Annals of Tropical Medicine and Parasitology* 17(3): 377–380.
Evans, A.M. 1924. Descriptions of new mosquitos from South America. *Annals of Tropical Medicine and Parasitology* 18(3): 363–375.

Evans, A.M. 1925a. Notes on Culicidae collected in Sierra Leone, with descriptions of a new species and a new variety. *Annals of Tropical Medicine and Parasitology* 19(1): 119–126, 1 pl.

Evans, A.M. 1925b. A new variety of *Anopheles marshalli* Theobald from the Belgian Congo. *Annals of Tropical Medicine and Parasitology* 19(2): 211–213.

Evans, A.M. 1925c. A new variety of *Anopheles marshallii*, from Sierra Leone. *Annals of Tropical Medicine and Parasitology* 19(4): 461–462.

Evans, A.M. 1926. Notes on Freetown mosquitos, with descriptions of new and little-known species. *Annals of Tropical Medicine and Parasitology* 20(1): 97–108, 1 pl.

Evans, A.M. 1927. A short illustrated guide to the anophelines of tropical and South America. *Memoirs of the Liverpool School of Tropical Medicine* 3: 1–79.

Evans, A.M. 1928. A new variety of *Armigeres* from Lagos, with descriptions of the larval and pupal stages. *Annals of Tropical Medicine and Parasitology* 22(1): 39–42.

Evans, A.M. 1929. *Aëdes (Aëdimorphus) apicoannulatus* Edwards and yellow fever: A correction. *Annals of Tropical Medicine and Parasitology* 23(4): 521–522.

Evans, A.M. 1931a. Notes on African anophelines. *Annals of Tropical Medicine and Parasitology* 25(1): 129–143.

Evans, A.M. 1931b. A new subspecies of *Anopheles funestus* Giles, from southern Rhodesia. *Annals of Tropical Medicine and Parasitology* 25(3–4): 545–549.

Evans, A.M. 1932. Notes on African mosquitoes. *Annals of Tropical Medicine and Parasitology* 26(1): 85–108.

Evans, A.M. 1933. *Anopheles demeilloni* sp. n., a new name for *Anopheles transvaalensis* of authors. *Annals of Tropical Medicine and Parasitology* 27(2): 265–269.

Evans, A.M. 1934. Further notes on African anophelines, with a description of a new group of *Myzomyia*. *Annals of Tropical Medicine and Parasitology* 28(4): 549–570.

Evans, A.M. 1935. Notes on anophelines. I.—Description of *Anopheles marshalli* [*sic*] var. *gibbinsi* from Uganda. II.—The characters of *A. maculipennis* var. *messeae* in Wirral, England. *Annals of Tropical Medicine and Parasitology* 29(4): 469–473.

Evans, A.M. 1936. Two new anopheline mosquitoes from Kenya. *Annals of Tropical Medicine and Parasitology* 30(4): 533–540.

Evans, A.M. 1938. *Mosquitoes of the Ethiopian Region. II.–Anophelini adults and early stages*. British Museum (Natural History), London. x, 404 pp.

Evans, A.M. & de Meillon, B. 1933. Notes on *Anopheles demeilloni (transvaalensis)* and *Anopheles garnhami* in South Africa, with descriptions of new varieties of these species. *Annals of Tropical Medicine and Parasitology* 27(2): 271–282.

Evans, A.M. & Garnham, P.C.C. 1936. The *funestus* series of *Anopheles* at Kisumu and a coastal locality in Kenya. *Annals of Tropical Medicine and Parasitology* 30(4): 511–520.

Evans, A.M. & Leeson, H.S. 1935. The *funestus* series of *Anopheles* in Southern Rhodesia, with description of a new variety. *Annals of Tropical Medicine and Parasitology* 29(1): 33–47.

Evans, A.M. & Leeson, H.S. 1937. Notes on variation in *Anopheles rivulorum* Leeson in East Africa, with description of a new variety. *Annals of Tropical Medicine and Parasitology* 31(3): 383–386.

Evenhuis, N.L. 1994. *Catalogue of the fossil flies of the World (Insecta: Diptera)*. Backhuys Publishers, Leiden. 600 pp.

Evenhuis, N.L. & Gon III, S.M. 1989. Family Culicidae, pp. 191–218 (Chapter 22). *In*: Evenhuis, N.L (ed.), *Catalog of the Diptera of the Australasian and Oceanian Regions*. Bishop Museum Special Publication 86, Bishop Museum Press and E.J. Brill, Honolulu.

Evenhuis, N.L. & Steffan, W.A. 1986. Classification of the subgenus *Toxorhynchites* (Diptera: Culicidae) II. Revision of the *Toxorhynchites acaudatus* group. *Journal of Medical Entomology* 23(5): 538–574.

Evenhuis, N.L., Pape, T. & Pont, A.C. 2016. Nomenclatural studies toward a world list of Diptera genus-group names. Part V: Pierre-Justin-Marie Macquart. *Zootaxa* 4172(1): 1–211.

Eversmann, E. 1834. Diptera. Wolgam fluvium inter et montes Uralenses observata. *Bulletin de la Société impériale des Naturalistes de Moscou* 7: 420–432.

Eysell, A. 1903. *Aëdes cinereus* (Hoffmgg.) und *Aëdes leucopygos* n. sp. *Abhandlungen und Bericht des Vereins für Naturkunde zu Kassel* 48: 285–306.

Eysell, A. 1905a. Sind die „Culiciden“ eine Familie? *Abhandlungen und Bericht des Vereins für Naturkunde zu Cassel* 49: 16–24.

Eysell, A. 1905b. Sind die „Culiciden“ eine Familie? *Archiv für Schiffs- und Tropen-Hygiene*. 9(2): 49–55.

Eysell, A. 1912. *Cyclophorus* (*Anopheles*) *nigripes* Staeger (nov. gen.). *Archiv für Schiffs- und Tropen-Hygiene* 16: 421–431.

Fabricius, J.C. 1775. *Systema entomologiae, sistens insectorum classes, ordines, genera, species adiectis synonymis, locis, descriptionibus, observationibus*. Flensburgi et Lipsiae. [xxxii], 832 pp.

Fabricius, J.C. 1787. *Mantissa Insectorum, sistens eorum species nuper detectas: adjectis synonymis, observationibus, descriptionibus, emendationibus*. Volume 2. Christ. Gottl. Proft, Hafni [= Copenhagen]. 382 pp.

Fabricius, J.C. 1794. *Entomologia systematica emendata et aucta: secundum classes, ordines, genera, species, adjectis synonimis, locis, observationibus, descriptionibus*. Volume 4. C.G. Proft, fil. & soc., Hafniae [=Copenhagen]. [vi], 472 pp., index generum (5 pp.).

Fabricius, J.C. 1805. *Systema antliatorum secundum ordines, genera, species, adiectis synonymis, locis, observationibus, descriptionibus*. C. Reichard, Brunsvigae [=Brunswick]. 372 pp., errata (1 p.), directory of published books (2 pp.), index + appendix (30 pp.).

Falleroni, D. 1926. Fauna anofelica italiana e suo "habitat„ (paludi, risaie, canali). Metodi di lotta contro la malaria. *Rivista di Malariologia* 5(5–6): 553–593.

Falleroni, D. 1932. *Zooprofilassi e sua applicazione in Ardea (Argo Romano): diverse razze di* Anopheles maculipennis. Cuggiani, Roma [Rome]. 48 pp.

Faran, M.E. 1979. *Anopheles (Nyssorhynchus) trinkae*, a new species in the Albimanus Section (Diptera: Culicidae). *Mosquito Systematics* 11(1): 26–39.

Faran, M.E. 1980. Mosquito studies (Diptera, Culicidae) XXXIV. A revision of the Albimanus Section of the subgenus *Nyssorhynchus* of *Anopheles*. *Contributions of the American Entomological Institute* 15(7): 1–215.

Faran, M.E. 1981. Synonymy of *Anopheles (Nyssorhynchus) noroestensis* with *An. (Nys.) evansi* [*sic*], with a description of the male genitalia of the lectotype of *An. (Nys.) evansi* [*sic*] (Diptera: Culicidae). *Mosquito Systematics* 13(1): 86–91.

Farner, D.S. 1945. A new species of *Aedes* from the Caroline Islands. (Diptera, Culicidae). *Proceedings of the Biological Society of Washington* 58: 59–61.

Farner, D.S. & Bohart, R.M. 1944. Three new species of Australasian *Aedes* (Diptera, Culicidae). *Proceedings of the Biological Society of Washington* 57: 117–122.

Fauran, P. 1961. Notes sur les moustiques de la Guyane. *Archives de l'Institut Pasteur de la Guyane Française l'Inini* 22(464): 1–15.

Fauran, P. & Pajot, F.X. 1974. *Wyeomyia ininicola* n. sp. *(Diptera, Culicidae)*, nouveau moustique de la Guyane française. *Cahiers ORSTOM. Série Entomologie médicale et Parasitologie* 12(2): 145–148.

Favre, V.V. [Фавр, В.В.] 1903. *An attempt in the study of malaria in Russia* [Попытка в изучении малярии в России]. Kharkov, Russia. 324 pp., map.

Felt, E.P. 1904. Mosquitos or Culicidae of New York State. *New York State Museum Bulletin* 79(323): 241–391, 57 pls, 391a–391f, 393–400.

Felt, E.P. 1905a. Studies in Culicidae. *New York State Museum Bulletin* 97(357): 442–497, 19 pls.

Felt, E.P. 1905b. *Culex brittoni* n. sp. *Entomological News and Proceedings of the Entomological Section of the Academy of Natural Sciences of Philadelphia* 16(3): 79–80.

Felt, E.P. & Young, D.B. 1904. Importance of isolated rearings from culicid larvae. *Science* 20: 312–313.

Feng, L.-C. 1938. The tree hole species of mosquitoes of Peiping, China. *Chinese Medical Journal, Supplement* 2: 503–525, 5 pls.

Feng, L.C. 1941. A new mosquito, *Topomyia houghtoni* sp. nov. from western Yunnan, China. *Bulletin of the Fan Memorial Institute of Biology, Peiping (Zoology Series)* 10: 243–254.

Ferrara, L. 1973. Une nouvelle espèce de moustique *(Diptera Culicidae)* capturée au Cameroun. *Orthopodomyia aureoantennata* n. sp. *Cahiers ORSTOM Série Entomologie médicale et Parasitologie* 11(3): 199–203.

Ferrara, L. 1974. Description d'un nouvel *Aedes* du sous genre *Diceromyia (Diptera, Culicidae)* découvert au Cameroun *Aedes (Diceromyia) mefouensis* n. sp. *Cahiers ORSTOM Série Entomologie médicale et Parasitologie* 12(1): 3–5.

Ferrara, L. & Eouzan, J.-P. 1974. Description de deux nouveaux *Eretmapodites*, du groupe *oedipodius* Graham, 1909 *(Diptera, Culicidae)* découverts au Cameroun. *Cahiers ORSTOM Série Entomologie médicale et Parasitologie* 12(4): 259–262.

Ficalbi, E. 1889a. Notizie preventive sulle zanzare italiane. I[a] Nota preventiva. Alcune generalità. – Descrizione di una specie nuova (*Culex hortensis*.). *Bullettino della Società Entomologica Italiana* 21: 20–30.

Ficalbi, E. 1889b. Notizie preventive sulle zanzare italiane. II[a] Nota preventiva (1). Descrizione di una specie nuova. *Bullettino della Società Entomologica Italiana* 21: 50–53.

Ficalbi, E. 1890a. Notizie preventive sulle zanzare italiane. VII[a] Nota preventiva (1). Descrizione di una specie nuova. *Bullettino della Società Entomologica Italiana* 22: 81–84.

Ficalbi, E. 1890b. Notizie preventive sulle zanzare italiane. IV. Nota preventiva. Descrizione di una specie nuova. Zanara di colorito modesto *Culex modestus*, n. sp. *Bullettino della Società Entomologica Italiana* (for 1889) 21: 93–94.

Ficalbi, E. 1890c. Notizie preventive sulle zanzare italiane. V[a] Nota preventiva (1). Descrizione di una specie nuova. Zanara elegante, *Culex elegans* n. sp. *Bullettino della Società Entomologica Italiana* (for 1889) 21(1–2): 95–100.

Ficalbi, E. 1890d. Notizie preventive sulle zanzare italiane. VI[a] Nota preventive (1). Quistioni zoologiche intorno al *Culex pipiens* e descrizione di una specie nuova *(Culex phytophagus)*. *Bullettino della Società Entomologica Italiana* (for 1889) 21(1–2): 124–131.

Ficalbi, E. 1893. Revisione delle specie Europee della famiglia delle zanzare. *Bullettino della Società Entomologica Italiana* 25: 48–61, 136–144.

Ficalbi, E. 1896. Notizie sulle Zanzare italiane. IX Nota. Descrizione di una specie nuova. Zanzara delle saline, *Culex salinus*, sp. n. *Bullettino della Società Entomologica Italiana* 28: 29–32.

Fitch, A. 1847. Winter insects of eastern New York. *American Journal of Agricultural Science* 5: 274–284.

Floch, H. & Abonnenc, E. 1945a. Les moustiques de la Guadeloupe (II). Les genres *Megarhinus*, *Aedes*, *Culex*, *Deinocerites*, *Mansonia* et *Wyeomyia*. *Publications / Institut Pasteur de la Guyane et du Territoire de l'Inini* 110: 1–48.

Floch, H. & Abonnenc, E. 1945b. Description de *Culex* nouveaux de la Guyane française. *Culex* (*Melanoconion*) *cauchensis* n. sp. *Culex* (*Melanoconion*) *cayennensis* n. sp. *Publications / Institut Pasteur de la Guyane et du Territoire de l'Inini* 112: 1–6.

Floch, H. & Abonnenc, E. 1945c. Description de *Culex* nouveaux de la Guyane Française [*sic*] (II). *Culex* (*Melanoconion*) *cavernicolus* n. sp. *Culex* (*Melanoconion*) *equinoxialis* n. sp. *Publications / Institut Pasteur de la Guyane et du Territoire de l'Inini* 114: 1–5.

Floch, H. & Abonnenc, E. 1945d. Description d'une nouvelle espèce d'anophèle: *A.* (*Stethomyia*) *canorii*. *Publications / Institut Pasteur de la Guyane et du Territoire de l'Inini* 116: 1– 3.

Floch, H. & Abonnenc, E. 1946a. Description de *Culex* nouveaux de la Guyane française (III). *Culex* (*Melanoconion*) *rabanicolus* n. sp. *Culex* (*Melanoconion*) *rorotaensis* n. sp. *Publications / Institut Pasteur de la Guyane et du Territoire de l'Inini* 120: 1–4.

Floch, H. & Abonnenc, E. 1946b. Description de *Culex* nouveaux de la Guyane Française [*sic*] (IV). *Culex* (*Melanoconion*) *punctiscapularis* n. sp. *Publications / Institut Pasteur de la Guyane et du Territoire de l'Inini* 122: 1–3.

Floch, H. & Abonnenc, E. 1947. Distribution des moustiques du genre *Culex* en Guyane française. *Publications / Institut Pasteur de la Guyane et du Territoire de l'Inini* 146: 1–9.

Floch, H. & Fauran, P. 1954. Nouveaux *Culex* en Guyane française: *Culex* (*Aedinus*) *amazonensis* Lutz et *Culex* (*Melnoconion*) *idottus* Dyar, *Culex* (*Melanoconion*) *vidali* n. sp. *Archives de Institut Pasteur de la Guyane Française* 15(333): 1–7.

Floch, H. & Fauran, P. 1955. Nouveaux *Culex* en Guyane française (VI). *Culex (Melanoconion) patientiae* n. sp. *Culex (Carrollia) manaensis* n. sp. *Culex (Microculex) reginae* n. sp. *Archives de l'Institut Pasteur de la Guyane Française et de l'Inini* 16(370): 1–7.

Fontenille, D. & Brunhes, J. 1985. Trois nouveaux culicidés de Madagascar: *Aedes (Aedimorphus) albodorsalis* n. sp., *Aedes (Aedimorphus) masoalensis* n. sp. et *Aedes (Aedimorphus) mathioti* n. sp. *Cahiers ORSTOM Série Entomologie médicale et Parasitologie* (for 1984) 22(2): 151–155.

Forattini, O.P. 1965. *Entomologia médica*. 2.° Volume. Culicini: *Culex*, *Aedes* e *Psorophora*. Universidade de São Paulo, São Paulo. 506 pp.

Forattini, O.P. & Sallum, M.A.M. 1985. A new species of *Culex (Melanoconion)* from southern Brazil (Diptera: Culicidae). *Revista de Saúde Pública* 19: 171–182.

Forattini, O.P. & Sallum, M.A.M. 1987a. Studies on some species of *Culex (Melanoconion)*, with the description of a new one from southern Brazil (Diptera: Culicidae). *Revista de Saúde Pública* 21(2): 123–156.

Forattini, O.P. & Sallum, M.A.M. 1987b. A new species of *Culex (Melanoconion)* from inland southern Brazil (Diptera: Culicidae). *Mosquito Systematics* 19(2): 167–172.

Forattini, O.P. & Sallum, M.A.M. 1989a. A new species of *Culex (Melanoconion)* from Atlantic tropical system of southern Brazil (Diptera: Culicidae). *Memórias do Instituto Oswaldo Cruz* 84(1): 1–4.

Forattini, O.P. & Sallum, M.A.M. 1989b. Taxonomic study and redescription of *Culex (Melanoconion) theobaldi* (Lutz, 1904) (Diptera: Culicidae). *Memórias do Instituto Oswaldo Cruz* 84(Supl. 4): 201–208.

Forattini, O.P. & Sallum, M.A.M. 1992. A new species of *Culex (Melanoconion)* from the Amazonian region (Diptera: Culicidae). *Memórias do Instituto Oswaldo Cruz* 87(2): 265–274.

Forattini, O.P. & Sallum, M.A.M. 1993a. Redescription of *Culex (Melanoconion) inadmirabilis* Dyar and synonymy of *Culex (Melanoconion) mistura* Komp and Rozeboom (Diptera: Culicidae). *Mosquito Systematics* 25(2): 77–88.

Forattini, O.P. & Sallum, M.A.M. 1993b. Taxonomic study of some species of the Educator Group of *Culex (Melanoconion)* (Diptera: Culicidae). *Mosquito Systematics* 25(2): 89–109.

Forattini, O.P. & Sallum, M.A.M. 1995. Two new species of the Spissipes Section of *Culex (Melanoconion)* (Diptera: Culicidae). *Mosquito Systematics* 27(2): 125–142.

Forattini, O.P. & Toda, A. 1966. Notas sôbre Culicidae (Diptera). 11. Algumas espécies amazônicas de *Microculex*. *Studia Entomologica* 9(1–4): 501–514.

Forattini, O.P., Rabello, E.X. & Souza Lopes, O. 1967. Notas sôbre *Culicidae (Diptera)*. 12- Nova espécie do subgênero *Culex*. *Revista Brasileira de Entomologia* 12: 12–16, 1 pl.

Forskål, P. 1775. *Descriptiones animalium, avium, amphibiorum, piscium, insectorum, vermium; quae in itinere orientali observavit Petrus Forskål. Post mortem auctoris edidit Carsten Niebuhr. Adjuncta est materia medica kahirina atque tabula maris Rubri geographica*. Ex officina Mölleri, Hauniae. 164 pp.

Foster, P.G., de Oliveira, T.M.P., Bergo, E.S., Conn, J.E., Sant'Ana, D.C., Nagaki, S.S., Nihei, S., Lamas, C.E., González, C., Moreira, C.C. & Sallum, M.A.M. 2017. Phylogeny of Anophelinae using mitochondrial protein coding genes. *Royal Society Open Science* 4: 170758.

Fox, I. 1953. Notes on Puerto Rican mosquitoes including a new species of *Culex*. *Journal of Parasitology* 39(2): 178–182.

Freeborn, S.B. 1949. Anophelines of the Nearctic Region, pp. 379–398. *In*: Boyd, M.F. (ed.), *Malariology. A comprehensive survey of all aspects of this group of diseases from a global standpoint*. Volume 1. W.B. Saunders Company, Philadelphia and London.

Froud, M.D. 1944. *Anopheles jebudensis* sp. nov., a new anopheline mosquito from southern Nigeria. *Annals of Tropical Medicine and Parasitology* 38(1): 73–77.

Gabaldon, A. 1940. Estudios sobre anofelinos. Serie I. 1. Descripción de *Anopheles (Nyssorhynchus) nuñez-tovari* [*sic*] n. sp. y consideraciones sobre una sub-division del grupo *Nyssorhynchus* (Diptera, Culicidae). *Publicaciones de la División de Malariología* 5: 3–7.

Gabaldon, A., Cova-Garcia, P. & Lopez, J.A. 1940. Estudios sobre anofelinos. Serie I. 2. *Anopheles (Nyssorhynchus) rangeli*, una nueva especie de la sub-serie *oswaldoi* (Diptera, Culicidae) de amplia distribucion en Venezuela. *Publicaciones de la Division de Malariologia* 5: 9–23.

Gabaldon, A., Cova-Garcia, P. & Lopez, J.A. 1941a. Estudios sobre anofelinos. Serie II. 1. *Anopheles (Nyssorhynchus) benarrochi*, una especie de la subserie *triannulatus*. *Publicaciones de la Division de Malariologia* 7: 3–24, 2 pls.

Gabaldon, A., Cova-Garcia, P. & Lopez, J.A. 1941b. Estudias sobre anofelines. Serie II. 2. *Anopheles (Arthuromyia) vargasi*, n. sp. y redescripcion del subgenero *Arthuromyia* Galvao [*sic*], 1941. *Publicaciones de la Division de Malariologia* 7: 25–56, 3 pls.

Gad, A.M., El Said, S., Hassan, A.N. & Shoukry, A. 1987. The distribution and ecology of the mosquitoes in the Red Sea Governorate, Egypt. *Journal of the Egyptian Society of Parasitology* 17(1): 207–221.

Gad, A., Harbach, R.E. & Harrison, B.A. 2006. *Anopheles (Cellia) ainshamsi*, n. sp. (Diptera: Culicidae), a saltwater species from the Red Sea coast of Egypt. *Proceedings of the Entomological Society of Washington* 108(2): 366–380.

Gaffigan, T.V. & Ward, R.A. 1985. Index to the second supplement to "A Catalog of the Mosquitoes of the World", with corrections and additions (Diptera: Culicidae). *Mosquito Systematics* 17(1): 52–63.

Galindo, P. & Blanton, F.S. 1954. Nine new species of Neotropical *Culex*, eight from Panama and one from Honduras (Diptera, Culicidae). *Annals of the Entomological Society of America* 47(2): 231–247.

Galindo, P. & Blanton, F.S. 1955. An annotated list of the *Culex* of Panama (Diptera, Culicidae). *Proceedings of the Entomological Society of Washington* 57(2): 68–74.

Galindo, P. & Mendez, E. 1961. Descriptions of four new species of *Culex* from Panama (Diptera: Culicidae). *Annals of the Entomological Society of America* 54(1): 1–4.

Galindo, P. & Trapido, H. 1956. Descriptions of two new subspecies of *Haemagogus mesodentatus* Komp and Kumm 1938, from Middle America (Diptera, Culicidae). *Proceedings of the Entomological Society of Washington* 58(4): 228–231.

Galindo, P. & Trapido, H. 1967. Description of *Haemagogus aeritinctus*, a new species from British Honduras, with a note on the validity of *Haemagogus lucifer* (H. D. and K.) (Diptera: Culicidae). *Proceedings of the Entomological Society of Washington* 69(2): 103–111.

Galindo, P., Blanton, F.S. & Peyton, E.L. 1954. A revision of the *Uranotaenia* of Panama with notes on other American species of the genus (Diptera, Culicidae). *Annals of the Entomological Society of America* 47(1): 107–177.

Galindo, P., Carpenter, S.J. & Trapido, H. 1951. Descriptions of two new species of *Wyeomyia* and the male of *Sabethes tarsopus* Dyar and Knab (Diptera, Culicidae). *Proceedings of the Entomological Society of Washington* 53(2): 86–96.

Galindo, P., Carpenter, S.J. & Trapido, H. 1953. The taxonomic status of the *Aedes leucocelaenus* complex with descriptions of two new forms (Diptera, Culicidae). *Annals of the Entomological Society of America* (for 1952) 45(4): 529–542.

Galliard, H. 1931a. *Culex brumpti*, n. sp. Moustique nouveau trouvé en Corse. *Annales de Parasitologie Humaine et Comparée* 9(2): 134–139.

Galliard, H. 1931b. Culicides du Gabon. I. – Culicinés, avec la description d'une espéce et de deux variétés nouvelles. *Annales de Parasitologie Humaine et Comparée* 9(3): 225–232.

Galliard, H. 1931c. Culicides du Gabon. II. Culicinés (*suite*). Remarques sur la biologie des *Mansonioides* et d'*Aedes* (*Stegomyia*) *argenteus* Poiret. *Annales de Parasitologie Humaine et Comparée* 9(6): 514–529.

Galliard, H. & Dang-van-Ngu [*sic*]. 1940. Une espèce nouvelle d'*Anophèles* du Tonkin, *A. tonkinensis* sp. nov. *Revue Médicale Française d'Extrême-Orient* 18(9–10): 595–598, 3 pls.

Galliard, H. & Ngu, D.-V. 1947. Culicinés du Tonkin. I. Liste des espèces et leur répartition. *Annales de Parasitologie Humaine et Comparée* 22(1–2): 75–80.

Galliard, H. & Ngu, D.V. 1950. Culicinés du Tonkin. III. – Culicidés de la Baie d'Along. *Annales de Parasitologie Humaine et Comparée* 25(1–2): 61–68.

Galvao, A.B. 1952. *Anopheles* (*Anopheles*) *rachoui*, uma nova espécie do grupo *Arribalzagia*. *Revista Brasileira de Malariologia e Doenças Tropicais* 4(1): 69–74.

Galvão, A.B. 1955. *Anopheles* (*Arribalzagia*) *bustamantei*, uma nova espécie *(Diptera, Culicidae)*. *Revista Brasileira de Malariologia e Doenças Tropicais* 7: 255–258.

Galvao, A.L.A. 1941a. Notas sôbre alguns anofelinos do sub-gênero *Nyssorhynchus* do norte do Brasil. *Revista de Biologia e Hygiene* 11(1): 92–96.

Galvao, A.L.A. 1941b. Contribuição ao conhecimento das especies de *Myzorhynchella (Diptera, Culicidae)*. *Arquivos de Zoologia* 2(18): 505–576, 13 pls.

Galvão, A.L.A. 1943. Chaves para a determinação das espécies do subgênero *Nyssorhynchus* do Brasil. *Arquivos de Higiene Saúde Pública* 8(19): 141–162.

Galvão, A.L.A. & Barretto, M.P. 1941. Sôbre um novo anofelino de São Paulo, *Anopheles* (*Anopheles*) *pseudotibiamaculatus* n. sp. (*Diptera, Culicidae*). *Revista de Biologia e Hygiene* 11(2): 65–70.

Galvão, A.L.A. & Damasceno, R.G. 1942a. *Anopheles* (*Nyssorhynchus*) *konderi* nova espécie de *Anopheles* do vale do Amazonas e considerações sôbre as espécies do complexo *tarsimaculatus* (Diptera, Culicidae). *Folia Clinica et Biologica* 14(5–6): 115–135.

Galvão, A.L.A. & Damasceno, R.G. 1942b. Sôbre um novo anofelino da Ilha de Marajó, *Anopheles* (*N.*) *Marajoara* N. Sp. *Anais Paulistas de Medicina e Cirurgia* 44(5): 424–427.

Galvão, A.L.A. & Damasceno, R.G. 1944. Observações sobre anofelinos do complexo *albitarsis* (Diptera Culicidae). *Anais da Faculdade de Medicina da Universidade de São Paulo* 20: 73–87.

Galvão, A.L.A. & Franco do Amaral, A.D. 1938. Sobre um novo anofelino de Campos do Jordão, Estado de São Paulo, *Anopheles* (*Nyssorhynchus*) *lanei* n. sp. (*Diptera, Culicidae*). *Revista de Biologia e Hygiene* 9(1): 8–16, 4 pls.

Galvão, A.L.A. & Franco do Amaral, A.D. 1940. Estudos sôbre os anofelinos do grupo *Myzorhynchella* com a descrição de uma espécie nova, *Anopheles* (*Nyssorhynchus*) *antunesi* n. sp. (Dipt. Culicidae). *Folia Clinica et Biologica* 12(5): 150–160.

Galvão, A.A. & Lane, J. 1937a. Notas sobre os *Nyssorhynchus* de S. Paulo. II. Descrição de uma nova especie *Anopheles* (*Nyssorhynchus*) *pessôai* (*Diptera, Culicidae*). *Revista de Biologia e Hygiene* (for 1936) 7(2): 67–79.

Galvão, A.L.A. & Lane, J. 1937b. Nota sobre os *Nyssorhynchus* de S. Paulo. VII. Estudo sobre as variedades deste grupo com a deserição de *Anopheles* (*Nyssorhynchus*) *albitarsis* Arrib., 1878 var. *limai* n. var. *Anais da Faculdade de Medicina da Universidade de São Paulo* 13: 211–238.

Galvão, A.A., Lane, J. & Corrêa, R. 1937. Notas sobre os *Nyssorhynchus* de S. Paulo. V. Sobre os *Nyssorhynchus* de Novo Oriente. *Revista de Biologia e Hygiene* 8(1): 37–45.

García, M. & Casal, O.H. 1967. *Toxorhynchites (Lynchiella) cavalierii*, nueva especie bromelícola de la República Argentina *(Diptera, Culicidae)*. *Physis* 26(73): 435–441.

Garcia, R., Jeffery, J. & Rudnick, A. 1969. *Culiseta marchettei*, a new species of the subgenus *Climacura* Howard, Dyar, and Knab from Malaysia, with notes on its biology (Diptera: Culicidae). *Journal of Medical Entomology* 6(3): 251–256.

Gaschen, H. 1934. Description d'un nouvel *Aedes* du Yunnan *Finlaya yunnanensis* sp. n. *Archives des Instituts Pasteur d'Indochine* 19: 331–335, 3 pls.

Gater, B.A.R. 1933. On *Anopheles baezai* n.sp., from the Malay Peninsula. *Bulletin of the Raffles Museum* 8: 162–168.

Gebert, S. 1948. Notes on a new species of *Aëdes*, sub-genus *Mucidus* (Diptera: Culicidae) found in Mauritius. *Entomologist* 81: 96–99.

Geoffroy, B. 1971a. Description du mâle de *Culex (Culiciomyia) eouzani* sp. n. *(Diptera Culicidae)* nouveau moustique de la République Centrafricaine. *Cahiers ORSTOM Série Entomologie médicale et Parasitologie* 9(3): 279–283.

Geoffroy, B. 1971b. Description d'un moustique nouveau de la République Centrafricaine *Aedes (Aedimorphus) adami* sp. n. *Cahiers ORSTOM Série Entomologie médicale et Parasitologie* 9(3): 273–277.

Geoffroy, B. 1971c. Description d'un nouveau moustique de la République Centrafricaine: *Culex (Neoculex) vattieri* sp. n. *Cahiers ORSTOM Série Entomologie médicale et Parasitologie* 9(3): 265–268.

Geoffroy, B. 1974. Description du mâle de *Culex (Mochthogenes) germaini* n. sp. *(Diptera, Culicidae)* moustique nouveau de République Centrafricaine. *Cahiers ORSTOM Série Entomologie médicale et Parasitologie* 12(2): 93–96.

Geoffroy, B. 1987. The *Aedes (Aedimorphus)* Domesticus Group (Diptera, Culicidae). I. New species, descriptions of *Aedes bambiotai* and *Aedes bancoi*. *Mosquito Systematics* 19(1): 100–110.

Geoffroy, B. & Hervé, J.-P. 1976. Description d'un nouveau moustique de République Centrafricaine, *Culex (Eumelanomyia) taufliebi* n. sp. *Cahiers ORSTOM Série Entomologie médicale et Parasitologie* 14(2): 151–154.

Germar, E.F. 1817. *Reise nach Dalmatien und in das Gebiet von Ragusa*. F.A. Brockhaus, Leipzig und Altenburg. xii, 325 pp., 11 pls.

Giaquinto Mira, M. 1931. Una nueva especie de *Anopheles* en Guatemala: *Anopheles hectoris* mihi. *Boletín de la Dirección General de Salubridad, Guatemala* 1(20/25): 606–615.

Giaquinto-Mira, M. 1950. Notes on the geographical distribution and biology of "Anophelinae" and "Culicinae" in Etiopia [*sic*]. *Revista di Malariologia* 29(5): 281–313.

Giebel, C. 1862. Wirbeltier und Insektenreste im Bernstein. *Zeitschrift für die Gesammten Naturwissenschaften* 20: 311–321.

Gil Collado, J. 1936. Culicidos de la Isla de Fernando Poo recogidos por la expedición J. Gil-F. Bonet. *EOS - Revista Espanola de Entomologia* 11(4): 311–329.

Giles, J[*sic*]. M. 1899. A description of the Culicidae employed by Major R. Ross, I.M.S., in his investigations on malaria. *Journal of Tropical Medicine* 2: 62–65.

Giles, G.M. 1900a. *A handbook of the gnats or mosquitoes, giving the anatomy and life history of the Culicidae*. John Bale, Sons & Danielsson, Ltd, London. viii, 374 pp., 8 pls.

Giles, G.M. 1900b. Addendum I. Description of two species of *Anopheles* from West Africa. *Memoirs of the Liverpool School of Tropical Medicine* 2: 49–51.

Giles, G.M. 1901a. A plea for the collective investigation of Indian Culicidae, with suggestions as to moot points for enquiry, and a prodromus of species known to the author. *Journal of the Bombay Natural History Society* 13(4): 592–610, 2 pls.

Giles, G.M. 1901b. Descriptions of four new species of *Anopheles* from India. *Entomologist's Monthly Magazine* 37: 196–198.

Giles, G.M. 1901c. Six new species of Culicidae from India. *Entomologist* 34(458): 192–197.

Giles, G.M. 1902. *A handbook of the gnats or mosquitoes giving the anatomy and life history of the Culicidae together with descriptions of all species noticed up to the present date*. Second edition. John Bale, Sons & Danielsson, Ltd, London. xi, [1], 530 pp., XVII pls.

Giles, G.M. 1904a. Notes on some collections of mosquitoes, &c., received from the Philippine Islands and Angola; with some incidental remarks upon classification. *Journal of Tropical Medicine* 7: 365–369.

Giles, G.M. 1904b. Notes on some collections of mosquitoes received from abroad. I.–A revised tabulation of the genus *Taeniorhynchus* Arribálzaga. *Journal of Tropical Medicine* 7: 381–384.

Giles, G.M. 1904c. *A revision of the Anophelinae, being a first supplement to the second edition of "A handbook of the gnats or mosquitoes"*. John Bale, Sons & Danielsson, Ltd, London. 47 pp.

Giles, G.M. 1906. Mosquito notes. I.–Note on a small collection of mosquitoes from Barhain [*sic*] in northern Arabia (shores of Persian Gulf). *Journal of Tropical Medicine* 9: 130–132.

Gillett, J.D. 1946. Notes on the subgenus *Coquillettidia* Dyar (Diptera, Culicidae). *Bulletin of Entomological Research* 36(4): 425–438.

Gillett, J.D. 1949. Further notes on the Ethiopian species of *Taeniorhynchus* Arribalzaga [*sic*] (Diptera, Culicidae). *Proceedings of the Royal Entomological Society of London, Series B: Taxonomy* 18(5–6): 97–102.

Gillett, J.D. 1975. The *Anopheles gambiae* complex; naming the sibling species. *Transactions of the Royal Society of Tropical Medicine and Hygiene* 69(3): 366–367.

Gillies, M.T. 1958. Notes on the biology of a new subspecies of *Anopheles wellcomei* (Diptera: Culicidae) from East Africa, and on the distribution of related forms. *Proceedings of the Royal Entomological Society of London, Series A: General Entomology* 33(1–3): 9–14.

Gillies, M.T. 1962. A new species of the *Anopheles funestus* complex (Diptera: Culicidae) from East Africa. *Proceedings of the Royal Entomological Society of London, Series B: Taxonomy* 31(7–8): 81–86.

Gillies, M.T. 1964. A new *Anopheles* (Diptera: Culicidae) from the East African coastal lowlands. *Proceedings of the Royal Entomological Society of London, Series B: Taxonomy* 33(3–4): 67–70.

Gillies, M.T. & Coetzee, M. 1987. A supplement to the Anophelinae of Africa south of the Sahara (Afrotropical Region). *Publications of the South African Institute for Medical Research* 55: 1–143.

Gillies, M.T. & de Meillon, B. 1968. The Anophelinae of Africa south of the Sahara (Ethiopian Zoogeographical Region). *Publications of the South African Institute for Medical Research* 54: 1–343.

Gimmerthal, B.A. 1845. Erster Beitrag zu einer künftig zu bearbeitenden Dipterologie Russlands. *Bulletin de la Société Impériale des Naturalistes de Moscou* 18(4): 287–331.

Girhe, B.E. & Sathe, T.V. 2001. On a new species of the genus *Aedes* Meigen (Diptera: Culicidae) from India. *Journal of Advanced Zoology* 22(1): 46–47.

Gistl, J.N.F.X. 1831. Kerfe in Copal eingeschlossen. *Isis (Oken's)* 24: 247–248.

Gjullin, C.M. 1946. A key to the *Aedes* females of America north of Mexico (Diptera, Culicidae). *Proceedings of the Entomological Society of Washington* 48(9): 215–236.

Glick, J.I. 1992. Illustrated key to the female *Anopheles* of southwestern Asia and Egypt (Diptera: Culicidae). *Mosquito Systematics* 24(2): 125–153.

Goeldi, E.A. 1902. *Os mosquitos no Pará. Encarados como uma calamidade publica*. Imprensa Official, Belem. 57 pp.

Goeldi, E.A. 1905. *Os mosquitos no Pará: Reunião de quatro trabalhos sobre os mosquitos indigenas, principalmente as especies que molestam o homem*. Memorias do Museu Goeldi (Museu Paraense) de Historia Natural e Ethnographia IV. C. Wiegandt, Pará, Brazil. 154 pp., 21 pls.

Gong, Z.-D. 1989a. A new species of genus *Topomyia* from Yunnan, China (Diptera: Culicidae) [in Chinese]. *Acta Entomologica Sinica* 32(1): 90–91.

Gong, Z.-D. 1989b. A new species of the genus *Topomyia* from Yunnan Province, China (Diptera: Culicidae) [in Chinese]. *Acta Zootaxonomica Sinica* 14(2): 220–223.

Gong, Z.-D. 1991. A new species of the subgenus *Topomyia* of genus *Topomyia* from Yunnan, China (Diptera: Culicidae) [in Chinese and English]. *Acta Zootaxonomica Sinica* 16(2): 228–231.

Gong, Z.-D. 1994. A new species of the genus *Topomyia* from West Yunnan, China (Diptera: Culicidae) [in Chinese]. *Acta Zootaxonomica Sinica* 19(2): 210–213.

Gong, Z.-D. 1996. A new species of *Topomyia* from the bamboo stumps in Yunnan, China (Diptera: Culicidae) [in Chinese]. *Acta Zootaxonomica Sinica* 21(3): 362–365.

Gong, Z.-D. & Ji, S.-H. 1989. A new species of *Tripteroides* from Yunnan (Diptera: Culicidae) [in Chinese]. *Acta Entomologica Sinica* 32(4): 482–484.

Gong, Z.D. & Li, Z.-H. 1999. A new species of *Heizmannia* from Yunnan, China (Diptera: Culicidae) [in Chinese]. *Acta Zootaxonomica Sinica* 24(1): 96–99.

Gong, Z. & Lu, B. 1986. A new *Aedes* of subgenus *Finlaya* from Yunnan, China [in Chinese]. *Zoological Research* 7(1): 1–4.

Gong, Z. & Lu, B. 1991. A new subgenus and species of the *Aedes* (Diptera: Culicidae) from Yunnan, China [in Chinese]. *Entomotaxonomia* 13(1): 55–59.

Gong, Z. & Lu, B. 1995. Studies on the genus *Topomyia* of China (Diptera: Culicidae) [in Chinese]. *Acta Entomologica Sinica* 38(4): 475–485.

González Broche, R. & Rodríguez Rodríguez, R. 2001. Nuevo subgénero de *Culex*, descripción de la pupa y de la larva y redescripción de la hembra de *Culex nicaroensis* Duret, (Díptera: Culicidae). *Avicennia* 14: 65–74.

González, C.R. & Sallum, M.A.M. 2010. *Anopheles (Nyssorhynchus) atacamensis* (Diptera: Culicidae), a new species from northern Chile. *Memórias do Instituto Oswaldo Cruz* 105(1): 13–24.

González, C.R., Reyes, C., Rada-Chaparro, V. & Saldarriaga-Córdoba, M. 2017. A new species of *Aedes* Meigen subgenus *Ochlerotatus* Lynch Arribálzaga (Diptera: Culicidae) from the coastal wetlands of the desert in northern Chile: morphological and molecular identification. *Zootaxa* 4273(1): 31–49.

Gonzalez-Rincones, R. & Guyon, L. 1953. *Clasificacion general de los Dipteros*. Universidad Central de Venezuela, Publicaciones del Instituto de Medicina Experimental "Jose Gregorio Hernandez", Caracas. 243 pp.

Gordeev, M.I., Zvantsov, A.B., Goriacheva, I.I., Sheĭkevich, E.V. & Ezhov, M.N. 2005. Description of the new species *Anopheles artemievi* sp.n. (Diptera, Culicidae) [in Russian]. *Meditsinskaia Parazitologiia i Parazitarnye Bolezni* 2005(2): 4–5.

Gordon, R.M. & Evans, A.M. 1922. Mosquitoes collected in the Manáos Region of the Amazon. *Annals of Tropical Medicine and Parasitology* 16(3): 315–338, 1 pl.

Gornostaeva, R.M. 2005. To the revision of the mosquitoes of the subgenus *Aedes* (Diptera, Culicidae) from the Palaearctic [in Russian]. *Parazitologiia* 39(6): 457–507.

Gough, L.H. 1910. On a collection of *Anopheles* made at Onderstepoort in the autumn of 1909. *Report of the Government Veterinary Bacteriologist / Union of South Africa* 1908–09: 115–121.

Grabham, M. 1905. Notes on some Jamaican Culicidae. *Canadian Entomologist* 37(12): 401–411.

Grabham, M. 1906a. Notes on some mosquitoes from Newcastle, Jamaica. *Canadian Entomologist* 38(5): 167–173.

Grabham, M. 1906b. Four new Culicidae from Jamaica, West Indies. *Canadian Entomologist* 38(11): 311–320.

Grabham, M. 1907. Notes on some new mosquitoes from Jamaica, West Indies. *Canadian Entomologist* 39(1): 25–26.

Graham, W.M. 1909. Four new species of the genus *Eretmapodites* (Theobald) from Ashanti. *Entomologist* 42(551, 553): 86–89, 157–159.

Graham, W.M. 1910. On new species of West-African *Culicidae*. *Annals and Magazine of Natural History, including Zoology, Botany, and Geology* (Eighth Series) 5: 264–273.

Grassi, B. 1898. Rapporti tra la malaria e peculiari insetti (zanzaroni e zanzare palustri). *Rendiconti delle seduti delle Reale Accademia dei Lincei. Classe de scienze fisiche, matematische e naturali* 7(2): 163–177.

Grassi, B. 1899. Ancora sulla malaria. *Rendiconti delle seduti delle Reale Accademia dei Lincei. Classe de scienze fisiche, matematische e naturali* 8: 559–561.

Grassi, B., Bignami, A. & Bastianelli, G. 1899. Resoconto degli studi fatti sulla malaria durante il mese di gennaio. *Reale Accademia dei Lincei. Rendiconti. Memorie della classe de scienze fisiche, matematische e naturali. Serie quinta* 8: 100–104.

Grassi, G.B. 1900. Studi di uno zoologo sulla marlaria. Memoria del Socio Battista Grassi. *Reale Accademia dei Lincei. Rendiconti. Classe de scienze fisiche, matematische e naturali. Memoirs* (5) 3: 299–516, 5 pls.

Green, E.E. 1901. *Mosquitoes and malaria*. Circular. Royal Botanic Gardens, Ceylon. Series I, No. 25: 345–368.

Griffith, E. & Pidgeon, E. 1832. Supplement on the order Diptera, pp. 721–780. *In*: Cuvier, G., Griffith, E., Pidgeon, E. & Gray, G.R. *The animal kingdom: arranged in conformity with its organisation. Volumes 14–15. The class Insecta*. Whittaker, Treacher, and Co., London. 793 pp.

Grjebine, A. 1950. Un nouveau moustique forestier de la basse Cote d'Ivoire *Eretmapodites pauliani* n. sp. *Bulletin de la Société de Pathologie exotique* 43(1–2): 45–50.

Grjebine, A. 1953a. Observations sur les nematoceres vulnerants de Madagascar, regions de Majunga et de la Mandraka. *Mémoires de l'Institut Scientifique de Madagascar Séries E* 4: 443–502.

Grjebine, A. 1953b. Une nouvelle espece d'anophele de Madagascar [Dipt. Culicidae]. *Le Naturaliste Malgache* 5(2): 203–209.
Grjebine, A. 1961. Un nouvel anophèle de Madagascar. *Bulletin de la Société de Pathologie exotique* (for 1960) 53(6): 939–942.
Grjebine, A. 1964. *Anopheles (Neomyzomyia) grenieri* n. sp. *(Diptera, Culicidae)*, nouvel anophèle de Madagascar. *Bulletin de la Société de Pathologie exotique* 57(1): 38–42.
Grjebine, A. 1966. *Faune de Madagascar. XXII. Insecies Diptéres Culicidae Anophelinae*. Centre National de la Recherche Scientifique, Office de la Recherche Scientifique et Technique Outre-Mer, Paris. 487 pp., table, map, 8 pls.
Grjebine, A. 1972. Une nouvelle espèce de moustique du Congo-Brazzaville: *Eretmapodites hamoni* n. sp. (Dipt. Culicidae). *Bulletin de la Société Entomologique de France* 77(9–10): 308–313.
Grjebine, A. 1979. Les moustiques des *Nepenthes* de Madagascar. Espèces nouvelles du genre *Uranotaenia. (Diptera, Culicidae). Annales de la Société Entomologique de France* 15(1): 53–74.
Grjebine, A. 1986. *Faune de Madagascar. 68. Insectes Diptères Culicidae Culicinae Ficalbiini*. Muséum nationale d'Histoire naturelle, Paris. 441 pp.
Grossbeck, J.A. 1904a. Description of a new *Culex*. *Canadian Entomologist* 36(11): 332.
Grossbeck, J.A. 1904b. Description of two new species of *Culex*. *Entomological News* 15(10): 332–333.
Grossbeck, J.A. 1905. New species of Culicidae. *Canadian Entomologist* 37(10): 359–360.
Grossbeck, J.A. 1906. Notes on *Culex squamiger*, Coq., with description of a closely-allied species. *Canadian Entomologist* 38(4): 129–131.
Grünberg, K. 1902. Ein neuer *Anopheles* aus Westafrika, *Anopheles Ziemanni* nov. spec. *Zoologischen Anzeiger* 25(677): 550–551.
Grünberg, K. 1905. Zur Kenntnis der Culicidenfauna von Kamerun und Togo. *Zoologischen Anzeiger* 29(12): 377–390.
Grünberg, K. 1907a. Ein neuer *Toxorhynchites* aus Kamerun. (Dipt.). *Deutsche entomologische Zeitschrift* 1907: 405–406.
Grünberg, K. 1907b. *Die blutsaugenden Dipteren. Leitfaden zur allgemeinen Orientierung, mit besonderer Berücksichtigung der in den Deutschen Kolonien lebenden Krankheitsüberträger*. Gustav Fischer, Jena. vi, 188 pp.
Grünberg, K. 1913. Zur Kenntnis der Culiciden-Fauna von Samoa. *Entomologische Rundschau* 30(22): 130–131.
Guenther, K. 1913. Die lebenden Bewohner der Kannen der insektenfressenden Pflanze *Nepenthes destillatoria* auf Ceylon. *Zeitschrift für wissenschaftliche Insektenbiologie* 9(6/7): 198–207.
Gutsevich, A.V. 1955. New and little known forms of mosquitoes (Diptera, Culicidae) [in Russian]. *Trudy Zoologicheskogo Instituta Akademii Nauk* 18: 320–324.
Gutsevich, A.V. 1962. A new species of the genus *Aedes* (Diptera, Culicidae) from Kazakhstan [in Russian]. *Entomologicheskoe Obozrenie* 41(4): 886–888.
Gutsevich, A.V. 1966. A new mosquito species from Uzbekistan, *Aedes* (*Ochlerotatus*) *simanini* (Culicidae) [in Russian]. *Zoologicheskii Zhurnal* 45(3): 457–459.
Hacker, H.P. 1921. Changes in nomenclature of species. *Report of the Malaria Bureau of the Federated Malay States* 2: 1.
Hackett, L.W. 1934. The present status of our knowledge of the sub-species of *Anopheles maculipennis*. *Transactions of the Royal Society of Tropical Medicine and Hygiene* 28(2): 109–128.
Hackett, L.W. & Lewis, D.J. 1935. A new variety of *Anopheles maculipennis* in southern Europe. *Rivista di Malariologia* 14(5): 377–383, 1 pl.
Hackett, L.W. & Missiroli, A. 1935. The varieties of *Anopheles maculipennis* and their relation to the distribution of malaria in Europe. *Rivista di Malariologia* 14(1): 45–109.
Haddow, A.J. 1946. The mosquitoes of Bwamba County, Uganda. IV.–Studies on the genus *Eretmapodites*, Theobald. *Bulletin of Entomological Research* 37(1): 57–82.
Haddow, A.J. & van Someren, E.C.C. 1950. A new species of *Stegomyia* Theobald from the Ruwenzori Range, Uganda. *Annals of Tropical Medicine and Parasitology* 44(3): 281–284.
Haddow, A.J., van Someren, E.C.C., Lumsden, W.H.R., Harper, J.O. & Gillett, J.D. 1951. The mosquitoes of Bwamba County, Uganda. VIII.—Records of occurrence, behaviour and habitat. *Bulletin of Entomological Research* 42(2): 207–238.
Haga, J. 1925. A new *Uranotaenia* (Culicidae, Diptera) from Alor, Lesser Sunda Islands. *Bulletin of Entomological Research* 16(1): 43–44.
Haga, J. 1930. Tabellen voor determinatie der in Nederländsch-Oost-Indië voorkomende Anophelinen. *Geneeskundig tijdschrift voor Nederlandsch-Indië* (for 1929) 70: 363–382.
Halcrow, J.G. 1957. A new sub-species of *Anopheles gambiae* Giles from Mauritius. *East African Medical Journal* 34(4): 133–135.
Haliday, A.H. 1833. Catalogue of Diptera occurring about Holywood in Downshire. *Entomological Magazine* 1: 147–180.
Hall, C.R., Howard, T.M. & Harbach, R.E. 1999. *Sabethes* (*Peytonulus*) *luxodens*, a new species of Sabethini (Diptera: Culicidae) from Ecuador. *Memórias do Instituto Oswaldo Cruz* 94(3): 329–338.
Hamon, J. 1953. Une nouvelle espece d'*Aedes* de l'ile de la Reunion (Diptere Culicidae). *Naturaliste Malgache* 5: 35–41.
Hamon, J. 1954a. Contribution a l'etude des culicides de la region de Bobo Dioulasso (Haute-Volta). *Annales de Parasitologie Humaine et Comparée* 29(5–6): 573–587.
Hamon, J. 1954b. Contribution a l'étude des culicidés d'Afrique Occidentale. Description de *Ficalbia (Ficalbia) grenieri* sp. n. *Bulletin de la Société de Pathologie exotique* 47(6): 835–839.
Hamon, J. 1955. Contribution a l'étude des culicides d'Afrique Occidentale. Description de *Uranotaenia devemyi* sp. n., *Culex grahami* var. *farakoensis* var. n. et de la larve de *Harpagomyia trichorostris* Theobald. *Bulletin de la Société de Pathologie exotique* (for 1954) 47(5): 672–678.
Hamon, J. 1957. Deux moustiques nouveaux de Cote-d'Ivoire: *Aedes* (*Aedimorphus*) *falabreguesi* sp. n., *Culex* (*Culex*) *castelli* sp. n. *Bulletin de la Société de Pathologie exotique* 50(5): 685–691.

Hamon, J. 1961. Les *Eretmapodites* du groupe *oedipodius* Graham. 1. Descriptions complémentaires et position systématique des six sous-espèces actuellement cannues. *Bulletin de la Société de Pathologie exotique* 54(4): 892–906.

Hamon, J. 1962. Description de deux nouveaux *Eretmapodites (Diptera, Culicidae)* capturés en Uganda: *E. vansomereni* sp. n. et *E. corbeti* sp. n. *Bulletin de la Société de Pathologie exotique* 55(2): 253–259.

Hamon, J. & Adam, J.P. 1955. Contribution a l'étude des culicidés (Diptères) d'Afrique. Description de *Harpagomyia moucheti* sp. n. *Bulletin de la Société de Pathologie exotique* 48(6): 882–885.

Hamon, J. & Adam, J.-P. 1959a. Description de deux nouveaux *Aedes* appartenant au groupe *tarsalis* Newstead en basse Cote-d'Ivoire (A. O. F.): *A.* (*Aedimorphus*) *reali* sp. n. et *A.* (*Aedimorphus*) *nyounae* sp. n. *Bulletin de la Société de Pathologie exotique* (for 1958) 51(1): 530–534.

Hamon, J. & Adam, J.P. 1959b. Description de deux nouveaux *Aedes* de Cote-d'Ivoire appartenant au sous-genre *Aedimorphus: A. ovazzai* sp. n. et *A. rickenbachi* sp. n. *Bulletin de la Société de Pathologie exotique* 52(2): 147–154.

Hamon, J. & Brengues, J. 1965. Observations sur les *Aedes (Aedimorphus)* d'Afrique avec description de deux nouvelles espèces: *Ae. lottei* n. sp. et *Ae. dialloi* n. sp. *Bulletin de la Société de Pathologie exotique* 58(1): 101–108.

Hamon, J. & Gandara, A. 1955. Contribution a l'étude des culicidés (Diptères) de la région Éthiopienne. Description de *Culex (Culiciomyia) cambournaci* n. sp. *Bulletin de la Société de Pathologie exotique* 48(6): 866–872.

Hamon, J. & Lambrecht, F.L. 1959. Description de *Culex (Culiciomyia) muspratti* n. sp. accompagnée de quelques notes sur les moustiques de la Forêt d'Irangi, Kivu, Congo Belge. *Bulletin de la Société de Pathologie exotique* 52(5): 582–587.

Hamon, J. & Mouchet, J. 1955. Contribution a l'étude des *Neoculex* (Diptères: Culicidés) de la région éthiopienne 2. – Description de *Neoculex adami* sp. n. *Bulletin de la Société de Pathologie exotique* 48(6): 860–862.

Hamon, J. & Ovazza, M. 1954. Une nouvelle espèce de moustique (*Culex shoae* n. sp.) vivant sur les bananiers du haut plateau d'Éthiopie. *Bulletin de la Société de Pathologie exotique* 47(3): 416–421.

Hamon, J. & Ovazza, M. 1956. Contribution a l'étude des culicidés d'Afrique. Observations sur le groupe *Culex annulioris* Theobald, 1901. *Bulletin de la Société de Pathologie exotique* 49(1): 89–99.

Hamon, J. & Rickenbach, A. 1954. Contribution a l'étude des culicidés d'Afrique occidentale. Description d'*Aedes (Aedimorphus) mattinglyi* sp. n., *Aedes (Banksinella) jamoti* sp. n. Notes complémentaires sur *Aedes (Aedimorphus) stokesi* Evans, *Aedes (Banksinella) bolensis* Edwards. *Bulletin de la Société de Pathologie exotique* 47(6): 930–941.

Hamon, J. & Rickenbach, A. 1955a. Contribution a l'étude des culicidés d'Afrique occidentale. Description d'*Anopheles brumpti* sp. n. *Bulletin de la Société de Pathologie exotique* 48(3): 342–344.

Hamon, J. & Rickenbach, A. 1955b. Contribution a l'etude des *Néoculex* (Diptères: Culicidés) de la Région éthiopienne. 1. – Corrections de quelques descriptions de terminalia mâles données par Edwards, avec étude d'une nouvelle variété. *Bulletin de la Société de Pathologie exotique* 48(6): 848–859.

Hamon, J. & van Someren, E.C.C. 1961a. Description de deux nouveaux moustiques du Tanganyika: *Culex (Culiciomyia) gilliesi* sp. n. et *Eretmapodites mattinglyi* sp. n. *Bulletin de la Société de Pathologie exotique* 54(4): 724–731.

Hamon, J. & van Someren, E.C.C. 1961b. Les *Eretmapodites* du groupe *oedipodius* Graham. II. Description de *E. grenieri* sp. n. et clé de détermination des espèces et sous-espèces du groupe *oedipodius*. *Bulletin de la Société de Pathologie exotique* 54(4): 907–913.

Hamon, J., Adam, J.P. & Mouchet, J. 1955. Contribution a l'étude des *Neoculex* (Diptères: Culicidés) de la région éthiopienne. 3. – Description de *Neoculex laplantei* sp. n. *Bulletin de la Société de Pathologie exotique* 48(6): 862–865.

Hamon, J., Holstein, M. & Rivola, E. 1957. Description d'un nouveau moustique du Congo Belge: *Culex (Neoculex) vinckei* sp. n. *Bulletin de la Société de Pathologie exotique* 50(5): 681–685.

Hamon, J., Service, M.W., Adam, J.P. & Taufflieb, R. 1961. Observations sur les *Aedes (Aedimorphus)* du groupe *tarsalis* Newstead avec description d'une nouvelle espèce: *Aedes grenieri* n. sp. *Bulletin de la Société de Pathologie exotique* 54(2): 375–388.

Hamon, J., Taufflieb, R. & Dyemkouma, A. 1961. Observations sur la variabilité d'*Anopheles rufipes*, Gough, 1910, avec description d'une nouvelle variété. *Bulletin de la Société de Pathologie exotique* 54(1): 24–28.

Hamon, J., Taufflieb, R. & Maillot, L. 1957. Description d'*Aedes (Aedimorphus) grjebinei* sp. n. accompagnée de quelques données nouvelles sur les moustiques d'A. E. F. *Bulletin de la Société de Pathologie exotique* 50(5): 692–696.

Handlirsch, A. 1924. Systematische Übersicht, pp. 377–1140. *In*: Schröder, C., *Handbuch der Entomologie* Band III. Gustav Fischer, Jena.

Harbach, R.E. 1985. A new species, *Culex (Culex) litwakae* (Diptera: Culicidae), from the coastal region of Kenya. *Mosquito Systematics* 17(3): 254–265.

Harbach, R.E. 1987. *Culex verutus,* a new species of the subgenus *Culex* (Diptera: Culicidae) from Sierra Leone. *Mosquito Systematics* (for 1986) 18(3–4): 216–229.

Harbach, R.E. 1988. The mosquitoes of the subgenus *Culex* in southwestern Asia and Egypt (Diptera: Culicidae). *Contributions of the American Entomological Institute* 24(1): vi, 1–240.

Harbach, R.E. 1991a. A new subgenus of the genus *Sabethes* (Diptera: Culicidae). *Mosquito Systematics* 23(1): 1–9.

Harbach, R.E. 1994. The subgenus *Sabethinus* of *Sabethes* (Diptera: Culicidae). *Systematic Entomology* 19(3): 207–234.

Harbach, R.E. 1995a. A new *Sabethes* of the subgenus *Peytonulus* (Diptera: Culicidae) with an unusual fourth-instar larva. *Entomologica scandinavica* 26(1): 87–96.

Harbach, R.E. 1995b. Two new species of the subgenus *Peytonulus* of *Sabethes* (Diptera: Culicidae) from Colombia. *Memórias do Instituto Oswaldo Cruz* 90(5): 583–587.

Harbach, R.E. 2012. *Culex pipiens* Linnaeus: Species versus species complex – taxonomic history and perspective. *Journal of the American Mosquito Control Association* 28(4s): 10–23.

Harbach, R.E. 2013. *Nyctomyia*, a replacement name for the preoccupied genus *Nyx* Harbach & Linton, 2013 (Diptera: Culicidae). *Zootaxa* 3691(1): 199.

Harbach, R.E. 2018. Two new species of the subgenus *Sabethinus* of *Sabethes* (Diptera: Culicidae) from Costa Rica, first confirmation of members of the subgenus in Central America. *Zootaxa* 4429(2): 269–280.

Harbach, R.E. & Culverwell, C.L. 2014. A new subgenus and species of *Topomyia* (Diptera: Culicidae: Sabethini) based on a remarkable male mosquito from Sabah, Malaysia. *Zootaxa* 3794(4): 575–580.

Harbach, R.E. & Greenwalt, D. 2012. Two Eocene species of *Culiseta* (Diptera: Culicidae) from the Kishenehn Formation in Montana. *Zootaxa* 3530: 25–34.

Harbach, R.E. & Howard, T.M. 2002. *Sabethes (Peytonulus) paradoxus*, a new species of Sabethini (Diptera: Culicidae) from Panama. *Proceedings of the Entomological Society of Washington* 104(2): 363–372.

Harbach, R.E. & Howard, T.M. 2007. Corrections in the status and rank of names used to denote varietal forms of mosquitoes (Diptera: Culicidae). *Zootaxa* 1542: 35–48.

Harbach, R.E. & Howard, T.M. 2009. Review of the genus *Chagasia* (Diptera: Culicidae: Anophelinae). *Zootaxa* 2210: 1–25.

Harbach, R.E. & Kitching, I.J. 1998. Phylogeny and classification of the Culicidae (Diptera). *Systematic Entomology* 23(4): 327–370.

Harbach, R.E. & Kitching, I.J. 2005. Reconsideration of anopheline phylogeny (Diptera: Culicidae: Anophelinae) based on morphological data. *Systematics and Biodiversity* 3(4): 345–374.

Harbach, R.E. & Kitching, I.J. 2015, 2016. The phylogeny of Anophelinae revisited: inferences about the origin and classification of *Anopheles* (Diptera: Culicidae). *Zoologica Scripta* (2015) doi:10.1111/zsc.12137; (2016) 45(1): 34–47.

Harbach, R.E. & Navarro, J.C. 1996. A new species of *Anopheles* subgenus *Kerteszia* (Diptera: Culicidae) from Venezuela. *Entomologica scandinavica* 27(2): 207–216.

Harbach, R.E. & Petersen, J.L. 1992. Two species previously confused under the concept of *Sabethes tarsopus* in Central America (Diptera: Culicidae). *Mosquito Systematics* 24(2): 102–123.

Harbach, R.E. & Peyton, E.L. 1990. A new subgenus in *Wyeomyia* (Diptera: Culicidae), with the reclassification and redescription of the type species, *Sabethes fernandezyepezi*. *Mosquito Systematics* 22(1): 15–23.

Harbach, R.E. & Peyton, E.L. 1991. Transfer of the subgenus *Davismyia* from *Wyeomyia* to *Sabethes* and description of the type species, *Miamyia petrocchiae* (Diptera: Culicidae). *Mosquito Systematics* (for 1990) 22(3): 149–159.

Harbach, R.E. & Peyton, E.L. 1992a. A new subgenus of *Wyeomyia* (Diptera: Culicidae), with the reclassification and redescription of *Wyeomyia (Davismyia) arborea*, *Wyeomyia (Dendromyia) tarsata* and *Sabethes (Sabethes) carrilloi*. *Mosquito Systematics* (for 1991) 23(2): 92–109.

Harbach, R.E. & Peyton, E.L. 1992b. A new subgenus of *Culex* in the Neotropical Region (Diptera: Culicidae). *Mosquito Systematics* 24(3): 242–252.

Harbach, R.E. & Peyton, E.L. 2000. Systematics of *Onirion*, a new genus of Sabethini (Diptera: Culicidae) from the Neotropical Region. *Bulletin of the Natural History Museum Entomology Series* 69(2): 115–169.

Harbach, R.E. & Rattanarithikul, R. 1988. A new species of the subgenus *Eumelanomyia* of *Culex* (Diptera: Culicidae) from Thailand. *Mosquito Systematics* 20(1): 69–76.

Harbach, R.E. & Taai, K. 2014. *Nyctomyia biunguiculata*, a new cavernicolous species of tribe Aedini (Diptera: Culicidae) from southern Thailand. *Zootaxa* 3895(3): 427–432.

Harbach, R.E. & Ward, R.A. 1994. Lectotype designation for *Anopheles (Cellia) rennellensis* Taylor and Maffi (Diptera: Culicidae). *Mosquito Systematics* 26(2): 73–74.

Harbach, R.E., Culverwell, C.L. & Kitching, I.J. 2017. Phylogeny of the nominotypical subgenus of *Culex* (Diptera: Culicidae): insights from analyses of anatomical data into interspecific relationships and species groups in an unresolved tree. *Systematics and Biodiversity* 15(4): 296–306.

Harbach, R.E., Dahl, C. & White, G.B. 1985. *Culex (Culex) pipiens* Linnaeus (Diptera: Culicidae): concepts, type designations, and description. *Proceedings of the Entomological Society of Washington* 87(1): 1–24.

Harbach, R.E., Garros, C., Nguyen, D.M. & Manguin, S. 2007. Formal taxonomy of species C of the *Anopheles minimus* sibling species complex (Diptera: Culicidae). *Zootaxa* 1654: 41–54.

Harbach, R.E., Harrison, B.A. & Gad, A.M. 1984. *Culex (Culex) molestus* Forskål (Diptera: Culicidae): neotype designation, description, variation, and taxonomic status. *Proceedings of the Entomological Society of Washington* 86(3): 521–542.

Harbach, R.E., Kitching, I.J., Culverwell, C.L., Dubois, J. & Linton, Y.-M. 2012. Phylogeny of mosquitoes of tribe Culicini (Diptera: Culicidae) based on morphological diversity. *Zoologica Scripta* 41(5): 499–514, 8 online figures and 2 appendices.

Harbach, R.E., Kitching, I.J., Culverwell, C.L., Howard, T.M. & Linton, Y.-M. 2013. *Nyx pholeocola*, a new genus and cavernicolous species of tribe Aedini (Diptera: Culicidae) from southern Thailand based on morphological and molecular data. *Zootaxa* 3683(2): 159–177.

Harbach, R.E., Parkin, E., Chen, B. & Butlin, R.K. 2006. *Anopheles (Cellia) minimus* Theobald (Diptera: Culicidae): neotype designation, characterization, and systematics. *Proceedings of the Entomological Society of Washington* 108(1): 198–209.

Harbach, R.E., Peyton, E.L. & Harrison, B.A. 1984. A new species of *Culex (Melanoconion)* from southern South America (Diptera: Culicidae). *Mosquito Systematics* 16(2): 185–200.

Harbach, R.E., Peyton, E.L. & Jakob, W.L. 1984. Synonymy of *Culex (Culex) oswaldoi* with *Culex (Culex) maxi* (Diptera, Culicidae). *Mosquito Systematics* (for 1983) 15(4): 310–317.

Harbach, R.E., Rattanarithikul, R. & Harrison, B.A. 2005. *Baimaia*, a new subgenus for *Anopheles kyondawensis* Abraham, a unique crabhole-breeding anopheline in southeastern Asia. *Proceedings of the Entomological Society of Washington* 107(4): 750–761.

Harbach R.E., Rattanarithikul, R. & Harrison, B.A. 2017. *Anopheles prachongae*, a new species of the Gigas Complex of subgenus *Anopheles* (Diptera: Culicidae) in Thailand, contrasted with known forms of the complex. *Zootaxa* 4236(1): 65–81.

Harbach, R.E., Rattanarithikul, R., Howard, T.M., Linton, Y.-M. & Kitching, I.J. 2007. Systematics of a new genus and cavernicolous species of the mosquito tribe Aedini (Diptera: Culicidae) from Thailand. *Proceedings of the Entomological Society of Washington* 109(2): 469–488.

Harbach, R.E., Vu, D.H. & Kitching, I.J. 2007. Systematics of *Kimia*, a new genus of Sabethini (Diptera: Culicidae) in the Oriental Region. *Proceedings of the Entomological Society of Washington* 109(1): 102–120.

Harris, T.W. 1835. VIII.—Insects, pp. 553–602. *In*: Hitchcock, E., *Report on the geology, mineralogy, botany, and zoology of Massachusetts*. Second edition. J.S. and C. Adams, Amherst, Massachusetts.

Harrison, B.A. 1973. *Anopheles (An.) reidi*, a new species of the *barbirostris* species complex from Sri Lanka (Diptera: Culicidae). *Proceedings of the Entomological Society of Washington* 75(3): 365–371.

Harrison, B.A. 1980. Medical entomology studies – XIII. The Myzomyia Series of *Anopheles (Cellia)* in Thailand, with emphasis on intra-interspecific variations (Diptera: Culicidae). *Contributions of the American Entomological Institute* 17(4): iv, 1-195.

Harrison, B.A. 1987. *Culex* subgenus *Thaiomyia* Bram, a synonym of *Culex* subgenus *Culiciomyia* Theobald (Diptera: Culicidae). *Mosquito Systematics* 19(1): 111–116.

Harrison, B.A. & Scanlon, J.E. 1974. *Anopheles (An.) pilinotum*, a new species name in the *aitkenii* complex for *An. insulaeflorum* from the Philippines and eastern Indonesia (Diptera: Culicidae). *Mosquito Systematics* 6(1): 32–40.

Harrison, B.A. & Scanlon, J.E. 1975. Medical entomology studies – II. The subgenus *Anopheles* in Thailand (Diptera: Culicidae). *Contributions of the American Entomological Institute* 12(1): iv, 1–307.

Harrison, B.A., Rattanarithikul, R., Peyton, E.L. & Mongkolpanya, K. 1991. Taxonomic changes, revised occurrence records and notes on the Culicidae of Thailand and neighboring countries. *Mosquito Systematics* (for 1990) 22(3): 196–227.

Harrison, B.A., Scanlon, J.E. & Reid, J.A. 1973. A new synonymy and new species name in the Southeast Asian *Anopheles hyrcanus* complex. *Mosquito Systematics* 5(4): 263–268.

Hasselquist, F. 1757. *Frederic Hasselquists ... Iter Palæstinum, eller Resa til Heliga Landet, förrättad ifrån År 1749 Til 1752, med beskrifningar, rön, anmårkningar, öfver de mårkvårdigaste Naturalier ... utgifven af Carl Linnæus*. Trykt på Lars Salvii, Stockholm. Foreword [14 pp.], 619 pp., content [1 p.].

Hatori, S. 1901. Species of *Anopheles* from Formosa. Kwampô (Official Reports of the Japanese Government) no. 5534, pp. 274–275.

Haworth, W.E. 1924. Mosquitoes and coconut palms. A mosquito survey of palm trees in East Africa and the problems resulting therefrom. *Transactions of the Royal Society of Tropical Medicine and Hygiene* 18(4): 162–198.

Hearle, E. 1923. A new mosquito from British Columbia (Culicidae, Diptera). *Canadian Entomologist* 55(1): 4–5.

Hearle, E. 1927. A new Canadian mosquito (Culicidae). *Canadian Entomologist* 59(5): 101–103.

Heinemann, S.J. & Belkin, J.N. 1978. Collection records of the project "Mosquitoes of Middle America" 11. Venezuela (VZ); Guianas: French Guiana (FG, FGC), Guyana (GUY), Surinam (SUR). *Mosquito Systematics* 10(3): 365–459.

Herrick, G.W. 1905. Notes on some Mississippi mosquitoes. *Entomological News and Proceedings of the Entomological Section of the Academy of Natural Sciences of Philadelphia* 16(9): 281–283, 1 pl.

Hervé, J.-P. & Geoffroy, B. 1974. Description d'un nouveau moustique de République Centrafricaine *Culex (Neoculex) subsalisburiensis* n.sp. *Cahiers ORSTOM Série Entomologie médicale et Parasitologie* 12(3): 191–193.

Hii, J.L.K., Peyton, E.L. & Shang, V.Y. 1988. Redescription of the adult and first descriptions of the larva and pupa of *Anopheles (Cellia) sulawesi* Waktoedi, a species of the Leucosphyrus Group from Sulawesi, Indonesia (Diptera: Culicidae). *Mosquito Systematics* 20(1): 41–54.

Hill, E. & Haydon, L.G. 1907. A contribution to the study of the characteristics of larvae of species of *Anophelina* in South Africa. *Annals of the Natal Government Museum* 1(2): 111–157, 12 pls.

Hill, R.B. & Hill, C.M. 1946. *Phoniomyia hirsuta*, a new sabethine from Jamaica (Diptera, Culicidae). *Proceedings of the Entomological Society of Washington* 48(2): 39–41.

Hitchcock, E. 1835. *Report on the geology, mineralogy, botany, and zoology of Massachusetts*. Second edition. J.S. and C. Adams, Amherst, Massachusetts. xii, 692 pp.

Ho, C. 1938. A new tree-hole breeding *Anopheles* from the island of Hainan. *Annals of Tropical Medicine and Parasitology* 32(3): 279–285.

Ho, K.-M. 1963. Report of a new mosquito, *Culex cheni* sp. nov. [in Chinese]. *Acta Entomologica Sinica* 12(3): 368–371.

Hoffmann, C.C. 1935. La formacion de razas en los *Anopheles* mexicanos. I. *A. maculipennis* y *A. quadrimaculatus* y una raza nueva del *maculipennis*. *Anales del Instituto de Biología, Universidad nacional Autónoma de México* 6: 3–22.

Hoffmann, C.C. 1938. La formacion de razas en los *Anopheles* mexicanus. II. *Anopheles albimanus* y sus variedades en la Republica Mexicana. *Anales del Instituto de Biología, Universidad nacional Autónoma de México* 9: 167–180.

Hogue, C.L. 1975. A new species of bromeliad-breeding *Culex (Culex)* from Cocos Island. *Mosquito Systematics* 7(4): 357–362.

Holstein, M. 1949. Le Problème d'Anophélès Gambiae [*sic*]. *Bulletin médical de l'Afrique occidentale française* Numéro Spécial: 155–160, 3 pls.

Holstein, M.H. 1951. Note sur l'épidémiologie du paludisme en Afrique-Occidentale française. *Bulletin of the World Health Organisation* 4: 463–473.

Hopkins, G.H.E. 1936. *Mosquitoes of the Ethiopian Region I.–Larval bionomics of mosquitoes and taxonomy of culicine larvae*. Printed by order of the Trustees, British Museum (Natural History), London. [vi], 250 pp.

Howard, L.O. 1901. *Mosquitoes: how they live; how they carry disease; how they are classified; how they may be destroyed*. McClure, Phillips & Co., New York. xv, 241 pp.

Howard, L.O., Dyar, H.G. & Knab, F. 1913. *The mosquitoes of North and Central America and the West Indies. Volumes 1 and 2*. Carnegie Institution of Washington Publication No. 159 [for 1912]. The Lord Baltimore Press, Baltimore. vii, 520 pp.; x, 150 pls.

Howard, L.O., Dyar, H.G. & Knab, F. 1915. *The mosquitoes of North and Central America and the West Indies. Volume three. Systematic description (in two parts). Part I*. Publication No. 159. Carnegie Institution of Washington, Washington, DC. vi, 523 pp.

Howard, L.O., Dyar, H.G. & Knab, F. 1917. *The mosquitoes of North and Central America and the West Indies. Volume four. Systematic description (in two parts). Part II*. Publication No. 159. Carnegie Institution of Washington, Washington, DC. pp. 525–1064.

Hsiao, C.-E. 1977. A new species of *Aedes* (Diptera: Culicidae) [in Chinese]. *Acta Entomologica Sinica* 20(4): 451–454.
Hsieh, L.-K. & Liao, T.-H. 1956. A list of Amoy mosquitoes with the description of a new species and a new variety [in Chinese]. *Acta Entomologica Sinica* 6(3): 123–127.
Hsü, C.-F. 1963. A new species of *Culex* from Hainan Island [in Chinese]. *Acta Entomologica Sinica* 12(2): 229–232.
Hsu, K.-C. 1964. Description of a new mosquito *Tripteroides* (*Tripteroides*) *szechwanensis* sp. nov. [in Chinese]. *Acta Entomologica Sinica* 13(2): 278–282.
Huang, Y.-M. 1968a. *Aedes* (*Verrallina*) of the Papuan Subregion (Diptera: Culicidae). *Pacific Insects Monograph* 17: 1–73.
Huang, Y.-M. 1968b. A new subgenus of *Aedes* (Diptera, Culicidae) with illustrated key to the subgenera of the Papuan Subregion (Diptera: Culicidae). *Journal of Medical Entomology* 5(2): 169–188.
Huang, Y.-M. 1969. A new species of *Aedes* (*Stegomyia*) from Thailand (Diptera: Culicidae). *Proceedings of the Entomological Society of Washington* 71(2): 234–239.
Huang, Y.-M. 1972. Contributions to the mosquito fauna of Southeast Asia. XIV. The subgenus *Stegomyia* of *Aedes* in Southeast Asia. I – The *scutellaris* group of species. *Contributions of the American Entomological Institute* 9(1): 1–109.
Huang, Y.-M. 1973. A new species of *Aedes* (*Stegomyia*) from Thailand and notes on the Mediopunctatus Subgroup (Diptera: Culicidae). *Proceedings of the Entomological Society of Washington* 75(2): 224–232.
Huang, Y.-M. 1974. A new species of *Aedes (Stegomyia)* from the Andaman Islands (Diptera: Culicidae). *Mosquito Systematics* 6(2): 137–141.
Huang, Y.-M. 1975. A new species of *Aedes (Stegomyia)* from Sri Lanka (Ceylon) (Diptera: Culicidae). *Mosquito Systematics* 7(4): 345–356.
Huang, Y.-M. 1977. Medical entomology studies – VII. The subgenus *Stegomyia* of *Aedes* in Southeast Asia. II – The *edwardsi* group of species. III – The *w-albus* group of species. (Diptera: Culicidae). *Contributions of the American Entomological Institute* 14(1): 1–111.
Huang, Y.-M. 1979. Medical entomology studies – XI. The subgenus *Stegomyia* of *Aedes* in the Oriental region with keys to the species (Diptera: Culicidae). *Contributions of the American Entomological Institute* 15(6): 1–79.
Huang, Y.-M. 1981. A redescription of *Aedes (Stegomyia) calceatus* Edwards and description of a new Afrotropical species, *Aedes (Stegomyia) ledgeri* (Diptera: Culicidae). *Mosquito Systematics* 13(1): 92–113.
Huang, Y.-M. 1985. A new African species of *Aedes* (Diptera: Culicidae). *Mosquito Systematics* 17(2): 108–120.
Huang, Y.-M. 1986a. Notes on the *Aedes* (*Diceromyia*) *furcifer* group, with a description of a new species (Diptera: Culicidae). *Proceedings of the Entomological Society of Washington* 88(4): 634–649.
Huang, Y.-M. 1986b. *Aedes* (*Stegomyia*) *corneti,* a new species of the *africanus* subgroup (Diptera: Culicidae). *Proceedings of the Entomological Society of Washington* 88(4): 764–776.
Huang, Y.-M. 1988a. *Aedes* (*Stegomyia*) *josiahae*, a new species of the *simpsoni* subgroup (Diptera: Culicidae). *Proceedings of the Entomological Society of Washington* 90(2): 155–163.
Huang, Y.-M. 1988b. The *Aedes (Stegomyia) pseudonigeria* group with emphasis on the species from the Afrotropical Region (Diptera: Culicidae). *Mosquito Systematics* 20(1): 1–20.
Huang, Y.-M. 1990. The subgenus *Stegomyia* of *Aedes* in the Afrotropical Region I. The *africanus* group of species (Diptera: Culicidae). *Contributions of the American Entomological Institute* 26(1): 1–90.
Huang, Y.-M. 1994. *Aedes* (*Stegomyia*) *mattinglyorum*, a new species of the *dendrophilus* group (Diptera: Culicidae). *Proceedings of the Entomological Society of Washington* 96(1): 11–21.
Huang, Y.-M. 1997. The subgenus *Stegomyia* of *Aedes* in the Afrotropical Region. II. The *dendrophilus* group of species (Diptera: Culicidae). *Contributions of the American Entomological Institute* 29(4): 1–104.
Huang, Y.-M. 2004. The subgenus *Stegomyia* of *Aedes* in the Afrotropical Region with keys to the species (Diptera: Culicidae). *Zootaxa* 700: 1–120.
Huang, Y.-M. 2005. *Cornetius*, a new subgenus of *Aedes*, and a redescription of *Aedes* (*Cornetius*) *cozi* Cornet (Diptera: Culicidae). *Proceedings of the Entomological Society of Washington* 107(3): 517–529.
Huang, Y.-M. & Hitchcock, J.C. 1980. Medical entomology studies – XII. A revision of the *Aedes scutellaris* group of Tonga (Diptera: Culicidae). *Contributions of the American Entomological Institute* 17(3): 1–107.
Huang, Y.-M. & Rueda, L.M. 2013. A new species of the genus *Orthopodomyia* from Ototomo, Cameroon, with notes on other *Orthopodomyia* group-species. *Journal of the American Mosquito Control Association* 29(3): 270–274.
Huang, Y.-M. & Rueda, L.M. 2015. *Pseudalbuginosus*, a new subgenus of *Aedes*, and a redescription of *Aedes* (*Pseudalbuginosus*) *grjebinei* Hamon, Taufflieb, and Maillot (Diptera: Culicidae). *Proceedings of the Entomological Society of Washington* 117(3): 381–388.
Huang, Y.-M., Mathis, W.N. & Wilkerson, R.C. 2010. *Coetzeemyia*, a new subgenus of *Aedes,* and a redescription of the holotype female of *Aedes* (*Coetzeemyia*) *fryeri* (Theobald) (Diptera: Culicidae). *Zootaxa* 2638: 1–24.
Hutchings, R.S.G. & Sallum, M.A.M. 2008. Two new species of *Culex* subgenus *Melanoconion* (Diptera: Culicidae) from the Amazon forest. *Zootaxa* 1920: 41–50.
Hutton, F.W. 1902. Additions to the *Diptera* fauna of New Zealand. *Transactions and Proceedings of the New Zealand Institute* (for 1901) 34: 179–196.
Iglisch, I. 1977. „Hausmücken" (*Culex-pipiens*-Gruppe) als mögliche Überträger des Serumhepatitis-Virus. Untersuchungen zur Genitalmorphologie. *Zeitschrift für angewandte Zoologie* 64(3): 257–290.
IMPE [Institute of Malariology, Parasitology and Entomology]. 1987. *Keys to the* Anopheles *in Vietnam (adults–pupae–larvae)* [in Vietnamese]. Department of Entomology, Institute of Malariology, Parasitology and Entomology, Hanoi. 64 pp.
Ingram, A. & de Meillon, B. 1927. A mosquito survey of certain parts of South Africa, with special reference to the carriers of malaria and their control. (Part I). *Publications of the South African Institute for Medical Research* 4(4): 1–81, 15 pls.
International Commission on Zoological Nomenclature. 1959. Opinion 548. Validation under the plenary powers of the generic name "*Toxorhynchites*" Theobald (Class Insecta, Order Diptera), as published in 1901 in the "Journal of Tropical Medicine". *Opinions and declarations rendered by the International Commission on Zoological Nomenclature* 20(14): 167–174.

International Commission on Zoological Nomenclature. 1964. Opinion 711. *Culex aegypti* Linnaeus, 1762 (Insecta, Diptera): validated and interpreted under the plenary powers. *Bulletin of Zoological Nomenclature* 21(4): 246–248.

International Commission on Zoological Nomenclature. 1965. Opinion 737. *Bironella gracilis* Theobald, 1905 (Insecta, Diptera): validated under the plenary powers. *Bulletin of Zoological Nomenclature* 22(3): 164–166.

International Commission on Zoological Nomenclature. 1968. Opinion 850. *Anopheles africanus* Theobald, 1901 (Insecta, Diptera): suppressed under the plenary powers. *Bulletin of Zoological Nomenclature* 25(1): 22–23.

International Commission on Zoological Nomenclature. 1969. Opinion 874. *Cellia errabunda* Swellengrebel, 1925 (Insecta, Diptera): suppressed under the plenary powers. *Bulletin of Zoological Nomenclature* 26(1): 10–11.

International Commission on Zoological Nomenclature. 1972. Opinion 979. *Culex albirostris* Macquart, 1850 (Insecta, Diptera): suppressed under the plenary powers. *Bulletin of Zoological Nomenclature* 29(1): 11–12.

International Commission on Zoological Nomenclature. 1982. Opinion 1213. *Toxorhynchites brevipalpis* Theobald, 1901 (Insecta, Diptera): conserved. *Bulletin of Zoological Nomenclature* 39(2): 122–123.

International Commission on Zoological Nomenclature. 1985. *International code of zoological nomenclature*. Third edition. International Trust for Zoological Nomenclature, London. xx, 338 pp.

International Commission on Zoological Nomenclature. 1991. Opinion 1644. *Culex stigmatosoma* Dyar, 1907 and *C. thriambus* Dyar, 1921 (Insecta, Diptera): specific names conserved. *Bulletin of Zoological Nomenclature* 48(2): 179–180.

International Commission on Zoological Nomenclature. 1999. *International code of zoological nomenclature*. Fourth Edition. International Trust for Zoological Nomenclature, London. xxix, 306 pp.

Iyengar, M.O.T. 1924. On the Indian anophelines of the *funestus* group and the description of a new species (Diptera, Culicidae). *Indian Journal of Medical Research* 12: 23–29, 1 pl.

Iyengar, M.O.T. 1929. Adult and larval stages of *Anopheles majidi*. *Indian Journal of Medical Research* 17(1): 1–10, 1 pl.

Iyengar, M.O.T. 1955. Distribution of mosquitoes in the South Pacific Region. *South Pacific Commission Technical Paper* 86: v, 47 pp.

Iyengar, M.O.T. & Menon, M.A.U. 1955. Mosquitos of the Maldive Islands. *Bulletin of Entomological Research* 46(1): 1–10.

Izquierdo, J.J. 1916. Investigaciones sobre paludismo en Puebla. PhD Tesis, Colegio del Estado de Puebla, México. 101 pp.

James, S.P. 1902. *Malaria in India*. Scientific Memoirs by Officers of the Medical and Sanitary Department of the Government of India. (New Series) No. 2. Office of the Superintendent of Government Printing, Calcutta. 106 pp.

James, S.P. 1910a. Notes on mosquitoes. *Paludism* 1: 33–38.

James, S.P. 1910b. A new arrangement of the Indian *Anophelinae*. *Records of the Indian Museum* 49(5): 95–109, 4 pls.

James, S.P. & Liston, W.G. 1904. *A monograph of the* Anopheles *mosquitoes of India*. Thacker, Spink & Co., Calcutta. [vi], 132 pp., 15 pls.

James, S.P. & Liston, W.G. 1911. *A monograph of the anopheline mosquitoes of India*. Second edition. Thacker, Spink and Co., Calcutta. 128 pp., 15 pls.

Jeffery, J., Oothuman, P. & Rudnick, A. 1987. *Culex (Eumelanomyia) richardgarciai*, a new species of mosquito from peninsular Malaysia (Diptera: Culicidae). *Mosquito Systematics* 19(2): 139–142.

Jenkins, D.W. & Carpenter, S.J. 1946. Ecology of the tree hole breeding mosquitoes of Nearctic North America. *Ecological Monographs* 16(1): 33–47.

Judd, D.D. 1996. Review of the systematics and phylogenetic relationships of the Sabethini (Diptera: Culicidae). *Systematic Entomology* 21(2): 129–150.

Judd, D.D. 1998. Review of a bromeliad-ovipositing lineage in *Wyeomyia* and the resurrection of *Hystatomyia* (Diptera: Culicidae). *Annals of the Entomological Society of America* 91(5): 572–589.

Kamimura, K. 1966. Comments on the distribution and taxonomy of *Anopheles* in Japan [in Japanese]. Abstracts of Speeches, 21st Annual Meeting, Western Japan Branch, Japan Society of Sanitary Zoology, p. 8.

Kamimura, K. 1968. The distribution and habit of medically important mosquitoes of Japan [in Japanese]. *Japanese Journal of Sanitary Zoology* 19(1): 15–34.

Kamimura, K. & Wada, Y. 1974. A new subspecies of *Culex* (*Barraudius*) *modestus* from Japan (Diptera, Culicidae). *Japanese Journal of Sanitary Zoology* 25(1): 13–20.

Kanda, T. & Oguma, Y. 1978. *Anopheles engarensis*, a new species related to *sinensis* from Hokkaido Island, Japan. *Mosquito Systematics* 10(1): 45–52.

Kang, W., Tan, J., Cao, C., Cheng, H., Yang, Y. & Huang, R. 1984. *Anopheles liangshanensis*, a new species related to *A. kweiyangensis* and *A. sinensis* from Sichuan, China (Diptera: Culicidae) [in Chinese]. *Sichuan Journal of Zoology* 3(2): 11–14.

Kano, R. & Hayashi, S. 1949. Studies on a mosquito, *Mansonia* (*Coquillettidia*) *ochracea* (Theobald). I. Notes on the adult and egg stages [in Japanese]. *Kontyû* 17(3): 23–26.

Kano, R., Nitahara, M. & Awaya, J. 1954. Description of a new mosquito, *Culex* (*Culiciomyia*) *sasai* n. sp., collected in the south-western part of Japan (Culicidae, Diptera) [in Japanese]. *Japanese Journal of Sanitary Zoology* 5(1–2): 14–20.

Karsch, F. 1887. Dipterologisches von der Delagoabei. *Entomologische Nachrichten* 13(2): 22–26.

Keating, W.H. (main author). 1824. *Narrative of an expedition to the source of St. Peter's River, Lake Winnepeek, Lake of the Woods, &c. &c. performed in the year1823, by order of the Hon. J. C. Calhoun, Secretary of War, under the command of Stephen H. Long, Major U.S. T.E.: compiled from the notes of Major Long, Messrs. Say, Keating, & Colhoun*. Volume II. H.C. Carey & I. Lea, Philadelphia. 459 pp.

Kengne, P., Antonio-Nkondjio, C., Awono-Ambene, H.P., Simard, F., Awolola, T.S. & Fontenille, D. 2007. Molecular differentiation of three closely related members of the mosquito species complex, *Anopheles moucheti*, by mitochondrial and ribosomal DNA polymorphism. *Medical and Veterinary Entomology* 21(2): 177–182.

Keshishian, M. 1938. A new species of the anopheline mosquito *A. sogdianus* sp. nov. in Tajikistan [in Russian]. *Meditsinskaya Parazitologiya i Parazitarnye Bolezni* 7(6): 888–896.

King, W.V. 1931. The Philippine varieties of *Anopheles gigas* and *Anopheles lindesayi*. *Philippine Journal of Science* 46(4): 751–757, 2 pls.

King, W.V. 1932a. The Philippine *Anopheles* of the *rossi-ludlowi* group. *Philippine Journal of Science* 47(3): 305–342, 4 pls.

King, W.V. 1932b. Three Philippine *Anopheles* of the *funestus-minimus* subgroup. *Philippine Journal of Science* 48(4): 485–523, 2 pls.

King, W.V. 1939. Varieties of *Anopheles crucians* Wied. *American Journal of Tropical Medicine* 19(5): 461–471.

King, W.V. & Baisas, F.E. 1936. A new species and a new variety of Philippine *Anopheles* related to *Anopheles leucosphyrus* Dönitz. *Proceedings of the Entomological Society of Washington* 38(5): 79–89.

King, W.V. & Bradley, G.H. 1937. Notes on *Culex erraticus* and related species in the United States (Diptera, Culicidae). *Annals of the Entomological Society of America* 30(2): 345–357.

King, W.V. & Bradley, G.H. 1941. General morphology of *Anopheles* and classification of the Nearctic species. *Publications of the American Association for the Advancement of Science* 15: 63–70.

King, W.V. & Hoogstraal, H. 1946a. The New Guinea species of *Culex* (*Culiciomyia*), with descriptions of two new species. *Proceedings of the Biological Society of Washington* 59: 143–153.

King, W.V. & Hoogstraal, H. 1946b. Two new species of *Aedes* (*Finlaya*) from New Guinea (Diptera: Culicidae). *Proceedings of the Entomological Society of Washington* 48(2): 37–38.

King, W.V. & Hoogstraal, H. 1946c. Three new species of *Aedes* from Netherlands New Guinea. *Proceedings of the Entomological Society of Washington* 48(4): 95–106.

King, W.V. & Hoogstraal, H. 1946d. Species of *Aedes* (*Finlaya*) of the *papuensis* group in the Australasian Region (Diptera, Culicidae). *Proceedings of the Entomological Society of Washington* 48(6): 135–157.

King, W.V. & Hoogstraal, H. 1946e. Two new species of mosquitoes of the genus *Ficalbia* from Netherlands New Guinea. *Proceedings of the Entomological Society of Washington* 48(7): 186–190.

King, W.V. & Hoogstraal, H. 1946f. Descriptions of three new species of mosquitoes of the genus *Aedes*, subgenus *Finlaya*, from New Guinea. *Journal of the Washington Academy of Sciences* 36(9): 305–313.

King, W.V. & Hoogstraal, H. 1947a. New species of New Guinea *Uranotaenia* of the *tibialis* group (Diptera: Culicidae). *Annals of the Entomological Society of America* (for 1946) 39(4): 585–596.

King, W.V. & Hoogstraal, H. 1947b. New Guinea species of mosquitoes of the genus *Aedes*, subgenus *Aedes*. *Journal of the Washington Academy of Sciences* 37(4): 113–134.

King, W.V. & Hoogstraal, H. 1947c. Two new species of *Culex* (*Neoculex*) from New Guinea (Diptera, Culicidae). *Proceedings of the Entomological Society of Washington* 49(2): 65–69.

King, W.V. & Hoogstraal, H. 1955. Three new species of New Guinea *Culex*, subgenus *Lophoceraomyia*, with notes on other species (Diptera, Culicidae). *Proceedings of the Entomological Society of Washington* 57(1): 1–11.

Kirby, W. 1837. Fauna Boreali-Americana. Part the fourth and last. The Insects, xxxix, 325 pp., 13 pls, errata. *In*: Richardson, J. (ed.), *Fauna Boreali-Americana; or the Zoology of the northern parts of British America: containing descriptions of the objects of natural history collected on the late Northern Land Expeditions, under command of Captain Sir John Franklin, R.N.* Josiah Fletcher, Norwich, U.K.

Kirkpatrick, T.W. 1925. Preliminary notes on the determination of Egyptian mosquitos, with descriptions of three species new to science. *Bulletin Société Entomologique d'Égypte* (for 1924) 8: 362–391.

Kirkpatrick, T.W. 1926. *Culex* (*Lasiosiphon*) *Adairi*, Nom. Nov. *Bulletin Société Entomologique d'Égypte* 10: 112.

Kitzmiller, J.B. 1982. *Anopheline names: their derivations and histories*. The Thomas Say Foundation. Volume VIII. Entomological Society of America, College Park, Maryland. vi, 639 pp.

Klein, J.M. 1969. A new species of *Ficalbia* (*Ravenalites*) from Cambodia (Diptera: Culicidae). *Proceedings of the Entomological Society of Washington* 71(2): 215–217.

Klein, J.M. 1970. A new species of *Culex* (*Neoculex*) from Cambodia (Diptera: Culicidae). *Proceedings of the Entomological Society of Washington* 72(4): 504–506.

Klein, J.M. 1973. Contributions to the mosquito fauna of Southeast Asia. XVII. The Cambodian *Aedes (Neomacleaya)* species with some new descriptions (Diptera: Culicidae). *Contributions of the American Entomological Institute* 10(1): 1–21.

Klein, J.-M. 1977. Deux nouvelles espèces de *Topomyia* du Cambodge *(Diptera, Culicidae)*. *Cahiers ORSTOM Série Entomologie médicale et Parasitologie* 15(2): 123–129.

Klein, J.-M. & Marks, E.N. 1960. Australian mosquitoes described by Macquart. I. Species in the Paris Museum, *Aëdes* (*Finlaya*) *alboannulatus* (Macquart), *Aëdes* (*Finlaya*) *rubrithorax* (Macquart), *Aëdes* (*Ochlerotatus*) *albirostris* (Macquart). New synonymy and a new species from New Zealand. *Proceedings of the Linnean Society of New South Wales* 85(1): 107–116.

Klein, J.M. & Sirivanakarn, S. 1970. Four new species of *Culex*, subgenus *Mochthogenes* from Southeast Asia (Diptera: Culicidae). *Proceedings of the Entomological Society of Washington* (for 1969) 71(4): 582–592.

Klein, J.-M., Rivière, F. & Séchan, Y. 1984. Une nouvelle espèce de moustique (Diptera: Culicidae) des îles Marquises (Polynésie Française): *Culex (Culex) toviiensis* n. sp. Description de la larve et de la nymphe. *Cahiers ORSTOM Série Entomologie médicale et Parasitologie* (for 1983) 21(2): 71–76.

Knab, F. 1907a. A new species of *Megarhinus*. *Canadian Entomologist* 39(2): 50–51.

Knab, F. 1907b. A new genus and species of sabethid mosquito. *Journal of the New York Entomological Society* 15(3): 120–121.

Knab, F. 1909. Two new Philippine Culicidae. *Entomological News* 20(9): 386–388.

Knab, F. 1913. A new bromelicolous *Megarhinus* (*Diptera; Culicidae*). *Insecutor Inscitiae Menstruus* 1(3): 35–36.

Knab, F. 1916. A new mosquito from the eastern United States. *Proceedings of the Biological Society of Washington* 29: 161–163.

Knight, K.L. 1946. The *Aedes* (*Finlaya*) *niveus* subgroup of Oriental mosquitoes. *Journal of the Washington Academy of Sciences* 36(8): 270–280.

Knight, K.L. 1947a. The *Aedes* (*Mucidus*) mosquitoes of the Pacific (Diptera: Culicidae). *Journal of the Washington Academy of Sciences* 37(9): 315–325.

Knight, K.L. 1947b. A new species of *Aedes* (*Christophersiomyia*) from the Philippines (Diptera: Culicidae). *Proceedings of the Biological Society of Washington* 60: 73–76.

Knight, K.L. 1948a. The *Aedes* (*Finlaya*) *chrysolineatus* group of mosquitoes (Diptera: Culicidae). *Annals of the Entomological Society of America* (for 1947) 40(4): 624–649.

Knight, K.L. 1948b. The *Aedes* (*Finlaya*) *albotaeniatus* group of mosquitoes (Diptera, Culicidae). *Proceedings of the Entomological Society of Washington* 50(1): 1–8.

Knight, K.L. 1948c. A new name for an *Aedes* (*Mucidus*) subspecies of the Solomon Islands, with two additional notes on *Aedes* (*Mucidus*) species (Diptera, Culicidae). *Proceedings of the Entomological Society of Washington* 50(3): 68–69.

Knight, K.L. 1953. Two new species of mosquitoes from the Yemen (Diptera: Culicidae). *Journal of the Washington Academy of Sciences* 43(10): 320–325.

Knight, K.L. 1968. Contributions to the mosquito fauna of Southeast Asia. – IV. Species of the subgroup *chrysolineatus* of group D, genus *Aedes*, subgenus *Finlaya* Theobald. *Contributions of the American Entomological Institute* 2(5): 1–45.

Knight, K.L. 1969. A new species of the genus *Aedes*, subgenus *Finlaya* Theobald, from India (Diptera: Culicidae). *Journal of the Kansas Entomological Society* 42(4): 382–386.

Knight, K.L. 1972. History of mosquito systematics. Part I. Eighteen century. *Mosquito Systematics* 4(1): 10–15.

Knight, K.L. 1978a. A new *Aedes (Finlaya)* mosquito from Thailand. *Mosquito Systematics* 10(1): 106–116.

Knight, K.L. 1978b. *Supplement to a catalog of the mosquitoes of the world (Diptera: Culicidae)*. The Thomas Say Foundation, Supplement to Volume VI. Entomological Society of America, College Park, Maryland. 107 pp.

Knight, K.L. & Abdel Malek, A.A. 1951. A morphological and biological study of *Culex pipiens* in the Cairo area of Egypt [Diptera-Culicidae]. *Bulletin de la Société Fouad 1er d'Entomologie* 35: 175–185.

Knight, K.L. & Farner, D.S. 1944. A correction in anopheline nomenclature (Diptera: Culicidae). *Proceedings of the Entomological Society of Washington* 46(5): 132–133.

Knight, K.L. & Harrison, B.A. 1988. A new *Aedes* (*Finlaya*) of the Niveus-Subgroup (Diptera: Culicidae). *Mosquito Systematics* (for 1987) 19(3): 212–236.

Knight, K.L. & Hull, W.B. 1951a. Three new species of *Aedes* from the Philippines (Diptera, Culicidae). *Pacific Science* 5(2): 197–203.

Knight, K.L. & Hull, W.B. 1951b. The *Aedes* mosquitoes of the Philippine Islands. I. Keys to species. Subgenera *Mucidus*, *Ochlerotatus,* and *Finlaya* (Diptera, Culicidae). *Pacific Science* 5(3): 211–251.

Knight, K.L. & Hull, W.B. 1952. The *Aedes* mosquitoes of the Philippine Islands. II. Subgenera *Skusea*, *Christophersiomyia*, *Geoskusea*, *Rhinoskusea*, and *Stegomyia* (Diptera, Culicidae). *Pacific Science* 6(2): 157–189.

Knight, K.L. & Hull, W.B. 1953. The *Aedes* mosquitoes of the Philippine Islands. III. Subgenera *Aedimorphus*, *Banksinella*, *Aedes*, and *Cancraëdes* (Diptera, Culicidae). *Pacific Science* 7(4): 453–481.

Knight, K.L. & Hurlbut, H.S. 1949. The mosquitoes of Ponape Island, eastern Carolines. *Journal of the Washington Academy of Sciences* 39(1): 20–34.

Knight, K.L. & Laffoon, J.L. 1946. The Oriental species of the *Aedes* (*Finlaya*) *kochi* group (Diptera: Culicidae). *Transactions of the American Entomological Society* 72(4): 203–225, 3 pls.

Knight, K.L. & Marks, E.N. 1952. An annotated checklist of the mosquitoes of the subgenus *Finlaya*, genus *Aedes*. *Proceedings of the United States National Museum* 101(3288): 513–574.

Knight, K.L. & Mattingly, P.F. 1950. The *Orthopodomyia anopheloides* subgroup of mosquitoes (Diptera, Culicidae). *Proceedings of the Entomological Society of Washington* 52(1): 1–20.

Knight, K.L. & Rozeboom, L. 1945. A new species of *Culex* from New Guinea (Diptera, Culicidae). *Proceedings of the Biological Society of Washington* 47(9): 289–295.

Knight, K.L. & Rozeboom, L.E. 1946. The *Aedes* (*Stegomyia*) *albolineatus* group (Diptera, Culicidae). *Proceedings of the Biological Society of Washington* 59: 83–98.

Knight, K.L. & Stone, A. 1977. *A catalog of the mosquitoes of the world (Diptera: Culicidae)*. Second edition. The Thomas Say Foundation. Volume VI. Entomological Society of America, College Park, Maryland. xi, 611 pp.

Koidzumi, M. 1917. On the anopheline mosquitoes from Formosa [in Japanese]. *Dobutsugaku Zassi* [*Zoological Magazine*] 29: 131–142, 351–355.

Koidzumi, M. 1918. On the *Stegomyia* mosquitoes of Taiwan [in Japanese]. *Transactions of the Natural History Society of Formosa* 8(38): 139–145.

Koidzumi, M. 1920. Preventive medical study of mosquitoes in Formosa [in Japanese]. *Daiwan Kenkyujo Hokoku* [*Report of the Formosa Government*] 8: 1–158.

Koidzumi, M. 1924. The anophelines of Formosa, pp. 96–101. *In*: Hoops, A.S. & Scharff, J.W. (eds), Far Eastern Association of Tropical Medicine–Transactions of the Fifth Biennial Congress held at Singapore, 1924. John Bale, Sons & Danielson, London.

Komp, W.H.W. 1926. A new *Culex* from Honduras. *Insecutor Inscitiae Menstruus* 14(1–3): 44–45.

Komp, W.H.W. 1932. A new *Culex*, *Culex vomerifer,* from Panama (Dipt., Culicidae). *Psyche* 39(3): 79–82.

Komp, W.H.W. 1935. Notes on the validity of the types of the species in the subgenera *Mochlostyrax* and *Melanoconion* in the U. S. National Museum. (Diptera, Culicidae.). *Proceedings of the Entomological Society of Washington* 37(1): 1–11.

Komp, W.H.W. 1936a. The male and larva of *Aedes dominicii* Rangel & Romero Sierra, and the male of *Aedes pseudodominicii* sp. nov., representatives of a new subgenus (*Soperia*) of the genus *Aedes*, from Colombia. *Proceedings of the Entomological Society of Washington* 38(4): 71–75.

Komp, W.H.W. 1936b. *Anopheles* (*Anopheles*) *chiriquiensis*, a new species of *Anopheles* from Panama (Diptera, Culicidae). *Proceedings of the Entomological Society of Washington* 38(7): 156–160.

Komp, W.H.W. 1936c. Description of nine new species of *Culex*, seven from Panama and two from Venezuela (Diptera, Culicidae). *Annals of the Entomological Society of America* 29(2): 319–334.
Komp, W.H.W. 1936d. *Anopheles* (*Nyssorhynchus*) *anomalophyllus*, a new species of *Anopheles* from Panama and Costa Rica (Dipt., Culicidae). *Proceedings of the Entomological Society of Washington* 38(7): 160–164.
Komp, W.H.W. 1937a. *Anopheles acanthotorynus*, a new species of the subgenus *Stethomyia* from Peru. (Diptera, Culicidae.). *Annals of the Entomological Society of America* 30(2): 358–360.
Komp, W.H.W. 1937b. The species of the subgenus *Kerteszia* of *Anopheles* (Diptera, Culicidae). *Annals of the Entomological Society of America* 30(3): 492–529.
Komp, W.H.W. 1938. *Aedes leucotaeniatus*, a new species of *Aedes* allied to *A. leucocelaenus* D. & S.; and descriptions of the male and larva of *A. leucocelaenus* D. & S. (Diptera, Culicidae). *Proceedings of the Entomological Society of Washington* 40(8): 260–266.
Komp, W.H.W. 1941. The species of *Nyssorhynchus* confused under *tarsimaculatus* Goeldi, and a new name, *A. emilianus*, for one species found in Para, Brazil (Diptera, Culicidae). *Annals of the Entomological Society of America* 34(4): 791–807.
Komp, W.H.W. 1943. *Anopheles clarki*, a new species of *Nyssorhynchus* of wide distribution in South America. (Diptera: Culicidae). *Proceedings of the Entomological Society of Washington* (for 1942) 44(9): 196–201.
Komp, W.H.W. 1949. *Aedes* (*Ochlerotatus*) *nubilus* Theobald, 1903 a synonym of *Aedes* (*Ochlerotatus*) *serratus* Theobald, 1901 (Diptera, Culicidae). *Proceedings of the Entomological Society of Washington* 51(3): 105–114.
Komp, W.H.W. & Brown, C.G. 1935. *Culex jubifer*, a new species of *Culex* from Panama (Diptera: Culicidae). *Annals of the Entomological Society of America* 28(2): 254–255.
Komp, W.H.W. & Curry, D.P. 1932. A new *Culex* from Panama (Dipt., Culicidae). *Psyche* 39(3): 82–84.
Komp, W.H.W. & Kumm, H.W. 1938. A new species of *Haemagogus*, *mesodentatus*, from Costa Rica, and a description of the larva of *Haemagogus anastasionis* Dyar (Diptera, Culicidae). *Proceedings of the Entomological Society of Washington* 40(9): 253–259.
Komp, W.H.W. & Rozeboom, L.E. 1951. Descriptions of eight new species of *Culex* subgenus *Melanoconion* (Diptera, Culicidae). *Proceedings of the Entomological Society of Washington* 53(3): 121–137.
Krishnamoorthy, R., Munirathinam, A., Dhananjeyan, K.J., Hiriyan, J., Mariappan, T., Samuel, P.P. & Venkatesh, A. 2013. Description of a new species, *Toxorhynchites (Toxorhynchites) tyagii* (Diptera: Culicidae), from Nilgiri hills, Western Ghats, southern India. *Zootaxa* 3701(4): 447–459.
Kuchartshuk, L.P. 1973. New species of mosquito from the genus *Aedes* (Diptera, Culicidae) [in Russian], pp. 98–101. *In*: *New and little-known species of Siberian fauna. Part 7* [in Russian]. Academy of Sciences, USSR, Siberian Branch, Institute of Biology, Publishing house 'Nauka', Novosibirsk.
Kuhlow, F. 1962. Beobachtungen und Experimente über den *Anopheles gambiae*-Komplex, Abtrennung von *Anopheles tangensis* n. sp. *Zeitschrift für Tropenmedizin und Parasitologie* 13(4): 442–449.
Kulasekera, V.L., Harrison, B.A. & Amerasinghe, F.P. 1989. *Anopheles* (*Anopheles*) *peytoni* new species, the "*An. insulaeflorum*" auct. from Sri Lanka (Diptera: Culicidae). *Mosquito Systematics* (for 1988) 20(3): 302–316.
Kulasekera, V., Knight, K.L. & Harbach, R.E. 1990. *Aedes (Finlaya) axitiosus*, new species of the Niveus Subgroup (Diptera: Culicidae) from East Malaysia. *Mosquito Systematics* 22(1): 26–33.
Kumm, H.W. 1933. Mosquitos breeding in bromeliads, at Bahia, Brazil. *Bulletin of Entomological Research* 24(4): 561–573, 2 pls.
Kumm, H.W., Komp, W.H.W. & Ruiz, H. 1940. The mosquitoes of Costa Rica. *American Journal of Tropical Medicine* 20(3): 385–422.
Kumm, H.W., Osorno-Mesa, E. & Boshell-Manrique, J. 1946. Studies on mosquitoes of the genus *Haemagogus* in Colombia (Diptera, Culicidae). *American Journal of Hygiene* 43(1): 13–28, 6 pls.
Laarman, J.J. 1958. *Anopheles Neomyzomia van thieli* n. sp. *Folia scientifica Africae centralis* 4: 18.
La Casse, W.J. & Yamaguti, S. 1948. *Mosquito fauna of Japan and Korea*. Office of the Surgeon, Headquarters I Corps, APO 301, U.S. Army, Japan. [viii], 273 pp.
La Casse, W.J. & Yamaguti, S. 1950. *Mosquito fauna of Japan and Korea*. Office of the Surgeon, Hq. 8th Army APO 343, Kyoto, Honshu. viii, 268 pp., Appendix I (7 pp.), Appendix II (213 pp.).
Laffoon, J. 1946. The Philippine mosquitoes of the genus *Aedes*, subgenus *Aedes*. *Journal of the Washington Academy of Sciences* 36(7): 228–245.
Lahille, F. 1904. Notes sur la classification des moustiques. *Actas y Trabajos del Segundo Congreso Médico Latino-Americano (Buenos Aires)* 2: 71–96.
Lahille, F. 1912. Sobre un *Anopheles*, una *Stegomyia* y la notacion de las nervaduras alares de los mosquitos. *Anales del Museo Nacional de Historia Natural de Buenos Aires* 23: 253–263, 5 pls.
Laird, M. 1957. A new mosquito from Fiji, *Aedes* (*Finlaya*) *freycinetiae* n.sp. (Diptera: Culicidae). *Pacific Science* 11(3): 342–351.
Laird, M. 1988. *The natural history of larval mosquito habitats*. Academic Press, London. xxvii, 555 pp.
Lambert, D.M. & Coetzee, M. 1982. A dual genetical and taxonomic approach to the resolution of the mosquito taxon, *Anopheles (Cellia) marshallii* (Culicidae). *Systematic Entomology* 7(3): 321–332.
Lane, C.J. 1992b. *Toxorhynchites auranticauda* sp.n., a new Indonesian mosquito and potential biocontrol agent. *Medical and Veterinary Entomology* 6(3): 301–305.
Lane, J. 1936a. II. Notas sobre mosquitos de São Paulo. *Boletim de Instituto Hygiene de São Paulo* 60: 3–15, 5 pls.
Lane, J. 1936b. Notas sobre culicideos de Matto Grasso. *Revista do Museu Paulista* 20: 173–206, 4 pls.
Lane, J. 1936c. I – Notas sobre mosquitos de São Paulo. *Revista do Museu Paulista* 20: 429–435.
Lane, J. 1939. Catálogo dos mosquitos neotrópicos.*Boletim Biológico Série Monográfica* 1: xi, 218 pp.
Lane, J. 1943. Sôbre o gênero *Uranotaenia* (Diptera, Culicidae, Culicini). *Revista de Entomologia* 14(1–2): 137–161.

Lane, J. 1944. A tribo Megarhinini no Brasil Meridional. *Revista de Entomologia* 15(1–2): 172–190.
Lane, J. 1945a. Os sabetineos da América. (Addenda e corrigenda). *Revista de Entomologia* 16(1–2): 132–157.
Lane, J. 1945b. Quatro espécies novas de *Culex* da Região Neotrópica. *Revista de Entomologia* 16(1–2): 204–209.
Lane, J. 1949. Anophelines of the Neotropical Region, pp. 399–418. *In*: Boyd, M.F. (ed.), *Malariology. A comprehensive survey of all aspects of this group of diseases from a global standpoint*. Volume 1. W.B. Saunders Company, Philadelphia and London.
Lane, J. 1951. Synonymy of Neotropical Culicidae (Diptera). *Proceedings of the Entomological Society of Washington* 53(6): 333–336.
Lane, J. 1953. *Neotropical Culicidae*. Volumes I and II. University of São Paulo, São Paulo. 1112 pp.
Lane, J. & Antunes, P.C.A. 1937. Nota sobre o genero *Mansonia*, sub-genero *Rhynchotaenia*, com a descripção de uma nova especie *(Diptera, Culicidae)*. *Revista do Museu Paulista* 23: 227–232.
Lane, J. & Causey, O.R. 1955. Additional data on Sabethini (Diptera, Culicidae). *Proceedings of the Entomological Society of Washington* 57(1): 11–17.
Lane, J. & Cerqueira, N.L. 1942. Os sabetíneos da América (Diptera, Culicidae). *Arquivos de Zoologia do Estudo de São Paulo* 3(9): 473–849.
Lane, J. & Cerqueira, N.L. 1957. The validity and change of name of two species of *Wyeomyia* (Diptera, Culicidae). *Proceedings of the Entomological Society of Washington* 59(5): 244.
Lane, J. & Coutinho, J.O. 1940. *Mansonia* subg. *Rhynchotaenia*: Descrição de duas espécies novaes dados sobre o subgênero (Dipt. Culicidae). *Revista de Entomologia* 11(1–2): 589–597.
Lane, J. & Forattini, O.P. 1952. New data on Sabethini (Diptera, Culicidae). *Proceedings of the Entomological Society of Washington* 54(5): 254–257.
Lane, J. & Guimarães, L.R. 1937. Notas sobre culicideos de Campos do Jordão. *Annaes Paulistas de Medicina e Cirugia* 33: 211–217.
Lane, J. & Ramalho, G.R. 1960. A new neotropical *Culex*. *Revista Brasileira de Entomologia* 9: 173–176.
Lane, J. & Whitman, L. 1943. Novas espécies de *Culex* do Brasil (Diptera, Culicidae). *Revista de Entomologia* 14(3): 389–408.
Lane, J. & Whitman, L. 1951. The subgenus "*Microculex*" in Brazil (Diptera, Culicidae). *Revista Brasileira de Biologia* 11(3): 341–366.
Lanzaro, G.C. & Eldridge, B.F. 1992. A classical and population genetic description of two new sibling species of *Aedes* (*Ochlerotatus*) *increpitus* Dyar. *Mosquito Systematics* 24(2): 85–101.
Lanzaro, G.C., Collier, T.C. & Lee, Y. 2015. Defining genetic, taxonomic, and geographic boundaries among species of the *Psorophora confinnis* (Diptera: Culicidae) complex in North and South America. *Journal of Medical Entomology* 52(5): 907–917.
Larrousse, F. 1925. Larve de *Culex* a branchies très développées nouvelle pour la faune Française *(Culex lavieri* n. sp.). *Annales de Parasitologie Humaine et Comparée* 3(1): 68–70.
Lassmann, G.W. 1944. Notas sobre una nueva especie de *Psorophora*. *Psorophora pisces* n. sp. (Culicidae, Diptera). *Boletín de Salubridad y Asistencia* 28/29: 3–9.
Lassmann, G.W. 1951. *Psorophora* (*Janthinosoma*) *totonaci*, a new species from Mexico (Diptera, Culicidae). *Proceedings of the Entomological Society of Washington* 53(5): 284–287.
Latreille, P.A. 1810. *Considérations générales sur l'ordre naturel des animaux composant les classes des Crustacés, des Arachnides et des Insectes; avec un tableau méthodique de leurs genres, disposés en familles*. F. Schoell, Paris. 444 pp.
Laurito, M., de Oliveira, T.M.P., Almirón, W.R. & Sallum, M.A.M. 2013. COI barcode versus morphological identification of *Culex* (*Culex*) (Diptera: Culicidae) species: a case study using samples from Argentina and Brazil. *Memórias do Instituto Oswaldo Cruz* 108(Suppl.1): 110–122.
Laveran, [A.] 1900. Sur un *Anopheles* provenant de Madagascar. *Comptes Rendus Hebdomadaires des Séances et Mémoires de la Société de Biologie* 52: 109–110.
Laveran, [A.] 1901a. Au sujet de culicides recueillis a Dijbouti et a la Nouvelle-Calédonie. *Comptes Rendus Hebdomadaires des Séances et Mémoires de la Société de Biologie* 53: 567–569.
Laveran, A. 1901b. Sur des culicides provenant de Hanoï (Tonkin). *Comptes Rendus Hebdomadaires des Séances et Mémoires de la Société de Biologie* 53: 991–993.
Laveran, A. 1901c. Sur des culicides provenant du Haut-Tonkin. *Comptes Rendus Hebdomadaires des Séances et Mémoires de la Société de Biologie* 53: 993–994.
Laveran, A. 1902a. Sur des culicides du Cambodge. *Comptes Rendus Hebdomadaires des Séances et Mémoires de la Société de Biologie* 54: 906–908.
Laveran, A. 1902b. Sur des culicides de Nouvelles-Hébrides. *Comptes Rendus Hebdomadaires des Séances et Mémoires de la Société de Biologie* 54: 908–910.
Laveran, [A.] 1902c. [Note]. *In*: Simond, P.L., Description d'un moustique dont le male possede un trompe faucille. *Comptes Rendus Hebdomadaires des Séances et Mémoires de la Société de Biologie* 54: 1158–1160.
Laveran, A. 1904. *Anopheles* et paludisme à Madagascar. Prophylaxie du paludisme. *Bulletin de l'Académie de Médecine Paris* 52: 197–218.
Leach, W.E. 1825. Descriptions of thirteen species of *Formica*, and three species of *Culex*, found in the environs of Nice. *Zoological Journal* 2(7): 289–293.
Lee, D.J. 1937. Notes on Australian mosquitoes (Diptera, Culicidae). Part IV. The genus *Theobaldia,* with description of a new species. *Proceedings of the Linnean Society of New South Wales* 62(5–6): 294–298.
Lee, D.J. 1944a. A new species of the genus *Anopheles* from northern Australia (Diptera, Culicidae). *Proceedings of the Linnean Society of New South Wales* 69(1–2): 21–25.

Lee, D.J. 1944b. Notes on Australian mosquitoes (Diptera, Culicidae). Part V. The genus *Armigeres* and new species of *Armigeres*, *Theobaldia* and *Culex*. *Proceedings of the Linnean Society of New South Wales* 69(5–6): 215–225.

Lee, D.J. 1946. Notes on Australian mosquitoes (Diptera, Culicidae). Part VI. The genus *Tripteroides* in the Australasian Region. *Proceedings of the Linnean Society of New South Wales* 70(5–6): 219–275, 2 pls.

Lee, D.J. & Woodhill, A.R. 1944. *The anopheline mosquitoes of the Australasian Region*. Publications of the University of Sydney, Department of Zoology, Monograph No. 2. Australian Medical Publishing Company Limited, Glebe, N.S.W. 29 pp.

Lee, D.J., Hicks, M.M., Debenham, M.L., Griffiths, M., Bryan, J.H. & Marks, E.N. 1988b. *The Culicidae of the Australasian Region. Volume 9. Nomenclature, synonymy, literature, distribution, biology and relation to disease. Genus* Culex *(Subgenera* Lutzia, Neoculex, *Subgenus undecided). Genera* Culiseta, Ficalbia, Heizmannia, Hodgesia, Malaya, Mansonia. Monograph Series, Entomology Monograph No. 2. Australian Government Publishing Service, Canberra. ix, 162 pp.

Lee, D.J., Hicks, M.M., Debenham, M.L., Griffiths, M., Marks, E.N. & Bryan, J.H. 1989b. *The Culicidae of the Australasian Region. Volume 8. Nomenclature, synonymy, literature, distribution, biology and relation to disease. Genus* Culex. *Subgenera* Culiciomyia, Eumelanomyia, Lophoceraomyia. Monograph Series, Entomology Monograph No. 2. Australian Government Publishing Service, Canberra. ix, 171 pp.

Lee, D.J., Hicks, M.M., Debenham, M.L., Griffiths, M., Marks, E.N., Bryan, J.H. & Russell, R.C. 1989a. *The Culicidae of the Australasian Region. Volume 7. Nomenclature, synonymy, literature, distribution, biology and relation to disease. Genus* Culex. *Subgenera* Acallyntrum, Culex. Monograph Series, Entomology Monograph No. 2. Australian Government Publishing Service, Canberra. ix, 281 pp.

Lee, D.J., Hicks, M.M., Griffiths, M., Debenham, M.L., Bryan, J.H., Russell, R.C., Geary, M. & Marks, E.N. 1987. *The Culicidae of the Australasian Region. Volume 4. Nomenclature, synonymy, literature, distribution, biology and relation to disease. Genus* Aedes. *Subgenera* Scutomyia, Stegomyia, Verrallina. Monograph Series, Entomology Monograph No. 2. Australian Government Publishing Service, Canberra. xiii, 324 pp.

Lee, D.J., Hicks, M.M., Griffiths, M., Debenham, M.L., Bryan, J.H., Russell, R.C., Geary, M. & Marks, E.N. 1988a. *The Culicidae of the Australasian Region. Volume 6. Nomenclature, synonymy, literature, distribution, biology and relation to disease. Genera* Armigeres, Bironella *and* Coquillettidia. Monograph Series, Entomology Monograph No. 2. Australian Government Publishing Service, Canberra. ix, 124 pp.

Lee, D.J., Hicks, M.M., Griffiths, M., Debenham, M.L., Marks, E.N., Bryan, J.H. & Russell, R.C. 1989c. *The Culicidae of the Australasian Region. Volume 11. Nomenclature, synonymy, literature, distribution, biology and relation to disease.Genera* Tripteroides, Uranotaenia, Wyeomyia, Zeugnomyia. Monograph Series, Entomology Monograph No. 2. Australian Government Publishing Service, Canberra. ix, 306 pp.

Lee, D.J., Hicks, M.M., Griffiths, M., Russell, R.C. & Marks, E.N. 1982. *The Culicidae of the Australasian Region. Volume 2. Nomenclature, synonymy, literature, distribution, biology and relation to disease. Genus* Aedeomyia. *Genus* Aedes *(subgenera* [Aedes] Aedimorphus, Chaetocruiomyia, Christophersiomyia, Edwardsaedes *and* Finlaya*)*. Monograph Series, Entomology Monograph No. 2. Australian Government Publishing Service, Canberra. v, 286 pp.

Lee, D.J., Hicks, M.M., Griffiths, M., Russell, R.C. & Marks, E.N. 1984. *The Culicidae of the Australasian Region. Volume 3. Nomenclature, synonymy, literature, distribution, biology and relation to disease. Genus* Aedes. *Subgenera* Geoskusea, Halaedes, Huaedes, Leptosomatomyia, Levua, Lorrainea, Macleaya, Mucidus, Neomelaniconion, Nothoskusea, Ochlerotatus, Paraedes, Pseudoskusea, Rhinoskusea. Monograph Series, Entomology Monograph No. 2. Australian Government Publishing Service, Canberra. v, 257 pp.

Leeson, H.S. 1930. A new variety of *Anopheles funestus*, Giles, from southern Rhodesia. *Bulletin of Entomological Research* 21(2): 125–126.

Leeson, H.S. 1935. Another anopheline of the *funestus* series from Southern Rhodesia. *Annals of Tropical Medicine and Parasitology* 29(1): 69–71.

Leeson, H.S. & Theodor, O. 1948. Mosquitos of Socotra. *Bulletin of Entomological Research* 39(2): 221–229.

Le Goff, G., Boussès, P. & Brunhes, J. 2007. Révision des *Neomelaniconion* Newstead (Diptera: Culicidae) de Madagascar: espèces présentes et description de cinq nouvelles espèces. *Bulletin de la Société Entomologiquede France* 43(2): 187–200.

Le Goff, G., Brengues, C. & Robert, V. 2013. *Stegomyia* mosquitoes in Mayotte, taxonomic study and description of *Stegomyia pia* n. sp. *Parasite* 20, Article no. 31. 18 pp.

Leguizamón, M.N. & Carpintero, D.J. 2004. Description of a new species of *Orthopodomyia* (Diptera: Culicidae) from the Neotropical region. *Studia dipterologia* 11(1): 203–205.

Lei, X. 1989a. A new species of *Aedes* (*Ochlerotatus*) (Diptera: Culicidae) [in Chinese]. *Sichuan Journal of Zoology* 8(1): 6–8.

Lei, X. 1989b. *The mosquito fauna of Sichuan, China* [in Chinese]. Chengdu University of Science and Technology Press, Chengdu, China. 292 pp.

Lei, X. 1996. A new species of *Anopheles* (Diptera: Culicidae) [in Chinese]. *Sichuan Journal of Zoology* 15(1): 1–4.

Lei, X. & Li, K. 1990. A new species of *Orthopodomyia* in Sichuan (Diptera, Culicidae) [in Chinese]. *Sichuan Journal of Zoology* 9(1): 8–10.

Leicester, G.F. 1903. A breeding place of certain forest mosquitoes in Malaya. *Journal of Tropical Medicine* 6: 291–292.

Leicester, G.F. 1908. The Culicidae of Malaya. *Studies from the Institute for Medical Research, Federated Malay States* 3(3): 18–261.

Leleup, N. 1952. Un Anophèle cavernicole nouveau du Kibali-Ituri: *Anopheles Faini* n. sp. *Revue de zoologie et de botanique Africaines* 46(1–2): 151–158.

Leleup, N. 1957. *Anopheles garnhami* sbsp. *basilewskyi*, sous-espece nouvelle de tres haute altitude (2.900 M.). *Folia Scientifica Africae Centralis* 3(3): 69.

Leleup, N. & Lips, M. 1950. Un Anophèle cavernicole nouveau du Katanga: *Anopheles Rodhaini* [*sic*] n. sp. *Revue de zoologie et de botanique Africaines* 43(4): 303–308.

Le Peletier, A.L.M. & Audinet-Serville, J.G. 1828. pp. 345–832, [1]. *In*: Latreille, P.A. (ed.), Histoire naturelle. Entomologie, ou histoire naturelle des crustacés, des arachnides et des insectes. *In*: Lareille, P.A., Le Peletier, A.L.M., Audinet-Serville, J.G. & Guérin-Méneville, F.É. (eds), *Encyclopédie methodique, ou par ordre de matieres; par une société de gens de lettres, de savans et d'artistes; Précédée d'un Vocabulaire universel, servant de Table pour tout l'Ouvrage, ornée des Portraits de MM. Diderot & D'Alembert, premiers Éditeurs de l'Encyclopédie.* Tome dixième. Mme Veuve Agasse, Paris.

Levi Castillo, R. 1944. *El complejo "Pseudopunctipennis" en el Ecuador (Diptera - Culicidae).* Imprenta de la Universidad, Guayaquil. 7 pp.

Leví Castillo, R. 1945. *Los anofelinos de la Republica del Ecuador.* Tomo primero. Artes Graficas Senefelder, Guayaquil, Ecuador. vi, 72 pp.

Levi-Castillo, R. 1951. *Los mosquitos del género* Haemagogus, *Williston, 1896 en América del Sur.* Editorial "Don Bosco", Cuenca, Ecuador. 77 pp.

Levi-Castillo, R. 1952. *Wyeomia* [*sic*] (*Wyeomyia*) *aphobema* var. *aequatorialis* var. n., a new sabethine mosquito from Ecuador (Diptera: Culicidae). *Proceedings of the Royal Entomological Society of London, Series B: Taxonomy* 21(9–10): 131–133.

Levi-Castillo, R. 1953a. Observations on the subgenus *Phalangomyia* of the genus *Culex* in Ecuador with description of a new species (Diptera: Culicidae). *Pacific Science* 7(2): 187–192.

Leví-Castillo, R. 1953b. Dos especies nuevas de mosquitos ecuatorianos (Diptera-culicidae). *Revista Ecuatoriana de Entomología y Parasitología* 1(1): 14–18, 2 pls.

Leví-Castillo, R. 1953c. Dos especies nuevas de mosquitos de la sierra ecuatoriana (Diptera-culicidae). *Revista Ecuatoriana de Entomología y Parasitología* 1(4): 63–70, 5 pls.

Levi-Castillo, R. 1953d. Una nueva especie de *Culex* procedente de la Provincia de Los Ríos, Ecuador: *Culex* (*Carrollia*) *Babahoyersis* N. Sp. (Diptera-Culicidae). *Revista Ecuatoriana de Entomología y Parasitología* 1(3): 91–94, 2 pls.

Leví-Castillo, R. 1953e. *Aedes garciai* n. sp., un Aedino nuevo para la fauna ecuatoriana.– *Lanesia*, nuevo subgénero de *Aedes. Revista Ecuatoriana de Entomología y Parasitología* 1(3): 101–106, 3 pls.

Leví-Castillo, R. 1953f. *Toxorhynchites aequatorianus* n. sp., especie neuva de *Toxorhynchites* hallada en Pichilingue, Los Ríos, Ecuador (Diptera-Culicidae). *Revista Ecuatoriana de Entomología y Parasitología* 1(4): 71–76, 3 pls.

Leví-Castillo, R. 1954a. Cuatro especies nuevas de Sabethini del Ecuador (Diptera–Culicidae). *Revista Ecuatoriana de Entomologia y Parasitología* 2(1–2): 247–255, 5 pls.

Leví-Castillo, R. 1954b. *Culex* (*Phalangomyia*) *azuayus* n. sp., un nuevo mosquito de altura del Ecuador (Diptera Culicidae). *Revista Ecuatoriana de Entomología y Parasitología* 2(1–2): 263–267, 4 pls.

Leví-Castillo, R. 1955a. Nueva división sistemática del Género *Haemagogus* Williston 1896 y una nueva especie ecuatoriana: *Haemagogus garciai*, N. Sp. *Revista Ecuatoriana de Entomología y Parasitología* 2(3–4): 359–364, 2 pls.

Levi-Castillo, R. 1955b. *Phoniomyia esmeraldasi*, a new mosquito from Ecuador. *Revista Ecuatoriana de Entomología y Parasitología* 2(3–4): 389–391, 1 pl.

Leví-Castillo, R. 1955c. *Haemagogus soperi* n. sp. mosquito trasmisor de fiebre amarilla selvática en el Ecuador (Diptera-Culicidae). *Revista Ecuatoriana de Entomología y Parasitología* 2(3–4): 479–484, 2 pls.

Leví-Castillo, R. 1955d. Un nuevo Anofelino de altura del Ecuador: *Anopheles gomezdelatorrei* N. Sp. *Revista Ecuatoriana de Entomología y Parasitología* 2(3–4): 509–514, 3 pls.

Lewis, D.J. 1937. A new species of *Anopheles* from the Anglo-Egyptian Sudan. *Proceedings of the Royal Entomological Society of London, Series B: Taxonomy* 6(9): 181–183, 1 pl.

Lewis, D.J. 1942. A new species of *Aëdes* (Dipt., Culicidae) from Eritrea. *Proceedings of the Royal Entomological Society of London, Series B: Taxonomy* 11(11): 165–166.

Lewis, D.J. 1944. A new subspecies of *Aëdes leesoni* Edwards (Dipt., Culicidae) from the Sudan. *Proceedings of the Royal Entomological Society of London, Series B: Taxonomy* 13(3–4): 27–29.

Lien, J.C. 1962. Non-anopheline mosquitoes of Taiwan: Annotated catalog and bibliography. *Pacific Insects* 4(3): 615–649.

Lien, J.C. 1967. New species of mosquitoes from Taiwan (Diptera: Culicidae). Part I. Two new species of *Aedes* (*Finlaya*). *Tropical Medicine* 9(4): 177–185.

Lien, J.C. 1968a. New species of mosquitoes from Taiwan (Diptera: Culicidae). Part II. New species of *Tripteroides*, *Orthopodomyia*, *Culiseta* and *Uranotaenia*. *Tropical Medicine* 10(1): 1–20.

Lien, J.C. 1968b. New species of mosquitoes from Taiwan (Diptera: Culicidae). Part III. Five new species of *Aedes*. *Tropical Medicine* 10(2): 95–115.

Lien, J.C. 1968c. New species of mosquitoes from Taiwan (Diptera: Culicidae). Part IV. *Mattinglyia catesi*, n. gen., n. sp. and four new species of genus *Heizmannia* Ludlow. *Tropical Medicine* 10(3): 127–153.

Lien, J.C. 1968d. New species of mosquitoes from Taiwan (Diptera: Culicidae). Part V. Three new subspecies of *Aedes* and seven new species of *Culex*. *Tropical Medicine* 10(4): 217–262.

Lindtrop, G.T. 1924. The white malarial mosquito – *Cellia* (*Anopheles*) *pulcherrima* Theob. in Azerbaijan [in Russian]. *Russkii Zhurnal Tropicheskoi Meditsiny* [*Russian Journal of Tropical Medicine*] 3: 38–40.

Linnaeus, C. 1758. *Systema naturae per regna tria naturae, secundum classes, ordines, genera, species, cum characteribus, differentiis, synonymis, locis.* Tomus 1, Editio Decima. Impensis Direct. Laurentii Salvii, Holmiae. 824 pp.

Linnaeus, C. 1762. Zweyter Theil, enthalt Beschreibungen verschiedener wichtiger Naturalien, pp. 267–606. *In*: Hasselquist, D.F. *Reise nach Palästina in den Jahren von 1749 bis 1752. Auf Befehl Ihro Majestat der Königinn von Schweden.* Johann Christian Koppe, Rostock, Germany.

Linnaeus, C. 1767. *Systema naturae per regna tria naturae, secundum classes, ordines, genera, species, cum characteribus, differentiis, synonymis, locis.* Twelfth edition, Volume 1, Part 2. Holmiae. 532 pp.

Linnaeus, C. [von Linné, C.] 1792. *Caroli Linnaei ... Flora Lapponica, exhibens plantas per Lapponiam crescentes, secundum systema sexuale, sollectas in itinere impensis Soc. Reg. Litter. et Scient. Sveciae A. CIↃ IↃ CC XXXII* [1732]. *Instituto. Additis synonymis, & locis natalibus omnium, descriptionibus & figuris rariorum, viribus medicatis & oeconomicis plurimarum. Amstelædami, Apud Salomonen Schouten CIƆ IƆ CC XXXVII* [1737]. [2], l, [1], 390, [40] pp., engr. front., 12 foldout pls.

Linthicum, K.J. 1988. A revision of the Argyritarsis Section of the subgenus *Nyssorhynchus* of *Anopheles* (Diptera: Culicidae). *Mosquito Systematics* 20(2): 98–271.

Linton, Y.-M., Dusfour, I., Howard, T.M., Ruiz, F., Manh, N.D., Dinh, T.H., Sochanta, T., Coosemans, M. & Harbach, R.E. 2005. *Anopheles (Cellia) epiroticus* (Diptera: Culicidae), a new malaria vector species in the Southeast Asian Sundaicus Complex. *Bulletin of Entomological Research* 95(4): 329–339.

Linton, Y.-M., Harbach, R.E., Chang, M.S., Anthony, T.G. & Matusop, A. 2001. Morphological and molecular identity of *Anopheles (Cellia) sundaicus* (Rodenwaldt) (Diptera: Culicidae), the nominotypical member of a malaria vector species complex in Southeast Asia. *Systematic Entomology* 26(3): 357–366.

Linton, Y.-M., Pecor, J.E., Porter, C.H., Mitchell, L.B., Garzón-Moreno, A., Foley, D.H., Pecor, D.B. & Wilkerson, R.C. 2013. Mosquitoes of eastern Amazonian Ecuador: biodiversity, bionomics and barcodes. *Memórias do Instituto Oswaldo Cruz* 108(Suppl. 1): 100–109.

Linton, Y.-M., Smith, L. & Harbach, R.E. 2002. Observations on the taxonomic status of *Anopheles subalpinus* Hackett & Lewis and *An. melanoon* Hackett. *European Mosquito Bulletin* 13: 1–7.

Lioy, P. 1863. I ditteri distribuiti secondo un nuovo metodo di classificazione natural [part]. *Atti del Reale Istituto Veneto di Scienze, Lettere ed Arti* (Series 3) 9(part 1): 187–236.

Lips, M. 1960a. *Anopheles* du Congo Belge. 2. Quelques especes des galeries forestieres. *Rivista di Parassitologia* 21(1): 39–64.

Lips, M. 1960b. *Anopheles* du Congo. 3. Faune des grottes et des anfractuosites. Références-récoltes-répartition et importance médicale actuelle. *Rivista di Parassitologia* 21(4): 289–306.

Lips, M. 1962. *Anopheles* du Congo (ex Belge). 7. Quelques especes des groupes *Cellia* et *Neocellia*. Références - récoltes - répartition et importance médicale actuelle. *Rivista di Parassitologia* 23(2): 107–134.

Liston, W.G. 1901. A year's experience of the habits of *Anopheles* in Ellichpur. The description of the species of *Anopheles* found in Ellichpur during the year. *Indian Medical Gazette* 36: 361–366, 441–443, 1 pl.

Liu, S.Z. & Lu, B.-L. 1984. A new *Aedes* from China (Diptera: Culicidae) [in Chinese]. *Acta Zootaxonomica Sinica* 9(1): 76–79.

Loew, H. 1845. *Dipterologische Beiträge*. W. Decker & Comp, Posen, Germany. 52 pp., 1 pl.

Loew, H. 1850. *Über den Bernstein und die Bernsteinfauna*. Programm der Koniglichen Realschule zu Meseritz. E.S. Mittler, Berlin. 48 pp.

Loew, [H.] 1866. Beschreibung einiger afrikanischen Diptera nemocera. *Berliner Entomologische Zeitschrift* 10(1–3): 55–62.

Loew, H. 1873. *Beschreibung europäischer Dipteren. Dritter Band. Systematische Beschreibung der bekannten europäischen zweiflügeligen Insecten. Von Johann Wilhelm Meigen. Zehnter Theil oder vierter Supplementband*. H.W. Schmidt, Halle, Germany. Viii, 320 pp.

Loew, H. 1874. Diptera nova a Hug. Theod. Christopho collecta. *Zeitschrift für die gesamte Naturwissenschaft* 43: 413–420.

Lounibos, L.P. 1979. Mosquitoes occurring in the axils of *Pandanus rabaiensis* Rendle on the Kenya coast. *Cahiers ORSTOM Série Entomologie médicale et Parasitologie* 17(1): 25–29.

Lourenço de Oliveira, R. 1983. Sobre uma nova espécie de *Phoniomyia* (Diptera: Culicidae) do Rio de Janeiro, Brasil. *Memórias do Instituto Oswaldo Cruz* 78(4): 501–505.

Lourenço-de-Oliveira, R. & da Silva, T.F. 1985. *Wyeomyia forcipenis*, new species of mosquito (Diptera: Culicidae) from Rio de Janeiro, Brazil. *Memórias do Instituto Oswaldo Cruz* 80(3): 321–326.

Lourenço-de-Oliveira, R. & da Silva, T.F. 1987. *Culex siphanulatus*, a new species of mosquito from the coast of Rio de Janeiro State, Brazil (Diptera: Culicidae). *Memórias do Instituto Oswaldo Cruz* 82(1): 101–110.

Lourenço-de-Oliveira, R. & Deane, L.M. 1984. What is *Anopheles allopha* (Lutz & Peryassú, 1921) (Diptera: Culicidae)? *Memórias do Instituto Oswaldo Cruz* 79(4): 509–510.

Lourenço-de-Oliveira, R., Harbach, R.E., Castro, M.G., Motta, M.A. & Peyton, E.L. 1999. *Wyeomyia (Prosopolepis) confusa* (Lutz): subgeneric validation, species description, and recognition of *Wyeomyia flui* (Bonne-Wepster and Bonne) as the senior synonym of *Wyeomyia kerri* del Ponte and Cerqueira. *Journal of the American Mosquito Control Association* 15(2): 200–212.

Lourenço-de-Oliveira, R., Motta, M.A. & de Castro, M.G. 1992. *Wyeomyia staminifera*, a new species of mosquito from Brazil (Diptera: Culicidae). *Memórias do Instituto Oswaldo Cruz* 87(1): 115–121.

Lu, B.-L. & Gong, Z.-D. 1986. Studies on the genus *Heizmannia* of China with descriptions of two new species (Diptera: Culicidae) [in Chinese]. *Acta Entomologica Sinica* 29(2): 181–189.

Lu, B. & Ji, S. 1997. Subgenus *Finlaya* Theobald, 1903, pp. 99–173 and figs 18–60. *In*: Lu, B., Li, B., Ji, S., Chen, H., Meng, Q., Su, L., Qu, F., Gong, Z. & Zhang, Z. (eds), *Diptera: Culicidae 1. Fauna Sinica, Insecta* [in Chinese]. Volume 8. Science Press, Beijing. xiii, 884 pp.

Lu, B.-L., Dong, X.-S. & Wang, X.-Z. 1986. A new species of subgenus *Topomyia* of genus *Topomyia* (Diptera: Culicidae) [in Chinese]. *Acta Zootaxonomica Sinica* 11(4): 406–408.

Ludlow, C.S. 1902a. Two Philippine mosquitoes. *Journal of the New York Entomological Society* 10(3): 127–131.

Ludlow, C.S. 1902b. Description of a new *Anopheles*. *Journal of the American Medical Association* 39(8): 426–427.

Ludlow, C.S. 1903. Some Philippine mosquitoes. *Journal of the New York Entomological Society* 11(3): 137–144.

Ludlow, C.S. 1904a. Concerning some Philippine mosquitoes. *Canadian Entomologist* 36(1): 69–72.

Ludlow, C.S. 1904b. Mosquito notes. *Canadian Entomologist* 36(8): 233–236.

Ludlow, C.S. 1904c. Mosquito notes.—No. 2. *Canadian Entomologist* 36(10): 297–301.

Ludlow, C.S. 1905a. Mosquito notes.—No. 3. *Canadian Entomologist* 37(3): 94–102, 129–135.

Ludlow, C.S. 1905b. A new North American *Taeniorhynchus*. *Canadian Entomologist* 37(3): 231–232.

Ludlow, C.S. 1905c. Mosquito notes—No. 4. *Canadian Entomologist* 37(11): 385–388.
Ludlow, C.S. 1906a. A new American mosquito. *George Washington University Bulletin* 5: 83–84.
Ludlow, C.S. 1906b. Mosquito notes.—No. 4. (Continued.) *Canadian Entomologist* 38(4): 132–134.
Ludlow, C.S. 1906c. An Alaskan mosquito. *Canadian Entomologist* 38(10): 326–328.
Ludlow, C.S. 1906d. Mosquito notes.—No. 5. *Canadian Entomologist* 38(11): 367–368.
Ludlow, C.S. 1907. Mosquito notes.—No. 5. —Continued. *Canadian Entomologist* 39(4, 8, 12): 129–131, 266–268, 413–414.
Ludlow, C.S. 1908a. Mosquito notes.—No. 6. *Canadian Entomologist* 40(1, 2, 9): 32–34, 50–52, 331–332.
Ludlow, C.S. 1908b. The mosquitos of the Philippine Islands; the distribution of certain species, and their occurrence in relation to the incidence of certain diseases. PhD Thesis, The George Washington University, Washington, D.C. 65 pp.
Ludlow, C.S. 1909a. Mosquito comment. *Canadian Entomologist* 41(1): 21–24.
Ludlow, C.S. 1909b. New Philippine mosquitoes. *Canadian Entomologist* 41(3): 97.
Ludlow, C.S. 1909c. Mosquito observations. *Canadian Entomologist* 41(7): 233–235.
Ludlow, C.S. 1910. Mosquito observations.—(continued). *Canadian Entomologist* 42(6): 193–196.
Ludlow, C.S. 1911a. A new Alaskan mosquito. *Canadian Entomologist* 43(5): 178–179.
Ludlow, C.S. 1911b. The Philippine mosquitoes. *Psyche* 18(4): 125–133.
Ludlow, C.S. 1914a. Philippine mosquitoes. *Psyche* 21(1): 30–32.
Ludlow, C.S. 1914b. A new anopheline. *Psyche* 21(4): 129–130.
Ludlow, C.S. 1914c. A new aëdine. *Psyche* 21(5): 159–160.
Ludlow, C.S. 1918. *Trichoprosopon* Theobald (Diptera; Culicidae). *Psyche* 25(3): 66–68.
Ludlow, C.S. 1919. New mosquitos from Panama. *Psyche* 26(6): 166–169.
Ludlow, C.S. 1920a. Siberian *Anopheles*. *Psyche* 27(4): 74–78.
Ludlow, C.S. 1920b. New Siberian Culicidae *(Diptera)*. *Insecutor Inscitiae Menstruus* (for 1919) 7(10–12): 151–161.
Ludlow, C.S. 1921. A new Philippine mosquito (*Diptera, Culicidae*). *Military Surgeon* 49: 690–691.
Luh, P.-L. & Lee, B.S. 1975. *Aedes* (subgenus *Ochlerotatus* Lynch Arribálzaga) – new species and a description of our country's [China] *Aedes dorsalis* group [in Chinese]. *Acta Entomologica Sinica* 18(4): 428–432.
Luh, P.-L. & Li, B.S. 1981. A new *Armigeres* from China (Diptera: Culicidae) [in Chinese]. *Acta Zootaxonomica Sinica* 6(2): 188–190.
Luh, P.-L. & Shih, J.-M. 1958. A new species of *Aedes* and a new subspecies of *Armigeres* from Kwangsi (Diptera: Culicidae) [in Chinese]. *Junshi Yixue Zazhi [Journal of Military Medicine]* 1(3): 222–226.
Luh, P.-L., Chao, K. & Xu, J.-M. 1974. A new *Culiseta* from China (Diptera: Culicidae) [in Chinese]. *Acta Entomologica Sinica* 17(4): 458–462.
Lutz, A. 1905. Novas especies de mosquitos do Brasil. *Imprensa Medica* 13(2, 3, 4, 5, 6, 7, 9, 11, 14, 15, 16, 18): 26–29, 48–52, 65–70, 81–84, 101–104, 125–128, 169–172, 202–204, 269–271, 287–290, 311–314, 347–350.
Lutz, A. 1921. Culicideos (systematica). Chave para os Culicideos. *Folha medica* 2: 161–163.
Lutz, A. 1928. *Estudios de zoología y parasitología venezolanas*. Rio de Janeiro. 133 pp., 1 table, 26 pls.
Lutz, A. & Neiva, A. 1911. Notas dipterolojicas. Dipterologische Mitteilungen. *Memórias do Instituto Oswaldo Cruz* 3(2): 295–300.
Lutz, A. & Neiva, A. 1913. Contribuições para a biolojia das megarininas com descrições de duas especies novas. *Memórias do Instituto Oswaldo Cruz* 5(2): 129–141.
Lynch Arribálzaga, F. 1878. Descripción de tres nuevos Culicidae de Buenos Aires. *El Naturalista Argentino* 1(5): 149–152.
Lynch Arribálzaga, F. 1891a. Dipterologia argentine. *Revista del Museo de La Plata* 1: 345–377.
Lynch Arribálzaga, F. 1891b. Dipterologia argentine. *Revista del Museo de La Plata* 2: 131–174, 5 pls.
Ma, S. 1981a. A new species of *Anopheles* (*Anopheles*) from China [in Chinese]. *Entomotaxonomia* 3(4): 277–279.
Ma, S.-F. 1981b. Studies on the *Anopheles* (*A.*) *sinensis* group of mosquitoes in China, including four new sibling species [in Chinese]. *Sinozoologia* 1(1): 59–70, 4 pls.
Ma, S.-F. 1982. Diptera: Culicidae [in Chinese], pp. 157–163. *In*: Qingzang Gaoyuan Zonghe Kexue Kaochadui (corporate author), *Xizang kunchong* [*Insects of Xizang (Tibet)*], Volume 2. Qingzang gaoyuan kexue kaocha congshu [Series of the Scientific Expedition to Qinghai-Xizang Plateau]. Science Press, Beijing.
Ma, Y.-J., Qu, F.-Y., Lei, X.-T. & Dong, X.-S. 2000. Comparison of rDNA-ITS2 sequences and morphological characters of *Anopheles kunmingensis* and *Anopheles liangshanensis* in China, with discussion on taxonomic status [in Chinese]. *China Journal of Parasitology and Parasitic Diseases* 18(2): 65–68.
Macdonald, W.W. 1958. Description of a new species of *Orthopodomyia* Theobald from Malaya, with a note on the genus in that country (Diptera: Culicidae). *Proceedings of the Royal Entomological Society of London, Series B: Taxonomy* 27(7–8): 121–126.
Macdonald, W.W. 1960. Malaysian parasites. XXXVIII. on the systematics and ecology of *Armigeres* subgenus *Leicesteria* (Diptera, Culicidae). *Studies from the Institute for Medical Research, Federated Malay States Malaya* 29: 110–153.
Macdonald, W.W. & Mattingly, P.F. 1960. A new species of *Udaya* from Malaya and a description of the early stages of *U. argyrurus* (Edwards, 1934) (Diptera: Culicidae). *Proceedings of the Royal Entomological Society of London, Series B: Taxonomy* 29(1–2): 22–28.
MacGregor, M.E. 1919. A new mosquito of the genus *Orthopodomyia* from a beech tree-hole in England. *Journal of the Royal Army Medical Corps* 33(6): 451–454.
MacGregor, M.E. 1924. *Aëdes* (*Stegomyia*) *mascarensis*, MacGregor: A new mosquito from Mauritius. *Bulletin of Entomological Research* 14(4): 409–412.

Mackerras, I.M. 1927. Notes on Australian mosquitoes (Diptera, Culicidae). Part ii. The zoogeography of the subgenus *Ochlerotatus*, with notes on the species. *Proceedings of the Linnean Society of New South Wales* 52(3): 284–298.

Macquart, J. 1834. *Histoire naturelle des Insectes. Diptères. Tome premier*. Librairie Encyclopédique de Roret, Paris. 576 pp., explanation of plates (8 pp.), 12 pls.

Macquart, J. 1838a. Diptères exotiques nouveaux ou peu connus. *Mémoires de la Société Royale des Sciences, de l'Agriculture et des Arts, de Lille* 2: 9–225, 25 pls.

Macquart, J. 1838b. *Diptères exotiques nouveaux ou peu connus*. Tome premier. – 1.re Partie. Libraire Encyclopédique de Roret, Paris. 323 pp., 14 pls.

Macquart, J. 1843. Diptères exotiques nouveaux ou peu connus. *Mémoires de la Société Royale des Sciences, de l'Agriculture et des Arts, de Lille* Année 1842: 162–460, 5 foldout tables, 36 pls.

Macquart, J. 1846a. Diptères exotiques nouveaux ou peu connus. Supplément. *Mémoires de la Société Royale des Sciences, de l'Agriculture et des Arts, de Lille* Année 1844: 133–364, 17 pls.

Macquart, J. 1846b. *Diptères exotiques nouveaux ou peu connus. Supplement. [I]*. (Extrait des *Mémoires de la Société Royale des Sciences, de l'Agriculture et des Arts, de Lille*. 1844.). Librairie Encyclopedique de Roret, Paris. 232 pp. (133–364), 20 pls.

Macquart, J. 1847. *Diptères exotiques nouveaux ou peu connus*. 2.e Supplément. Pp. 5–104. Paris. [Published separately].

Macquart, J. 1850. Diptères exotiques nouveaux ou peu connus. 4.e Supplément. *Mémoires de la Société des Sciences, de l'Agriculture et des Arts de Lille* Année 1849: 309–479, 14 pls.

Maffi, M. & Coluzzi, M. 1958. *Anopheles (Neocellia) salbaii* n. sp. Morfologia ed ambiente. *Rivista di Malariologia* 37(1–3): 45–55, 3 pls.

Maffi, M. & Tenorio, J.A. 1977. *Culex* (*Culex*) *banksensis*, a new species of mosquito (Diptera: Culicidae) from the Banks Islands, New Hebrides. *Pacific Insects* 17(4): 503–509.

Manalang, C. 1930. Morphology and classification of the Philippine variety of *Anopheles aconitus* Donitz, 1902, and *Anopheles minimus*, 1901. *Philippine Journal of Science* 43(2): 247–261, 1 pl.

Mangkoewinoto, R.M.M. 1919. Anophelines of West-Java. *Mededellingen van den Burgerlijken Geneeskundigen Dienst in Nederlandsch-Indië* 8(2): 41–82, 3 pls.

Mann F., G. 1950. Dos nuevas sub-especies del *Anopheles pseudopunctipennis* Th. 1901. *Biológica* 8–11: 33–42.

Marks, E.N. 1947a. Studies of Queensland mosquitoes. 1. The *Aëdes* (*Finlaya*) *kochi* group with descriptions of new species from Queensland, Bougainville and Fiji. *Papers / Department of Biology, University of Queensland* 2(5): 1–66.

Marks, E.N. 1947b. Studies of Queensland mosquitoes. 2. New species of *Aëdes* (subgenus *Finlaya*). *Papers / Department of Biology, University of Queensland* 2(6): 1–10.

Marks, E.N. 1948. Studies of Queensland mosquitoes. 3. The *Aëdes* (*Finlaya*) *australiensis* group. *Papers / Department of Biology, University of Queensland* 2(8): 1–42.

Marks, E.N. 1949. Studies of Queensland mosquitoes. 4. Some species of *Aëdes* (subgenus *Ochlerotatus*). *Papers / Department of Biology, University of Queensland* 2(11): 1–41.

Marks, E.N. 1951. The vector of filariasis in Polynesia: a change in nomenclature. *Annals of Tropical Medicine and Parasitology* 45(2): 137–140.

Mark, E.N. 1954. A review of the *Aedes scutellaris* subgroup with a study of variation in *Aedes pseudoscutellaris* (Theobald). (Diptera: Culicidae). *Bulletin of the British Museum (Natural History) Entomology* 3(10): 349–414, 1 pl.

Marks, E.N. 1955. Studies of Queensland mosquitoes. Part V—Some species of *Aedes* (subgenus *Finlaya*). *Papers / Department of Entomology, University of Queensland* 1(2): 11–29.

Marks, E.N. 1956. A new species of *Anopheles* from Queensland and notes on related species (Diptera: Culicidae). *Proceedings of the Royal Society of Queensland* 67(6): 41–52.

Marks, E.N. 1957a. The subgenus *Ochlerotatus* in the Australian Region (Diptera: Culicidae). I. Notes on classification, with the description of a new species. *Papers / Department of Entomology, University of Queensland* 1(5): 71–83.

Marks, E.N. 1957b. Some mosquitoes from Western Samoa, with a description of a new species of *Aëdes (Stegomyia)* (Diptera, Culicidae). *Annals of Tropical Medicine and Parasitology* 51(1): 50–57.

Marks, E.N. 1958. New species and records of the *Aedes kochi* group from eastern Australia (Diptera: Culicidae). *Proceedings of the Royal Society of Queensland* 69(6): 57–74.

Marks, E.N. 1959a. The subgenus *Ochlerotatus* in the Australian region (Diptera: Culicidae). II. Five new species from Western Australia. *Papers / Department of Entomology, University of Queensland* 1(8): 111–134.

Marks, E.N. 1959b. On two mosquitoes of the subgenus *Finlaya* (Diptera: Culicidae). *Proceedings of the Royal Society of Queensland* 70(4): 21–25.

Marks, E.N. 1960. Taxonomy and biology of some Papuan Culicidae. I. Genus *Topomyia*. *Pacific Insects* 2(1): 91–98.

Marks, E.N. 1963a. A revision of subgenus *Chaetocruiomyia* Theobald (Diptera: Culicidae). *Papers / Department of Entomology, University of Queensland* 1(13): 189–211.

Marks, E.N. 1963b. The subgenus *Ochlerotatus* in the Australian region (Diptera: Culicidae) V. The Stricklandi Section. *Journal of the Entomological Society of Queensland* 2(1): 31–47.

Marks, E.N. 1964a. The subgenus *Ochlerotatus* in the Australasian Region (Diptera: Culicidae). VII. Four new species. *Papers / Department of Entomology, University of Queensland* 2(3): 59–71.

Marks, E.N. 1964b. Notes on the subgenus *Chaetocruiomyia* Theobald (Diptera: Culicidae). *Proceedings of the Linnean Society of New South Wales* 89(1): 131–147.

Marks, E.N. 1968. Northern records of the genus *Culiseta* Felt in Australia with the description of a new species (Diptera: Culicidae). *Journal of the Australian Entomological Society* 7(1): 43–56.

Marks, E.N. 1977. *Aedes* (*Macleaya*) *stoneorum*, a new species from Queensland (Diptera: Culicidae). *Proceedings of the Entomological Society of Washington* 79(1): 33–37.

Marks, E.N. & Hodgkin, E.P. 1958. A new species of *Aedes* (*Finlaya*) from northern Australia (Diptera, Culicidae). *Proceedings of the Linnean Society of New South Wales* 83(1): 33–39.

Marlatt, C.L. 1903. Collecting notes on mosquitoes in Oriental countries. *Proceedings of the Entomological Society of Washington* 5(2): 111–123.

Marsh, F. 1933. A new species of *Anopheles* (*Myzomyia* group) from South-west Persia. *Stylops: Journal of Taxonomic Entomology* 2(9): 193–197, 1 pl.

Marshall, J.F. & Staley, J. 1933. *Theobaldia* (*Culicella*) *litorea* (Shute) n.sp. (Diptera, Culicidae). *Parasitology* 25(1): 119–126.

Marshall, J.F. & Staley, J. 1937. Some notes regarding the morphological and biological differentiation of *Culex pipiens* Linnaeus and *Culex molestus* Forskål (Diptera, Culicidae). *Proceedings of the Royal Entomological Society of London, Series A: General Entomology* 12(1–2): 17–26.

Martinez, A. & Prosen, A.F. 1953a. Una especie nueva de *Uranotaenia* (Dipt. Culic.) de Argentina. *Boletín de la Sociedad Entomológica Argentina* 2: 4.

Martínez, A. & Prosen, A.F. 1953b. Una nueva especie de mosquito de la Argentina. *Publicaciones / Misión de Estudios de Patología Regional Argentina, Jujuy, Universidad de Buenos Aires* 1: 1–2.

Martínez, A. & Prosen, A.F. 1955. Las especies de *Aedes* (*Howardina*) de Argentina y Bolivia (Diptera, Culicidae). *Publicaciones / Misión de Estudios de Patología Regional Argentina, Jujuy, Universidad de Buenos Aires* 26(85–86): 21–41, 2 pls.

Martínez, A. & Prosen, A.F. 1958. Una nueva especie de *Orthopodomyia* (Dipt. Culicidae). *Publicaciones / Misión de Estudios de Patología Regional Argentina, Jujuy, Universidad de Buenos Aires* 27(87–88): 17–22.

Martínez, A., Carcavallo, R.U. & Prosen, A.F. 1960. Una nueva especie de Culicidae de Bolivia (Díptera). *Revista de la Cátedra de Microbiología y Parasitología* 28(89): 25–29.

Martínez, A., Carcavallo, R.U. & Prosen, A.F. 1961. El genero *Haemagogus* Williston, 1896, en la Argentina (*Diptera – Culicidae*). *Anales del Instituto de Medicina Regional* (for 1960) 5(2): 63–86.

Martini, E. 1914. Some new American mosquitoes. *Insecutor Inscitiae Menstruus* 2(5): 65–76, 1 pl.

Martini, E. 1920a. Über Stechmücken besonders deren europäische Arten und ihre Bekämpfung veröffentlicht mit Unterstützung der Hamburgischen wissenschaftlichen Stiftung. *Archiv für Schiffs- und Tropenhygiene* 24(1): 1–267, 4 pls.

Martini, E. 1920b. Uber mecklenburgische Culicinen. *Sitzungsberichte und Abhandlungen der Naturforschenden Gesellschaft zu Rostock* 7: 203–209, 1 pl.

Martini, E. 1922. Bemerkungen zu einigen neueren Mückenarbeiten, welche auch für die deutsche Fauna wichtig sind. *Entomologische Mitteilungen* 11(4): 158–166.

Martini, E. 1924. Culiciden - Beobachtungen 1922, 1923. *Zeitschrift für angewandte Entomologie* 10(2): 436–447.

Martini, E. 1925. Zwei bemerkenswerte Culiciden von einem eigenartigen Biotop. *Internationale Revue der gesamte Hydrobiologie und Hydrographie* 12(5/6): 333–337.

Martini, E. 1926. Über die Stechmücken der Umgebung von Saratow. *Raboty Volzhskoi Biologicheskoi Stantsii* [*Arbeiten der Biologischen Wolga-Station*] 8: 189–228.

Martini, E. 1927. Über zwei neue Stechmückenarten aus Anatolien. *Archiv für Schiffs- und Tropenhygiene* 31: 386–390.

Martini, E. 1929. 11. u. 12. Culicidae, pp. 1–144. *In*: Lindner, E. (ed.), *Die Fliegen der palaerktischen Region. Band III*. E. Schweizerbart'sche Verlagsbuchhandlung (Erwin Nägele), Stuttgart.

Martini, E. 1930. 11. u. 12. Culicidae, pp. 145–320. *In*: Lindner, E. (ed.), *Die Fliegen der palaerktischen Region. Band III*. E. Schweizerbart'sche Verlagsbuchhandlung (Erwin Nägele), Stuttgart.

Martini, E. 1931a. Ueber einige südamerikanische Culiciden. *Revista de Entomologia Rio de Janeiro* 1(2): 199–219.

Martini, E. 1931b. Die Ausbeute der deutschen Chaco-Expedition 1925/26. – Diptera. XXV. Culicidae. *Konowia* 10(2): 116–120.

Martini, E. 1932a. *Chagasia fajardoi* Lutz var. *stigmopteryx*, nov. var. (Dipt. Culicidae). *Revista de Entomologia* 2(3): 276–277.

Martini, E. 1932b. Dos nuevos mosquitos *Anopheles*, procedentes del Estado de Chiapas, México. *Revista Mexicana de Biología* 12: 99–102.

Martini, E. 1935. *Los mosquitos de México*. Departamento de Salubridad Pública, Boletines Tecnicos, Serie A: Entomología Médica y Parasitología No. 1. Mexico, D.F. 65 pp., content [1 p.].

Maslov, A.V. 1964. On the systematics of bloodsucking mosquitoes of the group *Culiseta* (Diptera, Culicidae) [in Russian]. *Entomologicheskoe Obozrenie* 43(1): 193–217.

Maslov, A.V. 1967. *Krovososushchie komarȳ podtribȳ Culisetina (Diptera, Culicidae) mirovoĭ faunȳ* [*Bloodsucking mosquitoes of the subtribe Culisetina (Diptera, Culicidae) of the world*]. Académie Impériale des Sciences, Musée Zoologique, St Petersburg. Opredeliteli po faune SSSR, Vȳp. 93. 182 pp., 1 pl.

Matheson, R. 1933. A new species of mosquito from Colorado (Diptera, Culicidae). *Proceedings of the Entomological Society of Washington* 35(5): 69–71.

Matheson, R. 1934. Notes on mosquitoes from South America, with a description of a new species (Diptera, Culicidae). *Proceedings of the Entomological Society of Washington* 36(5): 120–122.

Matsumura, S. 1916. *Thousand insects of Japan* [Shin Nihon senchu zukai]. *Additamenta* [Shin Nihon senchu zukai]. Volume II. Keisei-sha, Tokyo [pp. 185–474, 25 pls, contents].

Mattingly, P.F. 1949a. *Ficalbia* (*Ficalbia*) *jacksoni* sp. n., a new species of mosquito (Diptera, Culicidae) from Hong-Kong. *Proceedings of the Royal Entomological Society of London, Series B: Taxonomy* 18(1–2): 9–11.

Mattingly, P.F. 1949b. Notes on some Oriental mosquitoes. *Proceedings of the Royal Entomological Society of London, Series B: Taxonomy* 18(11–12): 219–228.

Mattingly, P.F. 1951a. The *Culex pipiens* complex. Introduction. *Transactions of the Royal Entomological Society of London* 102(7): 331–342.
Mattingly, P.F. 1951b. *Culex (Culex) nakuruensis* sp. n. (Diptera, Culicidae), a remarkable new mosquito from the East African Highlands. *Proceedings of the Royal Entomological Society of London, Series B: Taxonomy* 20(3–4): 44–46.
Mattingly, P.F. 1952. The sub-genus *Stegomyia* (Diptera: Culicidae) in the Ethiopian Region. I. A preliminary study of the distribution of species occurring in the West African sub-region with notes on taxonomy and bionomics. *Bulletin of the British Museum (Natural History) Entomology* 2(5): 233–304.
Mattingly, P.F. 1953a. The sub-genus *Stegomyia* (Diptera: Culicidae) in the Ethiopian Region. II. Distribution of species confined to the East and South African sub-region. *Bulletin of the British Museum (Natural History) Entomology* 3(1): 1–65.
Mattingly, P.F. 1953b. A change of name among the British mosquitoes. (Diptera, Culicidae). *Proceedings of the Royal Entomological Society of London, Series B: Taxonomy* 22(5–6): 106–108.
Mattingly, P.F. 1953c. Notes on the Culicini of the Katanga (Diptera, Culicidae). Part I. Taxonomy. *Revue de zoologie et de botanique Africaines* 47(3–4): 311–343.
Mattingly, P.F. 1953d. New records and a new species of the subgenus *Stegomyia* (Diptera, Culicidae) from the Ethiopian Region. *Annals of Tropical Medicine and Parasitology* 47(3): 294–298.
Mattingly, P.F. 1954a. Notes on the subgenus *Stegomyia* (Diptera, Culicidae), with a description of a new species. *Annals of Tropical Medicine and Parasitology* 48(3): 259–270.
Mattingly, P.F. 1954b. Notes on Ethiopian *Uranotaenia* (Diptera: Culicidae) with a description of a new species. *Proceedings of the Royal Entomological Society of London, Series B: Taxonomy* 23(9–10): 167–171.
Mattingly, P.-F. 1955. Le sous-genre *Neoculex* (*Diptera, Culicidae*) dans la sous-région mediterraneenne. I. – Espèce, sous-espèce et synonymie nouvelles. *Annales de Parasitologie Humaine et Comparée* 30(4): 374–388.
Mattingly, P.F. 1957a. *The culicine mosquitoes of the Indomalayan Area. Part I: Genus* Ficalbia *Theobald*. British Museum (Natural History), London. 61 pp.
Mattingly, P.F. 1957b. Genetical aspects of the *Aëdes aegypti* problem. I.–Taxonomy and bionomics. *Annals of Tropical Medicine and Parasitology* 51(4): 392–408.
Mattingly, P.F. 1957c. *The culicine mosquitoes of the Indomalayan Area. Part II: Genus* Heizmannia *Ludlow*. British Museum (Natural History), London. 57 pp.
Mattingly, P.F. 1958a. *The culicine mosquitoes of the Indomalayan Area. Part III. Genus* Aëdes *Meigen, subgenera* Paraëdes *Edwards,* Rhinoskusea *Edwards and* Cancraëdes *Edwards*. British Museum (Natural History), London. 61 pp.
Mattingly, P.F. 1958b. A revision of *Paraëdes* Edwards and *Cancraëdes* Edwards (Diptera: Culicidae). *Proceedings of the Royal Entomological Society of London, Series B: Taxonomy* 27(5–6): 76–83.
Mattingly, P.F. 1959. *The culicine mosquitoes of the Indomalayan Area. Part IV. Genus* Aëdes *Meigen, subgenera* Skusea *Theobald,* Diceromyia *Theobald,* Geoskusea *Edwards and* Christophersiomyia *Barraud*. British Museum (Natural History), London. 61 pp.
Mattingly, P.F. 1960. *Aëdes* subgenus *Alanstonea* subgen. nov. (Diptera: Culicidae). *Proceedings of the Royal Entomological Society of London, Series B: Taxonomy* 29(11–12): 170–171.
Mattingly, P.F. 1961. *The culicine mosquitoes of the Indomalayan Area. Part V. Genus* Aëdes *Meigen, subgenera* Mucidus *Theobald,* Ochlerotatus *Lynch Arribalzaga and* Neomelaniconion *Newstead*. British Museum (Natural History), London. 62 pp.
Mattingly, P.F. 1963. New and remarkable *Aedes* (Diptera: Culicidae) from Africa. *Proceedings of the Royal Entomological Society of London, Series B: Taxonomy* 32(11–12): 165–170.
Mattingly, P.F. 1965. *The culicine mosquitoes of the Indomalayan area. Part VI. Genus* Aedes *Meigen, subgenus* Stegomyia *Theobald (Groups A, B and D)*. British Museum (Natural History), London. 67 pp.
Mattingly, P.F. 1970. Contributions to the mosquito fauna of Southeast Asia. - VI. The genus *Heizmannia* Ludlow in Southeast Asia. *Contributions of the American Entomological Institute* 5(7): 1–104.
Mattingly, P.F. 1971. Contributions to the mosquito fauna of Southeast Asia. – XII. Illustrated keys to the genera of mosquitoes (Diptera, Culicidae). *Contributions of the American Entomological Institute* 7(4): 1–84.
Mattingly, P.F. 1980. An interim reclassification of the genus *Tripteroides* with particular reference to the Australasian subgenera. *Mosquito Systematics* 12(2): 164–171.
Mattingly, P.F. 1981. Medical entomology studies – XIV. The subgenera *Rachionotomyia, Tricholeptomyia* and *Tripteroides* (Mabinii Group) of genus *Tripteroides* in the Oriental region (Diptera: Culicidae). *Contributions of the American Entomological Institute* 17(5): ii, 1–147.
Mattingly, P.F. & Adam, J.P. 1954. A new species of cave-dwelling anopheline from the French Cameroons. *Annals of Tropical Medicine and Parasitology* 48(1): 55–57.
Mattingly, P.F. & Brown, E.S. 1955. The mosquitos (Diptera: Culicidae) of the Seychelles. *Bulletin of Entomological Research* 46(1): 69–110, 3 pls.
Mattingly, P.F. & Bruce-Chwatt, L.J. 1954. Morphology and bionomics of *Aëdes (Stegomyia) pseudoafricanus* Chwatt (Diptera, Culicidae), with some notes on the distribution of the subgenus *Stegomyia* in Africa. *Annals of Tropical Medicine and Parasitology* 48(2): 183–193.
Mattingly, P.F. & Hamon, J. 1955. Position taxonomique et synonymie de quelques *Ficalbia* de la Région Ethiopienne (*Diptera, Culicidae*). *Annales de Parasitologie Humaine et Comparée* 30(5–6): 488–496.
Mattingly, P.F. & Knight, K.L. 1956. The mosquitoes of Arabia I. *Bulletin of the British Museum (Natural History) Entomology* 4(3): 91–141.
Mattingly, P.F. & Marks, E.N. 1955. Some Australian mosquitoes (Diptera, Culicidae) of the subgenera *Pseudoskusea* and *Neoculex*. *Proceedings of the Linnean Society of New South Wales* 80(2): 163–176.

Mattingly, P.F. & Rageau, J. 1958. *Culex* (*Culex*) *iyengari* n. sp., a new species of mosquito (Diptera, Culicidae) from the South Pacific. *Pacific Science* 12(3): 241–250.

Mattingly, P.F., Stone, A. & Knight, K.L. 1962. *Culex aegypti* Linnaeus, 1762 (Insect, Diptera); proposed validation and interpretation under the plenary powers of the species so named. *Bulletin of Zoological Nomenclature* 19(4): 208–219.

McArthur, J. 1949. A new variety of *Anopheles aitkeni* from Borneo. *Bulletin of Entomological Research* 40(1): 49–52.

McCormack, A.T. 1919. *Report of the Health Department of The Panama Canal for the calendar year 1918*. The Panama Canal Press, Mount Hope, Canal Zone. 120 pp.

McCracken, I. 1904. *Anopheles* in California, with description of a new species. *Entomological News* 15(1): 9–14, 1 pl.

McIntosh, B.M. 1971. The aedine subgenus *Neomelanoconion* Newstead (Culicidae, Diptera) in southern Africa with descriptions of two new species. *Journal of the Entomological Society of Southern Africa* 34(2): 319–333.

McIntosh, B.M. 1973. A taxonomic re-assessment of *Aedes* (*Ochlerotatus*) *caballus* (Theobald) (Diptera: Culicidae) including a description of a new species of *Ochlerotatus.Journal of the Entomological Society of Southern Africa* 36(2): 261–269.

McIntosh, B.M. 1975. A taxonomic revision of certain *Aedes* species (Diptera: Culicidae) of the subgenus *Aedimorphus* in southern Africa. *Journal of the Entomological Society of Southern Africa* 38(2): 251–287.

Medschid. 1928. Über *Aëdes lepidonotus* Edw. und *Aëdes Refiki* n. sp. *Archiv für Schiffs- und Tropen-Hygiene* 32: 306–315.

Medschid. 1930. *Culex martinii* n. sp. *Archiv für Schiffs- und Tropen-Hygiene* 34: 364–369.

Meigen, J.W. 1804. *Klassifikazion und Beschreibung der europäischen Zweiflügeligen Insekten. (Diptera Linn.)*. Erster Band, ester Abtheilung, mit VIII Kupfertafeln. Karl Reichard, Braunschweig. xxviii, 152 pp., 8 pls.

Meigen, J.W. 1818. *Systematische Beschreibung der bekannten europäischen zweiflügeligen Insekten*. Volume 1. Aachen. xxxvi, 332 pp., errata (1 p.), 11 pls.

Meigen, J.W. 1830a. *Abbildung aller bis jetzt bekannten zweiflügeligen Insekten*. I Heft. Schulzische Buchhandlung, Hamm. [4 pp.], 10 pls.

Meigen, J.W. 1830b. *Systematische Beschreibung der bekannten europäischen zweiflügeligen Insekten*. Volume 6. Hamm. iv, 401 pp., 12 pls.

Meigen, J.W. 1838. *Systematische Beschreibung der bekannten europaeischen zweifluegeligen Insekten*. Volume 7. Hamm. xii, 434 pp., 8 pls.

Mendoza, J.B. 1940. New *Anopheles* of the *Myzorhynchus* series. *Monthly Bulletin of the Bureau of Health* (*Manila*) 20(11): 353–369.

Mendoza, J.B. 1947. Two more Philippine *Anopheles* in the *Myzorhynchus* series. *Monthly Bulletin of the Bureau of Health* (*Manila*) 23(3): 171–184.

Meng, C.-H. 1957. *The taxonomy, biology and control of Chinese anopheline mosquitoes*. People's Hygiene Press, Peking. 177 pp.

Meng, C.H. 1958. A new species of *Culex* [in Chinese]. *Acta Entomologica Sinica* 8(4): 351–354.

Meng, C.H. 1962. A new species of *Aedes* (Diptera, Culicidae). *Aedes* (*Ochlerotatus*) *lasaensis* sp. nov. [in Chinese]. *Acta Entomologica Sinica* 11(2): 159–163.

Meng, C.-H. 1981. A proposed system of classification of *Aedes chrysolineatus* and related species [in Chinese]. *Entomotaxonomia* 3(4): 243–254.

Meng, C.H. & Wu, C.Y. 1962. A new species of *Theobaldia* (Diptera, Culicidae) [in Chinese]. *Acta Entomologica Sinica* 11(4): 382–387.

Meng, Q. & Chen, H. 1980. Arguments on the taxonomic positions of six species of *Culex* from China [in Chinese]. *Entomotaxonomia* 2(2): 161–164.

Menon, M.A.U. 1944. On a new species of the subgenus *Lophoceratomyia* Theobald 1905 (Diptera Culicidae). *Journal of the Malaria Institute of India* 5(3): 389–393.

Menon, M.A.U. 1950. The male of a new species of the subgenus *Aëdes* Meigen, 1818 (Diptera: Culicidae). *Proceedings of the Royal Entomological Society of London, Series B: Taxonomy* 19(9–10): 139–141.

Meunier, F. 1915. Nouvelles recherches sur quelques insectes des plâtrières d'Aix en Provence. *Verhandelingen der Koninklijke Akademie van Wetenschappen le Amsterdam (Tweede Sectie)* 18(5): 1–17, 5 pls.

Miao, J., Pan, J. & Jiang, W. 1988. Hybridization and chromosome observations on six species of the *Anopheles hyrcanus* group in China (Diptera: Culicidae) [in Chinese]. *Zoologica Research* 9(3): 231–237.

Middlekauff, W.W. 1944. A new species of *Aedes* from Florida (Diptera: Culicidae). *Proceedings of the Entomological Society of Washington* 46(2): 42–44.

Mihályi, F. 1955. *Aëdes hungaricus* n. sp. (Culicidae, Diptera). *Annales Historico-Naturales Musei Nationalis Hungarici* 47(series nova 6): 343–345.

Miller, D. 1922. A remarkable mosquito, *Opifex fuscus*, Hutton. *Bulletin of Entomological Research* 13(1): 115–126.

Missiroli, A. 1935. Nuova varietà di *Anopheles maculipennis. Annali d'Igiene Sperimentale* 45(5): 333.

Mitchell, A., Sperling, F.A.H. & Hickey, D.A. 2002. Higher-level phylogeny of mosquitoes (Diptera: Culicidae): mtDNA data support a derived placement for *Toxorhynchites. Insect Systematics and Evolution* 33(2): 163–174.

Mitchell, E.G. 1906. Mouth parts of mosquito larvae as indicative of habits. *Psyche* 13(1): 11–21.

Mitchell, E.G. 1907. Validity of the culicid subfamily Deinoceritinae. *Psyche* 14(1): 11–13.

Miyagi, I. 1976. Description of a new species of the genus *Topomyia* Leicester from the Ryukyu Islands, Japan (Diptera: Culicidae). *Tropical Medicine* 17(4): 201–210.

Miyagi, I. 1980. *Topomyia* (*Suaymyia*) *rausai*, a new species of mosquito from the Philippines (Diptera: Culicidae). *Journal of Medical Entomology* 17(2): 178–181.

Miyagi, I. & Toma, T. 1981. A new species of *Aedes (Stegomyia)* from Daito Islands, Ryukyu, Japan (Diptera: Culicidae). *Mosquito Systematics* (for 1980) 12(4): 428–440.

Miyagi, I. & Toma, T. 1989. A new species of *Topomyia (Suaymyia) suchariti* from Thailand (Diptera: Culicidae). *Mosquito Systematics* 21(1): 16–24.

Miyagi, I. & Toma, T. 1995. *Topomyia javaensis*: A new Indonesian species in the subgenus *Topomyia* (Diptera: Culicidae). *Mosquito Systematics* 27(3): 179–186.

Miyagi, I. & Toma, T. 1997. Description of *Topomyia irianensis* n. sp. and new records of *To. papuensis* from Maluku and Irian Jaya, Indonesia (Diptera: Culicidae). *Journal of the American Mosquito Control Association* 13(2): 134–139.

Miyagi, I. & Toma, T. 2001. *Tripteroides* (*Rachionotomyia*) *ponmeki* (Diptera: Culicidae): A new species from Khammouane Province, Lao PDR. *Medical Entomology and Zoology* 52(4): 269–277.

Miyagi, I. & Toma, T. 2005. *Topomyia roslihashimi*, a new species of the subgenus *Suaymyia* (Diptera: Culicidae) from Gombak, Peninsular Malaysia. *Medical Entomology and Zoology* 56(4): 275–282.

Miyagi, I. & Toma, T. 2007. A new mosquito of the genus *Topomyia* (Diptera, Culicidae) from a *Nepenthes* pitcher plant in a Bario highland of Sarawak, Malaysia. *Medical Entomology and Zoology* 58(3): 167–174.

Miyagi, I. & Toma, T. 2008. Description of a new species *Topomyia* (*Suaymyia*) *lehcharlesi* (Diptera: Culicidae) from Sarawak, Malaysia. *Medical Entomology and Zoology* 59(3): 163–170.

Miyagi, I. & Toma, T. 2010a. Descriptions of *Topomyia auriceps* Brug, 1939 and *Topomyia pseudoauriceps*, n. sp. from Sarawak, Malaysia (Diptera: Culicidae). *Medical Entomology and Zoology* 61(1): 27–38.

Miyagi, I. & Toma, T. 2010b. *Topomyia* (*Suaymyia*) *kelabitense* (Diptera: Culicidae), a new species from Sarawak, Malaysia. *Medical Entomology and Zoology* 61(4): 353–361.

Miyagi, I. & Toma, T. 2013. *Uranotaenia* (*Pseudoficalbia*) *tanakai* (Diptera, Culicidae), a new species from forest swamps, Iriomote Is., the Ryukyu Archipelago, Japan. *Medical Entomology and Zoology* 64(3): 167–174.

Miyagi, I., Toma, T. & Cabrera, B.D. 1983. Synonyms of *Topomyia rausai* with *Topomyia apsarae,* with description of the larva (Diptera: Culicidae). *Mosquito Systematics* 15(2): 111–116.

Miyagi, I., Toma, T. & Higa, Y. 2004. A new species of *Mimomyia* (*Ingramia*) from Indonesia (Diptera: Culicidae). *Medical Entomology and Zoology* 55(1): 11–20.

Miyagi, I., Toma, T. & Lien, J.C. 2004. *Ochlerotatus* (*Geoskusea*) *timorensis* (Culicidae: Diptera), a new species from crab-holes, West Timor, Indonesia. *Medical Entomology and Zoology* 55(2): 107–114.

Miyagi, I., Toma, T., Okazawa, T., Wong, S.F., Leh, M.U. & Yong, H.S. 2012. Three new phytotelma mosquitoes of the genus *Topomyia* (Diptera: Culicidae) from Katibas, Lanjak-Entimau, Sarawak, Malaysia. *Journal of Science and Technology in the Tropics* 8(2): 97–117.

Miyagi, I., Toma, T., Okazawa, T., Wong, S.F., Leh, M.U. & Yong, H.S. 2014. Description of a new species of *Topomyia* (*Topomyia*) *aliyusopi* (Diptera: Culicidae) from Kelabit highlands, Sarawak, Malaysia. *Journal of Science and Technology in the Tropics* 10(1): 27–38.

Miyagi, I., Toma, T. & Ramalingam, S. 1989. *Topomyia (Topomyia) hardini*, a new species from Sarawak Malaysia (Diptera: Culicidae). *Tropical Biomedicine* 6: 91–98.

Miyagi, I., Toma, T. & Ramalingam, S. 1991. *Topomyia* (*Topomyia*) *yongi*, a new species of mosquito from Peninsular Malaysia (Diptera: Culicidae). *Mosquito Systematics* (for 1990) 22(3): 185–191.

Miyagi, I., Toma, T. & Rivera, D.G. 1983. *Topomyia* (*Topomyia*) *cabrerai,* a new species from the Philippines (Diptera: Culicidae). *Mosquito Systematics* 15(1): 1–4.

Miyagi, I., Toma, T. & Tsukamoto, M. 1983. Description of a new species, *Tripteroides (Tripteroides) riverai* from the Philippines (Diptera: Culicidae). *Mosquito Systematics* 15(1): 5–17.

Miyazaki, I. 1951. On a new anopheline mosquito *Anopheles yatsushiroensis* n. sp. found in Kyushu, with some remarks on two related species of the genus. *Kyushu Memoirs of Medical Sciences* 2(3): 195–206, 2 pls.

Mochizuki, D. 1913. Culicidae in the district of Fukuoka [in Japanese]. *Fukuoka Ika Daigaku zasshi* 7(1): 24–32.

Mochtar, R. & Walandouw, E.K. 1934. Een variëteit van *A. gigas*, (Uit Sumatra's Westkust). *Geneeskundig Tijdschrift voor Nederlandsch-Indië* 74: 932–935.

Mogi, M. 1977. A new species of *Aedes* (Diptera: Culicidae) from Japan. *Tropical Medicine* 19(2): 129–140.

Mogi, M. & Toma, T. 1999. *Culex (Acallyntrum) miyagii* (Diptera: Culicidae): New species from Seram Island, Indonesia, with keys to the species of the subgenus. *Journal of Medical Entomology* 36(4): 473–480.

Monier, H. & Treillard, M. 1935. *Anopheles* (*Myzomyia*) *funestus* var. *imerinensis* var. n. de Madagascar. *Bulletin de la Société de Pathologie exotique* 28(7): 572–573.

Monchadskii, A.S. 1951. The larvae of bloodsucking mosquitoes of the USSR and adjoining countries (Subfm. Culicinae) [in Russian]. *Opredeliteli po faune SSSR* [*Tableaux analytiques de la faune de l'URSS*]. Tom 37. Akademiia nauk SSSR, Zoologicheskiĭ Institut, Moskva [Moscow]. 290 pp.

Montchadsky, A. 1936. Mosquito larvae (fam. Culicidae) of the USSR and adjacent countries [in Russian]. *Opredeliteli po faune SSSR* [*Analytical tables of the Fauna of the USSR*] No. 24. Zoological Institute of the Academy of Sciences, Moscow. 383 pp.

Montchadsky, A.S. & Garcia Avila, I. 1966. Las larvas de los mosquitos (Diptera: Culicidae) de Cuba. Su biología y determinación. *Poeyana* (*Serie A*) 28: 1–92, 1 Table.

Montchadsky, H.S. 1926. Larva of *Aedes* (*Ochlerotatus*) *pulchritarsis*, Rond., var. *stegomyina*, Stack. & Montch., nov., from Turkestan. *Bulletin of Entomological Research* 17(2): 151–157.

Morishita, K. 1946. A new variety of Formosan *Anopheles* – *Anopheles leucosphyrus* var. *takasagoensis* var. nov. [in Japanese]. *Journal of the Formosan Medical Association* 45(2): 21, 65–70.

Moses, D.A., Howard, T.M. & Harbach, R.E. 2000. A new species of the subgenus *Sabethoides* of *Sabethes* (Diptera: Culicidae) from Venezuela and Brazil. *Proceedings of the Entomological Society of Washington* 102(4): 991–1002.

Motoki, M.T., Wilkerson, R.C. & Sallum, M.A.M. 2009. The *Anopheles albitarsis* complex with the recognition of *Anopheles oryzalimnetes* Wilkerson and Motoki, n. sp. and *Anopheles janconnae* Wilkerson and Sallum, n. sp. (Diptera: Culicidae). *Memórias do Instituto Oswaldo Cruz* 104(6): 823–850.

Motschulsky, V. 1859. Catalogue des insectes rapportés des environs du fl. Amour, depuis la Schilka jusqu'à Nikolaëvsk, examinés et énumérés. *Bulletin de la Société impériale des Naturalistes de Moscou* 32(2): 487–507.

Motta, M.A. & Lourenço-de-Oliveira, R. 1995. *Wyeomyia luteoventralis* Theobald, the type species of the subgenus *Dendromyia* Theobald (Diptera: Culicidae). *Memórias do Instituto Oswaldo Cruz* 90(3): 375–385.

Motta, M.A. & Lourenço-de-Oliveira, R. 2000. The subgenus *Dendromyia* Theobald: a review with redescriptions of four species (Diptera: Culicidae). *Memórias do Instituto Oswaldo Cruz* 95(5): 649-683.

Motta, M.A. & Lourenço-de-Oliveira, R. 2005. *Spilonympha*, a new subgenus of *Wyeomyia* (Diptera: Culicidae) and description of a new species *Wyeomyia aningae*. *Annals of the Entomological Society of America* 98(6): 838–852.

Motta, M.A., Lourenço-de-Oliveira, R. & Sallum, M.A.M. 2007. Phylogeny of genus *Wyeomyia* (Diptera: Culicidae) inferred from morphological and allozyme data. *Canadian Entomologist* 139(5): 591–627.

Müller, O.F. 1764. *Fauna insectorum Fridrichsdalina, sive methodica descriptio insectorum agri Fridrichsdalensis, cum characteribus genericis et specificis, nominibus trivialibus, locis natalibus, iconibus allegatis, novisque pluribus speciebus additis*. Hafniae et Lipsiae. xxiv, 96 pp.

Mulligan, H.W. & Puri, I.M. 1936. Description of *Anopheles (Anopheles) habibi* n.sp. from Quetta, Baluchistan. *Records of the Malaria Survey of India* 6(1): 67–71, 1 pl.

Muspratt, J. 1950. Notes on *Aëdes* (Diptera, Culicidae) from Natal, with a description of a new species of the subgenus *Stegomyia*. *Journal of the Entomological Society of Southern Africa* 13(1): 73–79.

Muspratt, J. 1953. Research on South African Culicini (Diptera Culicidae) 1. Descriptions of two new *Aëdes* (*Ochlerotatus*) Lynch Arribalzaga. *Journal of the Entomological Society of Southern Africa* 16(1): 51–58.

Muspratt, J. 1955. Research on South African Culicini (Diptera, Culicidae). III. – A check-list of the species and their distribution, with notes on taxonomy, bionomics and identification. *Journal of the Entomological Society of Southern Africa* 18(2): 149–207.

Muspratt, J. 1956. Research on South African Culicini (Diptera, Culicidae). IV. – Additional distribution records, taxonomic descriptions and miscellaneous notes. *Journal of the Entomological Society of Southern Africa* 19(2): 37–46.

Muspratt, J. 1959. Research on South African Culicini (Diptera: Culicidae). V. A new species of and notes on the sub-genus *Mucidus*, with other observations and records. *Journal of the Entomological Society of Southern Africa* 22(1): 64–77.

Muspratt, J. 1961. Research on South African Culicini (Diptera: Culicidae). VI. Two new sub-species and notes on bionomics and distribution. *Journal of the Entomological Society of Southern Africa* 24(1): 92–103.

Myers, C.M. 1967. Identification and descriptions of *Aedes* eggs from California and Nevada (Diptera: Culicidae). *Canadian Entomologist* 99(8): 795–806.

Nagaki, S.S., da Silva, A.M. & Sallum, M.A.M. 2011. Redescription of *Anopheles (Nyssorhynchus) lutzii*, and resurrection of *Anopheles guarani* from synonymy with *An. lutzii* (Diptera: Culicidae). *Annals of the Entomological Society of America* 104(3): 374–388.

Nagaki, S.S., de Albuquerque Motta, M. & Sallum, M.A.M. 2010. Redescription of *Anopheles (Nyssorhynchus) antunesi* Galvão & Amaral and description of a new species of the Myzorhynchella Section (Diptera: Culicidae) from Serra da Mantiqueira, Brazil. *Memórias do Instituto Oswaldo Cruz* 105(3): 278–285.

Nainggolan, F.J. 1939. Over een variëteit van *Anopheles gigas* uit Oedjali Kalah (Noord Kerintji), en de variabiliteit harer vleugelteekening. *Geneeskundig Tijdschrift voor Nederlandsch-Indië* 79: 163–170, 1 pl.

Nakata, G. 1956. A new species of mosquito, *Aëdes (Finlaya) kobayashii* n. sp. from Kyoto, Japan [in Japanese]. *Japanese Journal of Sanitary Zoology* 7(2): 135.

Natarajan, R. & Rajavel, A.R. 2009. Description of a new species of *Culex* (*Lophoceraomyia*) from Orissa, India. *Journal of the American Mosquito Control Association* 25(4): 403–408.

Natarajan, R., Rajavel, A.R. & Jambulingam, P. 2016. Description of a new species of the genus *Hulecoeteomyia* (Diptera: Culicidae) from Meghalaya, India. *Zootaxa* 4137(3): 330–338.

Natarajan, R., Rajavel, A.R. & Jambulingam, P. 2017. Descriptions of three new species of *Uranotaenia* (*Pseudoficalbia*) (Diptera: Culicidae) from India. *Zootaxa* 4227(2): 251–265.

Natarajan, R., Rajavel, A.R. & Jambulingam, P. 2018. *Uranotaenia pseudostricklandi*, a new species in subgenus *Pseudoficalbia* (Diptera: Culicidae) from Kerala, India. *Zootaxa* 4429(1): 181–188.

Natvig, L.R. 1942. Über *Theobaldia bergrothi* Edw., *Th. borealis* Schingarew und *Th. glaphyroptera* Schiner. *Notulae Entomologicae* 22(1–2): 1–5.

Natvig, L.R. 1948. Contributions to the knowledge of the Danish and Fennoscandian mosquitoes: Culicini. *Norsk Entomologisk Tidsskrift Suppl.* 1: xxiii, 567 pp., 12 pls, map.

Navarro, J.-C. & Liria, J. 2007. *Wyeomyia trujilloi* Pulido y Sutil, 1981, nuevo sinónimo de *Wyeomyia bicornis* (Root, 1928) (Culicidae: Sabethini), con redescripción de la pupa y parte de la larva. *Boletín de Malariología y Salud Ambiental* 47(1): 89–102.

Neiva, A. 1906. Uma nova especie de anophelina brazileira. *Brazil-Medico* 20: 288–289.

Neiva, A. 1908a. Contribuição ao estudo dos dipteros brazileiros. Uma nova especie de *Sabethes*. *Brazil-Medico* 22(36): 351–352.

Neiva, A. 1908b. Das anophelinas Brazileiras. *Revista Medica de São Paulo* 11(22): 455–459.

Neiva, A. & Pinto, C. 1922a. Contribuição para o conhecimento das anophelinas do estado de Motto Grosso com a descripção de uma nova especie. *Brazil-Medico* 36: 321–322.

Neiva, A. & Pinto, C. 1922b. Considerações sobre o genero "*Cellia*" Theobald, com a deseripção de uma nova especie. *Brazil-Medico* 36(2): 355–357.

Neiva, A. & Pinto, C. 1922c. Commentarios sobre o genero *Uranotaenia* Arribalzaga, 1891 com descripção de uma nova especie. *Brazil-Medico* 36(2): 374.
Neiva, A. & Pinto, C. 1923. Sobre uma nova anophelina brazileira (*Cellia cuyabensis* nov. sp.). *Brazil-Medico* 37(1): 235–236.
Neveu-Lemaire, M. 1902a. Sur la classification des Culicides. *Comptes Rendus Hebdomadaires des Séances et Mémoires de la Société de Biologie* 54: 1329–1332.
Neveu-Lemaire, M. 1902b. Description de quelques moustiques de la Guyane. *Archives de Parasitologie* 6(1): 5–25.
Neveu-Lemaire, M. 1902c. Classification de la famille des Culicidae. *Mémoires de la Société Zoologique de France* 15: 195–227.
Neveu-Lemaire, M. 1905a. Sur un nouveau moustique appartenant a la sous-famille des *Anophelinae* (*Nyssorhynchus Bozasi* nov. sp.). *Comptes Rendus Hebdomadaires des Séances et Mémoires de la Société de Biologie* 59: 32–33.
Neveu-Lemaire, M. 1905b. Mission du Bourg de Bozas. Description d'une nouvelle espèce de *Stegomyia* recueillie par le Dr Brumpt a Harar. *Bulletin de la Société Zoologique de France* 30: 8–11.
Neveu-Lemaire, M. 1906. Étude des culicides africains. *Archives de Parasitologie* 10(2): 238–288.
Neveu-Lemaire, M. 1923. L'evolution de la classification des *Culicidae*. *Annales de Parasitologie Humaine et Comparée* 1(1): 90–107.
Newstead, R. & Carter, H.F. 1910. Descriptions of a new genus and three new species of anopheline mosquitos. *Annals of Tropical Medicine and Parasitology* 4(3): 377–383, 2 pls.
Newstead, R. & Carter, H.F. 1911a. On a new genus of Culicinae from the Amazon region. *Annals of Tropical Medicine and Parasitology* 4(4): 553–556.
Newstead, R. & Carter, H.F. 1911b. On some new species of African mosquitos (Culicidae). *Annals of Tropical Medicine and Parasitology* 5(2): 233–244, 1 pl.
Newstead, R. & Thomas, H.W. 1910. The mosquitos of the Amazon region. *Annals of Tropical Medicine and Parasitology* 4(1): 141–150, 1 pl.
Newstead, R., Dutton, J.E. & Todd, J.L. 1907. Insects and other Arthropoda collected in the Congo Free State. *Annals of Tropical Medicine and Parasitology* 1(1–5): 3–113.
Ngo, C.T., Harbach, R.E., Garros, C., Parzy, D., Le, H.Q. & Manguin, S. 2013. Taxonomic assessment of *Anopheles crawfordi* and *An. dangi* of the Hyrcanus Group of subgenus *Anopheles* in Vietnam. *Acta Tropica* 128(3): 623–629.
Nguyen, D.M., Tran, D.H., Harbach, R.E., Elphick, J. & Linton, Y.-M. 2000. A new species of the Hyrcanus Group of *Anopheles*, subgenus *Anopheles*, a secondary vector of malaria in coastal areas of southern Vietnam. *Journal of the American Mosquito Control Association* 16(3): 189–198.
Nguyen, D.M., Tran, D.H. & Nguyen, T.H. 1993. *Anopheles* (*Anopheles*) *vietnamensis*: A new species of the *Anopheles hyrcanus* group from Vietnam (Diptera: Culicidae). *Mosquito Systematics* 25(3): 215–221.
Nicolescu, G., Linton, Y.-M., Vladimirescu, A., Howard, T.M. & Harbach, R.E. 2004. Mosquitoes of the *Anopheles maculipennis* group (Diptera: Culicidae) in Romania, with the discovery and formal recognition of a new species based on molecular and morphological evidence. *Bulletin of Entomological Research* 95(6): 525–535.
Nielsen, L.T. 2009. *Ochlerotatus dahlae*, a new mosquito species from Utah. *Journal of the American Mosquito Control Association* 25(2): 123–125.
Noè, G. 1899. Contribuzione allo studio dei Culicidi. *Bollettino della Società Entomologica Italiana* 31: 235–262.
Nurul Huda, K.M. & Harrison, B.A. 1985. Priority of the name *Anopheles pseudojamesi* for the species previously called *An. ramsayi* (Diptera: Culicidae). *Mosquito Systematics* 17(1): 49–51.
Ogasawara, H. 1939a. On the genus *Megarhinus* in Japan, with description of a new variety [in Japanese]. *Entomological World, Tokyo* 7(63): 240–244.
Ogasawara, H. 1939b. A new mosquito from Shakujii (Tokyo) [in Japanese]. *Entomological World, Tokyo* 7(63): 237–239.
Ohama, S. 1947. Epidemiological study of malaria in Yaeyama. No. 2 report. On an anopheline mosquito, *Anopheles ohamai* (Ishigaki Island) 1947, which is newly found in Ishigaki Island [in Japanese]. *Records of the Public Health Department, Yaeyama Provincial Government* 4: 1–15, 2 pls.
Olivier, M. 1791. *Encyclopedique méthodique. Histoire naturelle. Insectes.* Volume 6. Panckoucke, Paris. [4], 704 pp.
Oliveira Coutinho, J. & Forattini, O.P. 1962. Notas sôbre Culicidae (Diptera). 4ª – Contribuição ao estudo do subgênero *Microculex* com a descriçao duma nova espécie. *Studia Entomologica* 5(1–4): 209–234.
O'Meara, G.F. & Craig, G.B. 1970. A new subspecies of *Aedes atropalpus* (Coquillett) from southwestern United States (Diptera: Culicidae). *Proceedings of the Entomological Society of Washington* 72(4): 475–479.
Osorno-Mesa, E. 1944. Two new species of *Haemagogus* from Colombia, *H. andinus* and *H. boshelli* (Diptera: Culicidae). *Proceedings of the Entomological Society of Washington* 46(7): 165–175.
Osorno-Mesa, E. 1947. Una nueva especie de *Anopheles* de Bogota, Colombia. *Caldasia* 4(20): 431–446.
Osorno-Mesa, E. & Muñoz-Sarmiento, F. 1948. Una nueva variedad de *Anopheles pseudopunctipennis*. *Caldasia* 5(21): 105–113, 3 pls.
Osten Sacken, [C.]R. 1868. Description of a new species of Culicidae. *Transactions of the American Entomological Society* 2: 47–48.
Osten Sacken, C.R. 1877. Western Diptera: Descriptions of new genera and species of Diptera from the region west of the Mississippi and especially from California. *Bulletin of the United States Geological and Geographical Survey of the Territories* 3(2): 189–354.
Otsuru, M. 1952. Epidemiological studies on post-war malaria [in Japanese]. *Japanese Medical Journal* 1470: 2109–2113.
Ôuchi, Y. 1939. A new species of *Megarhinus* mosquito from Amami-Oshiina, the southern Japan. *Journal of the Shanghai Science Institute, Section III* 4(21): 223–225, 1 pl.

Ovazza, M., Hamon, J. & Neri, P. 1956. Contribution a l'étude des diptères vulnérants de l'empire d'Éthiopie. I. Culicidae. *Bulletin de la Société de Pathologie exotique* 49(1): 151–182.

Owen, W.B. 1945. A new anopheline from the Solomon Islands with notes on its biology. *Journal of Parasitology* 31(4): 236–240.

Paine, R.W. & Edwards, F.W. 1929. Mosquitos from the Solomon Islands. *Bulletin of Entomological Research* 20(3): 303–316, 2 pls.

Pajot, F.-X. & Fauran, P. 1975. *Wyeomyia (Wyeomyia) pseudorobusta* n. sp. *(Diptera, Culicidae)*, nouveau moustique découvert en Guyane française. *Cahiers ORSTOM Série Entomologie médicale et Parasitologie* 13(3): 125–129.

Pajot, F.-X. & Geoffroy, B. 1971. *Aedes (Neomelanoconion) bergerardi* sp. n. une nouvelle espèce de *Culicidae* de la République Centrafricaine. *Cahiers ORSTOM Série Entomologie médicale et Parasitologie* 9(3): 269–272.

Pallas, P.S. 1771. *Reise durch versehiedene Provinzen des Russischen Reichs*. Volume 1. Kayserliche Academie der Wissenschaften, St Petersburg. [12], 504 pp., Tab. I–XI, Tab. A–L.

Pandazis, G. 1935. La faune des Culicides de Grèce. *Acta Instituti et Musei Zoologici Universitatis Atheniensis* 1(1/2): 1–27.

Paterson, G.C. & Shannon, R.C. 1927. Mosquitos de Embarcación (Salta) con notas sobre la zona biológica del Chaco (Chaco Life Zone). *Boletín del Instituto de Clínica Quirúrgica* 21–25: 1–10.

Patton, W.S. 1905. The culicid fauna of the Aden Hinterland, their haunts and habits. *Journal of the Bombay Natural History Society* 16(4): 623–637, 5 pls.

Pazos, J.H. 1908. Descripcion de nuevas especies de mosquitos de Cuba. *Anales de la Academia de Ciencias de la Habana* 45: 417–432, 4 pls.

Pearse, A.S. 1936. Parasites from Yucatan, pp. 45–59. *In*: Pearse, A.S., Creaser, E.P. & Hall, F.G. (eds), *The cenotes of Yucatan: a zoological and hydrographic survey*. Carnegie Institution of Washington Publication 457. Carnegie Institution of Washington, Washington, DC. [iv], 304 pp., 19 pls.

Pecor, J.E., Mallampalli, V.L., Harbach, R.E. & Peyton, E.L. 1992. Catalog and illustrated review of the subgenus *Melanoconion* of *Culex* (Diptera: Culicidae). *Contributions of the American Entomological Institute* 27(2): v, 1–228.

Perez Vigueras, I. 1956. *Los ixodidos y culicidos de Cuba. Su historia natural y medica*. La Habana. 579 pp.

Perty, M. 1841. *Allgemeine Naturgeschichte, als philosophische und humanitätswissenschaft für Naturforscher, Philosophen und das höher gebildete Publikum*. Volume 3. C. Fischer, Bern. xii, pp. 467–1119, i–xxxiv (index).

Peryassú, A.G. 1908. *Os culicideos do Brazil*. Instituto de Manguinhos, Rio de Janeiro, Brazil. vi, 407 pp., errata (1 p.), foldout map, 6 foldout tables, 27 pls.

Peryassú, A.G. 1921a. Os anophelineos do Brasil. *Archivos do Museo Nacional* 23: 9–101.

Peryassú, A. 1921b. Um novo Anophelineo Brasileiro. *Folha medica* 2(18): 141.

Peryassú, A. 1922a. Uma nova especie do Culicineo Brasileiro. *Folha medica* 3(15): 117–118.

Peryassú, A. 1922b. Duas novas especies de mosquitos do Brasil. *Folha medica* 3: 179.

Peryassú, A. 1923a. Uma nova especie de mosquito do Brasil. *Folha medica* 4: 2.

Peryassú, A. 1923b. Uma nova especie de anophelineo do genero *Cyclolepidopteron*. *Folha medica* 4: 68–69.

Peryassú, A. 1923c. Catalogo das subfamilias, generos, especies e synonimias de mosquitos pernilongos encontrados no Brasil. *Folha medica* 4: 74–76.

Peryassú, A. 1925. *Anopheles alagoanii* (n sp) Peryassú. *Folha medica* 6: 258–259.

Pessoa, S.B. & Galvão, A.A. 1936. Notes sobre algumas especies de mosquitos que se criam em bromelias (Diptera, Culicidae). *Revista de Biologia e Hygiene* 6(2): 79–90.

Peters, W. 1955a. The mosquitoes of Liberia (Diptera: Culicidae). *Proceedings of the Royal Entomological Society of London, Series B: Taxonomy* 24(5–6): 81–90.

Peters, W. 1955b. New Anophelini from the southern highlands of Tanganyika and notes on some other members of the genus *Anopheles* Meigen (Diptera: Culicidae). *Proceedings of the Royal Entomological Society of London, Series B: Taxonomy* 24(5–6): 95–103.

Peters, W. 1955c. The mosquitoes of Liberia (Diptera: Culicidae). *Proceedings of the Royal Entomological Society of London, Series B: Taxonomy* 24(7–8): 117–128.

Peters, W. 1958. Mosquitoes of New Guinea (Diptera: Culicidae). Part 1. *Proceedings of the Royal Entomological Society of London, Series B: Taxonomy* 27(5–6): 89–92.

Peters, W. 1959. Mosquitoes of New Guinea (Diptera: Culicidae). Part II.—Notes on *Tripteroides* Giles, 1904. *Proceedings of the Royal Entomological Society of London, Series B: Taxonomy* 28(9–10): 135–154.

Peters, W. 1963a. Mosquitoes of New Guinea (Diptera: Culicidae). IV.–Two new species, genera *Aedes* Meigen and *Armigeres* Theobald. *Proceedings of the Royal Entomological Society of London, Series B: Taxonomy* 32(1–2): 1–11.

Peters, W. 1963b. Mosquitoes of New Guinea (Diptera: Culicidae). Part V.—New species and previously undescribed pupae of the genus *Tripteroides* Giles. *Proceedings of the Royal Entomological Society of London, Series B: Taxonomy* 32(5–6): 63–71.

Peters, W. 1963c. Mosquitoes of New Guinea (Diptera: Culicidae). VI.—The genus *Tripteroides* Giles: a new species of the subgenus *Polylepidomyia* Theobald, notes on the taxonomy and distribution and a revised key to described pupae. *Proceedings of the Royal Entomological Society of London, Series B: Taxonomy* 32(7–8): 89–100.

Peters, W. 1963d. Mosquitoes of New Guinea (Diptera: Culicidae). VII.–A new species complex of *Uranotaenia* Lynch Arribálzaga, 1891. *Proceedings of the Royal Entomological Society of London, Series B: Taxonomy* 32(9–10): 135–144.

Peters, W. 1963e. Mosquitoes of New Guinea (Diptera: Culicidae). Part VIII.—New species of the genus *Uranotaenia* Lynch Arribálzaga, 1891. *Proceedings of the Royal Entomological Society of London, Series B: Taxonomy* 32(11–12): 203–209.

Peters, W. 1964. Mosquitoes of New Guinea (Diptera: Culicidae). Part IX.—Further new species of *Uranotaenia* Lynch Arribálzaga and notes on the genus. *Proceedings of the Royal Entomological Society of London, Series B: Taxonomy* 33(1–2): 19–29.

Petrocchi, J. 1923. Estado actual de la sistemática de los Culicinae (mosquitos) en la República Argentina. Enumeración de especies. *Revista del Instituto Bacteriológico del Departamento de Higiene* 3: 83–93.

Petrocchi [as Petrochi], J. 1925. Descripción de un nuevo *Anopheles*. *Revista del Instituto Bacteriológico del Departamento de Higiene* 4: 69–75.
Peus, F. 1930. Bemerkungen über *Theobaldia subochrea* Edw. (Dipt. Culicid.). *Mitteilungen der Deutschen Entomologischen Gesellschaft* 1: 52–60.
Peus, F. 1935. *Theobaldia* (Subg. *Culicella*) *ochroptera* sp. n., eine bisher unbekannte Stechmücke. *Markische Tierwelt: Zeitschrift für die faunistische Erforschung der Kurmark* 1: 113–121.
Peus, F. 1954. Über Stechmücken in Griechenland (Diptera, Culicidae). *Bonner zoologische Beiträge* 1: 73–86, 1 pl.
Peus, F. 1970. Bemerkenswerte Mücken am Tegeler Fliess. *Berliner Naturschutzblätter* Special issue (May): 18–26.
Peyton, E.L. 1972. A subgeneric classification of the genus *Uranotaenia* Lynch Arribalzaga, with a historical review and notes on other categories. *Mosquito Systematics* 4(2): 16–40.
Peyton, E.L. 1974. *Uranotaenia srilankensis*, a new species of the subgenus *Pseudoficalbia* from Sri Lanka (Diptera: Culicidae). *Mosquito Systematics* 6(3): 222–227.
Peyton, E.L. 1977. Medical entomology studies – X. A revision of the subgenus *Pseudoficalbia* of the genus *Uranotaenia* in Southeast Asia (Diptera: Culicidae). *Contributions of the American Entomological Institute* 14(3): iv, 1–273.
Peyton, E.L. & Harbach, R.E. 1991. A new species of *Culex* (*Melanoconion*) encountered during arbovirus surveillance in Mexico (Diptera: Culicidae). *Mosquito Systematics* 23(3): 199–204.
Peyton, E.L. & Harrison, B.A. 1979. *Anopheles (Cellia) dirus*, a new species of the Leucosphyrus Group from Thailand (Diptera: Culicidae). *Mosquito Systematics* 11(1): 40–52.
Peyton, E.L. & Harrison, B.A. 1980. *Anopheles (Cellia) takasagoensis* Morishita 1946, an additional species in the Balabacensis Complex of Southeast Asia (Diptera: Culicidae). *Mosquito Systematics* 12(3): 335–347.
Peyton, E.L. & Klein, J.M. 1970. Five new species of *Uranotaenia* from Southeast Asia (Diptera: Culicidae). *Proceedings of the Entomological Society of Washington* 72(2): 243–251.
Peyton, E.L. & Ramalingam, S. 1988. *Anopheles* (*Cellia*) *nemophilous*, a new species of the Leucosphyrus Group from Peninsular Malaysia and Thailand (Diptera: Culicidae). *Mosquito Systematics* 20(2): 272–299.
Peyton, E.L. & Rattanarithikul, R. 1970. Five additional new species of *Uranotaenia* from Southeast Asia (Diptera: Culicidae). *Proceedings of the Entomological Society of Washington* 72(3): 403–413.
Peyton, E.L., Roberts, D.R., Pinheiro, F.P., Vargas, R. & Balderama, F. 1983. Mosquito collections from a remote unstudied area of southeastern Bolivia. *Mosquito Systematics* 15(2): 61–89.
Phanitchakun, T., Wilai, P., Saingamsook, J., Namgay, R., Drukpa, T., Tsuda, Y., Walton, C., Harbach, R.E. & Somboon, P. 2017. *Culex* (*Culiciomyia*) *sasai* (Diptera: Culicidae), senior synonym of *Cx. spiculothorax* and a new country record for Bhutan. *Acta Tropica* 171: 194–108.
Philip, C.B. 1931. Two new species of *Uranotaenia* (Culicidae) from Nigeria, with notes on the genus in the Ethiopian Region. *Bulletin of Entomological Research* 22(2): 183–193.
Philippi, R.A. 1865. Aufzählung der chilenische Dipteren. *Verhandlungen der kaiserlich-königlichen zoologisch-botanischen Gesellschaft in Wien* 15: 595–782 + pls 23–29.
Pillai, J.S. 1966. *Culiseta novaezealandiae*, a new species of the subgenus *Climacura* Felt (Diptera: Culicidae: Culisetini), with notes on its ecology and development. *Transactions of the Royal Society of New Zealand, Zoology* 8(11): 125–133, 1 pl.
Pinto, C. 1932. Alguns mosquitos do Brasil e do oriente do Bolivia (Diptera. Culicidae). *Revista medico-cirurgica do Brasil* 40(10): 285–309.
Pinto, C. 1939. Disseminação do malaria pela aviação; biologia do *Anopheles gambiae* e outros anofelineos do Brasil. *Memórias do Instituto Oswaldo Cruz* 34(3): 293–430, 61 pls.
Piton, L. 1936. Addition a la faune entomologique des cinérites du Lac Chambon (Puy-de-Dome). *Revue Scientifique du Bourbonmais* 1936: 16–24.
Pittaluga, G. 1905. Estudios acerca de los Diptéros y de los parásitos que transmiten al hombre y á los animales domésticos. *Revista de la Real Academia de Ciencias Exactas, Físicas y Naturales de Madrid* 3(5): 402–504.
Podėnas, S. 1999. New species of fossil mosquitoes (Diptera, Culicidae) from the Baltic amber. *Acta Zoologica Lituanica* 9(3): 111–114.
Pohl, J.E. & Kollar, V. 1832. Brasiliens vorzüglich lästige Insecten. Besonderer Abdruck aus der Reise im Innern von Brasilien von Dr. Pohl [extracted from Pohl's "*Reise im Innern von Brasilien*"]. Wien. 20 pp., 1 pl.
Poinar, G., Jr. 2005. *Culex malariager*, n. sp. (Diptera: Culicidae) from Dominican amber: the first fossil mosquito vector of *Plasmodium*. *Proceedings of the Entomological Society of Washington* 107(3): 548–553.
Poinar, G.O., Jr., Zavortink, T.J., Pike, T. & Johnston, P.A. 2000. *Paleoculicis minutus* (Diptera: Culicidae) n. gen., n. sp., from Cretaceous Canadian amber, with a summary of described fossil mosquitoes. *Acta Geologica Hispanica* 35(1–2): 119–128.
Poiret, [J.-L.M.] 1787. Mémoire sur quelques insectes de Barbarie. *Journal de Physique, de Chimie, d'Histoire Naturelle et des Arts* 30(1): 241–245, 1 pl.
Porter, C.H. 2009. *Wyeomyia* (*Hystatomyia*) *baltae*, a new species of Sabethini (Diptera: Culicidae) from Peru. *Proceedings of the Entomological Society of Washington* 111(4): 807–825.
Porter, C.H. 2014. *Wyeomyia* (*Nunezia*) *paucartamboensis*, a new species of Sabethini (Diptera: Culicidae) from the Peruvian Andes with a diagnosis of the subgenus *Nunezia*. *Proceedings of the Entomological Society of Washington* 116(3): 311–338.
Porter, C.H. & Wolff E., M.I. 2004. A new species of *Wyeomyia* (*Hystatomyia*) (Diptera: Culicidae) from Colombia and a redescription of *Wy.* (*Hystatomyia*) *intonca* Dyar & Knab. *Zootaxa* 477: 1–31.
Portschinsky, J. 1884. Diptera europaea et asiatica nova aut minus cognita (Cum notis biologicis). *Trudy Russkago entomologicheskago obshchestva. Horae Societatis entomologicae rossicae, variis semonibus in Russia usitatis editae* 18: 122–134.
Portschinsky, J. 1910. Overview of the distribution of the most important pest animals in Russia in 1909 [in Russian]. *Ezhegodnik Glavnogo Upravleniya Zemleustroistva I Zemledeliya po Departamentu Zemledeliya* 3(3): 602–614.

Portschinsky, J. 1911. The malaria mosquito (*Anopheles claviger* F.) in connection with the malaria, their lives, their properties and circumvention [in Russian]. *Trudy Bīuro po ėntomologīi Uchenago Komiteta Glavnago Upravlenīia Zemleustroĭstva i Zemledīelīia* 5(1): 1–137, 3 pls.

Prado, A. 1931a. Contribuições ao conhecimento dos culicideos de São Paulo I. Notas sobre *Mansonia albifera* Prado e sobre o macho de *Mansonia albicosta* (Chagas). *Memorias do Instituto Butantan* 6: 193–198.

Prado, A. 1931b. Contribuições ao conhecimento dos culicideos de São Paulo IV. Uma nova especie de *Uranotaenia* (Diptera, Culicidae). *Memorias do Instituto Butantan* 6: 209–213.

Prashad, B. 1918. XVI. The description and life-history of a new species of *Anopheles* that breeds in holes in trees. *Records of the Indian Museum* 15(3): 123–127, 1 pl.

Pressat, A. 1905. *Le Paludisme et les Moustiques (Prophylaxie)*. Masson et C[ie], Paris. viii, 155 pp.

Pucat, A.M. 1965. The functional morphology of the mouthparts of some mosquito larvae. *Quaestiones Entomologicae* 1(2): 41–86.

Pulido F., J. & Sutil 0., E. 1981. *Wyeomyia (Wyeomyia) trujilloi* (Diptera, Culicidae) nueva especie de Venezuela. *Boletin de la Direccion de Malariologia y Saneamiento Ambiental* 21(3–4): 219–226.

Puri, I.M. 1928. On two species of Indian anopheline mosquitoes, *A. jeyporiensis* James and *A. moghulensis* Christophers. *Indian Journal of Medical Research* 16(2): 513–518, 2 pls.

Puri, I.M. 1929a. Description of the male, female, egg and larva of *Anopheles annandalei* var. *interruptus* nov. var., with corrections for the previous descriptions of the type species. *Indian Journal of Medical Research* 17(2): 385–395, 1 pl.

Puri, I.M. 1929b. A new tree-hole breeding *Anopheles* from South India—*Anopheles sintoni* sp. nov.—and a revised description of the larva *of A. culiciformis* Cogill. *Indian Journal of Medical Research* 17(2): 397–404, 1 pl.

Puri, I.M. 1930. A note on two species of Indian anopheline mosquitoes—*A. insulaeflorum* Swellengrebel and *A. aitkenii* James, with its variety *bengalensis* nov. var. *Indian Journal of Medical Research* 17(3): 953–956.

Puri, I.M. 1949. Anophelines of the Oriental Region, pp. 483–505. *In*: Boyd, M.F. (ed.), *Malariology. A comprehensive survey of all aspects of this group of diseases from a global standpoint*. Volume 1. W.B. Saunders Company, Philadelphia and London.

Qu, F.-Y. & Zhu, H.-M. 2008. On a new checklist of the anopheline mosquitoes in China with rectification for some specific names [in Chinese]. *Chinese Journal of Parasitology and Parasitological Diseases* 26(3): 210–216.

Qutubuddin, M. 1947. A new species of *Uranotaenia*, Culicidae-Diptera from Hyderabad (Deccan). *Indian Journal of Entomology* (for 1946) 8(1): 117–118.

Qutubuddin, M. 1951. *Uranotaenia mattinglyi* sp. n., a new species of mosquito (Diptera, Culicidae) from Hyderabad (Deccan), India. *Proceedings of the Royal Entomological Society of London, Series B: Taxonomy* 20(9–10): 107–109.

Qutubuddin, M. 1956. *Culex* (*Culex*) *afridii* sp. n. a new species of mosquito (Diptera, Culicidae) from Dacca, East Pakistan. *Mosquito News* 16(2): 140–142.

Qutubuddin, M. 1959. A new species of *Aëdimorphus* (Diptera: Culicidae) from Sudan Republic. *Annals and Magazine of Natural History* 2(13): 21–23.

Qutubuddin, M. 1972. A new species of *Finlaya* (Diptera: Culicidae) from Papua. *Pakistan Journal of Health* 22(1–4): 119–120.

Raffaele, G. 1928. Una nuova specie di "*Anopheles*„. *Rivista di Malariologia* 7: 11–17.

Rafinesque, C.S. 1820. Fishes of the River Ohio. *Western Review and Miscellaneous Magazine* 3(4): 244–252.

Rageau, J. & Hamon, J. 1957. *Aedes (Diptera Culicidae)* appartenant au sous-genre *Mucidus* en Nouvelle-Calédonie. *Bulletin de la Société de Pathologie exotique* 50(3): 372–378.

Rahman, S.J., Choudhury, D.S. & Kalra, N.L. 1969. Description of *Culex* (*Lophoceraomyia*) *raghavanii* sp. n. (Diptera: Culicidae) from Nilgiri Hills, South India. *Bulletin of the Indian Society for Malaria and Other Communicable Diseases* (for 1968) 5(1–2): 28–34.

Rahola, N., Makanga, B., Yangari, P., Jiolle, D., Fontenille, D., Renaud, F., Ollomo, B., Ayala, D., Prugnolle, F. & Paupy, C. 2014. Description of *Anopheles gabonensis*, a new species potentially involved in rodent malaria transmission in Gabon, Central Africa. *Infection, Genetics and Evolution* 28: 628–634.

Ramalingam, S. 1972. A new species of *Armigeres* from Sabah, Borneo (Diptera: Culicidae). *Proceedings of the Entomological Society of Washington* 74(4): 459–467.

Ramalingam, S. 1974. Some new records of *Anopheles* from Sabah, Malaysia. *Southeast Asian Journal of Tropical Medicine and Public Health* 5(1): 147–148.

Ramalingam, S. 1975. A new species of *Topomyia* from peninsular Malaysia (Diptera: Culicidae). *Mosquito Systematics* 7(3): 185–192.

Ramalingam, S. 1987. On the restriction of *Armigeres durhami* Edwards and the description of *Armigeres kesseli* n. sp. (Diptera: Culicidae). *Tropical Biomedicine* 4(1): 55–65.

Ramalingam, S. 1988. Description of a new species of *Topomyia* from peninsular Malaysia (Diptera: Culicidae). *Mosquito Systematics* (for 1987) 19(3): 245–250.

Ramalingam, S. & Banu, Q. 1987. Studies on the genus *Topomyia:* 2. Description of a new species from Sabah, Malaysia (Diptera: Culicidae). *Tropical Biomedicine* 4: 119–124.

Ramalingam, S. & Belkin, J.N. 1965. Mosquito studies (Diptera, Culicidae). III. Two new species of *Aedes* from Tonga and Samoa. *Contributions of the American Entomological Institute* 1(4): 1–10.

Ramalingam, S. & Pillai, A.G. 1972. A new species of *Malaya* from West Malaysia (Diptera: Culicidae). *Proceedings of the Entomological Society of Washington* 74(4): 451–459.

Ramalingam, S. & Ramakrishna, K. 1988. Studies on the genus *Topomyia*: 1. A new species from Sabah, Malaysia. *Mosquito Systematics* 20(1): 33–40.

Ramos, A.S. 1942. Sobre uma variedade nova de *Anopheles (Nyssorhynchus) oswaldoi* Peryassú, 1922 *(Diptera-Culicidae)*. *Arquivos de Higiene e Saúde Pública* 7(15): 61–71.

Rangel, R. & Romero Sierra, J.M. 1907. Mosquito número 9, pp. 45–48. *In*: Romero Sierra, J.M. 1907. Contribucion al estudio de los mosquitos de Caracas. Tesis do Doctorado No. 67, Facultad de Ciencias Medícas, Universidad Central de Venezuela, Caracas. [1–11]–55 pp.

Rattanarithikul, R. & Green, C.A. 1987. Formal recognition of the species of the *Anopheles maculatus* group (Diptera: Culicidae) occurring in Thailand, including the descriptions of two new species and a preliminary key to females. *Mosquito Systematics* (for 1986) 18(3,4): 246–278.

Rattanarithikul, R. & Harbach, R.E. 1991. *Anopheles maculatus* (Diptera: Culicidae) from the type locality of Hong Kong and two new species of the Maculatus Complex from the Philippines. *Mosquito Systematics* (for 1990) 22(3): 160–183.

Rattanarithikul, R. & Harrison, B.A. 1988. *Aedes (Finlaya) reinerti*, a new species from northern Thailand related to *Aedes (Finlaya) formosensis* Yamada (Diptera: Culicidae). *Mosquito Systematics* 20(1): 77–96.

Rattanarithikul, R., Harrison, B.A., Panthusiri, P., Peyton, E.L. & Coleman, R.E. 2006. Illustrated keys to the mosquitoes of Thailand. III. Genera *Aedeomyia*, *Ficalbia*, *Mimomyia*, *Hodgesia*, *Coquillettidia*, *Mansonia*, and *Uranotaenia*. *Southeast Asian Journal of Tropical Medicine and Public Health* 37(suppl. 1): 1–85.

Ravaonjanahary, C. & Bruhnes [*sic*], J. 1977. Un nouvel *Aedes* du genre *Skusea* découvert à Madagascar: *Aedes* (*Skusea*) *moucheti* sp. n. *Cahiers ORSTOM Série Entomologie médicale et Parasitologie* 15(3): 213–215.

Reid, J.A. 1947. Type specimens of Culicidae described by Laveran (Diptera: Culicidae). *Proceedings of the Royal Entomological Society of London, Series B: Taxonomy* 16(7–8): 86–91.

Reid, J.A. 1950. The *Anopheles umbrosus* group (Diptera: Culicidae). Part 1: systematics, with descriptions of two new species. *Transactions of the Royal Entomological Society of London* 101(9): 281–318.

Reid, J.A. 1953. The *Anopheles hyrcanus* group in south-east Asia (Diptera: Culicidae). *Bulletin of Entomological Research* 44(1): 5–76.

Reid, J.A. 1962. The *Anopheles barbirostris* group (Diptera, Culicidae). *Bulletin of Entomological Research* 53(1): 1–57.

Reid, J.A. 1963. Notes on anopheline mosquitoes from Malaya, with descriptions of three new species. *Annals of Tropical Medicine and Parasitology* 57(1): 97–116.

Reid, J.A. 1965. A revision of the *Anopheles aitkenii* group in Malaya and Borneo. *Annals of Tropical Medicine and Parasitology* 59(1): 106–125.

Reid, J.A. 1966. A note on *Anopheles subpictus* Grassi and *A. indefinitus* Ludlow (Diptera: Culicidae). *Journal of Medical Entomology* 3(3–4): 327–331.

Reid, J.A. 1968. *Anopheline mosquitoes of Malaya and Borneo*. Studies from the Institute for Medical Research Malaya, No. 31. Government of Malaysia, Kuala Lumpur. xiii, 520 pp.

Reinert, J.F. 1970. Contributions to the mosquito fauna of Southeast Asia. V. Genus *Aedes*, subgenus *Diceromyia* Theobald in Southeast Asia. *Contributions of the American Entomological Institute* 5(4): 1–43.

Reinert, J.F. 1972a. Contributions to the mosquito fauna of Southeast Asia. - XV. Genus *Aedes* Meigen, subgenus *Ayurakitia* Thurman. *Contributions of the American Entomological Institute* 9(2): 1–42.

Reinert, J.F. 1972b. *Aedes gouldi,* a new species of the subgenus *Aedimorphus* Theobald from West Pakistan (Diptera: Culicidae). *Proceedings of the Entomological Society of Washington* 74(4): 357–362.

Reinert, J.F. 1973a. *Aedes wainwrighti* Baisas, a synonym of *Aedes* (*Stegomyia*) *meronephada* (Dyar and Shannon), with notes on the subgenus *Stegomyia* Theobald (Diptera: Culicidae). *Mosquito Systematics* 5(1): 27–30.

Reinert, J.F. 1973b. Contributions to the mosquito fauna of Southeast Asia. – XVI. Genus *Aedes* Meigen, subgenus *Aedimorphus* Theobald in Southeast Asia. *Contributions of the American Entomological Institute* 9(5): 1–218.

Reinert, J.F. 1973c. *Aedes consonensis*, a new species of the subgenus *Neomacleaya* from South Vietnam (Diptera: Culicidae). *Mosquito Systematics* 5(3): 252–262.

Reinert, J.F. 1973d. Contributions to the mosquito fauna of Southeast Asia. -XVIII. A reconsideration of *Diceromyia* Theobald with the inclusion of *Aedes nummatus* Edwards and *Aedes pseudonummatus* new species (Diptera: Culicidae). *Contributions of the American Entomological Institute* 10(1): 22–40.

Reinert, J.F. 1973e. Contributions to the mosquito fauna of Southeast Asia. -XIX. *Bothaella,* a new subgenus of *Aedes* Meigen. *Contributions of the American Entomological Institute* 10(3): 1–51.

Reinert, J.F. 1974. Medical entomology studies – I. A new interpretation of the subgenus *Verrallina* of the genus *Aedes* (Diptera: Culicidae). *Contributions of the American Entomological Institute* 11(1): 1–249.

Reinert, J.F. 1976a. A new man-biting species of *Aedes (Paraedes)* from Southeast Asia (Diptera: Culicidae). *Mosquito Systematics* 8(3): 319–331.

Reinert, J.F. 1976b. Medical entomology studies – V. The subgenus *Rhinoskusea* of the genus *Aedes* (Diptera: Culicidae). *Contributions of the American Entomological Institute* 13(2): 1–60.

Reinert, J.F. 1977. Description of a new species of *Aedes (Verrallina)* from Sri Lanka (Diptera: Culicidae). *Mosquito Systematics* 9(3): 366–371.

Reinert, J.F. 1979. A description of *Isoaedes,* a new subgenus of *Aedes* Meigen, and its type-species, *Ae.* (*Isa.*) *cavaticus* new species (Diptera: Culicidae). *Mosquito Systematics* 11(2): 144–162.

Reinert, J.F. 1981. Medical entomology studies - XV. A revision of the subgenus *Paraedes* of the genus *Aedes* (Diptera: Culicidae). *Contributions of the American Entomological Institute* 18(4): 1–91.

Reinert, J.F. 1982. *Belkinius*, a new subgenus of *Aedes* and a description of *Ae.* (*Blk.*) *aurotaeniatus* Edwards (Diptera: Culicidae). *Mosquito Systematics* 14(2): 101–121.

Reinert, J.F. 1987. *Albuginosus*, a new subgenus of *Aedes* Meigen (Diptera: Culicidae) described from the Afrotropical region. *Mosquito Systematics* (for 1986) 18(3,4): 307–326.

Reinert, J.F. 1990. Medical entomology studies - XVII. Biosystematics of *Kenknightia*, a new subgenus of the mosquito genus *Aedes* from the Oriental Region (Diptera: Culicidae). *Contributions of the American Entomological Institute* 26(2): 1–119.

Reinert, J.F. 1993. Redescription of *Molpemyia*, and its revalidation as a subgenus of *Aedes* (Diptera: Culicidae). *Mosquito Systematics* 25(1): 41–63.

Reinert, J.F. 1999a. Descriptions of *Zavortinkius*, a new subgenus of *Aedes*, and the eleven included species from the Afrotropical Region (Diptera: Culicidae). *Contributions of the American Entomological Institute* 31(2): 1–105.

Reinert, J.F. 1999b. Restoration of *Verrallina* to generic rank in tribe Aedini (Diptera: Culicidae) and descriptions of the genus and three included subgenera. *Contributions of the American Entomological Institute* 31(3): 1–83.

Reinert, J.F. 2000a. Description of *Fredwardsius*, a new subgenus of *Aedes* (Diptera: Culicidae). *European Mosquito Bulletin* 6: 1–7.

Reinert, J.F. 2000b. Synonymy of subgenus *Sinoaedes* of genus *Aedes* with subgenus *Mattinglyia* of genus *Heizmannia*. *Journal of the American Mosquito Control Association* 16(1): 38–39.

Reinert, J.F. 2000c. Restoration of *Ayurakitia* to generic rank in tribe Aedini and a revised definition of the genus. *Journal of the American Mosquito Control Association* 16(2): 57–65.

Reinert, J.F. 2000d. New classification for the composite genus *Aedes* (Diptera: Culicidae: Aedini), elevation of subgenus *Ochlerotatus* to generic rank, reclassification of the other subgenera, and notes on certain subgenera and species. *Journal of the American Mosquito Control Association* 16(3): 175–188.

Reinert, J.F. 2003. Description of *Bruceharrisonius*, a new subgenus of *Ochlerotatus*, and a redescription of its type species *Oc. (Brh.) greenii*. *Journal of the American Mosquito Control Association* 19(4): 309–322.

Reinert, J.F. 2006a. *Tewarius*, a new genus of Aedini (Diptera: Culicidae). *Proceedings of the Entomological Society of Washington* 108(3): 639–645.

Reinert, J.F. 2006b. Description of subgenus *Pardomyia* Theobald and comparison with the nominotypical subgenus of *Mucidus* Theobald (Diptera: Culicidae: Aedini). *Journal of the American Mosquito Control Association* 22(4): 586–600.

Reinert, J.F. & Harbach, R.E. 2005. Corrections to the spelling of species names in genera *Psorophora* and *Verrallina*, with a note on the correct gender of *Udaya*. *Journal of the American Mosquito Control Association* 21(2): 227–228.

Reinert, J.F., Harbach, R.E. & Kitching, I.J. 2004. Phylogeny and classification of Aedini (Diptera: Culicidae) based on morphological characters of all life stages. *Zoological Journal of the Linnean Society* 142(3): 289–368.

Reinert, J.F., Harbach, R.E. & Kitching, I.J. 2006. Phylogeny and classification of *Finlaya* and allied taxa (Diptera: Culicidae: Aedini) based on morphological data from all life stages. *Zoological Journal of the Linnean Society* 148(1): 1–101.

Reinert, J.F., Harbach, R.E. & Kitching, I.J. 2008. Phylogeny and classification of *Ochlerotatus* and allied taxa (Diptera: Culicidae: Aedini) based on morphological data from all life stages. *Zoological Journal of the Linnean Society* 153(1): 29–114.

Reinert, J.F., Harbach, R.E. & Kitching, I.J. 2009. Phylogeny and classification of tribe Aedini (Diptera: Culicidae). *Zoological Journal of the Linnean Society* 157(4): 700–794 + 2 online appendices.

Reinert, J.F., Kaiser, P.E. & Seawright, J.A. 1997. Analysis of the *Anopheles* (*Anopheles*) *quadrimaculatus* complex of sibling species (Diptera: Culicidae) using morphological, cytological, molecular, genetic, biochemical, and ecological techniques in an integrated approach. *Journal of the American Mosquito Control Association* 13(Supplement): 1–102.

Reuben, R. 1967. *Aedes* (*Diceromyia*) *ramachandrai*, n. sp. (Diptera: Culicidae) from southern India. *Journal of Medical Entomology* 4(2): 234–236.

Ribbands, C.R. 1944a. Differences between *Anopheles melas (A. gambiae* var. *melas)* and *Anopheles gambiae*. *Annals of Tropical Medicine and Parasitology* 38(2): 85–86.

Ribbands, C.R. 1944b. Differences between *Anopheles melas* and *Anopheles gambiae* II.—Salinity relations of larvae and maxillary palp banding of adult females. *Annals of Tropical Medicine and Parasitology* 38(2): 87–99.

Ribeiro, H. 1969. Research on the mosquitoes of Angola. IV – Description of *Anopheles (Cellia) azevedoi* sp. nov. *(Diptera, Culicidae)*. *Anais da Escola Nacional de Saúde Pública e de Medicina Tropical* 3(1/4): 113–123.

Ribeiro, H. 1991. Research on the mosquito subfamily *Toxorhynchitinae (Diptera: Culicidae)*. I—The Afrotropical group *brevipalpis* (Adults) (1). *Arquivos do Museu Bocage* 2(3): 31–62.

Ribeiro, H., 1992a. Research on the mosquito subfamily Toxorhynchitinae (Diptera: Culicidae) II – description of the Afrotropical subgenus *Afrorhynchus* subg. nov. *Mosquito Systematics* (for 1991) 23(3): 195–198.

Ribeiro, H. 1992b. On the terrestrial and riverine fauna of the Democratic Republic of São Tomé and Príncipe. VII – Description of *Toxorhynchites capelai* sp. nov. (*Diptera, Culicidae*). *Garcia de Orta, Série de Zoologia* 19(1–2): 135–128.

Ribeiro, H. 1992c. Research on the mosquito subfamily Toxorhynchitinae (Diptera: Culicidae). III – Description of two new species from Angola. *Boletim da Sociedade Portuguesa de Entomologia* 2(Suplemento 3): 187–196.

Ribeiro, H. 1997. A new species of *Toxorhynchites* (Diptera: Culicidae) from Macau (China). *Journal of the American Mosquito Control Association* 13(3): 213–217.

Ribeiro, H. 2004. Les *Toxorhynchites* Theobald de Madagascar (Diptera: Culicidae). *Annales de la Société Entomologique de France* 40(3–4): 243–257.

Ribeiro, H. 2005. Three new species and one new subspecies of *Toxorhynchites* (Diptera: Culicidae) of the Afrotropical Region. *Journal of Medical Entomology* 42(3): 224–232.

Ribeiro, H. & Ramos, H.C. 1974. Research on the mosquitoes of Angola. VII – Redescription of the larva of *Aedes durbanensis durbanensis* (Theo., 1903) and description of *Aedes durbanensis angolae* subsp. nov. *Mosquito Systematics* 6(2): 93–98.

Ribeiro, H., da Cunha Ramos, H., Antunes Capela, R. & Alves Pires, C. 1977. Research on the mosquitoes of Portugal (*Diptera, Culicidae*) III – Further five new mosquito records. Garcia de Orta, *Série de Zoologia* 6(1–2): 51–60, 6 pls.

Ribeiro, H., da Cunha Ramos, H., Capela, R.A. & Pires, C.A. 1980. *Estudos, Ensaios e Documentos, N.° 135: Os mosquitos de Cabo Verde* (Diptera: Culicidae), *sistemática, distribuição, bioecologia e importância médica*. Junta de Investigações Científicas do Ultramar, Liboa. 141 pp., map, 80 figs.

Ribeiro, H., da Cunha Ramos, H., Capela, R.A. & Pires, C.A. 1998. Os mosquitos (Diptera: Culicidae) da Ilha de São Tomé. *Garcia de Orta, Série de Zoologia* 22(1–2): 1–20.

Ribeiro, H., da Cunha Ramos, H., Pires, C.A. & Antenues Capela, R. 1988. An annotated checklist of the mosquitoes of continental Portugal *(Diptera Culicidae)*. Actas, III Congreso Ibérico Entomologia, Granada, December 1988, pp. 233–253.

Ribeiro, H., da Cunha Ramos, H., Pires, C.A. & Capela, R.A. 1979. Description and biometric study of *Anopheles (Cellia) quadriannulatus davidsoni* ssp. n., a seventh member of the *Anopheles gambiae* Giles complex *(Diptera: Culicidae)* endemic to the Cape Verde archipelago. *Garcia de Orta, Série de Zoologia* 8(1–2): 75–88.

Ribeiro, H., Ramos, H.C., Pires, C.A. & Capela, R.A. 1987. Research of the mosquitoes of Portugal *(Diptera: Culicidae)* IX – A new anopheline record. *Garcia de Orta, Série de Zoologia* (for 1985) 12(1–2): 105–112, 1 pl.

Rickenbach, A. 1967. Description d'un nouvel *Eretmapodites (Diptera, Culicidae)* capturé au Cameroun. *Bulletin de la Société de Pathologie exotique* (for 1966) 59(3): 374–376.

Rickenbach, A. & Eouzan, J.-P. 1970. Description do quatre *Eretmapodites* nouveaux du groupe *plioleucus* Edwards 1941, capturés au Cameroun (Diptera, Culicidae). *Cahiers ORSTOM Série Entomologie médicale et Parasitologie* 8(2): 131–135.

Rickenbach, A. & Ferrara, L. 1965. Description de deux nouveaux *Aedes* du Cameroun appartenant au sous-genre *Aedimorphus* (Diptera, Culicidae). *Bulletin de la Société de Pathologie exotique* 58(1): 24–29.

Rickenbach, A. & Hamon, J. 1966. Description d'une nouvelle espèce d'*Orthopodomyia (Diptera, Culicidae)* capturée au Cameroun: *O. nkolbissonensis* n. sp. *Bulletin de la Société de Pathologie exotique* (for 1965) 58(6): 1112–1117.

Rickenbach, A. & Lombrici, G. 1974. Un nouvel *Eretmapodites* Theobald capturé au Cameroun: *E. eouzani* n. sp. [Dipt. Culicidae]. *Bulletin de la Société Entomologique de France* 79(7–8): 218–222.

Rickenbach, A. & Lombrici, G. 1975. Les *Eretmapodites* du groupe *leucopus* Graham, 1909. Position systématique des deux sous-espèces actuellement connues et description de deux espèces nouvelles. [Dipt. Culicidae]. *Bulletin de la Société Entomologique de France* 80(1–2): 39–42.

Rickenbach, A., Ferrara, L. & Eouzan, J.-P. 1968. Deux *Eretmapodites* nouveaux du Cameroun, appartenant au groupe *oedipodius* Graham, 1909 [Dipt. Culicidae]. *Bulletin de la Société Entomologique de France* (for 1967) 72(10): 297–300.

Rings, R.W. & Hill, S.O. 1948. The taxonomic status of *Aedes mathesoni* (Diptera, Culicidae). *Proceedings of the Entomological Society of Washington* 50(2): 41–49.

Rioux, J.-A. 1959. Liste sommaire des Culicides du Nord-Tchad. *Annales de Parasitologie Humaine et Comparée* 34(5–6): 733.

Rioux, J.-A. 1961 [for 1960]. Contribution a l'étude des culicides *(Diptera-Culicidae)* du Nord-Tchad, pp. 53–92. *In*: Rioux, J.A. (ed.), *Mission Épidémiologique au Nord-Tchad*. Arts et Métiers Graphiques, Paris. 132 pp. + foldout map.

Rioux, J.-A. & Arnold, M. 1955. Les culicides de Camargue (*étude systématique et écologique*). *Terre et la Vie* 4: 244–286, 2 pls.

Rioux, J.-A. & Pech, J.-M. 1959. Le biotype autogène de *Culex pipiens* L. ne doit pas être nommé *Culex molestus* Forskal *(Diptera Culicidae)*. *Cahiers des Naturalistes: Bulletin des Naturalistes Parisiens* 15: 115–117.

Rioux, J.A., Guilvard, E. & Pasteur, N. 1998. Description d'*Aedes* (*Ochlerotatus*) *coluzzii* n. sp. (Diptera, Culicidae), espèce jumelle A du complexe *detritus*. *Parassitologia* 40(3): 353–360.

Rivola, E. & Holstein, M.H. 1957. Note sur une variété d'*Anopheles nili* Theo. *Bulletin de la Société de Pathologie exotique* 50(3): 382–387.

Robineau-Desvoidy, J.-B. 1827. Essai sur la tribu des culicides. *Mémoires de la Société d'Histoire Naturelle de Paris* 3: 390–413, 1 pl.

Robinson, G.G. 1950. A new species of *Aedes* (*Finlaya*) from Northern Rhodesia. *Journal of the Entomological Society of Southern Africa* 13: 80–82.

Rocha, G.P., Lourenço-de-Oliveira, R. & Motta, M. de A. 2012. *Wyeomyia exallos*, a new species of sylvatic mosquito (Diptera: Culicidae) from Brazil. *Memórias do Instituto Oswaldo Cruz* 107(7): 928–934.

Rodenwaldt, E. 1925. Entomologische notities. III. *Geneeskundig Tijdschrift voor Nederlandsch-Indië* 65: 173–201, 3 pls.

Rodenwaldt, E. 1926. Entomologische notities. IV. *Geneeskundig Tijdschrift voor Nederlandsch-Indië* 66(6): 787–799.

Rodhain, F. & Boutonnier, A. 1983a. Description d'un nouvel *Aedes* du sous-genre *Ochlerotatus* (Diptera: Culicidae) de Madagascar: *Aedes ambreensis* nova species et considérations générales sur les femelles de sous-genre *Ochlerotatus* dans la région afro-tropicale. *Bulletin de la Société de Pathologie exotique* 76(5 Pt2): 825–833.

Rodhain, F. & Boutonnier, A. 1983b. *Aedes*, subgenus *Diceromyia* (Diptera: Culicidae). *Mosquito Systematics* 15(4): 330–336.

Rodhain, F. & Boutonnier, A. 1984. Le genre *Orthopodomyia* (Diptera: Culicidae) a Madagascar. *Archives de l'Institut Pasteur de Madagascar* (for 1983) 51(1): 203–248.

Rodhain, F. & Boutonnier, A. 1985. Checklist of the type specimens of mosquitoes (Diptera: Culicidae) in the Medical Entomology Collections of the Pasteur Institute in Paris. *Mosquito Systematics* (for 1984) 16(4): 271–281.

Romero Sierra, J.M. 1907. Contribucion al estudio de los mosquitos de Caracas. Tesis do Doctorado No. 67, Facultad de Ciencias Medícas, Universidad Central de Venezuela, Caracas. [1–11]–55 pp.

Romeo Viamonte, J.M. 1950. Los anofelinos de España y de la zona española del Protectorado de Marruecos. Su relación con la difusión del paludismo. *Revista de Sanidad e Higiene Pública* 24: 213–295, 17 pls.

Rondani, C. 1848. Esme di varie specie d'insetti ditteri brasiliani. *Studi entomologici (Turino)* 1: 63–112, 3 pls.

Rondani, C. 1872. Sulle specie Italiane del genere *Culex* Lin. *Bollettino della Societa Entomologica Italiana* 4: 29–31.

Ronderos, R.A. 1965. Notas sobre *Culicidae* Venezolanos (*Diptera*). *Physis* 25(70): 249–253.

Ronderos, R.A. & Bachmann, A.O. 1963a. A proposito del complejo *Mansonia (Diptera-Culicidae)*. *Revista de la Sociedad Entomológica Argentina* (for 1962) 25: 43–51.

Root, F.M. 1922. Notes on mosquitoes and other blood-sucking flies from Porto Rico. *American Journal of Hygiene* 2(4): 394–405.

Root, F.M. 1926. Studies on Brazilian mosquitoes. I. The anophelines of the *Nyssorhynchus* group. *American Journal of Hygiene* 6(5): 684–717, 9 pls.

Root, F.M. 1927a. Note on the mosquito fauna of the Republic of Haiti. *American Journal of Hygiene* 7(4): 463–469.

Root, F.M. 1927b. Studies on Brazilian mosquitoes. II. *Chagasia fajardoi*. *American Journal of Hygiene* 7(4): 470–480, 4 pls.

Root, F.M. 1927c. Studies on Brazilian mosquitoes. III. The genus *Culex.American Journal of Hygiene* 7(5): 574–598, 6 pls.

Root, F.M. 1927d. Studies on Brazilian mosquitoes. IV. Notes on some Brazilian species of *Anopheles. American Journal of Hygiene* 7(5): 599–605, 2 pls.

Root, F.M. 1937. A new species of *Uranotaenia* (Diptera, Culicidae) from Haiti. *Journal of Parasitology* 23(1): 98–100.

Roper, R. 1914. An account of some *Anopheline* mosquitos found in British North Borneo, with description of a new species. *Bulletin of Entomological Research* 5(2): 137–147, 2 maps.

Roque, A.B. 1903. Contribuição para o estudo da malaria e dos mosquitos de Angola. *Medicina Contemporanea* 6: 110–115.

Rosa-Freitas, M.G. 1989. *Anopheles (Nyssorhynchus) deaneorum:* A new species in the *albitarsis* complex (Diptera: Culicidae). *Memórias do Instituto Oswaldo Cruz* 84(4): 535–543.

Rosenberg, R. & Evenhuis, N.L. 1985. A new species of *Toxorhynchites* from Bangladesh (Diptera: Culicidae). *Mosquito Systematics* 17(1): 36–43.

Ross, E.S. 1943. The identity of *Aedes bimaculatus* (Coquillett) and a new subspecies of *Aedes fulvus* (Wiedemann) from the United States (Diptera, Culicidae). *Proceedings of the Entomological Society of Washington* 45(6): 143–151.

Ross, E.S. & Roberts, H.R. 1943. *Mosquito atlas. Part I. The Nearctic* Anopheles, *important malaria vectors of the Americas and* Aëdes aegypti, Culex quinquefasciatus. The American Entomological Society, the Academy of Natural Sciences, Philadelphia. iv, 44 pp.

Ross, R. 1898. Report on a preliminary investigation into malaria in the Sigur Ghat, Ootacamund. *Indian Medical Gazette* 33(4 & 5): 133–136, 173–175.

Rossi, G.C. & Harbach, R.E. 2008. *Phytotelmatomyia*, a new Neotropical subgenus of *Culex* (Diptera: Culicidae). *Zootaxa* 1879: 1–17.

Rossi, P. 1790. *Fauna Etrusca: Sistens insecta quae in provinciis Florentina et Pisana praesertim collegit Petrus Rossius.* Volume 2. Thomae Masi & Sociorum, Liburni. 348 pp., 10 pls.

Roth, L.M. 1945. The male and larva of *Psorophora (Janthinosoma) horrida* (Dyar and Knab) and a new species of *Psorophora* from the United States (Diptera: Culicidae). *Proceedings of the Entomological Society of Washington* 47(1): 1–23.

Rothwell, S. 1907. A new mosquito from India. *Entomologist* 40(525): 34–36.

Roubaud, E. 1934. Un type racial nouveau de l'*Anopheles maculipennis. Bulletin de la Société de Pathologie exotique* 27: 737–740, 1 pl.

Roubaud, E. 1935a. Variété nouvelle de l'*Anopheles maculipennis* au Maroc, *A. maculipennis sicaulti* (*n. var.*). *Bulletin de la Société de Pathologie exotique* 28(2): 107–111.

Roubaud, E. 1935b. La microstructure du flotteur de l'œuf dans les races biologiques de *Culex pipiens* L. *Bulletin de la Société de Pathologie exotique* 28: 443–445.

Roubaud, E. 1945. Le problème de l'espèce chez le moustique commun *Culex pipiens* L. *Bulletin de la Société de Pathologie exotique* 38(1–2): 47–60.

Roubaud, É. 1957. Biologie des moustiques. – Sur l'existence de deux entités biologiques distinctes dans la définition du moustique commun rural, *Culex pipiens pipiens. Compte Rendu de l'Académie des Sciences* 244: 3115–3116.

Roubaud, E. & Treillard, M. 1936. Sur une variété portugaise do l'*Anopheles maculipennis* (groupe *atroparvus*). *Bulletin de la Société de Pathologie exotique* 29: 726–730.

Rozeboom, L.E. 1935. *Culex rooti*, a new *Culex* from Panama (Diptera: Culicidae). *Annals of the Entomological Society of America* 28(2): 251–253.

Rozeboom, L.E. 1941. Distribution and ecology of the *Anopheles* mosquitoes of the Caribbean Region. *Publications of the American Association for the Advancement of Science* 15: 98–107.

Rozeboom, L.E. 1946. Three new *Finlayas* of the *dissimilis* group (Diptera: Culicidae) from the Philippine islands. *Journal of Parasitology* 32(6): 587–595.

Rozeboom, L.E. 1951. A new species of *Anopheles* from the Philippine islands (Diptera: Culicidae). *Journal of Parasitology* 37(5): 502–506.

Rozeboom, L.E. & Gabaldon, A. 1941. A summary of the *"tarsimaculatus"* complex of *Anopheles* (Diptera: Culicidae). *American Journal of Hygiene* 33(C): 88–100, 4 pls.

Rozeboom, L.E. & Knight, K.L. 1946. The *punctulatus* complex of *Anopheles* (Diptera: Culicidae). *Journal of Parasitology* 32(2): 95–131.

Rozeboom, L.E. & Komp, W.H.W. 1948. Three new species of *Culex* (Diptera: Culicidae) from Colombia. *Journal of Parasitology* 34(5): 396–406.

Rozeboom, L.E. & Komp, W.H.W. 1950a. A review of the species of *Culex* of the subgenus *Melanoconion* (Diptera, Culicidae). *Annals of the Entomological Society of America* 43(1): 75–114.

Rozeboom, L.E. & Komp, W.H.W. 1950b. A new *Microculex, elongatus*, from Colombia, with notes on the subgenus (Diptera, Culicidae). *Proceedings of the Entomological Society of Washington* 52(3): 147–157.

Rueda, L.M. 2005. Two new species of *Anopheles* (*Anopheles*) Hyrcanus Group (Diptera: Culicidae) from the Republic of South Korea. *Zootaxa* 941: 1–26.

Rueger, M.E. 1958. *Aedes (Ochlerotatus) barri,* a new species of mosquito from Minnesota (Diptera, Culicidae). *Journal of the Kansas Entomological Society* 31(1): 34–46.

Russell, P.F., Rozeboom, L.E. & Stone, A. 1943. *Keys to the anopheline mosquitoes of the world with notes on their identification, distribution, biology, and relation to malaria.* American Entomological Society; Academy of Natural Sciences, Philadelphia. 152 pp.

Rutledge, L.C., Ward, R.A. & Bickley, W.E. 1970. Experimental hybridization of geographic strains of *Anopheles stephensi* (Diptera: Culicidae). *Annals of the Entomological Society of America* 63(4): 1024–1030.

Sabrosky, C.W. 1999. *Family-group names in Diptera: an annotated catalog.* Backhuys Publishers, Leiden. [4], 576 pp.

Sakakibara, M. 1959. On *Anopheles* (*Anopheles*) *omorii* n. sp. *Endemic Diseases Bulletin of Nagasaki University* 1(3): 288–295, 3 pls.

Salem, H.H. 1938. The mosquito fauna of Sinai Peninsula (Egypt) with a description of two new species. *Publications of the Egyptian University Faculty of Medicine* 16: 1–31, 16 pls.

Saliternik, Z. & Theodor, O. 1942. On a new variety of *Anopheles turkhudi* from Palestine. *Journal of the Malaria Institute of India* 4(4): 429–434.

Sallum, M.A.M. & Forattini, O.P. 1996. Revision of the Spissipes Section of *Culex* (*Melanoconion*) (Diptera: Culicidae). *Journal of the American Mosquito Control Association* 12(3): 517–600.

Sallum, M.A.M. & Hutchings, R.S.G. 2003. Taxonomic studies on *Culex* (*Melanoconion*) *coppenamensis* Bonne-Wepster & Bonne (Diptera: Culicidae), and description of two new species from Brazil. *Memórias do Instituto Oswaldo Cruz* 98(5): 615–622.

Sallum, M.A.M., Foster, P.G., dos Santos, C.L.S., Flores, D.C., Motoki, M.T. & Bergo, E.S. 2010. Resurrection of two species from synonymy of *Anopheles* (*Nyssorhynchus*) *strodei* Root, and characterization of a distinct morphological form from the Strodei Complex (Diptera: Culicidae). *Journal of Medical Entomology* 47(4): 504–526.

Sallum, M.A.M., Hutchings, R.S.G., Leila, R. & Ferreira, R.L.M. 1997. *Culex gnomatos* a new species of the Spissipes Section of *Culex (Melanoconion)* (Diptera: Culicidae) from the Amazon region. *Memórias do Instituto Oswaldo Cruz* 92(2): 215–219.

Sallum, M.A.M., Peyton, E.L., Harrison, B.A. & Wilkerson, R.C. 2005a. Revision of the Leucosphyrus group of *Anopheles* (*Cellia*) (Diptera, Culicidae). *Revista Brasileira de Entomologia* 49(Supl. 1): 1–152.

Sallum, M.A.M., Peyton, E.L. & Wilkerson, R.C. 2005b. Six new species of the *Anopheles leucosphyrus* group, reinterpretation of *An. elegans* and vector implications. *Medical and Veterinary Entomology* 19(2): 158–199.

Sallum, M.A.M., Schultz, T.R. & Wilkerson, R.C. 2000. Phylogeny of Anophelinae (Diptera Culicidae) based on morphological characters. *Annals of the Entomological Society of America* 93(4): 745–775.

Sallum, M.A.M., Wilkerson, R.C. & Forattini, O.P. 1999. Taxonomic study of species formerly identified as *Anopheles mediopunctatus* and resurrection of *An. costai* (Diptera: Culicidae). *Journal of Medical Entomology* 36(3): 282–300.

Sanchez, J. 1885. Nota acerca de los moscos zancudos que han invadido la capital en el presente año. *La Naturaleza, Mexico* 7(13): 203–213.

Sanchez, J. 1886. Revista de historia natural. *La Naturaleza, Mexico* 7(20): 323–330.

Sánchez-Covisa Villa, A., Rodríguez Rodríguez, J.A. & Guillén Llera, J.L. 1985. Estudio de la larva de cuarto estado de las especies del subgenero *Finlaya (Diptera: Culicidae)* de España Peninsular. *Suplemento ao Boletim da Sociedade Portuguesa de Entomologia* 1 (Suplemento 1): 439–448.

Sandosham, A.A. 1944. *Malaria in Malai: a handbook for anti-malaria students, &c.* Syonan: Syonan Tokubetu-Si, Singapore. xvi, 288 pp., 8 pls.

Sandosham, A.A. 1959. *Malariology, with special reference to Malaya.* University of Malaya Press, Singapore. xix, 327 pp.

Sant'Ana, D.C. & Sallum, M.A.M. 2016, 2017. *Anopheles* (*Nyssorhynchus*) *striatus*, a new species of the Strodei Subgoup (Diptera, Culicidae). *Revista Brasileira de Entomologia* (2016) http://dx.doi.org/10.1016/j.rbe.2016.12.005; (2017) 61(2): 136–145.

Sasa, M. & Ishimura, K. 1951. A new species of mosquito, *Aedes* (*Finlaya*) *bunanoki*, n. sp: from Aomori, Japan. *Japanese Journal of Experimental Medicine* 21: 103–108.

Sasa, M. & Nakahashi, Y. 1952. A new species of the *Aedes* (*Finlaya*) *niveus* subgroup of mosquito from Japan. *Japanese Journal of Experimental Medicine* 22(3): 257–265.

Sasa, M. & Takahasi, H. 1948. Descriptions of *Culex rubensis* n. sp. and the adult male of *Theobaldia kanayamensis* Yamada [in Japanese]. *Japanese Journal of Bacteriology* 3(2): 51–52.

Sasa, M. & Tanimura, K. 1971. Index and consideration on taxonomy of the Japanese mosquitoes [in Japanese], pp. 1–47. *In*: Sasa, M. (ed.), *Progress in Medical Zoology I* [in Japanese]. Gakujutsusho Shuppankai, Tokyo.

Sasa, M., Kano, R. & Hayashi, S. 1950. A new species of mosquito, *Aedes koreicoides n.* [*sic*] sp., from Hokkaido. *Japanese Journal of Experimental Medicine* 20: 627–629.

Sasa, M., Kano, R. & Takahasi, H. 1950. A revision of the adult Japanese mosquitoes of the genus *Aedes*, subgenus *Aedes*, with description of two new species. *Japanese Journal of Experimental Medicine* 20: 631–640.

Sathe, T.V. & Girhe, B.E. 2002. Biodiversity of mosquitoes in Kolhapur District, Maharashtra. *Rivista di Parassitologia* 62(3): 193–198.

Sathe, T.V. & Jagtap, M. 2012. Taxonomy and molecular phylogeny of new species of the genus *Anopheles* (*Cellia*) Theobald (Culicidae: Diptera) from India. *Journal of Entomological Research* 36(1): 71–75.

Savage, H.M. 2005. Classification of mosquitoes in tribe Aedini (Diptera: Culicidae): paraphylyphobia, and classification versus cladistic analysis. *Journal of Medical Entomology* 42(6): 923–927.

Say, T. 1823. Descriptions of dipterous insects of the United States. *Journal of the Academy of Natural Sciences of Philadelphia* 3(1): 9–54.

Say, T. 1824. Appendix. Part 1.—Natural history. 1. Zoology, pp. 253–378. *In*: Keating, W.H. (main author), *Narrative of an expedition to the source of St. Peter's River, Lake Winnepeek, Lake of the Woods, &c. &c. performed in the year 1823, by order of the Hon. J. C. Calhoun, Secretary of War, under the command of Stephen H. Long, Major U. S. T.E.: compiled from the notes of Major Long, Messrs. Say, Keating, & Colhoun.* Volume II. H. C. Carey & I. Lea, Philadelphia.

Say, T. 1829. *Descriptions of North American* [*sic*] dipterous insects. *Journal of the Academy of Natural Sciences of Philadelphia* 6(1): 149–178.

Sazanova, O.N. 1955. Description of a new species of blood-sucking mosquito, *Aëdes* (*Ochlerotatus*) *grandilarva* sp. nova [in Russian]. Бюллетень Московского общества испытателей природы. Отдел биологический [*Bulletin of the Moscow Society of Naturalists. Biological Series*] 60(4): 99–102.

Scanlon, J.E. & Peyton, E.L. 1967. *Anopheles* (*Anopheles*) *tigertti,* a new species of the Aitkenii Group from Thailand. *Proceedings of the Entomological Society of Washington* 69(1): 18–23.

Schick, R.X. 1970a. Mosquito studies (Diptera, Culicidae). XX. The Terrens Group of *Aedes* (*Finlaya*). *Contributions of the American Entomological Institute* 5(3): 1–158.
Schick, R.X. 1970b. Mosquito studies (Diptera, Culicidae). XXIII. Additions and corrections to the revision of the *Aedes terrens* group. *Contributions of the American Entomological Institute* 7(1): 13–40.
Schiner, J.R. 1864. *Fauna austriaca. Die Fliegen (Diptera). Nach der analytischen Methode bearbeite, mit der Charakteristik sämmtlicher europäischer Gattungen, der Beschreibung aller in Deutschland vorkommenden Arten und der Aufzählung aller bisher beschriebenen europäischen Arten.* II Theil. Carl Gerold's Sohn, Wien. xxxii, 658 pp.
Schiner, J.R. 1868. *Reise der Österreichischen Fregatte Novara um die Erde in den Jahren 1857, 1858, 1859 unter den Befehln des Commodore B. von Wüllerstorf-Urbair.Zoologischer Theil. Zweiter Band. 1. Abtheilung. B. 1. Diptera.* Wien. vi, 388 pp., 4 pls.
Schmidt, E.R., Foley, D.H., Bugoro, H. & Bryan, J.H. 2003. A morphological study of the *Anopheles punctulatus* group (Diptera: Culicidae) in the Solomon Islands, with a description of *Anopheles* (*Cellia*) *irenicus* Schmidt, sp.n. *Bulletin of Entomological Research* 93(6): 515–526.
Schmidt, E.R., Foley, D.H., Hartel, G.F., Williams, G.M. & Bryan, J.H. 2001. Descriptions of the *Anopheles* (*Cellia*) *farauti* complex of sibling species (Diptera: Culicidae) in Australia. *Bulletin of Entomological Research* 91(5): 389–410.
Schneider, P. 1913. Beitrag zur Kenntnis der Culiciden in der Umgebung von Bonn. *Verhandlungen des Naturhistorischen Vereins der preufsischen Rheinlande und Westfalens* 70: 1–54.
Schouteden, H. 1912. Insectes recueillis au Congo au cours du voyage de S. A. R. le Prince Albert de Belgique. *Revue zoologique africaine* 2: 63–90.
Schrank, Franz von Paula. 1776. *Beyträge zur Naturgeschichte.* Caspar Fritsch, Leipzig. [6], 137 pp., [3 pp.], 7 foldout pls.
Schrank, Francisci de Pavla. 1781. *Envmeratio insectorm Avstriae indigenorum: cum figuris.* Avgvstae Vindelicorvm, Apvd vidvam E. Klett et Franck. 548 pp.
Scudder, S.H. 1882. *Nomenclator Zoologicus. An alphabetical list of all generic names that have been employed by naturalists for recent and fossil animals from the earliest times to the close of the year 1879. In two parts: I. Supplemental List. II. Universal Index.* Bulletin of the United States National Museum No.19. xix, 376 pp.
Scudder, S.H. 1890. *The tertiary insects of North America.* United States Geological Survey of the Territories. Volume 13. Government Printing Office, Washington, DC. 734 pp.
Sedaghat, M.M., Linton, Y.-M., Oshaghi, M.A., Vatandoost, H. & Harbach, R.E. 2003. The *Anopheles maculipennis* complex (Diptera: Culicidae) in Iran: molecular characterization and recognition of a new species. *Bulletin of Entomological Research* 93(6): 527–535.
Séguy, E. 1920. Les moustiques de France. *Bulletin du Muséum national d'histoire naturelle* 26(1, 2, 3, 4, 5, 60): 51–58, 141–147, 223–230, 322–329, 407–414, 512–519.
Séguy, E. 1921. Description d'un nouveau Moustique Français du groupe de l'*Aedes Mariae* et synopsis des espèces de ce groupe. *Bulletin de la Société Entomologique de France* 1921(13): 192–195.
Séguy, E. 1923. Remarques sur quelques moustiques. *Annales de la Société Entomologique de France* 92(1–2): 205–208.
Séguy, E. 1924. Les moustiques de l'Afrique Mineure de 1'Égypte et de la Syrie étude comparative des moustiques des régions méditerranéennes, de l'Europe centrale et septentrionale leurs parasites suivi du catalogue des culicides néarctiques et paléarctiques. *Encyclopédie Entomologique Série A. Travaux Généraux* 1: 1–257, 29 pls.
Séguy, E. 1925. Notes sur les moustiques de l'Afrique Mineure, de l'Égypte et de la Syrie. I. *Encyclopédie Entomologique Série B: Mémoires et Notes* 2(1): 13–21.
Séguy, E. 1927. Notes sur les moustiques de l'Afrique Mineure, de l'Égypte et de la Syrie. II. *Encyclopédie Entomologique Série B: Mémoires et Notes* 4(1): 27–28.
Senevet, G. 1932. A propos d'*Anopheles coustani* Laveran. *Archives de l'Institut Pasteur d'Algérie* 10(4): 478–484.
Senevet, G. 1936. Les moustiques de la Martinique. *Archives de l'Institut Pasteur d'Algérie* 14(2): 123–134.
Senevet, G. 1937. Les moustiques de la Guyane française (mission 1934). *Archives de l'Institut Pasteur d'Algérie* 15(3): 352–382.
Senevet, G. 1938. Les moustiques de la Guadeloupe (mission 1936). *Archives de l'Institut Pasteur d'Algérie* 16(2): 176–190.
Senevet, G. 1947. *Anopheles tonkinensis* synonyme de *A. alongensis*. *Archives de l'Institut Pasteur d'Algérie* 25(3–4): 214–217.
Senevet, G. & Abonnenc, E. 1938. Quelques anophélinés de la Guyane française. *Archives de l'Institut Pasteur d'Algérie* 16(4): 486–512.
Senevet, G. & Abonnenc, E. 1939a. Les moustiques de la Guyane française - II. Le genre *Culex*. *Archives de l'Institut Pasteur d'Algérie* 17(1): 62–134.
Senevet, G. & Abonnenc, E. 1939b. Les moustiques de la Guyane française. - III. Les sabéthinés. *Archives de l'Institut Pasteur d'Algérie* 17(2): 247–281.
Senevet, G. & Abonnenc, E. 1941. Les moustiques de la Guyane française. Le genre *Culex*. 2. Nouvelle espèce du sous-genre *Melanoconion*. *Archives de l'Institut Pasteur d'Algérie* 19(1): 41–44.
Senevet, G. & Abonnenc, E. 1946. Les moustiques de la Guyane française. X. Le genre *Culex* (3). Nouvelle espèce du sous-genre *Culex*. *Archives de l'Institut Pasteur d'Algérie* 24(2): 135–140.
Senevet, G. & Andarelli, L. 1955. Races et variétés de l'*Anopheles claviger* Meigen, 1804. *Archives de l'Institut Pasteur d'Algérie* 33(2): 128–137.
Senevet, G. & Prunnelle, M. 1927. Une nouvelle espèce d'anophèle en Algérie, *Anopheles marteri* n. sp. *Archives de l'Institut Pasteur d'Algérie* 5(4): 529–533.
Senevet, G. & Quiévreux, L. 1941. Les moustiques de la Martinique. (2e mémoire). *Archives de l'Institut Pasteur d'Algérie* 19(2): 248–264.
Senevet, G., Chabelard, R. & Abonnenc, E. 1942. Les moustiques de la Guyane. III. – Les sabéthinés (2). *Archives de l'Institut Pasteur d'Algérie* 20(4): 336–348.

Serafim, J., Jr. & Davis, N.C. 1933. Distribution of *Aëdes* (*Taeniorhynchus*) *taeniorhynchus* (Wiedemann). *Aëdes* (*Taeniorhynchus*) *jacobinae*, new species. *Annals of the Entomological Society of America* 26(1): 13–19.

Sergent, E. 1909. Liste des moustiques de l'Afrique du Nord (1re note). *Annales de la Société Entomologique de France* 78(3): 440–448.

Sergent, Ed. & Sergent, Ét. 1903. Observations sur les moustiques des environs d'Alger. *Annales de l'Institut Pasteur* 17(1): 60–67.

Sergent, Ed. & Sergent, Ét. 1905. Sur un Culicide nouveau, très commun a Biskra (*Grabhamia subtilis*). *Compte Rendu, Société de Biologie* 8: 673–674.

Service, M.W. 1958. A new *Anopheles* from the southern Cameroons, *Anopheles buxtoni* sp. nov. *Annals of Tropical Medicine and Parasitology* 52(2): 176–180.

Service, M.W. 1959a. A new species of the genus *Aëdes* from southern Nigeria. *Proceedings of the Royal Entomological Society of London, Series B: Taxonomy* 28(5–6): 73–76.

Service, M.W. 1959b. A new culicine mosquito, *Aëdes (Aëdimorphus) lokojensis* sp. nov., from northern Nigeria. *Revue de zoologie et de botanique Africaines* 59(3–4): 243–248.

Service, M.W. 1960. A new species and variety of *Anopheles* from Nigeria, with notes on *Anopheles flavicosta* Edwards and *Aëdes* (*Aë.*) *wendyae* Service. *Proceedings of the Royal Entomological Society of London, Series B: Taxonomy* 29(7–8): 85–90.

Service, M.W. 1970. A new *Anopheles* (Dipt., Culicidae) from Nigeria and a note on *A. pretoriensis* (Theobald). *Bulletin of Entomological Research* 59(4): 647–649.

Service, M.W. 1977. A new *Anopheles* (Dipt., Culicidae) from Nigeria together with a note on *An. brohieri* Edwards. *Annals of Tropical Medicine and Parasitology* 71(2): 233–236.

Service, M.W. 1993. Mosquitoes (Culicidae), pp. 120-240. *In*: Lane, R.P. & Crosskey, R.W. (eds), *Medical insects and arachnids.* Chapman & Hall, London.

Service, M.S. 2010. Sequel to Kitzmiller's *Anopheline Names: Their Derivations and Histories. Journal of Vector Ecology* 35(2): 213–266.

Shahgudian, E.R. 1956. Notes on *Anopheles marteri* Senevet and Prunnelle, 1927. *Proceedings of the Royal Entomological Society of London, Series A: General Entomology* 31(4–6): 71–75.

Shannon, R.C. 1927. Contribución a los estudios de las zonas biológicas de la República Argentina. *Revista de la Sociedad Entomológica Argentina* 1(4): 1–14.

Shannon, R.C. 1931. On the classification of Brazilian Culicidae with special reference to those capable of harboring the yellow fever virus. *Proceedings of the Entomological Society of Washington* 33(6): 125–164.

Shannon, R.C. 1933a. Correction. *Proceedings of the Entomological Society of Washington* 35(4): 58.

Shannon, R.C. 1933b. Anophelines of the Amazon Valley. *Proceedings of the Entomological Society of Washington* 35(7): 117–143.

Shannon, R.C. & Cerqueira, N.L. 1943. *Psorophora lanei*, um novo mosquito do Brasil e da Bolivia (Diptera, Culicidae). *Revista de Entomologia* 14: 135–137.

Shannon, R.C. & del Ponte, E. 1927. Cuatro notas sobre especies nuevas de Dípteros, Nematóceros, Hematófagos, o no, de la República Argentina. *Revista del Instituto Bacteriológico de Buenos Aires* 4(7): 724–736.

Shannon, R.C. & del Ponte, E. 1928. Los culicidos en la Argentina. *Revista del Instituto Bacteriológico de Buenos Aires* (for 1927) 5(1): 29–140.

Shevchenko, A.K. & Prudkina, N.S. 1973. On morphology of genitals in mosquito males from the *Aedes* genus [in Russian]. *Vestnik Zoologii [Zoological Record] / Institut Zoologii, Akademiya Nauk Ukrainskoi SSR* 6: 40–47.

Shidrawi, G.R. & Gillies, M.T. 1988. *Anopheles paltrinierii,* n. sp., (Culicidae: Diptera) from the Sultanate of Oman. *Mosquito Systematics* (for 1987) 19(3): 201–211.

Shingarev, N.L. 1926. New information on Culicidae of U.S.S.R. [in Russian]. *Russkiĭ Zhurnal Tropicheskoĭ Meditsiny* [*Russian Journal of Tropical Medicine*] 4(2): 47–48.

Shingarev, N.L. 1927. Notes on Culicidae. II. [in Russian]. *Russkiĭ Zhurnal Tropicheskoĭ Meditsiny* [*Russian Journal of Tropical Medicine*] 5: 545–550.

Shingarev, N.L. 1928. Notes on Culicidae. III. [in Russian]. *Russkiĭ Zhurnal Tropicheskoĭ Meditsiny* [*Russian Journal of Tropical Medicine*] 1: 47–53.

Shute, P.G. 1928. A new variety of *Culicella morsitans* (Theobald) (Diptera). *Entomologist* 61: 186.

Silva do Nascimento, T.F. & Lourenço-de-Oliveira, R. 2002. *Anopheles halophylus*, a new species of the subgenus *Nyssorhynchus* (Diptera: Culicidae) from Brazil. *Memórias do Instituto Oswaldo Cruz* 97(6): 801–811.

Singh, G. & Prakash, S. 2008. A new species of *Lutzia* (Diptera: Culicidae) comparable with *halifaxii* Theobald 1903 in a semi-arid zone of India, Chapter 24, pp. 195–204. *In*: Tyagi, B.K. (ed.), *Vector-borne diseases: Epidemiology and control.* Scientific Publishers (India), Jodhpur.

Sirivanakarn, S. 1968. The *Culex* subgenus *Lophoceraomyia in* New Guinea and Bismarck Archipelago (Diptera: Culicidae). *Pacific Insects Monographs* 17: 75–186.

Sirivanakarn, S. 1969. Note on the change of a specific name of New Guinea *Culex* (*Lophoceraomyia*) (Diptera: Culicidae). *Journal of Medical Entomology* 6(2): 119.

Sirivanakarn, S. 1971a. Contributions to the mosquito fauna of Southeast Asia. XI. Proposed reclassification of *Neoculex* Dyar based principally on the male terminalia. *Contributions of the American Entomological Institute* 7(3): 62–85.

Sirivanakarn, S. 1971b. *Culex* (*Neoculex*) *nematoides* (Dyar and Shannon 1925) an erroneous record from the Philippines (Diptera: Culicidae). *Mosquito Systematics Newsletter* 3(1): 1–3.

Sirivanakarn, S. 1972. Contributions to the mosquito fauna of Southeast Asia. XIII. The genus *Culex*, subgenus *Eumelanomyia* Theobald in Southeast Asia and adjacent areas. *Contributions of the American Entomological Institute* 8(6): i, 1–86.

Sirivanakarn, S. 1973a. Descriptions of two new species of *Culex* (*Lophoceraomyia*) with notes on three other species from the Papuan Subregion (Diptera: Culicidae). *Journal of Medical Entomology* 10(2): 212–216.

Sirivanakarn, S. 1973b. Three new species of *Culex* subgenus *Culiciomyia* Theobald from Southeast Asia and a redescription of the type of *C. tricuspis* Edwards from Alor, Lesser Sunda Islands, Indonesia (Diptera: Culicidae). *Proceedings of the Entomological Society of Washington* 75(1): 112–124.

Sirivanakarn, S. 1975. A new species of *Culex (Eumelanomyia)* Theobald from Manus Island, Papua-New Guinea (Diptera: Culicidae). *Mosquito Systematics* 7(4): 394–400.

Sirivanakarn, S. 1976. Medical entomology studies – III. A revision of the subgenus *Culex* in the Oriental Region (Diptera: Culicidae). *Contributions of the American Entomological Institute* 12(2): iii, 1–272.

Sirivanakarn, S. 1977a. Medical entomology studies – VI. A revision of the subgenus *Lophoceraomyia* of the genus *Culex* in the Oriental Region (Diptera: Culicidae). *Contributions of the American Entomological Institute* 13(4): iii, 1–245.

Sirivanakarn, S. 1977b. Additional descriptions of three species of *Culex (Eumelanomyia)* with the description of a new species from peninsular Malaysia (Diptera: Culicidae). *Mosquito Systematics* 9(1): 73–87.

Sirivanakarn, S. 1977c. Redescription of four Oriental species of *Culex* (*Culiciomyia*) and the description of a new species from Thailand (Diptera: Culicidae). *Mosquito Systematics* 9(2): 93–111.

Sirivanakarn, S. 1977d. A new species of *Culex (Eumelanomyia)* from India with descriptions of pupae and larvae of *Cx. pluvialis* Barraud and *Cx. iphis* Barraud (Diptera: Culicidae). *Mosquito Systematics* 9(4): 537–547.

Sirivanakarn, S. 1979. A new species of *Culex (Melanoconion)* from Bolivia and Ecuador (Diptera: Culicidae). *Mosquito Systematics* 11(2): 135–138.

Sirivanakarn, S. 1983. A review of the systematics and a proposed scheme of internal classification of the New World subgenus *Melanoconion* of *Culex* (Diptera, Culicidae). *Mosquito Systematics* (for 1982) 14(4): 265–333.

Sirivanakarn, S. & Belkin, J.N. 1980. The identity of *Culex (Melanoconion) taeniopus* Dyar and Knab and related species with notes on the synonymy and description of a new species (Diptera, Culicidae). *Mosquito Systematics* 12(1): 7–24.

Sirivanakarn, S. & Galindo, P. 1980. *Culex (Melanoconion) adamesi,* a new species from Panama (Diptera, Culicidae). *Mosquito Systematics* 12(1): 25–34.

Sirivanakarn, S. & Jakob, W.L. 1979. A new species of *Culex (Melanoconion)* from southern Brazil (Diptera: Culicidae). *Mosquito Systematics* 11(2): 139–143.

Sirivanakarn, S. & Jakob, W.L. 1982. *Culex (Melanoconion) sacchettae,* a new species from the state of São Paulo, Brazil (Diptera: Culicidae). *Mosquito Systematics* (for 1981) 13(2): 191–194.

Sirivanakarn, S. & Kurihara, I. 1973. A new species of *Culex,* subgenus *Culiciomyia* Theobald from Ceram, Indonesia (Diptera: Culicidae). *Proceedings of the Entomological Society of Washington* 75(2): 220–224.

Sirivanakarn, S. & Ramalingam, S. 1976. A new species of *Culex (Eumelanomyia)* Theobald with notes on three other species from Malaysia (Diptera: Culicidae). *Mosquito Systematics* 8(2): 209–216.

Sirivanakarn, S. & White, G.B. 1978. Neotype designation of *Culex quinquefasciatus* Say (Diptera: Culicidae). *Proceedings of the Entomological Society of Washington* 80(3): 360–372.

Skuse, F.A.A. 1889. Diptera of Australia. Part V—The Culicidae. *Proceedings of the Linnean Society of New South Wales* 3(4): 1717–1764, 1 pl.

Skuse, F.A.A. 1895. The banded mosquito of Bengal. *Indian Museum Notes* (for 1894) 3(5): 20.

Slooff, R. 1961. *Tripteroides* (*Rachisoura*) *tityae* (Diptera: Culicidae), a new species from Netherlands New Guinea. *Entomologische Berichten* 21(1): 22–23.

Slooff, R. 1963. *Uranotaenia gerdae,* a new species of the *tibialis* group, from West New Guinea (Diptera: Culicidae). *Annals of the Entomological Society of America* 56(1): 82–86.

Smith, M.E. 1952. A new northern *Aedes* mosquito, with notes on its close ally, *Aedes diantaeus* H., D., & K. (Diptera, Culicidae). *Bulletin of the Brooklyn Entomological Society* 47(2): 19–28.

Snow, K.R. 2001. The names of European mosquitoes: Part 9. *European Mosquito Bulletin* 11: 27–30.

Sobti, S.K. 1968. A new species of the *Anopheles funestus* complex (Diptera: Culicidae) from Zanzibar, United Republic of Tanzania. *Bulletin of the World Health Organisation* 38(3): 481–483.

Soesilo, R. 1932. *Anopheles* (*Bironella*) *walchi* n. sp. *Geneeskundig Tijdschrift voor Nederlandsch-Indië* 72: 1035–1036.

Soesilo, R. & van Slooten, J. 1931a. Verspreide aanteekeningen over Anophelinen in de Oost-Indischen Archipel. *Geneeskundig Tijdschrift voor Nederlandsch-Indië* 71(7–18): 1254–1259.

Soesilo, R. & van Slooten, J. 1931b. Miscellaneous notes on anopheline mosquitoes in the Dutch East Indies. *Mededeelingen van den Dienst Volksgezondheid in Nederlandsch-Indië* 20: 124–128.

Soghigian, J., Andreadis, T.G. & Livdahl, T.P. 2017. From ground pools to treeholes: convergent evolution of habitat and phenotype in *Aedes* mosquitoes. *BMC Evolutionary Biology* 17: 262.

Somboon, P., Rory, A., Tsuda, Y., Takagi, M. & Harbach, R.E. 2010. Systematics of *Anopheles* (*Cellia*) *yaeyamaensis* sp. n., alias species E of the *An. minimus* complex in southeastern Asia (Diptera: Culicidae). *Zootaxa* 2651: 43–51.

Somboon, P., Thongwat, D. & Harbach, R.E. 2011. *Anopheles* (*Cellia*) *rampae* n. sp., alias chromosomal form K of the Oriental Maculatus Group (Diptera: Culicidae) in Southeast Asia. *Zootaxa* 2810: 47–55.

Speiser, P. 1904. Zur Nomenclatur blutsaugender Dipteren Amerikas. *Insekten-Börse* 21(19): 148.

Speiser, P. 1909. 10. Diptera. 4. Orthorhapha, pp. 31–112. *In*: Sjöstedt, Y., *Wissenschaftliche ergebnisse der Schwedischen zoologischen expedition nach dem Kilimandjaro, dem Meru und den umgebenden Massaisteppen Deutsch-Ostafrikas 1905–1906*. 2. Band, Abteilung 8–14. P. Palmquists Aktiebolag, Stockholm.

Stackelberg, A.A. 1926. Materials for identification of Diptera Nematocera USSR. 2. Identification of families and genera in the adult stage [in Russian]. *Revue de Microbiologie, d'Epidémiologie et de Parasitologie. Saratov* [*Vestnik mikrobiologii, ėpidemiologii i parazitologii*] 5: 43–56, 24 figs.

Stackelberg, A.A. 1927. The Culicidae of the USSR and the countries bordering [in Russian]. *Analytical Tables of the Fauna of the USSR* 1: vi + 1–169.

Stackelberg, A.A. 1937. *Fauna de l'URSS. Insectes, Diptères. Vol. III, n⁰ 4. Fam. Culicidae (Subfam. Culicinae)* [in Russian]. Institut Zoologique de l'Académie des Sciences de l'URSS. Nouvelle Série nº 11. Moscow. x, 258 pp.

Stackelberg, A.A. 1943. A new species of *Finlaya*, Theo. (Diptera, Culicidae) from Ussuri Land. *Bulletin of Entomological Research* 34(4): 311.

Staeger, R.C. 1839. Systematik fortegnelse over de hidtil i Denmark fundne Diptera. *Naturhistorisk Tidsskrift (Series 1)* 2(6): 549–600.

Stanton, A.T. 1912. The *Anopheles* mosquitoes of Malaya and their larvae, with some notes on malaria-carrying species. *Journal of the London School of Tropical Medicine* 2(1): 3–11.

Stanton, A.T. 1915. A new anopheline mosquito from Sumatra. *Bulletin of Entomological Research* 5(4): 373–375.

Stanton, A.T. & Hacker, H.P. 1917. The *Anopheles* of Malaya—III. A new variety of *A. albotaeniatus*, Theo. *Bulletin of Entomological Research* 7(3): 273–275.

Starkey, G.S. & Webb, J.E., Jr. 1946. A new species of *Aedes* of the *Finlaya* group from Angaur Island. *Proceedings of the Entomological Society of Washington* 48(7): 179–182.

Statz, G. 1941. Plagegister aus dem Reinischen Braunkohlewnwalde. *Rheinischer Naturfreund* 5(1): 1–16.

Statz, G. 1944. Neue Dipteren (Nematocera) aus dem Oberoligocän von Rott. II. Teil. V. Familie Culicidae (Stechmücken). *Palaeontographica* 95(A): 108–120, 6 pls.

Steffan, W.A. 1977. Source of type material of *Toxorhynchites* (Diptera: Culicidae). *Mosquito Systematics* 9(1): 58–72.

Steffan, W.A. 1979. New *Culex* (*Acallyntrum*) from the Papuan Subregion (Diptera: Culicidae). *Journal of Medical Entomology* 16(2): 133–139.

Steffan, W.A. & Evenhuis, N.L. 1982. A new species of *Toxorhynchites* from Papua New Guinea (Diptera: Culicidae). *Mosquito Systematics* 14(1): 1–13.

Stegnii, V.N. & Kabanova, V.M. 1976. Cytoecological study of indigenous populations of the malaria mosquito in the territory of the U.S.S.R. I. Identification of a new species of *Anopheles* in the *maculipennis* complex by the cytodiagnostic method [in Russian]. *Meditsinskaya Parazitologiya i Parazitarnye Bolezni* 45(2): 192–198.

Stephens, J.F. 1825. Some observations on the British Tipulidae, together with descriptions of the species of *Culex* and *Anopheles* found in Britain. *Zoological Journal* 1(4): 448–457.

Stephens, J.F. 1828. Note on the foregoing paper, with a description of a new species of *Anopheles*. *Zoological Journal* 3(12): 502–504.

Stephens, J.F. 1829. *A systematic catalogue of British insects: being an attempt to arrange all the hitherto discovered indigenous insects in accordance with their natural affinities*. Baldwin and Cradock, London. xxxiv, 388 pp.

Stephens, J.F. 1846. *Illustrations of British entomology; or, a synopsis of indigenous insects: containing their generic and specific distinctions; with an account of their metamorphoses, times of appearance, localities, food, and economy, as far as practicable. Supplement*. Henry G. Bohn, London. vii, 32 pp.

Stephens, J.W.W. & Christophers, S.R. 1908. *The practical study of malaria and other blood parasites*. Third edition. Williams & Norgate, London. 414 pp.

Stoker, W.J. & Waktoedi Koesoemawinangoen, R. 1949. *Illustrated map of the anopheline imagines of Indonesia (English translation)*. Ministry of Health of Indonesia, Djakarta. 69 pp.

Stone, A. 1939. Two new *Aedes* from Guam (Diptera, Culicidae). *Proceedings of the Entomological Society of Washington* 41(5): 162–165.

Stone, A. 1944. Notes on the genus *Trichoprosopon* (Diptera, Culicidae). *Revista de Entomologia* 15(3): 335–341.

Stone, A. 1945a. A mosquito synonym (Diptera: Culicidae). *Proceedings of the Entomological Society of Washington* 47(2): 38–39.

Stone, A. 1945b. A new species of *Aedes* from Saipan and the larva of *Aedes pandani* (Diptera: Culicidae). *Proceedings of the Entomological Society of Washington* 47(3): 65–69.

Stone, A. 1957a. Corrections in the taxonomy and nomenclature of mosquitoes (Diptera, Culicidae). *Proceedings of the Entomological Society of Washington* (for 1956) 58(6): 333–344.

Stone, A. 1957b. Notes on types of mosquitoes in the Hungarian National Museum (Diptera, Culicidae). *Annals of the Entomological Society of America* 50(2): 171–174.

Stone, A. 1958a. The identity of *Culex aestuans* Wiedemann (Diptera, Culicidae). *Proceedings of the Entomological Society of Washington* 60(4): 186.

Stone, A. 1958b. Types of mosquitoes described by C. F. Adams in 1903 (Diptera, Culicidae). *Journal of the Kansas Entomological Society* 31(3): 235–237.

Stone, A. 1961a. A new subspecies of *Uranotaenia unguiculata* Edwards from Arabia (Diptera: Culicidae). *Proceedings of the Entomological Society of Washington* (for 1960) 62(4): 249.

Stone, A. 1961b. A synoptic catalog of the mosquitoes of the world, supplement I (Diptera: Culicidae). *Proceedings of the Entomological Society of Washington* 63(1): 29–52.

Stone, A. 1963. A synoptic catalog of the mosquitoes of the world, supplement II (Diptera: Culicidae). *Proceedings of the Entomological Society of Washington* 65(2): 117–140.

Stone, A. 1966. A new species of *Coquillettidia* from Samoa (Diptera: Culicidae). *Proceedings of the Entomological Society of Washington* 68(4): 331–333.

Stone, A. 1967. A synoptic catalog of the mosquitoes of the world, supplement III (Diptera: Culicidae). *Proceedings of the Entomological Society of Washington* 69(3): 197–224.

Stone, A. 1969. Bredin-Archbold-Smithsonian Biological Survey of Dominica: The mosquitoes of Dominica (Diptera: Culicidae). *Smithsonian Contributions to Zoology* 16: 1–8.

Stone, A. & Barreto, P. 1969. A new genus and species of mosquito from Colombia, *Galindomyia leei* (Diptera, Culicidae, Culicini). *Journal of Medical Entomology* 6(2): 143–146.

Stone, A. & Bohart, R.M. 1944. Studies on mosquitoes from the Philippine Islands and Australasia (Diptera: Culicidae). *Proceedings of the Entomological Society of Washington* 46(8): 205–225.

Stone, A. & Farner, D.S. 1945. Further notes on the *Aedes scutellaris* group (Diptera, Culicidae). *Proceedings of the Biological Society of Washington* 58: 155–162.

Stone, A. & Hair, J.A. 1968. A new *Culex (Melanoconion)* from Florida (Diptera, Culicidae). *Mosquito News* 28(1): 39–41.

Stone, A. & Knight, K.L. 1956. A new name in Culicidae (Diptera). *Proceedings of the Entomological Society of Washington* 58(1): 34.

Stone, A. & Knight, K.L. 1958. Two new names in mosquitoes (Diptera, Culicidae). *Proceedings of the Entomological Society of Washington* 60(2): 69.

Stone, A. & Penn, G.H. 1947. A new species of *Culex* from New Guinea (Diptera, Culicidae). *Journal of the Washington Academy of Sciences* 37(3): 89–91.

Stone, A. & Penn, G.H. 1948. A new subgenus and two new species of the genus *Culex* L. (Diptera, Culicidae). *Proceedings of the Entomological Society of Washington* 50(5): 109–120.

Stone, A. & Rosen, L. 1952. A new species of *Aedes* from Tahiti (Diptera: Culicidae). *Proceedings of the Hawaiian Entomological Society* 14(3): 425–428.

Stone, A. & Rosen, L. 1953. A new species of *Culex* from the Marquesas Islands and the larva of *Culex atriceps* Edwards (Diptera: Culicidae). *Journal of the Washington Academy of Sciences* 43(11): 354–358.

Stone, A. & Thurman, E.B. 1958. *Armigeres (Armigeres) baisasi,* a new mosquito from the Philippine Islands (Diptera: Culicidae). *Journal of the Washington Academy of Sciences* 48(7): 240–243.

Stone, A., Knight, K.L. & Starcke, H. 1959. *A synoptic catalog of the mosquitoes of the world (Diptera, Culicidae).* The Thomas Say Foundation. Volume VI. Entomological Society of America, College Park, Maryland. vi, 358 pp.

Strand, E. 1928. Miscellanea nomenclatorica zoologica et palaeontologica. *Archiv für Naaturgeschichte. Abteilung A* (for 1926) 92(8): 30–75.

Strickland, C. 1915a. Note on *Anopheles brevipalpis* Roper, and description of its egg and larva. *Indian Journal of Medical Research* 3(1): 201–204, 1 pl.

Strickland, C. 1915b. *Short key to the identification of the larvae of the common anopheline mosquitos of the Malay Peninsula.* Central Survey Office, Kuala Lumpur. 18 pp.

Strickland, C. 1916a. A new species of protanopheline from Malaya, *Myzorhynchus hunteri. Indian Journal of Medical Research* 4(2): 263–270.

Strickland, C. 1916b. An *umbrosus*-like anopheline from Malaya—*Myzorhynchus novumrosus* [*sic*]. *Indian Journal of Medical Research* 4(2): 271–273.

Strickland, [C.] 1917. A new species of anopheline, *Myzorhynchus similis* from the Malay Peninsula. *Indian Journal of Medical Research* 4(6): 611.

Strickland, C. & Chowdhury, K.L. 1927a. *An illustrated key to the identification of the anopheline larvae of India, Ceylon, and Malaya, west of Wallace's Line with practical notes on their collection.* Thacker, Spink & Co., Calcutta & Simla. xi, 67 pp.

Strickland, C. & Chowdhury, K.L. 1927b. A new species of anopheline *A. pseudojamesi* common in Bengal. *Indian Medical Gazette* 62(5): 240–243.

Strickland, C. & Chowdhury, K.L. 1931. *The anopheline larvae of the countries from India and the Orient to the Antipodes. A supplement (1931) to the anopheline larvae of India, Ceylon and Malaya.* Thacker, Spink & Co., Ltd., Calcutta & Simla. [i], 36 pp., 9 pls.

Strickland, E.H. 1911a. Some new Culicidae from Western Australia, South Queensland, and Tasmania. *Entomologist* 44(575): 130–134.

Strickland, E.H. 1911b. Some new Culicidae from Western Australia, South Queensland, and Tasmania. *Entomologist* 44(576, 577, 578): 179–182, 201–204, 249–251.

Strickland, E.H. 1911c. A new mosquito from Paraguay. *Entomologist* 44(579): 268–269.

Strickman, D. 1988. Redescription of the holotype of *Culex* (*Culex*) *peus* Speiser and taxonomy of *Culex* (*Culex*) *stigmatosoma* Dyar and *thriambus* Dyar (Diptera: Culicidae). *Proceedings of the Entomological Society of Washington* 90(4): 484–494.

Strickman, D. 1990. *Culex pseudostigmatosoma, Cx. yojoae* and *Cx. aquarius*: New Central American species in the subgenus *Culex* (Diptera: Culicidae). *Mosquito Systematics* (for 1989) 21(3): 143–177.

Stschelkanovzev, J. 1926. To the knowledge of the mosquito fauna of south-east RSFSR [in Russian]. *Bulletin Society Natural Voronezh* 1: 123–134.

Su, L. & Wang, C. 1981. A new species of the genus *Toxorhynchites* Theobald (Diptera: Culicidae) [in Chinese]. *Acta Entomologica Sinica* 24(3): 327–331.

Su, L. & Zhang, Y. 1987. A new species of the subgenus *Ochlerotatus* Lynch Arribalzaga (Diptera: Culicidae) [in Chinese]. *Acta Entomologica Sinica* 30(4): 450–452.

Su, L. & Zhang, Y. 1988. A new species of *Aedes* (Diptera: Culicidae) [in Chinese]. *Entomotaxonomia* 10(1–2): 11–13.

Su, L., Wang, C. & Li, F. 1978. A new species of *Aedes, Aedes* (*Edwardsaedes*) *antuensis* sp.nov. [in Chinese]. *Acta Entomologica Sinica* 21(4): 439–444.

Surcouf, J. 1912. Note sur les culicides. *Bulletin du Muséum national d'histoire naturelle* 18(2): 59–61.

Surcouf, J.M.R. & Gonzalez-Rincones, R. 1911. *Essai sur les Diptères vulnérants du Venezuela: Matériaux pour servir à l'étude des Diptères piqueurs et suceurs de sang de l'Amérique intertropicale. Première partie: Diptères nématocères vulnérants.* A. Maloine, Éditeur, Paris. v, 240 pp.

Surcouf, J.M.R. & Gonzalez-Rincones, R. 1912. Diptères piqueurs et suceurs de sang. Actuellement connus, de la République do Venezuela. *Archives de Parasitologie* 15(2): 248–314.

Sutil O., E., & Pulido F., J. 1978. *Sabethes* (*Sabethes*) *carrilloi* sp. n. de Venezuela. *Boletin de la Direccion de Malariologia y Saneamiento Ambiental* 18(3): 199–204.

Sutil Oramas, E. & Pulido F., J. 1974. *Wyeomyia cova-garciai* n. sp. *Boletín informativo de la Dirección de Malariología y Saneamiento Ambiental* 14(1–2): 24–26.

Sutil Oramas, E., Pulido Florenzano, J. & Amarista Meneses, J.R. 1987. Dos especies nuevas de *Culex* de Venezuela (Diptera: Culicidae). *Boletin de la Direccion de Malariologia y Saneamiento Ambiental* 27(1–4): 81–85.

Suzuki, Y., Tanimura, S., Miyagawa, M. & Murata, S. 1953. Species of mosquitoes collected in Tokushima City. III [in Japanese]. *Igaku to seibutsugaku* [*Medicine and Biology*] 28(1): 8–10.

Sweet, W.C. & Rao, B.A. 1937. Races of *A. stephensi* Liston, 1901. *Indian Medical Gazette* 72(11): 665–674.

Swellengrebel, N.H. 1914. Een nieuwe Anopheline veer Deli: *Myzorhynchus Argyropus* n. sp. (met autoreferaat.). *Geneeskundig Tijdschrift voor Nederlandsch-Indië* 54: 334–337, 1 pl.

Swellengrebel, N.H. 1917. *Myzomyia flava* n. sp., een nieuwe Anopheline veer Ned.-Indië. *Geneeskundig Tijdschrift voor Nederlandsch-Indië* 57: 807–809.

Swellengrebel, N.H. 1919. Eenige voor Nederl.-Indië nieuwe anophelinen. *Geneeskundig Tijdschrift voor Nederlandsch-Indië* 59: 1–12, 2 pls.

Swellengrebel, N.H. 1921. *Die anophelinen van Nederlandsch Oost-Indië*. Vereeniging Koloniaal Instituut te Amsterdam. Mededeeling N°. XV. Afdeeling Tropische Hygiene N°. 10. De Bussy, Amsterdam. x, 155 pp., 20 pls.

Swellengrebel, N.H. 1925. *Cellia errabunda* n. sp., een nieuwe anopheline voor Nederlandsch-Indié. *Geneeskundig Tijdschrift voor Nederlandsch-Indié* 69: 1913–1915.

Swellengrebel, N.H. & Rodenwaldt, E. 1932. *Die anophelen von Niederländisch-Ostindien*. 3te auflage. Gustav Fischer, Jena. viii, 242 + [2] pp., 24 pls.

Swellengrebel, N.H. & Swellengrebel de Graaf, J.M.H. 1920a. Addenda to bescription [*sic*] of larvea [*sic*] of Netherlands Indian anophelines. *Mededeelingen van den Burgerlijken Geneeskundigen Dienst in Nederlandsch-Indië* (for 1919) 9(Add.): 1–3, 2 pls.

Swellengrebel, N.H. & Swellengrebel de Graaf, J.M.H. 1920b. List of the anophelines of the Malay Archipelago with special reference to adults and larvae of new or incompletely described species or varieties. *Bulletin of Entomological Research* 11(1): 77–92.

Szadziewski, R. 1998. New mosquitoes from Baltic amber (Diptera: Culicidae). *Polskie Pismo Entomologiczne* 67(3–4): 233–244.

Szadziewski, R. & Gilka, W. 2011. A new fossil mosquito, with notes on the morphology and taxonomy of other species reported from Eocene Baltic amber (Diptera: Culicidae). *Polskie Pismo Entomologiczne* 80(4): 765–777.

Szadziewski, R. & Szadziewska, M.M. 1985. *Culex erikae* sp. n. (*Diptera, Culicidae*) from the Baltic amber. *Polskie Pismo Entomologiczne* 55: 513–518.

Taai, K. & Harbach, R.E. 2015. Systematics of the *Anopheles barbirostris* species complex (Diptera: Culicidae: Anophelinae) in Thailand. *Zoological Journal of the Linnean Society* 174(2): 244–264 + 15 online tables.

Taai, K., Baimai, V., Saeung, A., Thongsahuan, S., Min, G.-S., Otuska, Y., Park, M.-H., Fukuda, M., Somboon, P. & Choochote, W. 2013. Genetic compatibility between *Anopheles lesteri* from Korea and *Anopheles paraliae* from Thailand. *Memórias do Instituto Oswaldo Cruz* 108(3): 312–320.

Takahashi, M. 1968. Taxonomic and ecological notes on *Culex* (*Melanoconion*) *spissipes* (Theobald). *Journal of Medical Entomology* 5(3): 329–331.

Takei, H. 1941. Kinds and distribution of *Anopheles* in Hainan Island [in Japanese]. *Bulletin of the Naval Medical Association (Japan)* 30: 414–437, 541–555.

Tamayo, M.O. & García, C.A. 1907. Las aquas do Huacachina informe prosentado a la Sociedad Geografica de Lima. *Memoirs Municipal de Lima* 1906: 1–63. [Reproduced in part 30 September 1907 as 'Las Aquas do Huacachina' in *Boletin de la Sociedad Geográfica de Lima* 21: 241–251, including the description of *Anopheles peruvianus* Tamayo].

Tanaka, K. 1998. A new species of mosquitoes of *Aedes* (*Ochlerotatus*) (Diptera, Culicidae) from Japan. *Japanese Journal of Systematic Entomology* 4(2): 215–225.

Tanaka, K. 1999. Studies of the pupal mosquitoes (Diptera, Culicidae) of Japan. (1) *Aedes* (*Ochlerotatus*). *Japanese Journal of Systematic Entomology* 5(1): 105–124.

Tanaka, K. 2002. Studies on the pupal mosquitoes of Japan (6). *Aedes* (*Finlaya*) (Diptera, Culicidae). *Japanese Journal of Systematic Entomology* 8(2): 137–177.

Tanaka, K. 2003. Studies on the pupal mosquitoes of Japan (9). Genus *Lutzia*, with establishment of two new subgenera, *Metalutzia* and *Insulalutzia* (Diptera, Culicidae). *Japanese Journal of Systematic Entomology* 9(2): 159–169.

Tanaka, K. 2004. Studies on the pupal mosquitoes of Japan (11) subgenera *Oculeomyia* (stat. nov.) and *Sirivanakarnius* (nov.) of the genus *Culex*, with a key of pupal mosquitoes from Ogasawara-guntô (Diptera: Culicidae). *Medical Entomology and Zoology* 55(3): 217–231.

Tanaka, K. & Mizusawa, K. 1973. Two new species of the genus *Aedes* (*Neomacleaya*) from the Ryukyu Islands (Diptera, Culicidae). *Bulletin of the National Science Museum, Tokyo* 16(4): 625–638, 2 pls.

Tanaka, K., Mizusawa, K. & Saugstad, E.S. 1975a. Two new crab hole mosquitoes of the genus *Uranotaenia (Pseudoficalbia)* from the Ryukyu Islands (Diptera, Culicidae). *Mosquito Systematics* 7(1): 27–40.

Tanaka, K., Mizusawa, K. & Saugstad, E.S. 1975b. A new species of the genus *Aedes (Aedes)* from Japan, with synonymical notes on Japanese species of the subgenus *Aedes* (Diptera, Culicidae). *Mosquito Systematics* 7(3): 41–58, 174–177.

Tanaka, K., Mizusawa, K. & Saugstad, E.S. 1979. A revision of the adult and larval mosquitoes of Japan (including the Ryukyu Archipelago and the Ogasawara Islands) and Korea (Diptera: Culicidae). *Contributions of the American Entomological Institute* 16: vii, 1–987.

Tanaka, K., Saugstad, E.S. & Mizusawa, K. 1975. Mosquitoes of the Ryukyu Archipelago (Diptera: Culicidae). *Mosquito Systematics* 7(3): 207–233.

Tanimura, S. 1952. On the variation of the number of white bands of legs of *Aedes japonicus* Theobald collected in Tokushima City [in Japanese]. *Jikken Seibutsu Gakuhō* [*Bulletin of Experimental Biology*] 2(1): 173–178.

Taylor, B. 1972. A new species of *Aedes* from San Cristobal, British Solomon Islands Protectorate. *Journal of Medical Entomology* 9(4): 317–318.

Taylor, B. & Maffi, M. 1991. *Anopheles (Cellia) rennellensis*, a new species within the *punctulatus* complex of *Anopheles* (Diptera, Culicidae) from Rennell Island, pp. 195–198. *In*: Wolff, T. (ed), *Natural history of Rennell Island, British Solomon Islands. Volume 8 (Zoology). Scientific results of The Noona Dan Expedition (Rennell Section, 1962) and The Danish Rennell Expedition, 1965*. Danish Science Press, Copenhagen.

Taylor, B. & Tenorio, J.A. 1974. *Aedes (Finlaya) maffii*, a new species of mosquito from the British Solomon Islands (Diptera: Culicidae). *Journal of Medical Entomology* 11(5): 577–581.

Taylor, F.H. 1912a. Description of mosquitoes collected in the Northern Territory during the expedition, 1911. *Bulletin of the Northern Territory* la: 25–34, 1 pl.

Taylor, F.H. 1912b. Contribution to a knowledge of the Culicidae and Ixodidea of North Queensland and the Torres Strait. *Annual Report of Community Public Health Queensland* 1912: 27–29, 1 pl.

Taylor, F.H. 1913. Report of the entomologist. *Report of the Australian Institute of Tropical Medicine* 1911: 49–74, 3 pls.

Taylor, F.H. 1914a. A revision of the *Culicidae* in the Macleay Museum, Sydney. *Proceedings of the Linnean Society of New South Wales* (for 1913) 38(4): 747–760, 1 pl.

Taylor, F.H. 1914b. Contributions to a knowledge of Australian Culicidae. No. i. *Proceedings of the Linnean Society of New South Wales* 39(3): 454–468, 4 pls.

Taylor, F.H. 1914c. The *Culicidae* of Australia.—I. *Transactions of the Entomological Society of London* 1913: 683–708, 4 pls.

Taylor, F.H. 1914d. Culicidae from Papua. *Transactions of the Entomological Society of London* 1914: 185–205, 2 pls.

Taylor, F.H. 1915. Contributions to a knowledge of Australian Culicidae. No. ii. *Proceedings of the Linnean Society of New South Wales* 40(1): 176–184, 2 pls.

Taylor, F.H. 1916. Contributions to a knowledge of Australian *Culicidae*. No. iii. *Proceedings of the Linnean Society of New South Wales* 41(3): 564–574.

Taylor, F.H. 1919. Contributions to a knowledge of Australian Culicidae. No. iv. *Proceedings of the Linnean Society of New South Wales* (for 1918) 43(4): 826–843, 4 pls.

Taylor, F.H. 1920. Australian phlebotomic Diptera:— New Culicidae Tabanidae and synonymy. *Proceedings of the Royal Society of Victoria* 32(2): 164–167.

Taylor, F.H. 1927. Two new species of Australian Culicidae (Diptera). *Bulletin of Entomological Research* 18(1): 67–69.

Taylor, F.H. 1929. Notes on Australian Culicidae (Dipt.). *Bulletin of Entomological Research* 20(3): 271–277, 1 pl.

Taylor, F.H. 1933. A new species of *Finlaya* (order Diptera: family Culicidae) from Papua. *Australian Zoologist* 7(5): 370.

Taylor, F.H. 1934. The Diptera of the Territory of New Guinea. I. Family Culicidae. *Proceedings of the Linnean Society of New South Wales* 59(3–4): 229–236.

Taylor, F.H. 1943. Contributions to a knowledge of Australian Culicidae. No. VI. *Proceedings of the Linnean Society of New South Wales* 68(3–4): 153–157.

Taylor, F.H. 1944. Contributions to a knowledge of Australian Culicidae. No. VII. *Proceedings of the Linnean Society of New South Wales* 69(3–4): 120–128.

Taylor, F.H. 1946. Contributions to a knowledge of Australian Culicidae. No. VIII. *Proceedings of the Linnean Society of New South Wales* (for 1945) 70(5–6): 211–212.

Tennent, J.E. 1859. *Ceylon: an account of the island; physical, historical, and topographical; with notices of its natural history, antiquities and productions*. Second edition. Volume 1. Longman, Green, Longman, and Roberts, London. xxxix, 643 pp.

Tennent, J.E. 1861. *Sketches of the natural history of Ceylon with narratives and anecdotes illustrative of the habits and instincts of the mammalia, birds, reptiles, fishes, insects, &c. including a monograph of the elephant and a description of the modes of capturing and training it*. Longman, Green, Longman, and Roberts, London. xxiii, 500 pp.

Tenorio, J.A. 1975. *Bironella* (*Brugella*) *obscura*, a new species of mosquito from New Guinea (Diptera: Culicidae). *Journal of Medical Entomology* (for 1974) 11(6): 734–738.

Tenorio, J.A. 1977. Revision of the genus *Bironella* (Diptera: Culicidae). *Journal of Medical Entomology* 14(3): 317–361.

Tewari, S.C. & Hiriyan, J. 1992a. Description of a new species of *Aedes* (*Rhinoskusea*) from South India. *Mosquito Systematics* (for 1991) 23(2): 123–131.

Tewari, S.C. & Hiriyan, J. 1992b. Description of two new species of *Aedes* (*Diceromyia*) from South India (Diptera: Culicidae). *Mosquito Systematics* 24(2): 154–175.

Tewari, S.C. & Hiriyan, J. 1996. Revalidation and description of *Aedes* (*Finlaya*) *pallirostris* from northeastern India (Diptera: Culicidae). *Journal of the American Mosquito Control Association* 12(4): 713–720.

Theobald, F.V. 1900a. *Report on the collections of mosquitoes (Culicidae) received at the British Museum (Natural History) from various parts of the world in connection with the investigation into the causes of malaria, conducted by the Colonial Office and the Royal Society*. British Museum (Natural History), London. 12 pp.

Theobald, F.V. 1900b. A new *Anopheles* (*A. paludis*) from Sierra Leone. *Report of the Malaria Commission of the Royal Society* 1: 75–76.

Theobald, F.V. 1901a. *A monograph of the Culicidae or mosquitoes*. Volume 1. British Museum (Natural History), London. xviii, 424 pp. [viii, pls I–XXXVII and A–E under separate cover].

Theobald, F.V. 1901b. The classification of mosquitoes. *Journal of Tropical Medicine* 4: 229–235.

Theobald, F.V. 1901c. *A monograph of the Culicidae or mosquitoes*. Volume 2. British Museum (Natural History), London. viii, 391 pp.
Theobald, F.V. 1901d. Notes on a collection of mosquitoes from West Africa, and descriptions of new species. *Memoirs of the Liverpool School of Tropical Medicine* 4(Appendix): i–xiv, 3 pls.
Theobald, F.V. 1902a. A short description of the Culicidae of India, with descriptions of new species of *Anopheles*. *Proceedings of the Royal Society of London* 69(456): 367–394, 1 pl.
Theobald, F.V. 1902b. The classification of the Anophelina. *Journal of Tropical Medicine* 5: 181–183.
Theobald, F.V. 1903a. *A monograph of the Culicidae or mosquitoes*. Volume 3. British Museum (Natural History), London. xvii, errata [1 p.], 359 pp., foldout table, 17 pls.
Theobald, F.V. 1903b. Report on a collection of mosquitoes and other flies from Equatorial East Africa and the Nile provinces of Uganda. *Reports of the Sleeping Sickness Commission of the Royal Society* 7(3): 33–42, map.
Theobald, F.V. 1903c. Report on a collection of mosquitoes or Culicidae, etc, from Gambia, and descriptions of new species. *Memoirs of the Liverpool School of Tropical Medicine* 10(Appendix): i–xi, 2 pls.
Theobald, F.V. 1903d. A new culicid from Senegal and notes on the species of mosquitoes, etc. *Memoirs of the Liverpool School of Tropical Medicine* 11, Part 2, (Appendix): i–iii, 1 pl.
Theobald, F.V. 1903e. Two new Australian culicids. *Entomologist* 36(481): 154–157.
Theobald, F.V. 1903f. New Culicidae from the Federated Malay States. *Entomologist* 36(485): 256–259.
Theobald, F.V. 1903g. Two new Jamaican Culicidae. *Entomologist* 36(486): 281–283.
Theobald, F.V. 1903h. Notes on Culicidae and their larvae from Pecos, New Mexico, and description of a new *Grabhamia*. *Canadian Entomologist* 35(11): 311–316.
Theobald, F.V. 1903i. Description of a new North American *Culex*. *Canadian Entomologist* 35(8): 211–213.
Theobald, F.V. 1904a. A new culicid genus from Uganda. *Journal of Tropical Medicine* 7: 17–18.
Theobald, F.V. 1904b. The mosquitoes of Egypt, the Sudan and Abyssinia, pp. 62–83, 2 pls. *In*: Balfour, A. (director), *First Report of the Wellcome Research Laboratories at the Gordon Memorial College Khartoum*. Department of Education, Sudan Government, Khartoum.
Theobald, F.V. 1904c. New Culicidae from the Federated Malay States. *Entomologist* 37(488, 489, 490, 491, 493, 495, 496): 12–15, 36–39, 77–78, 111–113, 163–165, 211–213, 236–239.
Theobald, F.V. 1905a. A new genus of Culicidae. *Entomologist* 38(501): 52–56.
Theobald, F.V. 1905b. New Culicidae from the west coast of Africa. *Entomologist* 38(503): 101–104, 154–158.
Theobald, F.V. 1905c. A new *Stegomyia* from the Transvaal. *Entomologist* 38(508): 224–225.
Theobald, F.V. 1905d. *The mosquitoes or Culicidae of Jamaica*. Institute of Jamaica, Kingston. 40 pp., 9 pls.
Theobald, F.V. 1905e. *Genera Insectorum. Diptera, Fam. Culicidae*. Fascicle 26. V. Verteneuil & L. Desmet, Bruxelles. 50 pp., 2 pls.
Theobald, F.V. 1905f. A catalogue of the Culicidae in the Hungarian National Museum, with descriptions of new genera and species. *Annales Musei Nationalis Hungarici* 3(1): 61–119, figure legends, 4 pls.
Theobald, F.V. 1905g. New Culicidae from India, Africa, British Guiana, and Australia. *Journal of Economic Biology* 1: 17–36, 2 pls.
Theobald, F.V. 1905h. Some new mosquitoes from Ceylon. *Journal of the Bombay Natural History Society* 16(2): 237–250, 2 pls.
Theobald, F.V. 1905i. A new *Ficalbia* from West Africa. *Annals and Magazine of Natural History* 15(86): 199–200.
Theobald, F.V. 1906a. A new *Megarhinus*. *Entomologist* 39(522): 241.
Theobald, F.V. 1906b. Report on economic entomology. Part I. Second report on the mosquitoes or Culicidae of the Sudan, pp. 67–83. *In:* Balfour, A. (director), *Second Report of the Wellcome Research Laboratories at the Gordon Memorial College, Khartoum*. Department of Education, Sudan Government, Khartoum.
Theobald, F.V. 1907. *A monograph of the Culicidae or mosquitoes*. Volume 4. British Museum (Natural History), London. xix, 639 pp., 16 pls.
Theobald, F.V. 1908a. Notes on some Transvaal mosquitoes, including two new species and a new variety. *Entomologist* 41(540): 106–109.
Theobald, F.V. 1908b. First report on the collection of Culicidae and Corethridae in the Indian Museum, Calcutta, with descriptions of new genera and species. *Records of the Indian Museum* 2(3): 287–302.
Theobald, F.V. 1908c. New mosquitoes from the Sudan and list and synoptic table of all the known Sudanese species, pp. 249–267, 2 pls. *In*: Balfour, A. (director), *Third Report of the Wellcome Research Laboratories at the Gordon Memorial College, Khartoum*. Baillière, Tindall & Cox, London.
Theobald, F.V. 1909a. A new culicid genus. *Annals of Tropical Medicine and Parasitology* (for 1908) 2: 297.
Theobald, F.V. 1909b. *Descriptions of the new mosquitoes collected by Dr. Graham in Ashanti*. Gold Coast Miscellaneous Publication No. 237. Great Britain. Colonial Office, London. 31 pp., 2 pls.
Theobald, F.V. 1910a. Second report on the collection of Culicidae in the Indian Museum, Calcutta, with descriptions of new genera and species. *Records of the Indian Museum* 4(1): 1–33, 3 pls.
Theobald, F.V. 1910b. The Culicidae of Fiji, including two new species. *Entomologist* 43(565): 155–159.
Theobald, F.V. 1910c. *A monograph of the Culicidae or mosquitoes*. Volume 5. British Museum (Natural History), London. xiv, [1], 646 pp., 6 pls.
Theobald, F.V. 1911a. Uganda Culicidae, pp. 9–35. *In*: Theobald, F.V., *Novae Culicidae. Part 1*. Published by the author, Wye, Kent, UK.
Theobald, F.V. 1911b. The Culicidae or mosquitoes of the Transvaal. *Report of the Division of Entomology of the Department of Agriculture, Union of South Africa* 1: 232–272, 1 pl.

Theobald, F.V. 1911c. A new genus and two new species of Culicidae from the Sudan, pp. 151–15. *In*: Balfour, A. (director), *Fourth Report of the Wellcome Tropical Research Laboratories at the Gordon Memorial College, Khartoum. Volume B.—General Science*. Baillière, Tindall & Cox, London.

Theobald, F.V. 1911d. Culicidae of the R. Zool. Soc. "Natura Artis Magistra", Amsterdam and description of three new species. *Tijdschrift voor Entomologie* 54: 233–240.

Theobald, F.V. 1912a. Three new Culicidae from the Transvaal. *Entomologist* 45(586): 92–93.

Theobald, F.V. 1912b. No. V.—Diptera, Culicidae. *Transactions of the Linnean Society of London. Second Series: Zoology* 15(1): 81–94, 1 pl.

Theobald, F.V. 1913a. Second report on the mosquitoes of the Transvaal, pp. 315–342, 2 pls. *In*: *Union of South Africa, Department of Agriculture, Second Report of the Director of Veterinary Research*. Cape Times Limited, Government Printers, Cape Town.

Theobald, F.V. 1913b. Culicidae from New Caldeonia and the Loyalty Islands, pp. 163–164. *In*: Sarasin, F. & Roux, J., *Nova Caledonia. Forschungen in Neu-Caledonien und auf den Loyalty-Inseln. Recherches scientifiques en Nouvelle-Calédonie et aux iles Loyalty. A. Zoologie*. Vol. I, L. III. C.W. Kreidels Verlag, Wiesbaden.

Theobald, F.V. 1913c. A new mosquito from northern China. *Entomologist* 46(601): 179–180.

Theobald, F.V. 1913d. New Culicidae from the Sudan. *Annals of Tropical Medicine and Parasitology* 7(4): 591–602.

Theobald, F.V. 1914. A new mosquito from Samoa. *Entomologist* 47(608): 36–37.

Thompson, G.A. 1956. *Aedes stenei*, a new species of mosquito from Jamaica, B.W.I., with redescription of a closely related species (Diptera, Culicidae). *Mosquito News* 16(3): 196–200.

Thomson, C.G. 1869. Diptera. Species nova descripsit, pp. 443–614, 1 pl. *In*: K. Svenska Vtenskaps Akademien (corporate author), *Kongliga svenska Fregatten Eugenies resa omkring jorden. Vetenskapliga Iakitagelser. II. Zoologi. 1. Insecta*. P. A. Norstedt & Soner, Stockholm.

Thomson, F.W. 1903. Notes on the Culicidae of Dehra Dun. With a description of a new *Mansonia*, which mimics *Anopheles*. *Journal of Tropical Medicine* 6: 314–315.

Thurman, D.C., Jr 1954a. *Ayurakitia*, a new genus of mosquito from northern Thailand (Diptera: Culicidae). *Journal of the Washington Academy of Sciences* 44(6): 197–200.

Thurman, D.C. 1954b. The discovery of *Paraedes (Udaya) argyrurus* Edwards, 1934, in Thailand (Diptera: Culicidae). *Mosquito News* 14(2): 83–86.

Thurman, D.C. 1955. *Culex* (*Culiciomyia*) *termi*, an unusual new mosquito from Thailand (Diptera, Culicidae). *Proceedings of the Entomological Society of Washington* 57(1): 17–23.

Thurman, D.C. & Thurman, E.B. 1958. Two new species in the genus *Armigeres* Theobald, 1901, from northern Thailand (Diptera: Culicidae). *Journal of the Washington Academy of Sciences* 48(6): 186–187.

Thurman, E.B. 1959. A contribution to a revision of the Culicidae of northern Thailand. *University of Maryland Agricultural Experimental Station Bulletin* A-100: 1–182.

Thurman, E.B. & Winkler, E.C. 1950. A new species of mosquito in California, *Aedes* (*Ochlerotatus*) *bicristatus* (Diptera, Culicidae). *Proceedings of the Entomological Society of Washington* 52(5): 237–250.

Thurman, E.B., Haeger, J.S. & Mulrennan, J.A. 1949. The occurrence of *Aedes* (*Ochlerotatus*) *thelcter* Dyar in the Florida Keys. *Mosquito News* 9(4): 171–172.

Toma, T. & Higa, Y. 2004. A new species of *Ficalbia* (Diptera: Culicidae) from Iromote Island, Okinawa, Ryukyu Archipelago, Japan. *Medical Entomology and Zoology* 55(3): 195–199.

Toma, T. & Miyagi, I. 2002. *Armigeres* (*Armigeres*) *sembeli* (Diptera: Culicidae), a new species from Sulawesi, Indonesia and redescription of *Armigeres* (*Armigeres*) *giveni* Edwards. *Medical Entomology and Zoology* 53(4): 257–265.

Toma, T. & Miyagi, I. 2003. *Armigeres* (*Armigeres*) *laoensis* sp. nov. (Diptera: Culicidae) from Khammouane Province, Lao PDR. *Medical Entomology and Zoology* 54(2): 169–172.

Toma, T. & Mogi, M. 2003. *Topomyia* (*Suaymyia*) *miyagii* (Diptera: Culicidae): A new species from Flores Is., Indonesia. *Medical Entomology and Zoology* 54(1): 25–30.

Toma, T., Miyagi, I. & Cabrera, B.D. 1984. *Culex (Culiciomyia) azurini*, a new crab hole mosquito species from the Philippines. *Mosquito Systematics* 16(2): 172–182.

Toma, T., Miyagi, I. & Syafruddin. 1995. Description of *Armigeres (Armigeres) alkatirii*, a new species from Sulawesi, Indonesia (Diptera: Culicidae). *Mosquito Systematics* 27(2): 110–117.

Toma, T., Miyagi, I. & Tanaka, K. 1990. A new species of *Toxorhynchites* mosquito (Diptera: Culicidae) from the Ryukyu Archipelago, Japan. *Journal of Medical Entomology* 27(3): 344–355.

Torres Cañamares, F. 1945. Contribución al conocimiento del *Anopheles claviger* Mg. de España (Dip. *Cul.*). *Eos: Revista Española de Entomología* 20(3–4): 233–245.

Torres Cañamares, F. 1946. Nuevos datos sobre el *Anopheles marteri* Sen. y Pru. en España. ¿Se trata de una variedad? (Dipt. Cul.). *Eos: Revista Española de Entomología* 22(1–2): 47–58, 1 pl.

Torres Cañamares, F. 1951. Una nueva especie de *Aedes (Dipt. Cul.)*. *Eos: Revista Española de Entomología* 27(1): 79–92.

Toumanoff, C. 1931. Sur une variété nouvelle d'*Anopheles aconitus* Dönitz observée au Tonkin. *Compte Rendus des Séances de la Société de Biologie et de ses filiales* 107: 575–576.

Townsend, B.C. 1990. Culicidae, pp. 35–152. *In*: Townsend, B.C., Chainey, J.E., Crosskey, R.W., Pont, A.C., Lane, R.P., Boorman, J.P.T. & Crouch, C.A. (eds), *A catalogue of the types of bloodsucking flies in the British Museum (Natural History)*. (Occasional papers on systematic entomology British Museum (Natural History), no.7). British Museum (Natural History), London.

Townson, H., Dyer, N., McAlister, E., Satoto, T.B.T., Boewono, D.T., Bangs, M.J. & Harbach, R.E. 2013. Systematics of *Anopheles barbirostris* van der Wulp and a sibling species of the Barbirostris Complex (Diptera: Culicidae) in eastern Java, Indonesia. *Systematic Entomology* 38(1): 180–191.

Tsai, C. & Lien, J. 1950. A new species of *Aëdes* (*Finlaya*) found in Taiwan. *Journal of the Formosan Medical Association* 49(5): 177–183.

Tsuchimoto, S. 1948. On a variety of *Anopheles koreicus* [in Japanese]. *Medicine and Biology (Japan)* (for 1944) 6(4–6): 364–366.

Tsukamoto, M. 1989. Two new mosquito species from a pitcher plant of Mt. Kinabalu, Sabah, Malaysia: *Culex rajah* and *Toxorhynchites rajah* (Diptera: Culicidae). *Japanese Journal of Tropical Medicine and Hygiene* 17(3): 215–228.

Tsukamoto, M., Miyagi, I. & Toma, T. 1985. A revised checklist of Philippine mosquitoes. *Tropical Biomedicine* 2(2): 149–160.

Tsuzuki, J. 1902a. Malaria und ihre Vermittler in Japan. *Archiv für Schiffs- und Tropen-Hygiene* 6(9): 285–295.

Tsuzuki, Z. [*sic*] 1902b. Ueber die Ergebnisse meiner Malariaforschung in Hokkaido (Japan). *Zentralblatt für Bakteriologie, Parasitenkunde, Infektionskrankheiten (und Hygiene). Abteilung 1. Originale* 31(15): 763–768.

Tsuzuki, J. 1902c. A report of an investigation on malaria in Taiwan [in Japanese]. *Saikingaku Zasshi* 75: 25–51.

Tsuzuki, J. 1907. Über die *Anopheles*-Arten in Japan und einige Beiträge zur Kenntnis des Entwicklungsgangs der *Anopheles*-Larven. *Zoologische Jahrbücher* 25(5–6): 525–556, 4 pls.

Tung, L.H. 1955. On three new species of mosquitoes from Kiangsi Province, China [in Chinese]. *Acta Entomologica Sinica* 5(9): 309–315.

Turner, R.L. 1924. A new mosquito from Texas (*Diptera, Culicidae*). *Insecutor Inscitiae Menstruus* 12: 84.

Tyagi, B.K., Hiriyan, J., Tewari, S.C., Ayanar, K., Samuel, P.P., Arunachalam, N., Paramasivan, R., Krishnamoorthy, R., Dhananjeyan, K.J., Leo, S.V. & Rajendran, R. 2009. Description of a new species, *Anopheles pseudosundaicus* (Diptera: Culicidae) from Kerala, India. *Zootaxa* 2219: 49–60.

Ungureanu, E. 1956. Noi contributii la studiul geului *Theobaldia* in R.P.R. *Acaedmiă Republicii Populare Romîne Filială Lăşi, Studii şi Cercetări Ştinţifice Biologie şi Ştiinţe Agricole* 7(2): 1–9.

Unti, O. 1940a. Anofelinos do vale do Paraiba. Nova variedade e ciclo evolutivo do *Anófeles (Nyssorhynchus) Osvaldi Var. Ayrozai N. Var. Annaes Paulistas de Medicìna e Cirurgia* 40(11): 377–392.

Unti, O. 1940b. Anofelinos do Vale do Paraíba. Note *111.* - Biologia do *Anóféles (Nyssorhynchus) strodei Root, 1926* com a descrição d'uma variedade nova. *Anóféles (Nyssorhynchus) strodei ramosi* var. n. *Annaes Paulistas de Medicìna e Cirurgia* 40(12): 489–505.

Unti, O. 1941. Anofelinos do vale do Rio Paraíba, *Anopheles (Nyssorhynchus) strodei* Root, 1926 com a descrição de tres variedades novas. *Publica Serviço de Profilaxia da Malária* 33: 3–18.

Valencia, J.D. 1973. Mosquito studies (Diptera, Culicidae). XXXI. A revision of the subgenus *Carrollia* of *Culex*. *Contributions of the American Entomological Institute* 9(4): 1–134.

van den Assem, J. 1958. *Mansonia (Mansonioides) bonnewepsterae*, spec. nov. (Culicidae), with notes on habits and breeding place. *Tropical and Geographical Medicine* 10(3): 205–212.

van den Assem, J. 1959. Notes on New Guinean species of *Tripteroides*, subgenus *Rachisoura* (Diptera, Culicidae), with descriptions of two new species. *Tijdschrift voor Entomologie* 102(1): 35–55, 1 pl.

van der Wulp, F.M. 1867. Eenige Noord-Americaansche Diptera. *Tijdschrift voor Entomologie* 10(2): 125–164, 3 pls.

van der Wulp, F.M. 1881. Diptera, 63 pp., 3 pls. *In*: Veth, P.J., 1881–92. *Midden-Sumatra. Reizen en onderzoekingen der Sumatra-expeditie, uitgerust door het Aardrijkskundig genootschap, 1877–1879, beschreven door de leden der expeditie, onder toezicht van Prof. P.J. Veth.* Natuurlijke Historie. Negende afdeeling. E.J. Brill, Leiden.

van der Wulp, F.M. 1884. Note XXXVIII. On exotic Diptera. Part 1. *Notes from the Leyden Museum* 6(4): 248–256.

van Someren, E.C.C. 1945. Ethiopian Culicidae: Descriptions of two East African culicine larvae, not hitherto described, and a new subspecies from Abyssinia (Diptera). *Proceedings of the Royal Entomological Society of London, Series B: Taxonomy* 14(5–6): 77–80.

van Someren, E.C.C. 1946a. Ethiopian Culicidae: Notes and descriptions of some new species and hitherto unknown larvae and pupae (Diptera). *Transactions of the Royal Entomological Society of London* 96(7): 109–124.

van Someren, E.C.C. 1946b. Ethiopian Culicidae. Tribe Megarhinini–notes and descriptions. *Transactions of the Royal Entomological Society of London* 97(7): 177–186.

van Someren, E.C.C. 1946c. Ethiopian Culicidae—descriptions of the adults and pupae of two new species and two subspecies of the genus *Aëdes* from East Africa (Diptera). *Proceedings of the Royal Entomological Society of London, Series B: Taxonomy* 15(1–2): 1–6.

van Someren, E.C.C. 1947a. The description of a new mosquito from the Seychelles. *East African Medical Journal* 24(1): 29–35.

van Someren, E.C.C. 1947b. The description of a new *Anopheles* of the *Myzorhynchus* series from Madagascar with notes on its systematic position in relation to the Ethiopian species of this group. *East African Medical Journal* 24(1): 42–46.

van Someren, E.C.C. 1947c. Ethiopian Culicidae—the description of a new *Eretmapodites* Theobald. *Proceedings of the Royal Entomological Society of London, Series B: Taxonomy* 16(1–2): 13–15.

van Someren, E.C.C. 1948. Ethiopian Culicidae: Some new mosquitoes from Uganda. *Proceedings of the Royal Entomological Society of London, Series B: Taxonomy* 17(9–10): 128–132.

van Someren, E.C.C. 1949a. Ethiopian Culicidae—descriptions of four new mosquitoes from Madagascar. *Proceedings of the Royal Entomological Society of London, Series B: Taxonomy* 18(1–2): 3–8.

van Someren, E.C.C. 1949b. Ethiopian Culicidae—*Eretmapodites* Theobald: Description of four new species of the *chrysogaster* group with notes on the five known species of this group. *Proceedings of the Royal Entomological Society of London, Series B: Taxonomy* 18(7–8): 119–129.

van Someren, E.C.C. 1950a. Ethiopian Culicidae: Description of a new *Aëdes* from East Africa and notes on two known *Aëdes* with which it might be confused. *Proceedings of the Royal Entomological Society of London, Series B: Taxonomy* 19(1–2): 7–10.

van Someren, E.C.C. 1950b. Ethiopian Culicidae: A description of two new *Aëdes* of the subgenus *Aëdimorphus* Theobald, from Uganda. *Proceedings of the Royal Entomological Society of London, Series B: Taxonomy* 19(3–4): 39–41.

van Someren, E.C.C. 1950c. Ethiopian Culicidae: A description of a new sub-species of *Aëdes* (*Aëdimorphus*) *tricholabis* Edwards. *Proceedings of the Royal Entomological Society of London, Series B: Taxonomy* 19(5–6): 67–69.

van Someren, E.C.C. 1951. New Culicini from Kenya and Uganda. *Proceedings of the Royal Entomological Society of London, Series B: Taxonomy* 20(1–2): 1–9.

van Someren, E.C.C. 1954a. New Culicini from the Kenya Coast. *Proceedings of the Royal Entomological Society of London, Series B: Taxonomy* 23(5–6): 103–109.

van Someren, E.C.C. 1954b. Ethiopian Culicidae: descriptions of a new *Culex*, the female of *Eretmapodites tonsus* Edwards and the early stages of two *Aedes* of the subgenus *Banksinella* Theobald. *Proceedings of the Royal Entomological Society of London, Series B: Taxonomy* 23(7–8): 119–126.

van Someren, E.C.C. 1962. Ethiopian Culicidae: Three new *Aedes* from Tanganyika, with a description of the male of *Aedes usambara* Mattingly and the female of *Uranotaenia henrardi* Edwards. *Proceedings of the Royal Entomological Society of London, Series B: Taxonomy* 31(1–2): 19–26.

van Someren, E.C.C. 1967. The female and early stages of *Culex (Culex) nakuruensis* Mattingly, with a description of a new subspecies of *Culex (Culex) shoae* Hamon & Ovazza. *Proceedings of the Royal Entomological Society of London, Series B: Taxonomy* 36(1–2): 11–16.

van Someren, E.C.C. 1971. The description of a new *Aedes* mosquito of the subgenus *Skusea* from the Seychelles. *Journal of Entomology. Series B, Taxonomy* 40(1): 21–25.

van Someren, E.C.C. 1972. On the status of *Aedes* (*Ochlerotatus*) *fryeri* (Theobald) and *Aedes* (*Ochlerotatus*) *mombasaensis* Mattingly. *Mosquito Systematics* 4(3): 90.

van Someren, E.C.C. & Hamon, J. 1964. Ethiopian Culicidae (Diptera). A new species of *Culex* from Tanganyika, the description of the pupa of *Aedes usambara* Mattingly and the early stages of *Eretmapodites tonsus* Edwards. *Journal of the Entomological Society of Southern Africa* 27(1): 78–85.

van Thiel, P.H. 1927. Sur l'origine des variations de taille de l'*Anopheles maculipennis* dans les Pays-Bas. *Bulletin de la Société de Pathologie exotique* 20: 366–390.

Vargas, D.L. 1979. *Anopheles* (*Anopheles*) *veruslanei* Vargas, n. sp. (Diptera: Culicidae). *Gaceta Medica de Mexico* 115(2): 87–88.

Vargas, L. 1939. Datos acerca del *A. Pseudopunctipennis* y de un *Anopheles* nuevo de California. *Medicina: Revista Mexicana Suplemento* 19: 356–362.

Vargas, L. 1941a. El problema de las variedades do *A. punctipennis*. Descripción de *A. punctipennis stonei* n. var. *(Diptera. Culicidae)*. *Revista de la Sociedad Mexicana de Historia Natural* 2(2–3): 175–186.

Vargas, L. 1941b. *Aedes* (*Ochlerotatus*) *rozeboomi* nueva especie (Dipt. Culicidae). *Gaceta Medica de Mexico* 71(3): 393–395.

Vargas, L. 1941c. *Anopheles* (*Anopheles*) *Gabaldoni* n. sp. del Estado de Tabasco, México. (Dipt. Culicidae). *Medicina: Revista Mexicana* 21(395): 385–394.

Vargas, L. 1941d. New variety of *Anopheles pseudopunctipennis* (Diptera, Culicidae). *Bulletin of the Brooklyn Entomological Society* 36(2): 73–74.

Vargas, L. 1943a. Los subgeneros americanos de *Anopheles* (Diptera, Culicidae). *Anopheles* (*Russellia*) *xelajuensisde Leon* [*sic*], 1938 n. subgn. y *Anopheles* (*Coelodiazesis*) *fausti* n. sp. *Revista del Instituto de Salubridad y Enfermedades Tropicales* 4(1): 57–77.

Vargas, L. 1943b. *Anopheles earlei* Vargas, 1942, n. sp. norteamericana del grupo *maculipennis*. *Boletín de la Oficina Sanitaria Panamericana* 22(1): 8–12.

Vargas, L. 1950a. Note sobre la distribución holártica del grupo *maculipennis* y del subgénero *Coelodiazesis*. *Missirolimyia*, nov. subg. de *Anopheles*. (Diptera: Culicidae). *Rivista de Parassitologia* 11(2): 73–84.

Vargas, L. 1950b. Nota sobre *Aedes*. Revision de algunos *Ochlerotatus* americanos. *Kummyia* n. subgen. (Diptera Culicidae). *Revista de la Sociedad Mexicana de Historia Natural* 11(1–4): 149–159.

Vargas, L. 1950c. Los subgeneros de *Aëdes*. *Downsiomyia* n. subgen. (Diptera: Culicidae). *Revista del Instituto de Salubridad y Enfermedades Tropicales* 11(1): 61–69.

Vargas, L. 1956. *Psorophora (Psorophora) stonei* n. sp. *Revista del Instituto de Salubridad y Enfermedades Tropicales* 16(2): 15–16.

Vargas, L. & Diaz Nájera, A. 1961. Nueva especie de *Sabethes* de Mexico *(Diptera, Culicidae)*. *Revista del Instituto de Salubridad y Enfermedades Tropicales* 21(1–2): 73–77.

Vargas, L. & Downs, W.G. 1950. Tres especies nuevas de *Aedes* (Diptera, Culicidae). *Revista de la Sociedad Mexicana de Historia Natural* 11(1–4): 161–172, 5 pls.

Vargas, L. & Martínez Palacios, A. 1953. Descripcion de *Wyeomyia (Wyeomyia) stonei*, n. sp., y notas sobre otros Sabethini de Mexico. *Revista del Instituto de Salubridad y Enfermedades Tropicales* 13(4): 293–307.

Vargas, L. & Martínez Palacios, A. 1954. *Culex (Culex) coronator mooseri* n. subsp. de Mexico. *Revista del Instituto de Salubridad y Enfermedades Tropicales* 14(1): 33–37.

Venhuis, W.G. 1932. Een nieuwe anopheline, *Anopheles meraukensis*. *Geneeskundig Tijdschrift voor Nederlandsch-Indië* 72: 1040–1042, 1 pl.

Venhuis, W.G. 1933. Een nieuwe varieteit van de superspecies *An. punctulatus*. *Geneeskundig Tijdschrift voor Nederlandsch-Indië* 73: 203–206.

Venhuis, W.G. 1940. Een nieuwe *Anopheles* larve uit Indochina, *Anopheles alongensis* nov. sp. *Geneeskundig Tijdschrift voor Nederlandsch-Indië* 80: 173–178, 2 pls.

Ventrillon, E. 1904. Description de culicides de Madagascar. *Bulletin du Muséum National d'Histoire Naturelle* 10(8): 550–555.

Ventrillon, E. 1905a. Culicides nouveaux de Madagascar. *Archives de Parasitologie* 9(4): 441–450.

Ventrillon, E. 1905b. Culicides nouveaux de Madagascar. *Bulletin du Muséum National d'Histoire Naturelle* 11(6): 427–431.

Ventrillon, E. 1906a. *Culex* nouveaux de Madagascar. *Bulletin du Muséum National d'Histoire Naturelle* 12(2): 100–106.

Ventrillon, E. 1906b. *Stegomyia cartroni*, culicide nouveau de Madagascar. *Bulletin du Muséum National d'Histoire Naturelle* 12(3): 143–145.

Ventrillon, [E.] 1906c. *Cellia tananariviensis*. Culicide nouveau de Madagascar, 9e genre de la sous-famille des Anophelina. *Bulletin du Muséum National d'Histoire Naturelle* 12(4): 198–202.

Villeneuve, J. 1919. Deuxième note sur les Nématocères vulnérants [Dipt.] (Espèces françaises). *Bulletin de la Société Entomologique de France* 1919(2): 54–60.

Vincke, I. & Leleup, N. 1949. Un nouvel Anophèle dendrophile du Katanga: *Anopheles Berghei* sp. nov. *Revue de zoologie et de botanique Africaines* 42(3–4): 248–249.

Vockeroth, J.R. 1954. Notes on northern species of *Aedes*, with descriptions of two new species (Diptera: Culicidae). *Canadian Entomologist* 86(3): 109–116.

von Eichwald, E. 1837. *Reise auf dem Caspischen Meere, und in den Kaukasus: Unternommen in den jahren 1825–1826*. Esrter band. Zweite Abtheilung, den historishen Bericht der Reise in den Kaudasus enthaltend. J.G. Cotta'schen Buchhandlung, Stuttgart. x, 894 pp., 5 pls.

von Frauenfeld, G.R. 1867. Zoologische Miscellen. XI. *Verhandlungen der kaiserlich-königlichen zoologisch-botanischen Gesellschaft in Wien* 17(2–3): 425–502, 1 pl.

von Heyden, L. 1870. Fossile Dipteren aus der Braunkohle van Rott im Siebengebirge. *Palaeontographica* 17(6): 237–266, 2 pls.

von Humboldt, F.H.A. 1819. *Voyage aux régions équinoxiales du Nouveau Continent, fait en 1799, 1800, 1801, 1802, 1803 et 1804 par Al. de Humboldt et A. Bonpland; rédigé par Alexandre de Humboldt. Avec deux atlas, qui renferment, l'un les vues les cordillères et les monumens des peuples indigènes de l'amérique, et l'autre des cartes géographiques et physiques.* Tome Second. 722 pp. *In*: Humboldt, F.H.A. (ed.), *Voyage de Humboldt et Bonpland*. Premiére Partie. Relation historique, Paris.

von Röder, V. 1885. Dipteren von der Insel Portorico. *Stettiner Entomologische Zeitung* 46(7–9): 337–349.

Vu, T.P., Nguyen, D.M., Tran, D.H. & Nguyen, T.V. 1991. *Anopheles* (*Anopheles*) *cucphuongensis*: a new species from Vietnam (Diptera: Culicidae). *Mosquito Systematics* (for 1990) 22(3): 145–148.

Vythilingam, I., Jeffery, J., Harbach, R.E., Chan, S.T., Tan, C.H. & Asmad, M. 2007. *Anopheles* (*Cellia*) *kokhani*, n. sp. (Diptera: Culicidae) from Kapit, Sarawak, East Malaysia. *Proceedings of the Entomological Society of Washington* 109(2): 463–468.

Waktoedi Koesoemawinangoen, R. 1954. *Anophelini di Indonesia*. Kjilid 1. Kementerian Kesehatam Republik Indonesia, Djakarta. xiii, 192 pp.

Walker, F. 1848. *List of the specimens of dipterous insects in the collection of the British Museum. Part I*. British Museum, London. 229 pp.

Walker, F. 1856a. *Insecta Saundersiana: or characters of undescribed insects in the collection of William Wilson Saunders, Esq., F.R.S., F.L.S., &c. Vol. I. Diptera*. John van Voorst, London. 474 pp., 8 pls.

Walker, F. 1856b. Catalogue of the dipterous insects collected at Singapore and Malacca by Mr. A. R. Wallace, with descriptions of new species. *Journal of the Proceedings of the Linnean Society of London Zoology* 1(1): 4–39, 2 pls.

Walker, F. 1858. Catalogue of the dipterous insects collected in the Aru Islands by Mr. A. R. Wallace, with descriptions of new species. *Journal of the Proceedings of the Linnean Society of London Zoology* 3(10–11): 77–131.

Walker, F. 1859–1860. Catalogue of the dipterous insects collected at Makessar in Celebes, by Mr. A. R. Wallace, with descriptions of new species. *Journal of the Proceedings of the Linnean Society of London Zoology* 4(14–16): 90–144 [1859], 145–172 [1860].

Walker, F. 1860. Catalogue of the dipterous insects collected in Amboyna by Mr. A. R. Wallace, with descriptions of new species. *Journal of the Proceedings of the Linnean Society of London Zoology* 5(s1): 144–168.

Walker, F. 1861. Catalogue of the dipterous insects collected at Dorey, New Guinea, by Mr. A. R. Wallace, with descriptions of new species. *Journal of the Proceedings of the Linnean Society of London Zoology* 5(19): 229–257.

Walker, F. 1865. Descriptions of new species of the dipterous insects of New Guinea. *Journal of the Proceedings of the Linnean Society of London Zoology* 89(30–32): 102–130.

Walker, F. 1866. List of Diptera, pp. 337–340. *In*: Lord, J.K., *The naturalist in Vancouver Island and British Colombia*. Volume 2. Richard Bentley, London.

Walker, F. 1874. *Notes on Diptera and lists of species*. First series. E.W. Janson, London. 37 pp.

Waltl, J. 1835. *Reise durch Tyrol, Oberitalien und Piemont nach dem südlichen Spainen ... nebst einem Anhänge zoologischen Inhalts*. Zweiter Theil. Pustet'sche Buchhandlung, Passau. 120 pp.

Wang, H.-H. & Feng, C.-Y. 1964. Description of three new species of *Culex* (*Lophoceratomyia*) [in Chinese]. *Acta Zootaxonomica Sinica* 1(1): 37–41.

Wang, H.-H. & Feng, C.-Y. 1965. A new mosquito, *Uranotaenia lishanensis* sp. nov., with notes on the adult male and larva of *U. hebes* Barraud [in Chinese]. *Acta Zootaxonomica Sinica* 2(3): 206–210.

Wang, X.-X. 1984. Description of a new *Culex* species (Diptera: Culicidae) [in Chinese]. *Acta Entomologica Sinica* 27(2): 217–220.

Wanson, M. & Lebied, B. 1945. Un nouvel Anophèle cavernicole du Congo belge. *Anopheles* (*Myzomyia*) *Vanhoofi* spec. nov. *Revue de zoologie et de botanique Africaines* 39(1): 118–129.

Ward, R.A. 1979. Corrections in the nomenclature of three anophelines (Diptera: Culicidae). *Mosquito Systematics* 11(2): 130–134.

Ward, R.A. 1984. Second supplement to "A catalog of the mosquitoes of the world" (Diptera: Culicidae). *Mosquito Systematics* 16(3): 227–270.

Ward, R.A. 1992. Third supplement to "A catalog of the mosquitoes of the world" (Diptera: Culicidae). *Mosquito Systematics* 24(3): 177–230.

Waterhouse, C.O. 1905. Notes on some British *Culicidae. Annals and Magazine of Natural History* 16(96): 674–676.

Watson, M. 1910. A new anopheline from the Federated Malay States. *Annals of Tropical Medicine and Parasitology* 4: 251–253.

Wattal, B.L. & Kalra, N.L. 1965. An entomological survey of Dehra Dun Valley (Uttar Pradesh). Part II: Description of *Culex* (*Culiciomyia*) *ramakrishnii* sp. n., and allotype (♂) of *Uranotaenia maculipleura* Leicester, 1908. *Bulletin of the Indian Society for Malaria and Other Communicable Diseases* 2(4): 309–313.

Wattal, B.L., Bhatia, M.L. & Kalra, N.L. 1958. Some new records of culicines of Dehra Dun (Uttar Pradesh) with a description of a new variety. *Indian Journal of Malariology* 12(3): 217–230.

Wattal, B.L., Kalra, N.L. & Krishnan, K.S. 1966. Description of *Culex* (*Mochthogenes*) *shrivastavii Sp.* n. [*sic*] from Andaman Islands, India. *Bulletin of the Indian Society for Malaria and Other Communicable Diseases* 3(2): 159–161.

Weinbren, M.P. & Weinbren, B.M. 1970. Observations on the mosquito population in the irradiated forest at El Verde, pp. E-159–E-167. *In*: Odum, H.T. & Pigeon, K.F. (eds), *A Tropical Rain Forest: A Study Of Irradiation and Ecology at El Verde, Puerto Rico*. National Technical Information Service, Springfield, Virginia.

Wesenberg-Lund, C. 1921. Contributions to the biology of the Danish Culicidae. *Det Kongelige Danske Videnskabernes Selskabs Skrifter. Naturvidenskabelig og mathematisk Afdeling* (8) 7(1): 1– 210, 21 pls.

Westwood, D.O. 1835. Insectorum nonnullorum novum (ex ordine Dipterorum) descriptiones. *Annales de la Société Entomologique de France* 4: 681–685.

Weyenbergh, H. 1882. *Los habitantes del Río Primero*. Imprenta y Encuadernacion de Rius y Bechii, Montevideo, Uruguay. 27 pp.

Wharton, R.H. 1947. Notes on Australian mosquitoes (Diptera, Culicidae). Part VII. The genus *Harpagomyia. Proceedings of the Linnean Society of New South Wales* 72(1–2): 58–68.

Wharton, R.H. 1962. The biology of *Mansonia* mosquitoes in relation to the transmission of filariasis in Malaya. *Bulletin – Institute for Medical Research, Kuala Lumpur* 11: 1–114, 6 pls.

White, G.B. 1974. Priority of *Ingramia* Edwards, 1912, over *Ravenalites* Doucet, 1957 as the name for a subgenus of *Mimomyia* Theobald, 1903 (Diptera: Culicidae). *Mosquito Systematics* 6(4): 239–242.

White, G.B. 1975. Notes on a catalogue of Culicidae of the Ethiopian Region. *Mosquito Systematics* 7(4): 303–344.

White, G.B. 1978. Systematic reappraisal of the *Anopheles maculipennis* complex. *Mosquito Systematics* 10(1): 13–44.

White, G.B. 1980. Family Culicidae, pp. 114–148. *In*: Crosskey, R.W. (ed.), *Catalogue of the Diptera of the Afrotropical Region*. British Museum (Natural History), London.

White, G.B. 1985. *Anopheles bwambae* sp.n., a malaria vector in the Semliki Valley, Uganda, and its relationships with other sibling species of the *An. gambiae* complex (Diptera: Culicidae). *Systematic Entomology* 10(4): 501–522.

Wiedemann, C.R.G. 1819. Beschreibung neuer Zweiflügler aus Ostindien und Afrika. *Zoologisches Magazin* 1(3): 1–39.

Wiedemann, C.R.G. 1820. *Diptera exotica. Sectio I. antennis multiarticulatis*. Kiliae [Kiel, Germany]. xix, 38 pp.

Wiedemann, C.R.G. 1821. *Diptera exotica*. Pars I. Kiliae [Kiel, Germany]. xix, 244 pp., 2 pls.

Wiedemann, C.R.W. 1828. *Aussereuropäische zweiflügelige Insekten: als Fortsetzung des Meigenschen Werkes*. Volume 1. Hamm, Germany. xxxii, 608 pp.

Wijesundara, D.P. 1951. Notes on species of the subgenus *Aëdes* Meigen (sens. str.) in Ceylon, with descriptions of three new species. *Ceylon Journal of Science. Section B. Zoology and Geology* 24(3): 173–179, 2 pls.

Wilkerson, R.C. 1990. Elevation of *Anopheles chiriquiensis* from synonymy with *Anopheles parapunctipennis* and designation of name-bearing types for *Anopheles parapunctipennis* and *Anopheles parapunctipennis guatemalensis* (Diptera: Culicidae). *Mosquito Systematics* (for 1989) 21(3): 178–187.

Wilkerson, R.C. 1991. *Anopheles* (*Anopheles*) *calderoni* n.sp., a malaria vector of the Arribalzagia Series from Peru (Diptera: Culicidae). *Mosquito Systematics* 23(1): 25–38.

Wilkerson, R.C. & Linton, Y.-M. 2015. Elevation of *Pseudoskusea*, *Rusticoidus* and *Protomacleaya* to valid subgenera in the mosquito genus *Aedes* based on taxon naming criteria recently applied to other members of the Tribe [*sic*] Aedini (Diptera: Culicidae). *Parasites & Vectors* 8: 668.

Wilkerson, R.C. & Sallum, M.A.M. 1999. *Anopheles (Anopheles) forattinii:* a new species in Series Arribalzagia (Diptera: Culicidae). *Journal of Medical Entomology* 36(3): 345–354.

Wilkerson, R.C., Cong, L., Rueda, L.M., Kim, H.-C., Klein, T.A., Song, G.-H. & Strickman, D. 2003. Molecular confirmation of *Anopheles (Anopheles) lesteri* from the Republic of South Korea and its genetic identity with *An. (Ano.) anthropophagus* from China (Diptera: Culicidae). *Zootaxa* 378: 1–14.

Wilkerson, R.C., Linton, Y.-M., Fonseca, D.M., Schultz, T.R., Price, D.C. & Strickman, D.A. 2015. Making mosquito taxonomy useful: A stable classification of tribe Aedini that balances utility with current knowledge of evolutionary relationships. *Plos ONE* 10(7): e0133602.

Willcocks, F.C. 1910. A preliminary note on the prevalence of mosquitoes in Cairo and its environs. *Annals of Tropical Medicine and Parasitology* 3(5): 583–589.

Williams, D.M., Ebach, M.C. & Wheeler, Q.D. 2010. Beyond belief, the steady resurrection of phenetics, pp. 165–195. *In*: Williams, D.M. & Knapp, S. (eds), *Beyond cladistics: the branching of a paradigm*. University of California Press, Berkeley.

Williston, S.W. 1893. List of Diptera of the Death Valley Expedition, pp. 253–259. *In*: Riley, C.V., *Report on a small collection of insects made during the Death Valley Expedition* (pp. 235–268). [Series] *North American Fauna No. 7, The Death valley expedition. A biological survey of parts of California, Nevada, Arizona, and Utah*. Part 2. Government Printing Office, Washington.

Williston, S.W. 1896. On the Diptera of St. Vincent (West Indies). *Transactions of the Royal Entomological Society of London* 44: 253–446, 7 pls.

Williston, S.W. 1900. Culicidae, p. 224. *In*: Osten Sacken, C.R., 1886–1901. *Biologia Centrali-Americana. Insecta.Diptera*. Volume 1. The Supplement [by S.W. Williston, J.M. Aldrich, W.M Wheeler and A.L. Melander]. R. H. Porter, London.

Wirth, W.W. 1948. Diptera, pp. 86–136, 223–239. *In*: Usinger, R.L., La Rivers, I., Chandler, H.P. & Wirth, W.W., *Entomology 133: biology of aquatic and littoral insects*. Syllabus Series. University of California, Berkeley.

Wolfs, J. 1945a. *Anopheles* (*Anopheles*) *implexus* var. *henrardi* var. nova. *Recueil des Travaux des Sciences Medicales au Congo Belge* 3: 174–176.

Wolfs, J. 1945b. *Culex* (*Neoculex*) *Wansoni* sp. n. *Annales de la Société Belge de Médecine Tropicale* 25: 221–224, 2 pls.

Wolfs, J. 1947a. Un Culicide nouveau du Katanga, *Aedes* (*Banksinella*) *Bequaerti*, sp. n. *Revue de zoologie et de botanique Africaines* 40(1): 40–41.

Wolfs, J. 1947b. *Culex* (*Culex*) *bukavuensis* sp. n. *Revue de zoologie et de botanique Africaines* 40(2–3): 159–164.

Wolfs, J. 1947c. *Culex* (*Culex*) *Seldeslachtsi*, sp. nov. *Revue de zoologie et de botanique Africaines* 40(2–3): 224–228.

Wolfs, J. 1948a. *Taeniorhynchus* (*Coquillettidia*) *schoutedeni* n. sp. (Culic.). *Revue de zoologie et de botanique Africaines* 40(4): 244–247.

Wolfs, J. 1948b. *Taeniorhynchus* (*Coquillettidia*) *Vanoyei* sp. n. (Dipt. Culicidae). *Revue de zoologie et de botanique Africaines* 41(1): 83–86.

Wolfs, J. 1949. *Aedes* (*Stegomyia*) *apicoargenteus* var. *denderensis* var. n. (Culicidae). *Revue de zoologie et de botanique Africaines* 42(3–4): 190–192.

Wolfs, J. 1951. Un Culicide nouveau de la région éthiopienne: *Taeniorhynchus (Coquillettidia) karandalaensis* sp. n. *Revue de zoologie et de botanique Africaines* 45(1–2): 12–14.

Wolfs, J. 1958. Deux *Culicidae* nouveaux du Congo belge (Diptera Nematocera). *Revue de zoologie et de botanique Africaines* 58(3–4): 297–298.

Wood, D.M. 1977. Notes on the identities of some common Nearctic *Aedes* mosquitoes. *Mosquito News* 37(1): 71–81.

Wood, D.M., Dang, P.T. & Ellis, R.A. 1979. *The insects and arachnids of Canada. Part 6. The mosquitoes of Canada.* (*Diptera: Culicidae*). Canadian Government Publishing Centre, Supply and Services Canada, Hull, Quebec. 390 pp.

Woodhill, A.R. 1949. A new subspecies of *Aëdes* (*Stegomyia*) *scutellaris* Walker (Diptera, Culicidae) from northern Australia. *Proceedings of the Linnean Society of New South Wales* 74(3–4): 140–144.

Woodhill, A.R. & Lee, D.J. 1944a. The subspecies of *Anopheles amictus* Edwards (Diptera, Culicidae). *Proceedings of the Linnean Society of New South Wales* 69(1–2): 62–66.

Woodhill, A.R. & Lee, D.J. 1944b. Some new records and new synonymy of Australian species of *Anopheles* (Diptera, Culicidae). *Proceedings of the Linnean Society of New South Wales* 69(1–2): 67–72.

Worth, C.B. 1960. Description of a new species of the aedine subgenus *Neomelaniconion* from Tongaland, South Africa (Diptera: Culicidae). *Journal of the Entomological Society of Southern Africa* 23(2): 312–313.

Wu, S.-C. 1936. Further notes on the mosquitoes of Hangchow, Chekiang, with description of one new species. *Yearbook of the Bureau of Entomology of Chekiang Province* (for 1935) 5: 46–53.

Xavier, S.H. & da Silva Mattos, S. 1972. *Culex (Melanoconion) milwardi* sp. n. (Diptera, Culicidae). *Revista Brasileira de Biologia* 32(4): 569–571.

Xavier, S.H., da Silva, J.E. & da Silva Mattos, S. 1970a. *Culex (Melanoconion) olimpioi* sp. n. (Diptera, Culicidae). *Revista Brasileira de Malariologia e Doenças Tropicais* 22(1): 183–188.

Xavier, S.H., da Silva Mattos, S. & da Silva, J.E. 1970b. *Uranotaenia rachoui*, uma nova espécie do Brasil (Diptera, Culicidae). *Revista Brasileira de Malariologia e Doenças Tropicais* 22(2/4): 411–416.

Xu, J.-J. & Feng, L.-C. 1975. Studies on the *Anopheles hyrcanus* group of mosquitoes in China [in Chinese]. *Acta Entomologica Sinica* 18(1): 77–98, 6 pls.

Xu, J. & Luo, X. 1998. A new species of *Anopheles hyrcanus* group Nei Mongol Autonomous Region (Diptera: Culicidae). *Chinese Journal of Parasitology and Parasitic Diseases* 16(1): 45–52.

Yamada, M. 1937. A new species of *Anopheles* in Chosen (Korea). *Keijo Journal of Medicine* 8: 237–255, 5 pls.

Yamada, S. 1917. On two new species of mosquitoes from Japan [in Japanese]. *Dōbutsugaku zasshi* [*Zoological Magazine*] 29(341): 61–72.

Yamada, S. 1918. A new species of *Anopheles* collected from Hokkaido [in Japanese]. *Eiseigaku Densenbyogaku Zasshi* [*Journal of Hygiene and Infectious Diseases*] 13: 689–691.

Yamada, S. 1921. Description of ten new species of *Aedes* found in Japan, with notes on the relation between some of these mosquitoes and the larva of *Filaria bancrofti* Cobbold. *Annotationes Zoologicae Japonenses* 10(6): 45–81.

Yamada, S. 1924. A revision of the adult anopheline mosquitoes of Japan: systematic descriptions, their habits and their relations to human diseases, together with an account of three new species. *Scientific Reports from the Government Institute for Infectious Diseases, Tokyo Imperial University* 3: 215–241.

Yamada, S. 1925. A revision of the adult anopheline mosquitoes of Japan: systematic descriptions, their habits and their relations to human diseases. (Part II). *Scientific Reports from the Government Institute for Infectious Diseases, Tokyo Imperial University* 4: 447–493.

Yamada, S. 1927. An experimental study on twenty-four species of Japanese mosquitoes regarding their suitability as intermediate hosts for *Filaria bancrofti* Cobbold. *Scientific Reports from the Government Institute for Infectious Diseases, Tokyo Imperial University* 6: 559–622, 3 pls.

Yamada, S. 1932. Culicidae, pp. 210–235. *In*: Esaki, T. *et al*., *Nippon Konchu Zukan: Iconographia Insectorum Japonicorum*. Editio Prima. Hokuryukan, Tokyo.

Yamada, S. & Watanabe, P. 1918. A new species of Korean *Anopheles* [in Japanese]. *Japanese Journal of Experimental Medicine* 2: 206–209.

Yamaguti, S. & LaCasse, W.J. 1950. *Mosquito fauna of Guam*. Office of the Surgeon, Headquarters, 8th Army, APO 343. United States. 207th Malaria Survey Detachment. 101 pp.

Yamaguti, S. & LaCasse, W.J. 1952. *Culex* (*Culiciomyia*) *kyotoensis* n. sp. (Culicidae, Diptera) from Japan. Published by the authors, Kyoto, Japan. 5 pp., 2 pls.

Yamaguti, S. & Tamaboko, R. 1954. Two new species of *Aedes* (Culicidae, Diptera) from Japan. *Acta Medica Okayama* (for 1952) 8(4): 414–422, 4 pls.

Yang, J.-L., Li, F.-H. & Chen, H.-B. 1993. A new species of *Culex* from Yunnan (Diptera: Culicidae) [in Chinese]. *Acta Zootaxonomica Sinica* 18(2): 225–228.

Yao, Y.-C. & Wu, C.-C. 1944. A new anopheline mosquito, *Anopheles kweiyangensis*, sp. n. from Guiyang, Guizhou [in Chinese]. *National Medical Journal of China* 30(1): 1–10.

Yong, H.S., Cheong, W.H., Chiang, G.L., Dhaliwal, S.S., Loong, K.P. & Sarjan, R. 1983. Biochemical genetic relationship of three taxa of the malarial mosquito *Anopheles balabacensis* complex. *Comparative Biochemistry and Physiology Part B: Comparative Biochemistry* 76(3): 611–613.

Young, T.C.M. & Majid, S.A. 1928. A variety of Anopheles [*sic*] *karwari* collected in Coorg, S. India. *Indian Journal of Medical Research* 16(2): 469–471, 1 pl.

Zavortink, T.J. 1968. Mosquito studies (Diptera, Culicidae). VIII. A prodrome of the genus *Orthopodomyia*. *Contributions of the American Entomological Institute* 3(2): 1–221.

Zavortink, T.J. 1969. Mosquito studies (Diptera, Culicidae). XV. A new species of treehole breeding *Anopheles* from the southwestern United States. *Contributions of the American Entomological Institute* 4(4): 27–38.

Zavortink, T.J. 1970a. Mosquito studies (Diptera, Culicidae). XVI. A new species of treehole breeding *Aedes* (*Ochlerotatus*) from southern California. *Contributions of the American Entomological Institute* (for 1969) 5(1): 1–7.

Zavortink, T.J. 1970b. Mosquito studies (Diptera, Culicidae). XIX. The treehole *Anopheles* of the world. *Contributions of the American Entomological Institute* 5(2): 1–35.

Zavortink, T.J. 1970c. Mosquito studies (Diptera, Culicidae). XXII. A new subgenus and species of *Aedes* from Arizona. *Contributions of the American Entomological Institute* 7(1): 1–11.

Zavortink, T.J. 1972. Mosquito studies (Diptera, Culicidae). XXVIII. The New World species formerly placed in *Aedes* (*Finlaya*). *Contributions of the American Entomological Institute* 8(3): 1–206.

Zavortink, T.J. 1973. Mosquito studies (Diptera, Culicidae). XXIX. A review of the subgenus *Kerteszia* of *Anopheles*. *Contributions of the American Entomological Institute* 9(3): 1–54.

Zavortink, T.J. 1979. Mosquito studies (Diptera, Culicidae). XXXV. The new sabethine genus *Johnbelkinia* and a preliminary reclassification of the composite genus *Trichoprosopon*.*Contributions of the American Entomological Institute* 17(1): 1–61.

Zavortink, T.J. 1986. *Zinzala,* a new subgenus of *Wyeomyia* with two new species from pitcher-plants in Venezuela (Diptera, Culicidae, Sabethini). *Wasmann Journal of Biology* (for 1985) 43(1–2): 46–59.

Zavortink, T.J. & Chaverri, L.G. 2009. Resurrection of the names *Toxorhynchites moctezuma* (Dyar & Knab) and *Toxorhynchites hypoptes* (Knab) from synonymy with *Toxorhynchites theobaldi* (Dyar & Knab) (Diptera: Culicidae). *Proceedings of the Entomological Society of Washington* 111(4): 890–897.

Zavortink, T.J. & O'Meara, G.F. 1999. *Culex* (*Micraedes*) *biscaynensis* n. sp. from Florida (Diptera: Culicidae). *Journal of the American Mosquito Control Association* 15(3): 263–270.

Zavortink, T.J. & Poinar, G.O., Jr. 2000. *Anopheles* (*Nyssorhynchus*) *dominicanus* sp. n. (Diptera: Culicidae) from Dominican Amber. *Annals of the Entomological Society of America* 93(6): 1230–1235.

Zavortink, T.J. & Poinar, G.O., Jr. 2008. *Toxorhynchites* (*Toxorhynchites*) *mexicanus*, n. sp. (Diptera: Culicidae) from Mexican amber: a New World species with Old World affinities. *Proceedings of the Entomological Society of Washington* 110(1): 116–125.

Zetterstedt, J.W. 1838. *Insecta Lapponica descripta*. Voss, Lipsiae. 1139 pp.

Zetterstedt, J.W. 1850. *Diptera Scandinaviae disposita et descripta*. Volume 9, pp. 3367–4090. Officina Lundbergiana, sumtibus regiis, Lundae [Lund], Sweden.

Zhelochovtsev, A.N. 1937. Notes on the synonymy of the Palaearctic *Anopheles* [in Russian]. *Meditsinskaia Parazitologiia i Parazitarnye Bolezni* 6(5): 707–709.

Zhou, H., Zhu, G. & Lu, Y. 1999. One new species of the genus *Topomyia* (Diptera, Culicidae) [in Chinese]. *Acta Parasitology et Medica Entomologica Sinica* 6(2): 107–111.

Appendix 1 Current Higher-level Classification of the Culicidae

Subfamily	Tribe	Genus	Subgenus
Anophelinae		*Anopheles* Meigen	*Anopheles* Meigen
			Baimaia Harbach *et al.*
			Cellia Theobald
			Christya Theobald
			Kerteszia Theobald
			Lophopodomyia Antunes
			Nyssorhynchus Blanchard
			Stethomyia Theobald
		Bironella Theobald	*Bironella* Theobald
			Brugella Edwards
			Brugella Edwards
		Chagasia Cruz	–
Culicinae	Aedeomyiini	*Aedeomyia* Theobald	*Aedeomyia* Theobald
			Lepiothauma Enderlein
	Aedini	*Aedes* Meigen	*Abraedes* Zavortink
			Acartomyia Theobald
			Aedimorphus Theobald
			Alanstonea Mattingly
			Albuginosus Reinert
			Ayurakitia Thurman
			Aztecaedes Zavortink
			Belkinius Reinert
			Bifidistylus Reinert *et al.*
			Borichinda Har. & Rat.
			Bothaella Reinert
			Bruceharrisonius Reinert
			Cancraedes Edwards
			Catageiomyia Theobald
			Catatassomyia D. & Sh.
			Christophersiomyia Bar.
			Coetzeemyia Huang *et al.*
			Collessius Reinert *et al.*
			Cornetius Huang
			Dahliana Reinert *et al.*
			Danielsia Theobald
			Dendroskusea Edwards
			Diceromyia Theobald
			Dobrotworskyius Reinert *et al.*
			Downsiomyia Vargas
			Edwardsaedes Belkin
			Elpeytonius Reinert *et al.*
			Finlaya Theobald
			Fredwardsius Reinert
			Georgecraigius Reinert *et al.*
			Geoskusea Edwards
			Gilesius Reinert *et al.*

Continued

Subfamily	Tribe	Genus	Subgenus
			Gymnometopa Coquillett
			Halaedes Belkin
			Himalaius Reinert *et al*.
			Hopkinsius Reinert *et al*.
			Howardina Theobald
			Huaedes Huang
			Hulecoeteomyia Theobald
			Indusius Edwards
			Isoaedes Reinert
			Jarnellius Reinert *et al*.
			Jihlienius Reinert *et al*.
			Kenknightia Reinert
			Kompia Aitken
			Leptosomatomyia Theobald
			Levua Stone & Bohart
			Lewnielsenius Reinert *et al*.
			Lorrainea Belkin
			Luius Reinert *et al*.
			Macleaya Theobald
			Molpemyia Theobald
			Mucidus Theobald
			Neomelaniconion Newstead
			Nyctomyia Harbach
			Ochlerotatus Lynch Arr.
			Paraedes Edwards
			Patmarksia Reinert
			Paulianius Brunhes & Boussès
			Petermattinglyius Reinert *et al*.
			Phagomyia Theobald
			Polyleptiomyia Theobald
			Protomacleaya Theobald
			Pseudalbuginosus Hua. & Rue.
			Pseudarmigeres Stone & Knight
			Pseudoskusea Theobald
			Rampamyia Reinert *et al*.
			Rhinoskusea Edwards
			Rusticoidus Shev. & Prud.
			Sallumia Reinert *et al*.
			Scutomyia Theobald
			Skusea Theobald
			Stegomyia Theobald
			Tanakaius Reinert *et al*.
			Tewarius Reinert
			Vansomerenis Reinert *et al*.
			Zavortinkius Reinert
		Armigeres Theobald	*Armigeres* Theobald
			Leicesteria Theobald
		Eretmapodites Theobald	–
		Haemagogus Williston	*Conopostegus* Dyar
			Haemagogus Williston
		Heizmannia Ludlow	*Heizmannia* Ludlow
			Mattinglyia Lien
		Opifex Hutton	*Nothoskusea* Dumbleton
			Opifex Hutton
		Psorophora Rob.-Desvoidy	*Grabhamia* Theobald
			Janthinosoma Lynch Arr.
			Psorophora Rob.-Desvoidy

Continued

Subfamily	Tribe	Genus	Subgenus
		Udaya Thurman	–
		Verrallina Theobald	*Harbachius* Reinert
			Neomacleaya Theobald
			Verrallina Theobald
		Zeugnomyia Leicester	–
	Culicini	*Culex* Linnaeus	*Acalleomyia* Leicester
			Acallyntrum Stone & Penn
			Aedinus Lutz
			Afroculex Danilov
			Allimanta Casal & Garcia
			Anoedioporpa Dyar
			Barraudius Edwards
			Belkinomyia Adames & Galindo
			Carrollia Lutz
			Culex Linnaeus
			Culiciomyia Theobald
			Eumelanomyia Theobald
			Kitzmilleria Danilov
			Lasiosiphon Kirkpatrick
			Lophoceraomyia Theobald
			Maillotia Theobald
			Melanoconion Theobald
			Micraedes Coquillett
			Microculex Theobald
			Neoculex Dyar
			Nicaromyia Broche & Rod.
			Oculeomyia Theobald
			Phenacomyia Har. & Pey.
			Phytotelmatomyia Rossi & Har.
			Sirivanakarnius Tanaka
			Tinolestes Coquillett
		Deinocerites Theobald	–
		Galindomyia Stone & Bar.	–
		Lutzia Theobald	*Insulalutzia* Tanaka
			Metalutzia Tanaka
			Lutzia Theobald
	Culisetini	*Culiseta* Felt	*Allotheobaldia* Brolemann
			Austrotheobaldia Dobrotworsky
			Climacura Howard *et al.*
			Culicella Felt
			Culiseta Felt
			Neotheobaldia Dobrotworsky
			Theomyia Edwards
	Ficalbiini	*Ficalbia* Theobald	–
		Mimomyia Theobald	*Etorleptiomyia* Theobald
			Ingramia Edwards
			Mimomyia Theobald
	Hodgesiini	*Hodgesia* Theobald	–
	Mansoniini	*Coquillettidia* Dyar	*Austromansonia* Belkin
			Coquillettidia Dyar
			Rhynchotaenia Brèthes
		Mansonia Blanchard	*Mansonia* Blanchard
			Mansonioides Theobald
	Orthopodomyiini	*Orthopodomyia* Theobald	–
	Sabethini	*Isostomyia* Coquillett	–
		Johnbelkinia Zavortink	–

Continued

Subfamily	Tribe	Genus	Subgenus
		Kimia Vu & Harbach	–
		Limatus Theobald	–
		Malaya Leicester	–
		Maorigoeldia Edwards	–
		Onirion Harbach & Peyton	–
		Runchomyia Theobald	*Ctenogoeldia* Edwards
			Runchomyia Theobald
		Sabethes Rob.-Desvoidy	*Davismyia* Lane & Cerqueira
			Peytonulus Harbach
			Sabethes Rob.-Desvoidy
			Sabethinus Lutz
			Sabethoides Theobald
		Shannoniana Lane & Cer.	–
		Topomyia Leicester	*Miyagiella* Harbach
			Suaymyia Thurman
			Topomyia Leicester
		Trichoprosopon Theobald	–
		Tripteroides Giles	*Polylepidomyia* Theobald
			Rachionotomyia Theobald
			Rachisoura Theobald
			Tricholeptomyia D. & Sh.
			Tripteroides Giles
		Wyeomyia Theobald	*Antunesmyia* Lane & Cer.
			Caenomyiella Har. & Pey.
			Cruzmyia Lane & Cer.
			Decamyia Dyar
			Dendromyia Theobald
			Dodecamyia Dyar
			Exallomyia Har. & Pey.
			Hystatomyia Dyar
			Menolepis Lutz
			Miamyia Dyar
			Nunezia Dyar
			Phoniomyia Theobald
			Prosopolepis Lutz
			Spilonympha Motta & Lou.
			Triamyia Dyar
			Wyeomyia Theobald
			Zinzala Zavortink
	Toxorhynchitini	*Toxorhynchites* Theobald	*Afrorhynchus* Ribeiro
			Ankylorhynchus Lutz
			Lynchiella Lahille
			Toxorhynchites
	Uranotaeniini	*Uranotaenia* Lynch Arr.	*Pseudoficalbia* Theobald
			Uranotaenia Lynch Arr.

Appendix 2
Multi-generic and Traditional Classifications of the Genus *Aedes*

Multi-generic classification (*sensu* Reinert *et al*., 2009)		Composite classification (*sensu* Wilkerson *et al*., 2015)	
Genus	Subgenus	Genus	Subgenus
Abraedes		*Aedes*	*Abraedes*
Acartomyia		*Aedes*	*Acartomyia*
Aedes		*Aedes*	*Aedes*
'*Aedes*' *s. auct.*		*Aedes*	Subgenus uncertain
	'*Pseudalbuginosus*' *s. auct.*	*Aedes*	*Pseudalbuginosus*
Aedimorphus		*Aedes*	*Aedimorphus*
Alanstonea		*Aedes*	*Alanstonea*
Albuginosus		*Aedes*	*Albuginosus*
Ayurakitia		*Aedes*	*Ayurakitia*
Aztecaedes		*Aedes*	*Aztecaedes*
Belkinius		*Aedes*	*Belkinius*
Bifidistylus		*Aedes*	*Bifidistylus*
Borichinda		*Aedes*	*Borichinda*
Bothaella		*Aedes*	*Bothaella*
Bruceharrisonius		*Aedes*	*Bruceharrisonius*
Cancraedes		*Aedes*	*Cancraedes*
Catageiomyia		*Aedes*	*Catageiomyia*
Catatassomyia		*Aedes*	*Catatassomyia*
Christophersiomyia		*Aedes*	*Christophersiomyia*
Collessius	*Alloeomyia*	*Aedes*	*Collessius*
	Collessius	"	"
Cornetius		*Aedes*	*Cornetius*
Dahliana		*Aedes*	*Dahliana*
Danielsia		*Aedes*	*Danielsia*
Dendroskusea		*Aedes*	*Dendroskusea*
Diceromyia		*Aedes*	*Diceromyia*
Dobrotworskyius		*Aedes*	*Dobrotworskyius*
Downsiomyia		*Aedes*	*Downsiomyia*
Edwardsaedes		*Aedes*	*Edwardsaedes*
Elpeytonius		*Aedes*	*Elpeytonius*
Finlaya		*Aedes*	*Finlaya*
Fredwardsius		*Aedes*	*Fredwardsius*
Georgecraigius	*Georgecraigius*	*Aedes*	*Georgecraigius*
	Horsfallius	"	"
Geoskusea		*Aedes*	*Geoskusea*
Gilesius		*Aedes*	*Gilesius*
Gymnometopa		*Aedes*	*Gymnometopa*
Halaedes		*Aedes*	*Halaedes*
Himalaius		*Aedes*	*Himalaius*
Hopkinsius	*Hopkinsius*	*Aedes*	*Hopkinsius*
	Yamada	"	"
Howardina		*Aedes*	*Howardina*
Huaedes		*Aedes*	*Huaedes*

Continued

Multi-generic classification (*sensu* Reinert *et al*., 2009)		Composite classification (*sensu* Wilkerson *et al*., 2015)	
Genus	Subgenus	Genus	Subgenus
Hulecoeteomyia		*Aedes*	*Hulecoeteomyia*
Indusius		*Aedes*	*Indusius*
Isoaedes		*Aedes*	*Isoaedes*
Jarnellius		*Aedes*	*Jarnellius*
Jihlienius		*Aedes*	*Jihlienius*
Kenknightia		*Aedes*	*Kenknightia*
Kompia		*Aedes*	*Kompia*
Leptosomatomyia		*Aedes*	*Leptosomatomyia*
Levua		*Aedes*	*Levua*
Lewnielsenius		*Aedes*	*Lewnielsenius*
Lorrainea		*Aedes*	*Lorrainea*
Luius		*Aedes*	*Luius*
Macleaya	*Chaetocruiomyia*	*Aedes*	*Macleaya*
	Macleaya	"	"
Molpemyia		*Aedes*	*Molpemyia*
Mucidus	*Mucidus*	*Aedes*	*Mucidus*
	Pardomyia	"	"
Neomelaniconion		*Aedes*	*Neomelaniconion*
Nyctomyia		*Aedes*	*Nyctomyia*
Ochlerotatus	*Buvirilia*	*Aedes*	*Ochlerotatus*
	Chrysoconops	"	"
	Culicada	"	"
	Culicelsa	"	"
	Empihals	"	"
	Gilesia	"	"
	Juppius	"	"
	Lepidokeneon	"	"
	Ochlerotatus	"	"
	Pholeomyia	"	"
	Protoculex	"	"
	Pseudoskusea	"	*Pseudoskusea*
	Rusticoidus	"	*Rusticoidus*
	Woodius	"	*Ochlerotatus*
	Subgenus uncertain	"	"
'*Ochlerotatus*' *s. auct.*	'*Coetzeemyia*' *s. auct.*	"	*Coetzeemyia*
	'*Finlaya*' *s. auct.*	"	Subgenus uncertain
	'*Protomacleaya*' *s. auct.*	"	*Protomacleaya*
	Subgenus uncertain	"	*Ochlerotatus*
Paraedes		*Aedes*	*Paraedes*
Patmarksia		*Aedes*	*Patmarksia*
Paulianius		*Aedes*	*Paulianius*
Petermattinglyius	*Aglaonotus*	*Aedes*	*Petermattinglyius*
	Petermattinglyius	"	"
Phagomyia		*Aedes*	*Phagomyia*
Polyleptiomyia		*Aedes*	*Polyleptiomyia*
Pseudarmigeres		*Aedes*	*Pseudarmigeres*
Rampamyia		*Aedes*	*Rampamyia*
Rhinoskusea		*Aedes*	*Rhinoskusea*
Sallumia		*Aedes*	*Sallumia*
Scutomyia		*Aedes*	*Scutomyia*
Skusea		*Aedes*	*Skusea*

Continued

Multi-generic classification (*sensu* Reinert *et al*., 2009)		Composite classification (*sensu* Wilkerson *et al*., 2015)	
Genus	Subgenus	Genus	Subgenus
Stegomyia	*Actinothrix*	*Aedes*	*Stegomyia*
	Bohartius	"	"
	Heteraspidion	"	"
	Huangmyia	"	"
	Mukwaya	"	"
	Stegomyia	"	"
	Xyele	"	"
	Zoromorphus	"	"
	Subgenus uncertain	"	"
Tanakaius		*Aedes*	*Tanakaius*
Tewarius		*Aedes*	*Tewarius*
Vansomerenis		*Aedes*	*Vansomerenis*
Zavortinkius		*Aedes*	*Zavortinkius*

Appendix 3
Table of Changed Spellings of Species-group Names

The currently recognized extant species listed in the left column of the table below have had the spelling of their names changed to agree in gender with the name of the genus with which they are combined in the classifications listed in the centre and right columns. The 'Traditional classification' (centre column) includes the multi-subgeneric classification of *Aedes* (*sensu* Wilkerson *et al.*, 2015) and the 'Untraditional classification' (right column) includes the multi-generic classification of the composite *Aedes* (*sensu* Reinert *et al.*, 2009). The purpose of the table is to show how the original spelling of species names (left column, with their authorship and original generic placement) has changed when later combined with a genus name of a different gender (centre and right columns, with the generic name preceding the corrected spelling of the species name and the authorship in parentheses indicating that the name is now combined with a genus other than the one in which it was originally placed). Names that have retained their original spelling and generic placement are indicated by 'Unchanged'. Names spelled as originally proposed and transferred to a genus of the same gender are indicated as, for example, 'Unchanged (*Aedes*)', with the name in parentheses being the recipient (current) genus.

Original spelling and generic placement	Traditional classification	Untraditional classification
abditus Barraud, 1931: *Aedes*	*Verrallina abdita* (Barraud, 1931)	*Verrallina abdita* (Barraud, 1931)
abserratus Felt & Young, 1904: *Culex*	Unchanged (*Aedes*)	Unchanged (*Ochlerotatus*)
adustus Laffoon, 1946: *Aedes*	*Verrallina adusta* (Laffoon, 1946)	*Verrallina adusta* (Laffoon, 1946)
aenea Edwards, 1921: *Rachionotomyia*	*Tripteroides aeneus* (Edwards, 1921)	*Tripteroides aeneus* (Edwards, 1921)
aeneus Evans, 1926: *Megarhinus*	*Toxorhynchites aeneus* (Evans, 1926)	*Toxorhynchites aeneus* (Evans, 1926)
albolineata Theobald, 1904: *Scutomyia*	*Aedes albolineatus* (Theobald, 1904)	Unchanged
alboniveus Barraud, 1934: *Aedes*	Unchanged	*Downsiomyia albonivea* (Barraud, 1934)
albonotata Coquillett, 1906: *Gymnometopa*	*Aedes albonotatus* (Coquillett, 1906)	Unchanged (*Howardina*)
albopictus Skuse, 1895: *Culex*	Unchanged (*Aedes*)	*Stegomyia albopicta* (Skuse, 1895)
alboscutellata Theobald, 1905: *Lepidotomyia*	*Aedes alboscutellatus* (Theobald, 1905)	*Aedimorphus alboscutellatus* (Theobald, 1905)
albotaeniata Leicester, 1904: *Danielsia*	*Aedes albotaeniatus* (Leicester, 1904)	Unchanged
amaltheus de Meillon & Lavoipierre, 1944: *Aedes*	Unchanged	*Stegomyia amalthea* (de Meillon & Lavoipierre, 1944)
angustus Edwards, 1935: *Aedes*	Unchanged	*Stegomyia angusta* (Edwards, 1935)
annulatus Schrank, 1776: *Culex*	*Culiseta annulata* (Schrank, 1776)	*Culiseta annulata* (Schrank, 1776)
apicoargentea Theobald, 1909: *Stegomyia*	*Aedes apicoargenteus* (Theobald, 1909)	Unchanged
argenteopunctata Theobald, 1901: *Stegomyia*	*Aedes argenteopunctatus* (Theobald, 1901)	Unchanged (*Catageiomyia*)
aureus Edwards, 1915: *Taeniorhynchus*	*Coquillettidia aurea* (Edwards, 1915)	*Coquillettidia aurea* (Edwards, 1915)
aureolineatus Berlin, 1969: *Aedes*	Unchanged	*Howardina aureolineata* (Berlin, 1969)
aureosquammatus Ludlow, 1909: *Taeniorhynchus*	*Coquillettidia aureosquammata* (Ludlow, 1909)	*Coquillettidia aureosquammata* (Ludlow, 1909)
aurivittatus Cerqueira, 1943: *Aedes*	Unchanged	*Howardina aurivittata* (Cerqueira, 1943)
aurovenatus Worth, 1960: *Aedes*	Unchanged	*Neomelaniconion aurovenatum* (Worth, 1960)
axitiosus Kulasekera, Knight & Harbach, 1990: *Aedes*	Unchanged	*Downsiomyia axitiosa* (Kulasekera, Knight & Harbach, 1990)

Continued

Original spelling and generic placement	Traditional classification	Untraditional classification
azureosquamatus Bonne-Wepster, 1948: *Aedes*	*Verrallina azureosquamata* (Bonne-Wepster, 1948)	*Verrallina azureosquamata* (Bonne-Wepster, 1948)
bathanus Dyar, 1928: *Anopheles*	*Chagasia bathana* (Dyar, 1928)	*Chagasia bathana* (Dyar, 1928)
bicornuta Theobald, 1910: *Lophoceratomyia*	*Culex bicornutus* (Theobald, 1910)	*Culex bicornutus* (Theobald, 1910)
bifoliatus King & Hoogstraal, 1947: *Aedes*	*Verrallina bifoliata* (King & Hoogstraal, 1947)	*Verrallina bifoliata* (King & Hoogstraal, 1947)
bolensis Edwards, 1936: *Aedes*	Unchanged	*Neomelaniconion bolense* (Edwards, 1936)
brevivittatus Berlin, 1969: *Aedes*	Unchanged	*Howardina brevivittata* (Berlin, 1969)
brownscutumus Dong, Zhou & Dong, 1999: *Aedes*	Unchanged	*Bothaella brownscutuma* (Dong, Zhou & Dong, 1999)
caballa Theobald, 1912: *Grabhamia*	*Aedes caballus* (Theobald, 1912)	*Ochlerotatus caballus* (Theobald, 1912)
cacharana Barraud, 1923: *Finlaya*	*Aedes cacharanus* (Barraud, 1923)	Unchanged (*Phagomyia*)
caeruleocephala Leicester, 1908: *Colonemyia*	*Tripteroides caeruleocephalus* (Leicester, 1908)	*Tripteroides caeruleocephalus* (Leicester, 1908)
calceatus Edwards, 1924: *Aedes*	Unchanged	*Stegomyia calceata* (Edwards, 1924)
caledonica Edwards, 1922: *Rachionotomyia*	*Tripteroides caledonicus* (Edwards, 1922)	*Tripteroides caledonicus* (Edwards, 1922)
cautus Barraud, 1928: *Aedes*	*Verrallina cauta* (Barraud, 1928)	*Verrallina cauta* (Barraud, 1928)
celebicus Mattingly, 1959: *Aedes*	Unchanged	*Lorrainea celebica* (Mattingly, 1959)
ceylonicus Edwards, 1917: *Aedes*	*Verrallina ceylonica* (Edwards, 1917)	*Verrallina ceylonica* (Edwards, 1917)
chrysolineata Theobald, 1907: *Howardina*	*Aedes chrysolineatus* (Theobald, 1907)	Unchanged (*Hulecoeteomyia*)
cingulatus Fabricius, 1805: *Culex*	*Psorophora cingulata* (Fabricius, 1805)	*Psorophora cingulata* (Fabricius, 1805)
cingulata Leicester, 1908: *Leicesteria*	*Armigeres cingulatus* (Leicester, 1908)	*Armigeres cingulatus* (Leicester, 1908)
circumluteola Theobald, 1908: *Banksiella* [*sic*]	*Aedes circumluteolus* (Theobald, 1908)	*Neomelaniconion circumluteolum* (Theobald, 1908)
clavatus Barraud, 1931: *Aedes*	*Verrallina clavata* (Barraud, 1931)	*Verrallina clavata* (Barraud, 1931)
comatus Barraud, 1931: *Aedes*	*Verrallina comata* (Barraud, 1931)	*Verrallina comata* (Barraud, 1931)
comosus Reinert, 1974: *Aedes*	*Verrallina comosa* (Reinert, 1974)	*Verrallina comosa* (Reinert, 1974)
confusus Lutz, 1905: *Prosopolepis*	*Wyeomyia confusa* (Lutz, 1905)	*Wyeomyia confusa* (Lutz, 1905)
contiguus Edwards, 1936: *Aedes*	Unchanged	*Stegomyia contigua* (Edwards, 1936)
cretatus Delfinado, 1967: *Aedes*	*Verrallina cretata* (Delfinado, 1967)	*Verrallina cretata* (Delfinado, 1967)
cretinus Edwards, 1921: *Aedes*	Unchanged	*Stegomyia cretina* (Edwards, 1921)
cristatus Theobald, 1904: *Taeniorhynchus*	*Coquillettidia cristata* (Theobald, 1904)	*Coquillettidia cristata* (Theobald, 1904)
croceus Knight & Laffoon, 1946: *Aedes*	Unchanged	*Finlaya crocea* (Knight & Laffoon, 1946)
deccana Barraud, 1923: *Finlaya*	*Aedes deccanus* (Barraud, 1923)	Unchanged (*Phagomyia*)
digitatus Rondani, 1848: *Culex*	*Trichoprosopon digitatum* (Rondani, 1848)	*Trichoprosopon digitatum* (Rondani, 1848)
dobodurus King & Hoogstraal, 1946: *Aedes*	Unchanged	*Patmarksia dobodura* (King & Hoogstraal, 1946)
dolichocephala Leicester, 1908: *Leicesteria*	*Armigeres dolichocephalus* (Leicester, 1908)	*Armigeres dolichocephalus* (Leicester, 1908)
domestica Theobald, 1901: *Uranotaenia*	*Aedes domesticus* (Theobald, 1901)	*Aedimorphus domesticus* (Theobald, 1901)
fasciolatus Lynch Arribálzaga, 1891: *Taeniorhynchus*	*Coquillettidia fasciolata* (Lynch Arribálzaga, 1891)	*Coquillettidia fasciolata* (Lynch Arribálzaga, 1891)
flaveolus Coquillett, 1906: *Taeniorhynchus*	*Mansonia flaveola* (Coquillett, 1906)	*Mansonia flaveola* (Coquillett, 1906)
flavocinctus Edwards, 1936: *Taeniorhynchus*	*Coquillettidia flavocincta* (Edwards, 1936)	*Coquillettidia flavocincta* (Edwards, 1936)
flavopictus Yamada, 1921: *Aedes*	Unchanged	*Stegomyia flavopicta* (Yamada, 1921)

Continued

Original spelling and generic placement	Traditional classification	Untraditional classification
flava Leicester, 1908: *Chaetomyia*	*Armigeres flavus* (Leicester, 1908)	*Armigeres flavus* (Leicester, 1908)
fuscanus Wiedemann, 1820: *Culex*	*Lutzia fuscana* (Wiedemann, 1820)	*Lutzia fuscana* (Wiedemann, 1820)
fuscopennatus Theobald, 1903: *Taeniorhynchus*	*Coquillettidia fuscopennata* (Theobald, 1903)	*Coquillettidia fuscopennata* (Theobald, 1903)
geniculatus Olivier, 1791: *Culex*	Unchanged (*Aedes*)	*Dahliana geniculata* (Olivier, 1791)
gibbosus Delfinado, 1967: *Aedes*	*Verrallina gibbosa* (Delfinado, 1967)	*Verrallina gibbosa* (Delfinado, 1967)
grandiosa Williston, 1900: *Megarrhina*	*Toxorhynchites grandiosus* (Williston, 1900)	*Toxorhynchites grandiosus* (Williston, 1900)
harrisonicus Reinert, 1974: *Aedes*	*Verrallina harrisonica* (Reinert, 1974)	*Verrallina harrisonica* (Reinert, 1974)
hebrideus Edwards, 1926: *Aedes*	Unchanged	*Stegomyia hebridea* (Edwards, 1926)
hispidus Delfinado, 1967: *Aedes*	*Verrallina hispida* (Delfinado, 1967)	*Verrallina hispida* (Delfinado, 1967)
hollandius King & Hoogstraal, 1946: *Aedes*	Unchanged	*Patmarksia hollandia* (King & Hoogstraal, 1946)
horridus Dyar & Knab, 1908: *Aedes*	*Psorophora horrida* (Dyar & Knab, 1908)	*Psorophora horrida* (Dyar & Knab, 1908)
hybrida Leicester, 1908: *Colonemyia*	*Tripteroides hybridus* (Leicester, 1908)	*Tripteroides hybridus* (Leicester, 1908)
imitatus [*sic*] Baisas, 1946: *Topomyia*	*Kimia imitata* (Baisas, 1946)	*Kimia imitata* (Baisas, 1946)
indica Barraud, 1929: *Rachionotomyia*	*Tripteroides indicus* (Barraud, 1929)	*Tripteroides indicus* (Barraud, 1929)
inornata Theobald, 1905: *Gnophodeomyia*	*Culex inornatus* (Theobald, 1905)	*Culex inornatus* (Theobald, 1905)
inornata Walker, 1865: *Megarhina*	*Toxorhynchites inornatus* (Walker, 1865)	*Toxorhynchites inornatus* (Walker, 1865)
inquinatus Edwards, 1922: *Aedes*	Unchanged	*Phagomyia inquinata* (Edwards, 1922)
insolita Coquillett, 1906: *Verrallina*	*Aedes insolitus* (Coquillett, 1906)	'*Ochlerotatus*' *insolitus* (Coquillett, 1906)
iracundus Walker, 1848: *Culex*	*Coquillettidia iracunda* (Walker, 1848)	*Coquillettidia iracunda* (Walker, 1848)
iridescens Lutz, 1905: *Carrollia*	*Culex iridescens* (Lutz, 1905)	*Culex iridescens* (Lutz, 1905)
japonicus Theobald, 1901: *Culex*	Unchanged (*Aedes*)	*Hulecoeteomyia japonica* (Theobald, 1901)
kompongus Klein, 1973: *Aedes*	*Verrallina komponga* (Klein, 1973)	*Verrallina komponga* (Klein, 1973)
koreicus Edwards, 1917: *Ochlerotatus*	Unchanged (*Aedes*)	*Hulecoeteomyia koreica* (Edwards, 1917)
lacteus Knight, 1946: *Aedes*	Unchanged	*Downsiomyia lactea* (Knight, 1946)
lacustris Edwards, 1935: *Ficalbia*	*Mimomyia lacustris* (Edwards, 1935)	*Mimomyia lacustris* (Edwards, 1935)
lamelliferus Bohart & Ingram, 1946: *Aedes*	Unchanged	*Lorrainea lamellifera* (Bohart & Ingram, 1946)
lineatus [*sic*] Taylor, 1914: *Lepidotomyia*	*Verrallina lineata* (Taylor, 1914)	*Verrallina lineata* (Taylor, 1914)
lineatus von Humboldt, 1819: *Culex*	*Psorophora lineata* (von Humboldt, 1819)	*Psorophora lineata* (von Humboldt, 1819)
lineatopennis Ludlow, 1905: *Taeniorhynchus*	Unchanged (*Aedes*)	*Neomelaniconion lineatopenne* (Ludlow, 1905)
lineata Theobald, 1912: *Pseudohowardina*	*Culex lineatus* (Theobald, 1912)	*Culex lineatus* (Theobald, 1912)
litoreus Colless, 1958: *Aedes*	Unchanged	*Downsiomyia litorea* (Colless, 1958)
longiareolatus Macquart, 1838: *Culex*	*Culiseta longiareolata* (Macquart, 1838)	*Culiseta longiareolata* (Macquart, 1838)
luridus McIntosh, 1971: *Aedes*	Unchanged	*Neomelaniconion luridum* (McIntosh, 1971)
luteolateralis Theobald, 1901: *Culex*	Unchanged (*Aedes*)	*Neomelaniconion luteolaterale* (Theobald, 1901)
luteostriatus Robinson, 1950: *Aedes*	Unchanged	*Vansomerenis luteostriata* (Robinson, 1950)
magna Theobald, 1901: *Brachiomyia*	*Deinocerites magnus* (Theobald, 1901)	*Deinocerites magnus* (Theobald, 1901)
magnesiana Edwards, 1924: *Rachionotomyia*	*Tripteroides magnesianus* (Edwards, 1924)	*Tripteroides magnesianus* (Edwards, 1924)
mediopunctata Theobald, 1905: *Stegomyia*	*Aedes mediopunctatus* (Theobald, 1905)	Unchanged

Continued

Original spelling and generic placement	Traditional classification	Untraditional classification
mediovittata Coquillett, 1906: *Stegomyia*	*Aedes mediovittatus* (Coquillett, 1906)	Unchanged (*Gymnometopa*)
melanoptera Giles, 1904: *Phagomyia*	*Aedes melanopterus* (Giles, 1904)	Unchanged
melanurus Coquillett, 1902: *Culex*	*Culiseta melanura* (Coquillett, 1902)	*Culiseta melanura* (Coquillett, 1902)
metallica Edwards, 1912: *Stegomyia*	*Aedes metallicus* (Edwards, 1912)	Unchanged
metallicus Theobald, 1901: *Culex*	*Coquillettidia metallica* (Theobald, 1901)	*Coquillettidia metallica* (Theobald, 1901)
mexicanus Bellardi, 1859: *Culex*	*Psorophora mexicana* (Bellardi, 1859)	*Psorophora mexicana* (Bellardi, 1859)
microannulatus Theobald, 1911: *Chrysoconops*	*Coquillettidia microannulata* (Theobald, 1911)	*Coquillettidia microannulata* (Theobald, 1911)
micropterus Giles, 1901: *Culex*	Unchanged (*Aedes*)	*Dendroskusea microptera* (Giles, 1901)
microstictus Edwards, 1936: *Aedes*	Unchanged	*Catageiomyia microsticta* (Edwards, 1936)
minuta Theobald, 1901: *Stegomyia*	*Aedes minutus* (Theobald, 1901)	Unchanged (*Catageiomyia*)
minutissima Theobald, 1907: *Culiciomyia*	*Culex minutissimus* (Theobald, 1907)	*Culex minutissimus* (Theobald, 1907)
mixtus Edwards, 1936: *Aedes*	Unchanged	*Catageiomyia mixta* (Edwards, 1936)
moctezuma Dyar & Knab, 1906: *Megarhina* [*sic*]	*Toxorhynchites moctezuma* (Dyar & Knab, 1906)	*Toxorhynchites moctezuma* (Dyar & Knab, 1906)
neoafricanus Cornet, Valade & Dieng, 1978: *Aedes*	Unchanged	*Stegomyia neoafricana* (Cornet, Valade & Dieng, 1978)
neogeorgianus Belkin, 1962: *Aedes*	Unchanged	*Finlaya neogeorgiana* (Belkin, 1962)
nigricephala Theobald, 1901: *Stegomyia*	*Aedes nigricephalus* (Theobald, 1901)	*Aedimorphus nigricephalus* (Theobald, 1901)
nigrina Eckstein, 1918: *Culicada*	*Aedes nigrinus* (Eckstein, 1918)	*Ochlerotatus nigrinus* (Eckstein, 1918)
nigrochraceus Bonne-Wepster, 1930: *Taeniorhynchus*	*Coquillettidia nigrochracea* (Bonne-Wepster, 1930)	*Coquillettidia nigrochracea* (Bonne-Wepster, 1930)
nigropterum Le Goff, Boussès & Brunhes, 2007: *Neomelaniconion*	*Aedes nigropterus* (Le Goff, Boussès & Brunhes, 2007)	Unchanged
nigrosignatus Edwards, 1917: *Taeniorhynchus*	*Coquillettidia nigrosignata* (Edwards, 1917)	*Coquillettidia nigrosignata* (Edwards, 1917)
nippononiveus Sasa & Nakahashi, 1952: *Aedes*	Unchanged	*Downsiomyia nippononivea* (Sasa & Nakahashi, 1952)
nipponicus LaCasse & Yamaguti, 1948: *Aedes*	Unchanged	*Downsiomyia nipponica* (LaCasse & Yamaguti, 1948)
nivea Ludlow, 1903: *Stegomyia*	*Aedes niveus* (Ludlow, 1903)	Unchanged (*Downsiomyia*)
notoscriptus Skuse, 1889: *Culex*	Unchanged (*Aedes*)	*Rampamyia notoscripta* (Skuse, 1889)
novalbopictus Barraud, 1931: *Aedes*	Unchanged	*Stegomyia novalbopicta* (Barraud, 1931)
novochraceus Barraud, 1927: *Taeniorhynchus*	*Coquillettidia novochracea* (Barraud, 1927)	*Coquillettidia novochracea* (Barraud, 1927)
novoniveus Barraud, 1934: *Aedes*	Unchanged	*Downsiomyia novonivea* (Barraud, 1934)
nubicolus Laffoon, 1946: *Aedes*	*Verrallina nubicola* (Laffoon, 1946)	*Verrallina nubicola* (Laffoon, 1946)
obsoletus Huang, 1968: *Aedes*	*Verrallina obsoleta* (Huang, 1968)	*Verrallina obsoleta* (Huang, 1968)
occidentayunnanus Gong & Lu, 1991: *Aedes*	*Heizmannia occidentayunnana* (Gong & Lu, 1991)	*Heizmannia occidentayunnana* (Gong & Lu, 1991)
oceanicus Belkin, 1962: *Aedes*	Unchanged	*Finlaya oceanica* (Belkin, 1962)
ochraceus Theobald, 1903: *Taeniorhynchus*	*Coquillettidia ochracea* (Theobald, 1903)	*Coquillettidia ochracea* (Theobald, 1903)
omissa Edwards, 1914: *Leicesteria*	*Armigeres omissus* (Edwards, 1914)	*Armigeres omissus* (Edwards, 1914)
palmata Rodenwaldt, 1926: *Stethomyia*	*Anopheles palmatus* (Rodenwaldt, 1926)	*Anopheles palmatus* (Rodenwaldt, 1926)
palpale Newstead, 1907: *Neomelaniconion*	*Aedes palpalis* (Newstead, 1907)	Unchanged
pectinatus Edwards, 1914: *Leicesteria*	*Armigeres pectinatus* (Edwards, 1914)	*Armigeres pectinatus* (Edwards, 1914)
pecuniosus Edwards, 1922: *Aedes*	Unchanged	*Molpemyia pecuniosa* (Edwards, 1922)

Continued

Original spelling and generic placement	Traditional classification	Untraditional classification
pendula Edwards, 1914: *Leicesteria*	*Armigeres pendulus* (Edwards, 1914)	*Armigeres pendulus* (Edwards, 1914)
periskelata Giles, 1902: *Stegomyia*	*Aedes periskelatus* (Giles, 1902)	Unchanged (*Dendroskusea*)
personata Lutz, 1904: *Dendromyia*	*Onirion personatum* (Lutz, 1904)	*Onirion personatum* (Lutz, 1904)
pexus Colless, 1958: *Aedes*	Unchanged	*Downsiomyia pexa* (Colless, 1958):
phnomus Klein, 1973: *Aedes*	*Verrallina phnoma* (Klein, 1973)	*Verrallina phnoma* (Klein, 1973)
pipersalata Giles, 1902: *Stegomyia*	*Aedes pipersalatus* (Giles, 1902)	*Aedimorphus pipersalatus* (Giles, 1902)
platylepidus Knight & Hull, 1951: *Aedes*	Unchanged	*Scutomyia platylepida* (Knight & Hull, 1951)
plumiferus King & Hoogstraal, 1946: *Aedes*	Unchanged	*Phagomyia plumifera* (King & Hoogstraal, 1946)
plumosa Brug, 1931: *Rachionotomyia*	*Tripteroides plumosus* (Brug, 1931)	*Tripteroides plumosus* (Brug, 1931)
plumosus Theobald, 1901: *Culex*	*Mimomyia plumosa* (Theobald, 1901)	*Mimomyia plumosa* (Theobald, 1901)
pogonurus Edwards, 1936: *Aedes*	Unchanged	*Neomelaniconion pogonurum* (Edwards, 1936)
poicilia Theobald, 1903: *Finlaya*	*Aedes poicilius* (Theobald, 1903)	Unchanged
proxima Edwards, 1915: *Rachionotomyia*	*Tripteroides proximus* (Edwards, 1915)	*Tripteroides proximus* (Edwards, 1915)
pseudalbolineatus Brug, 1939: *Aedes*	Unchanged	*Scutomyia pseudalbolineata* (Brug, 1939)
pseudalbopicta Borel, 1928: *Stegomyia*	*Aedes pseudalbopictus* (Borel, 1928)	Unchanged
pseudoafricanus Chwatt, 1949: *Aedes*	Unchanged	*Stegomyia pseudoafricana* (Chwatt, 1949)
punctifemore [*sic*] Ludlow, 1921: *Stegomyia*	*Aedes punctifemoris* (Ludlow, 1921)	*Aedimorphus punctifemoris* (Ludlow, 1921)
punctocostalis Theobald, 1909: *Banksinella*	Unchanged (*Aedes*)	*Neomelaniconion punctocostale* (Theobald, 1909)
purpurata Edwards, 1921: *Rachionotomyia*	*Tripteroides purpuratus* (Edwards, 1921)	*Tripteroides purpuratus* (Edwards, 1921)
purpurea Theobald, 1910: *Molpemyia*	*Aedes purpureus* (Theobald, 1910)	Unchanged
quadrispinatus King & Hoogstraal, 1947: *Aedes*	*Verrallina quadrispinata* (King & Hoogstraal, 1947)	*Verrallina quadrispinata* (King & Hoogstraal, 1947)
quadrivittatus Coquillett, 1902: *Culex*	Unchanged (*Aedes*)	*Howardina quadrivittata* (Coquillett, 1902)
quinquelineatus Edwards, 1922: *Aedes*	Unchanged	*Rampamyia quinquelineata* (Edwards, 1922)
rarus Delfinado, 1968: *Aedes*	*Verrallina rara* (Delfinado, 1968)	*Verrallina rara* (Delfinado, 1968)
rutila [*sic*] Coquillett, 1896: *Megarhinus*	*Toxorhynchites rutilus* (Coquillett, 1896)	*Toxorhynchites rutilus* (Coquillett, 1896)
samoana Grünberg, 1913: *Finlaya*	*Aedes samoanus* (Grünberg, 1913)	Unchanged
sapphirinus Osten Sacken, 1868: *Aedes*	*Uranotaenia sapphirina* (Osten Sacken, 1868)	*Uranotaenia sapphirina* (Osten Sacken, 1868)
scutoscriptus Bohart & Ingram, 1946: *Aedes*	Unchanged	*Stegomyia scutoscripta* (Bohart & Ingram, 1946)
seculatus Menon, 1950: *Aedes*	*Verrallina seculata* (Menon, 1950)	*Verrallina seculata* (Menon, 1950)
sentanius King & Hoogstraal, 1947: *Aedes*	*Verrallina sentania* (King & Hoogstraal, 1947)	*Verrallina sentania* (King & Hoogstraal, 1947)
separata Leicester, 1908: *Megarhina*	*Anopheles separatus* (Leicester, 1908)	*Anopheles separatus* (Leicester, 1908)
septemstriatus Dyar & Knab, 1907: *Aedes*	Unchanged	*Howardina septemstriata* (Dyar & Knab, 1907)
serrata Barraud, 1929: *Rachionotomyia*	*Tripteroides serratus* (Barraud, 1929)	*Tripteroides serratus* (Barraud, 1929)
sexlineata Theobald, 1901: *Stegomyia*	*Aedes sexlineatus* (Theobald, 1901)	Unchanged (*Howardina*)
sibiricus Danilov & Filippova, 1978: *Aedes*	Unchanged	*Stegomyia sibirica* (Danilov & Filippova, 1978)
signifer Coquillett, 1896: *Culex*	*Orthopodomyia signifera* (Coquillett, 1896)	*Orthopodomyia signifera* (Coquillett, 1896)

Continued

Original spelling and generic placement	Traditional classification	Untraditional classification
simplus King & Hoogstraal, 1947: *Aedes*	*Verrallina simpla* (King & Hoogstraal, 1947)	*Verrallina simpla* (King & Hoogstraal, 1947)
soleatus Edwards, 1924: *Aedes*	Unchanged	*Stegomyia soleata* (Edwards, 1924)
speciosa Skuse, 1889: *Megarhina*	*Toxorhynchites speciosus* (Skuse, 1889)	*Toxorhynchites speciosus* (Skuse, 1889)
spinosus Berlin, 1969: *Aedes*	Unchanged	*Howardina spinosa* (Berlin, 1969)
spinosus Doucet, 1951: *Ravenalites*	*Mimomyia spinosa* (Doucet, 1951)	*Mimomyia spinosa* (Doucet, 1951)
subalbopictus Barraud, 1931: *Aedes*	Unchanged	*Stegomyia subalbopicta* (Barraud, 1931)
subargenteus Edwards, 1925: *Aedes*	Unchanged	*Stegomyia subargentea* (Edwards, 1925)
subniveus Edwards, 1922: *Aedes*	Unchanged	*Downsiomyia subnivea* (Edwards, 1922)
sundaica Rodenwaldt, 1925: *Myzomyia*	*Anopheles sundaicus* (Rodenwaldt, 1925)	*Anopheles sundaicus* (Rodenwaldt, 1925)
tibiamaculata Neiva, 1906: *Myzomyia*	*Anopheles tibiamaculatus* (Neiva, 1906)	*Anopheles tibiamaculatus* (Neiva, 1906)
tigripes de Grandpre & de Charmoy, 1901: *Culex*	*Lutzia tigripes* (de Grandpré & de Charmoy, 1901)	*Lutzia tigripes* (de Grandpré & de Charmoy, 1901)
tonsus Edwards, 1924: *Aedes*	Unchanged	*Geoskusea tonsa* (Edwards, 1924)
torosus Delfinado, 1967: *Aedes*	*Verrallina torosa* (Delfinado, 1967)	*Verrallina torosa* (Delfinado, 1967)
travestitus Brug, 1928: *Anopheles*	*Bironella travestita* (Brug, 1928)	*Bironella travestita* (Brug, 1928)
tremula Theobald, 1903: *Macleaya*	*Aedes tremulus* (Theobald, 1903)	Unchanged
triannulata Neiva & Pinto, 1922: *Cellia*	*Anopheles triannulatus* (Neiva & Pinto, 1922)	*Anopheles triannulatus* (Neiva & Pinto, 1922)
trispinatus King & Hoogstraal, 1947: *Aedes*	*Verrallina trispinata* (King & Hoogstraal, 1947)	*Verrallina trispinata* (King & Hoogstraal, 1947)
unidentatus McIntosh, 1971: *Aedes*	Unchanged	*Neomelaniconion unidentatum* (McIntosh, 1971)
unilineata Theobald, 1906: *Quasistegomyia*	*Aedes unilineatus* (Theobald, 1906)	Unchanged (*Stegomyia*)
vanus Colless, 1958: *Aedes*	Unchanged	*Downsiomyia vana* (Colless, 1958)
vanapus Huang, 1968: *Aedes*	*Verrallina vanapa* (Huang, 1968)	*Verrallina vanapa* (Huang, 1968)
vicina Edwards, 1914: *Rachionotomyia*	*Tripteroides vicinus* (Edwards, 1914)	*Tripteroides vicinus* (Edwards, 1914)
vicina Taylor, 1916: *Leucomyia*	*Culex vicinus* (Taylor, 1916)	*Culex vicinus* (Taylor, 1916)
w-alba Theobald, 1905: *Stegomyia*	*Aedes w-albus* (Theobald, 1905)	Unchanged

CABI – who we are and what we do

This book is published by **CABI**, an international not-for-profit organisation that improves people's lives worldwide by providing information and applying scientific expertise to solve problems in agriculture and the environment.

CABI is also a global publisher producing key scientific publications, including world renowned databases, as well as compendia, books, ebooks and full text electronic resources. We publish content in a wide range of subject areas including: agriculture and crop science / animal and veterinary sciences / ecology and conservation / environmental science / horticulture and plant sciences / human health, food science and nutrition / international development / leisure and tourism.

The profits from CABI's publishing activities enable us to work with farming communities around the world, supporting them as they battle with poor soil, invasive species and pests and diseases, to improve their livelihoods and help provide food for an ever growing population.

CABI is an international intergovernmental organisation, and we gratefully acknowledge the core financial support from our member countries (and lead agencies) including:

UKaid from the British people

Ministry of Agriculture People's Republic of China

Australian Government Australian Centre for International Agricultural Research

Agriculture and Agri-Food Canada

Ministry of Foreign Affairs of the Netherlands

Schweizerische Eidgenossenschaft Confédération suisse Confederazione Svizzera Confederaziun svizra Swiss Agency for Development and Cooperation SDC

Discover more

To read more about CABI's work, please visit: **www.cabi.org**

Browse our books at: **www.cabi.org/bookshop**,
or explore our online products at: **www.cabi.org/publishing-products**

Interested in writing for CABI? Find our author guidelines here:
www.cabi.org/publishing-products/information-for-authors/